Fourth SCAR Symposium on Antarctic Biology
Wilderness, South Africa, 12–16 September 1983

Sponsors

International Council for Scientific Unions (ICSU)
Scientific Committee on Antarctic Research (SCAR)
Scientific Committee on Oceanic Research (SCOR)
International Union of Biological Sciences (IUBS)
International Association of Biological Oceanography (IABO)
South African Scientific Committee on Antarctic Research (SASCAR)
South African Department of Transport

Organizers

Scientific Steering Committee, SCAR Working Group on Biology
Local Organizing Committee, SASCAR Subcommittee on Biological Sciences

In connection with the journal POLAR BIOLOGY.

Antarctic Nutrient Cycles and Food Webs

Edited by
W. R. Siegfried, P. R. Condy, and R. M. Laws

With 220 Tables and 390 Figures

Springer-Verlag
Berlin Heidelberg New York Tokyo

Prof. Walter R. SIEGFRIED

Percy FitzPatrick Institute
University of Cape Town
Rondebosch 7700, South Africa

Dr. Pat R. CONDY

Foundation for Research Development
C.S.I.R., P.O. Box 395
Pretoria 0001, South Africa

Dr. Richard M. LAWS

British Antarctic Survey
Natural Environment Research Council
High Cross, Madingley Road
Cambridge CB 3 OET, United Kingdom

ISBN 3-540-13417-4 Springer-Verlag Berlin Heidelberg New York Tokyo
ISBN 0-387-13417-4 Springer-Verlag New York Heidelberg Berlin Tokyo

Library of Congress Cataloging in Publication Data. SCAR Symposium on Antarctic Biology (4th : 1983 : Wilderness, South Africa) Antarctic nutrient cycles and food webs. Includes index. 1. Biogeochemical cycles – Antarctic regions – Congresses. 2. Food chains (Ecology) – Antarctic regions – Congresses. I. Siegfried, W. R. II. Condy, P. R. (Patrick Richard), 1948– . III. Laws, Richard M. IV. International Council of Scientific Unions. Scientific Committee on Antarctic Research. V. International Council of Scientific Unions. VI. SCAR Working Group on Biology. VII. Title. QH84.2.S33 1983 574.5'3 85-2796

Typesetting and bookbinding: Brühlsche Universitätsdruckerei, Giessen. Offsetprinting: Beltz Offsetdruck Hemsbach/Bergstraße
2131/3130-543210

Foreword

It is a pleasure and a distinct honour for me to greet the participants, guests and observers of this Fourth International Symposium on Antarctic Biology which has adopted nutrient cycles and food webs as its central theme.

On behalf of the Scientific Committee on Antarctic Research (SCAR) and other bodies of the International Council of Scientific Unions (ICSU), I bid you welcome. SCAR is pleased to acknowledge the role of the co-sponsors for this Symposium which include the Scientific Committee on Oceanic Research (SCOR), the International Association of Biological Oceanography (IABO), and the International Union of Biological Sciences (IUBS). In addition, SCAR and its co-sponsors wish to acknowledge the financial support of the Council for Scientific and Industrial Research (CSIR) and the Department of Transport (DOT) of the South African government. Nor should we forget to acknowledge also the role of the South African Scientific Committee on Antarctic Research (SASCAR) and one of its leaders and Vice President of SCAR, Mr. Jan de Wit, in arranging this charming venue for this Symposium.

The biological sciences in Antarctica are increasing in importance. SCAR's Working Group on Biology has been one of SCAR's most active sub-groups since SCAR was founded in 1958. Today, biologists have emerged as a major voice in SCAR through the Working Group on Biology, its Subcommittee on Conservation, the Group of Specialists on Seals, and the Group of Specialists on Southern Ocean Ecosystems and Their Living Resources. In a more immediate and pragmatic sense, the massive initiative of Antarctic marine biology through the SCAR-initiated Biological Investigations of Marine Antarctic Systems and Stocks (BIOMASS) research programme will be the major source of data than can provide a firm scientific basis for management policies eventually to be formulated by the Convention on the Conservation of Antarctic Marine Living Resources (CCAMLR). Moreover, if non-living resource exploitation should ever gain momentum in Antarctica, we will need the wisdom of Antarctic biologists to provide some major guidelines for preserving the unique ecology of the earth's only polar continent.

This Symposium was preceded by three others in the last 21 years. The first symposium on Antarctic biology was convened in Paris in 1962 and dealt with all aspects of biological research in Antarctica. The second met in Cambridge in 1968 and focused on Antarctic ecology. The third, in Washington, D.C., in 1974, selected as its main theme the adaptations within Antarctic ecosystems.

I hope that the publication of the proceedings of this particular Symposium will be given widespread distribution, and that many of you will seek a publication outlet for your papers in journals that are not specifically devoted to Antarctic research. It is evident to me that many biologists around the world are totally unaware of the level and sophistication of biological research in Antarctica.

In closing, let me say again how honoured I am to be present here with you to begin the Fourth Symposium on Antarctic Biology. It is an occasion for meeting old friends and making new friends and to exchange new ideas and discuss old concepts in an international setting. And above all, it is a time to tell the world once again that there exists a group of hardy men and women who have dedicated their careers to the understanding of the living world in one of nature's most fascinating and beguiling laboratories, Antarctica. I salute you for your efforts, and I wish for each of you a stimulating and rewarding experience at Wilderness. Thank you.

Wilderness, South Africa

J.H. ZUMBERGE*
President of SCAR

* University of Southern California, University Park, Los Angeles, California 90007, USA

Preface

The Scientific Committee on Antarctic Research (SCAR) arose during the International Geophysical Year (IGY) in 1957–58, which included an important Antarctic programme. This unprecedented international collaboration led to the recommendation to the International Council of Scientific Unions (ICSU) that a permanent organization should be established to sponsor a continuing Antarctic research programme after the end of IGY. In 1958 SCAR was created as a permanent Committee of ICSU, and is one of the more active ones. SCAR now has nine permanent Working Groups and like the other Working Groups, the main function of the Working Group on Biology is, by providing a regular forum for exchange of information, to foster and co-ordinate scientific activities in the Antarctic and to draw up agreed objectives that help to guide the various national programmes of research. It assists SCAR to respond to requests for information and advice on Antarctic matters, particularly from the Antarctic Treaty and other inter-governmental organizations. It has been particularly concerned with devising research programmes, giving advice and devising guidelines on conservation measures and on environmental issues. The first example was the Agreed Measures for the Conservation of Antarctic Flora and Fauna (1964). The most recent is the compilation of an annotated Atlas of Antarctic Protected Areas. Most business is expected to be conducted by correspondence but the Working Group meets as necessary in conjunction with the full SCAR meetings, held in alternate years, and organizes major symposia from time to time.

This book is a compilation of 92 of the 107 papers that were accepted for presentation at the Fourth SCAR Symposium on Antarctic Biology, held at Wilderness, South Africa, during 12–16 September 1983. The Symposium covered nutrient cycling and food webs in terrestrial, marine and freshwater ecosystems in the Antarctic and Sub-Antarctic regions, and the papers in the book are arranged accordingly. Each of the book's five main sections is introduced by a review paper which was solicited by the organizers of the Symposium. The balance of the papers consists of the full texts of 41 poster and 46 oral presentations at the Symposium.

The texts of all papers accepted for the Symposium were considered for publication in the proceedings, provided that they had been submitted at least six weeks before the Symposium. This arrangement allowed the editors to assess the quality of the contributions and to select appropriate reviewers from, in most cases, among the delegates attending the Symposium. All papers included in this book were reviewed, and authors of papers were required to resubmit revised manuscripts by the end of 1983. The editors are grateful to the many reviewers and to the contributors of papers for their assistance and co-operation which have greatly facilitated the relatively prompt publication of this volume. Although the entire volume is in English, approximately one-third of the papers were contributed by authors whose first language is not English. We thank them for their special efforts to reduce our editorial headaches. We also thank Nigel Bonner, Brenda Coetzee, Ione Davies and Margot Murray whose assistance in many ways allowed us to meet seemingly impossible deadlines.

In compliance with requirements set by the publishers, contributions comprising this volume accord with the language (English), style and conventions of the technical journal *Polar Biology*. The editors have, however, seen fit to depart in one instance from the practice of *Polar Biology*, in that the common names of species have been given initial capital letters. This has been done simply to reduce ambiguity in the prose. Many of us are familiar with the ambiguous: "The common grey whales swam slowly by". In a similar vein, there is the confusing and often conflicting use of the term "krill". This word, according to an array of dictionaries, is derived from the Norwegian *kril* (fry of fish) and is applied generally to planktonic crustaceans that constitute the principal food of baleen whales. In the context of modern literature dealing with the Antarctic, however, krill is used generally, but not exclusively, as the vernacular name of *Euphausia superba*. Hence, we favoured Antarctic Krill as the common specific name for this animal, and we have used krill as a collective name for all euphausiids (Euphausiacea). In other matters affecting the standardization of names of species and places, we have given preference to proper names that are in widespread use and that are comprehensible to a majority of international readers (e.g., cormorant *vice* shag, Antarctic Peninsula *vice* Graham Land). Similarly, we have accepted those scientific names of species that are recognized most widely, rather than follow slavishly the prescriptions of "official" or other "authoritative" checklists. We have omitted the authorities for Latin names. Widely recognized names that have been ruled as invalid have, of course, been rejected. Fortunately, we have had to deal with only one such case, regarding the nomenclature of the hyperiid amphipod known by most zoologists as *Parathemisto gaudichaudii* but now correctly named *Themisto gaudichaudii*. Trinomials have been used only when accurately known and essential to the text.

Finally, we hope that this book will convey some of the mood of excitement that pervades modern ecological research in the Antarctic and Sub-Antarctic regions. Old, well-worn ideas are toppling fast and many new, often revolutionary, concepts are creating expanding frontiers for scientific inquiry. Antarctic biology now has a number of hypotheses and sufficient empirical data to begin building syntheses. Young students as well as advanced scientists should be able to learn from this volume where both the progress and the gaps in our knowledge are most evident. The Antarctic is a unique region of peace and international co-operation where science is encouraged to prosper. The peace, the international co-operation and the conservation of the region's ecosystems are dependent to a considerable extent on the performance of scientists committed to Antarctic research.

Cape Town,
Pretoria, and
Cambridge

W.R. SIEGFRIED,
P.R. CONDY, and
R.M. LAWS

Contents

Introduction: Antarctic Biology, 1974–1983
R.M. Laws 1

Part 1. Marine Nutrient Cycles

Nutrient Cycles in Antarctic Marine Ecosystems
O. Holm-Hansen 6

Oceanic Frontal Systems and Biological Enhancement
J.R.E. Lutjeharms, N.M. Walters, and B.R. Allanson . (With 10 Figures) 11

The Distribution of Inorganic Nitrogen, Phosphorus, Silicon, and Dissolved Organic Matter in Surface and Deep Waters of the Southern Ocean
S. le Jehan and P. Tréguer . (With 6 Figures) 22

Trace Element Geochemistry in the Southern Ocean
M.J. Orren and P.M.S. Monteiro . (With 9 Figures) 30

A Contribution to the Oceanology of the Prince Edward Islands
B.R. Allanson, B. Boden, L. Parker, and C. Duncombe Rae . (With 11 Figures) . . 38

Nutrient Upwelling and Its Effects in the Lee of Marion Island
J.R. Grindley and P. David . (With 8 Figures) 46

Marine Biological and Oceanographical Investigations in Lützow-Holm-Bay, Antarctica
M. Fukuchi, A. Tanimura, and H. Ohtsuka . (With 4 Figures) 52

Light-Dependence of Phytoplankton Photosynthesis in the Antarctic Ocean: Implications for Regulating Productivity
M.M. Tilzer, B. von Bodungen, and V. Smetacek . (With 7 Figures) 60

Phytoplankton Biomass near a Receding Ice-Edge in the Ross Sea
W.O. Smith Jr. and D.M. Nelson . (With 9 Figures) 70

The Influence of Light on Growth and Development of the Sea-Ice Microbial Community of McMurdo Sound. (With 5 Figures)
C.W. Sullivan, A.C. Palmisano, S. Kottmeier, S. McGrath-Grossi, and R. Moe . . . 78

Physiological Response of Micro-Algae in the Ice-Platelet Layer to Low-Light Conditions
A.C. Palmisano and C.W. Sullivan . (With 2 Figures) 84

Autumnal Proliferation of Ice-Algae in Antarctic Sea-Ice
T. Hoshiai . (With 3 Figures) 89

Oceanographic Studies of Epi-Pelagic Ammonium Distributions: The Ross Sea NH_4^+ Flux Experiment
D.C. Biggs, A.F. Amos, and O. Holm-Hansen . (With 10 Figures) 93

Growth and Production of the Seaweed, *Himantothallus grandifolius*, at King George Island
G. Dieckmann, W. Reichardt, and K. Zieliński . (With 3 Figures) 104

Factors Influencing the Production, Decomposition and Distribution of Organic and Inorganic Matter in Admiralty Bay, King George Island
R. Dawson, W. Schramm, and M. Bölter . (With 5 Figures) 109

Kinetics and Trophic Role of Bacterial Degradation of Macro-Algae in Antarctic Coastal Waters
W. Reichardt and G. Dieckmann . (With 10 Figures) 115

The Role of Bacteria in the Cycling of Nutrients within the Maritime Antarctic Environment
A.C. Tanner 123

Heterotrophic Processes in a Kerguelen Mussel-Bed
D. Delille and G. Cahet . (With 6 Figures) 128

Part 2. Terrestrial and Freshwater Nutrient Cycles

Nutrient Cycling in Relation to Biological Productivity in Antarctic and Sub-Antarctic Terrestrial and Freshwater Ecosystems
R.I. Lewis Smith . (With 9 Figures) 138

The Role of Pygoscelid Penguin Rookeries in Nutrient Cycles at Admiralty Bay, King George Island
A. Myrcha, S.J. Pietr, and A. Tatur . (With 4 Figures) 156

Ornithogenic Phosphates on King George Island in the Maritime Antarctic
A. Tatur and A. Barczuk . (With 6 Figures) 163

Chemical Composition of Guano of Burrowing Petrel Chicks (Procellariidae) at Marion Island
S.R. Fugler 169

The Influence of Elephant Seals on the Terrestrial Ecosystem at Marion Island
K. Panagis . (With 1 Figure) 173

A Preliminary Study of the Action of Crustose Lichens on Rock Surfaces in Antarctica
D.W.H. Walton . (With 5 Figures) 180

Heterotrophic Acetylene Reduction in Soils at Marion Island
V.R. Smith 186

Cellulose Decomposition and Its Relationship to Nutrient Cycling at South Georgia
D.W.H. Walton . (With 6 Figures). 192

Bryophyte Production at Marion Island
S. Russell . (With 1 Figure) 200

Comparative Microbiology of Moss-Peat Decomposition on the Scotia Arc and Antarctic Peninsula
D.D. Wynn-Williams 204

Decomposition and Nutrient Cycling in *Rostkovia magellanica* from Two Contrasting Bogs on South Georgia
G.J. Lawson . (With 6 Figures) 211

Growth and Production of *Poa flabellata* in Relation to Soil Nutrient Status and Exposure at South Georgia
R.I. Lewis Smith . (With 6 Figures) 221

Methanogenesis and the Anaerobic Micro-Biology of a Wet Moss Community at Signy Island
M.R. Yarrington and D.D. Wynn-Williams . (With 3 Figures) 229

The Influence of Iron and Manganese on Nutrient Cycling in Shallow Freshwater Antarctic Lakes
J.B. Gallagher . (With 6 Figures) 234

Paucity of Nutrient Cycling and Absence of Food Chains in the Unique Lakes of Southern Victoria Land
B.C. Parker and G.M. Simmons, Jr 238

Factors Controlling Phytoplankton Populations in Maritime Antarctic Lakes
I. Hawes . (With 10 Figures) 245

Decomposition Processes in Maritime Antarctic Lakes
J.C. Ellis-Evans . (With 8 Figures) 253

Interactions of Bacterio- and Phyto-Plankton in Nutrient Cycling Within Eutrophic Heywood Lake, Signy Island
J.C. Ellis-Evans . (With 4 Figures) 261

Part 3. Marine Food Webs

Antarctic Marine Food Webs
G. Hempel . (With 3 Figures) 266

Choanoflagellates in the Antarctic Marine Food Chain
H.J. Marchant . (With 7 Figures) 271

Alternate Carbon Pathways at Lower Trophic Levels in the Antarctic Food Web
C.D. Hewes, O. Holm-Hansen, and E. Sakshaug . (With 6 Figures) 277

Spatial Variability of Phytoplankton and the Distribution and Abundance of Krill in the Indian Sector of the Southern Ocean
L.H. Weber and S.Z. El-Sayed . (With 3 Figures) 284

Abundance, Distribution and Behaviour of *Euphausia superba* in the Southern Ocean Between 15° and 30°E During FIBEX
I. Hampton. (With 10 Figures) 294

Variation in Geographical Distribution and Abundance of Larvae of Antarctic Krill, *Euphausia superba,* in the Southern Atlantic Ocean
I. Hempel . (With 3 Figures) 304

Vertical Distribution of Larvae of Antarctic Krill, *Euphausia superba*
I. Hempel . (With 3 Figures) 308

Feeding by *Euphausia superba* and Copepod Species in Response to Varying Concentrations of Phytoplankton
S.B. Schnack . (With 8 Figures) 311

Metabolism of Antarctic Krill, *Euphausia superba,* and Its Tropho-Dynamic Implications
R.Y. George . (With 2 Figures) 324

Aspects of the Nutritional Metabolism of Two Antarctic Euphausiids: *Euphausia superba* and *Thysanoessa macrura*
P. Mayzaud, J. Farber-Lorda, and M.C. Corre . (With 11 Figures) 330

Moult and Growth in Euphausiids
F. Buchholz . (With 6 Figures) 339

Structural and Functional Analyses of the Feeding Appendages of Krill Larvae
H-P. Marschall. (With 8 Figures) 346

Marine Macro-Plankton of Two Sub-Antarctic Islands
D.G.M. Miller . (With 5 Figures) 355

The Relationship Between Krill Food Requirements and Phytoplankton Production in a Sector of the Southern Indian Ocean . (With 5 Figures)
D.G.M. Miller, I. Hampton, J. Henry, R.W. Abrams, and J. Cooper 362

Feeding by Antarctic Krill, *Euphausia superba:* Does Size Matter?
L.B. Quetin and R.M. Ross . (With 3 Figures) 372

Nutrition of *Allotanais hirsutus* (Crustacea:Tanaidacea) at Kerguelen Island
D. Delille, L.D. Guidi, and J. Soyer . (With 2 Figures) 378

The Trophic Role of the Stone Crab, *Lithodes murrayi,* in the Benthic Ecosystem of the Crozet Islands
P.M. Arnaud and J.C. Miquel . (With 4 Figures) 381

The Energy Budget of a Population of the Antarctic Isopod *Serolis polita*
R.A. Luxmoore . (With 2 Figures) 389

Non-Cold-Adapted Metabolism in the Decapod *Chorismus antarcticus* and Other Sub-Antarctic Marine Crustaceans
J.G.H. Maxwell and R. Ralph . (With 7 Figures) 397

Food Webs and Functional Aspects of the Kelp, *Macrocystis pyrifera,* Community in the Beagle Channel, Chile
J.C. Castilla . (With 4 Figures) 407

Aspects of the Roles of Squid in Food Chains of Marine Antarctic Ecosystems
T. Nemoto, M. Okiyama, and M. Takahashi . (With 4 Figures) 415

The Role of Zooplankton in the Diets of Certain Sub-Antarctic Marine Fish
G. Duhamel and J.C. Hureau . (With 2 Figures) 421

The Evolution of Neutrally Buoyant Notothenioid Fishes: Their Specializations and Potential Interactions in the Antarctic Marine Food Web
J.T. Eastman . (With 1 Figure) 430

Krill Consumption by Antarctic Notothenioid Fish
K-H. Kock 437

The Early Life-History of the High-Antarctic Silverfish, *Pleuragramma antarcticum*
G. Hubold . (With 6 Figures) 445

Trophic Relationships Between Pelagic Fish and Euphausiids in Antarctic Waters
R. Williams . (With 2 Figures) 452

Chick Energy Balance in the Whitechinned Petrel, *Procellaria aequinoctialis*
A. Berruti, N.J. Adams, and C.R. Brown . (With 3 Figures) 460

Energy and Food Requirements of Pelagic Aerial Seabirds in Different Regions of the African Sector of the Southern Ocean
R.W. Abrams . (With 3 Figures) 466

Population and Energetic Aspects of the Relationship Between Blackbrowed and Greyheaded Albatrosses and the Southern Ocean Marine Environment
P.A. Prince 473

The Foraging Zones of Seabirds in the Crozet Islands Sector of the Southern Ocean. (With 8 Figures)
J.C. Stahl, P. Jouventin, J.L. Mougin, J.P. Roux, and H. Weimerskirch 478

A Preliminary Comparison of Marine Bird Biomass and Food Consumption Between the Southeastern Bering Sea and Parts of the Southern Ocean
G.L. Hunt, Jr . (With 1 Figure) 487

The Use of Lipid Profiles in Comparing the Diet of Seabirds
I.E. Horgan and J.A. Barrett 493

Biomass of Birds and Mammals in the Ross Sea
D.G. Ainley . (With 12 Figures) 498

Relationships Between Prey Life-Cycles and the Extent, Nature and Timing of Seal and Seabird Predation in the Scotia Sea
J.P. Croxall, P.A. Prince, and C. Ricketts . (With 15 Figures) 516

The Role of Giant Petrels in the Southern Ocean Ecosystem
S. Hunter 534

Diet and Energy Budget of the Antarctic Fur Seal, *Arctocephalus gazella*, at South Georgia
D.W. Doidge and J.P. Croxall . (With 3 Figures) 543

Cephalopod Prey of the Sub-Antarctic Fur Seal, *Arctocephalus tropicalis*, at Gough Island
M.N. Bester and P.A. Laycock 551

The Role of the Leopard Seal in the Tropho-Dynamics of the Antarctic Marine Ecosystem
D.B. Siniff and S. Stone . (With 3 Figures) 555

Weddell Seal–Antarctic Cod Interactions in McMurdo Sound, Antarctica
J.W. Testa, D.B. Siniff, M.J. Ross, and J.D. Winter . (With 4 Figures) 561

A Review of the Food Habits of the Small Cetaceans of the Antarctic and Sub-Antarctic
R.N.P. Goodall and A.R. Galeazzi 566

Energy Flow in the Southern Ocean Food Web
A. Clarke . (With 2 Figures) 573

Part 4. Terrestrial Food Webs

Terrestrial Food Webs in the Sub-Antarctic: Island Effects
A.E. Burger . (With 2 Figures) 582

Wind Transport of Electrostatically Charged Particles and Minute Organisms in Antarctica
W.S. Benninghoff and A.S. Benninghoff 592

Succession in the Microflora of Leaves and Litter of Three Plant Species at Sub-Antarctic Macquarie Island
E.J. Kerry and G.M. Weste . (With 3 Figures) 597

Organization and Dynamics of Oligochaeta and Diptera on Possession Island
P. Tréhen, M. Bouché, Ph. Vernon, and Y. Frenot . (With 7 Figures) 606

Arthropod Interactions in an Antarctic Terrestrial Community
W. Block . (With 4 Figures) 614

The Feeding Ecology of a Decreasing Feral House Cat, *Felis catus*, Population at Marion Island
P.J.J. van Rensburg 620

Part 5. Interactions Between Marine, Freshwater, and Terrestrial Systems

Interactions Between Antarctic and Sub-Antarctic Marine, Freshwater and Terrestrial Organisms
J.C. Hureau . (With 3 Figures) 626

The Intertidal and Shallow Subtidal Food Web at Marion Island
W.O. Blankley and J.R. Grindley. (With 1 Figure) 630

Durvillaea antarctica Production in Relation to Nutrient Cycling at Marion Island
P.G. Haxen and J.R. Grindley 637

Impact of Fur Seals on the Terrestrial Environment at South Georgia
W.N. Bonner. (With 4 Figures) 641

Chlorinated Hydrocarbons in Seabird Eggs from the Southern Atlantic and Indian Oceans
B.D. Gardner, W.R. Siegfried, and A.D. Connell 647

Organochlorines and Heavy Metals in Seals and Birds from the Weddell Sea
R. Schneider, G. Steinhagen-Schneider, and H-E. Drescher 652

Metal Concentrations in Antarctic Zooplankton Species
H.F.-K.O. Hennig, G.A. Eagle, C.D. McQuaid and L.H. Rickett. (With 4 Figures) 656

The Interaction of Soil and Lake Microflora at Signy Island
J.C. Ellis-Evans and D.D. Wynn-Williams. (With 1 Figure) 662

Trends in Crabeater Seal Age at Maturity: An Insight into Antarctic Marine Interactions
J.L. Bengtson and R.M. Laws. (With 3 Figures) 669

Summary and Conclusions

R.M. Laws 676

Subject Index 683

Introduction: Antarctic Biology, 1974–1983

R.M. LAWS[1]

The Antarctic, because of its remoteness, is still relatively unspoiled by human activity, although the nineteenth century sealing and twentieth century whaling industries caused an enormous perturbation of the marine ecosystem from which it is still recovering. Attention has recently turned to fish and krill (Euphausiacea) of which some 500,000 are now taken annually. Fortunately, the Antarctic Treaty nations have created a unique system of international agreements – the Antarctic Treaty itself, the Agreed Measures for the Conservation of Antarctic Flora and Fauna, the Convention for the Conservation of Antarctic Seals and the Convention for the Conservation of Antarctic Marine Living Resources – the last two of which came into force since the last SCAR biology Symposium in 1974. They require scientific advice in order to function and SCAR is well placed to provide this through its Working Group on Biology, its Group of Specialists on Seals and the Group of Specialists on Southern Ocean Ecosystems, which has initiated and co-ordinates the international BIOMASS Programme (Biological Investigations of Marine Antarctic Systems and Stocks).

The Antarctic and Sub-Antarctic together comprise one-eighth of the earth's surface; the area of the Southern Ocean alone south of the Antarctic Convergence is 36 million km^2, a tenth of the World Ocean. In contrast the land area of Antarctica is 14 million km^2, but only a quarter to a third of a million km^2 is ice free and supports life while the surface area of the lakes may be well under 5,000 km^2. The marine, terrestrial and freshwater ecosystems and the interactions between them correspond to the basic subdivisions of the papers in this Symposium.

This series of Symposia has been conceived and planned by the SCAR Working Group on Biology. The third Symposium was held in 1974; in his address G.A. Knox thought it would be the last to encompass the whole spectrum of Antarctic biological research and suggested that future symposia would be directed to more specific questions or more narrowly defined areas of interest.

For the fourth Symposium, Nutrient Cycles, and Food Webs was the theme chosen by the Working Group on Biology. It covers terrestrial, freshwater and marine ecosystems in the Antarctic and Sub-Antarctic regions and includes topics related to the international BIOMASS programme. It recognizes that understanding of processes and relationships is of very great scientific interest as well as being essential for management and conservation.

Nine years have passed since the last Symposium. This is much too long in my opinion – 4–6 years would be a more appropriate time span – and research in this interval has contributed greatly to understanding of the complexities of Antarctic ecosystems and to general theory over a wide range of sub-disciplines. There have also been far-reaching developments in related areas: computing and electronics; data-base management; sampling equipment (sonar, quantitative nets); automatic recording (micro-climates, nest recorders, activity recorders); telemetry (time-depth recorders, radio tracking, satellite imagery); and, increasing sophistication in field work (the application of labelled isotopes for studies of feeding activity, energetics, biochemistry, and hormone analysis, gel chromatography, and so on). Perhaps I may be permitted to give a personal view of developments in the last decade, starting with marine ecology, as the first topic in our programme.

Concerning the marine ecosystems, there has been, first, the development of satellite imagery. Since 1973 the pack-ice distribution has been described, including the identification of the important feature of the Weddell Polynya which was previously unknown. Gyres and eddies have been identified by the deployment of long-term recording current meters. Transport estimates have been published; the first accurate estimate for the Drake Passage flow is $139 \times 10^6\ m^3\ s^{-1}$ to the east, and there has been an active investigation of other current systems.

It is now recognized that nutrients are not limiting in the Antarctic seas as a whole and that wind-induced turbulence and consequent vertical mixing, by preventing phytoplankton staying in the euphotic zone has major effects on the phytoplankton and zooplankton. Remote sensing by the satellite-mounted Coastal Zone Colour Scanner will advance knowledge in this field. In terms of the flora the main advance is the recognition that the Southern Ocean phytoplankton is no more productive per unit area than in any other ocean except the Arctic Ocean. This is in total con-

1 British Antarctic Survey, NERC, High Cross, Madingley Road, Cambridge CB3 OET, England

Antarctic Nutrient Cycles and Food Webs
(ed. by W.R. Siegfried, P.R. Condy, and R.M. Laws)

trast with earlier ideas, but owing to its vast area the Southern Ocean is nonetheless in absolute terms very productive. Fluorimetric methods in vivo now make possible real-time continuous profiling. Other developments are: the recognition of the great importance of bacteria and decomposition processes; studies on the ice-associated algae; and, the investigation of macro-nutrient processes. The micro-nutrients are important, but still very little known – they include vitamins, especially B12, thiamine, biotin, trace elements, and chelating agents.

There have been few advances in our understanding of the marine benthos. However, more data have accumulated on densities and biogeography, there has been discussion of ecological strategies and it seems that the benthos have K-adapted qualities. An exciting development concerns information on life beneath the ice shelves, provided by the Ross Ice Shelf Project (RISP). Here there are low bacterial densities, sparse demersal Crustacea and no in-fauna in the sediments; in short, it is similar to the abyssal conditions elsewhere. RISP showed that basal metabolism of these organisms is low, but it must be stressed that the data refer to only one site. The past decade has seen no published community studies; autecological studies continue, but the major developments in this area concern the interplay between 'classical biology' and biochemistry-physiology. Precise measurements of resting oxygen consumption have led to a reappraisal of cold-adaptation, and we now have for fish some measures of growth efficiencies and production/biomass ratios. It would be nice to think that studies of communities might reappear (especially those in the deeper waters away from the continental shelf), and physiological approaches continue.

In zooplankton studies there have been great advances. The swarming behaviour of the Antarctic Krill, *Euphausia superba,* has attracted study: the diel cycle is now clearer and the size and morphology of swarms has been studied; super-swarms have been identified. The deployment of sonar in FIBEX (First International BIOMASS Experiment) has led to estimates of densities and the FIBEX survey in 4.5 million km^2 indicated a biomass of 78 x 10^6 t. Extrapolation to the overall range of Krill indicates a standing stock of 650 x 10^6 t. Understanding of growth rates of Krill is complicated by the experimental confirmation that starved individuals shrink in size, and the suggestion that this occurs naturally in the winter, influencing conclusions on growth rates and age classes. A further complication is the finding that female Krill undergo successive breeding cycles. Also, it has been established that Krill feeding is quite catholic – diatoms, dinoflagellates, other zooplankton, even cannibalism, and detritus. Very small phytoplankton organisms are very important primary producers and heterotrophic bacteria are important as well as micro-zooplankton. Physiology and biochemistry is providing insight into the zooplankton community, particularly how various organisms cope with a seasonal supply of food. It has been shown for example that *E. superba* lays down little or no wax ester reserves, an indication of year-round feeding, perhaps on detritus in winter; this is in contrast to *E. crystallorophias.* Recent research on lipofuscins (age pigments) as a means of ageing Krill shows great promise for cracking one of the most intransigent problems in marine invertebrate biology, that of ageing individuals.

Advances in the study of Antarctic fish encompass age determination, by scales and otoliths, and the confirmation that growth is slow. Reproduction is characterized by late sexual maturity, up to 5 years or more, large yolky eggs and low fecundity. The early life histories have been shown to fit in with the cycles of primary production and the abundance of copepod larvae as food. Studies on freezing avoidance and the anti-freeze (although how it acts is still a mystery), on metabolic adaptation at enzyme level, and on white-bloodedness have advanced. More results are available on biomass and production by species. There has been extensive progress in our understanding of how fish have evolved to maintain locomotor activity in cold water.

The BIOMASS Programme is important in unravelling marine interactions. It is beginning to put quantities into estimates of production at different levels. The identification of local interactions will give insight into ecosystem functioning and the refinement of models. Conversion efficiencies in the marine system have been shown to be low (1%), as between secondary production and tertiary consumers (seals, whales, birds), in contrast to the usual 10% value. SCUBA observations on Krill swarm behaviour have been enlightening but squid remain an enigma.

Seabird studies have advanced knowledge of pelagic distribution, population dynamics, and trophic relationships. Growth studies have involved sophisticated data gathering and biometrics treatment, and cross-fostering experiments have been enlightening. Growth strategies and the allocation of resources have received attention, and studies of food and foraging behaviour have become more quantitative. The questions are, first: what is eaten? and, secondly: how is food located and caught? To answer them, diets have been correlated with bird biology; for example, the lack of surface food in winter in the Antarctic enforces a northward migration or a change in diet. Age related changes in food preferences have been studied. Successful investigations have been made of foraging range (flight speeds and duration), and on the way in which variation in this characteristic of species might reduce competition.

Population structure and dynamics are better known from monitoring by means of long-term, known-age recaptures, leading also to evidence on sex- and age-specific mortality. Breeding success is age-related, and studied in known-age populations it is much better understood than before.

Energetics studies are dependent on activity patterns. These have established the pre-eminent role of the penguins, so we can assume that all birds in the Antarctic are effectively penguins and two-thirds are Adélie Penguins, *Pygoscelis adeliae.* In the Sub-Antarctic 80% of the birds are penguins and some 50% are Macaroni Penguins, *Eudyptes chrysolophus.*

The seals have also received more attention and they are estimated to consume ca. 130 x 10^6 t of Krill. Highlights of work in the last decade include the studies on Crabeater Seal, *Lobodon carcinophagus,* population dynamics; the establishment of the predator-prey relationship between Crabeaters and Leopard Seals, *Hydrurga leptonyx;* the finding that Leopard Seals, especially in winter, take large amounts of Krill; the demonstration of 4–5-yr cycles in age class strengths in Crabeater and Leopard Seals; and the continued growth of Antarctic Fur Seal, *Arctocephalus gazella,* stocks – soon to overtake the Northern Fur Seal, *Callorhinus ursinus,* as the world's most abundant eared seal. In contrast, there is the alarming demonstration of a large decrease in Indian Ocean Elephant Seal, *Mirounga leonina,* stocks. Other interesting results of fur seal research include the demonstration of density-dependent pup mortality, feeding behaviour related to Krill diurnal vertical migration, and studies of maternal and pup attendance-feeding behaviour.

Understanding of the ecology of whales appears to have made few advances, although some previously accepted concepts have been questioned, while the reduction in field studies has introduced more uncertainties into the understanding of the current status of whale populations.

Turning to freshwater research, the fastest growing areas have been in microbial ecology and in the chemistry and physics of natural waters (with reference to the biota). Up to now inadequate techniques have been limiting – but awareness of the significance of micro-organisms, especially Protozoa and bacteria, in nutrient cycling – and their interactions with larger organisms – is increasing as more appropriate techniques are applied. (This is true also of marine studies). With this greater emphasis on microbes and nutrient cycling, problems such as diffusion rates and the action of chelating agents, have led biologists to seek help from chemists and physicists, who can utilize the large body of their theory and apply it to biological problems. The emphasis is now more on physiological ecology and interactions of groups of organisms – synecology – but the study of organisms in isolation – autecology – is still important. The evolution of Antarctic lakes has been examined, as also the primary production by benthic algal felts.

In the terrestrial field soils have received more attention, particularly the saline soils in Dry Valleys, now dated and related to changes in the ice sheet. Work on Marion Island has been concerned with nutrient cycling, nitrogen fixation, and bacterial numbers. The weathering of rock and periglacial activity studies are worthy of note. Studies on lichen photosynthesis and physiology and cellulose breakdown also deserve mention. Micro-climate studies have received more attention with pioneering work at South Georgia on moss micro-climates, tussock grass, *Festuca* grassland, and *Acaena.* A fellfield study has begun on Signy Island, using solid-state logging equipment and Ambrose balls for measuring micro-habitat temperatures.

Most studies of soil-plant-water chemistry or nutrients have merely provided data to characterize a site with time restricted samples. Only very recently at a few sites have regular samples been taken to investigate seasonal trends. The only major study of nutrients has been on Marion Island, involving detailed investigations of primary and secondary production and the biogeochemical cycling of nutrients with particular reference to the interactions between birds and seals and the vegetation and primary production. Nutrient cycling on a seasonal basis has also been studied at Signy Island, but no relevant investigations have yet been undertaken on the Antarctic continent.

In the terrestrial environments micro-biological studies have been slow to develop – largely owing to the cryptic nature of microbes and the technical problems of Antarctic field work. An exciting finding is the occurrence of endolithic microbes (lichens, fungi, bacteria, and algae). It has been realized that the general absence of large above-ground herbivores has had a profound effect on the structure and functioning of terrestrial ecosystems. This shifts the emphasis of energy flux and nutrient cycling onto smaller organisms in the soil system, which is essentially a decomposer system with a few micro-predators.

The fauna has been studied in increasing detail – in terms of densities, life cycles, behaviour, biochemistry and physiology, and energetics. Aspects of freezing susceptibility and freezing tolerance have been explored – also respiration (as part of the energetic balance), supercooling and cryoprotectants. Finally, ecosystem studies at Marion Island and Signy Island have led to the production of models including the net primary production in quantitative terms – and calculation of ecological efficiencies.

These are personal views which it seemed worthwhile to mention in order to set the scene for the sections to come. It is intended that papers and discussions will: identify deficiencies in knowledge and evaluate problems in each field; consider multi-disciplinary approaches to the solution of these problems; and, identify future priorities for research.

The contributions which follow indicate in much greater detail what great progress has been made overall since 1974 and demonstrate that Antarctic biology is a thriving field of research.

Part 1
Marine Nutrient Cycles

Nutrient Cycles in Antarctic Marine Ecosystems

O. HOLM-HANSEN[1]

Summary. Studies of mineral cycling in the marine environment can be very informative in regard to understanding food web dynamics and general functioning of the ecosystem. Although emphasis is placed on the essential macro- and micro-elements, other mineral elements are also valuable tracers for various processes involving chemical transformations. Most chemical elements show a unique distribution pattern in the water column which reflects their biological and chemical reactivities and the rates at which they are resolubilized. Antarctic waters can be visualized as a giant chemostat, with nutrient rich water upwelling at the Divergence at high latitudes and the water ultimately downwelling at the Polar Front or the Sub-Tropical Convergence. In comparison with other major upwelling areas of the world, one would expect to find almost complete stripping of plant nutrients with time in the upper 50–100 m due to growth of phytoplankton. This condition of nutrient depletion would also be suggested by abundant data on rates of primary production in all sections of the Southern Ocean. Nitrate and phosphate, however, show only slight decreases in concentration in sections from close to the continent to close to the Sub-Tropical Convergence. This apparent contradiction has been explained by recent data which show that N is re-cycled 6–7 times in the euphotic zone before it settles out to deeper water as particulate organic N. These findings have focused attention on the nature of the food web which results in such very high rates of nutrient re-cycling in surface waters. Microbial populations seem to be the dominant organisms responsible for these transformations. With the exception of studies dealing with the distribution of silicon and related elements in the water column and sediments, little attention has been paid to questions of mineral cycling in Antarctic waters. Our understanding of the interaction between biological-physical-geochemical processes in the Antarctic is thus meagre in comparison with temperate and tropical waters.

1 Introduction

If one looks back at the previous SCAR (Scientific Committee on Antarctic Research) symposia on Antarctic biology, it is apparent that the "themes" are usually somewhat premature in that they deal with research problems for which there is not a large foundation of data. This is advantageous, however, in that they help to define important problems and to foster imaginative research in subsequent years. I suspect that this will also be true for the subject of marine nutrient cycles, which is one of the two major themes of this symposium.

What is meant by "nutrient cycles"? I interpret it to mean the sum total of all chemical transformations (both biotic and abiotic) of any element or compound through time and space. Although many biological oceanographers generally consider only N, P, and Si in regard to nutrient cycles, it is important to realize that the subject encompasses all the chemical elements, whether or not they are essential for biological systems. The chemical elements which are known to be essential, at least for some plant or animal life, and hence are of primary concern to us include the "macro" elements (C, H, O, N, S, P, Si, Mg, K, Ca, and Fe) and the "micro" elements (Mo, Mn, Co, Zn, Cu, V, B, Cl, and Na). There are data to suggest that some of the following elements may someday be included with the "essential" elements: F, Br, Cd, I, Ni, Se, and Sn. Considering the dearth of our knowledge concerning physiological requirements for trace elements, it seems appropriate in this symposium to consider all chemical elements which can tell us something about the functioning of the Antarctic ecosystem.

2 Nutrients in Antarctic Waters

If the oceans were devoid of biological activity, the chemcal composition of sea-salt would be almost uniform for nearly all elements. This is generally true for most of the major ions (Cl^-, SO_4^{-2}, Mg^+, K^+, Ca^{+2}, Na^+), but is not true for many of the minor constituents which include both essential and non-essential elements. These elements usually have distinctive distribution patterns which depend upon input source(s) and rates, physical mixing processes, and changes wrought by biological and chemical reactions. Thus, the distribution of such elements can be used as "chemical tracers" to tell us something about processes involving mixing, biological activity, and sedimentation rates. Distribution patterns of chemical elements in Antarctic waters are generally much less "structured" than at lower latitudes, in large part due to the dominant upwelling in

1 Scripps Institution of Oceanography, University of California, La Jolla, CA 92093, USA

Antarctic Nutrient Cycles and Food Webs
(ed. by W.R. Siegfried, P.R. Condy, and R.M. Laws)

Antarctic waters, the relative instability of the water column due to the temperature regime, and the relatively slow metabolic rates due to thermodynamic effects on enzyme activity rates. I do not know of a detailed study of any chemical element in Antarctic waters which would permit conclusions to be drawn regarding the rates of biological or physical processes which affect the distribution of that element. Much of the geochemical data known from Antarctic seas have been discussed recently by Broecker and Penge (1982). It is informative, however, to look at examples from other oceans to illustrate the kind of information which can be deduced from such studies. I will cite just a few examples. The distribution of Mn and Pb with depth shows highest concentration in surface waters, and relatively low values in deep water, indicating that the main input for these elements is from atmospheric fall-out (Schaule and Patterson 1981; Landing and Bruland 1980). In contrast to these profiles, the distribution patterns of Zn, Cd, Ni, and Cu all show relative depletion in surface waters and much higher concentrations in deep water. The rate of change in concentration with depth of these elements is very different, however, so that each element will usually show a "unique" or distinctive concentration profile. Cd, for instance, resembles the profiles of some plant nutrients (e.g., N and P) in having a maximal concentration at about 1 km, and decreasing concentrations below that depth. Cu, on the other hand, does not show such a maximum at 1 km, but instead the concentration often continues to increase down to 4–5 km depth. Other elements, such as Zn and Ni, show profiles somewhere in between those of Cd and Cu (Bruland 1980). These distinctive distribution patterns reflect differences in the chemical and biological reactivities of each element, including the rate at which they are remineralized throughout the water column.

3 Nutrient Cycles in Antarctic Waters

When considering nutrient cycling in a symposium such as this, I think one should include all waters south of the Sub-Tropical Convergence (S.T.C.) instead of the Antarctic Convergence (A.C.) My reasons for this are: when examining nutrient gradients in transects from tropical waters to the Antarctic continent, the most conspicuous change is often at the S.T.C., not at the A.C.; phytoplankton biomass also generally shows a more pronounced gradient at the S.T.C., than at the A.C.; and, many species of plankton are distributed both north and south of the A.C., with the result that the S.T.C. often represents the more distinct demarcation between warm-water and cold-water organisms.

I will generalize (and over-simplify) a great deal in an effort to introduce and discuss problems of nutrient cycling in Antarctic waters. In this context, one can view the Southern Ocean as a giant chemostat, with the following major components: injection and upwelling of nutrient-rich north Atlantic water, predominantly at the Antarctic Divergence (A.D.) which separates the West Wind Drift from the coastal East Wind Drift system (the waters of the West Wind Drift flow east-northeast, before ultimately submerging at the Antarctic Convergence and mixing with southward flowing water to form the Sub-Antarctic water); a corresponding "export" of Antarctic bottom water, which forms over relatively shallow areas, and flows north to mix with waters of the three major ocean basins; solar energy input, which drives primary production in addition to playing an important role in the overall heat budget; atmosphere-surface water exchange in regard to heat, CO_2, particulates, etc; and, cycling and transformations of elements and organic compounds throughout the water column, including some geological deposition.

Although the Antarctic Ocean is characterized by extensive upwelling, the interactions between the physical mixing processes and the biological processes are not nearly as well understood as in other upwelling areas. If one looks at the upwelling systems off Peru or South Africa (Barlow 1982; Andrews and Hutchings 1980), one can make some generalizations regarding nutrient transformations. As the newly upwelled water nears the surface and then flows horizontally, one finds that: the nutrients, such as nitrate, will initially be high (about 25 μM) and in time reach concentrations close to zero; phytoplankton biomass will be very low (< 0.5 μg chl-a l^{-1}) close to the source of upwelling, reach a maximum (often close to 25 μg chl-a l^{-1}), and then decrease with time as "grazing" pressure increases; and, that there is a concomitant elevation of water temperature during the same time period.

I would now like to try to relate such a series of events to the Antarctic ecosystem. The entire West Wind Drift zone may be likened to these upwelling plumes, where the waters close to the Antarctic Divergence represent the nutrient-rich waters, and the "aged" waters are found close to the A.C. Assuming an initial phytoplankton biomass of about 0.6 μg chl-a l^{-1}, a photosynthetic rate of 1.5 μg C/μg chl-a h^{-1}, a cellular C:N ratio of 5.0, 12 h of light d^{-1}, and a flow rate of 100 mm s^{-1}, one would expect that between 25 und 30 μ moles of nitrate l^{-1} would be consumed by the time that the upwelled water reached the zone of the A.C. (approximately 200 d in transit). Thus, one would expect to find almost complete stripping of nutrients in the aged water close to the A.C., a condition that is found for upwelling areas in other parts of the world. What is actually found when phytoplankton biomass and nutrient concentrations are measured in transects which extend from the A.D. to Sub-Antarctic water? Representative data are furnished by the recent work of Kuramoto and Koyama (1982) and Watanabe and Nakajima (1982), who made a long transect along 45°E longitude. Nitrate was 27 μM at 60°S, 22 μM at the A.C. (52°S), 20 μM a few degrees south of the S.T.C. (41°S), and < 1 μM at 39°S. Chl-a concentrations were irregular (between 0.1 and 0.5 μg l^{-1}) throughout the transect, with concentrations in Sub-Antarctic waters being about the same as those south of the A.C. Silicic acid, however, showed an interesting distribution pattern as it was high (57 μM) at 67° S, decreased sharply to 15 μM at 60°, and dropped to low values (approximately 1.0 μM) just north of the A.C. This unique distribution pattern for silicon, which is very different from most of the other plant nutrients, reflects its assimilation and polymerization by diatoms and silicoflagellates (both groups being abundant in Antarctic waters). This is combined with the relatively slow rate of regeneration of soluble silicic acid during downward transport of the particulate material in the water column.

Thus it is seen that the actual distribution of nutrients and plant biomass in the West Wind Drift is extremely different from the simplified model of upwelling plumes mentioned above. One can establish the "growth potential" of Antarctic waters by taking a jar of surface water anywhere south of the A.C. and incubating it on the deck of the ship under appropriate conditions of temperature and light. We have done this on many occasions, and found that within 2–3 weeks the chlorophyll-a concentrations increase to 25–50 μg l^{-1}, and nitrate concentrations approach zero. What, then, are the major processes which result in such a relatively small decrease in plant nutrients (e.g., nitrate) in Antarctic waters, and also in maintaining phytoplankton biomass at levels of only a few per cent of growth potential of the water? The answer must involve a combination of physical and biological processes. First, physical mixing processes will be of much importance in regard to: advection and eddy diffusion which will cause a continual injection of nutrients into the euphotic zone; and, the relative instability of the water column, which results in a loss of phytoplankton cells through settling into the aphotic zone. Secondly, the rate of cropping of phytoplankton biomass by herbivorous zooplankton has received relatively little attention in Antarctic waters. The magnitude of N recycling in the euphotic zone has been clearly demonstrated in recent studies utilizing ^{15}N-enriched substrates (nitrite, nitrate, ammonia, and urea). These N metabolism studies, which use the methodology described by Dugdale and Goering (1967), have been done by Slawyk (1979), Olson (1980), Glibert et al. (1982), Koike et al. (1981), and Rönner et al. (1983) in diverse areas of the Antarctic Ocean. These data conclude that: between 50 and 80% of all N assimilated by Antarctic phytoplankton is in the form of ammonia; and, the nanoplankton (< 20 μm in diameter) show a relatively higher rate of uptake of ammonia, as compared to nitrate, than do the micro-plankton (> 20 μM), which are predominantly diatoms.

What is the source of all this ammonia in Antarctic waters? Ammonia is not detectable in deep water (Biggs and Guffy, unpublished) and, hence, upwelling is of no consequence in this regard. Very small amounts originate from the atmosphere or from the melting of ice. It appears that nearly all the ammonia must originate from in situ processes occurring within the euphotic zone. The data of Biggs (1982) suggest that a maximum of 2% of the ammonia flux can be accounted for by grazing effects by net zooplankton (organisms > 200 μm), and so it appears that over 90% of the ammonia results from grazing by microbial organisms (< 200 μm). This conclusion has great import in regard to the structure and dynamics of the food web in Antarctic waters. Hitherto the prevailing view of the Antarctic "food chain" has been that of a fairly simple progression from micro-plankton-sized diatoms to krill (Euphausiacea) to whales, birds, seals, etc. This view is apparently in error. In order to have ammonia account for 75% of the total N assimilated by phytoplankton, 75% of primary production must be mineralized within the euphotic zone to maintain the observed ammonia assimilation rates. This has considerable significance in regard to the amount of primary production which is available for consumption by krill and higher trophic levels. As the nitrate-based primary production (approximately 25% of primary production) can be equated to the flux of photosynthate which can be "exported" to deeper waters (Eppley and Peterson 1979), this also limits the amount of organic C available for demersal and benthic organisms. This view of the Antarctic food web requires a re-orientation of our thinking in regard to modelling of the overall food web in Antarctic seas, to aid in resource evaluation and management.

These high rates of N mineralization within the euphotic zone can be extrapolated to all other chemical elements which react with living organisms. As mentioned previously, however, the rate of mineralization will be unique to each element. Certain elements, such as N, Mn, and P, can be expected to be mineralized rapidly, whereas others (e.g., Si, Fe, Ca, Cu) will have slower rates of solubilization. Theoretically, each chemical element thus would have a distinctive distribution pattern throughout the water column. The chances of documenting such detailed profiles for a variety of elements are maximized in highly stratified waters as found in tropical areas. This also results in primary production being limited by multiple nutrient deficiencies (e.g., N, P, Si, Fe). In the Antarctic, however, the relative instability of the water column will tend to minimize concentration gradients with depth, and thus make it more difficult to interpret distribution patterns for various elements as a function of rates of biological processes. Nutrient profiles in Antarctic waters generally show some loss of nitrate (from about 30 to 20 μM) and of reactive phosphate (from about 2.0 to 1.0 μM) in surface waters, but these lower concentrations are still in excess of those concentrations which would limit the rate of photosynthesis. Considerable attention has been given to the distribution of Si, which is commonly 10–20 μM near the surface and close to 100 μM toward the bottom of the euphotic zone. As noted previously, there is a much greater loss of Si from surface waters as compared to losses of nitrate or phosphate. There is a possibility that low Si concentrations may thus limit the biomass of diatoms or silicoflagellates in some areas of the

Antarctic seas (first discussed by Hart (1934)). What is the fate of this Si which is lost from the euphotic zone? DeMaster (1981) claims that the largest deposits of siliceous ooze in the world are found in Antarctic bottom sediments, which receive 60–80% of all the biogenic-Si deposited on the world's sea-floor. Such a large fraction of the total Si flux occurring in a relatively small fraction of the world's ocean is indicative of the unique features of the physical and biological regimes in the Antarctic. It could be expected that other elements which are also incorporated into diatoms (e.g., Ra, Ba, Ge) would show similarities to the distribution pattern shown by Si.

There has been very little work in the Antarctic where one attempts to relate the Redfield ratios of particulate organic material to the distribution of nutrients in the euphotic zone, and to analyze the results with measured rates of primary production, grazing, and physical mixing processes. During the *Vulcan* cruise (Holm-Hansen and Foster 1981) in the Scotia-Weddell seas in 1981 extensive data sets on these chemical and biological parameters were obtained on five north-south sections. These data are being analyzed at present.

It was seen above that our understanding of the Antarctic food web and of rates of nutrient regeneration were considerably advanced by utilization of ^{15}N-labelled substrates which permit one to determine individual rates of assimilation for different chemical species of the same element, as well as the rates of regeneration of that element by isotope dilution techniques. Two other isotopes (^{32}P and ^{30}Si) have been used for studies on rates of assimilation and regeneration in other oceans (Harrison 1980; Nelson and Goering 1977), but I am not aware of any published data using these isotopes in Antarctic waters.

Most studies on nutrient cycling in Antarctic waters have involved various components of the C cycle. With a few exceptions, these have dealt primarily with gross aspects of primary production, biomass and feeding preferences of net zooplankton and distribution of higher trophic levels. There has been little attention paid to the following aspects of the C cycle which are important in understanding the flow of energy and reduced C in Antarctic waters. Studies on inorganic C distribution and activity have been vigorously pursued in temperate and tropical waters, but have received almost no attention in the Antarctic. This includes the rate of exchange of CO_2 between the atmosphere and surface water, the rate of transfer of CO_2 from the upper mixed layer to deeper waters, activity of the inorganic C species with regard to various carboxylating reactions, and carbonate distribution relative to calcite and aragonite concentrations. The photo-autotrophic reduction of CO_2 by phytoplankton, benthic algae, and ice-associated algae will receive considerable attention during this symposium, but the distribution and activity of chemo-autotrophic organisms has scarcely been studied in the Antarctic. Of particular interest here are the nitrifying bacteria, which may play a significant role in N transformations throughout the entire water column, but particularly in the euphotic zone where ammonia concentrations often exceed 1.0 μM. Except for recent studies with sediment traps suspended in the water column, there is little information on the magnitude of the organic C flux to deeper water and to the sediments. The relative importance of physical and biological processes in this export of reduced C to the aphotic zone has not been ascertained.

The distribution and metabolic activity of heterotrophic microbial organisms is very poorly understood, in spite of a number of classical studies of bacteria in Antarctic waters. As new techniques and methodologies are now being applied to this problem, I am hopeful that our knowledge in this area will rapidly increase in the near future. This is particularly true for studying the chemical transformation undergone by organic matter during its downward transport in the water column, as well as its fate after deposition in the sediments. Recent studies demonstrate convincingly that well over 50% of phytoplankton biomass and primary production is accounted for by the nanoplankton populations (Hewes et al. 1983; Bröckel 1981). This is of much significance in regard to the complexity of the microbial food web, with concomitant implications in regard to rates of nutrient regeneration. The distribution and specific composition of dissolved organic C, as well as its importance in growth of heterotrophic organisms, is also poorly understood. It has been postulated that a significant amount of primary production in the Antarctic is lost to deeper waters and transported to tropical waters, where it supports high secondary production (Sorokin 1971). There is, however, little or no direct evidence to support this suggestion, which would have much impact on nutrient cycling and geochemistry in Antarctic waters.

The last subject I wish to comment on concerns the heterotrophic microbial organisms which are mentioned above in regard to ammonia production and to chemical transformations of organic material throughout the water column. Assuming that the ammonia results from grazing and subsequent excretion (of ammonia, urea, organic N compounds, etc.), the grazers must be largely restricted to the euphotic zone, or at least their excretory products must be predominantly limited to euphotic zone depths. As krill swarms commonly extend town to 200 m (Mathisen and Macaulay 1983), it is unlikely that krill are the major source of the ammonia wich drives primary production. It appears from studies of the past few years that nanoplankton-sized organisms account for much of this ammonia liberation. On-going research is concerned with trying to determine which components of the nano- or pico-plankton populations are most active in this generation of ammonia. Although bacterio-plankton historically have been assumed to be very important in such nutrient cycling, some researchers (Azam et al. 1983) question this assumption and in fact suggest that bacteria actually compete with autotrophic cells for inorganic nutrients such as P and N.

4 Conclusions

It is thus seen that our data and understanding of nutrient cycling in Antarctic waters are meagre, as compared to temperate or tropical waters. The Antarctic marine ecosystem, however, presents a unique environment in regard to many aspects of biological and geochemical oceanography, and hence much will be learned from the application of new techniques and methodologies. Hopefully, the scientific impetus afforded by bringing together at this symposium so many of the world's researchers interested in Antarctic waters will elucidate some of the questions I have discussed, and help to define the problems and the research needed to better understand the inter-relationships between biological processes and nutrient cycling.

References

Andrews WRH, Hutchings L (1980) Upwelling in the Southern Benguela current. Prog Oceanogr 9:1–81

Azam F, Fenchel T, Field JG, Gray JS, Meyer-Reil LA, Thingstad F (1983) The ecological role of water-column microbes in the sea. Mar Ecol Prog Ser 10:257–263

Barlow RG (1982) Phytoplankton ecology in the Southern Benguela Current. III. Dynamics of a bloom. J Exp Mar Biol Ecol 63: 239–248

Biggs DC (1982) Zooplankton excretion and NH_4^+ cycling in near surface waters of the Southern Ocean. I Ross Sea, Austral Summer 1977–1978. Polar Biol 1:55–67

Bröckel K von (1981) The importance of nanoplankton within the pelagic Antarctic ecosystem. Kiel Meeresforsch 5:61–67

Broecker WS, Penge TH (1982) Tracers in the sea. Lamont-Doherty Geological Observatory, Columbia University, Palisades, NY, 690 pp

Bruland KW (1980) Oceanographic distributions of cadmium, zinc, nickel, and copper in the North Pacific. Earth Planet Sci Lett 47:176–198

DeMaster DJ (1981) The supply and accumulation of silica in the marine environment. Geochim Cosmochim Acta 45:1715–1732

Dugdale RC, Goering JJ (1967) Uptake of new and regenerated nitrogen in primary productivity. Limnol Oceanogr 12:196–206

Eppley RW, Peterson BJ (1979) The flux of particulate matter to the deep ocean. Nature 282:677–680

Glibert PM, Biggs DC, McCarthy JJ (1982) Utilization of ammonium and nitrate during austral summer in the Scotia Sea. Deep-Sea Res 29:837–850

Harrison WG (1980) Nutrient regeneration and primary production in the sea. In: Falkowski PG (ed) Primary productivity in the sea. Plenum NY, pp 433–460

Hart TJ (1934) On the phytoplankton of the South-west Atlantic and the Bellingshausen Sea, 1929–31. Discovery Rep 8:1–268

Hewes CD, Holm-Hansen O, Sakshaug E (1983) Nanoplankton and microplankton studies during the circumnavigation cruise. Ant J US 18:169–171

Holm-Hansen O, Foster TD (1981) A multidisciplinary study of the eastern Scotia Sea. Ant J US 16:159–160

Koike I, Rönner U, Holm-Hansen O (1981) Microbial nitrogen metabolism in the Scotia Sea. Ant J US 16:165–166

Kuramoto S, Koyama K (1982) Preliminary report of the oceanographic observations in the 22nd Japanese Antarctic Research Expedition (1980–1981). Mem Natl Inst Polar Res Spec Issue 23:5–12

Landing WM, Bruland KW (1980) Manganese in the North Pacific. Earth Planet Sci Lett 49:45–56

Mathisen OA, Macaulay MC (1983) The morphological features of a super swarm of krill, *Euphausia superba* In: Nemoto T, Matsuda T (eds) Proceedings of the Biomass Colloquium in 1982, Tokyo, Tokyo, pp 153–164

Nelson DM, Goering JJ (1977) Near-surface silica dissolution in the upwelling region off northwest Africa. Deep-See Res 24:65–73

Olson RJ (1980) Nitrate and ammonium uptake in Antarctic waters. Limnol Oceanogr 25:1064–1074

Rönner U, Sörennson F, Holm-Hansen O (1983) Nitrogen assimilation by phytoplankton in the Scotia Sea. Polar Biol 2:137–147

Schaule BK, Patterson CC (1981) Lead concentrations in the northeast Pacific: Evidence for global anthropogenic perturbations. Earth Planet Sci Lett 54:97–116

Slawyk G (1979) ^{13}C and ^{15}N uptake by phytoplankton in the Antarctic upwelling area; Results from the *Antiprod I* cruise in the Indian Ocean sector. Aust J Mar Freshwater Res 30:431–448

Sorokin Yu I (1971) On the role of bacteria in the productivity of tropical oceanic waters. Int Rev Ges Hydrobiol 56:1–48

Watanabe K, Nakajima Y (1982) Vertical distribution of chlorophyll *a* along 45°E in the Southern Ocean. Mem Natl Inst Polar Res Spec Issue 23:73–86

Oceanic Frontal Systems and Biological Enhancement

J.R.E. LUTJEHARMS[1], N.M. WALTERS[2], and B.R. ALLANSON[3]

Summary. On a number of recent cruises between Africa and Antarctica closely-spaced measurements were made of hydrographic variables, as well as of primary productivity and chlorophyll-*a* concentration. We describe the occurrence of high levels of chlorophyll-*a* at certain specific locations, associated primarily with sea-surface fronts wherever such frontal systems have the characteristics of a convergence. Information gathered by the Coastal Zone Colour Scanner of the Nimbus-7 satellite shows a clear relationship between the horizontal shear edges of currents forming part of the southern Agulhas Current system and biological enhancement. Various possible physical mechanisms which may be responsible for biological enhancement at fronts with different characteristics are discussed.

1 Introduction

Since the publication of the seminal work on the physical characteristics of the Southern Ocean by Deacon (1937a, b), the primary importance of frontal systems in this ocean has been recognized. These fronts are also of importance to the biota of the area, since they represent jumps in the values of certain physical and chemical variables which determine species composition and productivity. Typical plankton communities are thus recognized for each individual zone between fronts (Hedgpeth 1969; Rakusa-Suszczewski 1978), dependent on the environmental factors prevalent in that zone (Holm-Hansen et al. 1977). These fronts are not, however, completely impervious and species may penetrate them in certain circumstances (Grindley and Lane 1979; Voronina 1962).

It is coming to be recognized that frontal systems do not act only as physical boundaries to biological communities but that they themselves may be areas of enhanced biological activity. For instance, Pingree and Mardell (1981) and Pingree et al. (1982) have studied biological enhancement at tidal fronts in the shelf seas around the British Isles, and other workers have investigated increases in the concentrations of surface chlorophyll-*a* at shelf-break fronts (Marra et al. 1982). Biological activity at fronts has furthermore been shown not to be restricted to lower trophic levels. Atkinson and Targett (1983) have demonstrated a relationship between fish distribution and upwelling along a front, and Ainley and Jacobs (1981) have reported affinities between seabirds and certain fronts in the Southern Ocean. Similar relationships have also been described by Abrams (to be publ.).

Less attention has, on the whole, been given to open-ocean fronts, particularly in the Southern Ocean, but Plancke (1977) has stated that increases in biomass and productivity occur at the northern edge of the Sub-Tropical Convergence area, and similar increases were found at the Antarctic Polar Front (Jacques and Minas 1981; Pomazanova 1980). On the basis of a preliminary study, using closely spaced stations between Africa and Antarctica, Allanson et al. (1981) demonstrated a correspondence between the geographical location of sea-surface fronts and high values in primary productivity. Thus, although not completely established, it would seem probable that sea-surface fronts play an important role in localized biological productivity.

2 Fronts in the Southern Ocean

The current perception of the frontal systems between South Africa and Antarctica is set out in Fig. 1. Proceeding southward from Africa, the first persistent frontal system is the Sub-Tropical Convergence. This front froms the boundary between Sub-Tropical Surface Water, thermally enhanced by Agulhas Current Water and Sub-Antarctic Surface Water. At the sea surface it is recognisable by an average decrease in temperature of 7.3°C (17.9–10.6°C) (Lutjeharms and Valentine, to be publ.). Average increases in reactive nitrate and phosphate of 8.7 μg at. 1^{-1} and 0.60 μg at. 1^{-1}, respectively, have also been observed (Allanson et al. 1981). The mean latitude of this frontal system is 41° 40′ S.

Embedded in the Sub-Antarctic Surface Water is the Sub-Antarctic Front. This front has also been observed south of Australia (Zillman 1970) and in the Drake Passage (Sievers and Emery 1978) and is defined as a sub-surface temperature gradient lying between the 3° and 5°C iso-

1 National Research Institute for Oceanology, CSIR, P.O. Box 320, Stellenbosch 7600, South Africa

2 National Physical Research Laboratory, CSIR, P.O. Box 395, Pretoria 0001, South Africa

3 Department of Zoology and Entomology, Rhodes University, P.O. Box 94, Grahamstown 6140, South Africa

Antarctic Nutrient Cycles and Food Webs (ed. by W.R. Siegfried, P.R. Condy, and R.M. Laws)

therms (Lutjeharms and Emery, to be publ.). No physical hypothesis has as yet been advanced to explain its existence. At the sea surface it may be recognized by an average drop of 3.9°C at a mean latitude of 46° 23′ S (Lutjeharms and Valentine, to be publ.).

South of Africa, the surface expression of the Antarctic Polar Front is found at an average latitude of 50° 18′ S. The sub-surface expression of this front is generally found about 50 km north of the surface expression (Lutjeharms and Valentine, to be publ.). Although generally considered to consist of a confluence of Antarctic Surface Water and Sub-Antarctic Surface Water (Fig. 1), the front's surface characteristics may depend on the variable wind stress and may consist of a combination of convergences and divergences (Wyrtki 1960; Ostapoff 1962, 1963).

The Antarctic Divergence, which is the presumed border between the Antarctic Circumpolar Current and the East Wind Drift, is not always clearly seen south of Africa. An upward bend in the isotherms at about 65° S (Lutjeharms and Emery, to be publ.) may represent this front or alternatively it may be interpreted as the southern inflow of the Weddell Gyre (A.L. Gordon et al. 1982). Close to the Antarctic continent, the Continental Water Boundary (not shown in Fig. 1) is usually observed by a decrease of about 2°C at 200 m (Lutjeharms et al. 1983).

3 Sea-Surface Temperature and Chlorophyll-a

Since 1981, a number of cruises in the Southern Ocean sector south of South Africa have been carried out with the express purpose of establishing relationships between physical fronts and peaks in primary productivity. Details of instrumentation and methods are given in Allanson et al. (1981). During one of these cruises (Fig. 2), the thermal characteristics of the upper 500 m were established by regular expendable bathythermograph (XBT) measurements. From this information, and from the sea-surface temperature distribution, the locations are shown (Fig. 3) of the five major fronts defined above. Complementary evidence is available from the distribution of nutrients (Fig. 3). A sharp increase in nitrate and also in phosphate (not shown in Fig. 3) occurs at the Sub-Tropical Convergence and an even sharper increase in silica defines the location of the Antarctic Polar Front.

Three major peaks in chlorophyll-a concentration were observed on this cruise track: at the Continental Water

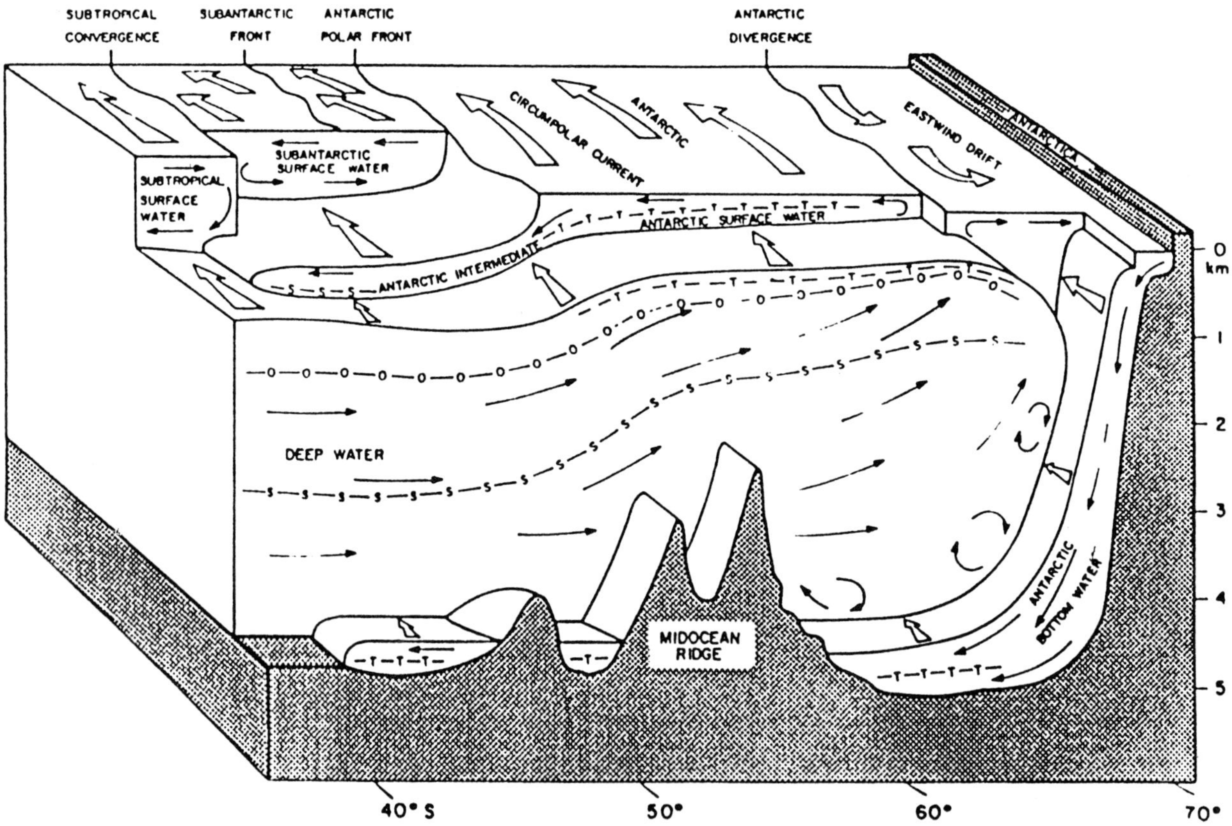

Fig. 1. Schematic presentation of water types and their movements, frontal zones and hydrographic variable maxima and minima in the Southern Ocean sector south of Africa. *Open arrows* indicate lateral movement; *solid arrows* inferred vertical movement. In the Deep Water the temperature maximum, the oxygen minimum and the salinity maximum layers are indicated by the *broken, lettered lines T, O,* and *S*, respectively. The temperature minimum of the Antarctic Surface Water and the salinity minimum layer of the Sub-Antarctic Intermediate Water are also shown by *lettered, broken lines*

Boundary, at the Antarctic Polar Front and directly to the north of the Sub-Tropical Convergence. Smaller increases which might be peaks were found at the Sub-Antarctic Front and at Bouvet Island, near 54° S, although their significance is open to question. An increase in chlorophyll-a over a wide latitudinal zone was also observed between 59° and 68° S, concomitant with a measurable reduction in nitrate and silica values (Fig. 3).

A second set of measurements along essentially the same track as above was made during 16–24 December 1982, but directed southward (Fig. 4). No XBT measurements were undertaken during this cruise, but regular sea-surface temperature measurements were used in conjunction with data in Lutjeharms and Valentine (to be publ.) to define with confidence the locations of the three major fronts (Fig. 5, page 15). The latitude of the Antarctic Polar Front on this occasion was nearly coincident with the sharpest slope in the silica increase, but this increase was spread over a relatively extensive latitudinal band. The distribution of nitrate (Fig. 5) indicates that the area to the south of the

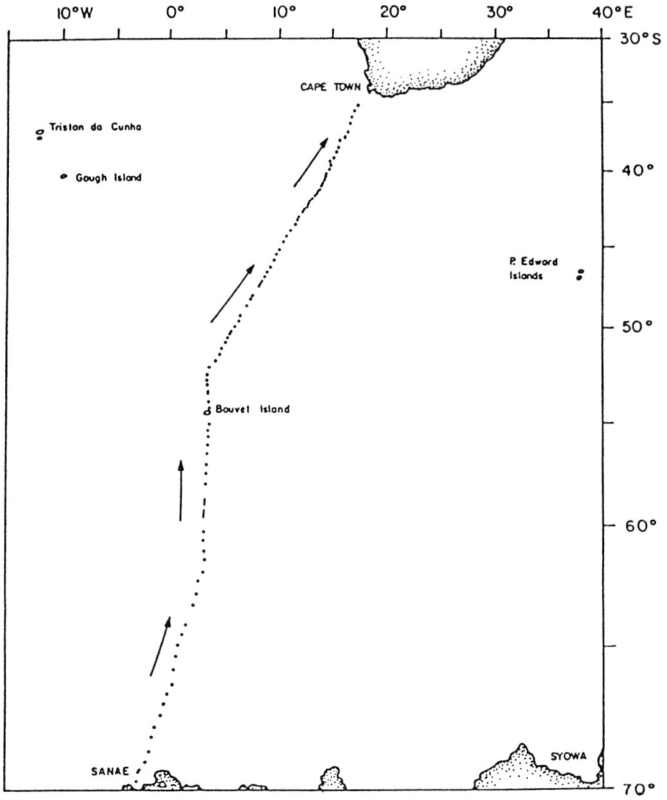

Fig. 2. Cruise track and chlorophyll-a stations operated by the M.V. *S A Agulhas* between the South African Antarctic base (SANAE) and Cape Town during January 1981

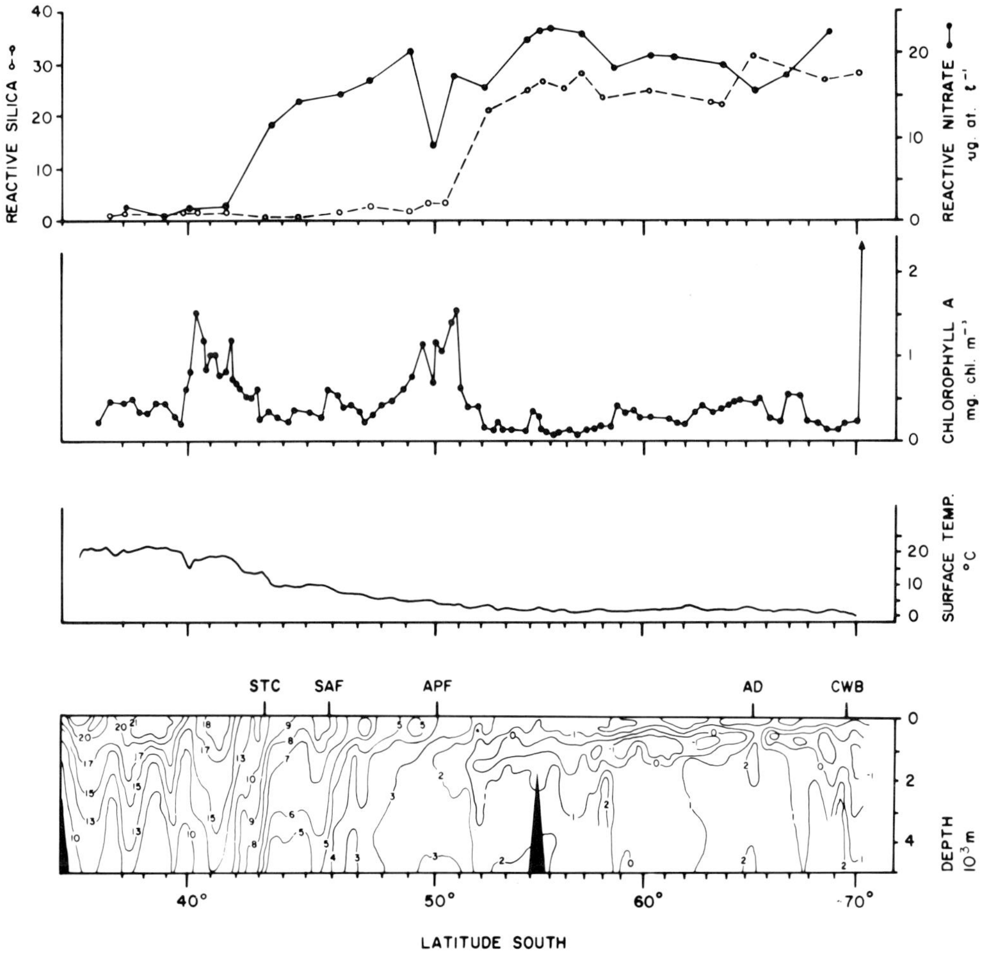

Fig. 3. Distribution of reactive silicate and reactive nitrate (*upper panel*), chlorophyll-a concentration (*second panel*) and temperature (*third panel*) at the sea surface along the cruise track shown in Fig. 2. The *bottom panel* portrays the concurrent isotherm distribution in the upper 500 m (after Allanson et al. 1981). The inferred locations of the Sub-Tropical Convergence (*STC*), the Sub-Antarctic Front (*SAF*), the Antarctic Polar Front (*APF*) the Antarctic Divergence (*AD*) and the Continental Water Boundary (*CWB*) are indicated

Sub-Tropical Convergence may be one where mixing occurs between Sub-Tropical Surface Water poor in nitrates and Sub-Antarctic Surface Water.

The distribution of chlorophyll-a during the 1982 cruise showed many of the characteristics observed in 1981. An elevation in chlorophyll-a values stretched in a band from 60° to 65° S. A significant peak also occurred just north of the Sub-Tropical Convergence, whereas peaks of lesser magnitude could be associated with the Sub-Antarctic Front and the border of the Agulhas Current in the extreme north. The strong peak between the Sub-Tropical Convergence and the Sub-Antarctic Front may be a function of the mixing across the Sub-Tropical Convergence, indicated by the distribution of nitrates (Fig. 5). The zone of greater concentration of chlorophyll-a south of 60° was again associated with a measurable reduction in nutrients.

From the above, and from knowledge obtained through a number of additional cruises that we have carried out, as well as from information in the literature, we assume with confidence that some of the features observed in the chlorophyll-a distribution are consistent ones which occur for significant periods. However, it has been shown (Lutjeharms et al. 1981; El-Sayed et al. 1979) that the frontal systems in this part of the Southern Ocean are subject to major changes and, in fact, the Sub-Tropical Convergence south of Africa has been described as one of dramatic dynamic variability (Lutjeharms and Baker 1980). It would thus be asking too much to expect features of the chlorophyll-a distribution to be present permanently. In this context, Voronina (1962) has shown how changes in even meteorological conditions may influence plankton distribution at the Antarctic Polar Front.

There are a number of possible explanations for some of the consistent features in the chlorophyll-a distribution. The slightly higher values in the zone between about 60° and 67° S were always present in summer, were always associated with slight decreases in nutrients and occurred only in that area where surface warming increased the stability in the upper 80 m of the water column significantly. Since primary productivity is not nutrient limited in this area (Riley 1946; Thomas and Dodson 1968; Eppley et al. 1969) and is not temperature or light limited (Holm-Hansen et al. 1977), the increase in thermal stratification may be the primary cause of the increase in chlorophyll-a. This conclusion was also reached by Witek et al. (1982) for the net-phytoplankton abundance in the southwestern Atlantic sector of the Southern Ocean. We assume that this is likely to be an early-summer phenomenon, since in winter there is very little thermal stratification in the upper layers. Although a small reduction in surface nutrients was noticeable, it is doubtful whether nutrients are consumed to below their limiting values. In the case of nitrates, for instance, this would be 2 μg at. l^{-1} (Eppley et al. 1969).

It has been suggested that bottom topographic effects may enhance productivity at certain locations in the Southern Ocean (Golubev et al. 1983) but these areas, if they exist, are most probably of geographically limited extent and can be ignored in the present context.

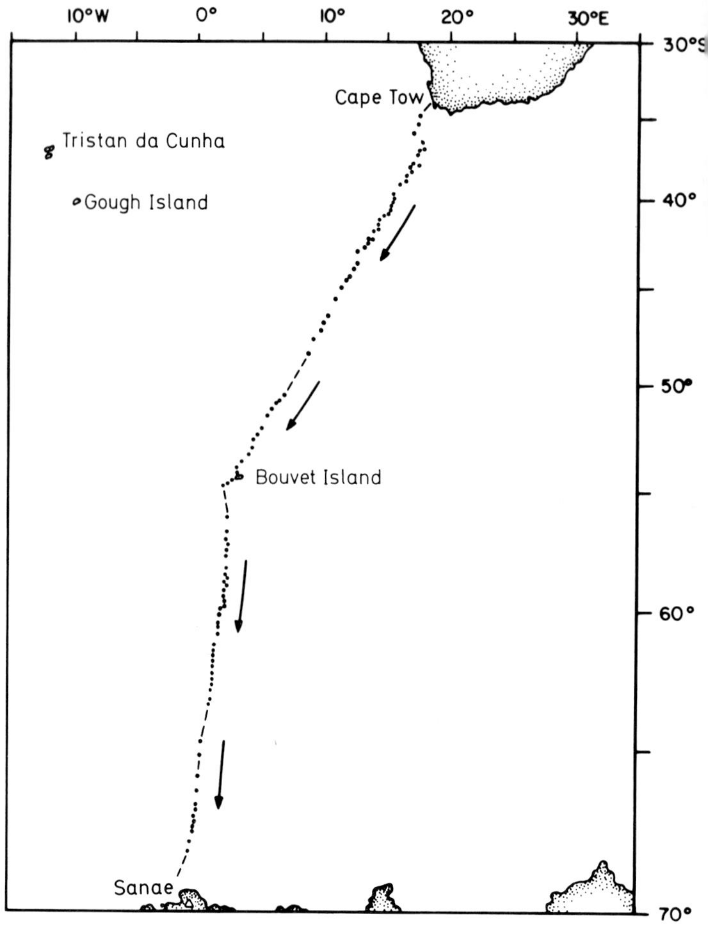

Fig. 4. Cruise track and chlorophyll-a stations operated by the M.V. *S A Agulhas* between Cape Town and the South African Antarctic base (*SANAE*) during December 1982

Peaks in the surface distribution of chlorophyll-a at the Antarctic Polar Front, at the Sub-Tropical Convergence and even perhaps to a lesser degree at the Sub-Antarctic Front may be due to one factor or a combination of different factors. First, the surface and the sub-surface expressions of these fronts hardly ever coincide. The average distance between the surface and sub-surface expressions of the Antarctic Polar Front, for instance, is about 50 km (Lutjeharms and Valentine, to be publ.). A sloping front of this kind could be responsible for the degree of thermal stability required to retain phytoplankton in the euphotic zone. This would be consistent with the results reported by Marra et al. (1982) who showed that, albeit on a smaller scale, the shelf-slope front, with its inclined isopycnals near the surface, provides stability and favours development of enhanced biomass levels.

Mixing of warm, nutrient-poor Sub-Tropical Surface Water across the Sub-Tropical Convergence may have the same stabilizing and, thus, biological enhancing effect, as was deduced from Fig. 5. Plancke (1977) found a second set of

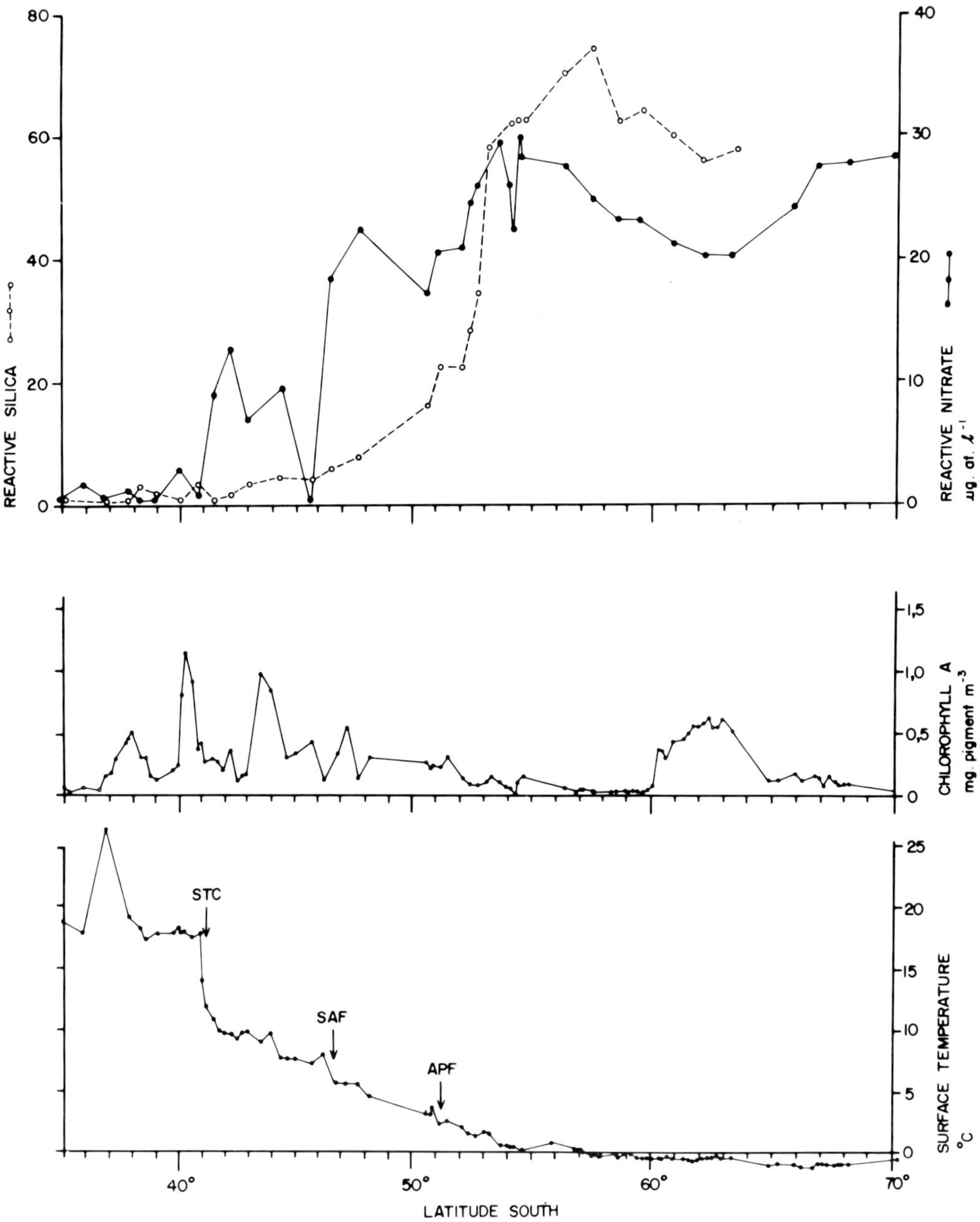

Fig. 5. Distribution of reactive silicate and reactive nitrate (*upper panel*), chlorophyll-a concentration (*middle panel*) and temperature (*lower panel*) at the sea surface along the cruise track shown in Fig. 4. The locations of the Sub-Tropical Convergence (*STC*), Sub-Antarctic Front (*SAF*) and Antarctic Polar Front (*APF*) are indicated according to their sea-surface temperature characteristics

surface chlorophyll-a maxima south of the Sub-Tropical Convergence at the northern limit of an area in which there was a second decrease in temperature; this was most probably at the boundary of cross-frontal mixing. Cross-frontal mixing may take the form of baroclinic eddies and has been observed to do so at upwelling fronts (Lutjeharms et al. 1980), at shelf-edge fronts (Pingree 1978) and across the Sub-Tropical Convergence (Lutjeharms 1981a). Pingree (1978) has suggested that such cross-frontal eddies will be important in the development of frontal phytoplankton blooms. Similarly, mixing in the opposite direction, taking nutrient-rich Sub-Antarctic Surface Water northwards across the Sub-Tropical Convergence, may also have biologically significant consequences in transporting nutrients from the well-mixed side of the frontal boundary to a more stratified side. This may explain why we, and others before us (Plancke 1977; Ichimura and Fukushima 1963; Steyaert 1973) have found major chlorophyll-a increases to the north of the Sub-Tropical Convergence.

Apart from stability considerations, enhanced chlorophyll-a concentrations at the major fronts may also arise from their confluence characteristics. Plankton with limited motility and other positively buoyant material would accumulate at such a convergence. Such aggregations of phytoplankton would lead to increases in the chlorophyll-a levels at the frontal boundary. The near-surface characteristics of the Antarctic Polar Front have been considered as a varying system of convergences and divergences (Wyrtki 1960; Ostapoff 1962, 1963), whereas the kinematic nature of the Sub-Antarctic Front is not known. Our sampling stations were probably too far apart to resolve fine detail of this nature. However, satellite imagery has the potential to resolve fine detail over a large geographical area instantaneously.

4 Satellite Remote Sensing

The advent of satellite instruments, such as the Coastal Zone Colour Scanner (CZCS) carried on the Nimbus-7 satellite (H.R. Gordon et al. 1982, 1983), that can measure ocean colour has facilitated study of chlorophyll concentrations by remote sensing. The CZCS is a scanning radiometer, recording radiances in six coregistered channels at 443, 520, 550, 670, and 750 nm as well as at 11.5 μm (Hovis et al. 1980).

In order to quantify the radiances observed, it is necessary to calibrate accurately the sensor and to remove the masking effect of the atmosphere due to reflection and scattering (Gordon 1978; Gordon et al. 1980). Various nonlinar, bio-optical algorithms have been constructed to achieve this (Clark 1981); the one developed for South African waters is described in detail by Walters (1983).

A selected bio-optical algorithm is applied to a particular water type according to its optical properties. The optical properties of water may be dominated by photoplankton and co-varying detrital material (case I) or by organic materials from terrigenous discharges and resuspension (case II) (Gordon et al. 1983). It was found that in the southern parts of the Agulhas Current system waters could be adequately described as case I (Walters 1982), but that on occasions some terrigenous material may enter the current causing complications (Walters and Neethling 1983). Since no adequate sea reference data are available for this area, the pigment concentration cannot be quantified accurately. Only the ratio of the sub-surface upwelled radiance (R_{31}) has therefore been computed and may be interpreted as semi-quantified chlorophyll or total suspended solid concentrations.

A severe limitation on the applicability of satellite remote sensing in the Southern Ocean is imposed by persistent cloud cover which obscures the radiances concerned. An analysis of cloud cover frequency south of Africa (Lutjeharms 1981a) has shown that the persistence of cloud cover south of 45° S is such as to make the acquisition of remotely sensed information impracticable. Thus, our studies have been concentrated on the Sub-Tropical Con-

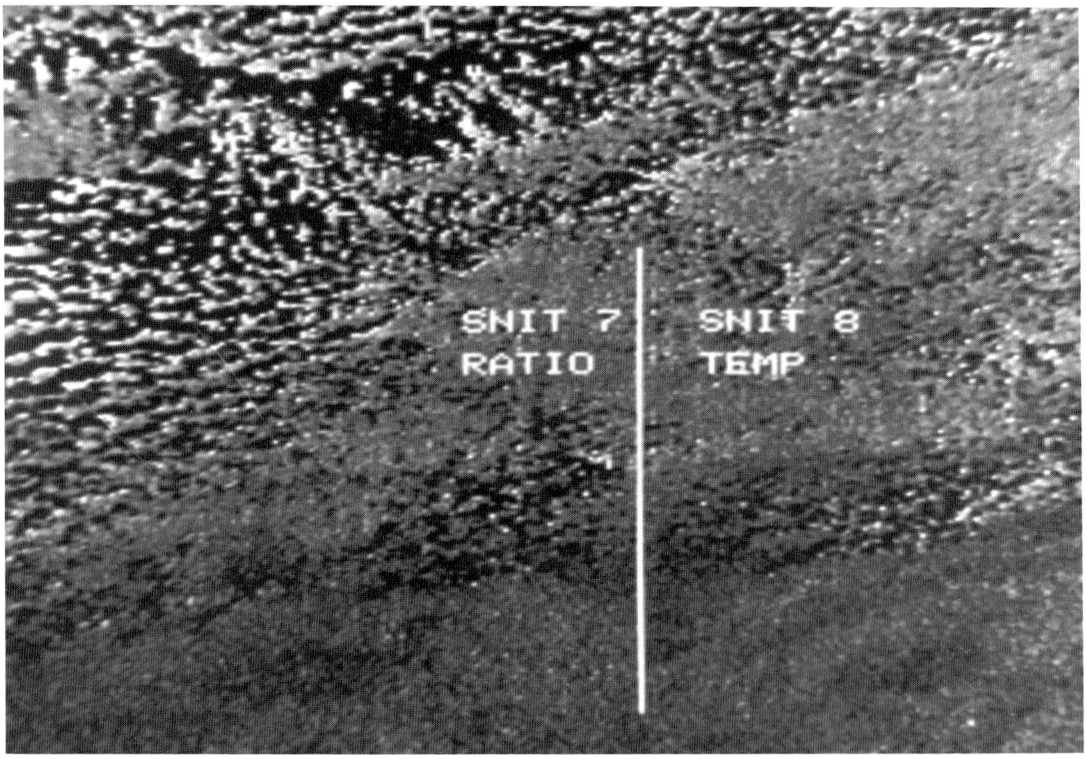

Fig. 6. Distribution of the radiance ratio R_{31}, representing a semi-quantitative measure of chlorophyll-a, astride the Sub-Tropical Convergence south of South Africa. The data are from the Coastal Zone Colour Scanner of the Nimbus-7 satellite for 16 June 1979. *Grainy areas* represent banks of cumulus clouds. *Snit 7/8* indicates the location of the temperature and chlorophyll-a sections shown in Fig. 7

vergence and its interaction with the southern Agulhas Current. A number of unusual and novel features have been found.

In Fig. 7 curves for the radiance ratio R_{31} are shown, as well as the estimated surface temperature along the same line. The location of the transect is shown in Fig. 6, superimposed on the chlorophyll-a field. Extensive cover by cumulus cloud complicates interpretation of the image. However, a sharp drop in temperature from 15°C to 10.8°C is evident (Fig. 7) and represents the position where the transect (Fig. 6) crosses the Sub-Tropical Convergence. A well defined, but narrow, peak in chlorophyll-a occurs at the thermal front. The resolution of the CZCS is about 1 km times 1 km and thus affords great detail. Although this peak might not be regarded as unusually salient on its own, it is consistent with our measurements at sea and with numerous visual interpretations of CZCS images not presented here. The thermal detail of mixing processes at this front is at times resolved by thermal infra-red satellite imagery. Combining the information from such imagery with CZCS-derived data for numerous occasions would aid in investigating the detail of biological enhancement at this front. Ideally, such measurements should be complemented by closely spaced measurements at sea (Mueller and LaViolette 1981).

A number of sections across elements of the southern Agulhas Current system, similar to the one described above, have been studied. With few exceptions, results support the thesis for strong biological enhancement at the surface fronts. One pertinent example involves a section across a warm Agulhas Current filament which had become detached from the current proper and was being advected into the South East Atlantic Ocean (Fig. 8, page 18). This phenomenon is known to occur in the area (Lutjeharms 1981b) and may be subsequently of crucial importance in determining the behaviour of the upwelling front off the southern coast of South Africa.

The temperature distribution across the warm-water filament is as one would expect. Outside the filament itself the sea-surface temperature was on average 13°C (Fig. 9, page 18) while inside the filament the temperature varied between 15° and 16°C. Borders between the two water-masses were sharp and well-defined. The chlorophyll-a distribution, as represented by the radiance ratio, did not resemble the temperature distribution. Values within the warm water were about the same as those outside. The chlorophyll-a peaks occurred just inside the boundaries of the water-masses. This simple portrayal is somewhat complicated by the fact that the traverse intersected the convoluted western border of the warm water twice, but reference to Fig. 8 resolves this complication. Such enhanced chlorophyll-a values at water-mass boundaries are not universal and may not occur at all times. Satellite measurements across a Gulf Stream ring did not show this (H.R. Gordon et al. 1982).

The strong, singular chlorophyll-a peaks, evident on the borders of Agulhas Current water, cannot be explained by

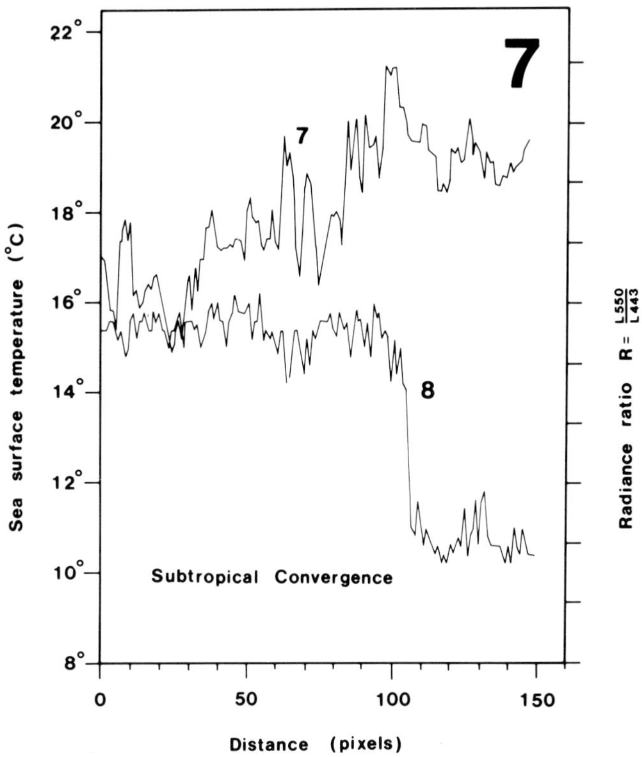

Fig. 7. The inferred sea-surface temperature (*8*) (from infra-red radiance values) and chlorophyll-a (*7*) (from the radiance ratio R_{31}) across the Sub-Tropical Convergence see (Fig. 6). Note the peak in chlorophyll-a concentration just north of the thermal front of the Sub-Tropical Convergence

the advection of nutrient-rich water, since nutrient values within the current resemble closely the values of water outside the current (Wyrtki 1971). While an increase in thermal stability may be caused at the edges of this warm-water intrusion, where warmer water overlies the colder South Atlantic water, this mechanism may be dismissed as one causing biological enhancement for two reasons. First, water in the surface layers of this area is already thermally stable and, secondly, filaments of this nature have been observed (Lutjeharms 1981b) to be advected rapidly and to disperse within a few weeks. A continuous plankton bloom in such a turbulent, fast-changing, environment is unlikely.

Two mechanisms are potential candidates for the biological enhancement observed. Atkinson (1977) has shown that when meanders occur on the Gulf Stream upwelling takes place on its landward side as it moves offshore. Such upwelling events within meanders may cause a large flux of nutrients into the euphotic zone, resulting in significant increases in phytoplankton production (Yoder et al. 1981). Dooley (1981) has described a similar situation where a shelf-current may entrain nutrient-rich deeper water and, thus, facilitate the occurrence of plankton blooms in areas of otherwise low primary productivity.

The Agulhas Current is known to exhibit meanders on various scales (Gründlingh 1979; Lutjeharms 1981c) and the presence of upwelling centres with enhanced produc-

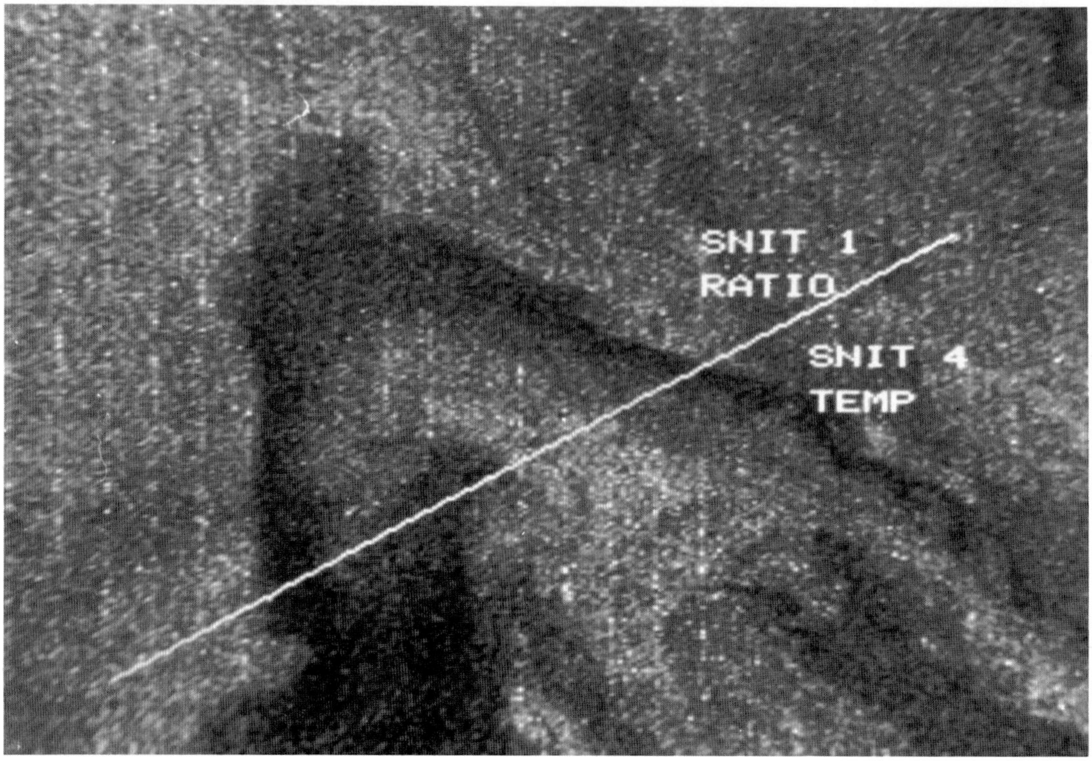

Fig. 8. Distribution of the radiance ratio R31, representing a semi-quantitative measure of chlorophyll-a, across a northward advecting filament of Agulhas Current water. Darker hues signify relatively high chlorophyll-a concentrations. The data are from the Coastal Zone Colour Scanner of the Nimbus-7 satellite for 16 June 1979. *Snit 1/4* indicates the location of the temperature and chlorophyll-a sections shown in Fig. 9

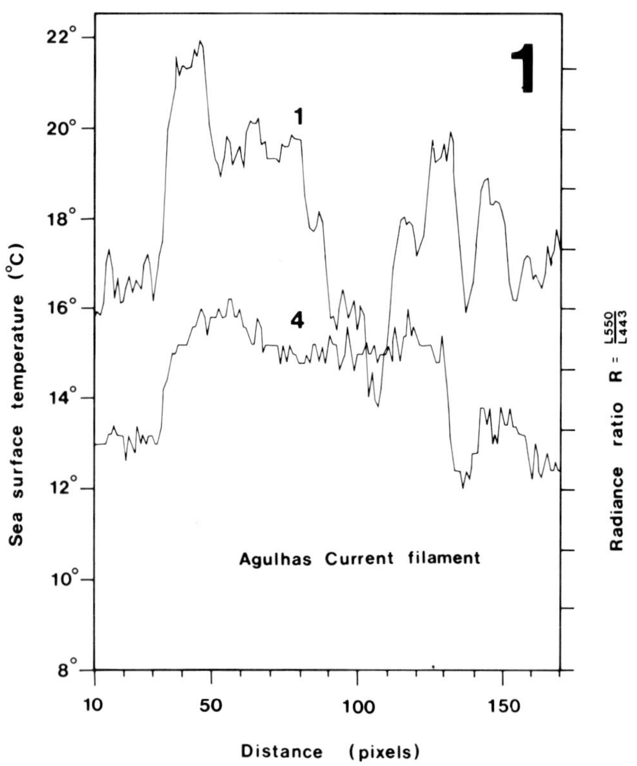

Fig. 9. The inferred sea-surface temperature (*4*) (from infra-red radiance values) and chlorophyll-a (*1*) (from the radiance ratio R31) across an Agulhas Current filament. Note the marked peaks in chlorophyll-a concentrations just inside the thermal borders of the warm water

tivity in the lee of such meanders is thus quite feasible. It has furthermore been shown (Walters and Neethling 1983) that plumes of sediment may be admixed onto the border of the Agulhas Current, creating a drawn-out ribbon of water with high sediment load bordering the current. An area of high productivity may undergo the same admixing and may form a drawn-out plume of high chlorophyll-a. This would not, however, explain the high chlorophyll-a values observed on the seaward border of the current which has no contact with the continental shelf.

The other mechanism we have considered takes into consideration the kinematic nature of the Agulhas Current. The growth of phytoplankton in this area is nutrient limited (i.e., the amount of nutrients available *per unit volume* to plankton being advected with the water-mass limit the primary productivity). However, organisms living in the border-mixing area of a fast-flowing current would experience a continuous replenishment of nutrients, albeit at a low level. Thus, the nutrients available *per unit time* might be much higher and might be sufficient to sustain a higher primary productivity. Lacking more detailed and more accurate data on the borders of the Agulhas Current, we propose this mechanism as one causing biological enhancement in areas of strong horizontal shear. Bearing in mind the high fish density observed in upwelling cells adjacent to the Gulf Stream (Atkinson and Targett 1983), it is instructive to note that tuna shoals are traditionally known by local fishermen to be found at the borders of the Agulhas Current (H.G. Hoy, pers. comm.).

5 Fronts and Biological Enhancement

We recognize two types of oceanic front, and for each type we offer a number of possible mechanisms to explain biological enhancement.

5.1 Horizontal Convergence Zone

In such a zone, portrayed schematically in Fig. 10a, a water-mass with a high nutrient concentration may supply the nutrients necessary to organisms in adjacent nutrient-poor water by cross-frontal mixing. Conditions favourable to increased primary productivity may also be created by the increase in density stratification in the upper layers, which enhances vertical stability and causes retention of phytoplankton in the euphotic zone. The continuous advection of immobile, "buoyant" material, such as phytoplankton, into the frontal zone may also be a mechanical means for increasing the chlorophyll-a concentration in a frontal region without necessarily increasing the primary productivity.

5.2 Horizontal Shearing Zone

If such a front (Fig. 10b) lies in an area of low vertical stability, a sloping frontal edge may fulfil this requirement for increased primary productivity. If one of the water-masses is nutrient-rich, bloom conditions may be created in the frontal zone. If the front forms part of a meandering shelf-current, bloom conditions may be created in upwelling cells formed in the lee of meanders. The high chlorophyll-a values in such cells may be admixed into the mixing zone of such a front and be advected far downstream of the original source. Organisms in nutrient-limited water may also profit by a steady influx of nutrients, even at a low level, into the mixing zone from a fast-flowing bordering current.

6 Conclusion

It is clear that frontal regions in the Southern Ocean and bordering areas may form unique areas of biological enhancement. Such enhancement may be intermittent, highly variable, and geographically restricted. A number of mechanisms or combinations of mechanisms may be responsible for an increase in productivity at any one front at any one time. Further investigations with finer resolutions are bound to produce instructive results for these poorly understood, important biogeographical, areas in the Southern Ocean (Mueller and LaViolette 1981).

Acknowledgements. We thank Prof. M.J. Orren for stimulating discussions, Messrs L. Parker, G.F.P. Rule and H.R. Valentine for collecting data at sea, and the captain and crew of the M.V. S A Agulhas for technical support. Mr. F.P. Anderson is the Principal Investigator for the South African CZCS/NIMBUS 7 Experiment on which some of our data were based.

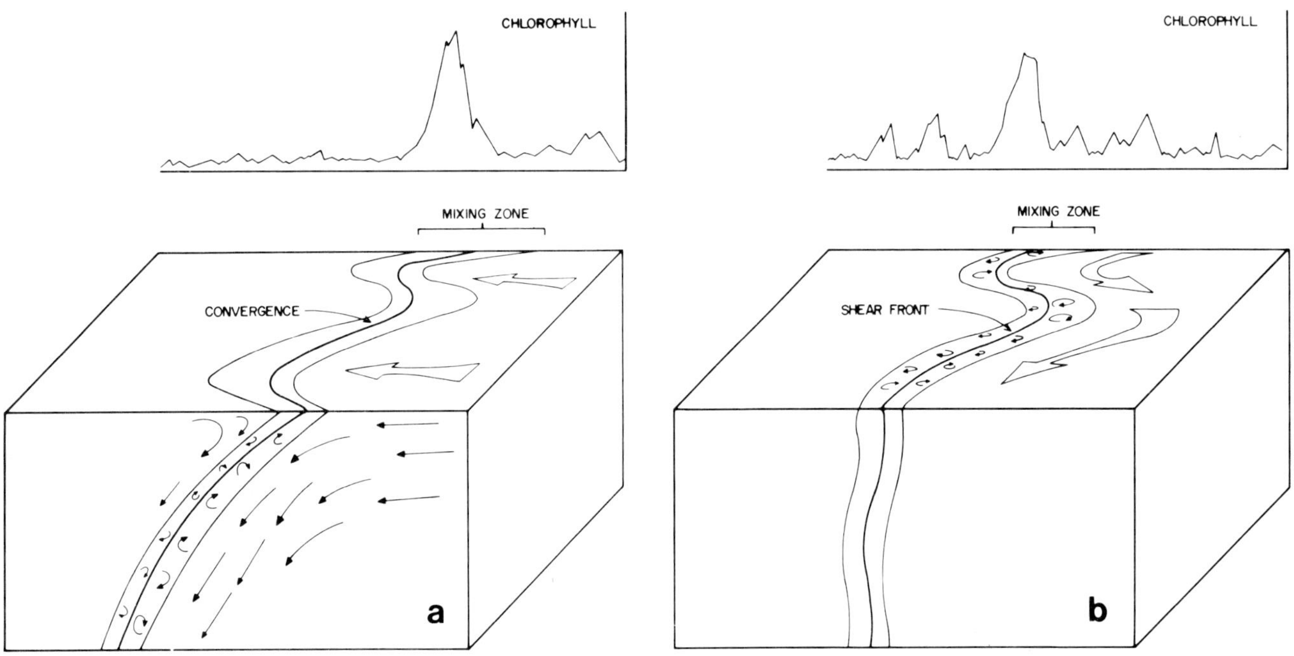

Fig. 10a, b. Schematic presentation of two types of dynamic ocean fronts found south of Africa and the distribution of chlorophyll-a concentration at the sea surface across them. **a** a vertical convergence zone, such as the Sub-Tropical Convergence and, under certain conditions, the Antarctic Polar Front. **b** a horizontal shear front, such as those found at the borders of the constituent parts of the southern Agulhas Current system

References

Abrams RW (to be publ.) Distribution of seabirds in the African sector of FIBEX. S Afr J Antarct Res

Ainley DG, Jacobs SS (1981) Sea-bird affinities for ocean and ice boundaries in the Antarctic. Deep-Sea Res 28(10A):1173–1185

Allanson BR, Hart RC, Lutjeharms JRE (1981) Observations on the nutrients, chlorophyll and primary production of the southern ocean south of Africa. S Afr J Antarct Res 10/11:3–14

Atkinson LP (1977) Modes of Gulf Stream intrusion into South Atlantic Bight shelf waters. Geophy Res Lett 4:583–586

Atkinson LP, Targett TE (1983) Upwelling along the 60-m isobath from Cape Canaveral to Cape Hatteras and its relationship to fish distribution. Deep-Sea Res 30(2A):221–226

Clark DK (1981) Phytoplankton pigment algorithms for the Nimbus-7 CZCS. In: Gower JRF (ed) Oceanography from Space. Mar Sci Vol 13 Plenum, New York pp 227–237

Deacon GER (1937a) The hydrology of the southern ocean. Discovery Reports Vol XV pp 1–124 + 44 plates

Deacon GER (1937b) Note on the dynamics of the southern ocean. Discovery Reports Vol XV pp 125–152

Dooley HD (1981) The role of axially varying vertical mixing along the path of a current in generating phytoplankton production. Philos Trans R Soc Lond A302:649–660

El-Sayed SZ, Benon P, David P, Grindley JR, Murail J-F (1979) Some aspects of the biology of the water column studied during the "Marion-Dufresne" cruise 08. CNFRA 44:127–134

Eppley RW, Rogers JN, McCarthy JJ (1969) Half saturation constants for uptake of nitrate and ammonia by marine phytoplankton. Limnol Oceanogr 14:912–920

Golubev YuN, Lomakin PD, Cherkesov LV (1983) Generation of topographic vortices and their possible effect on formation of productive zones in the ocean. Dopov Akad Nauk UKRSR Ser A(2):25–28

Gordon AL, Molinelli EJ, Baker TN (1982) Southern Ocean Atlas. Columbia University Press, New York 11 pp + 233 plates

Gordon HR (1978) Removal of atmospheric effects from satellite imagery of the oceans. Appl Optics 17:1631–1636

Gordon HR, Clark DK, Mueller JL, Hovis WA (1980) Phytoplankton pigments from the Nimbus-7 Coastal Zone Colour Scanner. Science 21:63–66

Gordon HR, Clark DK, Brown JW, Brown OB, Evans RH (1982) Satellite measurements of the phytoplankton pigment concentration in the surface waters of a warm core Gulf Stream ring. J Mar Res 40(2):491–502

Gordon HR, Clark DK, Brown JW, Brown OB, Evans RH, Broenkow WW (1983) Phytoplankton pigment concentrations in the Middle Atlantic Bight: comparison of ship determinations and CZCS estimates. Appl Optics 22(1):20–36

Grindley JR, Land SN (1979) Zooplankton around Marion and Prince Edward Islands. CNFRA Publ 44, Campagne MD.08/ Benthos-Premiers résultats pp 111–125

Gründlingh ML (1979) Observation of a large meander in the Agulhas Current. J Geophy Res 84(C7):3776–3778

Hedgpeth JW (1969) Introduction to Antarctic Zoogeography. In: Bushnell VD (ed) Distribution of selected groups of marine invertebrates in waters south of 35°S latitude. American Geographical Society, Antarctic Map Folio Series, Folio 11, 44 pp + 29 plates

Holm-Hansen O, El-Sayed SZ, Franceschini GA, Cuhel RL (1977) Primary production and the factors controlling phytoplankton growth in the Southern Ocean

Hovis WA, Clark DK, Anderson F, Austin RW, Wilson WH, Baker ET, Ball D, Gordon HR, Mueller JL, El-Sayed SZ, Sturm B, Wrigley RC, Yentsch CS (1980) Nimbus-7 Coastal Zone Colour Scanner: Systems description and initial imagery. Science 210: 80–83

Ichimura S, Fukushima H (1963) On the chlorophyll content of the surface water of the Indian and Antarctic Oceans. Bot Mag Tokyo 76:395–399

Jacques G, Minas M (1981) Production primaire dans le secteur indien de l'océan Antarctique en fin d'été. Oceanol Acta 4: 33–41

Lutjeharms JRE (1981a) Interaction between the Agulhas Current and the Subtropical Convergence. CSIR Res Rep 384, 39 pp

Lutjeharms JRE (1981b) Satellite studies on the South Atlantic upwelling system. In: Gower JFR (ed) Oceanography from Space. Mar Sci Vol 13 Plenum, New York pp 195–199

Lutjeharms JRE (1981c) Features of the southern Agulhas Current circulation from satellite remote sensing. S Afr J Sci 77(5): 231–236

Lutjeharms JRE, Baker DJ (1980) A statistical analysis of the mesoscale dynamics of the southern ocean. Deep-Sea Res 27A(2): 145–159

Lutjeharms JRE, Fromme GAW, Duncan CP (1980) Large scale motion in the SW Indian Ocean. CSIR Res Rep 375, 9 pp

Lutjeharms JRE, Fromme GAW, Valentine HR (1981) Oceanic frontal systems between Africa and Antarctica. EOS, Trans Am Geophys Union 62(45):924

Lutjeharms JRE, Emery WJ (to be publ.) The detailed thermal structure of the upper ocean layers between Cape Town and Antarctica during the period Jan–Feb 1978. S Afr J Antarct Res

Lutjeharms JRE, Fromme GAW, Valentine HR (1983) Die termiese oseaanstruktuur suid van Afrika; verslag oor sestien navorsingsvaarte. WNNR Navorsingsverslag 390, (to be publ.)

Lutjeharms JRE, Valentine HR (to be publ.) Southern Ocean thermal fronts south of Africa. Deep-Sea Res

Marra J, Houghton RW, Boardman DC, Neale PJ (1982) Variability in surface chlorophyll-a at a shelf-break front. J Mar Res 40(3): 575–591

Mueller JL, LaViolette PE (1981) Colour and temperature signatures of ocean fronts. In: Gower JFR (ed) Oceanography from Space. Mar Sci Vol 13 Plenum, New York pp 295–302

Ostapoff F (1962) On the frictionally induced transverse circulation of the Antarctic Circumpolar Current. Dtsch Hydrogr Z 15(3): 103–113

Ostapoff F (1963) A note on vertical motions in the region of the Antarctic Circumpolar Current. Monthly Weather Rev 91(10–12):727–729

Pingree RD (1978) Cyclonic eddies and cross-frontal mixing. J Mar Biol Assoc UK 58(4):955–963

Pingree RD, Mardell GT (1981) Slope turbulence, internal waves and phytoplankton growth at the Celtic Sea shelf-break. Philo Trans R Soc Lond A302:663–682

Pingree RD, Mardell GT, Holligan PM, Griffiths DK, Smithers J (1982) Celtic Sea and Armorican current structure and the vertical distributions of temperature and chlorophyll. Continental Shelf Res 1(1):99–116

Plancke J (1977) Phytoplankton biomass and productivity in the Subtropical Convergence area and shelves of the western Indian subantarctic islands. In: Llano GA (ed) Adaptations within Antarctic Ecosystems, Proceedings of the 3rd SCAR Symposium on Antarctic Biology pp 51–73

Pomazanova NP (1980) Divergence of drift currents in the Antarctic sector of the Indian Ocean. Tr Gidrometeorol Inst 80:110–115

Rakusa-Suszczewski S (1978) Environmental conditions within krill swarms. Pol Arch Hydrobiol 25(3):585–587

Riley GA (1946) Factors controlling phytoplankton populations on Georges Bank. J Mar Res 6:54–73

Sievers HA, Emery WJ (1978) Variability of the Antarctic Polar Frontal Zone in the Drake Passage – Summer 1976–1977. J Geophy Res 83:3010–3022

Steyaert J (1973) Distribution of plankton diatoms along an African-Antarctic transect. Invest Pesq 37(2):295–328

Thomas WH, Dodson AN (1968) Effects of phosphate concentration on cell division rates and yield of tropical oceanic diatom. Biol Bull Mar Lab, (Woods Hole) 134:199–208

Voronina NM (1962) On the dependence of the character of the boundary between Antarctic and Sub-Antarctic pelagic zones on the meteorological conditions. In: Antarctic Research, Maury Memorial Volume American Geophysical Union, Geophysical Memoir No 7 pp 160–162

Walters NM (1982) Satellietafstandwaarneming van oseaankleur. S Afr Tydskr Natuurwet Tegnol 1(3):156–161

Walters NM (1983) Verification of the utility of the Coastal Zone Colour Scanner in portraying pigment concentration along the Cape West Coast. S Afr J Physics (to be publ.)

Walters NM, Neethling DC (1983) A practical method to derive semi-quantitative total suspended solids distributions along the South African east coast from Nimbus CZCS data. CSIR Special Report, FIS 292, 18 pp

Witek Z, Pastuszak M, Grelowski A (1982) Net-phytoplankton abundance in western Antarctic and its relation to environmental conditions. Meeresforschung 29(3):166–180

Wyrtki K (1960) The Antarctic Convergence – and Divergence. Nature 187(4737):581–582

Wyrtki K (1971) Oceanographic atlas of the International Indian Ocean Expedition. National Science Foundation, Washington DC 531 pp

Yoder JA, Atkinson LP, Lee TN, Kim H, McClain C (1981) Role of Gulf Stream frontal eddies in forming phytoplankton patches on the outer southeastern US shelf. Limnol Oceanogr 26:1103–1110

Zillman JW (1970) Sea surface temperature gradients south of Australia. Aust Meteorol Mag 18:22–30

The Distribution of Inorganic Nitrogen, Phosphorus, Silicon and Dissolved Organic Matter in Surface and Deep Waters of the Southern Ocean

S. LE JEHAN and P. TRÉGUER[1]

Summary. Based on data collected during two cruises in the southern Indian Ocean, and by using hydrological models, uptake and regeneration ratios of nitrate, phosphate, and silicate are calculated and compared with those reported in the literature. The large gradient of silicate which occurs when the water-masses of the Antarctic Divergence drift northward, is mainly attributed to a biological phenomenon. In these waters of low primary productivity, phytoplankton growth needs large amounts of silicon, in comparison with nitrogen and phosphorus ($\Delta Si/\Delta N/\Delta P$ reach 150/15/1). In the Antarctic Intermediate Water and in the sub-surface, deep and bottom waters of the Antarctic Ocean, positive anomalies of silicate remain high beyond the maximum N and P layer. These results underline the unique nature of the biological cycle of Si in the Antarctic Ocean. The distribution of dissolved organic matter (total P and urea) is described for the oceanic water-masses and the more productive coastal waters of the Kerguelen Islands; concentrations being generally lower than for the World Ocean.

1 Introduction

Large amounts of nutrients are available in the near-surface waters of the Southern Ocean. However, the primary production remains low. Despite this low level of biological activity, large amounts of biogenic silica cover the Antarctic sediments. Thus, the distribution of nutrients in near-surface, deep, and bottom waters appears as a paradox.

Using data collected during Antiprod 1 (March 1977) and Antiprod 2 (March 1980) cruises, we describe the distributions of nitrate, phosphate, and silicate along a transect (66° 30′ E) south of the Polar Front Zone and along a coastal transect (north of Kerguelen archipelago). Data on dissolved organic matter (total phosphorus and urea) are also reported. By means of hydrological models, we calculate the uptake and regeneration ratios $\Delta Si/\Delta N/\Delta P$. The observed anomalies are explained from recent experimental work.

2 Materials and Methods

Sampling stations (Fig. 1) were located along a 66° 30′ E transect during March 1977 (Groupe Médiprod 1978) and March 1980 (Groupe Mediprod 1982). Samples were collected in 1.8–1 Niskin bottles. Nutrient concentrations were determined with an auto-analyzer (Technicon) according to Tréguer and Le Corre (1975). For total dissolved organic P (DOP) determination, samples were filtered through Whatman GF/C membranes (previously heated to 450°C); hydrogen peroxide (50 μl) was added to 80 ml of the seawater sample in a quartz tube and photo-oxidation was then carried out for 14 h with a 1,200 W ultraviolet source. Urea was analyzed automatically according to Aminot and Kerouel (1982).

1 Institut d'Etudes Marines, Brest, France

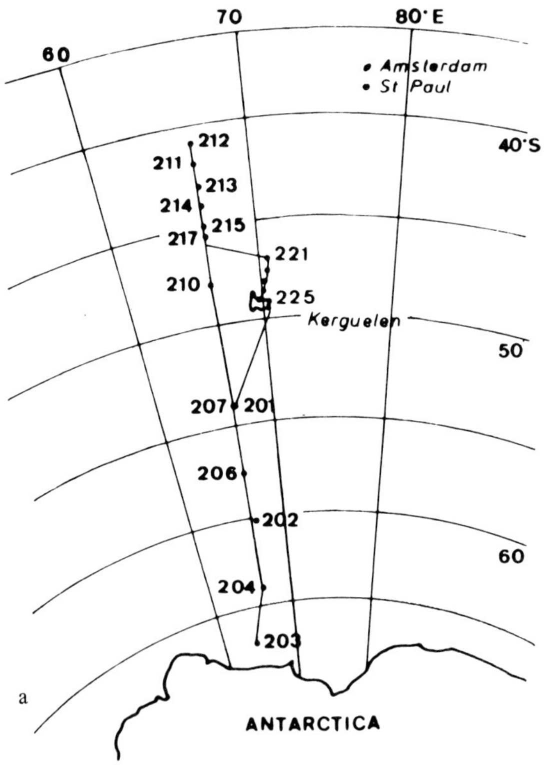

Fig. 1a. Antiprod 2 cruise track and stations occupied in the Southern Ocean (summer of 1980). The same transect was studied during March 1977 (Antiprod 1) when more stations were occupied

Antarctic Nutrient Cycles and Food Webs
(ed. by W.R. Siegfried, P.R. Condy, and R.M. Laws)

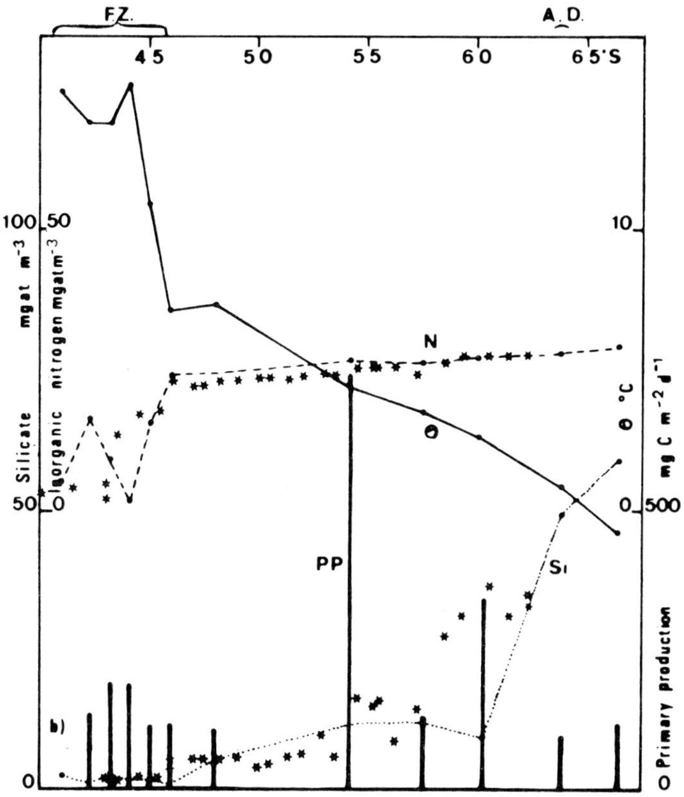

Fig. 1b. Distribution of mean temperature, total dissolved inorganic N, silicate and primary production in the photic zone of the surface waters between the frontal zone of convergences (*FZ*) and the Antarctic Divergence (*AD*) during Antiprod 2. Nitrate and nitrite and silicate values (*stars*) are from the Antiprod 1 cruise. (Groupe Médiprod 1978)

3 Results

3.1 Distribution of Inorganic and Organic Dissolved Nitrogen, Phosphorus and Silicon

3.1.1 Oceanic Area

The vertical distribution of the parameters in the whole of the water column is shown in Fig. 2. Our results agree with the general pattern observed in all the other sectors of the Southern Ocean (Deacon 1982). However, a unique surface Frontal Zone (FZ) occurs in our area. It results from the junction of the Antarctic and Sub-Tropical Convergences (Gamberoni et al. 1982), where eddies bring Indian Ocean Central Water (IOCW) into contact with Antarctic Surface Waters (ASW). By means of the Θ-S diagram and the multilinear regression method of Alvarez-Borrego et al. (1975), three water-masses are identified (Fig. 2b) south of the Frontal Zone: Antarctic Surface Water (ASW), Warm Deep Water (WDW), and Antarctic Bottom Water (AABW).

The Antarctic Surface Water is rich in nutrients (silicate = 59 mg at. m^{-3}; total dissolved inorganic N = 29.2 mg at. m^{-3}; phosphate = 1.92 mg at. m^{-3}). It is characterized by a rapid decrease of silicate from the Divergence to the Convergence (Fig. 1), compared to nitrate (and phosphate). Primary productivity is low, generally 350 mg C m^{-2} d^{-1} in the photic zone (90 m depth). Dissolved organic matter concentration occurs at usual levels in surface waters (0.10–0.15 mg at. P m^{-3}), whereas ammonium varies from 0.4 to 1.0 mg at. N m^{-3} with high concentrations near the Antarctic Divergence (60° S); urea varying little, from 0.13 to 0.20 in ASW (Fig. 3).

Two components are identified in the WDW (Fig. 2b). The main one corresponds to a salinity maximum (34,75‰) and is rich in nutrients (Fig. 2a). The second one, from 500 to 800 m is characterized by an oxygen minimum (< 4.3 ml l^{-1}) (Fig. 2a). the method of Alvarez-Borrego et al. (1975) shows that this results from conservative phenomena (Fig. 2b). Whereas the first component is mainly North Atlantic Deep Water, the second originates from the Indian Ocean (Callahan 1972). Nitrate and phosphate concentration maxima appear a few 100 m lower than the oxygen minimum, and are related to biological activity. In WDW, dissolved organic matter (Fig. 3) is low (0.01–0.02 mg at. P m^{-3}); being among the lowest concentrations in the World Ocean (Le Corre and Tréguer 1976). Antarctic Bottom Water is especially rich in silicate; concentrations reaching 150 mg at. m^{-3} in this area (Jacobs and Georgi 1977).

3.1.2 Coastal Area

Above the continental shelf, north of the Kerguelen archipelago, westerly winds prevail and generate upwelling (Murail et al. 1977). Phytoplankton biomass is higher than in the oceanic area, corresponding to a remarkable decrease in silicate concentrations (33–3 mg at. m^{-3}) compared with nitrate concentrations (31–15 mg at. m^{-3}) (Fig. 4). Ammonia and urea concentrations are high (2 and 0.8 mg at. N m^{-3}, respectively, in surface waters) and are correlated significantly; DOP is > 0.10 mg at. P m^{-3} in the surface waters, and is correlated significantly with urea. The changes in NH_4, DOP and urea are related to zooplankton which are especially active in this area.

3.2 Nitrogen, Phosphorus, and Silicon Anomalies

Non-conservative changes in total inorganic N, P, and Si were calculated by means of hydrological models.

3.2.1 Upper Layers

In the coastal area, the upwelling phenomenon generated a binary mixture of water-mass (end-members in Table 1), from which negative nutrient anomalies appear in the phot-

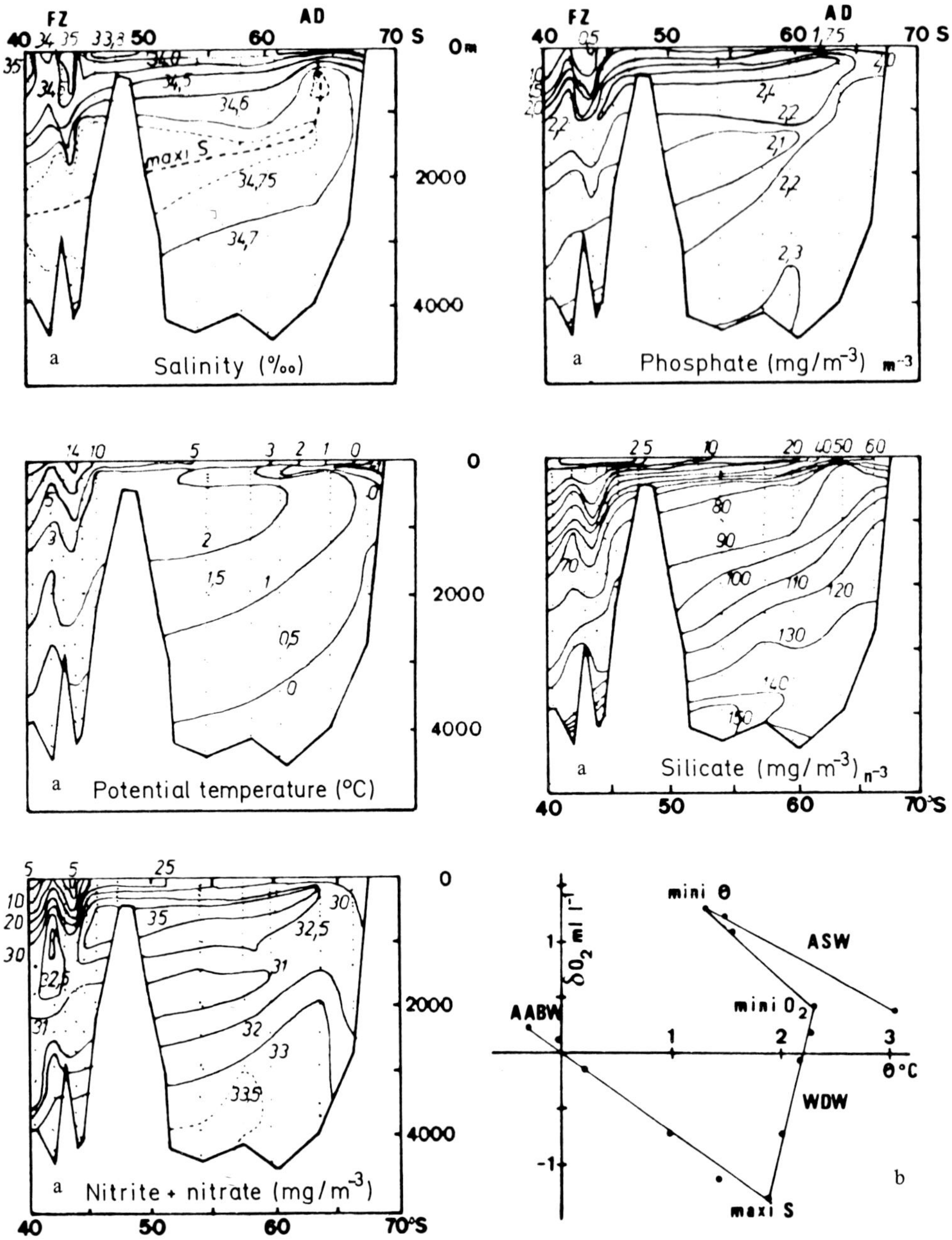

Fig. 2a,b. Vertical distribution of a salinity, temperature and nutrients, and b identification of water-masses (*ASW* Antarctic Surface Water; *WDW* Warm Deep Water; *AABW* Antarctic Bottom Water) at Station 207 (54° S, 66° 30′E) (see text for further particulars) during the Antiprod 2 cruise. (see Fig. 1)

ic zone (Fig. 5f). The correlation between the N and P anomalies is significant ($n = 32$, $r = 0.88$, $P < 0.05$) with mean ratios ($\Delta N/\Delta P = 12 \pm 0.4$) $\Delta Si/\Delta N$ ranging from 0.8 to 4.0 (Fig. 5).

In the oceanic area, during the ASW drift to the north, numerous phenomena (inputs at the Divergence, effects of melting ice, rain and evaporation) influence the distribution of hydrological parameters in near-surface waters. A general model to describe these effects is not attempted. However, from 47°–55° S, a linear gradient of physical parameters is observed, whereas the change in nutrients is non-conservative (Fig. 5). Negative anomalies of nutrients are calculated from a theoretical linear gradient (Table 1). The correlation between N and P anomalies is significant ($n = 41$, $r = 0{,}78$ $P < 0.05$) with a mean ratio $(\Delta N)/(\Delta P)$ of 14.0 ± 1.8. The $\Delta Si/\Delta N$ ratios vary from 4 to 13 (Fig. 5).

3.2.2 Sub-surface and Deep Waters

In sub-surface and deep waters, positive anomalies of nutrients, due to remineralization, are calculated, using hydrological models. In the Polar Front Zone, Antarctic Intermediate Water (AAIW), results from a binary mixture between IOCW and ASE (Fig. 6, Table 2). Positive anomalies of nutrients are calculated. The correlation between N and P is significant ($P < 0.05$). The mean regeneration $\Delta N/\Delta P$ ratios are 13.6 ± 1.4 ($n = 29$, $r = 0.89$) and $16.8 + 1.5$ ($n = 43$, $r = 0.86$), respectively, for the Antiprod 2 and 1 data (in agreement with Minas 1979). Figure 6 shows the rapid regeneration of N, which is active especially in the top 100 m; the recycling silicate is less obvious (some negative anomalies correspond to a probable underestimation of silicate concentration in the reference waters).

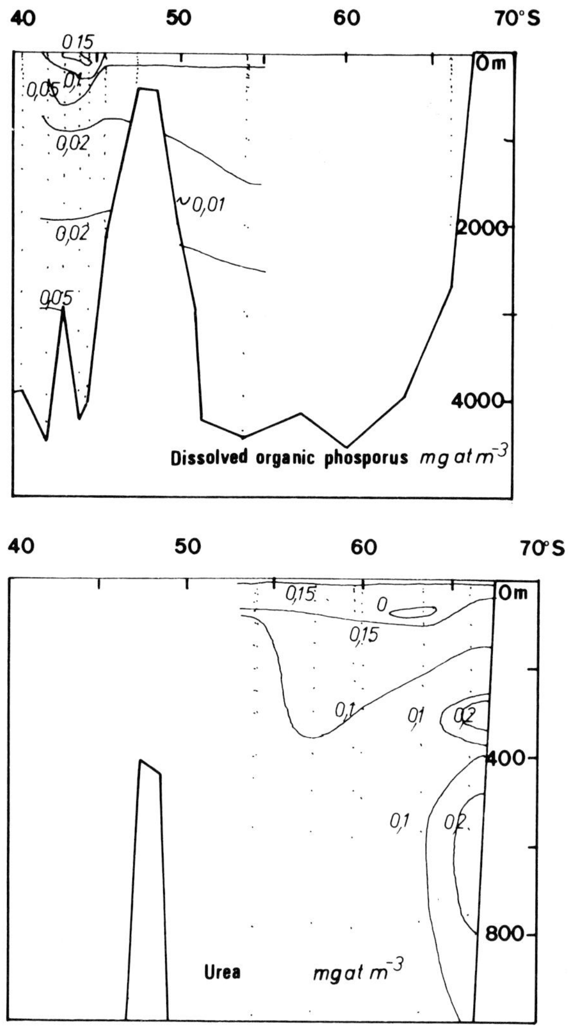

Fig. 3. Vertical distribution of dissolved organic phosphorus and urea along a transect (66° 30′E) during the Antiprod 2 cruise. (see Fig. 1)

The upwelling of WDW (south of 54° S) generates a ternary mixture in the sub-surface water, between ASW (T minimum) and WDW (2 components: S maximum; 0_2 minimum). In each sample the percentage of each end-member is calculated using a ternary diagram Θ-S (Fig. 6, Table 2). When comparing in situ concentrations and theoretical concentrations of nutrients, positive anomalies are calculated. The mean regeneration ratios $\Delta N/\Delta P$ are 14.3 ± 2.5 (n = 13, r = 0.87, $P < 0.05$) and 13.5 ± 0.9 (n = 30, r = 0.91, $P < 0.05$), respectively, for Antiprod 2 and 1 data. The distribution of silicate anomalies is analogous to the distribution observed for AAIW (Fig. 6). These anomalies increase with depth and are high below the depth of maximum regeneration of N. In the AABW, nutrient anomalies are calculated (Le Jehan 1982) using characteristics of reference waters: Foster and Carmack (1979) for Weddell Sea Bottom Water, and Jacobs and Georgi (1977) for WDW. Mean anomalies are, respectively, 2.5 mg at. N m^{-3}, 0.19 mg at. P m^{-3}, and 30 mg at. Si m^{-3}.

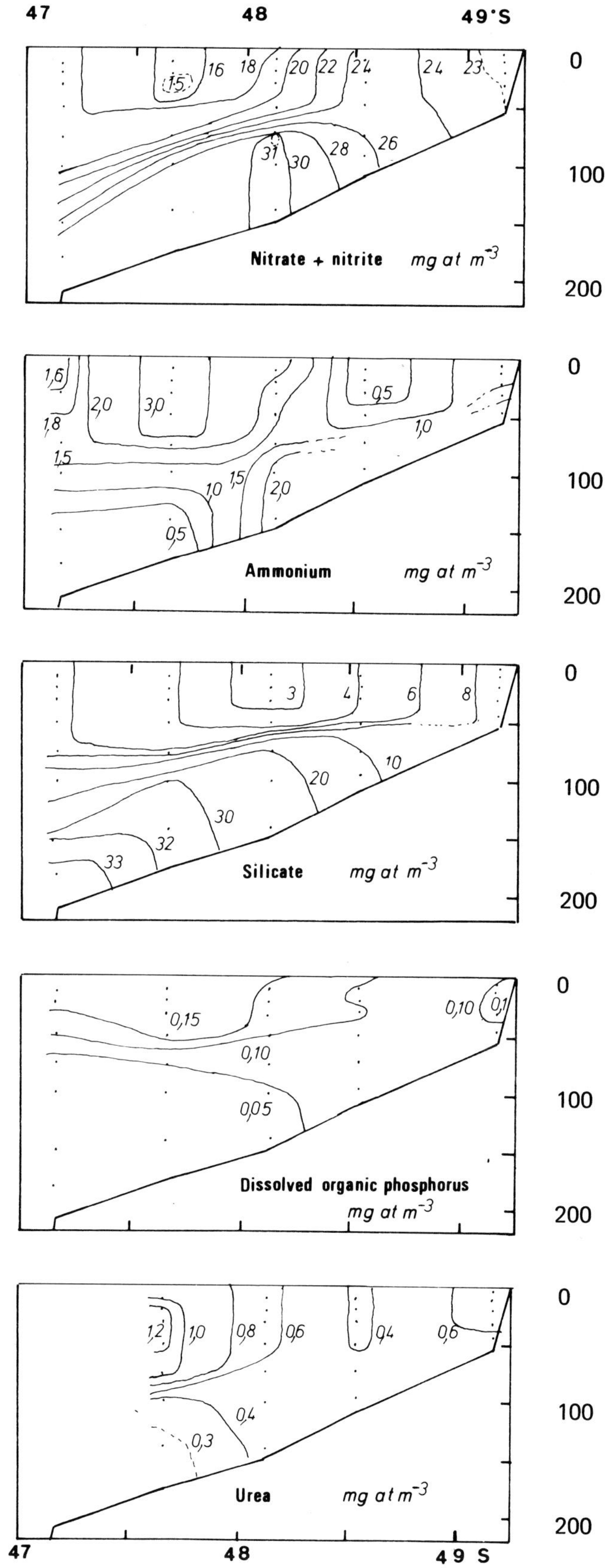

Fig. 4. Vertical distribution of nutrients and dissolved organic matter along a transect (see Le Jehan 1982) north of Kerguelen archipelago during the Antiprod 2 cruise. (see Fig. 1)

Table 1. Characteristics (salinity (‰)), temperature (°C) and inorganic N, P, and Si (mg at. m^3) of reference waters for calclulation of nutrient anomalies in deep waters. See text for explanation of abbreviations

	Salinity	Temperature	N	P	Si
Kerguelen archipelago					
Surface	33.68	6.5	25.64	1.75	8.8
Sub-surface	33.93	3.0	32.37	2.17	32.0
Oceanic area					
47°S	33.65	6.44	22.5	1.49	5.6
55°S	33.84	3.65	25.6	1.66	13.2

4 Discussion

In near surface waters our measurements of dissolved organic matter in the Southern Ocean are relatable to the low productivity of ASW. Dissolved organic P concentrations are minimal in WDW, the oldest water-mass of the Southern Ocean.

Recent observations and experimental work deal with the biological Si cycle in the Southern Ocean (reviewed in Le Jehan and Tréguer 1983). Jacques (1983) demonstrated the high demand of *Fragilariopsis kerguelensis* for silicate, the frequent thickness of the frustules, high sedimentation rates, and the presence of fluorescent cells at great depths

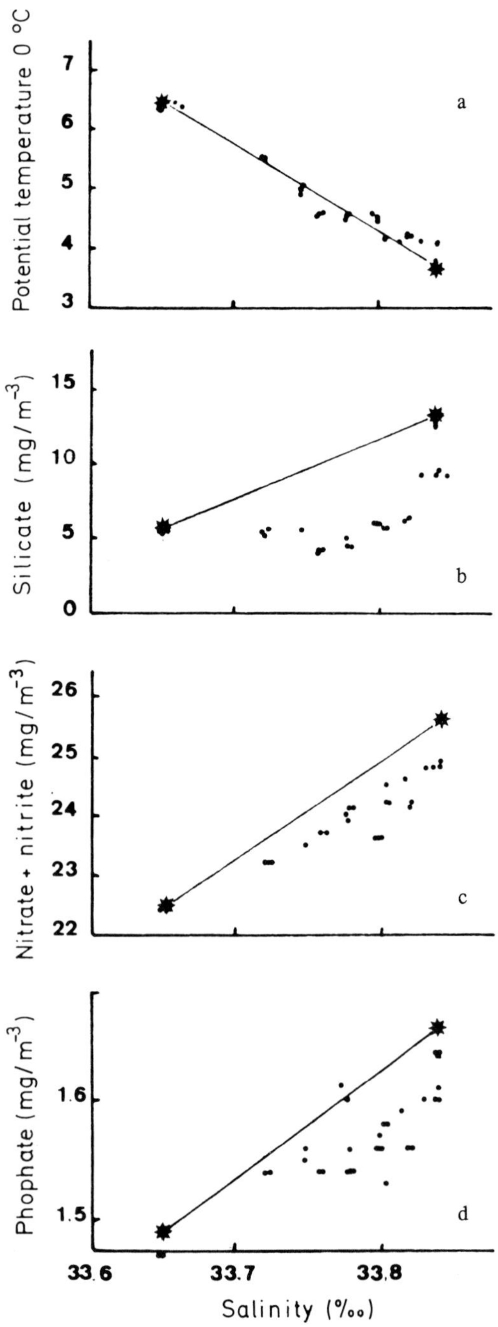

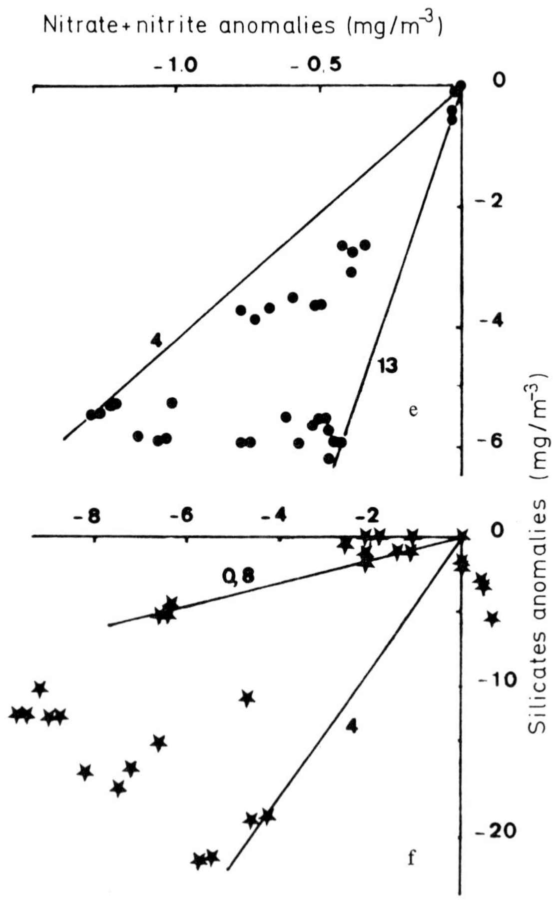

Fig. 5a–f. Negative anomalies of nutrients in the surface waters of a–e an oceanic area (47°–55°S, 66° 30′E) and f a coastal area north of Kerguelen archipelago. (After Groupe Médiprod 1978, 1982)

Table 2. Characteristics (salinity (°/₀₀), temperature (°C) and inorganic N, P, and Si (mg at. m^{-3}) of reference waters for the calculation of nutrient anomalies in deep waters. See text for explanation of abbreviations

	Salinity	Temperature	N	P	Si	Source
AAIW						
ASW	33.91	3.23	25.8	1.66	9.5	Groupe Médiprod (1978)
IOCW	35.43	14.97	4.9	0.40	4.0	Groupe Médiprod (1980)
Ternary model						
ASW (mini T)	33.98	−1.53	29.5	2.04	41.5	Groupe Médiprod (1978)
WDW (mini O_2)	34.60	3.23	33.5	2.28	70.0	Jacobs and Georgi (1977)
WDW (maxi S)	34.82	1.97	26.5	1.80	69.0	Jacobs and Georgi (1977)

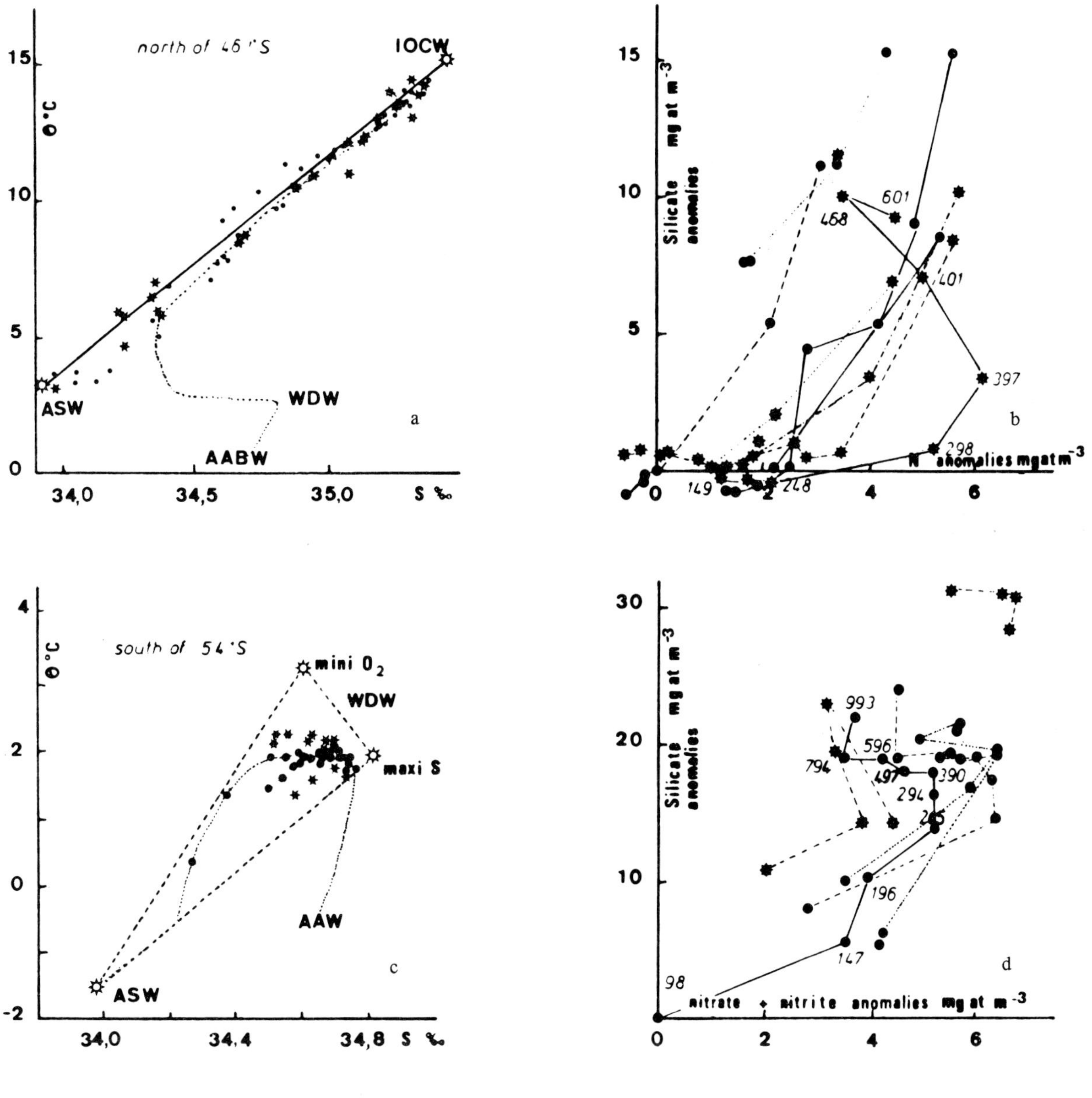

Fig. 6. Positive anomalies of nutrients in surface, deep and bottom waters of the Indian sector of the Southern Ocean, based on data from Groupe Médiprod [1978 (*dots*) and 1982 (*stars*)]. *AAIW* Antarctic Intermediate Water; *ASW* Antarctic Surface Water; *IOCW* Indian Ocean Central Water; *WDW* Warm Deep Water. (see text for further particulars)

together with the almost exclusive dominance of this species in the surface layer of the sediments. Nelson and Gordon (1982) showed that an average of only about one-third of the biogenic silica production is regenerated in the photic zone; two-thirds of the remaining silica being regenerated in sub-surface and deep water while the rest is exported towards the bottom layers, ultimately reaching sediments. These observations may explain the high positive silicate anomalies calculated by our model, and their presence in the whole of the water column. In contrast, N and P are recycled as usual, in the top 100 m, with $\Delta N/\Delta P$ ratios close to 16 (Redfield et al. 1963).

Negative anomalies calculated in ASW and in photic zones of coastal waters result from a balance between uptake and in situ regeneration. Whereas $\Delta N/\Delta P$ ratios, as calculated above, vary from 11 to 16 (not very different from the usual Redfield's uptake ratio), $\Delta Si/\Delta N$ ratios show abnormally high values (0.8–13) which are very different from the expected unity (see Richards 1958). Jacques (1983) demonstrated the need of diatoms for large amounts of silicate, whereas nitrate need is usual for eutrophic areas. Thus, high ratios of silicate anomalies may be forecast, related to phytoplankton activity. However, abnormal relative regeneration rates of N and Si are also observed in near-surface Antarctic waters. Nelson and Gordon (1982) measured low silicate recycling rates in the photic zone, compared with other eutrophic ecosystems (about 32% of the biogenic silica production is regenerated). Relative regeneration of N is high (Glibert 1982) and ammonium is available for phytoplankton growth, being assimilated preferentially to nitrate. Hence, the high demand of diatoms for silicate, on the one hand, and the low dissoluton of biogenic silica in the photic zone, on the other hand, compared with N, are convergent factors explaining the abnormally high $\Delta Si/\Delta N$ ratios calculated for surface water-masses.

5 Conclusion

The patterns of distribution of nutrients and dissolved organic matter in the southern Indian Ocean agree with previously reported observations in the Southern Ocean. In summer, the photic zone is poor in dissolved organic matter and in phytoplankton biomass, despite the richness in nutrients. The ratios of the N and P anomalies are close to Redfield's model, in near-surface (uptake) and deep waters (regeneration). The anomalies of the Si cycle in the Southern Ocean (high $\Delta Si/\Delta N$ in surface waters) and high positive ΔSi anomalies in deep and bottom waters, relative to N, are relatable to biological activity (high demand of Antarctic diatoms for silicate, high transport of biogenic silica toward the deep layers).

References

Alvarez-Borrego S, Guthrie D, Culberson CH, Park PK (1975) Test of Redfield's model for oxygen nutrient relationships using regression analysis. Limnol Oceanogr 20:795–805

Aminot A, Kerouel R (1982) Dosage automatique de l'urée dans l'eau de mer une méthode très sensible à la diacétylmonoxime. Can J Fish Aquat Sci 39:174–183

Callahan JE (1972) Structure and Circulation of deep water in the Antarctic. Deep-Sea Res 19:563–575

Deacon GER (1982) Physical and biological zonation in the Southern Ocean. Deep-Sea Res 29:1–10

Foster TD, Carmack EC (1979) Frontal zone mixing and Antarctic bottom water formation in the Southern Weddell Sea. Deep-Sea Res 23:301–307

Gamberoni L, Geronimi J, Jeannin PJ, Murail JF (1982) Section hydrologique le long du 66°30E entre les latitudes 40° et 62° Sud. Campagne ANTIPROD 1–MD 12 Oceanol Acata 5:289–299

Glibert PM (1982) Regional studies of daily, seasonal and size fractions variability in ammonium remineralization. Mar Biol 70: 209–222

Groupe Médiprod (1978) Résultats de la Campagne ANTIPROD 1 "Marion-Dufresne" 1–28 mars 1977, Res Camp Mer CNEXO 16 pp 94

Groupe Médiprod (1982) Résultats de la Campagne ANTIPROD 2 "Marion-Dufresne" 25 février – 18 mars 1980. CNFRA 53: 55–80

Jacobs SS, Georgi DI (1977) Observations on the Southwest Indian Antarctic Ocean. In: "A voyage of Discovery, G Deacon, 70th Anniversary". Deep-Sea Res vol suppl :43–84

Jacques G (1983) Some ecophysiological aspects of the antarctic phytoplankton. Polar Biol 2:27–33

Le Corre P, Tréguer P (1976) Contribution à l'étude de la matière organique dissoute et des sels nutritifs dans l'eau de mer. Caracteristiques chimiques du Golfe de Gascogne et des upwelling côtiers de l'Afrique du Nord-Ouest. Thèse Doctorat ès Sciences UBO Brest pp 490

Le Jehan S (1982) Contribution à l'étude des matières organiques et nutritives dans deux écosystèmes eutrophes: Cycles de l'azote, du phosphore et du silicum. Thèse de spécialité. UBO Brest pp 236

Le Jehan S, Tréguer P (1983) Uptake and regeneration $\Delta Si/\Delta N/\Delta P$ ratios in the indian sector of the Southern Ocean. Originality of the biological cycle of silicon. Polar Biol 2:127–136

Minas HJ (1979) Analyse de diagrammes de facteurs hydrologiques et chimiques (temperature, salinité, oxygène, sels nutritifs). Application à l'étude du systèmes production-régeneration dans les résurgences côtieres (côtes NW Africaines) et les zones à fort mélange vertical (Mediterranée Océan Antarctique). In: "Production Primaire et Secondaire, Colloque franco-soviétique". Publ CNEXO Actes du colloque 10:21–36

Murail JF, David P, Panouse M (1977) Résultats scientifiques de la campagne MD04/BENTHOS. Hydrologie du plateau continental des îles Kerguelen. CNFRA 42:41–64

Nelson DM, Gordon LI (1982) Production and pelagic dissolution of biogenic silicon in the Southern Ocean. Geochim Cosmochim Acta 46:491–501

Olson RI (1980) Nitrate and ammonium uptake in Antarctic waters. Limnol Oceanogr 25:1064–1074

Redfield AC, Ketchum BH, Richards FA (1963) The influence of organisms on the composition of seawater. In: Hill MN (ed) "The Sea". New York, London 2:26–77

Richards FA (1958) Dissolved silicate and related properties of some western north Atlantic and Carribean waters. J Mar Res 17:449–465

Tréguer P, Le Corre P (1975) Manuel d'analyses des sels nutritifs dans l'eau de mer. Utilisation de l'Auto Analyzer 11 Technicon. LOC-UBO. 2° edition Brest pp 110

Trace Element Geochemistry in the Southern Ocean

M.J. ORREN and P.M.S. MONTEIRO[1]

Summary. Water samples were collected on 2 surveys (1980, 1981) in the Southern Ocean south of South Africa, and these were analyzed for the trace elements copper, zinc and cadmium, in addition to the usual oceanographic parameters (temperature, salinity, dissolved oxygen, phosphate, nitrate and silicate). The results showed that, unlike other areas of the world ocean, there is little evidence that the trace elements follow the systematics of the nutrients. This is ascribed to the low productivity of the area. Lateral advection was found to affect distributions more than vertical processes. The oxygen minimum associated with Circumpolar Deep Water, advected into the region *via* the Drake Passage, was unexpectedly found to be a layer of minimum Cu concentration. Zinc and Cd showed little variation in the deep water, having values characteristic of those found elsewhere in the world ocean. The copper-silicate ratio was found to be a useful tracer, and was used to infer an input of dissolved silicate from the sediments into the bottom water.

1 Introduction

Recent advances in both the accuracy and the precision of methods used to determine trace metals in seawater have made it possible to study the systematics of their distribution in the ocean. These advances have been mainly concerned with realizing the extent, and, subsequently, minimizing the problem, of contamination at the sampling stage (Bewers and Windom 1982).

Transition-series metals are characterized by significant differences in structural and thermodynamic characteristics which give them a wide diversity of chemical properties. This makes them particularly useful as a means of studying the formation and destruction of particulates which act as chemical carriers in the biogeochemical cycling of the elements in the water column. Biogeochemical cycling of chemical components can be divided into two different categories: the shallow regenerative cycle, and the regenerative cycle. The former comprises the labile elements, such as P and N, which are fixed within organisms during growth and are redissolved rapidly by oxidation in the surface layers. The second category comprises the elements, such as Cu and Zn, which are fixed in more stable solid phases (opal, calcite) whose rate of regeneration is slower. The particulates of this cycle are responsible for the scavenging process which depletes the surface layers of biologically active elements.

Trace metals, being ubiquitous components of sea-water, can participate in the cycling process, either actively by biological uptake or passively through adsorption or coprecipitation, and it is through the distribution of an element in the water column that conclusions may be drawn as to the dominant biogeochemical process. Sampling density for trace metals in the Southern Ocean is low and generally restricted to surface waters (Boyle and Edmond 1975), with a few exceptions for Cu (Boyle et al. 1977), Cd (Boyle et al. 1976) in the South Pacific and particulate trace metals in the southeast Indian Ocean (Harris and Fabris 1979). This study has as its objective the identification of the most important biogeochemical processes which affect the concentration and distribution of Cu, Zn, and Cd in a large area of the Atlantic-Indian sector of the Southern Ocean (Fig. 1).

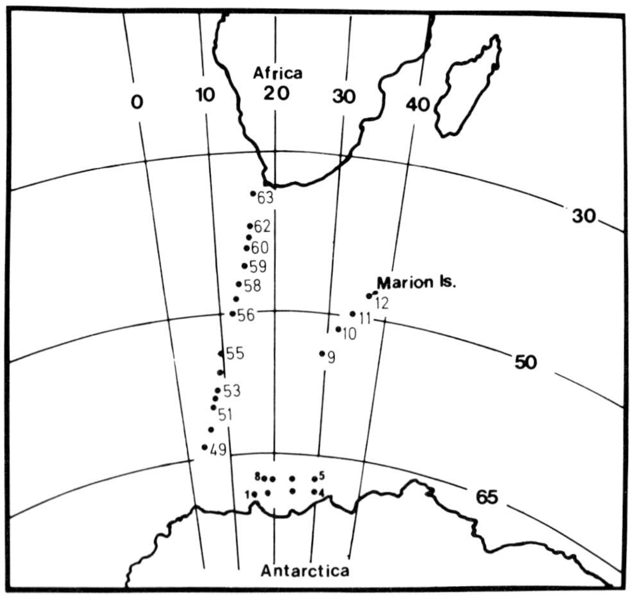

Fig. 1. Locations and numbers of selected stations sampled during February–March 1981

1 Department of Analytical Science, University of Cape Town, Rondebosch, 7700, South Africa

Antarctic Nutrient Cycles and Food Webs
(ed. by W.R. Siegfried, P.R. Condy, and R.M. Laws)

2 Materials and Methods

In addition to sampling for trace elements, at most stations measurements were made for temperature, salinity, silicate, phosphate, and primary production. Samples were collected in 8 l precleaned Niskin bottles suspended from a new stainless steel hydro-wire. The nutrients were determined ashore by standard auto-analyser techniques on unfiltered samples frozen at –15°C immediately after collection. Copper, Zn and Cd samples (unfiltered) of volume 1000 cm^3 were acidified to pH 1 with ARISTAR nitric acid and stored frozen at –20°C for shore analysis using atomic absorption-spectometry after a solvent extraction pre-concentration (Monteiro 1982). At all stages, the greatest care was taken to avoid metal contamination as outlined by Bewers and Windom (1982) and others. No direct intercalibration exercise for trace elements has been done, but the results show a good degree of oceanographic consistency and lie well within the ranges normally considered acceptable for marine trace analysis.

3 Results and Discussion

3.1 Nutrients

The vertical distribution of phosphate, a labile nutrient, is characterized by relatively high surface values (2 μmol kg^{-1}) increasing to a sharp mid-water maximum above 1000 m, below which there is minimal variability (Fig. 2). There is a deep minimum visible at Stations AG 9–11 which corresponds to the nutrient depleted core of North Atlantic Deep Water (NADW). The distribution of silicate is characterized by a gradual increase toward the bottom (Fig. 3), both as a result of its regenerative cycle and an input from

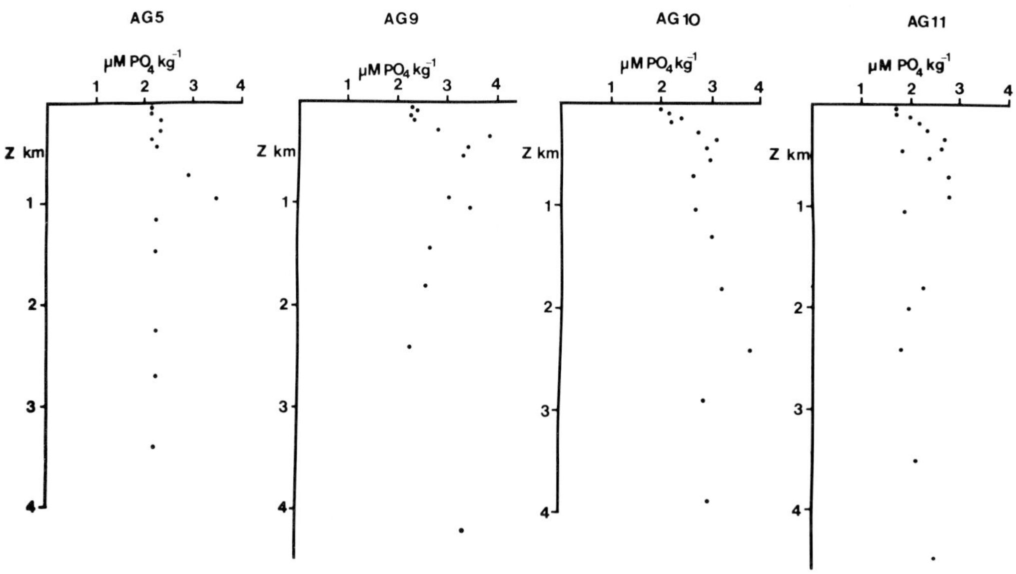

Fig. 2. Vertical distribution of phosphate at Stations AG 5, 9, 10, and 11 (see Fig. 1) during February–March 1981

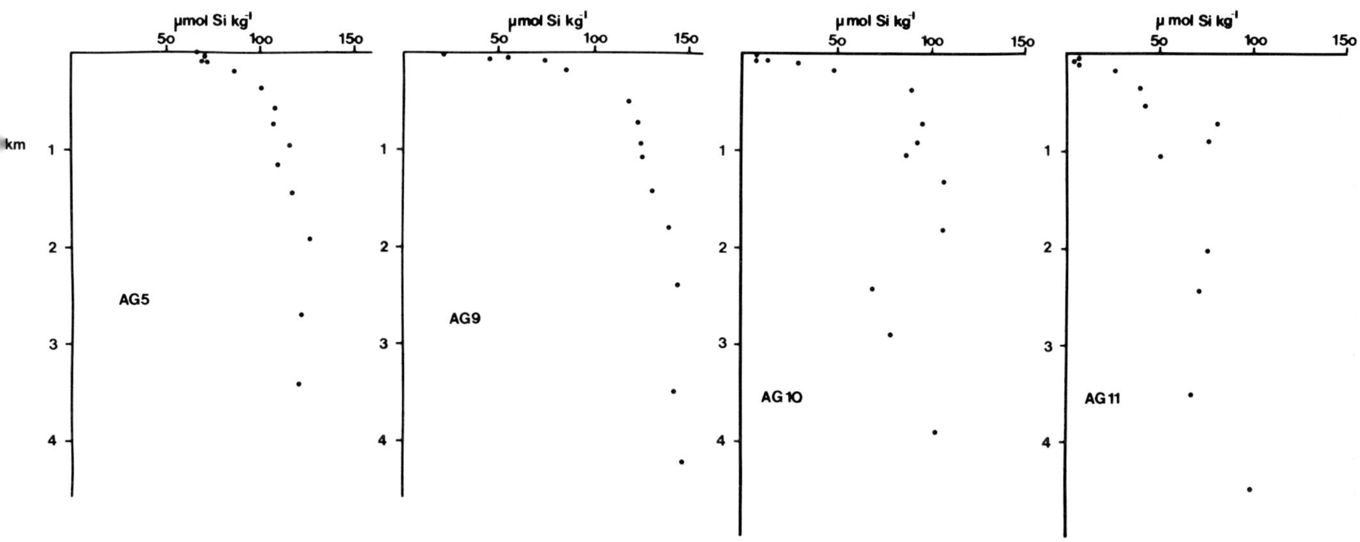

Fig. 3. Vertical distribution of silicate at Stations AG 5, 9, 10, and 11 (see Fig. 1) during February–March 1981

pelagic sediments in the Enderby basin (Edmond et al. 1979).

The concentrations of the chemical components in the surface layers were generally higher than those found in the North Atlantic (Yeats and Campbell 1983) and North Pacific (Bruland 1980). This is thought to reflect upwelling of enriched deep water, since the elemental ratios correspond closely to the ones found elsewhere.

3.2 Trace Elements

The three trace metals are dealt with separately below but, as a whole, the trend is that there is a surface depletion increasing to a mid-water maximum whose depth varies and below which the concentrations remain nearly constant. The exception to this is Cu at Station AG 9 where there is evidence of a significant bottom input. The near-surface concentrations show no evidence of any significant aeolian input.

3.2.1 Copper

The vertical profiles show that, to varying degrees, Cu is depleted near the surface and increases along a gradient to a sub-surface maximum layer in the upper thermocline (Fig. 4). The mean surface concentration at the southern group of stations (AG 1–8) is 0.98 nmol kg^{-1} and increases meridionally to a maximum of 1.73 nmol kg^{-1} at Station AG 10 which lies at the Polar Front zone (Table 1).

The surface Cu distribution is, as was suggested by Bruland (1980) and Yeats and Campbell (1983), governed by the processes of oxidative regeneration and scavenging. This was also found to be so in this study, on the basis of the relationship between Cu and nutrients and primary production. The Cu and nutrient data (Appendix 1) show that these chemical components are not growth limiting. These high surface concentrations are maintained by the upwelling of deep enriched water to the euphotic zone. It has been further observed that whereas central Pacific waters are more depleted in Cu (0.5 nmol kg^{-1}) those closer to the coastal upwelling areas have higher surface concentrations (1–2 nmol kg^{-1}) (Bruland 1980). Similarly, a transect of the Southern Ocean in the southwest Pacific showed surface concentrations to lie between 0.98 and 3.25 nmol kg^{-1} (Boyle and Edmond 1975).

Two important points should be noted from Fig. 4: the magnitude of the depletion of Cu at each station and how it changes meridionally; and, the depth of the sub-surface maximum which results from the regeneration-scavenging processes. Further, there are in this region two probable mechanisms which regulate this flux: the rate of primary production; and, the settling rate of the solid carrier phases. The lack of any particulate phase data precludes any calculations of fluxes of particulate Cu, but primary produc-

Table 1. Meridional changes in concentrations of copper and silicate (depth ≤ 50 m) in the Atlantic-Indian Sector of the Southern Ocean in February–March 1981

Station no.	Latitude (nominal) (°S)	Copper (nmol kg^{-1})	Silicate (μmol kg^{-1})	Cu/Si 10^{-3}
1	68	1.00	52.5	19.5
5	67	0.68	70.4	9.7
9	54	1.24	37.7	32.9
10	52	1.73	7.5	230.7
11	49	1.40	6.5	215.4

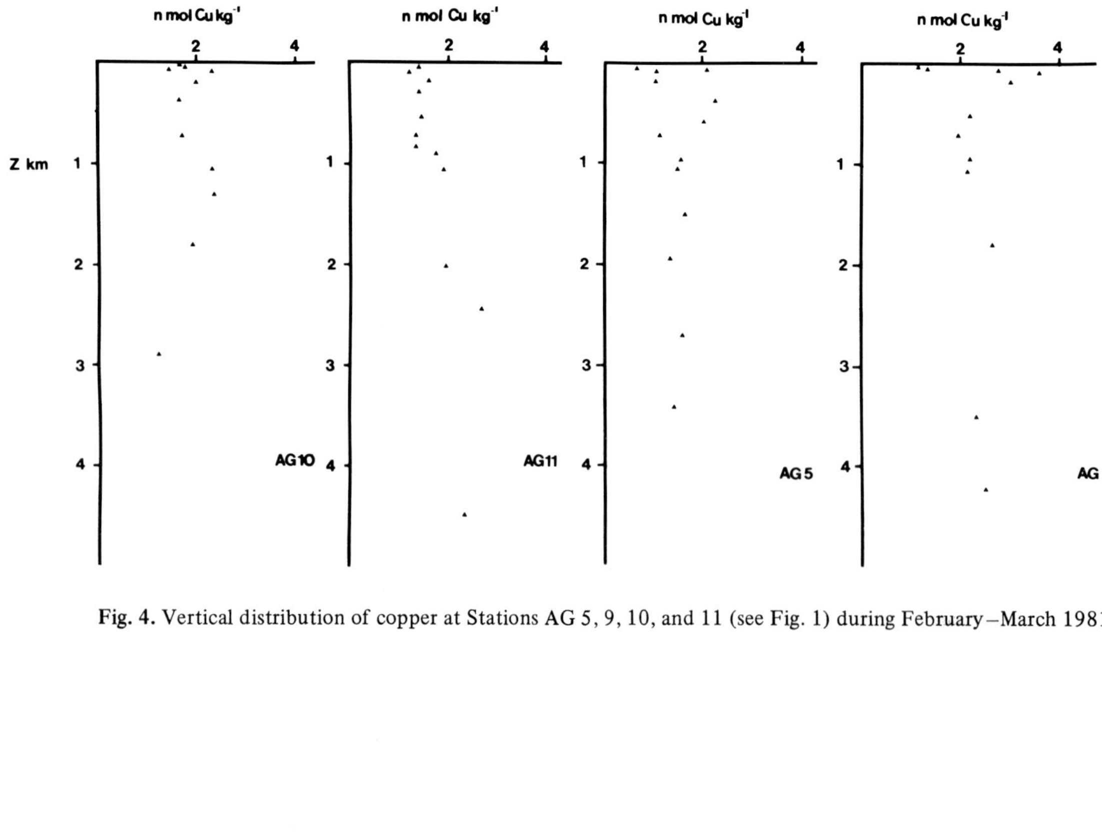

Fig. 4. Vertical distribution of copper at Stations AG 5, 9, 10, and 11 (see Fig. 1) during February–March 1981

Table 2. Meridional variability of primary production (x, y = intermediate production stations) in the Atlantic-Indian Sector of the Southern Ocean in February–March 1981

Station no.	Latitude (nominal) (°S)	Production mg C m^{-2} d^{-1}
2	68	157.9
3	68	181.9
4	68	244.9
5	67	227.9
7	67	80.1
8	67	129.4
x	61	143.5
y	58	204.9
9	54	212.6
11	49	305.5

tion data are used to link the observed distributions with biochemical processes.

Table 2 gives the measured primary production (PP) values integrated over depth to the 1% light level (Henry 1983, priv. comm.). Primary production rates vary considerably from station to station but on the meridional leg (Stations AG 8–11) there is a large northward increase in the productivity. If Cu distribution is indeed controlled by biological uptake, the dissolved concentration should be correlated negatively to the primary production. Most points fall within a straight line described by $Cu = -5 \times 10^3 \times P.P. + 2.00 (r^2 = 0.82)$. The points on the line all correspond to surface waters of Stations AG 1–9. The non-correlated points above the line (from Stations AG 10–11) cannot be explained by changes in upwelling or primary production rates.

The explanation is thought to lie in the fact that there is a progressive increase in the rate of regeneration of copper with the northward warming of the surface layers and its impact on rates of oxidation of organic matter. The net result of this effect is that there is a decoupling of the regeneration cycles of Cu and silicate, (Fig. 4) where the subsurface gradient for these elements correlates well at Stations 5–9 but not farther north at Stations 10–11. This raises the possiblity of using the different regeneration cycles for different trace elements as an indirect means to obtaining information on the regeneration/scavenging processes which dominate a given system. It also stresses the need for more data on the particulate phases prior to any serious modelling attempts being made.

3.2.2 Zinc

The Zn profiles in Fig. 5 show that this element is characterized by a surface minimum, whose value ranges from 15.5 nmol kg^{-1} at Station AG 5 to 5.38 nmol kg^{-1} at Station AG 11. At Stations AG 9–11 this depleted surface layer concentration increases over a sharp gradient to a mid-water maximum lying between 200 and 500 m. Below this depth the concentration falls close to the level of 15 nmol kg^{-1} with a slight increase with depth.

The vertical distribution shows close similarities with that of phosphate above 1000 m and silicate below this depth. Work carried out in the Pacific Ocean has shown a good correlation between Zn and silicate (Bruland 1980; Bruland et al. 1978). This result is verified both with surface data (Fig. 6) and with data for the whole of the water column (Fig. 7). The latter, however, only shows a good fit with data from Stations AG 9–11 which are used for the plot. This finding confirms the result obtained in the case of Cu, that oxidative regeneration releases the Zn into the dissolved state where it is scavenged by the predominantly silicate-based particulates settling down the water column.

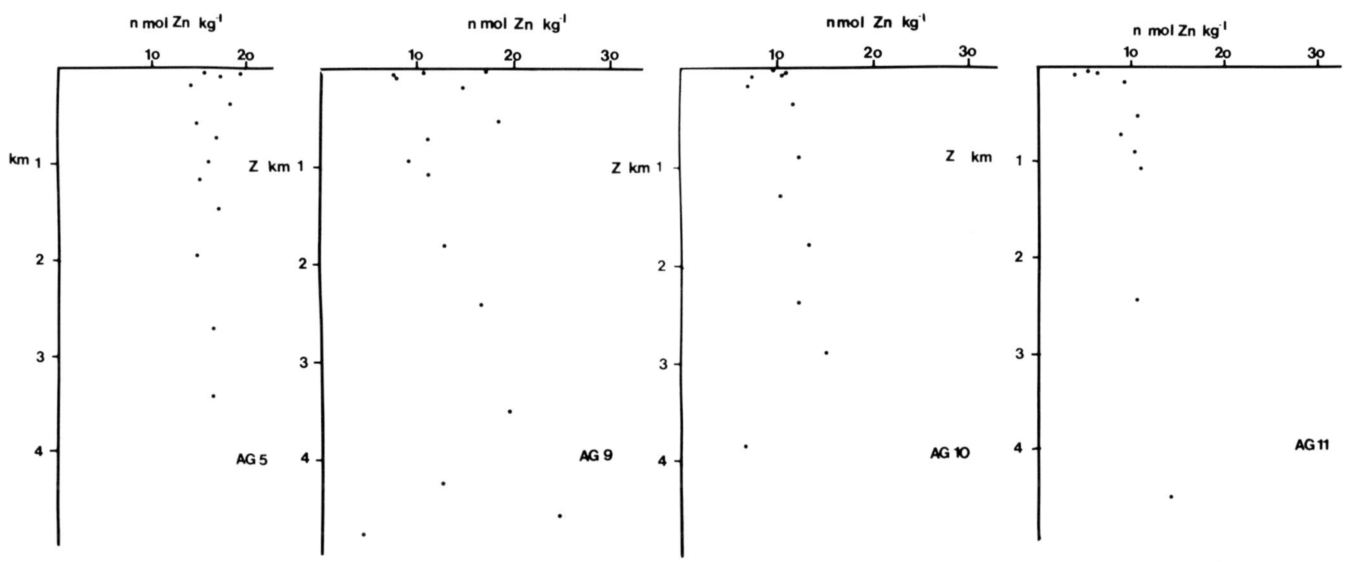

Fig. 5. Vertical distribution of zinc at Stations AG 5, 9, 10, and 11 (see Fig. 1) during February–March 1981

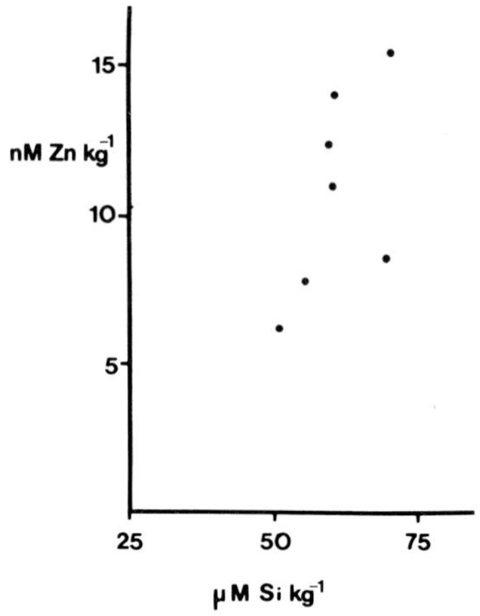

Fig. 6. The relationship between concentrations of zinc and silicate in surface waters (Z<50 m) at Stations AG 5, 9, 10, and 11 (see Fig. 1)

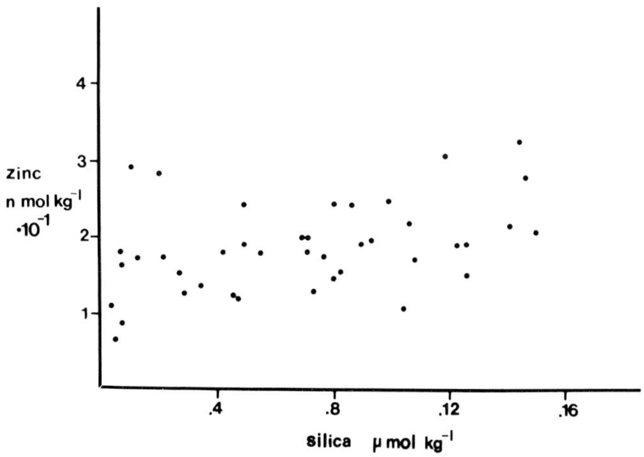

Fig. 7. The relationship between concentrations of zinc and silicate for the whole of the water column at Stations AG 9–11 (see Fig. 1)

This process is effective under the warmer surface water conditions north of Station AG 9. It is clear that due to these types of problems, it is difficult to define the lability of a metal only on the basis of the total dissolvable fraction and, in fact, particulate studies should be carried out should such results be required.

The mean Zn (nm):Si (μM) ration for the southerly group of stations is 1.6×10^{-4}, which is of the same order of magnitude as are the results from the Pacific where these values lie between 0.6 and 1.1×10^{-4} (Bruland 1980). The ratio increases northward where Stations AG 9–11 have a mean of 9.9×10^{-4}, as a result of the relative increase of Zn and the depletion of silicate in surface warmer waters. This pattern agrees with that of the Cu data.

3.2.3 Cadmium

The vertical distributions of Cd at Stations AG 5, 9, 10, 11 are shown in Fig. 8. These are characterized by weak subsurface gradients and a homogeneous distribution in deep water. The surface water values (Table 3) vary from 0.91–1.28 nmol kg^{-1} in the southern group of stations (AG 1–8) and decrease meridionally to a low value of 0.55 nmol kg^{-1} at Station AG 11.

As has been observed for silicate, phosphate, Cu, and Zn, the values for Cd are higher than those found in the North Atlantic (Yeats and Campbell 1983) and Pacific (Bruland 1980; Boyle et al. 1976), but this may again be accounted for by the upwelling of enriched deep water to the surface. This may be verified by comparing the ratio Cd:PO_4 from this study with those obtained elsewhere. A mean ratio of 4.0×10^{-4} (nmol : μmol) compares well with a mean of 2.7×10^{-4} in the North Atlantic (Yeats and Campbell 1983) and 3.5×10^{-4} in the Pacific (Boyle et al. 1976; Bruland 1980). Further, measurements made with *Sargas-*

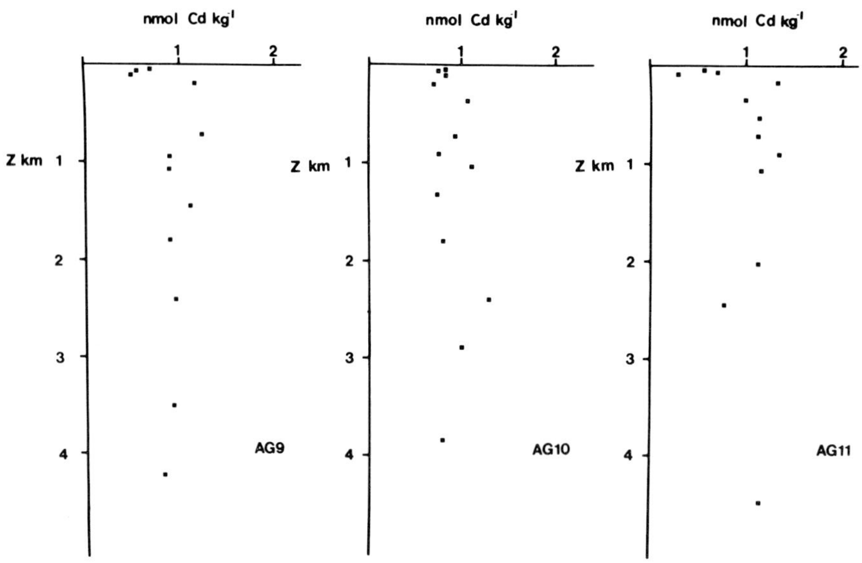

Fig. 8. Vertical distribution of cadmium at Stations AG 9–11 (see Fig. 1) during February–March 1981

Appendix 1. Chemical data for seawater sampled at four stations in the Southern Ocean during February–March 1981 in the Atlantic-Indian Sector of the Southern Ocean

Station no.	Depth (m)	Silicate (μmol kg^{-1})	Phosphate (nmol kg^{-1})	Copper (nmol kg^{-1})	Cadmium (nmol kg^{-1})	Zinc (nmol kg^{-1})
AG 5	50	70.40	2.16	0.68	1.10	15.50
	70	70.10	2.10	2.09	1.28	19.30
	90	71.30	2.15	1.07	1.10	17.20
	188	86.40	2.34	1.05	1.19	14.00
	377	100.40	2.17	2.27	1.01	18.30
67°20′S	568	107.20	2.12	2.01	0.82	14.80
28°00′E	720	106.90	2.92	1.15	1.10	16.80
	960	115.80	3.48	1.56	1.10	16.00
	1150	109.80	2.29	1.51	1.01	15.00
	1445	117.00	2.28	1.65	1.19	17.10
	1933	126.50	2.55	1.33	1.01	14.90
	2700	121.70	2.28	1.58	1.28	16.60
	3420	120.80	2.27	1.42	0.82	16.50
AG 9	20	20.70	2.63	1.14	1.47	17.10
	50	54.60	2.31	1.34	0.70	10.80
	70	45.50	2.30	2.78	0.56	7.73
	88	73.10	2.39	3.59	0.50	7.89
	181	84.90	2.56	3.12	1.17	14.80
54°03′S	520	118.20	3.12	2.17	–	18.40
29°22′E	720	122.50	4.18x	1.95	1.24	11.20
	941	124.80	3.03	2.22	0.90	9.08
	1080	125.80	3.55	2.14	0.89	11.20
	1421	130.60	2.67	4.26x	1.11	x23.70
	1800	139.30	2.52	2.65	0.89	12.90
	2394	144.70	2.73	3.83x	0.97	16.70
	3492	143.80	4.50x	2.32	0.93	19.70
	4230	148.60	3.39	2.54	0.82	12.60
AG 10	20	7.40	2.81	1.67	–	9.70
	50	7.50	1.99	1.79	0.83	10.90
	70	13.30	2.31	1.46	0.77	10.60
	87	29.10	2.16	2.33	0.83	7.69
	184	48.00	2.20	2.00	0.72	6.93
	366	89.00	3.10	1.65	1.06	11.70
51°40′S	720	95.00	2.66	1.71	0.95	x20.80
32°00′E	900	92.20	4.61x	4.14x	0.78	12.10
	1037	87.30	2.73	2.33	1.11	–
	1311	107.00	3.04	2.38	0.76	10.30
	1800	106.00	3.23	1.92	0.81	13.30
	2389	69.60	3.84	2.84x	1.30	12.20
	2889	79.40	2.35	1.21	1.03	15.00
	3845	103.00	3.01	–	0.81	6.77
AG 11	50	6.50	1.68	1.40	0.55	5.38
	70	4.20	2.02	3.06	0.71	6.43
	90	6.30	1.72	1.20	0.29	3.99
	170	26.60	2.15	1.62	1.33	9.41
	348	39.80	2.70	1.43	1.00	18.00
49°20′S	529	41.80	2.36	1.46	1.14	10.70
34°51′E	720	80.10	2.93	1.35	1.13	8.92
	889	76.80	2.77	1.73	1.34	10.40
	1060	50.00	1.86	1.90	1.18	11.20
	2019	75.40	1.96	1.97	1.15	18.30
	2430	71.10	1.81	2.69	0.80	10.80
	4493	98.70	2.53	2.34	1.18	14.70

Table 3. Cadmium concentrations in surface waters (depth ≤ 50 m) in the Atlantic-Indian Sector of the Southern Ocean in February–March 1981

Station no.	Latitude (nominal) (°S)	Cadmium nmol kg^{-1}	Phosphate μmol kg^{-1}	Cd/PO_4 10^{-4}
1	68	1.16	1.61	7.2
2	68	1.28	2.10	6.1
3	68	0.55	2.02	2.7
4	68	0.82	2.13	3.8
5	67	1.10	2.16	5.1
6	67	0.91	2.29	4.0
7	67	1.28	2.17	5.9
8	67	1.03	2.19	4.7
9	54	1.47	2.63	5.6
10	52	0.83	1.99	4.2
11	49	0.55	1.76	3.1
12	47	1.16	1.68	6.9

sum and planktonic animals (Boyle et al. 1976) put the ratio at 4.0×10^{-4} and 6.5×10^{-4}, respectively.

On the basis of two separate studies carried out in the Pacific Ocean (Boyle et al. 1976; Bruland 1980) it was found that the vertical distribution of Cd correlated well with phosphate. We plotted our surface Cd data against phosphate values to establish whether biological demand for Cd corresponded to that of phosphate. Figure 9 shows that most of the points do yield some correlation with the exception of five points with higher Cd values. These values were, however, obtained from Stations AG 3, 7, and 8, all of which are characterized by very low production levels. It is suggested, therefore, that Cd does indeed participate actively in the biogeochemical cycle and that it is regenerated rapidly near the surface at the senescent stage of a bloom. In deep water, the mean Cd concentration does not vary significantly from 1.00 nmol kg^{-1} which suggests that the vertical flux is very small and that there is no bottom input from interstitial waters.

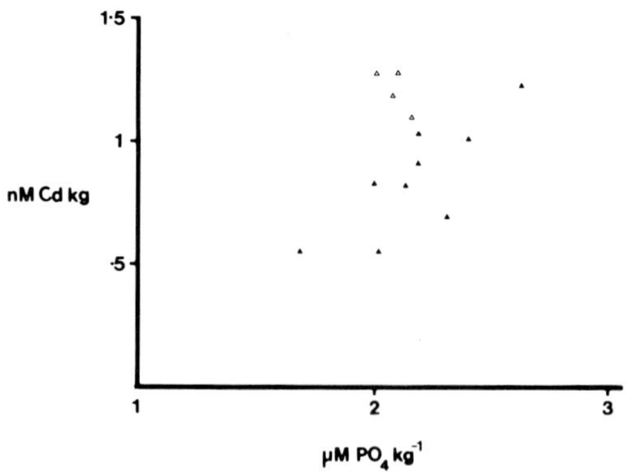

Fig. 9. The relationship between concentrations of cadmium and phosphate in surface waters (Z<50 m) at Stations AG 1–12 (see Fig. 1)

4 Conclusions

The main conclusion is that there is a close relationship between biochemical production in the surface layers of the Southern Ocean and the spatial variation of trace metal concentration. It was not established if any of these chemical components was growth limiting but due to the extent of upwelling, and the high micronutrient element concentrations, this is not thought to be a problem to production.

The trace metal concentrations (Cu, Zn, Cd) in surface waters are elevated with respect to those in the central gyres of both the Atlantic and Pacific oceans. This is thought to be a result of the upwelling of deep waters, as the aeolian component is minimal in this part of the Southern Ocean.

There is evidence that the depth of the regeneration cycle of each element changes as a function of temperature, which presumably alters the rates of dissolution and bacterially mediated oxidation. The rates of regeneration of chemical components have to be carefully examined to determine the precise relationship between chemical characteristics and local fluxes. The rates of regeneration are thought to be the key towards explaining the variation in rates of primary production in the euphotic layer of the Southern Ocean.

This study has set the groundwork for a future, more detailed, investigation of two specific aspects of the biogeochemistry of trace elements in the Southern Ocean: an investigation of trace metal concentrations above the main thermocline, including the dissolved and particulate phases together with measurements of primary production, POC, DOC, and photosynthetic pigments (this work would yield a better understanding of the biogeochemical systematics of trace elements with respect to local production levels and, hence, a more accurate mass balance can be calculated); and, the sediments underlying these waters are thought to be an important source of silicate and Cu, hence a closer look has to be taken at the composition of the sediments and interstitial waters, to derive a better understanding of the geochemical nature of the system.

Acknowledgements. This work was supported financially by the South African National Committee on Oceanographic Research (SANCOR) and the South African Scientific Committee on Antarctic Research (SASCAR).

References

Bewers JM, Windom HL (1982) Comparison of sampling devices for trace metal determination in sea water. Mar Chem 11(1):71–86

Boyle EA, Edmond JM (1975) Copper in surface waters south of New Zealand. Nature (Lond) 253:107–110

Boyle EA, Sclater FR, Edmond JM (1976) On the marine geochemistry of cadmium. Nature (Lond) 263–42–45

Boyle EA, Sclater FR, Edmond JM (1977) The distribution of dissolved copper in the Pacific. Earth Planet Sci Lett 37:38–54

Bruland KW (1980) Oceanographic distributions of cadmium, zinc, nickel and copper in the north Pacific. Earth Planet Sci Lett 47: 176–198

Bruland KW, Knauer GA, Martin JH (1978) Zinc in the north Pacific waters. Nature (Lond) 271:741–743

Edmond JM, Jacobs SS, Gordon AL, Mantyla AW (1979) Water column anomalies over opaline pelagic sediments and the origin of the deep silica maximum. J Geophys Res 84(12):7809–7826

Harris JE, Fabris GJ (1979) Concentrations of suspended matter and particulate cadmium, copper, lead and zinc in the Indian sector of the Antarctic ocean. Mar Chem 8(2):163–179

Monteiro PMS (1982) M Sc Dissertation

Yeats PA, Campbell JA (1983) Nickel, copper, cadmium and zinc in the north west Atlantic ocean. Mar Chem 12(1):43–58

A Contribution to the Oceanology of the Prince Edward Islands

B.R. ALLANSON, B. BODEN, L. PARKER, and C. DUNCOMBE RAE[1]

Summary. No evidence was found for upwelling of deep Antarctic water in the lee of the Prince Edward Islands in the southern Indian Ocean, notwithstanding the occurrence of strong westerly winds. This is emphasized by a comparison of thermoclinal slopes in the lee of Prince Edward Island and of Marion Island. Support also comes from surface contours of reactive nitrate, phosphorus, silica, and chlorophyll-*a* concentrations which decrease with distance from the islands. These contours suggest that primary productivity in the sea around the islands is locally enhanced through 'island mass effect'.

1 Introduction

The Prince Edward Islands, embracing Marion Island and Prince Edward Island (Fig. 1), are situated in the West Wind Drift. In common with many islands of the Southern Ocean, they are important breeding sites for marine birds and mammals (Van Zinderen Bakker 1971). The "island mass effect" (Raymont 1980) has often been used to explain the maintenance of the biomass of the birds and mammals and their production. However, there is no adequate explanation of either the validity of this view or of the mechanisms involved.

Hydrographic and biological studies in the lee of the Prince Edward Islands (Grindley and Lane 1979) suggest, on the basis of indicator zooplankton species, that upwelling of deep Antarctic water under the stress of strong predominantly north-westerly winds is the primary mechanism responsible for the productivity of the seas in the vicinity of the islands. However, El-Sayed et al. (1979) argued that the low silica concentrations in the surface water apparently militate against this hypothesis, notwithstanding a high concentration of phytoplankton at the time of sampling. These observations were made during a single visit to the islands in 1979. Since then, we have had the opportunity of increasing the grid of stations occupied and the periods when the islands were visited. Consequently, the seas surrounding the islands have been studied under varying meteorological conditions, including those which would have amply confirmed the hypothesis of Grindley and Lane (1979). However, we have been unable to do so to date and, consequently, present data here in support of an alternative explanation of the fertility of the sea around the P.E. Islands. Our evidence derives from the results of oceanographic studies designed, as far as time and equipment would permit, to describe physical, chemical and planktonic features of the seas surrounding the islands.

2 Methods

Data were collected at 26 oceanographic stations (Fig. 1) during voyages of the M.V. S.A. Agulhas in April–May 1982, and in May and November 1983. At each station, casts were made, using a Neil Brown conductivity temperature depth probe (CTD), and chemical samples were collected by Niskin bottles set at 50 m intervals down to

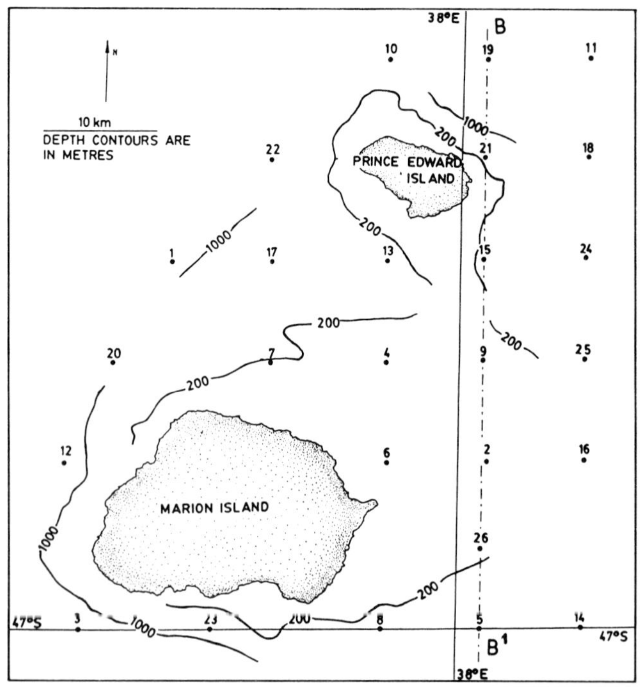

Fig. 1. Oceanographic Station (*1–26*) and transect *B–B*[1] at the Prince Edward Islands

1 Department of Zoology and Entomology, Rhodes University, Grahamstown 6140, South Africa

Antarctic Nutrient Cycles and Food Webs
(ed. by W.R. Siegfried, P.R. Condy, and R.M. Laws)

300 m. Phosphate as PO_4 (P), nitrate as NO_3 (N), nitrite as NO_2 (N), silicate as SiO_2 and chlorophyll-a were measured using the methods of Strickland and Parsons (1968). Ammonia as NH_4 (N) was measured according to Mackereth et al. (1978). Wind speeds were obtained from the South African Weather Bureau station on Marion Island.

Primary productivity of surface samples taken from each station was measured in an on-board incubator using the carbon-14 technique; specific details of our application of this technique are described by Allanson et al. (1981). The incubator maximum photosynthetic rate (P_{max}) was used to compare changes in the pattern of phytoplankton production throughout the study area. Unfortunately, no in situ incubations were possible. Zooplankton samples were collected by vertical hauls using a 0.6 m closing net (WP II, 200 μ mesh). Hauls were taken from 100 m to the surface, 300–100 m, and 500–300 m.

3 Results and Discussion

3.1 The Physical Environment

Based on vertical profiles of temperature and salinity (e.g., Fig. 2), a surface mixed layer occurred to 75 m at each station and a thermocline descended to about 200 m. Miller, Boden, and Parker (unpubl.) found isothermal conditions extending to 1,000 m in September 1982, confirming that stratification of the water column is seasonal.

Examination of surface isopycnals of sigma-t (Fig. 3) for May, 1983, shows a sloping isobaric surface in accordance with the simple rule that, in the southern hemisphere, the light water lies to the left of the denser waterr which is to the right of an observer looking in the direction of flow.

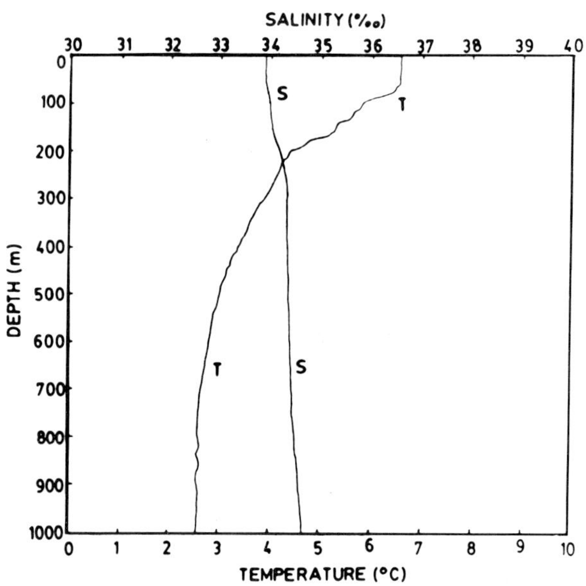

Fig. 2. Vertical profiles of temperature and salinity at Station 19 (see Fig. 1) in May 1983 and which are generally representative

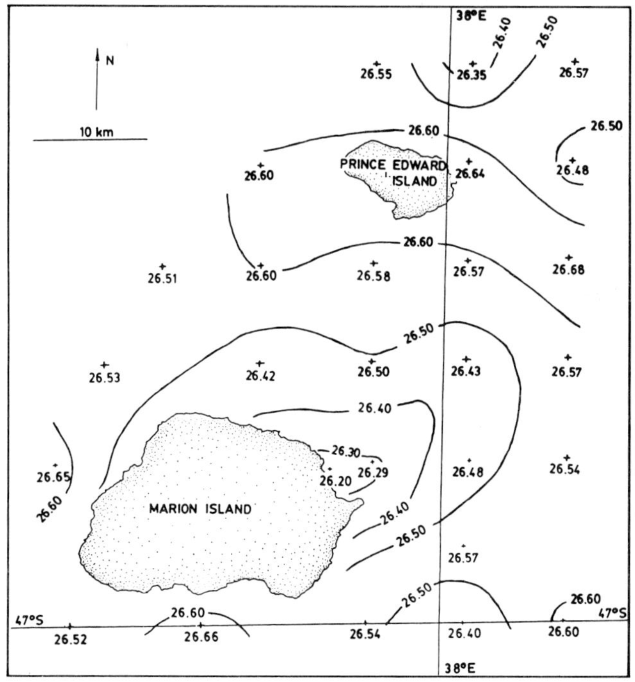

Fig. 3. Contoured surface values of sigma-t around the Prince Edward Islands in May 1983. Note plume of relatively low density water extending eastwards from Marion Island

Since isopycnals are assumed to be surfaces of equal potential density they may indicate direction of flow. Thus, because the isopycnals tend to be arranged circumferentially in the lee of Marion Island, it seems reasonable to expect the surface currents between the islands to match this pattern. An opportunity to test the likelihood occurred in November 1983. The currents observed then (by means of a Hydro Products current meter) are shown in Fig. 4, with those at 25 m being most strikingly correlated with the sigma-t pattern. The winds during this semi-synoptic survey were predominantly from the northwest at 37 km h^{-1}.

We argue that the eddy set up in the lee of Marion Island is the result of divergence of the mean easterly current flow around the island. Its existence depends not only on geostrophic effects, but also on the possibility that a major proportion of the kinetic energy of the easterly flow set up by the westerly winds is used to drive eddies of this kind. Wolanski and Imberger (1983) have demonstrated analytically how tidal currents passing islands on the Great Barrier Reef generate eddies in the horizontal velocity field in these coastal waters. That similar eddies, but of different origin, should occur in the lee of oceanic islands seems not unreasonable.

The origin of the decrease in density of the water towards the lee shore of Marion Island is uncertain, although the implication that it is due, at least in part, to run-off from the island is hard to dismiss, as the pattern of surface isohalines in Fig. 5 indicates. Alternatively, the reduced salinity in the immediate vicinity of the lee shore may represent containment of dilute seawater by the horizontal density gradient.

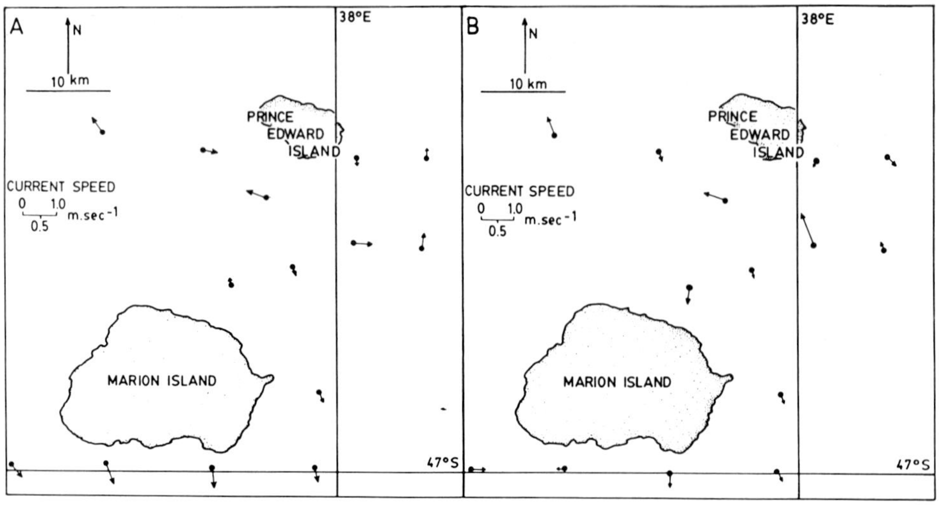

Fig. 4A, B. Current speeds and directions at two depths (surface **A**, 25 m **B**) around the Prince Edward Islands in November 1983

3.2. Nutrients

The surface conditions of PO_4 (P), NO_3 (N), NO_2 (N), and SiO_2 concentrations (Fig. 6) corroborate to some extent the sigma-t pattern. Good correlations were found between macro-nutrients and sigma-t for all depths and stations (Table 1).

The layer of buoyant water in the lee of Marion Island represents run-off that spreads out from the island to be incorporated in the main pattern of flow. This results in a tongue of nutrient-rich water to the east. The extent of this plume is not known, but its existence may be inferred from the elevated nutrient levels, and from the marked change in the surface sigma-t values, along the transect B–B[1] (Fig. 7). The ephemeral nature of the chemical structure of the shoal circuminsular waters is demonstrated in Table 2, showing that our values were consistently higher than those reported by Pascal (1979). This indicates that different watermasses were sampled. Vertical changes in concentrations throughout the depths sampled on both occasions were unremarkable (Table 2).

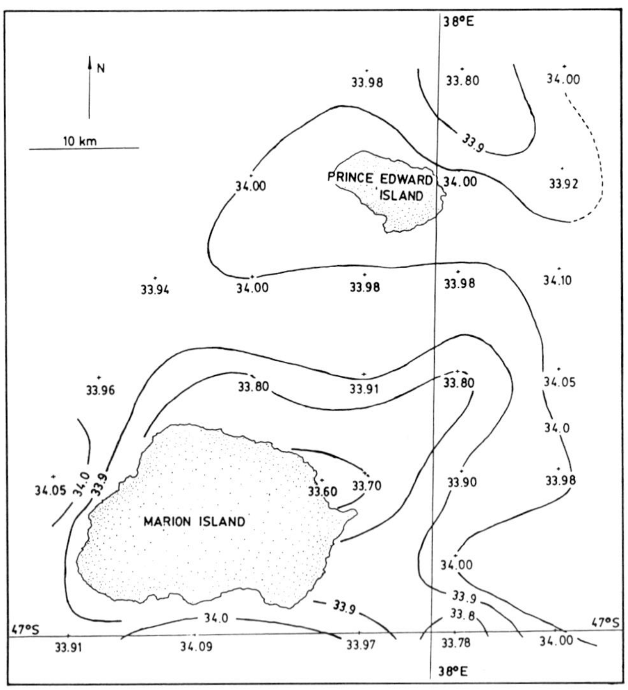

Fig. 5. Distribution of salinity (‰) at the surface around the Prince Edward Islands in May 1983

The ratio of NO_3 (N) to PO_4 (P) obtained in May, 1983, was 6.5:1 which is markedly lower than the ideal of 15:1 claimed by Cooper (1938) and later amended to 16.1 by Redfield et al. (1963). This may be attributed to a relatively high phosphate concentration in the surface waters, as a result of run-off from the islands. Pascal's (1979) data (Table 2) accord closely with the ideal value. The difference between two sets of values might be associated with the moulting of seabirds. According to Williams and Berruti (1978), an amount of 428,694 kg (dry wt.) of feathers is shed annually on Marion Island alone. About 25 % of the total feather mass is shed in March and this, allowing for decomposition, would leach a substantial nutrient contribution which could have affected our results in May. This does not negate Pascal's (1979) higher N:P ratio, obtained at roughly the same time of a different year; rather it emphasizes the fragility of the chemical structure of the system as noted above.

3.3 Phytoplankton and Primary Production

This account does not treat phytoplankton systematically or quantitatively. *Chaetoceros radicans* and *Rhizosolenia curvata* were in bloom at a few stations (notably Stations 1, 2 and 4) in May 1983 silicate values: 1.48–1.78 mmol m^3). El-Sayed et al. (1979) also found *C. radicans* in a bloom stage here, and speculated on its possible responsibility for a depleted silicate content of the water. The species was in bloom again in November, 1983. It is a temperate – sub-tropical form in the northern hemisphere, blooms

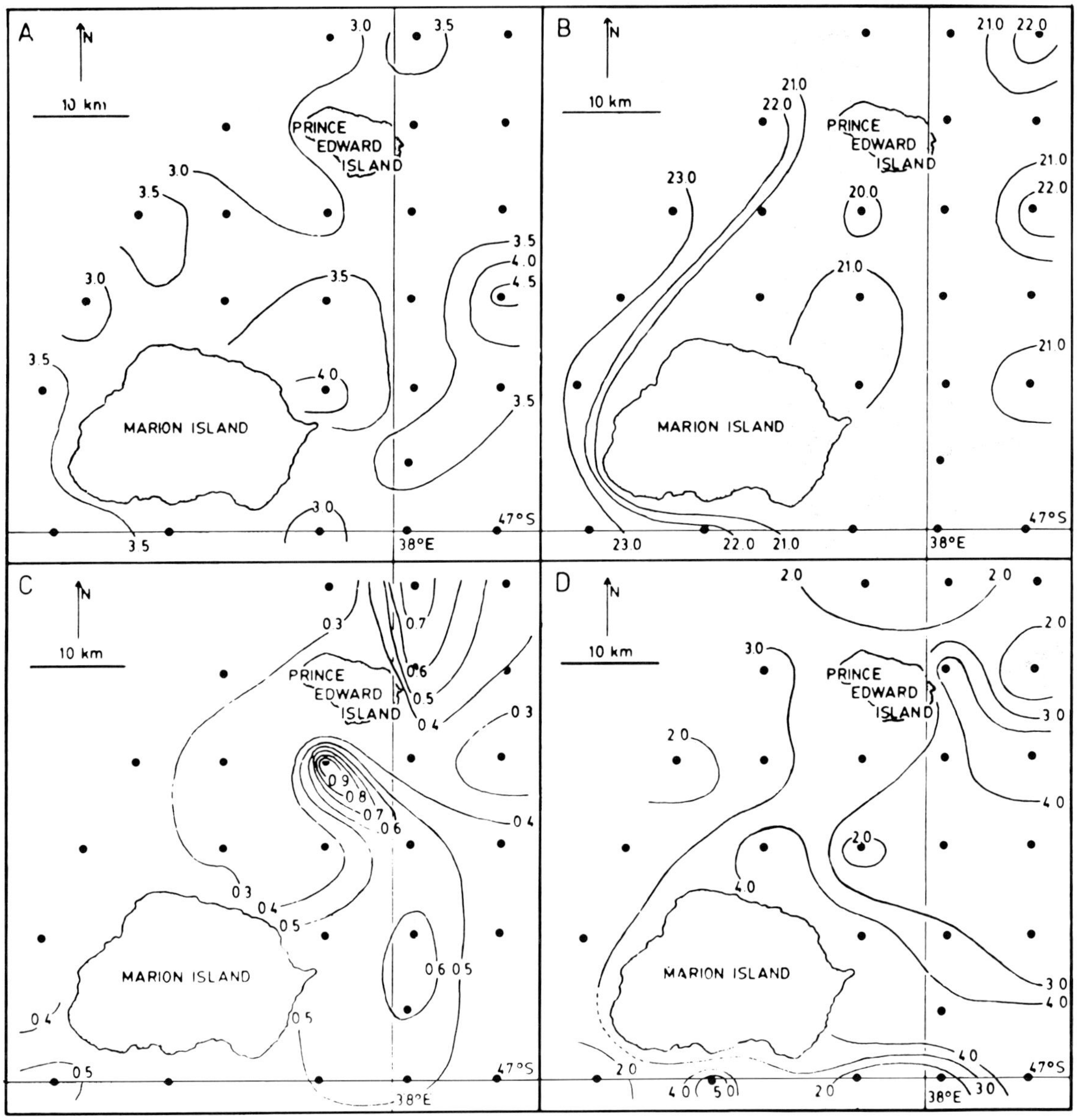

Fig. 6A–D. Contoured surface values of PO_4 (P), NO_3 (N), NO_2 (N) and SiO_2 in mmol m^3, indicated by **A, B, C** and **D**, respectively, around the Prince Edwards Islands in May 1983

readily, and is apparently bi-antitropical in temperate waters (i.e., it has a disjunct distribution on both sides of the tropics).

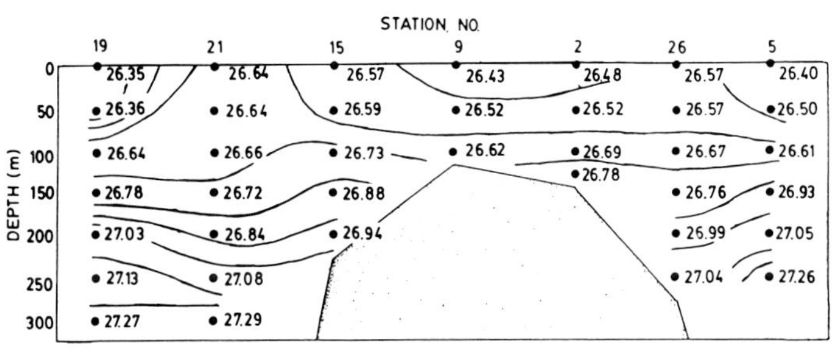

Fig. 7. Vertical sections of sigma-t along transect B–B[1] (see Fig. 1) in May 1983

Typical profiles of chlorophyll-a concentration are shown in Fig. 8A. The depth array revealed a maximum value at 50 m, decreasing from this depth to < 0.01 mg m^3 at 300 m. The real maximum is probably shallower and obscured by the wide sampling intervals. Maximum concentrations were found at Station 2, 4, and 5, in the lee of Marion Island (Fig. 9), and were associated with sharp gradients in isopycnals (Fig. 3).

The primary production potential (**PPP**) at each station suggests, with respect both to estimates of integral photosynthesis (Σ P) and to photosynthetic maximum (P_{max}), that the highest photosynthetic activity can be expected to occur in the lee of the islands (Fig. 10). The highest chlorophyll concentration is correlated with the highest estimate

Table 1. Pearson correlation coefficients for PO_4 (P), NO_3 (N), NO_2 (N), and SiO_2, and sigma-t at all depths and station around the Prince Edward Islands in May 1983

Sigma-t	PO_4 (P)	NO_3 (N)	NO_2 (N)	SiO_2 (N)	
Sigma-t	1.000 (n = 138) (p 0.001)	0.399 (n = 134) (p 0.001)	0.543 (n =132) (p 0.001)	– 0.520 (n = 132) (p 0.001)	0.896 (n = 133) (p 0.001)
PO_4 (P)	0.399 (n = 134) (p 0.001)	1.000 (n = 157) (p 0.001)	0.399 (n = 153) (p 0.001)	–0.273 (n = 153) (p 0.001)	0.366 (n = 154) (p 0.001)
NO_3 (N)	0.543 (n = 132) (p 0.001)	0.399 (n = 153) (p 0.001)	1.000 (n = 155) (p 0.001)	–0.389 (n = 155) (p 0.001)	0.530 (n = 153) (p 0.001)
NO_2 (N)	– 0.520 (n = 132) (p 0.001)	– 0.273 (n = 153) (p 0.001)	0.389 (n = 155) (p 0.001)	1.000 (n = 155) (p 0.001)	– 0.624 (n = 153) (p 0.001)
SiO_2	0.896 (n = 133) (p 0.001)	0.366 (n = 154) (p 0.001)	0.530 (n = 153) (p 0.001)	– 0.624 (n = 153) (p 0.001)	1.000 (n = 156) (p 0.001)

Table 2. Mean (± 1 S.D.) concentrations of macro-nutrients around the Prince Edward Islands in March 1976 (Pascal 1979) and May 1983 (this study)

Depth	PO_4	NO_3	NO_2	SiO_2	Year
5 – 87	1.19 ± 0.05	18.32 ± 0.66	0.26 ± 0.03	0.88 ± 0.44	1976
5 – 100	3.31 ± 0.57	21.62 ± 1.92	0.45 ± 0.16	3.95 ± 2.55	1983
125 – 275	1.50 ± 0.21	23.95 ± 3.93	0.23 ± 0.11	12.03 ± 0.89	1976
150 – 300	3.78 ± 0.50	24.47 ± 1.93	0.25 ± 0.12	13.19 ± 2.48	1983
5 – 275	1.32 ± 0.20	20.57 ± 3.72	0.24 ± 0.07	5.34 ± 7.00	1976
5 – 300	3.34 ± 0.60	22.93 ± 2.65	0.36 ± 0.18	8.43 ± 5.77	1983

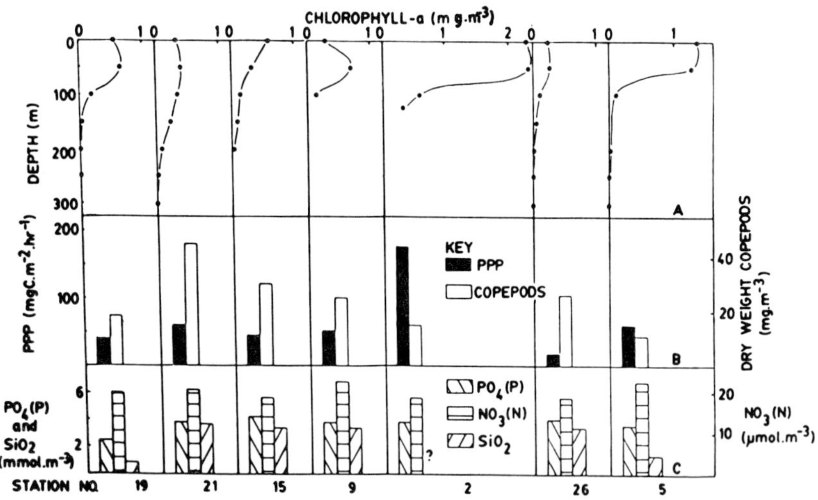

Fig. 8A–C. Vertical profiles of chlorophyll-a (*upper panel,* **A**), potential primary production and biomass of copepods (*middle panel,* **B**) and concentrations of PO_4 (P), NO_3 (N) and SiO_2 (*lower panel,* **C**) along transect B–B[1] (see Fig. 1) in May 1983

of integral production (Fig. 8A, B) but, although there may be hints, the data on reciprocal changes in PO_4 (P), NO_3 (N), and SiO_2 are equivocal (Fig. 8c). This is not surprising because the stations were occupied on different days under different meteorological conditions and, also, in view of the wind-field over the islands, modifications must be expected in the positions of the masses under wind stress. High PPP at two leeward sites (Fig. 10) requires explanation both as to the origin and the hydrodynamics maintaining its integrity.

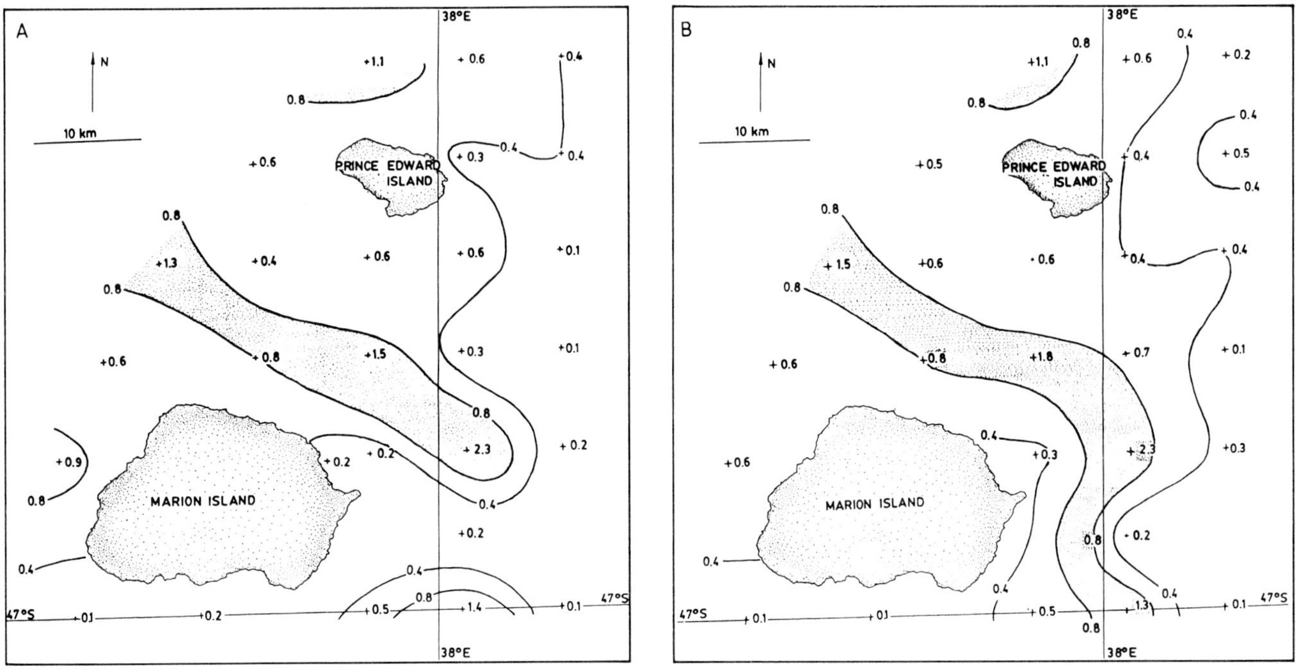

Fig. 9A, B. Contoured values of chlorophyll-a (mg m^3) at 2 depths (surface **A**; 50 m **B**) around the Prince Edward Island in May 1983

3.4 Zooplankton

Grindley and Lane (1979) reported the presence of a predominantly Antarctic copepod fauna in the sea surrounding the Prince Edward Islands, based largely on Vervoort's (1965) distributional study of "purely" Antarctic and Sub-Antarctic species. Although the ranges of these have since been extended their presence can be taken, by and large, as an indication of the presence of water of Antarctic origin either at present or in the near past.

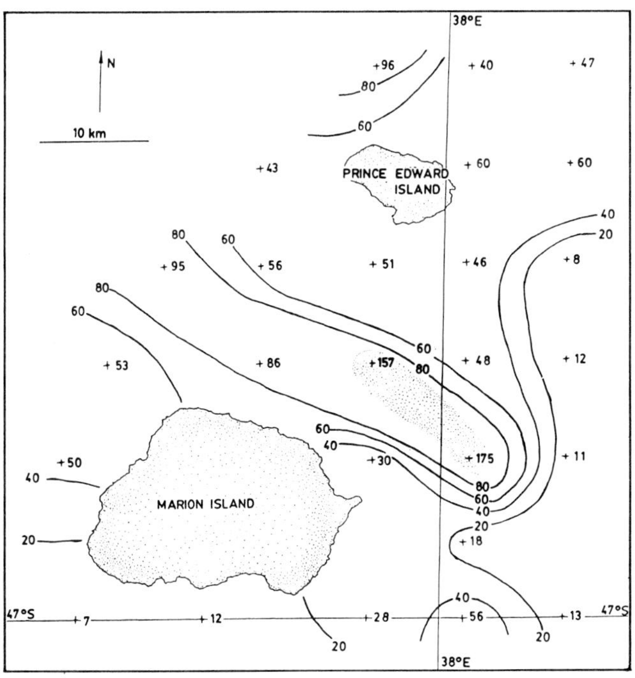

Fig. 10. Contoured values of potential primary production mg C m^2 h^{-1}) around the Prince Edward Islands in May 1983

A list of the zooplankton forms found by us at the Prince Edward Islands is contained in Appendix 1. The bulk and diversity of the zooplankton precluded detailed analyses for this report, but it is important to note that the collection is comprised of representatives from the Antarctic, sub-Antarctic, sub-tropical and even tropical zones. Hence, it is probable that the area surrounding the islands does not include an endemic holoplankton community, but one which is recruited from alien wandering water-masses. This suggestion is supported by the data on copepods in Fig. 11 which shows that their biomass in May 1983 was not matched in the preceding year at the same time.

4 Conclusions

Whereas the physical events and causes of upwelling along continental margins are comparatively well known (Schumann et al. 1972), understanding the physical properties of water flow around isolated islands has proved to be intractable, but no less important. We need to show how eddies can arise and act as biological containers (Wolanski et al., to be publ.). Complex current patterns in the vicinity of the Prince Edward Islands will arise not only from the effects on such pronounced planetary currents as the West Wind Drift, but also as a result of the stress of local wind-

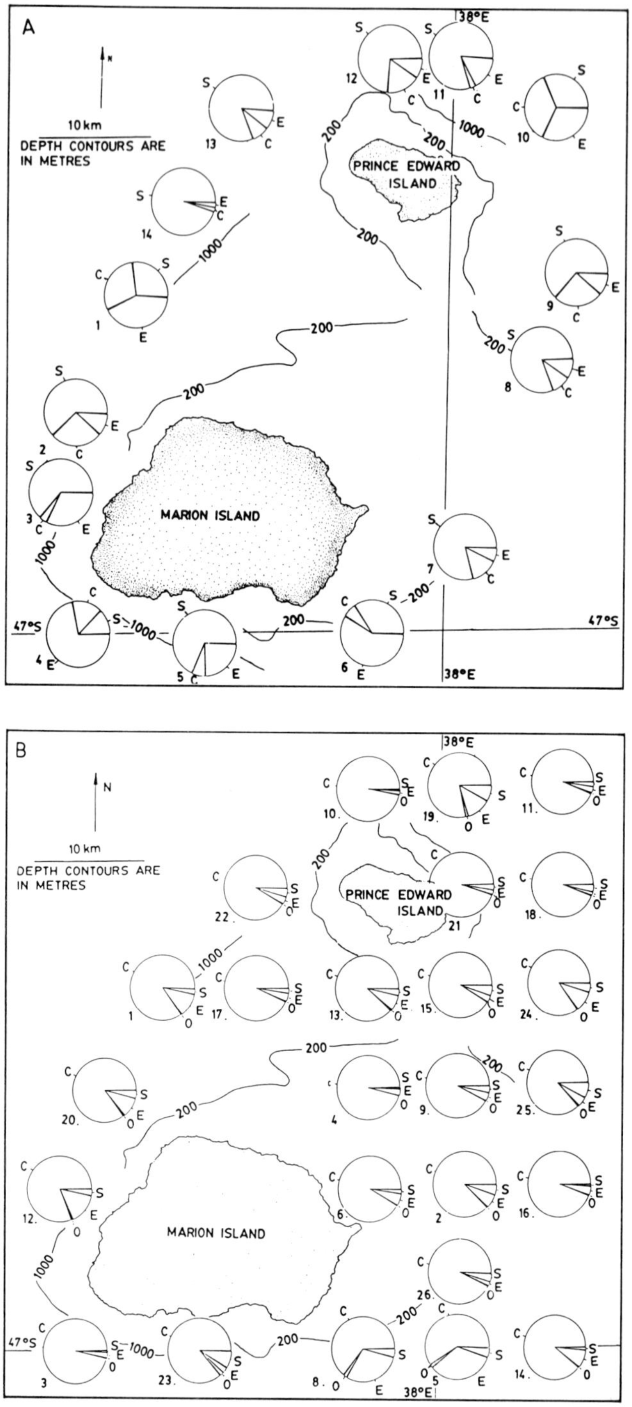

Fig. 11A, B. Four numerically dominant components of the zooplankton (expressed as % dry wt. per 1000 m^3) at stations (Station no. at *lower left* of each *pie-diagram*) around the Prince Edward Islands in May 1982 (**A**) and May 1983 (**B**). *C* copepods; *O* ostracods (insignificant in 1982); *S Sagitta; E* euphausiids

fields on the ocean surface in the immediate vicinity of the islands. Nor are we able to ignore the possible influence of the comparatively shallow saddle (150 m) which connects the islands in the Prince Edward group, and which lies obliquely across the path of the West Wind Drift.

We suggest, as an initial hypothesis, that where windshear is of sufficient magnitude, and when coupled with the effects of the island on the structure of the West Wind Drift, it causes vortex fields to be set up in the lee of Marion Island. The consequent downwelling cyclonic and upwelling anticyclonic eddies become foci of nutrient sequestration and biological production which form, decay and reform as wind-induced currents pass by the islands. These events constitute the hydrodynamic basis underlying the increased productivity of the seas surrounding islands and generally are known as the "island mass effect".

We suggest that the eddy fields contribute a succession of new interfaces at which the rate of delivery of nutrients normally in low concentration is increased. This is one of the mechanisms proposed by Lutjeharms et al. (1985) in the control of elevated production at the more permanent frontal features of the Southern Ocean. In addition, we suggest that nutrient delivery from the Prince Edward Islands is entrained in the illuminated buoyant surface waters in the lee of the islands, and under these conditions, algal blooms can be expected and do in fact occur.

Deacon (1983) appears to be unconvinced of the likelihood of upwelling, in the sense of Schumann et al. (1972), as the cause of production in the euphotic zone (50 m), since the surface silicate concentrations are low. He argues, furthermore, that while the 200 m Antarctic temperature minimum is a "useful indicator of the surface transition" in deep water, complex bottom topography in the vicinity of sub-Antarctic islands greatly diminishes the value of this marker. This argument reinforces our findings that the seas surrounding these islands are of complex origin and that permanent physical, chemical and biological facies cannot be expected. This complexity is exemplified by the apparent instability of the zooplanktonic fauna.

Certain changes in the species composition and biomass of the plankton may be attributed to vagrant eddies of alien origin or to frontal meanders. These expatriate water-masses probably transport their autochthonous plankton into the eddy field around the island; the plankton becoming entrained, whether the eddies are formally patterned or a random field of vorticity exists. The vertical motion in either case results in restricted horizontal recirculation, and the plankton will maintain its integrity until dissipated by physical processes or biological decay.

It is appreciated that more precisely determined hydrodynamic observations, with their subsequent analysis, are required before serious progress in the biological aspects of the oceanology at the Prince Edward Islands can be properly developed.

Acknowledgements. Financial support was provided by the South African Scientific Committee on Antarctic Research (SASCAR) and Rhodes University. The co-operation of Capt. W. Leith, his officers and crew, of the M.V. S.A. Agulhas is greatly appreciated, as is the editorial comment of Professor R. Siegfried.

Appendix 1. Zooplankton forms recorded at the Prince Edward Islands

PROTOZOA
- Foraminfera
  - *Globigerina* sp
  - *Globigerinoides* spd.
- Radiolaria
  - gen. et spp. non det

COELENTERATA
- Medusa spp.
- Siphonophora spp.

ANNELIDA
- Polychaeta
  - *Tomopteris* spp.
  - larvae gen. et spp. non det.

CHAETOGNATHA
- *Eukrohnia hamata*
- *Sagitta gazellae*
- *S. lyra*
- *S. maxima*
- *S. tasmanica*

MOLLUSCA
- Pteropoda
  - *Limacina* sp.
  - *Clione* sp.
  - *Clio* sp.
  - larvae sp.
- Gastropoda
  - gen. et spp. non det.
- Pleecypoda
  - Lamellibranch sp. – young

ARTHROPODA
- Crustacea
  - Ostracea
    - gen. et spp. non det.
  - Copepoda
    - *Aetidius armata*
    - *Calanus simillimus*
    - *Euchaeta* sp.
    - *Haloptilis* sp.
    - *Heterorhabdus* sp.
    - *Microcalanus* spp.
    - *Oithona* sp.
    - *Oncaea* sp.
    - *Metridia gerlachei*
    - *Pleuromamma robusta*
    - *Rhincalanus gigas*
    - *R. nasutus*
    - *Scaphocalanus magnus*
    - *Scolicithricella glacialis*
    - *Spinocalanus* sp.
  - Amphipoda
    - *Primno macropa*
    - *Phronimella elongata*
    - *Scina latifrons*
    - *Scina rattrayi kielhacki*
    - *Themisto gaudichaudii*
    - *Vibilia antarctica*
    - *V. armata*
  - Euphausiacea
    - *Euphausia lucens*
    - *E. similis*
    - *E. similis* v. *armata*
    - *Nematoscellis megalops*
    - *Stylocheiron elongatum*
    - *Thysanoessa macrura*
  - Isopoda
    - *Euridyce pulchra*

TUNICATA
- Appendiculata
  - *Oikopleura* and *Frittilaria*
- Thaliacea
  - *Salpa* sp.

VERTEBRATA
- Pisces
  - gen. et spp. non det.

References

Allanson BR, Hart RC, Lutjeharms JRE (1981) Observations on the nutrients, chlorophyll and primary production of the Southern Ocean south of South Africa. S Afr J Antarct Res 10(11):13

Cooper LHN (1938) Redefinition of the anomally of the nitrate-phosphate ratio. J M B AssocUK 23:178

Deacon GA (1983) Kerguelen, Antarctic and sub-Antarctic. Deep-Sea Res 30:77–81

El-Sayed SZ, Benon P, David P, Grindley JR, Murail J-f (1979) Some aspects of the biology of the water-column studied during the *Marion Dufresne* cruise 08. Comité National Francais des Recherches Antarctiques 44:127–134

Grindley JR, Lane SB (1979) Zooplankton around Marion and Prince Edward Island. Ibis 44:111–125

Lutjeharms JRE, Walters NM, Allanson BR (1985) Oceanic frontal systems and biological enhancement. In: Siegfried WR, Condy PR, Law RM (eds) Antarctic nutrient cycles and food webs (Proceedings of the 4th SCAR symposium on Antarctic biology). Springer, Berlin, Heidelberg, New York

Mackereth FJH, Heron J, Talling JF (1978) Water Analysis: Some revised methods for limnologists. Freshwater Biol AssocNo. 36: 120 pp

Pascal D (1979) Les éléments nutritifs des eaux bordants les isles Marion et Crozet (sud-ouest de l'océan indien). Comité National Francais des Recherches Antarctiques 44:61–78

Raymont JEG (1980) Plankton and productivity in the ocean. Pergamon, Oxford, 489 pp

Redfield AC, Ketchum BH, Richards FA (1963 In: The Sea, Ideas and Observations. Interscience, New York Vol II:26–77

Schumann EH, Perrins LA, Hunter II (1972) Upwelling along the South Coast of the Cape Province, South Africa. S Afr J Sci 78:238–242

Strickland JDH, Parsons TR (1968) A practical handbook of seawater analysis. Bull Fish Res Board Can 167 pp

Van Zinderen Bakker EM (1971) Marion and Prince Edward islands; report on the South African biological and geological expedition 1965–1966. Van Zinderen Bakker EM, Winterbottom JM, Dyer RA (eds). AA Balkema, Cape Town pp 1–15

Verwoort W (1965) Notes on the biogeography of free-living marine copepoda. Monogr Biol 15:381–400

Williams AJ, Berruti A (1978) Mineral and energy contributions of feathers moulted by penguins, gulls and cormorants to the Marion Island terrestrial ecosystem. S Afr J Antarct Res 8:71–74

Wolanski E, Imberger J (1983) Recirculating flows around islands in shallow coastal waters. MS Australian Institute of Marine Science, Townsville

Wolanski E, Pickard GL, Jupp DLB (1983) River plumes, coral reefs and mixing in the Gulf of Papua and the northern Great Barrier Reef. East Coast Shelf Sci (to be published)

Nutrient Upwelling and Its Effects in the Lee of Marion Island

J. R. GRINDLEY[1] and P. DAVID[2]

Summary. Investigations made in March 1976 revealed the presence of cold, nutrient-rich, upwelled water in the lee of Marion Island. It is tentatively suggested that this is wind-induced upwelling. In view of the persistent westerly winds in this region, this upwelling is probably a regular feature and must be of considerable significance to local ecosystems. Relatively low temperatures, high salinities and high values of nitrate, phosphate and silicate occurred in the lee of the island, at the 50 m depth level, where island surface run-off effects would be minimal. The patterns of distribution of these parameters indicate an upwelling process. This upwelling has been found to have a marked effect on the distribution of zooplankton in the area. The plankton community is remarkable in including many true Antarctic forms, normally found south of the Antarctic Convergence. Thus, 7 out of 8 characteristic Antarctic copepod species were present, while most of the typical Sub-Antarctic species were absent. *Oithona frigida, Ctenocalanus vanus* and *Calanus propinquus* were the most abundant Copepoda, while the only characteristic Sub-Antarctic species, *Clausocalanus laticeps,* occurred in much smaller numbers. The presence of certain bathy-pelagic species, such as *Aetidius armatus. Heterorhabdus austrinus* and *Haloptilus oxycephalus,* agrees with an upwelled origin of the water. The pattern of zooplankton distribution may be a result of the influence of the upwelling on phytoplankton production. The limited information that is available on phytoplankton production in the area appears to support this suggestion.

1 Introduction

During a cruise of the R.V. Marion Dufresne (MD-08) in March 1976, biological and hydrological investigations were carried out around the Prince Edward Islands (Frost et al. 1976). Zooplankton samples were collected at 12 stations. Additional samples were taken at 2 stations in the same area in November 1976, using the M.V. RSA. These samples were studied and 20 species of Copepoda and a number of species of other organisms were recorded. The copepod community was Antarctic rather than Sub-Antarctic in character in certain aspects. It was suggested that this peculiar species composition may have resulted from the upwelling of water from deeper levels in the lee of the Prince Edward Islands (Grindley and Lane 1979). This suggestion was offered purely on the basis of biological evidence, since the results of the hydrological investigations from the R.V. Marion Dufresne were not yet available. It has now been found that the hydrological findings do indeed reveal the presence of upwelled water in the lee of Marion Island and perhaps to some extent off Prince Edward Island as well. It is tentatively proposed that the pattern of this upwelling in the lee of the islands is wind-induced. In view of the persistent westerly winds in this region, this upwelling is probably a regular feature and must be of considerable significance to the local ecosystems.

The pattern of distribution of zooplankton recorded in 1976 indicated an area of relatively high zooplankton abundance in the lee of Marion Island (Grindley and Lane 1979). However, the limited number of sampling stations did not allow the distribution to be charted completely. Nevertheless, the plotted area of maximum quantitative zooplankton abundance corresponds closely to the site of nutrient upwelling revealed in the present investigation. It is suggested that the concentration of zooplankton may, therefore, be the result of enhanced primary productivity, resulting from the upwelling of nutrient-rich water. The limited information that is available on phytoplankton productivity in the area appears to support this suggestion (El-Sayed et al. 1979).

2 Materials and Methods

The materials and methods used in this study have been described by Benon and Murail (1979), El-Sayed et al. (1979b) and Grindley and Lane (1979).

3 Results

3.1 Zooplankton

The plankton community represented by the 1976 MD-08 series of samples is remarkable in including many true Antarctic elements normally found south of the Antarctic Con-

1 School of Environmental Studies, University of Cape Town, Rondebosch 7700, South Africa

2 Laboratoir d'Ecologie Numerique, 70 route Leon Lachamp, 13288 Marseille, Cedex 2, France

Antarctic Nutrient Cycles and Food Webs
(ed. by W.R. Siegfried, P.R. Condy, and R.M. Laws)

vergence. Vervoort (1965), in his review of the biogeography of Antarctic marine Copepoda, lists the following typical surface species: *Calanus propinquus, Calanoides acutus, Rhincalanus gigas, Metridia gerlachei, Scolithricella glacialis* and *Oithona frigida.* All but *M. gerlachei* are present in the plankton from around Marion Island. Vervoort (1965) lists the following three species as representing characteristic Sub-Antarctic surface species: *Calanus simillimus, Clausocalanus laticeps,* and *Eucalanus longiceps.* Of these, only *C. laticeps* is present in the samples and in smaller numbers than some of the typically Antarctic species. *Calanus simillimus,* which is completely absent from the samples is reported by Vervoort as being a most important species in the Sub-Antarctic. Whereas the distribution patterns of these species are not absolutely rigid, it is clear that the plankton community in the samples from near Marion Island is not a typical Sub-Antarctic community, although the island had been considered to lie north of the Antarctic Convergence (Van Zinderen Bakker et al. 1971). Recent work indicates that the Antarctic Convergence is complex near the Prince Edward Islands (Deacon 1983).

The seasonal movements of the most important Antarctic Copepoda provide clues to this anomalous distribution pattern. *Calanus propinquus,* which undertakes very marked vertical migrations, drifts gradually northwards in the Antarctic zone and is occasionally found far to the north of the Convergence in the Antarctic Intermediate Current (Vervoort 1965). *Rhincalanus gigas* is carried northward in Antarctic surface waters. Specimens may be found north of the Antarctic Convergence in summer (Ommaney 1936) and in surface waters even as far north as South Africa (Voronina 1970). The work of Ommaney and the more recent studies of Voronina have indicated that *R. gigas* descends into deeper water in the early autumn near the Antarctic Convergence. This would allow some northward transport in the Antarctic Intermediate Water as well as the well-documented southward drift in the Warm Deep Current. The vertical sections of Voronina (1970) show clearly that *R. gigas* is carried far north of the Antarctic Convergence in intermediate waters in summer and autumn. *Calanoides acutus* does not normally appear to survive the conditions north of the Antarctic Convergence. However, a few specimens have been captured in Sub-Antarctic midwater hauls (Vervoort 1965). Voronina (1970) shows that *C. acutus* is present in intermediate and even near-surface waters farther north than 49°S in February–March.

3.2 Hydrological Evidence for Upwelling

The hydrological data generated by the investigations carried out during MD-08 at the 13 stations indicated in Fig. 1 revealed cold, nutrient-rich, upwelled water in the lee of the islands. An account of aspects of the biology of the water column studied during the MD-08 cruise was published by El-Sayed et al. (1979). Mention was made of the suggestion made by Grindley and Lane (1979) regarding upwelling, and it was suggested that this was plausible and supportable in part by physical data. El-Sayed et al. (1979b), in their study of the productivity of the southwestern Indian Ocean, presented evidence of the high production and high levels of chlorophyll-a and phaeopigments during the MD-08 cruise which would have been influenced by upwelling of nutrient-rich water. In this account we attempt to document the form and extent of the upwelling from the MD-08 hydrological data.

Evidence of upwelling is apparent from six independent sets of data, including temperature, salinity, oxygen, phosphate, nitrate and silicate values. By systematic plotting of the vertical and horizontal distributions of each data set, generally consistent patterns of similar form were revealed in the area east and northeast of Marion Island. Temperature distribution at 100 m ranged from 5.12 to 6.41°C. The lowest values appear in the deep water northeast of the island. At the 75 m level, a core of even colder water appears at Station 12 (east of the island) and there is evidence of a pattern of concentric isotherms of higher temperature. The highest value (6.62°C) at this level lies south of the island at Station 31. At the 50 m level, the cold core has a minimum of 5.23°C at Station 12 and a comparable pattern of increasing temperatures is spread away from the core area.

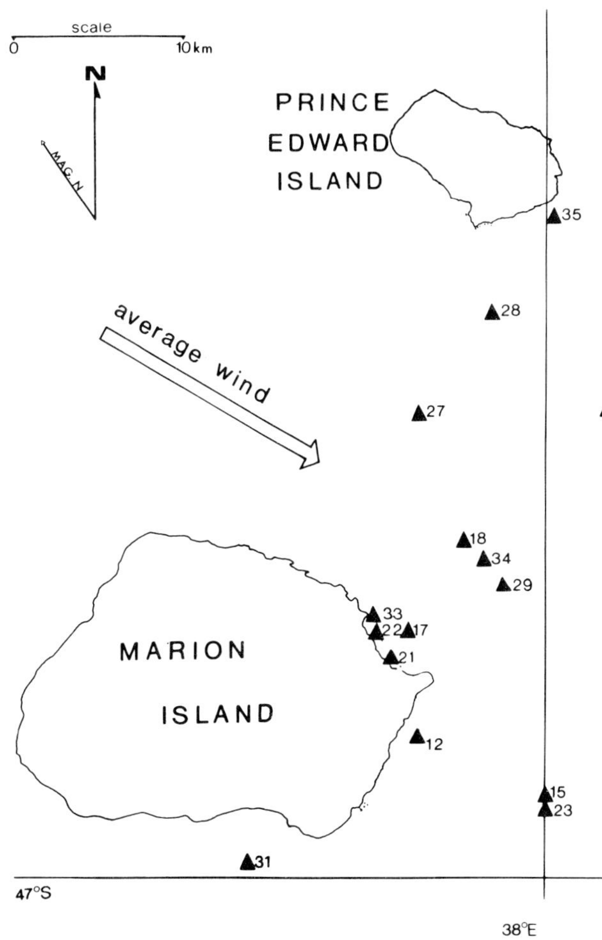

Fig. 1. Station numbers and positions, and average wind direction

At 30 m the minimum temperature is 5.80°C above the cold core. At higher levels the temperatures are higher and more dispersed. The highest temperature inshore on the surface was 7.12°C at Station 22. Figure 2 indicates the general pattern.

Salinity distribution at 100 m ranges from 33.826 to 33.899‰. At the 75 m level, at Station 12, the highest value of 33.928‰ appears to indicate a core of relatively

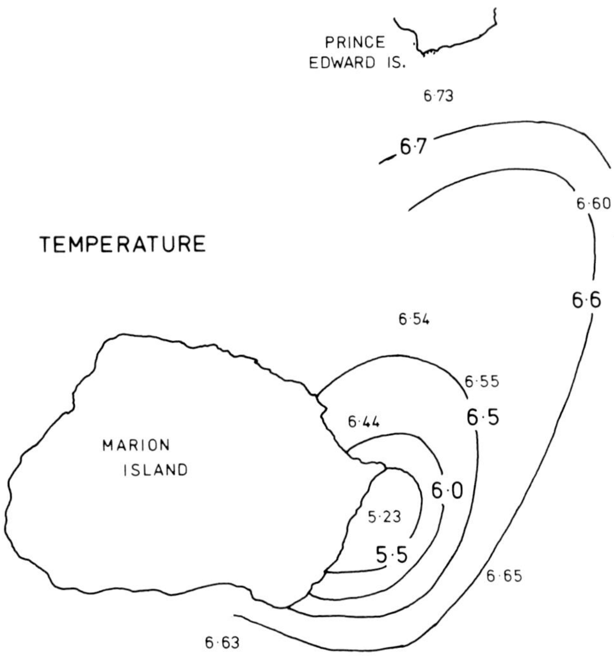

Fig. 2. Temperature (°C) at 50 m off Marion Island

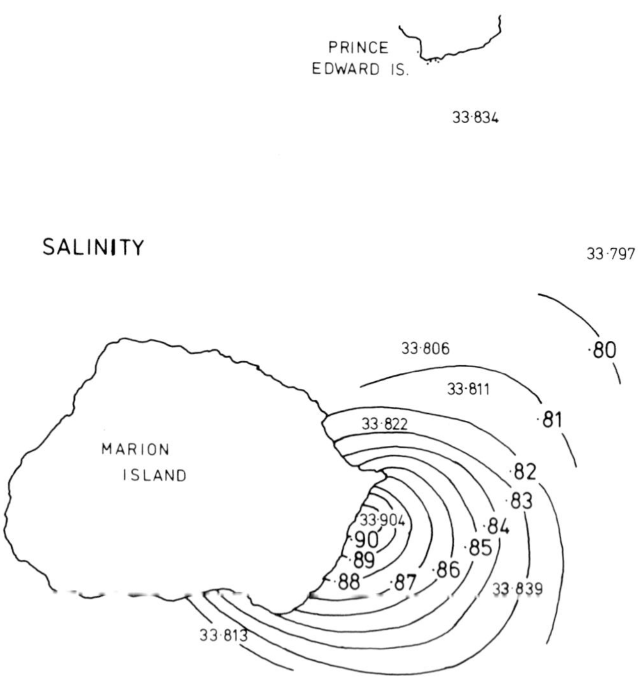

Fig. 3. Salinity (‰) at 50 m off Marion Island

high salinity water. At the 50 m level the maximum salinity in the core is 33.904‰ and all other values are lower with a concentric pattern of decrease (Fig. 3). A similar pattern appears at the 30 m level, but at the 0 m and even at the 10 m levels there is evidence of reduced salinity inshore, which is probably related to land run-off.

Oxygen concentrations are low at 100 m with a minimum value of 6.46 ml l^{-1} at Station 19. At the 75 m level, the lowest value east of the Island is 6.52 ml l^{-1} at Station 12 with higher values further out. At the 50 m level, a similar pattern appears with concentrations increasing in a concentric pattern from the core of low-oxygen water east of the island (Fig. 4). The 30, 20, 10 and 0 m levels show comparable patterns.

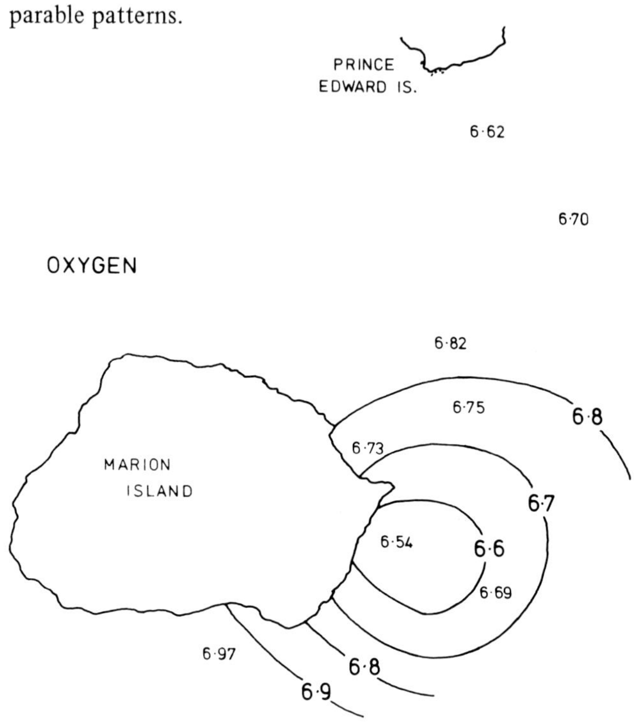

Fig. 4. Dissolved oxygen concentration (ml l^{-1}) at 50 m off Marion Island.

Phosphate concentrations at 100 m range from 1.06 to 1.41 μg at l^{-1}. At the 75 m level there is evidence of a core of higher concentrations with a maximum of 1.45 μg at l^{-1} at Station 17. At the 50 m level, the highest level is 1.39 μg at l^{-1} at Station 12 with lower values away from the core of concentration (Fig. 5). This pattern appears to be similar at shallower levels but anomalous high values probably related to land run-off appear inshore. A value of 1.47 μg at l^{-1} appears at Station 33 close inshore.

Nitrate concentrations at the 100 m level range from 9.4 to 22.4 μg at l^{-1}. At the 75 m level, the highest concentration of 20.4 μg at l^{-1} is at Station 12. At the 50 m level, the highest value of 20.6 μg at l^{-1}, also at Station 12, appears to represent a core of higher concentration with a concentric pattern of decrease (Fig. 6). A generally comparable pattern appears at shallower levels but some irregularities are present inshore.

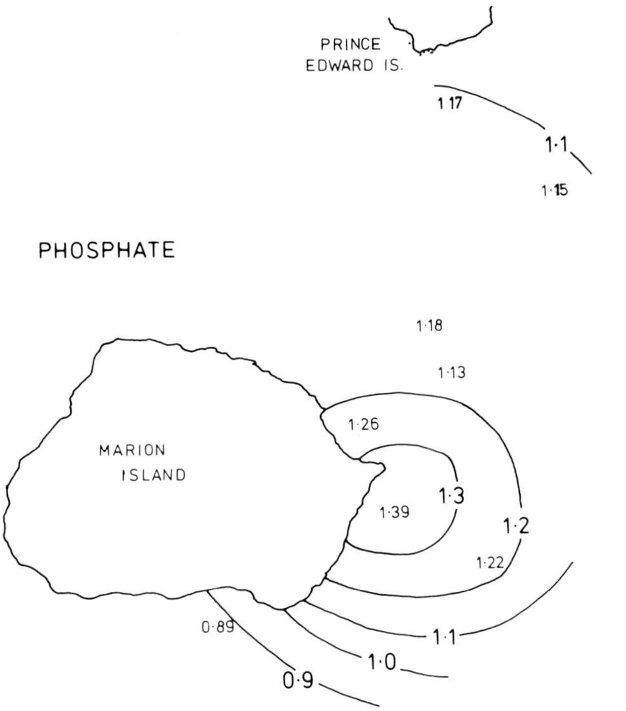

Fig. 5. Phosphate concentrations (μg at l^{-1}) at 50 m off Marion Island.

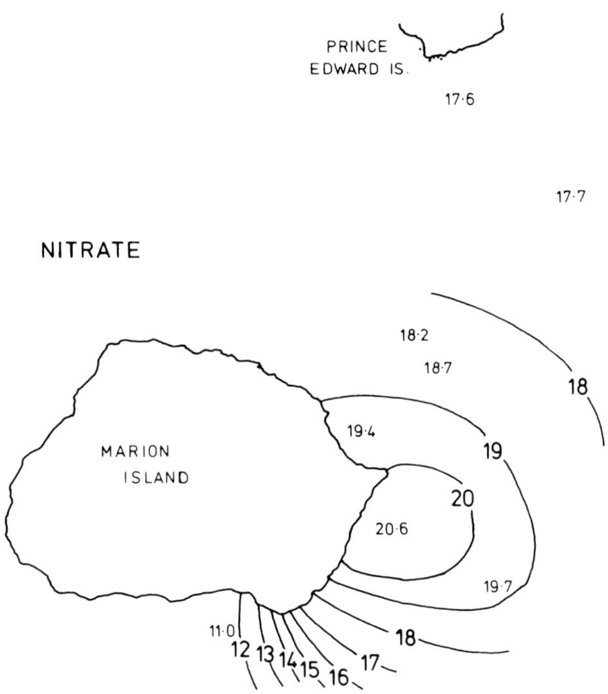

Fig. 6. Nitrate concentration (μg at l^{-1}) at 50 m off Marion Island

Silicate concentrations at the 100 m level range from 1.1 to 8.6 μg at l^{-1}. At the the 75 m level, the highest value of 4.5 μg at l^{-1} is at Station 12. At the 50 m level, values decrease from a maximum of 3.4 μg at l^{-1} at Station 12 in a concentric pattern (Fig. 7). Values are lower nearer the surface particularly inshore where phytoplankton produc-

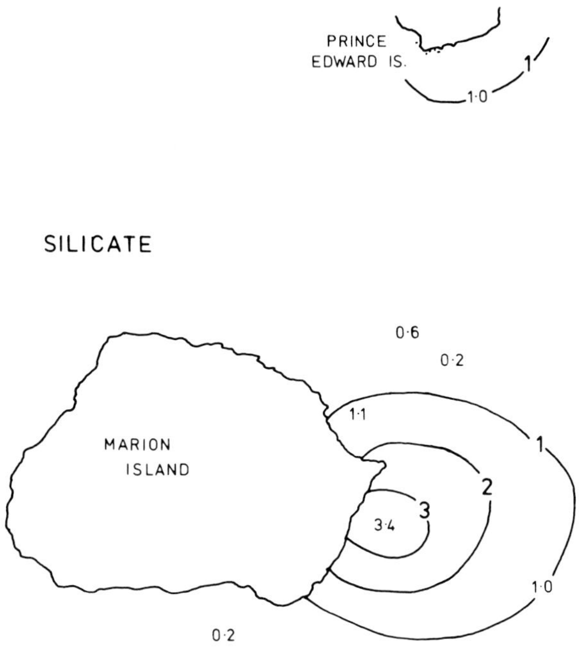

Fig. 7. Silicate concentration (μg l^{-1}) at 50 m off Marion Island

tion may be responsible for the reduction, as suggested by El-Sayed et al. (1979b).

The remarkably consistent pattern of distribution of these six independent parameters suggests that an upwelling mass of water with low temperature, high salinity, reduced oxygen concentration and increased phosphate, nitrate and silicate concentrations was present east of Marion Island in March 1976. All six sets of data suggest that similar anomalous concentrations were present southeast of Prince Edward Island as well, but the data are inadequate to determine patterns in that area.

4 Discussion

Marion Island and Prince Edward Island have been considered to lie more than 2° north of the Antarctic Convergence, and Van Zinderen Bakker et al. (1971) and other authors have stated that the islands are surrounded by Sub-Antarctic Surface Water. However, recent work has indicated that the Antarctic Convergence has complexities near these islands (Deacon 1983). The surface transition generally occurs where the temperature minimum of the Antarctic water sinks below 200 m on its descent below warmer water (Deacon 1934). However, the layer of cold, low-salinity water may be wide in this area where Benon and Murail (1979) described the sub-surface water as more Antarctic than Sub-Antarctic and suggested that the Antarctic Convergence was at about 46°S. The data provided by El-Sayed et al. (1979a) further supported this invasion of Antarctic water. Deacon (1983) notes that the Convention on Conservation of Antarctic Marine Living Resources "deems" the Antarctic Convergence to lie north of Marion Island. It appears that the upwelling east of Marion Island comes from a cold intermediate layer of Antarctic water, since the low silicate concentrations recorded indicate that it could not have come from any great depth.

In view of the persistent westerly gales that prevail at Marion Island and at Prince Edward Island, wind stress on the sea surface must be great. Surface water and floating algal fragments are observed to drift offshore. Schulze (1971), in his analysis of wind data from Marion Island, shows that westerly winds predominate and that winds from the northwest quadrant are the most frequent and strongest, often reaching gale force. Other wind directions or calm conditions are very infrequent. The westerly winds generate a surface boundary-layer flow generally referred to as the West Wind Drift. The velocity of such wind-driven currents is greatest at the surface. Here the stress acting on the surface of the water keeps it moving in the downwind direction at a fraction ($< 4\%$) of the wind speed. The surface current shear generates Reynolds stresses, transferring momentum downwards through the water but with velocity steadily decreasing with depth (Sverdrup et al. 1942). Surface water driven by wind stress streaming past an island thus generates weak sub-surface currents. The role of the Coriolis force in the lee of Marion Island cannot be established until further details of flows and densities become available, but it is perhaps significant that the areas of upwelling are situated to the left of the prevailing northwesterly winds.

Surface water in the lee of Marion Island, when exposed to wind stress, will tend to move offshore in a downwind direction perhaps with some Coriolis deflection as suggested above. Replacement of such wind drifted surface water can be achieved by advection of surface water from around the sides of the island, but the shear stresses can also cause the upwelling of sub-surface water from below against the density gradient. The data presented in Figs. 2–4 suggest that a dome-shaped vertical plume of upwelled water from a depth of approximately 100 m was drawn upwards towards the surface in this manner. Such a massive movement of water against density gradients requires substantial forcing and is thus likely to be due to strong surface wind drifts rather than weaker sub-surface currents. Similar patterns of purely wind-induced upwelling have been studied off the coast of South Africa by Grindley and Taylor (1970) and Cram (1970). The plume of water with anomalous values observed east of Marion Island (and perhaps also south-east of Prince Edward Island) can most readily be explained as wind-induced upwelling occurring in the lee of the islands. Such patterns of upwelling may be a regular feature and must be of considerable ecological significance locally.

The upwelling of nutrients would favour increased phytoplankton production, and there is some evidence of this. The primary productivity measured off Marion Island during cruise MD-08 was 17.95 $mgCm^3 d^{-1}$ at Station 29 (not 20.7 $mgCm^3d^{-1}$ at Station 59) (El-Sayed et al. 1979b). This high level of productivity was encountered at a depth of 6 m. Productivity values decreased rapidly to 2.9 $mgCm^3d^{-1}$ at 18 m and were negligible below the euphotic level. The abundant growth of phytoplankton off Marion Island was dominated by a bloom of *Chaetoceros radicans.* Only 21.5 ± 6.33% of the primary productivity off Marion Island was contributed by the nannoplankton. Chlorophyll-a concentrations of > 1.0 mg m^3 were occasionally observed at the stations located to the northeast of Marion Island (El-Sayed et al. 1979b). The productivity measured by El-Sayed off Marion Island was equivalent to 0.211 gCm^2d^{-1}. Mitchel-Innes (1967) obtained a productivity value of 0.135 $gCm^2 d^{-1}$ off Marion Island in July 1962. A comparable value of 0.110 $gCm^2 d^{-1}$ was recorded at Station 38 northeast of Prince Edward Island. In April 1963, Mitchell-Innes (1967) obtained an even higher figure of 0.266 $gCm^2 d^{-1}$ west of Marion Island.

The concentrations of zooplankton (Fig. 8) recorded northeast of Marion Island by Grindley and Lane (1979) may be related to the enhanced phytoplankton production in that area. The zooplankton samples included sufficient phytoplankton to colour the samples green. It seems unlikely that the northward drift of surface water from south of the

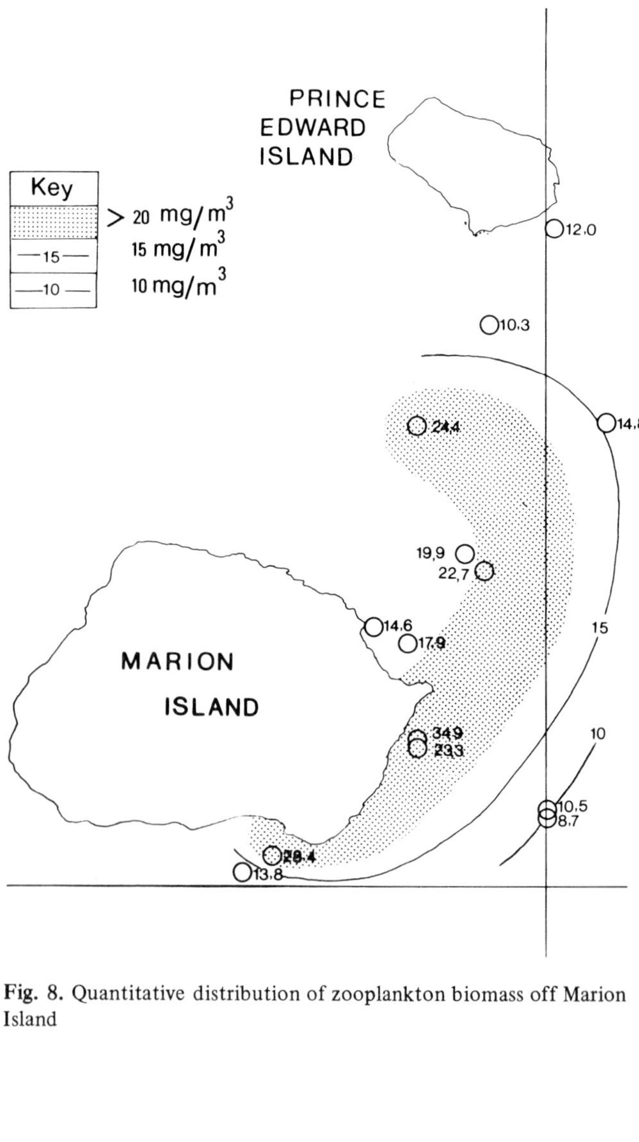

Fig. 8. Quantitative distribution of zooplankton biomass off Marion Island

Antarctic Convergence could explain the nature of the Marion Island samples. Although Antarctic species may be found north of the Convergence they are often at mid-water or intermediate levels (Voronina 1970). This suggests that the peculiar species composition of the Marion Island plankton community may result from the upwelling of water from intermediate levels in the lee of the islands. The strong and almost continuous westerly winds to which Marion Island and Prince Edward Island are exposed make wind-induced upwelling to the east of the islands probable through much of the year. The regular upwelling of cooler water from deeper levels might explain not only the 'Antarctic' character of much of the copepod community but also the presence of typically bathy-pelagic species such as *Aetidius armatus, Heterorhabdus austrinus* and *Haloptilus oxycephalus* (Vervoort 1965). The upwelling of nutrients would also account for the high rate of primary productivity that has been recorded by El-Sayed et al. (1979b).

Acknowledgements. We thank TAAF and SASCAR for making this study possible, the senior scientist Dr. P.M. Arnaud, and Dr. S.Z. El-Sayed, the Captain, Commandant G. Duranton, the officers, scientists and crew of the R.V. Marion Dufresne for their helpful co-operation at sea. Hydrological data were determined by P. Benon and M. Murail and nutrient determinations were by P. David.

References

Benon P, Murail JF (1979) Hydrologie des platformes peri-insulaires des iles Crozet, Marion et Prince Edward. CNFRA 44:39–60

Cram DL (1970) A suggested origin for the cold surface water in central False Bay. Trans Roy Soc S Afr 39:129–137

Deacon GER (1934) Die Nordgrenzen antarktischen und subantarktischen Wassers im Weltmeer. An Hydrogr Meteorol 42:129–136

Deacon GER (1983) Kerguelen, Antarctic and subantarctic. Deep-Sea Res 30:77–81

El-Sayed SZ, Benon P, David P, Grindley JR, Murail J-F (1979a) Some aspects of the biology of the water column studied during the 'Marion Dufresne' Cruise 08. CNFRA 44:127–134

El-Sayed SZ, Stockwell DA, Reheim HA, Taguchi S, Meyer MA (1979b) On the productivity of the southwestern Indian ocean. CNFRA 44:1–43

Frost PGH, Grindley JR, Wooldridge TH (1976) Report on South African participation in Cruise 08 of R.V. *Marion Dufresne* March–April 1976. S Afr J Antarct Res 6:28–29

Grindley JR (1978) Marine ecosystems of Marion Island. S Afr J Antarct Res 8:38–42

Grindley JR, Lane SB (1979) Zooplankton around Marion and Prince Edward Islands. CNFRA 44:111–125

Grindley JR, Taylor FJR (1970) Factors affecting plankton blooms in False Bay. Trans Roy Soc S Afr 39:201–210

Mitchell-Innes BA (1967) Primary production studies in the south-west Indian Ocean, 1961–1963. Invest Rep Oceanogr Res Inst (Durban) 14:1–20

Ommaney FD (1936) *Rhincalanus gigas* (Brady), a Copepod of the southern macroplankton. Discovery Rep 13:277–385

Schulze BR (1971) The climate of Marion Island. In: Marion and Prince Edward Islands. Report on the South African biological and ecological expeditions 1965–66. EM van Zinderen Bakker Sr, JM Winterbottom, RA Dyer (eds). AA Balkema, Cape Town: 16–31

Sverdrup HV, Johnson MW, Fleming RH (1942) The oceans their physics, chemistry and general biology. Prentice-Hall, Englewood Cliffs

Van Zinderen Bakker EM Sr (1971) Introduction. In: Marion and Prince Edward Islands. Report on the South African biological and geological expeditions 1965–66. EM van Zinderen Bakker Sr, JM Winterbottom, RA Dyer (eds). AA Balkema, Cape Town: 1–15

Vervoort W (1965) Notes on the biogeography and ecology of free-living marine Copepoda, pp 381–400 6 fig, In: Van Oye P and Van Mieghem J (eds) Biogeography and ecology in Antarctica. Monographiae biologicae 15. The Hague, Junk

Voronina VM (1970) Season cycles of some Antarctic Copepod species. 162–172 In: Holdgate MW (ed) Antarctic Ecology. Academic Press, London

Marine Biological and Oceanographical Investigations in Lützow-Holm Bay, Antarctica

M. FUKUCHI, A. TANIMURA, and H. OHTSUKA[1]

Summary. Routine samples of water and plankton were taken at 5 stations on fast-ice near Syowa Station (69° 00′ S, 39° 35′ E) in the eastern part of Lützow-Holm Bay, Antarctica. Sampling was carried out at 2–3 week intervals during 13 months between January 1982 and January 1983, excluding February 1982. Temperature, salinity, pH, dissolved oxygen, phosphate-P, silicate-Si, nitrite-N, nitrate-N, ammonium-N and chlorophyll-a, from just below the ice to near the bottom, were measured. Oceanographic conditions changed seasonally; the range of physical variables corresponding to that of the Antarctic Surface Water. Maximum chlorophyll-a concentration (11.30 $\mu g\ l^{-1}$) was observed at 5 m in late January of 1983. Integrated chlorophyll-a stock beneath the ice was much higher than has been reported from northern Antarctic open waters. Copepods, chaetognaths, ostracods, and euphausiids occurred throughout the year, and copepods were a major component of the zooplankton community. *Paralabidocera antarctica,* an Antarctic endemic copepod, formed a dense population just beneath the ice only in December and January. A preliminary account of this study is discussed.

1 Introduction

The biology of the ice-covered sea surrounding the Antarctic continent has received much less attention than that of open Antarctic seas, although the importance of such studies has been emphasized by Andriashev (1970), El-Sayed (1971), and Bradford (1978). Bunt (1960) investigated mainly hydrology and plankton near Mawson Station. Littlepage (1965) conducted oceanographic studies in McMurdo Sound. Near the Japanese Station, Syowa (69° 00′ S, 39° 35′ E), Hoshiai (1969) studied seasonal variation of chlorophyll-a and Wakatsuchi (1982) abserved seasonal oceanographic conditions.

The importance of shore-based studies has been stressed, as a part of the international BIOMASS (Biological Investigations of Marine Antarctic Systems and Stocks) programme. Within the Japanese BIOMASS project, the Japanese Antarctic Research Expedition planned a 3-yr programme of shore-based research between 1982 (JARE-23) and 1984 (JARE-25). In 1982, five stations were occupied near Syowa Station over depths ranging between 10 and 675 m. Lützow-Holm Bay is located between latitudes 68° 30′ S and 70° 00′ S and longitudes 34° 00′ E and 40° 00′ E, and its northern part deepens steeply once outside the 1,000 m isobath which is some 50 km from the coast. The Bay area south of the 1,000 m isobath usually remains ice-covered throughout the austral summer and the northern edge of the ice lies around 58–59° S during winter (Kusunoki 1981). Syowa Station is located on East Ongul Island in the eastern part of Lützow-Holm Bay. Routine observations were carried out between January 1982 and January 1983, each observation consisting of sampling water with Nansen and Van Dorn bottles, and sampling zooplankton by means of three methods. The results of a preliminary analysis of the data obtained from these observations are discussed in this paper.

2 Materials and Methods

The locations of the five oceanographic stations were chosen according to accessibility and sea depth (Fig. 1). Sea depths at Stations 1–5 were 10, 25, 50, 160 and 675 m, respectively. Station 1 was located in the narrow and shallow channel of Kita-no-seto Strait, and Station 5 was over the deep glacial trough in the Ongul Strait.

Seawater was sampled from just beneath the fast-ice to near the bottom. Temperature was measured, using a pair of protected reversing thermometers attached to each Nansen bottle. Nansen bottle samples were used for the measurement of salinity, pH, dissolved oxygen, phosphate-P, silicate-Si, nitrite-N, nitrate-N, and ammonium-N. Salinity was measured with an Auto-Lab Model 601 MkIII inductive salinometer, and was expressed on a practical salinity scale. Sigma-t was calculated from the data of temperature and salinity. pH was measured with a Denki Kagaku Keiki Co. Model HG-3 digital pH meter. Dissolved oxygen was determined by the Winkler method, as described by Strickland and Parsons (1968), nutrient salts were analyzed by the methods of Strickland and Parsons (1968), and ammonium-N was analyzed according to Solorzano (1969). Water samples from Van Dorn bottles were used for chlorophyll-a measurement. Chlorophyll-a concentration was determined by the fluorometric method (Saijo and Nishizawa 1969).

Zooplankton was sampled in three ways: a Norpac net, 450 mm in diameter, 1.80 m long and with a 0.11 mm

1 National Institute of Polar Research, 9–10 Kaga, 1-chome, Itabashi-ku, Tokyo 173, Japan

Antarctic Nutrient Cycles and Food Webs
(ed. by W.R. Siegfried, P.R. Condy, and R.M. Laws)

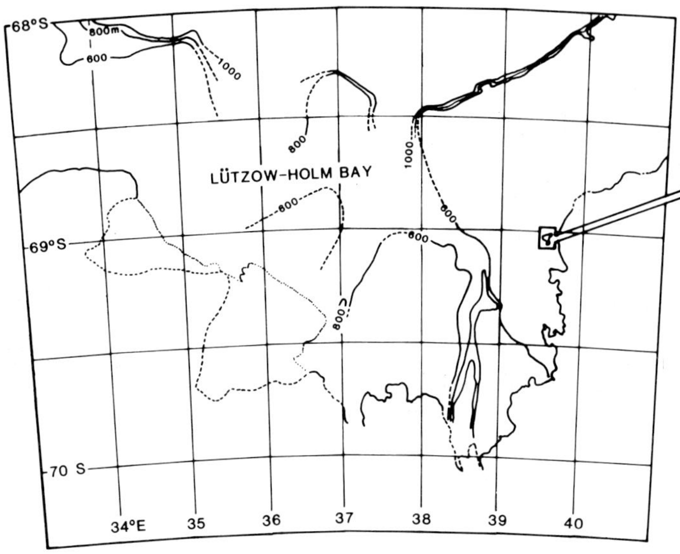

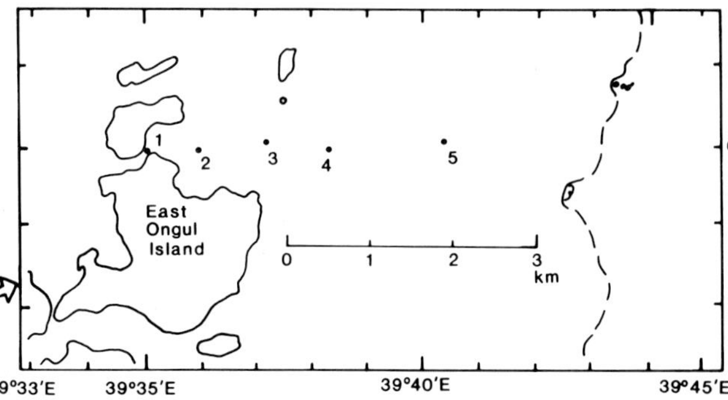

Station No	Position		Depth (m)
1	69°00'00"S	39°35'00"E	10
2	69°00'00"S,	39°36'00"E	25
3	68°59'57"S,	39°37'16"E	50
4	69°00'00"S,	39°38'20"E	160
5	68°59'57"S,	39°40'25"E	675

Fig. 1. Bathymetry of Lützow-Holm Bay, Antarctica (*left;* after Moriwaki and Yoshida 1983) and locations of stations in the eastern part of the Bay (*right*)

Table 1. Maximum and minimum values of oceanographic data at five stations throughout 21 routine observations. Extreme values recorded from 0–4 m depths in December and January are listed in parentheses; ranges exclude these extremes

Station no.		Temp. (°C)	Salinity	Sigma-t	pH	DO (ml l^{-1})	PO_4-P	SiO_3-Si	NO_2-N	NO_3-N	NH_4-N
							(μg at. l^{-1})				
1	Max.	−1.19 (−0.31)	34.241	27.560	8.12 (8.30)	8.85 (9.96)	2.48	84.0	0.82	27.0	1.7
	Min.	−1.87	33.746 (30.512)	27.157 (24.526)	7.78	6.93	1.10	34.9	0.00	11.6	0.0
2	Max.	−1.34 (−0.82)	34.196	27.525	8.23 (8.36)	9.14 (11.51)	3.02	82.0	0.82	29.2	1.4
	Min.	−1.88	33.525 (19.554)	26.978 (15.639)	7.89	6.95	0.96 (0.91)	36.7	0.02	8.4	0.0
3	Max.	−1.21 (−0.60)	34.216	27.535	8.29 (8.48)	10.01 (12.72)	2.50	81.6	0.67	42.8	2.3
	Min.	−1.89	33.772 (26.990)	27.183 (21.662)	7.88	6.61	0.56	27.7	0.04	8.2	0.0
4	Max.	−1.36 (+0.45)	34.333	27.627	8.13 (8.64)	8.12 (15.27)	2.40	83.4	0.54	38.9	1.5
	Min.	−1.87	33.693 (33.271)	27.119 (26.740)	7.86	6.60	0.95	33.7	0.00	8.4	0.0
5	Max.	−1.33	34.425	27.702	8.15	8.07	2.52	92.8	0.60	41.0	1.4
	Min.	−1.87	33.886	27.276	7.78	6.45	1.05	35.2	0.00	7.1	0.0
Range		−1.19 s −1.89	34.425 s 33.525	27.702 s 26.978	8.29 s 7.78	10.01 s 6.45	3.02 s 0.56	92.8 s 27.7	0.82 s 0.00	42.8 s 7.1	2.3 s 0.0

mesh (Motoda 1957), was hauled vertically from near the bottom to the surface at a speed of about 1 m s^{-1} at each station, a "NIPR-I" sampler, with a mesh size of 0.11 mm (Fukuchi et al. 1979) modified for operation down to 150 m, was used to take layered samples at between 5 and 10 selected depths, at each station (3 min of sampling was performed at each depth); and, at Stations 4 and 5 only, a parasol net was hauled vertically from near the bottom to the surface at a speed of about 1 m s^{-1} (the net having a hexagonal mouth with 0.5 m sides, a total length of 3.5 m and mesh sizes of 1.00 and 0.35 mm for the upper and lower parts, respectively). This net was designed for opening-closing operations. Net samples were preserved in ca. 5% buffered formalin-seawater, and were used for biomass measurements and microscopic examinations.

In January 1982, the fast-ice surrounding Syowa Station was in good condition for fieldwork and Stations 1 and 2 were occupied (routine observation no. 1). In February, only limited sampling could be done, due to bad weather and logistic arrangements for wintering. A complete set of routine observations at five stations was started in late March and was continued at intervals of 2–3 weeks until December (routine observation nos. 4–19). In January 1983, Station 5 was inaccessible, due to heavy melting of sea-ice in Ongul Strait, so that routine observations nos. 20 and 21 were carried out at Stations 1–4 only. A total of 21 sets of routine observations was completed, spanning 13 months.

3 Results and Discussion

3.1 Seasonal Variation in Physical and Chemical Parameters

Maximum and minimum values from all observations are summarized in Table 1. Seasonal variations in temperature and salinity at five stations are shown in Figs. 2 and 3, respectively.

At Stations 1 and 2, the water temperature in January was $>-1.70^{\circ}$C throughout the water column and exceeded -1.30°C at the surface. From March to June, the temperature decreased gradually and minimum temperatures below -1.85°C were recorded in May and June at all five stations. The lowest temperature occurred from surface to bottom at Stations 1–3, but was restricted to the upper 10 m at Stations 4 and 5. At Station 5, the -1.80°C isotherm deepened to about 250 m from April to the end of May. However, the -1.70°C isotherm moved upward from 550 m to 450 m during this period and had risen to 100 m by the end of July. Successively, isotherms of -1.60, -1.50, and 1.40°C moved upward during August and November. Seasonal change of temperature at Station 4 was similar to that occurring through the upper 150 m water column of Station 5. Also, the seasonal change of Station 3 resembled that in the 0–50 m column of Stations 4 and 5, except for the vertical range of the lowest temperature in June. Generally, temperature changed seasonally in a similar way at equivalent depths at all stations. Seasonal changes of isohalines are generally similar to those of temperature as seen in Figs. 2 and 3. However, the isohalines of 34.00 at Station 4, and 34.20 and 34.30 at Station 5 deepened between March and April. Generally, the salinity changes occurring at shallow stations were similar to those at corresponding depths at Station 5. The overall similarity between the temperature and salinity patterns at five stations suggests that they are the result of common processes, which presumably include both seasonal changes in incoming radiation and the effect of deeper water currents.

Wakatsuchi (1982) observed seasonal oceanographic conditions under fast-ice at scattered stations in the eastern part of Lützow-Holm Bay around Syowa Station. Temperature and salinity values reported by him generally correspond with the present data. Littlepage (1965) occupied a station on sea-ice over 290 m of water in McMurdo Sound. He showed that values of temperature and salinity throughout the water column there in 1961 ranged from -1.40 to -2.15°C and 33.96 to $34.99^{\circ}/_{oo}$, respectively. This range of temperature is about 0.30°C lower than the present data, and the salinity range is about $0.50^{\circ}/_{oo}$ higher. Ranges of temperature and salinity values in McMurdo Sound are within the ranges of Ross Sea and Ross Sea shelf water (see Fig. 5 in Gordon 1971). Similarly, the present data are within the ranges of the Antarctic Surface Water (see Fig. 6C in Houtman 1967). Different ranges of temperature and salinity values reflect regionality.

Concentrations of dissolved oxygen greater than 8.00 ml l^{-1} were recorded in April–May at all five stations. High values occurred through the entire water column at shallow stations (Stations 1–3), but these values were restricted to the upper 100 m at the deeper stations (Stations 4.5). Oxygen concentrations decreased gradually through winter to < 6.80 ml l^{-1}. Concentrations increased again in December–January when chlorophyll-a concentrations increased. pH values did not show any distinct seasonal variation but values greater than 8.00 were recorded in April–May and December–January when oxygen concentrations were high.

In contrast to the distinct seasonal variation observed for physical properties, levels of nutrient salts did not show clear seasonal trends. Phosphate-P, silicate-Si, and nitrate-N remained at high levels throughout. Concentrations of these salts were similar at all stations. Levels of ammonium-N and nitrite-N were generally low. However, ammonium-N concentrations exceeded 1.0 μg at. l^{-1} in late April and June at all five stations. Nitrite-N concentrations exceeded 0.4 μg at. l^{-1} in July and August at all five stations.

Fig. 2. Seasonal variation in temperature ($^{\circ}$C) at five stations. *H* and *L* indicate higher and lower temperature regimes than the surrounding water, respectively. *Dots* denote depths at which data were recorded

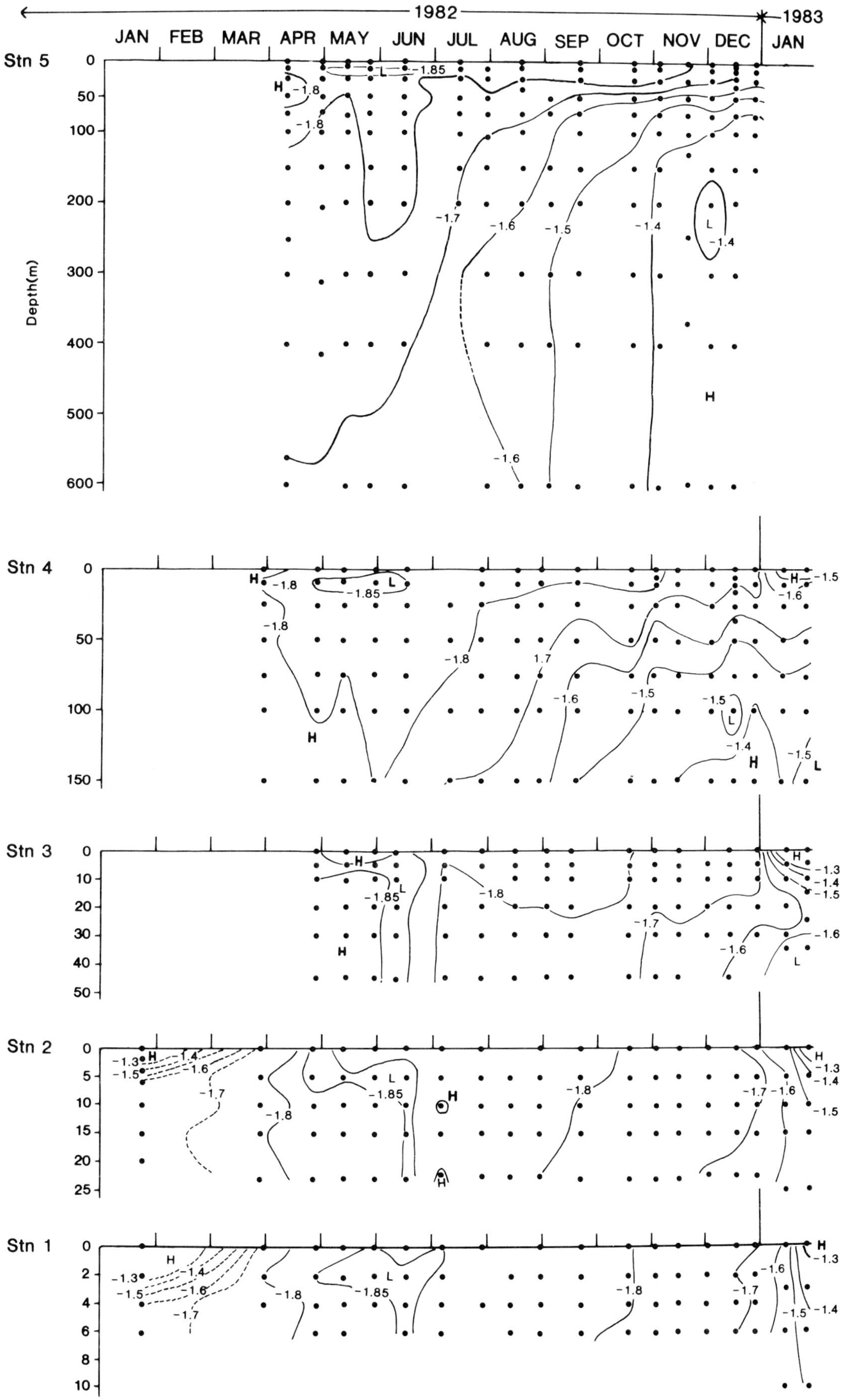

1982
1983
JAN FEB MAR APR MAY JUN JUL AUG SEP OCT NOV DEC JAN
Stn 5
Stn 4
Stn 3
Stn 2
Stn 1
Depth(m)
-1.85
-1.8
-1.7
-1.6
-1.5
-1.4
-1.3
H
L

Fig. 2

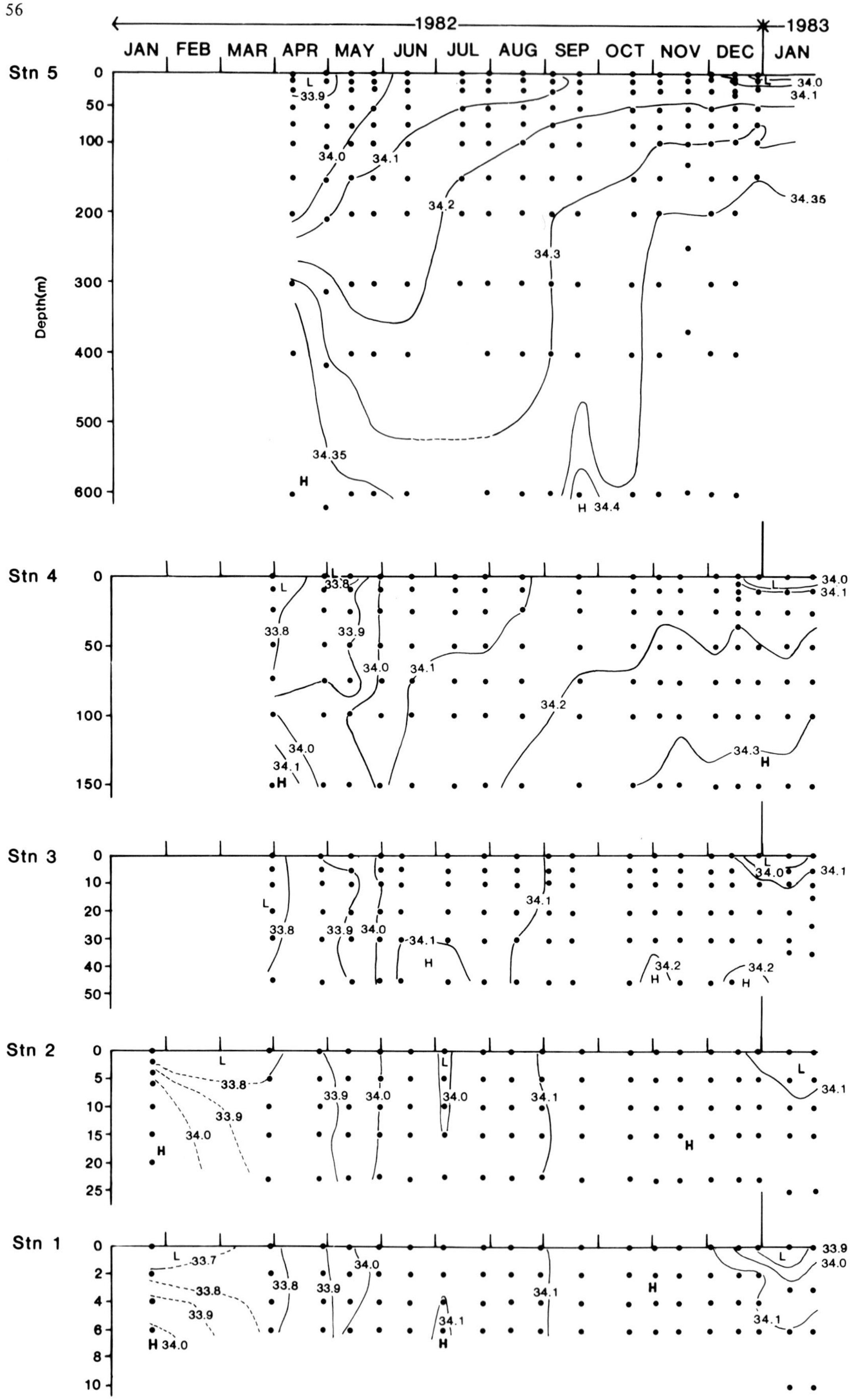

Fig. 3. Seasonal variation in salinity (practical salinity scale) at five stations. *H* and *L* indicate higher and lower salinity regimes than the surrounding water, respectively. *Dots* as in Fig. 2

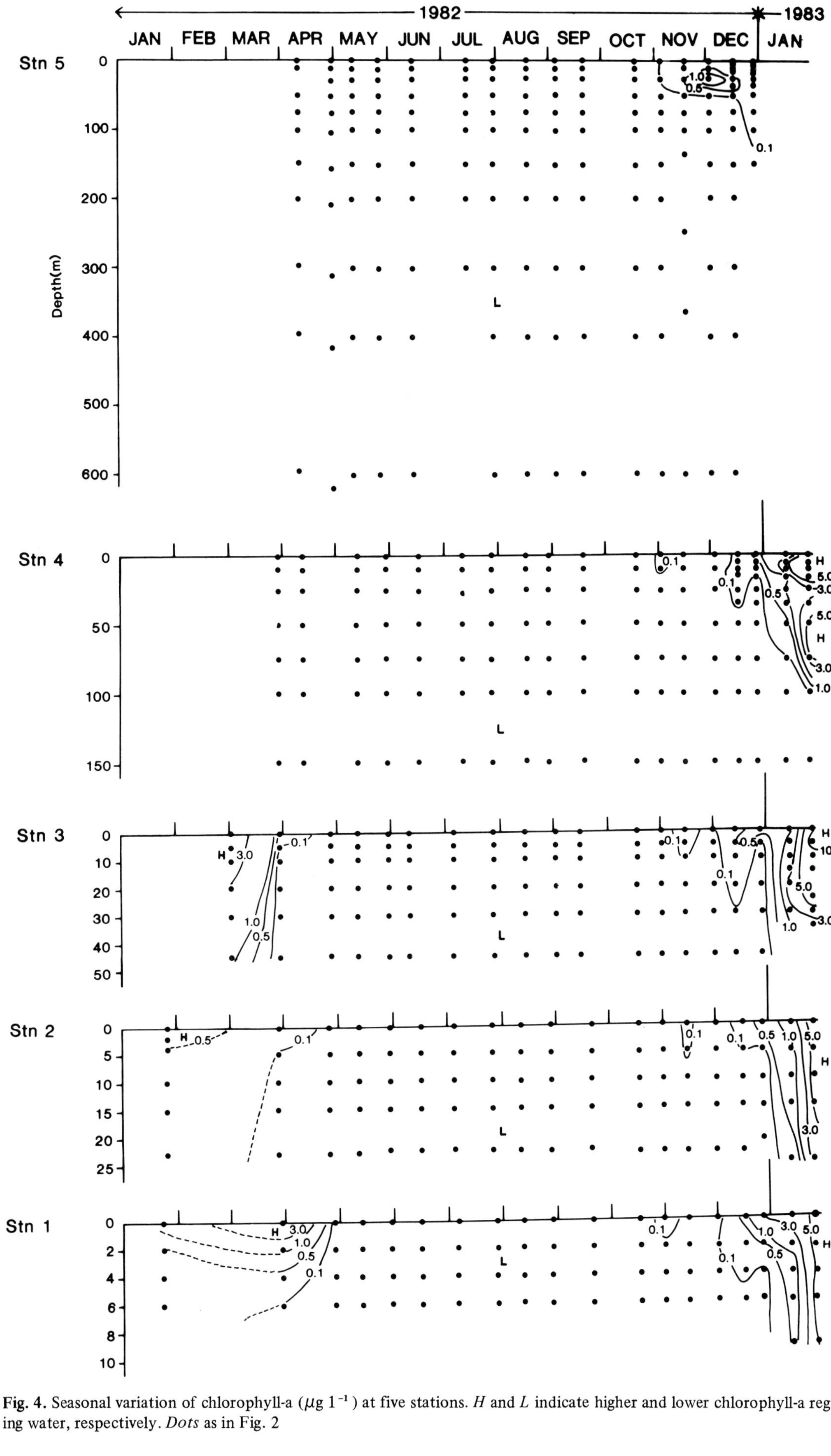

Fig. 4. Seasonal variation of chlorophyll-a (μg l^{-1}) at five stations. *H* and *L* indicate higher and lower chlorophyll-a regimes than the surrounding water, respectively. *Dots* as in Fig. 2

3.2 Seasonal Variation in Chlorophyll-a

Seasonal variation in chlorophyll-*a* at the five stations is shown in Fig. 4. The first measurements were made at Stations 1 and 2 in January 1982, when chlorophyll-*a* concentrations were high with a maximum of 4.70 $\mu g\ l^{-1}$ at the surface. By late March, concentrations were <0.50 $\mu g\ l^{-1}$ (excluding a high value > 3.00 $\mu g\ l^{-1}$ at Stn 1), and from April to November, they were < 0.10 $\mu g\ l^{-1}$. In early December, a value of 1.97 $\mu g\ l^{-1}$ was measured from 25 m at Station 5, but concentrations remained below 0.50 $\mu g\ l^{-1}$ at the other four stations. No observations were made at Station 5 in January 1983 but levels of chlorophyll-a concentrations increased at the other four stations. Maximum values were observed in late January: 7.01 at Station 1 (6 m), 9.74 at Station 2 (5 m), 11.30 at Station 3 (5 m) and 9.91 $\mu g\ l^{-1}$ at Station 4 (0 m). High concentrations were recorded for surface to bottom at Stations 1–3, but these were restricted to the upper 100 m at Station 4. Maximum integrated chlorophyll-a stocks for Stations 1, 2, 3, and 4 were 62.2 mg m^{-2} (0–1 m), 168.0 mg m^{-2} (0–25 m), 269.5 mg m^{-2} (0–50 m) and 464.2 mg m^{-2} (0–100 m), respectively.

Although no data were obtained between late January and late March 1982, it is likely that chlorophyll-a levels decreased during the period. Comparable measurements made in February and March of 1983 support this (Watanabe pers. comm.). Chlorophyll-a stocks under the fast-ice are very high when compared with data from open water (El-Sayed 1968; El-Sayed and Turner 1977; Kuroda and Fukuchi 1982). Chlorophyll-a concentrations near Syowa Station reported by Hoshiai (1969) agree with the present data.

3.3 Zooplankton Occurrence

Copepods, chaetognaths, ostracods and euphausiids occurred throughout the present series of collections, with copepods predominating. Euphausiids were collected by a parasol net most often at Station 5. Adult and juvenile stages of *Euphausia superba* and *E. crystallorophias* occurred, but only juvenile stages of *Thysanoessa macrura* and/or *T. vicina* (which cannot be distinguished) were collected.

At Station 5, integrated zooplankton biomass (0–600 m), as determined by settling volume from a Norpac net, was 176.5 ml m^{-2} in early April, 60 ml in late April–May, 80–90 ml in June–July, and then decreasing gradually to 30–40 ml between August and December. Biomass estimates from parasol net samples increased gradually during May and July but values were lower than those from samples caught by the Norpac net. Parasol net biomass continued to increase until early September, becoming more than twice as much as the biomass from the Norpac net. Between September and December, the parasol net value decreased slightly but it remained higher than the comparable Norpac figure. Because of the different mesh and mouth sizes of these nets, seasonal change of biomass might reflect the size spectrum of the zooplankton community. Therefore, it may be supposed that the abundance of small zooplankters decreased from autumn to spring, and that larger zooplankters became more numerous during winter.

Samples from modified "NIPR-I" sampler revealed that a calanoid copepod, *Paralabidocera antarctica,* endemic to Antarctic coastal waters, occurred very abundantly just beneath the under-surface of fast-ice in December–January, as had previously been noted in the region by Fukuchi and Sasaki (1981). Most individuals in this dense population were adult. This species disappeared by February and was not captured throughout autumn and winter. In late October–early November, young copepodite stages appeared in small numbers. Development progressed and adults occurred in January again. This species may play an important role in energy transfer from primary production in the ice and in water just beneath the ice during spring–summer.

4 Conclusions

The number of stations in the fast-ice area, covering a depth range of 10–675 m, and the frequency and total period of our observations, are remarkable in comparison with previous work reported from the Antarctic fast-ice area. Only preliminary accounts of the large body of data obtained by us are presented in this paper. However, it is clear that oceanographic conditions beneath the ice near Syowa Station differ from those in certain other Antarctic areas. Super-cooled water, below freezing point, was not recorded in our study. It is likely that the oceanographic conditions near Syowa Station are influenced by the Antarctic Surface Water to the north of Lützow-Holm Bay.

Acknowledgements. We express our gratitude to Prof. T. Hoshiai (of the National Institute of Polar Research, and leader of the JARE-23) for his valuable advice during this study. Thanks are also due to members of JARE-23 for their co-operation in fieldwork on the ice. Thanks are extended to Dr. G.W. Johnstone (of the Australian Antarctic Division) for his critical reading of the manuscript.

References

Andriashev A P (1970) Cryopelagic fishes of the Arctic and Antarctic and their significance in polar ecosystems. In: Holdgate M W (ed) Antarctic Ecology, Vol. 1. Academic, London, pp 297–304

Bradford J M (1978) Sea ice organisms and their importance to the Antarctic ecosystem (Review). N Z Antarct Rec 1(2): 43–50

Bunt J S (1960) Introductory studies of hydrology and plankton at Mawson, June 1956 to February 1957. ANARE Sci Rep Ser B 3: 1–135

El-Sayed S Z (1968) Primary productivity. In: Balech E, El-Sayed S Z, Hasle G, Neushul M, Zaneveld J S (eds) Antarctic Map Folio Ser 10. Primary productivity and benthic marine algae of the Antarctic and Subantarctic. American Geographical Society, New York, pp 1–6, pl 1

El-Sayed S Z (1971) Biological aspects of the pack ice ecosystem. In: Deacon G (ed) Symposium on Antarctic Ice and Water Masses, Tokyo 1970. SCAR, Cambridge, pp 35–54

El-Sayed S Z, Turner J T (1977) Productivity of the Antarctic and Tropical/Subtropical regions: A comparative study. In: Dunbar M J (ed) Polar Oceans, Arctic Institute of North America, Calgary, Alberta, pp 463–503

Fukuchi M, Sasaki H (1981) Phytoplankton and zooplankton standing stocks and downward flux of particulate material around fast ice edge of Lützow-Holm Bay, Antarctica. Mem Natl Inst Polar Res Ser E Biol Med Sci 34: 13–36

Fukuchi M, Tanimura A, Hoshiai T (1979) "NIPR-I", a new plankton sampler under sea ice. Bull Plankton Soc Japan 26: 104–109

Gordon A L (1971) Oceanography of Antarctic waters. In: Reid J L (ed) Antarctic research ser 15. Antarctic Oceanology I. American Geophysical Union, Washington DC, pp 169–203

Hoshiai T (1969) Seasonal variation of chlorophyll-*a* and hydrological conditions under sea ice at Syowa Station, Antarctica. Antarct Rec 35:52–67

Houtman T J (1967) Water masses and fronts in the Southern Ocean south of New Zealand. N Z Dep Sci Ind Res Bull 174: 1–40

Kuroda K, Fukuchi M (1982) Vertical distribution of chlorophyll-*a* in the Indian sector of the Antarctic Ocean in 1972–1973. Antarct Rec 74: 127–142

Kusunoki K (1981) Variations of sea ice conditions in Lützow-Holm Bay area, in Antarctica, in the last 20 years. In: Allison I (ed) Sea Level, Ice and Climatic Change. IAHS, Washington DC, pp 171–176 (IAHS Publ 131)

Littlepage J L (1965) Oceanographic investigations in McMurdo Sound, Antarctica. In: Llano G A (ed) Antarctic research ser 5. Biology of the Antarctic Seas II. American Geophysical Union, Washington DC, pp 1–37

Moriwaki K, Yoshida Y (1983) Submarine topography of Lützow-Holm Bay, Antarctica. Mem Natl Inst Polar Res Spec Issue 28: 247–258

Motoda S (1957) North Pacific Standard Plankton net. Inf Bull Planktol Jpn 4: 13–15

Saijo Y, Nishizawa S (1969) Excitation spectra in the fluorometric determination of chlorophyll-*a* and pheophytin-*a*. Mar Biol 2: 135–136

Solorzano J (1969) Determination of ammonia in natural waters by the phenolhypochlorite method. Limnol Oceanogr 14: 799–801

Strickland J D H, Parsons T R (1968) A practical handbook of sea water analysis. Fish Res Board Can Bull 167:1–311

Wakatsuchi M (1982) Seasonal variations in water structure under fast ice near Syowa Station. Antarctica, 1976. Antarct Rec 74: 85–108

Light-Dependence of Phytoplankton Photosynthesis in the Antarctic Ocean: Implications for Regulating Productivity

M.M. TILZER[1], B. von BODUNGEN[2], and V. SMETACEK[2]

Summary. The relationships of vertical profiles of phytoplankton photosynthesis to under-water light were studied at 23 stations in the South Scotia Sea and Bransfield Strait. On 3 occasions diurnal photosynthetic patterns were monitored. Chlorophyll-*a* concentrations varied by a factor of 16.5. Eighty per cent of the variations in light extinction could be explained by variations in chl-*a* concentration. Accordingly, the euphotic zone (1% surface light level) varied from 15 to 70 m. Photosynthetic profiles were studied in order to assess the production potential of Antarctic phytoplankton. The photosynthetic capacity (photosynthesis per chl-*a* at optimum light) and maximum quantum yield of photosynthesis (moles carbon dioxide assimilated per mole light quanta absorbed) on the average were smaller by a factor of 7 and 4, respectively, than in phytoplankton at lower latitudes. Diminished low-light photosynthesis suggests that in Antarctic waters temperature-controlled processes take over as rate-limiting steps in otherwise light-limited situations. The utilization efficiency of incident irradiance by phytoplankton at any level of chlorophyll concentration is diminished by both reductions in the photosynthetic capacity and lower light-limited quantum yields. By the combination of both effects, phytoplankton in Antarctic waters can utilize incident light only inefficiently even in situations where biomass accumulation is high.

1 Introduction

The contribution of phytoplankton as a primary source of organic C and energy to the subsequent food chain depends on both concentration of biomass and turnover rate of biomass C. The concentration of algal biomass strongly influences the efficiency at which phytoplankton can be utilized by herbivorous zooplankton (e.g., Frost 1972, 1975). The turnover rate of phytoplankton biomass ultimately determines the transfer of C and energy to the subsequent food chain (Tilzer 1984a).

The Antarctic phytoplankton is characterized by wide regional variations in both species composition and biomass. Maximum biomass can be far in excess of levels normally encountered in offshore waters (El-Sayed et al. 1964; El-Sayed and Weber 1982). Turnover rates of phytoplankton C which are reflected in relative photosynthetic rates (e.g., assimilation numbers) are generally low (Jacques 1983). On the other hand, N uptake studies have suggested rapid and extensive mineralization of organic matter within the euphotic zone (Rönner et al. 1983).

Antarctic phytoplankters live under a unique set of environmental conditions: consistently low water temperatures, small to negligible vertical density gradients, abundant nutrients, and ample light during summer (El-Sayed 1970). The question of whether the species have developed specific adaptations to the pervailing conditions is still unresolved (Neori and Holm-Hansen 1982; Jacques 1983). It is obvious, however, that phytoplankton productivity ultimately depends on the physiological responses of the algae to the extreme conditions under which they live.

This paper describes the photosynthetic responses of the algal community to underwater light. Photosynthetic profile parameters in Antarctic phytoplankton are compared with those of phytoplankton at lower latitudes. Special emphasis is placed on the efficiency at which light quanta are utilized for photosynthesis. The effect of varying concentrations of biomass on photosynthesis per unit of sea-surface area is examined, and estimates of upper limits to phytoplankton productivity are attempted. Our analysis shows that the restraints imposed by extremely low temperature ultimately restrict the efficiency of photosynthetic energy conversion and, thus, phytoplankton productivity in Antarctic waters.

2 Materials and Methods

This analysis is based on data of 23 vertical profiles and of three diurnal experiments (three profiles each) from the second leg of the 56th cruise of R.V. *Meteor* in November and December 1980 in the South Scotia Sea and the Bransfield Strait. All stations were located south of the Antarctic Convergence, ranging from latitudes of 54° 38′ (Station 19) to 64° 9′ (Station 139 off Anvers Island). The locations of all other stations are shown in Fig. 1.

Samples were collected by using 10–1 Niskin bottles from six to eight depths, ranging from the sea surface down to a depth corresponding to a light level of 0.1% of surface intensity. The sampling depths were determined by mea-

1 Limnological Institute, University of Constance, P.O. Box 5560, 7750 Konstanz, FRG

2 Institute for Marine Sciences, Division of Marine Planktology, University of Kiel, Düsternbrooker Weg 20, 2300 Kiel, FRG

Antarctic Nutrient Cycles and Food Webs
(ed. by W.R. Siegfried, P.R. Condy, and R.M. Laws)

surements of the underwater light gradient by an underwater probe ("Optiksonde"; Haardt and Maassen 1983). Photosynthetic rates were determined by simulated in situ incubations on deck at seawater temperatures by the radiocarbon technique. Jena glass bottles of 120 ml were used, shaded by gauze socks of varying light transmissions to simulate the light intensities at the depths of collection. Incubations lasted for 6 h beginning at ca. 13.30 hours local time. The incubation times during the diurnal experiments are given in Tables 3 and 4. The activity of the ^{14}C bicarbonate solution added to each sample was 10 μ Ci. Duplicate light and one dark bottle were used at each light level. Filtration was conducted at moderate vacuum onto 0.45 μM membrane filtres that were rinsed with seawater acidified to pH = 3 by HCl to remove remaining $^{14}CO_2$. In order to avoid any possible light shock, samples were processed inside the vessel.

Chlorophyll-a was determined spectro-photometrically from extracts in 90% acetone after Strickland and Parsons (1972) by using monochromatic equations. Nutrient analyses were performed in an auto-analyzer. Incident light was recorded by a pyrheliometer. Values in Joules were converted to quanta based on the assumption that 1 mol of PAR (photsynthetically active radiation) had an energy of 218 kJ and PAR was 46% of shortwave irradiance (Bannister 1974; Talling 1957; Tilzer 1983). Thus, 1 kJ was converted to 2.11 mmol of PAR.

2.1 Symbols, Units, and Definitions

C = Chlorophyll-a concentration, mg Chl-a m^{-3}.

I_o = Incident photosynthetically active radiation (PAR) mmol m^{-2} h^{-1}.

I'_o = Light intensity immediately below the water surface, reflection assumed 10%.

I_z = Light intensity at depth Z (m), μmol m^{-2} s^{-1} (= μ E m^{-2} s^{-1}) or mol m^{-2} h^{-1}.

I_k = Light intensity at onset of light saturation, units as above.

ϵ = Vertical light attentuation coefficient, ln-units m^{-1}, mean value over spectrum of PAR (400–700 m) and euphotic zone (0.01 I'_o).

$\bar{k}_c$ = Chlorophyll-specific light attentuation coefficient, mean value over spectrum of PAR and euphotic zone, ln-units m^{-1} [mg m^{-3} Chl-a$]^{-1}$ = mg^{-1} m^2 Chl-a.

k_w = Background light attenuation, units as ϵ.

P^B = Assimilation number, mg C [mg Chl-a$]^{-1}$ h^{-1}.

P^B_m = Maximum assimilation number at depth of light optimum ("photosynthetic capacity"). Units as above.

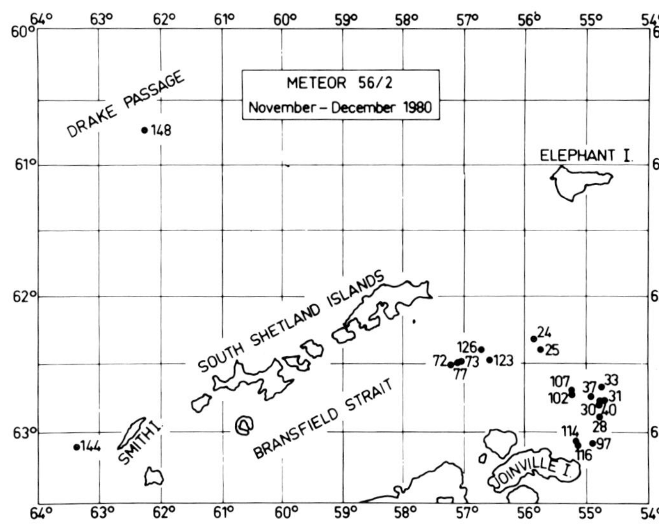

Fig. 1. Sampling stations during the second leg of 56th cruise of R.V. *Meteor.* Stations 19 and 20 are NNE of Elephant Island

a^B = Slope of the light-limited photosynthetic vs light intensity curve, mgC (Carbon) [mg Chl-a · mmol PAR$]^{-1}$ m^2. For conversion to the units used by Platt and Jassby (mg C [mg Chl-a$]^{-1}$ h^{-1} w^{-1} m^2) multiply by 16.5.

ϕ_z = Quantum yield of photosynthesis at depth z, mol C assimilated [mol PAR absorbed by Chl-a$]^{-1}$.

ϕ_m = Maximum quantum yield at limiting light intensities. Units as above.

π = Vertical integral of phytoplankton photosynthesis, mg C m^{-2} h^{-1}.

3 Results

3.1 Environmental Conditions Relevant to Phytoplankton Productivity

Only at the northernmost stations were water temperatures above 0°C. Vertical differences in general were extremely small (Table 1). Nutrient concentrations were in the range observed previously (e.g., Holm-Hansen et al. 1977; Rönner et al. 1983) with minimal vertical concentration gradients (Table 1). Regional variations in the observed nutrient concentrations followed different patterns for the nutrients measured. No statistically significant inverse relationship between concentrations of nutrients and biomass were apparent. This supports previous conclusions that phytoplankton growth was not nutrient-limited (Holm-Hansen et al. 1977; Paden et al. 1981). However, nitrate concentrations decreased slightly in areas of maximum phytoplankton biomass, but never to limiting levels. On the other hand, silica concentrations decreased sharply near the Antarctic Convergence. Jacques (1983) has attributed regional variations in silica levels to uptake by diatoms which require high concentrations for maximum growth. Concentrations of Chl-a were significantly correlated with POC ($r = 0.754$, $P < 0.001$). The regression analyses suggests a mean C: Chl-a ratio of 50 and a POC of roughly 50 $\mu g\ l^{-1}$ in the absence of phytoplankton. This analysis furthermore indicates low but variable concentrations of particulate organic matter other than phytoplankton (Tables 1 and 2). When assuming the suggested C:Chl-a ratio of 50, which is at the lower end of the range given by Strickland and Parsons (1972), phytoplankton C on the average comprised about 75% of total POC. Minima as low as 5% occurred in the Drake Passage and values approaching 100% coincided with high phytoplankton biomass (Table 1).

Phytoplankton species composition exhibited great regional differences. Nanoplankton was generally scarce (M. Elbrächter, pers. comm.). In the Southern Scotia Sea, where total biomass was low, *Corethron* was the dominant species. East of the Bransfield Strait (Station 25–46, 52–54) the assemblage was dominated by *Thalassiosira, Rhizosolenia, Chaetoceros* and *Nitzschia.* In the Bransfield Strait, where total biomass was at a maximum, *Phaeocystis* was dominant together with the four above-mentioned genera of diatoms (Elbrächter 1981).

The water transparency was highly variable. Euphotic depths as defined by the 1% light level ranged from 15 m to 70 m. Eighty per cent of the variations in the vertical light attenuation coefficient could be attributed to variations in chlorophyll concentrations. The regression analysis of vertical light extinction coefficients vs chlorophyll concentrations yielded the following equation:

$$\epsilon = 0.027\ C + 0.06;\ r = 0.89,\ P < 0.001 \qquad (1a)$$

This relationship is used below to estimate fractional light absorption by photosynthetic pigments (i.e., the proportions of underwater light that are absorbed by photosynthetic pigments and thus can be utilized for photosynthesis). In general terms Eq. (1a) can be written as

$$\epsilon = \bar{k}_c C + k_w \qquad (1b)$$

It is likely that chlorophyll-specific light attenuation ($\bar{k}_c$) is variable due to changing species composition. As a consequence, estimates of fractional light absorption probably are only approximate. In a water-layer of 1-m thickness the amount of light absorbed is $I_z - I_z + 1 = I_z\ (1-e^{-\epsilon})$ and the amount absorbed by algae is

$$PUR_{abs} = I_z\ (1-e^{-\epsilon})\ (\bar{k}_c C / a) \qquad (2)$$

(Tilzer 1983). Fractional light absorption varied highly between 5% and close to 100% (Table 1). The highest computed values should be over-estimates either due to underestimates of k_w or overestimates of $\bar{k}_c$, or both, because 100% algal light absorption is theoretically impossible.

However, fractional light absorption in Antarctic waters should be higher than in other systems at comparable chlorophyll concentrations, because of low background light attenuation (k_w). This could be explained by low dissolved colour and turbidity due to minimal terrestrial influence and small concentrations of non-algal POC (Table 1). Fractional light absorption by algae was significantly correlated with the proportions of algal C in total POC ($r^2 = 0.72$; $P < 0.001$, see Table 1). Moreover, chlorophyll – specific light attenuation ($\bar{k}_c$) is high as compared to more eutrophic systems and is typical for blue oligotrophic waters (Morel 1978; Tilzer 1978). Values of $\bar{k}_c$ should increase with depth in blue water (Atlas and Bannister 1980). Water was most transparent at wavelengths below 500 nm whenever Chl-a concentrations were low. Maximum transparency shifted to wavelengths greater than 550 nm only when phytoplankton was most abundant (Haardt and Maassen 1981).

3.2 Photosynthetic Profiles

The profiles in Fig. 2, over a chlorophyll concentration range of 16.5, show the effect of varying concentrations of biomass on both light-saturated maximum photosynthetic rates and the vertical extent of photosynthesis. During the diurnal experiments variable chlorophyll concentrations were recorded which were a consequence of patchiness and movements of the vessel relative to the water masses (Fig. 3). Talling (1971), for lakes, has used the equation

$$\pi = (P^B_m\ C/\epsilon)\ f\ (I'_o) \qquad (3)$$

to describe the factor-interaction controlling the vertical integral of phytoplankton photosynthesis and thus productivity per unit of water surface. This relationship can be applied whenever biomass is homogeneously distributed over the water column. This condition holds true in Antarctic waters (Table 1, Fig. 2). Talling (1957, 1971) has noted that integral photosynthesis is roughly equal to the product of light-saturated photosynthesis and the depth corresponding to a light-level of 50% of the intensity at onset of light saturation:

$$\pi = (P^B_m \cdot C/\epsilon) \cdot \ln\ (I'_o/0.5\ I_k) \qquad (4)$$

Integral photosynthesis is thus to a great extent controlled by the threshold intensity between light saturation and limitation. If assimilation numbers (P^B) are plotted against a depth-scale standardized with regard to the light intensity gradient, the light-dependence of the photosynthetic profiles can be demonstrated (Fig. 4).

Both the photosynthetic capacity (P^B_m) and the slope of the light-limited photosynthesis vs light intensity curve a^B, control I_k. The photosynthetic capacity depends on the number of photosynthetic units (n) and the photochemical turnover time: $P^B_m = n/T$, Falkowski (1981). The slope of

Table 1. Einvironmental parameters at stations where photosynthetic measurements were conducted (Fig. 1)

Station	°C mean euphotic	I_o PAR m mole m^{-2} h^{-1}	$\bar{\epsilon}$ m^{-1} euphotic	$\bar{k}_c C/\epsilon$ mean euphotic	PO_4 µg at l^{-1}	$NO_3 + NO_2$ µg at l^{-1}	SiO_4 µg at l^{-1}	POC µg l^{-1}	Phytoplank.[a] C % of POC
19	3.356 ± 0.051	2760	0.076	0.15	1.89 ± 0.07	14.03 ± 0.15	11.29 ± 0.46		
20	– 0.039 ± 0.111	2534	0.083	0.05	2.19 ± 0.05	8.40 ± 0.07	27.23 ± 0.89	149.0 ± 18.6	5.3
24	– 1.491 ± 0.096	2084	0.184	0.43	2.23 ± 0.65	7.53 ± 0.09	60.91 ± 0.63	261.7 ± 17.0	60.7
25	– 1.579 ± 0.027	1954	0.100	0.44	2.33 ± 0.06	8.62 ± 0.18	65.00 ± 1.27	142.9 ± 16.1	57.4
28	– 1.809 ± 0.014	1967	0.135	0.85	2.34 ± 0.02	10.32 ± 0.07	62.81 ± 1.02	196.1 ± 47.4	100
30	– 1.975 ± 0.000	1628	0.135	0.64	2.32 ± 0.07	11.04 ± 0.46	63.01 ± 1.18	228.1 ± 27.0	70.1
31	– 1.755 ± 0.009	3178	0.115	0.71	2.30 ± 0.04	10.56 ± 0.09	64.7 ± 0.67	202.8 ± 19.2	74.2
33	– 1.818 ± 0.004	1558	0.135	0.73	2.36 ± 0.06	10.53 ± 0.24	64.1 ± 1.36	219.4 ± 27.5	82.8
37	– 1.719 ± 0.017	3463	0.112	0.76	2.49 ± 0.01	12.39 ± 0.12	62.2 ± 1.07	206.3 ± 10.2	76.1
40	– 1.735 ± 0.025	2684	0.112	0.62	2.58 ± 0.26	13.55 ± 0.26	62.82 ± 0.36	178.7 ± 14.7	72.0
72	– 1.464 ± 0.004	2635	0.200	0.60	3.82 ± 0.58	21.44 ± 0.92	66.92 ± 2.05	249.6 ± 44.9	89.1
73	– 1.432 ± 0.014	3318	0.117	0.47	2.43 ± 0.10	14.52 ± 0.49	67.8 ± 1.48	148.16 ± 65.2	100
77	– 1.448 ± 0.003	2167	0.171	0.48	3.33 ± 0.13	25.44 ± 0.28	66.7 ± 0.74	176.7 ± 17.2	85.4
97	– 1.408 ± 0.865	1853	0.121	0.77	3.21 ± 0.18	24.10 ± 1.01	63.04 ± 1.07	209.6 ± 24.1	82.5
102	– 1.573 ± 0.018	3886	0.128	0.78	2.72 ± 0.12	19.57 ± 0.34	63.5 ± 0.39	197.2 ± 16.5	93.6
107	– 1.492 ± 0.026	2501	0.149	0.67	2.09 ± 0.32	24.40 ± 0.32	60.20 ± 0.80	227.7 ± 15.5	80.8
114	– 1.636 ± 0.003	2057	0.135	(1.00)	3.15 ± 0.08	30.01 ± 0.51	64.8 ± 1.12	287.7 ± 25.7	92.1
116	– 1.564 ± 0.058	1837	0.134	(1.00)	3.26 ± 0.84	27.75 ± 1.39	63.1 ± 1.05	194.6 ± 24.4	(141)
123	– 1.006 ± 0.187	2153	0.177	0.59	2.57 ± 0.14	21.71 ± 0.53	62.9 ± 0.76	286.7 ± 45.9	67.5
126	+ 2.000 ± 0.126	1279	0.307	0.64	2.02 ± 0.07	15.09 ± 0.41	64.2 ± 0.41	486.0 ± 23.1	90.0
139	– 1.386 ± 0.271	3202	0.067	0.37	2.42 ± 0.25	17.4 ± 1.16	65.9 ± 2.72	150.5 ± 45.1	30.2
144	– 0.756 ± 0.028	3401	0.154	0.60	2.24 ± 0.08	16.3 ± 0.52	51.1 ± 0.76	284.2 ± 15.2	61.6
148	+ 0.105 ± 0.551	1613	0.077	0.19	2.64 ± 0.30	19.4 ± 3.00	26.1 ± 1.81	242.4 ± 19.1	10.9

a Conversion of Chl a to phytoplankton carbon by assuming C:Chl a = 50

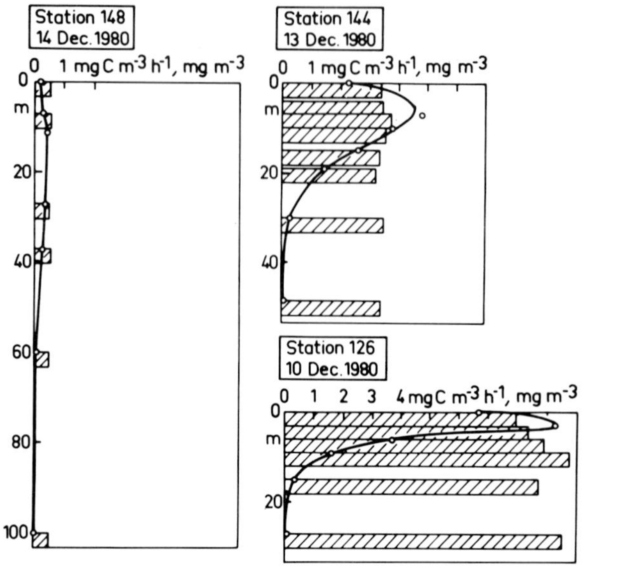

Fig. 2. Photosynthetic rates as measured by simulated in situ incubations (*solid lines, circles*) and chlorophyll-a concentrations (*hatched bars*) at three stations with increasing biomass. Further information is given in Tables 1 and 2

the light-limited P^B vs I-curve a^B is a function of the quantum yield of photosynthesis which reaches a saturation plateau at limiting light levels. The quantum yield of photosynthesis is a measure of the efficiency at which absorbed PAR is utilized for photosynthesis:

$$\phi = \frac{\text{mol C assimilated}}{\text{mol PAR absorbed}} \tag{5}$$

The slope a^B also depends on chlorophyll-specific light absorption ($\bar{k}_c$):

$$a^B = 12\,\phi_m \cdot \bar{k}_c \tag{6}$$

(Tilzer 1984b). It can be determined by regression analysis of light-limited values of P^B vs I_z. The threshold light intensity I_k then can be calculated by determination of the intersecting point of the light-limited slope and the saturation plateau.

Light-limited maximum quantum yields can be calculated from a^B thus obtained, by using Eq. (6). Independently, quantum yields can be calculated at any depth by

$$\phi_z = P_z^B / 12\, I_z\, \bar{k}_c \tag{7}$$

(Tilzer 1984b). At light-saturation the assimilation number should be constant if no light inhibition would occur.

$$P_m^B = 12\,\phi_z\,\bar{k}_c\,I_z = \text{const.}\ (I_z > I_k)$$

At light limitation the quantum yield is constant:

$$\phi_m = (1/12\,\bar{k}_c)\ \ (P_z^B / I_z) = \text{const.}\ (I_z < 0.5\ I_k)$$

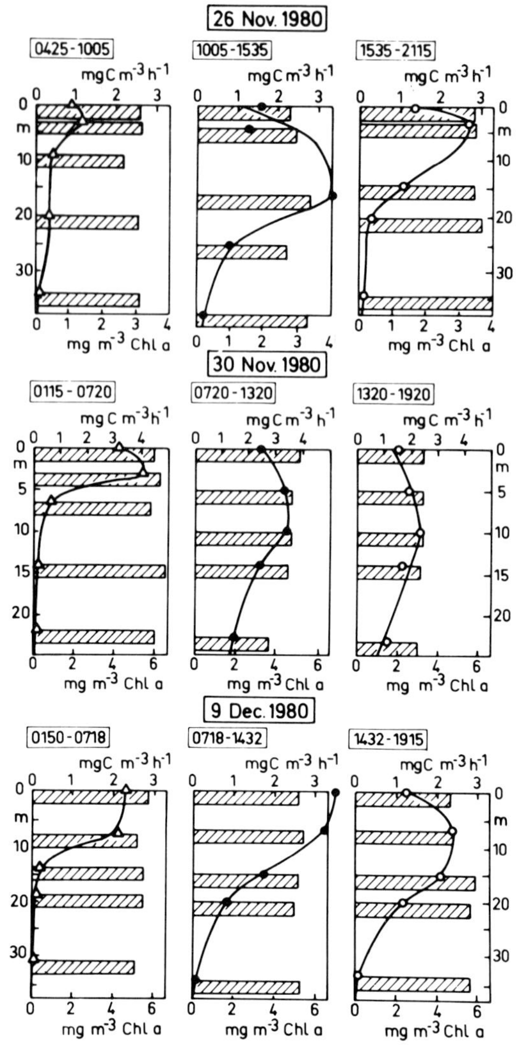

Fig. 3. Photosynthetic rates as measured by simulated in situ incubations (*solid lines*) and chlorophyll-a concentrations during diurnal experiments. Further information is given in Tables 3 and 4

The light intensity at onset of light saturation is defined by the validity of both equations:

$$I_k = (1/12\,\bar{k}_c) \cdot (P_m^B / \phi_m) = P_m^B / a^B \tag{8}$$

By substituting for I_k in Eq. (4) we obtain

$$\pi = (P_m^B \cdot C/\epsilon) \ln (2\, a^B\, I_o' / P_m^B), \text{ or} \tag{9a}$$

by solving Eq. (6),

$$\pi = (P_m^B\, C/\epsilon) \ln (24\, I_o'\, \Phi_m\, \bar{k}_c / P_m^B). \tag{9b}$$

These equations allow the description of the integral photosynthesis and thus phytoplankton productivity in terms of all important controlling parameters: incident irradiance, water transparency, chlorophyll concentration, photosynthetic capacity, and the slope of the light-limited P vs I

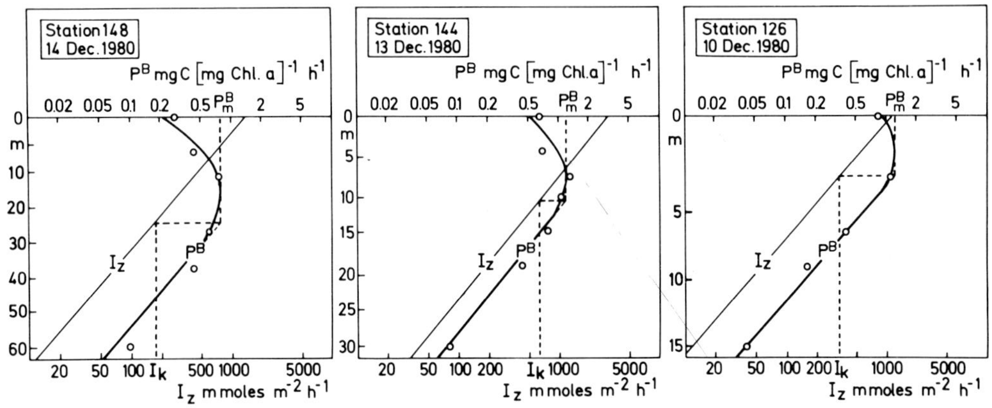

Fig. 4. Photosynthetic rates per unit of chlorophyll-a (P^B) plotted logarithmically against depth at the stations shown in Fig. 1. Depth scales have been standardized to a relative light gradient. Underwater light (I_z) is also given (*lower scale*)

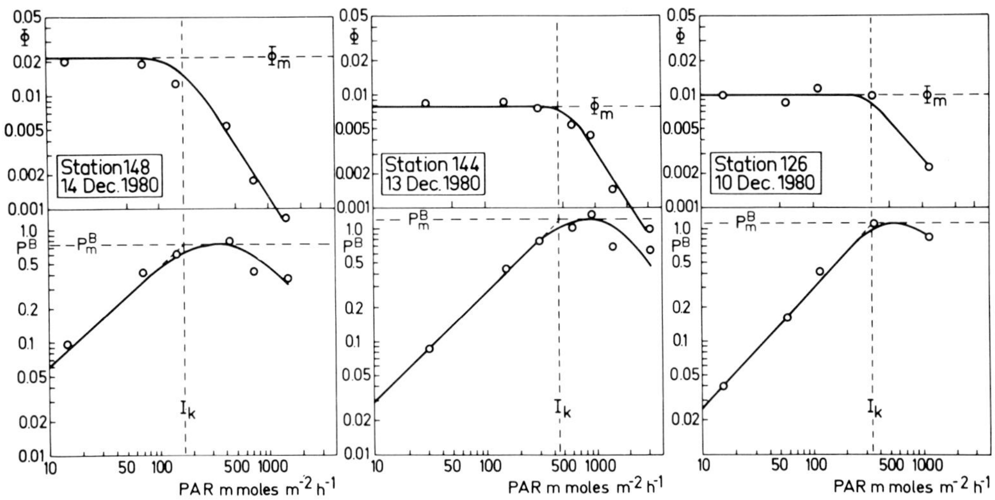

Fig. 5. Lightdependence of photosynthesis per unit of chlorophyll-a (P^B, *lower panels*) and of quantum yields (ϕ, *upper panels*). Quantum yields have been calculated by using Eq. 7 under the assumption of constant $\bar{k}_c = 0.027$. The plateau of mean ϕ_m was calculated from a^B by using Eqn. 6

curve which is a function of the maximum quantum yield and chlorophyll-specific light attenuation.

Figure 5 shows the relationship between assimilation numbers and quantum yields and the influence of both on I_k. Stations 126 and 144 are characterized by high photosynthetic capacities and low maximum quantum yields (and a^B). Therefore, I_k in both profiles is exceptionally high. At Station 148, low P^B_m is combined with high ϕ_m leading to low I_k.

4 Discussion

Photosynthetic capacities and the light-limited slopes of the photosynthesis vs light intensity curve were highly variable both between stations and at the same station within 1 d (Tables 2 and 3). However, values of P^B_m and, to a lesser extent of a^B, in general appear to be smaller in Antarctic phytoplankton than in algae from lower latitudes. This has been previously observed by Jacques (1983). Our field data for 1980 yield a mean photosynthetic capacity that is only about one-seventh of the values recorded at lower latitudes by Platt and Jassby (1976) and Côté und Platt (1983). The slope of the light-limited P^B vs. I curve in 1980 was smaller by a factor of 3.3 than the values of Côté and Platt (1983) and smaller by a factor of 4.7 than the values observed by Jassby and Platt (1976). Laboratory incubations in October and November 1983 at much lower concentrations of biomass (Chl-a concentrations 0.15–0.8 $mg\,m^3$) revealed the same basic trends (Tilzer, unpubl.).

The temperature dependence of light-saturated photosynthesis is well-established (e.g., Jewson 1977) and is to be expected because the photosynthetic capacity is controlled by enzymatic processes. The fact that the light-limited slope of the photosynthesis vs light intensity curve frequently is reduced as well, is more surprising. It is generally assumed that light-limited photosynthesis is

Table 2. Parameters characterizing photosynthetic profiles at stations shown in Fig. 1

Station	C mg m^{-3} Chl-a [a]	P^B_m	a^B [b]	ϕ_m [c]	I_k μE/m^{-2} S^{-1}	π mg cm^{-3} h^{-1}
19	0.43	0.50	0.0005	0.0015	175	8.9
20	0.16	0.53	0.0018	0.0055	55	15.2
24	3.18	0.87	0.0063	0.0194	37	52.7
25	1.64	0.81	0.0053	0.0164	22	54.0
28	4.27	1.02	0.0031	0.0094	92	78.9
30	3.20	0.36	0.0008	0.0024	118	78.2
31	3.01	0.80	0.0022	0.0067	103	61.8
33	3.63	0.70	0.0011	0.0032	172	34.3
37	3.14	0.41	0.0009	0.0029	98	36.5
40	2.57	0.50	0.0019	0.0059	63	30.3
72	4.45	0.68	0.0015	0.0045	63	77.0
73	3.11	0.70	0.0024	0.0075	74	43.5
77	3.02	0.80	0.0067	0.0207	78	45.9
97	3.46	0.80	0.0112	0.0346	19	57.8
102	3.69	0.70	0.0017	0.0052	109	57.1
107	3.68	0.41	0.0019	0.0050	42	41.0
114	5.30	0.54	0.0016	0.0051	87	56.8
116	5.50	0.45	0.0009	0.0027	110	47.5
123	3.87	0.74	0.0039	0.0121	55	43.7
126	8.75	1.10	0.0032	0.0099	84	58.6
139	0.91	0.45	0.0007	0.0029	151	13.9
144	3.47	1.19	0.0025	0.0078	129	66.7
148	0.53	0.75	0.0041	0.0127	47	15.3

a Mean value within euphotic zone
b a^B Determined by regression analysis
c Calculated from a^B by Eq. (5) assuming $\bar{k}_c = 0.027$

Table 3. Photosynthetic profile parameters during diurnal experiments (dimensions as in Table 2)

Date	Time[a]	P^B_m	a^B	ϕ_m	I_k
26 Nov. 1980	0425–1005	0.36	0.0008	0.0024	118
	1005–1535	0.80	0.0022	0.0067	103
	1535–2115	0.70	0.0011	0.0032	172
30 Nov. 1980	0115–0720	–	0.0093	0.0288	–
	0720–1320	0.68	0.0015	0.0048	63
	1320–1920	0.70	0.0024	0.0074	74
9 Dec. 1980	0150–0718	–	0.0049	0.0151	–
	0718–1432	0.64	0.0020	0.0060	95
	1432–1915	0.45	0.0008	0.0027	110

a No light saturation

controlled by temperature-independent photochemical reactions. Small light-limited slopes a^B would suggest diminished maximum quantum yields [Eq. (6)]. Neori and Holm-Hansen (1982) have observed that chlorophyll fluorescence in Antarctic phytoplankton increased between 4.5°C and in situ temperatures which should lead to decreasing light energy utilization at decreasing temperatures. One could speculate that in extremely cold water, temperature-dependent processes take over as rate-limiting reactions in otherwise light-limited situations. This concept can be reconciled with Liebig's Law of the Minimum.

Bannister and Weidemann (1984) have suggested that under light limitation quantum yields are in the range of 0.06–0.08. From our data a mean ϕ_m of 0.0085 (1980) and of 0.023 (1983) emerges which are significantly lower than maxima anticipated from theory. The mean values of

Table 4. Productivity-related parameters during diurnal experiments

Date	Time	Σ I_o mmol m^{-2} h^{-1}	Σ C mg Chl-*a* m^{-2}	π mg C m^{-2} h^{-1}	E^a TOT %
26 Nov. 1980	0425–1005	1628	103.2	14.62	0.08
	1005–1535	3178	120.5	61.78	0.17
	1535–2115	1558	123.5	34.33	0.19
30 Nov. 1980	0115–0720	168	136.8	22.80	1.20
	0720–1320	2635	105.5	77.02	0.26
	1320–1920	3318	82.3	43.54	0.12
9 Dec. 1980	0150–0718	326	261.9	26.35	0.71
	0718–1432	2057	281.1	56.83	0.24
	1432–1915	1837	279.8	47.53	0.23

a Light utilization efficiency : 100 π [Jm^{-2} h^{-1}] : I_o [JM^{-2} h^{-1}]

a^B and ϕ_m encountered in Antarctic waters are at the lower end of the range observed at lower latitudes (Table 2; Platt and Jassby 1976; Côté and Platt 1983; Tilzer 1984b).

Finally, we examine the implications of the photosynthetic characteristics of Antarctic phytoplankton for primary productivity. Equations 4 and 9 demonstrate how the photosynthetic parameters assessed in this study affect integral photosynthesis (π) and thus productivity. If both light-saturated and light-limited photosynthesis were affected in the same fashion, I_k would remain constant and integral photosynthesis would decrease linearly with P_m^B. Reduction of P_m^B alone would lead to little less than a linear decrease in integral photosynthesis because I_k would decrease and consequently the ratio in (I_o' / 0.5 I_k) would rise [Eq. (9)]. The weakest effect would have to be anticipated from a drop in ϕ_m, or a^B, respectively: integral photosynthesis should only decrease by the natural logarithms of a^B by lowering the ratio ln (I_o' /0.5 I_k) [Eq. (8)]. Unfortunately, no mean values of ln (I_o' / 0.5 I_k) from marine phytoplankton are available to us. Mean values for ln (I_o' / 0.5 I_k) in lakes vary from 1.91 (Megard et al. 1979) and 2.08 (Tilzer 1983), to 2.46 (Rodhe 1965). Therefore, I_k was at light levels of 29.6%, 25.0%, or 16.6% of surface intensity, respectively. The mean ratio of ln (I_o' / 0.5 I_k) from the data of this study is higher than observed previously (2.79 ± 0.58). Thus, I_k on the average is at a light level of only 12.6% of surface intensity. This supports the previous conclusion that the productivity of the Antarctic Ocean is to a greater extent affected by diminished light-saturated than by light-limited photosynthesis.

Our data can also be used to predict the hypothetical upper limit to integral photosynthesis (ψ) that would be attained if all light were absorbed by photosynthetic pigments. In this case, light absorption by non-photosynthetic material (k_w) would become negligible and Eq. (4) could be transformed to

$$\psi = (P_m^B / k_c) \ \ln (I_o' / 0.5 \ I_k) \quad (10)$$

(Bannister 1974; Megard et al. 1979; Tilzer 1984b). Actually this maximum can never be reached because there is always some background light attenuation (k_w). The degree to which the hypothetical upper limit to integral photosynthesis (ψ) is reached depends on fractional light absorption (i.e., the proportion of light that is absorbed by photosynthetic pigments relative to total light attenuation):

$$\pi / \psi = (\bar{k}_c C) / (\bar{k}_c C + k_w) \quad (11a)$$

This equation can be rewritten in analogy to the Michaelis-Menten equation (Megard et al. 1979):

$$\pi = \psi \cdot C / [(k_w / \bar{k}_c) + C] \quad (11b)$$

where $k_w / \bar{k}_c = C_o'$, that is the chlorophyll concentration where fractional light absorption is 0.5 (Fig. 6), comparable to the half-saturation constant in enzyme kinetics. From the mean values of P_m^B, $\bar{k}_c$ and ln (I_o' /0.5 I_k) based on the 1980 data a hypothetical upper limit to integral photosynthesis of 71 $mgCm^2 h^{-1}$ would emerge. This value appears to be unrealistically low because it is exceeded by observed data at fractional light absorptions ($\bar{k}_c C/\epsilon$) well below unity (Fig. 6). By using average photosynthetic profile parameters obtained in 1983 in a light-incubator, a hypothetical upper limit to integral photosynthesis of 194 $mgCm^2 h^{-1}$ is obtained. Maximum summer production rates in the range observed at high concentrations of biomass (El-Sayed 1970) then could be predicted. Given the variability of the relevant photosynthetic profile parameters both between stations and even within 1 day (Tables 2 and 3) and considering the possibility of wide variations in $\bar{k}_c$, the scatter in the measured data points relative to the theoretical saturation curve (Fig. 6) is not surprising.

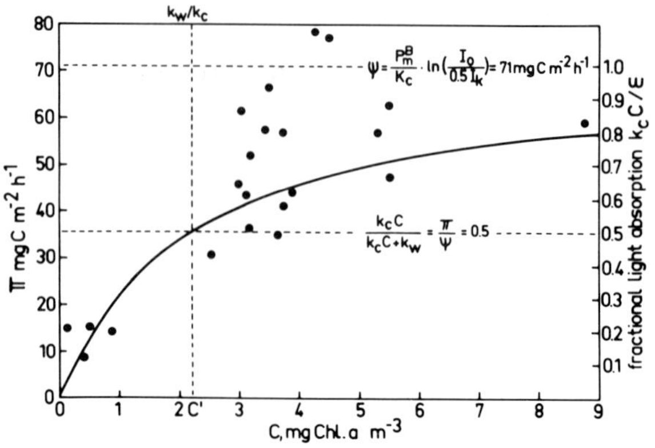

Fig. 6. Integral photosynthesis (π, *left scale*) plotted against mean euphotic Chl-a concentration (C, *lower scale*). The *upper broken line* gives the theoretical upper limit to productivity (ψ) which would be reached if all light were absorbed by photosynthetic pigments. Calculation by using Eq. (10) is based on mean values for relevant parameters observed during November–December 1980. *Solid line* gives the increase in fractional light absorption as calculated by using Eq. (11) which also gives information as to what degree the hypothetical upper limit to integral photosynthesis (ψ) is reached at any concentration of Chl-a. $C' = k_W/\bar{k}_c$ is the chlorophyll concentration at which fractional light absorption is 0.5 and consequently $\pi = 0.5\ \psi$

The efficiency at which light is converted into chemically bound energy is an integral measure of all factors controlling productivity (Dubinsky 1980). Because phytoplankters have to compete with their surrounding medium for available light, aquatic ecosystems in general are less productive than terrestrial communities (Tilzer et al. 1975). At any given level of chlorophyll concentration fractional light attenuation by Antarctic phytoplankton should be relatively high due to small non-algal turbidity of the water (k_W) and high chlorophyll-specific light-harvesting capabilities of the algae ($\bar{k}_c$). Both factors theoretically should enhance their light utilization efficiency relative to phytoplankton that is in an optically less favourable environment. However, this advantage is more than offset by the restraints imposed by consistently low water temperatures on both light-saturated and, to a lesser extent, light-limited photosynthesis. As a consequence, at any given fractional light absorption, the utilization efficiency of incident PAR should be smaller than at lower latitudes (Fig. 7). Moreover, the growing season is extremely short in the Antarctic Ocean. Therefore, the turnover rate of phytoplankton biomass is generally low and the transfer of C and energy to the subsequent consumer food chain is generally small despite sometimes high concentrations of biomass.

Acknowledgements. The first author (M.M.T.) thanks Professor Zeitzschel, Kiel, for his invitation to participate in the expedition. Dr. Haardt and Dr. Maassen provided light data, Dr. Elbrächter identified phytoplankton, P. Fritzsche, M. Mempel, L. Uhlman, R. Werner, and B. Beese provided technical assistance. The crew of R.V. *Meteor* rendered invaluable logistical support. The critical comments by Dr. W.W. Gieskes on an earlier draft of this manuscript are gratefully acknowledged. This research was funded by a grant from the Deutsche Forschungsgemeinschaft.

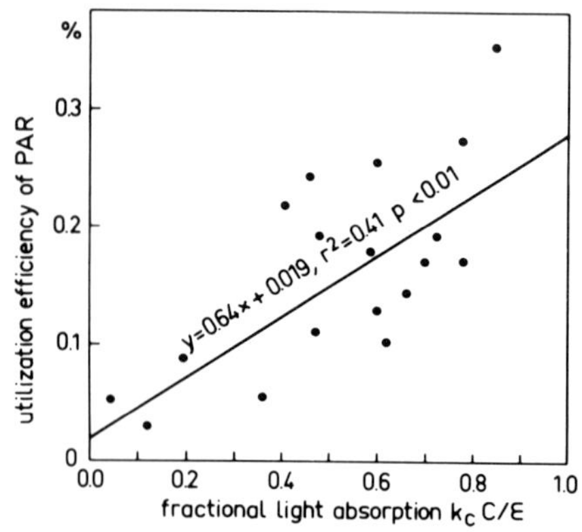

Fig. 7. Dependence of utilization efficiency of PAR (energy equivalents of C fixed photosynthetically per unit surface area divided by energy of PAR impinging at the same area) on fractional light absorption. Based on data from 1980

References

Atlas D, Bannister TT (1980) Dependence of mean spectral extinction coefficient of phytoplankton on depth, water colour, and species. Limnol Oceanogr 25:157–159

Bannister TT (1974) Production equations in terms of chlorophyll concentration, quantum yield, and upper limit to production. Limnol Oceanogr 19:1–12

Bannister TT, Weidemann AD (1984) The maximum quantum yield of phytoplankton photosynthesis in situ. J Plankton Res 6:275–294

Côté B, Platt T (1983) Day-to-day variations in the spring–summer photosynthetic parameters of coastal phytoplankton. Limnol Oceanogr 28:320–344

Dubinsky Z (1980) Light utilization efficiency in natural phytoplankton communities. In: Falkowski P (ed) Primary Productivity in the Sea, Plenum, New York, pp 83–97

Elbrächter M (1981) Arten und Größenspektrum des Mikroplanktons. In: Zeitzschel, Zenk BW, (eds) Beobachtungen und erste Ergebnisse der „Meteor" Reise 56 aus der Scotia See und der Bransfield Straße im November/Dezember 1980 (ANT I): Ein nautischer und wissenschaftlicher Bericht. Berichte aus dem Institut für Meereskunde, Kiel, Nr 80 pp 54–55

El-Sayed SZ (1970) On the productivity of the Southern Ocean (Atlantic and Pacific sectors). In: Holdgate A (ed) Antarctic Ecology, vol 1 Academic, New York, pp 119–135

El-Sayed SZ, Mandelli EF, Sugimura Y (1964) Primary organic production in the Drake Passage and Bransfield Strait. In: Lee M (ed) Biology of the Antarctic seas. I. Antarctic research ser. vol 1 American Geophysical Union, Washington DC, pp 1–11

El-Sayed SZ, Weber LH (1982) Spatial and temporal variations in phytoplankton biomass and primary productivity in the Southwest Atlantic and the Scotia Sea. Polar Biology 1:83–90

Falkowski PG (1981) Light-shade adaptation and assimilation numbers. J Plankton Res 3:203–216

Frost BW (1972) Effects of size and concentration of food particles on the feeding behaviour of *Clanus pacificus*. Limnol Oceangr 17:805–815

Frost BW (1975) A threshold feeding behaviour in *Calanus pacificus*. Limnol Oceanogr 20:263–266

Haardt H, Maassen R (1981) Optische Beobachtungen. In: Zeitzschel, BW Zenk (eds) Beobachtungen und erste Ergebnisse der „Meteor" Reise 56 aus der Scotia See und der Bransfield Straße im November/Dezember 1980 (ANT I): Ein nautischer und wissenschaftlicher Bericht. Berichte aus dem Institut für Meereskunde, Kiel Nr 80 pp 21–27

Haardt H, Maassen R (1983) CTD and optical data from the Antarctic-Meteor 56 ANT I-Part I: CTD and Chlorophyll profiles. Report 57, Sonderforschungsbereich 95, University of Kiel

Holm-Hansen O, El-Sayed SZ, Franzeschini G, Cukel R (1977) Primary production and the factors controlling phytoplankton growth in the Southern Ocean. In: Llano GA (ed) Adaptations within Antarctic Ecosystems Proc. 3rd SCAR Symposium Antarctic Biol. Smithsonian Institution pp 11–50

Jacques G (1983) Some ecophysiological aspects of the Antarctic phytoplankton. Polar Biology 2:27–33

Jewson DH (1977) The interaction of components controlling net phytoplankton photosynthesis in a well mixed lake (Lough Neagh, Northern Ireland). Freshwater Biol 6:551–576

Megard RO, Combs WS Jr, Smith PD, Knoll AS (1979) Attenuation of light and integral rates of photosynthesis attained by planktonic algae. Limnol Oceanogr 24:1038–1050

Morel A (1978) Available, usuable and stored radiant energy in relation to marine photosynthesis. Deep-Sea Res 25:673–688

Neori A, Holm-Hansen O (1982) Effect of temperature on rate of photosynthesis in Antarctic phytoplankton. Polar Biology 1:33–38

Paden CA, Hewes CD, Neori A, Holm-Hansen O, Weaver E, Kiefer DA, Sakshang E (1981) Phytoplankton studies in the Scotia Sea. Antarctic JUS 16:163–164

Platt T, Jassby AD (1976) The relationship between photosynthesis and light for natural assemblages of coastal marine phytoplankton. J Phycol 12:421–430

Rodhe W (1965) Standard correlations between pelagic photosynthesis and light. In: Goldman CR (ed) Primary Productivity in Aquatic Environments, Berkeley. Mem Ist Ital Idrobiol 18 (Suppl): 367–381

Rönner U, Sörensen F, Holm-Hansen O (1983) Nitrogen assimilation by phytoplankton in the Scotia Sea. Polar Biology 2:137–147

Strickland JDH, Parsons TR (1972) A practical handbook of seawater analyses. Bull Fish Res Board Can 167:1–311

Talling JF (1957) The phytoplankton population as a compound photosynthetic system. New Phytol 56:133–149

Talling JF (1971) The underwater light climate as a controlling factor in the production ecology of freshwater phytoplankton. Mitt Int Ver Limnol 19:214–143

Tilzer MM (1978) Prediction of productivity changes in Lake Tahoe at increasing phytoplankton biomass. Verh Int Limnol 2:407–413

Tilzer MM (1983) The importance of fractional light absorption by photosynthetic pigments for phytoplankton productivity in Lake Constance. Limnol Oceanogr 28:833–846

Tilzer MM (1984a) Estimation of phytoplankton loss rates from daily photosynthetic rates and observed biomass changes. J Plankton Res 6:309–324

Tilzer MM (1984b) The quantum yield as a fundamental parameter controlling vertical photosynthetic profiles of phytoplankton in Lake Constance. Arch Hydrobiol Suppl. 69:169–198

Tilzer MM, de Amezaga E, Goldman CR (1975) The efficiency of light energy utilization by lake phytoplankton. Verh Int Ver Limnol 19:800–807

Phytoplankton Biomass near a Receding Ice-Edge in the Ross Sea

W.O. SMITH Jr.[1] and D.M. NELSON[2]

Summary. Ice-edge zones have been hypothesized to be major sites of primary production and energy transfer within the Southern Ocean, and large biomass accumulations of upper-trophic-level organisms (birds and mammals) occur in these zones. However, few data exist to determine if these ice-edge regions support elevated levels of phytoplankton and enhanced rates of primary production. During 2 cruises on the USCGC Glacier in January–February, 1983, the chlorophyll distribution was measured in different areas of the Ross Sea. Our primary study area was located off the coast of southern Victorialand (76° 30′ S) in a region of receding pack-ice. We occupied 34 stations in a 100 x 300 km area of variable ice concentration. In comparison to control stations and previous data, chlorophyll levels were high, averaging 4.08 ± 1.46 mg chl-a m^{-3} at the depth of the chlorophyll maximum in the water column, and 128.2 ± 91.7 mg m^{-2} when integrated from the surface to 150 m. High surface chlorophyll levels appeared to be highly correlated with a stable surface layer at the edge of the receding ice-pack. At stations outside of the ice-edge bloom, stability at the surface was reduced and chlorophyll concentrations were markedly lower. Water column stability appeared to be a major factor in the initiation and maintenance of ice-edge phytoplankton blooms, and the roles of these blooms in the overall estimates of biogenic production and energy flux of the Southern Ocean need to be re-evaluated.

1 Introduction

The food webs of the Southern Ocean have been characterized as high biomass systems since the days of the great Antarctic whaling fleets. Numerous reports of high phytoplankton standing stocks in Antarctic waters have been reported, including the reports from the early expeditions of the Erebus, Terror and Discovery (e.g., Hart 1934). The Antarctic Krill, *Euphausia superba*, a dominant herbivore, exhibits schooling behaviour and forms locally dense swarms estimated to contain as many as 30,000 individuals m^{-3} or 33 kg m^{-3} (Everson 1977). The stocks of vertebrates are also enhanced. For example, baleen whales accounted for a biomass of 43.1×10^6 t (2.58 g m^{-2}) prior to intensive predation by man (Laws 1977), the biomass of Antarctic birds has been estimated to be 0.025 g m^{-2} (Laws 1977), and seal abundance has been estimated at 16 x 10^6 individuals. The standing stocks of the various trophic levels in the Southern Ocean initially appear to be large when compared to other pelagic systems.

The area near the edge of the pack-ice has also been shown to have elevated standing stocks of fauna and flora. El-Sayed (1970) observed a massive diatom bloom in the melting pack of the western Weddell Sea, and ice-edge phytoplankton blooms have been observed elsewhere (Marshall 1957; McRoy and Goering 1974; El-Sayed and Taguchi 1981). Marr (1962) also described an association of young Krill at the pack-ice – open-ocean interface. Higher trophic levels also tend to be concentrated in these regions. For example, bird populations, particularly Snow Petrels, *Pagodroma nivea,* and penguins (Spheniscidae), are strongly correlated with the presence of loose pack-ice (Ainley and Jacobs 1981), as are feeding baleen whales (Macintosh 1974). Although the spatial extent of the ice-edge system at any one time is small relative to the entire Southern Ocean, it expands and retreats over 20×10^6 km^2 each year (Zwalley et al. 1979). Consequently, even though it is probably restricted in space and time, the ice-edge may substantially influence the chemistry of the water column (Jennings et al. 1984) and the entire food web of the Antarctic marine ecosystem.

More recent quantitative estimates of phytoplankton biomass in the Antarctic have demonstrated the occurrence of large spatial and temporal variations. For example, surface chlorophyll concentrations in Antarctic and Sub-Antarctic waters averaged 0.29 mg m^{-3} but ranged from 0.01 to 118.4 mg m^{-3}, whereas vertically integrated (over the euphotic depths) chlorophyll concentrations averaged 17.40 mg m^{-2} (range 0.23–337.5 mg m^{-2}) (El-Sayed 1970; El-Sayed and Turner 1977). Holm-Hansen et al. (1977) found integrated chlorophyll levels in the upper 200 m of Antarctic waters to average 10.45 mg m^{-2} (range 0.73–30.8). Although the stations occupied do not necessarily represent an even sampling pattern in space or time, the observed range of chlorophyll concentrations emphasizes the variability to phytoplankton biomass within Antarctic waters.

We studied a receding ice-edge to assess the following properties of a marginal ice zone: the standing stock and primary production of phytoplankton in this region; the spatial extent of increased biogenic matter in the ice-edge

1 Botany Department and Graduate Program in Ecology, University of Tennessee, Knoxville, TN 37996, USA

2 College of Oceanography, Oregon State University, Corvallis, OR 97331, USA

Antarctic Nutrient Cycles and Food Webs
(ed. by W.R. Siegfried, P.R. Condy, and R.M. Laws)

region; and, the factors important in controlling the establishment and activity of the enhanced levels of phytoplankton biomass. This report describes the chlorophyll and temperature distributions occurring within a meso-scale region of the Ross Sea, and offers a preliminary interpretation of our findings.

2 Materials and Methods

Our study was conducted from 26 January–2 February, 1983, in the western Ross Sea near Granite Harbour and Oates Piedmont Glacier, Victorialand, from the USCGC Glacier (Fig. 1). A series of transects perpendicular to the ice-edge was occupied. Within these transects, 34 stations were completed in which the water column was sampled to a depth of 150 m. Additional locations were sampled for surface concentrations of biological parameters. Five stations were occupied along the Ross Ice-Shelf during 3–10 February, 1983.

During the study, the character of the ice-edge varied considerably as did overall ice concentrations. However, high concentrations of pack-ice (8/8) were generally found along the coast. In addition, the consolidated pack increased sharply to the south of our southernmost transect (Stations 14–23) at the entrance to McMurdo Sound. Variable concentrations of ice were encountered at stations within 100 km of the coast, but little or no pack-ice was observed in the outer half of our transects.

We measured temperature-conductivity profiles at each station with an Applied Microsystem Inc. CTD which was calibrated by discrete salinity and temperature measurements. XBT traces and temperatures of sea-surface samples also were recorded at each station. Water column samples were collected using 10–1 PVC samplers which were modified to ensure that rapidly sinking particles were sampled adequately. Samples for nutrients (nitrate, nitrite, phosphate, and silicon) were passed through polycarbonate filters and frozen for later analysis, but ammonium was determined at sea (Solorzano 1969). Chlorophyll was measured fluorometrically after extraction in 90% acetone. The fluorometer was standardized using known concentrations of purified chlorophyll (Sigma Chemical Inc) dissolved in 90% acetone. Additional data were also obtained on particulate nitrogen, carbon and silicon samples, but are not included in this report.

3 Results

3.1 Vertical Profiles

Of the 34 stations sampled, 18 exhibited profiles in which there was a chlorophyll maximum below 5 m. These maxima were not deep: the most frequently observed light depth at which the chlorophyll peaks occurred was 5%, which corresponded to an average depth of 12 m. The mean chlorophyll concentration at the depth of maximum pigment concentration within the bloom area (see below) was 4.22 ± 1.45 mg m^{-3}. Two representative profiles in which sub-surface chlorophyll maxima occurred are shown in Fig. 2.

Pigment values decreased rapidly below the euphotic zone, often by an order of magnitude or more. The integrated chlorophyll concentrations in the euphotic zone (through the 0.1% light depth) averaged 63.68 ± 19.52 mg m^{-3} in the bloom, whereas the chlorophyll concentrations integrated through 150 m averaged 116.55 ± 46.12 mg m^{-3}.

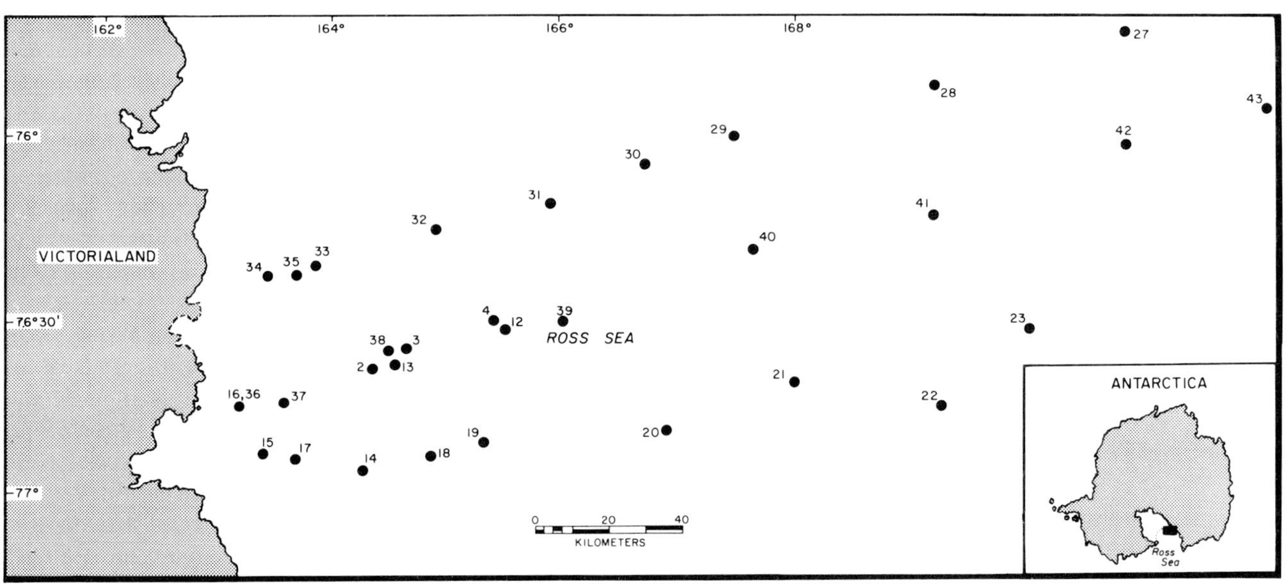

Fig. 1. Study area and stations

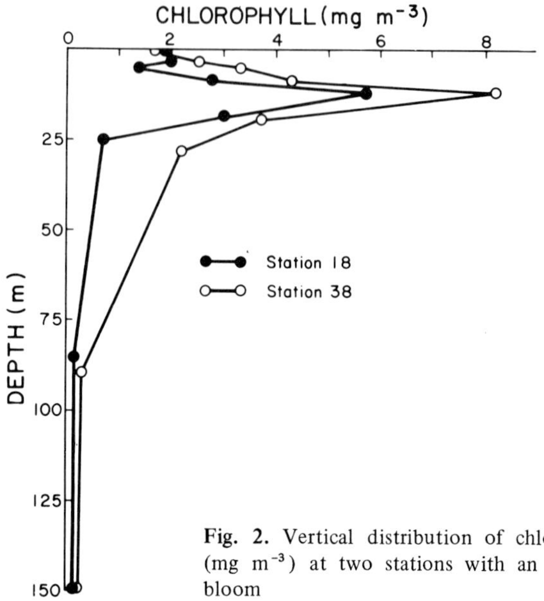

Fig. 2. Vertical distribution of chlorophyll (mg m^{-3}) at two stations with an ice-edge bloom

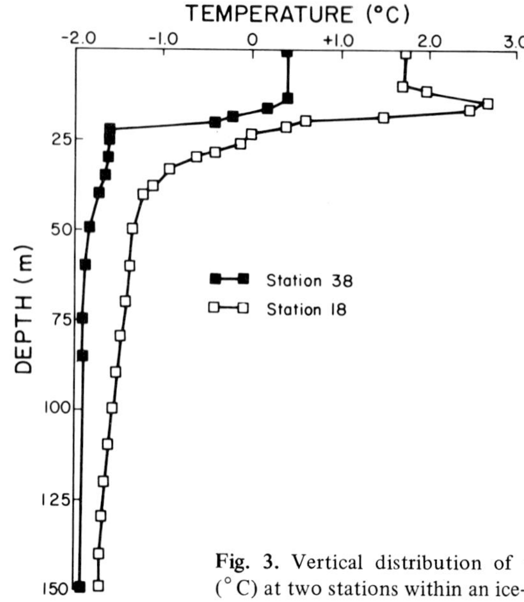

Fig. 3. Vertical distribution of temperature (° C) at two stations within an ice-edge bloom

3.2 Vertical Sections

During the cruise we were able to occupy three sections perpendicular to the ice-edge to determine the extent of the enhanced phytoplankton biomass levels. These sections were completed as rapidly as possible, with the longest transect (Stations 36–43) taking 30 h. This sampling regime allowed us to produce vertical transects of phytoplankton biomass which were quasi-synoptic in nature. We realize, however, that the marginal ice-zone is a dynamic region with a response time to physical forcing functions of the order of 1 d (Johannessen et al. 1983). Winds during the sections were relatively constant both in direction and magnitude (e.g., for Stations 36–43 wind direction ranged from 100°–178° and velocities from 4.2–11.4 m s^{-1}), so that we do not believe conditions changed significantly during the transect as a result of physical processes.

Chlorophyll distributions were similar for all transects (Fig. 4A, B, and C). Elevated concentrations (> 2 mg m^{-3}) were observed in the surface layers and extended seaward for approximately 200 km. The bloom was not restricted to the stations where ice was present but extended over large areas of the western Ross Sea. No information is available to assess the longshore variations in chlorophyll, but based on the data collected we did not observe the elevated biomass levels uncoupled from the presence of melting pack-ice.

3.3 Horizontal Distribution of Chlorophyll

The horizontal distribution of chlorophyll seemed to indicate two trends. The first was a decrease of chlorophyll away from the shore, and the second was a slight increase from south to north (Fig. 5). These patterns resulted in a central zone which was higher in surface chlorophyll than most of the rest of the study region. Station 30 had the highest surface value, 5.40 mg chl-a m^{-3}. Integrated euphotic zone (through the 0.1% light depth) chlorophyll values were in large part similar (Fig. 6), although the north-south trend was not as marked. Station 40 had the highest observed biomass, 102.2 mg chl-*a* m^{-2} (integrated from 0–28 m). A similar pattern was evident when chlorophyll values were integrated through 150 m.

3.4 Correlation of Temperature with Phytoplankton Biomass

Temperature is a key variable in ice-edge systems because it not only directly influences the rate of biological reactions, but it is a factor in controlling the physical regime in the water column (i.e., it alters density and controls ice formation). Surface water temperatures in our study region ranged from −1.11° to +1.94°C (Fig. 7), and regression analysis of both surface chlorophyll and euphotic zone chlorophyll concentrations with surface temperature showed no significant relationship. Therefore, it does not appear that temperature altered phytoplankton metabolic processes and directly resulted in the biomass variations observed in our study region.

Vertical profiles of temperature showed a strong correlation with chlorophyll within the ice-edge bloom (Fig. 3). This seemed to have resulted from stratification of the water column followed by phytoplankton growth. At stations outside the bloom, the correspondence of biomass and temperature was much less marked (Fig. 8). Vertical sections of temperature (Fig. 9) also showed that the surface layers with slightly elevated temperatures corresponded to the depths of increased phytoplankton biomass. The role of water column stability in promoting the ice-edge bloom appeared to be crucial during our study.

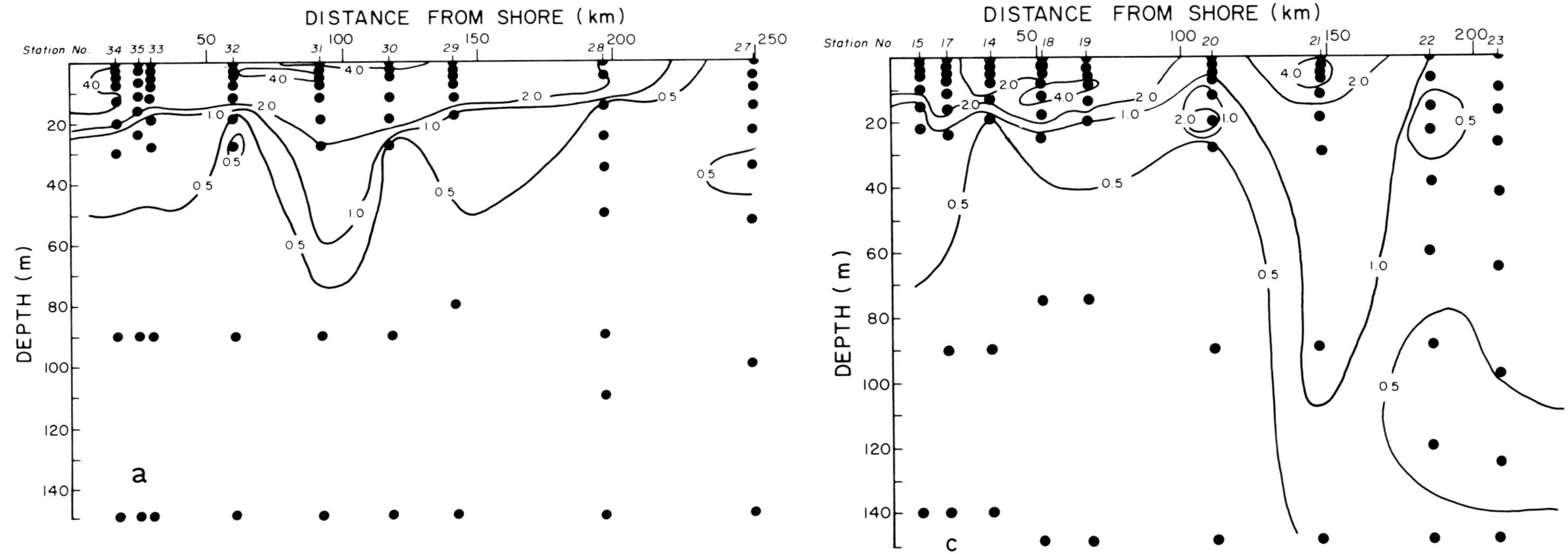

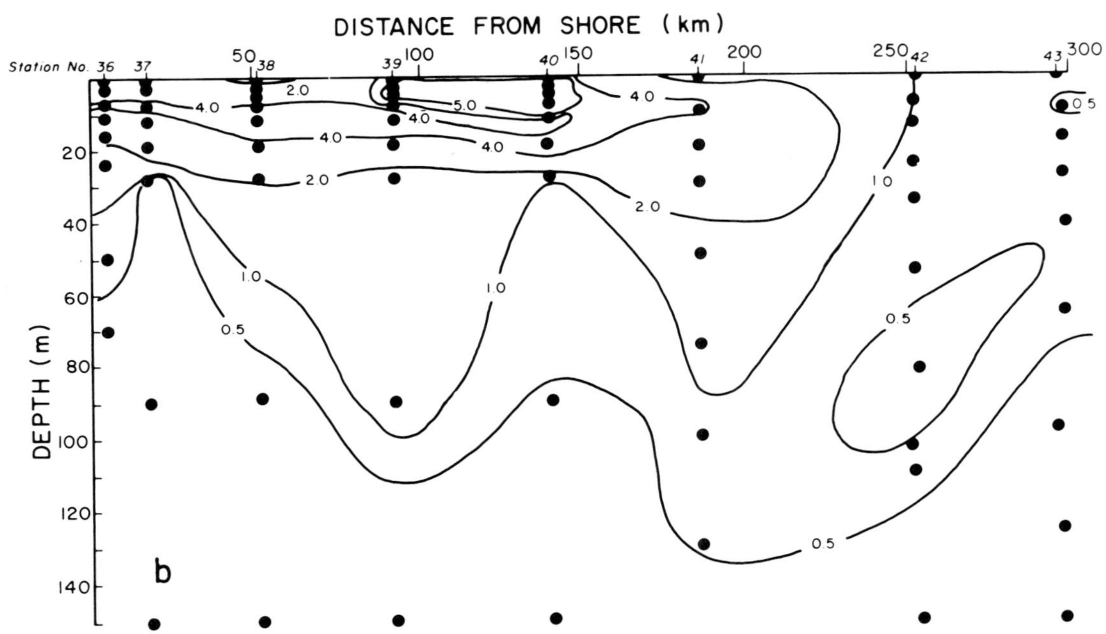

Fig. 4a–c. Vertical distribution of chlorophyll (mg m^{-3}) along three transects. **a** Stations 27–35; **b** Stations 36–43; **c** Stations 14–23. Depth of samples indicated by *dots*

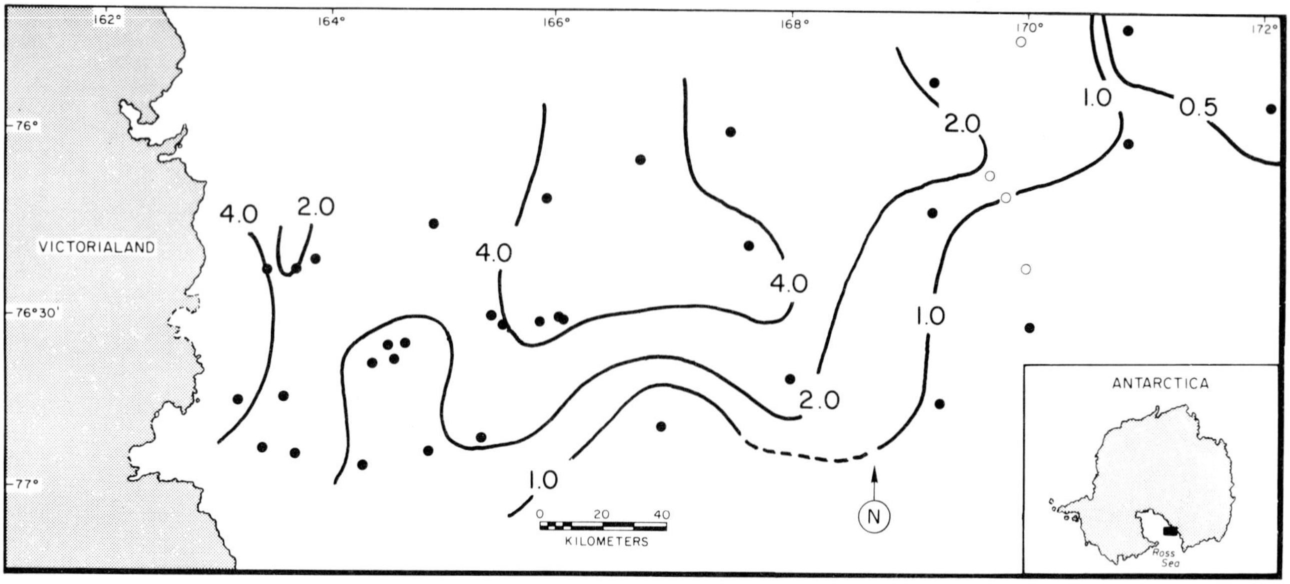

Fig. 5. Horizontal distribution of surface chlorophyll concentration (mg m^{-3}). Samples were collected over a 114-h period. Included are Stations 14–43) (Station 16 excluded)

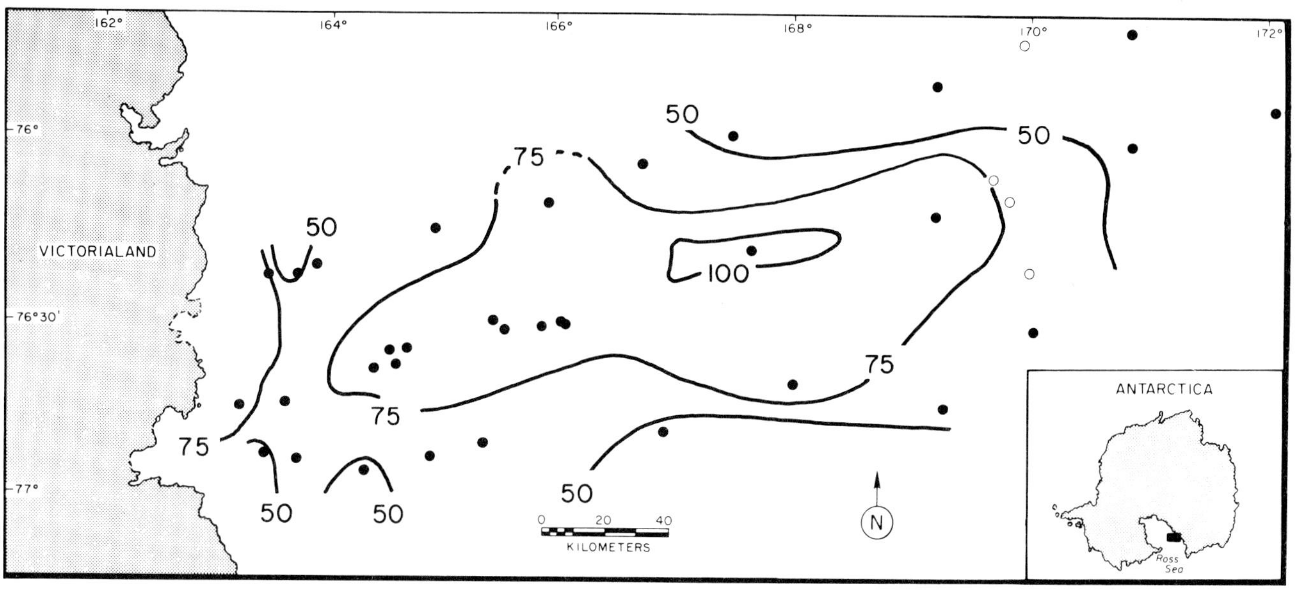

Fig. 6. Horizontal distribution of chlorophyll (mg m^{-2}) integrated from the surface to the depth of 0.1% light penetration. Included are data from Stations 14–43 (Station 16 excluded)

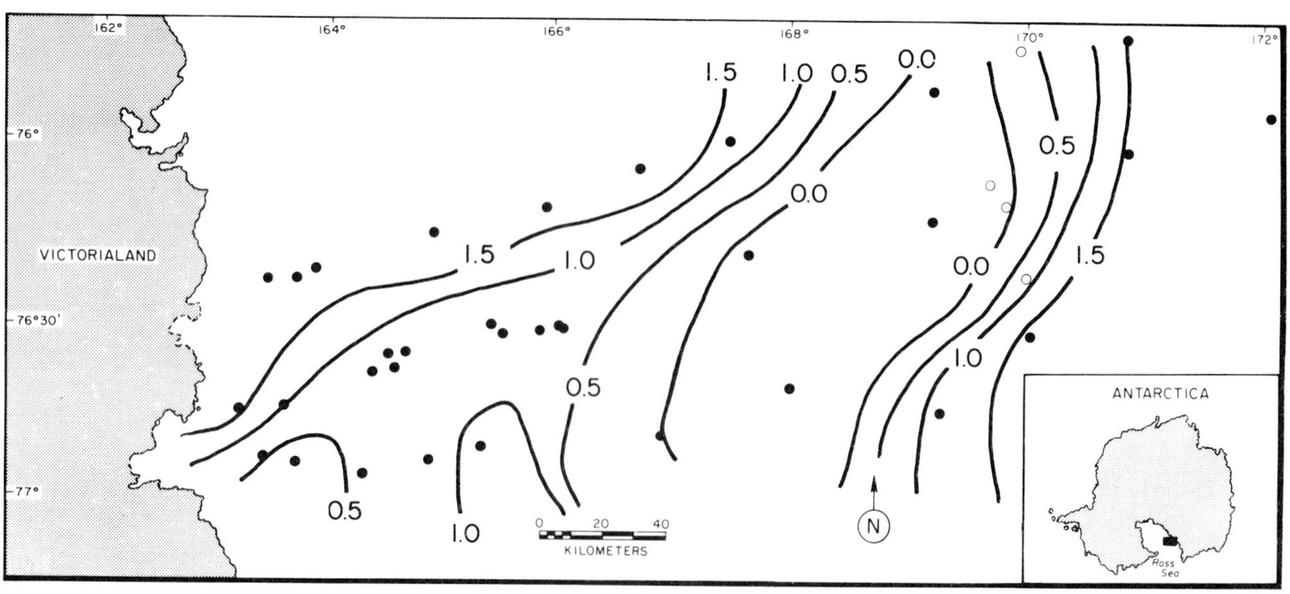

Fig. 7. Horizontal distribution of sea surface temperature (°C) as determined by bucket samples. Included are data from Stations 14–43 (Station 16 excluded)

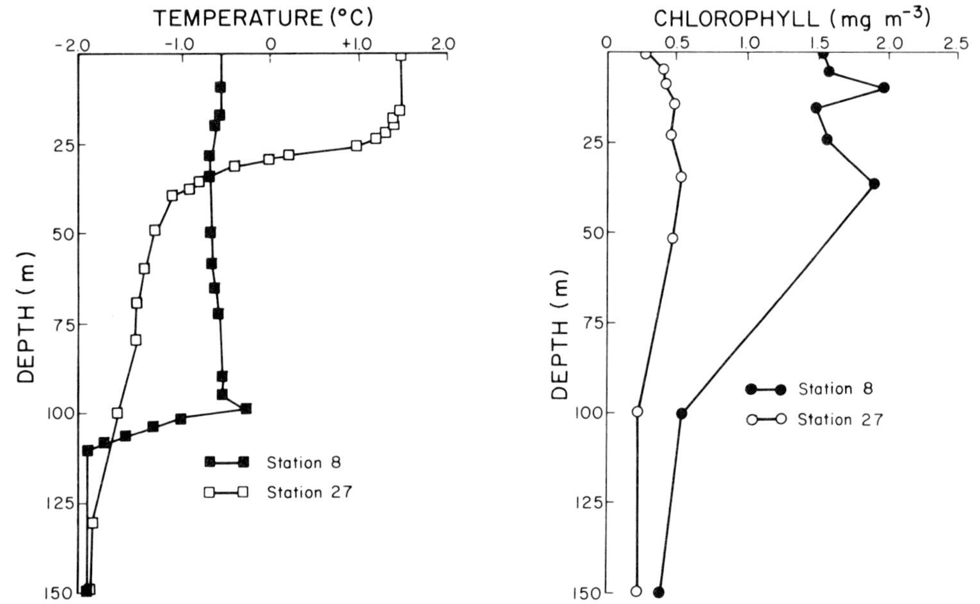

Fig. 8. Vertical distribution of chlorophyll (mg m^{-3}) and temperature (°C) at two stations outside the ice-edge bloom. Station 8 (Leg III) was located along the Ross Ice-Shelf (78° 06′ S, 178° 36′ E). Temperature data collected using a CTD unit. (data provided courtesy of Dr. S.S. Jacobs)

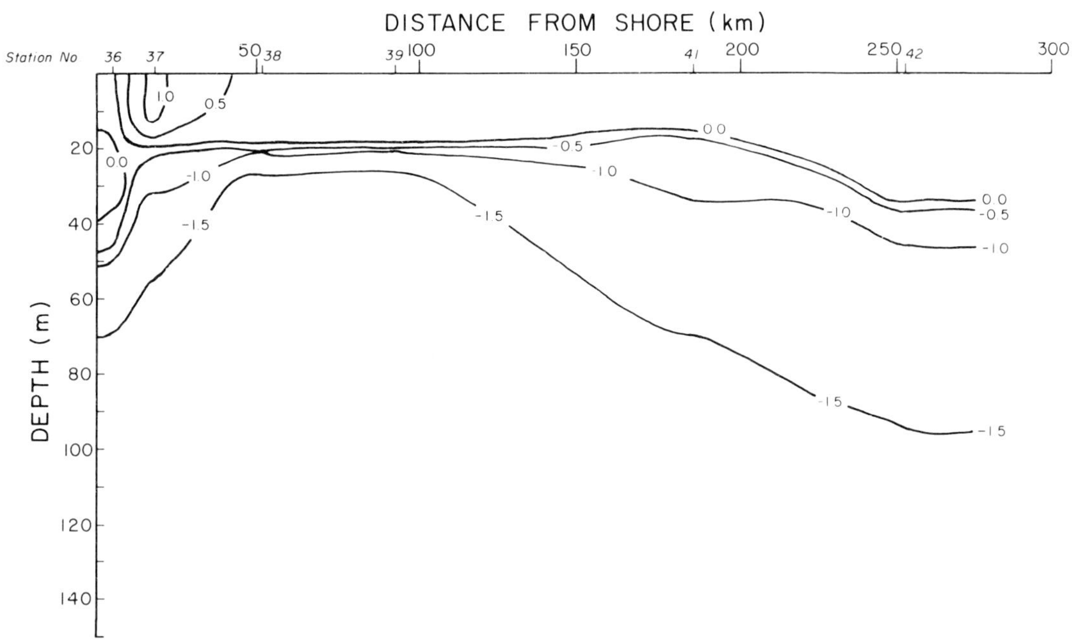

Fig. 9. Vertical distribution of temperature (°C) from Stations 36–43)

4 Discussion

Chlorophyll concentrations during our study were greater than most of those observed in other studies of the Southern Ocean, even those which have investigated the marginal ice-zone. For example, Holm-Hansen et al. (1977) found a mean chlorophyll value of 6.67 mg m^{-2} (n = 11; integrated from 0–200 m) for the Ross Sea Region, and El-Sayed and Taguchi (1981) found euphotic-zone chlorophyll to average ca. 1.65 mg m^{-3} in the southern Weddell Sea and 0.1 mg m^{-3} in the northern sector. An overall mean for the Pacific sector of 0.26 mg chl-a m^{-3} and 1.22 mg chl-a m^{-2} (surface and integrated euphotic zone values) for 723 stations has been reported (El-Sayed and Turner 1977), which indicates that the ice-edge bloom we observed was substantially elevated in phytoplankton biomass when compared to the Southern Ocean as a whole.

There are a number of possible causes for the observed bloom. One hypothesis which has been suggested previously is that meltwater from the receding ice-edge creates a low salinity (and hence low density), stable surface layer which decreases vertical mixing. Because Antarctic waters are high in nutrients, phytoplankton populations bloom in the stable, nutrient-replete environment (Marshall 1957; McRoy and Goering 1974). Another possibility is the production of a stable layer due to reduced turbulence, as a result of decreased wind stress caused by the physical presence of ice. A third hypothesis that has been suggested to explain the blooms is that upwelling could bring nutrients into the surface layers and stimulate phytoplankton growth. Finally, it has been hypothesized that epontic algae could be released from melting ice to create the high chlorophyll concentrations observed and/or seed a bloom near the ice-edge (Hameedi 1978).

Our temperature data indicated that a stable surface layer did exist within the marginal ice-zone. Meltwater per se would not create a layer of elevated temperature. However, we suggest that the low-salinity meltwater induced stability, and that the surface waters were then heated by 24-h solar radiation. By reducing the depth of mixing to less than that of the critical depth (Sverdrup 1953), net production could have occurred and formed the bloom we observed. We realize that density would be a much better indication of stability, but currently our salinity data are not fully analyzed. Our data also indicate that it is unlikely that a decreased mixed layer resulted from physical dampening of wind by ice. None of our stations along the Ross Ice-Shelf showed a surface bloom (Fig. 8), although meteorological conditions were similar. Ice-melt occurs predominantly at the bottom of the ice-shelf (Jacobs et al. 1979) so that surface stabilization via meltwater would not necessarily occur near an ice-shelf. The onshore-offshore pattern of biomass distribution was one which would be expected if stability were a major control mechanism of phytoplankton growth (i.e., growth within a stable layer and a rapid decrease in biomass when the stability is reduced). The weak north-south trend also may have resulted from ice effects, since dense pack-ice ocurred south of our southernmost transect. The distribution of biomass also no doubt reflects the general surface circulation of the region. Ainley and Jacobs (1981) reported that the overall movement of water off Victorialand is northward. Thus, the waters we studied presumably originated under the thick pack-ice and Ross Ice-Shelf, and were probably lower in phytoplankton biomass initially. The degree of influence of the area's circulation pattern on the distribution of chemical and biological variables is unknown.

The temperature sections (Fig. 9) showed no evidence for ice-edge upwelling. The process of upwelling has been modelled extensively (Neshyba 1977; Griesman 1979; Niebauer 1982; Roed and O'Brien 1983), and observed in the Arctic (Buckley et al. 1979; Johannessen et al. 1983). However, it has never been documented in the Antarctic marginal ice-zone. Johannessen et al. (1983) showed that upwelling occurred during off-ice wind conditions, due to an Ekman divergence caused by higher wind stress over ice than water. During our study the winds at stations within 50 km of the coast were usually from the north or longshore. Roed and O'Brien (1983) predicted that winds parallel to the ice-edge might also result in upwelling, so there is no clear consensus on the mechanisms operating in situ to produce upwelling. It is also possible that wind velocities were too low to result in upwelling. Johannessen et al. (1983) observed weak upwelling at winds of 5–10 m s^{-1}. During our study winds ranged between zero and 8.2 m s^{-1}, and averaged 4.6 m s^{-1} at stations within 50 km of the ice-edge. Even if the direction of the wind were favourable for upwelling, it is possible that its velocity was too low to produce upwelling. Finally, it is possible that the magnitude of any upwelling might have been too small to observe in our temperature or chlorophyll plots, and that the sampling regime was too coarse to detect any upwelling present. Nevertheless, it is by no means certain that ice-edge upwelling would facilitate a bloom in the Antarctic. Upwelling usually results in a nutrient flux into the euphotic zone and a stimulation of production. However, nutrient concentrations in surface waters were very high in the Southern Ocean during this study, and are so generally, and the addition of nutrients would not necessarily stimulate phytoplankton growth and may in fact result in a decrease in phytoplankton biomass by dilution of the bloom with low biomass sub-euphotic waters.

The degree to which epontic algae create or seed the ice-edge bloom remains unclear. Chlorophyll concentrations in the ice were not measured in our study, but a sample taken along the Ross Ice-Shelf contained 179 mg chl-a m^{-3}. The sample was semi-quantitative [i.e., an ice-floe was broken by the ship and a small (ca. 8000 cm^3) piece was melted and the chlorophyll analyzed quantitatively]. If this amount of chlorophyll were released as a point of source (i.e., if 179 mg chl were released regardless of the ice thickness) and if this were evenly distributed throughout a 30-m euphotic zone, the chlorophyll concentration of the water column would be increased for some finite period by 6 mg m^{-3}. If 10% of this chlorophyll were to remain in the euphotic zone, then a significant proportion of the bloom would have its origin as epontic forms. Because we presently do not know the activity of the ice algae or their sinking rates upon ice-melt, further research is needed to clarify the contribution of ice algae to the initiation and dynamics of ice-edge blooms.

The chlorophyll sections clearly show that the ice-edge bloom extended 200 km from the dense pack. This area is greater than other ice-edge blooms which have been reported. For example, the data of Alexander and Niebauer (1981) indicated that the extent of the bloom they studied in the Bering Sea was ca. 50 km. However, chlorophyll levels were extremely high throughout their entire study area, so that the influence of the ice-edge bloom is difficult to separate from the bloom caused by the shallow continental shelf of the Bering Sea. There are no data from the Antarctic with which we can compare our results. If the situation we observed occurs at all ice-edge regions in the Southern Ocean, then the amount of phytoplankton carbon produced in marginal ice zones is substantial, particularly in view of the large area over which the pack-ice recedes (up to 20 times 10^6 km^2). Ice-edge phytoplankton blooms may play a major role in biogeochemical cycles and trophic dynamics of the Antarctic, and we feel that an assessment of their importance may lead to a new understanding of Southern Ocean marine ecosystems.

Acknowledgements. This research was supported by grant DPP-8119572 from the Division of Polar Programs, National Science Foundation. We thank J. Elser, S. Moore, P. Whaling, M. Carbonell, D. Wilson, C. Weimer, and the officers and crew of the USCGC Glacier for assistance in the field, and Ms B. Walker for typing the draft of the manuscript.

References

Ainley DG, Jacobs SS (1981) Sea-bird affinities for ocean and ice boundaries in the Antarctic. Deep-Sea Res 28A:1173–1185

Alexander V, Niebauer HJ (1981) Oceanography of the Bering Sea ice-edge zone in spring. Limnol Oceanogr 26:1111–1125

Buckley JR, Gammelsrod T, Johannessen JA, Johannessen OM, Roed LP (1979) Upwelling: Oceanic structure at the edge of the arctic ice pack in winter. Science 203:165–167

El-Sayed SZ (1970) On the productivity of the Southern Ocean (Atlantic and Pacific sectors). In: Holdgate MW (ed) Antarctic Ecology, Academic, New York, pp 119–135

El-Sayed SZ, Taguchi S (1981) Primary production and standing crop of phytoplankton along the ice-edge in the Weddell Sea. Deep-Sea Res 28A:1017–1032

El-Sayed SZ, Turner JT (1977) Productivity of the Antarctic and tropical-subtropical regions: a comparative study. In: Dunbar M (ed), Polar Oceans, Proc SCOR/SCAR Polar Oceans Conf. Montreal, pp 463–504

Everson I (1977) The Living Resources of the Southern Ocean. FAO, Rome

Griesman P (1979) On upwelling driven by the melt of ice shelves and tidewater glaciers. Deep-Sea Res 26A:1051–1065

Hameedi MJ (1978) Aspects of water column primary productivity in the Chuckchi Sea during summer. Mar Biol 48:37–46

Hart TJ (1934) On the phytoplankton of the south-west Atlantic and Bellingshausen Sea. Discovery Rep 8:3–268

Holm-Hansen O, El-Sayed SZ; Franceschini GA, Cuhel RL (1977) Primary production and the factors controlling phytoplankton growth in the Southern Ocean. In: Llano GA (ed) Adaptations within Antarctic Ecosystems, Proceedings of the 3rd SCAR Symposium of Antarctic Biology, Smithsonian Inst, Washington, pp 11–50

Jacobs SS, Gordon AL, Ardai JL (1979) Circulation and melting beneath the Ross Ice Shelf. Science 203:439–444

Jennings JC, Gordon LI, Nelson DM (1984) Nutrient depletion indicates high primary production in the Weddell Sea. Nature 308:51–54

Johannessen OM, Johannessen JA, Morrison J, Farrelly BA, Svendsen EAS (1983) Oceanographic conditions of the marginal ice zone north of Svalbard in early fall 1979 with an emphasis on mesoscale processes. J Geophys Res 88:2755–2769

Laws RM (1977) The significance of vertebrates in the Antarctic marine ecosystem. In: Llano GA (ed) Adaptations within Antarctic Ecosystems, Proceedings of the 3rd SCAR Symposium of Antarctic Biology, Smithsonian Inst, Washington, pp 411–438

Mackintosh NA (1974) Sizes of krill eaten by whales in the Antarctic. Discovery Rep 36:157–178

Marr JWS (1962) The natural history and geography of the Antarctic krill (*Euphausia superba* Dana). Discovery Rep 32:33–464

Marshall PT (1957) Primary production in the Arctic. J Cons 23: 173–177

McRoy CP, Goering JJ (1974) The influence of ice on the primary productivity of the Bering Sea. In: Hood DW (ed) Oceangraphy of the Bering Sea pp 403–421

Neshyba S (1977) Upwelling by icebergs. Nature 267:507–508

Niebauer HJ (1982) Wind and melt driven ocean circulation in a marginal sea ice edge frontal system: a numerical model. Cont Shelf Res 1:49–98

Roed LP, O'Brien JJ (1983) A coupled ice-ocean model of upwelling in the marginal ice zone. J Geophys Res 88:2863–2872

Solorzano L (1969) Determination of ammonia in natural waters by the phenolhypochlorite method. Limnol Oceanogr 14:799–801

Sverdrup HU (1953) On conditions for the vernal blooming of phytoplankton. J Cons 18:287–295

Zwalley HJ, Parkinson C, Carsey F, Gloersen P, Campbell WJ, Ramseier RO (1979) Antarctic sea ice variations 1973–1975 NASA Weather Climate Rev Paper 56 Washington, DC

The Influence of Light on Growth and Development of the Sea-Ice Microbial Community of McMurdo Sound

C.W. SULLIVAN, A.C. PALMISANO, S. KOTTMEIER, S. McGRATH GROSSI, and R. MOE[1]

Summary. Microbial production in the Southern Ocean is not limited to benthic and planktonic habitats. The annual sea-ice provides a growth substratum and refugium for a complex microbial community composed primarily of micro-algae, bacteria and protozoans. We assessed growth and development of the sea-ice microbial community (SIMCO) in McMurdo Sound and obtained evidence that light was a major limiting factor. A light perturbation experiment was set up on the annual sea-ice of McMurdo Sound near Cape Armitage during October–December 1981 in which 2 experimental quadrats of 100 m^2 each were constructed. On 1 quadrat snow cover of 15–70 mm was maintained, while the adjacent quadrat received 0.7 m of snow to provide 2 different under-ice irradiances. Significant growth of ice algae occurred at irradiance of 0.2–2.9 $\mu E\ m^{-2}\ s^{-1}$. Estimates of *in situ* algal and bacterial growth rates indicated doubling times of 7 and 14 d, respectively. The growth of heterotrophic ice bacteria appeared to be coupled to growth of ice algae. At least 20 $\times$ 10^6 kg new carbon per yr is contributed to McMurdo Sound by SIMCOs. We conclude that ecosystem models of Southern Ocean food webs must consider not only total C input but also the dynamics of primary and secondary production derived from sea-ice microbial communities.

1 Introduction

Sea-ice in the Southern Ocean undergoes an annual cycle of growth and retreat during which time approximately 17 $\times$ 10^6 km^2 of the ice cover melts (Gordon 1981), releasing its contents to the water column. This cycle has an important influence on the physical and chemical characteristics of the region and, because of the biological community associated with sea-ice, it undoubtedly has a profound effect on the local biota as well.

Sea-ice provides a unique microhabitat with sufficient nutrients and light to allow high concentrations of microorganisms to build up within its boundaries. The physical structure of the ice also provides a safe refugium for the sea-ice microbial community (SIMCO), at least until melting occurs. Thus, the physics, chemistry, and biology of annual ice are very dynamic, being controlled by factors which regulate ice growth and retreat.

Few workers seem to fully appreciate the complex nature of the sea-ice microbial communities (SIMCOs) nor have many studied their dynamics. Despite their complexity we have begun to understand these systems by approaching the problems in a systematic, experimental fashion. Our initial approach has been to examine simultaneously ice algal and bacterial populations, the numerically (and biomass) dominant components of the SIMCO. We monitored algal and bacterial standing crop and activity by measuring the increase in cell number, carbon and chlorophyll-a concentrations and by performing metabolic rate experiments. The purpose of these investigations was to determine the factors that influence growth and development of SIMCOs; initially light, temperature and nutrients were considered to be the factors mostly likely to play an important role. In this report, findings are presented concerning the influence of light on the SIMCO in McMurdo Sound, Ross Sea, Antarctica. Preliminary reports of this work conducted during the austral summers of 1980 and 1981 appeared elsewhere (Sullivan and Palmisano 1981, 1982; Palmisano 1982; Sullivan et al. 1982, 1983). Aspects of this work are summarized here with an emphasis on the dynamics of the SIMCO spring bloom during 1981.

2 Materials and Methods

The net accumulation of the SIMCO was assessed in November 1981 at 9 stations with variable snow cover on the annual sea-ice of McMurdo Sound (Fig. 1). The ice at all sites was relatively uniform in depth (2.0 ± 0.2 m) except at Cape Royds where it was 0.8 m thick. Chlorophyll-a concentrations in the bottom 200 mm of ice varied inversely with the depth of snow on the ice when the sample was taken (Fig. 2). Ice thickness influences light penetration less than does snow thickness. This stems from the differing light extinction coefficients for ice and snow (Maykut and Grenfell 1975). Figure 2 shows the relationship of total downwelling irradiance as a function of snow depth over 2 m of Arctic annual sea-ice; the data were redrawn from Maykut and Grenfell (1975). Under-ice light falls off in a fashion similar to that of chlorophyll-a measured in our study, suggesting a cause-and-effect relationship between snow depth, under-ice irradiance and chlorophyll-a content of the ice.

1 Marine Biology Research Section, Department of Biological Sciences, University of Southern California, Los Angeles, CA 90089–0371, USA

Antarctic Nutrient Cycles and Food Webs (ed. by W.R. Siegfried, P.R. Condy, and R.M. Laws)

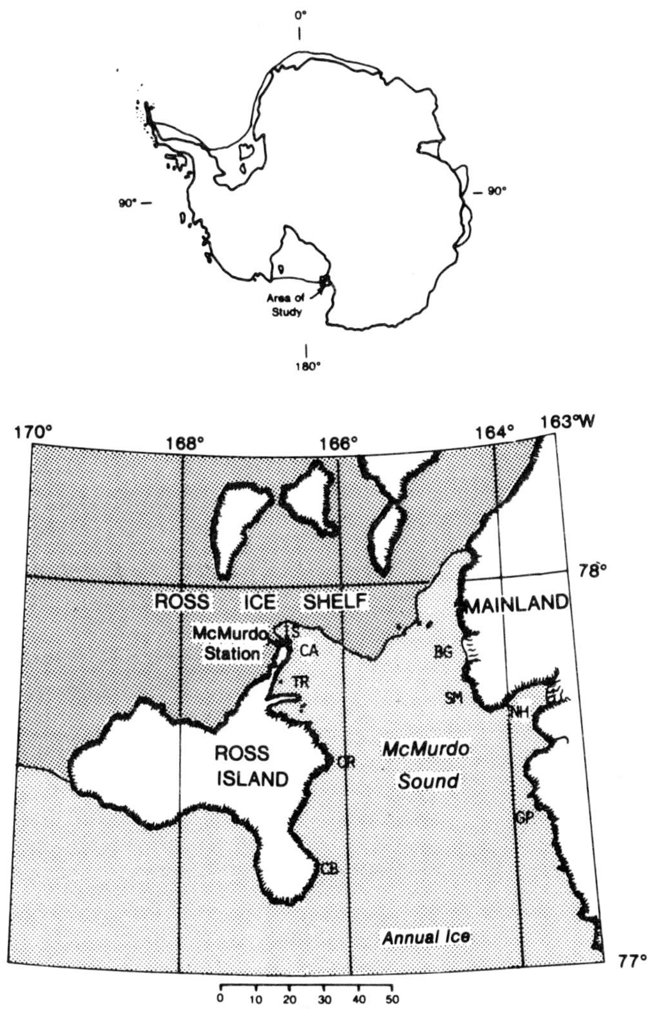

Fig. 1. Sample locations for sea-ice microbial community (SIMCO) studies conducted during 1981 in McMurdo Sound, Ross Sea, Antarctica. *IS* Ross Ice Shelf; *CA* Cape Armitage; *TR* Turtle Rock; *CR* Cape Royds; *CB* Cape Bird; *GP* Gneiss Point; *NH* New Harbor; *SM* Strand Moraines; *BG* Blue Glacier

To examine the effects of light on the in situ growth and development of SIMCOs, a perturbation experiment was set up on the annual sea-ice of McMurdo Sound just north of Cape Armitage during October–December 1981 (Palmisano 1982; Sullivan et al. 1982). Two experimental quadrats of 100 m^2 each were marked by placing 4 bamboo poles, with flags on both ends, through holes in the ice at each quadrat corner. After freezing in place, the underwater flags served to orientate SCUBA divers who sampled the undersurface of the ice in each quadrat. In one quadrat 15–70 mm of natural snow cover was maintained throughout the study, while the adjacent quadrat received 0.7 m of snow with the assistance of U.S. Navy snow scoops (Fig. 3). This treatment resulted in two different under-ice irradiances both of which were consistently $< 1\%$ surface irradiance throughout the 6-week study (Table 1). Because of the relative constancy of light and seawater temperature under the ice ($-1.86 \pm 0.2°C$; Littlepage 1965), and the high concentrations of nutrients in McMurdo Sound seawater (Littlepage 1965; Palmisano 1982), we could conduct a controlled light perturbation experiment in the natural environment.

Within each quadrat two to five ice cores were collected at random with a SIPRE ice auger at 19–25 day intervals. During the study period, the ice grew approximately 170 mm in thickness. Palmisano and Sullivan (1983) have shown that 99% of the chlorophyll-a and 90% of the bacterial biomass are located in the bottom 200 mm of the annual sea-ice in McMurdo Sound. Thus, the bottom 200 mm of ice cores provided source material for chemical analyses, enumeration of algal and bacterial cells, and determination of algal pigments and particulate organic carbon (POC).

Metabolic rate processes indicative of microbial growth were measured in samples from the ice-seawater interface. Samples from the undersurface of the ice were collected in a syringe-operated device by SCUBA divers, injected with suitable radio-chemicals into sealed Plexiglas chambers, and returned to the site of sampling. Divers fastened the chambers to the underside of the ice for 12–24 h in situ incubations. Throughout these procedures great care was taken not to subject samples to possible deleterious changes in temperature or light.

3 Results and Discussion

Accumulation of materials in the ice is the net result of biological and physical processes of addition and removal from the ice matrix. Important biological processes of addition include growth and metabolism, as well as cell adhesion to ice-crystal surfaces. However, physical entrapment of particles or living cells (especially from the organic enriched neuston) during the dendritic growth of ice-crystals at the advancing ice-seawater interface may also be important. Nucleation of ice-crystals by microbial cell surfaces could account for increases in the numbers of micro-organisms in the ice. A precedent for biologically mediated ice nucleation was established for the terrestrial bacterium *Pseudomonas syringae* (Maki et al. 1974). Biologically derived ice nuclei from living marine phytoplankton have also been reported (Vali and Schnell 1975). By analogy, similar phenomena might exist among polar algae and bacteria, which would cause their incorporation into the ice matrix during its growth (Ackley 1982).

Cells of the SIMCO might be removed by cryopelagic grazers such as copepods, amphipods, Krill (Euphausiacea) or larval fishes (Bradford 1978) at the ice-seawater interface, or by ice-associated protozoa and heterotrophic microflagellates within the ice matrix. Ice ablation from the lower surface during water column warming, heat absorption and drainage of freshwater through brine channels during ice melt cause loss to the water column. This primarily occurs in late December and in January, after the peak of the spring bloom.

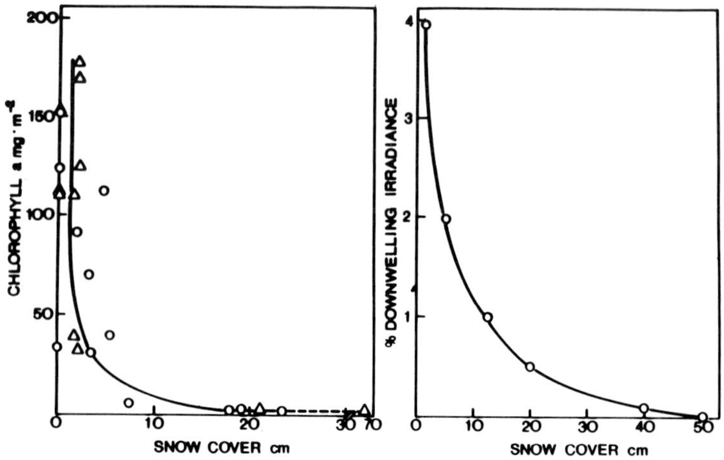

Fig. 2. Relationship between chlorophyll-a content, ice and downwelling irradiance measured under ice as a function of snow depth over 2 m of annual sea-ice. Data on downwelling irradiance under Arctic sea-ice redrawn from Maykut and Grenfell. (1975)

Fig. 3. Light perturbation experiment on the annual sea-ice near Cape Armitage. U.S. Navy snow scoop was used to construct a 100 m^2 quadrat with 0.7 m snow cover in order to darken the SIMCO at this site; adjacent control site was left undisturbed with 70–15 mm natural snow cover

Table 1. Snow and total under-ice irradiance (400–700 nm) in light perturbation experiment during October–December, 1981. (Data from Palmisano 1982)

	Snow depth (mm)	Irradiance under 2 m ice ($\mu E\ m^{-2}\ s^{-1}$)[a]
Control quadrat	72–15	0.20–2.90
Perturbed quadrat	700–670	0.06–0.11

a Four measurements were made by SCUBA divers using a scalar irradiance meter (QSI-140 Biospherical Instruments, Inc., San Diego, California)

With these considerations in mind, one must consider carefully the data presented in Fig. 4. In the control quadrat with natural snow cover, algal cells accumulated rapidly during the first 25 days of the experiment, increasing in concentration approximately 14-fold to 1.5×10^7 cells 1^1 (Fig. 4A). Chlorophyll-a concentration in the ice also increased during the experiment, but at a faster rate over the last 19 days when cell numbers increased only slightly (Fig. 4B). The chlorophyll-a content per cell increased after mid-November, possibly because of low light adaptation in response to self-shading resulting from high cell densities ($> 1.5 \times 10^7$ cells 1^{-1}) characteristic of this bottom-ice community. Floristic analysis revealed no shifts in species composition during this period, *Nitzschia stellata* and *Amphiprora* sp. accounting for 80% of the assemblage on both sampling dates. By contrast, there was no significant increase in either algal cell number or chlorophyll-a content of the ice in the perturbed quadrat where light levels never exceeded 0.11 $\mu E\ m^{-2}\ s^{-1}$. In situ incubations of the SIMCO beneath the perturbed quadrat revealed very low incorporation rates for $H^{14}CO_3$, suggesting that light was near compensation intensity for net C production.

Total numbers of ice bacteria, both free-living and those attached to particles, increased in the control quadrat

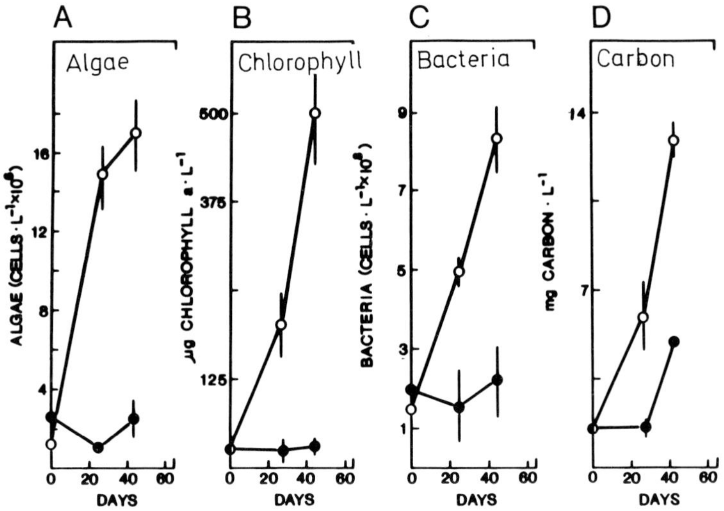

Fig. 4A–D. Accumulation of **A** algal cell number, **B** chlorophyll-a, **C** bacterial cell numbers, and **D** particulate organic carbon in the bottom 200 mm of annual sea-ice from a light perturbation experiment conducted on annual sea-ice north of Cape Armitage, McMurdo Sound. *Circles* represent data from control quadrat covered with 70–15 mm natural snow. *Dots* represent data from experimental quadrat covered with 0.7 m snow

almost six times during the experiment, whereas no net bacterial increase was observed in the perturbed quadrat. The available evidence indicates that sea-ice bacteria are largely heterotrophic, and their growth is closely coupled to the growth of ice algae. Micro-autoradiographs of the SIMCO following incubation with radio-labelled organic compounds showed that ice bacteria take up a variety of ^{3}H-amino acids, ^{3}H-glucose and ^{3}H-thymidine under in situ or simulated in situ conditions (Fig. 5). Also Dr. Hiromi Kobori, working in the authors' laboratory, has successfully isolated, from ice meltwater samples, more than 79 strains of psychrophilic/psychrotrophic bacteria which grow heterotrophically on marine agar media. Microscopic examination of ice meltwater samples failed to reveal autofluorescent cyanobacteria; nor did transmission electron micrographs show intracellular lamellar membranes typical of either cyanobacteria or phototrophic eubacteria.

The coupling of algal and heterotrophic bacterial growth is suggested by strong parallel increase in cell number in the control quadrat without net increase in the perturbed quadrat. Observations support this hypothesis: Sullivan and Palmisano (1981) presented evidence for a close physical coupling of bacteria and certain species of ice diatoms, especially of the genus *Amphiprora.* More recently, it was reported that the epibacteria comprised as much as 30% of the total bacterial assemblage in sea-ice (Grossi et al., to be publ.)

Particulate organic carbon (POC) concentrations in the unperturbed quadrat increased in a fashion similar to that observed for chlorophyll-a (Fig. 4D). Previous work (Palmisano and Sullivan 1983) suggests that primary production by ice algae far exceeds secondary production by ice bacteria. Thus the POC content, presumably an index of the ice algal assemblage or biogenic materials derived from them, is expected to follow the pattern of algal accumulation. In the unperturbed quadrat, the final C:chl-a ratio was only 26:1, suggesting that algae were a major source of POC, since C:chl-a ratios of 37:1 (Bunt and Lee 1970), 34:1 (Whitaker 1977), and 39:1 (Palmisano 1982) have been reported in SIMCOs by other investigators. In the perturbed quadrat, however, final C:chl-a ratios averaged 192, suggesting the presence of POC which was not associated with viable chlorophyll-containing cells.

Accumulation of POC in the perturbed quadrat (Fig. 4D) is not easily explained by the available data on microbial accumulation. After showing no increase during the first 25 days of the experiment, the POC content increased almost three-fold during the last 19 days. Extensive microscopic studies revealed no recognizable biological material to account for this increase. This C may be incorporated into the ice matrix from the water column by adsorption to ice surfaces or entrapment during dentritic growth of ice crystals at the ice-seawater interface. On the other hand, the fact that there is no net accumulation of algal or bacterial cells or chlorophyll-a in the perturbed quadrat suggests that physical processes are relatively less important to accumulation of biogenic materials in the ice. This idea should not be generalized, however, since significant differences may exist between biogenic accumulation mechanisms in frazil and congelation ice. Congelation ice predominates (Gow et al. 1981) in McMurdo Sound, whereas in the Weddell Sea frazil ice is the predominant type (Ackley 1982).

Primary productivity in the interface samples was measured in situ with $H^{14}CO_3$ at $-1.9°C$ and $\leqslant 2.90\ \mu E\ m^{-2}\ s^{-1}$ for 12 to 24 h. Ice algae fixed 0.31 mgC mg^{-1} chl-a h^{-1}; microautoradiographs showed that the codominant diatom species *Amphiprora* sp. and *N. stellata* were heavily labelled with ^{14}C. Micro-autoradiographic analyses of radiochemical incubations (Fig. 5) indicated that both diatoms and bacteria metabolized organic compounds, but that they did so selectively. *Amphiprora* specimens were labelled with ^{3}H-amino acids (diffuse grain development covering diatom surface) but apparently not ^{3}H-glucose or ^{3}H-thymidine, while epiphytic ice bacteria were labelled with all three (discrete clumps of grain development). Metabolic rate studies of micro-organisms from the ice-seawater inter-

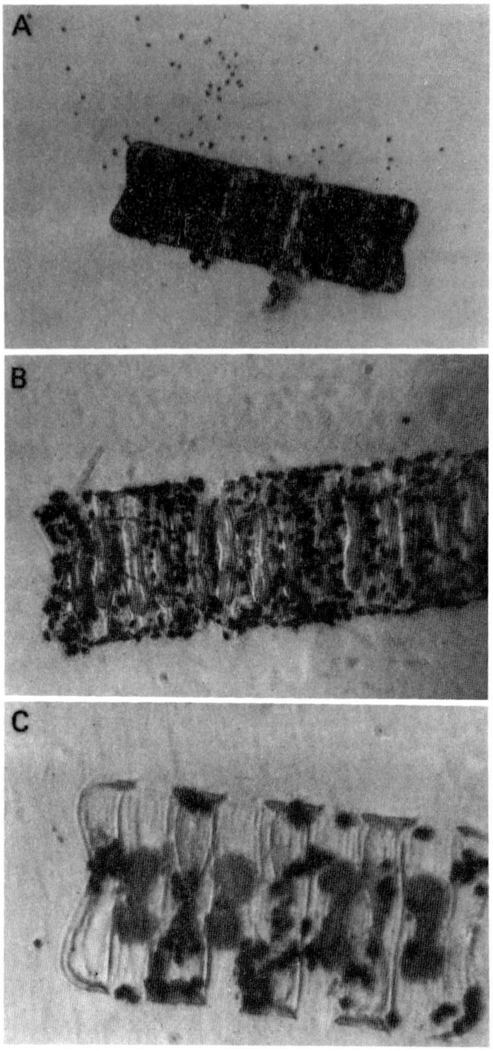

Fig. 5A–C. Micro-autoradiographs of SIMCO incubated under simulated in situ conditions with **A** ^{3}H-serine, **B** ^{3}H-glucose and **C** ^{3}H-thymidine. One ml of ice from the bottom 200 mm of ice cores was perfused with 0.2-μm filtered seawater containing radiochemicals (0.1 μCi ml^{-1} for 4 h at simulated in situ conditions −1.9 °C and 3–13 μE m^{-2} s^{-1}. Modified methods described by Hoppe (1976) were used except that stripping plates replaced liquid photographic emulsion

face or in small cubes of intact ice core perfused with seawater to which radiolabels were added, showed that the ice microflora were capable of energy-producing metabolism under simulated in situ conditions. The incorporation of ^{3}H-thymidine by bacteria demonstrated that nucleic acid synthesis, a prerequisite for cellular growth and division, takes place under in situ conditions in the ice.

Rigorous evaluation of microbial growth rates in the natural environment is difficult (Brock 1971; Hagstrom et al. 1979; Fuhrman and Azam 1980, 1982). However, because of the unique characteristics of the sea-ice habitat, we attempted a growth rate estimate for the two major components of the community: algae and bacteria. The estimate is based on evidence presented here which suggests that accumulation of micro-organisms results almost exclusively from their growth in the ice and at the ice-seawater interface except for initial seeding. This estimate is still conservative because, while the activity of herbivores and bactivores appears to be greatly limited early in the season by the physical structure of the ice habitat, we have not directly assessed the influence of "grazing" on the SIMCO standing crop.

Table 2 shows a comparison of growth rate estimates based on rates of accumulation of cells in the ice during the first half of the experiment before rates decreased. These rates are contrasted with "instantaneous" growth rate estimates for cells at the ice-seawater interface, using radio-chemical assay, after Eppley (1972). Algal growth rates are in good agreement whether calculated by changes in cell numbers or chlorophyll-*a* concentrations. The estimated doubling time of 7 days agrees remarkably well with the 6-day doubling time for the epontic algal community of McMurdo Sound reported by Bunt and Lee (1970). The bacterial assemblage appears to grow more slowly, with a 14-day doubling time. Ice bacteria when cultivated in the laboratory are capable of more rapid growth than diatoms (H. Kobori, pers. comm.). It can also be generalized that the inherent doubling time of most bacteria is shorter than that of most micro-algae. Thus, the observed relative in situ growth rates for algae and bacteria may reflect a dependence by the bacteria on algal photosynthetic products for their C and energy or inhibition of bacterial growth by algal antibiosis.

Less agreement was found between growth estimates based on accumulation and those obtained with radio-tracer methods designed to give an "instantaneous" growth rate (Table 2). Because of the complexity of the problem and its importance to microbial ecology, methods are currently being evaluated for assessing growth rates of the SIMCO, such as analysis of the frequency of dividing cells and dialysis chamber studies. Although difficult, the problem of making in situ growth rate estimates of microbial populations is tractable. Its solution is likely to require a good model system such as SIMCOs and a multifaceted approach.

An estimate of 4.9 g C m^{-2} for the annual production of SIMCOs to the ecosystem of McMurdo Sound was calculated by converting mean chlorophyll-a m^{-2} content of the ice at nine stations to C m^{-2} by the relationship 38 g C g^{-1} chlorophyll-a (Sullivan et al. 1982). This suggests an overall contribution of about 20 × 10^6 kg new C yr^{-1}. Such an estimate is reasonable, since production accumulates as the SIMCO is physically entrapped within the ice matrix, reducing losses from grazing and sinking. Nevertheless, this is an under-estimate of production for the following reasons: the spring bloom was not complete at sampling time; there may be a second bloom in autumn similar to that reported for Lütz-Holm Bay by Hoshiai (1977); losses due to "grazing" at the ice-seawater interface were not measured; and, secondary production by ice bacteria (Sullivan and Palmi-

Table 2. Growth rate estimates for the sea-ice microbial community during a light perturbation experiment

Basis for growth rate estimates	Control Quadrat		Perturbed Quadrat	
	Specific growth rate $\mu(d^{-1})$	Generation time g(d)	$\mu(d^{-1})$	g(d)
Algal cell numbers	0.099	7	0	no net growth
Chlorophyll-a	0.097	7	0	no net growth
Bacterial cell numbers	0.050	14		
$H^{14}CO_3$ primary production	0.014–0.086	8–50	0–0.27	26

sano 1981; Sullivan et al. 1982) was neglected. Because open water in McMurdo Sound is restricted at most to 1 or 2 months a year, in situ phytoplankton production is strongly limited. This leaves open the possibility that much of the new C input into the Sound, which supports rich and diverse benthic communities (Dayton and Oliver 1977), may be derived from phytoplankton advected from the Ross Sea to the north and in situ SIMCO production.

References

Ackley SF (1982) Ice scavenging and nucleation: The mechanisms for incorporation of algae into newly forming sea-ice. EOS Transactions, Am Geophys Union 63:65 (Abst)

Bradford JM (1978) Sea-ice organisms and their importance in the Antarctic ecosystem. NZ Antarct Res 1:43–50

Brock T (1971) Microbial growth rates in nature. Bacteriol Rev 35:39–58

Bunt JS, Wood EJF (1963) Microalgae and Antarctic sea-ice. Nature (Lond) 199:1254–1255

Bunt JS, Lee CC (1970) Seasonal primary production in Antarctic sea-ice at McMurdo Sound in 1967. J Mar Res 28:204–220

Dayton PK, Oliver JS (1977) Antarctic soft bottom benthos in oligotrophic and eutrophic environments. Science 197:55–58

Eppley RW (1972) Temperature and phytoplankton growth in the sea. Fish Bull 70(4):1063–1085

Fuhrman JA, Azam F (1980) Bacterioplankton secondary production estimates for coastal waters of British Columbia, Antarctica, and California. Appl Environ Microbiol 39(6):1085–1095

Fuhrman JA, Azam F (1982) Thymidine incorporation as a measure of heterotrophic bacterioplankton production in marine surface waters: Evaluation and field results. Mar Biol 66:109–120

Gordon AL (1981) Seasonality of Southern Ocean sea-ice. J Geophys Res 86:4193–4197

Gow AJ, Weeks WF, Goroni JW, Ackley SF (1981) Physical and structural characteristics of sea ice in McMurdo Sound. Antarct J US 16:94–95

Grossi S McGrath, Kottmeier ST; Sullivan CW (to be publ.) Sea-ice microbial communities III. Seasonal abundance of microalgae and associated bacteria, McMurdo Sound, Antarctica. Microb Ecol

Hagstrom A, Larsson U, Horsted P, Normark S (1979) Frequency of dividing cells, a new approach to the determination of bacterial growth rates in aquatic environments. Appl Environ Microbiol 37:805–812

Hoppe GH (1976) Determination of properties of actively metabolizing heterotrophic bacteria in the sea investigated by means of microautoradiography. Mar Biol 36:291–302

Hoshiai T (1977) Seasonal change of ice communities in sea-ice near Syowa Station, Antarctica. In: Dunbar MJ (ed) Polar Oceans, Arctic Institute of North America, Montreal pp 307–317

Littlepage JL (1965) Oceanographic investigations in McMurdo Sound, Antarctica. Antarct Res Ser 5:1–37

Maki LR, Galyan EL, Chang-Chen MM, Caldwell DR (1974) Ice nucleation induced by *Pseudomonas syringae*. Appl Microbiol 28:456–459

Maykut GA, Grenfell TC (1975) The spectral distribution of light beneath first-year sea-ice in the Arctic Ocean. Limnol Oceanogr 20:554–563

Palmisano AC (1982) Physiology of sea-ice diatoms: Strategies for winter survival. PhD thesis University of So. California, 118 pp

Palmisano AC, Sullivan CW (1983) Antarctic sea-ice microbial communities I. Distribution, abundance and primary production of ice microalgae in McMurdo Sound in 1980. Polar Biology 2:171–177

Sullivan CW, Grossi S, Kottmeier S, Palmisano A, Moe R (1983) Growth and interactions in Antarctic sea-ice. 3rd International Symposium on Microbial Ecology, Michigan State University, East Lansing, Michigan

Sullivan CW, Palmisano AC (1981) Sea-ice microbial communities in McMurdo Sound, Antarctica. Antarct J US 16(5):126–127

Sullivan CW, Palmisano AC (1982) Antarctic sea-ice microbial communities. EOS Transactions, Am Geophys Union 1984 63:47–48 (Abst)

Sullivan CW, Palmisano AC (1984) Antarctic sea-ice microbial communities II. Distribution, diversity and production of ice bacteria in McMurdo Sound. Appl Environ Microbiol 47:788–795

Sullivan CW, Palmisano AC, Kottmeier S, Moe R (1982) Development of the sea-ice microbial community (SIMCO) in McMurdo Sound, Antarctica. Antarct J US 17:155–157

Vali G, Schnell RC (1975) Biogenic sources of atmospheric ice nuclei: A review. EOS Transactions, Am Geophys Union 56:994 (Abst)

Whitaker TM (1977) Sea-ice habitats of Signy Island (South Orkneys) and their primary productivity. In: Llano GA (ed) Adaptations within Antarctic Ecosystems. Gulf Publishing Co, Houston pp 75–83

Physiological Response of Micro-Algae in the Ice-Platelet Layer to Low-Light Conditions

A.C. PALMISANO and C.W. SULLIVAN[1]

Summary. In McMurdo Sound, Antarctica, in 1982 a dense micro-algal community was associated with the ice-platelet layer of annual sea-ice. Under-ice irradiance was about 6 μE m^{-2} s^{-1}, less than 0.4% of surface downwelling irradiance. Two aspects of ice algal physiology appear to be important to growth under these conditions of low light. First, ice algae were shade-adapted, reaching a maximum photosynthetic rate of 0.6 mg C mg chl-a^{-1} h^{-1} at only 20 μE m^{-2} s^{-1}. Secondly, certain ice algae were found to take up ^{3}H-serine at natural substratum concentrations under both dark and low light conditions (15 and 48 μE m^{-2} s^{-1}, respectively). Since sea-ice micro-algae represent a significant fraction of primary production in some Antarctic pack-ice regions, knowledge of their physiology is crucial to understanding carbon flux and nutrient cycling in Antarctic marine ecosystems.

1 Introduction

Annual sea-ice provides a variety of micro-habitats for the growth of sea-ice microbial communities (SIMCOs) dominated by micro-algae. In polar oceans, micro-algae are associated with surface melt pools, floating brash ice in ice-edge zones, hard congelation ice, and, in the Antarctic, the underwater ice-platelet layer (Table 1). The physiology of sea-ice micro-algae is of special interest, because of the harsh conditions to which they are exposed. First, surface downwelling irradiance is episodic in the polar regions with continuous light in the summer and continuous darkness in the winter. Secondly, water temperatures in polar oceans rarely exceed 5°C and often approach the freezing point of seawater (−1.9°C). Finally, ice micro-algae are exposed to fluctuations in salinity and nutrients resulting from the yearly cycle of ice formation and deterioration. As ice forms in the autumn, salts are excluded from the ice matrix (Lake and Lewis 1970) creating brine pockets in which micro-algae may live and grow (Meguro et al. 1967). In the summer, however, sea-ice micro-algae are exposed to low salinity melt water from surface snow and ice (Bunt 1971).

Despite these conditions, relatively high standing crops of micro-algae have been reported in sea-ice (Table 1), particularly in the bottom of hard congelation ice in the Arctic and Antarctic, as well as the ice-platelet layer in the Antarctic. Maximum standing crops in McMurdo Sound in 1980 and 1981 were about 300 mg chl m^{-2} (Sullivan et al. 1982; Palmisano and Sullivan 1983), approaching the theoretical maximum diatom standing crop of 400 mg chl m^{-2} estimated by Steeman Nielsen (1962) for natural waters. Although the high standing crops may be partly a result of reduced "grazing" due to the inaccessibility of ice micro-algae to herbivores (Sullivan et al. 1982), we believe that these micro-algae are metabolically adapted for growth in the sea-ice micro-habitat. Here we describe aspects of the physiological ecology of micro-algae in the ice-platelet layer in McMurdo Sound, Antarctica, in 1982.

2 Materials and Methods

The study site was located about 200 m off Cape Armitage on annual sea-ice in McMurdo Sound. In 1982, a platelet-ice layer up to 1 m thick formed under 2 m of hard congelation ice. Micro-algae were associated with both the ice-platelet layer and the bottom of the congelation ice. Although both kinds of ice were sampled from under a variety of snow depths, we limit our discussion here to ice-platelet samples collected from an area with 50 mm snow cover.

Platelet-ice is very porous and consists of a mixture of congelation-type crystals and lamellar plates separated by brine layers (Lewis and Weeks 1970). Divers collected ice platelets and interstitial water lying between the ice platelets in 2-gallon Nalgene plastic jars. Under-ice irradiance was measured with a MER 1,000 spectro-radiometer with a cosine collector (Biospherical Instruments, San Diego, CA). Ice samples were taken to the Eklund Biological Centre at McMurdo Station where aliquots were melted into 0.2 μm filtered seawater at 0°C, then diluted in seawater to a final volume of less than 10^4 cells ml^{-1} to avoid self-shading in the following experiments.

To examine photosynthesis-irradiance relationships, 60 ml volumes were incubated with 0.1 μCi ml^{-1} ^{14}C-HCO$_3$ according to the methods of Steeman Nielsen (1952) as modified by Strickland and Parsons (1972). Samples were incubated at −1.9°C (the temperature of under-ice seawater

1 Marine Biology Research Section, Department of Biological Sciences, University of Southern California, Los Angeles, CA 90089-0371, USA

Antarctic Nutrient Cycles and Food Webs (ed. by W.R. Siegfried, P.R. Condy, and R.M. Laws)

Table 1. Maximum standing crops of micro-algae associated with annual sea-ice in the Antarctic and Arctic

Antarctic communities	Site	mg chl-a m^{-2}	Source
Surface. melt pool	Palmer Peninsula	117	Burkholder and Mandelli (1965)
Surface, melt pool	South Orkneys	244	Whitaker (1977)
Interior congelation ice	Weddell Sea	1.4	Ackley et al. (1979)
Bottom congelation ice	Syowa Station	30–65	Hoshiai (1981)
Bottom congelation ice	McMurdo Sound	294	Sullivan et al. (1982)
Bottom congelation ice	McMurdo Sound	309	Palmisano and Sullivan (1983)
Bottom platelet ice	McMurdo Sound	520 mg C m^2 1076 mg C m^2	Bunt and Lee (1970)
Arctic communities	**Site**	**mg chl-a m^{-2}**	**Source**
Surface, ice edge	Eastern Bering Sea	70	Niebauer et al. (1981)
Bottom congelation ice	Jones Sound	15.5, 10.7	Apollonio (1961, 1965)
Bottom congelation ice	Barrow	24	Meguro et al. (1967)
Bottom congelation ice	Beaufort Sea	30	Alexander (1974)
Bottom congelation ice	Barrow	30.5	Horner (1976)
Bottom congelation ice	Frobisher Bay	4.6	Grainger (1977)
Bottom congelation ice	Barrow Strait	11.9	Dunbar and Acreman (1980)
Bottom congelation ice	Frobisher Bay	7.6	Hsiao (1980)
Bottom congelation ice	Northeastern Bering Sea	64	Alexander and Chapman (1981)
Bottom congelation ice	Barrow, Narwhal Island	23, 26	Horner and Schrader (1982)

in McMurdo Sound) in an environmental chamber designed to mimic the quality and quantity of under-ice irradiance (Palmisano and Sullivan 1982). Fluorescent lights were fitted with blue-green filters, and total irradiance was adjusted with screens. Photosynthetic rate was determined at each of eight irradiances between 4 and 75 $\mu E\ m^{-2}\ s^{-1}$; two dark bottles served as controls. Samples were incubated for 24 h in constant light; preliminary studies showed that uptake of $^{14}C\text{-}HCO_3$ by ice micro-algae was linear for 30 h at –1.9°C.

Following incubation, samples were filtered onto 1 μM pore size Nuclepore filters at low vacuum (< 100 mm HG). Filters were rinsed with 0.2 μM filtered seawater, and fumed over concentrated HCl for 4 h to remove residual $^{14}CO_2$. Samples were placed in scintillation vials with 10 ml Aquasol II and counted in a Beckman LSC 100 scintillation counter. Counts were corrected to dpm's by the external standard ratio method. To measure extra-cellular organic C labelled with radio-carbon, 2 ml aliquots of the filtrate were acidified to pH2 and agitated on a shaker table for 12 h to purge residual $^{14}CO_2$. Previous work has shown that virtually all $^{14}CO_2$ is removed following 30 min of shaking (G. Taylor, pers. comm.).

Parallel samples of ice-algae diluted in filtered seawater, as described, were incubated with ^3H-serine in the dark and at 15 and 48 $\mu E\ m^{-2}\ s^{-1}$ to test the effect of light on heterotrophic uptake. Duplicate samples were incubated for 24 h at –1.9°C with 0.3 $\mu Ci\ ml^{-1}$ ^3H-serine; uptake was linear for 36 h at –1.9°C. About 19 nmol l^{-1} serine was added to the sample with the radio-label. Using high performance liquid chromatography, natural concentrations of serine in under-ice seawater at our study site in McMurdo Sound have been estimated at 385 ± 11 nmol l^{-1} (D. Manahan, pers. comm.). It is unlikely, therefore, that addition of the radio-label altered natural substratum concentration by more than 5%. A parallel set of 2% formalin-killed controls was incubated with the live samples. After 24 h, samples were filtered onto either 3.0 μm or 0.2 μm pore size filters, rinsed with seawater, and counted in the scintillation counter as previously described. Samples, preserved in 1% glutaraldehyde, were taken to the University of Southern California for micro-autoradiography (MAR) preparation by the methods of Hoppe (1976) as modified by Palmisano (1982).

3 Results

Irradiance under annual sea ice with 50 mm of snow cover was very low during the spring ice algal bloom with a mean of 6 $\mu E\ m^{-2}\ s^{-1}$. This represented only 0.4% of surface downwelling irradiance. It should be noted that this irradiance was measured at 1200 hours and, therefore, would probably overestimate daily irradiance because of the diurnal periodicity under the ice, as noted by Sullivan et al. (1982). Under-ice spectra showed a broad band in the blue-green region with a peak at 500 nm.

The photosynthesis-irradiance curve showed that a maximum photosynthetic rate was reached at about 20 $\mu E\ m^{-2}\ s^{-1}$ (Fig. 1). Extra-cellular organic C released by sea-ice diatoms was 0.2 mg C mg^{-1} chl-a h^{-1}. The assimilation number, adjusted for extracellular release, was 0.6 mg C mg^{-1} chl-a h^{-1}. Photosynthetic efficiency, determined by a linear regression

Table 2. Uptake of ^{3}H-serine (dpm ml^{-1}) at $-1.9°C$ by sea-ice microbial communities (SIMCOs) in the ice-platelet layer in three light treatments

Size fraction	Dark	15 μE m^{-2} s^{-1}	48 μE m^{-2} s^{-1}
$>$3.0 μm	3893 ± 715	4884 ± 1553	3668 ± 1037
$>$0.2 μm (total uptake)	4358 ± 551	5656 ± 480	3900 ± 464
% Total uptake in $>$3.0 μm fraction	89%	86%	94%

Figures are means ± S.D., based on 3 replicate samples which were melted into filtered seawater at 0°C prior to incubation

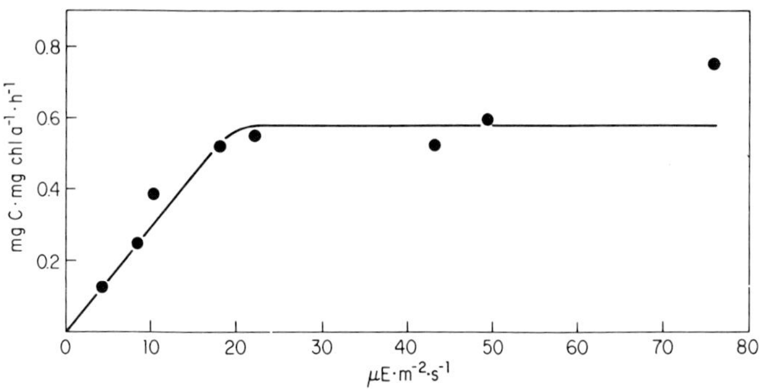

Fig. 1. Photosynthesis-irradiance relationship in micro-algae from the ice-platelet layer in McMurdo Sound, Antarctica, in 1982

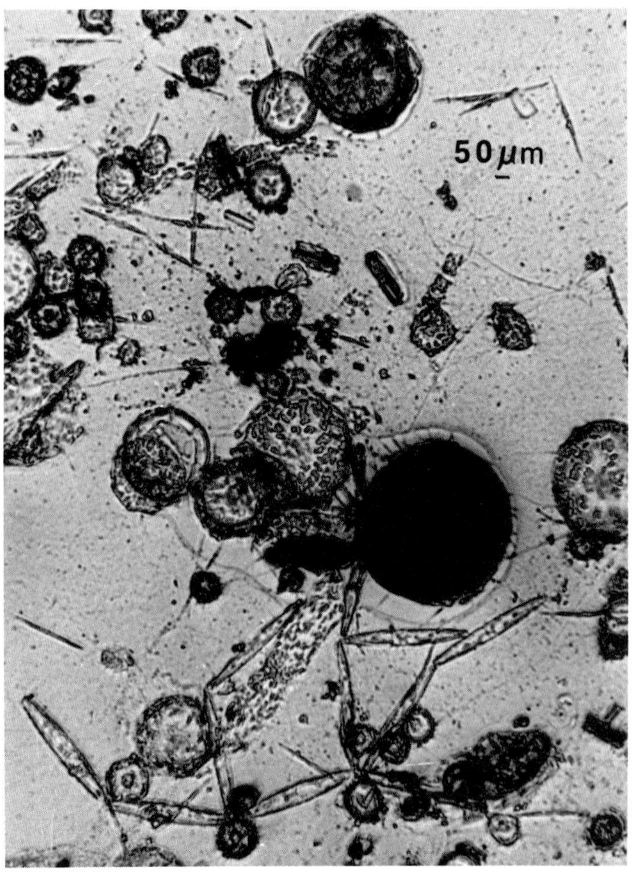

Fig. 2. Micro-autoradiograph of micro-algae from the ice-platelet layer which were incubated with ^{3}H-serine. A *heavily labelled Coscinodiscus* sp. appears *black*

on the slope of the initial part of the curve, was 0.026 mg C mg^{-1} chl-a h^{-1} (μE m^{-2} s^{-1})$^{-1}$.

^{3}H-serine was taken up by natural populations of ice-platelet micro-algae (Table 2). More than 85% of the total uptake ($>$ 0.2 μm fraction) was caught on a 3.0 μM pore size filter in each light treatment. There was no significant difference in the uptake of ^{3}H-serine in the fraction $>$ 3.0 μM among three light treatments by analysis of variance. This fraction contained virtually all the algal cells found in the platelet samples.

Unlike the core samples (congelation ice) in which pennate species dominated, centric diatoms were more abundant in the ice-platelet layer. MARs revealed that certain centric species including *Coscinodiscus* were heavily labelled with ^{3}H-serine, whereas others were not (Fig. 2).

4 Discussion

Despite an under-ice irradiance of only 6 μE m^{-2} s^{-1} ($<$0.4% of surface downwelling irradiance), micro-algae formed a spring bloom in the platelet layer of annual sea-ice under 50 mm of snow cover in McMurdo Sound in 1982. Irradiance under annual sea-ice in both the Antarctic and Arctic is typically $<$ 1% of surface irradiance during the spring ice diatom bloom (Sullivan and Palmisano 1981; Horner and Schrader 1982). In 1981, Sullivan et al. (1982) found a diurnal periodicity in under-ice irradiance during the spring bloom; while surface irradiance decreased from 1,500 to 1,300 μE m^{-2} s^{-1} between 1100 and 2300 hours, under-ice irradiance decreased from 2.9 to 0.5 μE m^{-2} s^{-1}. Two aspects of ice algal physiology may be important to their growth and survival under these low light conditions: (1) photo-adaptation and (2) heterotrophic nutrition.

Under-ice micro-algae appear to be shade-adapted, in that they reach their maximum photosynthetic rate at relatively low irradiance. In our study, photosynthesis by micro-algae in the ice-platelet layer was light saturated at only 20 μE m^{-2} s^{-1}. Similarly, Bunt (1964) found that micro-algae in the interstitial water between ice platelets were also light-saturated at 100 footcandles (about 22 μE m^{-2} s^{-1}). The assimilation number for the platelet community in 1982 was 0.6 mg C mg^{-1} chl-*a* h^{-1}; this estimate accounts for a loss of total fixed C by extra-cellular release. Bunt and Lee (1970) estimated an assimilation number of 0.4 for the interstitial micro-algae in the platelet-ice without accounting for extra-cellular loss. If their estimate is adjusted for a possible 33% loss, their assimilation number would be remarkably similar to ours. Assimilation values for sea-ice micro-algae are considerably lower than those reported for temperate phytoplankton, which typically range from 2 to 10 (Falkowski 1981).

The similarity between our observations and those of Bunt and Lee (1970) suggests that the micro-algae attached to platelets are physiologically similar to those found in the interstitial water between the platelets, and that our treatment of the ice-algae (melting platelets gradually into a larger volume of filtered seawater prior to experiment) did not adversely affect the algal cell physiology.

The proportion of extra-cellular release of radio-labelled organic carbon by sea-ice diatoms was higher than estimates reported recently for phytoplankton, which usually fall around 10% (Smith et al. 1977; Mague et al. 1980). The slow growth rates, low cell densities, low growth irradiance and long incubation times characteristic of studies of sea-ice micro-algae, however, would tend to produce lower, not higher rates of extra-cellular release or organic C (Sharp 1977). Because special care was taken to mimic under-ice conditions during incubation, to avoid cell damage during filtration, to purge residual $^{14}CO_2$ from samples, and because subtraction of dark values corrected for the presence of organic contaminants in the radio-label, other explanations must be sought for these high rates. It should be noted that the raw counts for each filtrate sample were always more than 500 cpm's above the dark background. A possible explanation is that the melting and dilution of samples may have increased extra-cellular release of organic C. However, sea-ice micro-algae are exposed to large changes in salinity during sea-ice formation and ablation. Moreover, sea-ice diatoms have been characterized as euryhaline, and Grant and Horner (1976) found growth rates of sea-ice diatom isolates to be constant over salinities ranging from 10 to 50%.

Certain micro-algae are facultatively heterotrophic in that they can use organic substrata, including sugars, amino acids, and organic acids, for C and energy (Hellebust and Lewin 1977). We reasoned that sea-ice micro-algae may be using heterotrophic nutrition to supplement photosynthesis during the low light conditions of the spring bloom, or to support maintenance metabolism during the dark polar winter. The possibility that algae under ice may be using heterotrophy for winter survival was suggested by Rodhe (1955) for Sub-Arctic lake phytoplankton, Wilce (1967) for Arctic seaweeds, and Allen (1971) for Arctic sea-ice diatoms. Using natural populations of Arctic sea-ice diatoms, however, Horner and Alexander (1972) concluded that heterotrophic uptake of ^{14}C-glucose, ^{14}C-glycine and ^{3}H-acetate by sea-ice micro-algae was negligible.

Palmisano and Sullivan (1982) tested two axenic Antarctic isolates of *Nitzschia cylindrus* (Grun.) Hasle and an unidentified axenic Arctic pennate diatom for facultative heterotrophy using ^{14}C-glucose. They found that dark incubation enhanced the incorporation and respiration of glucose as much as 60 X. In a previous study, Palmisano (1982) demonstrated the dark incorporation and respiration of ^{14}C-serine by natural populations of sea-ice micro-algae. Here we tested the effect of light on the heterotrophic uptake of ^{3}H-serine by natural ice algal populations. Based on a laboratory study, Palmisano and Sullivan (1982) hypothesized that heterotrophy in ice-algae may be induced in the dark. However, this was not the case for natural populations of ice-platelet algae; there was no significant difference in serine uptake among three light treatments. Some sea-ice algae, therefore, may use heterotrophic nutrition not only during the dark polar winter, but also during the spring bloom itself. Amino acids, probably derived in part from algal extra-cellular organic C, are found in relatively high concentration in ice-core melt and ice-seawater interface samples during the spring bloom (D. Manahan, pers. comm.).

Bacteria attached to micro-algae and detritus would be caught on the 3.0 μM porosity filter; therefore, their uptake of ^{3}H-serine would be included in this size fraction along with micro-algal uptake. However, centric diatoms dominating the ice-platelet layer showed a uniform labelling of the cell, unlike *Amphiprora* spp. which are colonized by attached bacteria and show a patchy label distribution. Thus, certain species of centric diatoms in the ice-platelet layer appeared to be able to take up ^{3}H-serine.

We conclude that micro-algae growing in the ice-platelet layers of annual sea-ice are physiologically adapted to grow at low ambient light levels. First, they reach maximum photosynthetic rates at the low irradiances typical under sea-ice. Although irradiance under annual sea-ice with a 50 mm snow cover was below saturation intensity, algae at the top of the platelet layer or under snow-free ice may experience light intensities that saturate photosynthesis. Secondly, certain species may supplement photosynthesis by heterotrophic uptake of organic substrata. Although algal heterotrophy does not provide new C for the micro-algal community, it could provide a means for more efficient utilization of the total C fixed. Thus, some portion of the fixed C that is lost by extra-cellular release may be re-assimilated by certain species and used as a source of C and energy. Efficient energy utilization is critical in SIMCOs where the annual growth period is limited to several months.

Acknowledgements. We thank R. Moe for diving support and S. McGrath Grossi for critically reviewing the manuscript. This study was supported by National Science Foundation grant DPP-81-17237.

References

Ackley SF, Buck KR, Taguchi S (1979) Standing crop of algae in the sea ice of the Weddell Sea region. Deep-Sea Res 26A:269–281

Alexander V (1974) Primary productivity regimes of the nearshore Beaufort Sea with reference to the potential role of ice biota. Symposium on the Beaufort Sea, Coastal and Shelf Research. Arct Inst N Am Arlington VA pp 609–632

Alexander V. Chapman T (1981) The role of epontic algal communities in Bering Sea Ice. In: Hood DW, Calder JA (eds) The Eastern Bering Sea Shelf: Oceanogr Res Vol 2. Inst Mar Sci U of Alaska Fairbanks pp 773–780

Allen MB (1971) High-latitude phytoplankton. Ann Rev Ecol Syst pp 261–276

Appolonio S (1961) The chlorophyll content of arctic sea ice. Arctic 14:197–200

Appolonio S (1965) Chlorophyll in actic sea ice. Arctic 18:118–122

Bunt JS (1964) Primary productivity of undersea ice in Antarctic waters. Influence of light and other factors on photosynthetic activities of Antarctic marine microalgae. Antarct Res Ser 1:27–31

Bunt JS (1971) Microbial productivity in the polar regions. Symposia of the Society for General Microbiology XXI. Microbes and Biological Productivity pp 333–353

Bunt JS, Lee CC (1970) Seasonal primary production in Antarctic sea ice at McMurdo Sound in 1967. J Mar Res 28:304–320

Burkholder PR, Mandelli EF (1965) Productivity of microalgae in Antarctic sea ice. Science (Wash DC) 148:872–874

Dunbar MJ, Acreman J (1980) Standing crops and species composition of diatoms in sea ice from Robeson Channel to the Gulf of St. Lawrence. Ophelia 19:61–72

Falkowski PG (1981) Light-shade adaptation and assimilation numbers. J Plankton Res 3:203–216

Grainger EH (1977) The annual nutrient cycle in sea ice. In: Dunbar MJ (ed) Polar Oceans. Arctic Institute of North America, Montreal pp 285–299

Grant WS, Horner RA (1976) Growth responses to salinity variation in four Arctic ice diatoms. J Phycol 12:180–185

Hellebust JA, Lewin J (1977) Heterotrophic nutrition. In: Werner D (ed) The Biology of Diatoms. Univ of California Press, Berkeley pp 169–197

Hoppe H-G (1976) Determination and properties of actively metabolizing heterotrophic bacteria in the sea, investigated by means of microautoradiography. Mar Biol 36:291–302

Horner RA (1976) Sea ice organisms. Oceanogr Mar Biol Ann Rev 14:167–182

Horner RA, Alexander V (1972) Algal populations in Arctic sea ice: An investigation of heterotrophy. Limnol Oceanogr 17:454–458

Horner RA, Schrader GC (1982) Relative contributions of ice algae, phytoplankton and benthic microalgae to primary production in nearshore regions of the Beaufort Sea. Arctic 35:485–503

Hoshiai T (1981) Proliferation of ice algae in the Syowa Station area, Antarctica. Mem Nat Inst Polar Res Ser E No 34 p 1–12

Hsiao SI (1980) Quantitative composition, distribution, community structure and standing stock of sea ice microalgae. Arctic 33:768–793

Lake RA, Lewis EL (1970) Salt rejection by sea ice during growth. J Geophys Res 75:583–597

Lewis EL, Weeks WF (1970) Sea ice: Some polar contrasts. In: Deacon G (ed) Symposium on Antarctic ice and water masses. Tokyo Japan pp 23–34

Mague TH, Friberg E, Hughes DJ, Morris I (1980) Extracellular release of carbon by marine phytoplankton: a physiological approach. Limnol Oceanogr 25:262–279

Meguro H, Ito K, Fukushima H (1967) Ice flora (bottom type): A mechanism of primary production in the polar seas and growth of diatoms in sea ice. Arctic 20:114–133

Niebauer HJ, Alexander V, Cooney RT (1981) Primary production at the Eastern Bering Sea ice edge: the physical and biological regimes. In: Hood DW, Calder JA (eds) The Eastern Bering Sea Shelf: Oceanogr Res Vol 2 Inst Mar Sci U Alaska pp 763–772

Palmisano AC (1982) Physiology of sea ice diatoms: Strategies for winter survival. PhD dissertation. Univ of Southern California, Los Angeles 118 pp

Palmisano AC, Sullivan CW (1982) Physiology of sea ice diatoms: I. Response of three polar diatoms to a simulated summer-winter transition. J Phycol 18:489–498

Palmisano AC, Sullivan CW (1983) Sea ice microbial communities (SIMCOs) I. Distribution, abundance and primary production of ice microalgae in McMurdo Sound in 1980. Polar Biology 2:171–177

Rodhe W (1955) Can plankton production proceed during winter darkness in subarctic lakes? Int Ver Theor Angew Limno Verh 12:117–122

Sharp JH (1977) Excretion of organic matter by marine phytoplankton: do healthy cells do it? Limnol Oceanogr 22:381–399

Smith WO, Barber RT, Huntsman SA (1977) Primary production off the coast of northwest Africa: excretion of dissolved organic matter and its heterotrophic uptake. Deep-Sea Res 24:36–47

Steeman Nielsen E (1952) The use of radioactive carbon (14C) for measuring organic production in the sea. Journal du Conseil. Conseil permanent international pour l'exploration de la mer 18:117–140

Steeman Nielsen E (1962) On the maximum quantity of plankton chlorophyll per surface unit of a lake or the sea. Int Rev Gesamten Hydrobiol 47:333–338

Strickland JDH, Parsons TR (1972) A practical handbook of sea water analysis. 2nd ed. Fish Res Board Can Ottawa Bull 167, 311 pp

Sullivan CW, Palmisano AC (1981) Sea ice microbial communities in McMurdo Sound, Antarctica. Antarct J US 16:126–127

Sullivan CW, Palmisano AC, Kottmeier S, Moe R (1982) Development of the sea ice microbial communities (SIMCO) in McMurdo Sound, Antarctica. Antarct J US 17:155–157

Whitaker TM (1977) Sea ice habitats of Signy Island (South Orkneys) and their primary productivity. In: Llano GA (ed) Adaptations within Antarctic Ecosystems. Gulf Publishing Co Houston pp 75–83

Wilce RT (1967) Heterotrophy in Arctic sublittoral seaweeds: An hypothesis. Bot Mar 10:185–197

Autumnal Proliferation of Ice-Algae in Antarctic Sea-Ice

T. HOSHIAI

Summary. Proliferation of ice-algae occurs during the austral spring-autumn period in ice-covered areas of the Antarctic Ocean. Autumnal proliferation occurs in limited areas, particularly in the new ice-fields, but in spring and summer proliferation occurs extensively in sea-ice. Results of plant pigment analyses, solar radiation measurements and observations of sea-ice growth in the autumn of 1982 showed that an increase of plant pigments occurred at the bottom of the sea-ice when the growth rate of the ice was low. A remarkable increase of ice-algae occurred in the autumn of 1970 when the growth rate of the sea-ice was extremely low; its thickness remaining at 0.3 m from late March until mid-April. New ice-fields are restricted in area, and the growth rate of sea-ice varies spatially and temporally. Hence, the autumnal enhancement of ice-algae is relatively limited. The growth rate of sea-ice is low in spring and summer, but the under-surface of the ice is stable for relatively long periods during which increases in solar radiation favour the proliferation of ice-algae in extensive areas of ice-covered sea.

1 Introduction

The importance of ice-algae as primary producers in polar marine ecosystems has been stressed since the 1960s, but assessments of the productivity and the contribution of ice-algae as primary producers to polar marine ecosystems are incomplete. Studies dealing with polar ice-algae have been made in mainly in spring and summer, but proliferation of ice-algae in the Antarctic occurs in the austral autumn as well (Hoshiai 1972, 1977, 1981a). Hence, it is necessary to investigate the mechanism of autumnal enhancement of ice-algae and to estimate the extent of areas in which such development occurs, in order to arrive at an estimate of the annual production of ice-algae in the Antarctic Ocean.

This paper deals with relationships between autumnal proliferation of ice-algae, solar radiation and the stability of the under-surface of sea-ice, in the vicinity of Syowa Station (69°00′S, 39°35′E), and compares the spring-summer proliferation of ice-algae with that observed in autumn.

National Institute of Polar Research, 9–10, Kaga 1-chome, Itabashi-ku, Tokyo 173, Japan

2 Materials and Methods

2.1 Autumnal Proliferation

Sea-ice cores were obtained at intervals between early March and mid-May 1982 at a fixed station in the Kita-no-seto Strait, about 500 m north of Syowa Station. The depth of snow and the thickness of sea-ice were measured. Each ice core was divided into a bottom section of 30 mm and 50–100 mm sections for the rest of the core. After melting, the plant pigments, chlorophyll-a and phaeopigments in each ice-core section were measured according to Saijo and Nishizawa (1969), with a fluorometer (Shimadzu FH 510). The plant pigments of the lowermost sections (30 mm) of the ice-cores are reported on in this paper, because ice-algae were most concentrated there (Hoshiai 1981a).

Continuous measurement of incident solar radiation on the sea-ice was carried out at the fixed station, using a pyranometer (MS 800 of Eko Seiki Co.) with a spectral filtering range of 305–2,800 nm. The results obtained on the sea-ice were correlated closely with global solar radiation data obtained by routine meteorological methods. Data obtained in 1970, 1975, and 1982 on autumnal development of ice-algae, solar radiation, growth of sea-ice, and depth of snow cover, were compared, to examine annual fluctuations in the relationships between these parameters. Plant pigments were measured fluorometrically (Hitachi FPL-2 fluorometer) in 1970 and 1975, according to the method of Yentsch and Menzel (1963). The data on solar radiation used in this comparison were those obtained by routine meteorological observations at Syowa Station.

2.2 Spring–Summer Proliferation

Ice-cores were obtained at five stations (A–E) near Syowa. Station A was in a snow-free area and the others (B–E) were in snow-covered areas. Samples were taken in the middle of each month from September to December 1982. Plant pigments, snow depth, sea-ice thickness, and solar radiation were measured as described above. Supplementary measurements were made at two localities (Y and Z) in snow-covered

Antarctic Nutrient Cycles and Food Webs
(ed. by W.R. Siegfried, P.R. Condy, and R.M. Laws)

areas on 6 and 7 October, and at three localities (X_1–X_3) in a snow-free area on 8 October 1982. These localities were more than 20 km south of Syowa Station.

3 Results and Discussion

3.1 Autumnal Proliferation in 1982

Changes in plant pigments, percentage of chlorophyll-a to total plant pigments, solar radiation, and sea-ice thickness are shown in Fig. 1. On 3 March, when the observations started, plant-pigment concentration was 8.72 mg m^{-3} in the bottom 30 mm section of the core. Concentrations began to increase in late March, and were higher than 14 mg m^{-3} between early and mid-April. Following a short and small decrease in late April, concentrations increased again to reach 17.68 mg m^{-3} during early May. Thereafter, an abrupt decrease occurred to 2.37 mg m^{-3} on 24 May (Fig. 1a) when coloration of sea-ice was not perceptible.

Solar irradiance on the surface of sea-ice varied from zero to more than 200 cal cm^{-2} d^{-1} during the observation period (Fig. 1c). Mean irradiance was 82.8 cal cm^{-2} d^{-1} between 21 March and 11 April when the first increase in plant pigments occurred. Hoshiai (1981b) speculated that coloration of sea-ice by ice-algae occurs when an amount of solar radiation between 150 and 250 cal cm^{-2} d^{-1} prevails, and the under-surface of the sea-ice is stable for about 2–3 weeks. The present results, however, show that Antarctic ice-algae can grow under lower irradiance. This was supported by the finding that the percentage of chlorophyll-a to total plant pigments increased from mid-March to mid-April and stayed at a high level thereafter (Fig. 1b). Chlorophyll-a percentage is an indicator of ice-algae activity (Yentsch 1965). Although an increase in sea-ice thickness began in early March 1982 and continued to increase gradually until mid-May (Fig. 1d), it should be noted that the increase of plant pigments occurred only during periods in which the growth of sea-ice was slow.

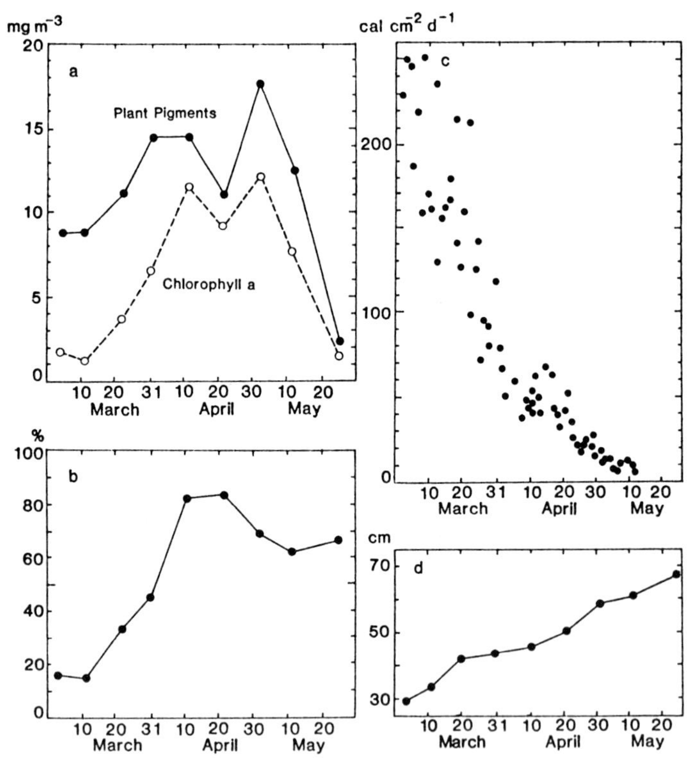

Fig. 1a–d. Changes in a plant pigments (sum of chlorophyll-a and phaeopigments) and chlorophyll-a b percentage of chlorophyll-a to plant pigments in the bottom layer of sea ice c solar irradiance on the sea-ice surface, and d sea-ice thickness at a fixed station in the Kita-no-seto Strait during March–May 1982

3.2 Annual Fluctuations in Autumnal Proliferation

A conspicuous coloration of sea-ice was observed in the autumn of 1970. The concentration of plant pigments was 6.35 mg m^{-3} in the bottom 10 mm of the ice on 27 March (Fig. 2). Subsequently, it increased to reach a peak of

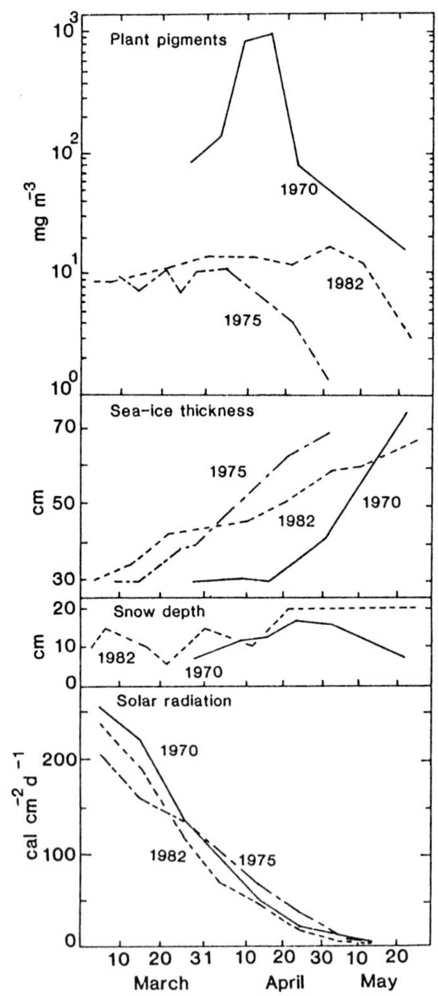

Fig. 2. Annual differences in change in concentrations of plant pigments in the bottom layer of sea-ice, incident radiation, snow depth and sea-ice thickness at a fixed station in the Kita-no-seto Strait during March–May 1970, 1975, and 1982

994.34 mg m^{-3} in the bottom 30 mm on 16 April. This increase between late March and mid-April occurred in both 1970 and 1982, though the concentrations in the 2 yrs were quite different. Moreover, the increase continued until later in the season in 1982 than in 1970. Coloration of sea-ice was not recognized in 1975 nor in 1982. In 1975 the chlorophyll-a concentration was less than that of 1982, and it began to decrease earlier in the season than in 1982.

Although annual fluctuation occurred in the amount of solar radiation supplied in the autumnal proliferation season, the amount of solar radiation in 1970, 1975, and 1982 seemed to be sufficient for proliferation of ice-algae, because the ice-algae grew during an irradiance as low as 12.1 cal cm^{-2} d^{-1} on the surface of the sea-ice in early May 1982.

Difference in ice-algal standing crop may be a function of irradiance reaching algal layers, rather than surface irradiance. The snow cover on the sea-ice reduces irradiance which penetrates into sea-ice. The recording of snow depth in 1975 was missed but, based on my experience, any differences between the 3 yrs in the snow depth were small at the observation site in Kita-no-seto Strait. The sea-ice thickness in the period from late March to early May in both 1975 and 1982 was 1.5 times that recorded in 1970. Although part of the irradiance is attenuated during passage through sea-ice, irradiance reaching the ice-algal layer is less influenced by sea-ice thickness than depth of snow cover (Bunt and Lee 1970; Maykut and Grenfell 1975).

A difference occurred in the growth rate of sea-ice between 1970 and the other years (Fig. 2). In 1970, the thickness of the ice hardly changed between late March and mid-April, remaining at about 0.3 m. During this period, a marked increase in ice-algae occurred. In contrast, in 1975 and 1982 the growth of sea-ice began in mid-March. However, in 1982 the rate of growth of the ice slowed between mid-March and early April, and also in late April, when the concentrations of plant pigments increased. The sea-ice grew relatively fast in 1975, and an increase in plant pigments was not observed. Thus it is likely that the increase of ice-algae was limited primarily by the growth rate of autumnal sea-ice in each of the 3 yrs considered here. Ice-algal communities develop especially in the bottom few centimetres of sea-ice (Hoshiai 1981a). When sea-ice grows, a new bottom layer is formed beneath the layer which was at the bottom. This change in sea-ice occurs successively. Hence, it is speculated that if the growth of the ice is rapid, ice-algae are unable to increase sufficiently to stain the ice, because of shortage of time for proliferation.

At the present stage of knowledge, the autumnal coloration of sea-ice is conspicuous in the new ice and the growth rate of new sea-ice fluctuates from year to year. Furthermore, the growth rate varies depending on topographical features, ocean and tidal currents, and other environmental factors. Spatial and temporal differences in the growth rate of sea-ice are remarkable in the new ice. Accordingly, the autumnal proliferation of ice-algae tends to occur in certain restricted areas and in particular years.

3.3 Spring–Summer Proliferation

The growth of sea-ice slows in mid-September, and stops after mid-October in the vicinity of Syowa Station (Naruse et al. 1971; Hoshiai 1981a). Hence it was suspected that the standing crop of ice-algae may relate to the snow cover on sea-ice during the spring–summer period.

The spring–summer change in plant pigments observed at five fixed stations (A–E) near Syowa Station, and the data obtained at five sites more than 20 km south of Syowa Station, are shown in Fig. 3. At Station A, in a snow-free area, a brown layer of ice-algae was observed and plant-pigment concentration was already high (34.80 mg m^{-3}) in the bottom 30 mm of ice in September. Plant-pigment levels increased subsequently, reaching a peak of 2829.71 mg m^{-3} in November. Thereafter, concentrations decreased to 283.29 mg m^{-3} in December. Plant pigment concentrations were low at the other stations in September, but started to increase in October (except at Station E) and continued to do so until December, when the observations were terminated. Coloration of sea-ice was observed at Station B in October and in November at Stations C, D, and E.

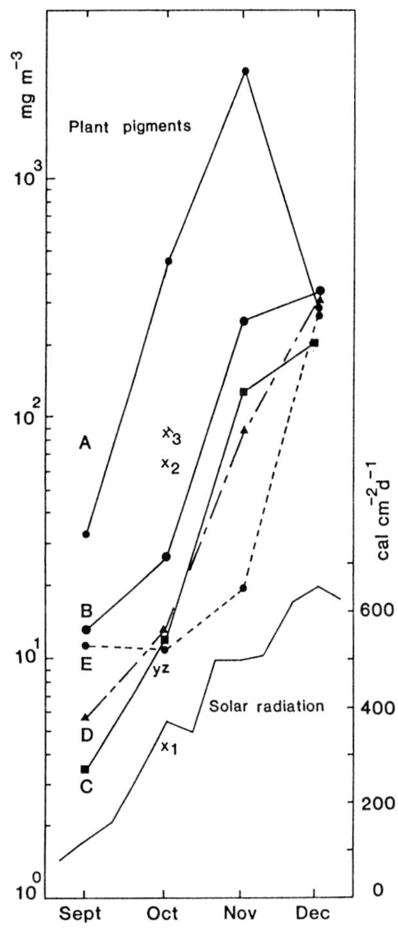

Fig. 3. Changes in concentrations of plant pigments in the bottom layer of sea-ice and incident solar radiation at five fixed stations (A–E) near Syowa Station during September–December 1982. See text for details of sampling sites **A–E, X_1–X_3** and **Y** and **Z**

Development of ice-algae seemed to start in late August or in the beginning of September at Station A when the level of solar radiation was still below 100 cal cm^{-2} d^{-1} (Fig. 3). However, this value equals that in mid-March–mid-April during which the autumnal proliferation was observed. According to Bunt and Lee (1970), about 50% of incident solar radiation penetrates to the bottom of sea-ice, through snow-free ice about 1 m thick. During the observation period, sea-ice thickness was 1.6–1.7 m at Station A and 1.07–1.23 m at Station D. It seemed that the irradiance reaching the sea-ice bottom was not much influenced by the thickness of the ice. On the contrary, it is likely that the snow cover affected the start of enhancement of the ice-algae. The snow depth changed from 150 to 50 mm at Station C, 200 to 50 mm at Station B, 300 to 150 mm at Station D, and 300 to 200 mm at Station E. Data obtained 20 km south of Syowa showed a similar relationship between depth of snow and ice-algal increase. Sullivan et al. (1985) also report a negative relationship between snow cover and ice-algal standing crop from McMurdo Sound.

This study suggests that a spring–summer proliferation of ice-algae occurs extensively in ice-covered areas of the Antarctic. Autumnal proliferation of ice-algae is much more restricted. However, in order to estimate primary production more accurately in the Antarctic Ocean, the autumnal proliferation of ice-algae should be considered and year-round studies of ice-algae at different localities in the ice-covered areas are required.

Acknowledgements. Mitsuo Fukuchi and Atsushi Tanimura, members of the 23rd Japanese Antarctic Expedition, measured plant pigments in 1982.

References

Bunt JS, Lee CC (1970) Seasonal primary production in Antarctic sea ice at McMurdo Sound in 1967 J Mar Res 28:304–230

Hoshiai T (1972) Diatom distribution in sea ice near McMurdo and Syowa Stations. Antarct J U S 7:84–85

Hoshiai T (1977) Seasonal change of ice communities in the sea ice near Syowa Station, Antarctica. In: Dunbar M J (ed) Polar Oceans Calgary Arctic Inst North Am pp 307–317

Hoshiai T (1981a) Proliferation of ice algae in the Syowa Station area Antarctica. Mem Natl Inst Polar Res Ser E (Biol Med) 34: 1–12

Hoshiai T (1981b) Solar radiation and stability of the undersurface of sea ice governing ice algal proliferation. Antarct Rec 73:23–29

Maykut GA, Grenfell TC (1975) The spectral distribution of light beneath first-year sea ice in the Arctic Ocean. Limnol Oceanogr 20:554–563

Naruse R, Endo Y, Ishida T, Ageta Y (1971) Observations of snow accumulation and sea ice at Syowa Station, Antarctica. Antarct Rec 40:57–64

Saijo Y, Nishizawa S (1969) Excitation spectra in the fluorometric determination of chlorophyll-*a* and phaeophytin-*a*. Mar Biol 2:135–136

Sullivan CW, Palmisano AC, Kottmeier S, McGrath-Grossi S, Moe R (1985) The influence of light on development of a sea-ice microbial community in McMurdo Sound. Siegfried WR, Condy PR, Laws RM (eds) In: Antarctic nutrient cycles and food webs (Proceedings of 4th SCAR Symposium on Antarctic biology). Springer Berlin Heidelberg New York

Yentsch CS (1965) Distribution of chlorophyll and phaeophytin in the open sea. Deep-Sea Res 12:653–666

Yentsch CS, Menzel DW (1963) A method for the determination of phytoplankton chlorophyll and phaeophytin by fluorescence. Deep-Sea Res 10:221–231

Oceanographic Studies of Epi-Pelagic Ammonium Distributions: The Ross Sea NH_4^+ Flux Experiment

D.C. BIGGS[1], A.F. AMOS[2], and O. HOLM-HANSEN[3]

Summary. During 16–23 January 1982, an interdisciplinary oceanographic study was carried out in the southwestern Ross Sea, supported by a grid of hydro-stations spaced along 175°E, 173°E, 171°E and 169°E at 5 min latitude intervals from 76° 50′S to the Ross Ice Shelf. To extend previous studies of oceanic nutrient dynamics and pathways of biogeochemical nutrient cycling we sought to define summer-time scales of spatial variability in water-column NH_4^+ in relation to water-mass characteristics, geostrophic flow, distance from ice edges, and plankton biomass. Near-surface NH_4^+ was correlated negatively with near-surface chlorophyll and with distance from ice edges. Lowest concentrations (0.05–0.35 μg at. NH_4^+ l^{-1}) occurred at open water stations north of 77° 10′S along transects 173°E and 175°E. Highest concentrations (0.50–1.35 μg at. NH_4^+ l^{-1}) occurred close along the Ross Ice Shelf, and at stations west of 171°E and within 50 km of a transition from open water to pack-ice. Ammonium maxima underlying the mixed layer clustered into 2 groups of sigma-theta density, reflecting local melting of continental, NH_4^+-rich ice of the Ross Ice Shelf and a net biological regeneration and accumulation of NH_4^+ under pack-ice. At locations and/or times where internal wave activity and the passage of meso-scale eddies flux this sub-surface NH_4^+ into the mixed layer, it is taken up preferentially to NO_3 and may help promote or maintain high chlorophyll standing stocks that are characteristic of mid-summer in this part of the Ross embayment.

1 Introduction

The southern Ross Sea along the barrier edge of the Ross Ice Shelf develops extensive summer blooms of planktonic algae. In the summer of 1978, for example, surface chlorophyll concentrations ≥ 1 mg m^{-3} extended for 100 s of km along the barrier and reached seaward for 10 s of km (El-Sayed et al. 1983). Measurements of nitrogenous nutrient concentrations within this 1978 phytoplankton bloom showed that NH_4^+ was depleted from near-surface waters, but reached concentrations of 0.4–1.4 μg at. NH_4^+ l^{-1} immediately underlying the algae. Because NH_4^+, when available, is taken up as a preferred nutrient by Southern Ocean plankton, Biggs (1982) hypothesized that summer-time algal blooms in the southern Ross Sea might be initiated, promoted, and maintained by this sub-surface NH_4^+ reservoir.

During 16–23 January 1982, an interdisciplinary oceanographic study of plankton, hydrography, and environmental NH_4^+ level was carried out in the southwestern corner of the Ross Sea (Biggs et al. 1982a). Working on board USCGC Glacier, we sought to define summer-time scales of spatial variability in water column NH_4^+ in relation to water-mass characteristics, geostrophic flow, distance from ice edges, and to plankton distribution, biomass and metabolic activity. Here we summarize the work dealing with nutrient chemistry and physical oceanography, and consider the formation of the sub-surface reservoir of NH_4^+ and certain physical processes which flux NH_4^+ to the overlying mixed layer. We present information on the general distribution and biomass of plankton, but the details will be reported elsewhere.

2 Methods

Twenty-six surface-bottom STD stations were spaced along 175°E, 173°E, 171°E, and 169°E at 5 min latitude intervals from 76° 50′S south to the Ross Ice Shelf (Fig. 1A). All stations were in open water, which in mid-January 1982 reached well beyond our study area northward to about 71°S and eastward along the barrier to about 165°W (see Antarctic Ice Analysis Charts, available from the US Navy-NOAA Joint Ice Center, Suitland, MD; in Biggs and Amos 1983). To the west of Stations 23–29, however, there was a sharp transition from open water to broken pack-ice (Fig. 1A). We used airborne radar profiles (from C-130 overflights) and high-resolution, local area coverage, satellite imagery of the McMurdo Station area (Nimbus and NOAA-6 VIS LAC, available through the US Navy Weather Office at McMurdo Station) to fix the position of this pack-ice edge between 13 and 23 January 1982. The position of the barrier edge of the Ross Ice Shelf was mapped by taking bearings with the ship's radar, co-ordinated with satellite navigation fixes on station at the southernmost hydro-stations.

1 Department of Oceanography, Texas A&M University, College Station, TX 77843, USA
2 Marine Science Institute, University of Texas, Port Aransas, TX 78373, USA
3 Scripps Institution of Oceanography, University of California, La Jolla, CA 92093, USA

Antarctic Nutrient Cycles and Food Webs
(ed. by W.R. Siegfried, P.R. Condy, and R.M. Laws)

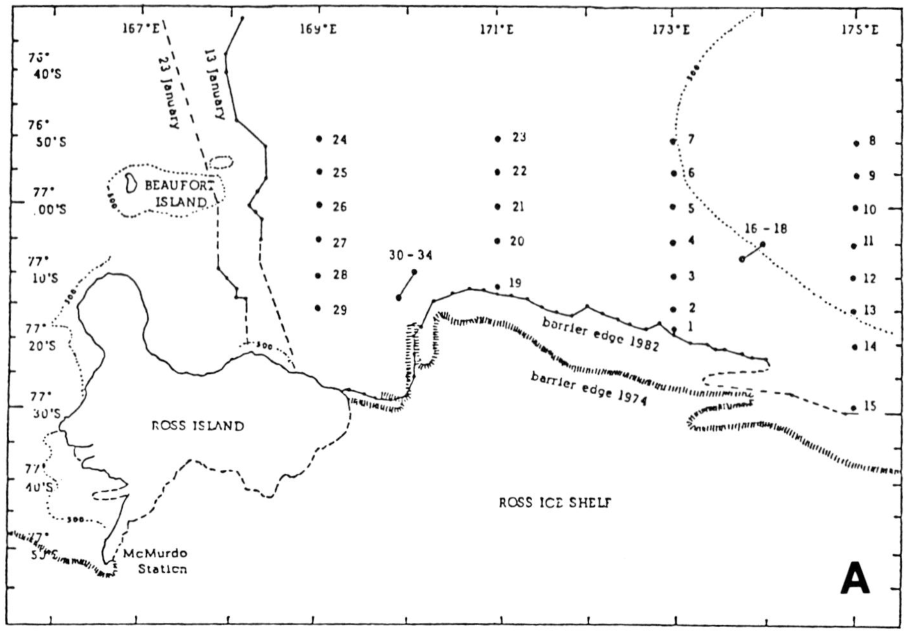

Fig. 1A. Oceanographic survey stations occupied by USCGC GLACIER, 16–23 January 1982. The survey area was in open water; the retreating edge of the sea-ice pack is indicated on 13 and 23 January

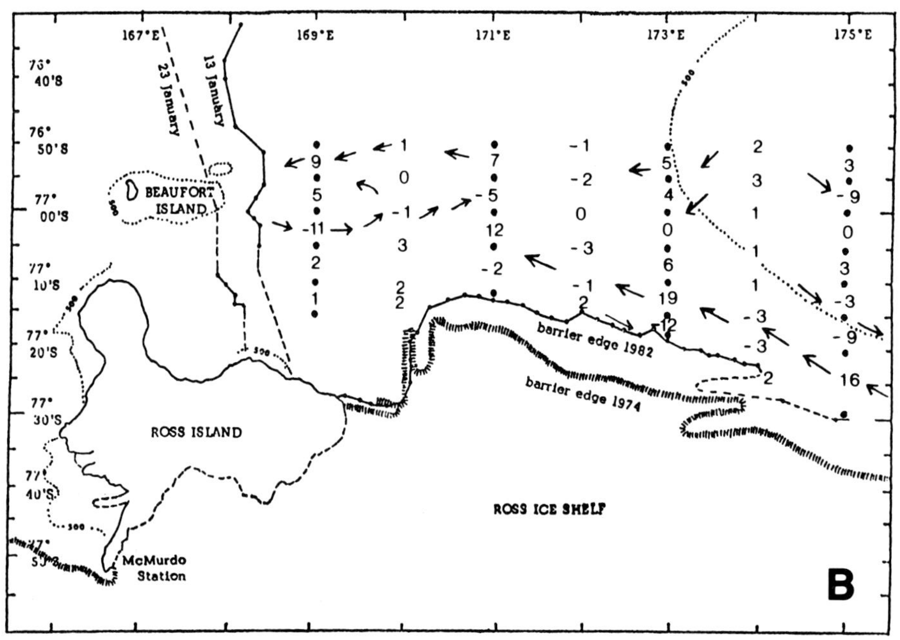

Fig. 1B. Dynamic chart of geostrophic shear between adjacent station pairs (cm s^{-1} surface current, relative to the 500 db surface or to the bottom if station depth was less than 500 m). *Negative numbers* between stations in north-south direction indicate flow to the east; *negative numbers* between stations in east-west direction indicate flow to the north

At each station, discrete water samples were collected in 12 2.5-l Go-Flo Niskin bottles mounted on a rosette multi-sampler attached to a Plessy 9040 STD. Ten bottles on the multi-sampler were tripped at standard levels within the upper 250 m (5, 20, 40, 60, 80, 100, 125, 150, 200, and 250 m); the remaining two bottles sampled the deep mid-water. Sub-samples for nutrient chemistry were drawn directly into 125-ml whirl-pac bags, which were analyzed on board the ship without filtration for ammonium, nitrate, nitrite, and silicate with a four-channel Technicon AA-II auto-analyzer. The whirl-pacs were refrigerated in the dark until analysis; most were analyzed within 12 h of collection and none was analyzed more than 48 h after collection. Nutrient chemistry analytical procedures followed those detailed previously for Southern Ocean shipboard auto-analyzer studies (Biggs et al. 1982b). Sub-samples for chlorophyll and phaeophytin were filtered through GF-C glass-fibre filters and processed on board the ship by fluorometric techniques (Holm-Hansen et al. 1965).

Zooplankton was collected at every other station with oblique hauls to 100 m, using 0.7 m bongo nets of 333-μm mesh. Flow through the nets was estimated from a frame-mounted TSK flowmeter; volumes filtered ranged from 70 to 230 m^3 per haul. After preservation in 4% buffered formaldehyde, zooplankton displacement volume was determined by the method of Ahlstrom and Thrailkill (1960). The flowmeter data indicated that the mesh of the net had not clogged, although the cod-end collections were frequently thick with diatoms, particularly along 175°E and 173°E. To separate the zooplankton from this slurry of phytoplankton, some collections were split 8 or 16 fold, using a Folsom plankton splitter. The sub-sample was then washed on a 210-μm mesh screen prior to the determination of displacement volume.

At two locations, the upper 100 m was sampled close along the drift track of a parachute drogued at a depth of 100 m in the sub-surface NH_4^+ reservoir. At these locations, we mated a submersible pump to the STD and, in co-operation with M.A. Johnson, A. Neori, J.D. Guffy, and M.A. Spears, sampled the continuous stream of water pumped aboard the ship, using an auto-analyzer, polarigraphic oxygen probe, and two Turner Designs model 10 flow-through fluorometers. Johnson (1981) has discussed the sampling and analytical capabilities of this system. These pumping stations correspond to STD casts 16–18 and 30–34, respectively, in Fig. 1A. Vertical pumped profiles afforded vertical resolution of ± 2 m; horizontal casts tracked time-series changes in NH_4^+ at selected depths for periods of 2–4 h.

A station-by-station archive of all the STD data, with the corresponding bottle-cast oxygen and bottle-cast nutrient data, is available as a technical report (Biggs and Amos 1983). A transect-by-transect summary of the vertical sections 0–250 m, including temperature, salinity, temperature-above-freezing, density, NH_4^+, NO_3^-, NO_2^-, $Si(OH)_4$ and geostrophic velocity, is included in the report.

3 Results

3.1 Water Types and General Circulation

Figure 1B charts the geostrophic surface current between adjacent station pairs, relative to 500 db (or to the bottom if station depth was < 500 m). No large-scale exchanges of water out from, or in, underneath the base of the Ross Ice Shelf are suggested from the thermohaline structure of this region. Instead, a westward component of the geostrophic shear predominated at most of the stations near the barrier edge. A westward-flowing surface current is in geostrophic balance with the curl of the January wind-stress, since surface winds blew from the S and SE between 16 and 23 January 1982 (Biggs and Amos 1983). Fig. 1B also indicates a local area of anticyclonic circulation to the east of Beaufort Island.

The temperature-salinity structure of this part of the Ross Sea has been described by Jacobs et al. (1970, 1979) and by Jacobs and Haines (1982). A T-S diagram of the 175°E section (Fig. 2) distinguishes three water types: Ross Sea Shelf Water (RSSW); a modified form of Circumpolar Deep Water (CDW); and, a modified form of Deep Ice Shelf Water (DISW). Temperature-salinity diagrams for the other three transects are given in Biggs and Amos (1983).

The RSSW comprised most of the deep water of the southwestern Ross Sea, except at stations on the relatively shallow (< 500 m) Ross Bank along 175°E. RSSW has salinity > 34.70‰ and temperature < –1.8°C. Along the barrier edge of the Ross Ice Shelf, the RSSW extended from the bottom to within 400 m of the surface. Farther off the edge, the RSSW reached to within 250 m–350 m of the surface. Maximum NH_4^+ concentrations in the RSSW were < 0.35 μg at. l^{-1}.

Modified CDW was found primarily along 175°E, where it formed a thick wedge of 60 m–250 m (Fig. 3). It was best developed at Stations 8–13 on Ross Bank, identified by the cluster of points midway between and roughly parallel to the 27.8 and 27.9 isopycnals on the T-S diagram (Fig. 2). This modified CDW, of origin far to the north of our study area, extended west to 173°E only at Station 5. Maximum NH_4^+ concentrations in this water type were < 0.40 μg at. l^{-1}.

The DISW is characterized by its proximity to the in situ freezing point, and is believed to be formed by melting-refreezing processes at the base of the Ross Ice Shelf (Jacobs et al. 1979). Modified DISW was present at depths of 200–400 m at Station 14 and 15, as a filament of water with temperatures near –2.0°C and salinities of 34.60–34.70‰. At these depths, water of this temperature and salinity lies within 0.2°C of the in situ freezing point (Fujino et al. 1974). From cross-sections of temperature-above-freezing (Fig. 4) and from cross-sections of salinity (Biggs and Amos 1983), the modified DISW can be traced west across 173°W (where it appears at Station 2, 3, and 4) to 171°E (where it appears at Stations 20 and 21). From our few bottle samples tripped within it, this DISW apparently had an NH_4^+ content of < 0.10 μg at. l^{-1}.

In summary, none of these three sub-surface water types was rich in NH_4^+, including the one formed from melting-refreezing interactions with the continental ice of the Ross Ice Shelf. However, a second more shallow-occurring Ice Shelf Water (SISW) that Jacobs et al. (1979) had frequently encountered during 1976–1979 at depths of 90–140 m along the barrier edge of the Ross Ice Shelf was conspicuous by its absence from our survey. Jacobs and Haines (1982) reported that this SISW had temperatures between –2.00 and –2.07°C, with salinity ranging from 34.35–34.38‰. This salinity suggests that, unlike DISW, SISW is modified more by ice-melt than by refreezing. Since the continental ice of the Ross Ice Shelf and of other non-marine ice sheets is high in NH_4^+ (Parker and Zeller 1979), this SISW may in fact represent an important source of NH_4+ to near-surface water and should be sampled intensively in time and space by future expeditions to the Ross Sea.

3.2 Pack-Ice, NH_4^+, and Phytoplankton Biomass

Near-surface NH_4^+ was correlated negatively with near-surface chlorophyll and with distance from ice edges. As shown in Fig. 1A, our westernmost line of stations along

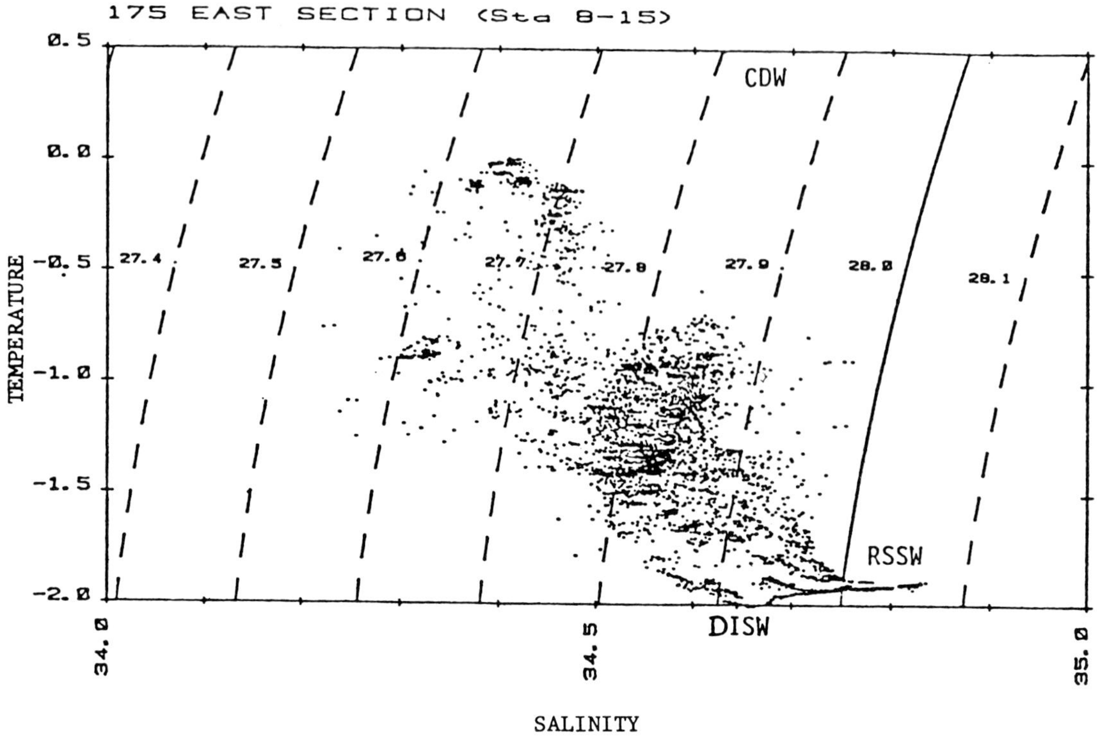

Fig. 2. Temperature-Salinity diagram of stations along 175°E, illustrating the thermohaline characteristics of modified Circumpolar Deep Water (*CDW*), Ross Sea Shelf Water (*RSSW*), and Deep Ice Shelf Water (*DISW*)

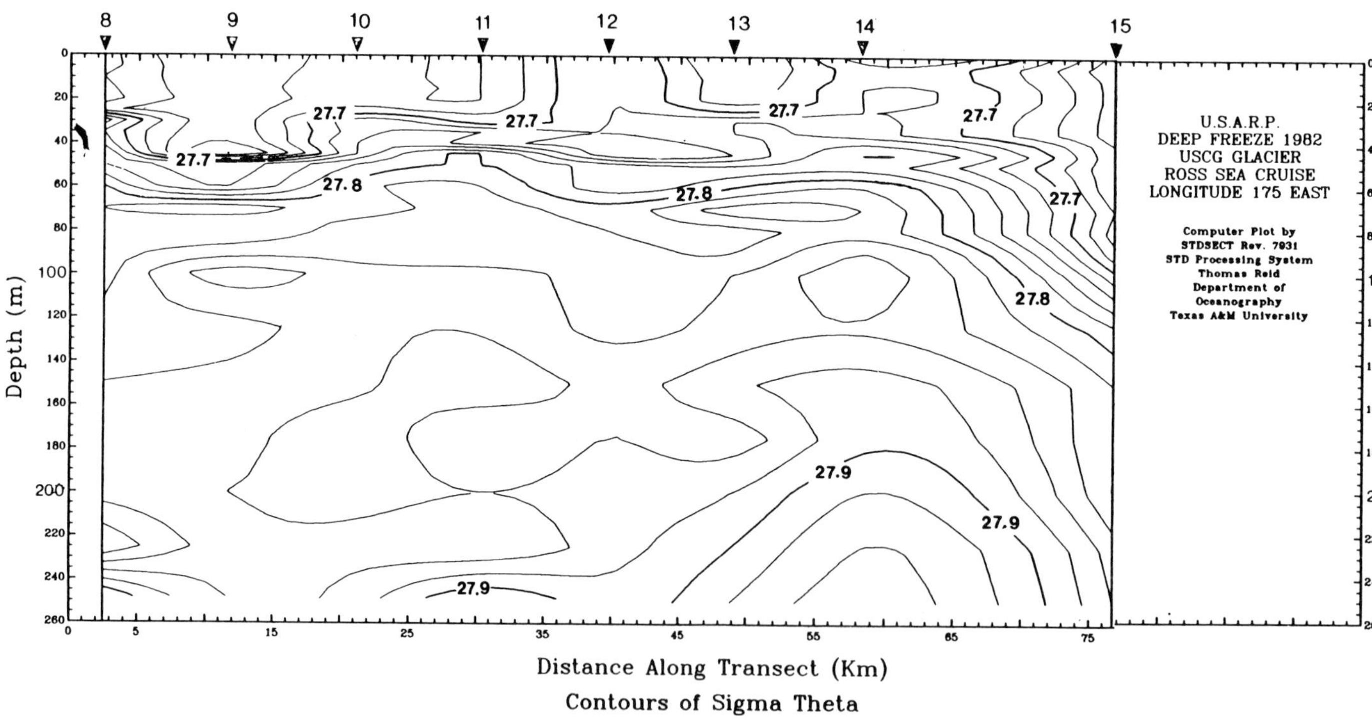

Fig. 3. Density field in the upper 250 m of stations along 175°E. The barrier edge of the Ross Ice Shelf is to the *right*, within 1–2 km of Station 15

169°E lay near a receding sea-ice edge. US Navy-NOAA records of environmental satellite imagery for the austral summer of 1982, which mapped the time transgressive behaviour of this and other ice edges, show that pack-ice had covered these westernmost stations until only about three weeks prior to our cruise (see time series of Antarctic Ice Analysis Charts, in Biggs and Amos 1983). Since mid-November 1981, when open water had first appeared at

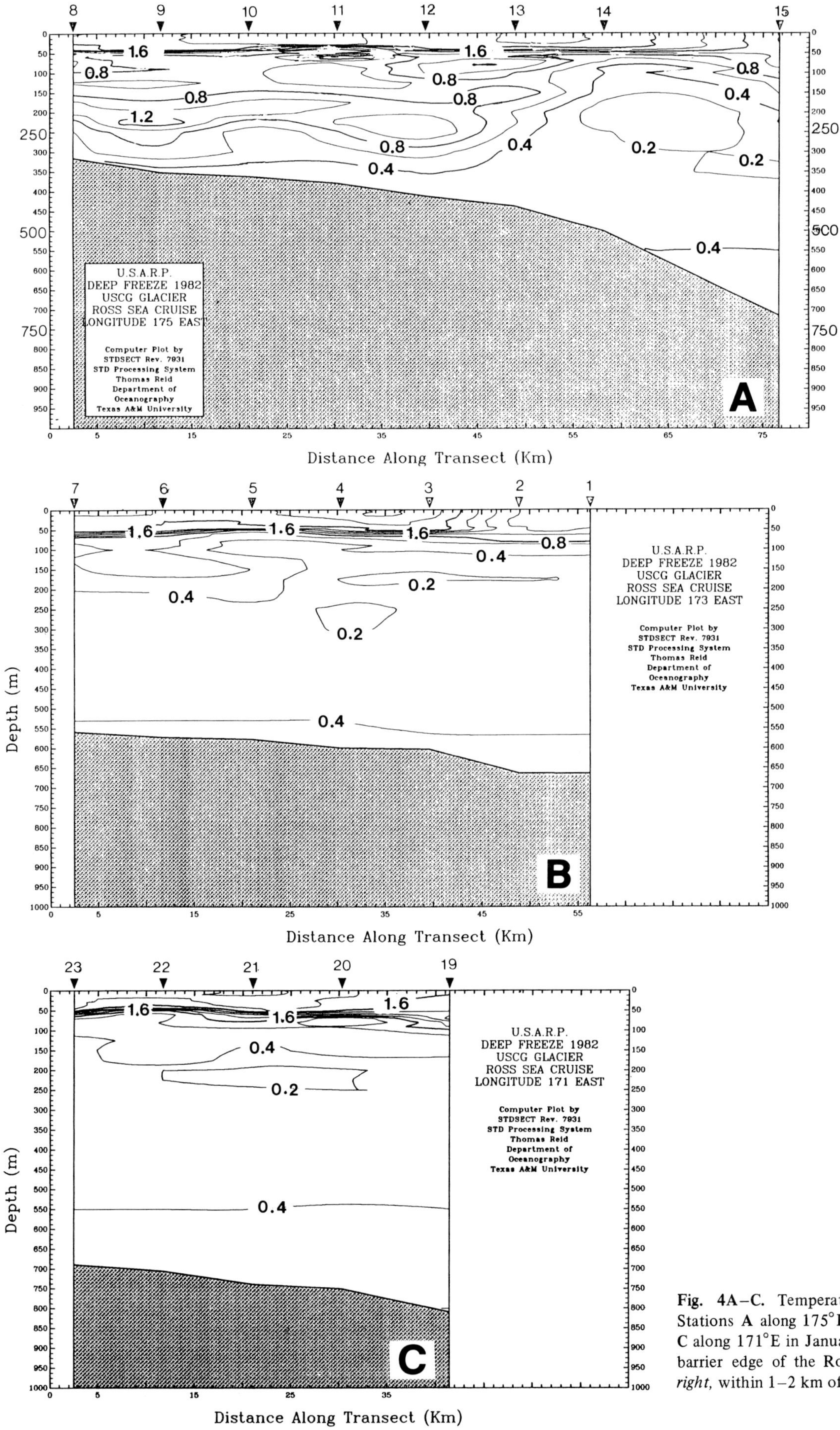

Fig. 4A–C. Temperature-above-freezing at Stations **A** along 175°E, **B** along 173°E, and **C** along 171°E in January 1982. As in Fig. 3, barrier edge of the Ross Ice Shelf is to the *right*, within 1–2 km of Station 15, 1 and 19

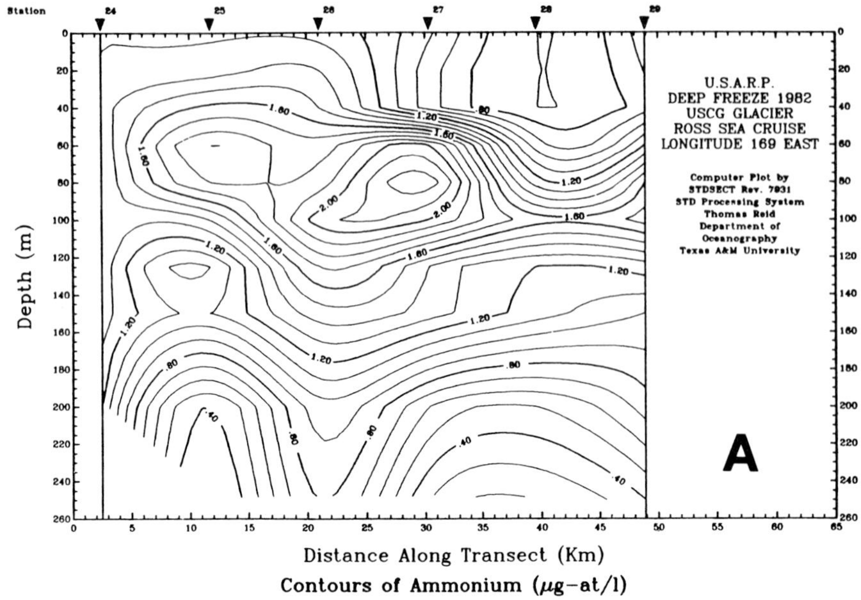

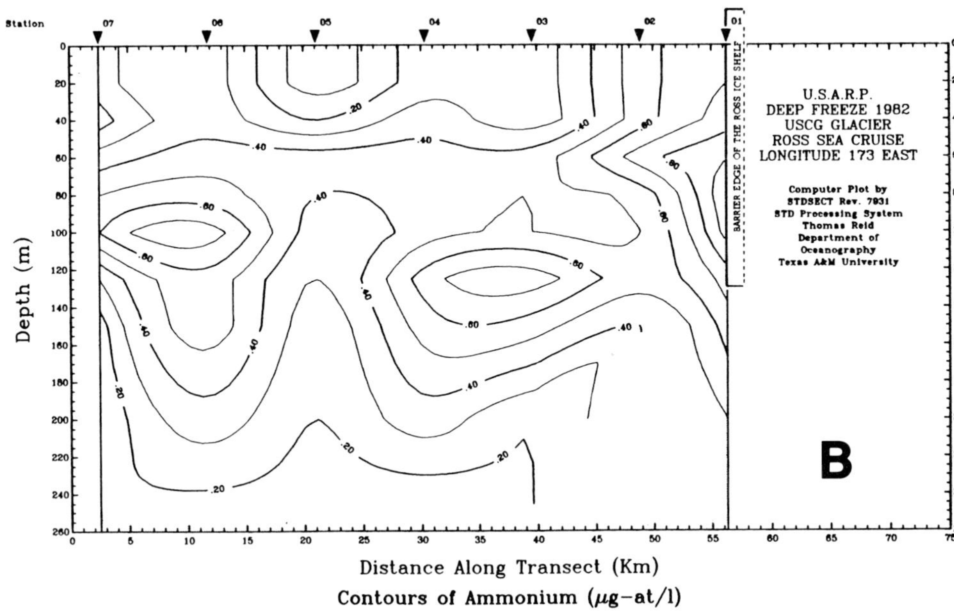

Fig. 5A, B. Ammonium concentration (μg at NH_4^+ l^{-1}) in the upper 250 m of stations **A** along 169°E, and **B** along 173°E. There were no NH_4^+ samples from 200 m and 250 m at Station 2, or from 250 m at Station 24

about 180°E in the southern Ross Sea, the pack-ice open water boundary retreated generally westward. Stations along 175°E and 173°E had been in open water for about 9 weeks, while those along 171°E had been in open water for about 7 weeks.

Although we carried out our survey in mid-summer, environmental conditions along 169°E near the pack-ice edge appeared far more analogous to those of early spring. The open water, although pre-charged with the high levels of nitrogenous nutrients, had low standing stocks of phytoplankton. Nitrate concentrations exceeded 22 μg at. l^{-1} throughout near-surface waters, and NH_4^+ reached 1.35 –2.15 μg at. l^{-1} (Fig. 5A). The highest NH_4^+ concentrations occurred at Stations 26 and 27, within a region of eastward flow in the local anticyclonic circulation (i.e. from the pack-ice zone towards the open water zone). Phytoplankton standing crop reached only 0.1–0.3 mg m^3 in near-surface waters, and ranged just 17–23 mg m^2 when integrated for the upper 100 m (Fig. 6). In summary, since the pack-ice had retreated only recently from 169°E, there seems to have been insufficient time for phytoplankton standing stocks to increase substantially.

Where open water had been present for 2 months or more, as along 173°E and 175°E, phytoplankton standing crop was high, ranging between 54 and 117 mg m^2 in the upper 100 m (Fig. 6). Here, NH_4^+ was markedly depleted, with near-surface concentrations as low as 0.05–0.10 μg at. l^{-1} (Fig. 7) and maximum concentrations nowhere > 1μg at. NH_4^+ l^{-1}. Nitrate, as well, was lower throughout the upper 250 m than it had been in the west, and near-surface

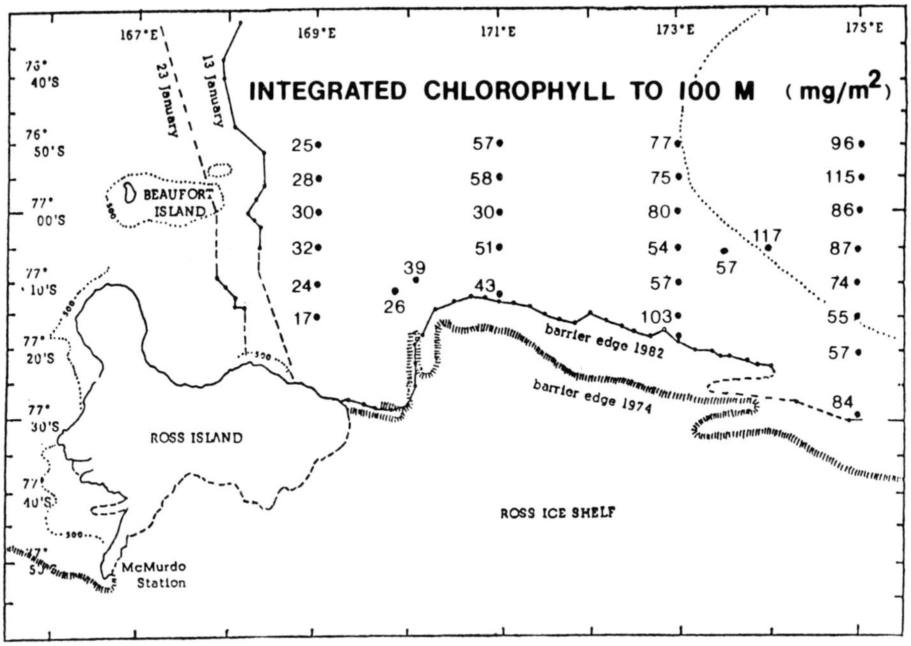

Fig. 6. Chlorophyll (mg m^{-2}) integrated from surface to 100 m

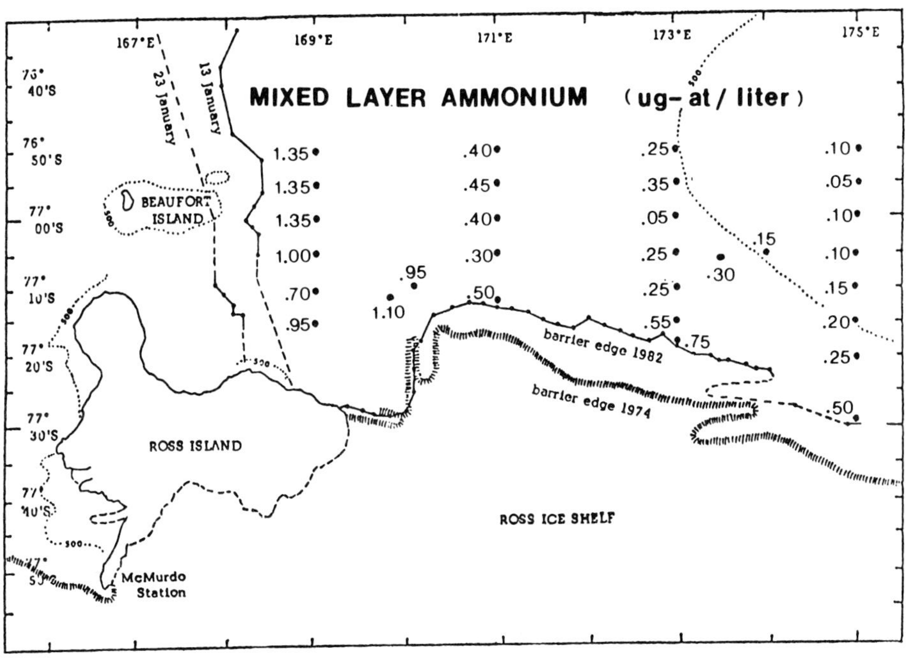

Fig. 7. Average NH_4^+ concentration within the mixed layer, from bottle samples at 5 m and 20 m

concentrations were as low as 15–17 μg at. NO_3^- l^{-1} (Biggs and Amos 1983).

Because phytoplankton takes up NH_4^+ as a preferred nutrient, NH_4^+ is depleted from the mixed layer in areas of high phytoplankton concentration. As the depletion of NH_4^+ proceeds, the algae shift from NH_4^+ to NO_3^- to meet their total daily nitrogenous requirement, and the net result is a highly significant negative correlation between near-surface NH_4^+ and near-surface chlorophyll. A non-parametric, pairwise ranked test of NH_4^+ concentration vs chlorophyll concentration for samples from 5 m and 20 m depths yielded a correlation of –0.77, significant at the P = 0.0001 level (Spearman Rank Test, n = 47).

Very close to the barrier edge of the Ross Ice Shelf, however, NH_4^+ chlorophyll relationships were more complicated. Here, locally high near-surface NH_4^+ concentrations occurred despite high standing stocks of phytoplankton (cf. Figs. 6 and 7). Since neither light, temperature, nor salinity were more extreme along the barrier than off the ice edge, the local barrier edge source of NH_4^+ seems likely.

3.3 Ross Ice Shelf, NH_4^+, and Zooplankton Biomass

Zooplankton, which regenerate NH_4^+ as a nitrogenous waste product of their respiratory metabolism, apparently are minor sources of NH_4^+ either along the barrier edge or

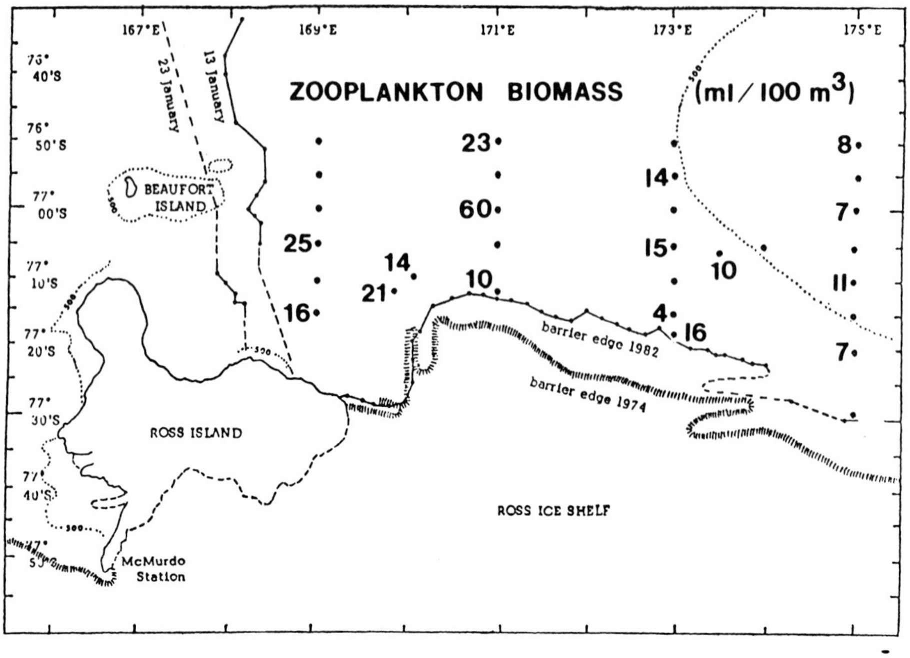

Fig. 8. Zooplankton displacement volume from oblique tows, upper 100 m

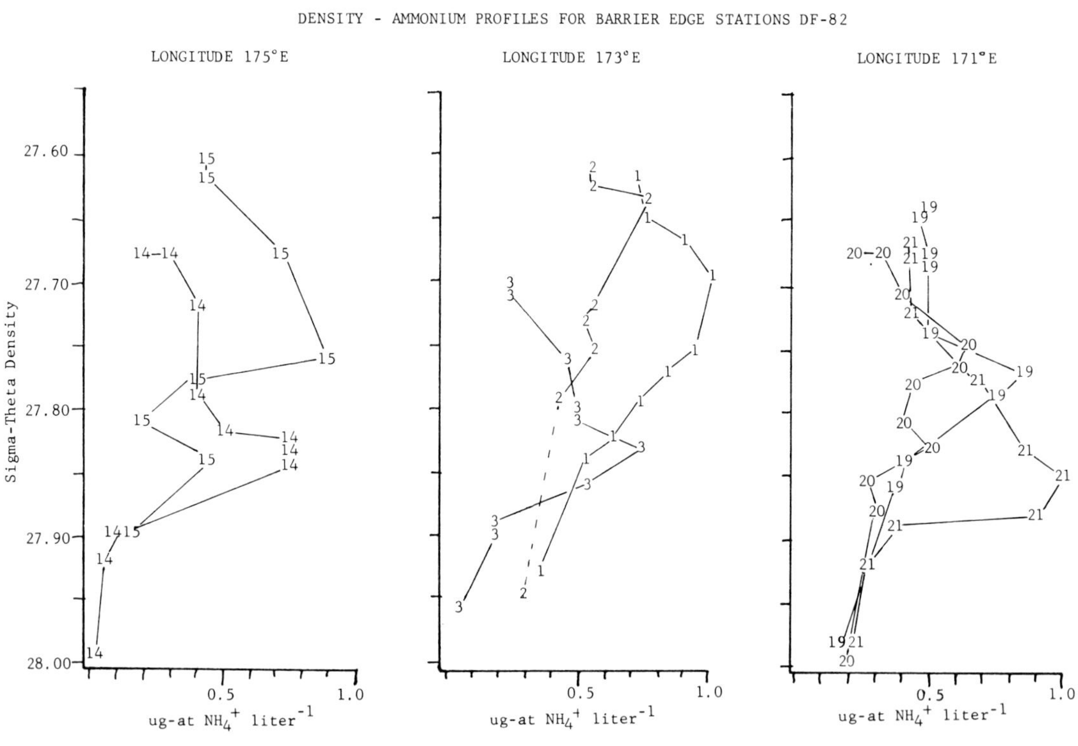

Fig. 9. NH_4^+ concentration as a function of density at stations within 18 km of the barrier edge of the Ross Ice Shelf. Note that NH_4^+ maxima cluster between 27.65 and 27.80 close in along the barrier, and between 27.81 and 27.90 farther offshore

elsewhere in the southwestern Ross Sea, since standing stocks were low throughout the upper 100 m. Most net-hauls caught < 20 ml displacement volume per 100 m^3 (Fig. 8). In fact, rather than being most abundant in areas of high chlorophyll, zooplankton were more abundant in the low chlorophyll, western part of the survey area. Hence, pairwise ranked correlation of zooplankton displacement volume (ml per 100 m^3) with chlorophyll integrated to 100 m was −0.67, significant at the P = 0.0044 level (Spearman Rank Test, n = 16).

In general, copepods were numerically the most abundant group within the zooplankton, averaging 48% of total numbers. Euphausiid calyptopis larvae were also common, averaging 26% of total numbers, and thecosome pteropods made up between 9 and 54% of total numbers. The latter showed a striking east-west gradient in age structure; being

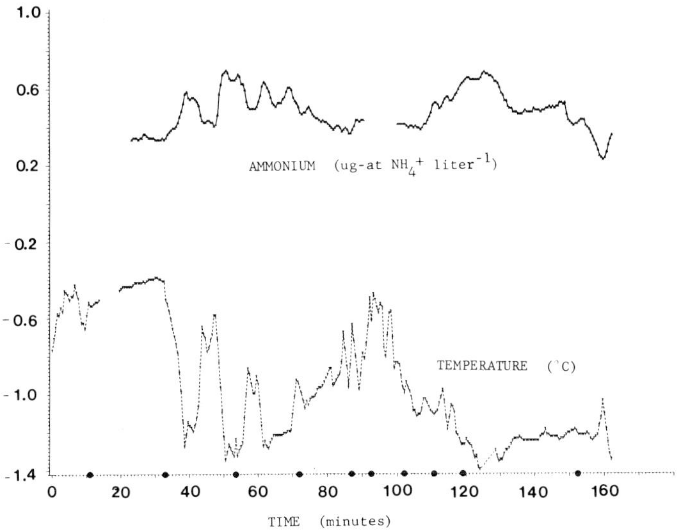

Fig. 10. Time-series of temperature and pumped NH_4^+ at a depth of 68 m at Station 17, plotted at 30 s intervals. Temperature was measured by STD and pumped NH_4^+ by flow-through auto-analyzer analysis. *Dots* along the x-axis indicate times when discrete samples were collected in rosette bottles hung on the STD, for calibration independent of the pumped sampling stream

represented mostly as veliger larvae in the east but as adults in the west.

3.4 NH_4^+ – Density Relationships

The influence of the barrier edge of the Ross Ice Shelf on NH_4^+ concentration is best visualized from NH_4^+ – density relationships. Figure 9 summarizes information for stations within 18 km of the barrier edge along 175°E, 173°E, and 171°E. Note that along all three lines, NH_4^+ maxima between densities of 27.65 and 27.80 were characteristics of stations closest to the barrier edge. Note also that the water with this NH_4^+-density signature disappeared rapidly with distance off the barrier edge, being replaced by water with a local NH_4^+ maximum of 27.81–27.90 units of sigma-theta. The latter signature was characteristic of the remaining stations farther off the edge, and those along 169°E (Biggs and Amos 1983). It was absent only over the Ross Bank, where most of the mid-water was comprised of low-NH_4^+ modified CDW.

The less dense of these two NH_4^+ sub-surface maxima suggests an association with local summer-time melting at the under-surface or along the terminal face of the Ross Ice Shelf. Its rapid disappearance off the edge simply reflects a northward shoaling isopycnal field (see Fig. 3). Whereas at the barrier edge, water of density 27.70 was at a depth of 60–100 m and water of density 27.80 was at a depth of 120–150 m, only 18 km north of the barrier edge water density of 27.70 was at the surface and water of density 27.80 reached the base of the mixed layer. This northward shoaling of isopycnal surfaces, with a geostrophic shear parallel to the barrier edge, should effectively mix NH_4^+ of melt-water origin relatively near the surface within just 5–10 km of the barrier edge.

3.5 NH_4^+ Variations Within the Pycnocline

Figure 10 presents a time series of temperature and NH_4^+ concentration pumped from a depth of 68 m within a gentle pycnocline at the base of the mixed layer at Station 17. In the upper 100 m at this station, NH_4^+ increased, while temperature decreased, with depth. Density at the surface was 27.66, and 27.80 at 100 m. Note that NH_4^+ concentration in this gentle pycnocline varied inversely with temperature, ranging from 0.2 μg at. l^{-1}, when temperature at 68 m was highest, to 0.7 μg at. l^{-1} when temperature at 68 m was lowest. This in-phase relationship between variations in NH_4^+ and variations in temperature is indicative of the passage of internal waves within the pycnocline. The magnitude of the temperature excursions during this time-series sampling, relative to vertical sections of temperature within the upper 100 m, suggests that the amplitude of these internal waves exceeded 20 m.

4 Discussion

4.1 Phytoplankton-NH_4^+-Zooplankton Relationships

It is now well documented that Southern Ocean phytoplankton meets most of its nitrogenous needs by taking up NH_4^+, despite high ambient concentrations of NO_3^- (Olson 1980; Koike et al. 1981; Glibert et al. 1982; Holm-Hansen 1985; Ronner et al., to be publ.). Since areas of high algal biomass can rapidly deplete NH_4^+ from near-surface waters, negative spatial correlations between near-surface NH_4^+ and chlorophyll are often reported (Biggs et al. 1982b). For an area of high algal biomass to be promoted or maintained by NH_4^+, mechanisms must exist to replace rapidly NH_4^+ depleted from the mixed layer by the daily photosynthetic activities of the phytoplankton.

Bidigare (1983) has recently reviewed the contribution of zooplankton NH_4^+ excretion to the N requirements for primary production in various marine environments. He states that the excretory contribution of macro-zooplankton, defined as plankters > 200 μm in size, ranges from about 10^{-1} to 1 μg at. NH_4^+ m^{-3} h^{-1} in oceanic-coastal areas of the world's oceans, or roughly 2–40% of the average local daily total phytoplankton N requirement (TPNR). According to Bidigare (1983), the remaining N deficit is probably supplied by micro-zooplankton excretion and/or by nitrate.

It is clear that the near-surface macro-zooplankton standing stock in the southern Ross Sea is especially low, relative to the average daily TPNR there. Biggs (1982) reported that Ross Sea macro-zooplankton biomass averaged only 0.06 ml m^{-3} in the upper 200 m in the summer of 1978 and, from measurements of NH_4^+ excretion by the principal taxa contributing to this biomass, judged that macro-zooplankton probably regenerate < 2% of the daily TPNR. Since the macro-zooplankton standing stock in summer 1982 averaged only 0.10 ml m^3 in the region of high chlorophyll concentration (Fig. 6), it seems highly unlikely that macro-zooplankton excretion was a major source of NH_4^+ for the TPNR in the summer of 1982.

Estimates of the abundance, population structure, and excretory impact of micro-zooplankton in the southern Ross Sea are not yet available. However, samples for the determination of micro-zooplankton abundance and population structure were collected by the method of Beers et al. (1975) from 0–100 m at the drogue buoy locations, and delivered to Dr. J.R. Beers for analysis. In addition, $^{15}NH_4^+$ isotope dilution experiments, to estimate the time course of remineralization of NH_4^+ by the micro-zooplankton – bacterial community, were carried out at these same locations by Dr. I. Koike. When these samples and experiments have been processed, they will allow a more complete comparison of the biogenic sources and sinks of NH_4^+ in the southern Ross Sea. However, we caution against reducing near-surface NH_4^+ dynamics strictly to biological terms. We believe that there are at least two non-biogenic processes which may be quite important in renewing NH_4^+ concentrations in the chlorophyll-rich, summer surface-waters of the southern Ross Sea. The first of these is a fine-scale process, which fluxes NH_4^+ vertically across the pycnocline. The second is a meso-scale process which arises as a consequence of the local circulation.

4.2 Renewal of NH_4^+ by Non-Biogenic Means

4.2.1 Internal Waves

The local temperature minima in Fig. 10 mark times and/or locations where water of high NH_4^+ content was in close contact with the base of the 30 m–50 m thick mixed layer in the chlorophyll-rich, eastern part of our survey area. If the passage of an internal wave-train(s) should increase the local vertical shear at wave crests and troughs or otherwise decrease the vertical stability within the gentle pycnocline, diffusive exchanges between the mixed layer and the underlying water will be enhanced. Should the shear increase to the point where the internal waves actually break, "patches" of high NH_4^+ concentration would be injected directly into the mixed layer. Given a mixed layer of 0.15 μg at. NH_4^+ l^{-1} overlying water of 0.70 μg at. NH_4^+ l^{-1}, such as existed at Station 17, internal wave activity holds the potential to be significant.

Whether the presence of the Ross Bank, rising to within 300 m–400 m of the surface just to the east of Station 17, exercises any topographic control over internal wave activity in the pycnocline is unknown. Smaller but quite regular excursions in temperature-NH_4^+ fields within the pycnocline were encountered at Station 33, over deeper water in the western part of our survey area. Here, the STD-pump combination was hung for 2 h at 65 m, and then lowered to 90 m for an additional 2 h (see Biggs and Amos 1983).

4.2.2 Eddies

Since high NH_4^+ concentrations seem to be associated with regions of pack-ice cover, processes which advect this water out from under the ice and into open water should also flux significant amounts of NH_4^+ to the zone of high phytoplankton growth. The anticyclonic circulation that we transected at northern stations along 169°E and 171°E is one example of this kind of meso-scale advective process. With NH_4^+ concentrations in excess of 1 μg at. l^{-1} from at least 160 m to the base of the mixed layer, this anticyclonic circulation represented the major reservoir of NH_4^+ within the open water area we surveyed. Note from a comparison of Fig. 1B and 6 that chlorophyll levels within the reservoir were locally higher than those outside its southern boundary. It is unknown whether this was a detached eddy, and whether its NH_4^+ load had originated biogenically or non-biogenically. Eddies of similar diameter have been observed in XBT surveys of the southern Ross Sea prior to 1982, although their NH_4^+ content was not sampled (S.S. Jacobs, pers. comm.).

4.3 The Retreating Ice Edge Environment

While there is danger in oversimplification, in the simplest case one might perhaps consider that the southwestern Ross Sea in the austral summer represents a region-wide retreating ice-edge environment. Whether a pre-charge of NH_4^+ in the water column underlying the retreating sea-ice may stimulate or accelerate the extensive development of the bloom conditions of phytoplankton which annually occur here is unknown. Whether the accumulation of NH_4^+ in the water column under the pack-ice results from the microbial decomposition of organic matter during the preceding austral winter is also unknown. Both of these possibilities should be investigated. However, pack-ice melt alone can represent but a minor source of NH_4^+ to the overlying water column, as Biggs (1982) has shown. For example if sea-ice averages 1.5 m thick and has an average NH_4^+ content of 1 μg at. l^{-1} (Oradovskiy 1974), each 1 m^2 of open water exposed by melt-down should receive 1.5 mg at NH_4^+. Distributed over a 25 m deep mixed layer, the average reservoir concentration would increase by only 0.06 μg at. NH_4^+ l^{-1}.

Finally, while ice-edges are widely acknowledged to be areas of locally high biological activity (Stirling 1982), our chlorophyll data emphasize the time-transgressive nature of algal growth there. Our understanding of productivity and NH_4^+ dynamics in the southwestern Ross Sea will benefit from additional study of the factors which structure primary productivity in retreating ice edge environments. It remains to be tested whether productivity is truly enhanced by the enhanced physical stability conferred on the near-surface density field by the lower than average salinities in zones of ice melt, or whether factors such as the degree to which the ice-edge zone is pre-charged with NH_4^+ offer a better means of explaining the time course of phytoplankton production there.

Acknowledgements. We thank M.A. Johnson, R.R. Bidigare, J.D. Guffy, M.A. Spears, J.L. Cox, C.D. Hewes, A. Neori, and I. Koike for their participation in the field programme, and we appreciate the keen interest and outstanding logistic support that the officers and marine science technicians of USCGC Glacier gave to us. Research support (DPP79-21355 and DPP80-20242) and travel grants from the US National Science Foundation for DCB and OHH to attend the Fourth SCAR Symposium on Antarctic Biology are gratefully acknowledged.

References

Ahlstrom EH, Thrailkill JR (1960) Plankton volume loss with time of preservation. CALCOFI Reports 9:57–73

Beers JR, Reid FMH, Stewart GL (1975) Microplankton of the North Pacific Central Gyre. Population structure and abundance, June 1973. Int Rev Gesamten Hydrobiol 60:607–638

Bidigare RR (1983) Nitrogen excretion by marine zooplankton. Chapter 13, In: Carpenter E and Capone D (eds) Nitrogen in the Marine Environment. Academic, London

Biggs DC (1982) Zooplankton excretion and NH_4^+ cycling in near-surface waters of the Southern Ocean. I. Ross Sea, austral summer 1977–1978. Polar Biology 1:55–67

Biggs DC, Amos AF (1983) Oceanographic data from the southwestern Ross Sea, January 1982: STD and nutrient chemistry programs. Tech Rep 83-3-T, Dept Oceanogr Texas A&M Univ pp 167

Biggs DC, Amos AF, Johnson MA, Holm-Hansen O, Neori A, Koike I, Bidigare RR, Cox JL (1982a) Ross Sea marine studies. Antarct J U S Ann. Rev Iss 17:144–154

Biggs DC, Johnson MA, Bidigare RR, Guffy JD, Holm-Hansen O (1982b) Shipboard autoanalyzer studies of nutrient chemistry, 0–200 m, in the Eastern Scotia Sea during FIBEX (January –March 1981). Tech Rep 82-11-T, Dept Oceanogr Texas A&M Univ pp 98

El-Sayed SZ, Biggs DC, Holm-Hansen O (1983) Phytoplankton standing crop, primary productivity and near-surface nitrogenous nutrient fields in the Ross Sea, Antarctica. Deep-Sea Res 30:871 –886

Fujino K, Lewis EL, Perkins RB (1974) The freezing point of seawater at pressures up to 100 bars. J Geophys Res 79:1792–1797

Glibert PM, Biggs DC, McCarthy JJ (1982) Utilization of ammonium and nitrate during austral summer in the Scotia Sea. Deep-Sea Res. 29:837–850

Holm-Hansen O (1985) Nutrient cycles in Antarctic marine ecosystems. In: Siegfried WR, Condy PR, Laws RM (eds) Antarctic nutrient cycles and food webs (Proceedings of the 4th SCAR symposium on Antarctic biology). Springer, Berlin Heidelberg New York

Holm-Hansen O, Lorenzen CJ, Holmes RW, Strickland JDH (1965) Fluorometric determination of chlorophyll. J Cons Int Explor Mer 30:3–15

Jacobs SS, Amos AF, Bruchhausen PM (1970) Ross Sea oceanography and Antarctic bottom water formation. Deep-Sea Res 17:935–962

Jacobs SS, Gordon AL, Ardai JL (1979) Circulation and melting beneath the Ross Ice Shelf. Science (Wash DC) 203:439–443

Jacobs SS, Haines WE (1982) Oceanographic data in the Ross Sea and along George-V Coast, 1976–1979. Tech Rep LDGO-82-1 Lamont-Doherty Geol Observ pp 505

Johnson MA (1981) Pumped profiles of ammonium and chlorophyll fluorescence from the upper 120 meters of the western Gulf of Mexico and the southwest Scotia Sea. Unpublished MS Thesis, Texas A&M University pp 100

Koike I, Ronner U, Holm-Hansen O (1981) Microbial nitrogen metabolism in the Scotia Sea. Antarctic J U.S Ann Rev Iss 16:165–166

Olson RJ (1980) Nitrate and ammonium uptake in Antarctic waters. Limnol Oceanogr 25:1064–1074

Oradovskiy SG (1974) Investigation of the chemical composition of antarctic sea ice. Oceanology (Moscow) 14:50–54 (English translation)

Parker BC, Zeller EJ (1979) Nitrogenous chemical composition of antarctic ice and snow. Antarctic J U.S. Ann Rev Iss 14:80–82

Ronner U, Sorensson F, Holm-Hansen O (to be published) Nitrogen assimilation by phytoplankton in the Scotia Sea. Polar Biology

Stirling I (1982) The biological importance of ice edges. EOS 63:46 –47

Growth and Production of the Seaweed, *Himantothallus grandifolius*, at King George Island

G. DIECKMANN[1], W. REICHARDT[2], and K. ZIELIŃSKI[3]

Summary. Growth of *Himantothallus grandifolius* was measured *in situ* for 32 d during late summer in Admiralty Bay, King Georg Island, Antarctica. Plants grew at a mean rate of 6.0 mm d^{-1} which is comparable with that of laminarians in southern temperate waters for the same time of the year. Conversion of length to biomass showed that the plants increased their biomass by approximately 22% during this period. This increase in biomass was almost negated by loss of frond material, due to constant erosion at the frond tips which are colonized by bacteria which promote erosion. Eroded material enters the detritus food chain and probably supports a rich and diverse coastal fauna.

1 Introduction

In contrast to Antarctica's sparse land flora, the region's coastal waters support large and spectacular submarine "forests" of brown and red algae which have only begun to be explored (Zanefeld 1968; Moe and Silva 1977; Ramirez 1982). *Himantothallus grandifolius* is a large kelp-like alga which appears to have a circumpolar distribution (Skottsberg 1964; Skottsberg and Neushul 1960), but about which otherwise little is known. The species, especially when represented by single bladed specimens, resembles *Laminaria longicruris* and *L. saccharina* in gross morphological appearance , but attains larger dimensions and also differs in other respects (Neushul 1963; Moe and Silva 1981). Until recently, its taxonomic status was unclear and it was assigned a provisional place among the Laminariaceae by Skottsberg and Neushul (1960). However, Moe and Silva (1981) have placed it definitively in the Desmarestiaceae (Desmarestiales).

In temperate waters throughout the world, large kelps have been recognized as important primary producers surpassing the production of phytoplankton in the areas where they occur (Mann 1973). Kelp-beds appear to function as distinctive ecosystems with recognizable food chains which in many cases lead to economically important animals, such as mussels and lobsters (Field et al. 1977). These systems also export large amounts of organic matter (detritus) which settles in deeper water and provides much of the energy-rich material on which benthic communities depend (Field et al. 1980a; b; Stuart et al. 1981). Colonization of algae by bacteria, and a rich and diverse benthic animal community observed in association with the algae in the Antarctic, leads to the speculation that the Antarctic coastal ecosystems may be structured similarly to kelp-bed ecosystems elsewhere.

The potential significance of macrophytic algae in the Antarctic has been pointed out by several (Neushul 1968; Zanefeld 1968; Dell 1972), but very little is known about their roles in the coastal ecosystems of the Antarctic and Sub-Antarctic. Previous reports dealing with algae in the Antarctic have been mainly of a descriptive nature (Papenfuss 1964; Delepine et al. 1965; Neushul 1968; Zanefeld 1966, 1968; Moe and Silva 1977, 1981), since ecological and physiological studies are usually handicapped by the severe conditions encountered in these regions. Our study, using rather simple techniques, attempted to determine the patterns of growth and production in *H. grandifolius,* in Admiralty Bay, and to establish a foundation for future investigations of the roles of macrophytic assemblages in Antarctic coastal ecosystems.

2 Study Area, Materials, and Methods

A detailed description of the hydrology of Admiralty Bay (62° 09′S, 58° 28′W), on the eastern side of King George Island, is given by Rakusa-Suszczewski (1980).

Himantothallus grandifolius is a perennial, sublittoral alga which grows at depths between 30 and 100 m in Admiralty Bay where it seems to form dense beds. Next to the dominant *Desmarestia* species it is probably one of the most important algae, as far as biomass is concerned (Moe and Silva 1977). Gepp and Gepp (1912), Skottsberg and Neushul (1960) and Moe and Silva (1977, 1981) give detailed descriptions of the plant. In contrast to their observations, however, and those of Zinova (1959) who found

1 Alfred-Wegener-Institut für Polarforschung, Columbus-Center, 2850 Bremerhaven, FRG
2 Institut für Meereskunde, Düsternbrooker Weg 20, 2300 Kiel, FRG
3 Department of Polar Research, Institute of Ecology, Polish Academy of Sciences, Dziekanów Leśny, 05–150 Lominanki, Poland

Antarctic Nutrient Cycles and Food Webs
(ed. by W.R. Siegfried, P.R. Condy, and R.M. Laws)

multiple branched specimens, most (69%) of the some 200 plants collected during the course of our investigation, were single bladed. Twenty-two per cent of the plants had two, and nine had three, blades. No plant with more than three blades was found. The largest plant dredged had two blades, 16.7 and 17.5 m long, with a maximum width of 0.90 m and a combined mass of 33 kg. These dimensions surpass those recorded in the literature. All large intact plants, longer than approximately 6 m, were fructifying at the distal ends of the blades.

Specimens of *H. grandifolius* were obtained from depths between 20 and 60 m using an anchor dredge. Twenty-five whole, undamaged plants, each with a single blade 0.2–10 m long, were placed in large barrels filled with seawater and immediately transported into shallow water for processing. The holdfasts were secured to a thick (80 mm) nylon rope 15 m long, using plastic-coated wire to which tags had been attached. Two anchors and a buoy were tied to the rope.

Measurements and punching of the plants for growth studies were done according to the method of Mann et al. (1979), which is based on that of Parke (1948). Holes (∅ = 5 mm) were punched with a cork-borer at 0.5 m intervals along the central axis of the blade, beginning at the junction with the stipe. Care was taken not to expose the plants to air during handling. The rope, together with the plants, was towed carefully into the bay and lowered to 25 m where it remained for 32 d between February and the beginning of March. Subsequently, the plants were taken into the laboratory where the distances moved by the punched holes along the length of the blades were recorded. Length increments were converted to biomass increments, using a regression of length to fresh mass obtained from 106 plants.

To determine percentage dry mass and organic content of different sections of a plant, an intact 15 m specimen was fractionated into holdfast, stipe and 1 m blade portions. The first 1 m was divided into two parts. Dry mass of three replicate subsamples per fraction was obtained by freeze drying, and the organic content was determined after combusting the dried material in a muffle furnace at 550°C for 4 h.

3 Results

Of the 25 plants punched and measured, 17 were recovered after 32 d in situ. All the blades showed substantial growth increments. Elongation occurred from the base to approximately half way up the length of the blades, with the rate of elongation decreasing progressively towards the tip (Fig. 1). The sum of the growth increments, as indicated by the distal displacement of the holes, gives the amount of new growth in terms of length. Although the largest plant had the greatest length increment, growth rate of the other plants could not be related to plant size (Table 1). The mean growth rate (increment in length) was 6.0 mm d^{-1},

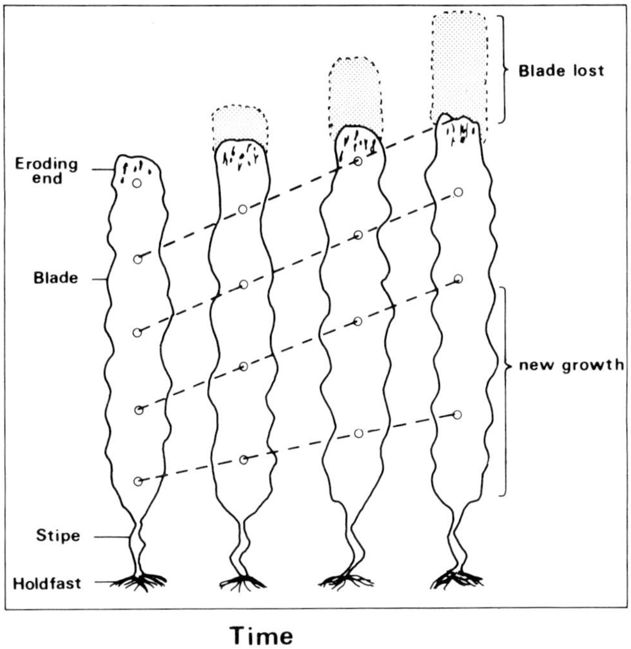

Fig. 1. General pattern of growth in mature *H. grandifolius*

which is the length that the plants would have grown had there been no erosion at the tips. Thus, it is possible to record the amount of tissue added, despite erosion of the ends having taken place. The amount of tissue lost can be determined by simple subtraction.

Changes in the percentage dry mass and organic matter in a mature plant give a further indication of the growth pattern observed (Fig. 2). Percentage dry mass and consequently organic matter were high in the stipe, as a result of its compact structure. It was not possible to determine the dry mass of the holdfast, because adhering particles could not be removed. The base of the blade showed a marked decrease in both dry mass and organic matter, compared to the stipe. This decrease in both parameters reached the lowest levels in the proximal 2–4 m of the blade. Both parameters increased up the blade, and with increasing age. The peak at the 5 m section (Fig. 2) is possibly due to some irregularity. Highest values were recorded near the distal end of the blade. At the tip, the area where the blade begins to senesce and disintegrate, dry mass and organic matter were again reduced; the organic matter decreasing prior to the dry mass.

As *H. grandifolius* blades increase in length, they become wider and thicker, so that the length to mass relationship is not linear. It is described by a power curve (Fig. 3). Using the equation

$$M = 0.0123L^{1.950}$$

(where M = Mass and L = Length), it is possible to convert each observation on initial length and subsequent length increment to a corresponding biomass increment. The amount of tissue lost by the plant can be quantified in a

Table 1. Length increments and calculated production of *H. grandifolius* plants over a period of 32 days during February and March 1983

Initial blade length mm	Length increment mm	Daily length increment mm	Calculated mass increment g	Calculated % increase in mass
200	50	2.0	2	50
400	70	2.0	6	38
860	240	8.0	45	62
1000	150	5.0	31	32
1030	160	5.0	34	55
1800	90	3.0	31	10
1830	120	4.0	42	13
1900	60	2.0	21	6
1930	140	4.0	51	14
2220	280	9.0	120	26
2600	90	3.0	43	7
3500	120	4.0	76	7
3600	280	9.0	187	16
3650	230	7.0	154	13
4000	100	3.0	72	5
4550	200	6.0	164	9
9100	640	20.0	1026	14
Mean	180	6.0	–	22

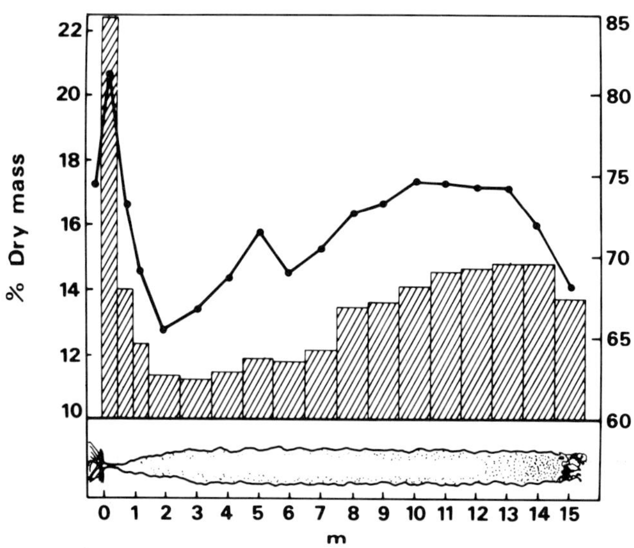

Fig. 2. Dry mass as a percentage of (wet wt.) mass (*histograms*), and organic matter as a percentage of (dry wt.) mass (*graph*) of a large *H. grandifolius* plant in relation to its length (*m*)

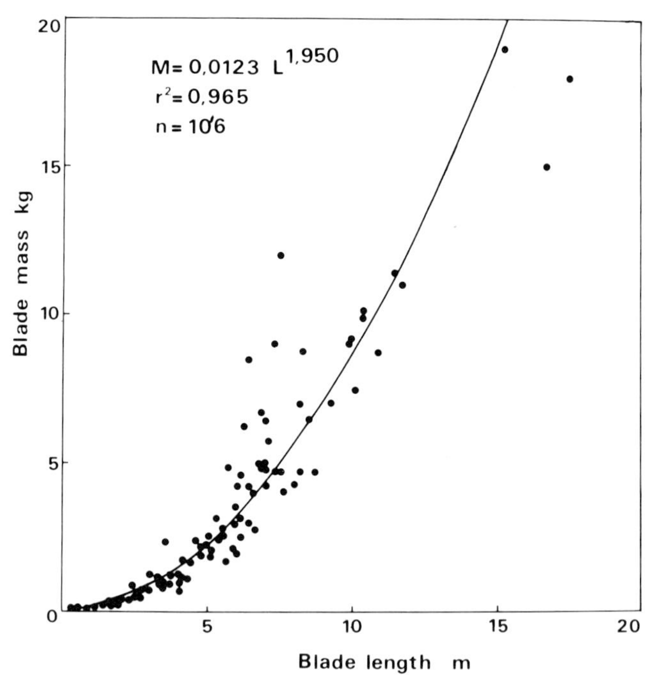

Fig. 3. Relationship between blade length (*m*) and (wet wt.) mass (*kg*) in *H. grandifolius*

similar way. Thus, the blades increased their biomass by a mean of 22% during the month of February (Table 1). The amount of tissue lost was not calculated.

4 Discussion

The pattern of growth in mature *H. grandifolius* plants is similar to that of laminarians, such as *L. digitata* and *L. longicruris* (Mann 1973) and *L. pallida* (Dieckmann 1980). There is a difference, however, in that the zone of elongation is not localized at the base of the blade, but is spread over approximately half of its length. Rate of elongation is fastest near the base and decreases progressively toward the tip. A similar growth pattern was observed in *Ecklonia maxima* by Mann et al. (1979).

The observed changes in percentage dry mass and organic matter in the blade are a clear reflection of the growth pattern observed. Thus, the base, because it has the intercalary meristem and is the area of active cell division and growth, has a higher organic content and, hence, higher dry mass than the adjacent blade tissue which represents the zone of cell enlargement and elongation. During this process water is taken up, resulting in a decrease in both parameters. The subsequent progressive increase in dry mass and organic matter towards the distal end of the blade is a result of cell maturation which is accompanied by accumulation of storage products and differentiation into reproductive tissue. The decrease at the tip is most likely caused by leaching of organic matter from senescing and disintegrating tissue (Newell et al. 1980; Stuart et al. 1981). Zieliński (1981) found a similar variation in the calorific value of different *H. grandifolius* sections.

Growth rates and production by *H. grandifolius* during the study period are surprisingly high, and are comparable with those of laminarians growing in temperate waters. *Laminaria pallida,* for example, has a frond elongation rate of 30–70 mm d^{-1} during February in South Africa, depending on depth of water (Dieckmann 1980). This represents a biomass increase of about 20–45%. However, February is late summer in the southern hemisphere, and for *L. pallida* late summer is a period of slow and decreasing growth rates when compared with those in the spring and early summer. It, therefore, is reasonable to assume that *H. grandifolius* also enters a late-summer phase of slow growth rates, in relation to a faster growth during spring and summer. While *L. pallida* grows under apparently optimum environmental conditions (Dieckmann 1980; Field et al. 1980a), *H. grandifolius* grows under harsh conditions which do not appear to be conducive to growth. However, judging by its luxuriant growth and large dimensions it seems to be perfectly adapted for a high growth rate. Furthermore, nutrients are not limiting in Antarctic waters (El-Sayed 1968; Rakusa-Suszczewski 1980) and light is available in excess during the summer. These conditions, together with the constantly low temperatures of the seawater, must favour a high ratio of photosynthesis to respiration during the spring and summer.

This investigation has answered certain questions attending the pattern of growth in *H. grandifolius,* but it has also revealed the paucity of information on environmental factors which influence algal growth in the Antarctic. Although the method applied to measure productivity of *H. grandifolius* is satisfactory, it is necessary to follow growth for at least 1 yr. Plants should be monitored continuously during the summer months and once before, and immediately after, winter. For data on standing stock, it will be necessary to carry out SCUBA surveys or ship-based underwater photographic surveys.

Despite the lack of information on the standing stock of *H. grandifolius,* it is interesting to calculate, as an example, the production by a single plant for the period of growth as measured. Since blades increased their biomass by 22% during the period of 32 d in late summer, a plant of 10 m length, with a mass of 10 kg (determined from the regression equation), increases its mass by 2.2 kg or 325.6 g (dry wt.) mass or 234.4 g organic matter with a calorific value of 4,000 kJ, as calculated from mean conversion factors given by Zieliński (1981). This gives an indication of the large production potential of these plants.

The occurrence of a rich and diverse benthic faunal community in association with algae in Admiralty Bay (Rakusa-Suszczewski 1980), and other coastal regions of the Antarctic (Dell 1972), points to the existence of a detritus-based food chain. Several of the macrophytic algae in the Antarctic are perennial. Thus, despite the probability of little or no growth occurring during winter, the plants still produce detritus due to constant erosion of the blade ends during months when little or no phytoplankton production occurs. Recent studies of the importance of detritus in Antarctic coastal waters (Dawson et al. 1985; Reichardt and Dieckmann 1985) have shown that the rate of bacterial decomposition of algal debris is comparable to that of temperate marine systems. Hence, it is probable that the Antarctic shallow coastal ecosystems is based at least in part on a detritus foodchain which is structured, and functions, similarly to those in temperate and sub-tropical coastal regions.

Acknowledgements. Two of us (G.D. and W.R.) are grateful to the Polish Academy of Sciences, and in particular Professor Rakusa-Suszczewski, for allowing us to work at "Arctowski" Station and for assistance. The work was supported by the Deutsche Forschungsgemeinschaft, Grant Nr. Schr. 143/7-1.

References

Dawson R, Schramm W, Bölter M (1985) Factors influencing the production and decomposition of organic matter in Admiralty Bay, King George Island. In: Siegfried WR, Condy PR, Laws RM (eds) Antarctic nutrient cycles and food webs (Proceedings of the 4th SCAR symposium on Antarctic biology. Springer, Berlin Heidelberg New York

Delepine R, Mackenzie-Lamb I, Zimmermann MH (1965) Preliminary report on the marine vegetation of the Antarctic Peninsula. In: Young C, Mclachlan TC (eds) Proceedings of the 5th Int Seaweed Symposium Pergamon, London pp 107–116

Dell RK (1972) Marine Algae. In: Russell FS, Young MS (eds) Advances in marine biology, 10 Academic, London pp 136–152

Dieckmann G (1980) Aspects of the Ecology of *Laminaria pallida* (Grev) J Ag. off the Cape Peninsula (South Africa) I. Seasonal Growth. Bot Mar 23:579–585

El-Sayed SZ (1968) Primary productivity. Am Geogr Soc Antarct Map Folio Series 10:1–6

Field JG, Jarmann NG, Dieckmann GS, Griffiths CL, Velimirov B, Zoutendyk P (1977) Sun, waves, seaweed and lobsters: The dynamics of a west-coast kelp bed. S Afr J Sci 73:7–10

Field JG, Griffiths CL, Griffiths RJ, Jarman N, Zoutendyk P, Velimirov B, Bowes A (1980a) Variation in structure and biomass of kelp communities along the south-west cape coast. Trans R Soc S Afr 44(2):145–203

Field JG, Griffiths CL, Linley EAS, Carter RA, Zoutendyk P (1980b) Upwelling in a nearshore marine ecosystem and its biological implications. Estuarine Coastal Mar Sci 11:133–150

Gepp A, and ES (1912) Marine algae of the Scottish National Antarctic Expedition. Rep Res Scott Nat Antarct Exped (Botany) 3:73–83

Mann KH (1973) Seaweeds, their productivity and strategy for growth. Science 182:975–981

Mann KH, Jarman N, Dieckmann G (1979) Development of a method for measuring the productivity of the kelp *Ecklonia maxima* (Osbeck) Papenf. Trans R Soc S Afr 44(1):27–41

Moe RL, Silva P (1977) Antarctic marine flora: Uniquely devoid of kelps. Science 196:1206–1208

Moe RL, Silva P (1981) Morphology and taxonomy of *Himantothallus* (including *Phaeoglossum* and *Phyllogigas*), an Antarctic member of the Desmarestiales (Phaeophyceae). J Phycol 17:15–29

Neushul M (1963) Reproductive morphology of Antarctic kelps. Bot Mar 5:19–24

Neushul M (1968) Benthic marine algae. Am Geogr Soc Antarct Map Folio Series 10:9–10

Newell RC, Lucas MI, Velimirov B, Seiderer LJ (1980) Quantitative significance of dissolved organic losses following fragmentation of kelp (*Ecklonia maxima* and *Laminaria pallida*). Mar Ecol Prog Ser 2:45–59

Papenfuss GF (1964) Catalogue and bibliography of Antarctic and subantarctic benthic marine algae. Am Geophys Union, Ant Res Ser 1:76 pp

Parke MW (1948) Studies in British laminariaceae. I. Growth in *Laminaria saccharina* (L) Lamour. J Mar Biol Assoc UK 27:651–109

Rakusa-Suszczewski S (1980) The role of near-shore research in gaining an understanding of the functioning of Antarctic Ecosystem. Pol Arch Hydrobiol 27(2):229–233

Ramirez ME (1982) Catalogo de las algas marinas del Territorio Chileno Antarctico. INACH, Ser Cient 29:39–67

Reichardt W, Dieckmann G (1985) Kinetics and trophic role of bacterial degradation of macroalgae from Antarctic coastal waters. In: Siegfried WR, Condy PR, Laws RM (eds) Antarctic nutrient cycles and food webs (Proceedings of the 4th SCAR symposium of Antarctic biology). Springer, Berlin Heidelberg New York

Skottsberg C (1964) Antarctic physiology. In: Carrick R, Holdgate M, Prevost J (eds) Biologie Antarctique. Hermann Paris pp 147–154

Skottsberg C, Neushul M (1960) *Phyllogigas* and *Himantothallus* Antarctic Phaeophyceae. Bot Mar 2:164–173

Stuart V, Lucas MI, Newell RC (1981) Heterotrophic utilization of particulate matter from the kelp *Laminaria pallida*. Mar Ecol Prog Ser 4:337–348

Zanefeld JS (1966) Vertical zonation of Antarctic and subantarctic benthic marine algae. Antarct J of US 1:211–213

Zanefeld JS (1968) Benthic marine algae, Ross Island to Balleny Islands. Am Geogr Soc Antarct Map Folio Series 10:10–12

Zieliński K (1981) Benthic macroalgae of Admiralty Bay (King George Island, South Shetland Islands) and circulation of algal matter between the water and the shore. Pol Polar Res 2(3–4): 71–94

Zinova AD (1959) On two brown algae from Antarctica, *Phyllogigas* and *Himantothallus*. Bot J Acad of Science USSR T XLIV

Factors Influencing the Production, Decomposition and Distribution of Organic and Inorganic Matter in Admiralty Bay, King George Island

R. DAWSON[1], W. SCHRAMM[2], and M. BÖLTER[3]

Summary. Four major sources of dissolved and particulate detritus interact to determine the composition of organic matter in Admiralty Bay: 1. production of dissolved and particulate matter (especially sugars, DOC), from extensive beds of benthic macrophytes via exudation and pulp formation from eroded intertidal zones during storms; 2. land run-off of nitrogencycle waste products from penguins and seals; 3. freshwater contributions from glacier-melts, carrying high loads of re-mineralized inorganic nutrients and particulate detritus; and, 4. in situ phytoplankton primary production and associated modification by secondary consumers. The parameters monitored under field conditions included inorganic nutrients, salinity, temperature, urea, DOC, POC, PON, chlorophyll-*a*, ^{14}C-primary production, dissolved mono- and polysaccharides, individual and combined amino acids, and particulate soluble and structural sugars. A detailed epi-fluorescence microscopical study of bacterial cell numbers, size classes and degree of attachment to particles, attempted to relate the heterotrophic interactions to the source and 'quality' of particulate and dissolved detritus.

1 Introduction

During January–February 1981, a study was made of the production of benthic macrophytes, and the release of dissolved and particulate material from these and other sources, in Admiralty Bay, King George Island, South Shetland Islands. The major sources of dissolved and particulate matter contributing to Admiralty Bay's inorganic and organic budget include the primary production of phytoplankton, land run-off from penguin colonies and seal wallows, infusion of glacier melts during summer, and contributions from extensive areas of macro-algal beds which fringe Admiralty Bay down to a depth of 90 m. The total algal cover accounts for 11.5% of the area, with maximal concentrations in an area of 3.5 km^2 close to Point Thomas (Fig. 1) (Furmanczyk and Zielinski 1982). During February–October 1979, the total quantity of algal matter washed ashore was estimated at 279 T dry wt. (Zielinski 1981).

1 IOC Assistant SEcretary, IOC/UNESCO, 7 Place de Fontenoy ·n-75700 Paris, France

2 Institut für Meereskunde an der Universität Kiel, Düsterbrookerweg 20, 2300 Kiel 1, FRG

3 Institut für Polarökologie der Universität Kiel, Olshausenstr. 40/60, 2300 Kiel 1, FRG

Laboratory experiments, involving exudates and pulps of the dominant algal species (Dawson and Schramm, unpubl.), showed that these species were potential suppliers of large amounts of dissolved sugars, particulate polysaccharides and DOC, in general, to secondary consumers in the food chain. The algae were found to be rich in mineral components and structural polysaccharides, but other organic compounds (e.g., proteins, lipids, pigments) occurred in quantities much smaller than in similar species in temperate marine environments (Czerpak et al. 1981).

The main aim of the investigation reported on here was to identify gradients in the biological and chemical properties of Admiralty Bay in relation to its different watermasses. Hence, we attempt to characterize the nutrient sources and their distributional gradients in this fjord. A distinction is made between the different classes of organic components in particulate and dissolved organic matter, and the results are discussed in relation to an investigation carried out 2 months earlier in the Bransfield Strait (Bölter and Dawson 1982).

2 Materials and Methods

All samples were taken at 1 m depth with 5 l (General Oceanics) water samplers; locations of sampling stations are shown in Fig. 1. The samplers were rinsed with 1 N HCl and copiously washed with distilled water. A detailed description of the analytical methods for the determination of carbohydrates is given by Bölter and Dawson (1982). The basic procedure followed the recommendations of Johnson and Sieburth (1977), Burney and Sieburth (1977) and Dawson and Liebezeit (1981, 1983). Based on these methods, we analyzed dissolved monocarbohydrates (DMCHO), dissolved polysaccharides (DPCHO), extractable monocarbohydrates of particulate matter (EMCHO) and extractable polysaccharides (EPCHO). Dissolved organic C (DOC) was analyzed according to the method of Schreurs (1978). Chlorophyll-a was determined by the trichromatic method of Jeffrey and Humphrey (1975). Inorganic nutrients and urea were analyzed according to the recommendations in Grasshoff et al. (1983). Salinity and temperature

Antarctic Nutrient Cycles and Food Webs
(ed. by W.R. Siegfried, P.R. Condy, and R.M. Laws)

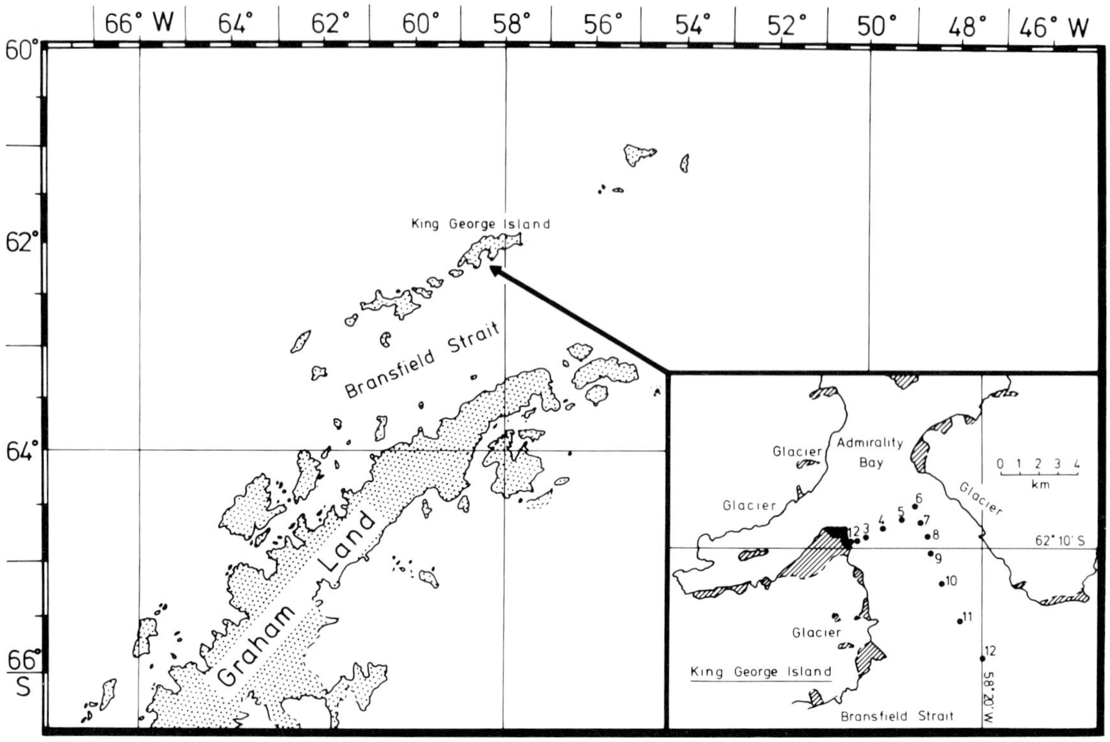

Fig. 1. Sampling Stations 1–12 in Admiralty Bay, King George Island. The area (shown in *black*) of dense macrophyte cover is close to Point Thomas. The *hatched areas* are ice-free zones supporting colonies of penguins and seal wallows

were measured, in situ, by probes. Particulate organic C (POC) and N (PON) were analyzed with a Hewlett-Packard CHN-analyzer (185B) (Ehrhatdt 1976). The estimation of primary production follows the method of Steemann-Nielsen (1952). Microbial cell counts were performed by the epifluorescence technique, according to Zimmermann et al. (1978). All chemical analyses (with the exception of the DOC analyses) were performed at the Polish Arctowski Station, within a few hours after sampling. The statistical analysis was carried out according to the scheme of Bölter et al. (1980), with refinements as given by Bölter and Meyer (1983).

3 Results and Discussion

A first impression of the nutrient load of the pelagic system in Admiralty Bay indicates that this system functions at high levels of both organic and inorganic nutrients. An investigation in the adjacent Bransfield Strait showed comparable values for inorganic nutrients (e.g., Bodungen et al. (1981) found concentrations of phosphate of 2.5 μg at. 1^{-1}, and most of the N in the form of nitrate). In the present study, we found ca. 80% of the inorganic N in the form of nitrate. The measurements of chlorophyll-a produced results similar to those reported by Bodungen et al. (1981) and Bölter and Dawson (1982) for the Bransfield Strait, some 2 months earlier, with values ranging between 5 and 11 μg 1^{-1} or even higher.

The corresponding values for dissolved and particulate organic C fractions are also comparable to the results obtained for the Bransfield Strait. Bölter and Dawson (1982) found the extractable carbohydrates of particulate matter to lie between 12 and 77 μg 1^{-1} (at Station 73, near King George Island) and, during intense phytoplankton blooms, values rose to over 300 μg 1^{-1} (EPCHO) in the surface waters. These agreements are also true for the determinations of DOC and dissolved monosaccharides (DMCHO). However, the ratios of extractable monosaccharides to polysaccharides (EMCHO:EPCHO) are much more constant within the pelagic system of the open Bransfield Strait than in Admiralty Bay (Table 1) and, thus, point to difference in the source material. A similar pattern in the ratios DMCHO:DPCHO is evident, with the exception of Station 4 and 5 (mean values are EMCHO:EPCHO = 0.74 ± 0.3; DMCHO:DPCHO = 0.9 ± 0.2). This indicates that a relatively higher proportion of the sugars in the dissolved pool are monomeric in character and, thus, potentially available to microheterotrophs.

3.1 Correlation Matrices

A correlation analysis of the individual chemical and biological determinations was made to provide insights into the structure of the Admiralty Bay ecosystem. The correlation matrix was subjected to a cluster analysis, which resulted in several groups of interacting parameters displaying similarities at the 1% and 5% levels. These clusters, describing bio-

Table 1. Inorganic and organic nutrients and primary production values at Stations 1–12 in Admiralty Bay

Station no.	Nitrite μmol l^{-1}	Urea μmol l^{-1}	DOC mgC l^{-1}	EMCHO μg l^{-1}	DMCHO μg l^{-1}	POC μg l^{-1}	PON μg l^{-1}	C/N	Prim. Prod. mgC l^{-1} h^{-1}	Chloroph. μg l^{-1}	EMCHO/ EPCHO	DMCHO/ DPCHO
1	0.2	1.09	2.01	29.3	257	790	105	7.5	0.026	1.3	0.48	0.80
2	0.23	1.64	5.76	35.0	300	770	125	6.2	–	8.0	0.51	0.80
3	0.2	1.36	1.20	19.4	219	1030	117	8.8	0.036	8.9	0.39	0.63
4	0.2	1.39	2.42	25.6	253	540	95	5.7	0.033	11.1	1.14	0.90
5	0.18	1.57	2.36	27.6	267	770	130	5.9	0.053	8.5	1.30	0.86
6	0.21	1.47	1.61	26.6	270	670	105	6.4	0.080	3.2	0.90	0.92
7	0.14	0.70	2.28	21.9	210	450	80	5.6	0.040	0.5	0.48	0.60
8	0.17	0.55	1.59	24.6	249	510	80	6.4	0.032	6.7	0.76	0.87
9	0.22	0.56	2.86	28.9	390	390	68	5.8	0.050	6.4	0.96	1.27
10	0.18	0.76	1.46	21.6	289	460	70	6.1	0.049	8.1	0.63	1.05
11	0.19	1.08	1.67	27.4	311	570	90	6.3	0.023	7.4	0.71	1.16
12	0.19	0.83	5.63	27.3	330	500	90	5.6	0.060	12.5	0.59	0.95

logical and chemical interactions, were examined in the light of geographical patterns and in relation to known water movements and sources of inorganic and organic nutrient contributions.

The first significant cluster existed between nitrate, total inorganic N, phosphate, temperature and the proportion of rod-shaped bacteria in the size class 1–3 μm. Figure 2 shows the course of nitrate and phosphate in relation to Stations 1–12: both nutrients initially have high levels which then decrease as the sea becomes more open. As mentioned above, nitrate represents, on average, about 80% of the total inorganic N. The ratio of the phosphate and the total inorganic N fluctuates around a value of 0.09. The two decreases for nitrate, as shown for Stations 1–3 and 11–12, respectively, had different origins: at Station 1–3 the decrease probably reflects the increasing distance of the terrestrial source, whereas the decrease at Stations 11–12 is thought to be due to different water-masses (discussed in detail below). The close correlation between nitrate and phosphate and their low variations indicates that they are not limiting for the plankton community. Nitrite and ammonia showed much higher fluctuations, indicating their rapid turnover.

The distribution of the fractions of the microbial population reflects the pattern described above. Concomitant with the decreasing nutrient levels at Stations 1–4, an apparent initial decrease occurs in the proportion of rod-shaped bacteria (1–3 μm) (49.5% at Station 1, 43.8% at Station 2), which decreases further to an average 20% at the other stations. Such correlations are less pronounced for the total cell number and the fraction attached to detritus (Fig. 3). An increase occurs in total cell numbers at Stations 7 and 8, which, however, remains unreflected in the course of other parameters.

A second group of correlating parameters comprises urea, POC, PON, and the ratio C:N (Table 1). This ratio remains relatively constant, with a mean value of 6.1 ± 0.6 (except at Station 3 where C:N = 8.77), suggesting a uniform distribution of the components of the seston in the environment. The urea profile points to the influence of penguin

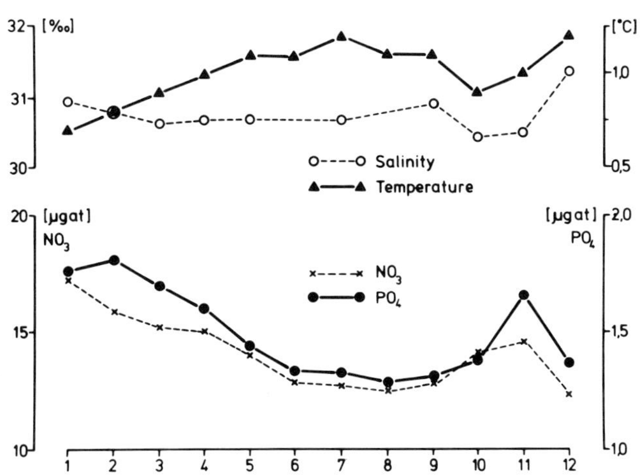

Fig. 2. Temperature, salinity, nitrate and phosphate at Stations 1–12 in Admiralty Bay

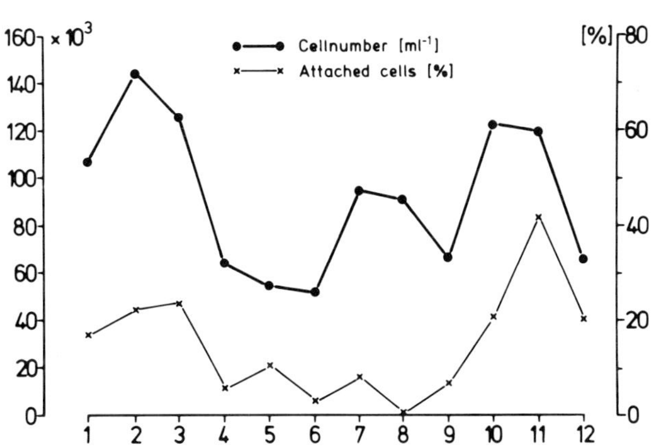

Fig. 3. Total cell number and the fraction of bacteria attached to particles at Stations 1–12 in Admiralty Bay

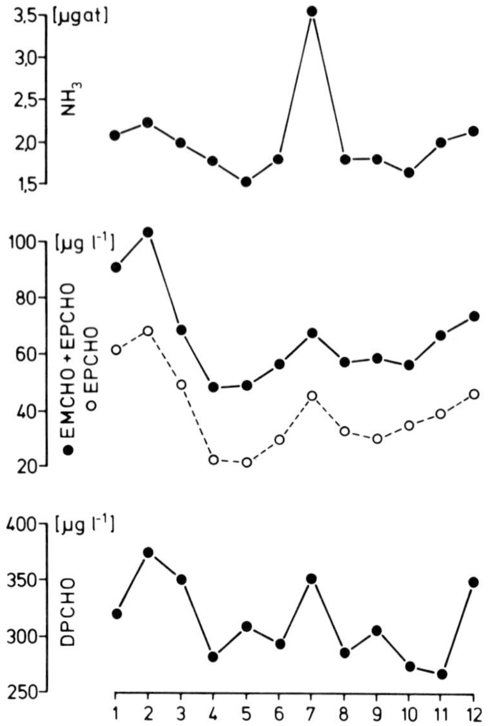

Fig. 4. Ammonia (NH_3) and different fractions of sugars (see text for abbreviations) at Stations 1–12 in Admiralty Bay

colonies (i.e., a terrestrial origin). However, the urea concentrations are not correlated with the trends of other inorganic nutrients.

Ammonia is clustered with the parameters describing the dissolved and particulate polysaccharide components (EPCHO, DPCHO, EPCHO + EMCHO) as is shown in Fig. 4. These carbohydrates display a distinct decrease with increasing distance from land, or, alternatively, from the macrophyte fringe at Station 1–3. At Station 7, ammonia increased to 3.53 μg at. l^{-1}, a peak which remains unreflected by other parameters or for which correlations are masked by other trends. Although increases were found in the total count of bacteria (Fig. 3), temperature, DOC, PCHO, DMCHO, nitrite and chlorophyll-a, their amplitudes reflect this poorly.

With comparatively low values of the ratio EMCHO: EPCHO at Stations 1–3, it can be argued that the particulate material at these locations contained only small amounts of extractable sugars (i.e., the organic nutrients especially glucose) which are directly assimilable by osmotrophic micro-organisms (Bölter 1981)). On the other hand, high amounts of dissolved polysaccharides and DOC were present at these stations, whereas dissolved monosaccharides (DMCHO) levels were minimal and, hence, are indicative of macrophyte detritus in an early stage of decomposition.

In general, the ratio of dissolved monosaccharides to polysaccharides (DMCHO:DPCHO) is in excess of 0.8:1 (1.27:1 at Station 9). Notable exceptions occurred at Stations 3 and 7 (0.63 and 0.60, respectively) where the dissolved polysaccharides dominated the sugar content of the seawater. The high value of DOC at Station 12 is also likely to be due to polysaccharides (EMCHO:EPCHO = 0.59). Exceptionally high values (> 7 mg DOC l^{-1}) were measured in the Bransfield Strait during intensive phytoplankton blooms. Much of this material, however, was considered to be colloidal in nature and originated from algal slimes (Bölter and Dawson 1982).

An unsatisfactory correlation appears to exist between the fraction of large rod-shaped bacteria (> 3 μm) and the course of nitrite: This finding is reminiscent of correlations between saprophytic bacteria, total microbial biomass and the number of active filamentous bacteria in a shallow water ecosystem (Bölter et al. 1981; Bölter and Meyer 1983), although both parameters represent only a small fraction of the total microbial population and the N-nutrients, respectively. Other fractions of the microbial population show only a few, if any, significant relationships with environmental conditions; the fluctuations being small (cocci < 0.5 μm = 2% ± 0.8 of the total number; cocci 0.5 –1.0 μm = 26% ± 4.5; rods < 1 μm: 33.4% ± 6).

Parameters describing phytoplankton activity, chlorophyll-a and primary production did not cluster during our analysis. The individual figures are given in Table 1. The trends for surface plankton run independently to those of any of the other parameters measured. Unfortunately, we have no information on zooplankters and their feeding activity which could be helpful in explaining the lack of correlations.

3.2 Zonation

The results generated by the transposed correlation matrix and the subsequent cluster analysis are summarized in Fig. 5 which partitions Admiralty Bay into five zones, these embrace Stations 1–3, 4–6, 7–8, 10–11 and 9, and 12, respectively. Stations 1 and 2 (and 3) are directly influenced by the phytobenthos close to the shore. They are characterized by high numbers of bacteria, a large fraction of bacteria attached to particles, and high proportions of rod-shaped bacteria (size class 1–3 μm) High levels of inorganic nutrients and organic matter accompany low ratios of EMCHO:EPCHO. Station 4 and 5 (and 6) appear to reflect the original planktonic system of the fjord. Bacterial cell numbers decrease concomitantly with the section content, as indicated by the lower amounts of POC as well as EPCHO and DPCHO. Cell numbers, phosphate and nitrate at Station 6 are similar to corresponding values in the outer part of the fjord, but the concentrations of urea and nitrite remain elevated. Stations 7 and 8 showed no correlation with each other or with any other station. An area of mixed water-masses, associated with changes in biological processes, could have accounted for effects such as high ammonia values at Station 7, and the maxima of DOC, DPCHO, EPCHO, and the minima of chlorophyll-a and nitrite. Stations 10 and 11 were correlated closely with Stations 1–3. Although this correlation is not significant at the 5 % level,

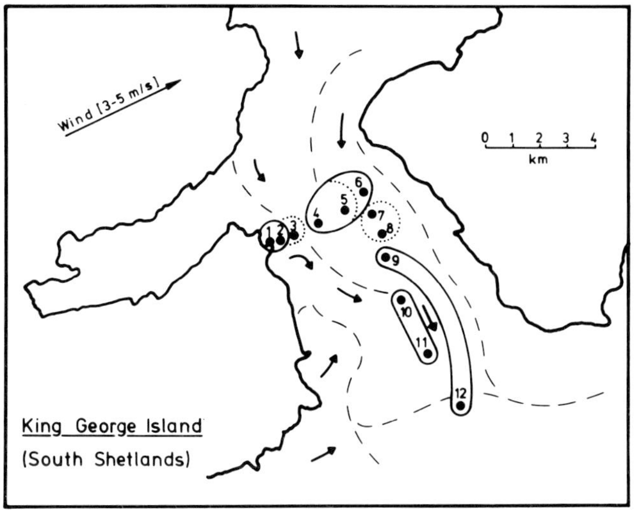

Fig. 5. Ordination of Stations 1–12 in Admiralty Bay, based on cluster analysis. Significant relationships between stations are indicated by *solid lines*; *broken lines* embrace associated stations (< 5% level of significance of association prevailing). Currents (from K. Furmanczyk, pers. comm.) are indicated by *arrows*

it indicates that these two sets of stations may be separated from all other stations by their negative correlation coefficient with them. The connection between Stations 1–3 and 10–11 may be due in part to the current system which transports native particulate (macro-phytobenthos) material. Relatively large amounts of slimy material were observed at Stations 1–3. The significant decrease in salinity at the outer stations points to a confluence with glacier-melts. Finally, Stations 9 and 12 reflect again the relatively stable condition of the open system of the fjord and/or pelagic system of the Bransfield Strait. The connections are documented further by increases in phosphate, nitrate and total bacterial number (Fig. 2 and 3).

4 Conclusions

The zonation presented above helps in understanding how the pelagic system of Admiralty Bay may be perturbed by certain coastal impacts. The influence of the benthic communities fringing Admiralty Bay (and presumably also other coastal areas of the Antarctic region) modifies both the quality and the quantity of organic and inorganic material. Thus, this system (and others similar to it) can be distinguished by its levels of organic and inorganic matter as well as its biological properties. However, the range over which the influence of the benthos occurs may be geographically relatively limited, and a proper estimate of the extent of the area of high benthic production requires a detailed description of the source material and its decomposition products. This rigid distinction between the different pools of organic material may alone be insufficient to trace the export and fate of other cell materials. Ratios of individual compounds (e.g., easily extractable vs. structural material, monosaccharides vs. polysaccharides) provide some holistic insight towards a description of the system, as opposed to mere estimations of the bulk of various compounds, such as DOC or POC. A more accurate "fingerprint" of detrital material may be gained from a detailed analysis of chlorophylls and carotenoid degradation products.

The high levels of nutrients at which ecosystems operate in the Antarctic detract from the usefulness of bulk parameters in modelling both phytoplankton and bacterial behaviour. During this study, the influence of inorganic nutrients on primary production, or on the heterotrophic (bacterial) population, is complicated and shows no clear causality. However, in spite of the levels of nutrients in this particular environment, additional contributions of organic and inorganic materials are reflected by sensitive sub-groups (e.g., the bacterial fraction represented by rods 1–3 μm) which could be used as indicators of such events.

Acknowledgements. We are grateful to E. Moczydlowski and Prof. Birkenmeyer for support during the summer and winter expeditions from Arctowski Station. Prof. G. Hempel made the logistical arrangements. The project was funded by the Deutsche Forschungsgemeinschaft (DFG Schr 143–6). Mrs. U. Oest conducted the bacterial counts, and Mr. W. Dzomla prepared the graphs.

References

Bodungen BV, Fritsche P, Mempel M, Schnack S, Smetacek V, Tilzer M, Uhlmann L, Werner R, Zeitzschel B (1981) Produktion, Akkumulation und Sedimentation von organischer Substanz in Beziehung zur Hydrographie. In: Zietzschel B and Zenk W (eds) Beobachtungen und erste Ergebnisse der "Meteor"-Reise 56 aus der Scotia-See und der Bransfield-Strasse im November/Dezember 1980 (ANT I): Ein nautischer und wissenschaftlicher Bericht. Der Inst Meeresk Kiel 80: pp 41–47

Bölter M (1981) DOC-turnover and microbial biomass production. Kiel Meeresforsch Sonderh 5:304–310

Bölter M, Dawson R (1982) Heterotrophic utilisation of biochemical compounds in Antarctic waters. Neth J Sea Res 16:315–332

Bölter M, Meyer M (1983) The sandy beach area of Kiel Fjord and Kiel Bight (Western Baltic Sea) – A structure analysis of shallow water ecosystems. In: McLachlan A, Erasmus T, (eds) Sandy beaches as ecosystems. Junk Publ Den Haag, pp 263–270

Bölter M, Meyer M, Probst B (1980) A statistical scheme for structural analysis in marine ecosystems. Ecol Model 9:143–151

Bölter M, Meyer-Reil L-A, Dawson R, Liebezeit G, Wolter K, Szwerinski H (1981) Structure analysis of shallow water ecosystems: Interactions of microbiological, chemical and physical characteristics measured in the overlying waters of sandy beach sediments. Estuarine Coastal Shelf Sci 13:579–589

Burney CM, Sieburth JMcN (1977) Dissolved carbohydrates in sea water. II. A spectrophotometric procedure for total carbohydrate analysis and polysaccharide estimation. Mar Chem 5:15–28

Czerpak R, Mical A, Gutkowski R, Siegien I, Jackiewicz I (1981) Chemism of some species of Antarctic macroalgae of the genera Adenocystis, Himantothallus, Leptosomia and Monostroma. Pol Polar Res 2:95–107

Dawson R, Liebezeit G (1981) The analytical methods for characterisation of organics in sea water. In: Duursma EK, Dawson R, (eds) Marine organic chemistry, Elsevier, Amsterdam pp 445–496

Dawson R, Liebezeit G (1983) The determination of amino acids and carbohydrates. In: Grasshoff K, Ehrhatdt M, Kremling K (eds) Methods of sea water analysis. Verlag Chemie, Weinheim, 2nd ed pp 319–345

EhrhatdtM (1976) Determination of particulate organic carbon and nitrogen. In: Grasshoff K (ed) Methods of seawater analysis. Verlag Chemie, Weinheim pp 215–219

Furmanczyk K, Zieliński K (1982) Distribution of macroalgae groupings in shallow waters of Admiralty Bay (King George Island, South Shetland Islands, Antarctic), plotted with the help of air photograph analysis. Pol Polar Res 3:41–47

Grasshoff K, Ehrhatdt M, Kremling K (eds) (1983) Methods of seawater analysis. Verlag Chemie, Weinheim 2nd ed

Jeffrey SW, Humphrey GF (1975) New spectrophotometric equations for determining chlorophylls a, b, c_1 and c_2 in higher plants, algae and natural phytoplankton. Biochem Physiol Pflanz 167: 191–194

Johnson KM, Sieburth JM (1977) Dissolved carbohydrates in seawater. I. A precise spectrophotometric analysis for monosaccharides. Mar Chem 5:1–13

Schreurs W (1978) An automated colorimetric method for the determination of dissolved organic carbon in seawater by UV destruction. Hydrobiol Bull 12:137–142

Steemann-Nielsen E (1952) The use of radio active (^{14}C) for measuring organic production in the sea. J Cons Perm Int Explor Mer 18:117–140

Zieliński K (1981) Benthic macroalgae of Admiralty Bay (King George Island, South Shetland Islands) and circulation of algal matter between the water and the shore. Pol Polar Res 2:71–94

Zimmermann R, Iturriga R, Becker-Birck J (1978) Simultaneous determination of total number of aquatic bacteria and the number thereof involved in respiration. Appl Environ Microbiol 36: 926–935

Kinetics and Trophic Role of Bacterial Degradation of Macro-Algae in Antarctic Coastal Waters

W. REICHARDT[1] and G. DIECKMANN[2]

Summary. Aerobic formation of detritus from debris of the Antarctic Kelp, *Himantothallus grandifolius,* yielded amounts of carbon and nitrogen 2 and 4 times higher, respectively, than fresh algae, and a concomitant decrease of the carbon:nitrogen ratio from 19 to approximately 10. These changes were achieved only in the presence of the indigenous Antarctic bacterio-flora. During formation of detritus from the Antarctic red alga, *Leptosomia simplex,* C:N ratios may drop even below the Redfield ratio reported for phytoplankton. The increases in the C and N contents of the debris and its colonization by epiphytic bacteria were characterized by saturation kinetics, which showed an initial lag, subsequent steep increment, and a final stationary phase. During its exponential phase, rates of bacterial colonization of debris from brown and red algae were comparable to those calculated for kelp in areas of temperate climates. There is preliminary evidence that some of the enzymes, which participate in the decomposition of the particulate detrital matter, are optimally adapted to the temperature regime of a permanently cold environment. The results obtained favour the idea that the formation of macro-algal detritus in Antarctic coastal waters is as efficient as in temperate regions. Preliminary assessments indicated that partly degraded thalli from both the brown and red algae were preferred over fresh thalli, as a food source, by intertidal amphipods.

1 Introduction

Primary productivity at the edge of the seas often reaches maximal values that are comparable to the productivity of tropical rain-forests (Mann 1973). In temperate latitudes, the total production of sub-littoral seaweeds, such as kelp, may be three times that of the total production of phytoplankton, in certain coastal areas. More than 90% of these benthic macrophytes enter detritus food chains. Despite the abundance of macro-algae in coastal Antarctic waters (Dieckmann et al. 1985), there seems to be no indication as to whether the Antarctic region provides conditions adequate for detritus food chains to play an efficient role as a major pathway for nutrient cycles.

Generally speaking, formation of detritus is triggered by disruption and ageing of the macrophyte tissues, followed by leaching and stimulation of microbial growth in the leachates as well as subsequent colonization of the fragments. The conversion of macrophyte debris to detritus is often characterized by: a succession of epiphytic bacteria which have the capacity to decompose detrital bio-polymers; a decrease of the C:N ratio; and, an increase of nutritive value of debris as food for macro-heterotrophic consumer organisms. In Antarctic waters, it can be anticipated that formation of detritus from macro-algae, the only benthic macrophytes in this environment, is further influenced by: the unique grinding forces exerted by pack-ice; adaptation of growth rates and enzyme activities of the epiphytic bacteria to a permanently cold temperature regime; and, the endemic nature and possibly different degradability of most of the predominant macro-algae.

Our investigation aimed to determine whether conditions for the formation of detritus in Antarctic coastal waters were adequate to postulate that detritus-based food chains may play a similarly efficient role as those in lower latitudes. Hence, experiments were designed to simulate certain aspects of detritus formation from artificially generated debris from two of the most abundant macro-algae in Admiralty Bay (South Shetland Islands). The species involved were *Himantothallus grandifolius* (a brown alga) and *Leptosomia simplex* (a red alga).

Time-course studies dealt with the C and N contents of the debris, colonization by epiphytic bacteria, and selected enzyme activities involved in the breakdown of particulate bio-polymers. As a biomass equivalent, the ATP content of the total suspensions was measured. Intertidal amphipods were used to assess the food value of the macro-algae.

2 Materials and Methods

2.1 Experimental Design

Five kg (wet wt) of *H. grandifolius* dredged from 30–60 m in Ezcurra Inlet (Admiralty Bay, King George Island) was cut and shredded, employing a high-speed blender (Ultraturrax) before final preparation. The debris was suspended in 6 l of 0°C Antarctic seawater and filtered through a 1.0 mm plankton net. The suspended matter in the filtrate was washed on a 200 μm-sieve and resuspended in 10 l of 0°C Antarctic seawater. Three kg (wet wt.) of *L. simplex,*

1 Institut für Meereskunde, Universität Kiel, 2300 Kiel, FRG
2 Alfred-Wegener-Institut für Polarforschung, Columbus-Center, 2850 Bremerhaven, FRG

Antarctic Nutrient Cycles and Food Webs
(ed. by W.R. Siegfried, P.R. Condy, and R.M. Laws)

collected from the upper sub-littoral of Admiralty Bay during low tide, was shredded, using an Ultraturrax blender for batches of 250 g in 500 ml 0°C Antarctic seawater, and subsequently filtered through 1.0 mm and 0.2 mm sieves to obtain the 0.2–1.0 mm particle fraction of suspension in 8 l of 0°C Antarctic seawater.

The suspended algal matter was incubated in the dark in aerated 3–l or 5–l flasks and 2–l (*L. simplex*) or 4.5–l (*H. grandifolius*) suspensions immersed in water baths at 0°C. At the beginning of each incubation a final blending (Ultraturrax) was carried out for 10 min. As controls, half of the flasks were irradiated (2 times for 1 h) with ultraviolet light using an immersible Hanau Sterisol lamp (TNN 15/35) to inactivate the micro-flora remaining on the debris and in suspension. The influence of dissolved nitrogenous organic matter was studied by adding Bacto-Peptone at a final concentration of 2.5 g l^{-1}.

The food value of the two algae was assessed by feeding 20 mm (diameter) discs, punched from fresh or partly degraded algal thalli, to 20 intertidal amphipods (Pontogeneiidae). These were kept in 250 ml perforated plastic containers in a 100 l tank with a continuous through-flow of fresh Antarctic seawater of ca. −1°C. At daily intervals, discs were taken out, blotted quickly between two layers of filter paper, and weighed in order to determine the daily consumption. At set time intervals, chosen according to the density of epiphytic bacteria in preliminary experiments, samples of the suspended debris were taken with a sterilized 50 ml syringe and processed immediately for ATP extraction, separation of the particulate fraction by centrifugation, and for enzyme assays.

2.2 Counting Epiphytic Bacteria

Five ml sub-samples of the suspension were poured into 10 ml centrifuge tubes, fixed with 0.5 ml of 38% formaldehyde and stored at 5°C. For acridine-orange (A.O.) epifluorescence microscopy, 0.5 ml of the staining solution (1,000 ppm acridine-orange in 0.15 M phosphate buffer, pH 4.7,) was added for 5 min, centrifuged (6,000 g), re-suspended in distilled water and again centrifuged. A drop of the re-suspended pellet (in 5 ml of distilled water) was used for counting the bacteria adhering to the debris, employing a Leitz Dialux 20 epi-fluorescence microscope with a 100 times NPL-Fluotar oil immersion objective. A 50 times 50 μm counting grid hat to cover algal surfaces in at least 20 different fields of vision. Results are given as numbers of adherent cells per 2,500 μm^2 unit surface area.

2.3 Chemical Analyses

Total adenosine triphosphate (ATP) was extracted immediately by heating 1 ml of a sub-sample and 1 ml of 0.4 M tris-HCL buffer, pH 7.7, for 5 min at 100°C. Extracts were stored frozen at −20°C. Assays based on the luciferin-luciferase reaction were carried out with a 50 μl sample volume and 2-ml 20 mM tris-HCL buffer, pH 7.7, containing 10 mM $MgSO_4$ and 1 mg ml^{-1} of FLE-250-luciferin-luciferase (Sigma), by measuring peak heights of bio-luminescence in a JRB-ATP-photometer.

The total particulate matter was collected by centrifugation, stored at −20°C and freeze-dried for the determination of dry wt., total C and N. Carbon and N were measured in triplicate sub-samples of approximately 100 μg in tin cups using a Carlo Erba elemental analyzer, model 1106, with automatic sampler and combustion at 1,020°C. Acetanilide was used as a standard.

2.4 Enzymatic Decomposition Potentials

To measure activities of bio-polymer–solubilizing enzymes, 5 ml sub-samples were extracted with 25 ml ice-cold Triton X100-buffer (0.02 M phosphate buffer, pH 7.0, in sterile seawater containing 2.5% Triton × 100 and 2 g l^{-1} polyvinylpyrrolidone as a stabilizer). Amylase, protease and cellulase activities were measured by incubating 3-ml extracts in 10-ml centrifuge tubes with 20 mg of the commercial particulate enzyme-substrates coupled to a blue dye (hide powder-, amylopectin-, and cellulose-azure, Sigma Chemicals). After 18-h incubation at 0°C (or employing a temperature gradient) the reactions were terminated by adding 1 ml of 16% formaldehyde in 0.1 M KH_2PO_4-Na_2HPO_4-buffer, pH 7.5. After 20 min of centrifugation (6,000 g) the absorbance was measured at 600 nm (10-mm path length using a Zeiss PMQ3 spectro-photometer. The absorbance was used as a relative unit.

3 Results and Discussion

3.1 Changes of Carbon and Nitrogen Contents in Macro-Algal Debris

Whereas the C:N atomic ratio of phytoplankton is 6.6, according to the Redfield ratio, the median C:N ratio of marine benthic macrophytes is as high as 18.3 (Atkinson and Smith 1983). Degradation of macrophyte debris is usually characterized by decreasing C:N ratios, which may be partly attributed to the build-up of micro-biota on the debris and subsequent immobilization of dissolved N (Thayer et al. 1977; Haines and Hanson 1979). On the other hand, decomposition experiments with particles derived from *Laminaria* suggest that N is conserved via formation of refractory polyphenol-complexed fractions in the plant debris (Robinson et al. 1982). Some authors even doubt that the contribution of micro-organisms to the N pools of detritus are significant (Christian and Wetzel 1978; Hanson 1982). Our experiments with detrital particles obtained from *H. grandifolius* showed a doubling of particulate C and a 4 times increase of particulate N after a week at 0°C (Fig. 1). In suspensions irradiated with UV light to inactivate epiphytic bacteria, this increase was markedly delayed, indicating that biological processes contributed considerably to the %C- and %N-increases, while the debris was converted to detritus.

Saturation type kinetics characterized the C- and N-fluxes of debris from both *H. grandifolius* and *L. simplex*, the most abundant macrophyte in the shallow sub-littoral. In contrast to *H. grandifolius*, only minor differences were found between the UV-irradiated and the untreated samples for *L. simplex* (Fig. 2). The further decrease of C and N after extended incubation (over 20 d) in debris amended with peptone may indicate that dissolved organic N plays a rate-limiting role for the decomposition of detritus (Robinson et al. 1982).

Enrichment of particulate N in debris from *L. simplex* started at relatively high levels of ca. 4%, which coincided with the maximum levels reached by mature detritus from *H. grandifolius*. Although certain red algae contain more N than kelp, contents as high as 3%–5% may indicate storage (Bird et al. 1982).

Can C- and N-fluxes in these experiments be considered as a specific achievement of the indigenous Antarctic micro-flora at ambient temperatures around 0°C? When debris of the Antarctic *H. grandifolius* was suspended in Baltic sea-water at 0°C, the increase of C and N again followed a saturation-type curve, yet the saturation value for N enrichment was only 50% of that noted under the influence of an Antarctic micro-flora (Figs. 1 and 3). This low N accumulation contributed to a final C:N ratio of 14 for mature kelp detritus. On the other hand, considerably lower C:N ratios were observed when the detritus had been allowed to form in water from the original habitat of the alga (Fig. 4). Thus, a direct relationship between the indigenous epiphytic micro-flora on the debris and its gain in N is indicated.

3.2 Growth of Bacteria on Algal Debris

Experimental manipulations (viz, UV-irradiation) causing temporary inactivation of part of the epiphytic micro-flora could not replace the use of analytical controls. Hence, conversion of the particulate algal biomass into bacterial biomass had to be confirmed by direct counting. There was no visible evidence of protozoans in any of these experiments.

Fresh debris of *H. grandifolius* and *L. simplex* was colonized by relatively small, mostly coccoid, bacterial cells (0.1–0.4μm), Similarly to kelp in temperate latitudes, the thalli of *H. grandifolius* were predominantly colonized along the junctions of the epidermal cell walls (Linley et al. 1981; Koop et al. 1982; Fig. 5). After a lag of 2 to 3 days dividing cells became more and more frequent, together with a predominance (95%) of relatively large rods (Fig. 5).

Bacterial cells counted on debris with a surface area of at least 2,500 μm^2 revealed exponential growth for 4–7 d following an initial adaptation phase of 1–3 d (Fig. 6). During exponential growth, the apparent doubling times

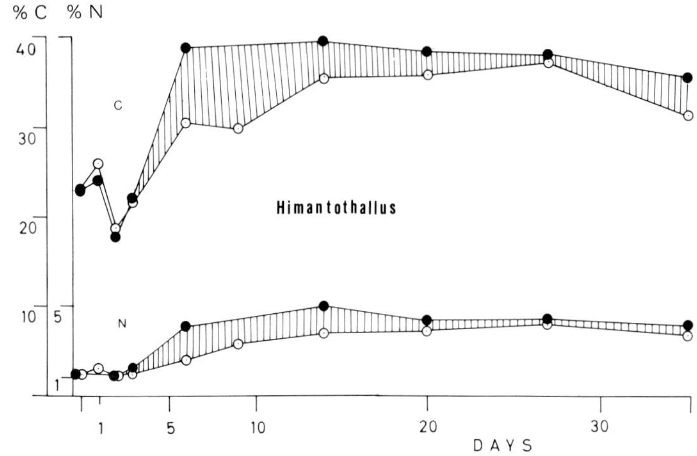

Fig. 1. Carbon and nitrogen during fluxes formation of detritus from *H. grandifolius* debris without (*dots*) and with UV-irradiation (*circles*)

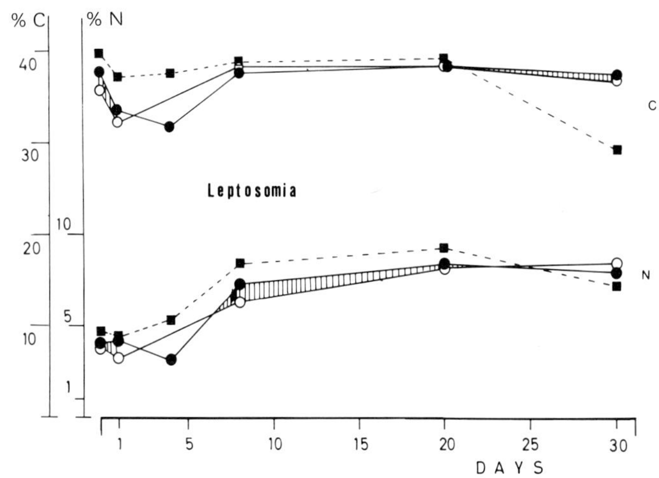

Fig. 2. Carbon and nitrogen fluxes during formation of detritus from *L. simplex* debris without (*dots*) and with (*circles*) UV-irradiation. *Broken line* traces results from a parallel experiment with the addition of peptone

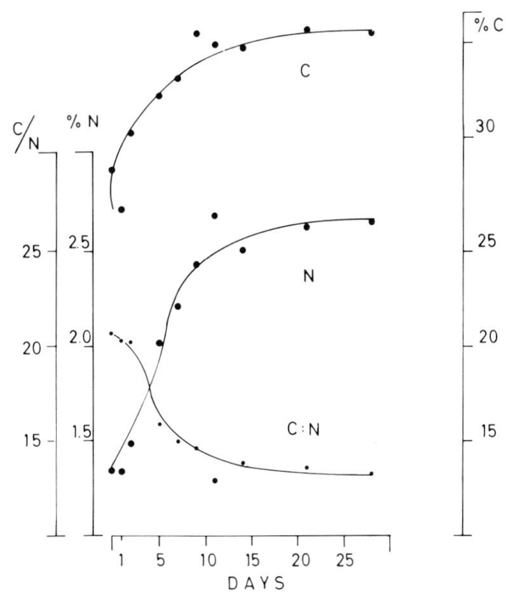

Fig. 3. Carbon and nitrogen in *H. grandifolius* debris during detritus formation in Baltic seawater. C:N ratios drawn on *left*

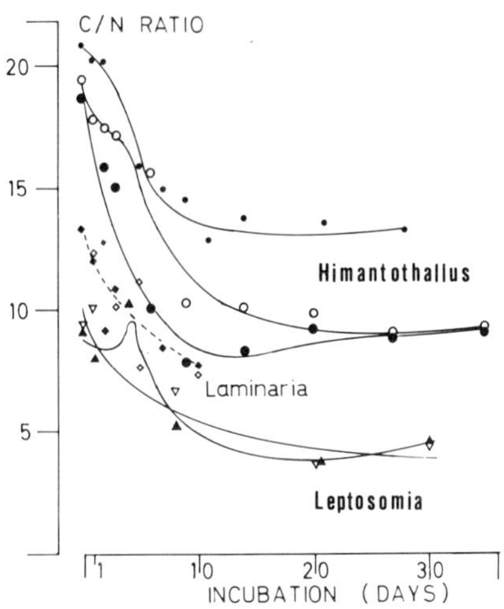

Fig. 4. Fluxes of C:N ratios in macro-algal debris under different treatments: *H. grandifolius*, in Antarctic seawater (●, o); *H. grandifolius* in Baltic Seawater (•); *L. simplex*, in Antarctic seawater (▲△); and, *Laminaria saccharina*, in Baltic seawater (■□). *Clear symbols* in all cases indicate UV-irradiation

ranged from 21.h for *L. simplex* to 27 h for *H. grandifolius*. These apparent doubling times must be viewed as a complex parameter based on the growth of adhering cells, as well as settling and detachment processes. Very similar values (24.8 h) were calculated for *Ecklonia maxima* debris from the South African coast, from data given by Koop et al. (1982). This indicates that the accumulation of a bacterial micro-flora on the surface of Antarctic macro-algal debris occurs at the same rate as in other parts of the world's oceans, in spite of the lower temperature regime.

Bacterial colonization of the debris correspond with the time course of %C and %N-increases (Figs. 1, 2, 6). Attached bacterial growth reached a maximum density of ca 400,000 cells mm^{-2} for detritus from both species. Densities of this order of magnitude also have been reported for other marine algae from other regions (Cundell et al. 1977; Koop et al. 1982), whereas bacterial densities on *Posidonia oceanica* apparently reach only 40,000 cells mm^{-2} (Velimirov et al. 1981). Bacterial counts indicated further that UV-irradiation had inactivated the epiphytic micro-flora only temporarily. The resulting time lag against the untreated sample disappeared after 1 day for *L. simplex* and 6 days for *H. grandifolius*, respectively.

So far only the particulate fraction of the experimental system has been considered. However, Newell et al. (1980) found that the conversion of kelp particles into bacterial biomass is much lower than bacterial growth on dissolved leachates. In our experiments, total dry wt. losses amounting to 37% in the first 6 days indicated intense leaching. To obtain a global biomass equivalent comprising both the algal debris and the suspended micro-organisms, ATP-concentrations were determined for the total suspensions. However, large variations of C:ATP ratios in kelp detritus limit the use of ATP analyses (Stuart 1982). In the case of *H. grandifolius*, this analysis was inapplicable, due to the masking effects with algal ATP. *Leptosomia simplex* debris, on the other hand, did not interfere with ATP measurement. A steep increase of ATP, after the bacterial colonization had already peaked, indicated a vigorous growth of suspended bacteria sustained by dissolved organic substrates (Fig. 7).

3.3 Food Value of Macro-Algal Debris

Despite the low temperature regime and the occurrence of largely endemic macro-algae in Antarctic coastal waters, conditions for the formation of macro-algal detritus are not more adverse than in the seas of certain other parts of the world. Nevertheless, the functional role of macro-algal detritus in Antarctic marine food chains has yet to be determined.

Preliminary tests to determine the nutritive value of macro-algal debris for Antarctic marine amphipods (Pontogeneiidae) indicated that the red alga, with its low C:N-ratio, was consumed more rapidly than the brown alga. Partly decomposed material was preferred over fresh thalli (Table 1, Fig. 8).

3.4 Decomposition Potential of Selected Enzymes

Bacteria involved in the process of detritus maturation are expected to produce exo-enzymes to depolymerize their particulate substrates. For detritus derived from *Spartina*, biphasic decomposition kinetics have been described (Marinucci 1982). Yet, as in most approaches to detritus biochemistry, it is extremely difficult to separate activities of the partly living algal cells from those of epiphytic bacteria. Ultra-violet irradiation showed only limited efficiency in helping to overcome this problem (Fig. 9). Nevertheless, at least at the beginning of an experiment, differences between the untreated and the UV-irradiated suspensions could be attributed to the attached micro-flora. In *H. grandifolius* debris, amylase activities maintained their initial level after detritus formation had entered its stationary phase. Protease activities, on the other hand, showed a considerable decrease toward the stationary phase in the untreated samples. Whether this can be seen as a regulatory phenomenon, due to the increment of nitrogenous compounds in the mature detritus, has yet to be investigated.

One of the most intriguing and partly unresolved questions concerns the intrinsic capacity of algal detritus to decompose efficiently at temperatures of 0°C and below, when amylase and protease have rate optima of around 25°C (Fig. 10). This is common among micro-organisms

Fig. 5A–E. Photomicrographs of epiphytic bacteria on debris of *L. simplex* and *H. grandifolius* at different stages of detritus maturation (epi-fluorescence microscopy). **A** Colonization of epidermis cell-wall junctions at the beginning of experiment with *H. grandifolius*. **B**, **C**, and **D** Colonization of *H. grandifolius* debris after 2, 6, and 28 days. **E** *Leptosomia simplex* debris after 20 days

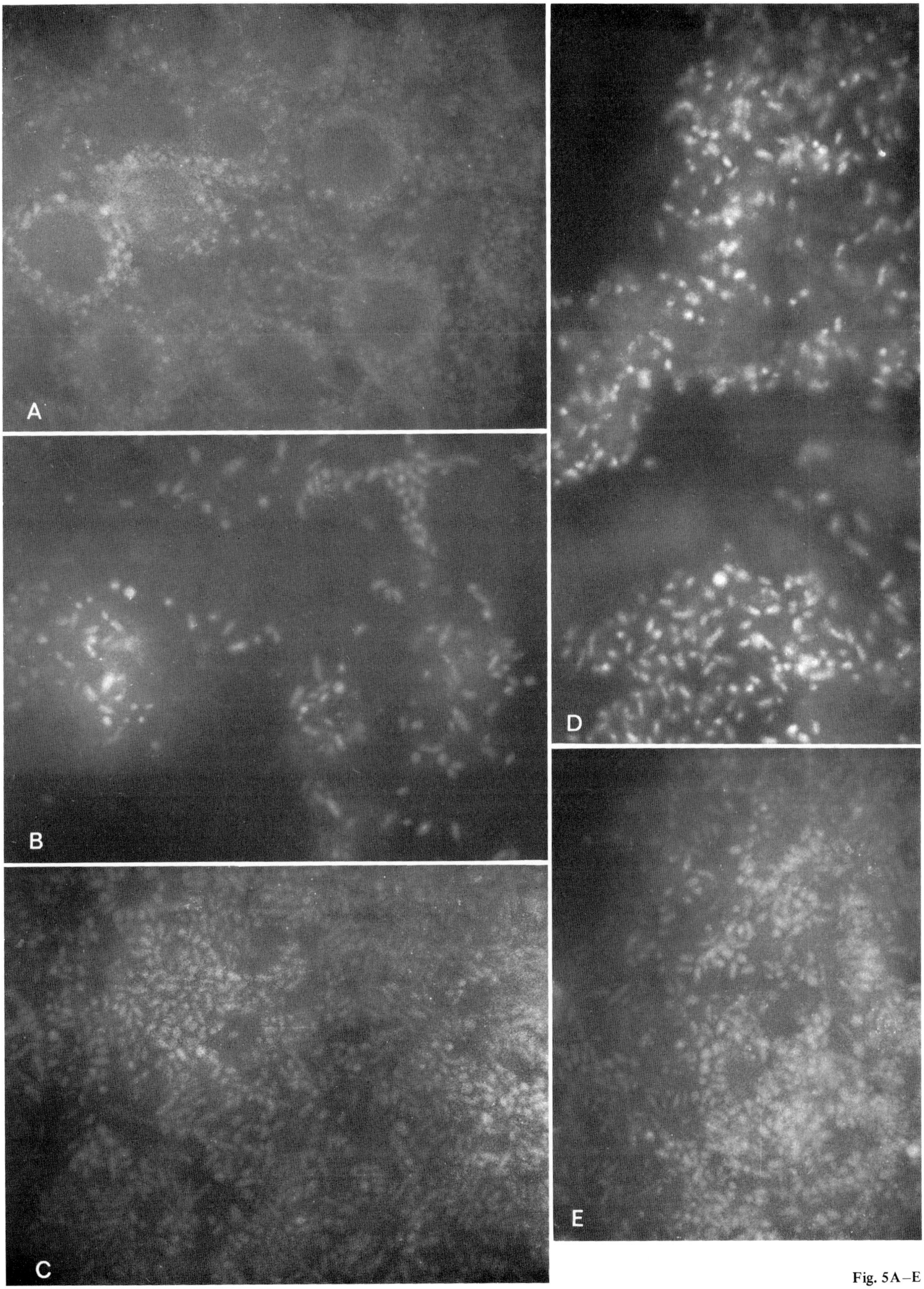

Fig. 5A–E

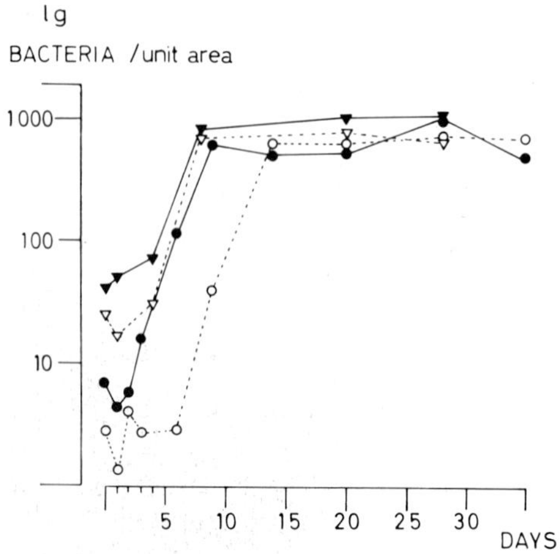

Fig. 6. Time-course of colonization of *H. grandifolius* (*dots*) and *L. simplex* (*triangles*) debris by epiphytic bacteria (log numbers per unit surface area of 2,500 μm²). *Clear symbols* in all cases indicate UV-irradiation

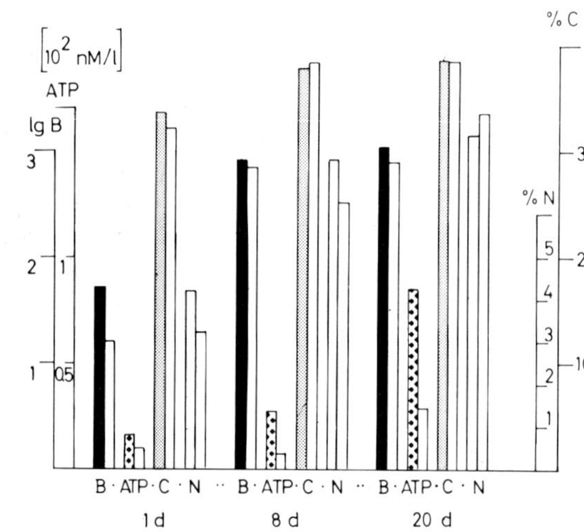

Fig. 7. Chemical and biological parameters for *L. simplex* during different phases of bacterial colonization. *lg B* log plot of bacterial cell densities per 2,500 μm² debris surface; nitrogen (*N*), carbon (*C*), and *ATP* concentrations. *Clear columns* in all cases indicate UV-irradiation

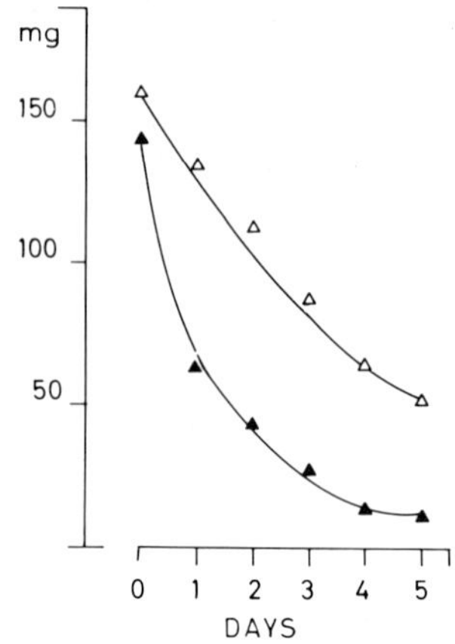

Fig. 8. Loss in fr. wt. of fresh (*clear triangles*) and partly degraded (*shaded triangles*) thallus discs from *L. simplex* due to consumption by intertidal amphipods

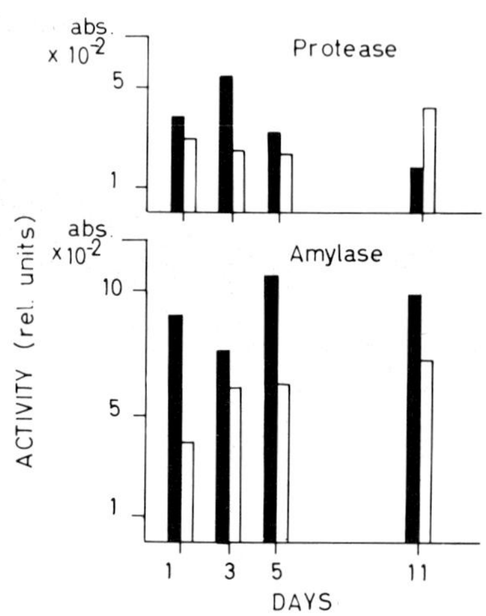

Fig. 9. Activities of protease and amylase in cultured suspensions of *H. grandifolius* debris at 0°C. Relative units based on absorbance (600 nm) of released dye after 18 h of Triton × 100 extract. Clear columns in all cases indicate UV-irradiation

Table 1. Elimination of fresh and partly degraded thallus discs of Antarctic macro-algae consumed by intertidal amphipods. Initial weight losses are after 2 days and 6 days for *L. simplex* and *H. grandifolius*, respectively

	Leptosomia simplex		*Himantothallus grandifolius*	
	fresh	partly degraded	fresh	partly degraded
Initial C:N	8.2	7.7	17.7	15.5
Initial % N	4.3	4.3	1.8	2.0
Initial fr. wt. (mg)	159 ± 10	144 ± 18	578 ± 22	317 ± 3
Weight loss d^{-1} (%)	14.5	30.6	0.9	5.4

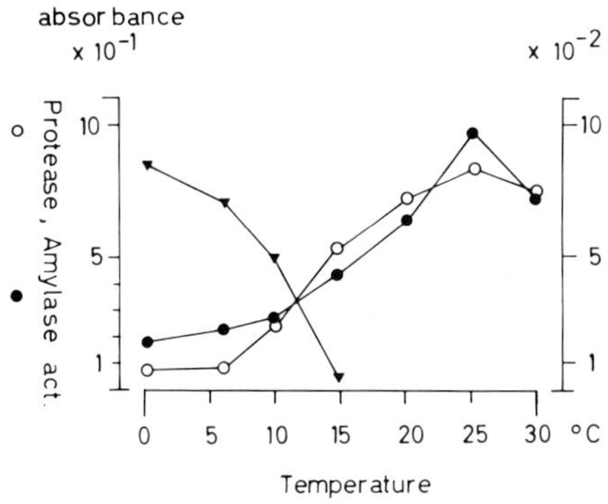

Fig. 10. Temperature characteristic of protease (*circles*), amylase (*dots*), and cellulase (*triangles*) in Triton X 100 extracts from *H. grandifolius* debris

which are adapted to permanently cold environments (Baross and Morita 1978). However, optimum adaptation to the ambient temperature regime of Antarctic waters is shown for cellulase activity from the same source (Fig. 10).

4 Conclusions

Our experiments, designed to simulate different aspects of the conversion of macro-algal fragments to a potential source of detrital food for macro-heterotrophic consumers, are the first of this type to be carried out in Antarctic waters; comparable research having been previously restricted to lower latitudes (Mann 1973; Newell and Lucas 1981). For simplification, the system chosen for time-course studies of ageing debris was confined to algal fragments and bacteria. This approach proved to be sufficient to confirm the potential effectiveness of detrital food chains in the coastal Antarctic environment with respect to the kinetics of attached bacterial growth, fluxes of particulate C and N, activities of selected depolymerizing enzymes and food selection by indigenous macro-heterotrophs.

A mass balance for the conversion of algal to bacterial substance was beyond the scope of this communication. It would have required additional information on the pool of DOM and densities of free supended bacteria, as well as an exact determination of the biomass volume of the total attached bacterio-flora. Since the debris represented living macro-algal fragements, parts of the increments of C% and N% were likely to have been caused by biological processes still associated with them. Physical processes related to mineralization and leaching were covered by the UV-irradiated blanks.

In temperate climates, conversion of kelp C to total bacterial C is strongly influenced by seasonal fluctuations of the predominant bacterial cell forms (i.e. rods or cocci) (Lucas et al. 1981; Newell and Lucas 1981). On the other hand, in our experiments with Antarctic macro-algae at 0°C, coccoid forms were confined exclusively to the surface of epidermal cells at the beginning of the experiment and did not contribute to the exponential phase of colonization of the debris. The majority of these forms may be considered as starved survival stages of rod-shaped cells (Amy et al. 1983; Reichardt and Morita 1982). Exponential colonization rates of the attached bacterio-flora, as well as temperature characteristics of selected enzymes involved in the decomposition of algal polymers, suggest that detritus formation in Antarctic coastal waters is as much adjusted to its environment as it is at lower latitudes. Relative enrichment of particulate N (as indicated by decreasing C:N ratios) may be an important fact for any detrital food chain, although the microbial contribution remains controversial. To help solve these questions, work is in progress to determine conversion efficiencies for attached fractions of detrital bacteria.

Acknowledgements. This investigation was supported financially by a grant (Schr. 143/7-1) from the Deutsche Forschungsgemeinschaft.

References

Amy PS, Pauling C, Morita RY (1983) Starvation-survival processes of a marine *Vibrio* sp. Appl Environ Microbiol 45:1041–1048

Atkinson MJ, Smith SV (1983) C:N:P ratios of benthic marine plants. Limnol Oceanogr 28:568–574

Baross JA, Morita RY (1978) Microbial life at low temperatures: Ecological aspects. In: Kushner DJ (ed) Microbial life in extreme environments. Academic, London, p 9–71

Bird KT, Habig C, DeBusk T (1982) Nitrogen allocation and storage patterns in *Gracilaria tikvahiae* (Rhodophyta). J Phycol 18:344–348

Christian RR, Wetzel RL (1978) Interaction between substrate, microbes and consumers of *Spartina* detritus in estuaries. In: Wiley ML (ed) Estuarine Interactions. Academic, New York pp 93–144

Cundell AM, Sleeter TD, Mitchell R (1977) Microbial populations associated with the surface of the brown alga *Ascophyllum nodosum.* Microb Ecol 4:81–91

Dieckmann G, Reichardt W, Zielinski K (1985) Growth and production of the seaweed, *Himantothallus grandifolius,* at King George Island. In: Siegfried WR, Condy PR, Laws RM (eds) Antarctic nutrient cycles and food webs (Proceedings of the 4th SCAR symposium on Antarctic Biology). Springer, Berlin Heidelberg New York

Haines EB, Hanson RB (1979) Experimental degradation of detritus made from the salt marsh plants *Spartina alterniflora* L., and *Juncus roemeriana* Scheele. J Exp Mar Biol Ecol 40:27–40

Hanson RB (1982) Organic nitrogen and caloric content of detritus. II Microbial biomass and activity. Estuarine Coastal Shelf Sci 14:325–336

Koop K, Newell RC, Lucas ML (1982) Biodegradation and carbon flow based on kelp (*Ecklonia maxima*) debris in a sandy beach. Mar Ecol Progr Ser 7:315–326

Linley EAS, Newell RC, Bosma SA (1981) Heterotrophic utilisation of mucilage released during fragmentation of kelp (*Ecklonia maxima* and *Laminaria pallida*). I. Mar Ecol Progr Ser 4:31–40

Lucas MI, Newell RC, Velimirov B (1981) Heterotrophic utilisation of mucilage released during fragmentation of kelp (*Ecklonia maxima* and *Laminaria pallida*). II. Mar Ecol Progr Ser 4:43–55

Mann KH (1973) Seaweeds: Their productivity and strategy for growth. Science 182:975–981

Marinucci AC (1982) Carbon and nitrogen fluxes during decomposition of *Spartina alterniflora* in a flow-through percolator. Biol Bull 162:53–69

Newell RC, Lucas MI (1981) The quantitative significance of dissolved and particulate organic matter released during fragmentation of kelp in coastal waters. Kieler Meeresforsch Sonderh 5: 336–369

Newell RC, Lucas MI, Velimirov B, Seiderer LJ (1980) The quantitative significance of dissolved organic losses following fragmentation of kelp (*Ecklonia maxima* and *Laminaria pallida*). Mar Ecol Progr Ser 2:45–59

Reichardt W, Morita RY (1982) Survival stages of a psychrotrophic *Cytophaga johnsonae* strain. Can J Microbiol 28:841–850

Robinson JD, Mann KH, Novitsky JA (1982) Conversion of the particulate fraction of seaweed detritus to bacterial biomass. Limnol Oceanogr 27:1072–1079

Stuart V (1982) Limitations of ATP as a measure of microbial biomass. S Afr Tydskr Dierkd 17:93–95

Thayer GW, Engel DW; Lacroix W (1977) Seasonal distribution and changes in the nutritive quality of living, dead and detrital fractions of *Zostera marina.* J E M Biol Ecol 30:109–127

Velimirov B, Ott JA, Novak R (1981) Microorganisms on macrophyte debris. Kieler Meeresforsch Sonderh 5:333–344

The Role of Bacteria in the Cycling of Nutrients within the Maritime Antarctic Environment

A.C. TANNER[1]

Summary. The Antarctic is an area of outstanding scientific interest. Early research into the role of micro-organisms in this unique environment centred on the terrestrial habitat, and it is only recently, with the upsurge of interest in the highly productive maritime Antarctic, that the role of micro-organisms in this environment has been considered. Studies of micro-organisms in the Southern Ocean have shown that bacteria exhibit a number of interesting properties, not least that of psychrophilic adaptation. In addition, these bacteria play a crucial role in the cycling of nutrients, and many scientists regard marine bacterial metabolism as a principal feature in the regeneration of primary nutrients. However, detailed, land-based studies have been necessary to elucidate further the complex seasonal interaction that occurs in the micro-flora. Benthic sediments have also proved to be important in mineralization processes in the near-shore maritime Antarctic, and studies in this area have shown that bacterial populations undergo cyclic variations related to a number of important environmental constraints. The taxonomy of these organisms is interesting, in that most of the bacteria isolated have not correlated well with previously described species. Numerical taxonomic methods are, however, proving to be useful. Parallels are emerging between data gathered from the Antarctic and from other environmentally similar areas. The association of bacteria with Krill, *Euphausia superba,* might be of great importance in nutrient cycling in the Southern Ocean.

1 Introduction

Antarctica can be conveniently divided into two distinct ecological zones: the maritime and the continental Antarctic, depending on the prevailing climatic conditions (Holdgate 1964). On the main Antarctic continent, temperatures rarely exceed 0°C, and the terrain is almost permanently frozen. In contrast, the maritime Antarctic, comprising the Antarctic coastal regions, the Antarctic Peninsula and its associated islands, has a much less severe climate. Here ice cover is seasonal and ambient temperatures generally are higher, ca. 0°C to +5°C. More lately, greater scientific interest has been shown in the maritime Antarctic, but early work concentrated on the micro-biology of the continent itself. The historical background of polar micro-biology is of interest, because it serves to illustrate some of the problems encountered in the study of micro-organisms occurring in these specialized environments.

1 Animal Health Development Department, Pfizer Central Research, Sandwich, Kent, CT13 9NJ, England

2 Historical Background

Bacteria were found in freshwater pools on ice floes in the North Polar Basin as early as 1897. Later, many bacterial types, including nitrogen-fixing species, were reported from this area by workers in Norway, Sweden, and Russia. During the many early Antarctic expeditions, scientists searched for, and in most cases found, micro-organisms: Ekelof (1908), with the Swedish Antarctic expedition (1901–1903), isolated bacteria from the soil of Grahamland, followed by Tsiklinsky (1908), with Charcot's French expedition (1903–1905), Gazert (1912), with the German Antarctic expedition (1901–1903), Pirie (1912), with the Scottish National Antarctic expedition (1902–1904) Atkinson (1914), with Scott's British Antarctic expedition (1910–1913), and McLean (1918) with Mawson's Australasian Antarctic expedition (1911–1914)).

Understandably, these early investigations were limited. Even Darling and Siple (1941), with the United States Antarctic expedition (1933–1935), could do little more than a qualitative survey of bacterial types present in ice, snow and mud. Two of the most significant problems facing the early workers were the lack of refrigerated incubators and the bacteriological techniques in common use at that time. The literature contains numerous examples of bacteria isolated from low-temperature habitats, being studied at temperatures far in excess of those ever likely to be encountered in situ. McLean (1918) isolated Antarctic micro-organisms from melt-water after incubation at 10°–15°C, but most isolations were from samples incubated at a much higher temperature. It has been shown that the past history of the source material is of great importance. If, for instance, the environment is exposed to sufficient radiant energy, this can cause inactivation of any psychrophiles present. Llano (1962) showed that for one location in the Horlick Mountains, Antarctica, the temperature of rock surfaces rose from –15°C to +27.8°C in 3 h due to incident solar radiation. This, and other examples, point to the high degree of thermolability demonstrated by true psychrophiles. Amongst the bacteriological techniques in common use at that time was the pour-plate method for the enumeration of micro-organisms. It is now known, however, that temperatures only some 3°–4°C above the gelling

Antarctic Nutrient Cycles and Food Webs
(ed. by W.R. Siegfried, P.R. Condy, and R.M. Laws)

temperature of agar (ca. 42°C) can be lethal to even facultative psychrophiles. A method of surface plating of samples on to pre-cooled agar plates or incubation on membrane filters at temperatures of < 10°C has been recommended for the successful isolation of psychrophiles.

3 The Role of Bacteria

The Antarctic marine environment can be divided into two main areas of study: the water column and the benthic sediment. The former is an area subject to change due to the effects of weather, radiation, tides, etc., but the latter is relatively unaffected by these changing conditions. The water column contains bacteria freely suspended in the water, attached to other organisms, to particles of organic debris and in the superficial layer above the sea-floor where they are abundant. In mud or sand, the numbers of bacteria tend to decrease with depth below the sediment surface. The population density of bacteria also decreases on passing from inshore waters to the open sea. Here the greatest population is found where phytoplankton are abundant and in the water-air and water-sediment interfaces.

Nutrient regeneration has been defined as the release of soluble organic or inorganic nutrients by, or from, living cells and their remains (Johannes 1968). The role of bacteria in the cycling of nutrients within temperate aquatic environments has been well documented. Seki et al. (1966) classed the action of bacteria on organic debris in sediments as secondary production, and other workers have concluded that, at times, benthic conversions by bacteria may well be the principal source of new organic material into the water column. Morita et al. (1971) considered bacteria to be the principal means of regeneration of primary nutrients by mineralization of organic matter. Stokes and Redmond (1966), after a quantitative survey of psychrophilic bacteria from a variety of environments, concluded that these bacteria probably play important roles in the cycling of matter.

Whilst it is widely accepted that Antarctic waters are highly productive, it has been suggested (Walsh 1969; El-Sayed 1973) that this may pertain more to coastal waters than to the whole Southern Ocean. Haeckel (1890) questioned the validity of the "high productivity" theory, and it has been suggested (Hart 1942) that the annual primary production of the Antarctic region might be similar to that of the English Channel. Other workers have also demonstrated a relatively low level of productivity in Antarctic regions (Burkholder and Burkholder 1967; Vargo 1968; Walsh 1969). It is probable that the intense summer productivity of shallow Antarctic sea areas is responsible for a misconception as to the productivity of these waters, for when the short duration of the bloom is considered (some 8–10 weeks) it can be seen that inshore production is comparable in magnitude to more temperate regions (Walsh 1969; Whitaker 1982). It has been calculated (Whitaker 1977) that, despite the environmental constraint of low temperature, primary production rates in the near-shore coastal waters surrounding Signy Island (60° 42′S, 45°C 35′W) are high during the algal bloom, amounting to between 80–290 g C m^2 yr^{-1}. It is interesting to note that these production rates are similar to those reported for Frobisher Bay (63° 40′N, 68° 27′W), in Canada (Grainger 1979).

Few detailed studies have been made of nutrient regeneration in Antarctic coastal waters, and information on the role of bacteria in the cycling of nutrients is sparse and almost entirely confined to preliminary investigations. Indeed, there are few data even on the occurrence of heterotrophic bacteria in Antarctic waters. Walls (1967) collected samples of water and sediment from various points along the Antarctic Convergence, and concluded that the bacteria isolated from these samples were similar to those found in other oceans. In a sampling transect along longitude 39°W, Kriss (1970) found the lowest concentrations of bacteria occurring in Sub-Arctic and Sub-Antarctic zones. He also reported considerable stratification of the water column in terms of bacterial numbers. Other workers have used the number of heterotrophs occurring in the water as an indication of the origin of that water-mass (Kriss et al. 1969; Kriss 1970, 1972a, 1972b, 1973; Mitsketvitch 1974), although the majority of these heterotrophic bacterial counts were performed at temperatures of ca. 20°–28°C (considerably higher than ambient), and this may explain the high percentage of apparently sterile samples obtained (Wiebe and Hendricks 1974). Investigations by Morita et al. (1971) have indicated that many psychrophilic bacteria occur in these regions, and that these thermal types of bacteria should receive primary consideration in any micro-biologically orientated research in Antarctic waters.

Whilst most of these preliminary investigations were conducted by ship-borne workers, it has needed detailed land-based studies to elucidate further the complex seasonal interactions that occur in the inshore marine micro-flora. Herbert and Bell (1974) examined samples of marine mud and seawater obtained from various points off the coast of Signy Island, using liquid culture enrichment techniques, for bacteria capable of cycling N and S compounds. They were able to show that with the exception of nitrifying bacteria, those capable of carrying out the specialized reactions of the S and N cycles were present in all the samples examined. In addition, they were able to demonstrate that the majority of these bacteria would grow, albeit slowly, and carry out these reactions at 5°C. However, they concluded that in situ fieldwork was required to determine the rates at which these nutrient transformations occurred in the natural Antarctic marine environment.

Tanner and Herbert (1981), in a more elaborate study, measured changes in phytoplankton populations, inorganic nutrient levels, total aerobic heterotrophic bacteria, denitrifying bacteria, proteolytic bacteria, and sulphate-reducing bacteria, together with rates of denitrification and heterotrophic N fixation. They were able to show that whilst levels of inorganic nutrients within the water column became depleted by the phytoplankton bloom, at no time did they become growth limiting. Total C–N analysis of the sediments showed that levels of these elements were low, and that there was no accumulation of moribund phytoplankton in the sediment. Populations of bacteria occurring in the sediments showed distinct cyclic variations in both type and number. Total viable counts of heterotrophic bacteria in the surface sediments were maximal during the phytoplankton bloom and then decreased in late autumn and winter. Maximum population densities of heterotrophs during the bloom reflected the increased organic input into the sediments, which were normally carbon-limited. Seasonal maxima in phosphate and ammonia levels within the water column followed the collapse of the bloom, and could be attributed to the action of proteolytic bacteria. Delille and Legarde (1974), investigating the activities of bacteria in the near-shore sediments of Kerguelen Island, showed that there were pronounced seasonal variations in the cycling of S, whereas no significant fluctuations were evident in the cycling of C.

Herbert and Bell (1974) and Tanner and Herbert (1981) demonstrated the presence of heterotrophic nitrogen-fixing bacteria in sediment samples from Signy Island. Low but detectable rates of N fixation, attributable to species of *Desulfovibrio* were recorded. However, the process was carbon-limited, and rates of N fixation could be increased by the addition of exogenous C. Whereas N fixation represents a potential input of new N into the sediements, denitrification results in its loss. There was considerable potential for denitrification in the Signy Island sediments but again the process was probably limited by C availability. Tanner (1981) showed that the formation of ammonia as an end product of nitrate reduction was more common than either nitrite or N gas in heterotrophic bacteria isolated from the Signy Island sediments.

Herbert (1981) and Herbert and Tanner (1977), showed that phototrophic bacteria occurred commonly in the water column and benthic sediments, but despite adaptation to the reduced light intensity and attenuated wavelength, no temperature adaptation was observed. Since the majority of these bacteria will only grow photo-autotrophically, their presence in this environment infers that they are physiologically active but the manner in which this is achieved has yet to be elucidated.

Rittenberg (1963), reviewing the role of bacteria in nutrient regeneration in the sea, concluded that almost any genera of bacteria can be isolated from the marine environment, but that this does not automatically infer that they play a significant role in mineralization. As such, a number of authors have studied the heterotrophic potential of the microbial flora of the Southern Ocean, using isotope-uptake experiments. Gillespie et al. (1976), studying the offshore microbial flora along a transect west of Heard Island, found that the uptake of ^{14}C-glucose and ^{14}C-acetate was low, and often could not be measured, especially in samples taken from below the euphotic zone. However, data obtained by these authors indicated that while activity was low, these oceanic microbial populations demonstrated considerable efficiency in their uptake of glucose and acetate. Tanner and Herbert (1982), investigating substrate uptake by heterotrophic microbial populations in the near-shore waters of Signy Island, showed a significantly lower efficiency of uptake by these littoral bacterial populations. Hodson et al. (1978, 1981), investigating the turnover of dissolved organic matter in McMurdo Sound, reported that the majority of dissolved adenosine triphosphate (DATP) uptake was attributable to free-living bacteria within the water column, and that rates of microbial utilization of DATP were inversely dependent upon water temperature, with maximal activity occurring at 0°C.

Currently, much interest is centred on Antarctic Krill, *Euphausia superba,* as a key organism in one of the world's largest marine ecosystems. The natural decomposition of Krill is therefore of considerable importance in the regeneration of nutrients within the Antarctic marine environment. Kelly et al. (1978) reported the presence of low numbers of *Pseudomonas, Flavobacterium, Moraxella* and *Alcaligenes* bacteria within fresh Krill. These isolates showed evidence of psychrophilic adaptation, and possessed active proteolytic and lipolytic enzymes. Locati et al. (1980), studying *post-mortem* changes in Krill, associated such changes with the multiplication of aerobic, psychrophilic bacteria (predominantly *Pseudomonas* species, although other genera were isolated as well). Zdanowski (1981) and Fevolden and Eidsa (1981), however, attributed the rapid decomposition of Krill more to autolysis than to bacterial activity, and suggested that the role of these micro-organisms might be more as saprophytes than as primary spoilage organisms.

4 Taxonomy

Interest has been shown in the taxonomy of bacteria isolated from the Antarctic environment. However, the classification of these bacteria presents considerable difficulties due to the lack of information concerning their basic physiology. Consequently, there is a lack of suitable determinative tests and a high proportion of these isolates do not identify fully when compared against currently available taxonomic keys. Much has been written on the advantages of numerical methods of taxonomy, and use has been made of computer-aided numerical methods for the classification of Antarctic bacteria. Pfister and Burkholder (1965), in a taxonomic study of Antarctic marine bacteria, found that most of their isolates were Gram-negative, heterotrophic, rod-shaped and mobile by means of flagellae. These and other workers have shown that bacteria belonging to the genera *Pseudomonas, Vibrio,* and *Flavobacterium* tend to predominate in low temperature environments such as those found in the maritime Antarctic. Tanner and Herbert (1982), in a numerical taxonomic study of Gram-negative bacteria isolated from the Antarctic marine environment, found that of 144 bacteria characterized, approximately 50% resembled members of the genera *Vibrio* and *Aeromonas*, 22% *Alcaligenes*, 2% *Pseudomonas*, and 2% *Flavobacterium*. However, some 24% of the isolates were not readily identifiable with previously described species listed in the eighth edition of Bergey's Manual (Buchanan and Gibbons 1974). Of these, a number showed considerable similarity with the recently described *Vibrio fluvialis* biotypes of Lee et al. (1981). It should be noted, however, that a large number of bacteria isolated from the Antarctic environment are biochemically relatively inactive, leading to a large proportion of negative similarities. This, together with the difficulty of finding suitable type strains for inclusion into numerical analyses of these bacteria, has hampered progress on their taxonomy. The use of more powerful resolving techniques such as measurement of guanine plus cytosine ratios, or nucleic-acid hybridization may well aid further investigations into the taxonomy of Antarctic marine bacteria.

5 Future Considerations

Whilst it is difficult to review the subject exhaustively within the constraints of this paper, a number of important points emerge concerning the role played by bacteria in the cycling of nutrients within the Antarctic marine environment. Bacteria are ubiquitous throughout the Southern Ocean, and in general demonstrate a high degree of adaptation to their environment. Temperature adaptation is widespread, with a high proportion of the bacteria isolated being classified as psychrophilic or psychrotrophic by presently accepted definition. The organisms show considerable nutritional versatility, although individual bacterial types are selective in their use of C sources. Most of the genera involved in the cycling of nutrients within more temperate waters have been isolated from the Antarctic marine environment, although their active participation in nutrient regeneration needs to be demonstrated. Uptake experiments have, however, shown that there is considerable potential for heterotrophic bacterial activity even considering the low in situ temperature of the Southern Ocean.

Further study is needed to elucidate the role of phototrophic bacteria, and to confirm the absence of various bacterial groups (e.g., sporeformers and aerobic nitrogen-fixing bacteria such as *Azotobacter* spp.). In addition, while it is clear that regeneration of nitrate does take place, the process by which this occurs has yet to be elucidated. It is probable that bacteria inhabiting the digestive tracts of certain gasteropods and related marine animals play a significant role in the regeneration of nutrients (e.g., in the degradation of chitin) and there is a requirement for further study to quantify this contribution. Similarly, the role of bacteria in the in situ decomposition of Krill poses some interesting questions.

Further investigations into the taxonomy and physiology of these highly adapted micro-organisms may well reveal the presence of previously undescribed species, and it is interesting to speculate on the similarities between Antarctic marine bacteria and those isolated from other environmentally similar areas, such as the Arctic and the ocean deeps. Perhaps the most important conclusion to be drawn from this brief review is that, whilst early researchers were hampered by limited facilities and a dearth of basic knowledge, with the increasingly sophisticated facilities available today, more detailed interdisciplinary studies can be instigated to elucidate further the role of bacteria in the complex processes that comprise nutrient regeneration within the Antarctic marine environment.

References

Atkinson EL (1914) In: Scott's Last Expedition, 1 London pp 211

Buchanan RE, Gibbons NE (1974) Bergey's Manual of Determinative Bacteriology. 8th Edition, Williams & Wilkins, Baltimore

Burkholder PR, Burkholder LM (1967) Primary productivity in surface waters of the South Pacific Ocean. Limnol Oceanogr 12:608–617

Darling CA, Siple PA (1941) Bacteria of Antarctica. J Bacteriol 42: 83–98

Delille D, Lagarde E (1974) Contribution a l'etude ecologique des Milieux subantarctiques, V Rev Inst Pasteur Lyon 7:149–165

Ekelof W (1908) Bakteriologische Studien während der Schwedischen Südpolar-Expedition (1901–1903), Stockholm 1908

El-Sayed SZ (1973) Biological Oceanography. Ant J US pp 93

Fevolden SE, Eidsa G (1981) Bacteriological characteristics of Antarctic krill. Sarsia 66:77–82

Gazert H (1912) Untersuchungen über Meeresbakterien und ihren Einfluß auf den Stoffwechsel im Meer. Deutsch Südpolar-Expedition 1901–1903.7 pp 268

Gillespie PA, Morita RY, Jones LP (1976) The heterotrophic activity for amino acids, glucose and acetate in Antarctic waters. J Oceanogr Soc Jpn 32:74–82

Grainger EH (1979) Primary production in Frobisher Bay, Arctic, Canada. In: Dunbar MJ (ed) Marine Production Mechanisms 9–30 EBP 20 CUP

Haeckel E (1890) Plankton Studien. (Transl by GW Field) Appendix 6 Rept Comm 1889–1891, US Comm Fish, Fisheries, Washington 1893 pp 565–641

Hart TJ (1942) Phytoplankton periodicity in Antarctic surface waters. Discovery Rep 21:261–365

Herbert RA (1981) Isolation and identification of photosynthetic bacteria (Rhodospirillaceae) from Antarctic marine and freshwater sediments. J Appl Bacteriol 41:75–80

Herbert RA, Bell CR (1974) Nutrient Cycling in the Antarctic marine environment. Br Antarct Surv Bull 39:7–11

Herbert RA, Tanner AC (1977) The Isolation and Characterization of photosynthetic bacteria from Antarctic marine sediments. J Appl Bacteriol 43:437–446

Hodson RE, Azam F, Holm-Hansen O, Karl D, Carlucci AF (1978) Bacterially medicated turnover of dissolved organic matter in Antarctic seawater. Abstr Annu Meet Am Soc Microbiol 78: pp 173

Hodson RE, Azam F, Carlucci AF, Fuhrman JA, Karl DM, Holm-Hansen O (1981) Microbiol uptake of dissolved organic matter in McMurdo Sound, Antarctica. Mar Biol 61:89–94

Holdgate MW (1964) Terrestrial ecology in the maritime antarctic. In: Carrick R, Holdgate MW, Prevost J (eds) Biologie Antarctique. Hermann, Paris pp 173–179

Johannes RE (1968) Nutrient regeneration in lakes and oceans. In: Droop MR, Ferguson Wood EJ (eds) Advances in Microbiology of the Sea. Vol 1 Academic, New York, pp 2–30–213

Kelly MD, Lukaschewsky S, Anderson CG (1978) Bacterial flora of Antarctic krill and some of their enzymatic properties. J Food Sci 43:1196–1197

Kriss AE (1970) Ecological-geographic patterns in the distribution of heterotrophic bacteria in the Atlantic Ocean. Mikrobiologiya 39:362–371

Kriss AE, Mishustina IE, Lebedeva MN (1969) Bacterial population densities (heterotrophs) in the water column of the Southern and Indian Oceans. Mikrobiologiya 38:511–517

Kriss AE (1972a) Quantitative distribution of heterotrophic bacteria in the Drake Passage. Mikrobiologiya 41:733–738

Kriss AE (1972b) Character of distribution of the heterotrophic microbial population in different regions of the Antarctic Ocean. Mikrobiologiya 46:1091–1098

Kriss AE (1973) Quantitative distribution of heterotrophic bacteria in the Southern Ocean between New Zealand and Antarctica. Mikrobiologiya 42:913–917

Lee VV, Shread P, Furniss AL, Bryant TN (1981) Taxonomy and description of *Vibrio fluvialis* sp. nov. J Appl Bacteriol 50:73–94

Llano GA (1962) The terrestrial life of the Antarctic. Sci Am 207: 212–230

Locati GA, Espeche ME, Fraile ER (1980) Changes in the bacteriological chemical and sensory characteristics of Antarctic krill during storage at 0–2°C. Rev Argent Microbiol 12:44–51

McLean AL (1918) Bacteria of ice and snow in Antarctica. Nature (Lond) 102:35–39

Mitsketvich IN (1974) Number and distribution of heterotrophic bacteria in South Atlantic and adjacent parts of the Southern Ocean. Mikrobiologiya 43:535–542

Morita RY, Gillespie PA, Jones LP (1971) Microbiology of Antarctic seawater. Ant J US 6:157

Pfister RM, Burkholder PR (1965) Numerical taxonomy of some bacteria isolated from Antarctic and tropical seawaters. J Bacteriol 90:863–872

Pirie JH (1912) Notes on Antarctic bacteriology. In: Bruce WS (ed) Scottish National Antarctic Expedition. Report on the scientific results of the voyage of SY "Scotia" 1902–1904 111. Scottish Oceanographical Laboratory Edinburgh pp 137–148

Rittenberg SC (1963) Marine bacteriology and the problem of mineralization. In: Oppenheimer CH (ed) Symposium on Marine Microbiology, Thomas, Springfield, Illinois pp 48–60

Seki H, Ohwada K, Tag N (1966) Utilization of carbon by microbial flora in stored sediments. Bull Nat Sci Mus (Tokyo) 9:387–412

Stokes JL, Redmond ML (1966) Quantitative ecology of psychrophilic microorganisms. Appl Microbiol 14:74–78

Tanner AC (1981) The microbiology of Antarctic marine sediments. PhD Thesis, University of Dundee

Tanner AC, Herbert RA (1981) Nutrient regeneration in Antarctic marine sediments. Kiel Meeresforsch Sonderh 5:390–395

Tanner AC, Herbert RA (1982) A numerical taxonomic study of Gram-negative bacteria isolated from the Antarctic marine environment. CNEXO (Actes Colloq) 14:31–38

Tsiklinsky M (1980) La flore microbienne dans les regions du Pole Sud Expedition antarctique francaise 1903–1905, Masson et Cie (eds). 3 Paris

Vargo GA (1968) Studies in phytoplankton ecology in tropical and subtropical environments of the Straits of Florida and its relation to physical factors. Bull Mar Sci Gulf Carib 18:5–60

Walls NW (1967) Bacteriology of Antarctic region waters and sediments. Ant J US 2:192–193

Walsh JJ (1969) Vertical distribution of Antarctic phytoplankton. Limnol Oceanogr 14:86–94

Wiebe WJ, Hendricks CW (1974) Distribution of heterotrophic bacteria in the transect of the Antarctic Ocean. In: Colwell RR, Morita RY (eds) Effects of the Ocean Environment on Microbiol Actitvity. University, Baltimore pp 524–535

Whitaker TM (1977) Plant production in inshore waters of Signy Island Antarctica. PhD Thesis, University of London

Whitaker TM (1982) Primary production of phytoplankton at Signy Island, South Orkneys. Proc R Soc Lond B Biol Sci 214:169–189

Zdanowski MK (1981) Growth of bacteria in the course of decomposition of *Euphausia superba* Dana. Bull Acad Pol Sci 29: 155–161

Heterotrophic Processes in a Kerguelen Mussel-Bed

D. DELILLE and G. CAHET[1]

Summary. A large mussel-bed occurs at Kerguelen Island, almost entirely blocking Bossière Fjord which is one of the longest inlets of the archipelago. This report describes the heterotrophic processes in the water and sediments attending the mussel-bed. Previous studies, involving physico-chemical parameters (temperature, salinity, mineral and organic C and N, dissolved oxygen, Eh, Ph and sulphides), failed to provide a satisfactory answer to the question of how the large mussel-bed is maintained. During the course of 1 yr, water and sediment samples were taken at 3 stations: 2 (at 25 and 9 m depth, respectively) on the 'downstream' side of the mussel-bed, and the third (at 2 m depth) in the 'upstream' part of the fjord close to 'continental' water inlets. Water and sediment samples were examined for their heterotrophic potential with ^{14}C compounds (glucose-glutamate), epifluorescence and MPN with Zobell's medium. Complementary tests were run with tracers to study assimilation dynamics (Vmax), using water samples. Sediment was analyzed for organic material (C and N), by gross methods or with organic fractions after chemical treatment.

1 Introduction

The Kerguelen archipelago, in the Sub-Antarctic region, is free of bacterial pollution (Delille 1977a). The natural bacterial populations react strongly to different environmental parameters (e.g., temperature, Eh, trophic source) in Morbihan Bay and particularly in Bossiere Fjord of the archipelago (Delille and Lagarde 1974; Delille 1977a,b; Delille et al. 1979). Access to the uppermost section of Bossiere Fjord is virtually blocked by the presence of a large bed of mussels, *Mytilus edulis kerguelensis.* In this paper we attempt to define the role that heterotrophic processes play in the functioning and the maintenance of this mussel-bed.

2 Methods and Materials

2.1 Sampling Stations

Kerguelen archipelago represents, with Heard Island, the only emerged part of the Kerguelen-Gaussberg Ridge. The archipelago covers some 6,500 km^2 and has a total coastline of some 2,800 km. In the south-east, Morbihan Bay (about 30 times 20 km) opens to the Indian Ocean through Royal Pass which is 12 km wide and about 40 m deep. To the northwest, there is a complex system of islands and a region of fjords which are natural collectors of rain- and melt-water. The largest of these fjords, Bossière Fjord (Fig. 1), is about 15 km long and on average more than 100 m wide but only a few 10's m wide in the narrowest channels. Access to the uppermost waters at Station 0 (2 m deep, rough sand) is made difficult by the mussel-bed, located between Station M (10 m deep, "fluid" mud, usually anoxic) and Station N (15 m deep, fine sand), leaving room for only a narrow, straight channel with significant tidal currents. The main rivers enter the fjord near Station 0, the only station where the water is relatively fresh (Delille 1977a).

2.2 Microbiology

Total bacterial abundance in the water was determined from 2-ml aseptic samples, fixed with filtered (0.2 μ) formol and kept at $-20°C$; for sediments, a pre-dilution was necessary. The microflora were counted in epifluorescence (Olympus BHA microscope) after filtration (0.2 μ Nuclepore) and an acridine orange stain (Hobbie et al. 1977, slightly modified). Counts of the viable aerobic-heterotrophic microflora were made by means of the MPN method, on Zobell 2216 medium (Oppenheimer and Zobell 1952).

Measurements of the heterotrophic potential of microflora in the water were made for Station M only. Estimates of Vmax were made following the Hobbie and Wright (1968) method adapted by Allen (1969). After addition of 0.10., 0.33, 0.65 and 1 31 μg ^{14}C U glucose per 1 (250 μ Ci mmole) or 0.09, 0.19, 0.38 and 0.76 μg ^{14}C U glutamic acid per 1 (250 μ Ci mmole), incubation was carried out in in situ temperature conditions; the incubation time being calculated in such a way that uptake was between 1% and 10% of the labelled substrates. For the sediments, the heterotrophic potential was evaluated by injecting a given quantity of (100 μl, 1 μ Ci ml) of ^{14}C U glucose or ^{14}C U glutamatic acid to 2-ml microcores. After incubation, a succession of differential extractions (HCl 2N, pyrophosphate + NaOH O, 1N, HCl 6N, combustion) allowed us to

[1] Laboratoire Arago, 66650 Banyuls-sur-Mer, France

Antarctic Nutrient Cycles and Food Webs
(ed. by W.R. Siegfried, P.R. Condy, and R.M. Laws)

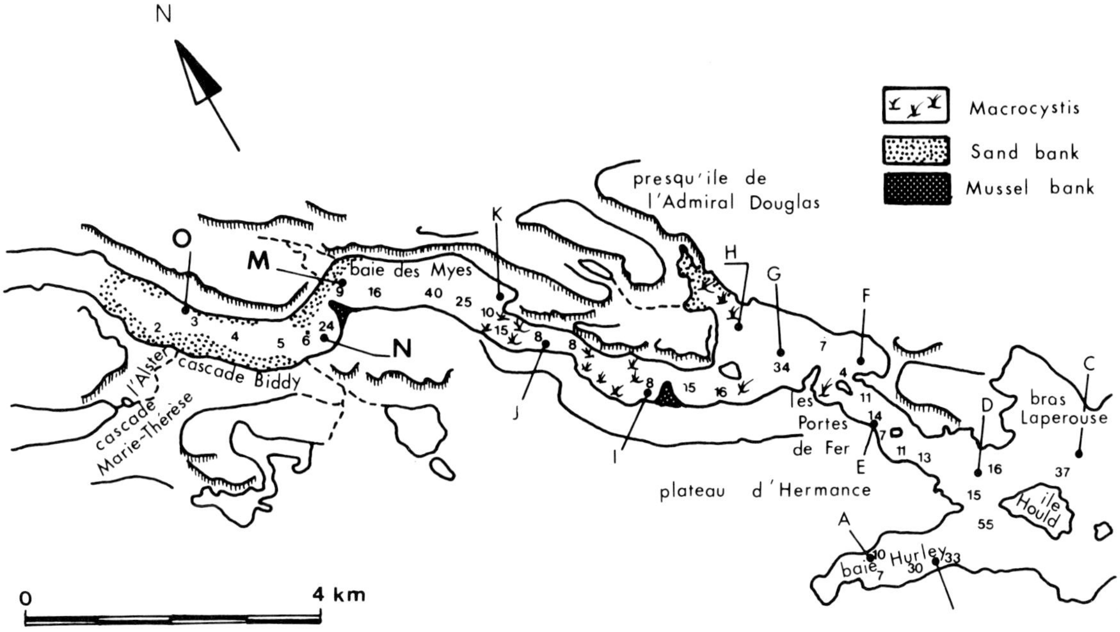

Fig. 1. Bossière Fjord (69° 40'S, 49° 20'W), showing locations of three sampling stations (*M*, *N*, and *O*) in relation to the mussel-bed

separate 5 fractions called: '^{14}C *respiré*, ^{14}C *acidosoluble*, ^{14}C *humus*, ^{14}C *hydrolyable de l'humine*, ^{14}C *residuel*' (Cahet and Delille 1982).

2.3 Chemistry

Carbon was oxydized (1200°C) in an induction furnace apparatus (LECO WR 12). The total organic carbon (C org. T) was estimated on separate aliquots of sediment treated with H_3PO_4 1M 60°C for decalcification and then by desiccation without washing. The amount of bound particulate carbon (CLP) was determined by adding HCl 10% and then washing with distilled water. The dissolved organic fraction obtain after decalcification (difference between C org. T and CLP) contains dissolved or weakly bound compounds (CH : acidosoluble carbon). The carbonate C corresponds to the difference between total C of the entire sample without treatment and total organic C.

Nitrogen was determined by the classical Kjeldahl method (vapour training and distilled NH_4^+ measurement), ammonium by direct distillation without early treatment, oxidized nitrogen (NO_2^- + NO_3^-) by second distillation of the same sample reduced by nascent hydrogen. Organic nitrogen corresponds to the difference between total nitrogen obtained after mineralization (H_2SO_4 + catalyzer at 400°C) and mineral nitrogen (NH_3 + NO_2^- + NO_3^-).

3 Results

3.1 Water

Water temperature varied seasonally, with a 9°C maximum in January–February and a 2°C minimum between June

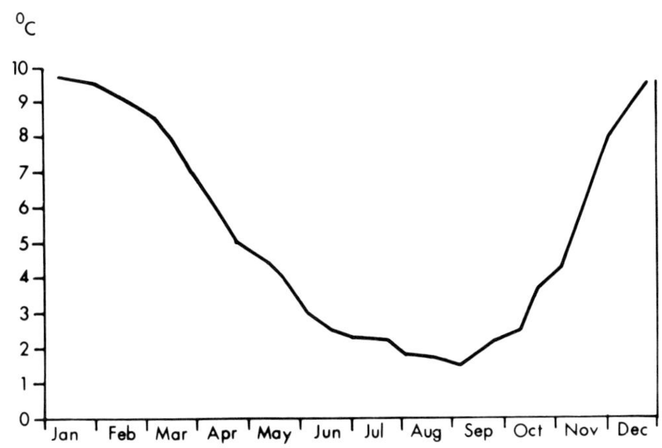

Fig. 2. Mean monthly temperatures of surface water in Bossière Fjord during January–December 1981

and October (Fig. 2). This is a typical pattern close to that observed during 1971–1972 (Delille 1977b).

The total microflora and abundance (direct counts) ranged between 10^5 and 10^7 bacteria ml^{-1}, with a mean value near 10^6 bact. ml^{-1}. The bacterial abundance showed a regular seasonal pattern. The distribution of values was the same for both of the upstream stations (N and O) (serial comparison, r = 0.84) with a single maximum in January, followed by a progressive decrease until August. The downstream station (M), with a higher average, showed a clear bimodal periodicity for the summer maximum (Fig. 3).

Counts of the viable aerobic-heterotrophic microflora ranged between 10^2 and 10^4 bact. ml^{-1} (Fig. 3), and showed a bimodal seasonal cycle with 2 maxima (January–February, and July–August). The upstream station (0) had an earlier maximum than the other two stations (serial comparison, r = 0.63). The heterotrophic potential confirms this general

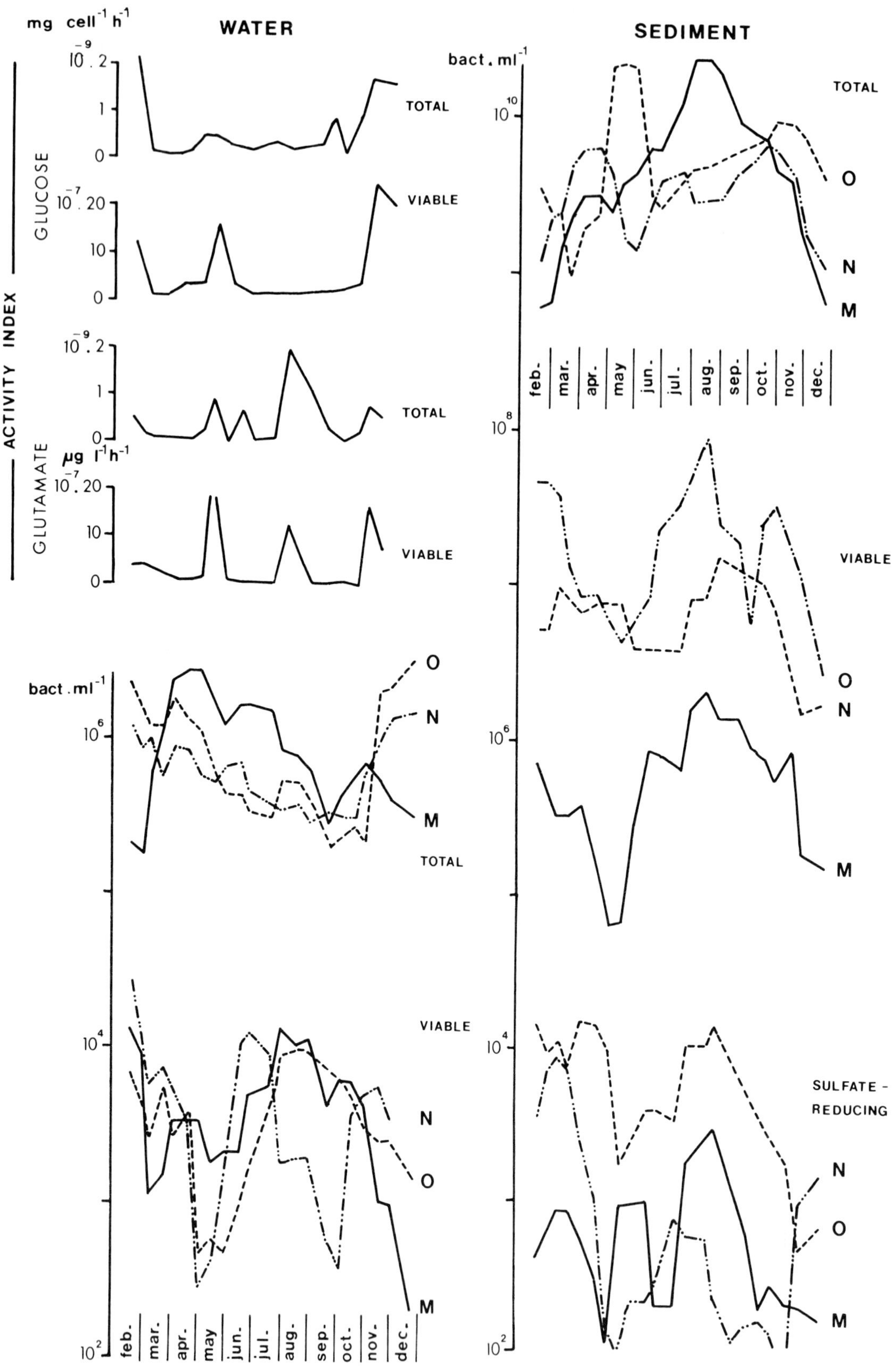

Fig. 3. Bacteriological parameters in water and sediments at three sampling stations (*M*, *N*, and *O*) in Bossière Fjord during February–December 1981

Table 1. Seasonal heterotrophic potential in surface water at Station M in Bossière Fjord during March–December 1981

	Dates																	
	27/3	8/4	22/4	7/5	20/5	4/6	18/6	2/7	22/7	2/8	19/8	3/9	8/10	21/10	4/11	18/11	2/12	26/12
Vmax ($\mu g\ l^{-1}\ h^{-1}$)																		
Glucose	1.9	Sat.	0.2	0.4	Sat.	0.4	0.4	0.9	Sat.	Sat.	0.5	Sat.	0.05	0.2	0.04	0.5	0.5	0.5
Glutamate	0.6	0.1	Sat.	0.09	0.2	0.2	0.8	0.2	0.4	0.05	0.04	0.9	0.2	0.2	0.03	0.1	0.25	0.2
Turnover time (h)																		
Glucose	1	11	3	2.3	–	0.5	4	1.5	–	–	2.5	–	7	2	23	13	6.5	2.5
Glutamate	1.5	8	–	13	4.5	1	2	0.2	3.5	3	15	3	2.5	6	40	6	3	4.5
Assimilation %																		
Glucose	44	47	45	40	73	20	52	13	73	86	46	79	40	60	71	87	86	76
Glutamate	24	47	49	40	78	87	87	2	82	32	72	33	81	21	40	74	88	77

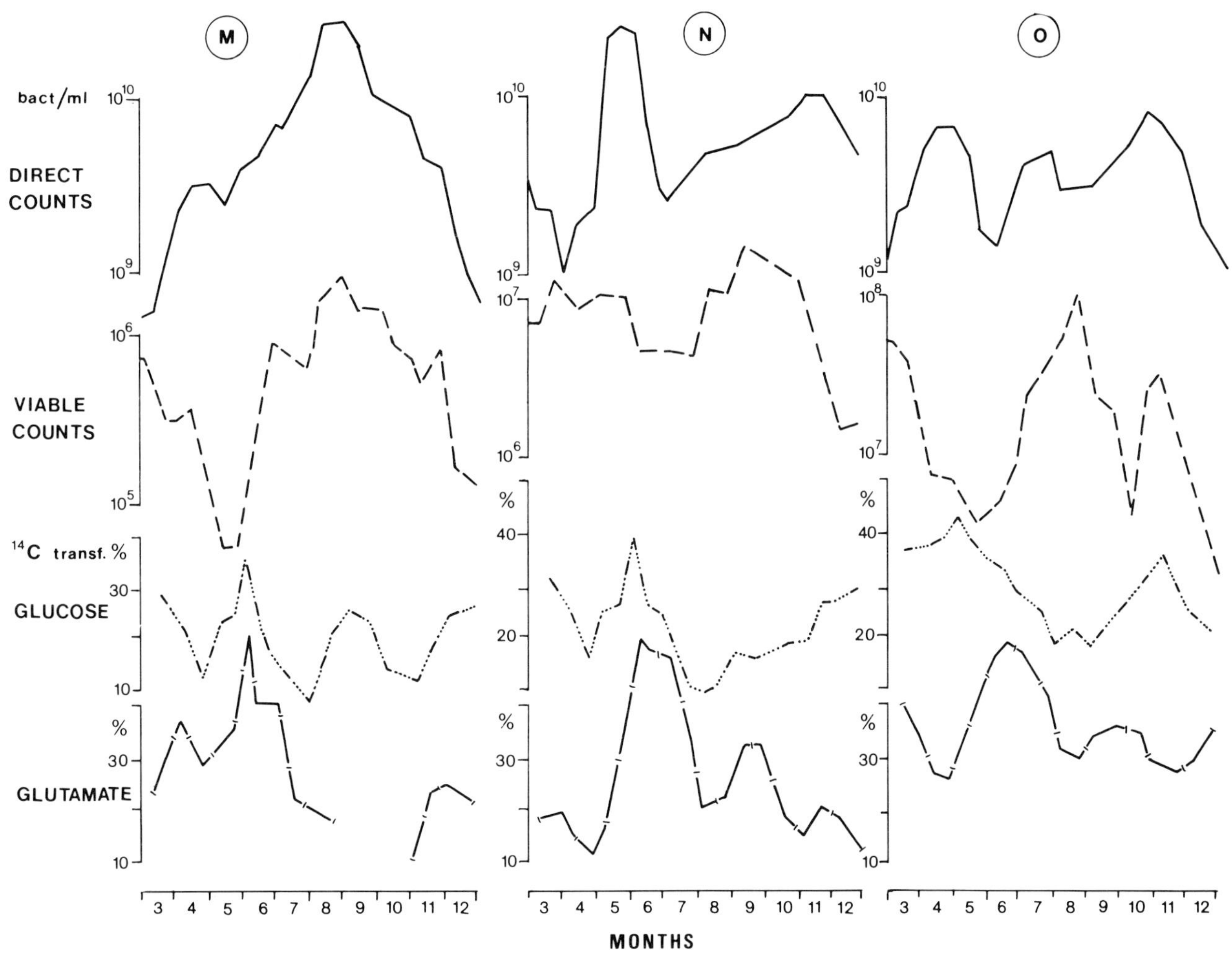

Fig. 4. Bacterial abundance and activity relationships in sediments at three sampling stations (*M*, *N*, and *O*) in Bossière Fjord during March–December 1981

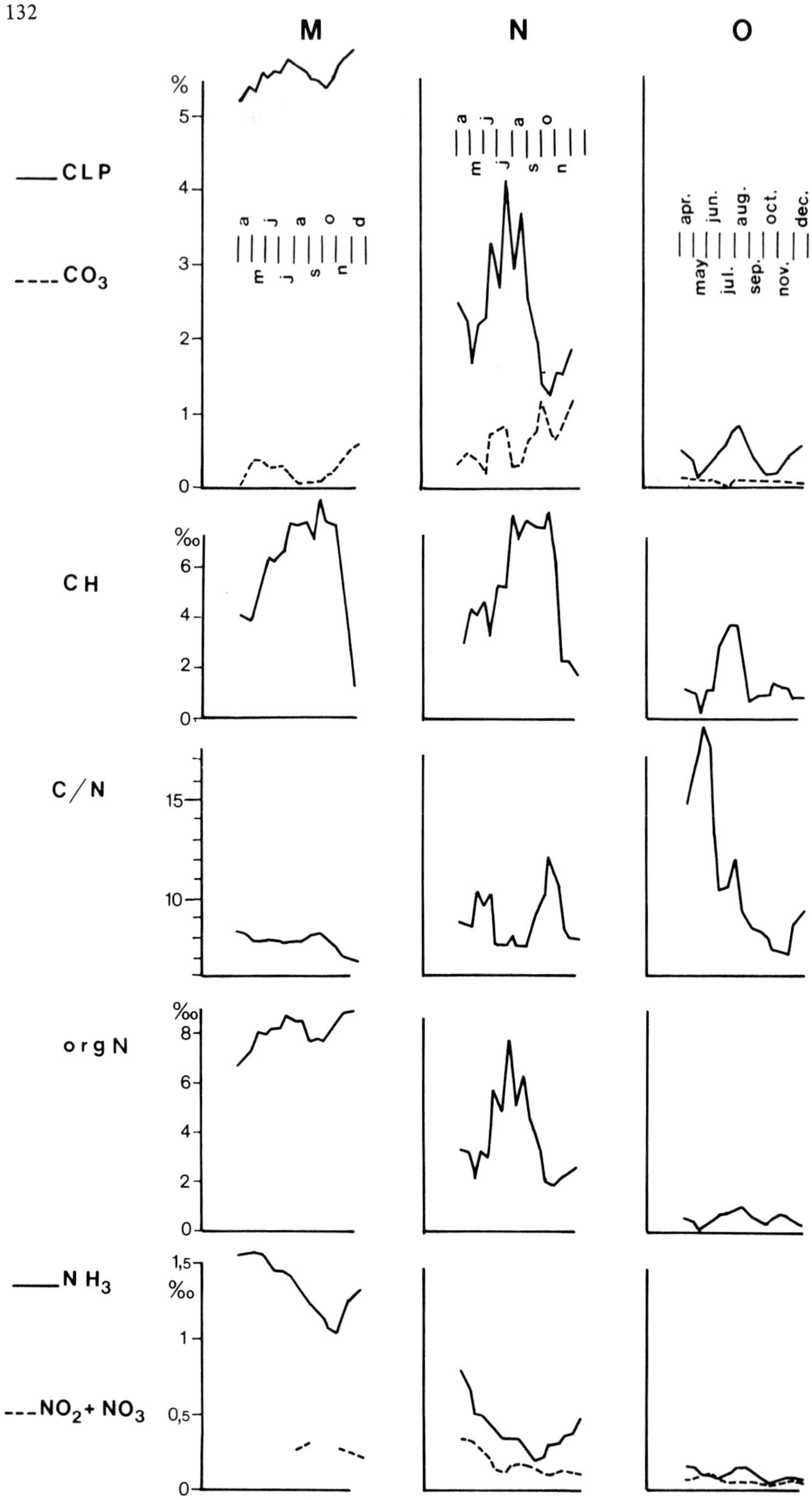

Fig. 5. Physicochemical parameters in sediments at three sampling stations (*M*, *N*, and *O*) in Bossière Fjord during March–December 1981

development (Table 1). Maximum Vmax values, close to 1 $\mu g\ l^{-1}\ h^{-1}$, were observed during bacterial growth periods (27 March, 18 June, 3 September). As Albright et al. (1980) showed for Sub-Arctic rivers, we found numerous saturation responses with glucose. This type of response is rare with glutamic acid, which permitted us to assess the chronological sequence of events. The turnover times were in general fairly uniform (shorter than 5 h) but exceptional events were recorded (19 August, 6 November), limited in time but with a strong amplitude (up to 23 h for glucose and 40 h for glutamate). These turnover times are shorter than those found by Albright et al. (1980) and Gillespie et al. (1976) at the same latitudes but in off-shore waters.

The percentage assimilation was very high in spring and autumn, and fell below 50% in summer. It is unstable in winter. In terms of biochemical systematics, we observed a

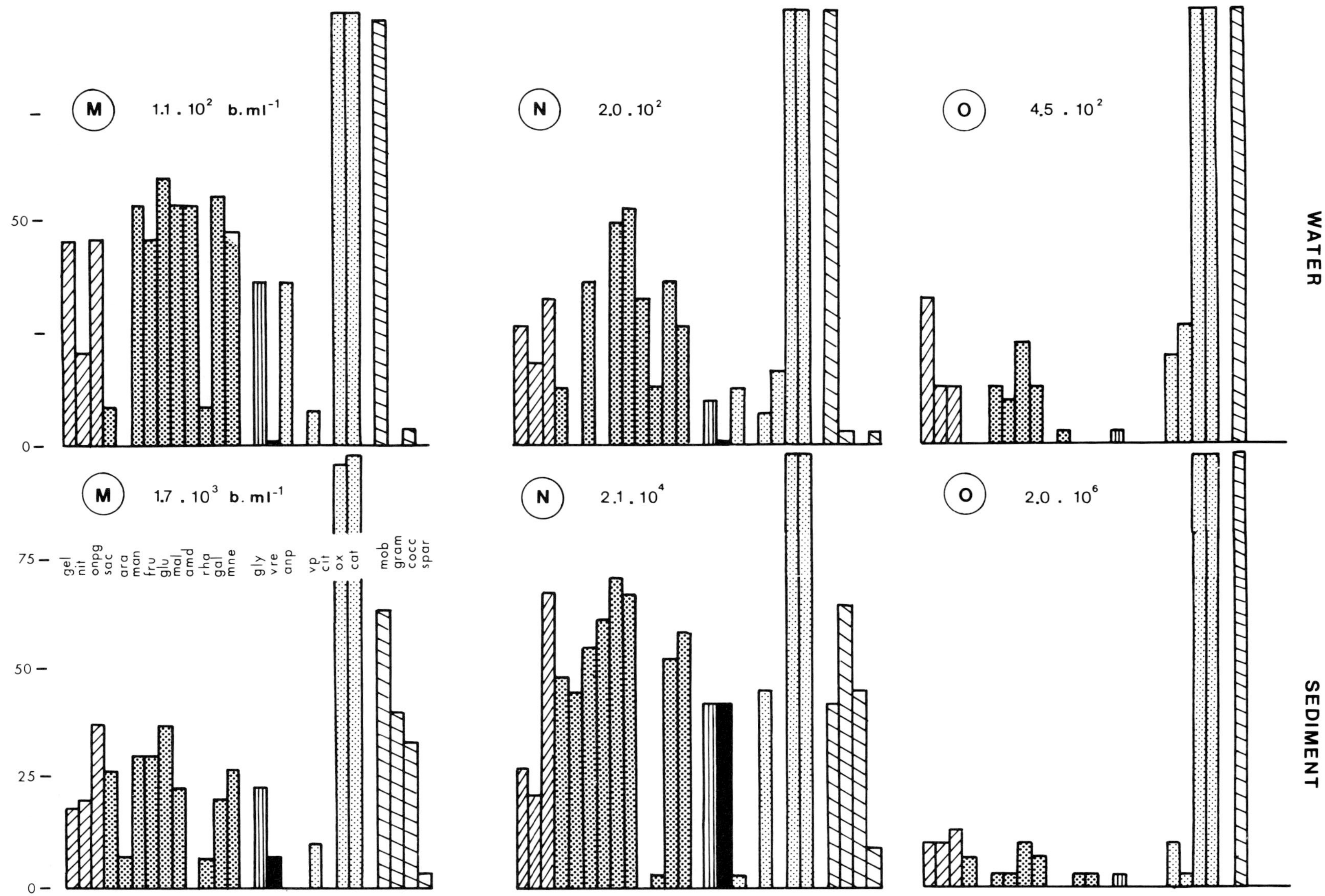

Fig. 6. Percentage positive response to different biochemical tests of bacteria isolated from water and sediments at three sampling stations (*M*, *N*, and *O*) in Bossière Fjord

regular distribution of the microflora as a function of the location of the sample stations. The diversity decreased towards the head of the fjord (Fig. 6).

3.2 Sediment

The total microflora abundance ranged from 10^9 to 10^{11} bact. ml^{-1} (Fig. 3). The seasonal variation is rather different from that of the water, and also differed between the three sampling stations. A first phase of general growth in March broke off in April at Station 0 and in May at Station N, but continued until August at Station M. A second growth phase in September–October occurred at Station N and O, but was indistinct at Station M. In November–December a uniform pattern of variation was found again for the three stations, corresponding to a general decrease in the microflora.

Counts of the viable heterotrophic microflora ranged between 10^5 and 10^8 bact. ml^{-1} for aerobic and 10^2 and 10^4 bact. ml^{-1} for sulphate-reducing bacteria (Fig. 3). As in the overlying water, the cycles were nearly bimodal with 2 maxima (February–March, and August–September) which is normal for the study area (Delille 1977a; Bovee et al. 1983). The mean phases of bacterial growth were preceded by a distinct renewal of heterotrophic activity in May–June (glucose and glutamate for Stations M and N, glutamate only for Station 0) (Fig. 4). At a taxonomic level, the general decrease in diversity in an up-fjord direction was masked partially by the very high diversity at Station N (Fig. 6).

The three stations were very different in terms of organic matter (Fig. 5). We observed a major decrease in CLP from Station M to Station O (annual means = 5.62%, 2.34%, and 0.39%); this difference being less marked when considering only the CH fraction. The seasonal variations were regular and uniform for the three stations, with enrichment occurring from May to July and reduction from August to November. The C:N ratio apparently indicates the presence of an increasing terrestrial influence with distance up the fjord. We also observed a succession of terrestrial inputs in May, July and September for Stations N and O.

4 Discussion

Our data serve to confirm several findings made previously (Delille and Cahet 1982). First, the variations in heterotrophic potential are relatively short-lived and tend to precede phases of viable heterotrophic growth. Inversely, turnover times, which are very long, may precede phases of decreasing micro-floral biomass. Secondly, the total counts reflect the actual abundance, but their variations in time are less pronounced than shown by the other microbiological parameters.

Temperature does not seem to play a very important role in these processes as long as it remains above 5°C. Below this value its influence on heterotrophic potential may become dominant (e.g., Gillespie et al. 1976), inducing saturation responses in water for glucose and low values of glutamate Vmax. In this study we were able to make use of a specific activity index (Wright 1978; Goulder 1979) which was used by Albright et al. (1980) in Sub-Arctic waters. We, similarly, observed very low index values per cell at the end of autumn and in winter. Albright et al. (1980) interpreted such responses as stress, due to the low quantity and quality of DOC and low temperatures. However, our results show that the response is better with glutamate during this cold period and is related to the viable microfloral growth. We found a very high activity index at the end of spring, during summer and in early autumn, an indication of increased organic supply (Wright 1978) in the water. This supply of organic matter contributes, together with relatively mild temperatures, to the observed bacterial bloom.

In the sediments, the influence of the enrichment in organic matter in June–August results in a significant growth of microflora at all stations. At Station O, the occurrence of two peaks of organic matter is reflected by two peaks of both total bacteria and viable microflora. The early autumn (March–April) maximum in viable heterotrophic microflora is not related to organic input, and does not involve an overall growth in the total bacterial biomass. This maximum appears to correspond to a re-activation of the cells present, rather than a real growth in the entire microflora. The episodic maxima observed in the distribution of the total microflora (May for Station N and April for Station O) are associated with an increase in heterotrophic potential (glucose and glutamate for Station N, glucose alone for Station O), but are not linked with any viable heterotrophic cell variation. These maxima are believed to be typical of opportunistic bacterial growth.

In the water, the anomaly observed in the total microflora from December to March at Station M, in contrast to the pattern at Station N and O, might interpreted as a deficit due to predation processes in the mussel-bed. This hypothesis is supported during this period by maxima of specific activity and bacterial assimilation (up to 90%). In terms of taxonomy, the mussel-bed does not seem to influence the microflora of the water. It may, however, play a major role in the microflora of the sediment at Station N. The presence of a positive response to sugars (46% vs. 20% for Station M and 4% for Station O) and of a positive urease response (42% vs. 7% for Station M and 0% for Station O) indicates a specialization of the microflora towards the degradation of zoological wastes. Accumulations of such products at Station N would also explain the relative enrichment in acid-soluble carbon compounds (CH), with a CH:CLP ratio twice as high as at Station M, and in carbonate carbon.

References

Albright LJ, Masuda KV, Ennis GL, Schreier H (1980) Microbial dynamics of two subarctic canadian rivers. Water Res 14:1353–1362

Allen HL (1969) Chemiorganotrophic utilization of dissolved organic compounds by planktonic algae and bacteria in a pond. Int Rev Gesamten Hydrobiol 54:1–33

Bovee F de, Cahet G, Delille D, Soyer J (1983) Bacteria meiofauna relationship in a subantarctic sand beach (Kerguelen). Proceed Intern Symp "Sandy beaches as ecosystems". Ed Junk Amsterdam

Cahet G, Delille D (1984) Disponibilité du carbone organique élaboré par voie photo, chemo et hétérotrophe dans divers sediments littoraux. Coll Intern Microbiologie Marine Marseille. 205–211

Delille D (1977a) Contribution à l'étude du role des populations bactérienne dans les principaux cycles biologiques (Cn, N, S, P) établis en milieu sédimentaire subantarctique (archipel de Kerguelen). These Doct Sci Nat Univ Claude Bernard, Lyon 286 pp

Delille D (1977b): Bacterial Sulfur and Nitrogen cycles in sedimentary deposits of Fjord Bossiere (Kerguelen Archipelago) In: Llano GA (ed) Adaptations within Antarctic Ecosystems, (Proc. 3th SCAR Symp. Antarctic Biology). Gulf Pub Comp pp 159–180

Delille D, Cahet G (1984) Croissance de populations hétérotrophes subantarctiques soumises a des enrichissements azotes. Effet des hydrocarbures. Coll Intern Microbiologie Marine Marseille. 213–219

Delille D, Lagarde E (1974) Recherches sur les cycles biologiques du soufre et du carbone dans les dépots sédimentaires de la baie due Morbihan (archipel de Kerguelen). Rev Inst Pasteur Lyon 7:149–165

Delille D Gadel F, Cahet G (1979) La matière organique dans les dépots de l'archipel de Kerguelen. Distribution spatiale et saisonnière. Ocenol Acta 2:181–193

Gillespie PA, Morita RY, Jones LP (1976) The heterotrophic activity for amino acids, glucose and acetate in antarctic waters. J Oceanogr Soc Jpn 32:74–82

Goulder R (1979) Vmax per bacterium and turnover rate per bacterium for glucose mineralization in natural waters. Curr Microbiol 2:365–368

Hobbie JE, Wright RT (1968) A new method for the study of bacteria in lakes: description and results. Mitt Int. Ver Theor Angew Limnol 14:64–71

Hobbie JE, Daley RJ, Jasper S (1977) Use of nuclepore filters for counting bacteria by fluorescence microscopy. Appl Environ Microbiol 33:1225–1228

Oppenheimer CH, Zobell CE (1952) The growth and viability of sixty three species of marine bacteria as influenced by hydrostatic pressure J Mar Res 11:10–18

Wright RT (1978) Measurement and significance of specific activity in the heterotrophic bacteria of natural waters. Appl Environ Microbiol 36:297–305

Part 2
Terrestrial and Freshwater Nutrient Cycles

Nutrient Cycling in Relation to Biological Productivity in Antarctic and Sub-Antarctic Terrestrial and Freshwater Ecosystems

R.I.Lewis SMITH[1]

Summary. Many terrestrial and freshwater ecosystems and their component ecological units in the Antarctic and Sub-Antarctic have been characterized in terms of single sets of chemical data. The biota and their growth, production and life cycles are often then related directly to the nutrient status of the system. However, such extensive studies provide no indication of the nutrient dynamics of the system. At only a few sites has intensive research involving frequent sampling and analysis on a seasonal or year-round basis been undertaken, mainly within the past 10–15 yr. Major advances have been made in recent years in investigating and understanding the processes involved in biogeochemical cycling, particularly of the major elements responsible for growth and production. Nutrient pathways through mineral and organic substrata, and contributions by precipitation, seaspray, fauna, etc., are gradually being quantified. The herbivore cycle in these relatively simple ecosystems is almost absent in terrestrial environments and much reduced in freshwater environments compared with corresponding ecosystems in other biomes. Food chains are short and virtually cease at the invertebrate herbivore level. Consequently, the decomposer cycle assumes the dominant role in the biological cycling of nutrients. The seasonal activity of critical groups of micro-organisms in relation to nutrient uptake and release is being studied experimentally both in the field and in the laboratory. The present state of this rapidly expanding area of research in the Antarctic and Sub-Antarctic is discussed, highlighting some of the more important findings. The need to develop more intensive long-term research programmes integrating geological, pedological, hydrological, micro-climatological and biological studies at selected ecologically important sites is emphasized. These regions offer ideal simple environments in which to test fundamental hypotheses concerning the functioning of ecosystems. The complex processes involved in nutrient cycling and their influence on the productivity of the associated biota may be more easily investigated in these south polar biomes than in most others.

1 Introduction

When the International Biological Programme (IBP) Tundra Biome studies of the structure and functioning of tundra ecosystems began in 1967, only three southern hemisphere study sites were selected: South Georgia, Macquarie Island and Signy Island. This systems approach was directed primarily towards assessing biomass, production and decomposition processes, carbon and energy transfer, and the synthesis of data to develop predictive models of energy flow within the ecosystems. Although a study of the main patterns of distribution of nutrients within sites was an integral part of this research, the much more complex study of nutrient cycling was attempted at only some of the more intensively worked Arctic sites, while for a few others only certain compartments within a nutrient cycle model have been evaluated. Most of these accounts have been collected and synthesized by Dowding et al. (1981).

Terrestrial and freshwater biological research in the Antarctic and Sub-Antarctic is much younger than in the corresponding northern biomes. Consequently, much of the current research is following the guidelines established by IBP in the Arctic and completed up to 10 yrs ago. Studies of nutrient cycling in the Antarctic and Sub-Antarctic are still in their infancy, and at only a very few sites has any concerted effort been made in this field in terrestrial and freshwater ecosystems. In this brief review the main areas of progress are highlighted and some suggestions are made on the future direction of research.

2 Terrestrial Ecosystems

Nutrient flux in Antarctic and Sub-Antarctic terrestrial ecosystems and within plants has not been extensively investigated, although occasional studies have been made for about 20 yrs. A brief review of this work has been given by R.I.L. Smith (1984), and most publications relating to soil chemistry and nutrients have been listed in an annotated bibliography by Walton (1980). However, the current interest in ecosystem development and functioning in these regions necessitates the monitoring and interpretation of seasonal changes in nutrient concentrations. The physical and biological processes involved in nutrient cycles are highly complex and consequently little progress has been made in identifying and quantifying the sources, inputs, transfers and losses within and between ecosystems or in assessing the nutrient requirements for the growth, production and survival of the biota.

Nutrient availability and turnover are limited by low metabolic activity and low rates of primary production, which are directly a response to the short growing season and low summer temperatures. With increasing latitude, the declining temperatures restrict the presence and availability

1 British Antarctic Survey, NERC, High Cross, Madingley Road, Cambridge CB3 OET, England

Antarctic Nutrient Cycles and Food Webs
(ed. by W.R. Siegfried, P.R. Condy, and R.M. Laws)

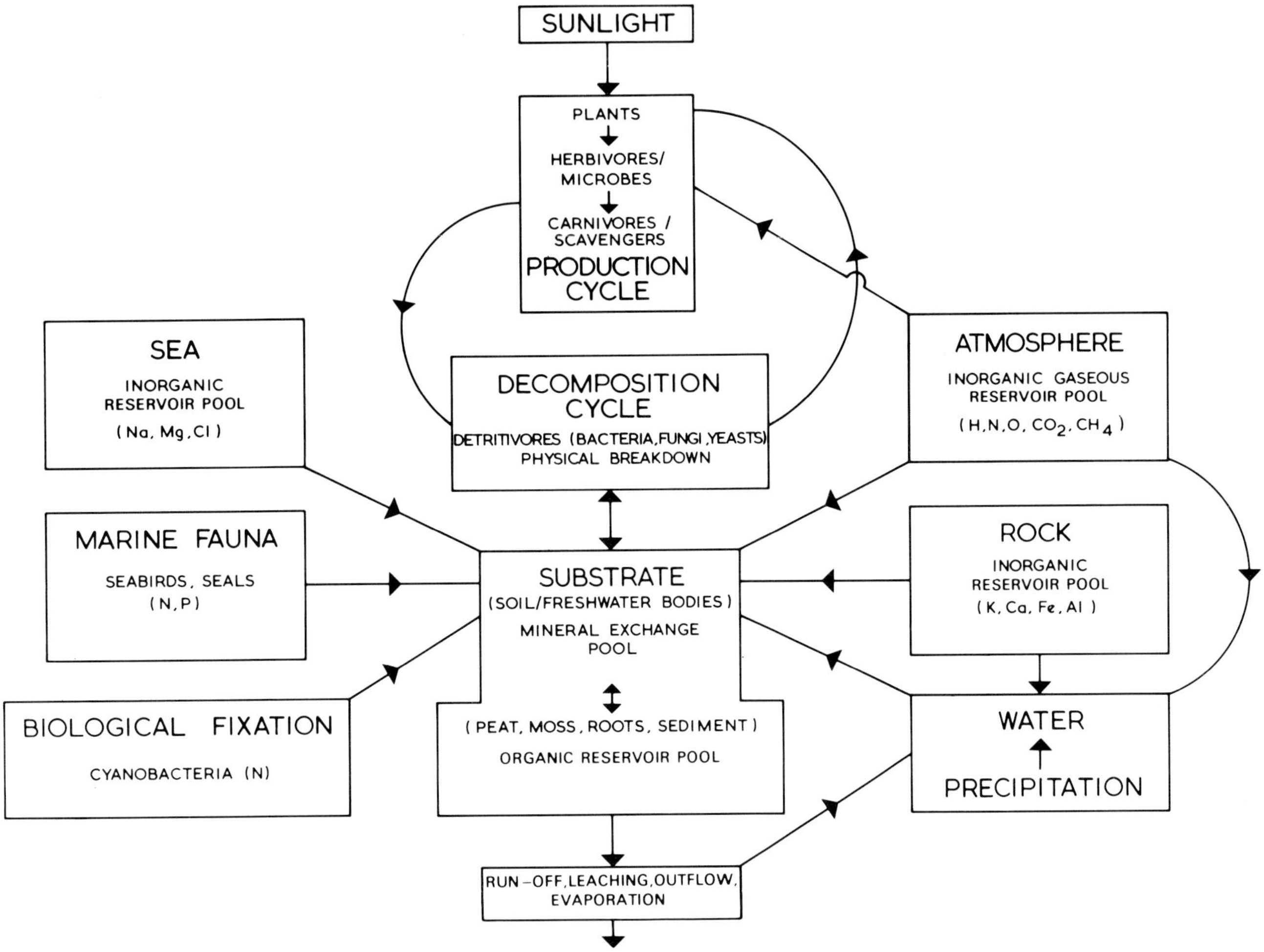

Fig. 1. Schematic representation of the major nutrient pathways in Antarctic and Sub-Antarctic terrestrial and freshwater ecosystems

of free water, which critically reduces biological composition, activity, production and energy flow. However, on several of the Sub-Antarctic islands, the growing season may be long but cool and some plant species continue C-fixation and dry-matter production throughout most of the year (Jenkin 1975); in some ecosystems microbial and detritivore populations and activity are considerable (M. Smith 1983). Consequently, in such systems nutrient turnover may be quite rapid, although in ecosystems with dry, mineral or frigic soils with an open or predominantly cryptogamic vegetation, turnover will be slow. The availability of nutrients, therefore, varies considerably with seasonal patterns of temperature, water and microbial decomposer activity.

A schematic representation of the major transfers of soil and lake nutrients in the polar biosphere is shown in Fig. 1. In a relatively mature ecosystem, nutrient gains and losses are more or less in equilibrium. Marine, precipitation and faunal contributions decrease with increasing distance from the sea and with increasing continentality.

In these polar environments, conditions for biological activity are more favourable below ground both before the beginning of melt and after the winter freeze-up, because of heat transfer through snow and ice and the high heat capacity of the substratum, especially if there is a high organic component. This reduces the number of freeze-thaw cycles at or near the surface and retains heat released when water does freeze (Walton 1982). This asynchrony of favourable conditions for biological activity above and below ground is important, and is responsible for much of the seasonal pattern of nutrient pool sizes. This, in turn, has a significant effect on the fluctuations in, and rate of, nutrient cycling, particularly in Sub-Antarctic ecosystems in which vascular plant root systems play an important role in nutrient absorption in summer and accumulation in winter.

The critical period for the release of most essential nutrients from their various sources is at snow-melt, when air temperatures rise above 0°C, but diel freeze-thaw cycles are also frequent. In the Sub-Antarctic, however, such cycles are prevalent only on South Georgia and Heard Island, and possibly to a lesser extent in high altitude fellfield (= *fjaeldmark*) on the other islands. Certain nutrients (e.g., P and K) may be released from organic matter before the melt and continue to be released after the commencement of the autumn freeze-up, since microbial activity and decomposition processes continue below 0°C (Fig. 2). Also,

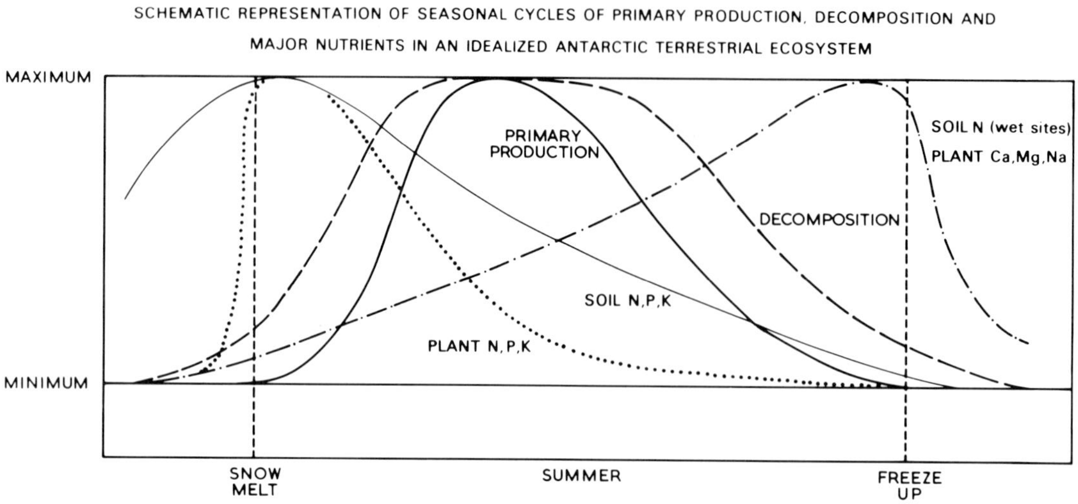

Fig. 2. Generalized seasonal cycles of primary production, decomposition and major nutrients in an idealized Antarctic or Sub-Antarctic terrestrial ecosystem

the thawing of frozen microbial and some plant and animal cells results in membrane damage and leakage of their contents, notably P and K as well as organic molecules such as carbohydrates and amino acids, into the soil solution. Thus, these organic and inorganic compounds reach maximum levels in the substratum immediately after the spring melt but decrease in concentration progressively throughout the growing season as they are utilized in primary production, metabolized or leached out of circulation. In contrast, exchangeable and soluble N (mainly in the form of ammonia) is in low concentration at melt and increases throughout the season. It is not clear why soil N should increase during this period when the increasing biomass of green assimilating material is utilizing large amounts of N, but some may be attributed to biological fixation and, in coastal areas, to contributions by seabirds and seals.

2.1 Marion Island

The only major study of nutrients in the Sub-Antarctic has been on Marion Island where the main objective of the biological programme has been the quantitative assessment of the food and energy cycles operative within the island's ecosystems. Since 1971, this has involved detailed investigations of primary and secondary production and the biogeochemical cycling of nutrients. Particular emphasis has been placed on the inter-relationships between the marine fauna and the vegetation and primary production (V. Smith 1977). More recently an intensive assessment of seasonal trends in soil and plant nutrients has been undertaken (V. Smith, to be publ.). Processes including patterns and rates of utilization, transformation, re-translocation, inputs and accumulation are included in the new long-term research strategy of 1981 as part of the ecological studies programme (SASCAR 1981).

Extensive studies of soil and plant nutrients in selected Marion Island ecosystems have provided an assessment of the nutrient status of the principal plants and vegetation types. Not surprisingly for an oceanic island with high rainfall and extensive organic deposits, the coastal soils have a high cation saturation and appreciable quantities of Na and Mg, although these decrease with increasing distance from the sea (V. Smith 1978a). The strong influence of seabirds and seals around the coasts and the cation absorbing capacity of the organic soils have caused substantial accumulation of organic N and P, which is largely responsible for the high productivity of some of the island's plant communities. However, the strongly acidic nature of these soils and the low temperatures inhibit microbial activity and, hence, organic matter decomposition and mineralization, so that levels of available N (as ammonia- and nitrate-N) are low and reduced by intense leaching and run-off, despite significant N-fixing by cyanobacteria associated with bryophytes in some mire communities. Available P is in low concentration but not deficient for plant growth. However, fellfield soils are low in all nutrients (V. Smith 1978b). Standing stocks of elements in the plant biomass may be high with means of 455 and 640 kg ha^{-1} in the above and below-ground components, respectively, for aggregated data from several low altitude vegetation types. Of the above-ground totals 45% is N and 14% K, whereas of the below-ground total 40% is N and 23% K. Because of the relatively low amount of below-ground standing crop and the accumulation of Na, Mg and K in the aerial parts only 58% of all elements are contained in below-ground plant parts compared with about 80% in comparable Arctic tundra vegetation (V. Smith 1976a). However, in various mire communities, despite large quantities of nutrients contained in the above-ground vascular plant necromass, over 60% of all elements occur in the below-ground component, with 90% (except K) occurring in open mire vegetation (V. Smith, to be publ.). Manuring by seabirds greatly enhances N and P

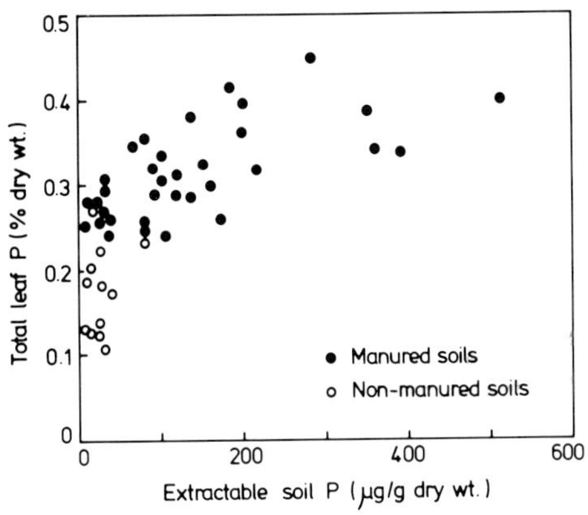

Fig. 3. Relationship between the concentration of extractable soil P and total leaf P in *Poa cookii* growing in bird-manured and non-manured soils on Marion Island. (From V. Smith 1979)

concentrations in the soil and plant leaves (Fig. 3), although it is the increase in N that causes a corresponding increase in P uptake (V. Smith 1978c, 1979); there are also higher concentrations of K, Fe, and Na in soil and plants associated with burrowing petrel colonies (V. Smith 1976b).

A major contribution to Sub-Antarctic nutrient cycling research has been in determining decomposition activity and N-fixation in Marion Island soils, peat and vegetation. Steyn and Smith (1981) found that mire and bog peat exhibit high numbers of aerobic micro-organisms, whereas fellfield microbial counts are low except beneath scattered cushion plants. They showed that sites affected by seabirds and seals possess up to 10 times the number of bacterial and twice the number of fungal colony-forming units than do uninfluenced soils. Lindeboom (1979) obtained similar results in the soil solution of biotically influenced areas. Many bacteria are capable of reducing nitrate to nitrite and also of forming ammonia from nitrate, but few organisms can reduce nitrate to gaseous N. The enhanced microbial population in biotically disturbed sites may be as much a response to the energy-rich proteins, lipids and uric acid excreted by the fauna as to the higher inorganic nutrient status of these sites (Smith and Steyn 1982).

Significant N-fixation has been demonstrated at temperatures between about 10 and 20°C, by free-living mats of cyanobacteria in mire communities and by epiphytic cyanobacteria associated with 14 out of 19 bryophyte species examined in a range of Marion Island communities (Croome 1973). Mire species, in particular *Brachythecium subplicatum,* had high nitrogenase activity (mean acetylene reduction 310 nmol g^{-1} dry wt. h^{-1}), but dry fellfield cushion mosses, despite possessing high densities of cyanobacteria, fixed only small amounts (1–5 nmol C_2H_2 g^{-1} dry wt. h^{-1}) (Smith and Russell 1982). However, only the often-spherical and mobile colonies of the fellfield moss *Ditrichum strictum* appeared capable of acetylene reduction in situ at an ambient temperature below 0°C, although the internal temperature of the colonies was considerably higher, whereas the looser wetter mire turf mosses remained at a low temperature, which inhibited reduction. *Ditrichum strictum* may therefore play an important role in the N-dynamics of nutrient-deficient fellfield habitats (Smith and Ashton 1981).

2.2 South Georgia

Nutrient cycling studies on South Georgia have been few and much of the work has not yet been published. However, in their studies on growth and production of selected plants and communities on the island, Walton and Smith (1980) assessed the seasonal change in the major elements in numerous species of plants. The more mobile elements (N, P, K, Na) tended to reach an early season peak and then decrease, while Ca and Mg tended to increase as the foliage matured. Although considerable differences in concentration were recorded in different species, the levels of an element in a particular life-form tended to be of the same magnitude. The highest levels of most elements were in the shrub-like suffruticose herb *Acaena magellanica* and also in several species growing in wet habitats where there is a small but continuous flow of nutrients through the substratum. Element contents were generally lowest in plants of dry or acid soils. The dense standing dead and litter components of *Poa flabellata* and *Festuca contracta* grasslands are important as a reservoir of nutrients that are released slowly, because of the low decomposition rates (Walton 1985a). This contrasts with *Acaena magellanica* herbfield where the deciduous leaves decay almost totally during winter and nutrient turnover is rapid (Walton 1977). Thus, whereas there is a substantial difference in the rate of recycling between the two ecosystem types, a relatively high proportion of the nutrient requirements of the grasses may be met by the standing dead component serving as a nutrient pool. Pratt and Smith (1982) obtained similar data and related changes in the diet of the introduced Reindeer, *Rangifer tarandus,* of South Georgia to seasonal trends in plant nutrient composition.

A major study by Lawson (1985, unpubl. data) of the production, decomposition and nutrient cycling in the rush *Rostkovia magellanica* growing in an ombrogenous bog and a soligenous mire has revealed the importance of leaf or organ age in relation to its nutrient content throughout the year. While the mean whole plant values indicated the overall seasonal trend, each leaf age class differs significantly in the level of nutrients contained therein (Fig. 4a,b). Also, it is only when individual age classes are analyzed that major differences between elements become apparent. For example, the concentrations of N, P, K, and Na decline as leaf age increases, whereas those of Ca and Fe increase with age; Mg and Mn increase up to the third oldest leaf then decrease in progressively older leaves. The Ca content of mire plants is much greater than in bog plants, whereas the converse is true for Mg and Mn; other elements (e.g., N, P, K, Na, Fe) are similar for corresponding age classes.

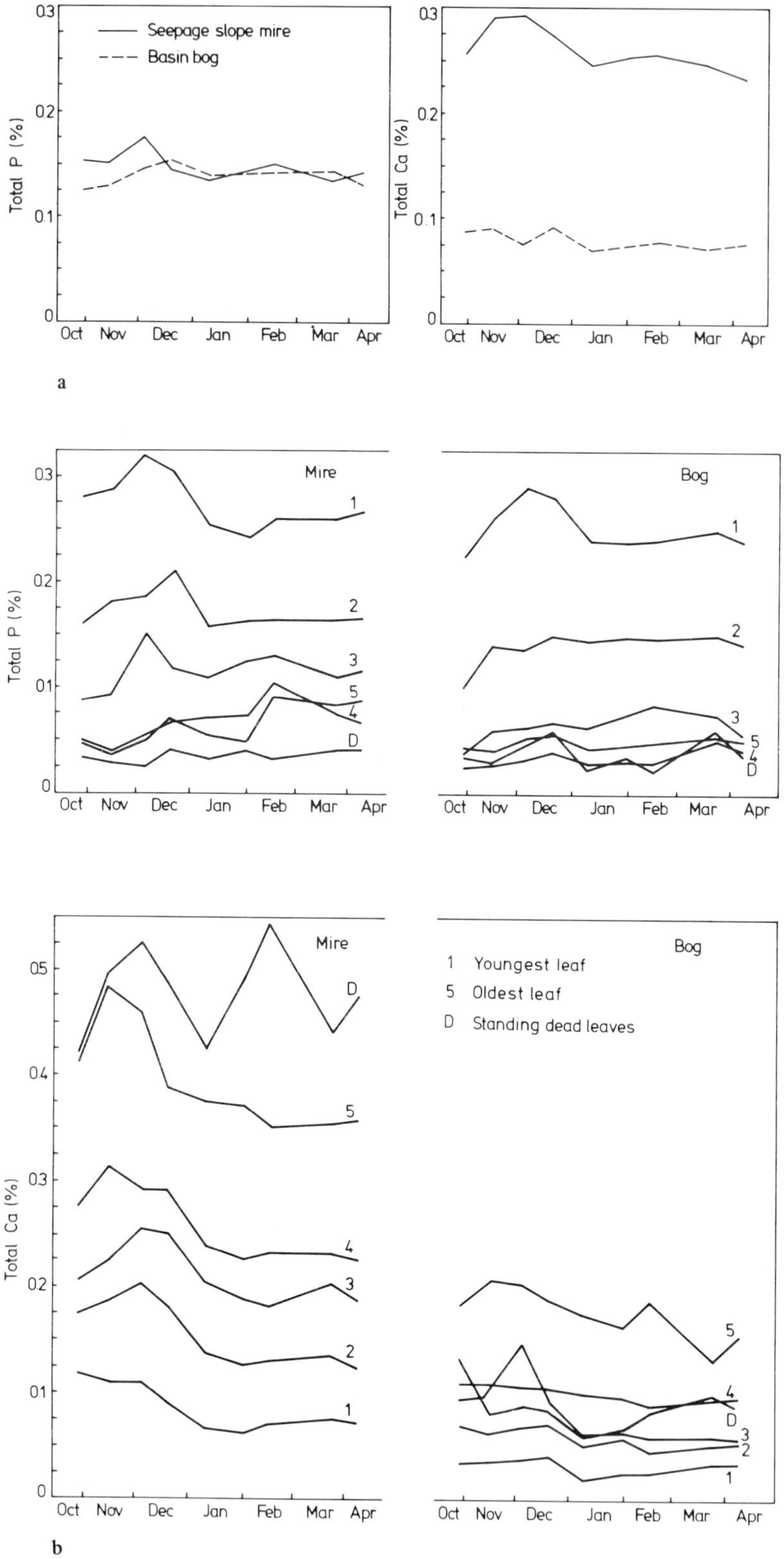

Fig. 4a,b. Seasonal change in total P and Ca in *Rostkovia magellanica* from bog and mire sites on South Georgia. **a** Mean concentrations for all aboveground parts. **b** concentrations in 6 leaf age classes

In his intensive study of the microbial ecology of four ecosystem soils on South Georgia, M. Smith (1983) developed a qualitative model of the major processes in soil nutrient cycling. He emphasized that the absence of native vertebrate herbivores and the relative paucity of the soil fauna must have a constraining effect on nutrient cycling by greatly reducing the degradation of standing material and the comminution of litter as well as by omitting the addition of nitrogenous waste products through faeces. The development and productivity of the soil depend on the current nutrient status and diversity of the flora and fauna as well as on the nutrient retention capacity and longevity

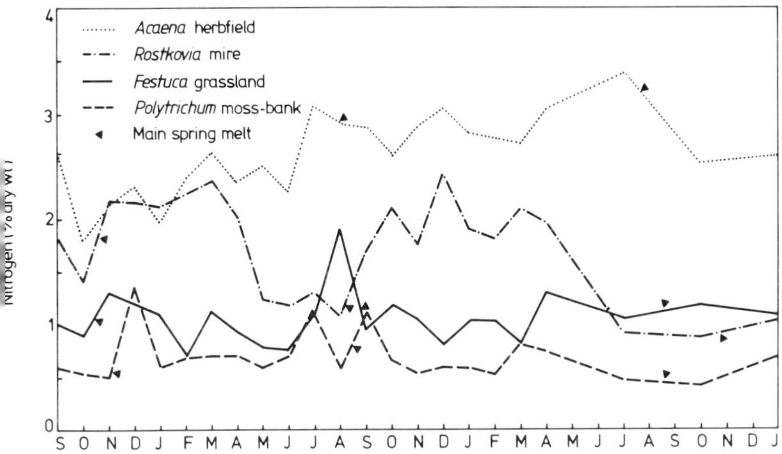

a

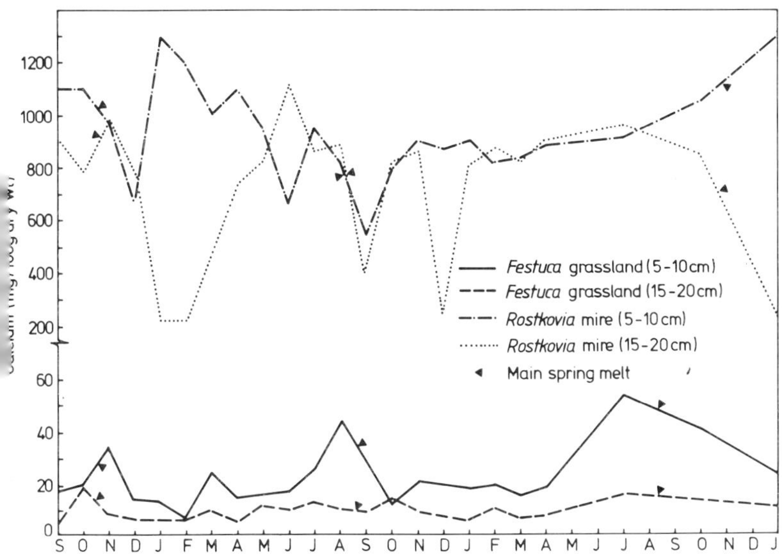

b

Fig. 5. a Seasonal change in total N at 5–10 cm depth in four ecosystem soils on South Georgia. **b** Seasonal change in extractable Ca at 2 depths in a dry grassland and a seepage slope mire on South Georgia

of that flora. This in turn determines the amounts of litter and nutrients available to the decomposer populations. Their response, affected by nutrients and population density, will determine the rate and efficiency of decomposition and hence the balance between nutrient cycling and nutrient accumulation.

Unlike the situation on Marion Island, epiphytic cyanobacteria are seldom prominent in bryophyte communities. Consequently, microbial fixation of atmospheric N is believed to be unimportant in the South Georgia ecosystems, although this has not been tested. At the study sites, seal, and bird influence was minimal and N and P are probably the major limiting factors affecting primary production and microbial biomass, although, as on the other Sub-Antarctic island, wherever there is biotic influence there is a marked increase in plant luxuriance and primary production. Vesicular-arbuscular mycorrhizal associations have recently been discovered in many of the South Georgia vascular flora (Christie and Nicolson 1983) and these may play an important role in the P cycle. Freeze-thaw cycles are suspected of being significant in the release of nutrients in the more severe Antarctic climate but, on South Georgia, M. Smith (1983) found no microbial bloom following freeze-thaw activity and Hurst et al. (to be publ.) demonstrated that repeated freezing and thawing had little effect on the release of K, phosphate or carbohydrates from mature leaves of *Acaena magellanica* or *Poa flabellata.* However, it is possible that freeze-thaw cycles affect only young active leaves.

An attempt to trace seasonal nutrient trends throughout the year in the soil in four ecosystems has indicated the need for replication of sample analysis rather than the bulking of samples to provide a single "average" datum per sample time. The data in Fig. 5a,b show an incoherent trend at two sample depths, but major differences between sites and soil types are reflected by inter-ecosystem differences in microbial biomass (M. Smith 1983) and net annual production (Walton 1976; Tallowin 1977; Smith and Stephenson 1975; Walton et al. 1975; R.I.L. Smith 1982; Lawson, unpubl.).

2.4 Signy Island

The only investigations of nutrient cycling on a seasonal basis in the Antarctic have been carried out at Signy Island, commencing in the early 1960s (Northover and Grimshaw 1967; Northover and Allen 1967), while Allen et al. (1967) and Holdgate et al. (1967) discussed the various sources of soil nutrients and the processes involved. In a *Chorisodontium aciphyllum-Polytrichum alpestre* peat and *Deschampsia antarctica* brown earth, ammonia-N and P, and to a lesser extent Na and K, increased steadily during summer and reached peak concentrations during winter, while Ca and Mg showed little seasonal change. Northover and Grimshaw (1967) demonstrated that repeated freezing and thawing, such as occur frequently in spring and late summer, increase the amount of extractable N and P only slightly in comparison with that released from comparable temperate soils. Holdgate et al. (1967) suggested that the more primitive and skeletal soils of Signy Island probably have fewer exchange sites for nutrient ions. They concluded that the nutrient regime of the Signy Island ecosystems is considerably influenced by the fauna, in particular bird excreta, and there is an accumulation of ions as winter progresses. Rainwater was found to contain large amounts of inorganic N and also Na, K, and Ca, but was relatively low in non-volatile phosphates (Allen et al. 1967; Allen and Northover 1967).

R.I.L. Smith (1978) reported that Na, K, and Ca are generally much higher in summer than in winter in a large number of Signy Island bryophytes and lichens. The difference was attributed to the probability of an annual cycle of elements within the plants, although the cycle may not be complementary with the seasonal trends in soil elements. However, the passive uptake of nutrients by non-vascular cryptogams may be simply a response to their availability in the soil. Wynn-Williams (1982), during the course of his intensive studies of microbial population dynamics in bryophyte communities, considered that N and P are unlikely to be limiting in Signy Island soils and that the nutrients most likely to regulate microbial activity, and hence decomposition and nutrient cycling to a large extent, are soluble carbohydrates and amino acids. These organic nutrients are probably released by membrane damage in moss cells during freeze-thaw action resulting in the rapid bloom of bacteria, yeasts and fungi at spring melt (Wynn-Williams 1980).

Anaerobic microbial activity has been demonstrated in wet moss carpets at Signy Island (Yarrington and Wynn-Williams 1985). Decomposition of carbohydrates, including cellulose, at this site (Wynn-Williams 1985) produces CO_2 and methane. Methanogenesis contributes 0.02 to 3.7% of the C loss from the permanently wet moss peat, probably reflecting the degree of anaerobiosis (Yarrington, pers. comm.). The immediate precursors of methane are a limited range of low molecular weight compounds such as formate, acetate, methylated amines and hydrogen, as well as CO_2. Some $^{14}CH_4$, but mainly $^{14}CO_2$, was produced from ^{14}C-labelled methionine or formate added to slurries of moss peat in vitro (Yarrington and Wynn-Williams 1984). The occasional blackening of silver stakes inserted into moss carpets (Wynn-Williams 1979; Davis 1981) was probably due to sulphide derived from amino acids rather than sulphate reduction.

Intra- and extra-cellular ion leakage from plants into the substratum may be an important source of nutrients, particularly in non-maritime influenced localities in the Antarctic. Hooker's (1977) investigation of cation and carbohydrate leakage in relation to duration of freezing and thallus turgidity demonstrated that a prolonged period of deep freezing (6 months or more at $-20°C$) can result in a cellular imbalance of cation concentration in various lichens. While short exposure (24 h) to deep freezing causes little damage to most species, prolonged desiccation is partly responsible for the imbalance due to freezing-induced dehydration. In general, plants in a desiccated state are resistant to membrane rupture but saturated thalli can suffer intracellular damage permitting cation and carbohydrate leakage to occur. However, much of this leakage is re-distributed to other sites within the thallus although a proportion is lost and may play an important part in the recycling of certain nutrient elements (especially Na, K, and Mg). Similar losses are caused by leaching, such as occurs during heavy rain or snow-melt by removal of extracellular nutrients, a proportion of which may be derived from intracellular leakage.

In his investigations of N cycling in two contrasting bryophyte ecosystems, Christie (unpubl.) could not detect autotrophic nitrifying bacteria despite the use of several isolation techniques. Two groups of heterotrophic anaerobic bacteria, which may be capable of fixing atmospheric N, clostridia and sulphate reducing bacteria occurred deep in the peat profile. However, they were present in extremely low numbers in a wet bryophyte carpet and were rarely detected in a drier moss turf. Epiphytic photosynthetic cyanobacteria (notably *Nostoc muscorum*) are much more abundant in the wet moss carpet than in the drier turf, but the thalli are not heterocystous throughout the summer (Broady 1979). Nitrogen input by biological dinitrogen fixation is therefore likely to be negligible, and the major source of inorganic N in both the hydric and mesic sites is probably precipitation, with an estimated 52 and 34 g ha^{-1} yr^{-1}, respectively, being deposited. Another possible source of N is ammonia volatilized from nearby penguin rookeries and absorbed from the atmosphere by the vegetation, but this source has not been quantified. The concentration of total N at consecutive 30 mm depths to 240 mm varies considerably in the wet moss carpet but to a much smaller degree in the drier moss turf. There is also a marked difference between the two sites, the former having much higher

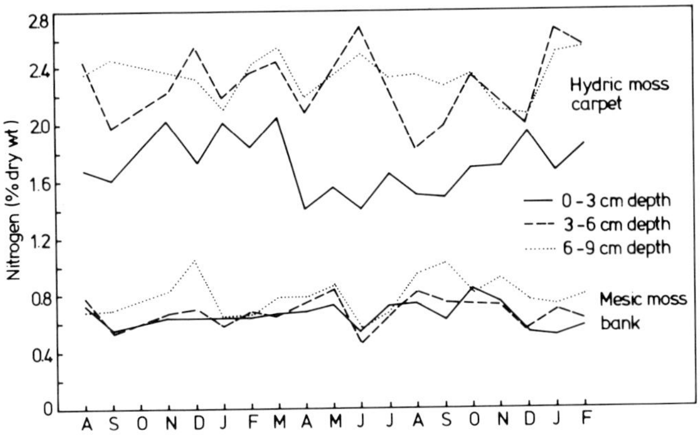

Fig. 6. Seasonal change in total N at various depths in a hydric moss carpet and a mesic turf on Signy Island

N levels (Fig. 6). Christie's (unpubl.) analysis illustrates the importance of sequential depth sampling, rather than sampling at a standard depth to provide a mean site value, if the interpretation of nutrient flux within a site is to be fully comprehended.

2.5 Continental Antarctica

No investigations of seasonal changes in soil or plant nutrients have yet been attempted on continental Antarctica, although various studies of weathering and soil chemical properties, and microbial, cryptogamic and vertebrate ecology have identified some of the processes involved in providing nutrients to terrestrial ecosystems.

The importance of birds as a major source of nutrients for Antarctic plants, which in turn serve as favourable habitats for invertebrates, has been referred to by most early biologists visiting the south polar regions. Indeed, the existence of much of the life of the interior of the continent owes its existence to N enrichment associated with isolated bird colonies (e.g., Siple 1938; Dalenius and Wilson 1958; Bowra et al. 1966). In inland mountain ranges there are locally adequate concentrations of inorganic N for moderate populations of heterotrophic bacteria, yeasts and algae to survive in the visually sterile primitive soils (Cameron et al. 1970; Parker et al. 1982a). However, the nitrate-N is of non-biogenic origin, being deposited by precipitation and accumulating as surface snow sublimes (Wilson and House 1965; Claridge and Campbell 1968; Parker et al. 1978). The levels of phosphate-P are negligible in the absence of marine bird populations. Davey (1983) has demonstrated significant diurnal variation in N fixation by freeliving cyanobacteria, especially *Nostoc commune,* with maximum acetylene reduction occurring shortly after mid-day. The rate at 0°C was 19% of the maximum rate (at 20°C). Dry foliose mats did not reduce acetylene, but activity was immediately resumed when the thalli were wetted. Nitrogenase activity of *Nostoc commune* growing epiphytically on mosses was enhanced with increasing temperature in moist samples but not in dry material.

The investigations of Friedmann's research group have revealed that endolithic cyanobacteria are widespread in rocks in the Dry Valleys. However, Friedmann and Kibler (1980) have shown them to be of little importance in fixing atmospheric N and, therefore, it is unlikely that these organisms serve as a source of N. A small amount of ammonia is released during decomposition of organic matter within the rock, although most of this is probably utilized again by the micro-organisms themselves, while partial loss of ammonia through volatilization is also likely.

3 Freshwater Ecosystems

Research on Antarctic and Sub-Antarctic freshwater ecosystems has developed mainly during the last 15 yrs. Comprehensive reviews of the morphology, chemistry and biota of Antarctic (but excluding Sub-Antarctic) freshwater lakes have been provided by Goldman (1970), Heywood (1972, 1977, 1978, 1984, to be publ.) and Priddle and Heywood (1980). Heywood's (1977) review summarized the physical and chemical properties of all Antarctic lakes for which data were available, while his 1984 review is particularly valuable in assessing lake geochemical dynamics in relation to the evolutionary status of the systems. Investigations of the kinetics of geochemical cycling are still in their infancy. The processes are complicated, involving interactions between catchment, lake morphometry, snow and ice cover, temperature, mixing processes, nutrient status and associated biological composition and activity, creating fundamental differences between shallow and deep nutrient-rich and nutrient-poor water-bodies. This is largely reflected in the level of primary production (Heywood, to be publ.).

The nutrient dynamics of inland waters are dependent on the proximity to the sea, the nature of the catchment, which may include bird and seal populations, and the temperature and precipitation regime of the region. The chemical status of the water-bodies is related to the age of the system, the local geology and the degree of colonization and development of the vegetation and organic matter accumulation in the catchment. Most inland lakes range from ultra-oligotrophic to oligotrophic, having originated relatively recently from proglacial systems. In contrast, in coastal lowland areas many lakes have progressed to mesotrophy or even eutrophy through transfer of nutrients via birds and seals present in the catchment.

The most extensive investigations on continental Antarctic lake systems have been those carried out in the Dry Valleys of south Victoria Land and, more recently, in the Vestfold Hills in Princess Elizabeth Land. From the point of view of freshwater limnology, few of these lakes may be classified as typical water-bodies since many are meromictic or saline, some of them have a marine origin, whereas others are permanently frozen, sometimes to a considerable depth. However, these latter types are ideal for sealed-system studies. Because of their unusual evolutionary history and resultant hypersalinity, saline lakes, even if they possess a temporary surface layer of fresh water, are not included in this account. The physical, chemical and biological characteristics of these lakes have been reviewed by Burton (1981), Wright and Burton (1981) and Heywood (1984).

Within the maritime Antarctic relatively large lakes are plentiful on several of the South Sandwich and South Shetland islands, on Signy Island (South Orkney Islands), and in the James Ross Island – Hope Bay region of the north-eastern Antarctic Peninsula; farther south lakes are infrequent and mostly small. Only those on Signy Island have been comprehensively studied and virtually nothing is known of the trophic status or biota of most of the others. The Signy Island lakes' system includes examples ranging from a permanently ice-covered lake to highly nutrient enriched lakes.

The only major investigations for Sub-Antarctic freshwater have been on Marion Island, although a wide range of lakes of various trophic status, and streams, exist on most of the other island groups.

3.1 Marion Island

The chemistry of the lentic and lotic waters of Marion Island has been studied in some detail by Grobbelaar (1974, 1975), although seasonal and vertical changes in the water columns have not been investigated.

The freshwater systems are acidic (mean pH 5.9) and dystrophic; Na and Cl concentrations are relatively high and in approximately the same proportions as in seawater. The ionic dominance of the major ions is Na > Mg > Ca > K > Cl > SO_4 > HCO_3, all but the latter being principally of oceanic rather than edaphic origin, deposited by sea spray and precipitation. V. Smith (to be publ.) calculated (from data provided in Grobbelaar 1978a) that the annual input of minerals by precipitation decreased from about 350 kg ha^{-1} at a distance of 200 m from the shore to 1 kg ha^{-1} on the highest peak 7 km inland. Appreciable amounts of NH_4-N and NO_2-N also occur in rain water. The high rainfall and relatively small volumes of freshwater bodies ensure a rapid replacement of the water mass and nutrient turnover. The leaching of iron-containing humic acids from the organic deposits beneath the mires and other well developed vegetation types in the catchment reduces alkalinity and hence pH of the lakes. There is consequently no carbonate and only small amounts of bicarbonate and CO_2 in the water. Lotic waters are of similar chemical composition but have slightly higher bicarbonate levels, probably due to the greater contribution by weathering and erosion of rock. The concentrations of Na and Cl and water conductivity are generally negatively correlated with the amount of rainfall (V. Smith, to be publ.) although a positive correlation occasionally has been demonstrated (Grobbelaar 1978b).

Nitrate- and ammonia-nitrogen and total dissolved P occur mainly in biotically influenced water. The seasonal variation of this enrichment causes corresponding changes in the physicochemical characteristics and in the populations and production of the biota of these contaminated waters. Elsewhere, non-biotically disturbed waters may be deficient in N and P, although some N may be leached from biologically fixed atmospheric N in mire catchments. Shallow freshwater bodies influenced by seabirds and seals exhibit high rates of both phytoplankton and benthic alga production. Primary production rates were much higher in contaminated eutrophic waters (Grobbelaar 1974) that support a benthic mat dominated by cyanophytes, whereas filamentous green algae predominate in uncontaminated waters. *Potamogeton* sp., the only oligate aquatic vascular plant in the island's flora, is also highly productive in the enriched lakes. Grobbelaar (1978c) demonstrated that no significant increase in growth response of benthic algal mats occurs when N, P or micro-nutrients were added individually to culture samples, but yield increased between 8 to 280 times when N and P were added simultaneously. The response was least in the most biotically enriched lake.

3.2 Macquarie Island

Of the other Sub-Antarctic island, the only limnological investigations involving lake chemistry have been preliminary studies of selected lakes on Macquarie Island (and to a very much smaller extent on Iles Kerguelen) by Evans (1970) and Tyler (1972). Most of the data presented are single-date analyses intended to differentiate the chemical properties of the water sampled near the lake shores, but Evans (1970) also determined lake phytoplankton production in relation to seasonal nutrient trends.

On Macquarie Island, as on Marion Island, the major source of ions is atmospheric precipitation, which is often heavily laden with sea-spray. Although all the major lakes are high on the island's plateau, the maritime influence is considerable. In very few lakes is the catchment suitable to allow geochemical processes to influence the water and change the normal ionic dominance from Na > Mg > Ca > K to Na > Ca > Mg > K. One of these, Square Lake, has a relatively high phosphate content, resembling the coastal, biotically-disturbed ponds, and is one of the few sites where the only aquatic vascular plant on the island, *Myriophyllum elatinoides,* grows in profusion. This is a similar situation to that of *Potamogeton* on Marion Island. Evans (1970) considered that it may play an important role in the re-cycling of epilimnetic P. He showed that a decrease in phosphate in early spring in this lake and in a coastal pond was correlated with the early phase of the spring phytoplankton bloom (detected as an increase in chlorophyll-a); the P level then increased rapidly as primary production decreased. Coastal ponds, which are organically rich and have large populations of heterotrophic bacteria, have much higher ammonium-nitrogen concentrations than the plateau lakes.

Nitrate-N and nitrite-N were generally negligible or absent in the inland waters, possibly because the nitrification process is inhibited by low temperatures and by the acidity of the water. Evans concluded that the concentrations of phosphate and nitrate have the greatest influence on primary production.

3.3 Signy Island

The physicochemical characteristics of all 17 lakes on Signy Island and the pattern of seasonal changes in selected lakes were initially investigated mainly by Heywood and his co-workers (Heywood 1968; Heywood et al. 1980). However, more recently a research team led by Ellis-Evans has been conducting a detailed integrated year-round programme of limnological research into biogeochemical cycling in the water and sediments of these lakes. Particular attention has been paid to the seasonality of and factors controlling phytoplankton, benthic flora, and bacterial activity in selected representatives of the various lake types (Ellis-Evans 1981a,b, 1982, to be publ.; Hawes 1983a,b). Additional studies of the role of the catchment in lake development and nutrient cycling are also underway. The small sizes and easy access of these lakes coupled with their great variety make Signy Island an ideal case-study site.

Low nutrient input (notably C and P) in run-off and leaching from the sparsely vegetated, often steep-sloped catchments around oligotrophic lakes contrast sharply with the substantial nutrient influx derived from the extensively moss covered catchments of the mesotrophic and eutrophic lakes. The gently sloping catchments of these latter lake systems result in increased chemical weathering, since there is a much longer retention time for ground water. Horne (1972) determined N-fixation rates in the foliose form of the blue-green alga *Nostoc commune* in the catchment of some of the lowland lakes and suggested that it may contribute a significant amount of the N recorded in runnels flowing through the vegetation into the lakes. Unicellular epiphytic cyanobacteria are also abundant in hydric moss communities (Broady 1979). such as occur in the wetter parts of these catchments and may also contribute to the N of the lakes as well as to the terrestrial ecosystems. A considerable amount of organic and inorganic particulate matter enters the mesotrophic and eutrophic lakes at spring melt and if not lost to the overflow, remains in suspension, due to wind-generated turbulence, until winter when it largely becomes incorporated into the sediments.

During summer, seals may congregate near the shores of some of these lakes and even swim in the lake water when it becomes ice-free. The degree of contamination by their excreta and moulted fur and skin and the re-suspension of sediments caused by their swimming is reflected in the major nutrient concentrations, which are very much higher than in the oligotrophic lakes (Table 1).

Following the formation of ice cover, microbiological activity results in a rapid decrease in oxygen concentration and an increase in carbon dioxide levels. For most of the winter, anaerobic conditions prevail in the lower levels of the mesotrophic and eutrophic lakes. Nutrient concentrations increase in all lake types to varying degrees after ice formation, reaching a peak in late winter due to mineralization and anoxic sediment release. Hawes' (1983a) detailed study of N and P cycling has shown that winter accumulation of these nutrients was substantial in mesotrophic Heywood Lake, whereas in oligotrophic Sombre Lake N barely doubled and P increased only three times against extremely low summer concentrations (Fig. 7). Even this increase was due in the main part to the presence (in winter) of a small anoxic sump in this lake. In more typical oligotrophic systems, far smaller seasonal changes occur (Heywood et al. 1980). The beginning of spring is characterized by the opening of inflows and outflows and the commencement of algal growth (Fig. 7). The influx of melt water and the breaking up of ice cover serves to destroy the vertical stratification present under winter-ice cover, resulting in a well-mixed water-body and even distribution of nutrients. Phytoplankton biomass and production are strongly correlated with the level of available P. Although P limitation is evident throughout the phytoplankton growing season in the oligotrophic lakes it is rather less important than N limitation in mesotrophic and eutrophic systems.

The oligotrophic lakes have poorly developed phyto- and bacterio-plankton populations compared to the nutrient enriched lakes, but substantial sediment microbial populations exist in both groups of lakes (Light et al. 1981; Ellis-Evans 1982). Herbert and Bell (1973) have isolated representatives of all the physiological groups of bacteria necessary to complete nutrient cycling in the various lake types. However, more recent work by Ellis-Evans (1985b) indicates that certain of these groups, notably sulphate reducers, are virtually inactive in the Signy Island lakes, due to the low temperatures and low concentrations of suitable electron acceptors. Most anaerobic activity would seem to be concerned with fermentation, but methanogenic bacteria have recently been found to be widespread, utilizing the products of fermentations to produce methane. Anaerobic bacteria are of significance in freshwater systems under winter-ice cover, but the majority of microbial decomposition appears to involve aerobic metabolism (Ellis-Evans 1985a).

Recent studies by Hawes (1985) have shown that chlorophyll-a increases rapidly in spring as soon as snow cover disappears, allowing a small percentage of the total irradiance to penetrate ice cover. The phytoplankton populations at this time have a high chlorophyll-a per cell content to maximize utilization of the low light levels in the presence of relatively abundant nutrients at the end of the winter, but C-fixation is limited. With the disappearance of the nutrients, the spring population crashes. As the ice cover disappears, a summer population capable of utilizing high irradiance levels develops and fixation rates increase, although the chlorophyll-a concentrations remain low. This sequence of

Table 1. Mean summer (S) and winter (W) water column data for three Signy Island lakes of different trophic status (from Heywood 1984)

	Cl (mg l^{-1})		NO_2-N ($\mu g\ l^{-1}$)		NO_3-N ($\mu g\ l^{-1}$)		orthoPO_4-P ($\mu g\ l^{-1}$)		SiO_2-Si ($\mu g\ l^{-1}$)	
	S	W	S	W	S	W	S	W	S	W
Moss Lake (oligotrophic)	23	49	1	4	21	50	0.5	0.4	143	267
Heywood Lake (mesotrophic)	35	56	5	21	39	128	16	58	174	368
Amos Lake (eutrophic)	37	95	204	85	321	721	597	261	192	657

events is most clearly seen in the enriched lakes where the greatest seasonal variation in nutrients occurs. The summer population in enriched lakes is subject to self-shading, so maximum C-fixation occurs just below the surface water layer. In the oligotrophic lakes, the greater penetration of light through the clear water column results in maximum fixation at some depth, with light inhibition evident in the surface-water populations.

In oligotrophic lakes, primary production is almost balanced by bacterial mineralization, whereas there appears to be a considerable imbalance in the enriched lakes. Total C gains by the planktonic and benthic biota in oligotrophic Sombre and mesotrophic Heywood Lake are roughly 21 and 179 g m^{-2} yr^{-1}, respectively, while losses approximate 19 and 68 g m^{-2} yr^{-1}, respectively. A similar pattern has been found for N and P budgets. As primary production and outflow losses increase there is a rapid reduction in nutrient concentrations. Phosphate and ammonia decrease to negligible levels in the oligotrophic lake, although concentrations at the base of the water column remain high until the ice cover disappears. In the mesotrophic lake, nitrate concentrations also fall although phosphate remains relatively high. The spring phytoplankton bloom consumes 72% of the P and 22% of the N in the oligotrophic lake and 40% of the P and 12% of the N in the mesotrophic lake, but much of this material is flushed from the systems as peak outflow rates coincide with maximum biomass. As the winter snow accumulation melts and the vegetation and underlying peat and soil thaw and faunal activity increases, the importance of run-off streams entering the lakes in supplying nutrients increases. During the melt-water inflow from the terrestrial catchment, the lakes not only receive an input of nutrients but also an inoculum of terrestrial microorganisms that have nutrient requirements different from those of the indigenous lake micro-flora (Ellis-Evans and Wynn-Williams 1985). The concentration of nitrate and phosphate in surface melt-water and groundwater is relatively high, even in areas with no biotic enrichment. Annual N and P fluxes (in mg m^{-2} yr^{-1}) for the meso- and oligotrophic lakes are estimated to be, respectively: N 660 (into the lake), 530 (out), and 1580 (in), 1017 (out), and P 350 (in), 140 (out), and 23 (in), 5 (out) (Hawes 1983a). The comparatively stable biomass and nutrient concentration associated with high photosynthetic rates suggest that recycling of nutrients is relatively rapid in both nutrient-poor and nutrient-rich lakes (Ellis-Evans 1985b).

3.4 Antarctic Peninsula

The only major study of nutrients in freshwater bodies in the Antarctic Peninsula region has been by Parker and his co-workers. Investigations on a eutrophic and mesotrophic (but described as oligotrophic) lake in the Arthur Harbour area of Anvers Island revealed, by simulated laboratory and field enrichment experiments, that ammonium ion is the major nutrient limiting the level of primary production in the mesotrophic lake. Phosphate was of lesser importance (Samsel and Parker 1972). The eutrophic lake is situated close to a large Adélie Penguin, *Pygoscelis adeliae,* rookery and several other bird colonies occur nearby. Their influence is the cause of the significantly higher total phosphate (50 times), ammonia-nitrogen (25 times), chloride (5 times), and dissolved organic carbon (7 times) than in the mesotrophic lake, which itself is influenced to some extent by a large population of skuas, (*Catharacta* spp.). These high concentrations are reflected in the primary productivity in the surface layers, being 20 times greater in the eutrophic lake than in the mesotrophic lake (Samsel and Parker 1971). Wodehouse and Parker (1981) demonstrated that gaseous ammonia volatilizing from microbiologically active guano in the penguin rookery to the windward of the enriched lake is carried by the prevailing winds over the lake which readily absorbs the ammonia. Airborne ammonia decreased progressively from 145 mg N m^{-2} d^{-1} at the edge of the rookery to 12 mg N m^{-2} d^{-1} at a point 200 m distant. Corresponding values for the vicinity of the other lake ca. 500 m to the windward of the rookery were 8–12 mg N m^{-2} d^{-1}.

3.5 Continental Antarctica

Many of the studies of continental freshwater lakes have been little more than preliminary surveys undertaken during relatively brief summer visits. The most intensive research on continental Antarctic lake systems has been that carried out in the permanently ice covered meromictic Lake Bonney and Lake Vanda. Here many limnologists, including Parker and Torii and their respective co-workers, have investigated and extensively published on the interactions between environmental factors and the microbial, planktonic and benthic algal mat communities with regard to nutrient turnover, production and decomposition processes

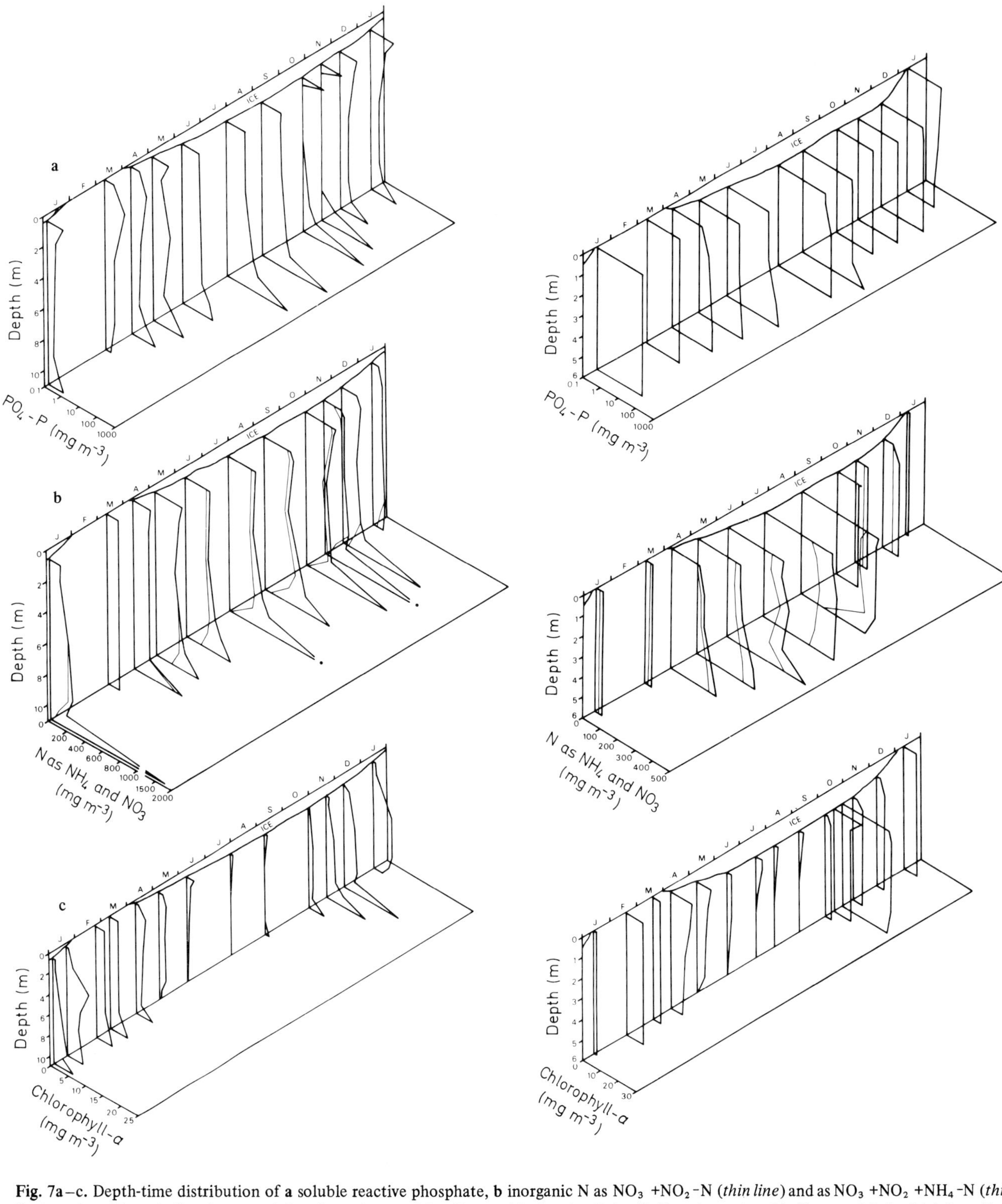

Fig. 7a–c. Depth-time distribution of **a** soluble reactive phosphate, **b** inorganic N as NO_3 +NO_2-N (*thin line*) and as NO_3 +NO_2 +NH_4-N (*thick line*), and **c** chlorophyll-*a* (corrected for phaeopigments) in oligotrophic Sombre Lake (*left*) and mesotrophic Heywood Lake (*right*) on Signy Island during 1980 (Hawes 1983a). Note log scale for phosphate axis

and energy flow through the system. Unfortunately, however, even these studies have not been of a continuous nature and the data obtained therefore require cautious interpretation in terms of annual cycles (e.g., Howell et al. 1978).

The catchment of Lake Bonney has active melt streams for only 2–4 weeks each year, commencing in late December. The main source of inorganic P is in the leachate from weathered rocks, soil, glacial flour, and sediment carried into the lake in these melt streams. These also contribute sig-

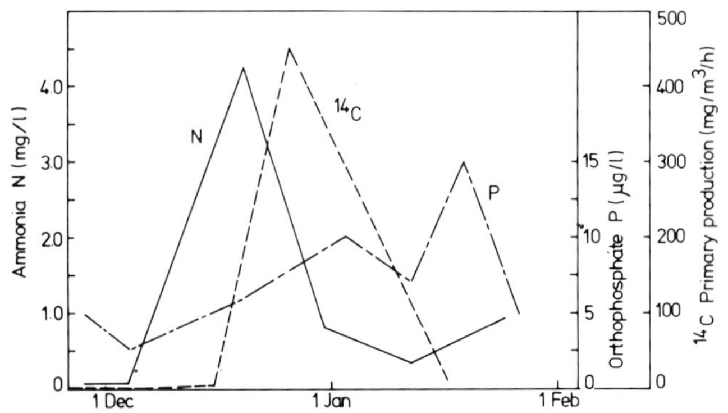

Fig. 8. Mean primary production in relation to mean N and P concentrations in the 4–10 m mixolimnion of Lake Bonney, south Victoria Land (From Hoehn et al. 1977; Parker et al. 1977)

nificant amounts of inorganic N, possibly derived from N-fixation. Benthic algal mats occur primarily around the points of such inflow where nutrient enrichment is greatest, but no evidence has been found to suggest that phytoplankton show similar horizontal distribution patterns. It is believed that subterranean flow of glacial melt-water occurs before surface flow and that the water rising from the lake bottom brings with it supplies of ammonia, nitrite, nitrate, organic C and P which have accumulated in the sediments due to settling of organic detritus in winter. The importance of ground-water as a source of nutrients and oxygen in the Dry Valleys has been discussed by Cartwright and Harris (1981).

Nitrogen, predominantly in the form of ammonia, increases rapidly early in the summer, reaching inexplicably high levels in the mixolimnion at mid-summer, then decrease coincides with the initiation of photosynthetic C-fixation, which reaches a maximum value about 1 week after the NH_4-N peak. As phytoplankton biomass increases, ammonia is utilized which leads to N-limitation and a subsequent decrease in C-fixation. A few days later, bacterial and yeast populations increase markedly (Parker et al. 1977). Maximum P concentrations do not coincide with maximum primary production, but slight peaks occur in early January, when P input from melt streams is greatest, and in late January, possibly when much of the surface P has reached the mixolimnion (Hoehn et al. 1977) (Fig. 8). However, maximum P input in the following season was recorded during the melt period, as might be predicted (Weand et al. 1977).

A survey by Allnutt et al. (1981) of in situ N fixation by benthic and moat blue-green algal mats in lakes in the Dry Valleys, and their inflow and outflow streams and adjacent soils indicated that measurable levels of fixation occur wherever the algae were in the presence of sufficient moisture, light, above-freezing temperatures and low concentrations of ammonium and nitrate ions. Fixation was detected between zero and 7°C in moat mats, but at higher temperatures in laboratory tests the rate increased severalfold. The positive correlation between N fixation and photosynthesis, and the presence of heterocystous blue-green algae (especially *Nostoc commune*), suggests that these are the principal fixing organisms.

In some permanently frozen lakes, benthic algal mats serve as biological sinks selectively accumulating elements in excess of concentrations in the water column. Parker et al. (1982b) estimated that substantial amounts of chemical constituents may then be removed from these lakes by mat lift-off and gradual escape through the ice. Such algal material takes 5–10 yrs to reach the lake surface and subsequently is lyophilized and dispersed by wind. This depletion of organic matter and nutrients helps to perpetuate their oligotrophic state.

High concentrations of total dissolved organic C (TOC) have been determined at various depths. Distinct seasonal changes in net photosynthetic production of phytoplankton and in TOC occur, with low levels of production in spring (early November) and a significant pulse of dissolved and extracellular organic matter around mid-summer (early December). This is possibly triggered by increased light intensity and duration and increased melt-water bringing inorganic nutrients, organic matter and cells into the mixolimnion. Parker et al. (1974) described this as implying an annual cycle of photosynthetic production and dark respiration by microbial communities. Changes in TOC in the highly saline deep water may be due to seepage input rather than in situ production, although the catchment area is almost totally devoid of biota that might serve as a nutrient source. The lake clearly possesses a large reservoir of organic matter, which may have accumulated over a long time-scale, but a significant fraction of the total is produced and recycled annually. Microbial decomposition is slow and the role of consumers is minimal.

Parker et al. (1977) suggested that up to 95% of the total photosynthetically fixed C may be released as extracellular products by phytoplankton, compared with only 43% for an eutrophic lake on Signy Island (Fogg and Horne 1970). However, both values are unrealistically high because of release of C from cells damaged by the techniques employed. More rigorous techniques used in studies on Signy Island have indicated that, in an oligotrophic lake, top and mid-water algal populations released 3%–17% of the total extracellular C, while benthic populations released 19%–39%, possibly indicative of unhealthy cells (Ellis-Evans 1981a); comparable data for a mesotrophic lake ranged from 10% to 30% (Light et al. 1981; Ellis-Evans 1981b). I. Hawes (pers. comm.), using even more sophisticated methods, has shown the range of values to be as low as 1%–10%. The organisms are well adapted to using low levels for metabolism and production. The high concentrations and rapid seasonal fluctuations of dissolved organic matter at various depths suggests the existence of metabolic pathways well adapted to recycling of organic matter within the simple microbial communities. As on Signy Island the primary controlling factor for phytoplankton and benthic algal production is light limitation of photosynthe-

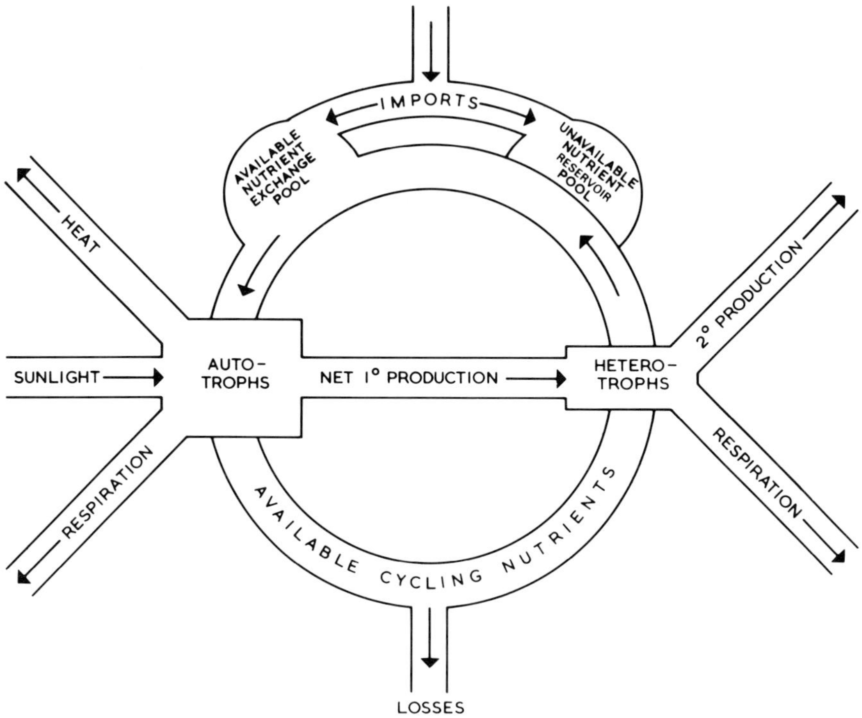

Fig. 9. Schematic relationship between C and nutrient cycles in the biosphere

sis; however, while N and P do not appear to be growth-limiting in continental lakes, there is evidence that these nutrients may be limiting in Signy Island lakes (Wootton, pers. comm.).

Elsewhere on the Antarctic continent few intensive studies have been made of freshwater lake systems. Accounts of lake and pond chemistry have been provided for the Schirmacher Oasis systems (Bardin and Leflat 1965) and the Ongul Islands lakes and ponds (Sugawara and Torii 1959; Murayama 1977), but give no indication of processes. However, MacNamara's (1972) work on the freshwater bodies of Molodezhnaya, Enderby Land, is of relevance. Chloride concentrations were found to be low (6–45 mg l^{-1}) in comparison with other coastal continental lakes, increasing with depth to 30 m. Sodium and Cl concentrations are high in temporary ponds and melt-pools due to evaporation. In contrast to the Lake Bonney situation, but similar to that on Signy Island, phosphate levels are quite high probably due to the use of these small water bodies by skuas. A seasonal phosphate cycle occurs with the highest levels in spring and the lowest in mid-summer following rapid metabolic uptake by the phytoplankton. As summer dilution of the lakes and pools occurs, ionic concentrations decrease. Unlike many Antarctic lake waters these are not highly mineralized and are slightly acidic to mildly alkaline due largely to the acidic gneisses of the region. Nitrogen occurs mainly in the form of ammonia, which most likely results from precipitation and other atmospheric processes, but the nitrate-N component is probably derived from local penguin rookeries. These low nutrient levels are reflected by the comparatively low primary production, although nothing is known of seasonal productivity. The region receives low precipitation, most of which falls when sea-ice is at a maximum during winter and highest ionic concentrations are associated with summer precipitation. Deposition of marine-derived elements is minimal and the degree of input by spray and precipitation decreases with increasing distance from the sea. Elsewhere, Brocas and Delwiche (1963) have also demonstrated decreases in Na, K, and Cl with increasing distance from the open sea.

In interior continental Antarctia Zeller and Parker (1981), Laird et al. (1982) and Parker et al. (1982c) found a strong correlation between nitrate ion concentration in firn and ice and solar activity, particularly aurorae. This suggests that the N levels in melt-water, and hence in continental lakes and terrestrial habitats, may exhibit an enhanced input every 11 yrs which may be reflected in inincreased primary production during sunspot maxima, but this has yet to be tested.

4 The Future Development of Ecosystem Research in South Polar Regions

The presence, concentration and availability of chemical elements in the biosphere, particularly those required as nutrients, are critical to the existence of life in any biome or habitat. They are of major importance in primary production and hence are key driving variables in energy circuits and food webs, both terrestrial and aquatic (Fig. 9). Most of the world's ecosystems have in some way been altered by man, and there is an increasing need to acquire baseline data from truly unperturbed habitats to generate predictive models of ecosystem functioning and to aid in the development of management policies for various natural resources. Such ecosystems prevail in the Sub-Antarctic and even more so in the Antarctic. These biomes offer ideal environments for case studies of ecosystem processes, for analyzing in detail fundamental biological problems and for testing ecological hypotheses. Such findings and principles may then be applied, at least in theory, to the more complex and economically productive systems of the world. Quantification of the input, accumulation, utilization, transfer, re-cycling and loss of chemical elements in the terrestrial and freshwater environments and the interactions with the marine environment is an essential requirement in the study of Antarctic and Sub-Antarctic ecosystem dynamics. Thus, to understand how ecosystems function and to assess the processes operating within and between each trophic level, it is advantageous to study simple ecosystems in which the biota comprise relatively few species, interactions are minimal and process rates reduced.

The development of nutrient dynamics research in the Antarctic and Sub-Antarctic can benefit greatly from the achievements of the I.B.P. studies at several Arctic tundra (notably by Chapin and his co-workers at Barrow, Alaska) and temperate moorland and grassland sites (see Dowding et al. 1981). These investigations have answered many questions about processes and interactions involved in ecosystem nutrient cycling and raised many others. Building on this experience, significant advances may be made in the understanding of ecosystem functioning, and in particular of nutrient and energy cycling, in the relatively simple systems of the south polar counterparts. For example, valuable comparisons can be made between the Sub-Antarctic ecosystems and the now well-documented research undertaken at Barrow, Alaska (see Tieszen 1978; Brown et al. 1980; Hobbie 1980), and between the Antarctic ecosystems and the high Arctic studies at Truelove Lowland, Devon Island (see Bliss 1977).

It is evident from reviewing the literature on nutrient cycling in the Antarctic and Sub-Antarctic biomes that this field of research is very restricted in location and in depth and intensity of investigations. Most of our current knowledge of ecosystem processes has been contributed by a few small groups of scientists working at a very few sites. There is an increasing awareness of the importance of biogeochemical cycling and of nutrients in general in the functioning of these stressed ecosystems. Perhaps the greatest advances in recent biological research have been made in microbiology and limnology. Numerous papers presented at this symposium have emphasized the important role of micro-organisms in the nutrient dynamics of both terrestrial and aquatic ecosystems. Associated studies of lake catchments serve to integrate the processes operating in terrestrial and inland aquatic ecosystems, while studies of the birds and seals provide a critical link between marine, terrestrial and freshwater environments. Also of particular interest are the investigations of various mechanisms operating in the derivation, transfer and loss of nutrients and of interactions between abiotic and biotic components, which permit the development of biological systems existing under some of the most stressful conditions on earth. However, most research has so far concentrated on biological processes, either in the production and release of nutrients or in their utilization in microbial and primary production. Relatively few studies (e.g. Gallagher 1985; Myrcha et al. 1985; Walton 1985b) have yet been directed towards physical processes concerning the release or transfer of nutrients as, for example, by various modes of weathering of rocks, by freeze-thaw action in rocks, soils and plants at both the community level and cytological level, or by precipitation and aerial transport. These should be important priorities for future research.

The organization of Antarctic biological research is controlled primarily by financial, logistic and manpower constraints and, unlike major field research programmes elsewhere in the world, it has not been possible to have teams of highly qualified scientists and technicians working throughout the year at various sites over several years. Only in the Dry Valley lakes programme has a relatively large team been able to operate for many years, but then only for short periods in summer since working throughout the year poses obvious difficulties. By contrast, the Signy Island terrestrial and freshwater ecosystems programmes have involved continuous year-round intensive studies by teams of only two or three scientists.

It is disappointing to see so little research on biological and environmental processes being carried out at many localities where research stations or other suitable habitation exists. For example, there is still a dearth of published information about the ecosystems of Îles Kerguelen, Îles Crozet, Macquarie Island, and the most extreme of all Sub-Antarctic islands, Heard Island. For some sites, volumes of data have emanated from several isolated or unrelated studies which have tended to characterize sites and situations rather than examine ecosystem dynamics or seasonal trends. At South Georgia, barely any freshwater studies have been attempted, yet the island possesses an exceptional sequence of easily accessible lakes ranging from a high altitude ultra-oligotrophic lake with surrounding late snow beds to a biotically influenced eutrophic lake surrounded by closed vegetation and peat deposits near sea level, the entire system being linked by a common stream. There is tremendous potential for valuable terrestrial and freshwater studies to be undertaken in the South Shetland Islands, where there is the greatest concentration of research stations anywhere in the Antarctic, yet very little has been achieved there. A similar situation exists on the Antarctic Peninsula. Perhaps the richest biological region on continental Antarctica exists at the Windmill Islands and adjacent mainland near Casey Station but no intensive studies have yet been undertaken there. Elsewhere, intensive studies have been conducted on abnormal or atypical systems such as at sites excessively influenced by birds or seals and in hypersaline lakes. It is crucial that we do not lose sight of the normal situation for it is the assessment and understanding of how such systems function that will have the most value in interpreting the more complex systems of the world.

No biological, physical or chemical process can be meaningfully interpreted from short-term studies and any predictive computer model is vulnerable to criticism when based on few data points. Field investigations and experiments should be complemented by adequately replicated laboratory experiments to simulate field situations under controlled conditions (e.g., Wynn-Williams 1982); these have the added advantage that processes can be accelerated. There are also many aspects of biogeochemical studies that lend themselves to manipulative experiments. For an accurate assessment of any study of ecosystem processes it is essential to continue investigations, sampling monitoring, analyses, etc., at regular short intervals throughout each of several years in order to determine within- and between-season and between-year variations. Many nations are currently involved in Antarctic and Sub-Antarctic biological research yet few have long-term research strategies involving an overlapping series of integrated exploratory or experimental projects relating to a major theme or aimed to answer specific ecological questions. It is the continuity of the component projects of such a programme that is so valuable in maintaining research of a high quality and which will lead to more highly acclaimed science. For the future development of Sub-Antarctic and Antarctic field research, more national organizations should be encouraged to establish finite study sites, representative of the type of ecosystem under examination, in which various interrelated studies are concentrated. Only in this way can a logical, meaningful and economical assessment of the overall functioning of the system, and of the processes operating within it, be achieved and modelled with accuracy.

Acknowledgements. I am grateful to Dr. J.C. Ellis-Evans for his critical revision of the freshwater section of this paper. I am also indebted to Drs. D.W.H. Walton and D.D. Wynn-Williams for their comments during the preparation of the manuscript.

References

Allen SE, Northover MJ (1967) Soil types and nutrients on Signy Island. In: Smith JE (organizer) A discussion on the Antarctic terrestrial ecosystem. Philos Trans R Soc, Ser B 252:179–185

Allen SE; Grimshaw HM; Holdgate MW (1967) Factors affecting the availability of plant nutrients on an Antarctic island. J Ecol 55: 381–396

Allnutt FCT, Parker BC, Seaburg KG, Simmons GM (1981) In situ nitrogen (C_2H_2) fixation in lakes of southern Victoria Land, Antarctica. Hydrobiol Bull 15:99–109

Bardin VI, Leflat ON (1965) Chemistry of waters in the Schirmacher ponds. Sov Antarct Exp Info Bull 5:361–363

Bliss LC (1977) Truelove Lowland, Devon Island, Canada: a high Arctic ecosystem. University of Alberta Press, Edmonton

Bowra GT, Holdgate MW, Tilbrook PJ (1966) Biological investigations in Tottanfjella and central Heimefrontfjella. Bull Br Antarct Surv 9:62–70

Broady PA (1979) The Signy Island terrestrial reference sites: IX. The ecology of the algae of site 2 – a moss carpet. Bull Br Antarct Surv 47:13–29

Brocas J, Delwiche R (1963) Cl, K, and Na concentrations in Antarctic snow and ice. J Geophys Res 68:3999–4000

Brown J, Miller PC, Tieszen LL, Bunnell FL (1980) An Arctic ecosystem: The coastal tundra at Barrow, Alaska. Dowden, Hutchinson & Ross Stroudsburg, Pennsylvania

Burton HR (1981) Chemistry, physics and evolution of Antarctic saline lakes – a review. Hydrobiol 82: 339–362

Cameron RE, King J, David CN (1970) Microbiology, ecology, and microclimatology of soil sites in dry valleys of southern Victoria Land, Antarctica. In: Holdgate MW (ed) Antarctic ecology Academic Press, London 2:702–716

Cartwright K, Harris KJH (1981) Hydrogeology of the Dry Valleys region, Antarctica. In: McGinnis LD (ed) Dry Valley Drilling Project. Antarctic Research Series 33:134–214. American Geophysical Union, Washington DC

Christie P, Nicolson TH (1983) Are mycorrhizas absent from the Antarctic? Trans Br Mycol Soc 80:557–560

Claridge GGC, Campbell IB (1968) Soils of the Shackleton Glacier region, Queen Maud Range, Antarctica. NZ J Sci 11:171–218

Croome RL (1973) Nitrogen fixation in the algal mats on Marion Island. S Afr J Antarct Res 3:64–67

Dalenius P, Wilson O (1958) On the soil fauna of the Antarctic and of the sub-Antarctic Island. The Oribatidae (Acari). Ark Zool 11: 393–425

Davey A (1983) Effects of abiotic factors on nitrogen fixation by bluegreen algae in Antarctica. Polar Biol 2:95–100

Davis RC (1981) Structure and function of two Antarctic terrestrial moss communities. Ecol Monogr 51:125–143

Dowding P, Chapin FS, Wielgolaski FE, Kilteather P (1981) Nutrients in tundra ecosystems. In: Bliss LC, Heal OW, Moore JJ (ed) Tundra ecosystems: a comparative analysis. Cambridge University Press, Cambridge , pp 647–683

Ellis-Evans JC (1981a) Freshwater microbiology in the Antarctic: I. Microbial numbers and activity in oligotrophic Moss Lake, Signy Island. Bull Br Antarct Surv 54:85–104

Ellis-Evans JC (1981b) Freshwater microbiology in the Antarctic: II. Microbial numbers and activity in nutrient-enriched Heywood Lake, Signy Island. Bull Br Antarct Surv 54:105–121

Ellis-Evans JC (1982) Seasonal microbial activity in Antarctic freshwater lake sediments. Polar Biol 1:29–140

Ellis-Evans JC (1985a) Decomposition processes in maritime Antarctic lakes. In: Siegfried WR, Condy PR, Laws RM (eds) Antarctic nutrient cycles and food webs (Proceedings of the 4th SCAR symposium on Antarctic biology). Springer, Berlin Heidelberg New York

Ellis-Evans JC (1984b) Methane in maritime Antarctic freshwater lakes. Polar Biol 3:63–71

Ellis-Evans JC (1985b) Interactions of bacterio- and phyto-plankton in nutrient cycling within eutrophic Heywood Lake, Signy Island. In: Siegfried WR, Condy PR, Laws RM (eds) Antarctic nutrient cycles and food webs (Proceedings of the 4th SCAR symposium on Antarctic biology). Springer, Berlin Heidelberg New York, pp

Ellis-Evans JC, Wynn-Williams DD (1985) The interaction of soil and lake micro-flora at Signy Island. In: Siegfried WR, Condy PR, Laws RM (eds) Antarctic nutrient cycles and food webs (Proceedings of the 4th SCAR symposium on Antarctic biology). Springer, Berlin Heidelberg New York

Evans AJ (1970) Some aspects of the ecology of a calanoid copepod, *Pseudoboeckella brevicaudata* Brady 1875, on a Sub-Antarctic island. ANARE Sci Rep Ser B (II) Zool Publication No. 110:1–100

Fogg GE, Horne AJ (1970) The physiology of Antarctic freshwater algae. In: Holdgate MW (ed) Antarctic ecology Academic Press, London 2:632–638

Friedmann EI, Kibler AP (1980) Nitrogen economy of endolothic microbial communities in hot and cold deserts. Microb Ecol 6: 95–108

Gallagher JB (1985) The influence of iron and manganese on nutrient cycling in shallow freshwater Antarctic lakes. In: Siegfried WR, Condy PR, Laws RM (eds) Antarctic nutrient cycles and food webs (Proceedings of the 4th SCAR symposium on Antarctic biology). Springer, Berlin Heidelberg New York

Goldman CR (1970) Antarctic freshwater ecosystems. In: Holdgate MW (ed) Antarctic ecology Academic Press, London 2:609–627

Grobbelaar JU (1974) Primary production in freshwater bodies of the sub-Antarctic island Marion. S Afr J Antarct Res 4:40–45

Grobbelaar JU (1975) The lentic and lotic freshwater types of Marion Island (sub-Antarctic): a limnological study. Ver Int Verein Limnol 19:1442–1449

Grobbelaar JU (1978a) Mechanisms controlling the composition of fresh waters on the sub-Antarctic island Marion. Arch Hydrobiol 83:145–157

Grobbelaar JU (1978b) The limnology of Marion Island; southern Indian Ocean. S Afr J Antarct Res 8:113–118

Grobbelaar JU (1978c) Factors limiting algal growth on the sub-Antarctic island Marion. Verh Int Verein Limnol 20:1159–1164

Hawes I (1983a) Nutrients and their effects on phytoplankton populations in lakes on Signy Island, Antarctica. Polar Biol 2: 115–126

Hawes I (1983b) Turbulent mixing and its consequence for phytoplankton development in two ice covered lakes. Bull Br Antarct Surv 60:69–82

Hawes I (1985) Factors controlling phytoplankton populations in maritime Antarctic lakes. In: Siegfried WR, Condy PR, Laws RM (eds) Antarctic nutrient cycles and food webs (Proceedings of the 4th SCAR symposium on Antarctic biology). Springer, Berlin Heidelberg New York

Herbert RA, Bell CR (1973) Nutrient cycling in fresh-water lakes on Signy Island, South Orkney Islands. Bull Br Antarct Surv 37:15–20

Heywood RB (1968) Ecology of the fresh-water lakes of Signy Island, South Orkney Islands: II. Physical and chemical properties of the lakes. Bull Br Antarct Surv 18:11–44

Heywood RB (1972) Antarctic limnology: A review. Bull Br Antarct Surv 29:34–65

Heywood RB (1977) Antarctic freshwater ecosystems: review and synthesis. In: Llano GA (ed) Adaptations within Antarctic ecosystems. Gulf, Washington DC pp 801–828

Heywood RB (1978) Maritime Antarctic lakes. Verh Int Verein Limnol 20:1210–1215

Heywood RB (1984) Antarctic inland waters. In: Laws RM (ed) Antarctic ecology Academic, London 1:279–344

Heywood RB (to be publ.) Limnological studies in the Antarctic Peninsula region [publication not yet known]

Heywood RB, Dartnell HJG, Priddle J (1980) Characteristics and classification of the lakes of Signy Island, South Orkney Island, Antarctica. Freshwater Biol 1):47–59

Hobbie JE (1980) Limnology of tundra ponds, Barrow, Alaska. Dowden, Hutchinson & Ross Inc, Stroudsburg, Pennsylvania

Hoehn RC, Parker BC, Fortner RD, Weand BL, Craft JA, Lane LS, Stavroe RW, Sugg HG, Whiteburst JT (1977) Nitrogen and phosphorus availability to plankton and benthic communities in Lake Bonney, southern Victoria Land, Antarctica. In: Llano GA (ed) Adaptations within Antarctic ecosystems. Gulf, Washington DC pp 859–872

Holdgate MW, Allen SE, Chambers MJG (1967) A preliminary investigation of the soils of Signy Island, South Orkney Islands. Bull Br Antarct Surv 12:53–71

Horne AJ (1972) The ecology of nitrogen fixation on Signy Island, South Orkney Islands. Bull Br Antarct Surv 27:1–18

Howell LW, Krutchkoff RG, Parker BC (1978) Development and use of a mathematical model for an Antarctic lake. Hydrobiol 60:129–134

Hurst JL, Pugh GJF, Walton DWH (to be publ.) The effects of freeze-thaw cycles and leaching on the loss of soluble carbohydrates from leaf material of two sub-Antarctic plants. Polar Biol

Jenkin JF (1975) Macquarie Island, Subantarctic. In: Rosswall T, Heal OW (eds) Structure and function of tundra ecosystems. Ecol Bull (Stockholm) 20:375–397

Laird CN, Zeller EJ, Armstrong TP (1982) Solar activity and nitrate deposition in South Pole snow. Geophys. Res Lett 9:1195–1198

Lawson GJ (1985) Decomposition and nutrient cycling in *Rostkovia magellanica* from two contrasting bogs on South Georgia. In: Siegfried WR, Condy PR, Laws RM (eds) Antarctic nutrient cycles and food webs (Proceedings of the 4th SCAR symposium on Antarctic biology). Springer, Berlin Heidelberg New York

Light JJ, Ellis-Evans JC, Priddle J (1981) Phytoplankton ecology in an Antarctic lake. Freshwater Biol 11:11–26

Lindeboom HJ (1979) Chemical and microbiological aspects of the nitrogen cycle on Marion Island (sub-Antarctic). PhD thesis, University of Groningen, 138 pp

MacNamara EE (1972) Some limnological observations from Enderby Land, Antarctica. Limnol Oceanogr 15:768–775

Murayama H (1977) General characteristics of the Antarctic lakes near Syowa Station. Antarct Rec 58:43–62

Myrcha A, Petri SJ, Tatur A (1985) The role of pygoscelid penguin rookeries in nutrient cycles at Admiralty Bay, King George Island. In: Siegfried WR, Condy PR, Laws RM (eds) Antarct nutrient cycles and food webs (Proceedings of the 4th SCAR symposium on Antarctic biology). Springer, Berlin Heidelberg New York

Northover MJ, Allen SE (1967) Seasonal availability of chemical nutrients on Signy Island. In: Smith JE (organizer) A discussion on the Antarctic terrestrial ecosystems. Philos Trans R Soc, Ser B 252:187–189

Northover MJ, Grimshaw HM (1967) Some seasonal trends in nutrient content of the soils of Signy Island, South Orkney Islands. Bull Br Antarct Surv 14:83–88

Parker BC, Whitehurst JT, Hoehn RC (1974) Observations of *in situ* concentrations and production of organic matter in an Antarctic meromectic lake. V J Sci 25:136–140

Parker BC, Hoehn RC, Paterson RA, Craft JA, Lane LS, Stavros RW, Sugg HG, Whitehurst JT, Fortner RD, Weand BL (1977) Changes in dissolved organic matter, photosynthetic production, and microbial community composition in Lake Bonney, southern Victoria Land, Antarctica. In: Llano GA (ed) Adaptations within Antarctic ecosystems. Gulf, Washington DC pp 873–890

Parker BC, Zeller EJ, Heiskell LE, Thomson WJ (1978) Non-biogenic fixed nitrogen in Antarctica and some ecological implications. Nature (Lond) 271:651–652

Parker BC, Boyer S, Allnutt FCT, Seaburg KG; Wharton RA, Simmons GM (1982a) Soils from the Pensacola Mountains, Anarctica: physical chemical and biological characteristics. Soil Biol Biochem 14:266–271

Parker BC, Simmons GM, Wharton RA, Seaburg KG, Love FG (1982b) Removal of organic and inorganic matter from Antarctic lakes by aerial escape of blue-green algal mats. J Phycol 18: 72–78

Parker BC, Zeller EJ, Gow AJ (1982c) Nitrate fluctuations in Antarctic snow and firn: potential sources and mechanisms of formation. Ann Glaciol 3:243–248

Pratt RM, Smith RI Lewis (1982) Seasonal trends in chemical composition of reindeer forage plants on South Georgia. Polar Biol 1:13–32

Priddle J, Heywood RB (1980) Evolution of Antarctic lake ecosystems. Biol J Linn Soc 14:51–66

Samsel GL, Parker BC (1971) Comparisons of two Antarctic lakes with different trophic states. V J Sci 22:177–182

Samsel GL, Parker BC (1972) Nutrient factors limiting primary productivity in simulated and field Antarctic microecosystems. V J Sci 23:64–71

SASCAR (1981) South African Antarctic Biological Research Programme: Aims, programme and organization. S Afr Nat Sci Progs Rep: 51 pp

Siple PA (1938) The second Byrd Antarctic Expedition – Botany. 1. Ecology and geographical distribution. Ann Mo Bot Gdn: 467–514

Smith RIL (1978) Summer and winter concentrations of sodium, potassium and calcium in some maritime Antarctic cryptogams. J Ecol 66:891–909

Smith RIL (1982) Growth and production in South Georgia bryophytes. CNFRA 51:229–239

Smith RIL (1984) Terrestrial plant biology of the sub-Antarctic and Antarctic. In: Laws RM (ed) Antarctic ecology Academic Press, London 1:61–162

Smith RIL, Stephenson C (1975) Preliminary growth studies on *Festuca contracta* T. Kirk and *Deschampsia antarctica* Desv. on South Georgia. Bull Br Antarct Surv 41/42:59–79

Smith VR (1976a) Standing crop and nutrient status of Marion Island (sub-Antarctic) vegetation. J S Afr Bot 42:231–263

Smith VR (1976b) The effect of burrowing species of Procellariidae on the nutrient status of inland tussock grasslands on Marion Island. J S Afr Bot 42:265–272

Smith VR (1977) A qualitative description of energy flow and nutrient cycling in the Marion Island terrestrial ecosystem. Pol Rec 18:361–370

Smith VR (1978a) Soil chemistry of Marion Island (Subantarctic). S Afr J Sci 74:174–175

Smith VR (1978b) Plant ecology of Marion Island: a review. S Afr J Antarct Res 8:21–30

Smith VR (1978c) Animal-plant-soil nutrient relationships on Marion Island (Subantarctic). Oecologia (Berl) 41:123–126

Smith VR (to be publ.) Seasonal dynamics of phytomass and chemical composition in Marion Island (sub-Antarctic) vegetation. III. Nutrient dynamics in mire-grassland communities. Polar Biol

Smith VR, Ashton PJ (1981) Bryophyte-cyanobacteria associations on sub-Antarctic Marion Island: are they important in nitrogen fixation? S Afr J Antarct Res 10/11:24–26

Smith VR, Russell S (1982) Acetylene reduction by bryophyte-cyanobacteria associations on a Subantarctic island. Polar Biol 1: 153–157

Smith VR, Steyn MG (1982) Soil microbial counts in relation to site characteristics at a Subantarctic island. Microb Ecol 8:253–266

Steyn MG, Smith VR (1981) Microbial populations in Marion Island soils. S Afr J Antarct Res 10/11:14–18

Sugawara K, Torii T (1959) Chemical composition of the waters of some ponds on East Ongul Island, Antarctica. Antarct Res 7:53 –55

Tallowin JRB (1977) The reproductive strategies of a Subantarctic grass, *Festuca contracta* T. Kirk. In: Llano GA (ed) Adaptations within Antarctica ecosystems. Gulf, Washington DC pp 967–980

Tieszen LL (1978) Vegetation and production ecology of an Alaskan Arctic tundra. Springer, Berlin Heidelberg New York

Tyler PA (1972) Reconnaissance limnology of sub-Antarctic islands I. Chemistry of lake waters from Macquarie Island and the Îles Kerguelen. Int Rev Gesamten Hydrobiol 57:579–778

Walton DWH (1976) Dry matter production in *Acaena* (Rosaceae) on a subantarctic island. J Ecol 64:399–415

Walton DWH (1977) Studies in *Acaena* (Rosaceae) II. Leaf production and senescence. Bull Br Antarct Surv:93–100

Walton DWH (1980) An annotated bibiography of Antarctic and sub-Antarctic pedology and periglacial processes. Br Antarct Surv Data 5:1–75

Walton DWH (1982) The Signy Island terrestrial reference sites: XV. Micro-climate monitoring. Bull Br Antarct Surv 55:111–126

Walton DWH (1985a) Cellulose decomposition and its relationship to nutrient cycling at South Georgia. In: Siegfried WR, Condy PR, Laws RM (eds) Antarctic nutrient cycles and food webs (Proceedings of the 4th SCAR symposium on Antarctic biology). Springer, Berlin Heidelberg New York

Walton DWH (1985b) A preliminary study of the action of crustose lichens on rock surfaces in Antarctica. In: Siegfried WR, Condy PR; Laws RM (eds) Antarctic nutrient cycles and food webs (Proceedings of the 4th SCAR symposium on Antarctic biology). Springer, Berlin Heidelberg New York

Walton DWH, Greene DM, Callaghan TV (1975) An assessment of primary production in a sub-Antarctic grassland on South Georgia. Bull Br Antarct Surv 41/42: 151–160

Walton DWH, Smith RIL (1980) Chemical composition of South Georgian vegetation. Bull Br Antarct Surv 49:117–135

Weand BL, Hoehn RC, Parker BC (1977) Nutrient fluxes in Lake Bonney – a meromictic Antarctic lake. Arch Hydrobiol 80:519 –530

Wilson AT, House DA (1965) Fixation of nitrogen by aurora and its contribution to the nitrogen balance of the earth. Nature (Lond) 205:793–794

Wodehouse EB, Parker BC (1981) Atmospheric ammonia nitrogen: a potential source of nitrogen eutrophication of freshwater Antarctic ecosystems. In: Parker BC (ed) Terrestrial Biology III. Antarctic Research Series 30:155–167. American Geophysical Union, Washington DC

Wright SW, Burton HR (1981) The biology of Antarctic saline lakes. Hydrobiol 82:319–338

Wynn-Williams DD (1979) Techniques used for studying terrestrial microbial ecology in the maritime Antarctic. In: Russell AD, Fuller R (ed) Cold tolerant microbes in spoilage and the environment. Soc Appl Bacteriol Tech Ser 13:67–81

Wynn-Williams DD (1980) Seasonal fluctuations in microbial activity in Antarctic moss peat. Biol J Linn Soc 14:11–28

Wynn-Williams DD (1982) Simulation of seasonal changes in microbial activity of maritime Antarctic peat. Soil Biol Biochem 14: 1–12

Wynn-Williams DD (1985) Comparative micro-biology of moss-peat decomposition on the Scotia arc and Antarctic Peninsula. In: Siegfried WR, Condy PR, Laws RM (eds) Antarctic nutrient cycles and food webs (Proceedings of the 4th SCAR symposium on Antarctic biology). Springer, Berlin Heidelberg New York

Yarrington MR, Wynn-Williams DD (1985) Methanogenesis and the anaerobic micro-biology of a wet moss community at Signy Island. In: Siegfried WR, Condy PR, Laws RM (eds) Antarctic nutrient cycles and food webs (Proceedings of the 4th SCAR symposium on Antarctic biology). Springer, Berlin Heidelberg New York

Zeller EJ, Parker BC (1981) Nitrate ion in Antarctic firn as a marker for solar activity. Geophys Res Lett 8:895–898

The Role of Pygoscelid Penguin Rookeries in Nutrient Cycles at Admiralty Bay, King George Island

A. MYRCHA[1], S.J. PIETR[2], and A. TATUR[1]

Summary. Bacterial, enzymatic and chemical analyses pointed to active micro-biological mineralization and transformation of penguin faeces in the maritime Antarctic. About 50% of C and N was volatilized during the first three weeks. Most of the N/NH_3 contained in percolated water was oxidized to N/NO_3 in the sub-surface layer of the soil. Uric acid and chitin were degraded totally in the second stage of mineralization. The chemical composition of the water flowing from the penguin rookeries depended on the rookery location and size and the amount of precipitation. Ornithogenic soils that developed around large rookeries of pygoscelid penguins were formed by the action of guano solutions on loams and gravels. The main soil-forming process was phosphatization. A description is given of the sequence of appearance of phosphatic minerals, in relation to changes in the chemical composition of guano leachates, during their soaking and reactions through the soils (struvite, hydroxylapatite, leukophosphite, minyulite, taranakite, viviante). A quantitative estimate was made of the ornithogenic P flow in the penguin rookery at Llano Point. We found that at least 11% of the P brought by penguins from the sea to the land is annually accumulated in the form of phosphatic deposits.

1 Introduction

Large rookeries of seabirds produce substantial amounts of organic matter (Siegfried 1982). This natural fertilization plays an important role in polar environments. Nutrients transported from decaying guano by water and air enrich the surrounds of penguin colonies and lead to the formation of specific ornithogenic soils (Tatur and Myrcha 1983, 1984). In the cold-temperate zone of the northern hemisphere, it has been found that run-off solutions from seabird colonies can enrich levels of nutrients in coastal waters, leading to significant increases in primary and secondary production (Golovkin and Gurevitch 1973; Golovkin and Garkovaja 1975; Bedard et al. 1980). There is a vast literature dealing with the processes of mineralization of guano, and phosphatization of rocks that accompany the washing out of guano deposits by rainfall, in warmer climatic zones (Hutchinson 1950; White and Warin 1964; Altschuler 1973).

Both micro-biological and chemical studies of the formation of ornithogenic soils have been largely restricted to the surface of the substratum of rookeries (Allen and Northover 1967; Boyd et al. 1970; Everett 1976; Wilson and Bain 1976; Smith 1978, 1979; Orchard and Corderoy 1983; Ramsay 1983). Changes in the chemical composition of run-off water in rookery areas have not get been reported. For these reasons, no accumulation of biophilous substances from the decomposition of bird faeces has been observed, nor have products of the reaction of faecal solutions with the parent rocks been found (Walton 1980). Because few data are available on this subject, it has usually been assumed that ornithogenic soils of the maritime Antarctic are similar to those described for continental Antarctica (Allen and Heal 1970; Ugolini 1972), and that guano produced by penguins is intensively and thoroughly washed out to sea. Climatic differences (primarily temperature and humidity) between the maritime Antarctic and continental Antarctica, prompted us to predict that the processes attending the decomposition of guano and the formation of ornithogenic soils might have different characteristics in the two regions.

In this paper we report, in part, on a complex study involving microbiological, hydrochemical and geochemical observations in areas surrounding large pygoscelid penguin rookeries in the maritime Antarctic. The study was designed to learn more about the formation of ornithogenic soils and to examine further the prediction made above.

2 Materials and Methods

Investigations were carried out in Admiralty Bay (62° 09′S, 58° 28′W), King George Island, during the austral summer of 1979–1980 when a total of some 45,000 pairs of Adélie, *Pygoscelis adeliae,* Chinstrap, *P. antarctica,* and Gentoo, *P. papua,* penguins bred on the seven ice-free strips of the coastline (Jabłonski 1984).

Rookeries of Adélie Penguins (at Point Thomas and Point Llano) usually were situated relatively far from the sea and nesting colonies in particular were localized on permeable ground (Fig. 1). Chemically aggressive guano solu-

1 Institute of Ecology, Polish Academy of Sciences, 05-150 Łomianki, Dziekanów Lesny n/Warsaw, Poland

2 Institute of Agricultural Chemistry, Soil Science and Microbiology, Agricultural Academy, Wrocław, Poland

Antarctic Nutrient Cycles and Food Webs
(ed. by W.R. Siegfried, P.R. Condy, and R.M. Laws)

◀ Fig. 1. The pygoscelid penguin rookery at Point Llano in Admiralty Bay. (Photo K. Furmanczyk)

Fig. 2. Chinstrap Penguin rookery on Chabrier Rock in Admiralty Bay. (Photo K. Furmanczyk) ▼

tions washed down from the rookeries reacted with the weathered cover of volcanic rock, during their long and complicated way down-slope to form a wide zone of alteration (Tatur and Myrcha 1983). Rookeries of Chinstrap Penguins (at Point Patelnia, Point Uchatka, Point Demay, Point Llano, Point Thomas, Shag Island, and Chabrier Rock) were localized mainly on steep, hard rocks near the sea (Fig. 2); waters washing down from these rookeries reach the sea relatively fast, and the zone of alteration, if it does exist, is smaller. Gentoo Penguins nested mainly in small colonies on stone beaches (at Point Llano and Point Thomas). Their total population was small and, therefore, of little significance here. Point Llano was our best study site, since it included some 18335 pairs of penguins in several colonies spread over an area of ca. 15 ha (Jabłonski 1984).

Micro-biological investigations were carried out, chemical compositions of running surface-water and water infiltrating the soil were analyzed, and soil profiles were examined with chemical and mineralogical analyses of subsequent layers in the penguin colonies and between the colonies and the coast. Micro-biological investigations were carried out in the laboratory, where the degree and type of mineralization of fresh penguin guano were examined; and in the field. Fresh guano, collected from the surface of the snow or stones, was incubated for three weeks at a room temperature of about 16°C, and in a shaded place outdoors with a temperature range of –1.5°C to 5.7°C (x = 2.3°C). In addition, samples were taken of a mixture of excreta from the nesting areas of different penguin species, together with water suspensions of organic-mineral sediments at different distances from the rookeries (i.e., direct vicinity, 10–15 m, 50–70 m, and in shallow water-bodies on the beach near the coast). Mineral and organic compounds (N/NH_3, N/NO_2, N/NO_3, total C, N, P, carbohydrates, chitin and uric acid) were determined in the samples used in the laboratory experiment and in those taken in the field. The activity of the following enzymes was also determined in these samples: gelatinase, asparaginase and uric acid hydrolases, and acid and alkaline phosphatases. Counts of different groups of bacteria were made after an incubation of 3–21 d, at a temperature of 18°C, on different substrates, using dilution-plate or tube methods (see Pietr et al. 1983).

Chemical analyses of run-off water from colonies were performed in analogous places under different meteorological conditions: just after heavy rainfall and in dry periods. During heavy rainfall the solutions with guano suspension were displaced down-slope, mainly onto the surface of the soil. During little rainfall, or melting of the snow cover, most of the solution percolated through the soil. Samples of water filtered and, after suitable dilution, the following determinations were made: pH, organic C, N/NH_3, N/NO_2, N/NO_3 and P, on the day of sampling, and Ca, Na, K, Mg, Al, and Fe after three months in the laboratory of the Institute of Ecology at Warsaw. Concurrently, similar analyses were performed on the suspensions gathered on the filters [detailed methods for these analyses are given in Tatur and Myrcha (1983)].

Specific features of the ornithogenic soils of the Admiralty Bay region were described, based on 44 soil profiles. Special attention was paid to the chemical and mineralogical analyses of ornithogenic remains, and to the products of the reaction between guano and rock. Soil samples were collected only from finer material, excluding fractions of stones whose numbers were estimated roughly. In the soil-containing parts, the following determinations were made: H_2O^+, pH C, N, H, P, Ca, Na, K, Mg, Al, Fe, Mn, and Sr, according to methods described in Tatur and Myrcha (1984). Using both chemical and physical methods, the mineralogical composition of phosphates occurring in subsequent layers of soil profiles was determined (Tatur and Barczuk 1984).

3 Results and Discussion

3.1 Qualitative Evaluation of the Decomposition Processes of Ornithogenic Matter, Accumulation in Soil and Return of Nutrients to the Sea

In the maritime Antarctic zone faeces of penguins undergo an intense mineralization (Pietr et al. 1983). Penguins excrete large amounts of faeces on land, mostly in regions occupied by breeding groups and also on routes connecting these areas with the sea. Penguin guano is heterogenous and contains three visible, separated fractions: a "white" fraction (27.7% N) consists of uric acid, a "red" fraction contains 65.8% of chitin from exoskeletons of Krill, *Euphausia superba,* and 2.45% of P. and a "green" fraction composed of undigested algal cells (the food of Krill) contains 9.90% N.

Mineralization of this diversified material proceeds in steps. A crucial role in this process is played by microbial activity. The high activity of micro-organisms observed in the maritime Antarctic, which promotes a fast rate of decomposition, is more typical for temperate-climate zones than for the much colder conditions found in contintental Antarctica (Boyd et al. 1970; Cameron et al. 1970; Cameron 1972; Orchard and Corderoy 1983; Ramsay 1983).

In an in situ experiment lasting three weeks it was found that under average climatic conditions during the austral summer at King George Island about 50% of the C and N had volatilized to the atmosphere from fresh guano, as a result of mineralization processes. Especially intense microbial processes occur during short periods of solar insolation which warm up the soil surface to temperature > 10°C. At a temperature of about 16°C the rate of mineralization increased four times. The mineralization of N compounds in fresh faeces was observed as an effect of activity of enzymes (proteases) from the digestive tract of penguins and under the influence of proteolytic and ammonifying groups of bacteria. The high numbers of saprophytic ($30–42 \times 10^6$ g^{-1} dry wt), proteolytic ($0.7–0.8 \times 10^6$ g^{-1} dry wt) and the calcium phosphate dissolving groups of bacteria, as well as gelatinase and phosphatase activities, were characteristic of organic matter from the rookery. Consequently, there is a degradation in the fresh guano of the organic compounds most vulnerable to decomposition, and this process is accompanied by a concentration of the more resistant compounds, such as chitin, uric acid and the phosphate minerals (hydroxylapatite and struvite). These minerals are formed from ions released from decomposing organic matter of guano enriched in marine aerosols. In partly decomposed and washed out guano, the rate of mineralization decreased and total microbial activity abruptly diminished. However, the activity of bacteria which decompose chitin, uric acid and hydroxylapatite, as well as that of the bacteria responsible for the ammonium nitrifying process, increased (Pietr et al. 1983; Pietr 1984).

Guano disappeared fast from the soil surface in the rookery. In cases where penguin rookeries are localized on steep coastal rocks, both guano and the products of its mineralization are easily and promptly washed off to the sea, even during a small amount of rainfall. Such a situation is characteristic of the nesting sites of Chinstrap Penguins (Fig. 2). When rookeries are located farther inland, on flat rocks surrounded by the weathered cover and separated from the sea by a broad beach, the flow of ornithogenic matter to the sea is more restrained. Such nesting places are typical of Adélie Penguins (Fig. 1).

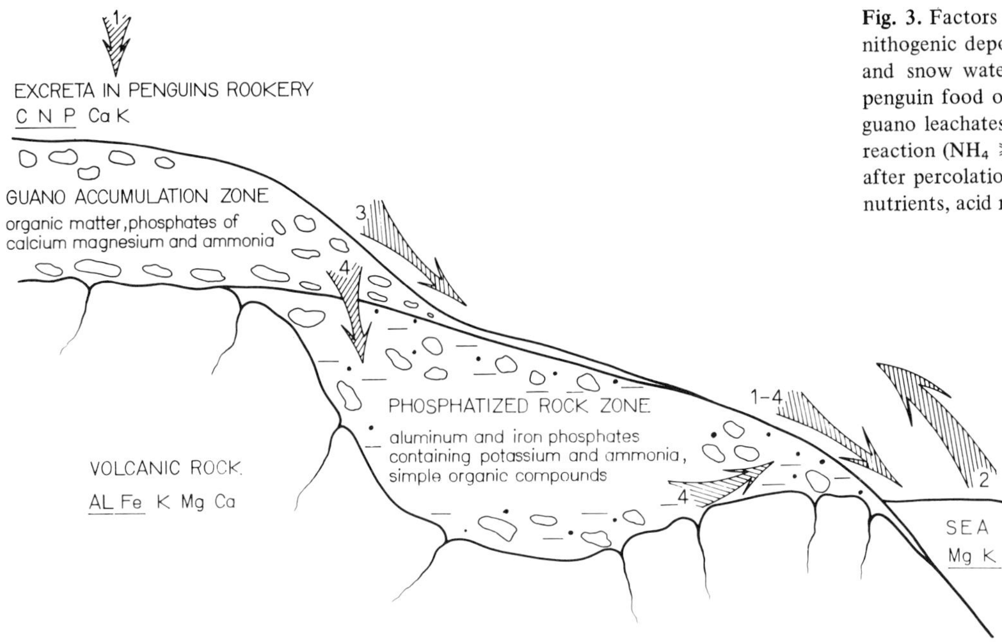

Fig. 3. Factors involved in the formation of ornithogenic deposits. Sources of water: (*1*) Rain and snow water; (*2*) Seawater, taken up with penguin food or by wind as aerosol; (*3*) Surface guano leachates, high level of nutrients, alkaline reaction ($NH_4 \gg NO_3$); and (*4*) Guano leachates after percolation through the soil, high level of nutrients, acid reaction ($NO_3 \geqslant NH_4$)

Delayed guano run-off facilitates mineralization which results in an increase in the ratio of simple ions to organic suspension in the waters reaching the sea. Among the mineral forms of N the concentration of nitrates can be as high as that of ammonium ion or sometimes even higher (Tatur and Myrcha 1983). Most of the C and N present in faeces can volatilize to the atmosphere due to prolonged processes of mineralization. In the situation just described the terrestrial deposits of ornithogenic materials can also become important (Tatur and Myrcha 1984).

Guano is washed partly into stony detritus in the rookery itself, and the remainder is smeared as a thin layer over the slopes around the nesting places of penguins. However, the guano leachates react with stony loams surrounding the rookeries, forming a large zone of phosphatized rock. The phosphatized zone shows a typical morphology for the soil and, consequently, was described as the "ornithogenic soils of the Maritime Antarctic Zone" (Tatur and Myrcha 1984). The division of the soil profile into horizons is easily noticeable in the field, and results from different mineral compositions involving phosphates in various layers. The phosphates play a key role in the genesis of the soil. They are formed either as a product of the precipitation from ornithogenic waters, or as a result of metasomatic reactions on the border of aquatic and silicate phases. The vertical distribution of these phosphates in the profile must remain in a close relationship with the chemical composition of the soil solution. Thus, changes in the chemical composition of the guano leachates percolating through and reacting with the soil proved to be a main factor controlling the vertical distribution of the phosphates in the profile of the ornithogenic soils.

The surface waters running off the penguins' breeding area are alkaline or neutral and contain changeable, often very highly concentrated, mineral forms of P and N. Among mineral forms of N, the ammonium form strongly predominates over the nitrate one. After percolation through the soil, the chemical composition of these waters undergoes stabilization and the reaction becomes acid (pH 4 or somewhat higher). Among mineral forms of N, the nitrates prevail over the ammonium ion (Fig. 3). Increase of acidity during percolation is probably due mainly to volatilization to the atmosphere, binding into phosphates and nitrification of the alkaline ammonium ion. The acid reaction is enhanced also by forming oxalic acid processes of organic matter mineralization. The final pH of the soil water results from the solubility equilibrium constant which is set during the reaction with phosphates occurring in the bottom of the ornithogenic soils.

The relationships between changes in chemical composition of guano leachates and the variability of mineral composition of the phosphates in the ornithogenic soils, presented in Fig. 4, have been analyzed in detail by Tatur and Barczuk (1984). Struvite is present exclusively on the surface of the soil. This mineral crystallizes from alkaline waters rich in ammonium ion and enriched with Mg. Hydroxylapatite is common in the whole surface layer of guano. It is the main mineral deriving there from the decomposing organic matter of penguin faeces. Leucophosphite occurs in the shallowest horizon of the phosphatized zone, often under a thin guano layer. This mineral is formed as the precipitate from neutral or slightly acid solutions still holding the ammonium ion. Periodical drying up and oxidation of this layer facilitates the occurrence of Fe^{III} built up in the crystalline structure of this mineral. Minyulite is typical for the deeper layer where it is formed as a product of metasomatic reactions between soil solution and the soil silicates. An adequately high concentration of nutrients and of F in the ornithogenic water S probably plays a significant role in the formation of this mineral. Tarankite is typical for the deepest, always wet, layer. It is formed as a precipitate from acid and most diluted solutions.

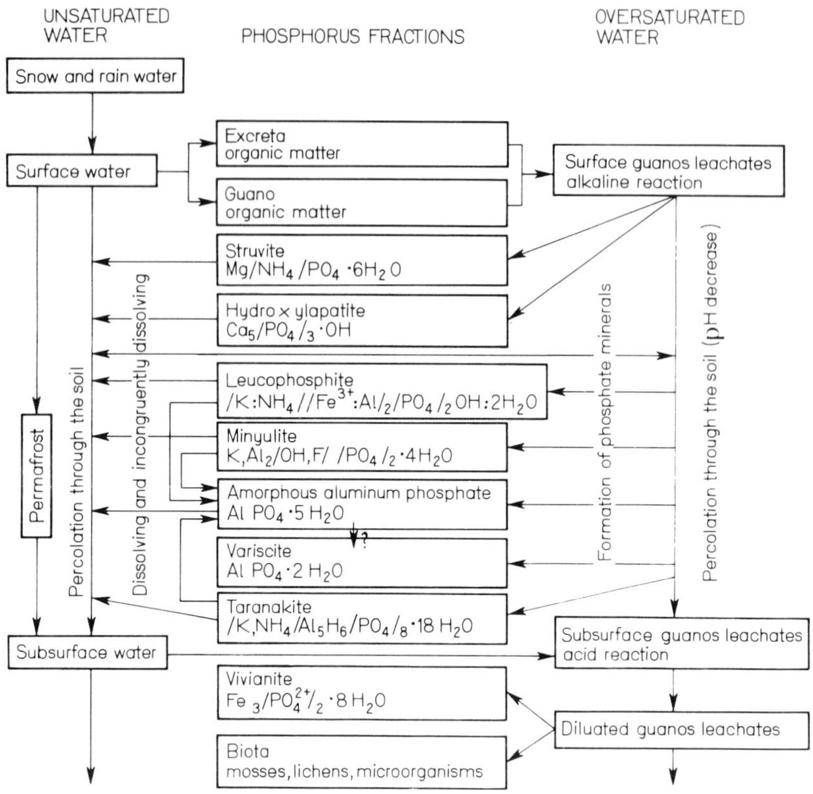

Fig. 4. Composition of the ornithogenic phosphorus deposits as affected by chemical changes in water

Leucophosphite, minyulite, and taranakite are formed and persist in the presence of excess alkali in ornithogenic solutions, but when solutions are poor in alkali (rain or melting water) a simple amorphous phosphate is formed. It can be formed as a product of direct metasomatic reaction between the soil solution and silicates, or it can result from the incongruent dissolving of aluminium phosphates containing the K and the ammonium ions. A small amount of a crystalline variscite-like form inside amorphous aluminium phosphate can be due to more acid conditions temporarily found in the percolating solutions. Vivianite occurs outside the zone of ornithogenic soils in a permanently wet, acidified, loamy layer under the turf banks where waters carrying ornithogenic P can reach periodically.

Phosphates occurring in ornithogenic soils are unstable in the high precipitation typical for the maritime Antarctic zone. The persistence of phosphates is possibly due to manuring by penguins which ensures the specific chemical conditions in the environment. A long interval in manuring must lead to a tendency common for this zone for acidification of the soils, which, coupled with leaching by the alkali-poor precipitation waters, should lead at first to dissolving guano minerals (struvite and hydroxylapatite) and then solution of the aluminium-iron phosphates containing K and the ammonia ion. The latter minerals, being more durable in an acid medium, become unstable in alkali-poor waters. The product of their incongruent solution will most likely be an amorphous aluminium phosphate, which either will dissolve or alter with age to a crystalline variscite-like form.

The paragenesis of phosphates observed in the ornithogenic soil results from the unstable equilibrium between opposite processes. The first one leads to formation of the minerals and their persistence during manuring. The second one leads to alteration and solution of these minerals when manuring ceases.

3.2 An Attempted Quantitative Estimate of Ornithogenic Phosphorus Flow through a Terrestrial Environment as Exemplified by the Rookery at Point Llano

In the conditions of the maritime Antarctic zone both guano and products of its mineralization are intensively washed to the sea. Thus, the discovery of the commonly observed deposition of P in soil raises the question of what is the rate of this process and, hence, what portion of the P carried from the sea by birds is retained on land each year. For a rough estimate of the rate of accumulation of P on land, the pygoscelid penguin rookeries isolated at Point Llano were selected for investigation. The run-off of guano and its mineralization products into the sea in this terrain is more difficult to trace and lasts longer than in the case of the Chinstrap Penguin rookeries which also occur in Admiralty Bay. Similar conditions occur in the region of Point Thomas where Adélie Penguins predominate in the rookeries. Thus, on the basis of our field studies, it seems that a direct, trustworthy measure of the amount of P returning to the sea is at present not possible. For this reason, the estimation of the accumulation rate of this element on land was based on data pertaining to: (1) the amount of P in ornithogenic soils; (2) the time of formation of these soils; and (3) the size of penguin populations that have a direct effect on ornithogenic soil formation.

1. From the analyses of amounts of P in numerous soil profiles investigated, it was estimated that in soils surrounding the Point Llano rookery at least 400 t of P is accumulated.

2. Based on the literature and our own observations concerning the modern geological history of this region, one can assume that adequate conditions enabling penguins to nest in places where contemporary rookeries occur have been in existence for about 500–700 yrs. These conditions appeared as a result of processes associated with the last retreat of the ice cover which brought about the emergence of the lowest parts of the sea coast (below 6 m).

3. In the breeding season 1979–80, 15013 pairs of Adélie, 3008 pairs of Gentoo and 314 pairs of Chinstrap penguins nested in the Point Llano rookery (Jabłonski 1984). From a review of the literature it is evident that there are large oscillations in the number of pygoscelid penguins nesting in the Admiralty Bay region (Croxall and Kirkwood 1979; Trivelpiece and Volkman 1979; Jabłonski 1984). The population census of 1979–80 was the highest since 1909. Almost all nesting places of penguins which had ever been used in the past were occupied and one can even suppose that there has been a terrestrial expansion during recent years. In the areas occupied by young nesting groups vegetation was still present, although considerably devastated. One can expect, then, that the population of penguins nesting in this region during the whole of the formation of P deposits, had not been larger than the population found in the season of 1979–80.

It is evident that at least 400 t of P has accumulated in the ornithogenic soils of Point Llano during a period probably not longer than 700 yrs. This P was derived from faeces excreted by penguins during their breeding seasons, based on an average of some 36700 adults and 24000 chicks per annum (Jabłonski, unpubl.). The cumulative energy requirement of the whole population can be calculated from bioenergetic data on the cost of maintenance of adult penguins (Le Resche and Boyd 1969; Bougaeff 1975) and of chicks during their whole development on land (Taylor 1982), as well as from the production value of young from hatching to entering the sea (Myrcha and Kaminski 1982; Taylor 1982). In these calculations, the duration of the breeding period and the growth periods of chicks of different species have been taken into consideration (Watson 1975; Taylor 1982; Jabłonski, unpubl.), as well as recent data from the region of Admiralty Bay on the frequency of chick-feeding and the period spent by adults at sea (Jabłonski, unpubl.; Trivelpiece and Volkman, pers. comm.).

Assuming that the pygoscelid penguins nesting on King George Island feed mainly on Antarctic Krill, *Euphausia superba,* and that the assimilation efficiency of this food is 75% which is similar to that ascertained for *Spheniscus demersus* (Cooper 1977), the relevant population might deposit on land slightly more than 200 t of faeces during one breeding season. Fresh faeces of penguins contain about 2.55% of P (Pietr et al. 1983). The population nesting at Point Llano thus deposits on land about 5.2 t of P each year. During 700 yrs, at least 0.57 t of P has been deposited in this area each year, which forms not less than 11% of the amount of this element carried onto the land by birds. The nesting penguins leave on land some P contained in egg-shells and feathers. The average weight of an egg-shell for the three species of penguins at Admiralty Bay is 15.5 g (Jabłonski, unpubl.) and the amount of P in the egg-shell is 3.316 ppm (Siegfried et al. 1978). Gentoo and Chinstrap penguins moulting, and young individuals of all three species, leave on land more than 3.5 t of feathers which contain not less than 120 ppm of P (Williams and Berruti 1978). In total, however, the amount of P left by penguins on land in egg-shells and feathers is negligible, and does not exceed 0.05% of its content in faeces.

The estimated rate of P accumulation on land may be a considerable underestimate, since substantial amounts of P that linger deep in stony soils of the phosphatized zone could have been omitted from the calculations. Moreover, the time of deposit formation could have been shorter than supposed (which was probably the maximum) and also the penguin population during the formation of the deposit could have been much smaller. The percentage amount of P accumulated on land each year, as compared with the amount brought in by birds from the sea, is likely to be higher than that estimated by us. This indicates that the processes taking place in ornithogenic soils may indeed slow down the rate of P cycling in the Antarctic environment, because P is being retained in the ornithogenic soils.

The fact that the P deposits on islands of the marine Antarctic zone are in the geological sense insignificant, results not only from the relatively slow rate of their accumulation, often discussed in the literature, but also from the relatively short time of their formation. In geological terms these forms are unstable. In the areas of their occurrence, processes of intense mechanical erosion exist, together with recurring phenomena of ice expansion, which must lead to a rapid and complete removal of any changes in the soil cover brought about by the activity of penguins. Therefore, it seems that the persistence of these deposits can be considered only in the historical sense.

Acknowledgements. This work was supported by the Polish Academy of Sciences within the MR-II-16 and MR-I-29 Projects carried out at the Arctowski Station during the Fourth Antarctic Expedition of 1979–80.

References

Allen SE, Heal OW (1970) Soils of the Maritime Antarctic Zone. In: Holdgate MW (ed) Antarctic Ecology, Academic Press, New York vol 2:693–701

Allen SE, Northover MJ (1967) Soil types and nutrients on Signy Island. Philos Trans R Soc Ser B 252:179–185

Altschuler ZS (1973) The weathering of phosphate deposits – geochemical and environmental aspects. In: Griffith EJ, Beeton A, Spencer IM, Mitchell DT (eds) Environmental phosphorus handbook. Wiley, New York, pp 33–96

Bedard J, Theriault IC, Berube J (1980) Assessment of the importance of nutrient recycling by seabirds in the St. Lawrence estuary. Canad J Fish Aquat Sci 37:583–588

Bougaeff S (1975) Variations ponderales et evaluation de la depense energetique chez le Manchot Adélie (*Pygoscelis adeliae*). C R Acad Sci 280:2373–2376

Boyd WL, Rothenburg J, Boyd JW (1970) Soil microorganisms at Paradise Harbour, Antarctica. Ecology 51:1040–1045

Cameron RE (1972) Microbial and ecologic investigations in Victoria Valley, Southern Victoria Land, Antarctica. Antarc Res Ser 20: 195–260

Cameron RE, King J, David CN (1970) Microbiology, ecology and microclimatology of soil sites in dry valleys of Southern Victoria Land, Antarctica. In: Holdgate MW (ed) Antarctic ecology. Academic Press, New York vol 2:702–716

Cooper J (1977) Energetic requirements for growth of the Jackass penguin. Zool Afr 12:201–213

Croxall JP, Kirkwood DK (1979) The breeding distribution of penguins on the Antarctic Peninsula and the islands of the Scotia Sea. Br Surv Publ Antarc 1–186

Everett KR (1976) A survey of the soils in the South Shetland Islands and adjacent parts of the Antarctic Peninsula. Ohio State Univ Inst Polar Stud Rep 85:1–44

Golovkin AN, Garkovaja JP (1975) Udobrenie pribereznyh vod murmanskogo poberezija ekskrementami ptic v blizi kolonij razlicnyh tipov. Biol Morya 5:49–57

Golovkin AN, Gurevitch VI (1973) Vydelenije i ocenka gidrohimiceskih anomalij v more metodom linijnyh diskriminantnyh funkcii (na primere barencovomorskih pticnyh bazarov). Okeanologiya 13:804–808

Hutchinson GE (1950) Biogeochemistry of vertebrate excretion. Bull Am Mus Nat Hist 96:1–554

Jabłonski B (1984) Distribution, numbers and breeding preferences of penguins at the region of the Admiralty Bay (King George Island, South Shetland Islands) in the season 1979/1980. Pol Polar Res 5:5–16

LeResche R, Boyd JC (1969) Response to acute hypothermia in Adélie penguin chicks. Commun Behav Biol 4:85–89

Myrcha A, Kaminski P (1982) Changes in body calorific value during nesting development of penguins of the genus *Pygoscelis*. Pol Polar Res 3:81–88

Orchard VA, Corderoy DM (1983) Influence of environmental factors on the decomposition of penguin guano in Antarctica. Polar Biol 1:199–204

Pietr SJ (1984) The physiological groups of bacteria in different soils at Admiralty Bay region (King George Island, South Shetland Islands, Antarctica). Pol Polar Res, in press

Pietr SJ, Tatur A, Myrcha A (1983) Mineralization of Penguin excrements at Admiralty Bay (King George Island, South Shetland Islands, Antarctica). Pol Polar Res 4:97–112

Ramsay AJ (1983) Bacterial biomass in ornithogenic soils of Antarctica. Polar Biol 1:221–225

Siegfried WR (1982) The roles of birds in ecological processes affecting the functioning of the terrestrial ecosystem at sub-Antarctic Marion Island. CNFRA 51:493–499

Siegfried WR, Williams AJ, Burger AE, Berruti A (1978) Mineral and energy contributions of eggs of selected species of seabirds to the Marion Island terrestrial ecosystem. S Afr J Antarct Res 8:75–87

Smith VR (1978) Animal-plant-soil nutrient relationship on Marion Island. Oecologia (Berl) 32:239–253

Smith VR (1979) The influence of seabird manuring on the phosphorus status at Marion Island (Subantarctic) soils. Oecologia (Berl) 41:123–126

Taylor JRE (1982) Bioenergetic of the development period of the pygoscelid penguin chicks. Ph.D. Thesis Institute of Ecology PAS pp 1–126

Tatur A, Myrcha A (1983) Changes in chemical composition of waters running off from the penguin rookeries at Admiralty Bay region (King George Island, South Shetland Islands, Antarctica). Pol Polar Res 4:113–128

Tatur A, Barczuk A (1984) Ornithogenic minerals around penguin rookeries on volcanic King George Island (Maritime Antarctic). Pol Polar Res 5:31–60

Tatur A, Myrcha A (1984) Ornithogenic soils on King George Island (Maritime Antarctic). Pol Polar Res 5:

Trivelpiece W, Volkman NJ (1979) Nest site competition between Adélie and chinstrap penguins: An ecological interpretation. Auk 96:675–681

Ugolini FC (1972) Ornithogenic soils of Antarctica. In: Llano GA (ed) Antarctic terrestrial biology. Antarct Res Ser 20:181–193

Walton DWH (1980) An annotated bibliography of Antarctic pedology and periglacial processes. Br Antarct Surv Bull 5

Watson GE (1975) Birds of the Antarctic and Subantarctic. Am Geophys Union, Washington pp 1–350

White WC, Warin ON (1964) A survey of phosphate deposits in the Southwest Pacific and Australian Waters. Bur Mineral Resour Aust Bull 69:1–173

Williams AJ, Berruti A (1978) Mineral and energy contributions of feathers moulted by penguins, gulls and cormorants to the Marion Island terrestrial ecosystem. S Afr J Antarct Res 8:71–74

Wilson MJ, Bain DC (1976) Occurrence of leucophosphite in a soil from Elephant Island, British Antarctic Territory. Am Mineral 61:1027–1028

Ornithogenic Phosphates on King George Island in the Maritime Antarctic

A. TATUR[1] and A. BARCZUK[2]

Summary. As a rule, the following sequence of phosphate minerals was found in ornithogenic soils around penguin rookeries: struvite and hydroxylapatite in the surface guano layer, leucophosphite in the upper part, of the phosphatized rock zone, with minyulite in the central part, and amorphous aluminium phosphate and taranakite in the deepest part. These minerals were identified on the basis of the X-ray diffraction patterns, chemical analysis and observation under petrographic and scanning-electron microscopes.

1 Introduction

Processes of rock phosphatization around bird colonies (causing the removal of phosphorus from the biogeochemical cycle) are common wherever high precipitation occurs. The abundance of water facilitates the reaction between guano leachates and the underlying rock material. Most descriptions of phosphatized rock are from the wet tropical zone (Hutchinson 1950; White and Warin 1964; Altschuler 1973). The few descriptions of such rocks in the temperate zone we can consider as the result of the lower economic significance of these deposits. In the maritime Antarctic, phosphatization has been recorded recently by Wilson and Bain (1976). They described the occurrence of the leucophosphite in the surface layer of the ornithogenic soils on Elephant Island, close to King George Island.

Several phosphates were found by us during examination of the ornithogenic soil around penguin rookeries on King George Island. Consideration of the vertical distribution of the phosphates in the profile proved to be more useful than chemical analysis in understanding the origin of these soils. The work extends our knowledge of P pathways in the biogeochemical cycle controlled by climatic conditions specific to the maritime Antarctic. Short descriptions of the phosphates occurring in the surface guano layer, as well as in the sub-surface zone of phosphatized volcanic rock, are given below.

1 Department of Biogeochemistry, Institute of Ecology, Polish Academy of Sciences, 05-092 Łomianki, Dziekanów Leśny n/ Warsaw, Poland

2 Institute of Geochemistry, Mineralogy and Petrography, University of Warsaw, 02-089 Warszawa, Zwirki i Wigury 93, Poland

2 Materials and Methods

The phosphates examined were taken from ornithogenic soils described by Tatur and Myrcha (1984). The mineral composition of these phosphates will be described in detail by Tatur and Barczuk (1984). Chemical analysis of the phosphates was carried out as described by Belopolskiy et al. (1974). Phosphate powder was X-rayed with a DRON-1 diffractometer. Visual morphological observations were performed under petrographic and scanning-electron microscopes (SEM).

3 Results

Struvite was widespread on the surface of the guano in small amounts. A thin layer of pure struvite crystals was found in one spot only. It was located in a place of intensive evaporation of guano leachates on the flat shelf of the rock (ca. 1 m^2). Struvite was identified on the basis of optical properties, chemical analysis and X-ray diffraction spacing (Table 1). Struvite forms well-defined crystals (approximate dimension = 1 mm, space group = Pmc 2) joined in greater aggregates (Fig. 1). Struvite has the following molecular formula: $Mg\ (NH_4)\ PO_4 \cdot 6H_2O$. Small amounts (0.37%) of Mn may substitute for Mg. An excess of Mg in relation to ammonium ions found during chemical analysis was probably due to mineral changes during storage of the sample under laboratory conditions.

Hydroxylapatite was the main phosphate of the guano layer. It was never found as a pure mineral, but always existed with organic matter detritus and silicate minerals. Hydroxylapatite was identified by means of X-rays in a sample containing 18% of P_2O_5 and 15% of CaO. The sample containing hydroxylapatite gave the following typical pattern: a wide and asymmetric peak (max. 2.8 Å) and weakly marked peaks (2.71 Å, 2.62 Å) on its left slope, and a distinctly sharp peak (3.44 Å).

Leucophosphite was the main mineral in the top layer of the phosphatized rock zone, where it was present together with silicates and oxides in plastic clay-like material among rubble and cobbles, probably as the stout prismatic crystals

Antarctic Nutrient Cycles and Food Webs
(ed. by W.R. Siegfried, P.R. Condy, and R.M. Laws)

Table 1. X-ray diffraction patterns for struvite, minyulite and taranakite in ornithogenic soils from King George Island. (All X-ray diffraction peaks found in the sample are given, with the more important of the complex ones marked "x")

Struvite

	d	$I/I_1 \cdot 100$
	6.12	9
	5.89	20
	5.61	68
	5.37	32
	4.58	4
	4.25	75
	4.14	45
	3.54	3
	3.46	4
x	3.28	25
	3.19	
	3.06	4
	3.01	7
	2.95	46
x	2.91	
	2.80	50
	2.71	
x	2.69	100
	2.65	
	2.54	4
	2.50	7
	2.39	4
	2.35	7
	2.25	3
	2.18	4
	2.13	7
	2.06	7
x	2.05	
	2.01	8
	1.981	16
	1.957	
	1.919	7
	1.863	5
	1.848	4
	1.792	21
	1.756	9
	1.735	11
	1.711	3
	1.678	3
	1.652	4
	1.591	12
	1.552	3

Minyulite

	d	$I/I_1 \cdot 100$
	6.73	36
	5.54	77
	4.86	6
	4.66	6
	4.23	10
	3.75	
	3.65	32
	3.67	
	3.39	188
	3.35	
	3.07	25
	2.97	17
	2.88	10
	2.76	9
	2.68	25
	2.61	19
	2.55	6
	2.43	
	2.40	
	2.38	32
	2.36	
x	2.33	
	2.27	14
	2.17	
x	2.14	41
	2.08	
	2.05	3
	2.02	9
x	1.918	16
	1.901	
	1.804	8
	1.778	5
	1.746	6
x	1.688	23
	1.667	
	1.613	7
	1.569	16

Taranakite

	d	$I/I_1 \cdot 100$
	15.92	100
x	7.92	
x	7.47	73
	7.18	
	5.87	24
	5.04	3
	4.66	4
x	4.31	30
	4.19	
•	4.01	10
x	3.81	68
	3.75	
	3.59	23
	3.35	29
	3.29	
x	3.17	54
x	3.14	
	2.95	17
	2.84	51
x	2.82	
x	2.73	21
	2.71	
	2.63	26
x	2.57	13
	2.54	
	2.48	3
	2.44	2
	2.39	16
	2.34	4
	2.26	8
	2.17	
	2.15	13
	2.12	
	2.09	23
x	2.07	
	1.963	13
	1.906	5
	1.866	5
	1.849	3
	1.820	
	1.800	19
	1.779	
	1.745	13
x	1.721	
	1.679	4
x	1.645	
	1.624	25
x	1.610	
	1.595	
	1.570	4

observed under the scanning microscope. This phosphatic clay-like material contained up to 22% of P_2O_5. Leucophosphite was never found as a pure mineral. It was identified, by means of X-ray data, in a mixture (with the other phosphates and silicates). All of the most important diffraction spacings typical of leucophosphite were found (among them 7.63 Å, 6.75 Å, 5.98 Å, and 3.06 Å). Chemical analysis suggested that, from the molecular formula (K, NH_4) (Fe^{III}, Al)$_2$ (PO_4)$_2$$\cdot_2$$H_2O$, potassium ions exceeded ammonium ions and the content of Al was close to that of Fe, or higher.

Minyulite was typical of the deeper part of the phosphatized rock zone, where it indurates the loamy material. Only in one profile did almost pure minyulite form a 1-m thick, hard, sponge-like layer. Slender rod prisms of minyulite (length = ca. 4.0 μm) had grown up from the massive wall of the pores and formed druse (Fig. 2). In some places, spheroidal agglomerations of needles were found (Fig. 3). Minyulite was identified on the basis of X-ray diffraction spacing (Table 1). Calculations, based on the chemical analysis of a mechanically enriched minyulite sample, gave the molecular formula $KAl_2(OH,F)_2\cdot_4H_2O$, with a small excess of elements connected with some admixture of silicates and organic matter. The optical properties of minyulite did not differ from those known for this mineral.

Aluminium phosphate amorphous to X-ray diffraction formed a hard brown coat on the phosphatized rock fragments in the deeper part of the soil profile (Figs. 4 and 5). At the bottom of one profile it was found as an almost pure mineral in a thin layer of gravel. After mechanical enrichment of this sample, analysis of amorphous aluminium phosphate gave the approximate formula $AlPO_4\cdot_5H_2O$, with a characteristically high concentration of Sr (0.22% SrO). Excess of elements connected with some admixture of silicates was noted. In some parts, the amorphous and isotropic aluminium phosphate becomes anisotropic. In the SEM-rounded plates of variscite-like crystals (diam. = ca. 1.0 μm) could be observed. Therefore, the possibility of small amounts of variscite occurring inside amorphous aluminium phosphate should not be excluded.

Taranakite always occurred in the deeper part of the phosphatized zone, either with other phosphates and silicates, or as a pure, white plastic clay-like material among rubble, below ground-water level. Taranakite was identified on the basis of X-ray diffraction spacing (Table 1). Crystals of taranakite occurred as typical pseudo-hexagonal plates of ca. 3.0 μm in diameter (Fig. 6). Calculation of its molecular formula gave: $K_{1.9}(NH_4)_{1.1}Al_5H_6(PO_4)_8\cdot 18\,H_2O$. Its optical properties did not differ from those typical of taranakite.

4 Discussion

The discovery of phosphatized rock around penguin rookeries on King George Island proves that ornithogenic mineralization in the maritime Antarctic zone is comparable with that in the temperate zone. This means that the suggestions of Allen and Heal (1970) and Ugolini (1972), regarding the method of ornithogenic soil formation on the islands of the maritime Antarctic zone and on the Antarctic continent, are not applicable to King George Island. The ornithogenic soils of continental Antarctica consist of guano lying directly on undecomposed rock, with no evidence of phosphatization of underlying rock (Syroechkovskiy 1959; Ugolini 1972). Any consideration of the annual biogeochemical cycle of materials affected by penguins should, therefore, take into account the P accumulated on the land.

The list of phosphate minerals in the ornithogenic soils on King George Island is surprisingly extensive. Struvite and hydroxylapatite are common everywhere in the guano, and taranakite and leucophosphite might be expected to occur in the phosphatized rock. However, the occurrence of amorphous aluminium phosphate and especially minyulite is more surprising here. Neither of these minerals have been found before in a site suggestive of an ornithogenic origin. The paragenesis of minyulite with amorphous aluminium phosphate has never before been reported in nature, although it has been obtained in a laboratory experiment (Cole and Jackson 1950).

Such a wide mineral differentiation as that found on King George Island has been obtained only in the laboratory under changeable conditions of formation (Haseman et al. 1950a,b; Lindsay et al. 1962; Hsu 1982). Phosphates resembling' those found in the phosphatized rock-zone studied by us were synthesized in the laboratory as products of reactions between acid soil silicates and phosphate fertilizers. Problems of the persistence and solubility of phosphates obtained in this way are basic ones for soil science. However, the synthesis of these minerals usually needs conditions quite different from those occurring in natural environments, using higher temperatures and ion concentrations unattainable in soil solutions. For this reason, we consider the area surrounding the penguin rookery studied by us to be an excellent natural laboratory where the phosphatization processes are apparently more effective than in laboratory simulations.

The wide variation of the minerals in the ornithogenic soil on King George Island is probably due to remarkable environmental conditions: high precipitation; continuous humidity; low temperature; specific chemical composition of penguin excreta (especially a high fluorine content); changes in composition of the soil water, as the result of periodical nesting of the penguins on the land; and, the lack of plants, and humus material derived from them. In our study area, these conditions have led to the development of a depth sequence of the phosphate minerals in ornithogenic soils. In extreme cases the pure monophosphate layers may be formed. Struvite and hydroxylapatite occurred in the surface guano layer, leucophosphite in the upper part of the phosphatized rock zone, minyulite in its central part, and amorphous aluminium phosphate and taranakite in its deepest part. The vertical sequence of phosphate distribution is probably the result of changes in the chemical com-

Fig. 1. Crystals of struvite (SEM × 130)

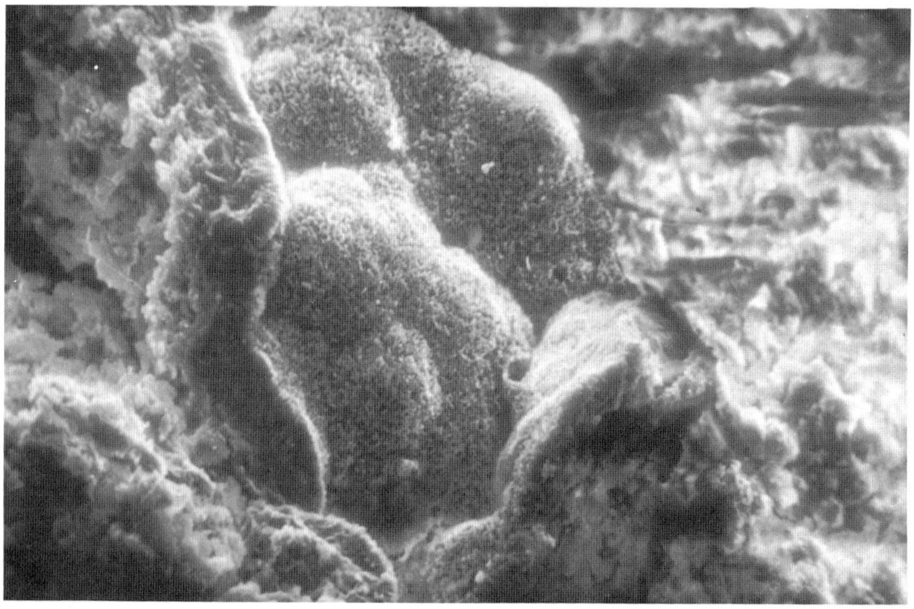

Fig. 2. Druse of minyulite (SEM × 330)

Fig. 3. Spheroidal aggregates of minyulite needles (SEM × 1500)

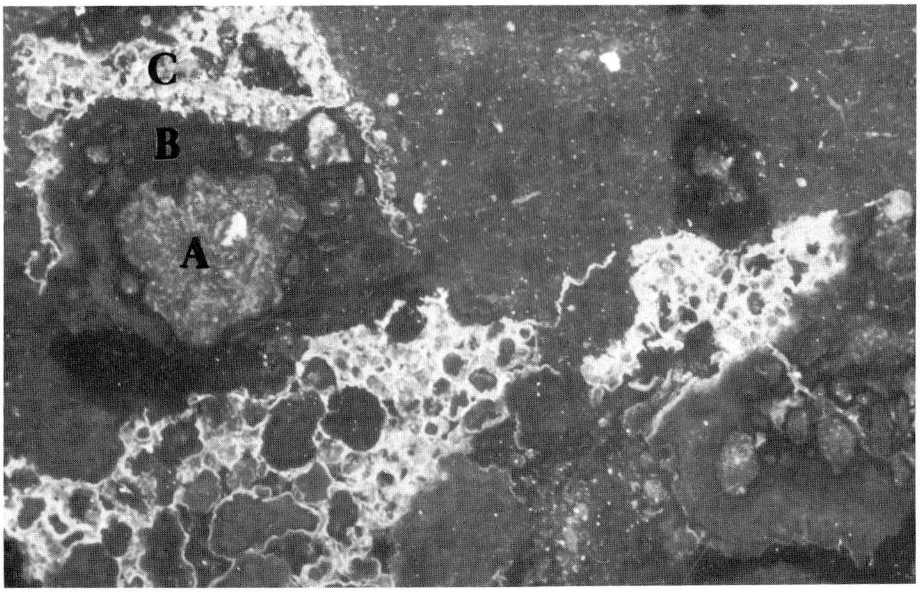

Fig. 4. Paragenesis of minyulite with amorphous aluminium phosphate. (*A*) clay pseudomorphs after fragement of rock (gray); (*B*) amorphous aluminium phosphate (dull brown); (*C*) minyulite (bright yellow). (Thin section under petrographic microscope × 20), reflected light)

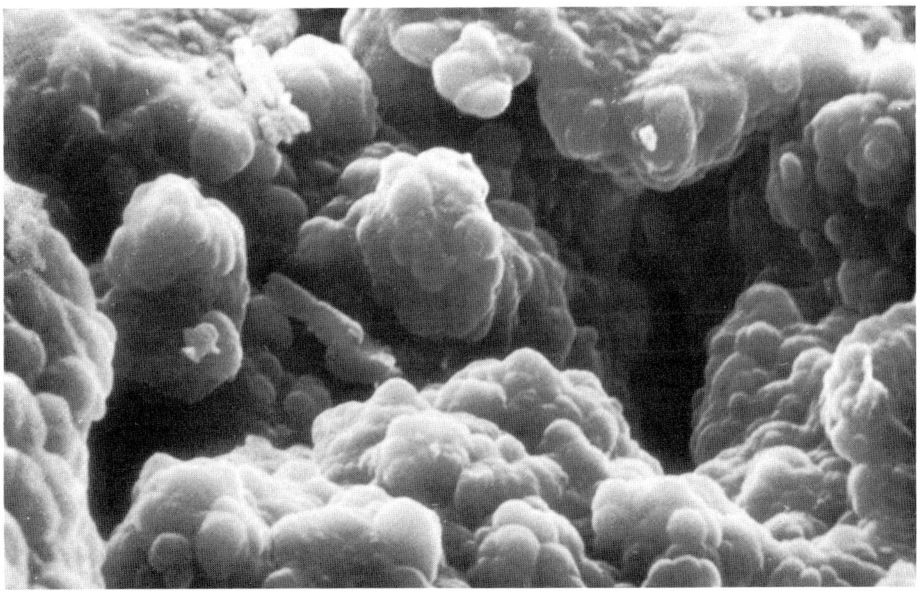

Fig. 5. Amorphous aluminium phosphate (SEM × 1250)

Fig. 6. Crystals of taranakite (SEM × 8600)

position of the guano leachates during their infiltration through the soil. This concept is described in detail by Myrcha et al. (1985).

Acknowledgements. This study was supported by the Polish Academy of Sciences, as part of projects MR-II-16 and MR-I-29, carried out at the Arctowski Station during the Fourth Antarctic Expedition of 1979–80. The authors are thankful to Dr. Andrzej Myrcha for presenting this paper at the Fourth SCAR Symposium on Antarctic Biology.

References

Allen SE, Heal OW (1970) Soils of the Maritime Antarctic Zone. Antarct Ecol 2:693–696

Altschuler ZS (1973) The weathering of phosphate deposits – geochemical and environmental aspects. In: Griffith EJ, Beaton A, Spencer JM, Mitchel DT (eds) Environmental Phosphorus Handbook. Wiley, New York, pp 33–96

Belopolskiy MP, Bunakowa NYu, Gumbar KK, Mikhaylova NA, Popov NP, Stolyarova MA (1974) Analiz fosfatov. In: Popov NP, Stolyarova MA (eds) Khimicheskiy analiz gornykh porod i mineralov, Nedra, Moskwa

Cole CV, Jackson ML (1950) Solubility equilibrium constants of dihydoxy – aluminium dihydrogen phosphate relating to a mechanism of phosphate fixation in soils. Soil Sci Soc Am Proc 15:84–89

Haseman JF, Brown EH, White CD (1950a) Some reaction of phosphate with clay and hydrous oxides of iron and aluminium. Soil Sci 70:257–271

Haseman JF, Lehr JR, Smith Jr and JP (1950b) Mineralogical character of some iron and aluminium phosphates containing potassium and ammonium. Soil Sci Soc Am Proc 15:76–84

Hsu PH (1982) Crystallization of variscite at room temperature. Soil Sci 133:3–13

Hutchinson GE (1950) Survey of contemporary knowledge of Biogeochemistry: 3. The Biogeochemistry of vertebrate excretion. Am Mus Nat Hist Bull 96:554 pp

Lindsay WL, Frazier AW, Stephenson HP (1962) Identification of reaction products from phosphate fertilizers in soil Soil Sci Soc Am Proc 26:446–452

Myrcha A, Pietr S, Tatur A (1985) The role of pygoscelid penguin rookeries in nutrient cycles at Admiralty Bay, King George Island. In: Siegfried WR, Condy PR, Laws RM (eds) Antarctic nutrient cycles and food webs (Proceedings of the 4th SCAR symposium on Antarctic Biology). Springer, Berlin Heidelberg New York

Syroechkovskiy EE (1959) The role of animals in primary soil formation in the conditions of the circumpolar regions of the earth (exemplified by the Antarctic). Zool Zh Ukr 38:1770–1775

Tatur A, Barczuk A (1984) Phosphates of ornithogenic soils on King George volcanic Island (Maritime Antarctic). Pol Polar Res 5 (to be published)

Tatur A, Myrcha A (1984) Ornithogenic soils on King George Island, South Shetland Islands (Maritime Antarctic). Pol Polar Res 5 (to be published)

Ugolini FC (1972) Ornithogenic soils of Antarctica. Antarct Res Ser 20:181–193

White WC, Warin ON (1964) A survey of phosphate deposits in the South-West Pacific and Australian Waters, Bur Miner Resour Aust, Bull 69:pp 173

Wilson MJ, Bain DD (1976) Occurrence of leucophosphite in a soil from Elephant Island. British Antarctic Territory. Am Mineral 61:1027–1028

Chemical Composition of Guano of Burrowing Petrel Chicks (Procellariidae) at Marion Island

S.R. FUGLER[1]

Summary. Guano was collected directly from burrowing petrel chicks (Procellariidae) in their nest chambers, throughout their growth periods at Marion Island (46° 54′S, 37° 45′E). Guano production was measured, and concentrations of selected chemical components were determined. Chemical composition of guano was similar to that previously obtained for a wide range of surface-nesting bird species at Marion Island.

1 Introduction

A research programme to determine the roles of marine birds in the transfer of nutrients from the sea to the terrestrial ecosystem at Sub-Antarctic Marion Island was initiated in 1973 (Siegfried 1978). As part of this programme, Burger et al. (1978) determined the chemical composition of guano produced by surface-nesting birds but, until now, no comparable data were available for burrowing petrels (Procellariidae) which are abundant at the island (Williams et al. 1979).

Smith (1976, 1978, 1979) showed that the nutrient contents of plants and soils were higher in areas supporting populations of burrowing petrels than in areas not greatly influenced by manuring by these birds, and that plant species composition was related to manuring. In this paper I describe a method for collecting quantitative samples of guano from burrowing petrel chicks of all ages within their nest chambers, as a preliminary to assessing the contribution of burrowing petrel guano to the terrestrial ecosystem of Marion Island. The chemical composition of the guano is described, emphasis on those elements considered to be of significance as nutrients for plant growth.

2 Methods

Guano was collected from chicks of five species of burrowing petrels at Marion Island (46° 54′S, 37° 45′E) between April 1982 and May 1983. These species were: Blue Petrel, *Halobaena caerulea,* Softplumaged Petrel, *Pterodroma mollis,* Greatwinged Petrel, *P. macroptera,* Grey Petrel, *Procellaria cinerea,* and Whitechinned Petrel, *P. aequinoctialis.*

2.1 Collection of Guano

Occupied burrows were prepared by making a hole into the nest chamber, which was plugged by a turf or stone so that chicks could be removed and replaced easily (Sinclair 1981). Guano was collected directly at regular intervals from chicks of known age throughout their growth periods in their nest chambers, by fitting them with commercially available rubber washing-up gloves with slits in the wrists. The chick was placed in the glove, with its head facing out of the neck of the glove, and its legs and wings were then pulled through the most convenient slits. Loose glove material was gathered up behind the chick's head into a loop, which was held in position by metal staples. For small chicks, the neck of the glove was turned back before stapling. For larger chicks, the glove was extended by material from another glove, and stapled into position. Condensation within the glove was not a serious problem. This method allowed the chick full mobility, and did not affect its being fed by the parents. Chicks were kept in gloves for 24 h periods.

2.2 Chemical Analysis of Guano

Inorganic (available) nitrogen (NH_4-N, NO_3-N, and NO_2-N) was determined by colorimetry (Bremner 1965a). Although these analyses were done at Marion Island, logistical reasons precluded the collection and analysis of guano samples within the 24 h time limit normally required. This problem was overcome by the addition of 5 ml of a streptomycin sulphate solution (0.006 g in 200 ml distilled water) to 0.2 g guano (weighed to an accuracy of 0.001 g) and made up to 500 ml with distilled water (VR Smith, pers comm). The dry wt. of guano collected was recorded after drying samples at 100°C for 72 h in a forced-draught oven. The dried guano samples were later digested (Allen et al. 1974) and total N was determined by the macro-Kjeldahl method (Bremner 1965b), total P by the molybdenum blue method (Murphy and Riley 1962) and Ca, Mg, K, and Na by atomic absorption spectroscopy (Pringle et al. 1968).

1 Fritzpatrick Institute, University of Cape Town, Rondebosch 7700, South Africa

Antarctic Nutrient Cycles and Food Webs
(ed. by W.R. Siegfried, P.R. Condy, and R.M. Laws)

Table 1. Production and chemical composition (selected elements) of guano from five species of burrowing petrel chicks at Marion Island

	Blue Petrel	Greatwinged Petrel	Softplumaged Petrel	Grey Petrel	Whitechinned Petrel
Guano production (g dry wt. d^{-1})	0.6 ± 0.5 n = 30	0.8 ± 0.7 n = 32	0.6 ± 0.5 n = 7	1.7 ± 1.3 n = 47	1.8 ± 1.5 n = 32
Water content (% dry wt.)	1000 ± 1316 n = 22	740 ± 496 n = 28	960 ± 450 n = 11	1254 ± 725 n = 14	2036 ± 3087 n = 30
NH_4-N ($mg\ g^{-1}$ dry wt.)	26.2 ± 18.6 n = 22	15.3 ± 19.9 n = 28	35.3 ± 20.6 n = 11	30.6 ± 30.9 n = 16	44.8 ± 21.8 n = 30
NO_3-N ($mg\ g^{-1}$ dry wt.)	1.1 ± 1.1 n = 22	0.3 ± 0.8 n = 28	0.3 ± 0.6 n = 11	< 0.1 n = 18	5.3 ± 9.8 n = 30
NO_2-N ($mg\ g^{-1}$ dry wt.)	0.1 ± 0.2 n = 22	2.6 ± 8.0 n = 28	< 0.1 n = 11	< 0.1 n = 18	< 0.1 n = 30
N ($mg\ g^{-1}$ dry wt.)	158.3 ± 56.7 n = 15	162.0 ± 53.1 n = 8	173.9 ± 67.8 n = 7	196.6 ± 49.1 n = 9	147.2 ± 74.8 n = 21
P ($mg\ g^{-1}$ dry wt.)	20.8 ± 12.5 n = 14	11.4 ± 4.8 n = 9	17.9 ± 8.1 n = 7	13.6 ± 5.5 n = 9	25.7 ± 14.7 n = 21
Ca ($mg\ g^{-1}$ dry wt.)	61.4 ± 56.8 n = 14	25.3 ± 32.9 n = 9	91.8 ± 89.3 n = 7	31.1 ± 32.6 n = 9	41.0 ± 42.0 n = 21
Mg ($mg\ g^{-1}$ dry wt.)	3.4 ± 1.1 n = 15	4.3 ± 1.6 n = 9	4.1 ± 2.2 n = 7	6.1 ± 2.8 n = 9	3.5 ± 1.5 n = 21
K ($mg\ g^{-1}$ dry wt.)	13.4 ± 8.9 n = 15	16.3 ± 11.0 n = 9	8.8 ± 13.2 n = 7	16.3 ± 4.7 n = 9	18.3 ± 14.2 n = 21
Na ($mg\ g^{-1}$ dry wt.)	8.2 ± 10.3 n = 15	30.0 ± 22.2 n = 9	23.2 ± 11.5 n = 7	17.8 ± 14.0 n = 9	20.7 ± 14.9 n = 21

Figures are means ± 1 S.D. integrated over the whole chick-growth period.

2.3 Statistical Methods

To identify possible relationships between age, amount of guano produced, and selected nutrients, regression analyses (linear, $\log_e$-linear and $\log_e$-$\log_e$, models) were done on all possible pair-wise combinations of variables (Dixon 1981).

3 Results

Large species produced most guano, ranging from a mean of 0.6 g guano (dry wt.) d^{-1} for the Blue Petrel and Softplumaged Petrel to 1.8 g d^{-1} for the Whitechinned Petrel (Table 1). In the Blue Petrel and Greatwinged Petrel, guano production increased with increasing age of the chicks (Table 2).

The guano of the 5 species was liquid in consistency (Table 1). Water content (expressed as % dry wt.) ranged from 740 (Greatwinged Petrel) to 2036 (Whitechinned Petrel). Most available N was in the form of ammonia (NH_4-N) (15.3–44.8 mg g^{-1}). Total N was higher (147.1–196 mg g^{-1}) than total P (11.4–25.7 mg g^{-1}) for each of the five species. Of the remaining selected elements, Ca was present in highest concentrations (25.3–91.8 mg g^{-1}) except for the Greatwinged Petrel; the concentration of Na being higher and Mg lower than for any other species (Table 1). Only two significant correlations between chick-age and a selected guano nutrient were found (Table 2).

4 Discussion

The chemical composition of burrowing petrel guano at Marion Island is essentially similar to that of surface-nesting seabirds breeding at the same locality (Table 1; Burger et al. 1978). However, the Na content of Greatwinged Petrel chick guano is the highest of all Marion Island seabirds.

Bedard et al. (1980), working in the St. Lawrence Estuary, Canada, studied soluble N content of guano of the

Table 2. Significant correlations from all possible pair-wise combinations of measured variables for guano of five species of burrowing petrel chicks

	Pair-wise combination	Correlation coefficient (r)	n	P
Blue Petrel	AGE vs GUANO	0.491	30	0.01
	AGE vs NO_3	0.560	22	0.01
	H_2O vs ln NO_3	0.855	15	0.001
	H_2O vs ln NH_4	0.596	22	0.01
	H_2O vs ln NO_2	0.795	12	0.01
	P vs Na	0.712	14	0.01
Greatwinged Petrel	ln AGE vs ln GUANO	0.477	37	0.01
	Na vs Mg	0.928	9	0.001
	ln K vs N	0.780	8	0.05
Softplumaged Petrel	H_2O vs ln N	0.903	7	0.01
	H_2O vs ln NH_4	0.689	11	0.05
Grey Petrel	AGE vs ln NH_4	0.532	18	0.05
Whitechinned Petrel	GUANO vs NO_2	0.547	30	0.01
	GUANO vs Ca	0.580	21	0.01
	H_2O vs NO_3	0.915	29	0.001
	Mg vs NO_3	0.817	9	0.01
	Mg vs Ca	0.634	21	0.01

Abbreviations: GUANO = guano production (g dry wt. d^{-1}); AGE = chicks in days; H_2O = water content (% dry wt.); for units of nutrients see Table 1.

Herring Gull, *Larus argentatus,* Great Blackbacked Gull, *L. marinus,* Ringbilled Gull, *L. delawarensis,* and the Common Eider Duck, *Somateria mollissima.* Bedard et al. (1980) suggested that seabird guano did not differ greatly in chemical composition with taxonomic grouping or with diet, although their data do not support this conclusion since the concentrations of soluble nutrients in Common Eider Duck guano were much lower than the concentrations of these nutrients in gull guano. This study and that of Burger et al. (1978) show that the chemical composition of seabird guano at Marion Island does not vary greatly between species or, at least in the case of burrowing petrels, with the age of chicks.

Most of the guano brought onto Marion Island by surface-nesting birds is deposited on bare rocks in penguin colonies where it is washed into the sea (Burger et al. 1978). Guano deposited in vegetated areas is more important to the terrestrial ecosystem of the Island, and burrowing petrels are the most important source of terrestrial manuring (Lindeboom 1979). Future reports will include seasonal estimates of the quantities of guano and nutrients deposited by burrowing petrels in different habitats at the Prince Edward Islands.

Acknowledgements. Scientific research at Marion Island is carried out under the auspices of the South African Scientific Committee for Antarctic Research. Financial and logistical support is provided by the Department of Transport. I thank Ian Newton for help in the field. Valdon Smith advised and helped with chemical analyses and loaned equipment, and Dr. T.M. Crowe advised on statistical methods.

References

Allen SE, Grimshaw HM, Parkinson JA, Quarmley C (1974) Chemical analysis of ecological materials. Blackwell, Oxford

Bedard J, Therriault JC, Berube J (1980) Assessment of the importance of nutrient recycling by seabirds in the St. Lawrence Estuary. Can J Fish Aquat Sci 37:583–588

Bremner JM (1965a) Inorganic forms of nitrogen. In: Black CA (ed) Methods of soil analysis. Am Soc Agron, Wisconsin (vol 2, pp 1178–1237)

Bremner JM (1965b) Total nitrogen. In: Black CA (ed) Methods of soil analysis. Am Soc Agron, Wisconsin (vol 2, pp 1149–1179)

Burger AE, Lindeboom HJ, Williams AJ (1978) The mineral and energy contributions of guano of selected species of birds to the Marion Island terrestrial ecosystem. S Afr J Antarct Res 8:59–70

Dixon WJ (ed) (1981) BMDP Statistical software. University of California Los Angeles

Lindeboom HJ (1979) Chemical and microbiological aspects of the nitrogen cycle on Marion Island (sub-Antarctic). PhD thesis, Univ Groningen

Murphy J, Riley JP (1962) A modified single solution method for the determination of phosphate in natural waters. Analytica Chim Acta 27:31–36

Pringle BH, Hissong DE, Katz B, Mulawka S (1968) Trace element accumulation by estuarine molluscs. J Sanit Eng Div Proc Am Soc Civ Eng 94:455–475

Siegfried WR (1978) Ornithological research at the Prince Edward Islands: a review of progress. S Afr J Antarct Res 8:30–33

Sinclair JC (1981) Techniques for observing Subantarctic burrowing petrels at the nest. Cormorant 9:67–71

Smith VR (1976) The effect of burrowing species of *Procellariidae* on the nutrient status of inland tussock grasslands on Marion Island. J S Afr Bot 42:265–272

Smith VR (1978) Animal-plant-soil nutrient relationships on Marion Island (Subantarctic). Oecologia (Berl) 32:239–253

Smith VR (1979) The influence of seabird manuring on the phosphorus status of Marion Island (Subantarctic) soils. Oecologia (Berl) 41:123–126

Williams AJ, Siegfried WR, Burger AE (1979) The Prince Edward Islands: a sanctuary for seabirds in the Southern Ocean. Biol Conserv 15:59–70

The Influence of Elephant Seals on the Terrestrial Ecosystem at Marion Island

K. PANAGIS[1]

Summary. The use of moulting sites by Elephant Seals, *Mirounga leonina,* at Marion Island has decreased by 13.9 % since 1974. Relatively few plant species are found at moulting sites that mostly are characterized by a *Poa cookii-Cotuletum plumosae* association, with *Poa cookii, P. annua* and *Cotula plumosa* occurring commonly. The behaviour of Elephant Seals results in a seasonally alternating degeneration and regeneration of the vegetation at moulting sites. Elephant Seals deposit substantial amounts of faeces, urine and moulted skin which, together with carcasses, results in elevated levels of nutrients in the soils at these sites, including the total forms of N, Ca, K, Mg and the available forms of Ca, Na, Mg and NH_4. High concentrations of total P and Ca, and available P, Ca and NH_4, reflect the combined effects of urination and defaecation. The relatively high levels of nutrients in areas used by Elephant Seals apparently constitute a principal factor affecting both plant and invertebrate distribution in parts of the coastal terrestrial ecosystem of Marion Island. Collembola and Acari species are the most prominent invertebrates in the soils at these sites. The total number of invertebrates is directly related to the effects of trampling by Elephant Seals.

1 Introduction

Concentrations of large numbers of moulting Elephant Seals, *Mirounga leonina* in vegetated areas have resulted in erosion at specific sites along the coastal plain of Marion Island (46°53′S, 37°52′E). The vegetation at these sites differs from that of the surrounding areas, and the minerals added to the sites through the actions of the Seals seem to be retained with little run-off to the sea. This differs from the effect of nesting seabirds along the coast, where manuring by penguins is associated with the production of peat which they later erode (Lindeboom 1979; Hall and Williams 1981). Trampling by Elephant Seals is thought to be the factor of greatest significance in the modification of the flora of Macquarie Island (Gillham 1961). Studies by Huntley (1971) on the vegetation of Marion Island have shown the influence of salt-spray on the distribution of plant communities, and Smith (1976, 1978) has emphasized the manuring influences of animals on the distribution of plant species. Changes occurring in plant communities within areas affected by seals at Marion Island have not been documented in detail before. The present paper reports on the dynamics of Elephant Seal moulting areas, emphasizing changes in soil properties, vegetation and invertebrate abundance.

1 Mammal Research Institute, University of Pretoria, Pretoria 0001, South Africa
Present address: Department of Agriculture and Nature Conservation, Private Bag 13306, Windhoek, South-West-Africa/Namibia

2 Materials and Methods

2.1 Soil Properties

Permanent quadrats were marked out at moulting sites (96 x 2 m at Boulder Beach, 76 x 2 m at Trypot Beach). Sampling points were 4 m apart throughout the quadrats. Cores of soil were taken in each quadrat for determination of percentage moisture, pH, bulk density, invertebrates, and mineral elements. Cores were 40 mm deep. Two soil samples were frozen for later determination of ammonia and nitrate. Soil depth and penetrometer measurements were made at each sampling point. An Elephant Seal utilization index was calculated for each 2 x 2 m quadrat, as the sum of the maximum number of animals for each day that they occurred in the area divided by the total number of observation days (97 days). Details of the methods used in the chemical analyses are given in Panagis (1984).

2.2 Nutrient and Energy Contents of Faeces and Urine

A re-inforced cage was used to contain sub-adult and adult female Elephant Seals in the field (Panagis 1984), in order to determine urine and faecal output and to obtain samples for chemical analyses. Faecal samples were air-dried; sub-samples being frozen and kept for ammonia and nitrate analyses. The energy content of fresh faeces (sub-samples of 0.3 g) was determined by means of a Gallenkamp ballistic bomb-calorimeter. Since priming with alcohol results in complete combustion and eliminates the need to dry samples, 0.2 g of 96% ethanol was burned with the sample. The energy value of the ethanol (6.9 kcal. g^{-1}) was substracted from the gross energy value. Faecal samples were collected every 12 h, frozen and kept for chemical analyses. pH readings were determined on sub-samples of freshly collected urine, using a portable pH meter.

Antarctic Nutrient Cycles and Food Webs
(ed. by W.R. Siegfried, P.R. Condy, and R.M. Laws)

2.3 Invertebrate Numbers

Soil samples were obtained with a core-sampler, 50 mm in diameter, 150 mm long and 215 cm^3 in volume. Arthropods were extracted using steep-sided Tullgren funnels. Intact soil samples were inverted on a mesh screen (1.5 mm), and the animals were collected directly in sampling bottles containing 70% alcohol. A light bulb (60 W) was used as a source of heat, and suspended directly above the soil samples. Samples were removed after 3 days and all the meso-fauna removed by hand. The efficiency of the method was not tested. Two species of Orabatei and two groups of Collembola were identified and the remaining species grouped under "others". The permanent quadrats were sampled during January, February and March, and the Boulder Beach site was also sampled in December (Panagis 1984).

3 Results

The location and relative size of currently occupied and unoccupied moulting sites are shown in Fig. 1. The plant communities at each site are listed in Appendix 1. The moulted areas comprise 36 separate sites ranging in size from 400 to 36,000 m^2.

Breeding and moulting sites at Kaalkoppie Beach and Mixed Pickle Cove were abandoned recently. The Kaalkoppie site was deserted as a result of volcanic activity in 1980, which blocked access to the beach and the adjacent vegetated moulting area, whereas the Mixed Pickle site seems to have been abandoned shortly after 1974; it is occupied at present by large numbers of Sub-Antarctic Fur Seals, *Arctocephalus tropicalis.* The five unoccupied sites (Fig. 1) indicate a decrease of 13.9% in site utilization since 1974. Decreases in utilization were also observed at the Boulder and Trypot sites. The general decrease in the Elephant Seal population at Marion Island is described by Skinner and van Aarde (1983).

More than one plant association occurred within a single moulting site, but the most important community was the *Callitriche antarcticae – Poa cookii* complex (Appendix 1). Salt-spray has been shown to be major factor influencing the coastal vegetation of Marion Island (Huntley 1968; Gremmen 1981). Huntley (1968) recognized two noda in a

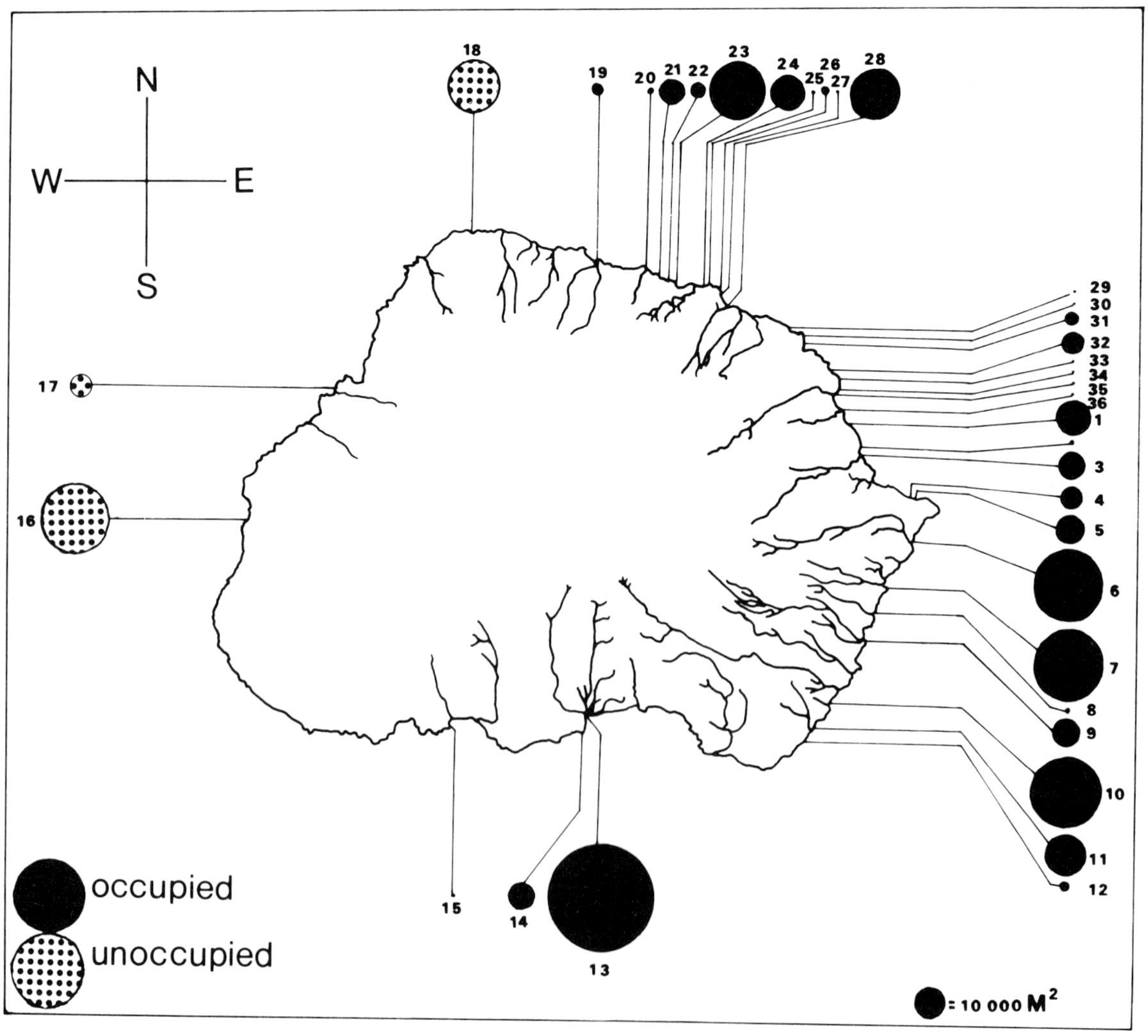

Fig. 1. Location and relative area of Elephant Seal moulting sites at Marion Island. Numbers correspond to those in Appendix 1

Appendix 1. Plant communities and their occurrence in Elephant Seal moulting sites at Marion Island

	Crassula moschata complex		*Callitriche antarcticae* – *Poa cookii* complex			
Site no. and name	*Cotula Crassuletum mochatae*	*Crassula moschatae-Azorelletum gelaginus*	*Montio fontanae-Callitrichetum antarcticae*	*Callitrichae antarcticae-Poetum annuae*	*Poa cookii Cotuletum plumosae*	*Leptodont proliferi Poetum cookii*
1. Boulder Beach				x	x	
2. Trypot North	x					
3. Trypot Beach			x	x	x	
4. Archway Beach				x	x	
5. Archway East	x					
6. Hansens Point					x	
7. Bullard West					x	x
8. Killerwhale Cove				x		
9. Waterfall Cove					x	
10. Kildalkey North				x	x	
11. Kildalkey Beach					x	
12. Hooker Cove					x	
13. Watertunnel	x			x	x	
14. Watertunnel Beach				x	x	
15. Goodhope West				x		
16. Kaalkoppie Beach	x					
17. Mixed Pickle		x				
18. Sealers Beach			x			
19. Storm Petrel	x					
20. Goney Beach	x					
21. Log Beach					x	
22. Log Beach South	x				x	
23. King Penguin West					x	
24. King Penguin				x		
25. King Penguin East					x	
26. Sea-elephant Point					x	
27. Sea-elephant Bay					x	
28. Sea-elephant Beach				x	x	
29. Sealers South					x	
30. Sealers Beach			x		x	
31. Ships Cove				x	x	
32. King Bird Head				x		
33. Duikers Point			x			
34. Rockhopper North	x					
35. Rockhopper Beach				x		
36. Rockhopper South				x		

salt spray complex, represented by a *Tillea moschata* halophytic herbfield (almost continuous along the west coast with the sea's influence being mainly mechanical and physiological) and a *Cotula plumosa* herbfield (occurring inland to the *Tillea moschata* herbfield) within the vicinity of an area used by moulting Elephant Seals. The plant communities occurring in areas influenced by seawater or salt-spray can be grouped together in the *Crassula moschata* complex (Gremmen 1981).

Mean values for soil properties at the Elephant Seal moulting sites at Boulder and Trypot beaches are given in Tables 1 and 2. The mean utilization index for the Boulder site is lower than that for the Trypot site. Higher values for available Ca, Mg, and Na as well as total N occurred at the Boulder site, whereas the Trypot site had greater values of NH_4, NO_3, chlorides, and total and available forms of K and P.

The nutrient content of Elephant Seal urine varied widely (Table 3). The urine of adult males had relatively high values of total N, Na, and NO_3, but lower values of P, K, Ca, Mg, and chlorides. The urine of adult females was relatively high in K, P, and chlorides. Total N ranged from 4.65 to 9.35, with a mean of 8.04% in faeces (Table 4). Values for P, K, Ca, Na, and NH_4 were high, and that of Mg, chlorides and NO_3 were low. Only a small proportion of the total N content was in the available form (1.62% wet sample); the greater part (8.04%) being in the total form.

The monthly densities, range and the coefficient of variation for all invertebrates for the months January to March 1981, at the two sites, are given in Table 5. No adult lepidopterans or slugs occurred at the Trypot site. The Oribatei (46,478 animals m^{-2}) and the Collembola (59,632 animals m^{-2}) were the most abundant groups at the Trypot and the Boulder sites, respectively. The average density for the Aphidae was highest at the Boulder site.

Table 1. Parameters of soil properties at the Boulder Beach Elephant Seal moulting site, December 1981

	Mean	Range	1 ± S.D.
Utilization index	0.29	0.02– 0.66	0.21
Distance to sea	110.20	66.20– 154.20	27.13
Aspect	112.70	00– 350.00	120.30
Slope	10.20	00– 32.00	9.00
Percentage moisture	736.60	155.11–1193.90	285.00
Temperature (°C)	5.70	4.00– 15.00	1.82
L.P.I.	– 1.04	– 23.90– 4.00	1.82
Bulk density	0.13	0.66– 0.21	0.41
pH	5.29	4.57– 7.03	0.50
Soil depth (cm)	46.20	7.80– 100.00	32.20
P-avail (ppm)	114.00	40.00– 338.00	96.12
K-avail (ppm)	685.30	206.00–2510.00	606.16
Ca-avail (ppm)	562.70	253.00– 771.00	137.54
Mg-avail (ppm)	1048.40	310.00–8793.00	4700.91
Na-avail (ppm)	337.30	49.00– 730.00	157.70
N-total (%)	3.19	2.55– 3.91	0.37
P-total (%)%)	0.20	0.13– 0.23	5.28
K-total (%)	0.15	0.13– 0.23	0.02
Ca-total (%)	0.14	0.09– 0.23	0.04
Mg-total (%)	0.10	0.01– 0.25	0.04
Na-total (%)	0.10	0.06– 0.19	0.03
Chlorides (%)	0.04	0.01– 0.15	0.03
Ash (%)	18.60	4.60– 44.00	13.37
NH_4^+ (%)	73.40	115.00– 189.00	50.54
NO_3^- (%)	0.64	00– 3.85	0.87

Table 2. Means, range and standard deviations for soil properties of samples collected at Trypot Beach, December 1980

	Mean	Range	1 ± S.D.
Utilization index	0.62	0 – 1.11	0.430
Distance to sea	34.40	27 – 42.70	5.220
Aspect	58.50	0 – 200.00	91.860
Slope	5.90	0 – 25.00	10.290
Percentage moisture	517.00	220.04– 851.84	181.190
Temperature (°C)	8.58	7.00– 12.00	1.470
L.P.I.	0.04	– 0.20– 0.40	0.158
Bulk density	0.14	0.09– 0.20	0.036
pH	5.88	4.70– 7.70	0.830
Soil depth (cm)	14.80	3.80– 38.30	9.490
P-avail (ppm)	644.20	176.00–1350.00	418.650
K-avail (ppm)	614.50	221.00–1168.00	293.940
Ca-avail (ppm)	1858.20	830.00–3320.00	801.460
Mg-avail (ppm)	595.85	195.00– 885.00	212.590
Na-avail (ppm)	200.60	23.00– 660.00	164.220
N-total (%)	2.86	2.00– 3.80	0.540
P-total (%)	0.34	0.15– 0.52	0.130
K-total (%)	0.16	0.11– 0.24	0.030
Ca-total (%)	0.44	0.24– 0.75	0.180
Mg-total (%)	0.12	0.06– 0.16	0.030
Na-total (%)	0.09	0.07– 0.11	0.020
Chlorides (%)	0.05	0.03– 0.08	0.010
Ash (%)	32.40	8.20– 47.10	12.060
NH_4^+ (%)	269.80	30.00–1307.00	344.450
NO_3^- (%)	4.55	0.00– 17.30	5.220

Table 3. Mean considerations of mineral elements (mg l^{-1}) in the urine of Elephant Seals

	pH	Total N	Total P	Total K	Total Ca	Total Mg	Total Na	Chlorides	NH_4	NO_3
Adult female	6.01	30,880	2,400	186	68	289	1,333	325	194	910
Adult female	6.09	31,290	1,800	86	63	316	1,295	248	317	1,071
Adult male	6.47	44,380	2,010	0	31	111	1,473	311	212	1,708
Sub-adult female	6.89	20,930	70	2	88	213	421	122	105	875

Table 4. Mean concentrations (%) of mineral elements in the faeces of adult Elephant Seals

	Total N	Total P	Total K	Total Ca	Total Mg	Total Na	Cl^-	Dry sample NH_4	Dry sample NO_3	Wet sample NH_4	Wet sample NO_3
Mean	8.04	0.45	0.53	0.39	0.04	1.21	0.36	0.91	0.05	1.62	0.26
Range	4.65–9.35	0.36–0.54	0.30–0.64	0.22–0.55	0.06–0.18	1.01–1.36	0.26–0.55	0.17–1.25	0–0.08	0.23–2.78	0.10–0.17
1 ± S.D.	2.26	0.10	0.15	0.18	0.06	0.16	0.13	0.50	0.04	1.05	0.30

Table 5 see page 178.

4 Discussion

Beaches consisting of flat, even profiles with smooth, rounded rocks and boulders on the east and north coasts of Marion Island are used extensively by breeding Elephant Seals (Condy 1977). Vegetated areas are, however, preferred during the moulting season when large numbers of animals concentrate along the drainage lines of rivers (Fig. 1). This seasonal change in habitat use applies to all age classes, and has been shown to occur at Gough Island (Bester 1980) and Macquarie Island (Carrick and Ingham 1962) as well.

Differences in plant communities and nutrient status occur, between sites used by moulting Elephant Seals and between such sites and adjacent areas. Previous workers (Huntley 1967, 1968, 1970, 1971; Smith 1976, 1977, 1978; Gremmen 1981) postulated that trampling by the animals and their contributions of nutrients were responsible for these differences. The regular seasonal use of moulting sites by Elephant Seals should eventually result in increases in the concentrations of many nutrients. However, the extent of leaching due to the high rainfall is not known. Defaecation and urination also influences the invertebrate fauna occurring at these sites. During the peak moult haul-out, invertebrate abundance is depressed by trampling.

Salt-spray is of particular importance in determining the distribution of certain plant species at certain moulting sites along the west coast of Marion Island. Where salt deposition is especially heavy, it over-rides the effects of the Elephant Seals and the resultant high nutrient levels. Plants such as *C. plumosa, T. moschata* and *A. selago* predominate. The latter two are eliminated from moulting sites where high levels of soil nutrients and low salt-spray deposition occur. Plants tolerant of high nutrient levels include *P. annua* and *A. stolinifera*, whereas *A. magellanica, R. biternatis* and *C. plumosa* are sensitive to high nutrient levels.

Acknowledgements. This paper forms part of a research project funded by the Department of Transport on the advice of the South African Scientific Committee for Antarctic Research. I am grateful for advice from Prof. J.D. Skinner, Dr. M.N. Bester and Messrs. V.R. Smith, R.J. van Aarde and P.J.J. van Rensburg.

Table 5. Mean density (no. animals.m^{-2}) range and coefficients of variation (Vx) for invertebrates at the Trypot and Boulder Elephant Seal moulting sites, January to March 1981

Species		Trypot								
		January			February			March		
		Mean	Range	Vx	Mean	Range	Vx	Mean	Range	Vx
Oligochaetae		474.3	0– 11,050	72.0	1,582.7	0– 7,310	116.5	654.5	0– 1,700	82.8
Lepidoptera										
Pringleophaga marioni	(adults)	–	–	–	–	–	–	–	–	–
	(larva)	365.5	0– 1,530	111,89	144.5	0– 680	158.9	91.8	0– 510	179.7
Embryonopsis halticella	(adults)	–	–	–	–	–	–	–	–	–
	(larva)	195.5	0– 340	77.9	25.5	0– 170	244.1	78.2	0– 340	143.1
Coleoptera										
Curculionidae	(adults)	39.1	0– 340	259.6	–	–	–	–	–	–
	(larva)	144.5	0– 850	172.9	130.9	0– 510	120.5	64.6	0– 340	199.7
Staphylinidae	(adults)	105.4	0– 340	105.7	117.3	0– 680	199.0	25.5	0– 170	244.1
Diptera	(larva)	248.2	0– 1,190	159.5	91.8	0– 510	180.0	195.5	0– 1,190	183.3
	(adults)	–	–	–	91.8	0– 510	179.7	39.1	0– 340	259.6
Arachnida		25.5	0– 170	244.1	52.7	0– 170	156.2	25.5	0– 340	360.6
Larval species[a]		13.6	0– 170	360.6	–	–	–	–	–	–
Hemiptera Aphidae		204.0	0– 1,530	220.5	195.5	0– 1,360	196.5	144.5	0– 340	116.7
Oribatei (Total)		46,478.0	680–269.110	151.9	14,201.5	170–39,780	90.0	21.537.3	850–119,000	151.9
Halozetes spp.		2,771.0	170– 7,310	88.0	1,137.3	0– 3,740	104.1	1,072.7	170– 4,080	116.4
Hyadesia spp.		36,159.0	850–252,450	186.6	8,853.6	0–33,320	112.6	8,277.3	170– 47,940	162.0
"Others"		7,548.0	340– 22,950	104.8	4,210.9	170–18,530	121.7	12,187.3	340–113,900	257.6
Collembola (Total)		3,617.0	510– 31,620	76.5	6,590	680–17,170	81.7	5,217.3	0– 27,200	135.6
Tullbergia antarctica		11,713.0	340– 31,620	81.9	5,701.8	680–13,260	72.7	2,104.6	0– 6,630	97.0
Isotomidae		13.6	0– 170	360.6	–	–	–	–	–	–
"Others"		1,922.7	0– 15,470	215.9	889.1	0– 8,160	252.4	3,190.9	0– 20,570	171.5
Slugs		–	–	–	–	–	–	–	–	–

Species		Boulder								
		January			February			March		
		Mean	Range	Vx	Mean	Range	Vx	Mean	Range	Vx
Oligochaetae		1,174.7	0– 3,570	91.8	821.1	0–3,060	104.7	746.3	0– 2,210	97.5
Lepidoptera										
Pringleophaga marioni	(adults)	–	–	–	–	–	–	6.8	0– 170	479.6
	(larva)	88.4	0– 340	113.7	73.1	0– 510	181.2	66.3	0– 340	184.6
Embryonopsis halticella	(adults)	–	–	–	–	–	–	–	–	–
	(larva)	96.9	0– 510	167.2	73.1	0– 680	217.4	119.0	0– 850	222.8
Coleoptera										
Curculionidae	(adults)	–	–	–	37.4	0– 340	238.5	6.8	0– 170	479.6
	(larva)	192.1	0– 1,190	128.8	185.3	0– 510	91.6	198.9	0– 510	104.8
Staphylinidae	(adults)	6.8	0– 170	479.6	22.1	0– 170	264.0	15.3	0– 340	479.6
Diptera	(larva)	494.7	0– 6,630	323.5	57.8	0– 510	264.0	147.9	0– 1,020	206.0
	(adults)	214.2	0– 1.870	200.4	73.1	0– 340	181.2	51.0	0– 340	183.6
Arachnida		66.3	0– 850	275.0	37.4	0– 340	275.9	59.5	0– 510	222.8
Larval species[a]		51.0	0– 510	231.0	44.2	0– 510	264.0	22.1	0– 340	350.9
Hemiptera Aphidae		243.1	0– 2,720	234.8	686.8	0– 6,800	207.0	2.927.4	0– 1,598	158.2
Oribatei (Total)		6,001.0	170– 18,020	89.0	3,666.9	0– 12,070	100.2	4,552.6	0– 19,550	122.4
Halozetes spp.		2,556.8	0– 9,690	112.1	1,479.0	0– 6,460	124.5	1,796.9	0– 13,430	171.8
Hyadesia spp.		1,898.9	0– 10,200	156.2	872.1	0– 3,910	141.2	516.8	0– 57,800	244.4
"Others"		1,545.3	0– 6,970	100.8	1,315.8	0– 8,670	164.4	2,150.5	0– 14,620	158.8
Collembola (Total)		59,632.6	5780–408,680	162.9	58,124.7	4590–206,550	84.6	107,128.9	12750–620,160	147.7
Tullbergia antarctica		554.2	0– 6,460	249.7	695.3	0– 5,270	186.3	229.5	0– 1,870	188.0
Isotomidae		50,304.7	510–389,130	187.4	48,716.9	0–189,210	91.7	88,576.8	2890–564,910	169.4
"Others"		8,773.7	680– 26,180	84.3	8,714.2	340– 32,130	95.2	19,397.4	510– 69,700	109.8
Slugs		22.1	0– 510	479.6	–	–	–	–	–	–

a Unidentified larval species

References

Bester MN (1980) The southern elephant seal *Mirounga leonina* at Gough Island. S Afr J Zool 15:238–239

Carrick R; Ingham SE (1962) Studies on the southern elephant seal *Mirounga leonina* (L). I Introduction to the series. CSIRO Wildl Res 7:89–101

Condy PR (1977) The ecology of the southern elephant seal *Mirounga leonina* (Linnaeus 1758), at Marion Island. Unpublished D.Sc thesis, University of Pretoria, Pretoria

Condy PR (1981) Annual food consumption and seasonal fluctuations in biomass of seals at Marion Island. Mammalia 45:23–30

Gillham ME (1961) Modification of sub-Antarctic flora on Macquarie Island by seals, birds and sea elephants. Proc R Soc Victoria 74: 1–11

Gremmen NJM (1981) The vegetation of the sub-Antarctic islands Marion and Prince Edward. W Junk Geobotany 3:149 pp

Hall KJ; Williams AJ (1981) Animals as agents of erosion at sub-Antarctic Marion Island. S Afr J Antarct Res 10/11:24–26

Huntley BJ (1967) A preliminary account of the vegetation of Marion and Prince Edward Islands. S Afr J Sci 63:235–241

Huntley BJ (1968) A floristic and ecological account of the vegetation of Marion and Prince Edward Islands, South Indian Ocean. Unpublished M Sc thesis, University of Pretoria, Pretoria

Huntley BJ (1970) Altitudinal distribution and phenology of Marion Island vascular plants. Tydskr Natuurwet 10:255–262

Huntley BJ (1971) Vegetation. In: Report on the South African biological and geological expedition 1965–1966. Van Zinderen Bakker E M, Winterbottom J M, Dyer R A (eds). Marion and Prince Edward islands. A A Balkema, Cape Town

Lindeboom HJ (1979) Chemical microbiological aspects of the nitrogen cycle on Marion Island (sub-Antarctic). PhD thesis, University of Groningen, Groningen, Netherlands

Panagis K (1984) The influence of southern elephant seals *Mirounga leonina* (Linnaeus) on the coastal terrestrial ecology of Marion Island. Unpublished M Sc thesis, University of Pretoria, Pretoria

Skinner JD, van Aarde RJ (1983) Observations on the trend of the breeding population of southern elephant seals *Mirounga leonina* at Marion Island. J Appl Ecol 20:707–712

Smith VR (1976) The effect of burrowing species of Procellariidae on the nutrient status of inland tussock grasslands on Marion Island. J S Afr Bot 42:265–272

Smith VR (1977) The chemical composition of Marion Island soils, plants and vegetation. S Afr J Antarct Res 7:28–39

Smith VR (1978) Animal-plant-soil nutrient relationships on Marion Island (sub-Antarctic). Oecologica (Berl) 32:239–253

A Preliminary Study of the Action of Crustose Lichens on Rock Surfaces in Antarctica

D.W.H. WALTON[1]

Summary. Initial investigations of the weathering effects of crustose lichens on mica-schist rock from Signy Island show considerable mechanical disruption of the rock surface, and penetration of the thallus into the rock. SEM studies show the presence of fungal propagules attached to the rock, and the growth form of the lichen thallus in colonizing bare rock.

1 Introduction

Biogeochemical weathering of minerals is much less well understood than rock weathering by physical and chemical agencies. Considerable discussion has taken place over many years on the role of lichens in mineral weathering and soil formation, and opinions have polarized into two camps. The traditional viewpoint is that lichens are important weathering agents, involved in secondary mineral formation, the release of mineral ions, decomposition of the rock surface and significant factors in the formation of a primitive soil (Schatz et al. 1956; Syers and Iskandar 1973; Jones et al. 1981). Other authors have, however, found no convincing evidence either of lichen-induced change in rock minerals or of their contribution to ion release and soil formation (Cooper and Rudolph 1953).

Much of this work was carried out before the development of the present sophisticated instrumentation. It has, however, recently been shown that studies using electron-microprobe analyses and scanning electron microscopy (SEM) on material from temperate regions (Ascaso et al. 1976; Jones et al. 1980; Wilson et al. 1981) can yield significant evidence of rock weathering by crustose lichens. Although Lindsay (1978) has suggested recently that lichens may play an insignificant role in the weathering of Antarctic rocks, Ugolini and Perdue (1968) have found evidence for chemical change, especially in ferric ion, phosphorus and aluminium. Further collections from Victoria Land (Ugolini and Grier 1969) appear not to have been analyzed. The work of E.L. Friedman (1982) on the ecology of endolithic lichens is also clearly relevant to this field.

1 British Antarctic Survey, NERC, High Cross, Madingley Road, Cambridge CB3 OET, England

As part of a long-term investigation into nutrient release and availability in Antarctic fellfields, preliminary studies are now underway on the role of lichens in the weathering of rocks on Signy Island, South Orkney Islands. As well as SEM and microprobe studies they will also involve optical examination of thin sections of lichen-encrusted boulders, and experimental treatment of mineral particles with lichen extracts. In this paper some initial results from thin sections and SEM studies are presented.

2 Optical Analysis of Thin Sections

2.1 Section Preparation

The preparation of thin sections of rock for optical examination under the microscope has involved extensive trials of various embedding resins. The most successful technique developed so far uses impregnation of the rock surface with Petropoxy 154 Resin (Palouse Petro Products, Rt 1, Box 92, Palouse, WA 99161, USA) without vacuum, and curing at ca. 120°C for 5 min. Slices are cut, using a diamond-tipped saw, and whenever possible these sections are large (ca. 50 x 50 mm). These slices are mounted on glass slides of selected thickness and lapped down to ca. 40 μ thick. They are then finished, using silicon carbide in paraffin as the abrasive, and finally set in clear resin beneath coverslips. Considerable problems were experienced in keeping the lichen thallus attached to the rock section. Prolonged curing of the resin sometimes increased the number of acceptable sections. In the specimens of weathered schist used for most sections stresses within existing fracture planes in the rock slices were increased during curing of the resin. This resulted in disintegration of both specimens and mounting slide. The heat used for curing generally changed the colour of the lichen thallus, whilst the resin had a slight solvent action on the pigments.

Most specimens successfully sectioned have been principally of micaschist, although some also have quartzite and garnet components. Examples of quartzite and calcite are also being examined. Minerals are identified in thin sections by standard mineralogical optical analysis. All sections illustrated here are of *Lecidella bullata.*

Antarctic Nutrient Cycles and Food Webs
(ed. by W.R. Siegfried, P.R. Condy, and R.M. Laws)

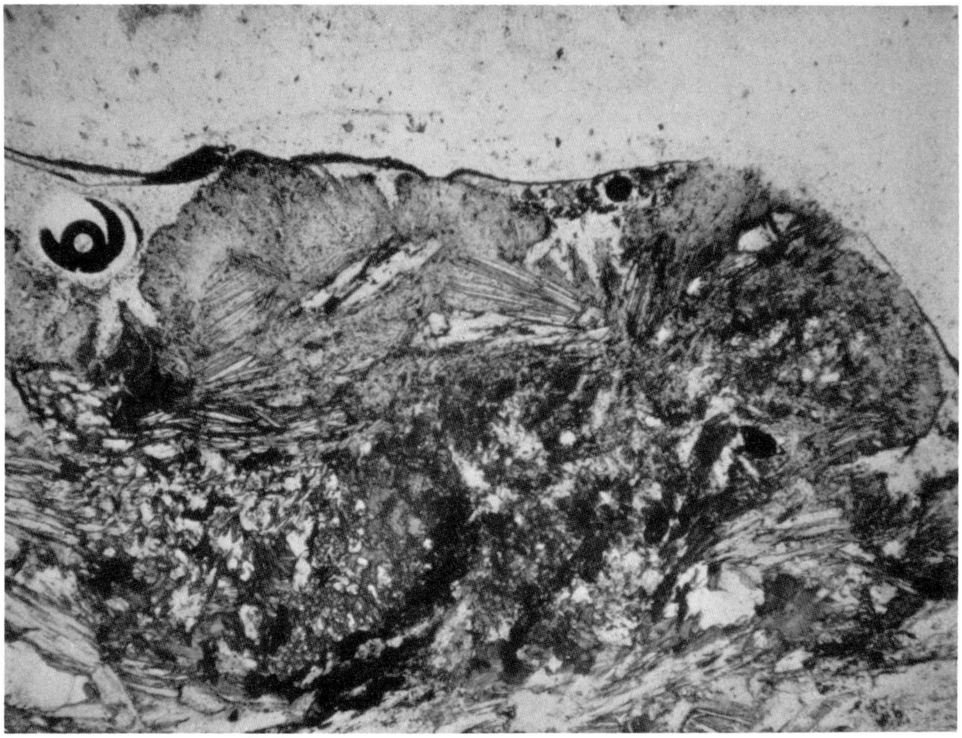

A

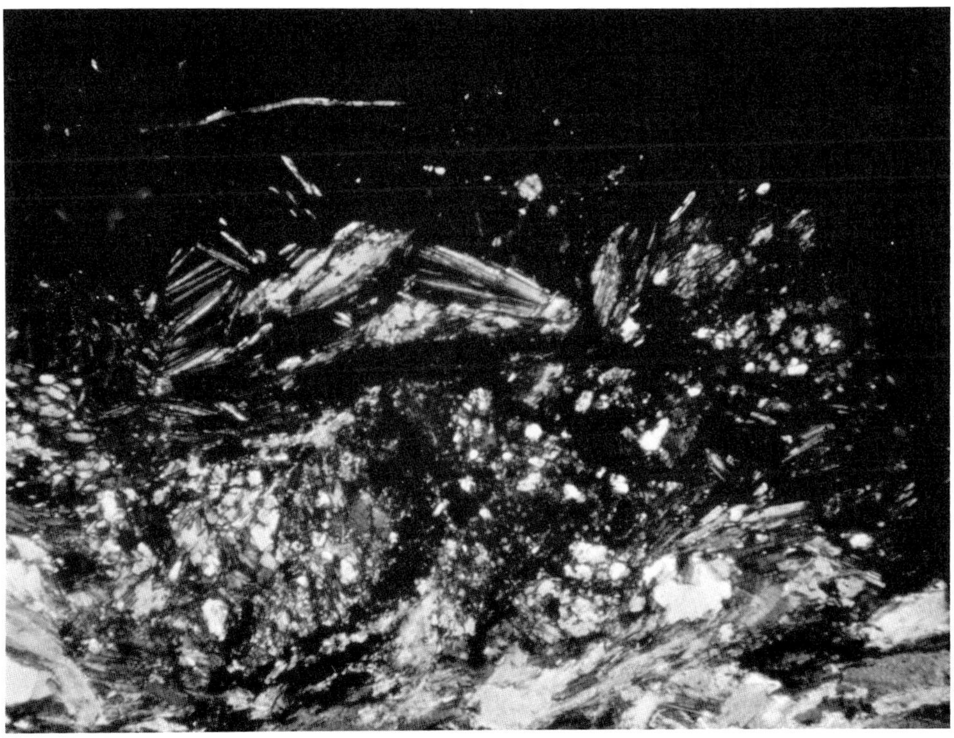

B

Fig. 1. A Transverse section through crustose lichen on quartz mica-schist. Under normal illumination, the edge of the thallus is clearly visible as is the rock substratum at the top. (x 55). **B** Same section under polarized light. Large pieces of mica have been removed from the basal rock and are splitting within the thallus into individual platelets

2.2 Section Analysis

In Fig. 1A the upper edge of the thallus is clearly visible in plane light, although it is not clear in places where the thallus ends and the rock surface begins. There is some evidence of rock material within the thallus. Fig. 1B, using polarized light, demonstrates clearly the nature and extent of the rock inclusions. Large pieces of mica have been taken up and are splitting along horizontal planes into individual sheets. On the right-hand side of this section numbers of small quartz crystals are scattered through the thallus. The whole of the central region of the section shows evidence of extensive penetration of the lichen into the rock surface.

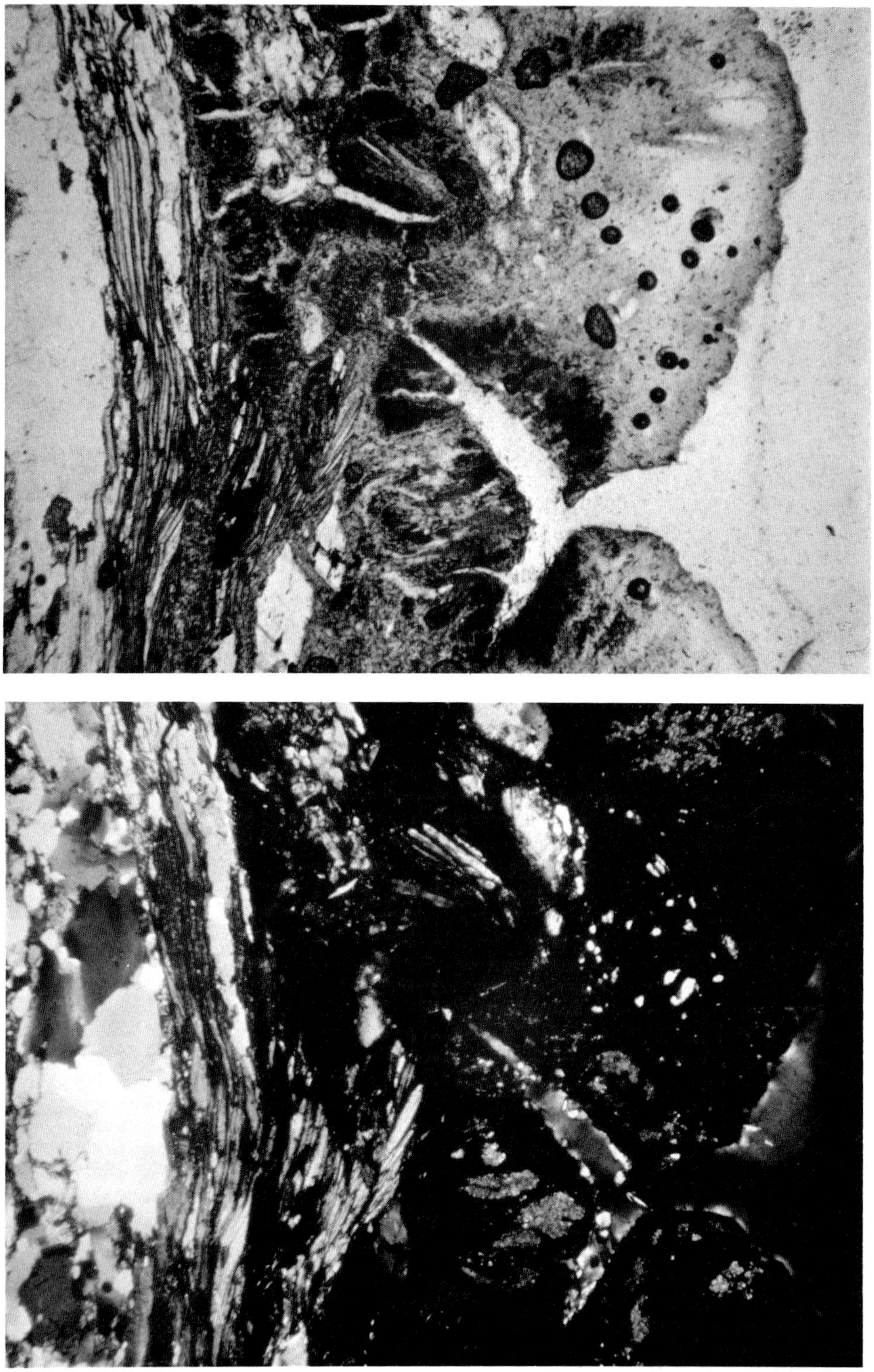

Fig. 2. A Transverse section through crustose lichen on quartz mica-schist. Under normal light both the thallus and the mica layers in the rock are clearly visible. Note how deeply the thallus itself is cracked by desiccation. (x 55). B Same section under polarized light. Mica platelets within the thallus are clearly visible. Small scattered quartz crystals can also be seen

A

B

Much larger quartz crystals are included within the thallus in Fig. 2. They are clearly visible in Fig. 2A, but the extent of crystal disintegration is only apparent in polarized light (Fig. 2B). The polarization also accentuates the lamellate stacks of mica at the rock surface and the disintegration plates within the thallus. There is clear evidence of lichen penetration into the mica at the rock surface, probably by rhizinae.

Mechanical weathering alone of mica sheets at the rock surface is shown in Fig. 3 where large pieces of mica are apparently peeling back from an area where there is no evidence for lichen growth into the rock.

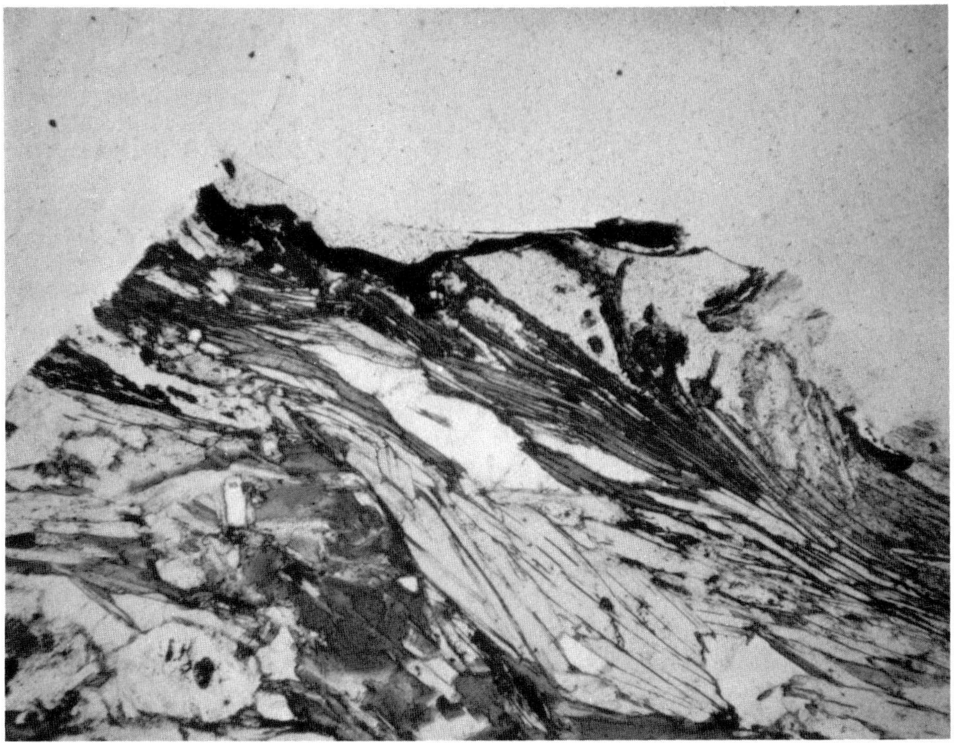

A

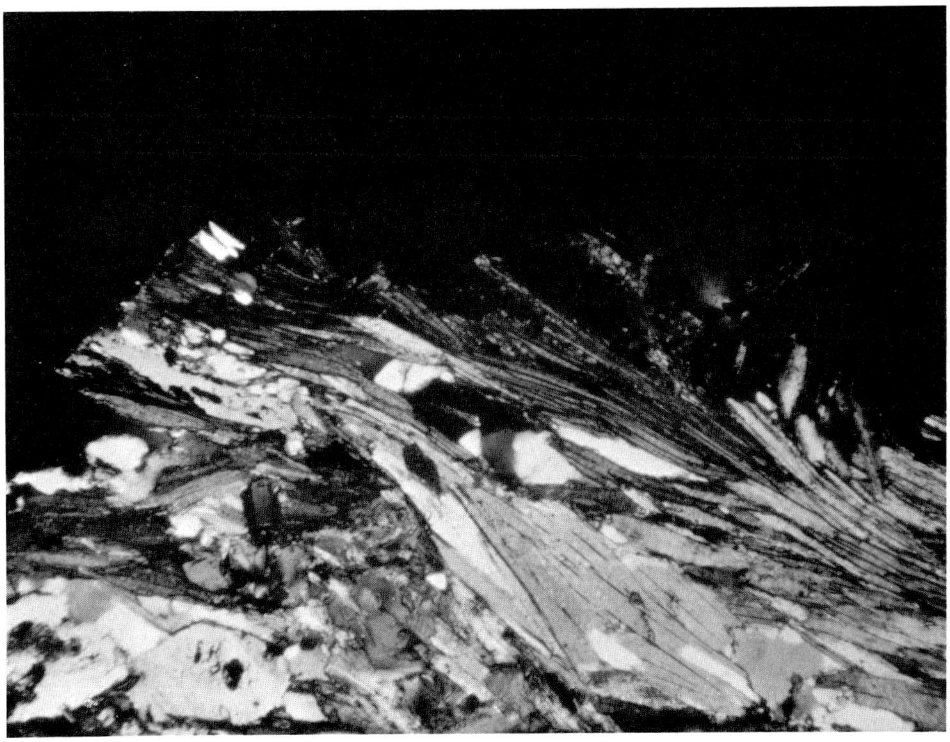

B

Fig. 3. A Transverse section through crustose lichen on quartz mica-schist. Weathering here may be due to mechanical penetration of water into the mica layers. Little lichen penetration is obvious. (x 55). **B** Same section under polarized light

3 SEM Rock-Surface Studies

3.1 Specimen Preparation

A range of lichen-encrusted rock specimens was collected on Signy Island and transported at -15°C to Cambridge. The specimens were allowed to thaw out, and dried at room temperature for several weeks. Small chips were taken from the surface, mounted on aluminium stubs and sputter-coated with gold. Specimens of *Lecidea* spp. were examined, using a Philipps Stereoscan at the Department of Earth Science, University of Cambridge.

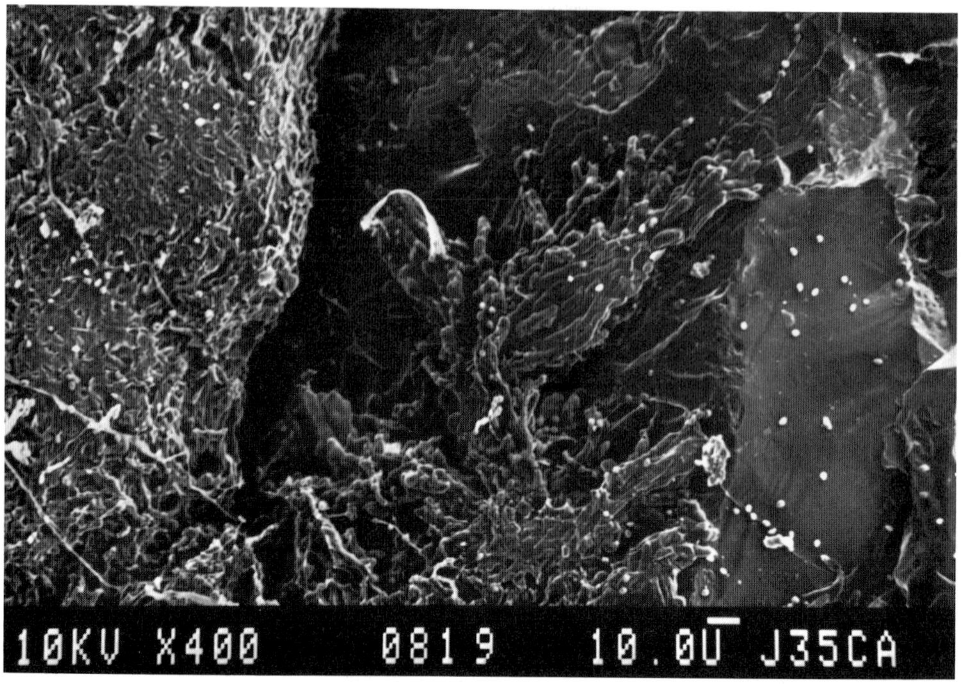

Fig. 4. SEM image of crustose lichens on mica-schist. Lobes of the lichen thallus are colonizing fresh rock surfaces. Note the number of propagules (probably fungal spores) on the rock surface

Fig. 5. SEM image of crustose lichen on mica-schist. A small flake of rock appears to be embedded in the thallus. Note the fungal hyphae on the lichen surface and the condiophore

3.2 Specimen Analysis

The lichen thallus favours the protection of surface discontinuities for growth, presumably since they offer both a roughened surface for attachment and a degree of protection from abrasion by wind-blown dust and ice particles (Fig. 4). In samples from more protected environments, fungal hyphae can be seen (Fig. 5) growing over the surface of the thallus. Hyphae have also been seen growing attached to bare rock surfaces and fungal spores have been seen on the surface of many specimens examined.

4 Discussion

Mechanical disintegration of rocks by lichens was first observed by Goeppert (1860). Systematic investigation of this phenomenon was undertaken by Fry (1924, 1927) who was able to show that the stress induced in the rock surface, during the drying of a model thallus (a gelatin film), was sufficient to cause fracturing of the surface of glass and shale. Experiments with *Xanthoria* showed that the haptera could detach surface fragments of rock during repeated wetting and drying cycles. Fry concluded (1924) that the mechanical disintegration of the rock surface preceded chemical degradation by the lichen.

Sections through *Lecidea lucida* (Fry 1924) showed plates of shale being lifted from the surface by end-penetration along horizontal lines of weakness. This appears very similar to the breakdown of the mica plates in Fig. 3, although there is little evidence of lichen penetration in that particular section. In Fry (1927), further sections through *Lecanora* spp. on shale demonstrated that greater stress is developed below the apothecia than below the vegetative thallus. The intercalated form of the shale fragments is similar to the arrangement of mica platelets within the Antarctic lichen thallus, although none of the Antarctic sections were made through apothecia. Fry's (1927) observations on *Rhizocarpon geographicum* on schist allowed her to conclude that temperate lichens on schist usually dislodge separate minerals or particles, rather than destroying the whole surface uniformly as on shale. The Antarctic material suggests that this also occurs there except where mica is found, and that it is preferentially eroded because of its lamellate structure and ease of penetration.

The preliminary SEM studies have demonstrated the frequency of propagules (fungal spores and possibly yeasts) on the rock specimens and indicated the type of micro-site in which germination and establishment might occur. Occasional crystals have also been seen associated with the edge of lichen thalli. In their initial survey of lichen–rock interactions, Ugolini and Perdue (1968) identified 11 elements below the lichen crust. Ferric iron was the most sensitive to mobilization, and alumina, silica and phosphorus also showed changes associated with lichen activity. Much further work, including examining rock from which the thallus has been removed, is now required to enable comparisons to be made between lichen weathering and physical weathering in both chemical and mineralogical terms.

The potential importance of other organisms in weathering must also be considered. Recent studies by Rades-Rohkohl et al. (1978) on bacterial weathering of quartz and by Rades-Rohkohl et al. (1979) on weathering by yeasts have provided preliminary data from laboratory experiments which are difficult to relate at the moment to ecological situations. The early work by Henderson and Duff (1963) on fungal activity on rocks also merits further investigation. Biogeochemical weathering on and in Antarctic rocks (see Friedmann 1982) is likely to be influenced by all these organisms, and as yet little is known about their importance and activity.

Acknowledgements. I am especially grateful to T. Bacon, Earth Sciences Division, British Antarctic Survey, for his work on the thin sections, and to the University of Cambridge for provision of SEM facilities.

References

Ascaso C, Galvan J, Ortega C (1976) The pedogenic action of *Parmelia conspersa, Rhizocarpon geographicum* and *Umbilicaria postulata.* Lichenologist (Oxf) 8:151–171

Cooper R, Rudolph ED (1953) The role of lichens in soil formation and plant succession. Ecology 34:805–807

Friedmann EI (1982) Endolithic microorganisms in the Antarctic Cold Desert. Science (Wash DC) 215:1045–1053

Fry EJ (1924) A suggested explanation of the mechanical action of lithophytic lichens on rocks (shale). Ann Bot (Lond) 38:175–196

Fry EJ (1927) The mechanical action of crustaceous lichens on substrata of shale, schist, gneiss, limestone and obsidian. Ann Bot (Lond) 41:437–460

Goeppert HR (1860) Über den Einfluß der Pflanzen auf felsiger Unterlage. Flora (Jena) 11:161–171

Henderson MEK, Duff RB (1963) The release of metallic and silicate ions from minerals, rocks and soils by fungal activity. J Soil Sci:14 236–246

Jones D, Wilson MJ, McHardy WJ (1981) Lichen weathering of rock-forming minerals: application of scanning electron microscopy and microprobe analysis. J Microsc 124:95–104

Jones D, Wilson MJ, Tort JM (1980). Weathering of a basalt by *Pertusaria corallina.* Lichenologist (Oxf) 12:277–290

Lindsay DC (1978) Role of lichens in Antarctic ecosystems. Bryologist 81:268–276

Rades-Rohkohl E, Franzle O, Hirsch P (1978) Behaviour, activities and effects of bacteria on synthetic quartz monocrystal surfaces. Microb Ecol 4:189–205

Rades-Rohkohl E, Hirsch P, Franzle O (1979) Neutron activation analysis for the demonstration of amphibolite rock-weathering activity of a yeast. Appl Environ Microbiol 38:1068

Schatz V, Schatz A, Trelawny GS, Barth K (1956) Significance of lichens as pedogenic (soil-forming) agents. Proc Pa Acad Sci 30:62–69

Syers JK, Iskandar IK (1973) Pedogenetic significance of lichens. In: Ahmadjian V, Hale ME (eds) The Lichens. Academic Press, New York, pp 225–248

Ugolini FC, Grier CC (1969) Biological weathering in Antarctica. Ant JUS 4:110–111

Ugolini FC, Perdue MJ (1968) Biological weathering in Antarctica. Ant JUS 3:166

Heterotrophic Acetylene Reduction in Soils at Marion Island

V.R. SMITH[1]

Summary. Peat samples from an oligotrophic mire on Marion Island (46° 54′S, 37° 45′E) exhibited little, if any, in situ heterotrophic acetylene reduction. Samples from this coastal minerotrophic mire and wallow (created by Elephant Seals, *Mirounga leonina*) showed high reduction rates when amended with glucose and phosphate and incubated anaerobically at 20°C, demonstrating that a potential for nitrogen fixation exists in all these soils. *Fjaeldmark* soil samples did not reduce acetylene under aerobic or anaerobic conditions, even if amended with glucose and phosphorus. Heterotrophic N fixation apparently does not contribute significantly to the reduced N status of the island's soils.

1 Introduction

A number of investigations into biological N fixation have been carried out in several Sub-Arctic Fennoscandian tundra sites (Granhall and Lid-Torsvik 1975), an Alaskan site (Alexander et al. 1978) and a Canadian high Arctic site (Jordan et al. 1978). Few comparable studies have been performed in the polar regions of the southern hemisphere. Fogg and Stewart (1968) and Horne (1972) demonstrated N fixation by cyanobacteria ("blue-green algae") and lichens on Signy Island in the maritime Antarctic. At Marion Island (46° 54′S, 37° 45′E), Croome (1973) and Lindeboom (1979) observed acetylene reduction by free-living cyanobacteria, occurring as mixed algal populations in a jelly-like mat on the surface of certain mire peats, and Smith and Ashton (1981) and Smith and Russell (1982) found that cyanobactera, epiphytic on bryophytes, exhibited rapid rates of acetylene reduction.

Situated in the Sub-Antarctic region, Marion Island experiences an oceanic climate dominated by high rainfall, high wind and low temperatures. Descriptions of the island's ecosystem are given in Van Zinderen Bakker et al. (1971) and Smith (1977a). The soils contain low amounts of inorganic N (Smith 1977b), so that any mechanism whereby reduced N is added to the ecosystem has important consequences in terms of primary production and decomposition. The soils are highly inorganic, water-logged and often anaerobic, and it was thought that under such conditions N fixation by heterotrophic bacteria might form an important contribution of reduced N to the inland's terrestrial ecosystem. Field and laboratory investigations into heterotrophic N fixation, using the acetylene-reduction assay, were therefore implemented and the results are reported here.

2 Materials and Methods

Acetylene-reduction incubations were performed in the field at an oligotrophic mire on the island's eastern coastal plain on 19 occasions between October 1979 and August 1980. The ca. 1.5 ha mire slopes gently (1:40) over its north-south length and has an average altitude of 30 m a.s.l. It receives most of its water directly from precipitation, but has a small *fjaeldmark* catchment on the eastern border. The water-table is rarely more than 200 mm from the surface. The main plant community in the mire is the association Lycopodio magellanicae–Jamesionelletum coloratae (sub-association Ranunculetosum biternati), dominated by bryophyte and graminoid species (Gremmen 1981). During October and November 1981 laboratory incubations were carried out on peat samples from this mire as well as from a coastal minerotrophic mire influenced by salt-spray and seabird manuring, an Elephant Seal, *Mirounga leonina,* wallow, and an inland *fjaeldmark* plateau with a skeletal soil consisting of volcanic ash mixed with organic matter.

2.1 In situ Acetylene-Reduction Incubations

Ten 100 mm long and 25 mm diameter peat cores were removed from the mire surface, using sterile stainless steel corers. Each core was extruded into an open-ended, 210 mm long, 25 mm i.d., glass incubation tube which was then stoppered at the bottom. The tubes were inserted into the holes made by the peat corer so that the internal peat surface coincided with that of the mire. After 12–18 h equilibration a silicone rubber stopper was fitted to the top of each tube. Five tubes were flushed with argon and 10% of the headspace of three of these replaced with acetylene which had been scrubbed twice through conc. H_2SO_4 and once through water. The other two argon-flushed tubes

1 Institute for Environmental Sciences, University of the Orange Free State, Bloemfontein 9301, South Africa

Antarctic Nutrient Cycles and Food Webs
(ed. by W.R. Siegfried, P.R. Condy, and R.M. Laws)

served as a control to assess endogenous ethylene production under anaerobic conditions. Of the remaining five tubes, two served as control incubations to assess aerobic endogenous ethylene production while 10% of the other three tubes was replaced with acetylene. At each incubation date a sample of the acetylene was taken to determine levels of ethylene contamination. All tubes were covered with aluminium foil for the duration of the incubations. After 24 h the headspace of each tube was sampled through the silicone stopper, using a double-sided needle into 3 ml Venoject® (Terumo Corporation, Japan) tubes which had been re-evacuated to a consistent vacuum. After a further 24 h the headspace was sampled again.

2.2 Laboratory Acetylene-Reduction Incubations

Three 50 mm long and 70 mm diameter peat cores were obtained from each of the four sites using a stainless steel corer. The cores were transported to the laboratory in new polyethylene bags. Half of one core was slurried with distilled water and the other half with an equal volume of glucose-potassium phosphate solution. The concentrations of glucose and phosphate were sufficient to amend the soils to 1% glucose and 50 ppm P (wet soil weight basis). Sub-samples (3 g–7 g, accurately weighed) of the fortified slurry were placed in 20 100 mm long, 9 mm i.d., cylindrical glass tubes open at the bottom and possessing a short, open side-arm near the top. A large sub-sample was weighed, dried at 105°C for 48 h and reweighed to assess the moisture content. The unfortified slurry was treated in a similar manner. All the flasks were stoppered at the bottom, and the side arms of 10 tubes of each slurry type were fitted with a serum stopper and evacuated to 0.01 atmospheres before flushing with argon. The evacuation-flushing cycle was repeated 3 times, establishing an anaerobic series of unfortified and fortified sub-samples. All tubes were then placed in a dark incubator at 20°C. After 36 h, serum stoppers were fitted to the remaining tubes (aerobic series) which were then placed on a manifold connected to a supply of air containing 10% scrubbed acetylene. The other end of the manifold was connected via a manometer to a vacuum pump. The procedure recommended by Matsuguchi et al. (1979) was then used to introduce the assay gas to the samples. The flasks were slowly evacuated to 0.01 atmospheres before relieving the vacuum with the air-acetylene mixture. The procedure was repeated 4 times at intervals of 1 min. One tube each of the aerobic fortified and unfortified series was subjected to the same evacuation-flushing procedure, using air without acetylene to assess endogenous ethylene production during the incubation. Argon containing 10% acetylene was introduced to the argon-flushed tubes in a similar manner. Pure argon was added to 1 tube each of the fortified and unfortified slurry series to establish controls for assessing endogenous ethylene production under anaerobic conditions. Following the gas introduction procedure, all tubes were incubated in the dark at 20°C for 3 h. The tubes were vibrated strongly on a vortex stirrer for 30 s before sampling the headspace, using 3-ml Venoject tubes. Sub-samples of each batch of acetylene used were taken to determine levels of ethylene contamination.

2.3 Gas-Chromatographic Determination of Ethylene in the Gas Samples

Concentrations of ethylene in the incubation headspace samples were determined using a Hewlett Packard 5830A gas-chromatograph with a built-in integrator and fitted with a Poropak N column. The separation was run at 50°C using N_2 carrier gas at 17 ml min^{-1} and detection was by flame ionization. C_2H_4 concentrations in the headspace were corrected for the solubility of C_2H_4 in the sample water content at the temperature of the sample.

2.4 Bacterial Counts

A longitudinal sub-sample was aseptically sliced from the centre of the second peat core from each site, placed in sterile quarter-strength Ringer's solution and shaken vigorously for 2 min. Serial dilutions of the suspension were spread-plated onto sodium albuminate agar (Matsuguchi et al. 1979). Triplicate platings were made for both aerobic and anaerobic incubations. After 8 (aerobic) or 10 (anaerobic) days of incubation at 20°C visible colonies were counted. Anaerobic incubations (under H_2/CO_2) were carried out in a BBL (Becton, Dickinson & Co. USA) anaerobic jar.

2.5 Soil Chemical Analyses

The remainder of the peat core used for bacterial enumeration was utilized to determine pH (combination electrode immersed in a slurry of 1 part soil: 2 parts water), NH_4-N [on 0.5M NaCl extracts by the indophenol-blue reaction (Solorzano 1969)] and NO_2 and NO_3-N [Greiss-Ilosvay reaction on the same extracts, NO_3 being determined as NO_2 after reduction with spongy cadmium (Mackereth et al. 1978)]. Determinations of inorganic forms of N, pH, moisture content and bacterial numbers were all commenced within 1 h of sampling. The third peat core was crumbled, air-dried and used to determine organic C by the Walkley-Black method (Allison 1965), total N (H_2SO_4-K_2SO_4-$CuSO_4$-Se digestion followed by steam distillation determination of NH_4-N), total P (mixed acid digestion followed by phosphomolybdate-blue colour development) and "available" P by anion-resin extraction (Smith 1979).

3 Results and Discussion

All the peats possessed high moisture contents, were acid, highly organic and contained an appreciable amount of organ-

Table 1. Soil chemical properties and bacterial counts at four sites on Marion Island. All values are on a dry soil weight basis

	Moisture %	pH	Organic C %	Total N %	Total P %	NH_4-N ($\mu g\ g^{-1}$)	NO_2-N ($\mu g\ g^{-1}$)	NO_3-N ($\mu g\ g^{-1}$)	Available "P" ($\mu g\ g^{-1}$)	Bacterial counts (colony-forming units times $10^5\ g^{-1}$)	
										Aerobic	Anaerobic
Oligotrophic mire	1241	4.9	48	2.1	0.11	29	0	0	40	39.1	8.9
Minerotrophic mire	2014	5.2	54	1.8	0.13	45	0	0	231	282.1	87.1
Fjaeldmark	507	5.2	24	1.1	0.20	0	0	0	0	5.4	0.2
Elephant Seal wallow	1259	5.6	45	4.7	0.20	65	1.7	0.9	190	115.0	107.7

Table 2. Acetylene reduction rates (nmol $C_2H_2\ g^{-1}$ dry peat h^{-1}) during incubation at an oligotrophic mire site on Marion Island

Date	Aerobic				Aerobic			
	1st incubation		2nd incubation		1st incubation		2nd incubation	
3,10.– 5.10.79	0.06±0.03	(0.04–0.09)	0.04±0.03	(0 –0.07)	0.01±0.01	(0 –0.02)	0	
18.10.–20.10.79	0.04±0.03	(0.01–0.07)	0.02±0.04	(0 –0.07)	0		0.02±0.03	(0 –0.06)
1.11.– 3.11.79	0.07±0.02	(0.05–0.09)	0.11±0.10	(0 –0.20)	0.02±0.02	(0 –0.03)	0.01±0.02	(0 –0.03)
19.11.–21.11.79	0.11±0.09	(0 –0.17)	0.06±0.08	(0.01–0.15)	0.05±0.02	(0.03–0.07)	0 ±0.01	(0 –0.01)
29.11.– 1.12.79	0.05±0.08	(0 –0.15)			0.05±0.04	(0.02–0.10)	0.06±0.06	(0 –0.12)
17. 1.–19. 1.80	0.01±0.01	(0 –0.02)	0.02±0.02	(0.01–0.04)	0		0	
28. 1.–30. 1.80	0				0.06±0.05	(0 –0.10)	0	
8. 2.–10. 2.80	0.03±0.02	(0 –0.05)	0.04±0.03	(0.01–0.07)	0.04±0.04	(0 –0.08)	0.01±0.02	(0 –0.03)
21. 2.–23. 2.80	0.05±0.06	(0 –0.12)	0.03±0.02	(0.01–0.06)	0.03±0.05	(0 –0.09)	0 ±0.01	(0 –0.01)
10. 3.–12. 3.80	0.01±0.02	(0 –0.04)			0		0	
27. 3.–29. 3.80	0		0.07±0.08	(0 –0.16)	0		0 ±0.01	(0 –0.01)
7. 4.– 9. 4.80	0.05±0.03	(0.02–0.07)	0.05±0.04	(0.02–0.09)	0.06±0.09	(0 –0.17)	0	
22. 4.–24. 4.80	0.04±0.03	(0.02–0.07)	0.02±0.03	(0 –0.06)	0.02±0.03	(0 –0.06)	0 ±0.01	(0 –0.01)
6. 5.– 9. 5.80	0.03±0.02	(0.01–0.06)	0.02±0.03	(0 –0.05)	0.03±0.04	(0 –0.07)	0.01±0.01	(0 –0.02)
22. 5.–24. 5.80	0.03±0.02	(0.01–0.04)	0.01±0.01	(0 –0.02)	0 ±0.01	(0 –0.01)	0 ±0.01	(0 –0.01)
27. 6.–29. 6.80	0.04±0.01	(0.03–0.05)	0.02±0.01	(0.02–0.03)	0.04±0.01	(0.03–0.05)	0.03±0.01	(0.02–0.04)
14. 7.–16. 7.80	0.04±0.01	(0.03–0.06)	0.04±0.01	(0.03–0.05)	0.05±0.02	(0.04–0.06)	0.04±0.01	(0.03–0.04)
31. 7.– 2. 8.80	0.04 ±0.01	(0.03–0.04)			0.03±0.01	(0.03–0.04)	0.01±0.02	(0 –0.03)
12. 8.–14. 8.80	0.11±0.16	(0 –0.29)	0.04±0.01	(0.03–0.04)	0.03±0.01	(0.03–0.03)	0.03±0.01	(0.03–0.03)

Values are means ± S.D. Ranges in parentheses. N = 3.

ic N (Table 1). As expected for the acid, water-logged, low-temperature soil regime, NH_4^+ was the predominant form of inorganic N. However, amounts of NH_4-N were low. Even the Elephant Seal wallow mud value of 65 $\mu g\ g^{-1}$ (dry wt.) soil represents a concentration of 4.8 $\mu g\ cm^{-3}$ fresh soil volume, compared with 32 $\mu g\ cm^3$ in a silty peat of a wet tundra meadow at Barrow, Alaska (Flint and Gersper 1974) and 17 $\mu g\ cm^3$ in an ombrotrophic Sub-Arctic mire at Stordalen, Sweden (Rosswall and Granhall 1980). Compared with total P contents, "available" P levels were also low; most of the P in the island's soils being present as unavailable sesquioxide complexes or in the organic fraction (Smith 1979). The *fjaeldmark* soil possessed especially low contents of "available" nutrients.

The minerotrophic mire peat and the Elephant Seal wallow mud yielded high aerobic and anaerobic bacteria counts (Table 1). This agrees with the observation by Steyn and Smith (1981) that manuring by animals enhances microbial numbers in the island's soil. The *fjaeldmark* soil exhibited low viable counts, especially of anaerobic bacteria.

Acetylene-reduction rates for the oligotrophic mire samples during the in situ incubations are presented in Table 2. The values were corrected for endogenous ethylene production. Ethylene contamination of the acetylene was not found. The gas-chromatographic technique employed could detect ethylene concentrations down to 1 x 10^{11} mol cm^{-3} gas sample, corresponding to approximately 5 x 10^{12} mol C_2H_4 produced g^{-1} dry peat h^{-1} under the particular incubation conditions. However, several errors and uncertainties which substantially modify the confidence which can be accorded to this detection limit are introduced during the sample processing and general assay procedure. The most important of these are those associated with applying the fresh:dry weight ratio determined on a separate sub-sample to the sub-samples used in the incubations; the measurement of the head space volumes in the incubation tubes; dilution and leakage during gas sub-sampling; ethylene carry-over contamination of the gas syringe between sample injections; and, errors in determining background levels of ethylene in the acetylene and endogenous ethylene production. In addition, the solubility product of ethylene in the peat solution was assumed to be the same as in pure water and absorption of the gas onto reactive particles was assumed not to have occurred. Because of these uncertainties, a positive result cannot be ascribed confidently to acetylene reduction values below 50 x 10^{12} mol g^{-1} h^{-1} (i.e. 0.05 nmol g^{-1} h^{-1}). The mean acetylene reduction rates presented in Table 2 represent 228 incubations, of which 186 (88 aerobic, 98 anaerobic) gave estimates below this value. Considering the mean estimates for each incubation date, 59 (28 aerobic, 31 anaerobic) out of 76 were below 0.05 nmol C_2H_2 reduced g^{-1} h^{-1}. The highest mean estimate (0.11 nmol g^{-1} h^{-1}) is very low compared with values (4.5–310.6 nmol g^{-1} moss tissue h^{-1}) exhibited by cyanobacteria-bryophyte associations in the island's mires (Smith and Russell 1982).

Temperatures in the tubes during in situ incubations were not measured. However, it is likely that they correspond closely to the temperature of the soil in which the tubes were inserted. Monthly soil temperature data for the oligotrophic mire between October 1979 and September 1980 are given in Table 3. The monthly mean value was calculated from the daily averages of the maximum and minimum temperatures. The min. and max. values given in Table 3 are the absolute values for each month. Mean soil temperature was low throughout the year with a small annual range (4.4°–10.3°C). The difference between the highest and lowest recorded temperatures was 11.1°C. It is not known whether these temperatures were sufficiently low enough to suppress heterotrophic nitrogenase activity and therefore account for the very low to zero production rates in Table 2. Alexander (1974) stated that the optimum temperature for N fixation in tundra organisms is around 15°C. It is likely that this temperature was attained in the top of the peat several times during the year; the data in Table 3 representing integrated values over the top 100 mm depth.

Results of the acetylene-ethylene reduction assay depend markedly on the assay procedure applied. Matsuguchi et al. (1979) demonstrated that in water-logged soils, especially, gas diffusion proceeds so gradually that precautions must be taken to ensure adequate movement of acetylene into the soil and ethylene back into the headspace. A short incubation is also preferable because the growth of N-fixing bacteria may be inhibited by acetylene (Brouzes and Knowles 1971) and the ethylene produced may be decomposed during incubation (De Bont and Mulder 1974; Cornforth 1975; Yoshida and Suzuki 1975). Based on extensive

Table 3. Soil temperatures (0–200 mm depth) at an oligotrophic mire site on Marion Island during October 1979–September 1980

Month	Min.	Max.	Mean[a]
October	3.3	10.0	5.4
November	5.0	10.5	7.4
December	5.3	13.0	8.4
January	7.2	12.2	9.7
February	7.5	13.6	10.3
March	7.0	12.4	9.7
April	5.0	9.5	7.9
May	5.5	10.8	7.4
June	3.5	8.3	5.7
July	2.5	6.6	4.4
August	3.4	6.0	4.5
September	3.1	10.0	5.8

a Calculated as the monthly mean of the $\frac{max + min}{2}$ values for

experimentation, Matsuguchi et al. (1979) proposed that the acetylene reduction assay procedure (see under Materials and Methods above) for the laboratory incubations be applied to water-logged soils, since the evacuation-flushing cycles ensure saturation of the peat with the assay gas and the vigorous agitation before headspace sampling maximizes ethylene diffusion out of the peat solution. Short incubation times at temperatures somewhat higher than those in the field were also employed. These conditions yield acetylene reduction estimates several times larger than the usual assay techniques (acetylene injection into the headspace or single evacuation followed by addition of acetylene-containing assay gas, long incubations, direct sampling of headspace without agitation) and almost 80% of the V_{max} of water-logged samples can be measured in this way. Fortification of the sub-samples with glucose and phosphate also enhances their acetylene reduction capability, especially after the 36 h pre-incubation period.

Using the laboratory assay conditions described, peats from the four sites exhibited the acetylene reduction rates presented in Table 4. Endogenous ethylene production did not occur under these conditions. Except for one anaerobic replicate of the wallow mud, samples not amended with glucose and phosphate did not reduce acetylene under anaerobic or aerobic conditions. Aerobically incubated fortified sub-samples of the mire peats and the *fjaeldmark* soil also exhibited no reduction, except for replicates one of the two mire peats. The low values for these two replicates, as well as that for the corresponding anaerobic, fortified *fjaeldmark* replicate, are probably due to carry-over contamination of C_2H_4 in the sample injection syringe, since these samples were each chromatographed immediately after a calibration standard containing a high concentration of C_2H_4.

Fortified peats from the two mires and the wallow exhibited marked reduction rates under anaerobic conditions. In addition, the fortified wallow replicates incubated aerobically also showed substantial nitrogenase activity. The

Table 4. Acetylene reduction rates (nmol C_2H_2 g^{-1} dry wt. h^{-1}) during laboratory incubations of peat samples from four sites on Marion Island

Site	Replicate	Aerobic		Anaerobic	
		Unamended	+ Glucose, P	Unamended	+ Glucose, P
Oligotrophic mire	1	0	2.2	0	498.2
	2	0	0	0	404.9
	3	0	0	0	932.4
	4	0	0	0	712.5
			0.6± 1.1[a]		637 ±235.3[a]
Coastal minerotrophic mire	1	0	2.5	0	198.0
	2	0	0	0	206.5
	3	0	0	0	159.8
	4	0	0	0	143.5
			0.6± 1.3[a]		177 ± 30.1[a]
Fjaeldmark	1	0	0	0	1.2
	2	0	0	0	0
	3	0	0	0	0
	4	0	0	0	0
					0.3± 0.6[a]
Elephant Seal wallow	1	0	61.3	0	331.4
	2	0	149.4	0	467.3
	3	0	195.6	0	280.7
	4	0	201.6	5.4	219.6
			152.0± 64.8[a]	1.4± 2.7[a]	324.8±105.5[a]

a Mean ± S.D.

wallow samples were anaerobic on collection and possessed high numbers of aerobic and anaerobic bacteria. It is likely that, despite the prior gas-flushing procedure adopted, anaerobic conditions were established rapidly during the "aerobic" incubations. Fortified *fjaeldmark* samples did not reduce acetylene under anaerobic or aerobic conditions, except for the apparent exception noted above. This may be related to the low numbers of bacteria in the *fjaeldmark* soil.

In order to assess whether glucose or P (or both) were responsible for the enhanced C_2H_2 reduction under anaerobic conditions, 24 sub-samples of the oligotrophic mire peat were fortified with either glucose (1% wet peat weight), PO_4-P (50 ppm wet peat weight) or glucose + P (8 sub-samples each treatment). A further eight sub-samples were not fortified. All the sub-samples were pre-incubated for 24 h at 20°C under argon in the dark. Using the Matsuguchi (et al. 1979) procedure, 10% C_2H_2 in argon was then introduced to six sub-samples (and pure argon to the remaining two sub-samples) of each treatment. After 3 h dark incubation at 20°C, the presence of C_2H_4 in the headspace of each sub-sample was determined using the gas-chromatograph. Concentrations of C_2H_4 were not quantified. Endogenous C_2H_4 production did not occur, and C_2H_2 reduction was shown only by the sub-samples which had been amended with glucose (Table 5). This suggests that energy sources, rather than nutrients, limit nitrogenase activity in the mire peats. However, further investigations into the influence of elements other than P are necessary before the importance of nutrients for heterotrophic N fixation in the peats of Marion Island can be evaluated.

Table 5. Presence (+) or absence (–) of C_2H_2 reduction by peat amended with glucose and/or phosphate and incubated in the dark for 3 h at 20°C

Amendment	Replicate 1	2	3	4	5	6
Nil	–	–	–	–	–	–
+P	–	–	–	–	–	–
+G	+	+	+	+	+	+
+GP	+	+	+	+	+	+

4 Conclusion

The results suggest that heterotrophic bacteria potentially able to fix N are present in at least some Marion Island soils. However, because of a lack of readily utilizable energy sources (and perhaps nutrients), they rarely fix in situ. Low temperatures may also reduce nitrogenase activity in these organisms, although direct evidence for this is lacking. Photo-autotrophic fixation is probably more important than heterotrophic fixation, since free-living cyanobacteria and, especially, those associated with bryophytes rapidly reduce acetylene under field conditions. This is similar to observations at most northern hemisphere tundra sites (Alexander 1974) where heterotrophic fixation, especially by anaerobes, is not regarded as a significant source of reduced N.

Acknowledgements. This work was supported financially and logistically by the South African Department of Transport, on the advice of the South African Scientific Committee on Antarctic Research. MG Steyn carried out the field incubations and assisted in the laboratory incubations. Professor DF Toerien, Institute for Environmental Sciences, assisted in the preparation of the manuscript.

References

Alexander V (1974) A synthesis of the IBP Tundra Biome circumpolar study of nitrogen fixation. In: Holding AJ, Heal OW, MacLean SF, Flanagan PW (eds) Soil organisms and decomposition in tundra. Tundra Biome Steering Committee, Stockholm pp 109–121

Alexander V, Billington M, Schell DM (1978) Nitrogen fixation in Arctic and Alpine tundra. In: Tieszen L (ed) Vegetation and production ecology of an Alaskan Arctic tundra. (Ecol Studies 29). Springer, Heidelberg New York Berlin, pp 539–558

Allison LE (1965) Organic carbon. In: Black CA (ed) Methods of soil analysis, vol 2. American Society of Agronomy, Madison, pp 1367–1378

Brouzes R, Knowles R (1971) Inhibition of growth of *Clostridium pasteurianum* by acetylene: Implication for nitrogen fixation assay. Can J Microbiol 17:1483–1490

Cornforth IS (1975) The persistence of ethylene in aerobic soils. Plant Soil 42:85–96

Croome RL (1973) Nitrogen fixation in the algal mats on Marion Island. S Afr J Antarct Res 3:64–67

De Bont JMA, Mulder BG (1974) Nitrogen fixation and cooxidation of ethylene by a methane-utilizing bacterium. J Gen Microbiol 83:113–121

Flint PS, Gersper PL (1974) Nitrogen nutrient levels in Arctic tundra soils. In: Holding AJ, Heal OW, MacLean SF, Flanagan PW (eds) Soil organisms and decomposition in tundra. Tundra Biome Steering Committee, Stockholm, pp 375–387

Fogg GE, Stewart WDP (1968) *In situ* determinations of biological nitrogen fixation in Antarctica. Br Antarct Surv Bull 15:39–46

Granhall U, Lid-Torsvik V (1975) Nitrogen fixation by bacteria and freeliving blue-green algae in tundra areas. In: Wielgolaski FE (ed) Fennoscandian tundra ecosystems. Part 1, Plants and microorganisms. (Ecological Studies 16). Springer, Berlin Heidelberg New York, pp 305–315

Gremmen NJM (1981) The vegetation of the sub-Antarctic islands Marion and Prince Edward. Geobotany 3

Horne AJ (1972) The ecology of nitrogen fixation on Signy Island, South Orkney Islands. Br Antarct Surv Bull 27:1–18

Jordan DC, McNicol PJ, Marshall MR (1978) Biological nitrogen fixation in the terrestrial environment of a high Arctic ecosystem (Truelove Lowland, Devon Island, N.W.T.). Can J Microbiol 14: 643–649

Lindeboom HJ (1979) Chemical and microbiological aspects of the nitrogen cycle on Marion Island (sub-Antarctic). PhD thesis. Rijksuniversiteit te Groningen, The Netherlands

Mackereth FJH, Heron J, Talling JF (1978) Water analysis: Some revised methods for limnologists. Freshwater Biol Assoc Sci Publ No. 36:71–73

Matsuguchi T, Shimomura T, Lee SK (1979) Factors regulating acetylene reduction assay for measuring heterotrophic nitrogen fixation in water-logged soils. Soil Sci Plant Nutr 25:323–336

Rosswall T, Granhall U (1980) Nitrogen cycling in a sub-Arctic Ombrotrophic mire. In: Sonesson M (ed) Ecology of a Subarctic mire (Ecological Bulletins 30). Swedish Natural Science Research Council, Stockholm, pp 209–234

Smith VR (1977a) A qualitative description of energy flow and nutrient cycling in the Marion Island terrestrial ecosystem. Polar Record 18:361–370

Smith VR (1977b) The chemical composition of Marion Island soils, plants and vegetation. S Afr J Antarct Res 7:28–39

Smith VR (1979) Evaluation of a resin-bag procedure for determining plant-available P in organic, volcanic soils. Plant Soil 53: 245–249

Smith VR, Ashton PJ (1981) Bryophyte-cyanobacteria associations on Subantarctic Marion Island: are they important in nitrogen fixation? S Afr J Antarct Res 10/11:24–26

Smith VR, Russell S (1982) Acetylene reduction by bryophyte-cyanobacteria associations on a Subantarctic island. Polar Biology 1:153–157

Solorzano L (1969) Determination of ammonia in natural waters by the phenol-hypochlorite method. Limnol Oceanogr 14:799 –801

Steyn MG, Smith VR (1981) Microbial populations in Marion Island soils. S Afr J Antarct Res 10/11:14–18

Van Zinderen Bakker EM, Sr, Winterbottom JM, Dyer RA (1971) Marion and Prince Edward Islands; Report on the South African biological and geological expedition 1965–1966. AA Balkema, Cape Town

Yoshida T, Suzuki M (1975) Formation and degradation of ethylene in submerged rice soils. Soil Sci Plant Nutr 21:129–135

Cellulose Decomposition and Its Relationship to Nutrient Cycling at South Georgia

D.W.H. WALTON[1]

Summary. Data for cellulose decomposition at South Georgia in litter bags and cotton strips are combined with information on the cellulolytic activity of fungi to show that major differences in the rate of cellulose breakdown are possible. An explanation for this is offered in terms of plant growth form, micro-climate and substratum quality. Cellulose break-down is linked to nutrient cycling for particular species via the percentage element content in standing dead organic matter. It is suggested that at South Georgia decomposition is mediated via a different system to that described for Arctic communities.

1 Introduction

South Georgia, a Sub-Antarctic island lying well within the Antarctic Convergence and with extensive permanent snow and ice (Smith and Walton 1975), falls within the broad definition of tundra. It was the most important southern hemisphere site incorporated within the International Biological Programme Tundra Biome and the results of those investigations (Callaghan et al. 1976) were included in various Biome Syntheses (Bliss et al. 1981; Rosswall and Heal 1975; Bliss and Wielgolaski 1973). Yet, when detailed comparisons have been made between Arctic and Antarctic plant communities in attempts to find polar analogues (Aleksandrova 1980; Bliss 1979), it has become clear that the limited species diversity, the lack of particular growth forms and the absence of indigenous grazing animals common to all Antarctic communities does not allow the easy categorization of the southern plant stands into classes defined for northern vegetation. An inadequate understanding of Antarctic and Sub-Antarctic vegetation by Arctic biologists has been due largely to the very limited published data, and it is only recently that a workable first classification of the southern vegetation has been produced (R.I.L. Smith 1984).

Within the confines of these Sub-Antarctic and Antarctic plant stands, can we then apply conclusions drawn from work in other tundra areas? Although the answer must be "yes" for some processes it is far from clear that this can be easily done for nutrient cycling and plant decomposition.

1 British Antarctic Survey, NERC, High Cross, Madingley Road, Cambridge CB3 OET, Great Britain

Data comparisons in the Tundra Biome synthesis volume (Bliss et al. 1981) show that little information was then available on litter break-down and nutrient release in Sub-Antarctic ecosystems. This is still true for Îles Kerguelen, Îles Crozet and Heard Island but not for South Georgia, Marion Island and Macquarie Island. This paper will be limited primarily to South Georgian communities, but the occurrence of the same or similar plant communities throughout the Sub-Antarctic should make it possible to test the conclusions elsewhere.

Where conditions favour primary production, tundra ecosystems in general tend to have proportionally larger pools of dead organic matter than temperate and tropical ecosystems. This accumulation of energy and nutrients in the litter and standing dead categories inhibits the rate of mineral nutrient cycling. The reduced decomposition rate is usually ascribed to short summers, low temperatures and moisture stresses (too much or too little water) (Heal et al. 1981). Attention has also been directed to litter quality (Cleve 1974).

It is the intention in this paper to examine the decomposition of cellulose at South Georgia, to suggest which factors may control it and to relate this process to nutrient cycling in different community types.

2 Data Sources

Data on rates of cellulose breakdown in the soil will be drawn mainly from M.J. Smith (1983) and will be of two types :(1) decomposition of natural plant litter in bags at different sites (*Festuca contracta* grassland, *Polytrichum alpestre* mossbank, *Acaena magellanica* dwarf shrub stand, *Poa flabellata* grassland); (2) break-down of a standard cellulose substratum (cotton cloth) at different sites (*Festuca* grassland, *Polytrichum* mossbank, *Acaena* dwarf shrub stand, *Rostkovia magellanica* mire).

Data on the temperature optima of cellulolytic fungi and their occurrence on South Georgian plants will be taken from Hurst (1982). Micro-climate information is from the British Antarctic Survey (BAS) computer data files (Walton, unpubl.). Information on community biomass for the grassland is from Walton et al. (1975) and for the *A. magellanica*

Antarctic Nutrient Cycles and Food Webs
(ed. by W.R. Siegfried, P.R. Condy, and R.M. Laws)

stand from Walton (1973), whilst nutrient data are taken from Pratt and Smith (1982) and Walton and Smith (1979). These data sources do not all refer to the same sites but all sites used were situated in Cumberland Bay, South Georgia.

3 Cellulose Decomposition

3.1 Litter Bags

To compare rates of litter break-down at two sites (dried grass culms at a *Festuca* grassland and moss stems in *Polytrichum* mossbank), constant fraction loss rates were estimated from logarithmic plots of weight loss against time (Fig. 1). The rate for the moss bank (10 mg g^{-1} yr^{-1}) was an order of magnitude less than for the grass site (140 mg g^{-1} yr^{-1}). These estimates were for material incorporated within the top layer of the soil rather than for fresh litter on the surface, and any constraints due to seasonal surface drying are likely to be under-represented by the bag data.

Analyses of the change in *α*-cellulose and holocellulose of the material within the bags (Fig. 2) shows that, although the original cellulose levels were much higher in grass litter than in moss, percentage levels after 2.5 years were similar for both species. Thus, most of the loss of grass material from the bags must have been due to cellulose break-down, whereas the lack of break-down of moss cellulose must account for the very low constant fraction loss rates

Acaena litter bags in an *Acaena* community (Walton 1973) showed a very rapid rate of leaf break-down with over 90% loss occurring in 14 weeks, the remaining 10% consisting of petioles. Weight loss for standing dead of *P.*

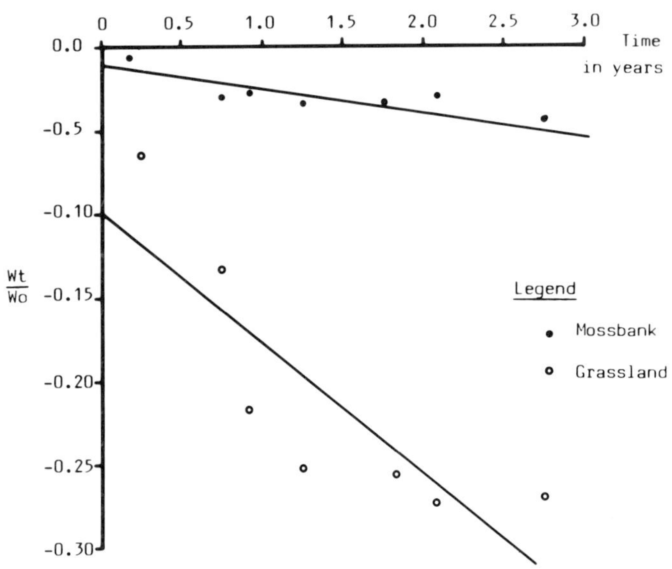

Fig. 1. A comparison of litter decomposition at 50 mm–100 mm depth at two sites, exemplified by their constant fraction loss rate ($\log_e \frac{Wt}{Wo}$). (Data from M.J. Smith 1983)

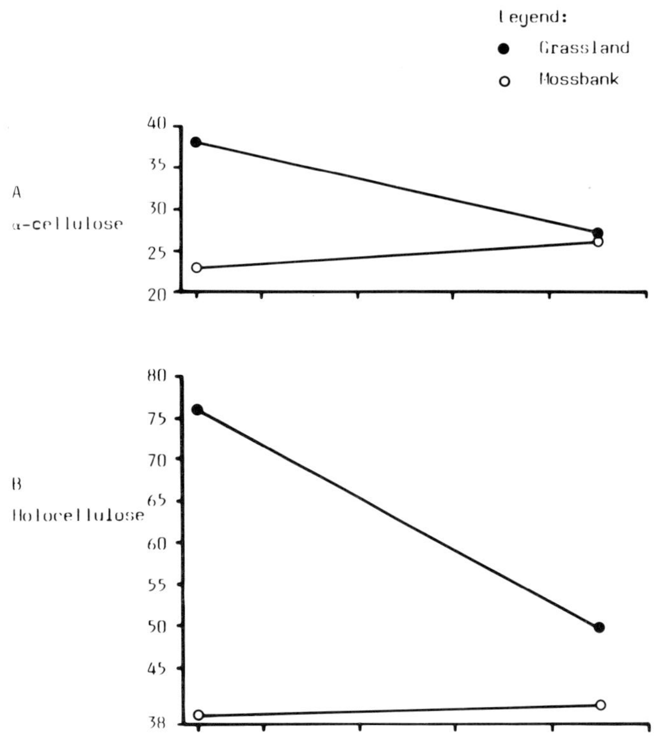

Fig. 2. Cellulose analyses of decomposing grass and moss litter over 2 years (all results as % dry wt. (Data from M.J. Smith 1983)

flabellata in litter bags at a grassland site ranged from 12% to 30% yr^{-1}, depending on the degree of exposure of the site (Smith and Walton 1975).

3.2 Cotton Strips

In order to compare site potential for degradation of cellulose alone, without the complexities inherent in natural litter, it is necessary to use a standard substratum. The use of standardized cotton strips for studying decomposition has been described in detail elsewhere (Latter and Howson 1977; Walton and Allsopp 1977). The advantage of using the change in tensile strength of the cotton fibres, rather than absolute weight loss, is that the earliest stages of enzymatic attack can be detected, before any glucose units are actually lost from the cellulose chains.

Comparisons of loss of tensile strength between three sites (grassland, mossbank and mire) can be made by plotting percentage loss per 100 d (Fig. 3). The mossbank showed the least ability to degrade cotton cellulose, although it does not appear to have been remarkably different from the grassland in the overall profile response at the end of this period. Despite very similar rates of degradation at the surface of both grassland and mire, the latter showed a markedly different profile response. Peak decomposition was at 120 mm–150 mm, with a rate at 260 mm which was 2 times as high as that for the mossbank at the same depth. At the mire, moisture was never limiting and anaerobic conditions could occur (as recorded by silver rods) below

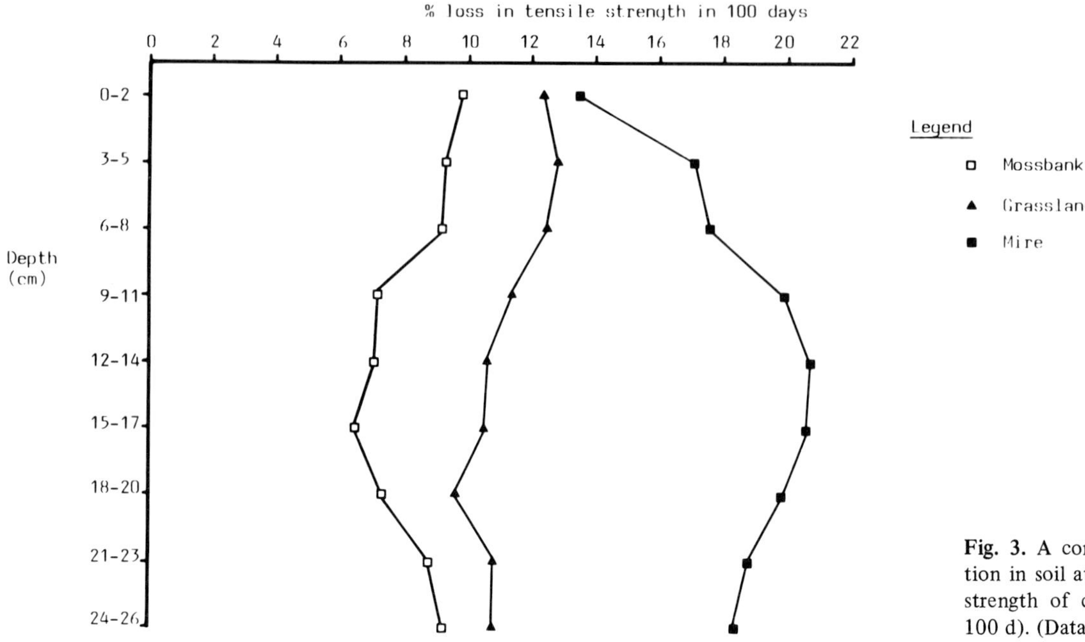

Fig. 3. A comparison of cellulose degradation in soil at three sites, in terms of tensile strength of cotton fabric (standardized to 100 d). (Data from M.J. Smith 1983)

160 mm. This site, with a continuous through-flow of drainage water, is likely to have had a considerably higher nutrient status than either the grassland or the mossbank.

Percentage rates of cellulose break-down (as measured by tensile strength loss) can be plotted to give three-dimensional representation of decomposition with depth and time (Fig. 4). Figure 4A again shows the marked activity peak for the mire at 120 mm–150 mm. This occurred shortly after insertion of the strips and subsequent decomposition remained most active at this level in the profile. In contrast, cellulolytic activity at both the grassland (Fig. 4B) and mossbank (Fig. 4C) was mainly restricted to near the top and bottom of the profile with, in the case of the mossbank, a very slow rate of break-down in the centre of the profile. Peak cellulolytic activity for both of these sites occurred generally in the latter part of the summer. Although at times very slow, none of the figures shows a nil rate of break-down over the winter period or under possible anaerobic conditions.

By contrast, Fig. 5 illustrates the extremely fast rate of decay at the *A. magellanica* site where, down to 140 mm, almost complete loss of tensile strength occurred over the summer (November–February). Over the winter period substantial cellulose break-down took place, even down to 260 mm depth. At the grassland, total tensile strength loss reached over 90% close to the surface after 2 years, and at the mossbank break-down in 2 years over much of the profile was only ca. 55%.

There are, thus, very marked differences in the ability to degrade standard cellulose in the soils underlying different vegetation types. The differences parallel those shown by litter bags at those sites.

3.3 Microbial Physiology

M.J. Smith (1983) has shown that the cellulolytic bacteria present at the South Georgian sites are only very weakly active, and that cellulose degradation is principally due to fungal activity. During more detailed studies on litter and phylloplane fungi several species were isolated for physiological testing (Hurst 1982). *Chyrosporium pannorum* and *Botrytis cinerea* both showed peak cellulolytic activity at 20°C, whereas *Chaetophoma* spp. and *Cladosporium sphaerospermum* both showed cellulose degradation with maximum activity at ca. 25°C. All showed significant activity (usually ca. 60% of maximum) at 1°C. Linear extension rate of these species also peaked at 20°–25°C with slow but measurable growth at 1°C. Experimental incubation at 5°C of *P. flabellata* litter with *C. pannorum* showed a significant increase ($P > 0.05$) in percentage loss of dry weight over the sterile controls.

3.4 Leaf Morphology and Growth Form

All South Georgian phanerorganic species are perennial and several have winter-green leaves. *Poa flabellata, F. contracta* and *R. magellanica* all retain some green leaves throughout the winter period. Conversely, *A. magellanica* shows autumnal senescence of all of its leaves and, although they do not actually fall off as in a truly deciduous species, this species is effectively deciduous. Environmental action quickly destroys the integrity of the leaf blade and at the end of winter only a few petioles remain attached to the stem.

Poa flabellata, F. contracta and *R. magellanica* all have tussock growth forms. In these graminoid species leaves die

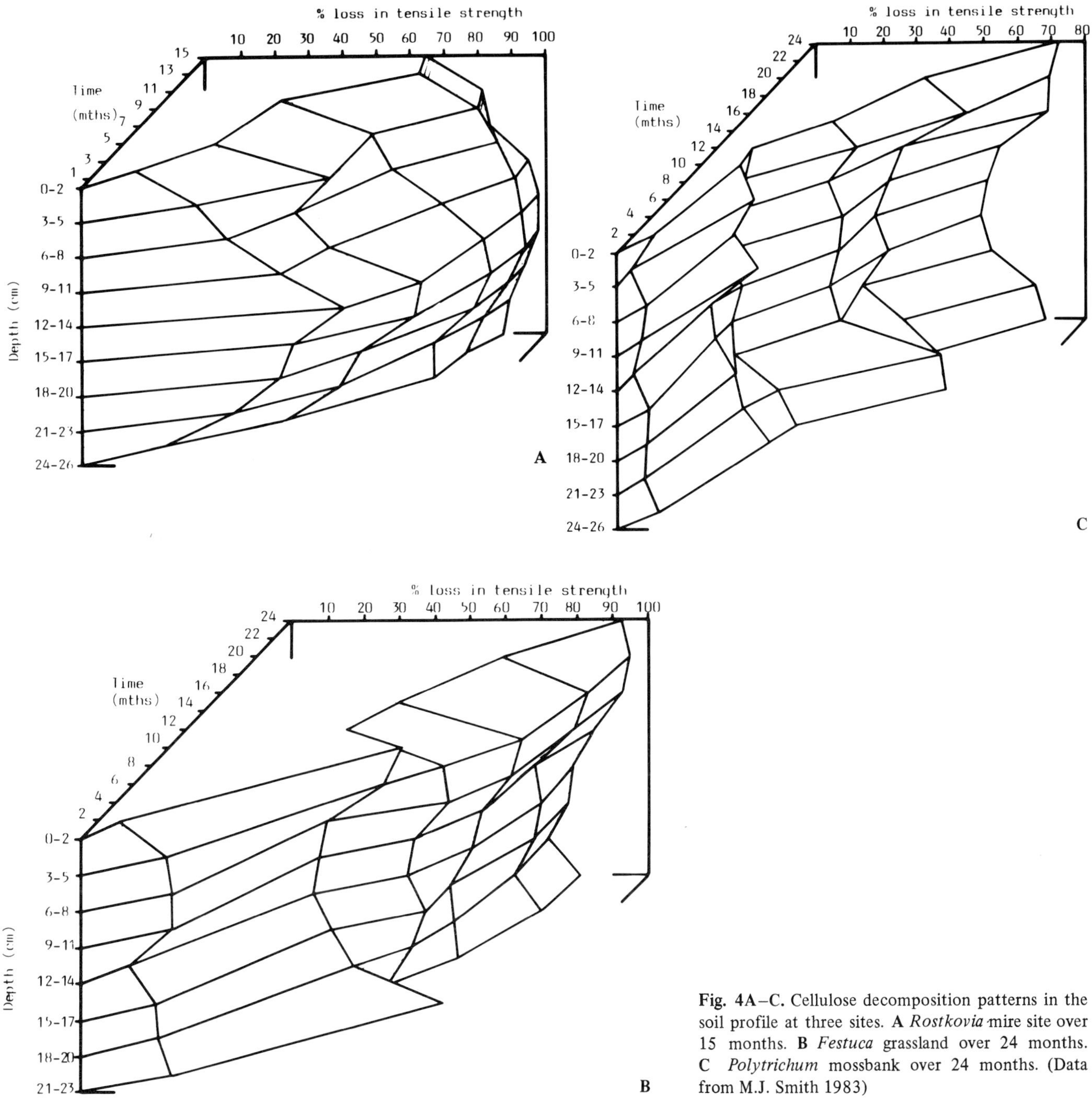

Fig. 4A–C. Cellulose decomposition patterns in the soil profile at three sites. **A** *Rostkovia* mire site over 15 months. **B** *Festuca* grassland over 24 months. **C** *Polytrichum* mossbank over 24 months. (Data from M.J. Smith 1983)

back progressively from the tip and remain upright within the tussock for up to 2 years after death. All three species have xeric characteristics in their leaf anatomy (e.g., thick cuticles, sunken stomata and inrolled lamina). Once the leaf has died and all mobile material has been translocated out, the lamina remains as a dry cellulose shell, little wetted by rain, subject to continual desiccation by wind and radiative heating and possibly exposed to low temperatures above the snow cover during winter. This constitutes an unpromising and difficult micro-habitat, both for fungal colonization and for growth; none of the cellulolytic phylloplane species tested by Hurst (1982) showed growth below water activity 0.8. Fragmentation and loss of the leaf tip can occur within a few months of death but the whole lamina is very unlikely to enter the litter layer at this time. Thus, each of these species has a significant percentage of above-ground material classifiable as standing dead. *Acaena magellanica* leaves (comprising mainly the softened leaf pinna) are quickly transferred into the litter layer. The main litter fall begins in late February.

In the mossbank, death of leaves occurs gradually upwards along the stem, the apex remaining green throughout the winter. No litter layer develops, since the lower leaf material turns into peat often without complete decompo-

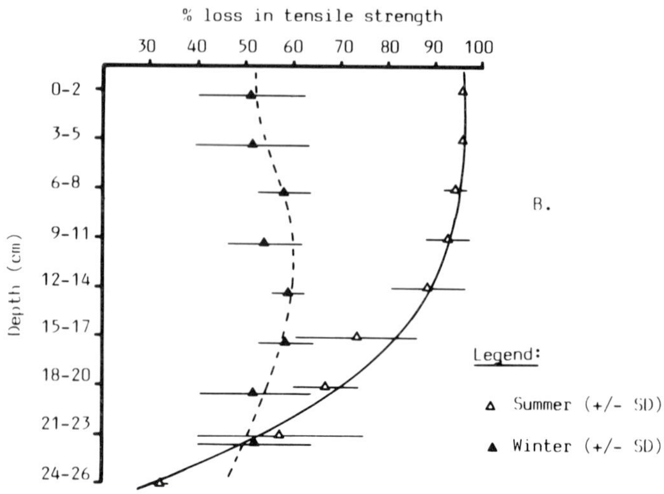

Fig. 5. Cellulose decomposition patterns for winter and summer periods in an *Acaena* stand. (standardized to 100 d.) (Data from M.J. Smith 1983)

sition. The pattern of growth and population dynamics of shoots for *Polytrichum alpestre* are described in detail in Collins (1976).

3.5 Chemical Composition

Although the majority of data available are for living leaves, they can be used to demonstrate the wide range of chemical substratum potential available for initial microbial colonization (Table 1). *Acaena* has substantially less fibre content than the other species, whereas *Poa, Festuca* and *Polytrichum* all have high fibre contents. *Polytrichum* has more than 2 times the holocellulose content of *Acaena* leaves, with that of the graminoid species high throughout the season.

The most striking features of the mineral contents are the low levels of all nutrients in *Polytrichum* and the high levels in *Acaena.* Especially important is the high N content of *Acaena* leaves (up to 4.6%). The only data for dead material are for *Acaena* attached dead leaves and *Festuca* standing dead leaves. Even in the dead leaves of *Acaena,* nutrient levels are still often higher than in living *Polytrichum* moss; *Festuca* dead leaves are very similar to *Polytrichum* in nutrient content.

3.6 Micro-Climate

For present purposes, temperature data from a *Festuca* grassland site will be used as representative of South Georgian micro-climates (Fig. 6). Comparisons with data from other South Georgian communities show that litter and upper-soil temperatures are very similar during winter periods with snow cover. Summer temperatures are cooler in the bog site and *Acaena* site and often warmer in the *Poa* site (Walton, unpubl.).

Maximum litter temperatures reached ca. 17°C in January 1977, with frosts beginning in early April in 1977 and mid-March in 1978. With the exception of a short period in May 1977, when the litter temperatures reached –8° due to a short period of snow-melt, temperatures throughout both winters were very close to zero both in the litter and in the soil. Thus, decomposer organisms are unlikely to have been subjected to lethal cold temperatures but may have suffered stress in summer from high temperatures for limited periods.

The moisture content of the litter layermoss and upper soil varied (Table 2) with the season. At the grassland, the litter layer was always wetter than the soil (although subject to much greater short-term fluctuations), whereas at the mossbank the surface layer was always drier than the peat beneath. The mire was always wet (80–90%), more so than the peat underlying the *Acaena* stand (20–50%). Limited evaporimeter measurements within vegetation have established that, during the summer, humidity remains high within the closed *Acaena* canopy but fluctuates considerably in the *Festuca* grassland. Only within the shaded litter close to the tussock does humidity remain high throughout the summer in the *Poa* community.

Table 1. Seasonal range of percentage chemical content of leaves of South Georgian plants (Pratt and Smith 1982)

	Crude fibre	Holocellulose	Total N	Total P	Total K
Poa flabellata – living	18–32	35–45	1.2–2.4	0.1–0.3	0.6–0.9
Festuca contracta – living	20–30	36–53	1.0–2.0	0.1–0.3	1.0–1.9
– dead			0.5	0.1	0.2
Acaena magellanica – living	7–13	5– 2.6	1.6–4.6	0.1–0.5	0.6–2.6
– dead			1.2–1.5	0.2–0.3	0.1–0.5
Rostkovia magellanica – living	23–35	45–52	1.7–2.1	0.1–0.2	0.7–1.1
Polytrichum alpestre – living	23–37	50–64	0.5–1.0	0.1	0.4

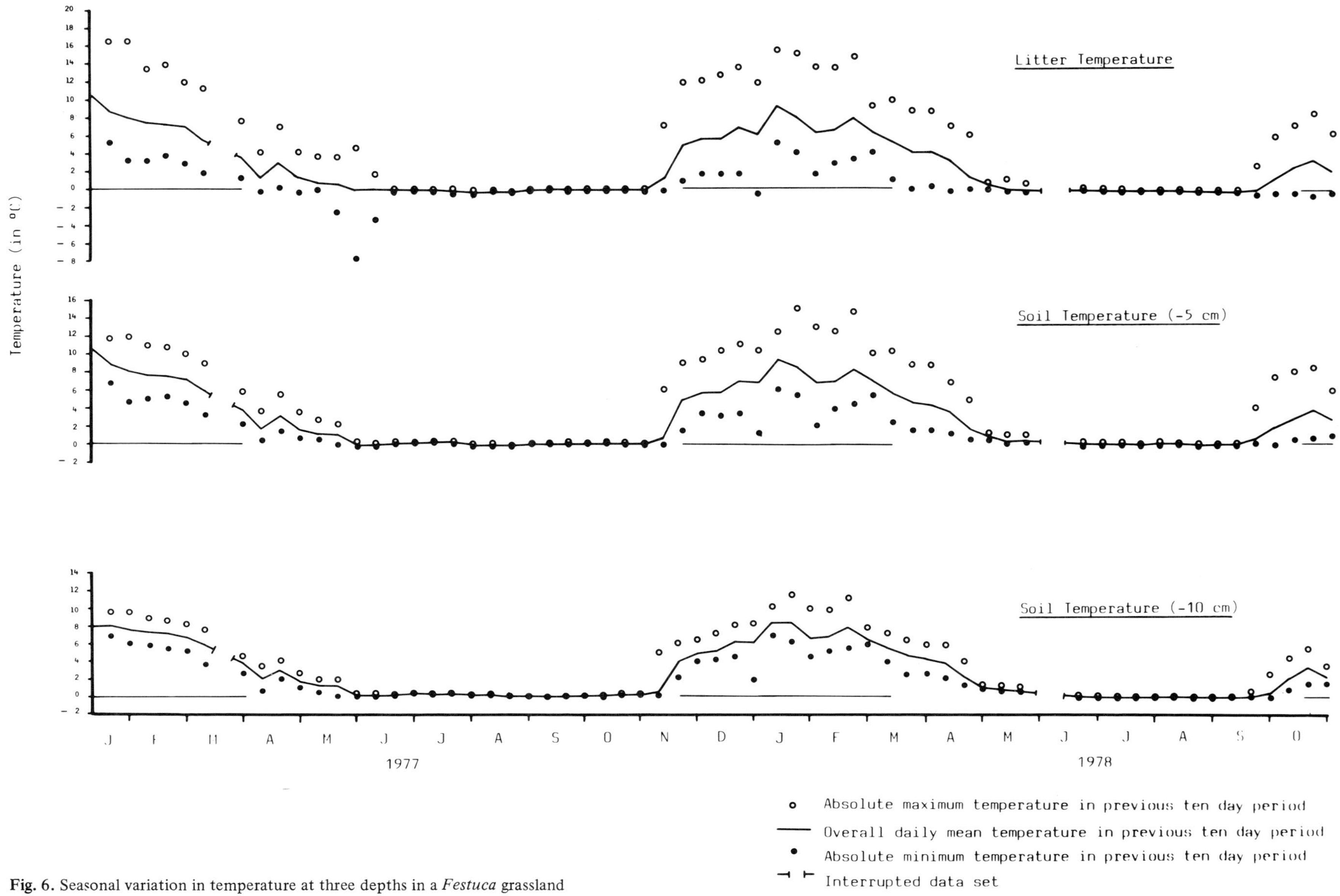

Fig. 6. Seasonal variation in temperature at three depths in a *Festuca* grassland

Table 2. Moisture levels for litter and soil in two South Georgian communities. Values as percentage of fr. wt. (M.J. Smith 1983)

		Spring	Summer	Autumn	Winter
Festuca grassland	plant litter	77–79	71–78	76–79	88–91
	0–20 mm soil	56–61	48–63	51–60	78–83
	60–80 mm soil	28–30	30–38	35–40	56–59
Moss bank	0–30 mm	77–84	70–73	80–81	87–92
	60–80 mm	88–92	78–80	82–87	91–92

4 Synthesis

In general terms the rates of decomposition of plant material can be summarized as rapid (temperatures high, humidity high, tissues soft, and tissues nutrient rich) or slow (temperatures low, humidity low, tissues fibrous, and tissues nutrient poor). When applied to South Georgian plants the following tentative explanations apply to particular species:

Acaena magellanica – rapid decomposition of leaf material due to high N content of leaves, high humidity within canopy, low fibre content in lamina and short leaf life;
Festuca contracta – slow decomposition of leaves due to retention of leaves in an upright position for long periods when dead, difficulties of fungal penetration due to leaf morphology, low leaf water content, extended leaf life and low nutrient status;
Poa flabellata – slow decomposition of leaves due to high fibre content, low leaf water content when in standing dead, low nutrient status and extended leaf life;
Rostkovia magellanica – slow decomposition of leaves due to high fibre content, retention of leaves in an upright position when dead, extended leaf life and poor nutrient status; and,
Polytrichum alpestre – low decomposition of plants due to high holocellulose and crude fibre, very low nutrient content and low microbial populations (especially of fungi, see M.J. Smith 1983).

Thus, the generalizations reflect the differences in decomposition between plant species. There are, however, clear distinctions from the Arctic situation. There, where the large pools of N, P, and C bound in dead soil organic matter clearly implicate decomposers as the major bottleneck in nutrient flow (Dowding et al. 1981), the low soil oxygen and resultant anaerobic conditions allow bacteria to dominate the micro-flora. These bacteria appear to lack the ability to decompose many of the more complex substrata at low temperature (Flanagan and Bunnell 1980), unlike the fungi which are the dominant decomposers on South Georgia.

On South Georgia, the winter temperatures under snow are rarely very low, whereas in summer high temperatures are not uncommon in several community types. The important cellulolytic fungi have been shown to function over a temperature range from 1°–30°C allowing cellulose decomposition to occur at the soil surface, given adequate water and nutrients, over at least 7 months of the year. The cotton-strip data have established that cellulose degradation within the soil profile can continue throughout the year. The differences in rates of cotton-strip break-down and weight loss in litter bag experiments agree closely with what might have been predicted from chemical and morphological evidence for leaves of particular species. It would appear that generally South Georgian winter temperatures in the litter and soil layers are not low enough to stop cellulose degradation by fungi, as long as adequate nutrients are available. Thus, the factors restricting decomposition are not principally temperature, bacteria and oxygen availability which limit decomposition in the Arctic, but are more specifically related to plant form, substratum potential and dominance of fungi.

The effect of decomposition on nutrient cycling in the South Georgian communities depends on the vegation type (Table 3). In *Festuca* grassland, where 40–50% of the above-ground mineral content is in standing dead, nutrient

Table 3. Mineral content as a percentage of dry wt. In certain South Georgian plant communities. (Data calculated from Walton and Smith 1979)

		January	March
Festuca contracta grassland	Live	60	47
	Dead	40	53
Acaena magellanica herbfield	Live	69	80
	Dead	31	20
		January	February
Poa flabellata grassland	Live	56	64
	Dead	44	36

cycling is slow, and this must also apply to those *P. flabellata* stands where mineral contributions from birds or seals is negligible. *Rostkovia* communities are often on very wet cold soils with low oxygen tensions, and low decomposition potential makes nutrient recycling within the plant, with minimal loss of minerals to the soil system, highly advantageous. On the other hand, the lower percentage of minerals within dead leaves of *Acaena* is presumably related to the deciduous nature of the leaves and the need for rapid recycling and nutrient availability in spring for the swift production of new leaves.

It is accepted that the arguments presented here are relatively simplistic, limited to only a few of the South Georgian species and based on incomplete data. Yet, it is suggested that they do demonstrate at least some of the principal restraints on nutrient cycling in important South Georgian communities. More detailed future work should be able to relate weight loss to microbial respiration, assess the ability of a more diverse micro-flora to break down South Georgian litter, quantify nutrient contribution to specific communities from animal activities and estimate nutrient limitation as a factor in dry matter production, as R.I.L. Smith (1984b) has already attempted for *P. flabellata*.

Acknowledgements. I am very grateful to Dr. M.J. Smith for the use of his data and his comments on this paper.

References

Aleksandrova VD (1980) The Arctic and Antarctic: Their division into geobotanical areas. Cambridge University Press, Cambridge

Bliss LC, Heal OW, Moore JJ (1981) Tundra ecosystems: a comparative analysis. Cambridge University Press, Cambridge

Bliss LC (1979) Vascular plant vegetation of the Southern Circumpolar Region in relation to antarctic, alpine and arctic vegetation. Can J Bot 57:2167–2178

Bliss LC, Wielgolaski FE (eds) (1973) Proceedings of the conference on primary production and production processes. Tundra Biome Steering Committee, Edmonton and Oslo

Callaghan TV, Smith RIL, Walton DWH (1976) The I.B.P. Bipolar Botanical Project. Philos Trans R Soc Lond B 274:315–319

Cleve KV (1974) Organic matter quality in relation to decomposition. In: Holding AJ, Heal OW, MacLean SF, Flanagan PW (eds) Soil Organisms and Decomposition in Tundra. Tundra Biome Steering Committee, Stockholm pp 311–324

Collins NJ (1976) Growth and population dynamics of the moss *Polytrichum alpestre* in the Maritime Antarctic. Oikos 27:389–401

Dowding P, Chapin F, Wielgolaski FE, Kirfeather P 1981) Nutrients in tundra ecosystems. In: Bliss LC, Heal OW, Moore JJ (eds) Tundra ecosystems: a comparative analysis. Cambridge University Press, Cambridge pp 647–683

Flanagan PW, Bunnell FL (1980) Microflora activities and decomposition. In: Brown PC, Bunnell FL (eds) An arctic ecosystem: The coastal tundra at Barrow, Alaska. Dowden Hutchinson and Ross, Stroudsburg pp 291–334

Heal OW, Flanagan PW, French DD, MacLean SF (1981) Decomposition and accumulation of organic matter. In: Bliss LC, Heal OW, Moore JJ (eds) Tundra ecosystems: A comparative analysis. Cambridge University Press, Cambridge

Hurst JL (1982) Biology of fungi on plants in the sub-Antarctic. Ph.D. thesis. University of Aston in Birmingham. 236 pp (unpublished)

Latter PM, Howson G (1977) The use of cotton strips to indicate cellulose decomposition in the field. Pedobiologia 17:145–155

Pratt RM, Smith RIL (1982) Seasonal trends in chemical composition of reindeer forage plants on South Georgia. Pol Biol 1:13–32

Rosswall T, Heal OW (eds) (1975) Structure and function of tundra ecosystems. Ecol Bull (Stockholm) 20

Smith MJ (1983) The microbial ecology of Sub-Antarctic tundra soils. Ph.D. thesis. University of Surrey. 256 pp (unpublished)

Smith RIL (1984) Vegetation. In: Laws RM (ed) Antarctic Ecology. Vol 1. Academic Press, London pp 61–161

Smith RIL (1985) Growth and production of *Poa flabellata* of South Georgia in relation to soil nutrient status and exposure. In: Siegfried WR, Condy PR, Laws RM (eds) Antarctic nutrient cycles and food webs (Proceedings of the 4th SCAR symposium on Antarctic biology). Springer, Berlin Heidelberg New York pp

Smith RIL, Walton DWH (1975) South Georgia, Subantarctic. Ecol Bull (Stockholm) 20:399–424

Walton DWH (1973) Changes in standing crop and dry matter production in an *Acaena* community on South Georgia. In: Bliss LC, Wielgolaski FE (eds) Primary production and production processes, Tundra Biome. Proceedings of the Conference, Dublin, Ireland, April 1973. University of Alberta Printing Services, Edmonton pp 185–190

Walton DWH, Allsopp D (1977) A new test cloth for soil burial trials and other studies on cellulose decomposition. Int Biodeterior Bull 13:112–115

Walton DWH, Greene DM, Callaghan TV (1975) An assessment of primary production in a sub-Antarctic grassland on South Georgia. Bull Br Antarct Surv 41/42:151–160

Walton DWH, Smith RIL (1979) The chemical composition of South Georgian vegetation. Bull Br Antarc Surv 49:117–135

Bryophyte Production at Marion Island

S. RUSSELL[1]

Summary. Annual shoot production of bryophytes at Marion Island (45° 54′S, 37° 45′E) ranges from 21 g m² in fellfield to 754 g m² in drainage lines. This trend is correlated with soil-water content on a xerix-hydric gradient. Plant-water content, energy values, ash content and, to a lesser extent, chlorophyll content show a similar relationship with production trends in a restricted sample of species from each major habitat type. However, further studies are expected to reveal a more complex relationship of variables associated with growth, reflecting diversity in micro-habitat and growth form.

1 Introduction

Standing crop and production data have been reported for the vascular plants of Marion Island (46° 54′S, 37° 45′E) (Huntley 1972). However, the bryophyte component of the island's vegetation has been neglected. Bryophytes contribute approximately 25% of the total lowland plant biomass, increasing locally to 60% in *Agrostis* mires (Smith 1978), and become progressively more dominant in the fellfield habitat at altitudes in excess of 300 m a.s.l. (pers. observ.).

A 2-yr study of net annual production in 16 bryophyte species in the island's major habitats was carried out during 1980–1982. Studies of water, energy, ash and chlorophyll contents were also made to examine the hypothesis that these variables are related in the same way as plant community pattern to the xerix-hydric habitat moisture

Table 1. Characteristics of bryophyte species from the major habitats at Marion Island

	Ditrichum strictum	*Brachythecium rutabulum*	*Blepharidophyllum densifolium*	*Brachythecium subplicatum*	*Clasmotocolea vermicularis*
	Fellfield	Slopes	Mires	Drainage lines	Biotic/Salt
Mean soil water content (% dry wt.)[a]	85	588	1160	2896	1081
Growth form	Large cushion	Weft	Carpet	Carpet	Mat, Thread-like form
% Cover in growth plot	13	45	95	90	80
Mean stem density (Shoots cm^{-2})	23.7	11.0	42.0	23.2	106.5
Annual shoot production (g m^{-2})	21	113	326	754	409
Plant tissue water content (% dry wt.)	228	689	1378	2200	854
Chlorophyll content (mg g^{-1})	0.143	1.375	0.868	1.534	no data
Energy value (kJ g^{-1})	16.276	16.025	15.606	14.740	15.703
Ash content (% dry wt.)	0.515	2.743	3.954	5.253	1.290

a Figures for soil water content are averaged from the multiple relevée data of Gremmen (1982) for each habitat.

1 Department of Plant Sciences, Fort Hare University, Alice, Ciskei

Antarctic Nutrient Cycles and Food Webs
(ed. by W.R. Siegfried, P.R. Condy, and R.M. Laws)

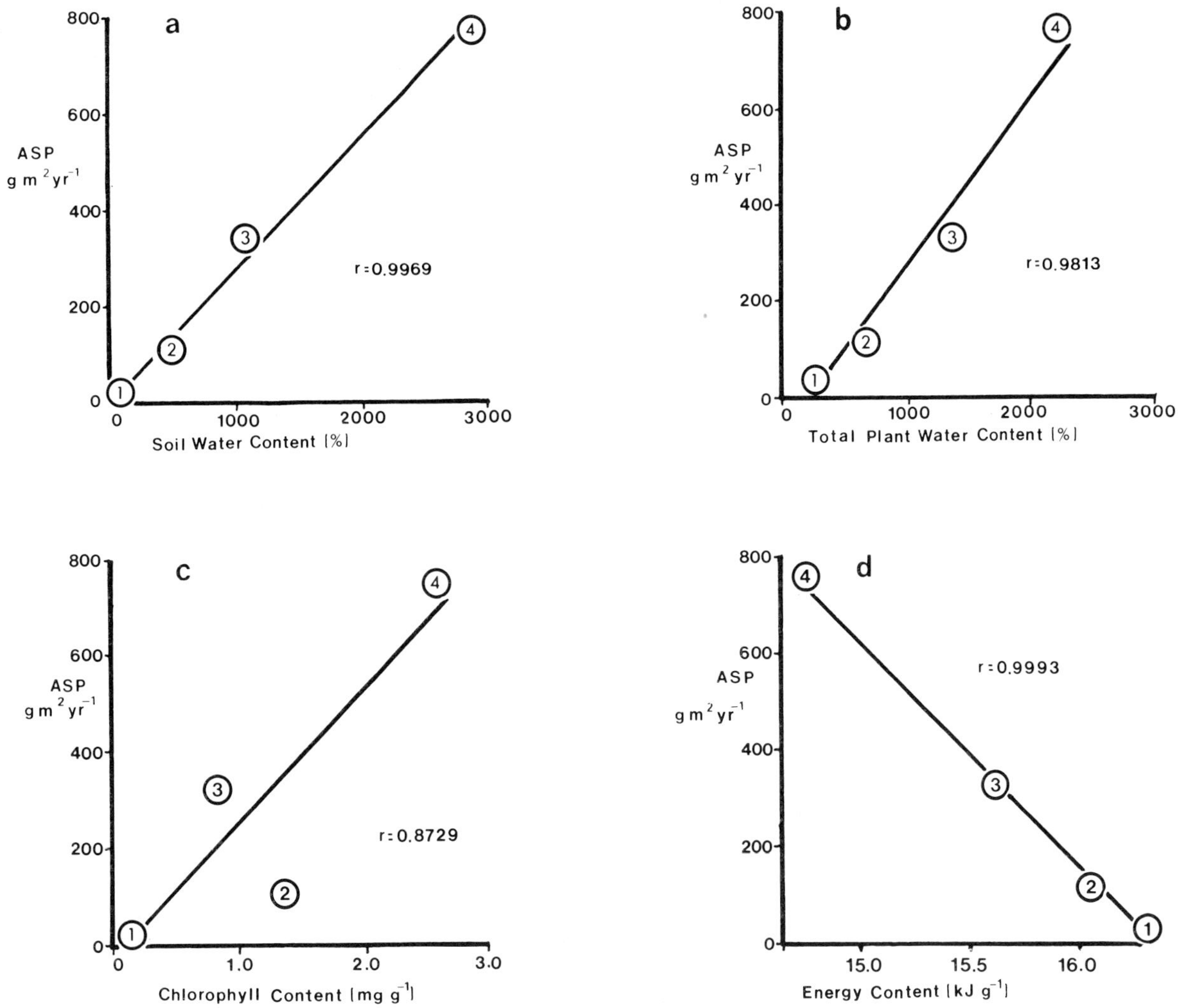

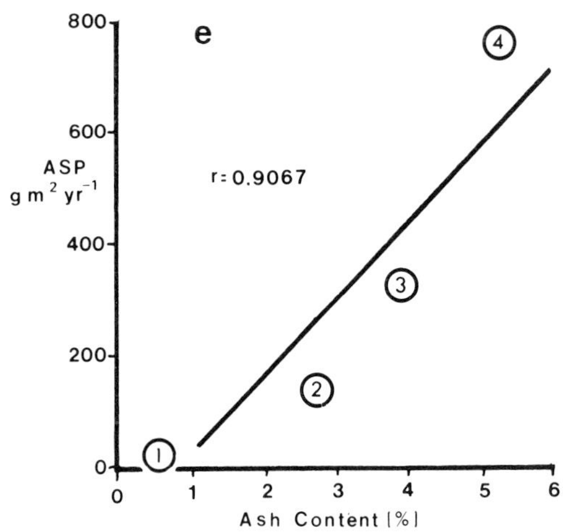

Fig. 1a–e. Annual shoot production (ASP) of four species of bryophytes as a function of: **a** Soil-water. **b** Total plant water. **c** Chlorophyll **d** Energy. **e** Ash contents, respectively, in relation to four habitat types at Marion Island. *1 Ditrichum strictum* – fellfield; *2 Brachythecium rutabulum* – slopes; *3 Blepharidophyllum densifolium* – mires; *4 Brachythecium subplicatum* – drainage lines

gradient between fellfield and drainage lines (Gremmen 1975).

This paper presents preliminary results for selected species representing the range of production values obtained at Marion Island. Some general conclusions can be drawn that support the hypothesis. The results of an analysis of plant nutrient status for the species in question were not available in time for inclusion in the paper.

2 Materials and Methods

Characteristics of five bryophyte species typical of the island's major habitats are shown in Table 1. The first four habitats are arranged in series from dry to wet according to Gremmen 1982). The influence of salt-spray and manuring by seabirds and seals is predominant close to the coast, and, for comparative purposes, data for *Clasmatocolea vermicularis,* a liverwort species from this habitat, are given in Table 1 as well.

Annual shoot production was measured by marking and harvesting a minimum of 280 bryophyte stems per species from growth plots in each habitat type. Plots ranged in size from 3 to 30 m^2 depending on the number of methods used to measure growth (Russell 1984). Tissue water content was averaged from 24 replicates per species by weighing field-fresh samples within a few hours of collection and after oven drying to constant weight at 90°C. Samples consisted of 20-mm diameter cores taken from the top 20 mm of the colony where the depth of stems allowed.

Energy contents were determined from triplicated 1 g samples of each bryophyte species, free from soil particles, in a Gallenkamp adiabatic bomb calorimeter with corrections for fuse wire oxidation and acid production. The sample included the entire standing crop (i.e. green and brown portions of the stems above soil level or above the point where decomposition rendered extraction of entire stems impossible). Ash contents were calculated from muffle furnace residues after heating at 500°C for 5 h. Chlorophyll was determined with a Spectronic 20 spectrophotometer, using standard extraction methods and the absorption coefficient of Arnon, as given in Sestak (1971).

3 Results and Discussion

Annual shoot production data are plotted against soil-water content, plant-tissue water content, energy value, ash content, and chlorophyll value, in Fig. 1. The production value for *Ditrichum strictum,* from the fellfield habitat, was corrected after ashing, for wind-blown dust contamination.

The strong correlation between soil-water regime and vegetation pattern noted by Gremmen (1982) is reflected also in the comparison of soil-water content with production values in Fig. 1a. Lowest production values are found in the exposed fellfield habitat and the greatest occurred in the drainage line situation with its constant minerotrophic water flow. These production values are comparable with those found by other workers in the Sub-Antarctic (Jenkin 1975; Smith and Walton 1975; Smith 1982). The values for plant-tissue water content (Fig. 1b) follow a similar trend. The chlorophyll contents are low in fellfield and high in drainage lines but the values for slope and mire species are transposed giving a correspondingly lower correlation coefficient (Fig. 1c). Bryophyte energy values show a remarkably strong inverse relationship with production when expressed on a dw basis (Fig. 1d). In common with the findings of Vitt and Pakarinen (1977), however, the trend is less clear when these values are expressed on an ash-free basis. Ash content is lowest in the fellfield habitat and highest in the drainage line species (Fig. 1e). Absolute values compare well with the data of Forman (1968) for alpine bryophytes, but are lower than the figures obtained, for example, by Vitt and Pakarinen (1977) and Smith and Walton (1975). The latter authors analyzed the uppermost 20 mm of the shoot, representing mainly the current year's growth rather than the entire standing crop as in the present study.

It is probable that relatively high nutrient contributions and production-decomposition rates in the biotic/salt-spray zone, influenced the growth-associated variables for *Clasmatocolea vermicularis* which diverges from the trends indicated by species on the xeric-hydric gradient (Table 1).

Differences in growth form and micro-habitat diversity contribute to a high degree of variability in results from bryophyte production studies (Wielgolaski et al. 1981). Experience has shown that high correlations between annual production values and growth-associated variables, such as those studied here, are rarely obtained. The correlations reported here should, therefore, be viewed with circumspection. The species studied can be regarded as "mid-point" representatives of their habitat types. Within the mire habitat, for example, there is often a wide range of micro-topographic types and a "mosaic" community pattern (Gremmen 1982). It is expected that the results of a statistical analysis of the complete data set will reflect this complexity and shed light on the reasons for variation in plant growth within each major habitat.

Acknowledgements. Thanks are due to the Institute for Environmental Sciences, University of the Orange Free State, Bloemfontein, the South African Department of Transport, and the South African Scientific Committee on Antarctic Research for sponsoring this study.

References

Forman RTT (1968) Caloric values of bryophytes. Bryologist 71: 344–347

Gremmen NJM (1975) Preliminary report on the 5th expedition to Marion Island, 1964/75. Internal Report (typescript), Institute for Environmental Sciences, University of the Orange Free State, Bloemfontein, S Africa

Gremmen NJM (1982) The vegetation of the subantarctic islands Marion and Prince Edward. Junk, The Hague

Huntley BJ (1972) Aerial standing crop of Marion Island plant communities. J S Afr Bot 38:115–119

Jenkin JF (1975) Macquarie Island, Subantarctic. In: Rosswall T, Heal OW (eds). Structure and function of tundra ecosystems. Ecological Bulletins 20. Swedish Natural Science Research Council, Stockholm. pp 375–397

Russell S (1984) Growth measurement in bryophytes: a case study. J Hattori Bot Lab 55:147–157

Sestak Z (1971) Determination of chlorophylls a and b. In: Sestak Z, Catsky J, Jarvis PG (eds). Plant photosynthetic production manual methods. Junk, The Hague

Smith RIL (1982) Growth and production in South Georgia bryophytes. Com Nat Franc Recherch Antarct 51:229–239

Smith RIL, Walton DWH (1975) South Georgia, Subantarctic. In: Rosswall T, Heal OW (eds) Structure and function of tundra ecosystems. Ecological Bulletins 20. Swedish Natural Science Research Council, Stockholm pp 399–423

Smith VR (1978) Animal-plant-soil nutrient relationships on Marion Island (Subantarctic). Oecologia (Berl) 32:239–253

Vitt DH, Pakarinen P (1977) The bryophyte vegetation, production and organic compounds of Truelove Lowland. In: Bliss LC (ed) Truelove Lowland, Devon Island, Canada: a high Arctic ecosystem. University of Alberta, Edmonton pp 225–244

Wielgolaski FE, Bliss LC, Svoboda J, Doyle G (1981) Primary production of tundra. In: Bliss, Heal OW, Moore D (eds) Tundra ecosystems. Cambridge University pp 187–225

Comparative Microbiology of Moss-Peat Decomposition on the Scotia Arc and Antarctic Peninsula

D.D. WYNN-WILLIAMS[1]

Summary. The hypothesis that microbial activity and abundance in Antarctic peat are regulated more by substrate and moisture than directly by temperature or latitude was broadly supported throughout a 51° to 68° S transect. A consistent difference between respiration in *Polytrichum*- and *Drepanocladus*-peat supported the substrate-dependence notion. A correlation between respiration and a broad mixture range in *Drepanocladus* was precluded by a narrower range in *Polytrichum*. The absence of correlation between O_2 uptake and CO_2 release suggested anaerobic CO_2-production in wet conditions. Corrections using Q_{10} coefficients indicated limited temperature responses. Moisture-dependent decomposition gave significant loss in tensile strength (LTS) of cotton strips nearer the surface in wet sites than dry ones. Ground isolation partly accounted for LTS at high latitude. Strip discoloration suggested cellulose decomposition by chromogens found at the sites. Viable counts did not correlate with respiration. Counts consistently different between fellfield and moss communities reflected substrate characteristics.

1 Introduction

The decomposition of dissolved organic carbon (DOC) and cellulose by micro-organisms in a moss turf and a moss carpet site at Signy Island, South Orkney Islands, has been investigated since 1975 (Wynn-Williams 1980). The importance of phyllosphere yeasts in the decomposition of DOC was reported by Wynn-Williams (1980, 1982). The latter studies revealed groups of micro-organisms which were particularly associated with the decomposition of DOC and macromolecules. The origins and translocation of microbial groups from one niche into another according to edaphic and environmental conditions was also demonstrated (Ellis-Evans and Wynn-Williams 1985).

The present paper extends these studies geographically to a transect from Stanley (51°45′S) in the Falkland Islands to Refuge Islands (68°22′S), including Signy Island (60°44′) and other islands on the Scotia Arc. It also broadens the investigation into a comparison of oxygen uptake and carbon dioxide release by decomposing moss peat at these sites. It also related these to the breakdown of cellulose, and the micro-organisms associated with these activities. An account of such a broad comparison will necessarily be simplistic. However, it will focus on the hypothesis that the composition and decomposer activity of the terrestrial microflora of the maritime Antarctic is dependent more on the physical and chemical nature of the substrate and its moisture content than on the direct effect of temperature or of increasing climatic severity with latitude.

2 Study Sites

For locations, climate and edaphic details of selected sites where measurements of respiratory gas exchange were made in the 1980–81 season see Table 1. Additional sites where microbiological samples were collected included Stanley, Falkland Islands (51°45′S), Beachêne Island (52°55′S), Bird Island, South Georgia (54°00′S), Deception Island (62°57′S), Rothera Point (67°34′S) and Adelaide Island (67°46′S). Where possible, samples were taken from *Polytrichum alpestre* moss turf, *Drepanocladus uncinatus* moss carpet and adjoining fellfield areas of mainly bare mineral debris.

3 Methods

Sampling procedures were as described in Wynn-Williams (1979). Homogenization of samples for microbiological analysis and spread plating for colony-forming units (CFU) were as in Wynn-Williams (1983). Although it is appreciated that viable counts greatly underestimate the size of the microflora (Martin et al. 1982), they are justified for detecting relative changes in the opportunistic fast-growing microbial populations prevailing in moss peats and fellfields which are subject to freeze-thaw stress. Casein peptone starch agar (CPSA) amended with Actidone (Upjohn) at 50 mg l^{-1} was the primary medium for bacteria, and Sabouraud dextrose agar amended with Aureomycin (Lederle) at 30 mg l^{-1} and sodium propionate (250 mg l^{-1}) was the primary medium for yeasts and filamentous fungi. Distinctive micro-organisms which were noted as indicators of decomposer activity, edaphic features or distribution patterns included *Chromobacterium*, gold chromogenic bacteria in general, *Cytophaga* in particular, and *Mortierella* and *Mucor*.

1 British Antarctic Survey, NERC, High Cross, Madingley Road, Cambridge CB3 OET, Great Britain

Antarctic Nutrient Cycles and Food Webs
(ed. by W.R. Siegfried, P.R. Condy, and R.M. Laws)

Respiratory activity of field-fresh moss peat at ca. 10–30 mm depth immediately beneath the green shoot layer was measured as oxygen-uptake by 4–8 replicate cores (27 mm diam., ca. 11.5 cm^3). These were inserted in the removable bases of special flasks for use in a Gilson differential respirometer (Wynn-Williams 1979). Carbon dioxide evolution was measured concurrently on a further 4–8 replicate cores in identical flasks fitted with Suba Seal bungs in the apertures. Reagents were added by syringe via the side arm. Five ml of 0.05 N KOH containing 0.22% Teepol and 3% stock indicator solution were injected into the annular well before incubation in a cooled incubator at peat ambient temperature. The stock indicator solution consisted of 1 part phenolphthalein solution (1% in 70% ethanol) plus 1 part thymolphthalein solution (1% in absolute ethanol). After incubation, 0.5 ml saturated barium chloride was added to precipitate potentially-interfering potassium carbonate. The solution was then back-titrated with 0.025 N HCl to a colourless end point.

Respiration data obtained at ambient peat temperature were corrected for Q_{10} using the logarithmic response e^{qt}, where $q = (\ln Q_{10})/10$ and t is the range of the temperature adjusted to a standard temperature of 5°C, ($t = t_5 - t_{ambient}$).

Decomposition of cellulose was measured at Faraday and Rothera by inserting cotton strips into the peat profile for a season (1981–82) and measuring the loss in tensile strength (LTS) at 100% RH as described by Walton and Allsopp (1977).

4 Results and Discussion

4.1 Respiration as a Measure of Decomposition

The hypothesis suggests that because of the different substrate quality and micro-environments within moss turf (e.g., *Polytrichum*) and moss carpets (e.g., *Drepanocladus*), their respiratory gas exchange may differ. No attempt was made to measure the respiratory gas exchange of fellfield soil samples as the apparatus available was insufficiently sensitive for such low activity.

It is generally assumed that the majority of respiration in such peat is due to microbial activity (Davis 1981). A preliminary assay of O_2-uptake by excised rhizoids from *Polytrichum* at Signy Island suggested that the contribution of below-ground moss tissue to total peat respiration was 3%–14%. This assumed an active rhizoid biomass of 1.3 g m^2 (dry wt.) at the 10–30 mm depth (after Davis 1981). These data were corrected to a standard 5°C using a Q_{10} coefficient (change in rate per 10°C) which compensates for the exponential metabolic response to temperature. At Signy Island in spring 1975, these were Q_{10} = 2.95 for *Polytrichum* peat and 2.50 for *Drepanocladus* peat in the range of 0 to 5°C. These values were slightly lower than those generally found for tundra, Q_{10} = 3.5–4.0 (Heal et al. 1982). Such experiments in vitro should be treated with caution. Nevertheless, although regression analysis of data from the present transect failed to show any correlation between viable microbial counts and gas exchange (probably due to small values of n), the results presented here represent essentially microbial decomposer activity. Respiration probably reflected decomposition mainly of DOC exuded and flushed from frost-damaged cells (Hurst 1982).

The object of a transect rather than a single-site study was to detect general trends for the maritime Antarctic. Peat respiration at the selected transect sites (Table 1) was tested by Analysis of Variance (ANOVA) of the data as follows: Between the two communities (n = 54 and 64 for O_2 uptake and CO_2 release respectively); between the 5 sites sampled for *Polytrichum* (n = 23 and 26 respectively) and *Drepanocladus* (n = 23 and 30 respectively). All variance ratios were significant at $P < 0.01$ except for insignificant variation in *Polytrichum* peat O_2 uptake between the sites. This latter consistency was eliminated by correcting the data to the standard 5°C. The significant differences in both O_2-uptake and CO_2-evolution between the two communities were consistent with the hypothesis of substrate-dependence. This may include an element of pH-dependence since *Polytrichum* peat was significantly more acid (mean ± SE = pH 4.1 ± 0.1) than *Drepanocladus* (pH 5.2 ± 1.0). The hypothesis of substrate-dependence was further supported by apparent consistency of O_2-uptake at ambient temperature within *Polytrichum* communities between all sites. However, this was also partly due to their relatively constant moisture content.

Attempts were made to determine which edaphic and microclimatic factors (Table 1) were primary regulators of O_2-uptake and CO_2-evolution, and if the two measures of respiration differed in their response to them.

The main physical difference between the two moss communities throughout the transect was their moisture content (Table 1). The carpets were much wetter and often waterlogged. This restricts the diffusion of oxygen and other gases and may establish anaerobic conditions. These will affect O_2 uptake more than CO_2 evolution. There was no direct relationship between O_2 uptake and the narrow range of peat moisture at the drier *Polytrichum* sites (Table 1). However, the moisture content of *Drepanocladus* peat varied greatly between sites, giving predictably correlated O_2 uptake ($r = -0.927$, $n = 5$, $P < 0.05$). This correlation was improved by correction to the 5°C standard temperature.

Respiration was not directly related to the mean annual air temperature (Table 1). Under the conditions prevailing ath the five selected locations at the time of sampling, neither O_2 uptake nor CO_2 evolution correlated with the ambient temperature of *Polytrichum* peat, despite its known Q_{10} coefficient. This contrasted with the importance of peat temperature in comparable tundra at Barrow, Alaska, where it accounted for 40% of the variation in res-

Table 1. Characteristics of sites sampled for respiration and microbiological studies of moss peat

Location	Latitude	Mean (± SD) annual insolation (h)	Mean annual air temperature (°C) (1971–75)	Peat at 10–30 mm depth					
				Sampling date	Temperature (°C)	*Polytrichum* %H_2O (dry wt.)	pH	*Drepanocladus* %H_2O (dry wt.)	pH
South Georgia	54° 16′S	1216 ± 124	2.08	5 February 1981	5.0	390	3.8	1409	3.9
Signy Island	60° 44′S	562 ± 78	−4.22	2 February 1981	3.0	367	4.1	1570	5.2
Almirante Brown	64° 53′S	NA	−2.44	19 December 1980	1.0	399	4.0	574	6.8
Faraday	65° 15′S	840 ± 122	−2.64	12 December 1980	1.0	471	4.1	498	5.1
Jenny Island	67° 44′S	944 ± 147[a]	−3.72[a]	14 March 1981	4.0	309	4.7	NP	NP
Refuge Islands	68° 22′S	NA	−5.22[b]	11 March 1981	2.0	NP	NP	700	4.9

Abbreviations: NA, Not available; NP, Not present

a Data for Adelaide Island (67° 46′S)
b Data for San Martin Base (68° 07′S)

piration (63% in combination with moisture), and at Hardangervidda, Norway, where it accounted for 81% (Flanagan and Veum 1974). However both these sites were wet and more similar to the wetter *Drepanocladus* sites here where temperature affected CO_2 evolution ($r = 0.936$, $n = 5$, $P < 0.02$) unlike the drier *Polytrichum* turf. Nevertheless, correction of *Polytrichum* peat O_2 uptake to the standard 5°C eliminated the intersite consistency, implying some measure of temperature-dependence.

CO_2-evolution at ambient temperature in both communities varied significantly between the sites. This contrast with the consistency of O_2-uptake by *Polytrichum* reflects CO_2-generating pathways other than aerobic respiration. Carbon flux may be rate-limited by temperature in wetter communities, as was shown for a similar ombrotrophic mire at Abisko (Sweden) by Svensson (1980) and for methane-flux at Signy Island by Yarrington and Wynn-Williams (1985). However, standardization to 5°C to compare potential rather than actual activity may be artificial as optimal temperatures rarely occur in these regions.

There was no correlation between O_2-uptake and CO_2-evolution in either community, probably because of the influence of anaerobic conditions in wet moss carpets which rarely prevail in moss turf (Wynn-Williams 1980). This emphasized the value of using both methods of respiration analysis simultaneously.

Full details of the respiration studies will be published elsewhere, but their similarity to the findings for the International Biological Programme tundra sites should be mentioned. The potential respiration curve for moss turf peat at Signy Island, derived from the respiration-temperature curve at optimum moisture for total litter at Barrow, predicts a CO_2-release of 22.1 $\mu l\ g^{-1}$ (dry wt.) h^{-1} for early February (Heal et al. 1982). The corresponding respiration rate on 2 February 1981 was 20.0 $\mu l\ g^{-1}\ h^{-1}$. This similarity supports the general relevance of the present transect data.

4.2 Decomposition of Macromolecules

Although decomposition is generally slow at prevailing maritime Antarctic temperatures (Table 1), microbial respiratory enzymes are active as long as free water is present (Wynn-Williams 1982). Decomposition of macromolecular components of moss and peat, such as cellulose, may be significant in a single growing season. This is illustrated by results (Table 2) relating the 1981–82 season to a detailed study at Signy Island in 1977–78 (Wynn-Williams 1980). The LTS was corrected for a standard 100-day growing season. The pattern of cellulose decomposition in the peat profiles varied both between sites and communities. It was significant nearer the surface in wet moss carpets than in drier moss turf, suggesting limitation by moisture. However, other regulatory factors include anaerobic conditions which do not necessarily preclude cellulose breakdown. Minerals may also be more readily available at lower depths and may have a dramatic effect on peat decomposition (Svensson 1980). One season's LTS at depth 10–30 mm at Rothera (50%) was much greater than at Faraday (17%) despite its more southerly latitude. However, the mean annual sunshine at Rothera is ca. 1150 h compared with only ca. 840 h at Faraday and ca. 560 h at Signy Island (D.W.S. Limbert, pers. comm.), so that ground heating at this shallow depth will have a marked influence irrespective of air temperature.

During the study at Signy Island in 1977–78, Wynn-Williams (1980) detected a highly significant correlation between LTS (kg cm^{-1}) and gold discoloration (expressed on a scale of 1 to 3) where $r = 0.754$, $P < 0.001$, $n = 23$. This was described by the regression $y = 2.874x + 2.448$ where y = LTS and x = discoloration. Gold chromogens were common in peat communities and spreading colonies of *Cytophaga* were frequently seen. Such observations

Table 2. Significance (*t*-Test) of loss in tensile strength (LTS) of cotton strips buried in moss peat at Signy Island (1977–78) and at Faraday and Rothera (1980–81) relative to insertion controls

Depth	Moss turf (*Polytrichum*)		Moss carpet (*Drepanocladus*)		
(mm)	Signy I. (60° S)	Faraday (65° S)	Signy I. (60° S)	Faraday (65° S)	Rothera (67° 30′S)
10– 30	–	–	–	b	b
40– 60	–	–	b	b	NP
70– 90	–	–	b	b	NP
100–120	b	a	b	b	NP
130–150	b	–	b	–	NP
160–180	b	–	b	–	NP

a Significant at $P < 0.05$
b Significant at $P < 0.01$
NP: Not present

led to the monitoring of ecologically-significant microorganisms such as *Cytophaga* and *Chromobacterium* amongst the general population.

The hypothesis of substrate-dependent decomposition, partly confirmed here, is consistent with the conclusions for IBP tundra sites (Heal et al. 1982), suggesting a bipolar trend. However, the apparently greater importance of moisture over field temperature for decomposition in the maritime Antarctic (particularly at cloudy sites) was the converse of IBP findings. This may be due partly to the low summer temperatures at the Antarctic sites (Limbert 1979) relative to Arctic ones at similar latitudes (Rosswall and Heal 1975).

4.3 Composition of the Microflora

The main viable microbial populations of the three terrestrial communities compared in the transect are summarized in Table 3. The often large standard errors reflect variation between sites. This necessitated ANOVA to test the significance of difference between communities and within them (between sites). This validated testing of the hypothesis which predicted consistency of the microflora within each community irrespective of its sampling site.

The absence of correlation between viable counts and respiration suggested that the prevailing edaphic and microclimatic conditions were influencing respiration differently

Table 3. Mean viable microbial counts (Colony-forming units, CFU x 10^4 g^{-1} dry wt.) in three dominant communities on a latitude transect including South Georgia, Signy Island, Almirante Brown and Rothera

Microbial	Depth	Community								
		Polytrichum			*Drepanocladus*			Fellfield		
group	(mm)	Mean	SE	n[a]	Mean	SE	n	Mean	SE	n[b]
Heterotrophic	0–10	3,020	1,188	11	46,211	12,413	15	170.0	64.0	14
bacteria	10–30	7,613	180	9	137,300	3,401	16	"	"	"
Gold	0–10	1,872	1,018	11	3,667	1,102	15	23.7	9.4	14
bacteria	10–30	420	175	8	40	12	14	"	"	"
Yeasts	0–10	128	62	12	441	250	16	0.31	0.16	15
	10–30	19	5	12	39	13	16	"	"	"
Filamentous	0–10	86	65	12	25	7	15	0.86	0.71	15
microfungi	10–30	24	4	12	115	52	16	"	"	"

a Total replicate cores
b Samples from 0–30 mm were bulked for analysis

Table 4. Variance Ratios (F) for viable microbial counts (CFU g^{-1} dry wt.) between sites dominated by *Polytrichum*, *Drepanocladus* and fellfield communities

Community	Microbial group	Depth (mm)	F^{1}_{n-2}	Total replicate cores (n)
Polytrichum	Heterotrophic	0–10	2.2	11
(k = 3)[a]	bacteria	10–30	9.1[b]	9
	Yeasts	0–10	11.2[c]	12
		10–30	0.4	12
Drepanocladus	Heterotrophic	0–10	26.8[c]	15
(k = 4)	bacteria	10–30	12.2[c]	16
	Yeasts	0–10	1.3	16
		10–30	5.0[b]	16
Fellfield	Heterotrophic			
(k = 4)	bacteria	0–30	2.5	15
	Yeasts	0–30	4.3[b]	15

a Comparisons were made between microbial populations in *Polytrichum* at South Georgia, Signy Island and Almirante Brown. Comparisons between *Drepanocladus* and fellfield populations also included Rothera
b Significant at $P < 0.05$
c Significant at $P < 0.01$

Table 5. Comparative heterotrophic bacterial counts (g^{-1} dry wt.) in *Drepanocladus* communities before and after construction of the British Antarctic Research Station at Rothera Point, Adelaide Island

Site	Depth (mm)	Mean (± SE) colony-forming units x 10^4 February 1976		n	February 1981		n	1981/1976 ratio
Rothera Site 1	0–10	307.8	153.0	3	101,420	16,320	4	330
	10–30	133.6	20.5	4	28,280	7,360	4	212
Rothera Site 2	0–10	123.8	34.9	4	11,810	775	4	95
	10–30	15.7	3.5	3	4,900	2,030	4	312
Signy, SIRS 2	0–10	4,130	–	1	79,420	700	3	19
	10–30	33.8	6.4	2	313.4	37.9	2	9

from growth and reproduction. In some cases there was significant variation in microbial populations between sites, as illustrated by total heterotrophic bacteria and yeasts (Table 4). A major source of variation within the transect was the exceptionally high total heterotrophic bacterial CFU count at Rothera. This provoked re-examination of comparable counts made on the same medium at the same time of the year in 1976 before the building of the BAS station, ca. 100 m from site 1 and ca. 250 m from site 2. The comparisons in Table 5 include concomitant data from Signy Island as a control. The ratio of CFU before and after the construction showed a 20-fold greater increase at Rothera than at Signy Island. Many factors may be responsible for this, but a distribution of nutrients or bacterial contaminants from the station may have been contributory. It may therefore have been inappropriate to use CFU data from Rothera site 1 to represent the microflora of *Drepanocladus* in this area. In contrast, the consistency of the bacterial flora of *Polytrichum* between sites was remarkable, considering their range of latitude and climates.

Table 6. Variance Ratios (F) for viable microbial counts (CFU g^{-1} dry wt.) between three dominant communities on a latitude transect including South Georgia, Signy Island, Almirante Brown and Rothera

Microbial group	Depth (mm)	Communities compared: *Polytrichum*/ *Drepanocladus* F^{1}_{n-2}	n^{c}	*Polytrichum*/ fellfield F^{1}_{n-2}	n	*Drepanocladus*/ fellfield F^{1}_{n-2}	n
Heterotrophic	0–10	8.7[b]	26	7.4[b]	25	12.8[b]	29
bacteria	10–30	2.9	25	17.1[b]	23	5.3[a]	30
Gold	0–10	1.3	26	4.3[a]	25	10.2[b]	29
bacteria	10–30	8.2[a]	22	9.2[a]	22	2.0	28
Yeasts	0–10	1.1	28	5.4[a]	27	2.9	31
	10–30	1.6	28	19.3[b]	27	7.9[b]	31
Filamentous	0–10	1.1	27	2.1	27	12.0[b]	30
microfungi	10–30	2.3	28	33.6[b]	27	4.5[a]	31

a Significant at $P < 0.05$
b Significant at $P < 0.01$
c Sum of replicate cores from the 4 sites

The demonstration of significant differences in microflora between the communities of each site (Table 6) was perhaps more important as this was positive support of the hypothesis of a substrate-associated microbial population. When the data from individual replicate cores at selected sites were compared, only the total heterotrophic bacteria of the phyllosphere differed significantly between *Polytrichum* and *Drepanocladus.* This suggested that their bacterial communities were essentially similar. These comprised mainly decomposer organisms in contrast to the colonizer-dominated microflora of fellfield which differed from both moss communities.

The difference between the heterotrophic bacterial microfloras of both *Polytrichum* and *Drepanocladus* sites at the selected locations and those of adjacent fellfield soils was consistently significant over the whole transect (F^{1}_{12} = 6.5 and F^{1}_{13} = 4.8 respectively, $P < 0.05$). When all 11 transect locations were compared, the difference in mean microbial counts between the communities was not as marked as in the selected sites but was still significant.

No correlation was apparent here between counts of bacteria, yeasts or microfungi and either moisture or pH. This was probably due to interactions and variations in substrate quality from site to site, although Smith and Steyn (1982) found that pH could account for 48% of fungal variation in the Sub-Antarctic soils of Marion Island. The pH range throughout the transect communities was probably too restricted for such a general trend to emerge. However, individual groups of micro-organisms may show specific trends better than complex populations.

4.4 Distribution of Ecologically Significant Micro-Organisms

During this study, *Chromobacterium* emerged as a valuable indicator of edaphic and biogeographical influences (e.g., pH and moisture) on maritime Antarctic bacteria (Wynn-Williams 1982). Gold chromogenic bacteria were prevalent in all three communities in similar proportion to the total count (Tables 3 and 6). Their prevalence in peat communities was of significance relative to the discoloration of cotton strips, with which potentially-cellulolytic *Cytophaga* may be associated.

Yeasts have been shown to be exploiters of DOC in moss peat ecosystems (Wynn-Williams 1980, 1982) so their limited occurrence in fellfields (Table 3) in the absence of developed cryptogamic communities, was predicable.

Microfungi such as *Chrysosporium pannorum* in the Antarctic utilize DOC but can also attack macromolecules such as cellulose (Pugh and Allsopp 1982) and are therefore more adaptable than yeasts and "sugar fungi". These authors described the widespread distribution of such a "sugar fungus", *Mortierella,* at Signy Island. It was observed on the present transect in moss peat only at South Georgia and Signy Island, but it was also found in fellfield soils with discontinuous moss and lichen cover at Signy Island. Physiologically similar strains of *Mucro* were also found in organic soils and peat in the sites at Stanley, Beauchêne Island, Bird Island and Almirante Brown. It was isolated unexpectedly from 1 yr-exposure glacial moraine inland of Husvik, South Georgia. This was probably wind-blown from local soil to become part of the morainic "propagule bank". Pugh and Allsopp (1982) considered the abundance of yeasts and *Mortierella* (characteristically fast growing and high sporulating) in the Antarctic to be anomalous. This is true considering that most ice-free land is bare fellfield, but the moss peat communities described in this paper are important components of the Antarctic ecosystems in which such ruderal organisms flourish.

5 Conclusions

The hypothesis of the regulation of microbial activity and abundance in Antarctic moss peat by substrate quality and moisture was broadly supported by analyses of respiratory activity, cellulose decomposition and viable microbial counts. The ability to test such a hypothesis was due to a large extent to the relative simplicity of terrestrial Antarctic ecosystems, which enables interacting processes to be more easily separated than in more complicated and less isolated environments.

Acknowledgements. I thank the officers and crew of R.R.S. Bransfield for the shipborne help during the 1980–81 transect, and the Base Commanders and personnel of Grytviken, Bird Island, Signy Island, Almirante Brown, Faraday and Rothera for their help ashore. I also thank Drs R.G. Booth, P. Griffiths, R.I. Lewis Smith and M.B. Usher, and W.N. Bonner and T. Heilbronn for their assistance with sampling. I am grateful to Dr W. Block and K. Richard for experimental work on rhizoids at Signy Island. Miss M.C. Vivian-Neal and P.V. Tearle helped with bacterial classification. D.W.S. Limbert and J.P. Corkill kindly provided data from the records of the B.A.S. meteorological section. I thank Dr Block for helpful criticism of the manuscript.

References

Davis RC (1981) Structure and function of two Antarctic terrestrial moss communities. Ecol Monogr 5:125–143

Ellis-Evans JC, Wynn-Williams DD (1985) The interaction of soil and lake microflora at Signy Island. In: Siegfried WR, Condy PR, Laws RM (eds) Antarctic nutrient cycles and food webs (Proceedings of the fourth SCAR symposium on Antarctic biology). Springer, Berlin Heidelberg New York

Flanagan PW; and Veum AK (1974) Relationships between respiration, weight loss, temperature and moisture in organic residues in tundra. In: Holding AJ, Heal OW, Maclean SF, Flanagan PW (eds) Soil Organisms and Decomposition in Tundra. IBP Tundra Biome Steering Committee, Stockholm pp 249–277

Heal OW; Flanagan PW; French DD; Maclean SF (1982) Decomposition and accumulation of organic matter. In: Bliss LC, Heal OW, Moore JJ (eds) Tundra Ecosystems: A comparative Analysis. Cambridge University Press, Cambridge pp 587–633

Hurst JL (1982) Biology of fungi on plants in the Sub-Antarctic. Ph.D. thesis. University of Aston in Birmingham

Limbert DWS (1979) Climatological summary for 1975. Br Antarct Surv Bull 47:129–135

Martin NJ; Siwasin J; Holding AJ (1982) The bacterial population of a blanket peat. J Appl Bacteriol 53:35–48

Pugh GHF, Allsopp D (1982) Microfungi on Signy Island, South Orkney Islands. Br Antarct Surv Bull 57:56–68

Rosswall T, Heal OW (1975) The IBP Tundra Biome – an introduction. In Rosswall T and Heal OW (eds) Structure and Function of Tundra Ecosystems. Ecol Bull (Stockholm) 20:7–16

Smith VR; Steyn MG (1982) Soil microbial counts in relation to site characteristics at a subantarctic island. Microb Ecol 8:253–266

Svensson BH (1980) Carbon dioxide and methane fluxes from the ombrotrophic parts of a subarctic mire. In: Sonessen M (ed) Ecology of a Subarctic Mire. Ecol Bull (Stockholm) 30:235–250

Walton DWH, Allsopp D (1977) A new test cloth for soil burial trials and other studies on cellulose decomposition. Int Biodeterior Bull 13:112–115

Wynn-Williams DD (1979) Techniques for studying terrestrial microbial ecology in the maritime Antarctic. In: Russell AD, Fuller R (eds) Cold Tolerant Organisms in Spoilage and the Environment. Academic, London pp 67–82

Wynn-Williams DD (1980) Seasonal fluctuations in microbial activity in Antarctic moss peat. Biol J Linn Soc 14:11–18

Wynn-Williams DD (1982) Simulation of seasonal changes in microbial activity of maritime Antarctic peat. Soil Biol Biochem 14:1–12

Wynn-Williams DD (1983) Distribution and characteristics of *Chromobacterium* in the Maritime and Sub-Antarctic. Polar Biol. 2:101–108

Decomposition and Nutrient Cycling in *Rostkovia magellanica* from Two Contrasting Bogs on South Georgia

G.J. LAWSON[1]

Summary. The chemical composition of *Rostkovia magellanica* tillers was contrasted between a eutrophic seepage slope-mire and a mesotrophic basin-bog. Tillers in the slope-mire have higher concentrations of Ca (+214%), K (+17%), N (+13%) and Fe (+6%) but lower concentrations of Na (−22%), Mg (−54%) and Mn (−24%) than those in the basin-bog. These differences generally reflect the higher levels of exchangeable nutrients and pH in peat from the seepage-slope. Five elements are removed from senescing leaves (in the order K > P > N > Na > Mg), whereas 3 are accumulated (Fe > Ca > Mn). At both sites, initially rapid loss of weight of litterbag content was halted over winter. Weight loss in the second summer was much faster on the seepage-slope. The content of mobile elements (K, P, N, Na) decreased considerably overall through leaching and decomposition, but N and P contents increased during the second summer at the slope and bog sites, respectively, and this may represent fungal activity. Litterbags composed of different age classes of leaves remained chemically distinct from each other (with the exception of K) during the 15-month trial.

1 Introduction

It is well established that, as plant materials senesce and subsequently decompose, certain elements will be rapidly removed by translocation, leaching or microbial action, whilst others remain locked within structural tissue or are even absorbed from the environment (Swift et al. 1979). Many studies have compared the nutrient composition of living and dead plant parts, and in a deciduous species this may be sufficient to characterize the annual nutrient cycle. The leaves of many perennial species, however, may live for more than one season and dead graminaceous leaves can remain attached to the parent tiller for several years. This may complicate the calculation of annual nutrient budgets, but it does provide the opportunity to quantify the rate of chemical change associated with leaf senescence and death. Callaghan (1980), for example, has been able to determine the pattern of chemical change in successive annual segments of the slow growing clubmoss *Lycopodium annotinium,* dating back for up to 25 yrs. Another method of quantifying the rate of chemical change during decomposition is to enclose dead leaves in mesh bags and leave them in place amongst the litter layer on the soil surface (Bocock and Gilbert 1957).

In situ chemical changes, as standing dead leaves senesce, may be different to the more rapid decomposition of fallen leaf litter or material in litterbags. Nevertheless, it can be useful to compare the two processes by enclosing excised leaves of known age in litterbags and contrasting decomposition with that of leaves still attached to the plant. It is somewhat artificial to use litterbags of living leaves in this way, and the work of, for example, Coulson and Butterfield (1978) has been criticized for this reason. However, with long-lived graminoid tillers typical of tundra regions, it is often impossible to define the precise point of leaf death. Hence, the aim of this study was to use plant material of different ages and chemical compositions, to relate the speed and nature of decomposition to the quality of the decomposing substratum and the environmental conditions.

The Sub-Antarctic island of South Georgia has a flora of only 24 native vascular species, yet many habitats are available for comparing productive plant growth. *Rostkovia magellanica* shows considerable phenotypic plasticity, growing on a range of bogs, mires, flushes and fellfields. Thus, it provides an excellent opportunity for quantifying the decomposition rate of leaves from a single species in its different native habitats.

2 Study Area

Two sites were employed within the International Biological Programme (IBP) study area at King Edward Cove on South Georgia (Smith and Walton 1975). The one site is a seepage-slope (Fig. 1), dominated by the loose carpet-forming moss *Tortula robusta,* occasionally associated with *Drepanocladus uncinatus* and *Calliergon sarmentosum.* Vascular plant cover is principally sparse *Rostkovia magellanica* with some *Juncus scheuchzerioides* and stunted *Acaena magellanica.* The water table is continuous at the peat surface and mean pH at the base of the peat layer ($\bar{x}$ = 47 mm thickness) is comparatively high at 5.97. Peat depth is up to 1.95 m and the peat is marked by a very high extractable Ca content (Table 1), indicating eutrophic conditions.

1 Institute of Terrestrial Ecology, Merlwood Research Station, Grange-over-Sands, Cumbria LA 11 6JU, United Kingdom

Antarctic Nutrient Cycles and Food Webs
(ed. by W.R. Siegfried, P.R. Condy, and R.M. Laws)

Fig. 1. The seepage-slope site: a soligenous eutrophic mire at South Georgia

Table 1. Physical and chemical features of two study sites at South Georgia

	Seepage-Slope	Basin-Bog
Altitude (a.s.l.)	85 m	60 m
Distance from sea	0.75 km	0.25 km
Average snow melt	mid-November	mid-October
Maximum snow depth	1.05 m	0.40 m
Depth of moss layer	80 mm	10–150 mm
January mean surface temp.	0.1°C	– 0.1°C
July mean surface temp.	12.3°C	17.7°C
Mean pH	5.97	4.72
Field moisture content (%)	670	408
Total N (% dry wt.)	2.49	2.23
Extractable Na (mg 100 g dry wt.)	32.1	23.2
Extractable K (")	61.7	25.0
Extractable Ca (")	1053.1	142.0
Extractable Mg (")	67.8	52.6
Extractable P (")	9.1	3.8

The second site is a basin-bog (Fig. 2) dominated by dense *Rostkovia magellanica* with an understory of tall turf-forming mosses (*Chorisodontium aciphyllum* and *Polytrichum alpestre*). On the upper portions of the turves it is sufficiently dry to permit the growth of lichens, such as *Cladonia rangeferina,* other cladonias, and *Pseudocyphellaria freycinetii,* together with foliose liverworts, particularly *Blepharidophyllum densifolium.* The water-table depth fluctuates between zero (i.e., level with the base of the moss turf, which may be up to 150 mm in height) and 220 mm. Mean pH at the peat surface is 4.72, and the site can be classified as mesotrophic (Table 1). Annual above-ground production of *Rostkovia* averages 170 g m^{-2}, whereas at the seepage-slope it is only 33 g m^{-2} (Lawson, unpubl.). Maps of both sites are presented in Smith (1981), together with a detailed account of the types of peat and peat-forming vegetation on South Georgia. This vegetation is important in the coastal zones, and these two sites are typical of many throughout the island.

3 Materials and Methods

3.1 Harvests

Nine harvests of 100 *Rostkovia* tillers were taken from both sites at intervals through the summer of 1975–76 (6 months). Concurrent measurements of tiller density enabled the rate of biomass production to be calculated. Laminas

Fig. 2. The basin-bog site: an ombrogenous mesotrophic bog at South Georgia

were separated from sheaths in the five youngest leaves (sheaths from the first four leaves being combined), and the laminas were weighed and analyzed in five separate age classes. Observations on tagged tillers indicated that two leaves are produced per year at both sites. Old leaves (>3–4 yrs old), living rhizomes and stem bases, buds and roots were also separated and analyzed.

3.2 Litterbags

Onehundred tillers were collected in late December and randomly divided into 3 groups of 20. These were subsequently sub-divided into the following plant fractions: laminas 1, 2, 3, and 4, respectively, sheaths from leaves 1 to 4 combined, and old leaves. Dry weights were obtained after freeze-drying, and the plant material was wrapped in 2 mm terylene gauze and heat sealed. Long brass skewers were then used to secure the bags to the moss surface.

Two series of 144 bags each were placed on the basin-bog and seepage-slope sites on 2 January, 1976, using material collected from respective sites; 54 bags per site were harvested on 1.4.76, 36 on 4.11.76 and the remaining 54 bags on 29.3.77.

3.3 Chemical Analysis

All samples were either freeze-dried to constant weight on South Georgia or taken to England in deep-frozen storage for oven-drying at 60°C. Samples from the production harvests (9 dates x 9 plant parts) and the old leaf and sheath litterbags were ground through a 1 mm mesh in a Glen Creston hammer mill. Prior to digestion they were divided into two replicate samples, weighing approximately 1 g. Litterbags containing laminas were of insufficient mass to be ground in the hammer mill, and they were finely chopped by hand and thoroughly mixed prior to division into two replicates.

Mixed perchloric acid – hydrogen peroxide digestion (Allen et al. 1974) allowed N and metal cation determinations to be performed on a single sample. Calcium, Mg, Mn, and Fe were determined, using a Baird-Atomic MKIII atomic-absorption spectro-photometer. Potassium and Na determinations were made using a flame emission head on the same instrument. Nitrogen and P were determined using an auto-analyzer, after the methods of Crooke and Simpson (1971) and Murphy and Riley (1962).

4 Results

4.1 Harvests

4.1.1 Between Site Differences

Considerable differences were found in the nutrient concentrations between tillers from the two sites. Concentrations of Ca (+214%,***), K (+17%,***), N (+13%,***), P (+7%, n.s.), and Fe (+6%, n.s.) are relatively high at the seepage-slope, whereas concentrations of Na (–22%,**), Ms (–54%,***) and Mn (–214%,***) are relatively low (Table 2). However, since tillers from the basin-bog averaged 20% heavier than those from the seepage-slope, when % concentrations are converted to contents per tiller (Fig. 3) the seepage-slope is higher only for Ca and K.

Table 2. A comparison of average weight (mg tiller) and chemical composition, based on nine harvests of plant material taken from *Rostkovia magellanica* between 28 October 1975 and 8 April 1976 at South Georgia. All figures for chemical elements are percentages

	Weight	P	N	K	Ca	Mg	Fe	Na	Mn
Seepage-Slope									
Lamina 1	10.84	0.274	2.01	1.24	0.084	0.082	0.007	0.149	0.004
Lamina 2	12.23	0.173	1.77	1.05	0.154	0.089	0.007	0.135	0.007
Lamina 3	9.62	0.114	1.33	0.71	0.215	0.110	0.018	0.143	0.013
Lamina 4	7.56	0.069	0.99	0.24	0.259	0.093	0.021	0.095	0.013
Lamina 5	6.11	0.064	0.86	0.11	0.381	0.070	0.029	0.071	0.010
Sheaths (1–4)	22.31	0.115	0.70	0.95	0.304	0.069	0.010	0.216	0.004
Old leaves	57.65	0.034	0.59	0.15	0.474	0.043	0.084	0.078	0.003
Rhizomes + Stems	26.14	0.204	1.03	0.94	0.220	0.070	0.031	0.207	0.005
Buds	2.54	0.245	0.82	0.21	0.227	0.091	0.082	0.242	0.004
Roots	5.36								
Average for Laminas	46.36	0.139	1.39	0.67	0.219	0.089	0.017	0.119	0.009
Average for Laminas	160.36	0.145	1.13	0.63	0.258	0.079	0.031	0.145	0.007
Basin-Bog									
Lamina 1	15.66	0.248	1.74	1.08	0.030	0.106	0.006	0.212	0.015
Lamina 2	15.95	0.142	1.52	1.00	0.059	0.129	0.010	0.168	0.027
Lamina 3	11.98	0.064	0.93	0.58	0.083	0.160	0.025	0.155	0.044
Lamina 4	9.28	0.036	0.75	0.17	0.101	0.163	0.027	0.097	0.043
Lamina 5	7.15	0.050	0.68	0.09	0.175	0.140	0.041	0.081	0.033
Sheaths (1–4)	28.34	0.185	0.70	0.63	0.051	0.100	0.014	0.228	0.012
Old leaves	77.48	0.048	0.50	0.17	0.082	0.084	0.060	0.089	0.011
Rhizomes + Stems	18.13	0.170	1.26	0.91	0.039	0.107	0.032	0.282	0.007
Buds	4.23	0.295	0.94	0.19	0.124	0.108	0.053	0.226	0.007
Roots	13.89								
Average for Laminas	60.02	0.108	1.12	0.58	0.090	0.140	0.022	0.143	0.033
Average for Tillers	202.09	0.135	1.00	0.54	0.082	0.122	0.029	0.177	0.022

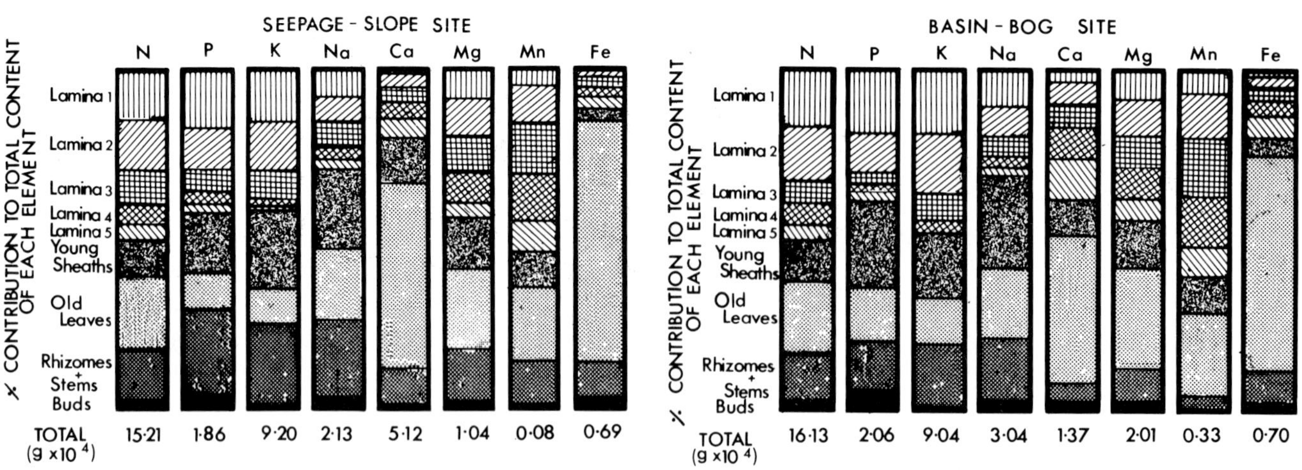

Fig. 3. The contributions of different parts of *Rostkovia magellanica* to the total content of each macro-element in an average tiller at a seepage-slope site (*above*) and a basin-bog site (*below*) at South Georgia

Table 3. Changes in chemical content (g x 10^6 tiller) associated with the ageing of laminae of *Rostkovia* at South Georgia

	P	N	K	Ca	Mg	Fe	Na	Mn
Seepage-Slope								
Content in Lamina 1	30.6	227	140	9.22	9.22	0.77	16.9	0.43
% Remaining in Lamina 2	71.2	98.7	96.1	209	122	119	99.8	200
% Remaining in Lamina 3	38.9	62.0	55.6	239	125	243	91.0	305
% Remaining in Lamina 4	19.6	38.0	15.0	241	88	230	48.9	263
% Remaining in Lamina 5	15.3	28.2	5.8	314	55	288	31.2	179
Basin-Bog								
Content in Lamina 1	38.5	271	167	4.70	16.6	0.92	33.2	2.32
% Remaining in Lamina 2	58.0	88.3	94.5	199	124	176	79.6	187
% Remaining in Lamina 3	21.3	44.0	44.7	224	123	342	59.7	243
% Remaining in Lamina 4	10.2	30.4	10.7	237	108	332	31.7	202
% Remaining in Lamina 5	12.6	24.5	5.3	367	82	443	23.9	134

Table 4. The comparative importance of nitrogen, phosphorus and potassium in different plant parts of *Rostkovia magellanica.* The ratios express % concentrations

	Seepage-Slope			Basin-Bog		
	N/P	N/K	K/P	N/P	N/K	K/P
Buds	3.3	3.9	0.9	3.2	4.3	0.6
Rhizomes and stems	5.0	1.1	4.6	7.4	1.4	5.4
Young sheaths	6.1	0.7	8.3	3.8	1.1	3.4
Lamina 1 (youngest)	7.3	1.6	4.5	7.0	1.6	4.4
Lamina 5	13.4	7.9	1.7	13.6	7.6	1.8
Old leaves	17.4	3.9	4.4	10.4	2.9	3.5

4.1.2 Differences Between Plant Parts

Comparisons of nutrient concentrations in different plant parts at the sites generally mirror the trends described above for the tiller as a whole. Exceptions to the general trend that the basin-bog has relatively low concentrations of the more mobile elements are the high levels of: P in buds, sheaths and old leaves, N in buds and rhizomes + stems, Fe in laminas and K in old leaves (Table 2).

Data on the rate of change in relative nutrient content within the five lamina age classes (Table 3) demonstrate faster decreases of the more mobile elements at the basin-bog site. At both sites, the ease of removal of different elements is in the order K > P > N > Na > Mg. The other three nutrients were accumulated in the order Fe > Ca > Mn.

The relative proportions of N, P, and K differ widely within the three types of young tissue (Table 4). Buds show high P and low K concentrations, sheaths have a particularly high K concentration, and the youngest lamina contains a high level of N.

4.1.3 Seasonal Trends

The age classes shown in Fig. 4 represent cohorts through which individual laminas can progress: as new leaves are produced during the growing season laminas will move into successively older age classes, and the nutrient content of each class therefore represents an interplay between the production of new leaves and any leaching, translocation or decomposition which may take place. The general trend is for concentrations of N, P, and K in the youngest lamina to increase for a short period in spring and then decrease through the remainder of the season. Older laminas, on the other hand, show a slight increase through the season, which is most marked for K in lamina 3.

Negligible or very modest seasonal trends were observed in the other plant parts (e.g., seasonal decrease in the N, P, and K concentration of sheaths), but they are of doubtful significance and are not presented here.

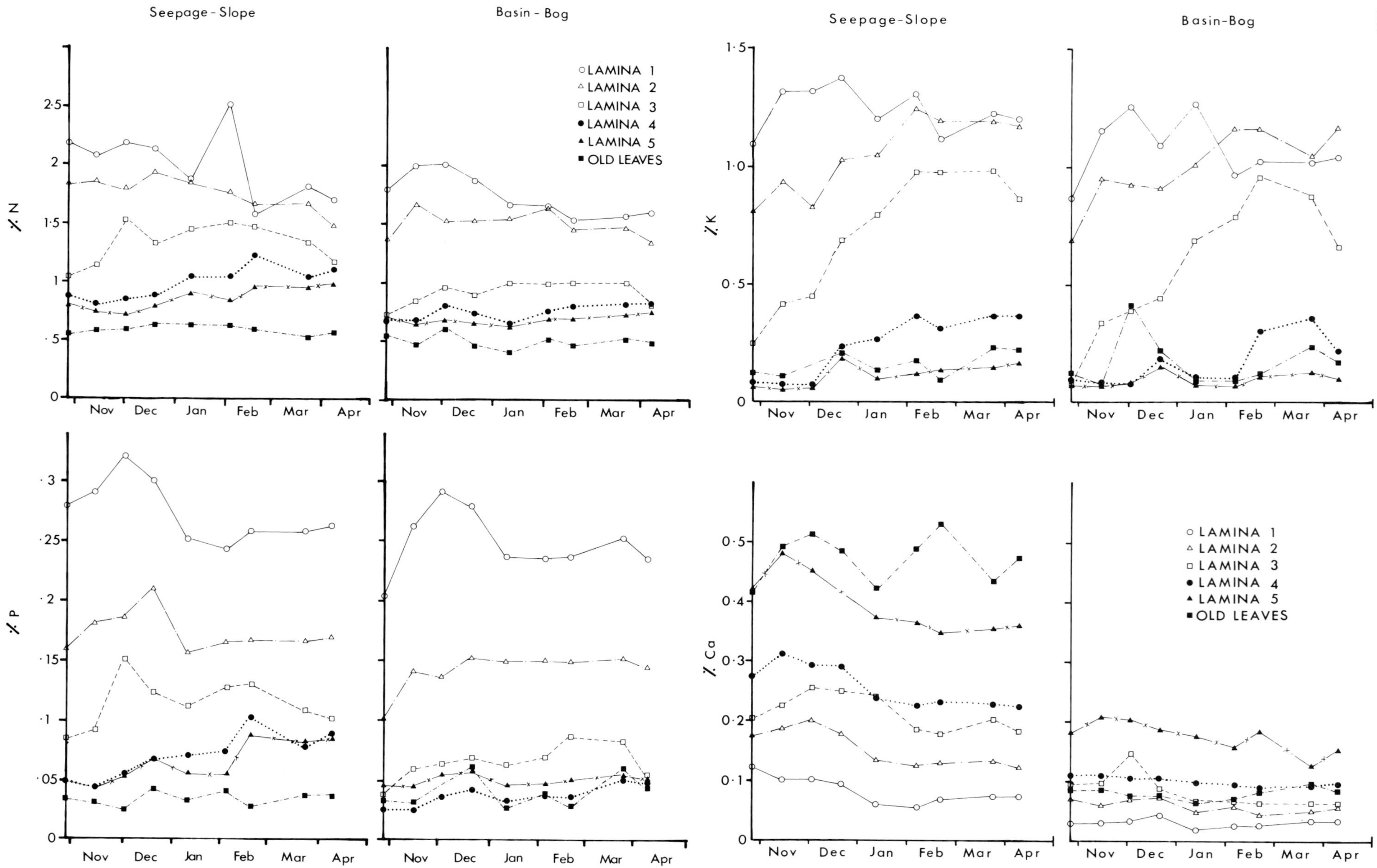

Fig. 4. Seasonal changes in the concentration of nitrogen, phosphorus, potassium and calcium in different parts of *Rostkovia magellanica* at a seepage-slope site and a basin-bog site at South Georgia

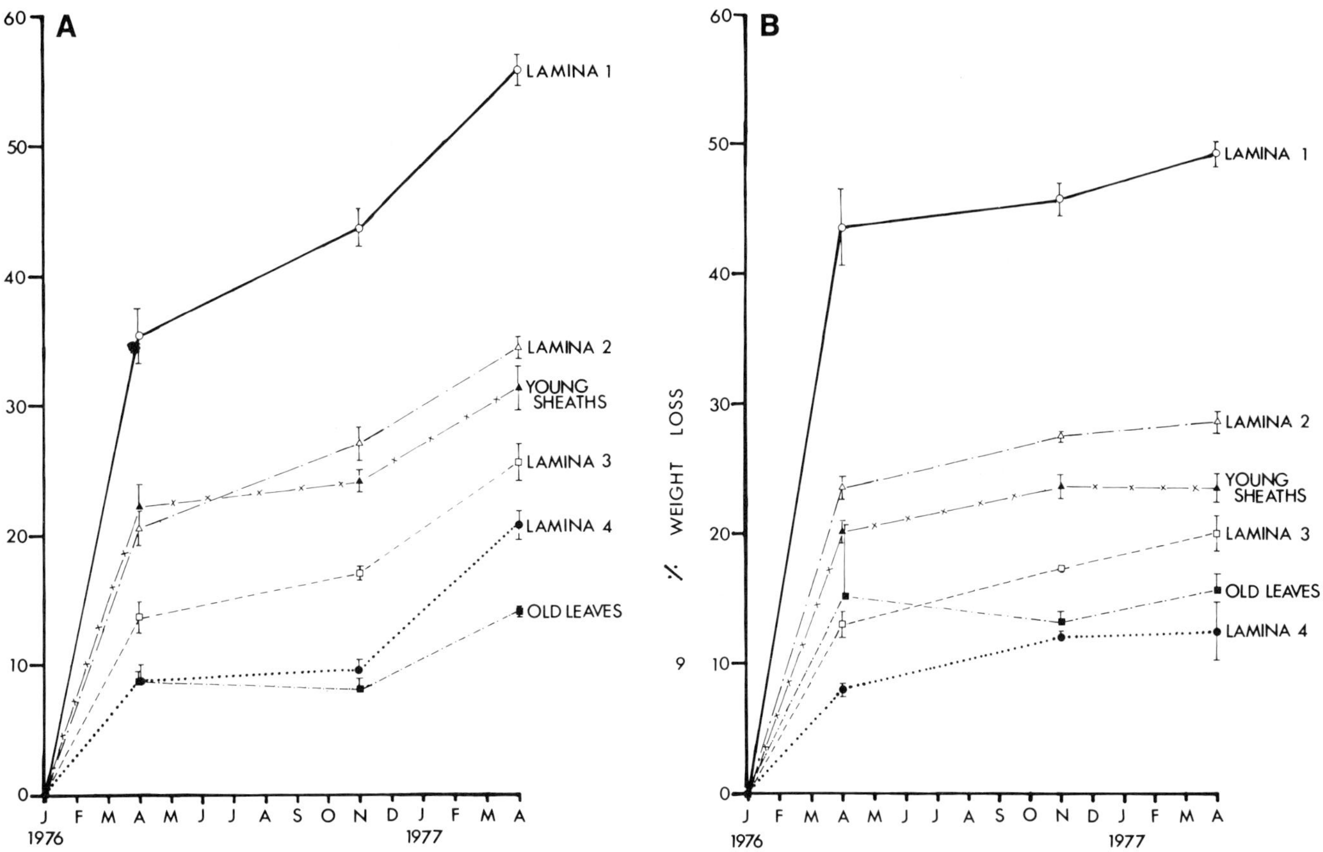

Fig. 5A,B. Weight losses of *Rostkovia magellanica* material in litterbags from a seepage-slope site (*A*) and a basin-bog site (*B*) at South Georgia

4.2 Litterbags

Weight losses during the first 3 months showed a direct correlation with initial nutrient concentration, varying between 43% for the youngest lamina and 8% for old leaves (Fig. 5). Losses of weight are related more closely to litter quality than to site conditions, although the maintenance of stable weights at the basin-bog, after the first 3 months of rapid loss, is likely to be explained by the fact that this site has colder winter temperatures, a drier moss surface and higher acidity. Natural decay is most realistically shown by the loss rate of old leaves, which in both sites was around 14%–15% over the 15-month period.

Phosphorus levels decreased throughout the entire period at the seepage-slope, but at the basin-bog the older laminas and leaves showed a considerable increase in P content (192% for lamina 4) during the first summer. In the second summer, the percentage concentration increased considerably in all plant factions at this site. The absolute content of P also increased during this period; combined sets of litterbags containing the 6 different plant fractions increasing from 6.7 mg P tiller to 13.0 mg P tiller. However, during the same 170-day period in summer, the average content of P in the seepage-slope litterbags decreased from 15.3 mg tiller to 10.0 mg tiller (Figs. 5 and 6).

Nitrogen levels decreased continuously at the basin-bog, but at the seepage-slope the percentage concentration (Fig. 6) increased considerably during the final summer. This pattern was also evident with absolute content of N in the litterbags: increasing from 146.8 to 168.5 mg tiller at the seepage-slope and decreasing from 155.4 mg to 110.0 mg tiller at the basin-bog (Figs. 5 and 6).

Potassium rapidly decreased to a stable level of around 0.1% which appears to be little affected by the origin of the litterbag material. Calcium and Fe increased with leaf and bag age, but markedly less so at the basin-bog. Magnesium concentration increased with leaf age, but levels in litterbags changed relatively little during the study period. However, contents of Mg decreased by 30% at the seepage-slope and 10% at the basin-bog sites. Sodium levels at both sites decreased rapidly during the first 3 months of placement and then remained comparatively stable (Figs. 5 and 6).

4.3 Differences Between Litterbag and In Situ Decomposition

Since approximately two leaves are produced per annum at both sites, it can be assumed that laminas 4 and 5 have undergone a period of in situ decomposition similar to the 15 months span of the litterbag experiment for laminas 1 and 2. Comparisons of the nutrient concentrations of these plant fractions (Table 5) indicates that litterbag decomposi-

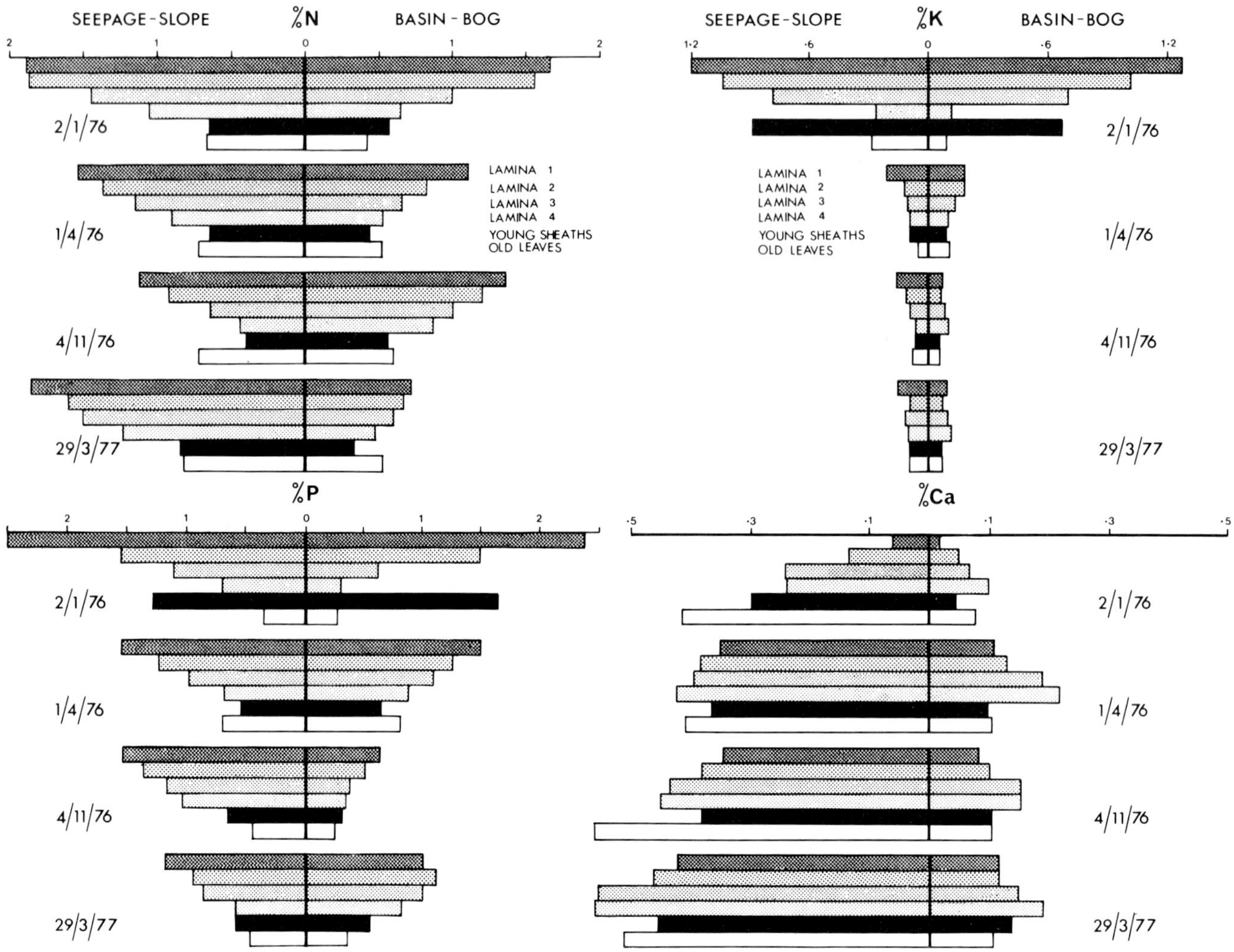

Fig. 6. Changes in nutrient concentration of *Rostkovia magellanica* material in litterbags at a seepage-slope site and a basin-bog site at South Georgia

Table 5. Percentage concentration of nutrients in the two youngest laminas after 15 months of exposure in litterbags and approximately the same period of in situ senescence at two study sites on South Georgia

	Seepage-Slope				Basin-Bog			
	Last litterbag harvest		In situ tillers		Last litterbag harvest		In situ tillers	
	Lamina 1	Lamina 2	Lamina 4 (Old lamina 1)	Lamina 5 (Old lamina 2)	Lamina 1	Lamina 2	Lamina 4 (Old lamina 1)	Lamina 5 (Old lamina 2)
N	1.860	1.611	0.994	0.857	0.723	0.673	0.750	0.684
P	0.119	0.096	0.069	0.064	0.102	0.112	0.036	0.050
K	0.155	0.096	0.238	0.113	0.087	0.064	0.169	0.092
Ca	0.421	0.463	0.259	0.381	0.113	0.115	0.101	0.175
Mg	0.066	0.071	0.093	0.070	0.102	0.115	0.163	0.140
Fe	0.192	0.173	0.021	0.029	0.118	0.102	0.027	0.041

tion causes higher relative concentrations of N (+87%), P (+62%), Ca (+38%) and Fe (+628%) and lower concentrations of K (−30%), and Mg (−15%). The pattern is similar at the basin-bog, with the exception that P rather than N is retained within litterbags: N (−3%), P (+149%), Ca (−17%), Fe (+224%), K (−62%) and Mg (−29%).

5 Discussion

The chemical composition of South Georgian vegetation is described in Walton and Smith (1979) and Pratt and Smith (1982). In higher plants generally, these workers found a decrease in N, P and K through the summer, matched by an increase in Na, Ca and Mg. Leaves of *Rostkovia magellanica,* on the other hand, did not display a seasonal trend in the concentration of any element. Walton and Smith (1979) quote higher N, P and K, and lower Ca than do Pratt and Smith (1982) and this may be due to the fact that the former analyzed "living leaves" (and sheaths), wherears the latter selected leaves which may have been up to 50% dead. This demonstrates the difficulty in deriving a meaningful average chemical content for a winter-green species such as *Rostkovia,* with widely divergent contents in distinct plant parts.

Considering the above *proviso,* there is no evidence that the chemical composition of *Rostkovia* at South Georgia is substantially different to that of other rushes and sedges in tundra bogs (Tamm 1954; Chapin et al. 1975; Smith 1977; Malmer and Nihlgard 1980). The only anomalies are the high concentrations of Na, Mg and Fe in the South Georgia bogs, and the first two of these are adequately explained by the influence of sea-spray.

Although it is often the case that the chemical composition of mire waters has little effect on the chemical composition within plant species (Malmer 1958), it seems in this case that the higher levels of extractable P, K, and Ca observed at the seepage-slope are reflected by higher concentrations of these elements in the tillers. The higher content of Mg in tillers from the basin bog may be in compensation for the low levels of Ca at this site. The higher concentration of Fe and Mg in tillers from the basin-bog (pH 4.72), compared with the seepage-slope (pH 5.97), confirms the observation of Malmer (1958) that these elements are preferentially absorbed on acid sites.

The chemical differences reported here between plant parts confirms those of other studies (Williams et al. 1977; Malmer and Nihlgard 1980). The ease of removal of elements in ageing leaves is in the order K > P > N > Mg, with Fe, Ca and Mn being accumulated in old tissues. Whilst N, P, and K are largely concentrated in young tissues, it is interesting that young laminas, young sheaths and buds should display such different proportions of the 3 elements (Table 5). However, it is understandable that K concentrations should be lower in below-ground buds, because this element is particularly susceptible to leaching, and it is normal for P to be concentrated in the active meristems of buds.

Table 3 apparently shows that the mobile nutrients are removed more rapidly from the basin-bog, and this may mean that internal translocation and conservation of nutrients is relatively great at this site. It is indeed generally the case that internal conservation of nutrients is greater at infertile sites (Goodman and Perkins 1959), but the basin-bog does not conform entirely to this thesis because of the comparatively high P concentration in the old-leaf and lamina 5 fractions. Certainly the level of extractable P at this site is especially low (Table 1) and it may be that old leaves, which decompose very slowly in situ and can remain attached to the tiller for up to 8 yr, are being used as a store of P.

The increase in content of P during the second summer in all plant fractions collected from the basin-bog offers substantiation to the observation that this element appears to be accumulated in old tissue. Yet, in this case, there is no storage function apparent. It is uncertain why such an increase should be observed at the one site and not the other, but a similar increase in P has been observed in deciduous leaf litter (Gosz et al. 1973), and during the first phase of decomposition in pine needles (Staff and Berg 1982) and blanket-bog vegetation (Heal et al. 1978). It is generally attributed to an increasing population of decomposer microorganisms.

Nitrogen concentration increased at the seepage-slope in all bags collected at the end of the second summer. An initial increase in N, followed by a gradual decrease, can be a common occurrence in litterbags (Swift et al. 1979). It is assumed to be associated with an increased biomass of fungi and bacteria, which have a C:N ratio of 4–9:1 as opposed to around 90:1 in plant tissues (Buckman and Brady 1960). This explanation conforms with considerable evidence of fungal activity in the last collection of bags from this site, where up to 20 perithecia of *Microthyrium fuegianum* per 10 mm of lamina were observed in the bags containing younger laminas. No such infestation was evident at the basin-bog, presumably because of the much drier conditions, and no increase in N concentration was observed at this site.

Considerable losses by leaching are implicated in the very rapid initial decrease of most elements in the litterbags. There is a suspicion that freeze-drying, even if carried out quickly, may cause the subsequent leaching of an unusually high amount of nutrient (Timmons et al. 1970). Further answers will be provided when a third litterbags series, which did not use freeze-drying, has been analyzed.

This paper presents preliminary results from two alternative methods of studying the chemical changes associated with senescence and decomposition at contrasting sites. At both sites, bags containing leaves of different ages remain chemically distinct for at least 15 months, demonstrating that initial nutrient concentration is more important than site condition in determining the pattern of decomposition. The considerable differences in nutrient concentration between leaves which have aged in situ and those which were excised prematurely and decomposed within litterbags are reflections of the complex interactions between leaching, translocation and decomposition. Questions attend the balance between physical and biological decomposition processes. At the basin-bog, for example, there is little difference in N concentration after similar periods of in situ or litterbag decomposition, whereas decomposing micro-organisms appear to have increased N levels at the eutrophic seepage-slope site. At both sites, P is lost much more rapidly whilst leaves remain attached to the tiller, presumably because of translocation. Potassium is lost less rapidly, however, from in situ leaves, reflecting the fact that it is more likely to be removed by leaching than by translocation. Iron is accumulated rapidly in litterbag material of all ages but has a low concentration in in situ leaves. These patterns of chemical change reaffirm the complexities of decomposition processes, and emphasize the need to conduct microbiological and radiochemical tracer studies in conjunction with standard investigations of weight loss and nutrient content.

Acknowledgements. This work was carried out while I was employed by the British Antarctic Survey, and I am particularly grateful to Dr. R.I.L. Smith for encouragement and for commenting on this manuscript.

References

Allen SE; Grimshaw HM; Parkinson JA; Quarmby C (1974) Chemical analysis of ecological materials. Blackwell, Oxford 563 pp

Buckman HO; Brady NC (1960) The nature of properties of soils. Macmillan, New York, 653 pp

Bocock KL; Gilbert OJW (1957) The disappearance of leaf litter under different woodland conditions. Plant Soil 9:179–185

Callaghan TV (1980) Age-related patterns of nutrient allocation in *Lycopodium annotinum* from Swedish Lapland. Oikos 35:373–386

Chapin FS; van Cleve K; Tieszen LL (1975) Seasonal nutrient dynamics of tundra vegetation at Barrow, Alaska. Arct Alp Res 7:209–226

Coulson JC; Butterfield J (1978) An investigation of the biotic factors determining the rates of plant decomposition on blanket bog. J Ecol 66:631–650

Crooke WM; Simpson WC (1971) Determination of ammonium in Keldahl digests of crops by an automated procedure. J Sci Food Agric 22:9–10

Goodman GT; Perkins DF (1959) Mineral uptake and retention in cottongrass (*Eriophorum vaginatum* L.). Nature 184:467–468

Gosz JR; Likens GE; Borman FH (1973) Nutrient release from decomposition leaf and branch litter in the Hubbard Brook forest, New Hampshire. Ecol Monogr 47:173–191

Heal OW; Latter PM; Howson G (1978) A study of the rates of decomposition of organic matter. In: Heal OW, Perkins DF (eds) Production ecology of British moors and montane grasslands. Springer, Berlin Heidelberg New York, pp 113–135

Malmer M (1958) Chemical composition of mire plants and peat. Bot Not 11:274–288

Malmer N; Nihlgard B (1980) Supply and transport of mineral nutrients. Ecol Bull (Stockholm) 30:63–95

Murphy J; Riley JP (1962) A modified soil solution method for the determination of phosphate in waters. An Chim Acta 27:31–36

Pratt RM; Smith RIL (1982) Seasonal trends in chemical composition of reindeer forage plants on South Georgia. Polar Biol 1: 13–32

Smith RIL (1981) Types of peat and peat-forming vegetation on South Georgia. Br Antarct Surv Bull 53:119–139

Smith RIL; Walton DWH (1975) South Georgia, Subantarctic. Ecol Bull (Stockholm) 20:399–423

Smith VR (1977) The chemical composition of Marion Island soils, plants and vegetation. S Afr J Antarct Res 7:28–39

Staff H; Berg (1982) Accumulation and release of plant nutrients in decomposing Scots pine needle litter. Long term decomposition in a Scots pine forest II. Can J Bot 60:1561–1568

Swift MJ; Heal OW; Anderson JM (1979) Decomposition in terrestrial ecosystems. Blackwell, Oxford, 372 pp

Tamm CO (1954) Some observations on the nutrient turn-over in a bog community dominated by *Eriophorum vaginatum L.* Oikos 5:189–194

Timmons DR; Holt RF; Latterell JJ (1970) Leaching of crop residues as a source of nutrients in surface runoff water. Water Resour Res 6:1367–1375

Walton DWH; Smith RIL (1979) The chemical composition of South Georgian vegetation. Br Antarct Surv Bull 49:117–135

Williams PA; Nes P; O'Connor KF (1977) Macro-element pools and fluxes in tall-tussock (*Chinochloa*) grasslands, Canterbury, New Zealand. NZJ Bot 15:443–476

Growth and Production of *Poa flabellata* in Relation to Soil Nutrient Status and Exposure at South Georgia

R.I. LEWIS SMITH[1]

Summary. Tall tussock grass, *Poa flabellata*, is the dominant and most widespread phanerogam on South Georgia. Development of tussock grasslands and luxuriance of individual plants are greatly enhanced in areas influenced by seals and seabirds. Such stands are easily distinguished from nonbiotically influenced ones by the large stature of the plants, because of increased annual productivity of aerial parts, and by the deep green colour of the leaves resulting from increased chlorophyll content and chloroplast size, both factors being associated with nitrogen enrichment. The wide ecological amplitude of the grass in relation to edaphic, moisture and microclimatic factors, its ability to tolerate high nutrient levels, deposition of wind-blown sea-spray and disturbance by seals, the high energy reserves, and its rapid growth, all contribute to its success in this harsh environment. The grass shows a preference, in terms of growth and productivity, for an organic soil with a high moisture content and nutrient status, factors which prevail in the biotically influenced coastal stands of dense luxuriant tussock grass. Tussock is capable of comparable growth and production in both sheltered and exposed habitats so long as nutrients and water are not limiting, although there are important differences in growth-forms and tiller-production. High growth rates and productivity are also maintained as plant density increases, although tiller production and growth-form vary.

1 Introduction

Poa flabellata (Lam.) Rasp. [= *Parodiochloa flabellata* (Lam.) CE Hubbard (see Groves 1981)] is a large winter-green tussock-forming grass occurring on Tierra del Fuego, Islas Diego Ramirez, the Falkland Islands, Gough Island and the Sub-Antarctic island of South Georgia which represents the climatic limit of its geographical range. On South Georgia it is the most abundant and widespread vascular plant, forming extensive almost monospecific stands on coastal plains, raised beaches, hillsides and scree slopes from high-water mark to an altitude of ca. 250 m (Greene 1964; Smith and Walton 1975a). It favours organic soils and is highly tolerant of sea-spray deposition and biotic disturbance. Its luxuriance and productivity are enhanced wherever there are breeding populations of Elephant and Fur seals, penguins, albatrosses and various petrels which enrich the substratum. Here, the grass may reach a height of 2 m, although in the Falkland Islands it can exceed 3 m. Gunn (1976) found that mature plants have an above-ground biomass of up to 16 kg m^{-2} (live above ground) + 10 kg m^{-2} (standing dead) + 10 kg m^{-2} (total roots). Net dry production may be as high as 10 kg m^{-2} yr^{-1} (above ground) and 2 kg m^{-2} yr^{-1} (below ground). However, these exceptional data refer to individual plants of about 1 m^{2} and not to the grassland community as a whole; a subjective estimate for a typical dense tussock grassland may reduce values to about 25% (Smith 1984).

2 Material and Methods

Field trials were conducted at King Edward Cove, Cumberland East Bay, South Georgia, to determine the effects of natural soil type, nutrient status and exposure on the growth and dry weight production of *P. flabellata* tillers and seedlings during summer. Matched tillers, about 120–150 mm tall and with four or five visible leaves, derived from the same clones, and two-leaved seedlings were planted in perforated 100 or 150 mm diameter polythene pots containing the soils or directly into the soil in 1 m^{2} cleared plots amongst the natural vegetation. Most trials commenced about 20 December and the plants were harvested after about 3–3.5 months.

3 Effect of Soil Type and Soil Chemistry on the Growth of *Poa flabellata* Tillers

3.1 Experimental Procedure

Five-leaf tillers (n = 20 plants per treatment) were planted in: (1) the soil in situ in six different ecosystems (Elephant Seal, *Mirounga leonina,* wallow peat amongst tussock grass (W), *Rostkovia magellanica-Tortula robusta* seepage slope mire peat (M), *Polytrichum alpestre* moss bank peat (B), *Festuca contracta* grassland loam (G), low altitude (70 m a.s.l.) sparsely vegetated exposed fellfield (Fl), and high altitude (300 m a.s.l.) sparsely vegetated exposed fellfield (Fh)) which were watered only by rainfall; (2) the same soils in 150 mm diameter pots (4 tillers per pot) in a

1 British Antarctic Survey, NERC, High Cross, Madingley Road, Cambridge CB3 OET, Great Britain

Antarctic Nutrient Cycles and Food Webs
(ed. by W.R. Siegfried, P.R. Condy, and R.M. Laws)

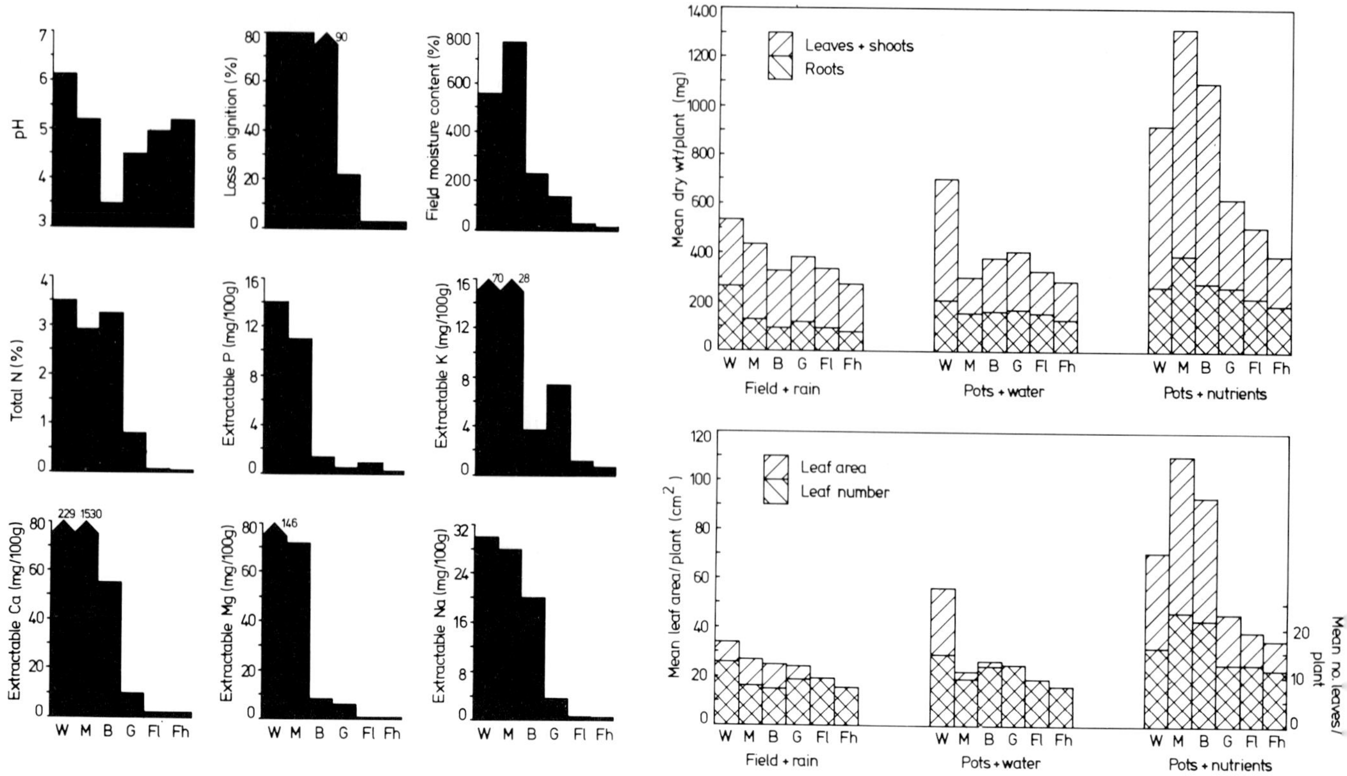

Fig. 1. Early-summer chemical and physical characteristics of soils from six ecosystem types, and growth of *Poa flabellata* tillers in the same untreated soil types in situ and in pots with two treatments at a sheltered site

sheltered site amongst tussock grass at 5 m a.s.l., and watered every 2–3 days with rainwater; and (4 tillers per pot) in a sheltered site amongst tussock grass at 5 m a.s.l., and watered every 2–3 days with rainwater; and (3), as (2) but watered with a non-limiting nutrient medium (modified Hoagland's solution, see Lewis and Greene 1970). The concentrations of nutrients in the soils at the commencement of the trial (22 December) were determined (Fig. 1).

3.2 Results

The results (Fig. 1) after 90 days show:

1. Growth and production are greatly enhanced by the addition of nutrients. This is primarily because of an increase in leaf area, although in the mire and moss bank peats there was also a significant increase in leaf number.

2. Growth and production are slightly enhanced in most soils by the regular addition of rainwater, whereas natural irregular precipitation creates periods of desiccation which cause physiological stress and inhibit growth.

3. Production is generally higher in the organic soils than in the mineral soils but in the field plots this difference is only slight. The nutrient-rich wallow peat treated with rainwater may release absorbed nutrients to a greater extent than the other soils. The higher productivity in the mire and moss bank soils treated with nutrients may be due to the high retention capacity of the decaying mosses forming these peats; the wallow soil is composed mainly of tussock grass peat.

4. The slightly greater growth in the field wallow plot may be a response to the high concentrations of nutrients. The level of ammonium nitrogen is also much higher in this soil than in others. The mineral soils are deficient in all nutrients; this is reflected in the inferior growth and production in these soils.

5. Production in the field plots appears to follow most closely the trend in potassium concentrations. However, it is likely that growth is affected by a combination of nutrients rather than primarily by one.

4 Effect of Soil Type, Nutrient Status and Exposure on the Growth of *Poa flabellata* Tillers

4.1 Experimental Procedure

Four-leaf tillers (n = 20 plants per treatment) were planted in 150 mm diameter pots (4 tillers per pot) containing an organic tussock wallow peat and a mineral fellfield soil at: a sheltered tussock grass site at ca. 35 m a.s.l. (Sh), an intermediate *Festuca* grassland site at ca. 50 m (Int), and an expo-

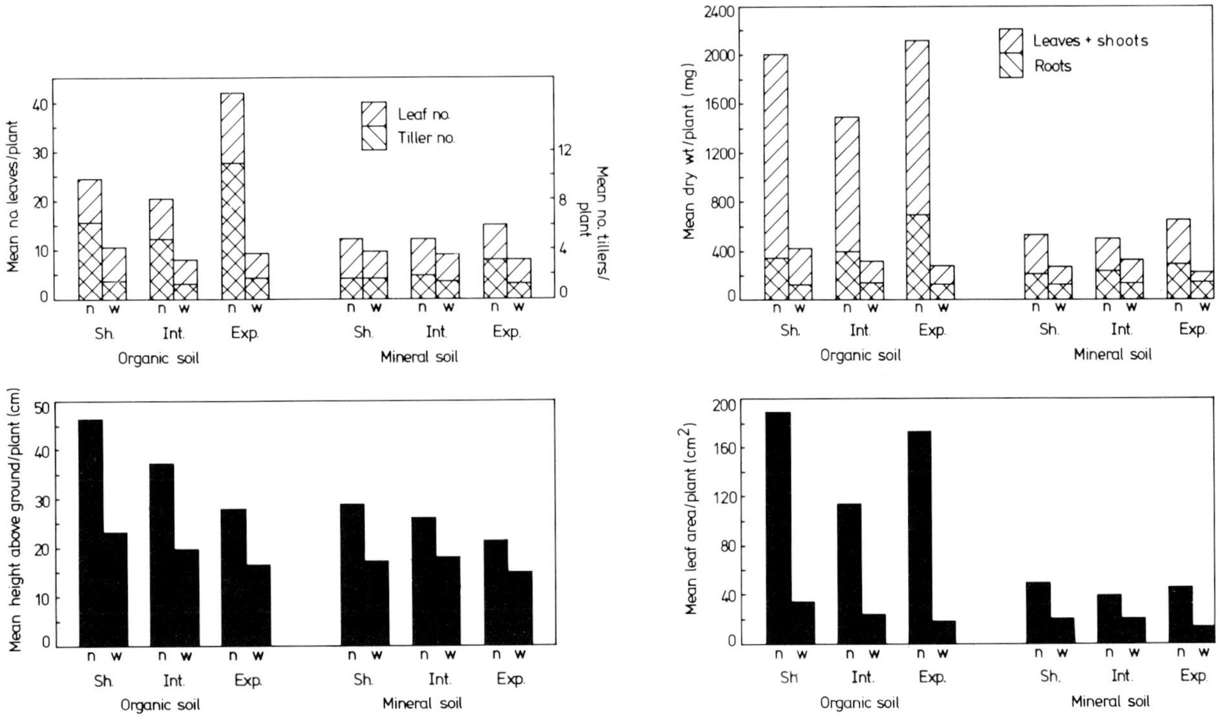

Fig. 2. Growth of *Poa flabellata* tillers in pots in two soil types with two treatments in three ecosystems of varying exposure

sed fellfield site at ca. 85 m (Exp.). Replicate sets of pots at each site were treated with oligotrophic stream water (w) and a non-limiting nutrient medium (n).

4.2 Results

The results (Fig. 2) after 100 days show:

1. Growth and production are greatly enhanced by the regular addition of nutrients, particularly in organic soil, irrespective of exposure.

2. Growth and production do not differ significantly either between soil type or exposure factor when treated with stream water.

3. Plant stature decreases with increasing exposure but production and leaf area increase considerably at the most exposed site, due to greater development of a shorter and denser growth form producing a greater number of daughter tillers and leaves.

5 Effect of Soil Type and Nutrient Status on the Growth of *Poa flabellata* Seedings in Different Ecosystems

5.1 Experimental Procedure

Two-leaf, single-root seedlings (n = 50 plants per treatment) were planted in 100 mm diameter pots (5 seedlings per pot) containing an organic tussock wallow peat or a mineral fellfield soil at: a sheltered tussock grass site at ca. 35 m a.s.l. (Poa), a moderately exposed *Festuca* grassland at ca. 50 m (Fest), a relatively sheltered permanently wet *Rostkovia* mire at ca. 60 m (Rost), and an exposed fellfield site at ca. 85 m (Fe l). Replicate sets of pots at each site were treated with oligotrophic stream water (w) and a non-limiting nutrient medium (n). Some additional plants were grown in pots containing *Festuca* grassland loam-like brown soil.

5.2 Results

The results (Figs. 3 and 4) after 80 days show:

1. Growth and production are greatly enhanced by the regular addition of nutrients, particularly in the organic soil, irrespective of exposure or the natural community.

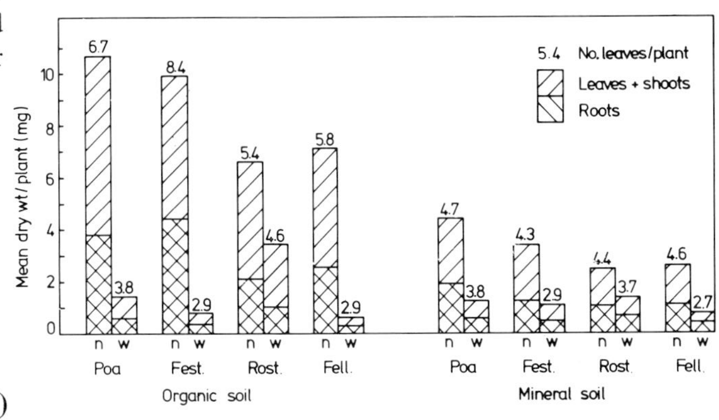

Fig. 3. Growth of *Poa flabellata* seedlings in pots in two soil types with two treatments in four ecosystems of varying exposure

Fig. 4. Diazo prints of largest *Poa flabellata* seedlings after 80 days in pots in three soil types with two treatments at three sites of varying exposure. *Upper* and *lower* figures are above and below ground dry wt. (mg), respectively. *Top row Poa* site; *centre row Festuca* site; *bottom row* Fellfield site. *Left-hand column* organic soil; *centre column* loam; *right-hand column* mineral soil. *Left-hand plant of each pair* nutrient treated; *right-hand plant of each pair* water treated

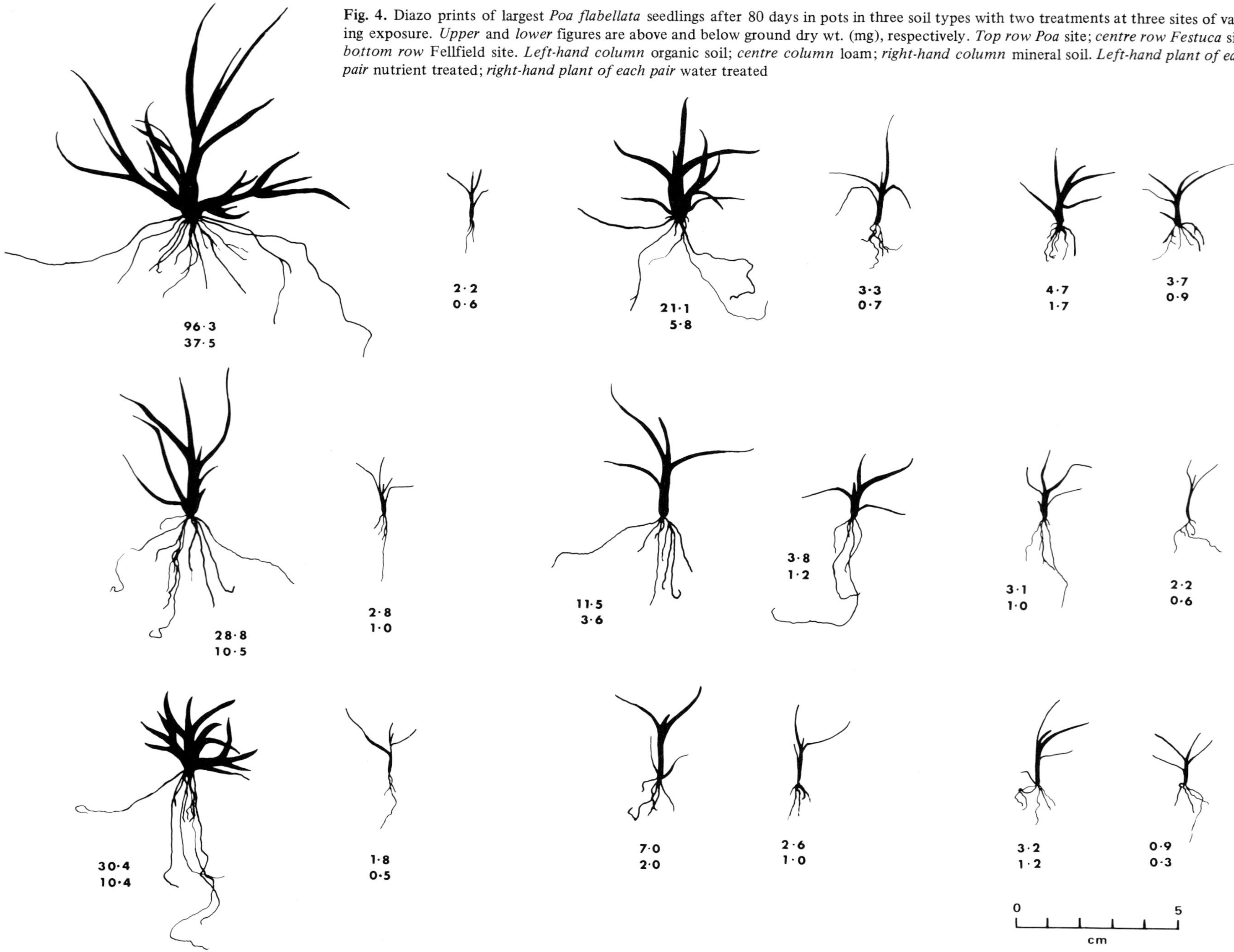

Table 1. Mean relative growth rate (g g^{-1} wk^{-1}) for *Poa flabellata* seedlings after 80 days

Soil and treatment		Site			
		Poa	Festuca	Rostkovia	Fellfield
Peat	water	0.08	0.03	0.15	0.01
	nutrients	0.26	0.25	0.22	0.21
Mineral	water	0.04	0.05	0.08	0.03
	nutrients	0.20	0.15	0.13	0.13
Loam	water	0.07	–	–	–
	nutrients	0.18	–	–	–

2. Growth and production do not differ significantly between soil type or exposure factor when treated with water, except for pots in the mire where the soil received a continual supply of eutrophic ground water rich in NH_4-N, Ca, Mg, K and Fe.

3. The slightly higher growth and production values at the grassland and fellfield sites than at the mire site when treated with nutrients is caused by increased daughter tiller and leaf production at these more exposed sites.

4. The above ground: below ground dry weight ratio is about 3:1 irrespective of soil type, treatment or exposure. Relative growth rates (RGRs) for the whole plants (Table 1) were similar in peat at each site, when treated with nutrients; RGRs for nutrient-treated mineral soil were rather less, whereas those for both soil types treated only with water were considerably lower, with the exception of plants grown in peat at the wet site. Values were consistently lowest for plants grown at the exposed fellfield site. However, seedlings grown in peat at the margin of an Elephant Seal wallow had RGRs of 0.42 g g^{-1} wk^{-1}, indicating the highly favourable growing conditions afforded by this natural medium although survival is largely dependent on the degree of perturbation by the seals (Smith 1971). Higher RGRs may be expected over shorter periods of favourable weather within the duration of the experiment, particularly around mid-summer, as has been shown for Oats, *Avena sativa,* grown as phytometers on South Georgia (Smith and Walton 1975b). RGRs for Oats grown in untreated *Festuca* soil were depressed in relation to those for plants grown in nutrient-treated vermiculite, indicating edaphic and nutrient limitations (Walton and Smith 1976). Comparable RGRs were obtained for *Phleum alpinum* tillers grown in vermiculite at sites of varying exposure (Callaghan and Lewis 1971), but tillers of *Deschampsia antarctica* and *Festuca contracta* grown in the same medium had slightly higher growth rates, although they decreased later in the summer (Smith and Stephenson 1975).

6 Growth of *Poa flabellata* Seedlings and Tillers over Two Seasons

6.1 Experimental Procedure

Two-leaf seedlings (three per pot) and four leaf tillers (two per pot) were planted in pots (150 mm diameter) of vermiculite at a sheltered low altitude (5 m a.s.l.) tussock grass site, and tillers only were grown similarly at a relatively exposed *Festuca* grassland site at ca. 40 m a.s.l. The pots were treated regularly with a non-limiting nutrient medium and harvests of 30 plants were taken at intervals over a 16-month period.

6.2 Results

The results (means of 30 whole plants per harvest plotted as cumulative fitted curvilinear regressions) show (Fig. 5) that the rate of development during the first summer is slow, possibly as the plants require this period to become established. However, immediately after the disappearance of the winter snow, growth and production increase rapidly. Relative growth rates of seedlings become negative during winter as some leaf tissue dies. The difference between tillers grown at the sheltered and exposed sites increases towards the end of each summer.

Although RGRs are most valid for the exponential phase of growth, they can be used over greater periods to interpret a longer-term growth strategy in a long-lived plant (Table 2). Growth rates are low compared with those for temperate species, 0.36 g g^{-1} wk^{-1} being the maximum determined for seedlings and 0.19 g g^{-1} wk^{-1} for tillers. The mean rates over the first growing season were slightly higher for seedlings of *P. flabellata* than those of *P. foliosa* grown under similar conditions on Macquarie Island (0.31 compared with 0.23 g g^{-1} wk^{-1} in nutrient treated vermiculite), but tiller growth appeared almost identical (0.10 and 0.13 g

Table 2. Inter-harvest and long-term production (Pn g wk^{-1}) and relative growth rate (Rw g g^{-1} wk^{-1}) values for *Poa flabellata* seedlings and tillers (data from Gunn 1976)

	Seedlings		Tillers			
Weeks after planting	Sheltered site Pn	Rw	Sheltered site Pn	Rw	Exposed site Pn	Rw
5	0.0003	0.36	0.06	0.13	0.04	0.10
9	0.001	0.34	0.17	0.19	0.11	0.15
13	0.003	0.21	0.12	0.29	0.09	0.09
17	–	–	0.27	0.12	0.34	0.18
24	–	–	0.63	0.02	–	–
33	–0.0002	–0.01	–	–	–	–
41	–	–	–	–	–0.01	–0.002
43	0.01	0.18	0.006	0.001	–	–
51	–	–	0.10	0.03	0.24	0.06
54	0.26	0.29	–	–	–	–
59	0.66	0.16	–	–	–	–
61	–	–	1.12	0.13	0.30	0.05
65	–	–	0.79	0.05	0.65	0.07
1st summer	0.001	0.31	0.13	0.10	0.08	0.13
2nd summer	0.25	0.24	0.70	0.08	0.16	0.06
summer–winter–summer	0.11	0.17	0.28	0.06	0.15	0.05

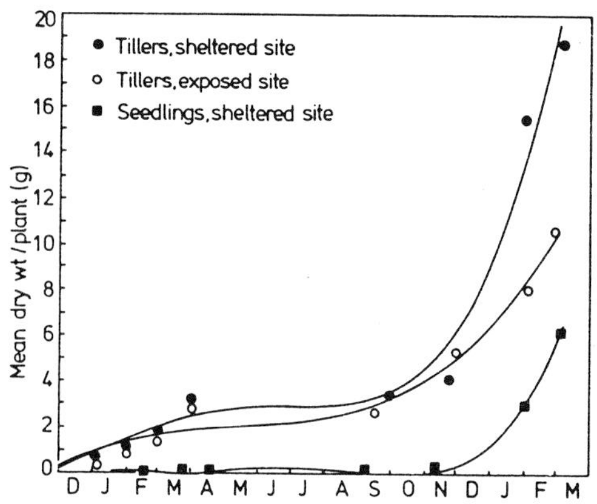

Fig. 5. Growth of *Poa flabellata* seedlings and tillers over two summers. (data from Gunn 1976)

g^{-1} wk^{-1} for *P. flabellata* at sheltered and exposed sites, respectively, compared with 0.13 g g^{-1} wk^{-1} for *P. foliosa* at a sheltered site). There was no significant difference between *P. foliosa* seedlings or tiller growth in nutrient-treated soil and in vermiculite (Jenkin 1972). Dry weight production in *P. flabellata* was greater during the second summer with seedling growth being proportionately more rapid than tiller growth. However, RGRs were lower, particularly in the tillers, in the second summer suggesting a decreasing growth rate with age. Over the 65-week growing period, production of tillers at the sheltered site was twice that at the exposed site although there was negligible or negative growth during winter. Variation in growth rate during a summer is strongly correlated with environmental variables, especially temperature and amount of sunshine.

7 Effect of Plant Density on the Growth of *Poa flabellata* Tillers

7.1 Experimental Procedure

Four-leaf tillers were planted in 150 mm diameter pots of vermiculite treated regularly with a non-limiting medium. Replicate sets of pots contained tillers in densities of 1, 2, 4 (n = 20 plants per density), 8 (n = 32 plants) and 16 (n = 48 plants) per pot. The pots were treated regularly with a non-limiting nutrient medium.

7.2 Results

The results (Fig. 6) after 100 days show a steady decrease in growth and production as density, root interaction and competition for water and nutrients increase; however, plant stature increases producing narrower leaves and less robust plants.

8 Discussion

The trials clearly indicate that above and below ground growth in *Poa flabellata* is greatly enhanced in soils supplemented with nutrients and also in soils with a high organic content. A combination of these two factors, such as exist in biotically disturbed areas, results in the highest production rates. However, growth and development in mineral soils, albeit at a low rate of production, has been

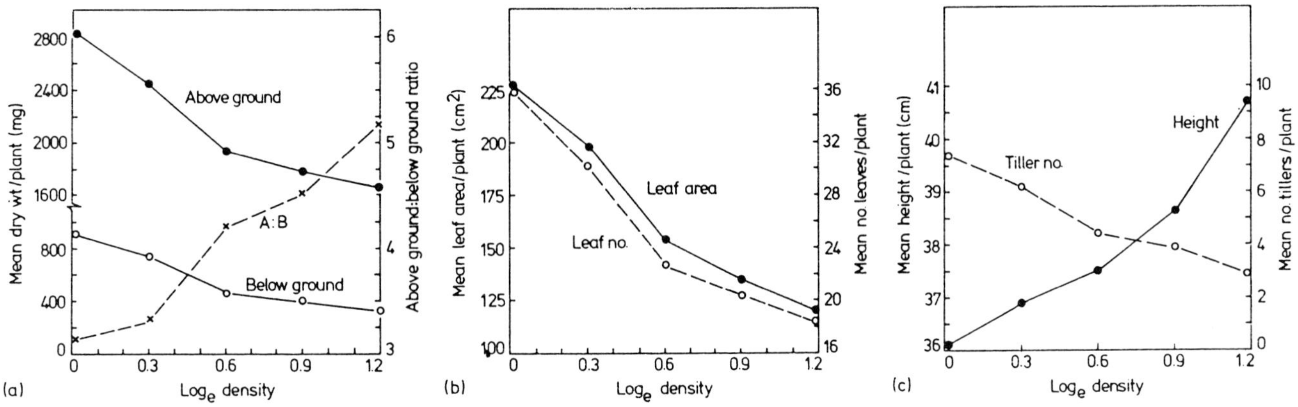

Fig. 6a–c. Growth of *Poa flabellata* tillers in pots of nutrient-treated vermiculite at various plant densities

shown to occur, thereby confirming the grasses' ability to colonize relatively nutrient deficient scree slopes.

Plants growing in sites of natural soil enrichment (i.e., associated with seal and seabird colonies, and following burning of a tussock stand) have increased levels of most nutrients in the foliage (Walton and Smith 1980).

However, in these trials it has not been possible to test the actual nutrient requirements of the grass; it obviously favours a high nitrogen concentration (see Davies 1939), but high levels of other elements (e.g., P, Na) may also be important for maximum production. Jellings et al. (1983) have shown that the leaves of *Poa flabellata* plants associated with seal wallows have larger chloroplasts and fewer mesophyll cells per unit area than plants in non-biotically disturbed sites. In the enriched sites the chloroplast compartment is 18% larger. This is largely a response to the high nitrogen concentration in the substratum which creates the dark green coloration in the luxuriant plants growing in such habitats. Regular treatment with water improves development only slightly compared with irregular irrigation by precipitation alone, except for the wallow peat which may release a greater proportion of its relatively high concentrations of nutrients while in a wet state, and the plants grown at a soligenous mire site where the experimental soils are continually supplied with fairly eutrophic seepage water. Growth in the naturally nutrient-deficient mineral fellfield soil is exceedingly poor but increases considerably when nutrients are added, although dry wt. production is generally several-fold less than in the corresponding organic soils. Water-treated plants grown at an exposed site show depressed growth but when treated with nutrients dry wt. production exceeds that in the much more favourable sheltered site in both organic and mineral soil types. This is due to the response of plants to wind and possibly lower temperatures which reduce plant stature but stimulate the development of an increased number of daughter tillers, thus considerably increasing the leaf number and total dry wt. As with other grasses on South Georgia, with ecological amplitudes ranging from sheltered to exposed habitats, plants become shorter and more compact with a higher density of tillers and leaves in the latter situations (Callaghan 1974). This effect is not so marked in seedlings (see also Smith 1971). However, all these field trials were undertaken during one summer. When extended over two summers, growth rates and production appear to be low during the first year but increase dramatically during the second (Gunn 1976). Examination of relatively large mature plants in the field suggests that the rate of tiller production must be quite rapid over a period of a decade or more, particularly in low altitude biotically enriched organic soils. Its rapid rate of development was first noted by Moody (1845) who reported that "it seems to take three years (from the seed) to arrive at perfection" (i.e., to reach maturity and produce seed).

A relatively high proportion of *Poa flabellata* seed is viable with about 35% of unvernalized seed germinating after 40 days in laboratory tests; a brief (2-day) period at -2°C does not affect this percentage although a longer period (9 days) at this temperature reduces germination to 10%–20% (Gunn 1976). Germination in the field often occurs in great abundance but establishment is possible by only a few plants growing on moist peat. Large numbers of seed germinate amongst turf-forming mosses (especially *Chorisodontium aciphyllum* and *Polytrichum alpestre*) in tussock grassland, but the rapid and frequent fluctuations of temperature and moisture in the apical zone of these mosses causes the death of most of the seedlings before the root system can penetrate to a more permanently moist depth. The ability of *P. flabellata* to readily germinate and become established in moist nutrient-rich organic soil and to tolerate trampling by birds and seals must be highly advantageous in the recolonization of sites of severe erosion. Therefore, as long as there is not complete removal of the organic soil and given adequate protection from the rapidly increasing Antarctic Fur Seal, *Arctocephalus gazella*, population in areas where they have eradicated large expanses of living tussock grass (Bonner 1985), recolonization either by natural dispersal of seed or by artificial reseeding should be relatively rapid and successful. Even regeneration of recently killed grass from the tussock pedestals can proceed quickly when protected from the impact of fauna, as has been shown in exclosures in tussock grassland severely grazed by Reindeer, *Rangifier tarandus* (Smith 1982).

References

Bonner WN (1985) Impact of fur seals on the terrestrial environment at South Georgia. In: Siegfried WR, Condy PR, Laws RM (eds) Antarctic nutrient cycles and food webs (Proceedings of the fourth SCAR symposium on Antarctic biology). Springer, Berlin Heidelberg New York pp

Callaghan TV (1974) Intraspecific variation in *Phleum alpinum* L. with specific reference to polar populations. Arct Alp Res 6:361–401

Callaghan TV; Lewis MC (1971) The growth of *Phleum alpinum* L. in contrasting habitats at a sub-Antarctic station. New Phytol 70:1143–1154

Davies W (1939) The grassland of the Falkland Islands. Government Printer, Stanley, Falkland Islands

Greene SW (1964) The vascular flora of South Georgia. Br Antarct Surv Sci Rep 45:1–58

Groves EW (1981) Vascular plant collections from the Tristan da Cunha group of islands. Bull Br Mus (Nat Hist), Bot series 8: 333–420

Gunn TC (1976) The autecology of *Poa flabellata* (Lam.) Hook. f. Ph.D. thesis, University of Manchester:166 pp (Unpublished)

Jellings AJ, Usher MB, Leech RM (1983) Chloroplast size in tall and short phenotypes of *Poa flabellata* on South Georgia. Br Antarct Surv Bull 59:41–46

Jenkin JF (1972) Studies on plant growth in a sub-Antarctic environment. Ph.D. thesis, University of Melbourne:297 pp + appendices (Unpublished)

Lewis MC; Greene SW (1970) A comparison of plant growth at an Arctic and Antarctic station. In: Holdgate MW (ed) Antarctic ecology. Academic Press, London 1:838–850

Moody RC (1845) Further account of the tussock grass. J Royal Agric Soc 5:50–52

Smith RI Lewis (1971) An outline of the Antarctic programme of the Bipolar Botanical Project. In: Heal OW (ed) International Biological Programme Tundra Biome Working Meeting on Analyses of Ecosystems, Kevo, Finland, September 1970. IBP Tundra Biome Steering Committee and Atlantic Richfield Company, London, pp 50–71

Smith RI Lewis (1982) The impact of introduced reindeer on the vegetation of South Georgia. Comité National français des Recherces Antarctiques 51:531

Smith RI Lewis (1984) Terrestrial Plant Biology of the Sub-Antarctic and Antarctic. In: Laws RM (ed) Antarctic Ecology Academic Press, London 1:61–162

Smith RI Lewis; Stephenson C (1975) Preliminary growth studies on *Festuca contracta.* T. Kirk and *Deschampsia antarctica* Desv. on South Georgia. Br Antarc Surv Bull 41/42:59–75

Smith RI Lewis; Walton DWH (1975a) South Georgia, Subantarctic. In: Rosswall T, Heal OW (ed). Structure and function of tundra ecosystems. Ecol Bull (Stockholm) 20:399–423

Smith RI Lewis; Walton DWH (1975b) A growth analysis technique for assessing habitat severity in tundra regions. Ann of Bot 39: 831–843

Walton DWH; Smith RI Lewis (1976) Some limitations on plant growth and development in tundra regions – an investigation using phytometers. New Phytol 76:501–510

Walton DWH; Smith RI Lewis (1980) Chemical composition of South Georgian plants. Br Antarct Surv Bull 49:117–135

Methanogenesis and the Anaerobic Micro-Biology of a Wet Moss Community at Signy Island

M.R. YARRINGTON[1] and D.D. WYNN-WILLIAMS[2]

Summary. A wet moss-carpet site on Signy Island, South Orkney Islands, has the physico-chemical properties necessary for the existence of an anaerobic micro-flora. It has been shown to possess such a micro-flora, capable of actively metabolizing under the prevailing climatic conditions, where low temperatures and prolonged periods of freezing occur. The biological production of methane is a terminal step in the mineralization of organic carbon by anaerobic micro-organisms, requiring consistently low redox potentials (−330 mV), and the availability of a restricted range of low molecular weight C compounds as substrates. Thus, the methanogenesis measured in situ and in the laboratory peat slurry experiments, is a reliable indicator of anaerobic microbial metabolism. The moss-carpet exhibited redox potentials as low as −400 mV (corrected to pH 7.0) and released an average of 1.24 mg C m^2 d^{-1} as methane during the Antarctic summer of December 1981 to March 1982. Laboratory experiments on homogenized moss-peat samples maintained anaerobically, showed that methane production was influenced by environmental temperature and pH, and that under natural environmental conditions in the moss-carpet, the methanogens were probably metabolizing sub-optimally. Inconclusive results were obtained from experiments designed to test substrate specificity for methanogenesis; however, they suggested that methionine and formate could be at least partially metabolized with the formation of methane.

1 Introduction

The moss carpet study site on Signy Island, South Orkney Islands, is a waterlogged soligenous moss-peat, with a vegetation type commonly found in the maritime Antarctic (Tillbrook 1973). A contrasting semi-ombrogenous moss-turf site was also selected, and both have been subjected to a series of biological investigations in an attempt to describe and quantify the floral, faunal and microbial components to construct an ecosystem model. This quantifies the cycling of organic matter in terms of standing crops and flows of organic C through the systems (Davis 1980, 1981). The model showed that for the wetter moss-carpet site there was a higher rate of decomposition of organic matter than could be accounted for by the components examined initially (Davis 1981). This was in contrast to the drier, aerobic moss-turf site, where the sum of component contributions balanced the measured overall decomposition rate. It was postulated that in the moss-carpet site the aerobic microbial component was insufficient to account for all the microbial decomposition, and that chemical and environmental conditions at this site were suitable for significant anaerobic microbial activity (Wynn-Williams 1980). Some of the preliminary results of the subsequent investigation of this anaerobic microbial component are described here.

The complete mineralization of organic compounds such as cellulose within an anaerobic microbial community leads to the production of carbon dioxide and methane, which are subsequently largely returned to the atmosphere (Fenchel and Jorganssen 1977). The production of methane, by a very specialized group of bacteria, is a terminal process in the anaerobic carbon "food-chain", the immediate precursors of methane being a limited range of low molecular weight organic C compounds produced by other anaerobes. Known substrates utilized as both C and energy sources by methanogenic bacteria include formate, acetate, methylated amines and hydrogen plus carbon dioxide (Anthony 1982). In order for methanogenesis to occur the environment must have a redox potential of less than −330 mV, and all oxygen must be excluded (Anthony 1982). The biological production of methane is therefore a reliable indicator of both the necessary conditions for, and the activity of anaerobic micro-organisms.

The present investigation at the Signy Island moss-carpet study site aimed to demonstrate and quantify C cycling by an anaerobic micro-flora tolerant of low temperature and pH, and capable of generating significant amounts of methane. Such activity would help to quantify discrepancies in the aerobic ecosystem model for decomposition at the site.

2 Site Characteristics

The wet moss-carpet consisted of the three mosses *Calliergidium austrostramineum, Calliergon sarmentosum* and *Drepenocladus uncinatus* with the associated leafy liverwort *Cephaloziella varians.* The top 5–10 mm of the profile con-

1 Marine Science Laboratories, University College of North Wales, Menai Bridge, Anglesey LL59 5EH, Wales, (Formerly: Dept. Botany and Microbiology, University College of Wales, Aberystwyth) Great Britain

2 British Antarctic Survey, NERC, High Cross, Madingley Road, Cambridge CB3 OET, Great Britain

Antarctic Nutrient Cycles and Food Webs
(ed. by W.R. Siegfried, P.R. Condy, and R.M. Laws)

sisted of green moss shoots with intact moss stems usually visible for 30 mm–40 mm. Below the intact stems the peat was more amorphous, and was usually a yellow-brown colour which changed to dark brown on exposure to air.

For 8 months of the year the site is completely frozen, but after the main thaw in December 1981 until the freeze in May 1982 the site was thawed and permanently wet, with open pools of water present due to its location in a shallow depression on the Gourlay Peninsula at Signy Island.

The mean depth of the peat in the study was 130 mm, with a maximum depth of 300 mm. The moss-carpet had a mean pH value of approximately 5.2, and the mean summer temperature at 15 mm depth (for the 3-year period 1972–74) was 2.1°C; the mean monthly maximum and minimum temperatures over the same period being 3.5°C and –9.1°C (Davis 1981). Further details of the micro-climate are given by Walton (1977, 1982).

3 Materials and Methods

3.1 Redox Potential

Redox potentials were measured in the field using platinum wire electrodes (tip diameter 0.376 mm, length 4 mm) constructed as in Jones (1979), in circuit with a calomel reference electrode and a millivolt meter (CD64, Waldon Precision Apparatus, UK). Measurements of redox potentials through the peat profile were made at 5 mm intervals. All readings were corrected to give equivalent values at pH 7.0. Equilibration times of up to 5 min for each reading were commonly necessary.

3.2 Methane Gas Collection and Measurement

Svensson (1980) used closed containers and gas-chromatography to collect and measure the methane released from a Sub-Arctic mire, and a similar method was used at the Signy Island site. One-litre Kilner jars with their bases removed were inserted into the intact moss-carpet to a depth of 50 mm. A total of 16 jars was used in the study area; these were sampled on 13 occasions during the period 22 December 1981 to 18 March 1982. The jars, which were located permanently, were sealed with a lid containing recessed rubber stoppers to allow sampling of the gas phase. At the end of each 24 h period of sealed incubation, samples of the gases in the head spaces of the jars were removed in 1-ml plastic syringes. Mixing of the gases prior to sampling was achieved by repeatedly pumping the syringe; after sampling, the syringes were inserted into rubber stoppers to prevent gas loss before analysis. Three replicate samples were taken from each jar. Gas samples were returned to the laboratory at the Signy Island research station and analyzed on a PU4500 gas-chromatograph (Pye Unicam Ltd., UK) fitted with a flame ionization detector. Peak heights were measured on a 3390A electronic integrator (Hewlett-Packard Ltd., UK). Separation of gases in the sample was achieved on a 1.6 m X 4 mm Pyrex column containing Chromasorb 102 (Phase Separations Ltd., UK) with an OFN carrier gas flow rate of 30 ml min^{-1}, oven temperature 105°C, injector temperature 100°C and detector temperature 150°C. Standard gas mixtures of 100 ppm methane in N_2 (Phase Separations Ltd., UK) were used to calibrate the instrument.

3.3 Experiments with Homogenized Peat Samples

Deep-frozen peat cores collected from the study site and returned frozen to Britain were thawed at 5°C under a $H_2:CO_2:N_2$ (10:10:80) gas mixture in an anaerobic jar. When thawed, they were homogenized in an MSE ATO-MIX blender (MSE Scientific Instruments, UK) under a $H_2:CO_2:N_2$ (10:10:80) gas phase within an anaerobic cabinet (Forma Scientific, USA). Aseptic techniques were used whenever possible. The only addition to the peat homogenate at this stage was sterile distilled water. Ten-ml portions of the resultant peat slurry were distributed aseptically into sterile 30-ml bottles. The bottles were then sealed with recessed rubber stoppers (Astell Hearson, UK).

3.3.1 Effect of Temperature and pH upon Methane Production

A peat slurry was produced by the method described above. 300 ml of the slurry was placed in a sterile beaker under a $H_2:CO_2N_2$ (10:10:80) gas phase. The original pH of the slurry was 5.43. Sufficient sterile 1.0M hydrochloric acid was added to change the pH value to 5.40; 10-ml portions were distributed to three sterile 30-ml bottles as above. More acid was added and the process repeated to generate a series of slurries in the range of pH 4.5–5.4. The procedure was repeated with another 300-ml portion of the original slurry using sterile 1.0M sodium hydroxide to generate a series of slurries in the pH range of 5.5–9.0. The resultant three sets of slurries covering the pH range 4.5–9.0 were incubated at 2°, 11° and 27°C, respectively.

3.3.2 Substrate Utilization by Methanogens

0.1M solutions of sodium acetate, sodium formate, methanol and methionine were made up in 0.2M phosphate buffer solution (pH 7.0) and autoclaved in sealed bottles (121 °C for 15 min). The substrate solutions were added to separate bottles of peat slurry to give a final concentration of 10 mM substrate. Six replicate bottles of slurry were used for each substrate; 2 bottles of each series contained Durham tubes holding 0.5 ml of 5.0 M sodium hydroxide solution, and 4 had no such addition. The bottles were incubated at 25°C. After 114 h incubation a solution of 2-

bromethanosulfonic acid, an inhibitor of methanogenesis, was added to one bottle in each substrate series to give final concentration of 10 mM.

Methane production was measured at various times during the incubation of the amended peat slurries. One-ml samples of the gas phase were withdrawn into plastic syringes and analyzed as above. Results were expressed as volumes of methane per million volumes of the gas phase (v.p.m.).

4 Results and Discussion

When the redox potentials at various depths in the moss-carpet were measured in situ, the values decreased with depth so that the lowest redox potential was found at the base (ca. 150 mm) of the peat profile. Fig. 1 shows some typical redox potential profiles (determined at 5 mm intervals). The zone of rapid change in redox potential, although at different depths in different parts of the site, was a constant feature of such profiles. It was often associated with increased physical resistance to probe insertion. This zone correspond with the transition from intact moss stems to amorphous peat beneath. The lowest redox potential recorded for each peat profile varied over the site, reaching values as low as –400 mV (adjusted to pH 7.0), suitable for methanogenesis.

Methane production rates in situ varied both with the location of the incubation chamber and with sampling date to give a range of values from 0.02–15.2 mg C m^2 d^{-1} as methane. The mean methane production rate for the 16 incubation chambers and the summer season from December 1981 to March 1982 (13 samples taken over 85 days) was 1.24 mg C m^2 d^{-1}. The methane concentrations in the

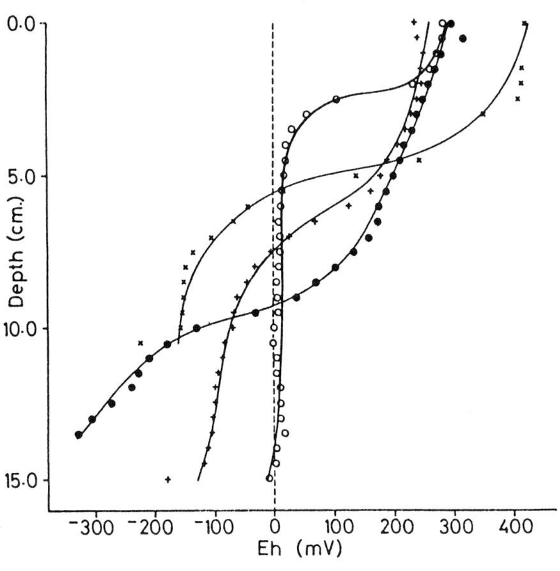

Fig. 1. Examples of typical redox potential profiles recorded from several sampling areas in the moss carpet on Signy Island, ●, o, x and + indicate four distinct profiles

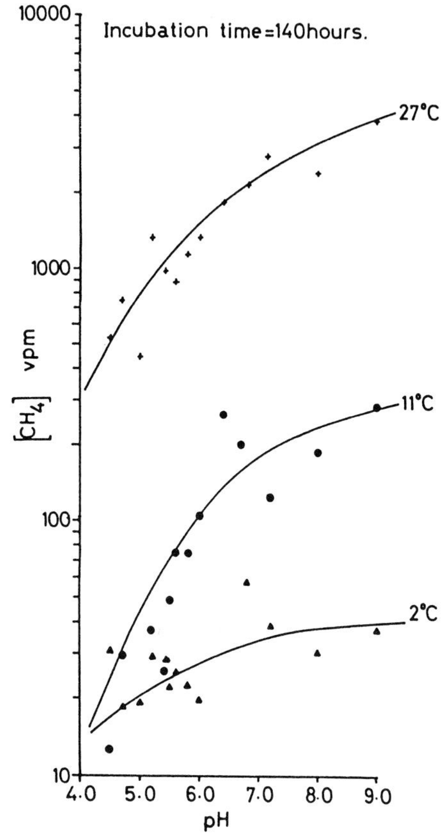

Fig. 2. Methane concentrations (calculated as volumes of methane per million volumes of the gas sample, v.p.m.) found in the gaseous head-space above samples of homogenized moss-peat slurries, maintained anaerobically in sealed containers. The initial pH values of the slurries covered the pH range between 4.5 and 9.0, and replicate series were incubated for 140 h at 2, 11 and 27°C

gas phase of some of the incubation chambers were sometimes seen to increase and then decrease over a 24-h period when samples were taken at 2-h intervals. This was possibly caused by the action of methylotrophic micro-organisms growing aerobically at the moss-peat surface.

The complete inhibition of methanogenesis by the addition of 2-bromoethanesulfonic acid to the anaerobically-maintained peat slurries confirmed the occurrence of biogenic methane production. Further laboratory experiments on such slurries showed that both pH and temperature affected the rate of methane production (Fig. 2). When acid or alkaline additions were used to manipulate the initial pH value of the slurry, in the absence of any additional buffering capacity, it was found that higher pH and temperature values (within the ranges tested) individually and collectively favoured methane production. Examination of the pH of the peat slurries at the start of the experiments and after 430 h incubation, indicated that when the initial pH was > 5.2, acid was produced which caused the pH to fall (see Fig. 3). This was probably due to the production of fatty acids by fermentative micro-organisms involved in anaerobic decomposition of the organic matter. A greater pH change occurred at the higher incubation temperatures used,

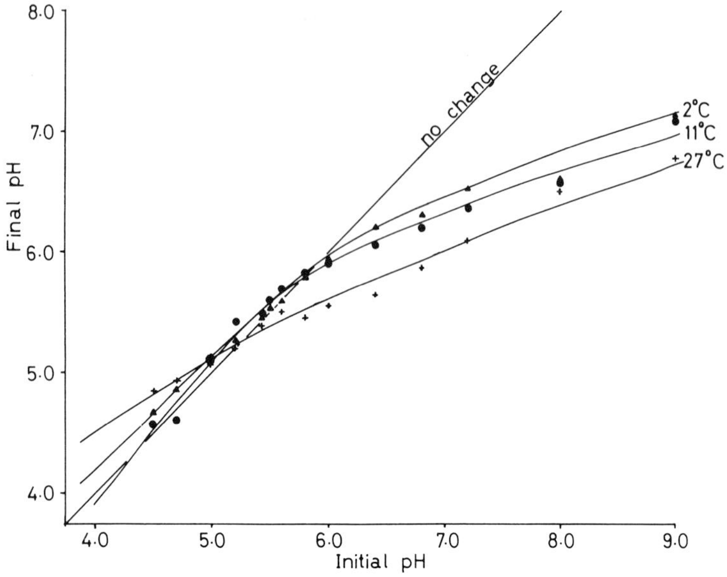

Fig. 3. pH values which developed in sealed anaerobic containers of homogenized peat slurries, which were adjusted to various pH values and then incubated for 430 h. Replicate series were incubated at 2, 11 and 27°C. The base line representing no change in pH during the incubation period is included for comparative purposes

Table 1. Methane concentrations (v.p.m.) found in the gaseous headspace above samples of homogenized moss-peat slurries, maintained anaerobically in sealed containers, amended with carbon substrates

	Methane concentration	
Substrate	+NaOH	−NaOH
Control	4,120	2,740
Acetate	4,470	2,490
Methanol	2,830	1,450
Formate	4,540	3,780
Methionine	10,100	7,100

5.0 M sodium hydroxide (NaOH), used in some cases, reduced the carbon dioxide concentration in the head-space gases from approximately 10.0% to 0.1%. Methane concentrations measured after 431 h incubation

implying increased acid production, and therefore increased general anaerobic activity than at the lower incubation temperatures. A less marked difference was apparent at initial pH values < 5.2, when pH increased over the period of incubation, implying utilization of acids. At a pH of < 5.2, an inhibition of the acid-producing fermentative microorganisms seemed to occur. Again, higher temperatures favoured a larger change in pH.

The amendment of peat slurries with potential carbon-energy sources gave inconclusive results, possibly due to the disruption of not only physical niches, but also syntrophic associations, and nutrient rich cells. However, substrates which directly or indirectly stimulated methanogenesis included formate and, unexpectedly, methionine (Table 1). When 5.0 M sodium hydroxide solution was used, to remove carbon dioxide from the headspace gases, only methionine caused stimulation of methane production above control levels; however, in the absence of sodium hydroxide solution, both formate and methionine stimulated methane production. Addition of (methyl-14C) methionine, 14C formate or 14C carbon dioxide to active methane producing peat slurry did not yield measurable amounts of 14C methane, although labelled carbon dioxide was rapidly detected in headspace gases for both formate and methionine addition (M Yarrington, unpubl.). It seems probable that methionine and formate indirectly stimulated methanogenensis in the peat slurry, and the major substrates for methanogenesis in the moss-carpet have yet to be recognized. This merits further study, since Lovely and Klug (1983) suggested that methylamine and trimethylamine may be important substrates for methanogenesis in eutrophic lake sediments.

Attempts have been made to culture the anaerobic micro-organisms in the moss-carpet which are responsible for the anaerobic decomposition of high molecular weight C sources such as cellulose, giving carbon dioxide and methane as the end products. Methane-producing cultures have been isolated but not characterized. Cellulolytic, fermentative and photosynthetic bacteria have likewise been obtained, although not yet purified and identified.

5 Conclusions

An anaerobic micro-flora capable of significant methanogenesis in wet moss peat at low temperature and pH has been demonstrated at Signy Island. Heterogenous distribution of micro-environments in vivo of sufficiently low E_h compelled the use of peat slurries to show that methanogenesis in field conditions was probably sub-optimal. Preliminary studies in vitro suggested that formate and methionine were amongst carbon-energy sources stimulatory for methanogenesis in this peat.

Acknowledgements. The studies reported here were supported by a Science and Engineering Research Council CASE Research Studentship awarded to M.R. Yarrington, who is also grateful to the British Antarctic Survey for the opportunity to visit the Antarctic, and to the men of the Signy Island base for their encouragement and assistance. We would like to thank Dr. W.C. Block for critical review of the manuscript and Dr. M.E. Rhodes-Roberts for supervision and advice.

References

Anthony C (1982) The biochemistry of methylotrophs. Academic, London

Davis RC (1980) Peat respiration and decomposition in Antarctic terrestrial moss communities. Biol J Linn Soc 14:39–49

Davis RC (1981) Structure and function of two Antarctic terrestrial moss communities. Ecol Monogr 51:125–143

Fenchel TM, Jorgansen BB (1977) Detritus food chains of aquatic ecosystems: the role of bacteria. Adv Microb Ecol 1:1–58

Jones JG (1979) Microbial activity in lake sediments with particular reference to electrode potential gradients. J Gen Microbiol 115: 19–26

Lovely DR, Klug MJ (1983) Methanogenesis from methanol and methylamine and acetogenesis from hydrogen and carbon dioxide in the sediments of a eutrophic lake. Appl Environ Microbiol 45:1310–1315

Svensson BH (1980) Carbon dioxide and methane fluxes from the ombrotrophic parts of a subarctic mire. In: Sonesson M (ed) Ecology of a subarctic mire. Ecol Bull (Stockholm) 30:235–250

Tillbrook PJ (1973) The Signy Island terrestrial reference site: 1. An introduction. Br Antarct Surv Bull 33–34:65–76

Walton DWH (1977) Radiation and soil temperatures 1972–74: Signy Island terrestrial reference sites. Br Antarct Surv Data 1:1–49

Walton DWH (1982) The Signy Island terrestrial reference sites: XV Microclimate monitoring, 1972–74. Br Antarct Surv Bull 55:111–126

Wynn-Williams DD (1980) Seasonal fluctuations in microbial activity in Antarctic moss peat. Biol J Linn Soc 14:11–28

The Influence of Iron and Manganese on Nutrient Cycling in Shallow Freshwater Antarctic Lakes

J.B. GALLAGHER[1]

Summary. Two lakes, mesotrophic Heywood Lake and oligotrophic Sombre Lake, at Signy Island, have been studied over 1 full yr (1980–1981). The development of deep water anoxia in Sombre Lake is accompanied by release, from the sediments, of nitrogen, silicon, phosphorus, iron, and manganese. Iron is released from anoxic sediments in massive amounts, but is quickly returned through oxidation to iron oxides, near the anoxic-oxic boundary and through possible ferrous carbonate precipitation, within the anoxic region. Phosphorus is also quickly returned to sediments, at the oxic-anoxic boundary, which suggests efficient co-precipitation-sorption by the recently oxidized ferrous. Phosphorus is therefore unable to enter the main waterbody. Melt water inflows are the source of P supply to the surface waters. Nitrogen and Si appear to have little association with axides and, therefore, freely pass into the oxic main water-body of the oligotrophic system. There are strong seasonal correlations between the vertical profiles of Fe and P, under the ice cover in Heywood Lake. Nitrogen profiles appear independent of the other determinants, Si, P, Fe, and Mn. Towards the end of the period of bottom-water anoxia, during the spring melt, high concentrations of ferrous, released from the anoxic sediments, have been detected in oxygenated waters. This apparent stability, in the presence of oxygen, may be attributable to a combination of low pH and the possible influence of high concentrations of humic acids from the melt inflows. Sulphide concentrations may become sufficiently high to influence Fe transport significantly, probably through ferrous sulphide precipitation, within the anoxic region.

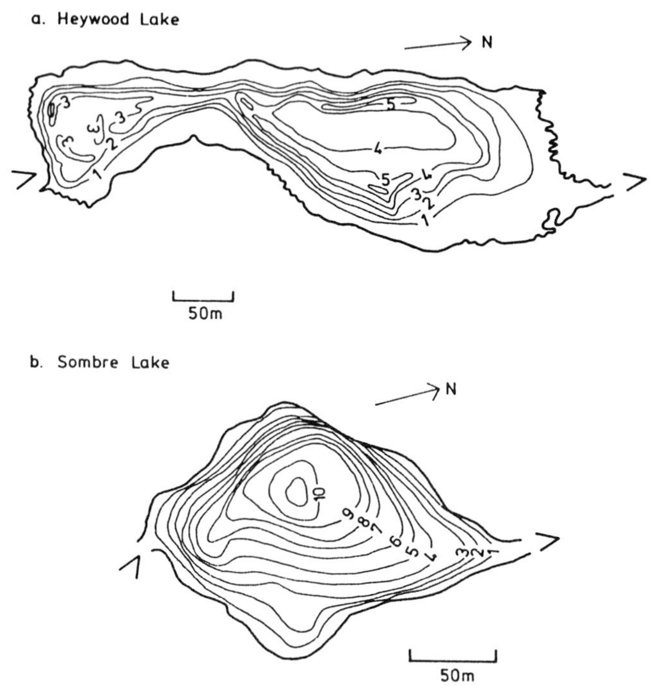

Fig. 1. Bathymetric maps of Heywood and Sombre lakes, with 1 m interval contours. Inflow and outflow streams shown by >

Introduction

The study of redox active elements, such as iron and manganese, facilitates a detailed description of a lake's physico-chemical status and may offer explanations for otherwise obscure patterns of nutrient cycling and nutrient availability (Sholkovitz and Copeland 1982). In this paper I describe how water-column movements of Fe and Mn and their redox changes may control the movements and speciation of nutrients in oligotrophic and eutrophic freshwater Antarctic lakes. The paper is based on work carried out in Heywood and Sombre lakes (Fig. 1) at Signy Island, South Orkney Islands.

1 British Antarctic Survey, NERC, High Cross, Madingley Road, Cambridge CB3 OET, Great Britain

2 Methods

Water was sampled with the exclusion of air, using a device (Fig. 2) modified from that design by Magnuson and Stuntz (1970). A sediment detector was attached to the sampling head (Cunningham and Davison 1980), permitting water samples to be taken from the vicinity of the sediment-water interface without the bottom sediments. Water from pH, ferrous Fe, sulphide, total carbon dioxide, and particulate carbonate was collected in glass stoppered B.O.D. bottles. Samples for total and total filterable P, N, Si, Fe and, Mn were collected in polypropylene bottles, and filtered within a few h of collection. In some instances, an in situ filtration apparatus, in conjunction with the sampling device, was used (Fig. 2). Methods for the analysis of the above determinants are detailed in Gallagher (in press).

Antarctic Nutrient Cycles and Food Webs
(ed. by W.R. Siegfried, P.R. Condy, and R.M. Laws)

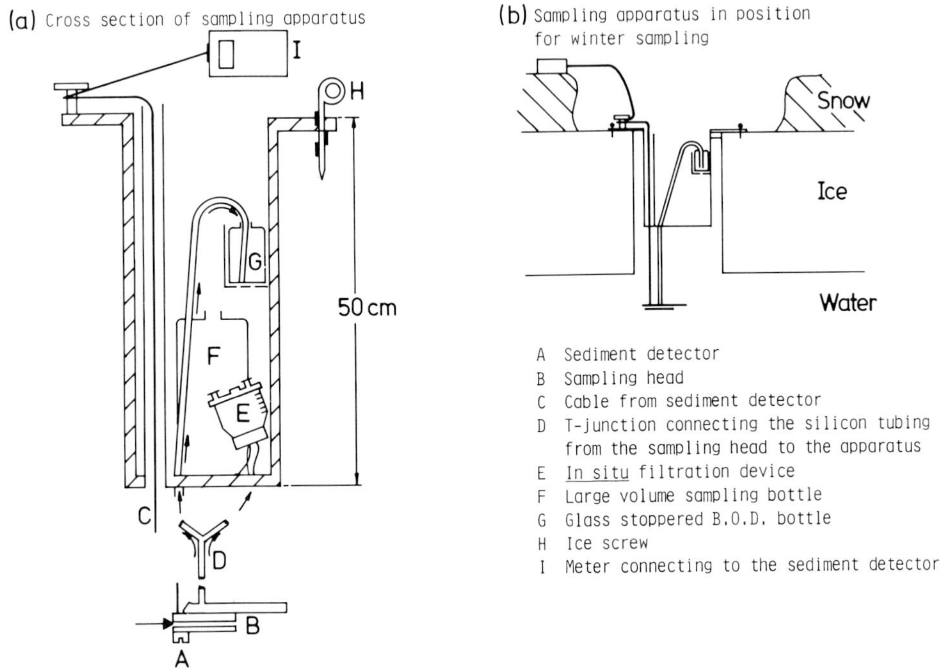

Fig. 2a,b. Apparatus for taking water samples, anoxically, from under ice

3 Results

3.1. Oligotrophic Sombre Lake

The development of deep-water anoxia is accompanied by the release of Si, P, N and Mn (as manganous) from the anoxic sediments, together with massive amounts of Fe (as ferrous) (Fig. 3).

Figures 3 and 4 show that ferrous and P concentrations drop markedly at the anoxic–oxic boundary. This indicates rapid ferrous oxidation to ferric oxides together with sorption co-precipitation of P on to the ferric oxides, with subsequent sedimentation of the ferric oxide-phosphorus "complex" back to the sediments. Thus, the major P contribution to the surface waters is derived from inflows during the spring–melt period. Silicon and N vertical profiles (Fig. 3) indicate that these elements are free to enter the main oxic water–body and, therefore, have little association with Fe. Similarly, Mn vertical profiles (Fig. 3) indicate that Mn is free to enter the main oxic water-body, through its small association with Fe and its greater tolerance to oxidation and precipatation in oxygenated waters (Stumm and Morgan 1981).

Sulphide concentrations are never sufficiently high enough (40 $\mu g\ l^{-1}$) to influence Fe movements within

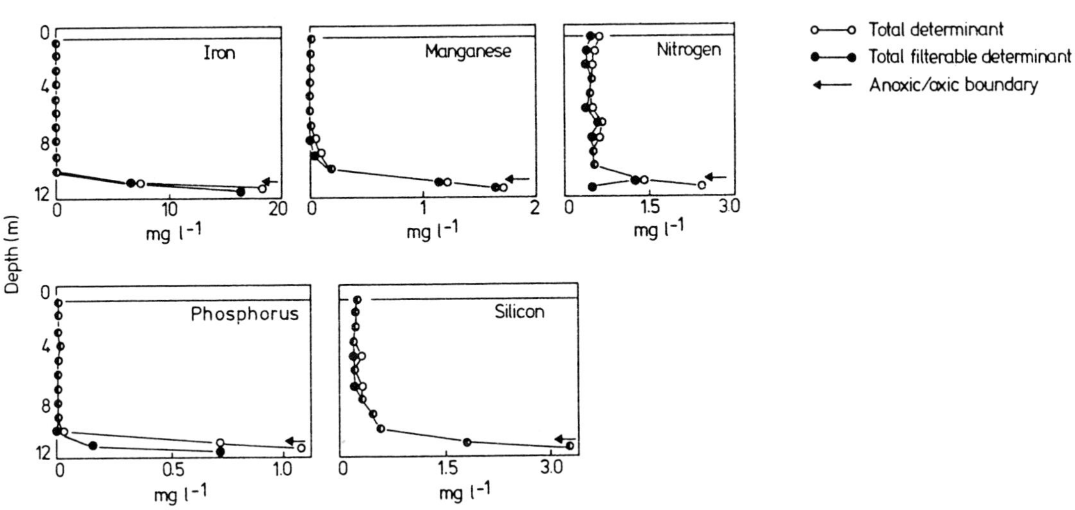

Fig. 3. Total and total filterable profiles of N, Si, P, Fe, and Mn in Sombre Lake, sampled on 21 October 1981

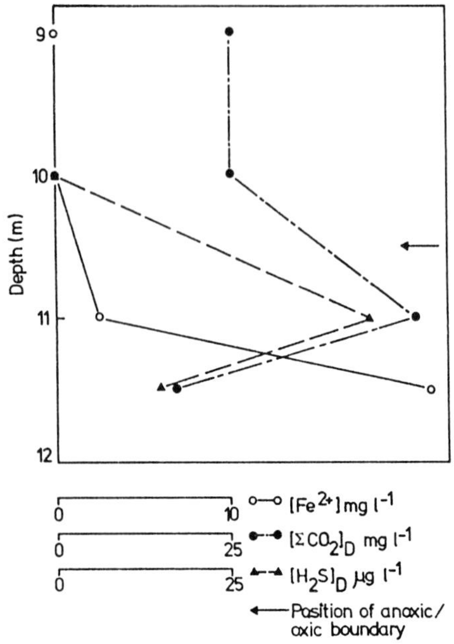

Fig. 4. Ferrous Fe, total inorganic C and total dissolved sulphide profiles in Sombre Lake bottom water, sampled on 21 October 1981

the water column. However, carbonate supply, from the total carbon dioxide, is sufficient to affect Fe movements through ferrous carbonate precipitation within the anoxic region. The fact that the products of ferrous carbonate ion activity often exceed the ferrous carbonate solubility product provides indirect evidence for this mechanism. Further evidence is provided by direct measurement of particulate carbonate at the anoxic sediment-water interface (0.2 $\mu g\ l^{-1}$). All of this indicates the likelihood of ferrous carbonate precipitation and its importance to the movement of Fe in the water column in Sombre Lake.

3.2. Eutrophic Heywood Lake

Under ice cover, there are strong correlations between profiles of Fe and P, indicating that Fe may control P speciation. Nitrogen profiles are independent of the other determinants, Si, P, Fe, and Mn (Fig. 5), and can therefore be used as a biotic control in studies of the movements of other nutrients (e.g., P movements which are under the influence of Fe).

Sulphide and ferrous ion products are in agreement with the ferrous solubility product, indicating ferrous sulphide precipitation within the anoxic region. Sulphide concentrations here, unlike in Sombre Lake, are sufficiently great to affect the Fe movements, as shown by the convex nature of the Fe and sulphide profiles towards the bottom of the water column (Fig. 6).

Towards the end of the period of bottom–water anoxia, during the spring melt, significant concentrations of ferrous, released from the sediments, are detected in oxygenated surface waters. This apparent stability is prob-

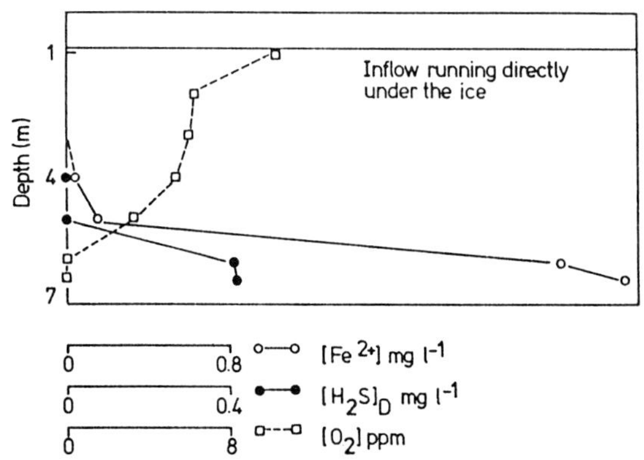

Fig. 6. Ferrous Fe, total dissolved sulphide and dissolved oxygen profiles in Heywood Lake, sample on 17 November 1981

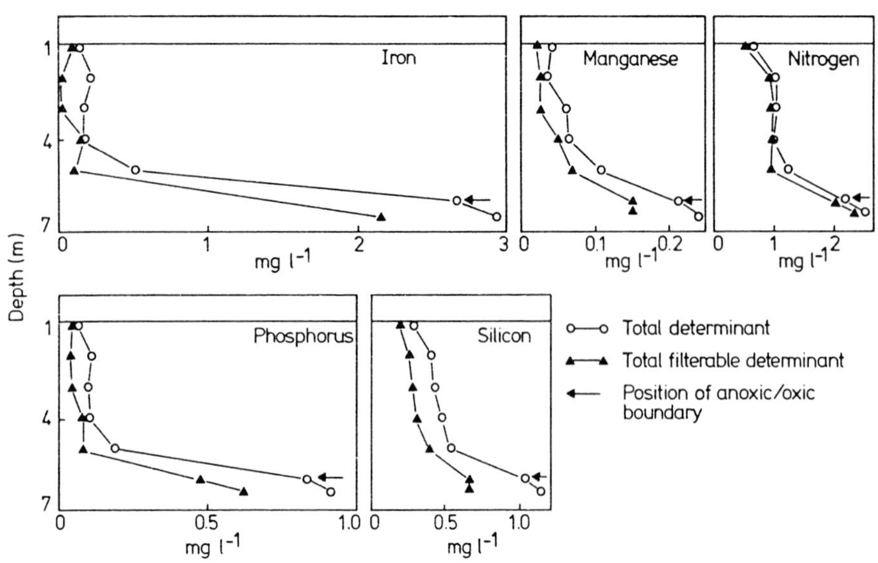

Fig. 5. Total and total filterable profiles of N, Si, P, Fe, and Mn in Heywood Lake, sampled on 17 November 1981

ably a consequence of the low pH (5.96–6.13), which markedly reduces the oxid tion rate (Stumm and Morgan 1981). It is also possible that high concentrations of humic acids, leaching from the catchment moss-banks to the surface water via the melt-water inflows, may have stabilized the ferrous by a catalytic oxidation mechanism (Stumm and Morgan 1981).

4 Conclusions

The movements and speciation of Fe quickly respond to changing redox conditions, and control the movements and speciation of P, but have little effect on the movements of other nutrients such as Si and N. Hawes (1983) has shown that P is the major limiting factor of phytoplankton growth. In Sombre Lake, the anoxic sump is a small component of the total lake volume and, therefore, represents a small component of the total P supply to the phytoplankton. In other similar Antarctic lakes where the sump is more substantial, the absence of a P supply from the anoxic sump to the surface waters, by Fe movements, will be an important consideration. In lakes similar to Heywood Lake, in which N, and not P, is the limiting nutrient (Hawes 1983), the different redox behaviour of Fe and, thus, P also becomes an important consideration in reasons for nutrient limitation. Hence, an understanding of the cycling of Fe is important for gaining insights into nutrient cycling in Antarctic lakes.

References

Cunningham CR, Davison W (1980) An opto-electronic sediment detector and its use in the chemical micro-profiling of lakes. Freshwater Biol 10:413–418

Gallagher JB (in press) Interactions of iron and manganese with nutrients in an Antarctic freshwater lake. Bull Br Antarct Surv

Hawes I (1983) Nutrients and their effects on phytoplankton populations in lakes on Signy Island, Antarctica. Polar Biol 2:115–126

Magnuson JJ, Stuntz WE (1970) A syphon water sampler for use through ice. Limnol Oceanogr 15:156–158

Sholkovitz ER, Copeland D (1982) The chemistry of suspended matter in Esthwaite Water, a biologically productive lake with seasonally anoxic hypolimnion. Geochim Cosmochim Acta 46: 393–410

Stumm W, Morgan JJ (1981) Aquatic chemistry. Wiley, New York pp 464–469

Paucity of Nutrient Cycling and Absence of Food Chains in the Unique Lakes of Southern Victoria Land

B.C. PARKER and G.M. SIMMONS, Jr[1]

Summary. Some of the lakes of southern Victoria Land, representing the highest latitude (77° S) lakes investigated in the southern hemisphere, are older than most glacially carved lakes. Features which combine to make these lakes unique include the 3.50–6.0 m thick, poorly transparent, turbulence-inhibiting perennial ice covers which overlie strongly physically and chemically stratified water columns. Specific non-biological factors influencing the structure and functioning of the relatively simple planktonic and benthic communities of these lakes are: a widely variable but normally limiting photosynthetically available radiation (PAR); stable low temperatures; high salinities (at greater depths of some lakes); supersaturated oxygen (at shallower depths); low turbulence and a virtual lack of vertical mixing; and frequent phosphorus and sometimes carbon dioxide limitations. Although a few unique, endemic algal and bacterial species occur within these lakes, a more striking feature is the low diversity of physiologically adapted strains of quite cosmopolitan taxa comprising the bulk of the planktonic and benthic communities of all the lakes. Also striking is the virtual absence of consumer animals, except for a few protozoa. The dominant plankton consists of a sparse assemblage of continuously swimming eukaryote micro-algal flagellates and non-swimming heterotrophic bacteria. The benthos of much of these lakes is covered with a great biomass of bluegreen algal-diatomaceous mats. The combined limitation of light, low temperatures, and low nutrients maintain the oligotrophic state of these lakes, and inhibit both decomposition with recycling of nutrients and the development of a significant food chain. Losses of nutrients to the accumulating sediments and losses by the algal mats, which escape through the perennial lake-ice and are blown away, combine to perpetuate the conditions of low productivity and low decomposition.

1 Introduction

Since 1972 , members of the Virginia Polytechnic Institute and State University have investigated one or more of 12 lakes located at about 77°S in the dry valley oasis of southern Victoria Land, Antarctica. A number of earlier workers investigating primarily two of these lakes contributed significantly to the identification of problems which have subsequently pursued (e.g., Armitage and House 1962; Angino and Armitage 1963; Goldman et al. 1963; Goldman 1964). This paper reviews information for 7 lakes within this dry valley oasis with particular attention to those physical, chemical and biological features which contribute to the paucity of nutrient cycling and relative absence of food chains. Because the materials and methods applied to these investigations have been described in detail previously, we have summarized this information in Table 1.

1 Department of Biology, Virginia Polytechnic Institute and State University, Blacksburg, VA 24061, USA

2 Select Physical and Chemical Variables

Table 2 gives the location, surface area, maximum depth, and average perennial ice thickness for the seven lakes, all of which occur within a 10,000 km^2 grid in a region of similar climate and geomorphology. Although these lakes probably do not share identical ages and geological histories, most are older than glacially carved lakes in the northern hemisphere. For example, Hendy et al. (1977) placed the origin of Lake Bonney at > 100,000 yrs and claimed that Taylor Valley has been at least partially ice-free even longer. We have obtained a 2.8 m length of sediment core from Lake Fryxell which may represent a 50,000-yr accumulation, based on a ^{14}C date (unpublished data). An important feature of the 7 lakes is the insulation of their water colums by perennial ice-covers ranging in thickness from 3.5 –6.0 m (Table 2). This feature is at least partially responsible for a number of peculiar properties of these lakes.

2.1 Wind-Generated Turbulence or Mixing

The ice-covers of the lakes all but eliminate wind-generated turbulence or mixing, which in turn perpetuates the strong physical and chemical stratification and selects for the plankton micro-organisms. Parker et al. (1980) attributed the presence of a select few species of strong, perpetually swimming, planktonic algae (e.g., *Chroomonas lacustris, Chlamydomonas subcaudata*) and the striking lack of planktonic diatoms in these lakes to this lack of turbulence. Heterotrophic bacterio-plankton present in large numbers are of extremely small size and thus will sink only slowly if at all (Mikell et al. 1984). These bacterioplankton are unattached and, thus, resemble the "indigenous of autochthonous plankto-bacteria" of the oceanic environment which have no requirement for solid surfaces (Sieburth 1979).

Antarctic Nutrient Cycles and Food Webs
(ed. by W.R. Siegfried, P.R. Condy, and R.M. Laws)

Table 1. Key to methods elaborated in previously published papers for Antarctic lakes

Method	Major References
Melting access holes/coring	Love et al. (1982)
Turbulence/mixing	Parker et al. (1980)
Photosynthetically available radiation (PAR, $\mu Em^{-2} s^{-1}$)	Allnutt et al. (1981); Parker et al. (1982a); Seaburg et al. (1983)
Temperature (°C)	Hoehn et al. (1977); Parker et al. (1982a); Wharton et al. (1983)
Conductivity ($\mu mhoscm^{-1}$)	Parker et al. (1982a)
Alkalinity/free CO_2	Hoehn et al. (1977); Parker et al. (1982a)
Dissolved oxygen	Hoehn et al. (1977); Parker et al. (1982a); Wharton et al. (1983)
^{33}P uptake kinetics	Seaburg et al. (to be published)
Nitrate-N, nitrite-N, ammonium-N	Fortner et al. (1976); Hoehn et al. (1977); Allnutt et al. (1981)
Nitrogen (C_2H_2)-fixation	Allnutt et al. (1981)
Orthophosphorus-P	Hoehn et al. (1977); Seaburg et al. (to be publsihed)
Condensed phosphorus	Hoehn et al. (1977)
Organic carbon	Parker et al. (1974, 1977); Allnutt et al. (1981)
Primary production	Parker et al. (1977, 1982a); Sugg (1977)
Macro-/micro-elements	Weand et al. (1976); Parker et al. (1982b)
Particulate extractable ATP	Simmons et al. (1983)
Particulate extractable chlorophyll a	Parker et al. (1982a); Simmons et al. (1983)
Algal/protozoan taxa	Cathey et al. (1981, 1982); Seaburg et al. (1979); Parker et al. (1982a)

Table 2. Seven lakes of southern Victoria Land with their approximate locations, surface areas (km^2), maximum piezometric depths (m), average ice thickness (m)

Lake	Coordinates	Surface Area	Depth	Ice Thickness
Bonney-East	77° 43′S, 162° 23′E	3.90	33	4.5
Chad	77° 38′S, 162° 45′E	0.17	7	5.5
Fryxell	77° 37′S, 163° 07′E	5.40	19	4.5
Hoare	77° 38′S, 162° 53′E	1.80	32	5.5
Joyce	77° 43′S, 161° 33′E	0.80	30	5.0
Miers	78° 07′S, 163° 54′E	1.10	20	6.0
Vanda	77° 32′S, 161° 33′E	5.00	70	3.5[a]

[a] Ice cover contains cracks

2.2 Light Transmission Through Ice and Water

The perennial ice-covers have varying quantities of soil blown onto their surfaces and varying quantities of gas bubbles and algal mat pieces frozen in ice, all of which reduce the total photosynthetically available radiation (PAR, 400–700 nm) reaching the water (Parker et al. 1981, 1982a, b). Table 3 shows that Lake Vanda's ice-cover, which is the thinnest and is cracked with fewest trapped gas bubbles or algal mat pieces, allows transmission of up to 18% of surface PAR. On the brightest austral summer days, PAR penetrating Lake Vanda ice has reached 234 μE m^{-2} s^{-1}. Transmission through Lake Vanda water also is greatest as a result of its very low extinction coefficient of 0.055 (Kaspar et al. 1982). Lake Bonney ice, by contrast, allows transmission of only 7.0% of surface PAR, and has an extinction coefficient of 0.138. Lake Miers has a nearly opaque ice-cover, allowing 0.35 of surface PAR to penetrate. No measurable PAR occurred below certain depths in Lakes Fryxell and Hoare which have extinctions of 1.002 and 0.164, respectively, (Parker et al. 1982a).

2.3 Stable Low Temperatures

The perennial ice-covers thermally insulate the underlying water from the wide ranging ambient surface temperature changes. Thus, relatively constant temperatures occur at any given depth in a lake (Table 3). With respect to the heat balances of Lakes Vanda and Bonney, which have respectively warm bottom and intermediate layers, some solar heating has been proposed (Ragotzkie and Likens 1964). Wharton et al. (to be publ.) report on temperature data for five of these lakes, and Seaburg et al. (1981) have shown that a significant number of planktonic algae in these lakes are obligately or facultatively adapted to growth at low temperatures.

2.4 Salinity Variations

Table 3 shows the striking variation in salinity with depth in several of these lakes, expressed as specific conductance (see Parker et al. 1982a). All lakes are fresh beneath the perennial ice. Lakes Chad, Hoare and Miers are essentially fresh throughout their depths, whereas Lakes Bonney and Vanda have hypersaline deep waters. Craig et al. (1975) reported the discovery of the unusual metastable crystal, dihydrohalite, in Lake Bonney.

2.5 Oxygen Supersaturation

Apparently unique to these perennially ice-covered lakes are the dissolved oxygen levels, which are 2– > 3 times nor-

Table 3. Ranges in important physical and chemical features for seven southern Victoria Land lakes at various depths throughout the austral summer: Photosynthetically Available Radiation (PAR, as % of surface PAR penetrating ice; range in μEm^{-2} s^{-1}); Temperature (°C); Conductivity (μmhoscm^{-1}); Dissolved O_2 (mgD0 l^{-1}); Available Inorganic Carbon (AIC, mg CO_2 +HCO_3 l^{-1}); Nitrate- + Nitrite-N (μg NO_3 +NO_2-N l^{-1}); Ammonium-N (μg NH_4-N l^{-1}); Orthophosphate-P (μgo-PO_4-P l^{-1}). Ranges are for the austral summer

	Bonney-East	Chad	Fryxell	Hoare	Joyce	Miers	Vanda
% PAR	7.0 – 0.20	0.8 –0.60	1.5 – 0	0.8 – 0	1.2 – 0.02	0.30– 0.02	18.0 – 0.01
μEm^2 s^{-1}	91.0 – 2.6	10.4 –7.8	19.5 – 0	10.4 – 0	15.6 – 0.26	3.9 – 0.26	234.0 – 0.13
°C	– 2.0 – +7.0	0.00–2.00	0.00– 4.00	– 0.30 – 1.00	0.00– 2.50	0.00– 5.80	0 – 25.0
μmhoscm^{-1}	500 – 156,000	400	500 – 8,600	400 – 800	500 –3,400	500 –610	400 –123,000
DO	2.0 – 45.8	?40.0	0 – 41.9	0 – 52.1	0 – 42.0	0 – 40.0	0 – 29.6
AIC	10.0 – 734.0	? 8.0	80.0 – 722	8.0 – 78.0	9.0 – 130.0	10.0 – 65.0	6.0 – 140
pH	5.3 – 8.4	7.8 –8.2	7.0 – 8.4	7.3 – 8.9	7.3 – 8.2	6.8 – 8.4	5.4 – 8.2
NH_4-N	0 – 4,200	6.86	1.4 – >100	0 – >100			1.4 – 24,463
NO_3+NO_2-N	63.2 – 2,500	4.90	0 – 4.48	0 – 49.4			15.4 – 3,275
o-PO_4-P	1.395– 11.811		1,302 – 39.711	0.775– 3.906			0.30– 252.7

mal saturation in the shallower more brightly illuminated waters (Table 3) (Parker et al. 1982a; Wharton et al. 1983). Apparently the annual O_2 production by photosynthesis exceeds the annual respiratory O_2 consumption, and equilibration of gases with the atmosphere is inhibited or entirely prevented by the perennial ice-covers. Lake Vanda is an exception with cracks in its ice-cover through which at least some gas has been observed to escape (Kasper et al. 1982). Note that 5 of the lakes have anaerobic bottom water beneath the super-saturated oxygen layer (Table 3).

2.6 Inorganic Carbon, Nitrogen and Orthophosphorus

Table 3 summarizes the ranges for available inorganic carbon, ammonium-nitrogen, nitrate- and nitrite-nitrogen, and orthophosphorus determined at various depths in these lakes over several austral summers. Inorganic C values in the shallower parts of these lakes are low, relative to the deeper waters (Parker et al. 1982a). In fact, with the progress of the austral summer inorganic C often drops, pH rises, and free CO_2 is depleted. Probably this is due to increased photosynthesis at these shallower depths. We have noted this especially in Lakes Fryxell and Hoare (Parker et al. 1980). Inorganic N and orthophosphorus levels show generally similar patterns with lower concentrations at the shallower more brightly illuminated depths. Anaerobic bottom waters are, however, the exception in that they contain high concentrations of ammonium-nitrogen, orthophosphorus and CO_2 without detectable nitrate or nitrite (Fortner et al. 1976; Hoehn et al. 1977; Weand et al. 1977; Allnutt et al. 1981; Seaburg et al. to be publ.).

The above eight physical or chemical variables have been listed, because we believe that they are primarily influential in regulating the biological community structure and metabolic processes important to nutrient cycling and the development of food chains in these lakes. In summary, these eight variables are:

1. lack of wind-generated turbulence or mixing;
2. much reduced light transmission through ice and water;
3. stable low temperatures;
4. salinity differences;
5. the unusual oxygen regime, especially the oxygen supersaturation;
6. available inorganic carbon;
7. available inorganic nitrogen; and,
8. available orthophosphate-phosphorus.

3 Community Structure

Two major communities occur within the seven lakes: a planktonic and a benthic mat community. The distribution of algal, protozoan, and metazoan taxa and their distribution within these two communities is given in Appendix 1. Both communities also contain bacteria and yeasts, about which little is presently known. Flagellates dominate amongst the planktonic algae. Six of the seven lakes have the cryptophycean *Chroomonas lacustris* with the chlorophycean *Chlamydomonas subcaudata* or *Chloromonas alpina* as the most abundant species. Other algal taxa are considerably less common. Notably rare in the plankton are diatoms and a few protozoa.

Benthic algal-bacterial mats occur abundantly attached to the gravelly substrata in all seven lakes. The most conspicuous mats are the thin moat mats that grow in the ice-free areas which develop around the lake edges during most austral summers. These are easily seen from the surface, whereas the mats beneath perennial ice have been studied more effectively by SCUBA diving (Love et al. 1982). The macro-morphological types of these below-ice mats and their development have been described in detail by Wharton et al. (1983). They include (2) liftoff, (3) pinnacle, (4) prostrate aerobic, and (5) prostrate anaerobic mats. The dominant algal taxa in the first four below-ice mats are the cyanophycean *Phormidium frigidum* and *Lyngbya martensiana* together with various pennate diatoms. Anaerobic mats lack diatoms but have *Phormidium frigidum* with photosynthetic and heterotrophic bacteria. Protozoa and the metazoan rotifers, tardigrades and nematodes are not very common in below-ice aerobic mats, but usually can be found relatively frequently in the moat mats. Hence, consumer animals are practically absent from the plankton and are rare in the perennial benthic mats. This fact has been clearly documented by Cathey et al. (1981, 1982), especially at Lakes Fryxell and Hoare.

Another unusual feature of these lakes is the fate of some of the lift-off mat. Parker et al. (1982b) pointed out that the columnar upward growth of lift-off mats was the result of oxygen-producing photosynthesis with entrapment of oxygen bubbles up to ca. 10 mm in diameter within the bluegreen algal matrix of the mats. The entrapped O_2 also facilitates a tearing loose of mat pieces from the gravelly substratum during the austral summer. Thus, mat clumps lift off, float to beneath the perennial ice and at the austral summer's end, become frozen into the new ice formed. Over a period of 10–12 yrs, through annual ice ablation loss at the surface and new ice formed below, the frozen mat clumps reach the upper lake ice surface. They are lyophilized and are blown away. This lift-off, freeze-in, and escape process is so extensive within these lakes that a significant quantity of nutrient salts, sediments, and organic matter is removed from the shallower areas of these lakes. This unusual process, thus, appears to substitute at least partly for the absence of out-flow streams from these lakes (Parker et al. 1982b).

3.1 Metabolic Studies in Review

In situ studies conducted at these lakes, which address at least some aspects of nutrient cycling, have included pri-

Appendix 1. Algal (1–48), achlorophyllous protozoan (A–F), and metazoan (I–III) taxa common to seven southern Victoria Land lake plankton and one or more attached moat or benthic algal mats, listed in order of approximately descending importance

Lake	Plankton	Attached
Bonney-East	1, 4, 8, 16, 9, 6, 5, 12, 14, A, B	22, 23, 4, 35, 14, 39–48, A, B, I, II, III
Chad	Not adequately surveyed	22, 21, 20, 31, 29, 23, 36, 26, A, I, II, III
Fryxell	1, 4, 9, 10, 5, 6, A, D, E, F	22, 29, 23, 27, 33, 35, 36, 14, 39–46, 15, 48, A, D, E, F, I, II
Hoare	1, 6, 18, A, C, D	22, 23, 28, 29, 24, 25, 30, 32, 4, 34, 33, 14, 35, 36, 38, 39, 41–43, 45, 46, A, C, D, I, II, III
Joyce	1, 4, 18, A, C, D	22, etc., but not adequately surveyed
Miers	1, 18, 17, 9	22, etc., but not adequately surveyed
Vanda	2, 3, 12, 11, 17, 15, 18, 7, 19, A, D	22, 23, 33, 14, 15, 35–39, 41–43, 45, 47, 48, A, D, I

Taxa: Predominantly planktonic

1. *Chroomonas lacustris* Pascher and Rutner
2. *Ochromonas* sp. Wyssotzki
3. *Ochromonas minuscula* Conrad
4. *Chlamydomonas subcaudata* Wille
5. *Chlamydomonas intermedia* Chodat
6. *Chloromonas alpina* Wille
7. *Brachiomonas submarina* Hazen
8. *Thorakomonas feldmannii* Bourelly
9. *Chlorella vulgaris* Beyerinck
10. *Furcilia lobosa* Stokes
11. *Monodus coccomyxa* Pascher
12. *Nitzschia sublinearis* Hustedt
13. *Achnanthes brevipes* (Kützing) Cleve
14. *Navicula cryptocephala* Kützing
15. *Hantzschia amphioxys* var. *maior* Grunow
16. *Dinoflagellate* (colorless)
17. *Oscillatoria retzii* Agardh
18. *Oscillatoria* sp. (2–3 μm diam) Vaucher
19. *Lyngbya* sp. (2–3 μm diam) Agardh

Predominantly benthic or non-planktonic

20. *Oscillatoria limosa* (Roth) Agardh
21. *Oscillatoria priestleyi* West and West
22. *Phormidium frigidum* Fritsch
23. *Lyngbya martensiana* Meneghini
24. *Microcoleus paludosus* var. *acuminatus* Gardner
25. *Porphyrosiphon notarisii* (Menehg.) Kutz
26. *Schizothrix antarctica* Fritsch
27. *Anabaena cylindrica* Lemm.
28. *Anabaena* sp. Bory
29. *Nostoc commune* Vaucher
30. *Nostoc longistaffi* Fritsch
31. *Calothrix gypsophila* Bornet and Flahault
32. *Synechococcus aeruginosus* Nag.
33. *Tetracystis* sp. Brown and Bold
34. *Chlorococcum* sp. Rabenhorst emend Starr
35. *Stauroneis anceps* Ehrenberg
36. *Caloneis ventricosa* var. *truncatula* (Grunow) Meister
37. *Cymbella* sp. Agardh
38. *Navicula fragilarioides* Krasske
39. *Navicula gausii* (Heiden) Hustedt
40. *Navicula litos* Hohn and Hellerman
41. *Navicula muticopsis* Van Huerck
42. *Navicula muticopsis* f. *murrayi* (West and West) Ko-Bayashi
43. *Navicula muticopsis* f. *reducta* West and West
44. *Navicula seminulum* Grunow
45. *Navicula shackletoni* West and West
46. *Nitzschia angustata* (W. Smith) Grunow
47. *Nitzschia frustulum* (Kutz) Grunow
48. *Pinnularia cymatopleura* West and West

A. *Stombidium* sp. Claparede and Lachmann
B. *Sphaerophyra* sp. Claparede and Lachmann
C. Peritrich
D. Ciliate
E. Actinobolinidae
F. *Collodictyon* sp.

I. Rotifera
II. Tardigrada
III. Nematoda

mary production by the ^{14}C method, mostly with the plankton (Parker et al. 1977, 1982a; Seaburg et al. 1983), extractable chlorophyll-*a* and adenosine triphosphate (Parker et al. 1982a; Simmons et al. 1983), measurement of nitrogenase activity using acetylene-reduction (Allnutt et al. 1981), and ^{33}P uptake experiments (Parker et al. 1980; Seaburg et al. to be publ.).

Based on primary production rates during the austral summer, dominant phytoplankton, phytoplankton volume, and to a lesser extent chlorophyll-*a* and extinction coefficients for PAR through water, Parker et al. (1982a) showed that these lakes were oligotrophic. Based on acetylene-reduction experiments in three lakes (Chad, Fryxell, Hoare), Allnutt et al. (1981) showed that N was not severely limiting in benthic habitats and that only in shallow, more brightly illuminated, warmer moat, melt-stream and peripheral lake areas, in which heterocystous bluegreen algae grew, was significant nitrogenase activity measureable. Heterotrophic bacterial nitrogenase was ruled out (Allnutt et al. 1981). Based upon 40 ^{33}P tracer experiments (adding small amounts of high activity ^{33}P) and 20 non-tracer experiments (adding 4–5 concentrations of unlabelled P salt + ^{33}P), Seaburg et al. (to be publ.) concluded that P was one major limiting nutrient to the phytoplankton of these lakes, especially in shallow benthic waters. A scarcity of P, as well as possibly other nutrient elements, also has been supported by the demonstration that large benthic algal mat pieces with bound minerals and nutrient salts escape through the perennial lake-ice (Parker et al. 1982b). The lift-off mats, thus, constitute a sink for the differential removal of nutrient salts. A second sink for nutrient salt removal are the prostrate aerobic and anaerobic (perhaps pinnacle) mats, most of which remain in place and accumulate sediments as modern stromatolites (Parker et al. 1981; Wharton et al. 1982, 1983; Love et al. 1983). Anaerobic stromatolitic cores up to 2.8 m long, estimated to represent up to 50,000 yrs accumulation, have been retrieved from Lake Fryxell (unpublished data). Simmons et al. (1983) have observed, using considerably shorter sediment cores, that undegraded chlorophyll-a but not ATP can be extracted at considerable depth from a number of Antarctic lake cores. In contrast, degraded chlorophyll-a and ATP were readily detectable in temperate lake cores. Finally, chloroform-extractable lipids from an anaerobic core from Lake Hoare reveal a remarkable preservation of free fatty acids, hydrocarbons, triglycerides, alkanes, sterols, and steroids throughout the core length (Orcutt and Parker, unpubl.) The above three points, together with the dissolved oxygen super-saturation of shallower lake waters, constitute strong evidence to support the view that decomposition rates by heterotrophic microorganisms are attenuated in these lakes. Add to this the essential lack of a food chain, and it seems evident that nutrient cycling also is at a minimal level. The lack of nutrient cycling and losses to the sediments, or from the lake-ice, through algal mat lift-off and escape apparently have combined to perpetuate their oligotrophic state and essentially arrest the more typical process of eutrophication.

4 Conclusions

Data published and on hand for one or more of the seven dry valley oasis lakes indicate the following:

1. Extreme lack of turbulence or mixing partially selects for the plankton and ensures almost no resuspension of matter lost to the sediments.
2. Light is limiting to photosynthesis throughout most depths below perennial ice, even during the austral summer.
3. Despite light limitation, oxygen-producing photosynthesis can occur and at shallower depths has induced perpetual super-saturation of oxygen, such that algal mats lift-off and escape through ice, removing nutrients and trapped sediments from the lakes.
4. Cold temperatures throughout most depths also limit metabolic processes, such as photosynthetic C fixation, nitrogen fixation, P uptake, organic C uptake and assimiliation – respiration, and growth.
5. Hypersalinity of deeper waters in Lakes Bonney and Vanda, together with low light, apparently inhibits growth and development of a significant algal-microbial mat community.
6. O_2/CO_2 which can be very high in parts of these lakes may inhibit photosynthesis, enhance algal photorespiration, and both enhance and inhibit bacterial growth and metabolism.
7. Nitrogen fixation appears largely confined to shallow ice-free moat and peripheral areas, where nitrogenase activity remains very temperature restricted even though abundant heterocystous bluegreen algae are present.
8. ^{33}P uptake nontracer [$^{33P+P}o$] level in situ experiments suggest that while turnover rates are very low, orthophosphorus concentrations also are low such that P limitation is likely.
9. An extremely low diversity and low biomass of consumer animals occurs in both planktonic and benthic communities of these oligotrophic lakes.

These nine points combined help confirm and at least partially explain two rather unusual features for lakes in general, namely a much reduced or nearly absent food chain and nutrient cycle. We make no claim for having all of the details or numbers to describe quantitatively the nutrient sources, fates, and rates of flow within these lakes. Such respresents the next logical goal in research, for us or others.

Acknowledgements. We are grateful to the National Science Foundation's Division of Polar Programs for grant nos 76-23996 and 80-12988 which supported this research. We thank numerous field and laboratory personnel for assistance in the collecting and processing of data.

References

Allnut FTC; Parker BC; Seaburg KG; Simmons Jr GM (1981) In situ nitrogen (C_2H_2)-fixation in lakes of southern Victoria Land, Antarctica. Hydrobiol Bull 15(3):99–109

Angino EE; Armitage KB (1963) A geochemical study of lakes Bonney and Vanda Victoria Land, Antarctica. J Geol 71(1):89–95

Armitage KB; House HB (1962) A limnological reconnaissance in the area of McMurdo Sound, Antarctica. Limnol Oceanogr 7: 36–41

Cathey DD; Parker BC; Simmons Jr GM; Yongue WH; Van Brunt MR (1981) The Microfauna of algal mats and artificial substrates in southern Victoria Land lakes of Antarctica. Hydrobiologia 85: 3–15

Cathey DD; Simmons Jr GM; Parker BC; Yongue WH; VanBrunt MR (1982) Protozoan colonization of artificial substrates in two Antarctic lakes. Trans Am Microsc Soc 101(4):353–367

Craig Jr JR; Light JF; Parker BC; Mudrey MG (1975) Identification of hydrohalite. Antarct JUS X:178–179

Fortner RD; Weand BL; Hoehn RC; Parker BC (1976) Revaluation of orthophosphorus and inorganic nitrogen levels in an Antarctic meromictic lake. Hydrobiologia 49:229–232

Goldman CR (1964) Primary productivity studies in Antarctic lakes. In: Carrick R, Holdgate MW, Prevost M (eds) Biologique antarctique, Hermann, Paris, 291–299

Goldman CR; Mason DT; Wood JB (1963) Light injury and inhibition in Antarctic freshwater phytoplankton. Limnol Oceanogr 8(3):313–322

Hendy CH; Wilson AT; Popplewell KB; House JE (1977) Dating of geochemical events in Lake Bonney, Antarctica, and their relation to glacial and climate change. N Z J Geol Geophys 20:1103–1122

Hoehn RC; Parker BC; Fortner RD; Weand BL; Craft J; Lane L; Stavros R; Sugg H; Whitehurst J (1977) Nitrogen and phosphorus availability to plankton and benthic communities in Lake Bonney, southern Victoria Land, Antarctica. In: Llano GA (ed) Adaptations Within Antarctic Ecosystems, Gulf Publ. Co., Houston

Kaspar M; Simmons Jr GM; Parker BC; Seaburg KG; Wharton Jr RA; Lewis-Smith RI (1982) A species of *Bryum* Hedw. from Lake Vanda, Antarctica. Bryologist 84(4):424–430

Love FG; Simmons Jr GM; Wharton Jr RA; Parker BC (1982) Methods for melting dive holes in thick ice and vibracoring beneath ice. J Sediment Petrol 43:644–647

Love FG; Simmons Jr GM; Wharton Jr RA; Parker BC; Seaburg KG (1983) Modern *Conophyton*-like algal mats discovered in Lake Vanda, Antarctica. Geomicrobiol J 3(1):33–48

Mikell AT; Parker BC (1984) Response of the heterotrophic community to high dissolved oxygen in Lake Hoare, Antarctica. Appl Envir Microbiol 47(5):1062–1066

Parker BC; Whitehurst JT; Hoehn RC (1974) Observations of in situ concentrations and production of organic matter in an Antarctic meromictic lake. Virginia J Sci 25:136–140

Parker BC; Simmons Jr GM; Seaburg KG; Wharton Jr RA (1980) Ecosystem comparison of oasis lakes and soils (ECOLS). Antarct J US XV:167–170

Parker BC; Simmons Jr GM; Love FG; Seaburg KG; Wharton Jr RA (1981) Modern stromatolites in Antarctic dry valley lakes. BioScience 31:656–661

Parker BC; Simmons Jr GM; Seaburg KG; Cathey DD; Allnutt FCT (1982a) Comparative ecology of plankton communities in seven Antarctic oasis lakes. J Plankton Res 4(2):271–286

Parker BC; Simmons Jr GM; Wharton Jr RA; Seaburg KG; Love FG (1982b) Removal of organic and inorganic matter from Antarctic lakes by aerial escape of bluegreen algal mats. J Phycol 18(1): 72–78

Ragotzkie RA; Likens GE (1964) The heat balance of two Antarctic lakes. Limnol Oceanogr 9(3):412–425

Seaburg KG; Kaspar M; Parker BC (1983) Photosynthetic quantum efficiencies of phytoplankton from perennially ice covered Antarctic lakes. J Phycol 19(4):

Seaburg KG; Parker BC; Prescott GW; Whitford LA (1979) The algae of southern Victoria Land, Antarctica: A taxonomic and distributional study. Bibliothecia Phycologia (Cramer J, ed) pp 1–166

Seaburg KG; Parker BC; Simmons Jr GM (1981) Temperature-growth responses of algal isolates from Antarctic oases. J Phycol 17(4): 353–360

Seaburg KG; Parker BC; Simmons Jr GM; Wharton Jr RA (to be published) Phosphorus – a major limiting nutrient for Antarctic dry valley lakes. Polar Biol

Sieburth J McN (1979) Sea microbes. Oxford University Press, NY, pp 1–491

Simmons Jr GM, Wharton Jr RA, Parker BC, Andersen D (1983) Chlorophyll and adenosine triphosphate levels in Antarctic and temperate lake sediments. Microb Ecol 9:123–135

Sugg Jr HG (1977) Distribution, development, internal structure, productivity and respiration rates of benthic algal mats in Lake Bonney. MS thesis, Virginia Polytechnic Institute & State University, Blacksburg, VA pp 1–122

Weand BL, Hoehn RC, Parker BC (1976) Trace element distribution in an Antarctic meromictic lake. Hydrobiol Bull 10(2):104–114

Weand BL; Hoehn RC; Parker BC (1977) Nutrient fluxes in Lake Bonney – a meromictic Antarctic lake. Arch Hydrobiol 80(4): 519–530

Wharton Jr RA; Parker BC; Simmons Jr GM (1983) Distribution, species composition, and morphology of algal mats (stromalites) in Antarctic dry valley lakes. Phycologia 22(4):355–365

Wharton Jr RA, Parker BC, Simmons Jr GM, Seaburg KG, Love FG (1982) Biogenic calcite structures forming in Lake Fryxell, Antarctica. Nature 295:403–405

Factors Controlling Phytoplankton Populations in Maritime Antarctic Lakes

I. HAWES[1]

Summary. The 17 lakes on Signy Island, South Orkney Islands, provide examples of most lake types found in the maritime Antarctic. Continuous monitoring of physico-chemical parameters in these lakes has revealed considerable differences in nutrient and phytoplankton concentrations. These can be linked to the level of natural enrichment (by birds and seals) occurring on the catchment areas. The interdependence of nutrient loadings, nutrient concentrations and standing crop of phytoplankton produced is a consistent feature of phytoplankton ecology in both temperate and high-Arctic lake systems, despite the restrictions imposed by extreme environmental conditions in the latter case. Under winter snow and ice cover, photosynthesis was undetectable in the lakes studied and phytoplankton populations decreased. Inorganic nutrient regeneration occurred in both oligotrophic and mesotrophic lakes but was much more extensive in the latter. The algae responded rapidly to increased light levels in spring and a shade-adapted population developed. In the oligotrophic lake, phosphate rapidly became limiting, whereas the winter accumulation permitted a dense population to develop in the mesotrophic lake despite low irradiance. Turbulent mixing under the ice cover facilitated growth in the latter. Major losses to the outflow and grazers as well as the onset of nutrient limitation terminated this spring bloom. During summer, the oligotrophic lake remained phosphate limited, while the high P load from the animal-enriched catchment resulted in nitrogen limitation in the mesotrophic lake. Relatively high production rates were maintained and may have been supported by rapid internal nutrient recycling.

1 Introduction

Signy Island (60° 43′S, 45° 38′W) lies at the northern edge of the Weddell Sea (Fig. 1) and falls within the maritime Antarctic climatic zone (Holdgate 1964). Mean monthly air temperatures in 1980 varied between +1.7 and –17.5°C. There are 17 lakes on the island, all of which are small, with a maximum recorded depth of 15 m. Water temperatures in these lakes are low, seldom rising above 5°C in summer, and the strong, virtually continuous, winds ensure that no vertical stratification develops during the open-water period. The lakes are ice-covered for 9–12 months each year but phytoplankton develops under the ice in spring and during the summer open-water period (Light et al. 1981). Hawes (1983a) has shown that this phytoplankton is capable of high rates of photosynthesis, but that its density in a range of these lakes is largely determined by the availability of nutrients.

1 British Antarctic Survey, NERC, High Cross, Madingley Road, Cambridge CB3 OET, Great Britain

Among other environmental variables, light may be of particular relevance in affecting the growth of the planktonic algae. The combined effects of seasonal changes, together with winter snow and ice cover, can lead to fluctuations over three orders of magnitude in Signy Island lakes (Priddle 1980). Changes of comparable magnitudes have been reported from northern polar regions and, together with the ability of algae to adapt to them, are of great im-

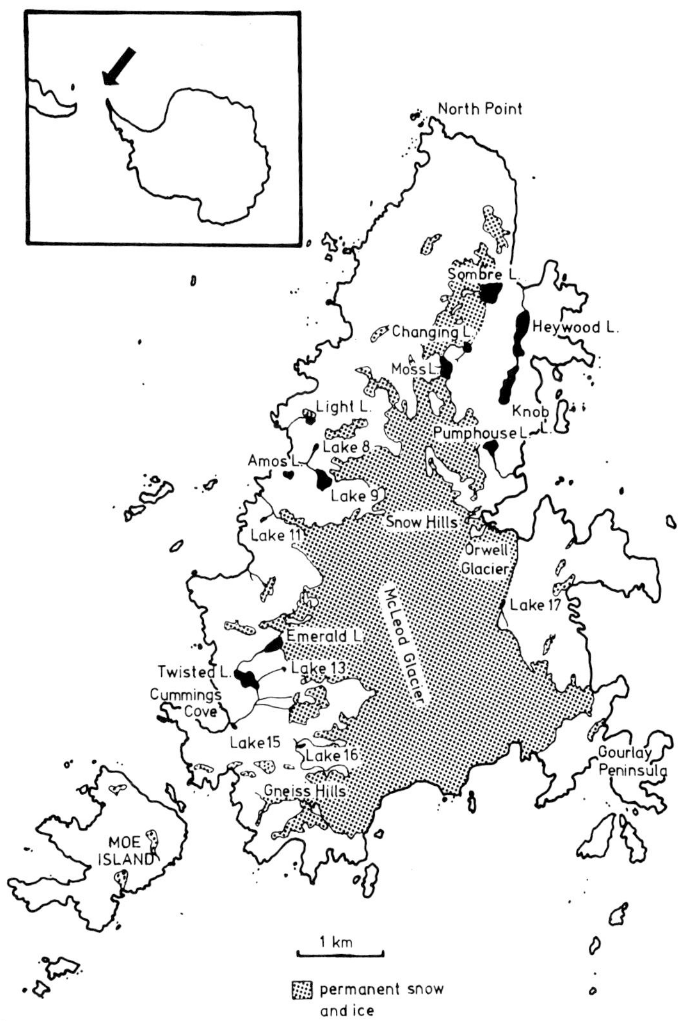

Fig. 1. Map of Signy Island, showing lakes (*black*) and areas of permanent ice cocer (*stippled*)

Antarctic Nutrient Cycles and Food Webs
(ed. by W.R. Siegfried, P.R. Condy, and R.M. Laws)

portance in determining phytoplankton production, although algal biomass frequently has been found to be nutrient limited (Hobbie 1973; Kalff and Welch 1974). Vincent (1981) attributed low rates of photosynthesis in permanently ice covered lakes in the dry valleys of southern Victoria Land to a combination of low radiation and strong shade adaptation of the algae, although here, also, nutrients appeared to play an important role in determining the standing crop.

There have been few comprehensive ecological studies of maritime Antarctic lakes, but the limited information available tends to confirm the importance of a high nutrient loading for extensive summer phytoplankton development (Goldman et al. 1972; Samsel and Parker 1972). In this respect, Antarctic lakes do not appear to differ significantly from lakes in other geographical locations. In this paper, I summarize current knowledge of factors which are important in controlling the seasonal cycle of phytoplankton growth in two lakes at Signy Island, South Orkney Islands.

2 The Lakes

The lakes (Heywood and Sombre) discussed here lie close together in a single geological formation (Fig. 2), but are of contrasting character. The catchment of Heywood Lake contains extensive areas of moss carpet, is frequented by large numbers of birds and seals during summer and is free of permanent ice cover. The catchment of Sombre Lake is more typically Antarctic, consisting predominantly of bare rock, scree and permanent snowfields. Moss cover is scant, and few seabirds and no seals encroach the catchment. Estimates of phosphorus loading (as inorganic phosphate) reflect these differences: Heywood Lake receiving 350 mg m^{-2} yr^{-1}, whereas Sombre Lake receives < 10% of this at 23 mg m^{-2} yr^{-1} (Hawes 1983a). Radiation incident at the surface of the two lakes is essentially the same. Longer duration of ice cover on Sombre Lake, however, means that the high irradiance within the water column associated with the ice-free summer period are delayed by approximately 1 month.

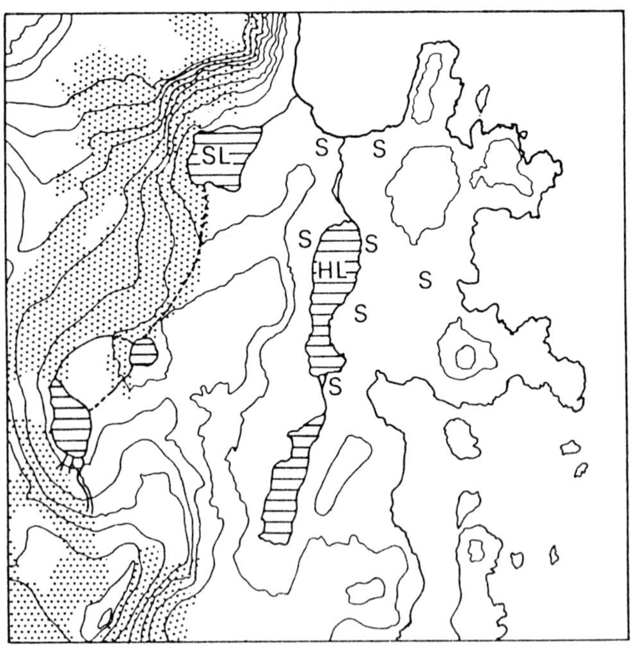

Fig. 2. Location of Heywood (*HL*) and Sombre (*SL*) lakes, showing permanent snow and ice (*stippled*) and moss banks frequented in summer by seals (*S*)

3 Materials and Methods

Full details of sampling techniques, analytical procedures and the ^{14}C uptake method for estimating photosynthesis are given elsewhere (Hawes 1983a,b).

4 Results and Discussion

4.1 Seasonal Phytoplankton Development

Chlorophyll-a measurements have been made routinely in both lakes for a number of years, and a regular pattern of development is apparent (Fig. 3).

In both lakes, winter concentrations are low and in situ photosynthesis is undetectable. The loss of the overlying snow layer in September–October increased the penetration of incident radiation to the water column from < 1% to > 20%. Since incident radiation is itself increasing at this time, a substantial rise in the availability of radiation within the lake ensues. Mean daily surface-penetrating radiation in Sombre Lake increased from 0.4 to 35 $\mu E\ m^{-2}\ s^{-1}$ from 12 September to mid-October 1980. After a short lag period, the concentration of chlorophyll-a increases and in situ photosynthesis becomes detectable (Hawes 1983a). Since water temperatures do not increase until well after spring growth has commenced, initiation is entirely a light induced process which is a consistent feature of ice-covered lakes (Pechlaner 1971; Hobbie 1973; Ramberg 1979).

Chlorophyll-a concentrations continue to increase in Sombre Lake until a maximum is reached during the short, ice-free summer, which then decreased with the reformation of ice cover. In Heywood Lake, the spring increase is much more rapid and often rises to extremely high concentrations, before falling equally quickly 2–3 weeks later. A second peak then develops in summer, and is considerably denser than in Sombre Lake. This summer peak contains less chlorophyll-a than is present during spring but the particulate carbon:chlorophyll-a ratio is three times greater (Table 1), and a similar phytoplankton biomass in terms of organic C may be present.

Although photosynthesis remains undetectable during winter, the phytoplankton does retain a photosynthetic

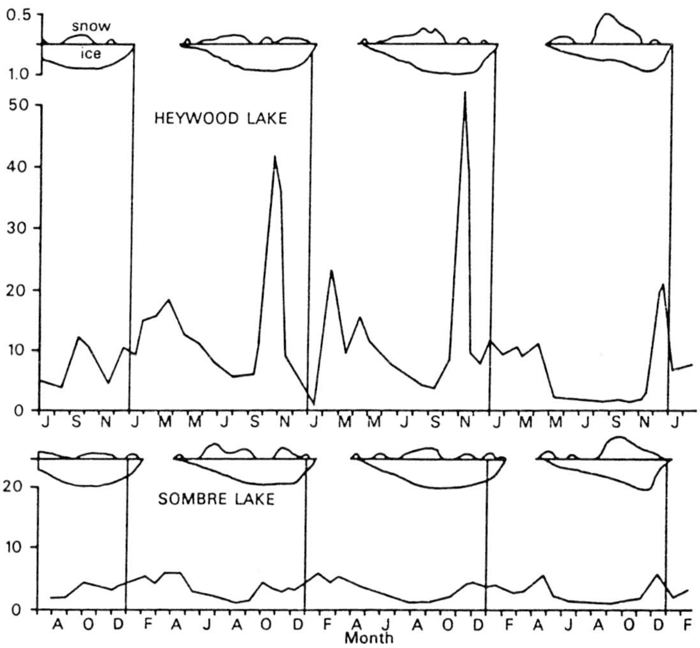

Fig. 3. Mean chlorophyll-a concentrations (uncorrected for phaeopigments) in Heywood and Sombre lakes from mid-1978 to 1981; depth of snow and ice cover shown for each month. (Mean concentrations calculated for deep-spot sampling sites)

capacity. Photosynthesis-radiation curves constructed for these populations (Fig. 4) show characteristics of extreme shade adaptation (see Beardall and Morris 1976) with a low radiation-saturated rate and one of the lowest light-saturated points on record at 15 μE m^{-2} s^{-1} (Harris 1978, Fig. 5). Similar curves were constructed monthly throughout the seasonal cycle, and changes in the important characteristics of saturating radiation flux (I_k), saturated-photosynthetic rate (P_{max}) and slope of the initial portion of the curve (see Talling 1957) are given in Table 2.

Adaptation to increased radiation in spring involves increases in both maximum rate of photosynthesis and saturating-radiation level. A consequence of this is that at the low spring radiation levels, photosynthesis proceeds below its saturated rate for much of the day. Under these unusual conditions, photosynthesis is directly limited by the level of surface-penetrating radiation, together with the photosynthetic efficiency of the phytoplankton (Talling 1971). Since this efficiency is unlikely to be limited by the relatively high concentrations of nutrients in the early stages of spring growth (Hawes 1983a), the initial rate of increase is effectively dependent on the availability of radiation. The pattern of subsequent development, however, reflects the abundance and distribution of nutrients at the end of winter.

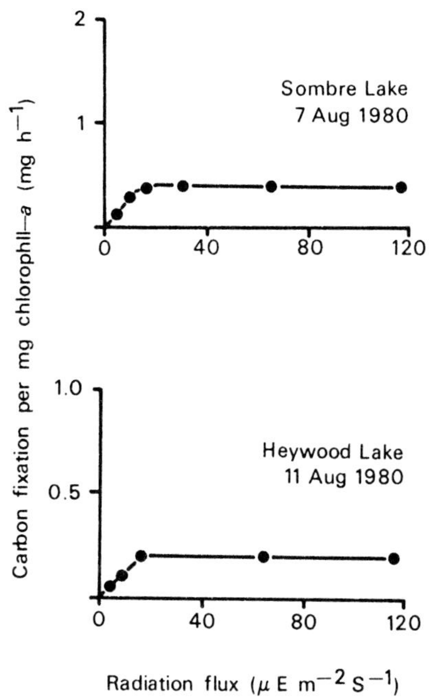

Fig. 4. Photosynthesis-radiation curves for over-wintering phytoplankton in Sombre and Heywood lakes

Table 1. Typical elemental ratios of seston in winter, spring, and summer in Heywood and Sombre lakes. All samples from mid-depth

	Heywood Lake				Sombre Lake			
	C:Chl.a	N:P	C:P	C:N	C:Chl.a	N:P	C:P	C:N
Winter								
15/09/80	29	5.3	30	5.7	36	10.0	84	13.3
Spring								
25/10/80	39	5.6	37	6.5	42	10.0	93	9.3
Summer								
15/01/81	122	4.4	57	12.8	47	11.4	127	11.2

Table 2. Photosynthetic characteristics of phytoplankton from Heywood and Sombre lakes, 1980. See text for explanation of symbols, I_k is expressed as $\mu E\ m^{-2}\ s^{-1}$ and P_{max} as mg C (mg chlorophyll-a)$^{-1}$ h^{-1}

	Sombre Lake			Heywood Lake		
Month	I_k	P_{max}	P_{max}/I_k	I_k	P_{max}	P_{max}/I_k
Jan.	38	1.8	0.04	35	2.8	0.08
Feb.	55	6.1	0.11	–	–	–
March	34	11.2	0.33	31	9.2	0.30
April	20	5.6	0.29	26	7.3	0.28
May	15	4.3	0.29	20	3.0	0.15
June	14	3.7	0.26	15	2.2	0.15
July	16	2.4	0.15	15	1.3	0.09
Aug.	16	1.8	0.11	11	0.4	0.04
Sept.	–	–	–	16	1.1	0.07
Oct.	40	4.6	0.08	35	4.4	0.13
Nov.	35	3.2	0.09	34	3.3	0.10
Dec.	–	–	–	40	1.8	0.05

4.2 Nutrient Regulation of Spring Growth

In oligotrophic Sombre Lake, P concentrations are low at the commencement of spring growth except at the base of the water column (Fig. 5), where anoxic winter conditions result in the dissolution of sediment-bound phosphates (Hawes 1983b; Gallagher 1985). The depth gradient of temperature in this lake confers a stability on the water column which effectively prevents movement of water or solutes between depths (Gallagher 1985). As the concentration of chlorophyll-a rises, P falls to undetectable levels in surface and mid-water samples; bio-assays indicate a P deficiency, and further increases in standing crops are nutrient limited. Phosphorus is more readily available in the deeper water and the majority of the spring increase occurs here, despite the lower radiation levels (Fig. 5). Phosphorus limitation is known to limit growth rates (Senft 1978; Gotham and Rhee 1981) and, consistent with this, the shallow, P-limited populations showed lower light-saturated rates of photosynthesis than those in deeper water (Fig. 6). Where adaptation is to radiation levels only, populations at depth show lower satu-

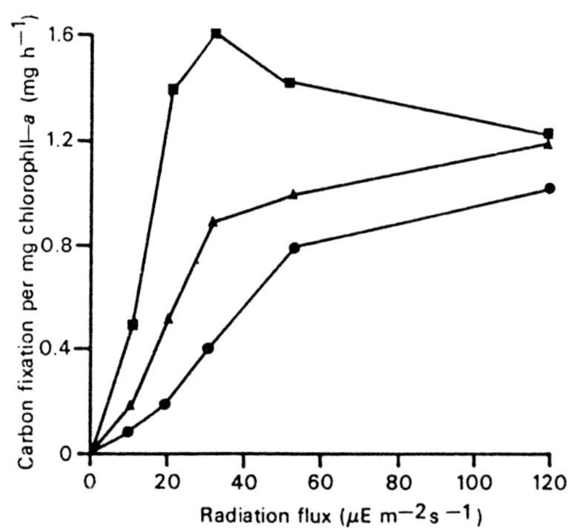

Fig. 6. Relationships between radiation flux and phytoplankton photosynthesis in samples from different depths in Sombre Lake under ice cover in spring. *Dots, triangles* and *rectangles* represent values from samples taken at the bottom, mid-water and surface, respectively

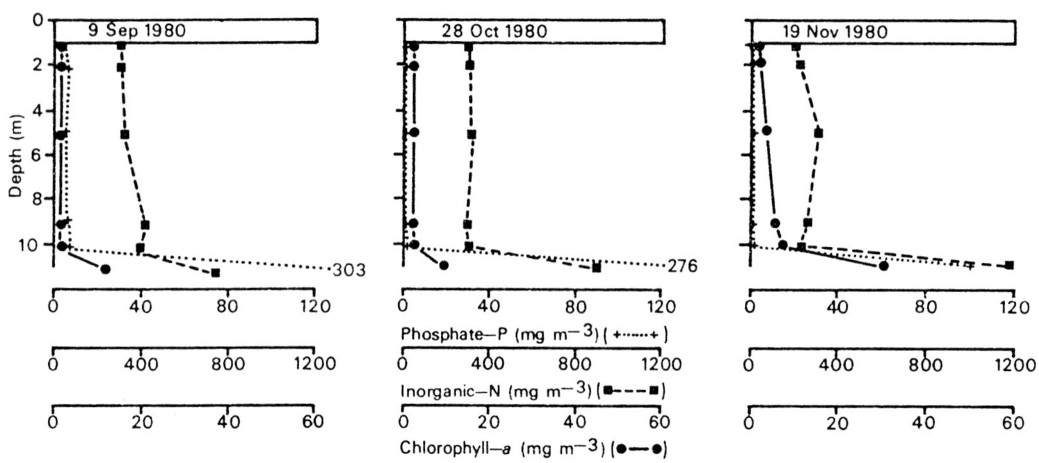

Fig. 5. Development of vertical profiles of phosphate, inorganic nitrogen and chlorophyll-a from late-winter through spring in Sombre Lake

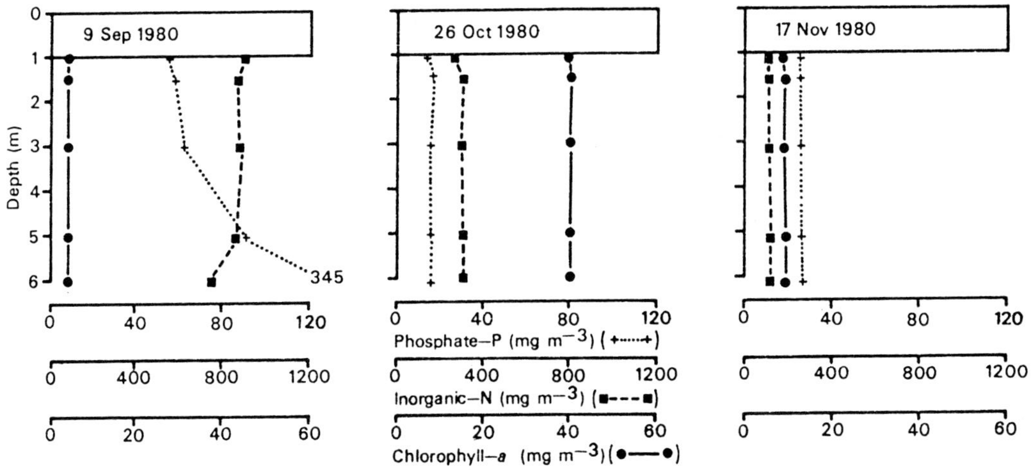

Fig. 7. Development of vertical profiles of phosphate, inorganic nitrogen and chlorophyll-a from late-winter through spring in Heywood Lake

rated rates of photosynthesis than shallow ones (Tilzer and Schwarz (1976). Clearly, low nutrient concentrations have an overriding effect on potential growth rates in mid-water, whereas at 10- and 11-m depths, where radiation levels are only 10–15% of that penetrating the snow-ice layer, radiation becomes more important and limits the rate at which available nutrients can be incorporated.

In the more eutrophic Heywood Lake, nutrient concentrations are high throughout the water column at the end of winter (Fig. 7). This permits high growth rates to be maintained, and a considerably denser algal population develops than that in Sombre Lake, before first P and later N become limiting, as the winter accumulation is depleted. The high P concentrations in this lake permit N to be depleted much more severely than in Sombre Lake, and N becomes the single most limiting nutrient. As limitation becomes more acute, the light-saturated rate of photosynthesis (hence potential growth rate) falls (Table 2) and, as in Sombre Lake, the productivity of the lake is limited both by the low radiation levels and the inability of the plankton to effectively utilize them. Light levels at depth are particularly low during the bloom, because of the rapid vertical extinction of light by the dense phytoplankton. In

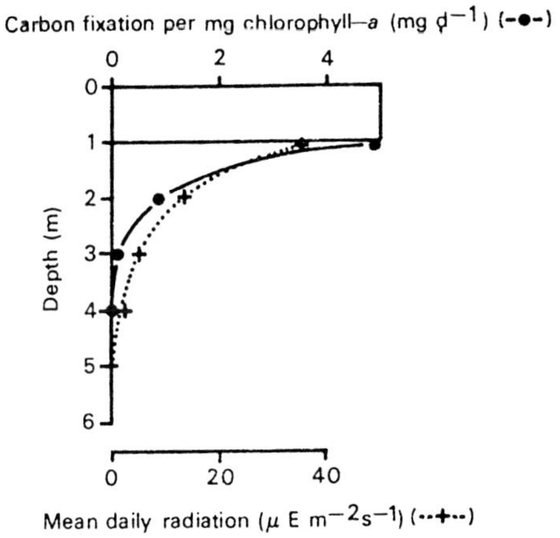

Fig. 8. In situ photosynthesis and mean daily radiation under ice in spring in Heywood Lake

situ ^{14}C incubations show the euphotic zone to be restricted to the upper 2 m (Fig. 8). It is possible that further population increases are limited by the phenomenon of self-shading (Talling 1971), and that the maximum biomass possible under prevailing radiation levels has been reached.

The homogenous vertical distribution of nutrients and chlorophyll-a in Heywood Lake appears to be maintained by under-ice turbulent mixing (Hawes 1983b). Thus, there is a temporal, rather than spatial, distinction between surface and bottom samples and light and nutrients are equally important in determining growth rates at all depths. The spring population crashes when the light and nutrient-limited growth rate fails to exceed losses from the water column. In Heywood Lake, these losses can be attributed to two main factors, outflow and grazing. Recordings of discharge show high outflow rates to occur shortly after spring growth reaches a maximum (Fig. 9) and, in Heywood Lake, because of mixing, this outflow has a similar chlorophyll content to the lake itself. If outflow loss is assumed to be drawn evenly from throughout the lake, up to 18 mg chlorophyll-a m^{-3} would have been lost by the end of November 1980, representing 35% of the bloom. Nutrients tied up in this material will also be lost from the system.

Unlike the permanently ice-capped lakes of the southern Victoria Land dry valleys (Goldman et al. 1972; Vincent 1981), herbivorous zooplankton occur at Signy Island and grazing appears to represent the second main source of loss of bloom material. Of particular importance in Heywood Lake are the larvae of the anostracan *Branchinecta gaini.* These hatch almost synchronously just after the bloom has begun and feed in the plankton, where they may attain densities of 40,000 individuals m^{-3} before adopting the benthic adult habit. Preliminary studies of "grazing" rates using radio-labelled algae, indicate a consumption of up to 0.08 mg chlorophyll-a per animal d^{-1} during the decrease of the spring maximum. At this rate, a further 40% of the bloom may be lost, leaving only a small proportion for which the fate is not known. Possibly because of the low standing crops of phytoplankton during spring, Sombre Lake does not support a large population of *B. gaini* and

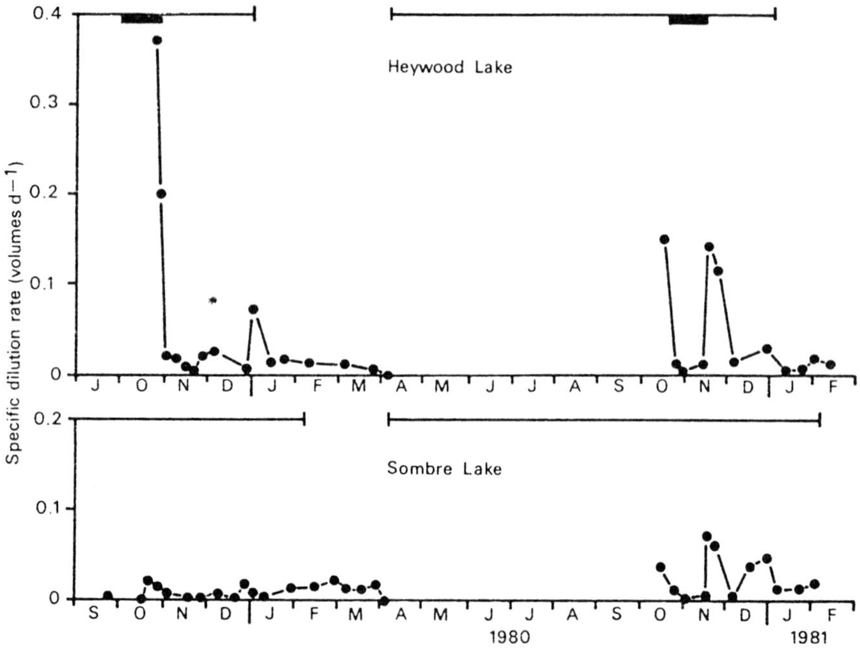

Fig. 9. Seasonal variations in specific dilution rates of Heywood and Sombre lakes. *Line* at the *top* indicates duration of ice cover, *black bar* indicates duration of spring bloom in Heywood Lake

grazing pressure is considerably less. In this lake, too, inflowing melt-water forms a relatively discrete, low density layer immediately beneath the ice in spring which does not mix with the resident lake water (Hawes 1983b). This is comparable to that described by Rigler (1978) in the High Arctic Char Lake. Since most of the algal material and dissolved nutrients lie well below this layer, little is lost via this route either. Hence, any loss of algae from Sombre Lake is likely to be due to the death of cells within the water column or sedimentation and, consequently, the chlorophyll-a concentration does not fluctuate as dramatically as in Heywood Lake.

At the end of the spring growth phase, the conditions in these two lakes are quite different. In eutrophic Heywood Lake, the high P concentrations have permitted the supply of inorganic N to be largely bound up as algal material, much of which has been subsequently lost from the system. Further increases in standing crop will be dependent on remineralization of this nutrient from organic remnants (grazers, dissolved organic material and sediment), of the spring bloom, the significance of which is at present unknown, and influx from the catchment.

In Sombre Lake, N concentrations remain high, while P is depleted both by algal activity and precipitation as particulate-bound phosphates when the bottom water becomes oxic (Gallagher 1985). Increases in algal biomass will again depend on resupply and/or recycling processes.

4.3 Open Water

During the ice-free period, bio-assays show the phytoplankton to remain nutrient limited. The high concentrations of P in streams entering Heywood Lake maintain the low N:P ratio and N remains the most limiting nutrient. The unenriched Sombre Lake catchment provides little P, and this remains the limiting nutrient throughout the summer (Hawes 1983a). Radiation levels are, however, greater and the summer rate of photosynthesis is high with maximum radiation-saturated rates of 9–10 mg C (mg chlorophyll-a)$^{-1}$ h^{-1} in vitro (Table 2) and 80 to 60 mg C (mg chlorophyll-a)$^{-1}$ d^{-1} in situ in Heywood and Sombre Lakes, respectively (Fig. 10). These rates confirm high values reported by Light et al. (1981) and, as these authors point out, are higher than many lakes at comparable latitudes with similar temperature regimes.

The small net input of nutrients calculated by Hawes (1983a) is inadequate to permit all of this carbon fixation to result in the formation of algal material. Using the C:N and C:P ratios from Table 1, the contribution of N to Heywood Lake would provide for only 1.7 g algal C m^{-2} yr^{-1} to be produced and the contribution of P to Sombre Lake is sufficient for 2.3 g m^{-2} yr^{-1}. Daily rates of photosynthesis reported by Light et al. (1981) and Hawes (1983a) approach or exceed these calculated annual values. There is clearly a severe discrepancy, and several observations suggest that this is attributable to recycling of fixed material within the water column.

First, there is no evidence for extensive production of algal material during summer. Chlorophyll concentrations are relatively stable, and losses to outflow and grazers are small. Similarly, the rapid changes in nutrient concentrations seen during the spring increases are not seen in summer, and uptake would appear to balance supply. Ellis-Evans (1981a,b) has demonstrated the occurrence of heterotrophic bacteria in Signy Island lakes, with an activity maximum during the open-water period, a prerequisite for effective cycling. Finally, Light et al. (1981) and Hawes (1983a)

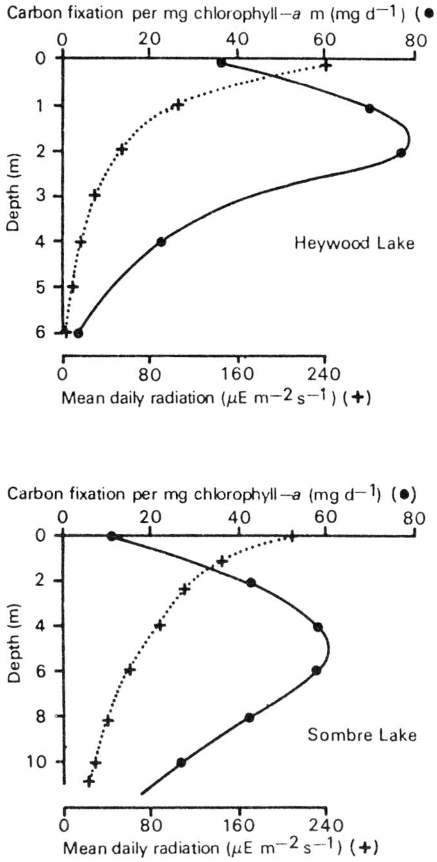

Fig. 10. In situ photosynthesis and mean daily P.A.R. levels in Heywood and Sombre lakes during the open-water period in summer

have shown that with the cessation of algal photosynthesis when ice-cover reforms, inorganic nutrients accumulate rapidly, indicating a continuous process of mineralization.

Rates of nutrient uptake and release have not been measured in these lakes, and the nature of the recycled material is not known. It may represent small molecules released by photosynthesizing algae (Fogg 1977), in which case C will be cycled faster than other nutrients or may result from the remineralization of complete algal cells, when all nutrients will be recycled. The short food chains in maritime Antarctic lakes, where planktonic predators are virtually absent may assist in this rapid turnover; herbivorous zooplankton having been shown elsewhere to contribute substantially to rapid recycling (Larow and McNaught 1978). Although biomass remains nutrient limited, recycling enables the lakes to attain high photosynthetic rates (hence absorption of available solar energy), a situation reported from other apparently nutrient-poor areas (Schindler and Nighswander 1970; Morris et al. 1971).

Acknowledgements. Thanks are due to the British Antarctic Survey for providing facilities for this work and access to unpublished information. Dr. J.C. Ellis-Evans commented on drafts of the manuscript.

References

Beardall J; Morris I (1976) The concept of light intensity adaptation in marine phytoplankton. Some experiments with *Phaeodactylum tricornutum.* Mar Biol 6:377–387

Ellis-Evans JC (1981a) Freshwater microbiology in the Antarctic I. Microbial numbers and activity in oligotrophic Moss Lake. Bull Br Antarct Surv 54:85–104

Ellis-Evans JC (1981b) Freshwater microbiology in the Antarctic II. Microbial numbers and activity in nutrient-enriched Heywood Lake. Bull Br Antarct Surv 54:105–121

Fogg GE (1977) Aquatic primary productivity in the Antarctic. Philos Trans R Soc Lond Series B 279:27–38

Gallagher JB (1985) The influence of iron and manganese on nutrient cylcing in shallow freshwater Antarctic lakes. In: Siegfried WR, Condy PR, Laws RM (eds). Antarctic nutrient cycles and food webs (Proceedings of the Fourth SCAR symposium on Antarctic Biology). Springer, Berlin Heidelberg New York

Goldman CR, Mason DT, Wood BJB (1972) Comparative study of the limnology of two small lakes on Ross Island, Antarctica. In: Llano GA (ed) Antarctic Terrestrial Biology, American Geophysical Union Antarctic Research Series 20, Washington 1–50

Gotham IJ, Rhee G-Y (1981) Comparative kinetic studies of phosphate-limited growth and phosphate uptake in phytoplankton in continuous culture. J Phycol 17:257–265

Harris GP (1978) Photosynthesis, productivity and growth: The physiological ecology of phytoplankton. Ergb Limnol 10:1–171

Hawes I (1983a) Nutrients and their effects on phytoplankton populations in lakes on Signy Island, Antarctica. Polar Biol 2: 115–126

Hawes I (1983b) Turbulence and its consequences for phytoplankton development in two ice-covered Antarctic lakes. Bull Br Antarct Surv 60:69–82

Hobbie JE (1973) Arctic limnology: A review. In: Britton ME (ed) Alaskan Arctic Tundra. Arctic Institute of North America Technical Paper No 25

Holdgate MW (1964) Terrestrial ecology in the maritime Antarctic. In: Carrick R; Holdgate M; Prevost J (eds) Biologie Antarctique. Paris, Hermann pp 173–179

Larow EJ, McNaught DC (1978) Systems and organismal aspects of phosphorus re-mineralization. Hydrobiologia 59:151–154

Kalff J, Welch HE (1974) Phytoplankton production in Char Lake, a natural polar lake and in Meretta Lake, a polluted polar lake, Cornwallis Island, Northwest Territories. J Fish Res Board Can 31:621–636

Light JJ, Ellis-Evans JC, Priddle J (1981) Phytoplankton ecology in an Antarctic lake. Freshwater Biol 11:11–26

Morris I, Yentsch CM, Yentsch CS (1971) The physiological state with respect to nitrogen of phytoplankton from low nutrient sub-tropical waters as measured by the effect of ammonium ion on dark carbon dioxide fixation. Limnol Oceanogr 16:859–868

Pechlaner R (1971) The phytoplankton spring outburst and its conditions in Lake Erken (Sweden). Limnol Oceanogr 15:113–130

Priddle J (1980) The production ecology of benthic plants in some Antarctic lakes. I. *In situ* production studies. J Ecol 68:141–166

Ramberg B (1979) Relations between phytoplankton and light climate in 2 Swedish forest lakes. Int Rev Gesamten Hydrobiol 64:749–782

Rigler FH (1978) Limnology in the High Arctic: A case study of Char Lake. Verh Int Ver Limnol 20:127–140

Samsel GL, Parker BC (1972) Limnological investigations in the area of Anvers Island, Antarctica. Hydrobiologia 40:505–511

Schindler DW and Nighswander JE (1970) Nutrient supply and primary production in Clear Lake, eastern Ontario. J Fish Res Board Can 27:2009–2036

Senft WH (1978) Dependence of light-saturated rates of algal photosynthesis on intracellular concentrations of phosphorus. Limnol Oceanogr 23:709–718

Talling JF (1957) Photosynthetic characteristics of some freshwater plankton diatoms in relation to underwater radiation. New Phytol 56:29–50

Talling JF (1971) The underwater light climate as a controlling factor in the production ecology of freshwater phytoplankton. Mitt Int Ver Theor Angew Limnol 19:214–243

Tilzer MM, Schwarz K (1976) Seasonal and vertical patterns of phytoplankton light adaptation in a high mountain lake. Arch Hydrobiol 77:488–504

Vincent WF (1981) Production strategies in Antarctic inland waters: phytoplankton ecology in a permanently ice-covered lake. Ecology 61:1215–1224

Decomposition Processes in Maritime Antarctic Lakes

J.C. ELLIS-EVANS[1]

Summary. The freshwater lakes on Signy Island have been the focus of an intensive research programme for almost 12 years. In this time, considerable information has been accumulated on planktonic and benthic decomposition. This paper brings together the above data and indicates the effectiveness with which organic material is decomposed by micro-flora in both oligotrophic and eutrophic systems. The majority of mineralization in both lake types is aerobic and concerned with readily assimilable organic material. Degradation of phytoplankton debris is slow and mainly occurs in surface sediments. Proteins, lipids, starchy materials, pectin and cellulose are degraded in the top 20 mm of sediment by aerobic and anaerobic heterotrophs. Although sulphate reducers may be present in substantial numbers, data from seasonal sulphate measurements and laboratory experiments indicate that organic sulphur decomposers are probably more significant in Antarctic lake sediments. Recently, methanogenic bacteria have been found in both oligotrophic and eutrophic lakes on Signy Island. High levels of methane build-up under winter ice cover, indicating significant activity even at such consistently low temperatures. From such information it is possible to assess the significance of decomposition by bacteria within Antarctic lakes of different nutrient status.

1 Introduction

During the past 12 years, the British Antarctic Survey has investigated the seasonal fluctuations in biomass and activity of both primary producers and decomposers in a wide range of lake types at Signy Island (Priddle 1980a,b; Light et al. 1981; Ellis-Evans 1981a,b,c, 1982). The factors influencing seasonality amongst the primary producers have been studied in some detail, as Hawes (1985) has shown. Comparable studies of the factors controlling decomposition in the lakes are as yet incomplete, but sufficient information has been accumulated to support a description of the range of microbial decomposer activities in Signy Island lakes.

2 Study Area

The lakes are described in detail by Heywood et al. (1980). The work presented here refers to five of these systems which conveniently span the full range of lake types. Ultra-oligotrophic Gneiss Lake is permanently ice-covered and has a poorly developed benthic community. Oligotrophic Moss Lake and Sombre Lake have well developed benthic plant communities, but Sombre Lake also develops a small anoxic sump in winter. Heywood Lake is meso-eutrophic with a dense phytoplankton in summer. Both Heywood Lake and eutrophic Amos Lake receive N and P enrichment from the presence of seals and birds in the catchments, but the effect is more pronounced in the latter system because of its smaller volume. All Signy Island lakes, with the exception of Gneiss Lake, are ice-free for 2–4 months each year and the annual temperature range in these lakes is 0°–6°C.

3 Methods

Most of the techniques employed here have been fully described elsewhere (Ellis-Evans 1981b, c, 1982, 1984); cotton-strip experiments have been described in Ellis-Evans (1981a) and the enzyme assays are described in Jones (1979). Chemical analyses were outlined in Table 1 of Heywood et al. (1980), with the exception of sulphate (Tabatabai 1974) and sulphide (Golterman 1969) analyses.

4 Results

4.1 Aerobic Decomposition

The South Orkney region is moderately windy (annual mean speed at Signy Research Station, 6 m s^{-1} and all the Signy Island lakes are relatively shallow (16 m). During the summer period when ice cover is absent, the lakes are subject to complete wind-induced mixing. Under winter ice and snow cover, such mixing cannot occur and relatively stable vertical gradients may develop (but see Hawes (1985)). During this time, snow cover markedly reduces light penetration of the ice cover and reduces primary production, but respiration continues, leading to oxygen depletion, particularly in the more enriched lakes (Fig. 1). In Moss

1 British Antarctic Survey, NERC, High Cross, Madingley Road, Cambridge CB3 OET, Great Britain

Antarctic Nutrient Cycles and Food Webs
(ed. by W.R. Siegfried, P.R. Condy, and R.M. Laws)

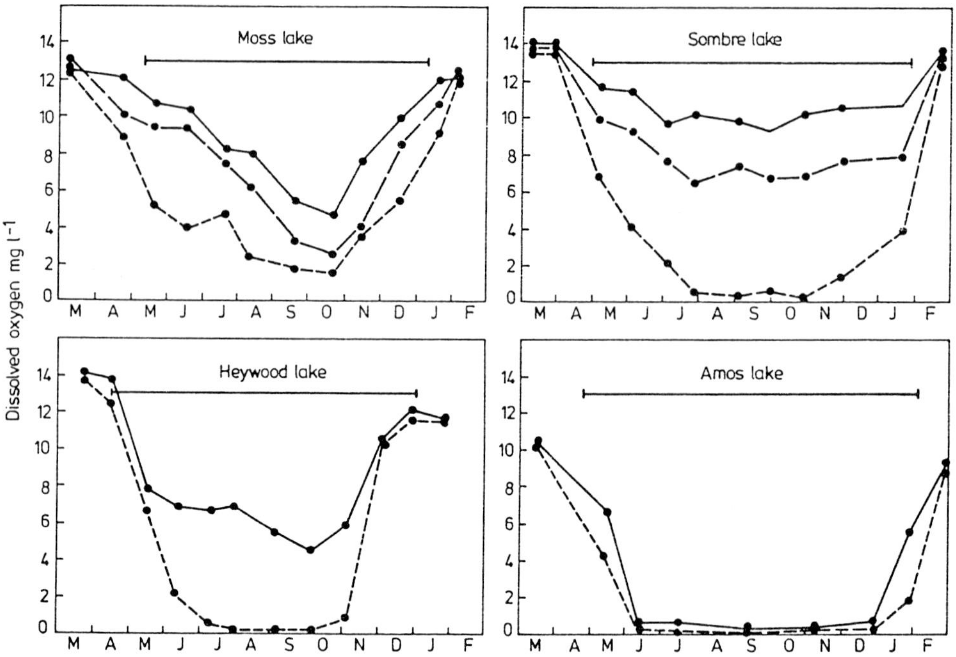

Fig. 1. Seasonal fluctuations in dissolved oxygen concentrations for 4 Signy Island lakes. —— surface waters, — — mid depth, – – – – bottom water, *Horizontal bars* indicate duration of ice cover

Lake and Gneiss Lake (not illustrated) oxygen concentrations never attain limiting levels (< 2 mg l^{-1} Ellis-Evans 1982)), but limitation invariably occurs throughout much of the water column of Heywood and Amos Lakes and in the sump region of Sombre Lake. It therefore follows that in summer, water column decomposition is concerned with aerobic processes, whereas in winter the trend of increasing anoxia with increase in trophic status suggests a potentially significant role for anaerobic bacteria in surface sediments and in bottom waters of the enriched lakes.

4.2 Uptake and Mineralization of Small DOC Molecules

In all lakes so far studied, both bacterio-plankton numbers (direct microscopic and plate-count techniques) and activity (uptake of ^{14}C uniformly labelled glucose or acetate) essentially follow seasonal fluctuations in phytoplankton numbers and activity (Fig. 2). More recent studies using a broader range of carboxylic and amino acids with ^{3}H labelling (to increase sensitivity) have confirmed that lake bacteria utilize a broad spectrum of organic compounds. Seasonal rates of acetate uptake are slow, ranging from 2–26 ng l^{-1} h^{-1} for Moss Lake to 19–101 ng l^{-1} h^{-1} in enriched Heywood Lake and 32–442 ng l^{-1} h^{-1} in grossly enriched Amos Lake. These data show a direct correlation with population size (direct counts) and indicate that, given the cold temperature regime, it is probably substrate availability that governs the amount of bacterio-plankton. Turnover times (Fig. 2), in the range 204–347 h for acetate in Moss Lake, are indicative of low physiological activity in the bacterio-plankton year-round. In contrast, both Heywood Lake and Amos Lake show considerable seasonal variation in acetate turnover times (range 27–550 h),

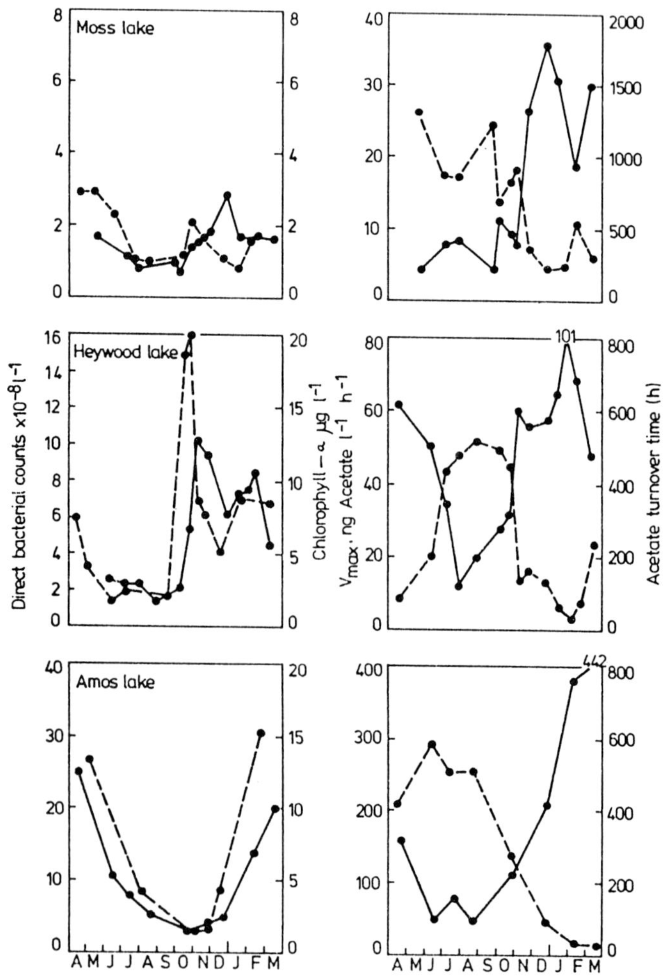

Fig. 2. Seasonal fluctuations in chlorophyll-a (– –), direct bacterial counts (——), heterotrophic potential (——) expressed as V_{max} and acetate turnover (– –) in three lakes of differing trophic status. Acetate V_{max} values and turnover times were calculated as described in Ellis-Evans (1981a)

suggesting low activity in winter and relatively high activity in summer.

Uptake of small DOC molecules by aerobic sediment populations proceeds at rates up to 100 times faster than in the overlying water-body. Sediment bacteria populations are also several orders of magnitude higher than planktonic populations, but preliminary work using micro-autoradiographic techniques indicate that only a small proportion of the population is actively metabolizing. Calculations show that activity per cell is higher in the sediments than in the overlying water. Laboratory experiments indicate that uptake of most readily assimilable organic substances is optimal in the range 16°–22°C, so the process is always temperature-restricted in Signy Island lakes.

4.3 Macro-Molecule Utilization

A major proportion of aerobic bacteria from Signy Island lakes are capable of proteolysis and amylolysis (Ellis-Evans 1981b,c, 1982). A smaller, but still significant proportion has the potential for lipolysis. Proteolytic and amylolytic bacteria were a greater proportion of the total population in winter than in summer. This may represent a switch in utilization pattern from uptake of small DOC molecules to macro-molecule degradation.

Water samples from Moss Lake and Heywood Lake have been filtered through 0.2 μm membrane filter and Whatman No. 1 paper filters to separate free enzyme activity from membrane-bound activity. The results have shown that virtually all amylase and proteinase activity is membrane-bound, since no detectable free enzyme activity was found in any ot the samples. Data for amylase activity are presented in Table 1. It can be inferred from the results that hydrolytic enzyme activity is located in the attached bacteria component of the bacterio-plankton and this is supported by the finding that attached bacteria populations have a higher percentage of proteolytic and amylolytic bacteria than the free living component. The temperature optimum for the amylase complex is around 20°C and maximum temperatures around 30°–37°C, which may reflect a close association with the temperature-sensitive cell membrane of these essentially psychotrophic bacteria (Herbert and Bhakoo 1979; Ellis-Evans 1981a).

In addition to macro-molecules, such as proteins, large carbohydrates and lipids, considerable amounts of cellulose, chitin and pectins are available for degradation in the lakes, and Table 2 indicates the occurrence of bacteria capable of utilizing macro-molecules in a range of lake types. The results indicate that whereas proteolytic, amylolytic and lipolytic organisms are of widespread occurrence, pectin utilization is virtually restricted to sediment samples; cellulolysis occurs only under anaerobic conditions and chitin is apparently not degraded.

In an attempt to assess cellulolytic activity, the cotton-strip method of Latter and Howson (1977) has been modified for use in lake sediment studies (Ellis-Evans 1981a). The results for Moss Lake shown in Fig. 3 demonstrate the virtual absence of cellulose breakdown, assessed as loss of tensile strength with time, in water overlying the sediments and slow but measurable degradation of the cotton strips within the sediments. There is often some carry-down during strip insertion, so it has proved difficult to establish the major site of cellulolysis within the sediment profile. However, whenever degradation was allowed to continue to fibre destruction, holes always occurred in the 10–60 mm strata. Both algal felt and moss-dominated sites were investigated, and no significant differences between sites was

Table 1. Enzyme activity in filtered lake-water samples

	(μg starch utilized l^{-1} d^{-1})							
	Membrane-filtered				Paper-filtered			
Month	Top-Water		Bottom-Water		Top-Water		Bottom-Water	
	SL	*HL*	*SL*	*HL*	*SL*	*HL*	*SL*	*HL*
August	ND	ND	ND	ND	ND	140 ± 81	160 ± 30	225 ± 49
September	ND	ND	ND	ND	ND	206 ± 92	280 ± 25	190 ± 84
October	ND	110[a]	140[a]	ND	ND	170 ± 75	170 ± 34	235 ± 80
November	ND	ND	ND	ND	190 ± 31	490 ± 247	260 ± 19	645 ± 259
December	ND	ND	ND	200[a]	220 ± 38	870 ± 283	180 ± 32	445 ± 176
January	ND	ND	ND	ND	240 ± 39	510 ± 155	310 ± 25	620 ± 311
February	ND	ND	ND	ND	180 ± 52	490 ± 203	260 ± 22	545 ± 292

All values are means of 10 replicates ± 1 S.D. ND = not detectable
Abbreviations: SL = Sombre Lake, HL = Heywood Lake
a Suspected contamination

Table 2. Occurrence of heterotrophic bacteria capable of utilizing macro-molecules in the four study lakes. Clearing of the agar medium constituted a positive reaction in all cases. Clearing on starch plates had to be detected by addition of half strength Lugols Iodine

		Casein		Starch		Tributyrin		Cellulose		Chitin		Pectin	
		w	m	w	m	w	m	w	m	w	m	w	m
Upland	Gneiss Lake	+	+	+	+	+	+	–	–	–	–	–	–
	Moss Lake	+	+	+	+	+	+	–	a	–	–	–	+
Coastal	Heywood Lake	+	+	+	+	+	+	–	a	–	–	–	+
	Amos Lake	+	+	+	+	+	+	–	–	–	–	+	+

Abbreviations: w = water, m = mud. + = positive reaction, – = negative reaction. a = positive under anaerobic conditions

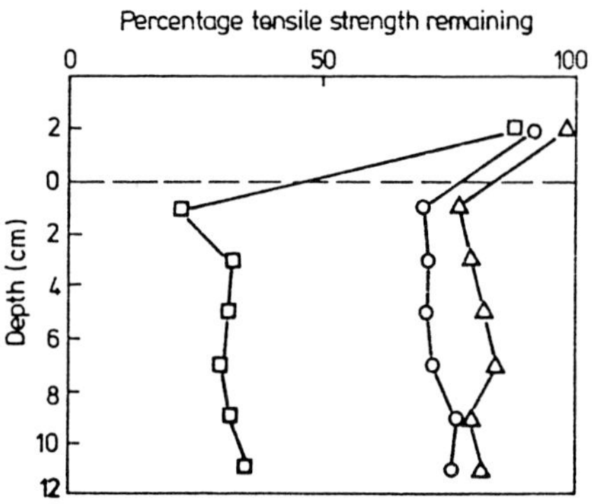

Fig. 3. Vertical distribution of cellulolytic activity in Moss Lake sediments expressed as percentage tensile strength remaining in cotton strips (compared to intact control strips) after 3 months (△), 6 months (○) and 9 months (□) incubation

established. The method could not, unfortunately, be employed in turbid lakes as visibility was inadequate for divers to operate.

4.4 Anaerobic Decomposition

From vertical sediment profiles, aerobes appear to be largely restricted to the water column and the top few centimetres of the sediments. Vertical Eh profiles of upland and coastal lake sediments support this finding, and indicate that anoxic conditions exist year-round just below the sediment surface in all three lakes studied (Fig. 4). In Moss Lake and the littoral zone of Sombre Lake, Eh profiles are extremely stable and Ellis-Evans (1984) has suggested that it is the benthic algal felt that endows this seasonal stability. In contrast, coastal lake sediments, which have no algal felt, are characterized by considerable seasonal variation in Eh profile. The presence of a more open sediment particle structure also undoubtedly contributes to the variations in Eh observed in the latter lakes.

Quantitative assessment of process rates by monitoring the appearance of ammonia, nitrogen, sulphide and methane, coupled with the disappearance of nitrate and sulphate, has proved extremely difficult. Free sulphide is rarely detected in Signy Island lakes, because it is rapidly precipitated as FeS due to the presence of excess Fe^{3+} ions for reduction (Gallagher 1985). The appearance of ammonia in water overlying anoxic sediments is complicated by the fact that chemical binding of ammonia occurs in freshwater sediments (Jones et al. 1982). An obvious solution to such problems is to utilize labelled substances (^{35}S, ^{15}N) in short-term incubations and this approach is currently under investigation. For now, the relative importance of fermentative and redox-related processes to decomposition in Signy Island has to be elucidated from population size and distribution data coupled with laboratory experimentation.

4.5 Fermentative Processes

From work with bacterial isolates, it is clear that lake sediments do contain bacterial populations capable of degrading cellulose, starch and proteins anaerobically at environmental temperatures. Relatively few occur in ultra-oligotrophic Gneiss Lake, but substantial numbers are present in the other lake systems, particularly in the enriched lakes in winter. The distribution of fermentative organisms is not necessarily redox-related, but experiments with sediment cores using the enzyme assay methods described in Jones (1979) do indicate a transition from a tricarboxylic acid cycle-based metabolism to a fermentative one, with malate dehydrogenase and lactate dehydrogenase increasing with increasing depth (Fig. 5a). In contrast, estimates of protease activity in relation to depth (Heywood Lake sediment samples) indicate that peak activity is always associated with the sediment-water interface and decreases rapidly with depth (Fig. 5b). This shows hydrolytic enzyme activity is associated with the region of highest substrate concen-

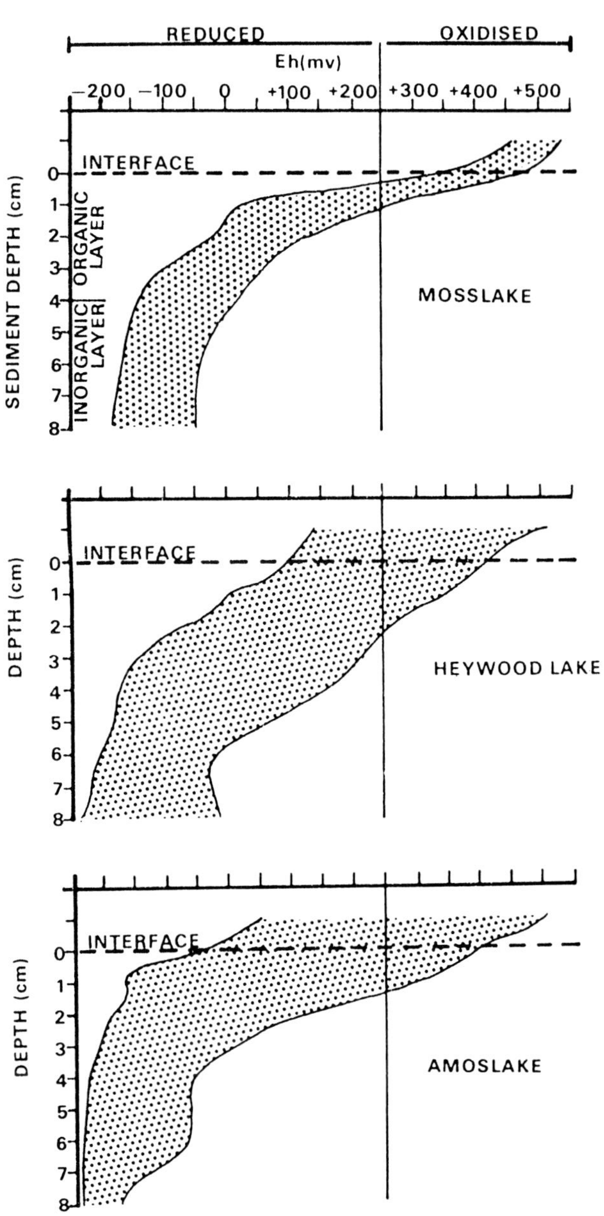

Fig. 4. Limits of seasonal Eh fluctuations in depth profiles of three lake sediments. For the purposes of this study +250 mV was taken as the nominal division between oxidizing and reducing conditions. Approximate depth of the brown organic layer is marked on the Moss Lake Figure

trations (sedimenting detritus) and is not subject to Eh status.

4.6 Nitrate Reduction

MPN counts (Fig. 6a) of denitrifying bacteria indicate that numbers peak in summer when surface sediments are once more aerobic. On the other hand, chemical data for bottom waters of Sombre Lake and Heywood Lake indicate winter disappearance of nitrate coupled with the appearance (Figs. 6b,c) of ammonia. Laboratory experiments indicate that although a large number of bacterial isolates can reduce nitrate to nitrite, relatively few can take nitrite through to N and the majority, under laboratory conditions, ferment nitrate to ammonia. This has considerable ecological relevance since, in the absence of substantial N-fixation, N will be lost from the ecosystem, whereas ammonia can be readily assimilated. Preliminary field data indicate that in Sombre Lake, nitrate fermenters ($NO_3^- \rightarrow NH_3^+$) dominate early winter populations of nitrate reducers, but that denitrifiers ($NO_3^- \rightarrow N_2$) become increasingly important as winter progresses. In Heywood Lake, nitrate fermenters are in the majority all winter. Data presented in Figs. 5b, 5c provide some support for these findings, but clearly the application of more sophisticated techniques (gas-chromatography, stable-isotope studies) are required to quantify the contribution of nitrate fermentation and denitrification. A further consideration is the significance of de-amination in ammonia production and particularly the degradation of urea, which is an important component of phytoplankton debris. Evidence from experiments with bacterial isolates indicate that nitrate reduction and urea breakdown are unimportant in oligotrophic lakes, but that de-amination may be of considerable significance in algal mats. Urea degradation has been detected in bacterial cultures from eutrophic lake sediments.

4.7 Sulphate Reduction

Dissimilatory sulphate reduction is a major anaerobic process in marine environments (Jorgensen 1980), but in Signy

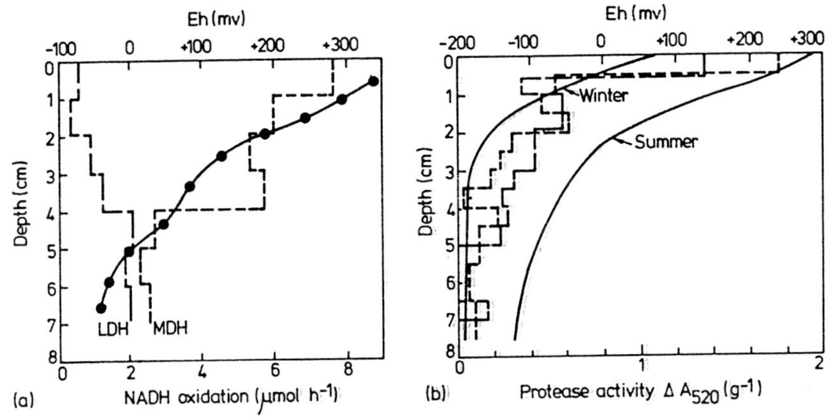

Fig. 5.a The vertical distribution of lactate dehydrogenase (*LDH*) and malate dehydrogenase (*MDH*) with respect to *Eh* (•—•) in a sediment profile. **b** Vertical distribution of protease activity in winter (— —) and summer (– – –) core profiles. *Eh* profiles for the same cores are also indicated

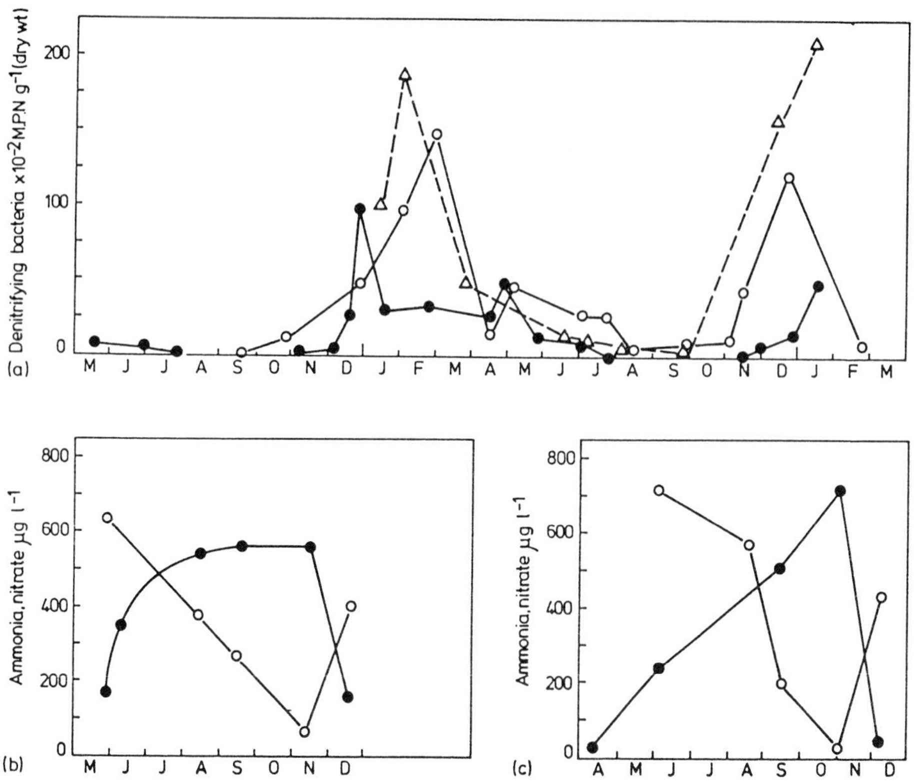

Fig. 6.a Seasonal fluctuations in numbers of denitrifying bacteria in the sediments of Moss Lake (●), Heywood Lake (○) and Amos lake (△). **b** Changes in ammonia (●) and nitrate (○) concentrations in bottom waters of oligotrophic Sombre Lake during the period of ice cover. **c** Changes in ammonia (●) and nitrate (○) concentrations in bottom waters of nutrient-enriched Heywood Lake

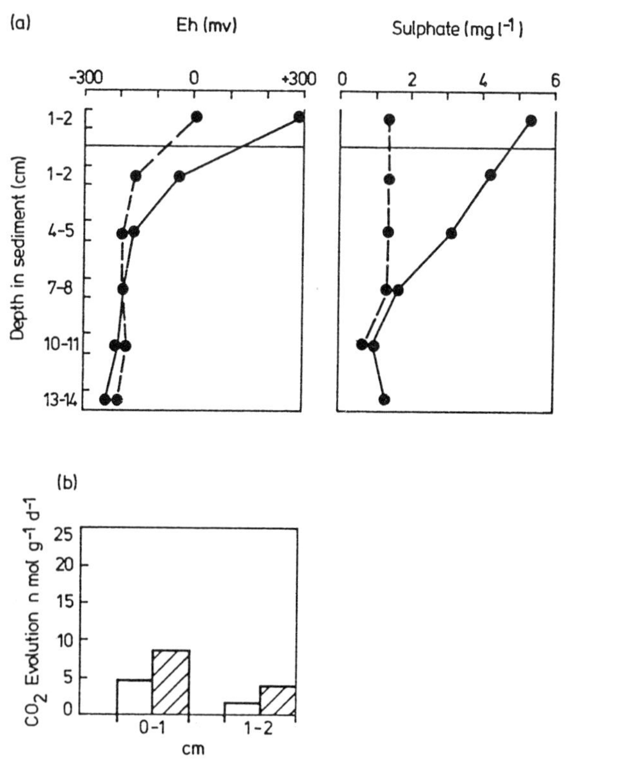

Fig. 7.a Data obtained with an interstitial sampler, showing changes in vertical distribution of sulphate and Eh between summer (——) and winter (– – –). **b** Differences in $^{14}CO_2$ evolution rate in the presence (□) and absence (▨) of sulphate (500 μM) at 2 depths in Sombre Lake sediment inoculated with ^{14}C-acetate

Island freshwater systems sulphate concentrations rarely exceed 60 μmol l^{-1} in the water column, and sediment interstitial concentrations (Fig. 7a) are invariably near the limits of detection in winter when anoxic conditions are most suitable for sulphate reduction (B. Gallagher, pers. comm.; Ellis-Evans 1984). Addition of sulphate to ^{14}C-acetate labelled cores did result in increased $^{14}CO_2$ production (Fig. 7b), but rates nevertheless were extremely low. Temperature optima for sulphate reducers from Antarctic lakes were similar to those previously reported from temperate systems (Jones et al. 1982), being around 32°C with very low activity at ambient lake temperatures. Since sulphate reducers are clearly not suited to the low temperature regime of the Signy Island lakes and permanently sulphate limited, it seems unlikely that sulphate reduction has any great significance in these systems.

Although sulphate concentrations change relatively little seasonally, sulphide has been found to accumulate in certain lakes (notably Amos Lake) under winter ice cover (Ellis-Evans 1982). This can probably be attributed to decomposition of sulphur-containing proteins by putrefactive bacteria. This has previously been proposed for certain other freshwater environments (Molongoski and Klug 1980; Jones et al. 1982), and the presence of substantial numbers of such bacteria in Signy Island lake sediments (Herbert and Bell 1973) suggest a similar situation. Forsberg (1980) has shown that dissimilatory sulphate reducers can produce sulphide from sulphur amino acids, so the two processes of sulphide production are clearly not mutually exclusive with respect to the organisms involved.

4.8 Methane Production

Substantial quantities of dissolved methane have been found in all the lakes studied, including ultra-oligotrophic Gneiss Lake (Ellis-Evans 1984). Methane production is a strictly anaerobic process and Gneiss Lake sediments are essentially aerobic throughout the year. Nevertheless, biogenic methane production has been confirmed to occur in these sediments. On this basis, methanogenesis is potentially of widespread occurrence in the maritime Antarctic.

The addition of sulphate to lake sediments markedly reduces conversion of $^{14}CO_2$ to methane (Fig. 8a) indicating an interaction between sulphate reducers and methanogens. Bryant et al. (1977) have shown with anaerobic mixed cultures that sulphate reducing bacteria can function as fermentative organisms and that in the absence of sufficient sulphate, certain genera may become net hydrogen donors to those methanogens utilizing CO_2 and H_2 as methane precursors. Jones et al. (1982) have produced evidence to indicate it also occurs in a eutrophic temperate lake. Comparison of acetate and CO_2/H_2 as methane precursors (Fig. 8b) indicates that the latter is probably the more important pathway in Antarctic lakes (Ellis-Evans 1984), though acetate is clearly also used to a significant extent.

4.9 Significance of Anaerobic Processes

Recent work by Ellis-Evans (1982) has suggested that virtually all autochthonous production in oligotrophic system at Signy Island can probably be accounted for in terms of aerobic decomposition. Recent work in Sombre Lake indicates that methanogens utilize approximately 13% of the annual sedimentation component (Ellis-Evans 1984), so even in oligotrophic systems there exist a measurable anaerobic decomposition component. In the eutrophic lakes, a significant proportion of autochthonous production could not be accounted for by aerobic decomposition, "grazing", outflow losses or sedimentation (Ellis-Evans 1982; Hawes 1984). In addition, there is a substantial allochthonous input from the catchment and this, coupled with the extent of anoxia which develops in eutrophic lakes at Signy Island in winter, suggests that anaerobic processes could have greater significance in these systems. However, micro-biological evidence to date suggests that anaerobes are unlikely to be capable of degrading more than a small proportion of the total organic input to enriched systems and that much of the material is instead ultimately incorporated into permanent deposits. The presence of virtually intact material of terrestrial origin over 200 mm beneath the surface of Heywood Lake sediments supports this conclusion.

Acknowledgements. I am grateful to the British Antarctic Survey for providing facilities to undertake this work and particularly the Signy Island Base members for their assistance in the field.

References

Bryant MP, Campbell LL, Reddy CA, Carbill MR (1977) Growth of *Desulfovibrio* in lactate or ethanol media low in sulfate in association with H_2-utilizing methanogenic bacteria. Appl Environ Microbiol 33:1162–1169

Ellis-Evans JC (1981a) Freshwater microbiology in the Antarctic I- Microbial numbers and activity in oligotrophic Moss Lake. Bull Br Antarct Surv 54:85–104

Ellis-Evans JC (1981b) Freshwater microbiology in the Antarctic II- Microbial numbers and activity in nutrient-enriched Heywood Lake. Bull Br Antarct Surv 54:105–121

Ellis-Evans JC (1981c) Freshwater microbiology at Signy Island, South Orkney Islands, Antarctica. PhD thesis (CNAA) 283 pp

Ellis-Evans JC (1982) Seasonal microbial activity in Antarctic freshwater lake sediments. Polar Biology 1:129–140

Ellis-Evans JC (1984) Methane in maritime Antarctic freshwater lakes. Polar Biology 3:63–71

Forsberg CW (1980) Sulphide production from cysteine by *Desulfovibrio desulfuricans.* Appl Environ Microbiol 39:453–455

Gallagher BJ (1985) The influence of iron and manganese on nutrient cycling in shallow freshwater Antarctic lakes. In: Siegfried WR, Condy PR, Laws RM (eds) Antarctic nutrient cycles and food webs (Proceedings of the fourth SCAR symposium on Antarctic biology). Springer, Berlin Heidelberg New York

Golterman HL (1969) Methods for Chemical Analyses of Fresh Water IBP Handbook No 8. Blackwell Oxford, Edinburgh 188 pp

Hawes I (1985) Factors controlling phytoplankton populations in maritime Antarctic lakes. In: Siegfried WR, Condy PR, Laws RM (eds) Antarctic nutrient cycles and food webs (Proceedings of the fourth SCAR symposium on Antarctic biology). Springer, Berlin Heidelberg New York

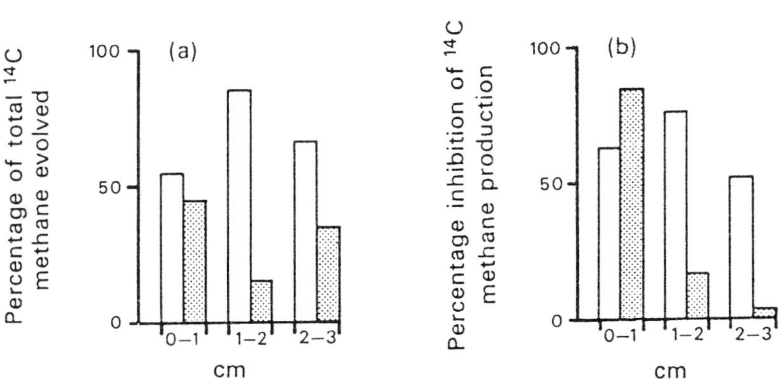

Fig. 8.a ^{14}C-methane evolved from (□) $^{14}CO_2$ and from (▨) ^{14}C-acetate at 3 depths in the sediment profile. Results expressed as a percentage of total ^{14}C-methane production. **b** Percentage inhibition of ^{14}C methane evolution from (□) $^{14}CO_2$ and (▨) ^{14}C-acetate in the presence of 500 μM sulphate at three depths in the sediment profile

Herbert RA, Bell CR (1973) Nutrient cycling in freshwater lakes on Signy Island, South Orkney Islands. Bull Br Antarct Surv 37: 15–20

Herbert RA, Bhakoo M (1979) Microbial growth at low temperatures. In Russell AD, Fuller R (eds) Cold tolerant microbes in spoilage and the environment. Academic, London pp 1–16

Heywood RB, Dartnall HJG, Priddle J (1980) Characteristics and classification of the lakes of Signy Island, South Orkney Islands, Antarctica. Freshwater Biol 10:47–59

Jones JG (1971) Studies on freshwater bacteria – factors which influence the population and its activity. J Ecol 59:593–613

Jones JG (1979) Microbial activity in lake sediments with particular reference to electrode potential gradients. J Gen Microbiol 115: 19–26

Jones JG, Simon BM, Horsley RW (1982) Microbiological sources of ammonia in freshwater lake sediments. J Gen Microbiol 128: 2823–2831

Jorgensen BB (1980) Mineralization and the bacterial cycling of carbon, nitrogen and sulphur in marine sediments. In Ellwood DC *et al.* (eds) Contemporary Microbial Ecology. Academic, London, pp 239–25

Latter PM, Howson G (1977) The use of cotton strips to indicate cellulose decomposition in the field. Pedobiologia 17:145–155

Light JJ, Ellis-Evans JC, Priddle J (1981) Phytoplankton ecology in an Antarctic lake. Freshwater Biol 11:11–16

Molongoski JJ, Klug MJ (1980) Anaerobic metabolism of particulate organic matter in the sediments of a hypereutrophic lake. Freshwater Biol 10:507–518

Priddle J (1980a) The production ecology of benthic plants in some Antarctic lakes. I *in situ* production studies. J Ecol 68:141–153

Priddle J (1980b) The production ecology of benthic plants in some Antarctic lakes. II laboratory physiology studies. J Ecol 68:155–166

Tabatabai MA (1974) Determination of sulphate in water samples. Sulphur Inst J 10:11–12

Interactions of Bacterio- and Phyto-Plankton in Nutrient Cycling Within Eutrophic Heywood Lake, Signy Island

J.C. ELLIS-EVANS[1]

Summary. Recent estimates of whole lake carbon budgets within eutrophic Heywood Lake and oligotrophic Sombre Lake, have revealed a significant disparity between production (as measured by ^{14}C) and turnover values in the eutrophic system. Whereas the turnover of photosynthetically fixed C in Sombre Lake could be adequately explained in terms of bacterial decomposition, grazing, sedimentation and loss to the outflow, up to 70% of Heywood Lake production remained unexplained by the above processes. In Heywood Lake, carbon dioxide can become limiting to phytoplankton populations at certain periods of the production season. Bacterial breakdown of algal debris and extra-cellular products of photosynthesis (detected by ^{14}C method) is too slow to generate the necessary levels of carbon dioxide, so a second extra-cellular DOC pool comprising small readily assimilable substrates was postulated. Using a variety of techniques, evidence was found to support this proposal. The possibility that N and P are also rapidly recycled was investigated using axenic algal cultures. Preliminary experiments indicated that bacteria were having a beneficial effect on algal production in both N and P-depleted environments. It is suggested that such mechanisms could explain the high production values recorded from enriched Antarctic lakes.

1 Introduction

It has been evident for some time that whereas turnover of photosynthetic production in oligotrophic maritime Antarctic lakes could be readily accounted for by bacterial decomposition, grazing, sedimentation and loss to the outflow, a large proportion of eutrophic Heywood Lake production (up to 70%) remained unexplained by the above processes (Ellis-Evans 1981a,b; Hawes 1983a). Recent work by Hawes (1983b) has shown that despite the high photosynthetic rate, fluctuations both in biomass and nutrient concentrations were relatively small. It was concluded that the available pool of nutrients restricted the standing crop, but that rapid recycling and the net nutrient input permitted high growth rates, as previously suggested by Kalff and Welch (1974) for high latitude Arctic lakes. Only one instance of direct nutrient limitation on photosynthetic rate was observed in Heywood Lake and this occurred under spring ice cover when demand for inorganic C by the developing phytoplankton outstripped the rate of supply. The work described here was initially concerned with the possible role of bacterio-plankton in supplying inorganic C to spring phytoplankton populations, but was subsequently broadened to include the supply of N and P to summer populations in Heywood Lake and oligotrophic Sombre Lake. Hawes (1983b) has established that N is the major limitation to summer phytoplankton development in Heywood Lake, whereas P is limiting in Sombre Lake.

1 British Antarctic Survey, NERC, High Cross, Madingley Road, Cambridge CB3 OET, Great Britain

2 Methods

Initial experiments were undertaken to assess the rate at which labelled phytoplankton debris and extra-cellular products were mineralized by natural populations. Uniformly ^{14}C-labelled algal debris and exudates were prepared as described by Cole and Likens (1979). Quintiplicate 500 ml quantities of 1.0 μm Nucleopore-filtered lake water, to which known quantities of labelled debris or exudates had been added, were incubated in opaque glass bottles at 3° or 10°C for 72 h. Preliminary experiments indicated that 1.0 μm Nucleopore filters removed all algal cells whilst permitting over 87% of lake bacteria to pass through. At the end of incubation, each water sample was acidified, and $^{14}CO_2$ degassed by a N stream to be trapped in an ethanolamine-cellosolve mixture as described by Cole and Likens (1979). Following addition of Picofluor-15(Packard) scintillation cocktail, each sample was counted by liquid scintillation.

Axenic cultures were prepared by standard sub-culturing techniques and two essentially pure cultures, tentatively identified as *Chlorella* sp. and *Rhodomonas palustris,* were obtained. These were employed to investigate the relationship between algae and bacteria under defined incubation conditions. Micro-autoradiography was used to establish the distribution of ^{14}C label amongst planktonic populations and the "stripping" technique of Hoppe (1977) was employed in all instances.

Differential filtration was always undertaken with Nucleopore filters and wherever possible gravity filtration was used. Slight vacuum was applied in the case of 0.2 μm filters which were found to remove all bacterial cells.

Antarctic Nutrient Cycles and Food Webs
(ed. by W.R. Siegfried, P.R. Condy, and R.M. Laws)

3 Results and Discussion

Figure 1a indicates that mineralization of ^{14}C-labelled phytoplankton debris was essentially linear over 72 h incubation period at both 3° and 10° C. Approximately 1.3% of the total label added (ca. 5×10^6 dpm) was mineralized during the incubation at 3°C, whereas 2.4%, was mineralized at 10°C. Clearly, the process is temperature-limited over the annual temperature range of 0°–5°C experienced in Heywood Lake. Control bottles contained 500 ml of heat-sterilized lake water plus a filter bearing the labelled algal debris, and the results indicated virtually no microbial or $^{14}CO_2$ contamination. Long-term incubations (Fig. 1 b) show that linearity continued up to ca. 50 days at 3°C and ca. 40 days at 10°C. Neither dissolved oxygen nor dissolved inorganic C concentrations changed markedly during the 72 h incubations, but changes were evident in the more lengthy incubations. Such changes probably alter rates of metabolism (Ellis-Evans 1982); microbial growth on the walls of the bottles may also be significant. Nevertheless, the percentage total label added which had been mineralized by the end of the linear phase was remarkably similar for both incubation temperatures (31–35%).

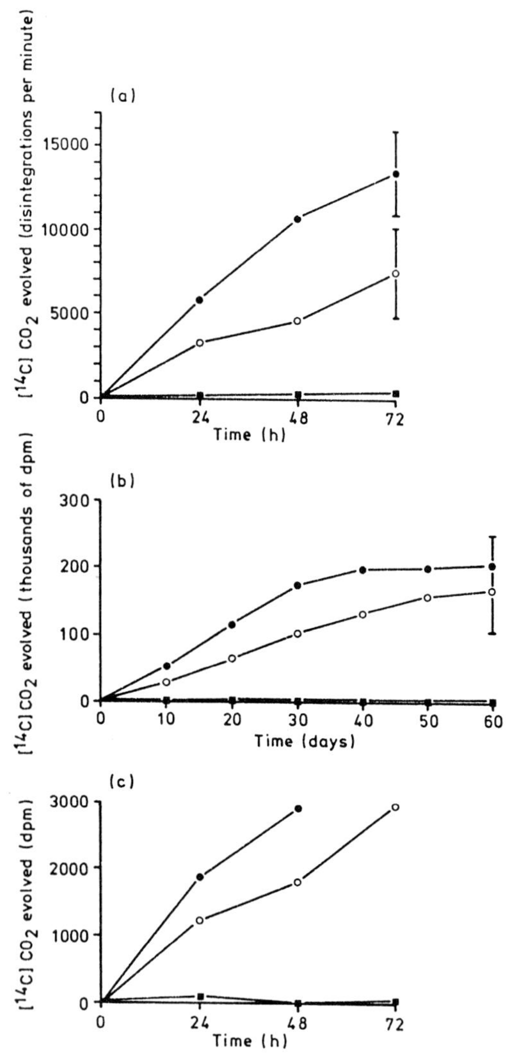

Fig. 1.a Mineralization of ^{14}C-labelled phytoplankton debris with time at 3°C (○) and 10°C (●). The heat sterilized control (■) was incubated at 10°C. Results are expressed as disintegrations per min^{-1} per sample. **b** Long-term experiment to demonstrate the relationship between mineralization and incubation time at different temperatures; 3°C (○), 10°C (●), control (■). **c** Mineralization of ^{14}C-labelled algal exudates at 3°C (○) and 10°C (●) over a 72 h period; control (■). Vertical bars represent ± 1 S.E.

Breakdown of ^{14}C-labelled algal exudate at 3° and 10°C was also linear over the incubation period (Fig. 1c) and approximately 10% of the total label was mineralized in 72 h. Long-term incubations at 3°C indicated that 57–71% of the labelled exudates were mineralized during the linear phase (ca. 20 days). The data indicate that exudates are mineralized more quickly and more completely than algal debris.

Conversion of the above data to CO_2 units yielded rates of 0.88 mg CO_2 m^3 d^{-1} and 0.35 mg CO_2 m^3 d^{-1}, respectively, for mineralization of algal debris and algal exudates by summer bacterial populations from Heywood Lake. The total (1.23 mg CO_2 m^3 d^{-1}) is equivalent to $<$0.1% of the daily phytoplankton requirement for CO_2 and significantly less than the daily deficit of 50 mmol m^3 for spring phytoplankton populations developing under thick ice cover (Hawes 1983b). Since microbial populations in early spring (Oct.–Nov.) are smaller than in summer (Ellis-Evans 1981b), it would seem unlikely that bacteria make a significant contribution to the CO_2 supply during the spring period of CO_2-limitation. However, non-axenic algae were used to prepare the labelled exudates and it is well established that the ^{14}C-labelling method employed here would only yield the more resistant exudate components, since the most readily assimilable substrata would have already been incorporated into bacterial POC or even mineralized.

A further series of experiments was therefore undertaken with axenic algal cultures. When a time-course experiment was carried out using exudates from these latter sources, a 2-stage process was invariably observed with both cultures (Fig. 2). The initial linear phase was comparatively rapid but short lived (ca. 15 h at 3°C), whereas the second stage was relatively slower and remained linear for the duration of the experiment. Controls using heat-sterilized water indicated some contamination but nothing significant. Micro-autoradiography indicated that labelled bacteria were present within 1 h of the experiment being initiated when axenic algal exudates were used, but appeared only after 2.5–3.0 h when non-axenic algal exudates were employed. The above data suggest that axenic culture exudates contain quantities of rapidly assimilable organic substances which are not detected in experiments utilizing non-axenic natural populations, because the attached bacteria remove these substances too quickly to be measured as DOC by conventional ^{14}C production methods.

The potential influence of CO_2 production by bacteria populations on phytoplankton activity can be assessed with

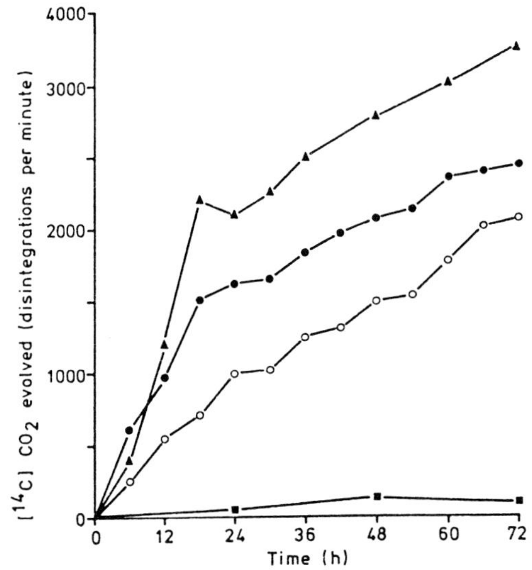

Fig. 2. Mineralization at 3°C of ^{14}C-labelled exudates from axenic algal cultures. *Chlorella* sp. (▲) and *Rhodomanas palustris* (●); non-axenic *Chlorella* sp. exudates (○) were also run for comparison

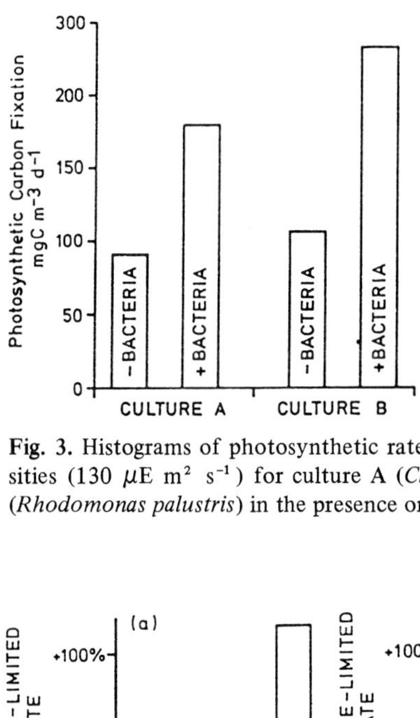

Fig. 3. Histograms of photosynthetic rates at saturating light intensities (130 μE m^2 s^{-1}) for culture A (*Chlorella* sp.) and culture B (*Rhodomonas palustris*) in the presence or absence of bacteria

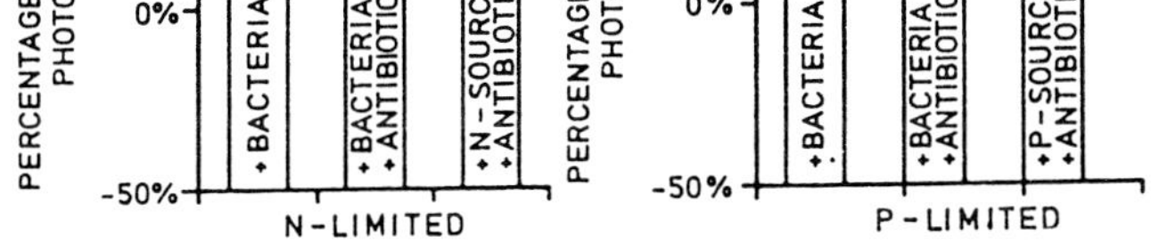

Fig. 4.a Percentage change in photosynthetic rate of nitrogen-limited *Chlorella* sp. in the presence of bacteria, bacteria plus antibiotics, or an N-source plus antibiotics. **b** Percentage change in photosynthetic rate of phosphorus-limited *Chlorella* sp. in the presence of bacteria, bacteria plus antibiotics, or a P-source plus antibiotics

axenic cultures. Both test cultures were incubated in 300 ml bottles (with no headspace) under constant saturating illumination (130 μE m^2 s^{-1}) and the photosynthetic rate estimated over the second 24 h period. Results presented in Fig. 3 demonstrate that the presence of bacteria markedly increased the photosynthetic rate, which suggests that rapid recycling of C was providing the necessary carbon dioxide. Further confirmation was provided by micro-autoradiography. Axenic algae were cultured under CO_2 limitation in the presence of labelled exudates only or exudates and bacteria. After 72 h, micro-autoradiographs were prepared and these revealed that ca. 27% of the algae were labelled in the presence of bacteria, whereas consistently $< 5\%$ were labelled in their absence. Although Satoh and Hanya (1976) have shown that algae can themselves bring about some re-mineralization of organics, it is more likely that the small numbers of algae labelled here in the absence of bacteria reflect $^{14}CO_2$ contamination.

These results indicate that two DOC pools exist in the Heywood Lake water column. The one comprises small readily assimilable compounds which are rapidly recycled between algae and bacteria and a second group of substances which are degraded more slowly. Particulate organic material constitutes a third source of organic C and has a very slow turnover time. It is not possible to quantify the importance of rapid recycling of C to Heywood Lake planktonic populations with the methods exployed here, but rapid recycling implies a rapid response and certainly laboratory cultures were shown to positively benefit from such a mechanism.

If C is rapidly recycled, it seems reasonable to suggest that N and P may also be recycled in this fashion. Water samples were therefore taken from Heywood Lake during the summer period when inorganic N was limiting (Hawes 1983b) and filtered through 0.2 μm filters to remove planktonic organisms. Axenic algal cultures were added to 300 ml aliquots and incubated under natural light conditions for a further 24 h–48 h. One set of bottles was left unamended to provide an estimate of photosynthetic activity under N limited conditions. Other sets were amended with the natural bacteria population, bacteria plus antibiotics (35 units penicillin per ml; 4 μg chloramphenicol per ml) or antibiotics plus N (250 μg NH_4 Cl l^{-1}). The bottles were incubated for 72 h at 5°C under saturating light intensities and then aliquots removed for ^{14}C analysis. Results presented in Fig. 4a indicate that the presence of bacteria enhanced the photosynthetic rate, whereas addition of antibiotics and bacteria resulted in negligible change. Addition of antibiotics and a N source again resulted in enhanced rates which show that the antibiotics were not affecting the algae. Repeating the experiment with Sombre Lake water which is P-deficient (Hawes 1983 b) revealed a similar pattern (Fig. 4b) with the presence of bacteria enhancing production, bacteria and antibiotics having negligible effect and antibiotics plus a P source (30 μg phosphate l^{-1}) again enhancing production.

4 Conclusion

The evidence presented here indicates that rapid recycling of nutrients is probably occurring in Heywood Lake. Since factors such as "grazing" outflow losses and sedimentation do not adequately explain the turnover of photosynthetically derived organic material, the process of rapid exchange proposed here may indeed be of major importance in enriched Antarctic lakes. Similarly, the process may have relevance to oligotrophic systems where phosphate is the major limiting factor, since results presented here indirectly show an exchange of phosphate between bacteria and algae which may permit greater photosynthetic activity than would otherwise be predicted. This could reasonably explain the high photosynthetic efficiency found in oligotrophic lake algae (Hawes 1983b).

Acknowledgements. I am grateful to the British Antarctic Survey for co-operation whilst undertaking this work, particularly members of the BAS base at Signy Island during summer 1981–82.

References

Cole JJ, Likens GE (1979) Measurements of mineralisation of phytoplankton detritus in an oligotrophic lake. Limnol Oceanogr 24:541–547

Ellis-Evans JC (1981a) Freshwater microbiology in the Antarctic. 1. Microbial numbers and activity in oligotrophic Moss Lake. Bull Br Antarct Surv 54:85–104

Ellis-Evans JC (1981b) Freshwater microbiology in the Antarctic. 2. Microbial numbers and activity in nutrient-enriched Heywood Lake. Bull Br Antarct Surv 54:105–121

Ellis-Evans JC (1982) Seasonal microbial activity in Antarctic freshwater lake sediments. Polar Biology 1:129–140

Hawes I (1983a) Turbulent mixing and its consequences for phytoplankton development in two ice-covered Antarctic lakes. Bull Br Antarct Surv 60:69–82

Hawes I (1983b) Nutrients and their effects on phytoplankton populations in lakes on Signy Island, Antarctica. Polar Biology 2:115–226

Hoppe H-G (1977) Analysis of actively metabolising bacterial populations with the autoradiographic technique. In: Microbial Ecology of a Brackish Water Enviroment (Rheinheimer G., ed):171–196 Springer, Berlin Heidelberg New York

Kalff J, Welch HE (1974) Phytoplankton production in Char Lake, a natural polar lake and in Meretta Lake, a polluted polar lake, Cornwallis Island, Northwest Territories. J Fish Res Board Can 31:621–636

Satoh Y, Hanya T (1976) Decomposition of urea by larger particulate fraction and the free bacteria fraction in a pond water. Int Rev Gesamten Hyrobiol 61:799–806

Part 3
Marine Food Webs

Antarctic Marine Food Webs

G. HEMPEL[1]

Summary. Traditional views on the productivity of the Southern Ocean have changed greatly. High primary production in the shelf region and near the pack-ice and the short diatom-Krill-whale food chain were first taken as indications of a very large renewable resource. However, it is now known that primary production in the open parts of the Southern Ocean is relatively low, and Soviet estimates of Krill, *Euphausia superba,* stocks have decreased by an order of magnitude. The pelagic Southern Ocean contains 3 major zones which, although they interact, may for convenience be considered as having distinct food webs. These zones are: the ice-free zone of the West Wind Drift dominated by herbivorous copepods, salps and small euphausiids; the seasonal pack-ice zone of the East Wind Drift and adjacent fronts and eddies in which Krill is the dominant element; and the permanent sea-ice zone near the Antarctic continent and its ice shelves where zooplankton biomass is low and, thus, predators are scarce. The Krill system is a product of a long evolutionary process in isolation. It is characterized by high seasonality in light and ice cover but otherwise stable conditions of low temperature and high nutrient supply. Within the Krill-dominated food chain unusually large size steps, and high energy demands for maintenance and swimming, require large concentrations of food to keep the effort for searching and feeding low during the short summer season. This pelagic system running on high energy costs is in contrast to the low energy system of the rich Antarctic benthos, particularly in pack-ice regions. Because survival in the Antarctic is energetically expensive, there is little net gain in the Krill system, despite its short food chain. However, the large biomass stored in Krill swarms and in large vertebrates has made the Krill system attractive to exploitation by man. Antarctic communities typically exhibit K-strategies but paradoxically there is great environmental variability in the pack-ice zone. Any model of food chain dynamics, or any management strategy, must take into account not only the low efficiency of the food chain but also the variability of its key elements.

1 Introduction

The previous Antarctic biology symposia, organized by SCAR (Scientific Committee on Antarctic Research), dealt with aspects of evolution, stability, endemism and diversity of the Antarctic marine ecosystem, as well as its biogeography, productivity and living resources. Food webs were described in general, and a few crude attempts were made to quantify Antarctic Krill, *Euphausia superba,* production. The discussions were based on a wealth of data taken during summer in the ice-free zone of the Southern Ocean.

Views on the productivity of the Southern Ocean have changed greatly over the years. High primary production in the shelf region and near the pack-ice, together with the short food chain (diatoms-Krill-whales) in the same areas, were taken first as indications of a very high, potentially exploitable, resource. This optimism has faded away. We now know that overall primary production in the oceanic parts of the Southern Ocean is much lower than previously supposed. The estimates by Soviet scientists of Krill stocks and their annual production have decreased by an order of magnitude (Lubimova et al. 1980). BIOMASS (Biological Investigations of Marine Antarctic Systems and Stocks) is now yielding a more detailed and realistic picture of the distribution and abundance of some of the key elements of Southern Ocean food webs (Hempel 1983). However, actual annual production at any trophic level and the quantitative interactions within the food webs are still largely unknown.

The Krill-dominated food web is unique in the World Ocean and has the potential of being a very important resource. Therefore, most research has focused on this component of the Antarctic marine ecosystem, ignoring the fact that the larger part of the Southern Ocean is occupied by food webs without much Krill.

In this paper I will deal briefly with certain key features of the Southern Ocean and its food webs in general, and then go on to review the Krill system in particular.

2 General Features

The Antarctic marine ecosystem is a semi-closed unit. The northern boundary is fairly well defined by the Antarctic Convergence in the upper 500–1000 m (Fig. 1). Whales and birds cross the Convergence regularly, and are responsible for the export of a certain amount of organic matter. In the deeper layers, there is a meridional transport of nutrients and zooplankton by the relatively warm Deep Water moving southward and by the Antarctic Bottom Water moving northward. At its continental boundary, the Southern Ocean is not influenced by terrestrial biota and river input.

1 Alfred-Wegener-Institut für Polarforschung, Columbus Center, 2850 Bremerhaven, FRG

Antarctic Nutrient Cycles and Food Webs
(ed. by W.R. Siegfried, P.R. Condy, and R.M. Laws)

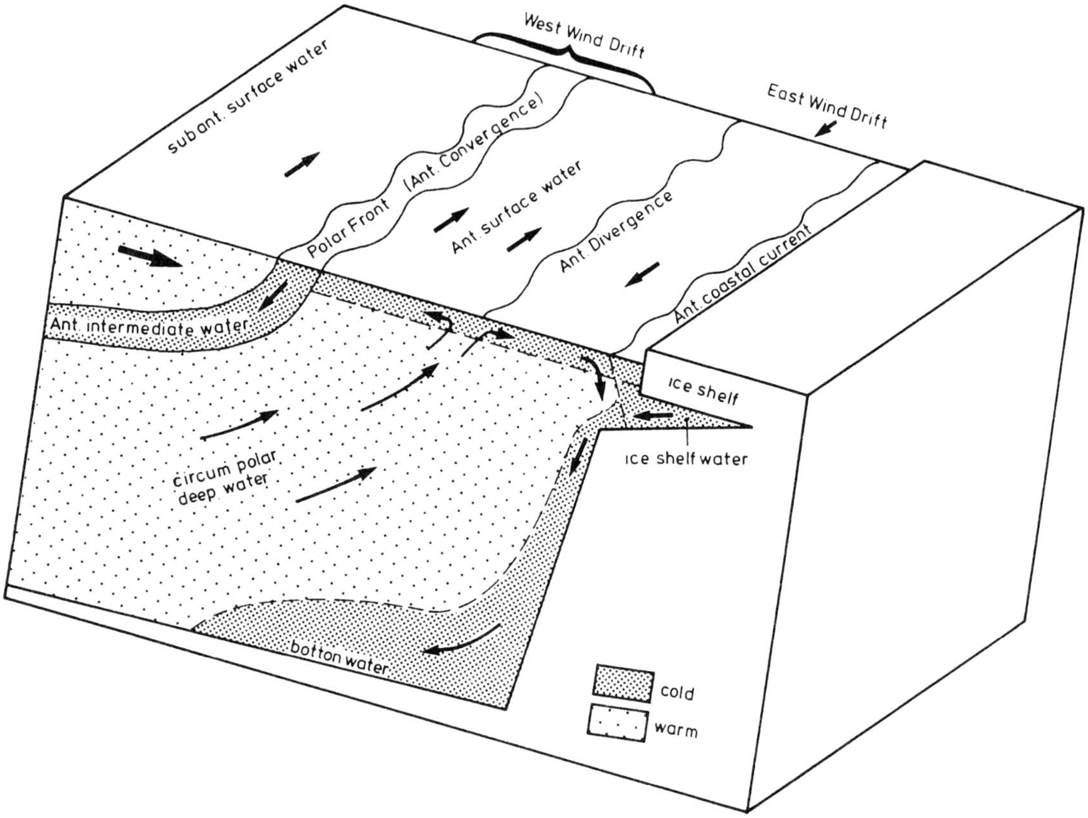

Fig. 1. Structure of Antarctic water-masses and current systems (modified after Gordon and Goldberg 1970)

The high degree of endemism and the eco-physiological adaptations of many of the species of the Southern Ocean are indications of a long evolutionary process in isolation, linked with rather unique community structures.

Primary production in the Southern Ocean seems to be limited mainly by light conditions. Vertical instability of the water column reduces the light exposure of algal cells, preventing the cells from making full use of the rich nutrient supply. Furthermore, about 50% of the total area of the Southern Ocean is covered by pack-ice at the end of the winter; this ice shrinks to a 10% coverage in the austral summer. Important effects of the ice cover are the shading of the water underneath, the layers of melt-water in spring and summer, and the liberation of large amounts of ice-algae at the break-up and melting of the pack-ice. Year-to-year variations in extent and timing of the ice cover must have a very large effect on the otherwise stable life conditions of the Southern Ocean.

3 Zonation

The Southern Ocean is not as ecologically uniform as is implied by the general term "Antarctic ecosystem". Topographically, it is a ring with the bottom generally deepening to the north. However, the Scotia Arc and the Kerguelen-Gaussberg Ridge and the large shallow bights of the Weddell and Ross Seas, as well as more local features, such as the Queen Maud Rise and Prydz Bay, cause eddies in the otherwise circular current system of the West Wind Drift and the East Wind Drift plus the Weddell Drift.

The Antarctic Convergence and the Weddell-Scotia Confluence tend to meander, and the Antarctic Divergence might not always be well defined. Satellite-tracked buoys, remote sensing of surface temperature, and frequent measurements by ships of opportunity, under the programme of the IOC Southern Ocean group and the SCAR/SCOR WG 74, aim at facilitating an understanding of the structure and variability of the Southern Ocean current system and its interaction with the ice cover. Recently, Deacon (1982) has discussed the various Antarctic boundary zones and their possible biological significance.

Biogeographical sub-divisions of the Southern Ocean vary with the taxonomic or ecological group they were made for. Obviously any scheme related to benthos, including demersal fish, is largely dependent on bottom topography and on the evolutionary immigration routes. In the pelagic zone, the relationship with water-masses, fronts and seasonal ice cover becomes more obvious. Thus, Hart (1942), for phytoplankton, and Voronina (1966) and Lubimova et al. (1980), for zooplankton communities, distinguished rather similar zones (Fig. 2). Only in areas of major oceanographic disturbance, such as in the vicinity of the Antarctic Peninsula and the Scotia Sea-South Georgia region, did Hart (1942) and subsequent authors find deviations from the zonal distribution pattern.

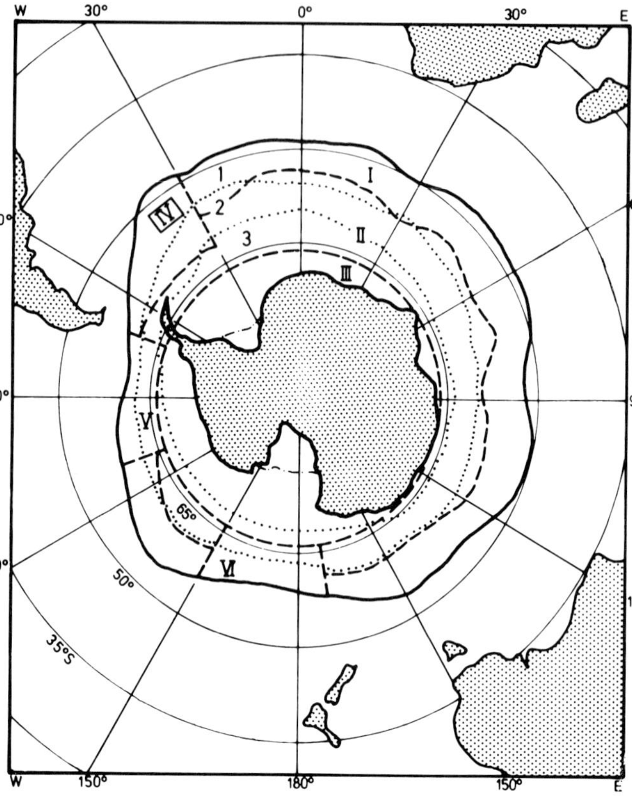

Fig. 2. Zonation of phytoplankton (*Roman numerals* and *dashed lines;* after Hart 1942) and zooplankton (*Arabic numerals* and *dotted lines;* after Voronina 1971) in the Southern Ocean

The three zones and their pelagic communities can be briefly described as follows:

3.1 Ice-Free Zone

The ice-free zone of the oceanic West Wind Drift is rich in nutrients but relatively poor in primary production and phytoplankton biomass. The zooplankton is rather similar to that of the northern North Atlantic. Herbivorous copepods, salps and small euphausiids are dominant. Except for South Georgia, Krill is mostly absent. Squid, myctophids and some juveniles of benthic fish species are present, but play a minor role in the West Wind Drift compared to the shoaling fish of the boreal zones of the North Atlantic and North Pacific. Flying birds tend to concentrate in the region of the Antarctic Convergence.

3.2 Seasonal Pack-Ice Zone

The seasonal pack-ice zone is covered by ice in winter and spring, but it is mainly ice-free in summer and autumn. It occupies most of the East Wind Drift and its large eddies along the Antarctic Divergence, and includes the northern branch of the Weddell Gyre and the waters off the Antarctic Peninsula. Large parts of the zone are comparatively shallow, particularly in the Atlantic sector. Possibly a certain stabilization of the upper layers in the vicinity of the pack-ice and the seeding by ice-algae make this zone the most productive one on a yearly basis. With the breaking up and retreat of the pack-ice in spring and summer, a series of phytoplankton blooms may proceed from north to south. Large amounts of phytoplankton sink to the bottom, out of this system. In spring, Schnack et al. (1984) measured a sedimentation rate of live *Corethron* and *Thalassiosira* aggregates amounting to one-third of the daily production in the euphotic zone above. The system comprises a complex food web with salps, copepods, euphausiids, fish larvae, and chaetognaths. However, Krill is the dominant element, which has opened the way for successful pelagic exploitation of the high primary production of the seasonal pack-ice zone. Shoals of Krill are sufficiently large and dense to provide the food base for baleen whales, Crabeater Seals, *Lobodon carcinophagus*, and Adélie Penguins, *Pygoscelis adeliae*, at low energy cost. The highly mobile mammals and birds are obviously better adapted for exploiting Krill than are most fish.

3.3 Permanent Pack-Ice Zone

The permanent pack-ice zone extends over the area of the cold near-shore water-masses (ice-shelf water) which are partly separated from the East Wind Drift by the Continental Convergence. These water-masses are particularly well developed in the shallow parts of the inner Weddell Sea and in the Ross Sea. They are the habitat of a biotic community with a very intense, but brief, production period. In January 1983, von Bröckel (1984) measured a maximum primary production of 1.7 g C m^2 d^{-1}. Zooplankton biomass is low. Krill is replaced by the small *Euphausia crystallorophias.* Accordingly, krill-eating mammals and birds are scarce. Pelagic fish, particularly *Pleuragramma antarcticum*, seem to be more abundant than elsewhere. Their adults feed mainly on *E. crystallorophias* (Hubold 1984), as do the few Crabeater Seals (Drescher, pers. comm.). Most of the primary production by phytoplankton and ice-algae goes to the extremely rich epi-benthic fauna of suspension feeders such as sponges and echinoderms. They are the food source of certain crustaceans, cephalopods and of a large number of fish species which are taken by Emperor Penguins, *Aptenodytes forsteri*, and Weddell Seals, *Leptonychotes weddelli.* It seems that systems of this kind, having brief and irregular outbursts of primary productivity, cannot sustain a rich pelagic food web but only a community of benthic feeders which exploit the detritus concentrations at the water-seabed interface. They do not use up energy for swimming, and can survive long periods of starvation at negligible metabolic cost. Their growth is presumably very slow most of the time.

Very little is known about the biological processes in the permanent pack-ice zone itself, particularly in winter. Epontic ice-algae have been suggested as a major source of food for herbivorous zooplankton and Krill. Detritus feeders at the sea-floor may depend largely on sinking ice-algae. Non-feeding resting stages, found in calanoid copepods, might be common in many animals (Hirche 1983).

4 The Krill System

The Krill system is only one of several systems in the Southern Ocean, and the short food chain determined by Krill is only part of a more complex food web. However, the Krill system is the most outstanding product of a long evolutionary process in isolation, characterized by high seasonality in light and ice cover under otherwise stable conditions of low temperature and high nutrient supply.

Euphausia superba is well adapted to the seasonal pack-ice zone, with its phytoplankton supply being highly variable in terms of species, size of cells and in temporal and spatial distribution. Individual Krill are fast swimmers, attaining speeds up to 600 mm s^{-1} with quick reaction times and a high endurance in permanently swimming and filtering at 100–150 mm s^{-1} (Kils 1981). Adults are very large in relation to other pelagic herbivorous crustaceans. The filtering basket and mouth parts of Krill can cope with a food spectrum of 10 orders of magnitude in weight, ranging from nanoplankton to adult Krill, but Krill can also scrape off ice-algae and heterotrophs from underneath the pack-ice (Kils 1983). In fact, the feeding apparatus of Krill is one of the most efficient energy-capturing systems in the animal kingdom.

It is typical for highly productive pelagic areas to host short food chains having relatively large herbivores (e.g., anchoveta, sardines, red crabs, etc.). The trophic step from Krill to whales is about 8 orders of magnitude which is much larger than the comparable step in a "normal" food chain (Fig. 3). It is worth noting that most consumers of Krill are specialized to feed only on this single food species, while Krill itself has a wide food spectrum.

In Krill, as well as in marine mammals and birds, energy demands for metabolism are extremely high. In swimming Krill, the energy demand per unit body weight increases

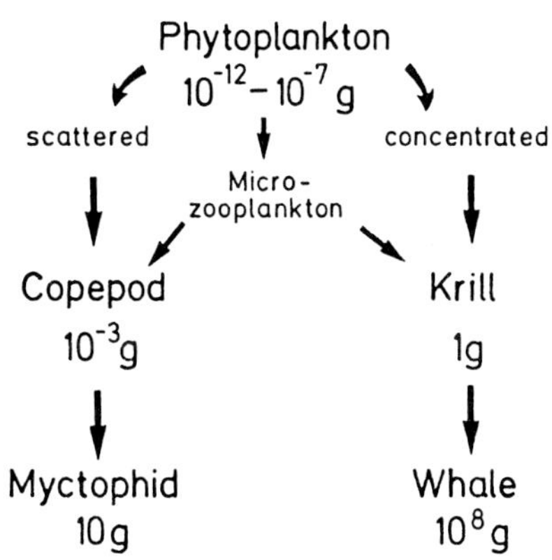

Fig. 3. Size steps (weight units) in two avenues of Antarctic food webs

with body weight, in contrast to the situation, for instance, in whales. The energy budget of Krill during the long period of low primary production is not known. Adult whales and penguins may lose half their body weight in winter. There are some indications that Krill might shrink during the winter period (Ikeda and Dixon 1982). Reproduction in Krill and whales is low, and starts relatively late in life. Krill and whales are representatives of the high-energy cost strategy of survival under highly seasonal conditions, whereas sponges and echinoderms represent the low-energy cost strategy which is also adopted by benthic crustaceans (Clarke 1985, Luxmoore 1985).

Currently we have no indication that the Krill system is food limited at the first trophic step (i.e., between phytoplankton and Krill), although large concentrations of Krill must have an appreciable "grazing" effect on phytoplankton blooms, at least on a local scale. Possible food limitations in the second trophic step (i.e. from Krill to its warm-blooded consumers) have been suggested by reports of an increase in growth and reproduction in seals and whales over the 1960's and 1970's. This needs to be confirmed and monitored very closely. The decrease in the guano-producing birds off Peru after the reduction of the Anchoveta, *Engraulis ringens,* is a negative example of food limitation in a similar system (Murphy 1972). Unfortunately, we do not know how the Krill population itself reacted to the decrease in whale stocks. Such information would have helped in the formulation of realistic management models for the Convention on the Conservation of Antarctic Marine Living Resources (CCAMLR).

Within the Antarctic ecosystem it is only the seasonal pack-ice zone and its northern fringes and branches which hold some promise for continuous exploitation by man. However, the Krill system migth not be tuned to allow a very high fishing pressure. In spite of a rich nutrient supply, phytoplankton production is limited by the seasonal incidence of solar radiation and ice cover, and by low temperature. The rich patches of phytoplankton near and in the pack-ice are only partially exploited by Krill and other pelagic herbivores. Much of the phytoplankton goes to the benthos which acts largely as a sink. Secondary production by Krill is limited by the high energy costs of swimming, etc., and the long period of low food supply in winter. On the tertiary level, food limitations again occur and the ecological efficiency of warm-blooded vertebrates is comparatively low.

Mere survival in the Antarctic is very expensive – at least in the pelagic zone – so there is little net gain in the Krill system compared with the short length of its food chain. A sustainable yield of 2 × 10^6 of whales after restoration of the stocks would mean only ca. 0.03% of the primary production of the Southern Ocean and even a Krill fishery of 50 × 10^6 t yr^{-1} would not take more than 0.5% of the carbon produced by phytoplankton in the

Southern Ocean. This ratio would be similar to the fishing yield obtained from the much longer food chain in the North Sea. Admittedly these estimates are not well-founded, since we have no good figures for overall primary production in the Southern Ocean in general or in the Krill system in particular. The given figure for the potential yield of Krill is also doubtful. The abundance estimates of Krill obtained from acoustic surveys are much higher than those from experimental fishing. On the other hand, Croxall et al. (1985) found that, around South Georgia, Krill biomass estimates determined from predator consumption were much higher than those from acoustic surveys. The conversion of biomass figures into estimates of annual production is rather uncertain too. Growth rate and energy requirements in situ as well as the figures on consumption of Krill by the main consumers are far from being well established. In many cases laboratory data and field data do not agree. A multi-laboratory project under BIOMASS is just developing which will aim at obtaining more accurate life-history data for Krill.

The large biomass which is stored in Krill swarms and in the large individuals, schools and colonies of marine birds and mammals (and to a lesser extent in fish and squid) have made the Krill system attractive to exploitation by man since the onset of sealing in the late 18th century. History has, however, shown that none of the Antarctic's living resources (seals, whales, and fish) has been able to sustain a long-lasting exploitation after the depletion of the virgin stock. Krill is different on account of its somewhat shorter life span and higher reproduction rate, but considerably more groundwork is needed before we can forecast somewhat safely how to exploit this important source of protein.

We get used to thinking of Antarctic communities as comprising systems of organisms with late and low reproduction, long life spans, large body size and high total biomass with low net production. This K-strategy is typical under stable or regularly changing environmental conditions. However, conditions in the seasonal pack-ice zone are highly variable from year to year and from month to month. The position of the oceanic fronts and of the outflow of water from the Weddell Sea into the West Wind Drift are also variable. The large fluctuations in the occurrence of Krill larvae, with the remarkably strong year-class of 1981 (Hempel 1983), and in the distribution and abundance of adolescent and adult Krill in the open Scotia Sea and around South Georgia (e.g., Hempel et al. 1979) seem to reflect these changes in the oceanic and ice regimes. Salps seem to undergo short-term shifts in distribution and abundance, possibly in reaction to the variability in occurrence and composition of phytoplankton blooms.

Any realistic model of food-chain dynamics in the Southern Ocean and any management of the exploitation of the Antarctic living resources has to take into account not only the low efficiency of the food chain in the pelagic system but also the variability of its key elements.

References

Bröckel K von (1984) Primary production in the Weddell Sea. Submitted to Polar Biology

Clarke A (1985) Energy flow in the Southern Ocean food web. In: Siegfried WR, Condy PR, Laws RM (eds) Antarctic nutrient cycles and food webs (Proceedings of the 4th SCAR symposium of Antarctic biology). Springer, Berlin Heidelberg New York

Croxall JP, Prince PA, Ricketts C (1985) Relationships between prey life-cycles and the extent nature and timing of seal and seabird predation in the Scotia Sea. In: Siegfried WR, Condy PR, Laws RM (eds) Antarctic nutrient cycles and food webs (Proceedings of the 4th SCAR symposium on Antarctic biology). Springer, Berlin Heidelberg New York

Deacon GER (1982) Physical and biological zonation in the Southern Ocean. Deep-Sea Res 19:1–15

Gordon AL, Goldberg RD (1970) Circumpolar characteristics of Antarctic waters. Antarctic Map Folio Series 13, Am Geograph Soc New York

Hart TJ (1942) Phytoplankton periodicity in Antarctic surface waters. Discovery Rep 21:261–356

Hempel G (1983) FIBEX – An international survey in the Southern Ocean: Review and outlook. Mem Nat Inst Polar Res Tokyo. Special Issue 27:1–15

Hempel I (1983) Studies in eggs and larvae of *Euphausia superba* and *Euphausia crystallorophias* in the Atlantic sector of the Southern Ocean. Ber Polarforsch Sonderh 4:30–46

Hempel I, Hempel G, Baker A de C (1979) Early life history stages of krill (*Euphausia superba*) in Bransfield Strait and Weddell Sea. Meeresforschung 27:267–281

Hirche HJ (1983) Overwintering of *Calanus finmarchicus* and *Calanus helgolandicus*. Mar Ecol Prog Ser 11:281–290

Hubold G (1985) The early life-history of the high-Antarctic Silverfish, *Pleuragramma antarcticum*. In: Siegfried WR, Condy PR, Laws RM (eds) Antarctic nutrient cycles and food webs (Proceedings of the 4th SCAR symposium on Antarctic biology). Springer, Berlin Heidelberg New York

Ikeda T, Dixon P (1982) Observations on moulting in Antarctic krill (*Euphausia superba* Dana). Aust J Mar Freshwater Res 33:71–76

Kils U (1981) The swimming behaviour, swimming performance and energy balance of Antarctic krill, *Euphausia superba*. BIOMASS Scientific Series No 3, 122 pp

Kils U (1983) Swimming and feeding of Antarctic krill, *Euphausia superba* – some outstanding energetics and dynamics, some unique morphological details. Ber Polarforsch Sonderh 4:130–155

Lubimova et al. (1980) Results of the Soviet research into assessment of stocks, ecology, and the role of krill in the Antarctic ecosystem. Published as manuscript

Luxmoore RA (1985) The energy budget of a population of the Antarctic isopod *Serolis polita*. In: Siegfried WR, Condy PR, Laws RM (eds) Antarctic nutrient cycles and food webs (Proceedings of the 4th SCAR symposium on Antarctic biology). Springer, Berlin Heidelberg New York

Murphy GI (1972) Fisheries in upwelling regions with special reference to Peruvian waters. Geoforum 11:63–71

Schnack SB, Smetacek V, von Bodungen B, Stegmann P (1984) Utilization of phytoplankton by copepods in Antarctic waters during spring. In: Gray J (ed) Proc 18th Europ Mar Biol Symp, Oslo. John Wiley, London, New York (to be published)

Voronina NM (1966) The distribution of zooplankton biomass in the Southern Ocean. Deep-Sea Res 29:1–15

Voronina NM (1971) Annual cycle of Antarctic plankton. In: Beklemishev (ed) Fundamentals of the biological productivity of the ocean and its exploitation. Nauka, Moscow, pp 64–71

Choanoflagellates in the Antarctic Marine Food Chain

H.J. MARCHANT[1]

Summary. Choanoflagellates (collared heterotrophic flagellates) bearing an extra-cellular siliceous lorica, comprise 10–40% of the total number of living Antarctic marine nanoplankton. Here I report on the species composition, distribution and abundance of choanoflagellates in Prydz Bay, as well as their presence under the sea-ice at an inshore site near Davis station. Studies on freshly collected and cultured material indicate that, as well as bacteria, the food of choanoflagellates includes nanoplanktonic primary producers, particulate extra-cellular products and cellular debris from larger phytoplankton.

1 Introduction

The larger components of the Antarctic marine plankton, particularly the diatoms, have received considerable attention. Their taxonomy and distribution have been studied in detail (e.g., Zernova 1970), while more recent investigations have also included measurements of their productivity (e.g., Holm-Hansen et al. 1977). In comparison, nanoplanktonic organisms, those in the size range 2–20 μm (sensu Sieburth et al. 1978), have received much less attention. This is surprising, in that the widespread abundance of these minute organisms in the Southern Ocean has been known for some time (e.g., Hasle 1969). Studies during the summers of 1980–81 and 1982–83, as well as an investigation at a coastal station throughout 1982, have confirmed Hasle's findings and the more recent reports of abundant Antarctic marine nanoplankton (El-Sayed and Turner 1977; von Bröckel 1981; Buck and Garrison 1983). These studies indicate that, as has been found elsewhere, nanoplanktonic organisms are of great importance both as primary producers (Hallegraeff 1981) and as heterotrophs (Azam et al. 1983).

Heterotrophic micro-flagellates play an important role in marine ecosystems in various parts of the world (Fenchel 1982a). One group of these protozoa, the choanoflagellates, especially those which have a siliceous lorica (Acantheocidae), are a highly conspicuous and abundant component of the Antarctic marine nanonplankton. In the Weddell Sea their abundance averaged 3.6 $\times$ 10^5 cells l^{-1} (Silver et al. 1980; Buck and Garrison 1983). Although the few quantitative accounts that have been published on choanoflagellates are variable, it is clear that this group of organisms is one of the more abundant components of the world's oceans. That they occur in such relatively large numbers suggests that they play an important role in the pelagic food web. Here I report studies on their species composition and abundance in the water column of Prydz Bay, as well as optical and electron microscopical observations on their ingestion of food as part of an investigation of their role in the Antarctic marine ecosystem.

1 Antarctic Division, Department of Science and Technology, Channel Highway, 7150 Kingston, Tasmania, Australia

2 Materials and Methods

2.1 Collection of Samples

Samples of choanoflagellates were taken from Prydz Bay, (68°S, 75°E), Antarctica, during the First International BIOMASS Experiment (FIBEX) in January–March 1981 and on another cruise during November–December 1982 in the same area.

Plankton samples were concentrated while at sea, by using a continuous flow centrifuge the cup of which was constructed to the dimensions given in Davis (1957). The centrifuge was powered by a modified compressed air driven die grinder. Seawater was led to the centrifuge through plastic plumbing from a centrifugal pump, the intake of which was near the stern of the vessel at a depth of 5 m. At hydrographic stations, water samples were taken at various depths in 5 l Niskin bottles on a rosette sampler attached to a Mark III Neil Brown conductivity, temperature, depth (CTD) unit. In addition, samples were taken throughout 1982 at an inshore site near the Australian Antarctic station of Davis (68° 35′S, 77° 58′E). Here samples were collected through a hole in the ice with a Kammerer bottle.

2.2 Processing of Samples

Plankton was concentrated by slow filtration of 1–2 l to a final volume of 5 ml, using a 0.8 μm pore-sized Millipore filter. Both on board ship and at the shore station, the con-

Antarctic Nutrient Cycles and Food Webs
(ed. by W.R. Siegfried, P.R. Condy, and R.M. Laws)

centrated micro- and nano-plankton were treated by a number of methods.

At sea, planktonic organisms were identified and enumerated using a Zeiss WL optical microscope equipped with phase-contrast optics. The instrument was isolated from the vibrations of the ship's engines sufficiently well to permit the use of the highest magnification of an optical microscope. Organisms were photographed using a Zeiss MC63 camera and a Zeiss microflash unit. Other samples of the concentrate were placed on glass coverslips and formvar-coated electron microscope grids which had been treated with polylysine (Marchant and Thomas 1983), fixed by 30s exposure to OsO_4 vapour and rinsed with distilled water after allowing the organisms to settle and attach to the polylysine for 20 min. On returning to Australia, the grids were shadow-cast with platinum-palladium and examined by transmission electron microscopy (TEM). For scanning electron microscopy (SEM), a thin layer of gold was applied to the coverslips with a sputter coater.

Other sub-samples of the plankton were transferred to enrichment cultures and returned to Australia in growth cabinets maintained at 1°C with continuous illumination. Cultures of *Diäphanoeca multiannulata,* growing with *Phaeocystis pouchetii* and *D. grandis,* which was isolated from Clear Lake in the Vestfold Hills, were prepared for SEM on polylysine-coated coverslips, as described above. Observations on living organisms were made using a Zeiss Photomicroscope fitted with Nomarski differential interference contrast optics, an incident light fluorescent attachment and a cooling stage which was maintained at 0°–3°C by 40% methanol pumped from a Julabo recirculating water bath.

3 Results and Discussion

3.1 Composition of the Micro- and Nano-plankton

The proportions of the major components of micro- and nano-plankton from various oceanic stations in Prydz Bay are shown in Table 1. These values are derived from counts of the numbers of living cells. The heterotrophic micro-flagellates occurring in Antarctic waters include choano-flagellates, non-pigmented dinoflagellates and tintinnids, of which choanoflagellates are the most numerous. At oceanic stations the mean abundance of choanoflagellates was $2.7 \times 10^4 \pm 1.7 \times 10^4$ cells l^{-1}.

Table 1. Percentage composition of living micro- and nanoplankton based on cell numbers

Date	Local time	Position	Diatoms > 20 μm	Diatoms < 20 μm	Dinoflagellates	Flagellates	Choanoflagellates
29.01.81	1000 h	68° 35.8′S, 71° 26.4′E	10	81	1	7	1
14.02.81	1100 h	63° 12.2′S, 71° 24.2′E	51	36	3	8	3
15.02.81	1040 h	63° 14.7′S, 77° 43.8′E	0	72	3	16	8
16.02.81	1100 h	63° 12.6′S, 83° 00.6′E	7	67	0	23	3
21.02.81	1035 h	64° 59.7′S, 67° 04.2′E	32	39	3	22	4
23.02.81	1100 h	66° 19.4′S, 78° 33.1′E	5	48	< 1	46	< 1
27.02.81	1030 h	Davis	26	47	< 1	16	11
01.03.81	1600 h	66° 40.7′S, 78° 31.4′E	16	43	1	38	2
03.03.81	1545 h	67° 43.2′S, 73° 47.9′E	1	63	2	30	5
10.03.81	0600 h	62° 13.0′S, 68° 57.0′E	30	39	< 1	27	4
20.11.82	1100 h	61° 53.9′S, 88° 50.1′E	12	54	2	21	11
20.11.82	1400 h	61° 13.5′S, 88° 29.7′E	13	59	1	18	8
21.11.82	0900 h	63° 03.5′S, 88° 29.9′E	14	58	1	22	4
21.11.82	2030 h	63° 12.5′S, 85° 15.2′E	7	49	1	41	2
23.11.82	1000 h	61° 47.5′S, 81° 26.7′E	11	49	5	19	16
26.11.82	1030 h	62° 10.1′S, 63° 01.1′E	9	61	1	28	2
02.12.82	1000 h	66° 10.8′S, 62° 55.2′E	5	53	0	37	5
03.12.82	1000 h	67° 19.6′S, 65° 34.0′E	13	26	9	52	0
08.12.82	1100 h	68° 35.4′S, 72° 04.2′E	49	16	5	32	0
17.12.82	1000 h	65° 47.5′S, 73° 27.2′E	1	31	1	64	2

At the inshore site, choanoflagellates were found to be common in the summer when their numbers were in the order of 5×10^5 cells l^{-1}. Between April and December they were scarce under the ice where the phytoplankton population was also very low. Despite extensive searching, no living choanoflagellates were found within the sea-ice at the coastal station. Both Buck (1981) and Takahashi (1981) report choanoflagellates from ice samples, but did not ascertain whether they were living.

Because of the thermal lability and susceptibility of many dinoflagellates and flagellates to distortion by fixatives, and the fact that choanoflagellates have a resistant lorica, there is a tendency to systematically underestimate the number of dinoflagellates and flagellates and to overestimate the number of choanoflagellates. The counts of organisms presented here were made on living material immediately after concentration. Buck and Garrison (1983), in their study of the protists from the ice-edge of the Weddell Sea, counted only intact specimens. Despite variations in the methodology of counting choanoflagellates and other heterotrophic micro-flagellates, it is clear that they are abundant in Antarctic waters and are therefore likely to be playing a significant role in the marine ecosystem.

3.2 Species Composition and Distribution of Antarctic Choanoflagellates

Choanoflagellates have been reported from the Antarctic waters of Terre Adélie (Deflandre 1960), the Weddell Sea (Silver et al. 1980; Buck 1981) and Lützow-Holm Bay (Takahashi 1981). The 18 species of choanoflagellates identified in coastal and oceanic samples from Prydz Bay, listed in Table 2, differ markedly in size and in the detailed morphology of their feeding apparatus. The consequences of these differences are considered in the next section.

3.3 Ecological Role of Choanoflagellates

Choanoflagellates are considered to be bacterivorous. While there is good evidence that they are consumers of bacteria (Leadbeater and Morton 1974; Fenchel 1982b) and cyanobacteria (Laval 1971), my observations on fixed and living Antarctic marine choanoflagellates suggest that they ingest material in addition to bacteria. Particulate organic matter, including extra-cellular products of phytoplankton, which occurs abundantly in the Antarctic marine ecosystem has

Table 2. List of choanoflagellate species from Prydz Bay

Species	Oceanic	Inshore
Acanthoeca spectabilis, Ellis (1930)		+
Acanthoecopsis unguiculata, Thomsen (1973)	+	+
Bicosta spinifera, (Thronsden) Leadbeater (1978)	+	+
Calliacantha natans, (Grøntved) Leadbeater (1978)	+	+
Calliacantha simplex, Manton and Oates (1979b)	+	+
Crinolina aperta (Leadbeater) Thomson (1976)	+	+
Diaphanoeca grandis, Ellis (1930)		+
Diaphanoeca multiannulata, Buck (1980)	+	+
Diaphanoeca pedicellata, Leadbeater (1972)	+	+
Parvicorbicula circularis, Thomsen (1976)	+	+
Parvicorbicula socialis, (Meunier) Deflandre (1960)	+	+
Pleurasiga reynoldsii, Thronsden (1970)	+	
Pleurasiga sphyrelata, Thomsen 1973	+	
Pleurasiga orculaeformis, Schiller (1925) sensu Leadbeater (1973)		+
Pleurasiga sp.	+	+
Saepicula leadbeateri, Takahashi (1981)		+
Stephanoeca diplocostata paucicostata, Thronsden (1969)		+

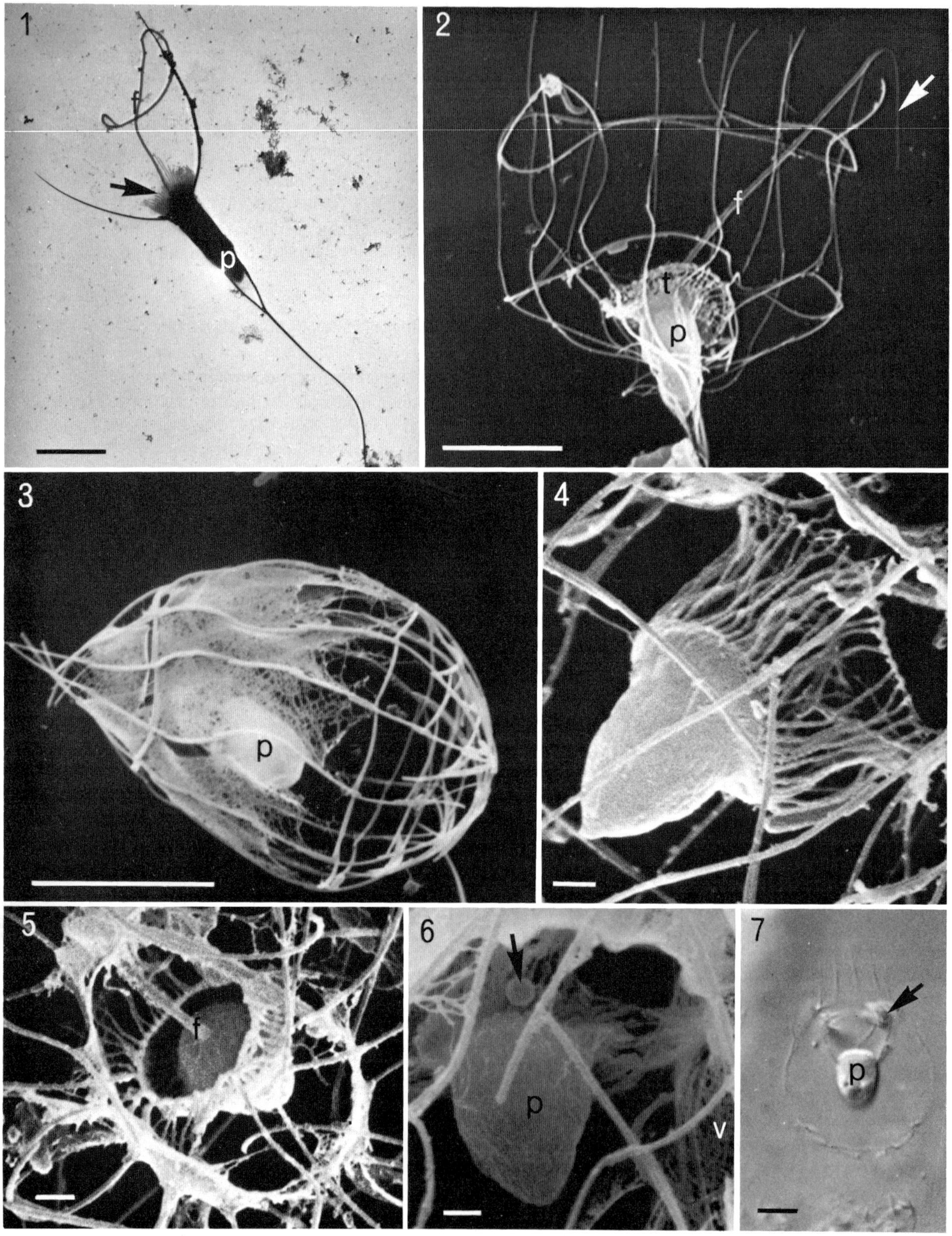

Fig. 1. *Bicosta spinifera.* A ring of tentacles (*arrow*) encircles the single flagellum emerging from the protoplast (*p*) suspended from the lorica. Shadow cast specimen, TEM. *Scale bar* 5 μm

Fig. 2. *Diaphanocea pedicellata.* The protoplast (*p*) is located at the base of the large lorica. A ring of tentacles (*t*) encircles the flagellum (*f*) which bears a long hair-point (*arrow*). SEM. *Scale marker* 10 μm

Fig. 3. A veil of fine filaments covers much of the anterior part of the lorica of *Diaphanoeca grandis* obscuring the protoplast (*p*). The collapsed anterior end of the lorica is an artifact of specimen preparation. SEM. *Scale marker* 10 μm to be considered as a potential food source for choanoflagellates.

Phaeocystis pouchetii, a prymnesiophyte, occurs extensively in Antarctic waters (Kashkin 1963) and can be so abundant as to clog plankton nets. Cell counts of $> 10^6$ cells l^{-1} are not uncommon. This colonial alga secretes 16%–64% of its photo-assimilated C, largely in the form of mucilage (Guillard and Hellebust 1971). Many other phytoplankton members produce large quantities of particulate organic material (Bölter and Dawson 1982). Observations made both at sea and on laboratory cultures indicate that dense bacterial growths use the mucilage of senescent *Phaeocystis* as substrates. Few bacteria are associated with actively growing colonies, probably due to the occurrence of acrylic acid, an inhibitor of both gram-positive and negative bacteria, in the mucilage (Sieburth 1960). I have found that *Diaphanoeca multiannulata,* one of the largest Antarctic choanoflagellates, grew prolifically with actively growing cultures of *Phaeocystis* with which extremely few bacteria are associated.

Loricate choanoflagellates exhibit considerable variation in morphology. Organisms such as *Biocosta* (Fig. 1) *Calliacantha, Pleurasiga, Parvicorbicula,* and *Diaphanoeca pedicellata* (Fig. 2) have more or less exposed feeding baskets, whereas the feeding baskets of others, such as *Diaphanoeca grandis* and *D. multiannulata,* are largely surrounded by the lorica (Fig. 3). Scanning electron microscopy reveals that a veil of fine fibrils lies on the inside of the anterior part of the lorica of both *D. grandis* and *D. multiannulata* (Figs. 3 and 6). The extent of the veil varies considerably between individuals in the same culture. This delicate veil, or membrane, lying within the lorica has been reported from a number of choanoflagellates and has been considered to be involved in supporting the protoplast within the lorica (Leadbeater 1975). Its position in the anterior part of the lorica of *D. grandis* and *D. multiannulata,* and my finding of particles of detritus, nanoplanktonic phytoplankton, as well as bacteria on its inner surface, suggest that it may also function in trapping food particles either directly or by influencing the flow of water past the feeding basket. Whether such a veil enables these relatively large choanoflagellates to utilize a wider range of material for food than species without the veil and whether it increases their feeding efficiency are being investigated. The feeding basket of tentacles that encircles the flagellum is extensive in *D. multiannulata* (Figs. 4 and 5). The protoplast of this choanoflagellate has not been previously illustrated.

It has been established that bacteria utilize the extracellular products secreted by phytoplankton (Fuhrman and Azam 1980; Larsson and Hagstrom 1979; Azam and Hodson 1977; Iturriaga and Zsolnay 1981). Grazing by heterotrophic micro-flagellates has been suggested as the way in which marine bacterial populations are kept at a relatively low, constant concentration (Lightheart 1969; Sorokin 1981; Banse 1982; Fenchel 1982b). Whereas it has been well established that marine choanoflagellates consume bacteria, apparently little work has been done to investigate what else they ingest. My studies carried out on both freshly collected, as well as cultured material indicate that, in addition to bacteria, they consume nanoplankton and mucilage secreted by phytoplankton (Figs. 6 and 7). Differences in the morphology of the feeding apparatus as well as in the size of the protoplast of the various species of choanoflagellates indicate the likelihood of dietary differences between the various species. Thus, the diet of small species would consist of bacteria and particulate organic matter, whereas the larger organisms can potentially graze on larger organisms and particles.

Acknowledgements. I sincerely thank Lu Peiding, National Bureau of Oceanography, Qingdao, Peoples' Republic of China, and Mr. Rick Perrin for winter observations on choanoflagellates.

References

Azam F, Fenchel T, Field JG, Gray JS, Meyer-Reil LA, Thingstad F (1983) The ecological role of water-column microbes in the sea. Mar Ecol Prog Ser 10:257–263

Azam F, Hodson RE (1977) Size distribution and activity of marine microheterotrophs. Limnol Oceanogr 22:492–501

Banse K (1982) Cell volumes, maximal growth rates of unicellular algae and ciliates and the role of ciliates in the marine pelagial. Limnol Oceanogr 27:1059–1071

Bölter M, Dawson R (1982) Heterotrophic utilization of biochemical components in Antarctic waters. Neth J Sea Res 16:315–332

Bröckel K von (1981) The importance of nanoplankton within the pelagic Antarctic ecosystem. Kiel Meeresforsch Sonderh 5:61–67

Buck KR (1981) A study of choanoflagellates (Acanthoecidae) from the Weddell Sea, including a description of *Diaphanoeca multiannulata* n. sp. J Protozool 28:47–54

Buck KR, Garrison DL (1983) Protists from the ice-edge region of the Weddell Sea. Deep-Sea Res 30:1261–1277

Davis PS (1957) A method for the determination of chlorophyll in seawater. CSIRO Division of Fisheries, Report No. 7 CSIRO Sydney. pp 1–8

Fig. 4. Protoplast of *Diaphanoeca multiannulata* with a ring of tentacles fused at their tips, making up the feeding basket. SEM. *Scale marker* 1 μm

Fig. 5. Protoplast of *Diaphanoeca multiannulata* showing the flagellum (*f*) and surrounding feeding basket. SEM. *Scale marker* 1 μm

Fig. 6. A blue-green algal cell (*arrow*) attached to the feeding basket of *Diaphanoeca grandis.* SEM. *Scale marker* 1 μm

Fig. 7. *Diaphanoeca grandis* with a nanoplanktonic flagellate (*arrow*) attached to its feeding basket. Nomarski differential interference contrast optics. *Scale marker* 5 μm

Deflandre G (1960) Sur la présence de *Parvicorbicula* n.g. *socialis* (Meunier) dan le plancton de l'Antarctique (Terre Adélie) Rev Algol 5:183–189

El-Sayed SZ, Turner JT (1977) Productivity of the Antarctic and tropical /subtropical regions : a comparative study In: Dunbar MJ (ed) Polar Oceans, Proceedings of the SCAR Polar Oceans Conference, Montreal 1974. Arctic Institute of North America, Calgary, pp 463–503

Fenchel T (1982a) Ecology of heterotrophic micro-flagellates. I. Some important forms and their functional morphology. Mar Ecol Prog Ser 8:211–223

Fenchel T (1982b) Ecology of heterotrophic micro-flagellates. IV. Quantitative occurrence and importance as bacterial consumers. Mar Ecol Prog Ser 9:35–42

Fuhrman JA, Azam F (1980) Bacterioplankton secondary production estimates for coastal waters of British Columbia, Antarctica and California. Appl Environ Microbiol 39:1085–1095

Guillard RRL, Hellebust JA (1971) Growth and the production of extracellular substances by two strains of *Phaeocystis poucheti.* J Phycol 7:330–338

Hallegraeff GM (1981) Seasonal study of phytoplankton pigments and species at a coastal station off Sydney : Importance of diatoms and the nanoplankton. Mar Biol 61:107–118

Hasle GR (1969) An analysis of the phytoplankton of the Pacific Southern Ocean : abundance, composition, and distribution during the Brategg Expedition, 1947–1948. Hvalradets Skr 52: 1–168

Holm-Hansen O, El-Sayed SZ, Franceschini GA, Cuhel RL (1977) Primary production and the factors controlling phytoplankton growth in the Southern Ocean. In: Llano GA (ed) Adaptations within Antarctic Ecosystems. Gulf Publishing Co Houston pp 11 –50

Iturriaga R, Zsolnay A (1981) Transformation of some dissolved organic components by a natural heterotrophic population. Mar Biol 62:125–129

Kashkin NI (1963) Data on the ecology of *Phaeocystis pouchetii* (Hariot) Lagerheim 1893 (Chrysophyceae) II Habitat and specification of biogeographical characteristics. Okeanologiya 3:697 –705

Larsson U, Hagström A (1979) Phytoplankton exudate release as an energy source for the growth of pelagic bacteria. Mar Biol 52: 199–206

Laval M (1971) Ultrastructure et mode de nutrition du choanoflagelle *Salpingoeca pelagica* sp. nov. Comparaison avec les choanocytes des spongaires. Protistologica 7:325–336

Leadbeater BSC (1975) A microscopical study of the marine choanoflagellate *Savillea micropora* (Norris) comb. nov., and preliminary observations on lorica development in *S. micropora* and *Stephanocea diplocostata* Ellis. Protoplasma 83:111–129

Leadbeater BSC (1979) Developmental studies on the loricate choanoflagellate *Stephanoeca diplocostata* Ellis I. Ultrastructure of the non-dividing cell and costal strip production. Protoplasma 98:241–262

Leadbeater BSC, Morton C (1974) A microscopical study of marine species of *Codosiga* James-Clark (Choanoflagellata) with special reference to the ingestion of bacteria. Biol J Linn Soc 6:337 –347

Lightheart B (1969) Planktonic and benthic bacteriorvorous protozoa at eleven stations in Puget Sound and adjacent Pacific ocean. J Fish Res Board Can 26:299–304

Marchant HJ, Thomas DP (1983) Polylysine as an adhesive for the attachment of nanoplankton to substrates for electron microscopy. J Microsc (Oxf) 131:127–129

Sieburth J McN (1960) Acrylic acid an "antibiotic" principle in *Phaeocystis* blooms. Science (Wash DC) 132:676–677

Sieburth J McN, Smetacek V, Lenz J (1978) Pelagic ecosystem structure : heterotrophic compartments of the plankton and their relationship to plankton size fractions. Limnol Oceanogr 23:1256–1263

Silver MW, Mitchell JG, Ringo DL (1980) Siliceous nanoplankton. II. Newly discovered cysts and abundant choanoflagellates from the Weddell Sea, Antarctica. Mar Biol 58:211–217

Sorokin YI (1981) Microheterotrophic organisms in marine ecosystems. In: Longhurst AR (ed) Analysis of marine ecosystems. Academic, London, pp 293–342

Takahashi E (1981) Loricate and scale-bearing protists from Lützow-Holm Bay, Antarctica. I. Species of the Acanthoecidae and the Centrohelida found at a site selected on the fast ice. Antarct Rec 73:1–22

Zernova VV (1970) Phytoplankton of the Southern Ocean In: Holdgate MW (ed) Antarctic Ecology I. Academic, London, pp 136 –142

Alternate Carbon Pathways at Lower Trophic Levels in the Antarctic Food Web

C.D. HEWES[1], O. HOLM-HANSEN[1], and E. SAKSHAUG[2]

Summary. A large percentage (> 50%) of total chlorophyll found in waters surrounding Antarctica is contained in the pico- and nanoplankton (< 20 μm) size-fraction. About 30% of the total nanoplankton biomass consists of heterotrophic flagellates. The total phytoplankton standing stock is low, averaging < 0.6 μg Chl 1^{-1} in the East Wind Drift area. Higher concentrations are found in the shallow portions of the Weddell Sea and over the Scotia Ridge. In routine incubations pre-screened with a 200 μm Nitex net there is a measurable loss of chlorophyll and cells in the nanoplankton size category. This decrease in autotrophic biomass, as compared to the control incubations, apparently results from feeding by heterotrophic microplankton. Ciliate populations in Antarctic environments can attain high celldensities, which suggests they may be an important food reserve for animals requiring microplankton sized food particles. This new information requires that we revise the classical concept of the Antarctic 'diatom-krill' food chain to one which incorporates the feeding by Krill (Euphausiacea) on many large-volumed particles (protozoans as well as diatoms), with protozoans acting as a link which couples pico- and nano-plankton production to higher trophic levels.

1 Introduction

The Antarctic Ocean traditionally has been regarded as very rich in biological production (Fricker 1900; Lillie 1913). This was due primarily to the high standing stock of Krill (Euphausiacea), seals and whales. It was assumed by most investigators that primary production would have to be very high in order to support the biomass of these organisms of higher trophic levels. This concept for the Southern Ocean helped to establish the model of the Antarctic food chain as one of a simple trophic progression from large-volumed diatoms to Krill to whales (Hart 1934, 1942). Whereas very rich phytoplankton concentrations (> 150 μg chlorophyll-a 1^{-1}) have been reported by El-Sayed (1967) and Hewes et al. (1983) for the southern Weddell Sea, extensive studies during the past decade have indicated that primary production in the Southern Ocean is moderate to low as compared to that of other oceanic areas.

There is, of course, much temporal and geographical variation in primary production, but representative values (in mg carbon m^2 d^{-1}) for the Southern Ocean are 134 (Holm-Hansen et al. 1977), 265 (Bröckel 1981), 104–410 (El-Sayed and Taguchi 1981) and 544 (El-Sayed 1967). All but the last of these daily production rates are lower than those reported for the southern California Bight (Eppley et al. 1979) and, because of the short productive season, when converted to annual rates are comparable to those of oligotrophic tropical gyres (Eppley and Peterson 1979). Growth rates measured for Antarctic phytoplankton are generally low (generation times of several days; see Holm-Hansen et al. 1977) and are less than half the rate expected by the growth-temperature models of Eppley (1972) and Goldman and Carpenter (1974). The impression of high primary production in the Antarctic, therefore, appears to reflect the great biomass that can accumulate at times more any actual high rates of growth. Previous studies (Holm-Hansen et al. 1977; Smith and Morris 1980; Paden et al. 1981; Fogg and Hayes 1982) furnish strong evidence that overall phytoplankton growth rates are not limited by an inorganic or organic nutrient deficiency. Phytoplankton growth rates are thought to be limited by a combination of effects from light (El-Sayed and Mandelli 1965) and temperature (Neori and Holm-Hansen 1982). Therefore, the patchiness of phytoplankton biomass in Antarctic waters is usually considered to be a result of feeding by net-zooplankton (> 200 μm) and physical mixing processes (El-Sayed 1971; Paden et al. 1981; Meyer and El-Sayed 1983).

Most of the early investigations dealing with phytoplankton in the Antarctic, including many of the studies done on the R.V. Eltanin, were based on samples obtained with phytoplankton nets (usually about 35 μm mesh). This may partially explain why the Southern Ocean is often thought to be dominated by the relatively large-volumed (20–200 μm equivalent spherical diameter, ESD) micro-diatoms (Hart 1934, 1942; Hasle 1970; Nemoto and Harrison 1981). This has also resulted in a lack of interest and, hence, paucity of data for members of the non-mineralized phytoplankton component (Balech 1970). These include the Cryptophyceae, Crysophyceae, and Dinophyceae, many of which have relatively small cells. In water samples from the Scotia and Weddell seas (Holm-Hansen and Foster 1981), nano-

1 Polar Research Program, Marine Biology Building, A-002, Scripps Institution of Oceanography, University of California, La Jolla, CA 92093, USA

2 Biological States, University of Trondheim, N 7000 Trondheim, Norway

Antarctic Nutrient Cycles and Food Webs
(ed. by W.R. Siegfried, P.R. Condy, and R.M. Laws)

plankton (2–20 μm ESD) generally accounted for between 25 and 75% of the total estimated phytoplankton carbon (Reid and Sakshaug, unpubl. floristic data). This is in agreement with data from other investigators (El-Sayed 1971; Fay 1973; Horner 1976; Silver et al. 1980; Bröckel 1981; El-Sayed and Taguchi 1981).

An evaluation recently made of the food sources available to dense Krill, *Euphausia superba,* aggregations in the Southern Ocean (Holm-Hansen and Huntley 1984) concluded that, based on chlorophyll concentrations and zooplankton estimates, the phytoplankton standing stock was sufficient to sustain "maintenance" metabolism, but was not adequate to provide for normal growth of the euphausiids. However, estimates of Krill abundance, based on samples obtained by nets, are much lower than those obtained from acoustic surveys (Macaulay 1981), and the possibility thus exists that the phytoplankton stocks would not support even the "maintenance" requirements of the euphausiids. It is also important to consider the feeding preference and filtering efficiency of Krill in regard to the demonstrated dominance of nanoplankton in Antarctic waters. Krill have maximal filtering rates which are achieved with particles 30 μm or larger; half-maximal rates have been found for 6–10 μm particles (Boyd 1982; McClatchie and Boyd 1983; Quetin and Ross 1985). Additionally, preferential feeding on micro-sized particles by Krill has been substantiated by Meyer and El-Sayed (1983). Thus, it would appear that a large portion of primary production may not be suitable for efficient grazing directly by Krill. However, other biological mechanisms could exist to "concentrate" this production into suitable sized particles, thus acting as an intermediate link in the pathway of carbon from nanoplankton to the large net-zooplankton. Therefore, one of the goals of the current study was to investigate the relative size and productivity of the nanoplankton standing stocks and to better understand the ultimate fate of the photosynthate in these small cells.

2 Materials and Methods

Water samples were obtained during the circumnavigation cruise of the USCGC Polar Star (Holm-Hansen and Chapman 1983), the cruise track of which is shown in Fig. 1. Water was obtained either with 5.0 1 PVC Niskin bottles when the ship was stopped at stations or with a 2.0 1 PVC sampler when the ship was underway. In the latter case, water samples were from the upper 1 m. For most routine stations, part of each water sample was filtered first through a 20-μm Nitex net to separate the nanoplankton from the microplankton. At major stations, where microscopical analyses or feeding experiments were performed, either a 10-μm or a 20-μm Nitex mesh was used to obtain a "nanoplankton" fraction, and the "microplankton" sample was pre-filtered through a 202-μm Nitex net. The fractionated water samples where then used for the following determinations.

Samples for measurement of chlorophyll-a were passed through glass-fibre filters (Whatman GF/C, 25 mm), extracted in absolute methanol (Holm-Hansen and Riemann 1978), and the chlorophyll concentrations determined by

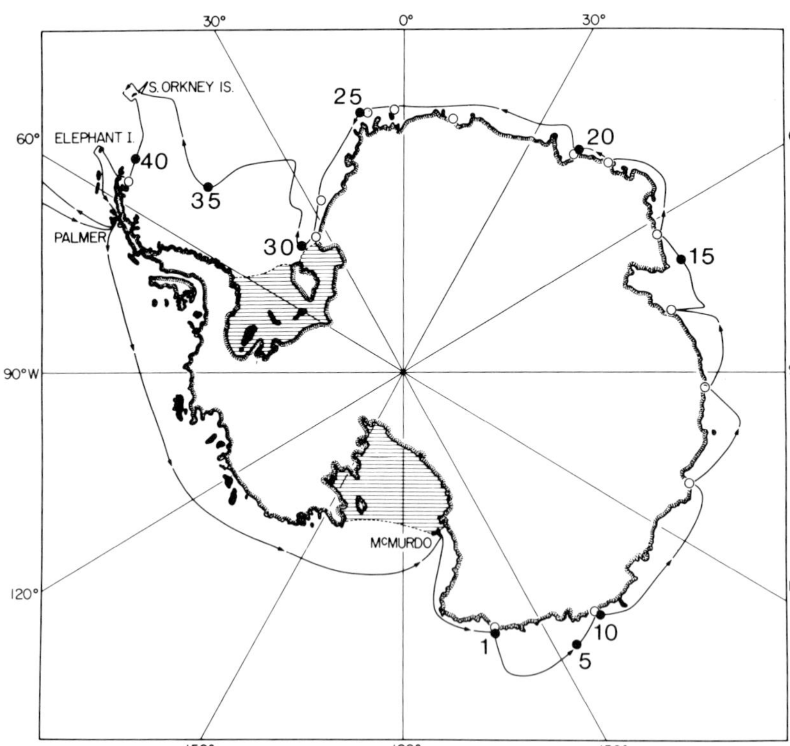

Fig. 1. Cruise track of USCGC Polar Star which departed Palmer Station on 27 December 1982 and returned on 6 March 1983. *Dots* show every fifth sampling station (Stations 1–40). *Open circles* indicate locations of 14 research bases visited during the cruise

measurement of fluorescence (Holm-Hansen et al. 1965). Microscopical analysis was made of nanoplankton in less than 200-μm size-fractions onboard ship, using the Filter-Transfer-Freeze (FTF) technique (Hewes and Holm-Hansen 1983). This method, in brief, was carried out by having predetermined volumes of water samples fixed with 1% gluteraldehyde (total dilution), and concentrated by filtration onto 0.4 μm Nuclepore filters. The filters were transferred carefully to microscope slides and, with the particulates sandwiched between the filter and glass slide, were frozen with a canned freon gas (Histo-spray, Fisher Scientific). All cells are thus frozen in an even layer of ice from which the filter is peeled off. The frozen cells are then embedded in a glycerin-gel mounting medium and allowed to thaw. This type of preparation can be immediately examined under the microscope. Autotrophic flagellates were distinguished from heterotrophic flagellates by using an epi-fluorescence unit to excite chlorophyll to fluoresce, so that it could be observed with the microscope. The methodology and the quantitative and qualitative aspects of the techniques are discussed more thoroughly in other papers (Hewes and Holm-Hansen 1983; Hewes et al. 1984). Cell size and density were converted to cell C estimates for diatoms and non-diatoms using the Strathman (1967) equation. The Strathman equation tends to overestimate C for small cells (< ca. 3 μm ESD), so that an upper value of 2.0 was given to the product of the conversions between the wet-weight (wet wt.) to dry-weight (dry wt.) and dry wt. to C.

Near Stations 36 and 37 (Fig. 1), experiments were conducted to measure the significance of feeding by micro-zooplankton on cells in the < 10-μm size fraction. The experimental design consisted of monitoring, by microscopical analysis as described above, changes through time in the cell numbers of nanoplankton in both of these size-fractioned populations. The water containing the experimental ("grazed") group was the < 200-μm size-fraction; that of the control ("ungrazed") group was pre-screened with the 10-μm Nitex. These samples were incubated on deck as "cultures" with 50% attenuated light and maintained at surface seawater temperature. The cultures contained ciliates and large heterotrophic dinoflagellates actively feeding on autotrophic nanoplankton (as seen using epi-fluorescence microscopical methods). Meta-zooplankton were not observed in any of these samples. Concurrently with obtaining most of the samples for microscopical preparation, chlorophyll was also determined for the cultures as described above.

3 Results and Discussion

Figure 2 shows the concentration of chlorophyll-a in the pico- and nanoplankton (< 20 μm) and microplankton (20–200 μm) in all water samples collected. The < 20-μm group shows smaller fluctuations in chlorophyll concentration than the microplankton populations. Three major biologically active areas can be distinguished from these data: the East Wind Drift (Stations 1–25); the Weddell Sea (Stations 26–36); and the Scotia Ridge area (Stations 37–46). The East Wind Drift and Weddell Sea areas had approximately the same proportions of nanoplankton to microplankton populations, as 63% and 53%, respectively, of the total chlorophyll passed through a 20-μm screen. The Scotia Ridge area had a significantly higher proportionate concentration (86%) of nanoplankton. The Weddell Sea, which is characterized by extensive shallow areas, had much higher standing stocks of both nano- and microplankton (Fig. 2 and 3). This is not simply a result of shallow (< 1000 m) waters, since comparable depths in the East Wind Drift generally showed low phytoplankton biomass. The shallow waters surrounding the South Orkney Islands had high chlorophyll concentrations consisting predominantly of nano-flagellates, as determined by microscopical analysis.

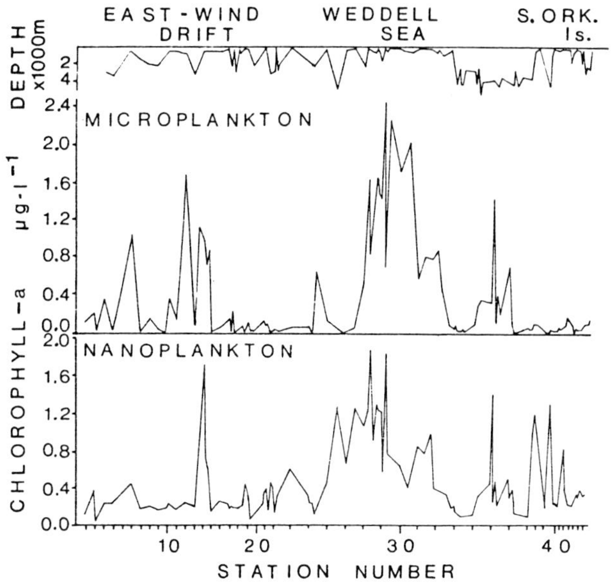

Fig. 2. Chlorophyll-a content in nanoplankton (< 20 μm) and in microplankton (> 20 μm), together with water depth, during the circumnavigation cruise. Station 1 was at approximately 159° E and Station 42 at 56° W. All water samples were taken from the surface (upper 1 m). Chlorophyll concentrations in the microplankton were obtained by subtracting the nanoplankton value from the "total" chlorophyll concentration

We suspect that much of the fluctuation in microplankton biomass is a reflection of feeding pressure from net-zooplankton (> 200 μm). Some evidence for this has been presented by Holm-Hansen and Huntley (1984). Thus, we could characterize most of the East Wind Drift zone as a "blue-water" environment, in which chlorophyll-a concentrations are low (< 0.6 μg l^{-1}), not correlated with depth, and the phytoplankton biomass is dominated by nanoplankton. Microscopical examinations (using the FTF method, Hewes et al. 1984) revealed that *Fragilariopsis nana* (a pennate diatom, 1–2 times 5 μm) was often the dominant phytoplankton species in terms of biomass and cell abundance in waters of the East Wind Drift.

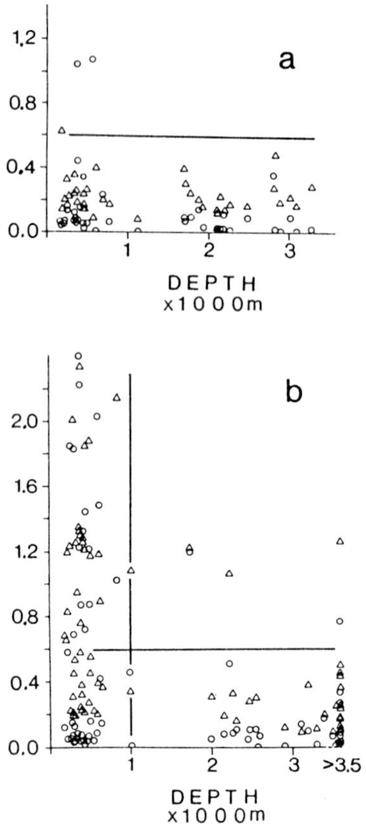

Fig. 3a,b. Concentration of chlorophyll-a in nanoplankton (*triangles*) and microplankton (*circles*) from surface waters in relation to the depth of the total water column. **a** in East Wind Drift (Stations 1–15). **b** in Weddell Sea and Scotia Ridge area (Stations 26–46). Lines were drawn arbitrarily to emphasize differences and similarities between areas of the East Wind Drift and the Scotia Ridge Weddell Sea areas

The average nanoplankton biomass, as obtained by on-board microscopical analysis, for 15 stations in the East Wind Drift, was 48 μg C 1^{-1} for autotrophs with a range of 10–303 μg C 1^{-1} (in *F. nana* the range of C was 0.6–126 μg 1^{-1}). In contrast, the average for heterotrophic nano-flagellates was 24 μg C 1^{-1}, with a range of 7°–60°C 1^{-1}. Preliminary estimates for micro-heterotrophic (ciliate and flagellate) biomass at some stations was in excess of 25 μg C 1^{-1}.

The few samples for which both phytoplankton and ciliate populations were analyzed from the Scotia and Weddell seas during our cruise in 1981 (Holm-Hansen and Foster 1981) have shown that the protozoan component is significant when compared with comparably sized autotrophs. Floristic biomass estimates for these stations were obtained with inverted microscope procedures (Hasle 1978), using formalin-preserved water samples, and cell volumes and population size converted to biomass, using the equations developed by Strathman (1967). Ciliate biomass estimates were obtained using similar methods (Beers et al. 1975) and were made available to us through the courtesy of J.R. Beers. At five stations (two stations included euphotic zone depth profiles, with one of these stations operating in a diatom bloom) in the Scotia Sea in 1981, tintinnid biomass ranged from 0.024–3.54 μg C 1^{-1} ($\bar{x}$ = 1.2 ± 1.3), and total ciliate biomass ranged from 1.9–7.3 μg C 1^{-1} ($\bar{x}$ = 3.6 ± 2.4), which can be compared to > 20 μm diatom biomass of 0.7–42.6 ($\bar{x}$ = 13.9 ± 17.8 μg C 1^{-1}). Tintinnid and total ciliate biomass are compared with the "monad-flagellate" biomass in Fig. 4a. Six of the 7 points (see regression line) suggest that the biomass of ciliates was proportional to that of flagellates. This is consistent with the idea that nano-flagellates can be food for ciliates. The trophic mode of flagellates could not be determined for these samples (i.e. the Utermöhl (1958) method was used), and all flagellates were classified as "phytoplankton". Nevertheless, the mean total ciliate biomass was 17% of the mean total phytoplankton biomass and 40% of the total micro-plankton biomass. For these data, diatoms are the only planktonic group which are unquestionably autotrophic, and for which the proportion of phytoplankton biomass may be compared with ciliates. The proportion of ciliates to centric diatoms and to all diatoms > 20 μm (*Synedra reinboldii,* a pennate diatom, was a dominant species at one station) for non-bloom waters is shown in Fig. 4b; There did not appear to be any correlation between ciliates and centric diatoms or micro-sized diatoms. However, ciliates at times did comprise a significant fraction of the total micro-plankton biomass.

At two geographically different places (deep Weddell Sea and shallow Scotia Ridge), it was found that protozoans (ciliates) could effectively control population growth of the

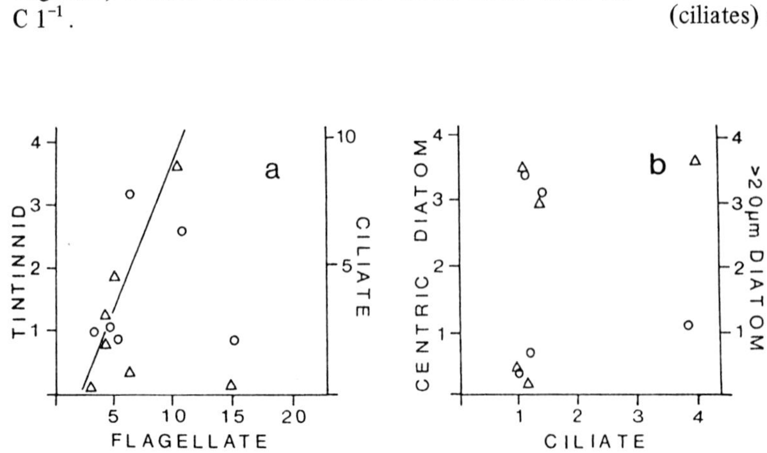

Fig. 4a,b. Ciliate and "phytoplankton" biomass data (in μg C 1^{-1}) obtained from waters around the South Orkney Islands in 1981. **a** shows the relationships between total ciliate (*triangles*) and tintinnid (*circles*) biomass to the biomass of "monads and flagellates". **b** shows the relative biomass of ciliates to centric diatoms (*circles*) and to 20-μm ESD diatoms (*triangles*) for non-bloom waters. (tintinnid and ciliate biomass estimates by courtesy of J.R. Beers)

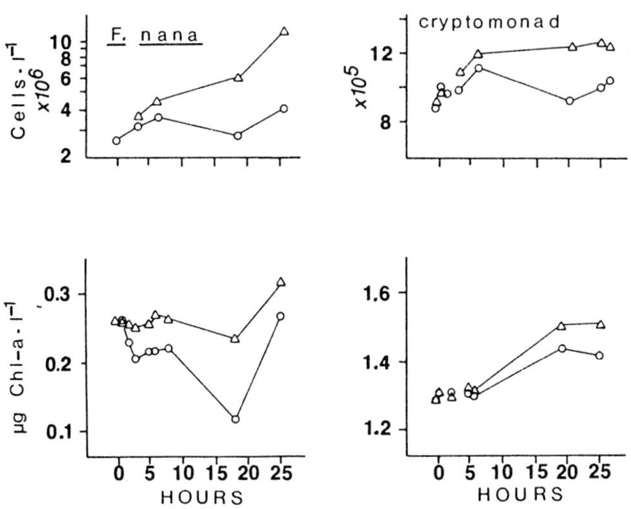

Fig. 5. The results of two experiments measuring the effect of feeding by micro-zooplankton on nanoplankton populations in natural water samples taken from the Weddell Sea. *Fragilariopsis nana* was the dominating phytoplankton species in the sample from Station 36; a 10-μm diameter cryptomonad was dominant at Station 37. *Top* sets of graphs show cell density in the cultures; *bottom* graphs show changes in chlorophyll content of the culture. In the *top* graphs, cell counts of the species are compared between the 10 μm pre-screened (*triangles*) and 202 μm pre-screened (*circles*) cultures incubated for 1 day. In the *bottom* graphs, chlorophyll in the "ungrazed" (< 10-μm pre-screened) culture is indicated by *triangles*, and the "grazed" fraction (< 10 μm post-screen of the < 202-μm culture) is indicated by *circles*

dominant phytoplankton species and that the less fed upon population had a much higher measured rate of growth (Fig. 5). Also, one "ungrazed" phytoplankton species (*F. nana*) attained rates of growth in excess of one doubling per day, close to the theoretical rate predicted from Eppley's (19772) model. Evidence for "grazing" was also reflected in the chlorophyll concentrations of the control group, in relation to those of the 10-μm size-fractioned experimental group.

This was, however, a preliminary study, and much more intensive work will be necessary before we know the geographical extent and the seasonal importance of such feeding pressure. The significance of the data presented above is that: nanoplankton in Antarctic waters might have rates of growth much greater than previously suspected; feeding by protozoans on pico- and nanoplankton may be extensive enough to regulate the size of their populations to fairly uniform concentrations; and, as suggested by the amount of phytoplankton consumed and because the assimilation efficiency for ciliates is presumed to be high (Fenchel 1980a–c), a rapid protozoan growth rate (and hence high heterotrophic production) may occur. Thus, a large flux of C from rapidly growing autotrophic cells might go directly into the micro-zooplankton pool, creating an efficient transfer of autotrophic C to larger particles. It is reasonable to suggest that protozoans, feeding on nanoplankton, thus play a major link in the transformation of "inaccessible" (to Krill) primary and secondary production into "packages" of C more suitably sized for consumption by net-zooplankton.

The relative proportions of autrophic and heterotrophic biomasses in micro- and nanoplankton as reported above are consistent with other data in the literature. Bröckel (1981) found that "protozooplankton" (e.g., ciliate) biomass ranged from 8–110 mg C m^{-2}, in the northern part of the Weddell Sea, to 325 mg C m^{-2} near South Georgia. He found that this component averaged 16% of the total "phytoplankton" biomass. In the Weddell Sea, choanoflagellates have been reported to average 10^5 cells l^{-1} (approximately 5 mg C m^{-3}) (Silver et al. 1980; Marchant 1984), which compares with a total phytoplankton biomass of 1.9–3.3 g C m^{-2} (Holm-Hansen, unpublished). That ciliates contain significant amounts of C in much of the Southern Ocean argues against the idea that they could not exist in open waters for lack of food (e.g., Banse 1982). The sizable amount of autotrophic plus heterotrophic C of nanoplankton populations would seemingly be sufficient to support such micro-heterotrophic organisms (see Fenchel 1980c). Thus, a reasonable hypothesis is that extensive feeding on pico- and nanoplankton by nano- and microplankton (i.e. ciliates and phagotrophic flagellates) takes place in the euphotic zone of Antarctic waters, and this heterotrophic microbial process may account for a significant portion of the total Antarctic heterotrophic production. This, then, may be comparable to the large-volumed diatom production at many times of the year in terms of providing food for the larger zooplankton such as Krill.

A more realistic view of microbial dynamics than that presented by Hart (1934, 1942), and others since, might be depicted diagrammatically in Fig. 6. Within reasonable limits, feeding by protozoan populations may effectively regulate the size of pico- and nanoplankton populations, keeping them at fairly uniform levels. This would imply that a large flux of DOC would result through microbial feeding, as compared with models which suggest that DOC occurs predominantly through exudation by phytoplankton (see Azam et al. 1983). Micro-phytoplankton cells (i.e. diatoms such as *Rhizosolenia* or *Corethron*), due to their large size, would be little fed upon by this heterotrophic community. If protozoans were to control cell densities, net nanoplankton population growth could approach zero, and measured increases (as by floristics, ^{14}C, chlorophyll, ATP, etc.) would reflect lower rate estimates than those which really occur. Additionally, as eukaryotic microbial heterotrophs have a significant population size, inclusion of these as "phytoplankton" would increase initial C values. This would reduce the phytoplankton specific growth rates estimated from the increase of photosynthetically assimilated new C. All these factors added together could tend to give a false impression of actual phytoplankton growth. On the other hand, the large phytoplankton species may have much slower growth rates than their nanoplankton counterparts at a given temperature (Eppley 1972; Banse 1976,

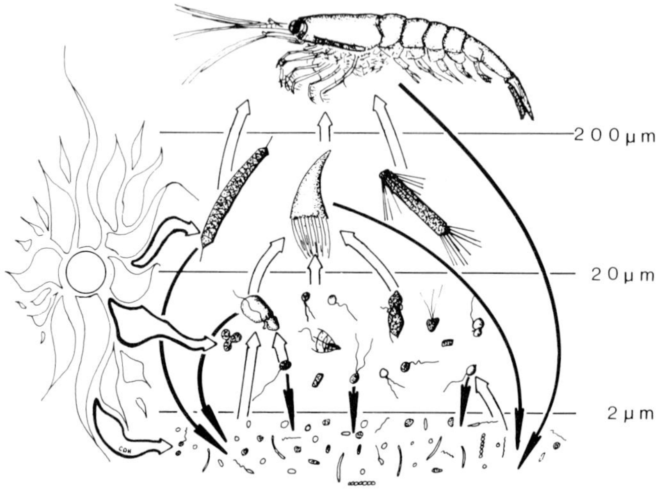

Fig. 6. Part of the Antarctic food web, which might account for some of the discrepancies apparent in the classically conceived linear food chain. Both autotrophic (*dark shading*) and heterotrophic (*clear*) organisms occur in the same size categories represented as micro- (20–200 µm), nano- (2–20 µm) and pico-plankton (< 2 µm). Within each category, a contribution of energy is required either from sunlight (*clear arrows*) or from organic material for heterotrophic assimilation (*dark arrows*). In general, protozoans will feed rapidly and efficiently on cells smaller than themselves. Pico- and nanoplanton are thought not to be efficiently fed on by net-zooplankton, such as Krill, but rather this "inaccessible" carbon is obtained indirectly by feeding on micro-zooplankton. Thus, the protozoan community may act as a major trophic link by concentrating small cells into larger particulates within the Antarctic food web

1982; Malone 1980). If so, these cells might significantly bias measured rates of growth of the total population below the values expected in this cold environment. Thus, reported slow rates for the Southern Ocean (e.g., Holm-Hansen et al. 1977) might have reflected that of the more classical (e.g., diatom) communities. Krill are much more efficient at feeding on microplankton than nanoplankton (Quetin and Ross 1985) and, since the potential micro-heterotrophic production is high with respect to that of microplankton, it is reasonable to propose that protozoans, at times, constitute a major source of food for Krill in the Southern Ocean.

Acknowledgements. We thank R.W. Eppley and J.R. Beers for reviewing the manuscript, J.R. Beers and F.M.H. Reid for providing ciliate and floristic data, respectively, from the Vulcan expeditions, and Ms D. Eck for manuscript preparation. We also thank the crews of the R.V. *Melville* (1981) and USCGC *Polar Star* (1983) for their help and co-operation. This work was supported by National Science Foundation Grants DPP79-21295 and DPP82-12362.

References

Azam F; Fenchel T; Field JG, Gray JS; Meyer-Reil LA; Thingstad F (1983) The ecological role of water-column microbes in the sea. Mar Ecol Prog Ser 10:257–263

Balech E (1970) The distribution and endemism of some Antarctic micro-plankters. In: Holdgate, MW (ed) Antarctic ecology, vol I. Academic, New York, pp 143–147

Banse K (1976) Rates of growth, respiration and photosynthesis of unicellular algae as related to cell size – A review. J Phycol 12: 135–140

Banse K (1982) Cell volumes, maximal growth rates of unicellular algae and ciliates, and the role of ciliates in the marine pelagial. Limnol Oceanogr 27(6):1059–1071

Beers JR; Reid FMH; Stewart G (1975) Microplankton of the North Pacific Central Gyre. Population structure and abundance, June 1973. Int Rev Gesamten Hydrobiol 60:607–638

Boyd CM (1982) Feeding rates and mechanisms of the Antarctic krill, *Euphausia superba.* Trans, Am Geophys Union 63:939

Bröckel K von (1981) The importance of nanoplankton within the pelagic Antarctic ecosystem. Kiel Meeresforsch 5:61–67

El-Sayed SZ (1967) On the productivity of the southwest Atlantic Ocean and the water west of the Antarctic peninsula. In: Llano, G and Schmitt, W (eds) Biology of the Antarctic seas. III. Antarctic Research Ser, vol II. American Geographical Union, Washington DC, pp 15–47

El-Sayed SZ (1971) Observations on phytoplankton bloom in the Weddell Sea. In: Llano GA, Wallen IE (eds) Biology of the Antarctic Seas, Am Geophys Union 17:301–312

El-Sayed SZ; Mandelli EP (1965) Primary production and standing crop of phytoplankton in the Weddell Sea and Drake Passage. In: Biology of the Antarctic Seas II, American Geophysical Union, Washington, D.C., 5:87–106

El-Sayed SZ; Taguchi S (1981) Primary production and standing crop of phytoplankton along the ice-edge in the Weddell Sea. Deep-Sea Res 28:1017–1032

Eppley RW (1972) Temperature and phytoplankton growth in the sea. Fish Bull 70(4):1063–1085

Eppley RW; Peterson BJ (1979) Particulate organic matter flux and planktonic new production in the deep ocean. Nature 282 (5740): 677–680

Eppley RW; Renger EH; Harrison WG (1979) Nitrate and phytoplankton production in Southern California coastal waters. Limnol Oceanogr 24:483–494

Fay R (1973) Significance of nanoplankton in primary production of the Ross Sea, Antarctica, during the 1972 austral summer. PhD thesis, Texas AM University, 184 pp

Fenchel T (1980a) Suspension feeding in ciliated protozoa: Functional response and particle size selection. Microb Ecol 6:1–11

Fenchel T (1980b) Suspension feeding in ciliated protozoa: Feeding rates and their ecological significance. Microb Ecol 6:13–25

Fenchel T (1980c) Relation between particle size selection and clearance in suspension-feeding ciliates. Limnol Oceanogr 25(4): 733–738

Fogg GE; Hayes PK (1982) The relative importance of nutrients and hydrographic features for the growth of Antarctic phytoplankton. Joint Oceanographic Assembly, Halifax, Nova Scotia, August 1982, pp 60–61

Fricker K (1900) The Antarctic regions. Sonneschein, A (trans) MacMillan, New York, 327 pp

Goldman JC; Carpenter EJ (1974) A kinetic approach to the effect of temperature on algal growth. Limnol Oceanogr 19:756–766

Hart TJ (1934) On the phytoplankton of the South-west Atlantic and the Bellingshausen Sea, 1929–31. Discovery Rep 8:1–268

Hart TJ (1942) Phytoplankton periodicity in Antarctic surface waters. Discovery Rep 21:261–356

Hasle GR (1970) Discussion section. In: Holdgate, MW (ed) Antarctic ecology. Academic, New York, p 187

Hasle GR (1978) The inverted microscope method. In: Sournia, A (ed) Phytoplankton manual, UNESCO, U.K., pp 88–96

Hewes CD; Holm-Hansen O (1983) A method for recovering nanoplankton from filters for identification with the microscope: The filter-transfer-freeze (FTF) technique. Limnol Oceanogr 28:389–394

Hewes CD; Holm-Hansen O; Sakshaug E (1983) Nanoplankton and microplankton studies during the circumnavigation cruise. Ant J US 18:169–171

Hewes CD; Reid FMH; Holm-Hansen O (1984) The quantitative analysis of nanoplankton: A study of methods. J Plankton Res 6:601–613

Holm-Hansen O; Lorenzen CJ; Holmes RW; Strickland JDH (1965) Fluorometric determination of chlorophyll. J Cons perm int Explor Mer 30:3–15

Holm-Hansen O; El-Sayed SZ; Franceschini GA; Cuhel RL (1977) Primary production and the factors controlling phytoplankton growth in the Southern Ocean. In: Llano, GA (ed) Adaptations within Antarctic ecosystems. Gulf, Houston, Texas, pp 11–50

Holm-Hansen O; Riemann B (1978) Chlorophyll *a* determination: Improvements in methodology. Oikos 30:438–447

Holm-Hansen O; Foster TD (1981) A multidisciplinary study of the eastern Scotia Sea. Ant J US 16:159–160

Holm-Hansen O; Chapman AS (1983) Antarctic circumnavigation cruise, 1983, Ant J US 18:162–163

Holm-Hansen O; Huntley M (1984) Feeding requirements of krill in relation to food sources. J Crust Biol 4(special 1):156–173

Horner RA (1976) Sea ice organisms. Oceanogr Mar Biol Annu Rev 14:167–182

Lillie DG (1913) Summary of biological work carried out on board the *Terra Nova,* 1910–1913. In: Huxley, L (ed) Scott's last expedition, vol II. Smith Elder, London, p 475

Macaulay MC (1981) Distribution and abundance of krill in the Scotia Sea as observed acoustically, Ant J US 16:166–167

Malone T (1980) Algal size. In: Morris, I (ed) The physiological ecology of phytoplankton. University Calif Press, pp 433–463

Marchant HJ (1985) Choanoflagellates in the Antarctic marine food chain. In: Siegfried WR, Condy PR, Laws RM (eds) Antarctic nutrient cycles and food webs (Proceedings of the 4th SCAR symposium on Antarctic biology). Springer, Berlin Heidelberg New York

McClatchie S; Boyd CM (1983) A morphological study of sieve efficiencies and mandibular surfaces in the Antarctic krill *Euphausia superba.* Can J Fish Aquat Sci 40:955–967

Meyer MA; El-Sayed SZ (1983) Grazing of *Euphausia superba* Dana on natural phytoplankton populations. Polar Biol 1:193–197

Nemoto T; Harrison G (1981) High latitude ecosystems. In: Longhurst, AR (ed) Analysis of marine ecosystems. Academic, New York, pp 95–126

Neori A; Holm-Hansen O (1982) Effect of temperature on rate of photosynthesis in Antarctic phytoplankton. Polar Biol 1:33–38

Paden CA; Hewes CD; Neori A; Holm-Hansen O; Weaver E; Kiefer DA; Sakshaug E (1981) Phytoplankton studies in the Scotia Sea. Ant J US 16:163–164

Quetin LB; Ross RM (1985) Feeding by Antarctic krill, *Euphausia superba:* Does size matter? In: Siegfried WR, Condy PR, Laws RM (eds) Antarctic nutrient cycles and food webs (Proceedings of the 4th SCAR symposium on Antarctic biology). Springer, Berlin Heidelberg New York

Silver MW; Mitchell JG; Ringo DL (1980) Siliceous nanoplankton. II. Newly discovered cysts and abundant choanoflagellates from the Weddell Sea, Antarctica. Mar Biol 58:211–217

Smith AE; Morris I (1980) Pathways of carbon assimilation in phytoplankton from the Antarctic Ocean. Limnol 25:865–872

Strathman RR (1967) Estimating the organic carbon content of phytoplankton from cell volume or plasma volume. Limnol Oceanogr 12:411–418

Utermöhl H (1958) Zur Vervollkommnung der quantitativen phytoplankton-Metodik. Mitt Int Ver Theor Agnew Limnol Verh 9:1–38

Spatial Variability of Phytoplankton and the Distribution and Abundance of Krill in the Indian Sector of the Southern Ocean

L.H. WEBER and S.Z. EL-SAYED[1]

Summary. In conjunction with an acoustic survey of Krill, *Euphausia superba,* carried out during the 1981 FIBEX cruise of the M.V. S.A. Agulhas, water samples were collected for the analysis of phytoplankton biomass, productivity and metabolic state. Samples were size-fractionated to determine what portions of the biomass and productivity were in the nano (< 20 μm) and net-plankton (> 20 μm) fractions. Inter-relationships between the phytoplankton, Krill and physico-chemical variables were analyzed with the aid of multiple regression and canonical correlation techniques. With other variables held constant, phytoplankton biomass was found to be negatively correlated with mean integrated Krill density and number of Krill swarms, but positively correlated with Krill swarm biomass. Preliminary analysis of continuous integrated Krill and surface in vivo fluorescence profiles indicated strong day-night periodicities. For Krill, this was a sampling artifact. Day-night periodicity in the fluorescence profiles appears to have been real, and must be accounted for if these records are to be used in determining the spatial scale of variability in phytoplankton standing stocks.

1 Introduction

Constituting perhaps 50% of the Antarctic zooplankton biomass, Krill, *Euphausia superba,* is the dominant herbivorous species and serves as a key link between marine primary producers and top predators (Holdgate 1967). The broad distributional patterns of Krill are known (Marr 1962; Mackintosh 1973; Voronina 1974; Everson 1977). In general, *E. superba* has a circumpolar, but asymmetrical, distribution in waters south of the Polar Front, which is believed by certain authors to be due largely to physical constraints on the population (Deacon 1979, 1982; Deacon and Baker 1980). However, numerous behavioural characteristics, including breeding, feeding, predator avoidance and chemo-reception, have also been suggested as factors which govern the distribution of Krill and affect their swarming-schooling incidence (Hamner et al. 1983).

During the First International BIOMASS Investigation (FIBEX) investigators working the Bransfield Strait noted that areas of high Krill concentration usually had relatively low standing crops of phytoplankton (Uribe 1982; Rakusa-Suszczewski 1982). A similar inverse relationship was observed in an area north of Elephant Island, which was sampled prior to and during the occurrence of a dense swarm of Krill furcilia (Holm-Hansen and Huntley, to be publ.). Based on studies of Krill feeding and egestion behaviour, pigment content of faeces, Krill densities and ambient chlorophyll concentration, Antezana and Ray (to be publ.) concluded that for this swarm, feeding and swarming were co-occurring events. This study on the behaviour of a swarm in nature substantiates laboratory experiments which suggested that Krill feed while swarming (Antezana et al. 1982).

Meyer and El-Sayed (1983) reported that phytoplankton cells > 20 μm were fed on more efficiently by Krill than smaller cells. McClatchie and Boyd (1983), using electron micrographs of Krill feeding appendages, suggested that particles > 60 μm would be retained with maximum efficiency, those between 6–20 μm in diameter with 30–40% efficiency, and that for particles < 6 μm retention efficiency drops off rapidly. It appears that much of the selectivity in feeding by *E. superba* is a result of mechanical limitations of the filtering structure (Marschall 1985; Quetin and Ross 1985). These studies suggest a need to re-evaluate the conclusion of Kawamura (1981) that *E. superba* preferentially occupies waters where small diatoms are prevalent. Ligowski (1982) reported the presence of fragments of *Corethron criophilum* and the dominance of *Chaetoceros* sp. in certain samples, and the occurrence of other large forms in the stomach contents of *E. superba.* The presence or absence of large diatom species in Krill-rich as contrasted to Krill-poor waters most probably has a temporal component.

The purpose of our study was to examine the distributional patterns of *E. superba* in relation to its primary food source, the phytoplankton. Inter-relationships between Krill and phytoplankton are expressed as: correlations between Krill density-aggregations and phytoplankton biomass; shifts in the size distribution of the phytoplankton prey community; and, spatial patterns of predator and prey biomass distribution. Each of these expressions is analyzed in the context of the physico-chemical environment, using multivariate and spectral analysis techniques.

2 Materials and Methods

Data were collected during the FIBEX cruise (10 February–20 March, 1981) of the M.V. S.A. Agulhas, in the general

1 Department of Oceanography, Texas A&M University, College Station, TX 77843, USA

Antarctic Nutrient Cycles and Food Webs
(ed. by W.R. Siegfried, P.R. Condy, and R.M. Laws)

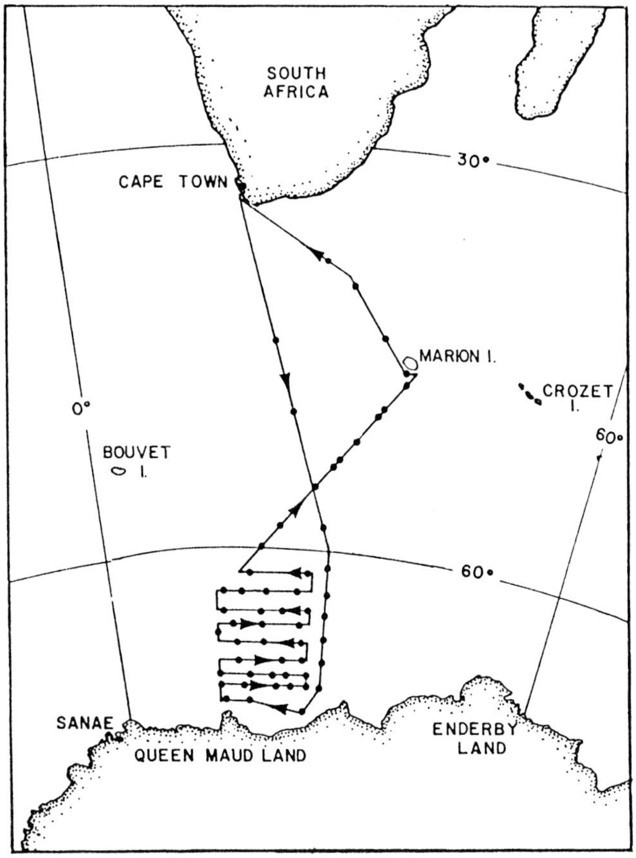

Fig. 1. Cruise track of the M.V. S.A. Agulhas during FIBEX, February–March 1981

area between 15° and 30°E and 50° and 70°S (Fig. 1). The data reported here represent only a small part of a much larger suite lodged in the South African Sea Fisheries Research Institute (Hampton 1985).

2.1 Discrete Samples

Stations were occupied daily at approximately 1000 and 2400 h local time. Eight-litre Niskin samples were collected from depths corresponding to the penetration of 100, 40, 18, 8, 5, 1.25, and 0.25% of surface irradiance, as determined by use of a Secchi disc. Aliquots from each Niskin sampler were analyzed for ATP, total adenylates and adenylate energy charge (Karl and Holm-Hansen 1978), chlorophyll-a and phaeopigment concentrations (Holm-Hansen et al. 1965; Strickland and Parsons 1972) and primary productivity (Steeman Nielsen 1952; O'Reilly and Thomas 1979). Aliquots used in the ^{14}C incorporation experiments were placed in 100-ml Whirl-Pak bags (NASCO, Inc.) and contained in a deck incubator. The various light regimes of the sampled depths were simulated by using various combinations of spectral and neutral density filters (Schott Optical Glass Co.). Surface irradiance was measured by a Biospherical Instruments QSI-140 quantum meter which integrated photosynthetically active radiation (PAR) over the 4-h incubation period. The relative proportions of each of the above parameters attributable to the net (> 20-μm) and the nano (< 20-μm) size fractions of the plankton community were determined by serial filtration through a 20-μm Nitex mesh and a Toyo Type GB100 R glass-fibre filter. Station data were analyzed statistically using SAS (Statistical Analysis System, Ray 1982) procedures. Variables are as defined in Table 1.

2.2 Continuous Samples

Acoustic estimates of Krill biomass were given as mean densities, averaged in most cases over 30-min intervals (see Hampton 1985). Since the distance travelled during each time interval was also given, data points equidistant in space could be extracted from the record. An interval of 10 km was chosen, since this was the approximate modal value of the distance travelled per 30 min. In vivo fluorescence of near-surface waters was monitored continuously by diverting flow from the ship's seawater pump through the flow-through cell of a Turner Designs, Model 10 fluorometer. In order to correlate in vivo fluorescence with concentration of extracted chlorophyll, samples were taken from the fluorometer's out-flow every 6 h for in vitro analysis. Continuous profiles of in vivo fluorescence and surface seawater temperature were recorded with a strip-chart recorder. These records of in vivo fluorescence and temperature were sub-sampled with a Numonics Model 1224 electronic digitizer at 1 and 2-km intervals, respectively. Since the acoustic log yielded average ship speed over 30-min intervals, variable ship speed was accounted for by changing the distance between points collected from the strip charts. Auto-correlations for the three data sets were computed using FESTSA (Fast and EaSy Time Series Analysis) procedures (Brooks 1976).

3 Results

3.1 Analysis of Discrete Data

Correlation coefficients (r) between pairs of variables are given in Table 2. In general, bivariate correlations are weak. Of the physical, chemical and Krill variables, Max Tons is most highly correlated with the measured phytoplankton parameters, with correlation coefficients of ≥ 0.60 between Max Tons and Nano and Net AT, Nano and Net Chl, and Nano Phaeo. Because of the very high correlation between salinity and sigma-t, salinity was not included in subsequent multiple regression or canonical correlation analyses in order to avoid multi-colinearity. Similarly, due to correlation between Phaeo and pigment ratio values, and because the latter are ecologically more interpretable, Nano Phaeo and Net Phaeo were also omitted from subsequent analyses.

Table 1. Definitions of coded variables used in statistical analyses

Nano-Net Prod	=	Primary productivity
Nano-Net AT	=	Total adenylate (ATP+ADP+AMP) concentration
Nano-Net EC	=	Adenylate energy charge
Nano-Net Chl	=	Chlorophyll-a concentration
Nano-Net Phaeo	=	Phaeopigment concentration
Nano-Net AN	=	Assimilation number (Prod/Chl)
Nano-Net PigRat	=	Pigment ratio (Chl/Chl + Phaeo)
Integ Krill[a]	=	Mean Krill density (g m^{-2}) integrated to 100 m depth and averaged over 10 km before and after each station
No. Swarms[a]	=	Number of Krill swarms detected within 10 km before and after each station
Mn Radius[a]	=	Mean radius of Krill swarms detected within 10 km before and after each station
Max Tons[a]	=	Biomass of largest Krill swarm detected within 10 km before and after each station
PAR[a]	=	Mean quanta cm^{-2} s^{-1} during each ^{14}C incubation
Stability[a]	=	(1/P) * (△P/△Z) where P = density, △Z = difference in depth between Om and the depth of 0.25% light penetration

a For varaibles with only 1 value per station, the same value was assigned to all depths

Using each of the phytoplankton and Krill parameters as dependent variables, coefficients of determination (r^2) for all possible multiple regression equations with from 1 to 6 independent variables were computed. Mallows' C_p statistic aided in the selection of models with the appropriate number of independent variables (Freund and Minton 1979; Ray 1982). Selected models were analyzed to yield standardized partial regression coefficients, P values and r^2 (Tables 3–4).

The importance of Krill as a factor in regulating phytoplankton biomass-distribution is suggested by the entrance of one or more Krill parameters into each of the models for predicting adenylates, chlorophyll, and the amount of net chlorophyll relative to phaeopigment . In general, the calculated phytoplankton models predict that low phytoplankton biomass and relatively high concentrations of phaeopigments will be associated with high integrated Krill densities and increased numbers of Krill swarms. On the other hand, increased phytoplankton biomass is predicted for areas characterized by individual Krill swarms of high biomass.

The possible significance of silicate in relation to phytoplankton is indicated by its positive correlation with Nano AN and its negative correlation with Nano and Net Chl. Similarly, both Nano and Net AN are positive functions of temperature, while Nano and Net Chl and Nano AT each shows an inverse correlation with this physical variable. Stability and latitude also enter as predictor variables in several of the phytoplankton models. Stability is positively correlated with Nano AT, but negatively correlated with Nano Chl. Although stability appears as a predictor of Net AN and Net PigRat, it does not enter into the models for Nano AN or Nano PigRat. Latitude may represent a significant variable which was not included in this analysis, or combinations of variables which could not be adequately represented with simple multiple regression techniques.

In each of the models for the four Krill parameters, Net PigRat has a negative coefficient. If a low pigment ratio (relatively high proportion of phaeopigments) is taken as an indication of "grazing" (Currie 1962), these models suggest that: Krill feed preferentially on net-plankton; and, feeding pressure is a positive function of integrated Krill density, the number of swarms, the size of the swarms, and the biomass of swarms. The apparent inconsistency in the relationship between Mn Radius and Net PigRat is worth noting. As a predictor variable of Net PigRat, Mn Radius has a positive coefficient, whereas in the model for Mn Radius, Net PigRat has a negative coefficient. This disagreement is resolved when Max Tons is added as an independent variable to the model for Mn Radius. With Max Tons held constant, the partial regression coefficient of Net PigRat reverses its sign, while coefficients of the other independent variables retain their sign. A possible interpretation is that feeding occurs most actively in small swarms of increased density.

able 2. Simple correlation coefficients for all pairs of phytoplankton, Krill and physico-chemical variables

	Te,ü	Salinity	Sigma-t	DO	Silicate	Nitrate	Latitude	Stability	Max Tons	Nano Prod	Net Prod	Nano AT	Nano Chl	Net Chl	Nano Phago	Net Phago	Nano AN	Net AN	Net AN	
ight	0.38	−0.40	−0.43							0.30							0.39			
emp		−0.62	−0.74		−0.44	−0.31	−0.83	−0.44							−0.36	−0.45			0.36	
alinity			*0.98*	−0.57	0.53	0.55		−0.38		−0.25					0.58	0.50			−0.45	
igma-t				−0.53	0.56	0.54	0.39								0.57	0.53			−0.44	
O					−0.32	−0.34									−0.58	−0.48			0.38	
hosphate					0.30	0.29	−0.52	−0.35		−0.26									−0.25	
ilicate						0.60	0.44								0.42	0.36			−0.29	
atitude								0.39											0.39	
teg Krill					−0.49				0.66					0.48	0.47					
n Radius									0.50			0.50								
ano AT					−0.31			0.46	*0.60*		0.25		0.49	0.41					0.34	0.41
et AT		0.41	0.39	−0.26		0.26			*0.60*			0.32	0.58	0.60	0.70	0.30	−0.37	−0.31	−0.29	0.33
ano EC																			−0.26	−0.26
ano Chl	−0.35	0.44	0.52			0.31			*0.73*					0.63	0.40		−0.39	−0.28		0.28
et Chl		0.26	0.26						*0.80*						0.36		−0.28	−0.33		0.47
ano Phaeo						0.48			*0.60*							0.67	−0.28	−0.26	*−0.76*	
et Phaeo						0.32													0.58	*0.39*
ano AN		−0.36	−0.36							0.69								0.36		
et AN											0.59									0.35
ano PigRat						−0.30														0.35

nly r values ≥ 0.25 (with P-values ≤ 0.05) are included; n varied between 60 and 286. For comparisons involving variables having only 1 value per station ee Table 1), n varied from 16–37. The values of correlations referred to in the text appear in italics

In addition to Net PigRat, Nano AT appears as a predictor variable in three of the Krill models, stability in two of them, and nine other variables in only one of the models.

Canonical correlation, unlike multiple regression, allows the inclusion of more than one variable on both sides of a model equation (Gittins 1979). From the two sets of variables, the procedure finds pairs of linear combinations (canonical variables) which are maximally correlated. Interrelationships between phytoplankton parameters and physical, chemical and Krill variables are summarized by the two significant ($P < 0.05$) pairs of canonical variables given in Table 5. The ecologically interpretable variables are those with both large coefficients and large values of r. If a variable's canonical coefficient and the given r value are opposite in sign, the weight assigned to that coefficient is a result of correlations between that variable and other variables from the same set (Cooley and Lohnes 1971; Ray 1982).

Both pairs of canonical variables (Table 5) identify relationships characterized by a positive correlation between Max Tons and chlorophyll. For the first pair, Nano Chl has the largest coefficient, while Net Chl has the greatest loading in the second pair of canonical variables. The two dimensions described by the two pairs of canonical variables may represent differing relationships in separate geographical regions. For the first pair, temperature and stability are loaded negatively, whereas stability has a positive coefficient in the second dimension. This may be a reflection of latitudinal differences.

Although the r^2 values for both pairs of canonical variables are reasonably high, redundancy analysis shows that only 30% of the variance in the phytoplankton data set is explained by the two significant canonical variables of the environmental data set. This low percentage is not atypical of canonical correlation analysis and results, in part, from the fact that the procedure searches for pairs of canonical variables which are maximally correlated with each other. The computed linear combinations do not necessarily account for maximum variance in their own respective data sets (Gittins 1979). Thus, this analysis is

Table 3. Selected multiple regression models for phytoplankton variables. Only models for which P values of all coefficients are < 0.10 are given (n = 44)

Dependent variable	Independent variables	Standardized coefficients	r^2
Nano AT	Stability	0.94	0.57
	Max Tons	0.57	
	Latitude	–0.56	
	No. Swarms	–0.45	
	Mn Radius	0.32	
	Integ Krill	–0.23	
Net AT	Latitude	–0.76	0.48
	Temp	–0.58	
	Max Tons	0.42	
	Integ Krill	–0.25	
	No. Swarms	–0.23	
Nano Chl	Temp	–0.82	0.64
	Max Tons	0.78	
	Phosphate	0.28	
	Silicate	–0.27	
	Stability	–0.20	
Net Chl	Max Tons	0.85	0.70
	Silicate	–0.50	
	Temp	–0.33	
	Nitrate	0.26	
	Integ Krill	–0.22	
	Mn Radius	–0.19	
	No. Swarms	–0.16	
Nano AN	Silicate	0.54	0.40
	Phosphate	–0.40	
	Sigma-t	–0.38	
	Temp	0.35	
Net AN	Latitude	1.10	0.40
	No. Swarms	0.97	
	Stability	–0.86	
	PAR	0.63	
	Temp	0.48	
Net PigRat	No. Swarms	–0.61	0.52
	Latitude	–0.47	
	Mn Radius	0.38	
	Stability	0.36	
	Integ Krill	–0.28	
	Nitrate	0.19	

extremely useful in revealing associations but often has little predictive value.

The results of the analysis in which No. Swarms and Max Tons were related simultaneously to temperature, stability and phytoplankton variables are summarized in Table 6. In the first pair of canonical variables, a small number of swarms having a large biomass is associated with increased phytoplankton concentrations and low water column stability. In the second pair of canonical variables, an increased number of swarms is associated with increased phaeopigments relative to net chlorophyll and with increased stability. As in the previous phytoplankton analysis, the fact that the sign of the coefficient for stability is opposite in the two dimensions suggests that differing phytoplankton-Krill inter-relationships may be associated with different physical regimes. Redundancy analysis shows that, taken together, the two canonical variables of the environment account for 62% of the variance in the two Krill parameters.

The vertical migration of Krill to very near the surface at night frequently prevented their detection acoustically (Hampton 1985), so that reported night-time values are

Table 4. Selected multiple regression models for Krill variables. Only models for which the P values of all coefficients are < 0.10 are given (n = 13)

Dependent variable	Independent variables	Standardized coefficients	r^2
Integ Krill	Nano Prod	2.83	0.81
	Nano AN	−1.96	
	Temp	0.64	
	Net PigRat	−0.60	
No. Swarms	Net PigRat	−1.28	0.94
	Nano AT	0.94	
	Net EC	−0.79	
	Latitude	−0.43	
	PAR	−0.31	
Mn Radius	Nano AT	2.09	0.80
	Stability	−1.49	
	Nano PigRat	1.19	
	Net PigRat	−1.00	
	Net Chl	−0.64	
Max Tons	Nano AT	1.32	0.86
	Net PigRat	−0.86	
	Nano Chl	0.77	
	Stability	−0.69	

Table 5. Canonical correlation analysis of phytoplankton with environmental (Krill and physico-chemical) variables for day-time stations

Variable	Canonical variable			
	I		II	
	Coeff.	r	Coeff.	r
Phytoplankton				
Nano AT			0.79	0.73
Net AT				
Nano Chl	0.83	0.82	−0.34	0.30
Net Chl	0.21	0.72	0.88	0.49
Nano AN				
Net AN				
Nano PigRat				
Net PigRat	0.25	0.33		
Environmental				
Temp	−0.72	−0.48		
Sigma-t				
Silicate			−0.67	−0.47
Nitrate	0.41	0.58	0.19	−0.26
Stability	−0.55	−0.43	0.74	0.68
Integ Krill				
Max Tons	0.41	0.43	0.57	0.36
Canonical r^2	0.78		0.73	
Redundancy	0.21		0.09	

Values have been omitted where the standardized canonical coefficients are < 0.10 and/or the correlation coefficient between a variable and its canonical variable is < 0.25 (n = 45).

Table 6. Canonical correlation analysis of No. Swarms and Max Tons with Temp, Stability and phytoplankton variables for day-time stations

Variable	Canonical variable			
	I		II	
	Coeff.	r	Coeff.	r
Krill				
No. Swarms	−0.49	−0.30	0.89	0.95
Max Tons	0.97	0.87	0.31	0.48
Environmental				
Temp				
Stability	−0.52	−0.44	0.66	0.50
Nano AT	0.42	0.50	0.11	0.35
Net AT				
Nano Chl	0.11	0.59		
Net Chl	0.66	0.84		
Nano AN				
Net AN	−0.15	−0.33	0.22	0.39
Nano PigRat				
Net PigRat			−0.66	−0.50
Canonical r^2	0.68		0.57	
Redundancy	0.29		0.33	

Values have been omitted where the standardized canonical coefficients are < 0.10 and/or the correlation coefficient between a variable and its canonical variable is < 0.25 (n = 53)

generally low. To ensure that this pattern was strictly a sampling artifact, a t-test was used to discern differences between the day-time and night-time means of each variable. Integ Krill has a mean of 2.40 g m^{-2} for day-time stations and 0.71 g m^{-2} for night-time stations (P = 0.064). The only other variable with significantly different means was Nano Chl which had a night-time mean of 0.098 mg m^{-3} and a day-time mean of 0.079 mg m^{-3} (P = 0.075). Although integrated Krill density was not adequately sampled at night, it appears that Krill swarming characteristics (No. Swarms, Mn Radius, and Max Tons) do not display any day-night differences.

In an effort to test the validity of the multi-variate relationships, separate analyses were performed on water column means and integrated values of the various parameters. In all cases, results were similar to those reported. It is emphasized that the above statistical exercises have served to generate, rather than to test, hypotheses. Although calculated levels of significance have been used as general guides, their meaningfulness is questionable because many of the variables have non-normal frequency distributions, and because multiple, rather than single, tests have been run in choosing the selected models.

3.2 Analysis of Continuous Data

Continuous profiles of integrated Krill biomass, surface in vivo fluorescence and surface seawater temperature, with the cruise track stretched into a continuous straight line, are shown in Fig. 2. The indicated day-night cycles appear to be of different lengths due to variable ship speed and length of time at each station. This graphical representation indicates several problems which must be addressed prior to the application of spectral analysis techniques. As noted earlier, night-time acoustic estimates of Krill biomass were only about 30% of the day-time values. The result is an artifactual, low-frequency periodicity which tends to overshadow shorter scales of variability that may be of ecological significance. Because the ship traversed a different distance during each day-night cycle, this low-frequency period does not correspond to a particular spatial scale. Auto-correlation analysis of a similar profile, from the March, 1980, M.V. S.A. Agulhas pre-FIBEX cruise, showed a weak positive correlation between points 300–400 km apart, a distance generally corresponding to that covered during 1 day-night period. More significantly, maximum correlation occurred between points separated by 60–70 km or less, with near zero correlation at distances of 70–300 km.

In vivo fluorescence also displays strong diel periodicity. Results are similar to those reported for the central equatorial Pacific (Setser et al. 1982) and for the Southern Ocean region south of Australia (Yamaguchi and Shibata 1982). A plot of in vivo fluorescence, normalized against corresponding values of extracted chlorophyll, and time of day (Fig. 3), suggests that the diel periodicity in fluorescence may have a physiological basis (Kiefer 1973; Loftus and

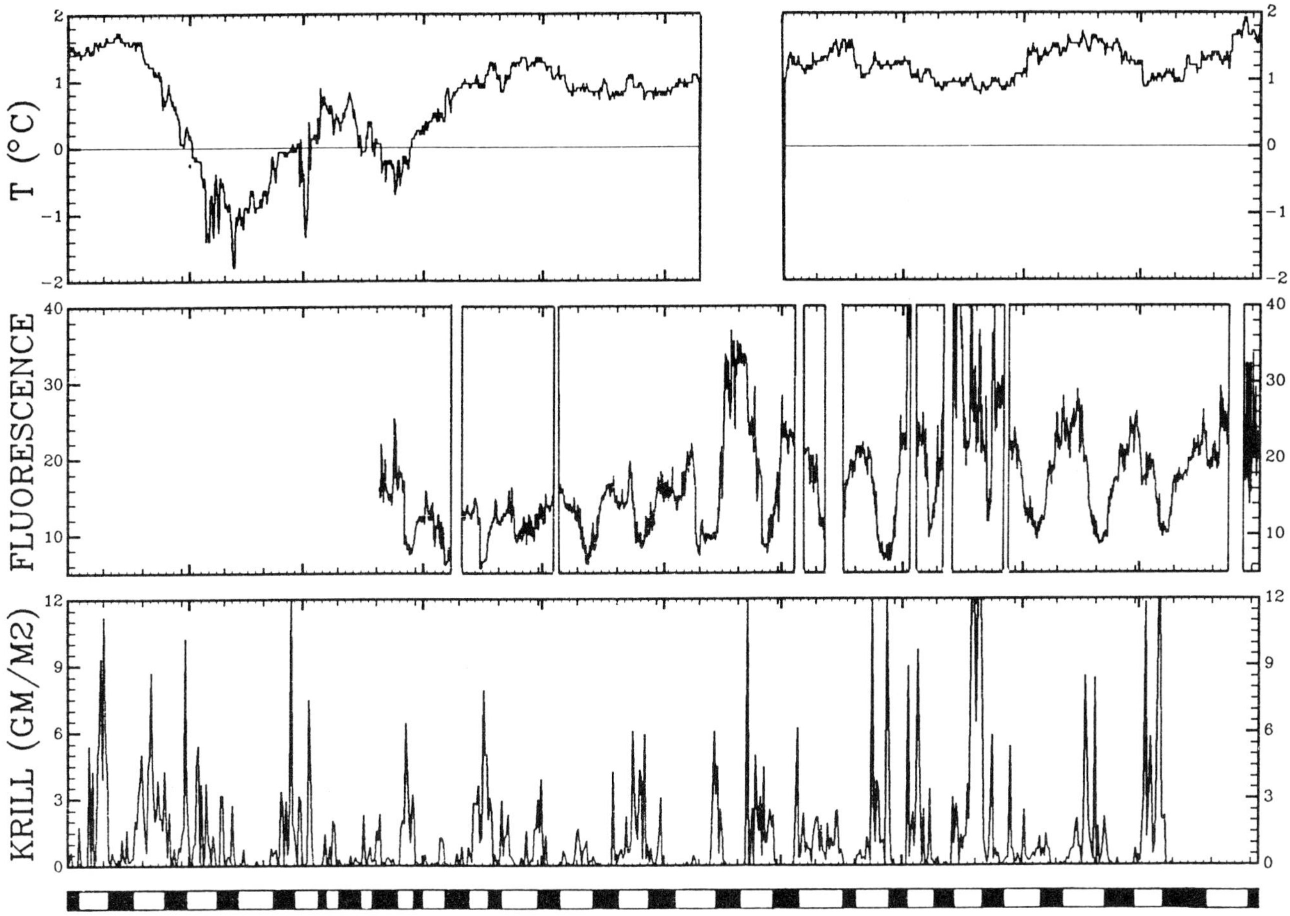

Fig. 2. Profiles of temperature, in vivo fluorescence and integrated Krill density during the FIBEX cruise of the M.V. S.A. Agulhas in February–March 1981. The x axes represent distance (10 mm = 300 km). The day (*clear*)-night (*shaded*) cycle is shown in the *horizontal bar*

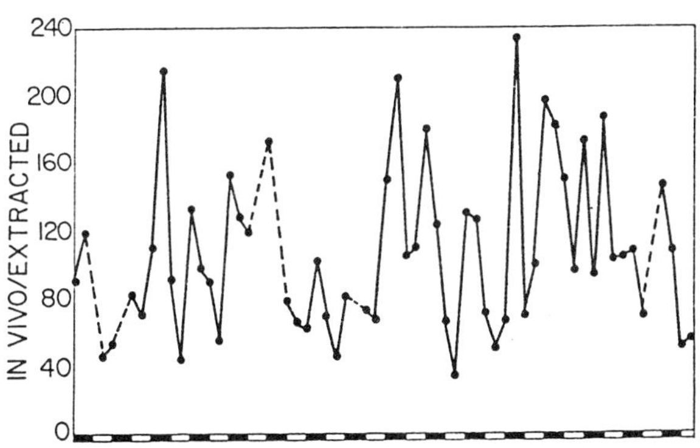

Fig. 3. Day-night differences in the amount of in vivo fluorescence per unit chlorophyll during the FIBEX cruise of the M.V. S.A. Agulhas in February–March 1981. The day (*clear*)-night (*shaded*) cycle is shown in the *horizontal bar*

Seliger 1975). Maximum fluorescence per unit chlorophyll occurs at night, when minimal amounts of excitation light would be expected to be utilized by the phytoplankton's photosynthetic apparatus. Again drawing from the 1980 M.V. S.A. Agulhas data set, auto-correlation analysis of the fluorescence profile revealed positive correlations at a scale of 300–400 km and at scales of < 110 km.

4 Discussion

There are two analytically different approaches to studying the correlations between groups of variables. Classical regression-type analyses quantify relationships between variables given their mean values at discrete points. No account is taken of the spatial or temporal distribution of the variables. Spectral analysis, on the other hand, evaluates the distribution of a parameter's variance as a function of the distance or time between sampling points (Platt and Denman 1975). Both approaches have been attempted here in an effort to define relationships between phytoplankton and Krill and between these two populations and their physico-chemical environment.

Although feeding pressure has been proposed as a factor which controls the biomass and productivity of Antarctic phytoplankton (e.g., Holm-Hansen et al. 1977; Meyer and El-Sayed 1983), the availability of phytoplankton data taken concurrently with the acoustic estimation of Krill biomass was allowed, perhaps for the first time, a statistical examination of this hypothesis. Inter-relationships among the suite of phytoplankton, Krill and physico-chemical variables are observable at several levels of complexity through the use of simple bivariate correlations, multiple regression and canonical correlation analysis.

In the simple correlation matrix (Table 2), both adenylates and chlorophyll are correlated more strongly with maximum swarm biomass than with any other variable. In both the multiple regression and canonical correlation analysis, the correlation coefficient between phytoplankton biomass and Max Tons remains strongly positive. Although the total absence of any time course measurements makes interpretation questionable, it appears that Krill are somehow cueing in on areas of local phytoplankton abundance. Additionally, in support of the conclusion of Meyer and El-Sayed (1983) and McClatchie and Boyd (1983) concerning the size-dependent selectivity of Krill feeding, the amount of phaeopigments relative to net chlorophyll concentration is correlated positively with each of the four Krill parameters in the multiple regression models.

Physico-chemical factors which display a significant relationship with the measured phytoplankton variables include temperature, stability and silicate. The controlling influence of each of these factors has been examined previously (e.g., El-Sayed and Mandelli 1965; Holm-Hansen et al. 1977; Allanson et al. 1981), mostly through the use of bivariate correlation analysis. It is interesting to consider the various dimensions of interaction between these variables with the aid of multiple regression and canonical correlation analysis. The possible importance of silicate as a limiting nutrient is noteworthy. This is evidenced by silicate's negative partial correlation with phytoplankton biomass and, more convincingly, by its positive partial correlation with assimilation number.

The problem of day-night periodicity in the Krill and fluorescence records has prevented completion of the proposed spectral analysis. Our plans for further analysis of these continuous profiles include examination of individual day-time sections and then the combination of results to obtain an average picture. The fact that important spatial scales appear to be less than the distance covered during an average day-time period justifies this approach. Additionally, since in vivo fluorescence does not display a step-like response to changes between day and night, the day-time sections of fluorescence will be normalized to a mean irradiance curve.

The application of various statistical procedures to the suite of data obtained during the M.V. S.A. Agulhas FIBEX cruise has enabled us to generate several hypotheses concerning phytoplankton-Krill inter-relationships. In order to test these hypotheses, future studies must resolve finer spatial scales and include time-course measurements in order to examine the underlying mechanisms.

Acknowledgements. The excellent co-operation of I. Hampton, D. Miller, J. Henry, and their associates at the South African Sea Fisheries Research Institute is gratefully acknowledged. We thank Dr. E. Hofmann for advice on data analysis, and Dr. S. Worley for guidance in using FESTSA subroutines. This study was supported, in part, by a grant (DPP-8111107) from the Division of Polar Programs, National Science Foundation.

References

Allanson BR; Hart RC; Lutjeharms JRE (1981) Observations on nutrients, chlorophyll and primary production of the Southern Ocean south of Africa. S. Afr J Antarct Res 10:3–14

Antezana T; Ray K (to be published) Feeding of *Euphausia superba* in a swarm north of Elephant Island. J Crust Biol

Antezana T; Ray K; Melo C (1982) Trophic behaviour of *Euphausia superba* Dana in laboratory conditions. Polar Biol 1:77–82

Brooks DA (1976) FESTSA (Fast and EaSy Time Series Analysis) Center for Marine and Coastal Studies, North Carolina State Univ., Raleigh, North Carolina

Cooley WW; Lohnes PR (1971) Multivariate data analysis. Wiley, New York

Currie RI (1962) Pigments in zooplankton faeces. Nature 193:956–957

Deacon GER (1979) The Weddell gyre. Deep-Sea Res 26:981–995

Deacon GER (1982) Physical and biological zonation in the Southern Ocean. Deep-Sea Res 29:1–15

Deacon GER, Baker A (1980) Note on the distributional range of *Euphausia superba.* BIOMASS Newsletter 2:16–19

El-Sayed SZ; Mandelli EF (1965) Primary production and standing crop of phytoplankton in the Weddell Sea and Drake Passage. In: Llano GA (ed) Biology of the Antarctic Seas II. American Geophysical Union, Washington D.C. (Antarctic Research Series, Vol 5, pp 87–106)

Everson I (1977) The living resources of the Southern Ocean. Food and Agricultural Organization of the United Nations, Rome (Southern Ocean Fisheries Survey Programme GLO/SO/77/1)

Freund RJ; Minton PD (1979) Regression methods: a tool for data analysis. Marcel Dekker, New York Basel

Gittins R (1979) Ecological applications of canonical analysis. In: Orloci L, Rao CR, Stiteler WM (eds) Multivariate methods in ecological work. International Co-operative Publ House, Fairland, Maryland, pp 309–535

Hamner WM; Hamner PP; Strand SW, Gilmer RW (1983) Behaviour of Antarctic krill, *Euphausia superba:* Chemoreception, feeding, schooling, and molting. Science 220:433–435

Hampton I (1985) Abundance, distribution and behaviour of *Euphausia superba* in the Southern Ocean between 15°E and 30°E during FIBEX. In: Siegfried WR, Condy PR, Laws RM (eds) Antarctic nutrient cycles and food webs (Proceedings of the 4th SCAR symposium on Antarctic Biology). Springer, Berlin Heidelberg New York

Holdgate MW (1967) The Antarctic ecosystem. Philos Trans R Soc Lond 252:363–383

Holm-Hansen O; Lorenzen CJ; Holmes RW; Strickland JDH (1965) Fluorometric determination of chlorophyll. J Cons perm int Explor Mer 30:3–15

Holm-Hansen O; El-Sayed SZ; Franceschini GA; Cuhel R (1977) Primary production and the factors controlling phytoplankton growth in the Antarctic seas. In: Llano GA (ed) Adaptations within Antarctic Ecosystems (Proceedings of the 3rd SCAR symposium on Antarctic Biology). Gulf, Houston, pp 11–50

Holm-Hansen O; Huntley M (to be published) Feeding requirements of krill in relation to food sources. J Crust Biol

Karl DM; Holm-Hansen O (1978) Methodology and measurement of adenylate energy charge ratios in environmental samples. Mar Biol 48:185–197

Kawamura A (1981) Food habits of *Euphausia superba* and the diatom community. In: El-Sayed SZ (ed) Selected contributions to the Woods Hole Conference on Living Resources of the Southern Ocean 1976. SCAR/SCOR, Cambridge (BIOMASS, Vol II, pp 65–68)

Kiefer DA (1973) The *in vivo* measurement of chlorophyll by fluorometry. In: Stevenson LH, Colwell RR (eds) Esturarine microbial ecology. University of S.C. Press, Columbia, pp 421–430

Ligowski R (1982) Phytogenic food of *Euphausia superba* Dana caught in the southern Drake Passage and the Bransfield Strait, February–March 1981 (BIOMASS-FIBEX) Pol Polar Res 3: 281–288

Loftus ME, Seliger HH (1975) Some limitations of the *in vivo* fluorescence technique. Chesapeake Science 16:79–92

Mackintosh NA (1973) Distribution of post-larval krill in the Antarctic. Discovery Rep 36:95–156

Marr JWS (1962) The natural history and geography of the Antarctic krill (Euphausia superba) Discovery Rep 32:33–464

Marschall HP (1985) Structural and functional analyses of the feeding appendages of krill larvae. In: Siegfried WR, Condy PR, Laws RM (eds) Antarctic nutrient cycles and food webs (Proceedings of the 4th SCAR symposium on Antarctic Biology). Springer, Berlin Heidelberg New York

McClatchie S; Boyd CM (1983) A morphological study of sieve efficiencies and mandibular surfaces in the Antarctic krill, *Euphausia superba.* Can J Fish Aquatic Sci 40:955–967

Meyer MA; El-Sayed SZ (1983) Grazing of *Euphausia superba* Dana on natural phytoplankton populations. Polar Biol 1:193–197

O'Reilly JE; Thomas JP (1979) A manual for the measurement of total daily primary productivity on MARMAP and OCEAN PULSE cruises using ^{14}C simulated *in situ* sunlight incubation. NOAA Northeast Fisheries Center, Highlands, New Jersey (Ocean Pulse tech man no 1)

Platt T; Denman KL (1975) Spectral analysis in ecology. Annu Rev Ecol Syst 6:189–210

Quetin LB, Ross RM (1985) Feeding by Antarctic krill: Does size matter? In: Siegfried WR, Condy PR, Laws RM (eds) Antarctic nutrient cycles and food webs (Proceedings of the 4th SCAR symposium on Antarctic Biology). Springer, Berlin Heidelberg New York

Rakusa-Suszczewski S (1982) Feeding of *Euphausia superba* Dana under natural conditions. Pol Polar Res 3:289–297

Ray AA (1982) SAS user's guide: statistics. SAS Institute Inc., Cary, North Carolina

Setser PJ; Guinasso NL; Schink DR (1982) Daily patterns of fluorescence *in vivo* in the central equatorial Pacific. J Mar Res 40:453–471

Steeman Nielsen E (1952) The use of radioactive carbon for measuring organic production in the sea. J Cons 18:117–140

Strickland JDH; Parsons TR (1972) A practical handbook of seawater analysis. Bull Fish Res Board Can 167:201–203

Uribe E (1982) Influence of the phytoplankton and primary production of the Antarctic waters in relationship with the distribution and behaviour of krill. Instituto Antarctico Chileno, Scientific Series, No 28, pp 147–163

Voronina NM (1974) An attempt at a functional analysis of the distributional range of *Euphausia superba.* Mar Biol 24:347–352

Yamaguchi Y; Shibata Y (1982) Standing stock and distribution of phytoplankton chlorophyll in the Southern Ocean south of Australia. Trans Tokyo Univ Fish 5:111–128

Abundance, Distribution and Behaviour of *Euphausia superba* in the Southern Ocean Between 15° and 30° E During FIBEX

I. HAMPTON[1]

Summary. The abundance, distribution and behaviour of the Antarctic Krill, *Euphausia superba,* between 15° E and 30° E, and south of 62° S, was studied acoustically during FIBEX (First International BIOMASS Experiment) in February–March 1981. The biomass is estimated at 9.05×10^5 t. The Krill were spread relatively uniformly over the survey area, probably because of an absence of large-scale hydrodynamic features to concentrate them in any part of the region, and were almost invariably found in swarms. The mean biomass and density of the 1304 swarms detected were estimated by multi-channel echo integration at 0.139 t and 31 g m^{-2} respectively. Ninety-eight per cent of the swarms were estimated to have weighed < 1 t. Most of the biomass, however, was contained within a few large swarms. From relationships between swarm size and concurrent grazing indices, it is suggested tentatively that the large swarms were migrating or foraging and the small swarms were feeding or digesting. The distributions of swarm depth, biomass and density at night and by day indicate that swarms which did not migrate to the surface at night behaved almost identically to those detected by day. The depth distribution of Krill in the vicinity of 13 day-time stations indicates that Krill tended to be concentrated near the top of the thermocline. It is suggested that this was due to lateral current shear in the boundary region creating favourable conditions for feeding and aggregation. About 35% of the Krill in the vicinity of the stations were found at oxygen levels below reported lethal limits, suggesting that these limits may be too high or that a measure of adaptation to low oxygen concentrations may have occurred.

1 Introduction

Because of its aggregating habits and general shallow depth distribution, Antarctic Krill, *Euphausia superba,* can be surveyed and studied acoustically. In recent years there have been several large-scale echo integrator surveys of Krill (e.g., Kalinowski 1982; Doi 1982; Lillo and Guzman 1982), as well as a number of small-scale integrator studies of Krill swarms (e.g., Everson 1982; Mathisen and Macaulay 1983). During the FIBEX (First International BIOMASS Experiment) survey in 1981 (Hempel 1983), Krill abundance and distribution in three large sections of the Southern Ocean were determined from echo integrator data collected on 11 ships during a 3-month period (Hampton 1983).

This paper describes an acoustic study of Krill which was conducted from M.V. S.A. Agulhas during FIBEX. The work was carried out between 15°E and 30°E from 16 February to 10 March 1981. The total abundance of Krill in the survey area is estimated and the sampling error assessed. Acoustic data on the depths, sizes and densities of 1304 individual Krill swarms, obtained using a multi-channel integrator, are presented and related as far as possible to vertical thermal structure and vertical distributions of chlorophyll-a and dissolved oxygen. It is emphasized that the survey was designed primarily to estimate the abundance of Krill and not to study Krill behaviour. Consequently, hypotheses presented here on the reasons for certain observed behavioural features are of necessity highly tentative, and need to be substantiated by intensive acoustic and biological studies of Krill swarms in small areas over extended periods.

Phytoplankton biomass, productivity and metabolic state were assessed at 22 stations during the cruise (Weber and El-Sayed 1985). These authors have used Krill parameters from the present study to examine relationships between Krill and phytoplankton in the survey area.

2 Materials and Methods

The survey track is shown in Fig. 1. Apart from time spent on station or in towing nets, the survey ran continuously. Krill were detected and assessed with a Simrad EKS 120-kHz echo sounder interfaced to a custom-built digital echo integrator. Echoes from the top 100 m of the water column were integrated in 20 5-m channels, and the integrals logged on magnetic tape. In the few cases where Krill were found deeper than 100 m, the number of channels was increased to 40, giving a 200-m integration range, The echo sounder was calibrated immediately before the cruise, using a remotely-controlled hydrophone system (Hampton et al. 1981). Parameters measured were source level (SL), voltage receiving sensitivity (VRS), effective pulse length, and beam pattern in a horizontal plane parallel to the transducer face, from which the transducer's directivity index was computed. The measured values are shown in Table 1. Integrated echo intensities in each channel were corrected at processing time for time-varied-gain (TVG) errors in the receiver by applying a correction factor appropriate to the TVG error at the mid-point of the channel. Echo integrals recorded

1 Sea Fisheries Research Institute, Private Bag X2, Roggebaai 8012, South Africa

Antarctic Nutrient Cycles and Food Webs
(ed. by W.R. Siegfried, P.R. Condy, and R.M. Laws)

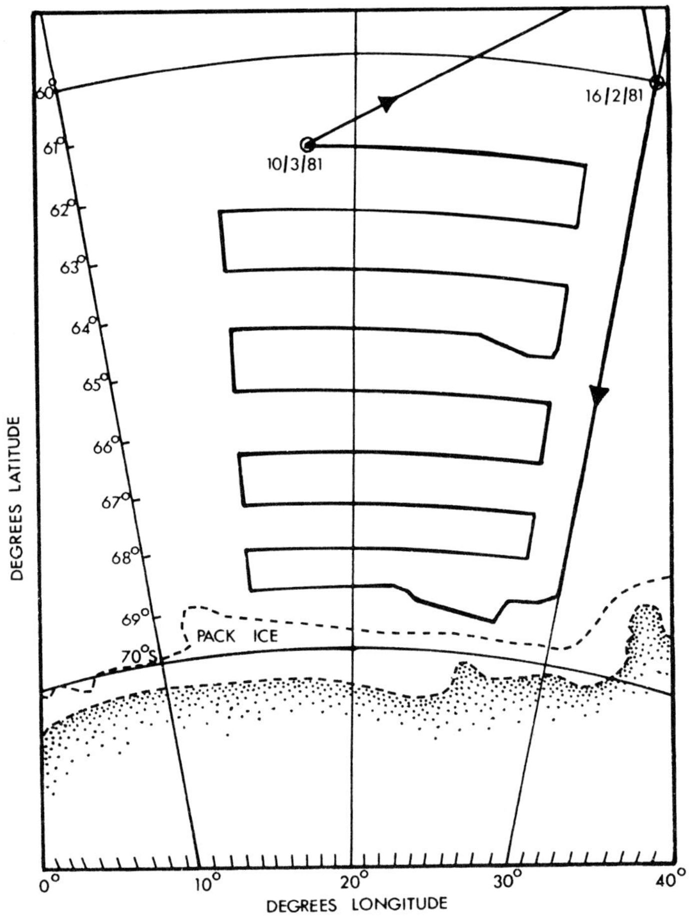

Fig. 1. Cruise track of M.V. S.A. Agulhas during FIBEX

from individual aggregations could be selected by software, allowing the densities and intercepted lengths of individual Krill swarms to be computed. Swarm depths and heights were measured off the echo start. Acoustic targets were identified by aimed fishing with an opening-closing RMT-8 or RMT-2 net (Robertson et al. 1981). Krill on the surface were captured with a Neuston net weighted to tow with the frame about one-third submerged. Stations were occupied every day at about 1000 and 2400 h local time. At each station, a CTD cast was made to at least 100 m. Water samples were collected at discrete depths between the surface and the 0.25% light level for the analysis of oxygen, nutrient and chlorophyll content. Aliquots were also taken from each sample for the analysis of phytoplankton biomass, productivity and metabolic state. This work is described by Weber and El-Sayed (1985).

Table 1. Calibration parameters for 120-kHz echo sounder

SL (dB re. 1 μPa at 1 m)	221.0
VRS (dB re. 1 Volt μPa^{-1} at max. TVG)	–104.7
Directivity index (dB)	– 18.3
Effective pulse length (ms)	0.58

3 Results

3.1 Biomass Estimation

The primary parameter on which the biomass estimation was based is ρ_i, the mean weight of Krill per unit area (i.e., the mean surface density) for the ith integration period.

Echoes were usually integrated over a 30-min period, corresponding to a steaming distance of about 9 km on average. The values were estimated from

$$\rho_i = \frac{c\,\Delta R\,10^{0.1(\bar{S}_v)_i}}{b} \; \frac{\sum_{j=1}^{M} n_j\, l_j^{\,d}}{\sum_{j=1}^{M} n_j\, l_j^{\,a}} \quad , \qquad (1)$$

where $(\bar{S}_v)_i$ is the mean volume back-scattering strength (MVBS) for interval i, a and b are constants in the Krill target strength-length relationship $t_s = al^b$, and c and d are constants in the Krill length-weight relationship $w = cl^d$. n_j is the number of Krill in length class l_j for the biological sample attributed to the interval, and M is the number of length classes in the sample. ΔR is the vertical extent of the integration channel (usually 100 m).

The target strength constants were taken from the following expression, presented by K.I. Yudanov at the Post-FIBEX Data Interpretation Workshop in 1981 (Hampton 1983),

$$TS = 10 \text{ Log } t_s = 20 \text{ Log } 1 - 77.2$$

(TS in dB, l in cm). This gives a = 1.90 times 10^{-10} and b = 2.0. Yudanov's expression was chosen, since his measurements were made in situ at 120 kHz over a reasonable range of animal sizes (35–50 mm), and because it was adopted as the common expression for conversion of the FIBEX MVBS values to density. Yudanov obtained the expression by regressing the maximum echoes received from single dispersed Krill at sea against the log of the maximum length of animals in nearby catches.

The constants c and d were obtained from Jazdzewski et al. (1978)

$$w = 0.0018\, 1^{3.383} \quad ,$$

where w is in mg (wet wt.) and l in mm. All length-weight conversions in this study were made by means of the above expression.

The distribution of Krill over the survey region is shown in Fig. 2, which was obtained by applying Eq. (1) to all intervals. For clarity, the values are grouped into 2-h intervals. In Fig. 3 all values are plotted in a continuous time sequence. This figure is reproduced from Fig. 2 in Weber and El-Sayed (1985).

Since there appeared to be a strong periodicity in the daytime and nighttime density estimates, the significance of the differences between day and night means was tested

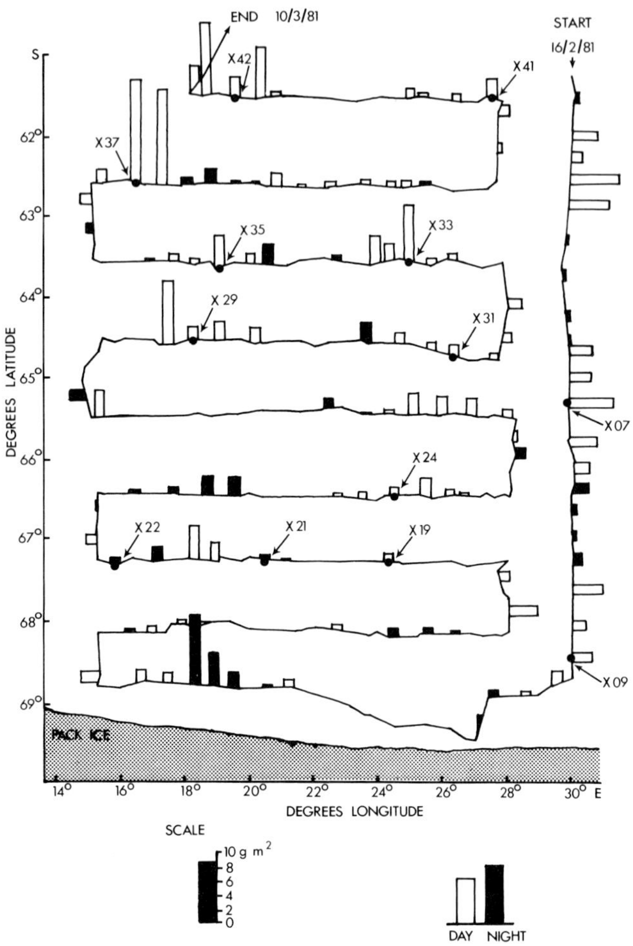

Fig. 2. Horizontal distribution of Krill in the survey area. 30-min integrator readings grouped into 2-h intervals

using both paired (day with following night) and pooled data. The null hypothesis that there was no difference in population density by day and night was rejected at the 0.01% level in both cases. The respective overall means were 1.45 and 0.59 g m^{-2}. In view of the known tendency of *E. superba* to migrate to the surface at night (e.g., Marr 1962; Pavlov 1969; Everson 1977) where it is undetectable by echo sounders, the difference was attributed to sampling bias, rather than to any real day-night difference in population abundance. All night-time values of ρ_i were therefore discarded for the purpose of biomass estimation.

$\bar{\varphi}_D$, the mean density of Krill in the survey area using only the day-time measurements, was estimated from

$$\bar{\rho}_D = \frac{\sum_{i=1}^{N} \rho_i \, L_i}{L_D} \quad , \qquad (2)$$

where L_i is the distance steamed during the i th interval, L_D is the total distance steamed by day and N is the number of daytime intervals. The L_i/L_D values are weighting factors which account for intervals of unequal distance. Eq. (2) gave $\bar{\rho}_D$ = 1.46 g m^{-2}, which applied over the survey area of 6.2 x 10^5 km^2, gives a biomass estimate, $\hat{B}$, of 9.05 x , 10^5 t. Note that this is an unstratified estimate. Because of the absence of any clear trends in abundance over the survey area (see Fig. 2), there was no basis for post-survey stratification of the area.

3.2 Estimation of Variance in $\hat{B}$

The variance in $\hat{B}$ is a function of many factors, such as sampling error, errors in the target strength expression used, and calibration errors. The sampling error was estimated from the variance in the daily estimates of mean density, on the assumption that each of these was an independent estimate of the population mean. Since successive daytime transects were on average about 150 km apart, the serial correlation between the daily estimates should be negligible, so that the assumption of independence should be valid. The overall variance $\hat{B}$ was estimated from

$$\sigma_{\hat{B}}^2 = \frac{K A^2}{K-1} \sum_{k=1}^{K} \frac{(\bar{\rho}_k - \bar{\rho}_D)^2 \, D_k^2}{D_T^2} \quad , \qquad (3)$$

where $\bar{\rho}_k$ is the estimate of mean density for day k, D_k is the distance steamed on day k, and D_T the total distance steamed for all days. K is the number of days, and A is the survey area, which was assumed to have been measured with negligible error. The D_k^2/D_T^2 factors are weighting factors, introduced to allow for differences in the distance steamed per day. From Eq. (3) σ_B was estimated as 1.58 x 10^5 t, giving a coefficient of variation ($\sigma_{\hat{B}}/\hat{B}$) of 0.175.

Systematic errors are incurred predominantly in determining the target strength from the length distribution and in converting echo-integrator readings to mean volume back-scattering strength. Because of large uncertainties in the constants a and b in particular, and the difficulties in calibrating echo-sounders accurately with hydrophones, these errors could well exceed the sampling error. Unfortunately they cannot be estimated formally at present as there are insufficient data to carry out the necessary error analysis.

The absence of confidence limits on the estimate of absolute standing stock prohibits the use of the estimate for much beyond the crude trophodynamic calculations attempted by Miller et al. (1985). However, where abundance estimates are compared within the same survey, such as when analyzing distributional features, only the relative error is important and useful comparisons can be made in the presence of large systematic errors.

3.3 Horizontal Distribution

Figures 2 and 3 indicate no clear large-scale trends in Krill distribution within the survey area. This is confirmed in part by Weber and El-Sayed's (1985) finding of no signif-

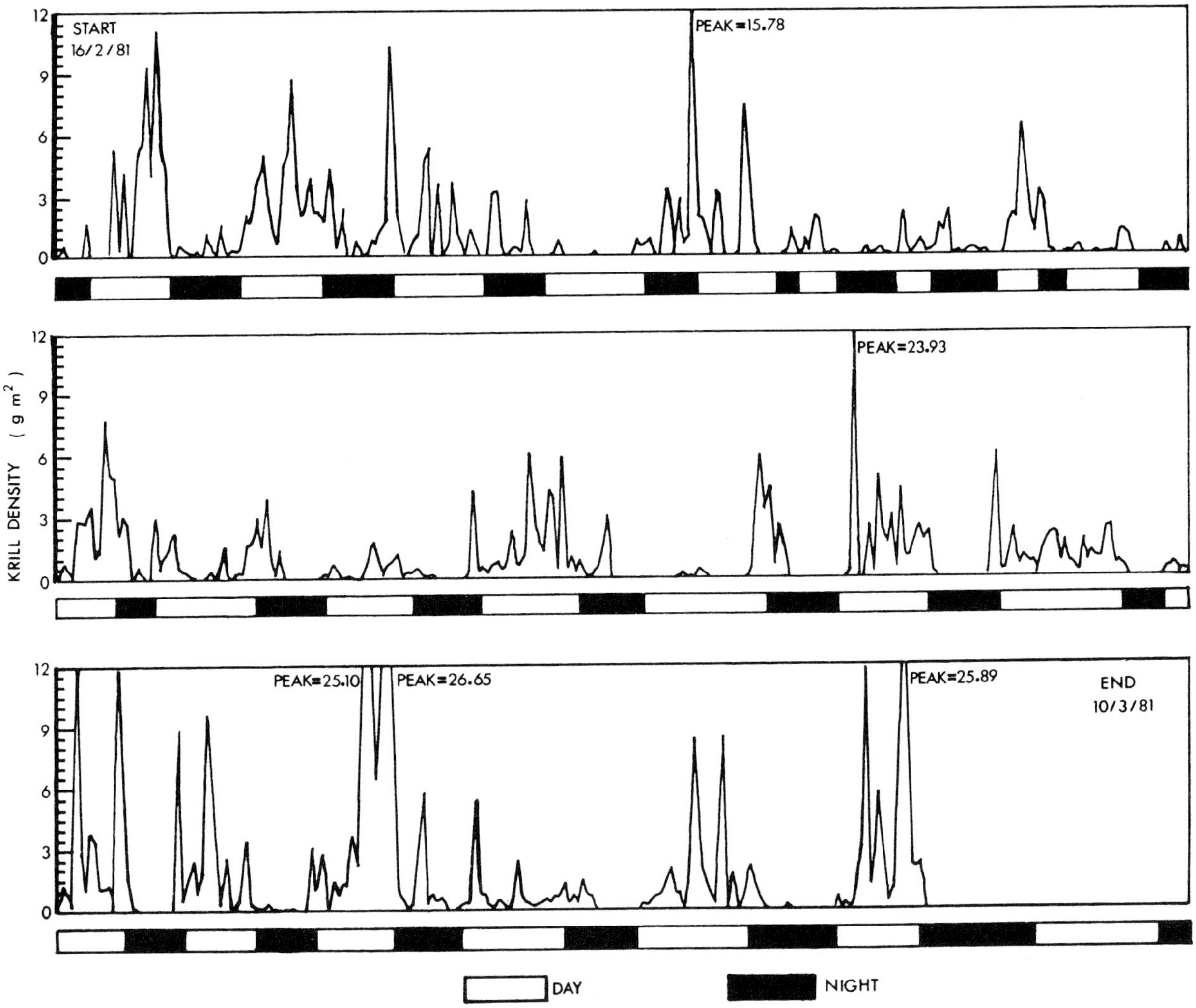

Fig. 3. Continuous time sequence of Krill density plotted from 30-min integrator readings for the whole survey area. (After Weber and El-Sayed 1984, Fig. 2)

icant correlation between Krill abundance and latitude in their bi-variate correlation analysis. Horizontal patchiness on a smaller scale is evident from the variation in 30-min density values shown in Fig. 3. Daytime values ranged from 0 to 26.65 g m^{-2} (18 times the daytime mean) and the nighttime values from 0 to 15.78 g m^{-2} (26 times the nighttime mean). The diurnal trends in estimated density reported above are clearly evident.

Within the 30 min (± 9 km) intervals, further patchiness was observed, the Krill invariability being aggregated into well-defined cohesive formations fitting the description of swarms or schools in Kalinowski and Witek's (1982) classification scheme. Krill were rarely detected in any of the other formations described by Kalinowski and Witek (irregular or scattered formations, layers or super-patches).

The sizes of individual swarms were estimated from their intercepted lengths along the acoustic transects, with the standard correction for finite beam-width (e.g., Forbes and Nakken 1972). It was assumed for simplicity that all swarms were circular in cross-section, from which it follows that the intercepted length underestimates the true diameter by a factor of $\pi/4$ on average. Correcting for this gives the following estimator of swarm biomass

$$\hat{b}_s = \frac{4}{\pi} d_s^2 h_s \rho_s \quad , \tag{4}$$

where d_s is the intercepted length, h_s is the mean height, measured off the echo-chart, and ρ_s is the mean volume density of the swarm calculated from its mean volume back-scattering strength.

The day and night distributions of $\hat{b}_s$ are shown in Figs. 4a and 4b, respectively, and the combined distribution in Fig. 4c. All have been logarithmically transformed to cope with the wide range of values. Daytime values ranged from < 0.01 t (10 kg) to 22.3 t with a mean of 0.138 t, and nighttime values from < 0.01 t to 6.63 t, with a mean of 0.146 t. The overall mean was 0.139 t (139 kg). An accumulated frequency plot for day and night values combined (Fig. 5) shows that a large proportion of the biomass was contained within a very small proportion of the swarms.

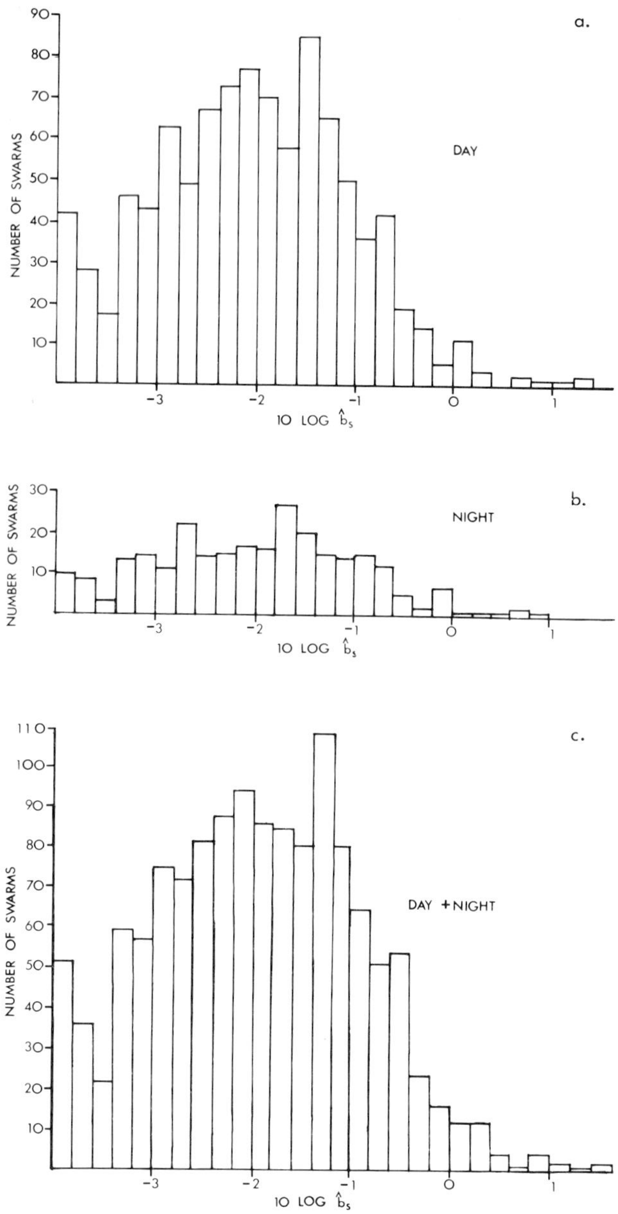

Fig. 4a–c. Distribution of Krill swarm biomass **a** by day, **b** at night and **c** for day and night combined

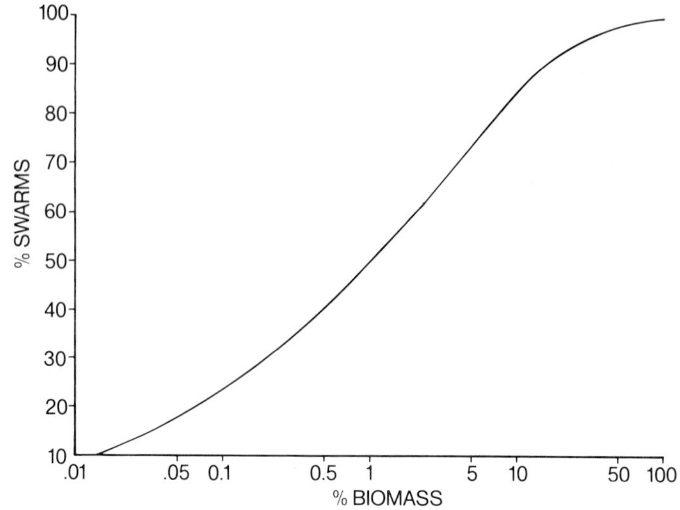

Fig. 5. Accumulated frequencies of Krill swarm biomass

Note, however, that for quantitative purposes, Figs. 4 and 5 should be treated with caution, since the line-intercept method of estimating size distribution is biased towards large swarms because of the greater probability of intercepting them. The assumption of circular swarms can also introduce distortions, particularly if the swarms tend to be considerably longer in one direction than another. Conformations such as this (i.e., ellipses, "streamers" or "ribbons") have been reported from visual observations (e.g., Marr 1962; Nemoto et al. 1981; Hamner et al. 1983) and were common among our side-scan sonar images of Krill swarms on previous cruises (e.g. Fig. 6). Randomly orientated transects will have the greatest probability of intersecting such swarms parallel to the short axis, in which case Eq. (4) severely underestimates the biomass. Conversely, the probability of interception is lowest in a direction parallel to the long axis, in which case use of the intercept in Eq. (4) is likely to over-estimate the biomass severely. Consequently, the distribution of $\hat{b}_s$ will be characterized by a comparatively high proportion of small swarms and a few exceptionally large ones. These characteristics are evident in all the distributions in Fig. 4.

In Figs. 7a und 7b the day- and nighttime distributions of swarm mean volume back-scattering strength (MVBS) are shown. The distributions are remarkably similar and the means almost identical (–46.2 dB and –46.3 dB, respectively), as are the standard deviations (3.40 dB and 3.53 dB, respectively). For a target strength of –66.5 dB (appropriate for a length of 34.0 mm, the mean length of all animals captured during the survey), an MVBS of –46 dB gives a density of 114 individuals m^{-3}. The corresponding weight density for animals of this size is 31 g m^{-3}.

3.4 Vertical Distribution

In Fig. 8 the mean depth of detected swarms at each hour of the day throughout the survey is shown. No pronounced diel trends are evident except for the period 2000–2200 h local time when swarms tended to be about 20 m shallower than the 24-h average (47.4 m). Note, however, the marked difference in the number of swarms detected by day and night (1025 vs. 279).

Figures 9a and 9b show the depth distributions of swarms detected by day and night, respectively. The respective means are 47.9 and 45.6 m, with standard errors 4.4 and 4.6 m. A t-test using pooled data showed no significant difference in day and night means ($P > 0.5$).

3.5 Relationships between Krill and the Environment

Weber and El-Sayed (1985) found Krill abundance during the survey to be negatively correlated with integrated

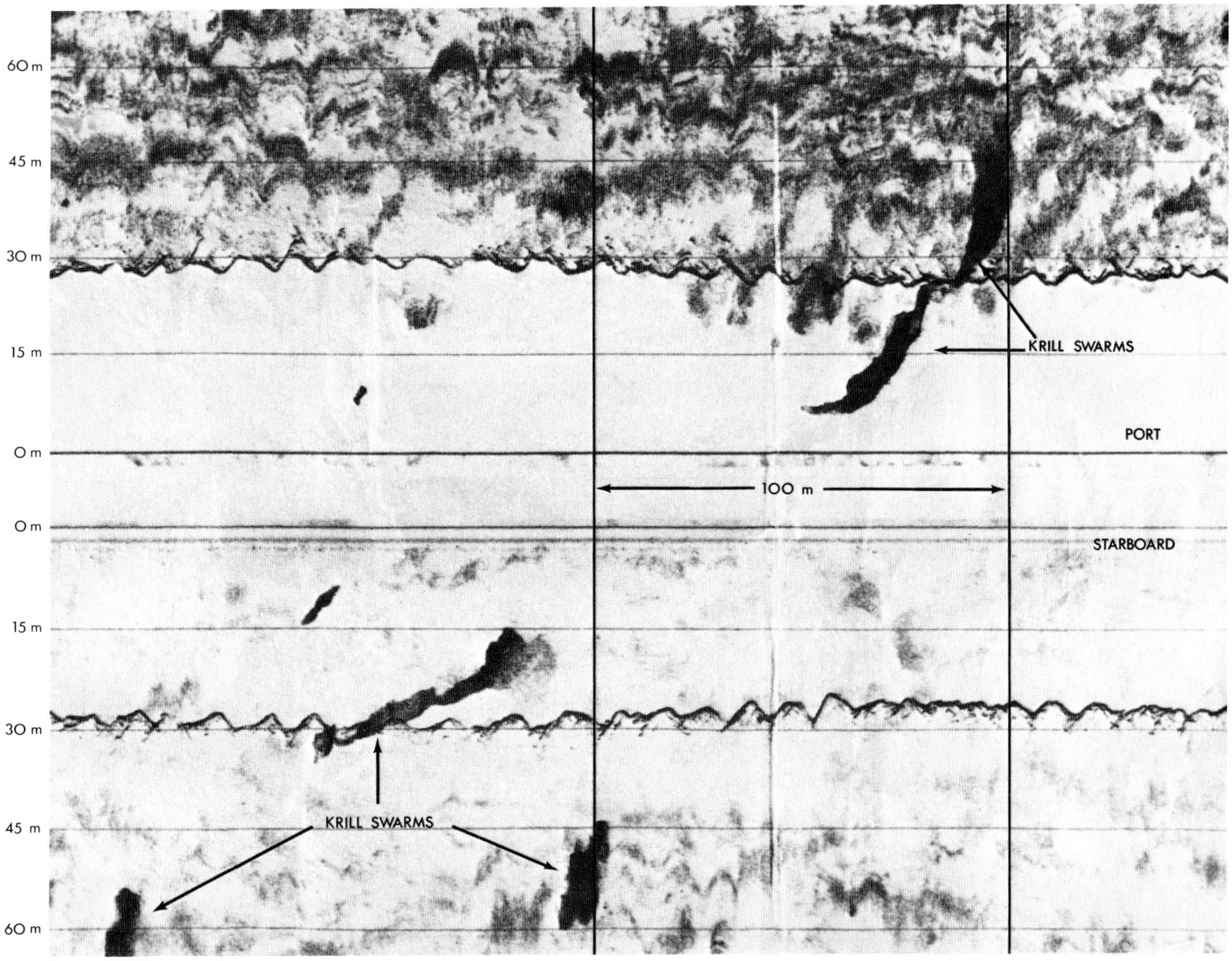

Fig. 6. 100-kHz side-scan sonar images of Krill swarms detected from M.V. *S.A. Agulhas* in March 1980. The images have not been corrected for geometric distortion but at the towing speed used (1.5 m s^{-1}), the distortion is minor

phytoplankton biomass, but could find no correlation with any physical or chemical variable. Variables tested were water column averages of temperature, salinity, dissolved oxygen, nitrate and silicate. In this section an attempt is made to investigate relationships between the vertical distribution of Krill in the vicinity of stations and the vertical distributions of temperature, Chlorophyll-a and dissolved oxygen at the stations.

Krill vertical distribution profiles were obtained by multichannel integration, in 10-m depth intervals, of all Krill echoes received 50 km on either side of the station positions. The analysis was restricted to the 21 daytime stations, 8 of which were discarded since there were no significant amounts of Krill in the vicinity. The positions of the 13 remaining stations are shown in Fig. 2. Vertical profiles of Krill abundance, temperature, chlorophyll-a and dissolved oxygen at one typical station (Station 42), are shown in Fig. 10.

4 Temperature

It appeared that Krill tended to be most abundant near the top of the thermocline (e.g., Fig. 10). To test for this, the following binomial test for co-incidence of peaks was devised. The water column between 0 and 100 m was subdivided into 10 equal intervals and the frequency with which the Krill peak coincided with different parts of the thermocline was noted. It was found that at 6 of the 13 stations, the Krill peak coincided with the channel in which the top of the thermocline lay. If the Krill were randomly distributed between 0 and 100 m, the probability of the peak falling within any particular channel would be 0.1, and r the number of coincidences, would have a binomial distribution with $p = 0.1$, $q = 0.9$ and $n = 13$. The normal deviate is approximately $(|r - np| - 1/2)/\sqrt{npq}$, which for $r \doteq 6$, is 3.89. The hypothesis that $p = 0.1$ is therefore rejected at the 0.1% level, implying that peaks in Krill abundance

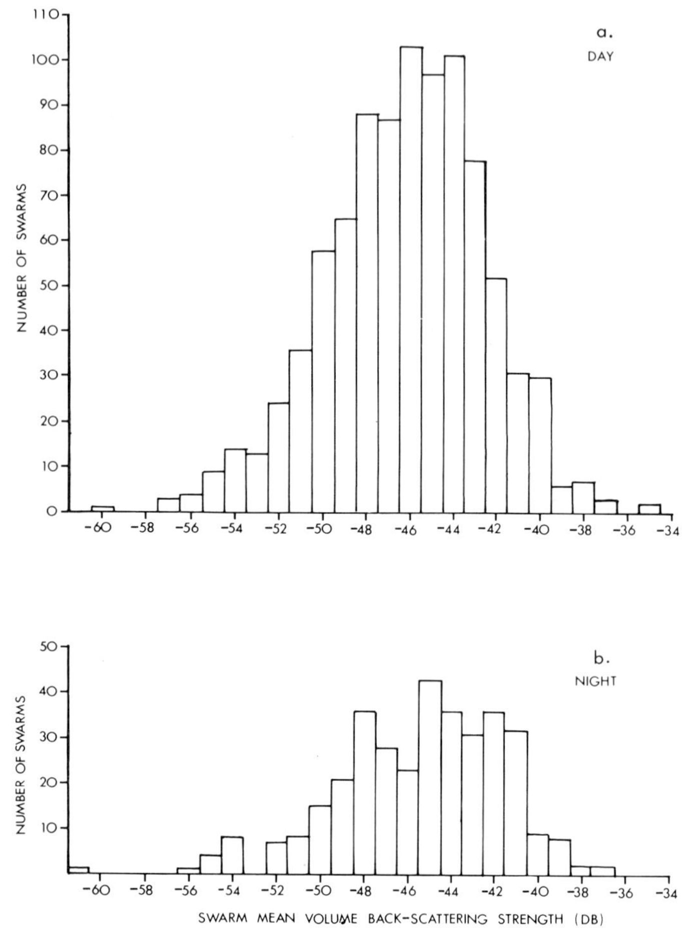

Fig. 7a, b. Distribution of mean volume back-scattering strength (MVBS) of Krill swarms **a** by day, **b** at night

were found close to the top of the thermocline far more frequently than would have been expected by chance. No significant coincidence with any other feature of the vertical thermal structure could be found.

5 Chlorophyll-a

Correlation between Krill abundance and chlorophyll-a concentration in each of the 10 depth channels was tested through a Spearman rank test as it was clear that neither variable was normally distributed. The rank correlation coefficients, r_s, for the 13 stations ranged between –0.49 and 0.79, with only one (r_s = 0.79 at Station 41) being significant at the 5% level. A pooled estimate of the population correlation coefficient was obtained by combining the values for the 13 stations through Fisher's z-transformation (Snedecor and Cochran 1967). This gave $\bar{z} = 0.273$ and $\sigma_{\bar{z}} = 0.165$, with 98 d.f. The corresponding value of r_s is 0.267 with 95% confidence limits –0.050 and 0.535, respectively. Assuming $\bar{z}$ to be normally distributed, the null hypothesis that the population correlation coefficient is zero, is not rejected at the 95% level.

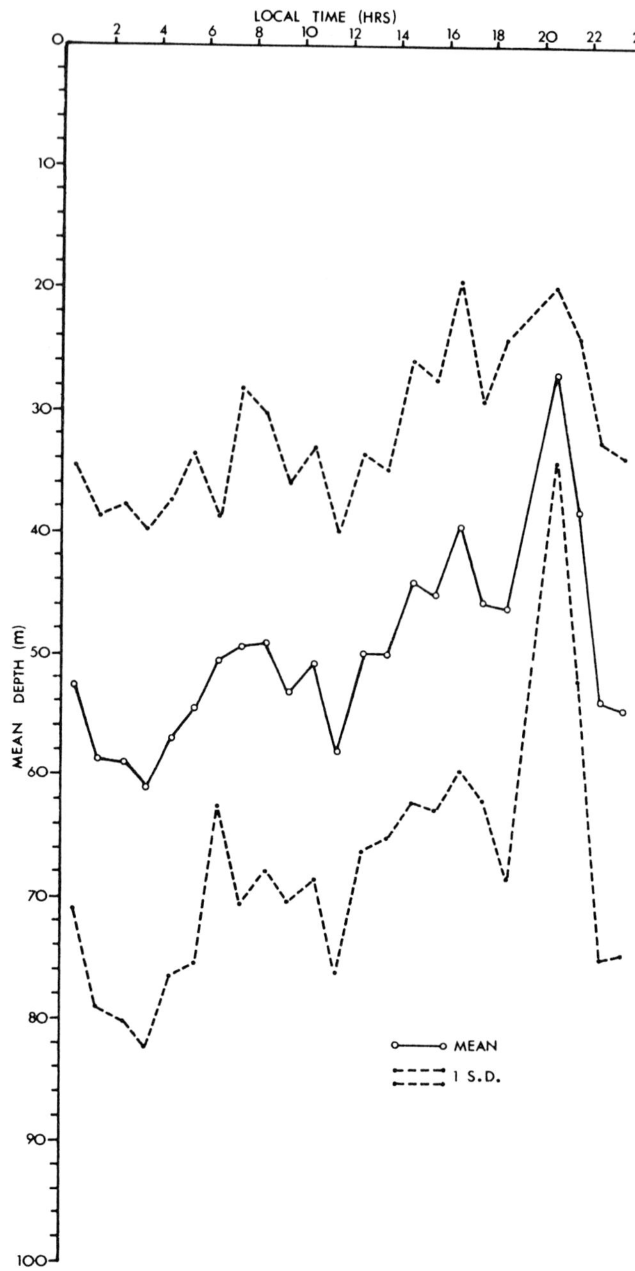

Fig. 8. Mean depth of detected Krill swarms at each hour of the day

6 Dissolved Oxygen

Dissolved oxygen levels in the upper 100 m at the 13 stations were generally close to saturation (approximately 7.9 ml l^{-1}), with only a slight decrease with depth in most cases. Nonetheless, at 9 of the stations, oxygen levels at 100 m were below 85% saturation, the critical level for Krill survival according to Kils' (1979) laboratory experiments. About 35% of the Krill in the vicinity of the stations were found at depths where the oxygen level was below this. The implications of this are discussed below.

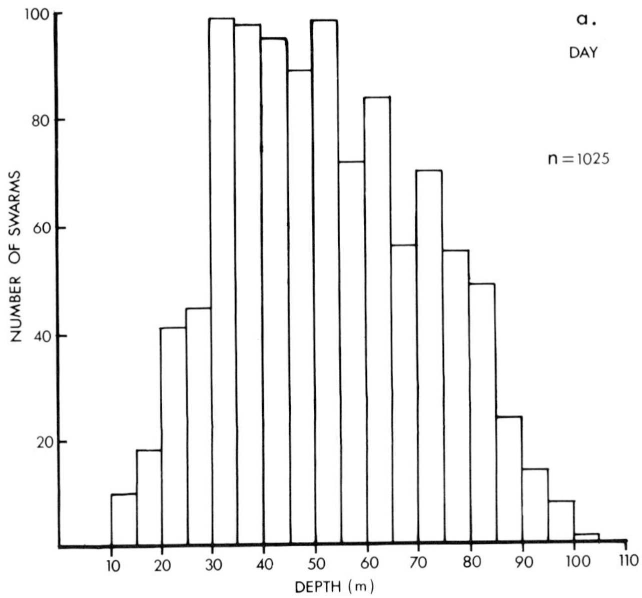

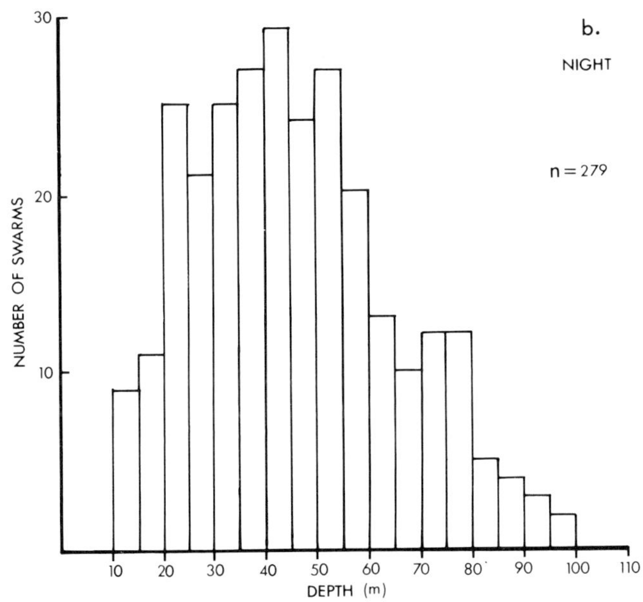

Fig. 9a, b. Depth distribution of detected Krill swarms **a** by day and **b** at night

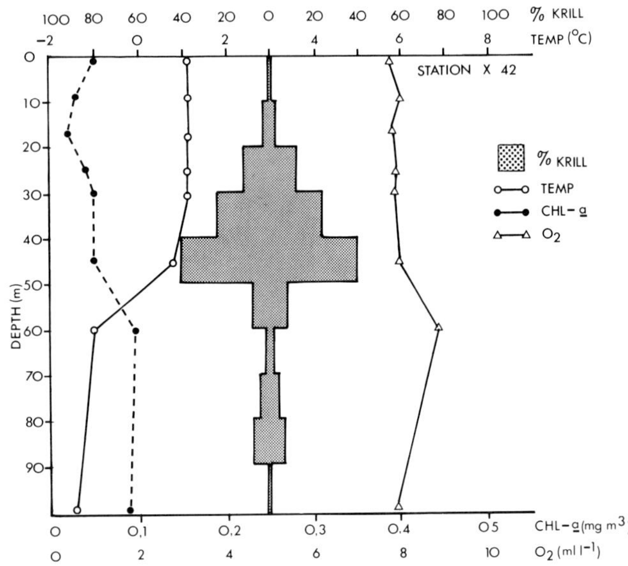

Fig. 10. Vertical profiles of Krill abundance, temperature, chlorophyll-a and dissolved oxygen at Station 42. Krill abundance profile constructed from underway integrator data 50 km on either side of station position

7 Discussion

The density of Krill in the survey area was low compared to the estimated mean density for all the FIBEX areas (17.0 g m^{-2}, Hampton 1983), and falls below the range estimated by Marr (1962) for the surface density of Krill in the Weddell and East Wind Zones (2.5–29 g m^{-2}). Also notable was the absence of any strong distributional features on a large scale, in contrast to certain other areas surveyed during FIBEX (e.g., the Antarctic Peninsula region and Prydz Bay) where a high proportion of the biomass was concentrated into relatively small areas (Hampton 1983). The widespread distribution and low abundance are consistent with the conclusions of Witek et al. (1982) that Krill tend to be concentrated in regions of high current shear (i.e., in frontal zones, meanders and eddies, and regions of sharp change in bottom topography). Our survey lay totally within oceanic waters, with no pronounced frontal interactions, even in the region of the Antarctic Divergence (M.J. Orren, pers. comm.). In such areas, Witek et al. (1982) postulated that Krill are dispersed in the laminar-like flow of the water and tend to form widely-spaced small swarms.

Despite the possibility of bias in the estimates of swarm biomass for reasons already discussed, it is clear from Fig. 5 that the great majority of the swarms were small, with the biomass of most probably amounting to < 1 t. J. Kalinowski (pers. comm.) found a similar distribution in an acoustic study of 23,000 Krill aggregations in the Antarctic Peninsula and Scotia Sea regions between 1976 and 1979. He found that approximately 66% of all swarms detected had a biomass < 1 t, as against 98% in the present study. He also found that the biomass of swarms tended to be greater in neritic regions (particularly around South Georgia). If, as is probably more appropriate, the comparison between Kalinowski's distribution and ours could be restricted to swarms found in more oceanic areas, such as the Scotia Sea and Drake Passage, the distributions would be even more similar.

It must, therefore, be concluded that Krill show a general tendency to form small swarms in the summer months, although many workers have reported finding swarms that are far larger than average. The present survey was not designed to investigate swarming processes, but Weber and El-Sayed's (1985) concurrent investigation of correlations between Krill aggregation parameters and chlorophyll-phaeophytin ratios offers one possible explanation for the observed size distribution. Their analysis suggests that for a given abundance of Krill, "grazing" pressure (indicated by a relatively high proportion of phaeopigments) tended to be lower where the Krill were aggregated into a few large swarms, and higher when they were dispersed into a large number of small swarms. This suggests that the rare large swarms could be non-feeding aggregations, perhaps migrating or foraging in the manner described by Hamner et al. (1983), who found rapid feeding and orientated swimming (which is necessary for migration) to be incompatible. The common small swarms could be aggregations of non-orientated individuals engaged in feeding or digestion. The qualitative features of the swarm biomass distributions shown in Fig. 4, and Weber and El-Sayed's (1985) correlations, are consistent with this behavioural pattern, but it is emphasized that other interpretations are possible. Concurrent acoustic and feeding studies in small areas over periods of at least a few days are necessary to test an hypothesis such as this.

The estimates of mean swarm biomass and density obtained in this study (139 kg and 31 g m^{-3}) are roughly comparable with overall mean values of 416 kg and 72 g m^{-3}, obtained by J. Kalinowski (pers. comm.) and with an acoustic estimate by Mathisen and Macaulay (1983) of the mean density of a super-swarm near Elephant Island (approximately 80 g m^{-3}). These densities are consistent with photographic observations of Krill swarms by Guzman and Marin (1983), who reported a maximum density of 550 animals m^{-3} (approximately 400 g m^{-3}) from 700 exposures of swarms at sea. The density estimates are, however, very much lower than those reported from commercial catches and sampling of visible swarms, which range from about 1 kg m^{-3} to 33 kg m^{-3} (Everson 1977). They are also in major disagreement with SCUBA diver reports (Hamner et al. 1983) of Krill swarms containing 20000–50000 individuals m^{-3} (20–50 kg m^{-3} for animals weighing 1 g). No conclusive explanation for these discrepancies can be given at this stage, but it is noted that all the non-acoustic estimates are based on comparatively few observations which could have been heavily biased towards larger and denser swarms. The acoustic studies indicate that swarms do occasionally attain such high densities, so the discrepancies may not be as puzzling as they at first appear.

The difference in the day and night surface density estimates could be due either to Krill becoming less detectable at night, or to a diel change in the mean target strength of individuals, such as was suggested by Everson (1982) to account for diel changes in mean volume back-scattering strengths within a Krill patch studied continuously for 7 days. Krill could become undetectable acoustically by migrating to within about 10 m of the surface, or by dispersing so widely that their echo returns become indistinguishable against background noise levels. Evidence for both forms of behaviour has been widely reported (e.g., Marr 1962; Pavlov 1969; Kalinowski and Witek 1982; Yudanov 1982), but in this survey the frequency with which Krill were found in Neuston nets at night (D.G.M. Miller, pers. comm.) and the scarcity of dispersed sound-scattering layers at any time of day or night, favour the former explanation. The possibility of a general diel change in target strength cannot be ruled out entirely, but appears unlikely from the similarity between the day and night-time distributions of swarm scattering strength (Fig. 7). This similarity is more likely to be due to similar volume densities (and therefore similar target strengths) than to changes in target strength compensating for changes in density to keep the scattering strength constant. It is concluded, therefore, that upward migration into the acoustically blind surface zone was probably responsible for the lower density estimates at night.

Figures 4, 7, 8, and 9 indicate that swarms which did not migrate to the surface at night behaved in much the same way as those detected during the day, having similar depth, biomass and density distributions. It appears, therefore, that Krill at night either migrated to the surface or did not migrate at all, with about 60% of the population on average falling into the former category (on the basis of the overall difference in daytime and nighttime integrator readings). If, as has been suggested by Pavlov (1969) and Mezykowski and Rakusa-Suszczewski (1979), Krill migrate to the surface at night to feed, the migrating proportion could have been composed predominantly of feeding animals, and the non-migrating portion of swarms consisting of replete animals. Alternative explanations for the separation involving differential migration rates for different size classes (e.g., Everson and Ward 1980) were ruled out, since the mean length of animals caught on the surface at night (30.1 mm) was not significantly different ($t = 1.5, P = 0.171$) from that of Krill taken deeper down during the day (33.9 mm).

The apparent tendency for Krill to concentrate near the top of the thermocline (at least by day), is of interest. Since chlorophyll-a concentrations were always greatest below the thermocline, it seems unlikely that this behaviour was related directly to food density. It is tentatively suggested instead that lateral velocity shear between the surface water and the cooler, more saline winter water beneath it, could have created conditions favourable to Krill near the boundary region. A.J. Boyd (pers. comm.) measured shear velocities between 20 m and 100 m (i.e., above and below the thermocline) at 22 stations on the cruise, using drogues, and found relative speeds of about 0.1 m s^{-1} on average. Although small, relative motion such as this, if confined to the region of the thermocline, could be advantageous for Krill swarms situated there, since the swarms could easily move into new water by undertaking small upward and downward migrations. Lateral shear near the thermocline could also increase the probability of swarms meeting and amalgamating there, which would tend to increase the abundance of Krill in this region. This kind of mechanism, in which Krill are brought into contact by hydrodynamic forces and then amalgamate through social behaviour was proposed by Witek et al. (1982) to explain the formation of high concentrations of swarms on a large scale.

In spite of the deep chlorophyll maxima, Krill encountered during the survey were almost totally confined to the upper 100 m. The general decrease in oxygen concentrations with depth and the experimentally-demonstrated sensitivity of Krill to low oxygen tensions (Kils 1979), suggests that the depth range may have been primarily limited by oxygen deficiencies. In laboratory experiments, Kils found that non-adapted Krill were unable to survive at oxygen concentrations below 85% saturation (approximately 6.7 ml l^{-1} in this case). About 65% of the Krill around the 13 selected stations were situated at depths where oxygen levels were above this. The fact that 35% of the Krill were found at depths where there was apparently insufficient oxygen to meet their respiratory demand, may be support for Clarke and Morris' view (1983) that Kils' (1979) critical oxygen level is too high. Alternatively, a measure of adaptation to low oxygen concentrations may have taken place (Kils found that Krill have the ability to adapt to levels as low as 50% saturation over a period of about a day). Despite these uncertainties as to the effective critical oxygen level under natural conditions, Kils' (1979) view of Krill as an animal confined to well-oxygenated surface waters is generally supported by the current study.

Acknowledgements. Mr. J. Henry is thanked for providing the physical, chemical and chlorophyll-a data at the stations, and Mr. D.G.M. Miller for providing Krill biological data. I also thank Mr. Miller for helpful discussions on aspects of Krill biology, and Miss D.A. Leach for preparing the figures. I am grateful to Mr. L.H. Weber and Professor S.Z. El-Sayed of Texas A and M University, for agreeing to their Fig. 2 being reproduced here. The scientific complement and officers and crew of M.V. S.A. Agulhas during FIBEX are thanked for their assistance in collecting the data. The project was funded by the South African National Committee for Oceanographic Research and the Sea Fisheries Research Institute, Cape Town. The South African Department of Transport Affairs made M.V. S.A. Agulhas available for the work.

References

Clarke A; Morris DJ (1983) Towards an energy budget for Krill: the physiology and biochemistry of *Euphausia superba* Dana. Polar Biol 2:69–86

Doi T (1982) Target strength, distribution pattern and abundance of Krill, *Euphausia superba,* in the Antarctic observed by *Kaiyo Maru* in the 1981 season. Contribution No. 26, Symposium on Fisheries Acoustics, Bergen, Norway, 21–24 June

Everson I (1977) The living resources of the Southern Ocean. Rome FAO United Nations Development Programme, Southern Ocean Fisheries Survey Programme GLO/SO/77/1:156 pp

Everson I (1982) Diurnal variations in mean volume backscattering strength of an Antarctic Krill (*Euphausia superba*) patch. J Plankton Res 4(1):155–162

Everson I; Ward P (1980) Aspects of Scotia Sea zooplankton. Biol J Linn Soc Lond 14:93–101

Forbes ST; Nakken O (1972) Manual of Methods for fisheries resource survey and appraisal. FAO Rome FIRM/M5:138 pp

Guzman O; Marin B (1983) Hydroacoustic and photographic techniques applied to study the behaviour of Krill (*Euphausia superba*). In: Proceedings of the Biomass Colloquium in 1982. Mem Natl Inst Polar Res Spec Issue 27:129–152

Hamner WM; Hamner PP; Strand SW; Gilmer RW (1983) Behaviour of Antarctic Krill, *Euphausia superba:* chemo-reception, feeding, schooling and moulting. Science 220:433–435

Hampton I (1983) Preliminary report on the FIBEX acoustic work to estimate the abundance of *Euphausia superba.* In: Proceedings of the BIOMASS Colloquium in 1982. Mem Natl Inst Polar Res Spec Issue 27:165–175

Hampton I; de Bruyn P; Leonard WIR; Paterson MJ (1981) Remotely controlled hydrophone system for calibration of hull-mounted echo-sounder transducers. Fish Bull S Afr 14:115–121

Hempel G (1983) FIBEX – an international survey in the Southern Ocean: review and outlook. In: Proceedings of the BIOMASS Colloquium in 1982. Mem Natl Inst Polar Res Spec Issue 27:1–15

Jazdzewski K; Porebski J; Rakusa-Suszczewski S; Witek Z; Wolnomiejski N (1978) Biological and populational studies on Krill near South Georgia in the summer 1976. Pol Arch Hydrobiol 25: 607–631

Kalinowski J; Witek Z (1982) Forms of Antarctic Krill aggregations. ICES Biological Oceanography Committee C.M. 1982/L:60:9 pp

Kils U (1979) Performance of Antarctic Krill *Euphausia superba,* at different levels of oxygen saturation. Meeresforschung 27:35–47

Lillo S; Guzman O (1982) Study of the abundance, distribution and behaviour of Krill at the Bransfield Strait and Drake Passage by means of hydroacoustic techniques. Institute Antarctico Chileno (INACH) Sci Series 28:17–45

Marr JWS (1962) The natural history and geography of the Antarctic Krill (*Euphausia superba* Dana). Discovery Rep 32:33–464

Mathisen OA; Macaulay MC (1983) The morphological features of a super swarm of Krill. In: Proceedings of the BIOMASS Colloquium in 1982. Mem Natl Inst Polar Res Spec Issue 27:153–164

Mezykowski T; Rakusa-Suszczewski S (1979) The circadian rhythms in *Euphausia superba* Dana and its carbohydrate metabolism. Meeresforschung 27:124–129

Miller DGM; Hampton I; Henry J; Abrams RW; Cooper J (1985) The relationship between Krill food requirements and phytoplankton production in a sector of the Southern Indian Ocean. In: Siegfried WR, Condy PR, Laws RM (eds) Antarctic nutrient cycles and food webs (Proceedings of the 4th SCAR symposium on Antarctic biology). Springer, Berlin Heidelberg New York

Nemoto T; Doi T; Nasu K (1981) Biological characteristics of krill caught in the Southern Ocean. BIOMASS 2:47–63

Pavlov VYa (1969) The feeding of krill and some features of its behaviour. Tr VNIRO 66:207–222 (issued as MAFF Transl ns 94)

Robertson AA; Alexander DGW; Miller DGM (1981) Modified collapsible opening and closing midwater trawls (RMT-8 and RMT-2). Fish Bull S Afr 14:103–113

Snedecor GW; Cochran WG (1967) Statistical Methods. Iowa State University Press, Iowa (6th ed)

Weber LH; El-sayed SZ (1985) Spatial variability of phytoplankton and Krill distribution and abundance in the Indian sector of the Southern Ocean. In: Siegfried WR, Condy PR, Laws RM (eds) Antarctic nutrient cycles and food webs (Proceedings of the 4th SCAR symposium on Antarctic biology). Springer, Berlin Heidelberg New York

Witek Z; Grelowski A; Kalinowski J (1982) Formation of Antarctic Krill concentrations in relation to hydrodynamic processes and social behaviour. ICES Biological Oceanography Committee C.M. 1982/L:59:22 pp

Yudanov KI (1982) Finding and quantitative estimation of Krill concentrations by hydroacoustic instruments. Contribution No. 70, Symposium on Fisheries Acoustics, Bergen, Norway, 21–24 June

Variation in Geographical Distribution and Abundance of Larvae of Antarctic Krill, *Euphausia superba*, in the Southern Atlantic Ocean

I. HEMPEL[1]

Summary. The geographical distribution and annual fluctuations in abundance of *Euphausia superba* larvae, mainly calyptopis stages, were studied in 1976, 1978 and 1981 in the central Scotia Sea and in the Bransfield Strait region. Comparisons are restricted to January-March, since previous studies have shown that relatively few larvae occur earlier in the season. In 1981, larval abundance was highest in the open ocean of the central Scotia Sea and in the shelf region close to Elephant Island. Abundance was significantly lower in 1978 when only a few 'rich' catches were made northeast of the South Orkney Islands and in the Bransfield Strait. In 1976, Krill larvae occurred in fair numbers in both the Bransfield Strait and the northern Scotia Sea. The Bransfield Strait region appears to be a major spawning area for Krill. Spawning in the open ocean apparently varies from year to year, in both space and time, for which explanations are not yet available. The same holds for large fluctuations in the abundance of larval Krill.

1 Introduction

Quantitative accounts of annual fluctuations attending the early life stages of the Antarctic Krill, *Euphausia superba*, are still largely lacking (Hempel et al. 1979).

During West German expeditions to the Scotia Sea and the Bransfield Strait in 1975–76, 1977–78 and 1980–81, primarily aimed at a survey of the spawning areas of Krill, the distribution of Krill larvae was studied in conjunction with micro-nekton trawling for adult Krill. In this paper, I use data from the three expeditions to describe annual variation in the geographical distribution and abundance of Krill larvae.

2 Materials and Methods

The sampling programmes, methods, and main results of the 3 expeditions have been described by Hempel and Hempel (1978), Hempel et al. (1979) and Hempel (1982), and station and sample lists have been compiled by Wörner (1978), Wörner and Kühn (1978) and Klages (1982). Larval samples were collected with a 1 m^2 rectangular mid-water trawl, fitted with 330 μm mesh, that formed part of RMT 1 + 8 assembly (Baker et al. 1973).

1 Institut für Polarökologie, Universität Kiel, Olshausenstr. 40, 2300 Kiel, FRG

Larvae were counted either in whole samples or in quantitative sub-samples, and larval abundance was expressed in numbers per 1000 m^3 of filtered volume of water. Normally, the gear was equipped with a flow-meter. When the flow-meter did not work, the path of the net was estimated from the duration of the haul and the ship's speed. This figure was multiplied by the cross section of the mouth of the net to obtain an estimate of the volume of water filtered.

Krill larvae were rare during November and December. Thus, I report here on samples taken during January-March only. The following numbers of samples were collected: 58 (1976), 127 (1978) and 114 (1981). Sampling was spread fairly evenly over the central Scotia Sea, the waters around Elephant Island and the Bransfield Strait.

3 Results

In 1976, 64% of the hauls were positive in the Bransfield Strait and its western approaches (Fig. 1). Larval abundance reached 2×10^4 animals 1000 m^{-3} in the upper 140–200 m of the water column. In February, most of the larvae were calyptopis stages I and II. No positive hauls for larvae were obtained between the South Orkney Islands and a position at 67°S in the central Weddell Sea. In the central Scotia Sea (between 56°S and 60°S), 7 out of 15 samples contained Krill larvae (maximum 1.1×10^3 animals 1000 m^{-3}, mainly calyptopis I) with older larvae, up to early furcilia, occurring in late February.

Krill larvae were less abundant in 1978 than in 1976. Out of 127 hauls, only 18 yielded larvae; larval densities exceeding 1000 animals 1000 m^{-3} occurring in only three hauls. Mainly eggs and nauplii were caught in January and February 1978, indicating a delay in spawning compared with 1976. The geographical distribution of the few positive stations was similar to that in 1976, being concentrated south of Elephant Island and northwest of South Orkney Islands (Fig. 2).

Figure 3 shows the generally higher abundance and wider geographical distribution of Krill larvae (again mainly calyptopis I) in 1981. Taken together, 80% of the hauls were positive. In the Bransfield Strait and on the shelf and

Antarctic Nutrient Cycles and Food Webs
(ed. by W.R. Siegfried, P.R. Condy, and R.M. Laws)

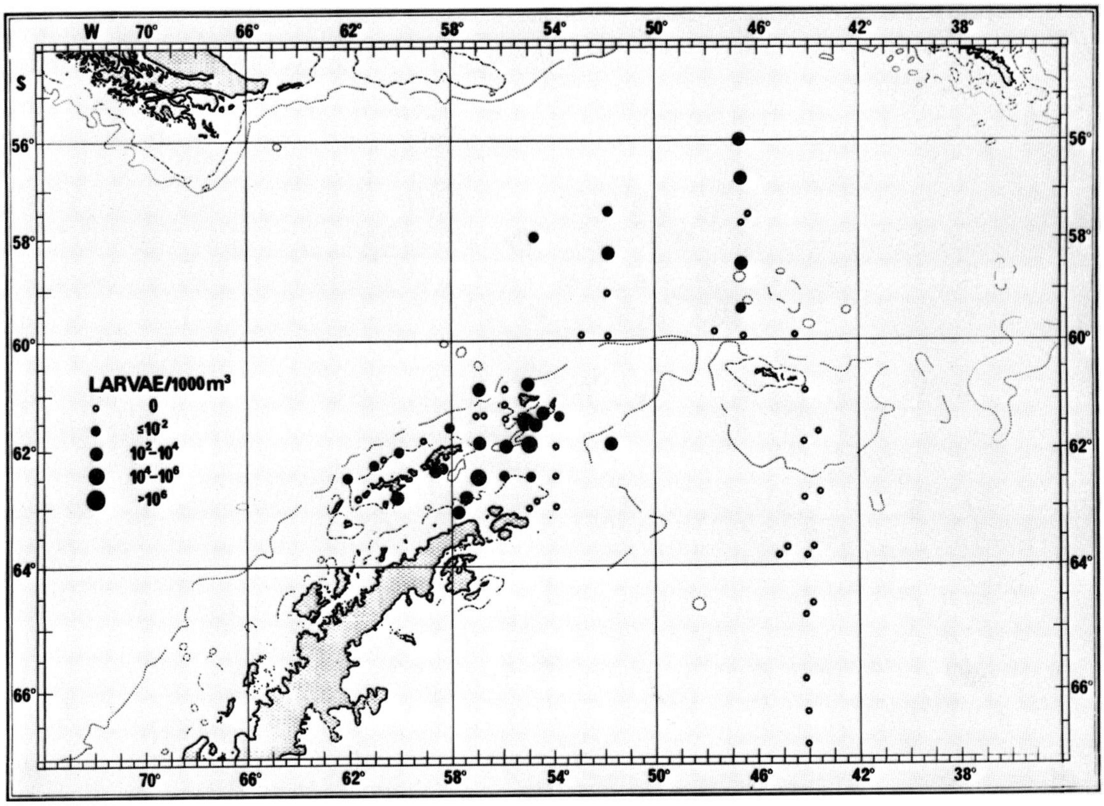

Fig. 1. Distribution and abundance of Krill larvae taken by RMT 8 + 1 in the upper 140–200 m, in January–March 1976

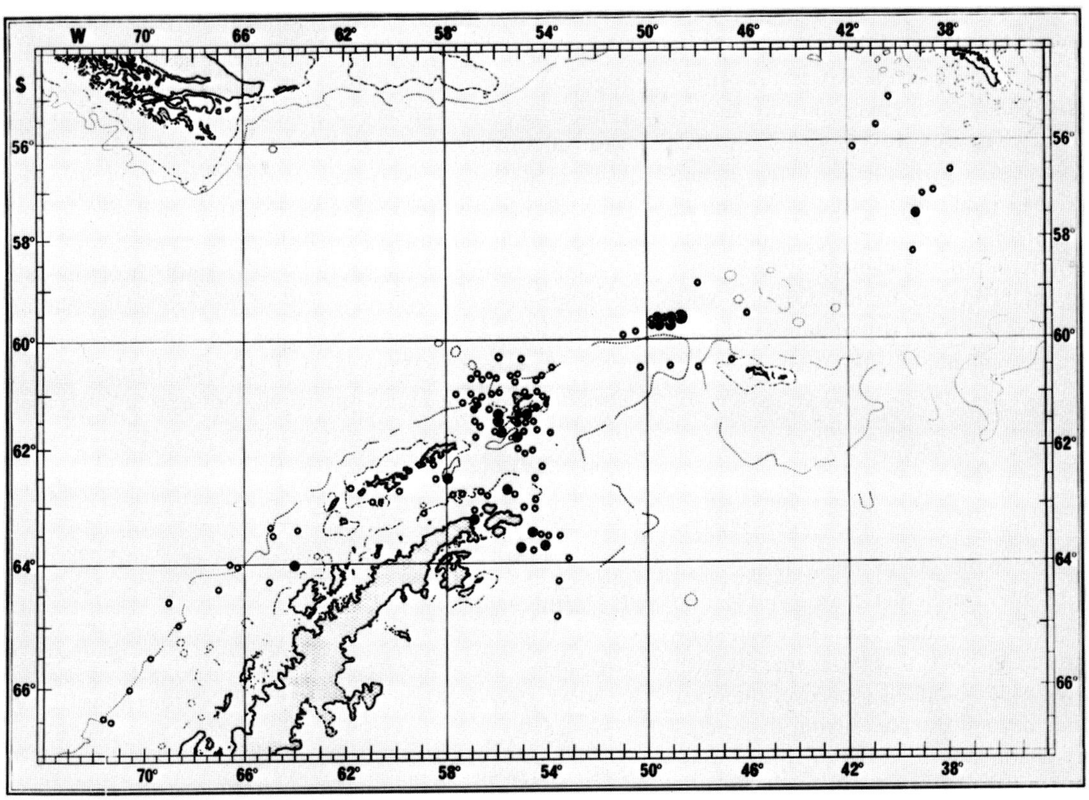

Fig. 2. Distribution and abundance of Krill larvae taken by RMT 8 + 1 in the upper 140–200 m, in January–March 1978. See Fig. 1 for explanation of symbols

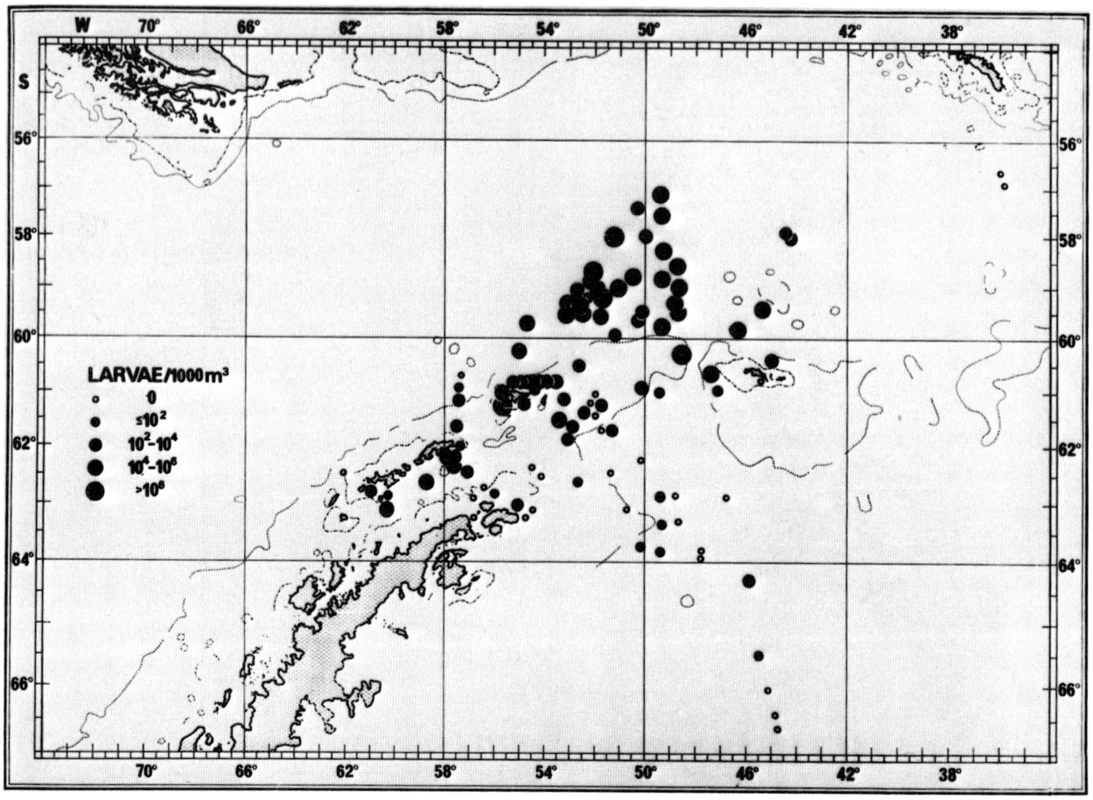

Fig. 3. Distribution and abundance of Krill larvae taken by RMT 8 + 1 in the upper 140–200 m, in January–March 1981

slope between Elephant Island and the South Orkney Islands, Krill larvae occurred all over the area and some larvae were found even in the inner Weddell Sea. More important is the fact that in 1981 all stations in the central Scotia Sea yielded large numbers of larvae and the highest abundances were found there (up to 1.2×10^6 animals 1000 m^{-3}). This area was characterized by low larval abundance during the previous expeditions, with a maximum of 1082 larvae 1000 m^{-3} in 1976. Calyptopis I predominated in late January and early February, whereas 3 weeks later mainly calyptopis II and III were found, together with furcilia stages. This was similar to the situation in 1976, and much earlier than when comparable stages were found in 1978.

4 Discussion

It is clear that Krill larval abundance in the western part of the Atlantic Sector of the Southern Ocean is subject to marked annual fluctuation. Maximum abundance varied by three orders of magnitude. In 1981, a year of generally high larval abundance, larvae were abundant even over the open ocean in the Scotia Sea beyond the continental shelf edge. These changes are very hard to interpret. The effects of atmospheric and oceanic circulation, and of oceanic frontal systems on the distribution of adult Krill in the southern Atlantic Ocean have been suggested especially by Soviet authors since the days of commercial whaling (e.g., Beklemishev 1960; Makarow 1972; Maslennikov 1972; Maslennikov and Solyankin 1980). These reports, and much of the literature since Marr's (1962) publication, which have been summarized by Everson (1977), point to a relationship between the distribution and abundance of Krill and particular hydrographic regimes. However, variability in the spawning of Krill has not been studied in detail, and we do not know which environmental factors determine time and area of spawning and possibly affect fecundity of individual Krill. Moreover, in spite of studies of the vertical distribution of the eggs and nauplii of Krill (e.g., Marr 1962; Voronina 1974; Makarow 1975; Hempel et al. 1979; Hempel 1984), nothing is known about the survival of eggs and larvae in the sea. Hence, it would be premature to speculate on possible causes of early mortality in Krill.

References

Beklemishev KW (1960) Southern atmospheric cyclones and the whale feeding grounds in the Antarctic. Nature (Lond) 187: 530–531

Everson I (1977) The living resources of the Southern Ocean. FAO, Rome 156 pp

Hempel I (1982) Zur Verbreitung der Brut von *Euphausia superba* in der mittleren Scotia Sea während FIBEX 1981. Arch Fisch Wiss 33:89–95

Hempel I (1984) Vertical distribution of larvae of Antarctic Krill, *Euphausia superba.* In: Siegfried WR, Condy PR, Laws RM, (eds) Antarctic nutrient cycles and food webs (Proceedings of the 4th SCAR symposium on Antarctic biology). Springer, Berlin Heidelberg New York

Hempel I, Hempel G (1978) Larval krill (*Euphausia superba*) in the plankton and neuston samples of the German Antarctic Expedition 1975/76. Meeresforschung 26:206–216

Hempel I, Hempel G, Baker A de C (1979) Early life history stages of krill (*Euphausia superba*) in Bransfield Strait and Weddell Sea. Meeresforschung 27:267–281

Klages N (1982) Deutsche Antarktis Expedition 1980/81 mit FS "Meteor". Liste der Zooplankton-Mikronektonnetzfänge. Ber Polarforsch 2:29 pp

Makarow RR (1972) Life cycle and distribution pattern of *Euphausia superba* Dana. (in Russian). Tr Vses Nauchno-Issled Inst Morsk Khoz Okeanogr 77:85–92

Makarow RR (1975) Vertical distribution of euphausiid eggs and larvae of the northeastern coast of South Georgia Island, Oceanology 15:708–711

Marr JWS (1962) The natural history and geography of the Antarctic krill (*Euphausia superba* Dana). Discovery Rep 32:33–464

Maslennikov VV (1972) On the effect of water dynamics on the distribution of *Euphausia superba* Dana in the area of South Georgia. Tr Vses Nauchno-Issled Inst Morsk Rybn Khoz Okeanogr 75:107–117

Maslennikow VV, Solyankin EV (1980) The role of water dynamics in the maintenance of *Euphausia superba* Dana population in the Weddell Sea. Okeanologiya 20:295–300

Voronina NM (1974) An attempt at a functional analysis of the distributional range of *Euphausia superba.* Mar Biol 24:347–352

Wörner FG (1978) Liste der Mikronekton- und Zooplankonfänge der 2. Deutschen Antarktis-Expedition. Ber Inst f Meeresk Univ Kiel 60, 82 pp

Wörner FG, Kühn A (1978) Liste der Mikronekton- und Zooplanktonfänge der deutschen Antarktis-Expedition 1975/76. Ber Inst f Meeresk Univ Kiel 48, 55 pp

Vertical Distribution of Larvae of Antarctic Krill, *Euphausia superba*

I. HEMPEL[1]

Summary. The vertical distribution of eggs and larvae of the Antarctic Krill, *Euphausia superba*, was studied during January 1981 in the Scotia Sea. Nauplii were scattered over a wide range of depths, whereas metanauplii were found concentrated between 1000 and 500 m. Calyptopes were not found below 1000 m but large numbers were present in the upper 500 m, predominantly in the 100–30 m layer. The herbivorous copepod, *Calanoides acutus*, was confined closer to the surface. The omnivorous copepod, *Metridia gerlachei*, occurred at depths > 100 m, down to 500 m. It was the most abundant copepod found in the central Scotia Sea. In the eastern Weddell Sea, *E. superba* larvae occurred together with larvae of *Euphausia crystallorophias* which were more abundant. *Euphausia superba* larvae tended to congregate at greater depths (between 200 and 300 m) than those of *E. crystallorophias* which were most abundant near the surface.

1 Introduction

The vertical distribution of Antarctic Krill, *Euphausia superba*, larvae has been discussed by Marr (1962) who postulated that Krill eggs sink from near-surface waters to the sea bed on the continental shelf and to a depth of ca. 2000–3000 m in the open ocean. The nauplii and metanauplii return to surface waters after hatching during a hypothesized "developmental ascent". In recent years, this hypothesis has been examined by a number of authors (Nast 1979; Hempel et al. 1979), on the basis of a limited number of samples from mid-water and surface layers only; no samples having been considered from depths > 1000 m.

To date, net samples have contained very few nauplii and metanauplii, but calyptopis stages of Krill have been readily apparent in the upper 200 m. It is possible that in this layer Krill larvae may compete for food with other herbivorous zooplankton and that they are subject to predation by chaetognaths and other carnivorous members of the zooplankton. It is likely that the survival of Krill larvae is influenced by the extent to which competitors for food and potential predators remain at the same depths as the larvae themselves. Thus, it seemed worthwhile to analyze the vertical distribution and abundance of Krill larvae in relation to associated elements of the accompanying zooplankton collected during discrete depth sampling.

1 Institut für Polarökologie, Universität Kiel, Olshausenstr. 40, 2300 Kiel, FRG

2 Materials and Methods

In January 1981, Nansen closing nets (NCN) (mouth diameter = 600 mm; 300 μm mesh) were operated from the R.V. Meteor in the Scotia Sea, Bransfield Strait and Elephant Island regions, as follows: four hauls between 2000 and 1000 m, seven hauls (1000 m–500 m) and seven hauls (500 m–140 m). In February, a multiple opening-closing net (MOCNESS) (mouth diameter = 1.0 m; eight nets with 330 μm mesh) was employed to sample in eight depth strata between 500 m and the surface. The results obtained at a single station on 10 February 1981, are presented here. This station was located at 60°20′S, 48°23′W, in the centre of an area with a high abundance of Krill larvae in deep water west of the South Orkney Islands.

During the Weddell Sea expedition of the R.V. Polarsirkel in 1979–80, discrete depth samples were also taken by a vertical closing net (mouth diameter = 700 mm; 250 μm mesh) similar to the NCN used in 1981. Data from Station 95, occupied on 14 February 1980, at 72°42′S, 19°18′W, off Cape Norwegia in 2300 m of water, are reported here.

Euphausiid larvae were counted in either the whole of a net-sample or in quantitative sub-samples. In the MOCNESS samples, copepods were counted as well. Larval stages and copepods were identified, using a dissecting microscope. The volume of water filtered was determined by a flow-meter in the MOCNESS, or estimated by multiplying the vertical path of the net by its mouth opening.

3 Results and Discussion

The season 1980–81 was unusually rich in Krill larvae in the Scotia Sea. Figure 1 summarizes the abundance of Krill larvae caught by 18 hauls with the NCN for different strata between 2000 m and 140 m. The average frequency of nauplii remained relatively constant irrespective of the depth sampled. Metanauplii were virtually absent below 1000 m. They were dominant between 500 and 1000 m, and their frequency decreased closer to the surface. Calyptopis I were found mainly in the upper 500 m. No calyptopis was found deeper than 1000 m. Table 1 gives relative abundances of the various developmental stages in each

Antarctic Nutrient Cycles and Food Webs
(ed. by W.R. Siegfried, P.R. Condy, and R.M. Laws)

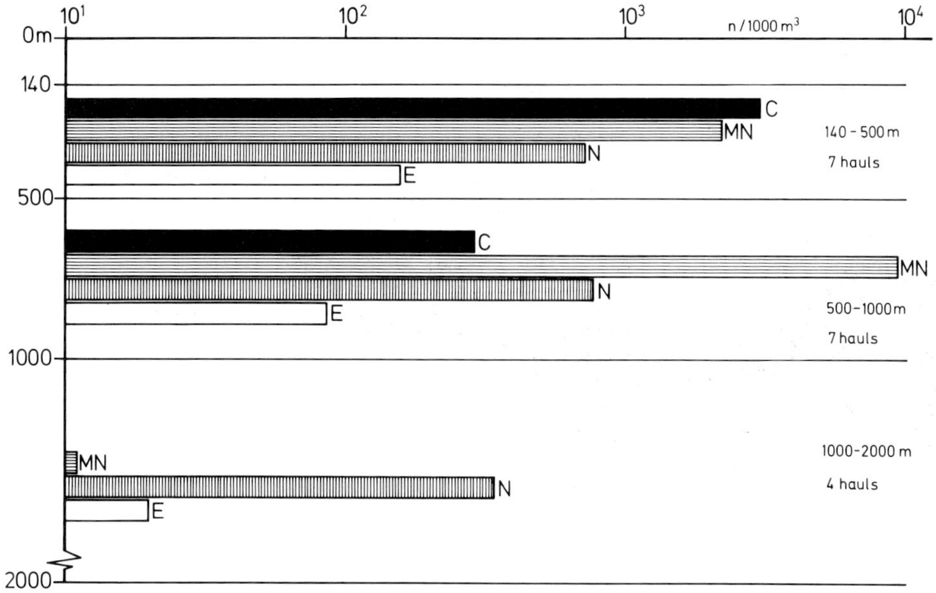

Fig. 1. Abundance of *E. superba* eggs (*E*), nauplii (*N*), metanauplii (*MN*) and calyptopis stages (*C*) in relation to depth in the Bransfield Strait and Scotia Sea, January 1981

depth layer. These figures indicate a vertical succession of larval stages, which supports Marr's (1962) hypothetical "developmental ascent". It also confirms his view that, in deep water, hatching of the nauplii may occur below 1000 m.

Table 1. Relative abundance (%) of different developmental stages of *E. superba* in the Scotia Sea and Bransfield Strait, January 1981

Depth (m)	Eggs	Nauplii	Metanauplii	Calyptopes
140– 500	2.5	11.9	36.0	49.3
500–1000	0.8	6.9	89.0	2.7
1000–2000	5.4	91.5	3.0	0

As in earlier studies, very few Krill eggs and relatively low numbers of nauplii were recorded. This could be attributed to destruction of the very fragile eggs and nauplii during hauling, and also during the handling of samples on deck (Marschall and Hirche 1984). It is, therefore, not surprising that deep hauls produced (in absolute numbers) fewer eggs and nauplii than shorter, shallower hauls.

A more detailed analysis of the vertical distribution of calyptopis larvae (mainly CI) in the upper 500 m is presented in the left-hand part of Fig. 2. Abundance of larvae was highest between 150 and 30 m, with a maximum of 2×10^6 larva 1000 m^{-3} at 60–30 m. In the surface layer (30–0 m) and at depths of 150 m, larval abundances were about 10^4 larva 1000 m^{-3}. A similar analysis for the upper 325 m off Cape Norwegia (Fig. 3) showed decreasing numbers of Krill calyptopis per 1000 m^3 with decreasing depth: 10^4

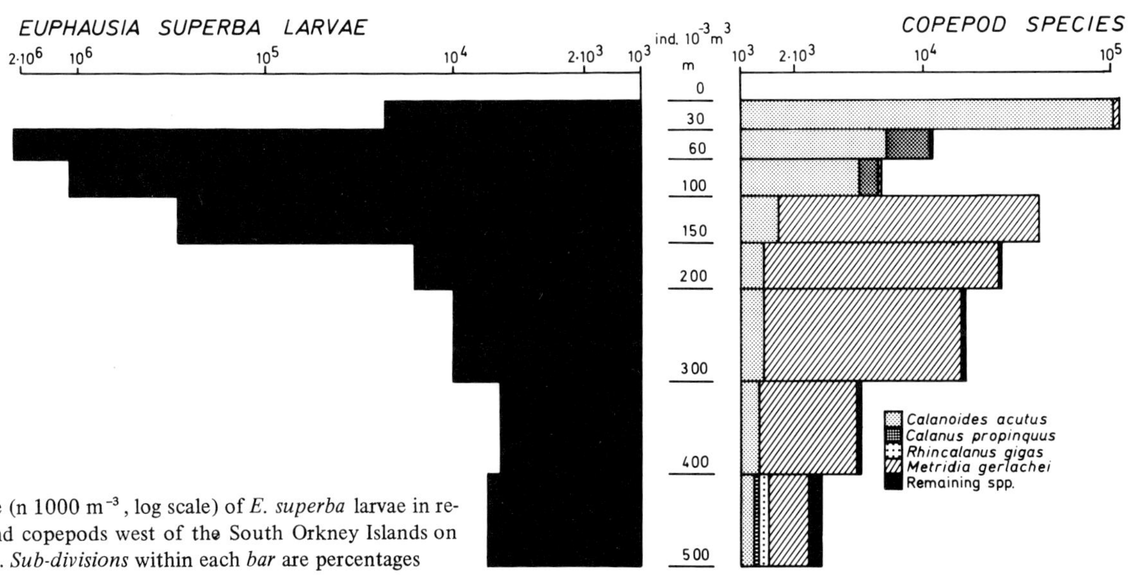

Fig. 2. Abundance (n 1000 m^{-3}, log scale) of *E. superba* larvae in relation to depth and copepods west of the South Orkney Islands on 10 February 1981. *Sub-divisions* within each *bar* are percentages

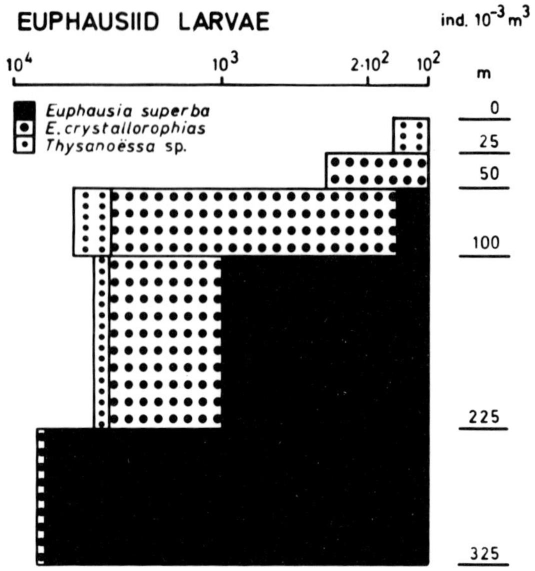

Fig. 3. Abundance (n 1000 m^{-3}, log scale) of various euphausiid larvae in relation to depth off Cape Norwegia, on 14 February 1980. *Sub-divisions* within each *bar* are percentages

larvae at 325–225 m, 10^3 larvae at 225–100 m and 10^2 larvae between 100 and 50 m. Krill larvae were not found in the near-surface layer above 50 m. At the same station (Station 95) larvae of *E. crystallorophias* occurred in substantial numbers, mainly in the two layers between 100 and 25 m. Larvae of *Thyssanoessa* sp. were found in most depths between 225 m and the surface, although numbers decreased with depth. The reasons for the deeper occurrence of *E. superba* larvae off Cape Norwegia compared to the Scotia Sea are not yet understood. It should be mentioned that Cape Norwegia marked the southern edge of the area of distribution of *E. superba* larvae during 1980–81 (Hempel and Hempel 1982; Hempel et al. 1983). Any current oceanographic interpretation (e.g. Voronina 1979) of the occurrence and vertical distribution of Krill larvae seems premature and highly speculative.

Results from the MOCNESS haul west of the South Orkney Islands indicated that copepod numbers were low in the 100-60-30 m depth strata (Fig. 2). Krill larvae attained maximum abundance in these layers, which was at least two orders of magnitude greater than that of the copepods. *Calanoides acutus,* together with *Calanus propinquus,* predominated in the upper 100 m. These species, being herbivorous, conceivably could compete with Krill larvae for food. A drastic shift in the copepod fauna was observed between 150–100 m. In this layer, as in all deeper ones down to 500 m, the omnivorous copepod *Metridia gerlachei* was dominant. The total number of copepods decreased between 100 and 500 m, but diversity was highest at maximum depths sampled.

Acknowledgements. The NCN samples in 1981 were taken by Dr. J. Kinzer and Mr. D. Jarosch. The MOCNESS was operated by Dr. N. Klages. Mrs. C. Dieckmann and E. Mizdalski (Alfred-Wegener-Institut für Polarforschung, Bremerhaven) prepared the figures. The work was supported by a grant of the Federal Ministry of Science and Technology.

References

Hempel I, Hempel G, Baker A de C (1979) Early life history stages of krill (*Euphausia superba*) in Bransfield Strait and Weddell Sea. Meeresforschung 27: 267–281

Hempel I, Hempel G (1982) Distribution of euphausiid larvae in the southern Weddell Sea. Meeresforschung 29: 253–266

Hempel I, Hubold G, Kaczmaruk B, Keller R. Weigmann-Haass R (1983) Distribution of some groups of zooplankton in the inner Weddell Sea in summer 1979/80. Ber Polarforsch 9:35 pp

Marr J (1962) The natural history and geography of the Antarctic krill (*Euphausia superba* Dana). Discovery Rep. 32: 33–464

Marschall HP, Hirche HJ (1984) Development of eggs and nauplii of (*Euphausia superba*). Polar Biol 2:245–250

Nast F (1979) The vertical distribution of larval and adult krill (*Euphausia superba*) on a time station south of Elephant Island, South Shetlands, Meeresforschung 27: 103–118

Voronina NM (1974) An attempt at a functional analysis of the distributional range of *Euphausia superba.* Mar Biol 24:324–352

Feeding by *Euphausia superba* and Copepod Species in Response to Varying Concentrations of Phytoplankton

S.B. SCHNACK[1]

Summary. The feeding behaviour of dominant copepod species (*Calanus propinquus, Calanoides acutus, Rhincalanus gigas, Metridia gerlachei* and *M. curticauda*) and of the Antarctic Krill, *Euphausia superba,* was studied during 2 cruises in the Atlantic sector of the Southern Ocean. The phytoplankton composition differed markedly between the 2 cruises. In November-December 1980, *Thalassiosira* spp. dominated close to the ice-edge, whereas *Corethron criophilum* was predominant in the Drake Passage. In February 1982, the diatoms, *C. criophilum, Chaetoceros* spp. and *Biddulphia* spp., as well as μ-flagellates were most abundant. However, all copepod species, as well as Krill, fed efficiently on the most abundant food item and, on both cruises, these animals fed on the same food organisms. However, feeding rates differed between the species; *Metridia gerlachei* showing higher feeding rates (in terms of food ingested on a weight-specific basis) than the other animals studied.

1 Introduction

Interest in the feeding behaviour of Antarctic marine animals has increased considerably during recent years. However, studies have been focused mainly on the Antarctic Krill, *Euphausia superba* (Antezana et al. 1982; Meyer and El-Sayed 1983; Kato et al. 1979, 1982; Boyd et al. to be publ.; Morris et al. 1983; Morris to be publ.), while Antarctic copepod species have been neglected. One reason for this is probably the remarkable position of *E. superba* as an important filter feeder in the Antarctic system. This animal makes up a large part of the total zooplankton biomass. However, *E. superba* does not comprise the major fraction of the primary consumers in all areas of the Antarctic where copepods are also important (Mackintosh 1934). Thus, in order to understand the functioning of the food web in the Antarctic, it is important to study any potential competition between *E. superba* and abundant copepod species.

This report deals with the feeding behaviour of abundant, large copepod species (*Rhincalanus gigas, Calanoides acutus, Calanus propinquus, Metridia gerlachei* and *M. curticauda*) in comparison with *E. superba.* The study focused on the feeding responses of individual copepod species and Krill to changes in phytoplankton composition and abundance in the south-western sector of the Atlantic Ocean.

1 Alfred-Wegener-Institut für Polarforschung, Columbus-Center, 2850 Bremerhaven, FRG

2 Materials and Methods

Copepods were collected with a Nansen net (200 μm mesh) hauled vertically in the upper 150–200 m. Healthy copepods were sorted from the catches and placed in 5-1 bottles containing ambient seawater. Adult and sub-adult *E. superba* were selected from specimens obtained with a RMT net. The euphausiids were treated in the same way as the copepods. Due to the low concentrations of copepods and euphausiids at many stations, the captured animals were kept until a sufficient number was available for filling three sets of experimental bottles for each species. The animals were allowed 1-3-d periods to adapt to the composition and the concentration of the food supply.

Plankton samples were collected with Niskin bottles within the upper 100 m (November–Deecember 1980) or from the surface with a ship-pump (February 1982), and were stored in a 50-l polyethylene carboy. The experiments with captive animals were run with food at ambient concentration, and also with an enriched supply to obtain information on the animals' responses to changes in the density of food. Phytoplankton for this purpose was obtained from the surface of the sea, using a net with a 20-μm mesh. The phytoplankton suspension was then added to the carboy and thoroughly mixed. After mixing, 5-1 bottles were filled with the phytoplankton, and the experimental animals were added.

Adult females and copepodids of stage V or *R. gigas, C. propinquus, C. acutus* and *M. gerlachei* were used in the experiments. In February 1982, experiments were also run with *M. curticauda.* Copepodite stages I–IV were available for the two *Metridia* species only. The concentration of the copepods was between 20 and 80 individuals per 5 l, and that of the euphausiids between 1 and 5 individuals per 5 l, depending on the size of individuals. One control without animals was run for each experimental set.

The bottles used in the experiments were stoppered with silicon plugs. Circulation of water within the bottles was maintained at 200 ml h^{-1}, by way of a peristaltic pump. All experiments were run at 0°C in the dark and lasted between 8 and 9.5 h. Phytoplankton (= food) samples taken at the beginning and at the end of each experiment were preserved with Lugol's solution (November–December 1980) or

Antarctic Nutrient Cycles and Food Webs
(ed. by W.R. Siegfried, P.R. Condy, and R.M. Laws)

with 1% formalin (February 1982). Phytoplankton concentration was determined from sub-samples of 100–200 ml, using an inverted microscope. Only phytoplankton species or groups with numbers of individuals > 30 per 200 ml were included in the calculations. Cell volume and carbon content were calculated according to Strathmann (1967). Mean concentration for each phytoplankton species or group available as food, as well as copepod and euphausiid filtration and ingestion rates for each phytoplankton species or group, were calculated according to Frost (1972). Selection of food items was quantified, using the selectivity index of Jacobs (1974). At the end of the experiments, the content of the experimental bottles was concentrated with a 55-μm sieve, and faecal pellets and animals were removed by pipette. Fifty per cent of the total number of animals were deep-frozen, and the others were preserved with 4% formalin for later identification. In the laboratory, the deep-frozen animals were dried at 60°C for 24 h (copepods) or for 48 h (euphausiids), and weighed individually on a Sartorius electro-balance. The C content of the copepods was analyzed with a Carlo-Erba or a Perkin Elmer CN-analyzer. For *E. superba,* C content was taken to be 40% of the dry wt. mass (Lenz 1974). Assimilation efficiency was determined from the ratio of ash-free dry wt. to dry wt. for samples of food and faecal pellets (Conover 1966).

3 Results

Growth of phytoplankton did not exceed 1% of the initial food concentration in the control bottles during the experimental period. The mean dry wt., C content, and C:N ratio per individual animal of the copepod species studied in November–December 1980 and in February 1982 are given in Table 1. The mean dry wt. of *E. superba,* in various size-classes, was 30, 140, and 200 mg for 30–35, 40–45 and 50–55 mm individuals, respectively. A difference occurred in the weight ratios between the copepods. In spring the ratio was about 5.7 and in late summer it was 6.6 (i.e. the largest copepod, *R. gigas,* was about 5.7 and 6.6 heavier than the copepodids of *Metridia*).

Different ecological zones with distinct phytoplankton compositions were found in November–December 1980 (Schnack et al. to be publ.). The experiments reported here were carried out either in a "*Thalassiosira*" zone or in a "*Corethron*" dominated zone. In February 1982, the diatoms, *Corethron criophilum, Chaetoceros* spp., and *Biddulphia* spp., as well as μ-flagellates, were most abundant.

Figures 1 and 2 give the results of the experiments carried out in November–December 1980 when "centric diatoms" dominated. Here, "centric diatoms" are taken to include all organisms of the "*Thalassiosira*" type. In Fig. 1, column A represents the ambient food supply experiment and column B the enriched one. The initial food concentrations were 77.3 and 189 μg C 1^{-1}, respectively. The bottom diagram shows the food density (μg C 1^{-1}) in the control in relation to the cell volume of the food organisms.

There was a distinct difference in the filtration rate between the copepod species. *Rhincalanus gigas* and *C. acutus* had higher filtration rates on small phytoplankton particles (μ-flagellates) than on the phytoplankton biomass peak (most dense concentration of "centric diatoms"), whereas *C. propinquus, M. gerlachei* and the copepodids showed highest filtration rates on the most abundant food particle ("centric diatoms") and relative low filtration rates on small particles (μ-flagellates). All three size-classes of *E. su-*

Table 1. Mean (± 1 S.D.) dry wt. (μg ind.) and C content and C:N ratio of five copepod species

Nov.–Dec. 1980	n	dry wt.	Carbon content	C:N
Rhincalanus gigas	12	1,102 ± 30.1	505 ± 26.2	4.1
Calanus propinquus	9	442 ± 16.4	213 ± 18.5	4.3
Calanoides acutus	15	300 ± 16.9	139 ± 15.5	5.0
Metridia gerlachei	11	341 ± 17.2	160 ± 8.6	4.0
Copepodids	23	185 ± 19.7	89 ± 13.5	4.7
February 1982				
Rhincalanus gigas	9	980 ± 52.0	392 ± 10.4	3.9
Calanus propinquus	13	424 ± 21.5	189 ± 16.5	4.1
Calanoides acutus	12	270 ± 21.7	108 ± 11.5	4.7
Metridia gerlachei	10	260 ± 14.4	101 ± 9.6	3.6
Metridia curticauda	9	227 ± 18.5	98 ± 10.0	4.3
Copepodids	16	144 ± 12.2	59 ± 9.8	4.9

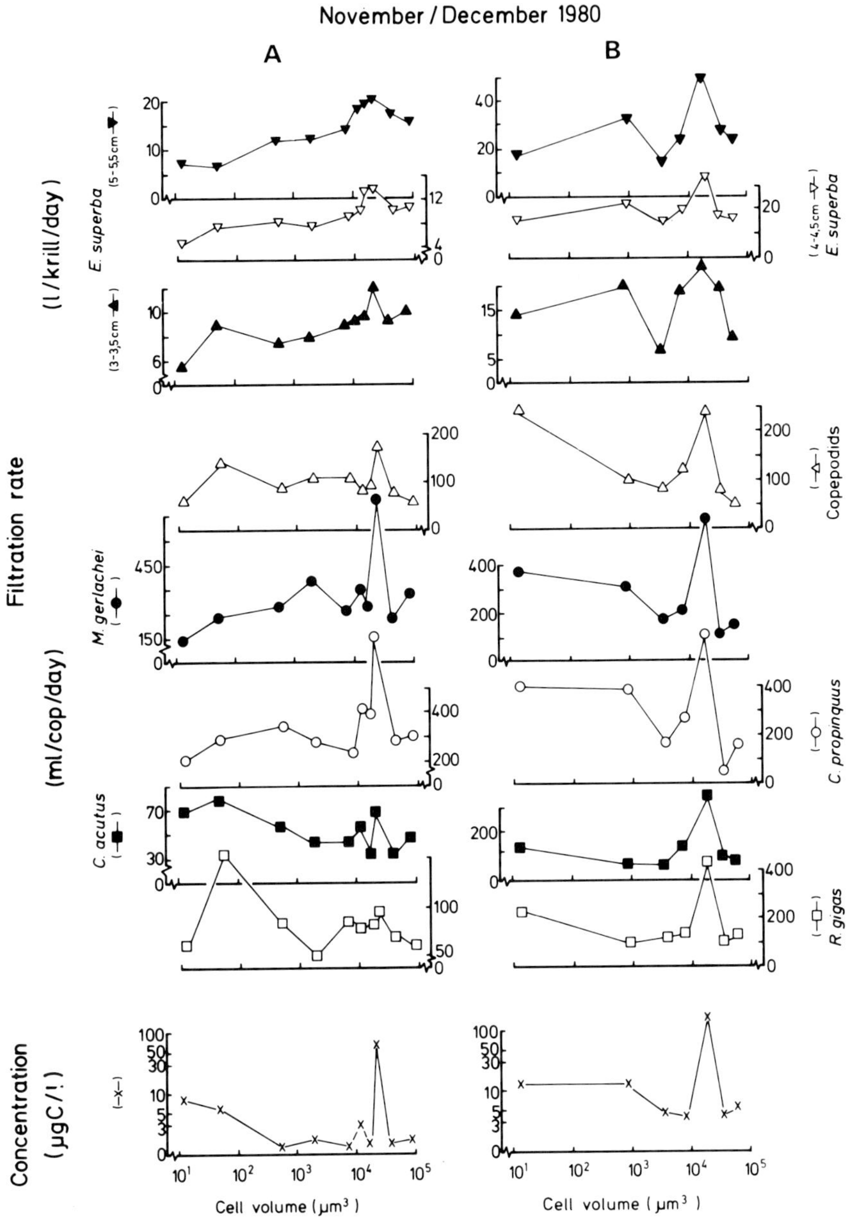

Fig. 1A, B. Filtration rates of *E. superba* and copepods in relation to two concentrations of a "centric diatom" food supply. Graphs **A** reflect an ambient food supply (77 μg C 1^{-1}) and **B** an enriched food supply (189 μg C 1^{-1})

perba had highest filtration rates on the biomass peak, but not as distinct as for *M. gerlachei* and *C. propinquus.* No significant correlation between filtration rates and concentration of the food organisms could be detected (Spearman rank correlation).

These results are different from those obtained in the enriched-food experiment (column B) in which all filtration rates had a similar trend. The animals had highest filtration rates on the biomass peak ("centric diatoms"), except for copepodids which had slightly higher rates on small organisms (μ-flagellates). The higher filtration rates, relative to ambient food supply, for *R. gigas* and *C. acutus* were surprising, as were also the higher filtration rates on small particles for *M. gerlachei, C. propinquus* and copepodids. All three size-classes of *E. superba* had similar filtration rates. There was no significant correlation between the concentration and the filtration rates for *R. gigas* and *C. acutus* and for all size-classes of *E. superba.* For *M. gerlachei* and the copepodids there was a significant correlation at the 5% level, and similarly in *C. propinquus* at the 1% level (Spearman rank correlation).

The relative proportions of the dominant phytoplankton groups in the food suspension as well as in the diet of the animals are shown in Fig. 2. The top diagram (A) gives the results of the experiments with an ambient food supply. "Centric diatoms" made up 79% of the initial phytoplankton biomass in terms of C, and μ-flagellates 15%. The balance consisted of "pennate diatoms" (*Rhizosolenia* spp.), *C. criophilum, Biddulphia* spp. and the silico-flagellate *Distephanus speculum. Rhincalanus gigas* and *C. acutus,* as well as the smallest size-class of *E. superba,* ingested the three phytoplankton groups in proportions similar to those

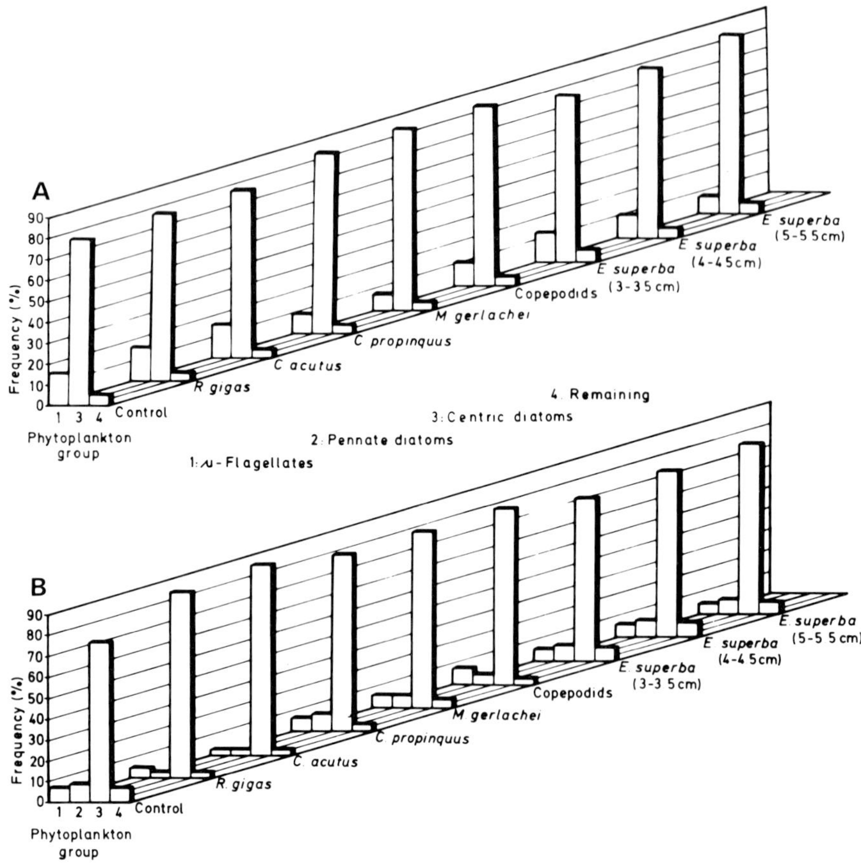

Fig. 2. Relative proportions of four different phytoplankton groups (*1* μ-flagellates; *2* "pennate diatoms"; *3* "centric diatoms"; *4* remainder) in relation to ingestion by *E. superba* and copepods which fed either on an ambient food supply (77 μg C l^{-1}) (*top diagram*) or an enriched food supply (189 μg C l^{-1}) (*bottom diagram*)

for their availability. *Calanus propinquus*, copepodids and *M. gerlachei* and the two larger size-classes of *E. superba* ingested slightly higher relative proportions of "centric diatoms" (between 84% and 87%). The selectivity index showed positive values for all animals on "centric diatoms". *Rhincalanus gigas* and *C. acutus* also had positive values on nano-flagellates.

In the enriched food experiment (Fig. 2B), the remaining group consisted of *C. criophilum, Biddulphia* spp. and *D. speculum;* "pennate diatoms" accounting for 9%. The highest relative proportion (77%) was again recorded for "centric diatoms"; nano-flagellates accounting for 7%. All copepods had higher relative proportions on "centric diatoms" (between 84% and 91%). The size-classes of *E. superba* ingested similar relative proportions to those available in the suspension. The selectivity index was positive only on "centric diatoms" for all animals, with the exception of copepodids for which μ-flagellates also had a positive index.

The results of the experiments in November–December 1980, in which *C. criophilum* was the most dominant food particle, are given in Figs. 3 and 4. In Fig. 3, column C represents the ambient food-supply experiment with an initial concentration of 72 μ-g C l^{-1}. Apart from the copepods, only the smallest size-class (30–35 mm) of *E. superba* was studied in these experiments. All animals had highest filtrations rates on the biomass peak (*C. criophilum* cell volume = 7700 μm^3), followed by high filtration rates on the largest food particle ("centric diatoms"; cell volume = 20800 μm^3). The relatively low filtration rates on particles of about 600 μm^3 (*Nitzschia* spp.) and 9600 μm^3 (dinoflagellates) in *R. gigas, C. acutus, C. propinquus* and in *M. gerlachei* and copepodids are interesting. Dinoflagellates were represented mainly by *Gyrodinium* spp. and *Protoperidinium* sp. No significant correlation between concentration and filtration rates of the single food organisms could be found (Spearman rank correlation).

The right-hand column (D) in Fig. 3 represents the enriched food experiment (initial food concentrations = 177 μg C l^{-1}). In these experiments, *C. criophilum* was again the most abundant, and also the largest, organism, possibly due to the 20 μm plankton net retaining large cells more efficiently than smaller ones. All animals filtered at highest rates on the biomass peak (i.e. on *C. criophilum*), and secondly on "centric diatoms" (cell volume = 13800 μm^3). A relatively low filtration rate was found for all animals on *Chaetoceros* spp. (cell volume = 240 μm^3). No significant correlation between concentration and filtration rates was obtained for *R. gigas, C. acutus* and *E. superba.* In *C. propinquus* there was a significant correlation at the 5% level, and in *M. gerlachei* and in copepodids at the 1% level (Spearman rank correlation).

In Fig. 4 the top diagram (C) gives the results of the ambient-food experiment in which *C. criophilum* made up

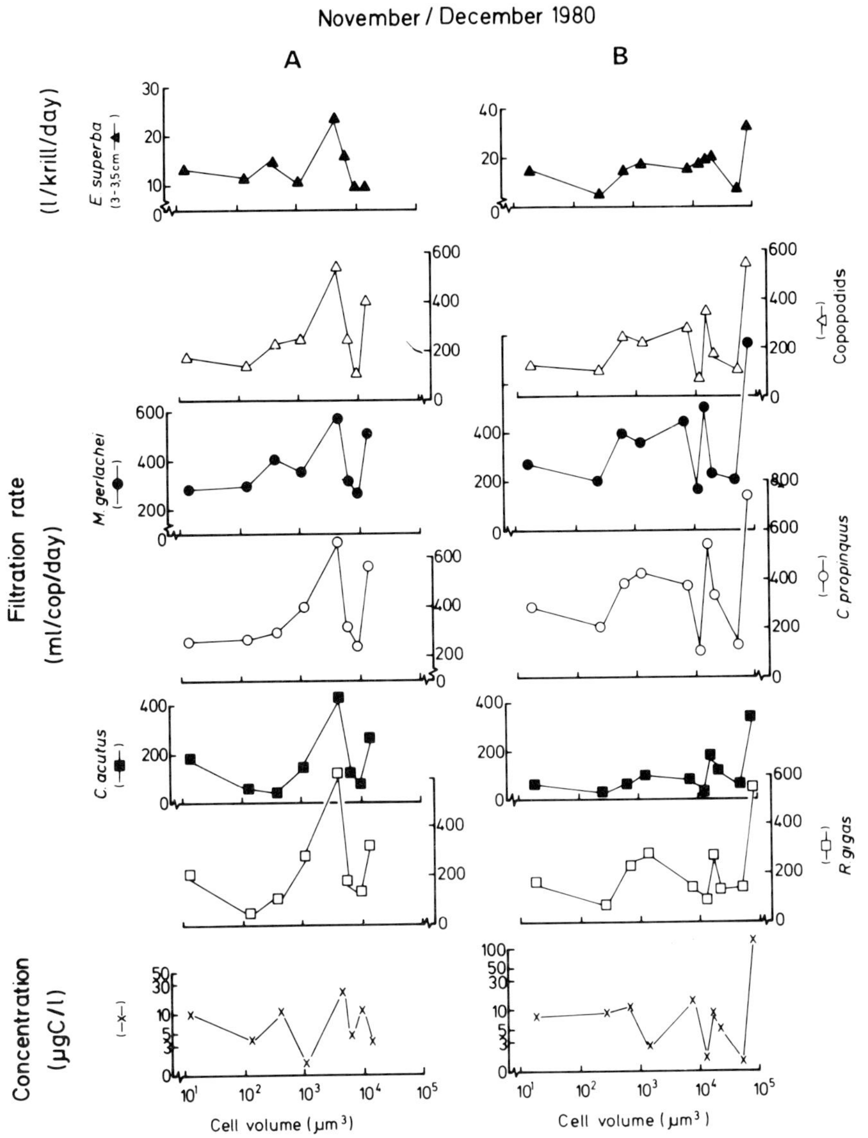

Fig. 3A, B. Filtration rates of *E. superba* and copepods in relation to two concentrations of a *"Corethron"* food supply. Graphs **A** reflect an ambient food supply (72 µg C l⁻¹) and **B** an enriched food supply (177 µg C l⁻¹)

28% (µ-flagellates 25%, pennate diatoms 16%, centric diatoms 5%), and *D. speculum* 6% of the initial food biomass. The remainder, consisting of *Chaetoceros* spp. and dinoflagellates, made up 21%. The relative proportions appear rather similar to those of the control, with a slightly higher proportion of feeding on *C. criophilum.* All animals had a positive selectivity index on *C. criophilum,* and all copepods also on "centric diatoms". The bottom diagram (D) gives the results of the enriched food experiment in which *C. criophilum* made up 67% of the initial food biomass. The relative proportion of *C. criophilum* in the animals' diet varied between 73% and 85%. All animals had a positive selectivity index only on *C. chriophilum.* The results of the experiments from February 1982 are summarized in Figs. 5 and 6. In the ambient food experiment (column E), µ-flagellates and the diatoms *Chaetoceros* spp., *C. criophilum* and *Biddulphia* spp. were most abundant. The filtration rates for all animals showed a zig-zag pattern similar to that of the control, with higher filtration rates on the biomass peaks (i.e. on µ-flagellates, *Chaetoceros* spp., *C. criophilum* and *Biddulphia* spp. with cell volumes of 70, 2050, 26 × 10^3 and 13 × 10^4 µm³, respectively). All animals had a positive selectivity index on *C. criophilum,* and also for copepodids on *Chaetoceros* spp. and *E. superba* on µ-flagellates. There was a significant correlation between the filtration rates and the concentrations at the 1% level for all animals (Spearman rank correlation). Column F (Fig. 5) gives the result of the enriched food experiment. A similarity between the filtration rate curves and the concentration curve of the control is apparent. The filtration rates in *E. superba* in these experiments (February 1982) were about 5–6 times lower than those recorded in November–December 1980.

There was a significant correlation between the concentration and the filtration rates at the 5% level for *C. propinquus,* and at the 1% level for all other animals (Spearman

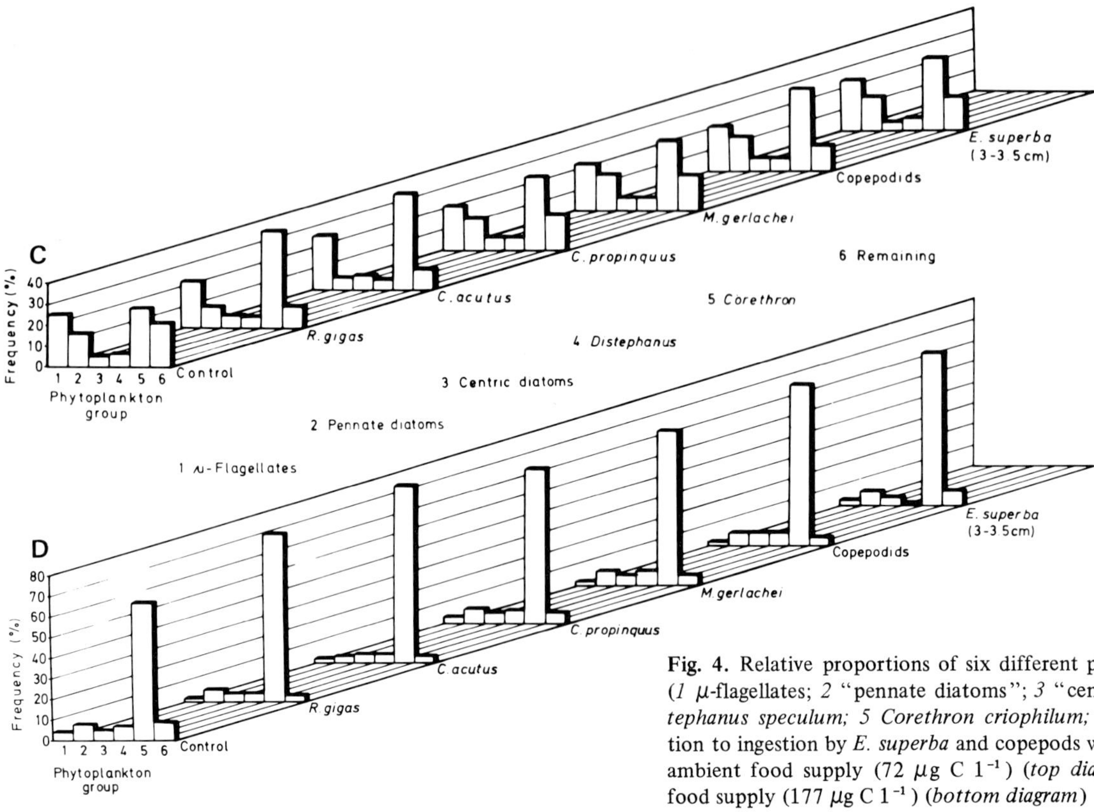

Fig. 4. Relative proportions of six different phytoplankton groups (*1* μ-flagellates; *2* "pennate diatoms"; *3* "centric diatoms"; *4 Distephanus speculum; 5 Corethron criophilum; 6* remainder) in relation to ingestion by *E. superba* and copepods which fed either on an ambient food supply (72 μg C 1^{-1}) (*top diagram*) or an enriched food supply (177 μg C 1^{-1}) (*bottom diagram*)

rank correlation). The relative proportions of the phytoplankton groups in the food and in the animals' ration are shown in Fig. 6. The top diagram (E) again gives the result of the ambient concentration experiment. *Corethron criophilum* made up 34%, *Chaetoceros* spp. 23%, *Biddulphia* spp. 17% and μ-flagellates 21% of the initial food biomass. The remaining group consisted of "pennate diatoms", "centric diatoms" and of *Rhizosolenia* spp. The relative proportion of *C. criophilum* was higher in the animals' diet than in the control. The selectivity index is positive for all animals on *C. criophilum* and on *Chaetoceros* spp.. *Euphausia superba* also had a higher ratio on μ-flagellates, which is also reflected by a positive selectivity index. In the concentrated food experiment (Fig. 5F), the relative proportion of *C. criophilum* increased to 43% of the initial food concentration and that of μ-flagellates decreased to 13%. *Chaetoceros* spp. constituted 23% and *Biddulphia* spp. 16%. The remaining group included *Rhizosolenia* spp. "centric diatoms", "pennate diatoms", *D. speculum* and dinoflagellates. Higher proportions of *C. criophilum* were again found in the animals' diet, with a positive selectivity index. The relative proportions for μ-flagellates and for the remaining group were lower than in the control. There were positive selectivity indices for *M. gerlachei, M. curticauda, C. propinquus* and *R. gigas* on *Chaetoceros* spp. and for copepodids on μ-flagellates.

Table 2 gives the mean weight-specific filtration rates for the different copepods and Krill feeding on the dominant phytoplankton organisms in all experiments. Adult females of *M. gerlachei* had the highest values, apart from two exceptions in which the copepodids of *Metridia* registered higher ones. Figure 7 summarizes the results of all experiments. Filtration rates and the log ingestion rates are plotted against the log mean C concentration for all copepods (except *M. curticauda*) and for the smallest size-class of *E. superba*. Each point represents a single value for a single phytoplankton species. The November–December 1980 and the February 1982 data for *E. superba* were calculated separately. For all animals, a linear relationship between the log ingestion rate and the log mean C concentration was significant (0.1% level). A linear relationship between the filtration rate and the log mean C concentration was significant (5% level) for *M. gerlachei, R. gigas, C. acutus* and *E. superba*.

The weight specific ingestion rate is plotted against body C in Fig. 8. The variation between the ingestion rates obtained from the experiments with lower and higher food concentrations differ between the animals. The lowest variation occurred in *M. gerlachei* and *C. propinquus. Metridia curticauda* and the two larger size-classes of *E. superba* were not taken into account, because only two values were available. There is a decrease in weight-specific ingestion rate with increasing body weight. However, the large *M. gerlachei* showed higher values than the smaller *C. acutus* and, in all experiments carried out in November–December 1980, *E. superba* (smallest size-class) had a higher weight-specific ingestion rate than the much smaller *R. gigas*. This is contrary to the results of the experiments conducted in Febru-

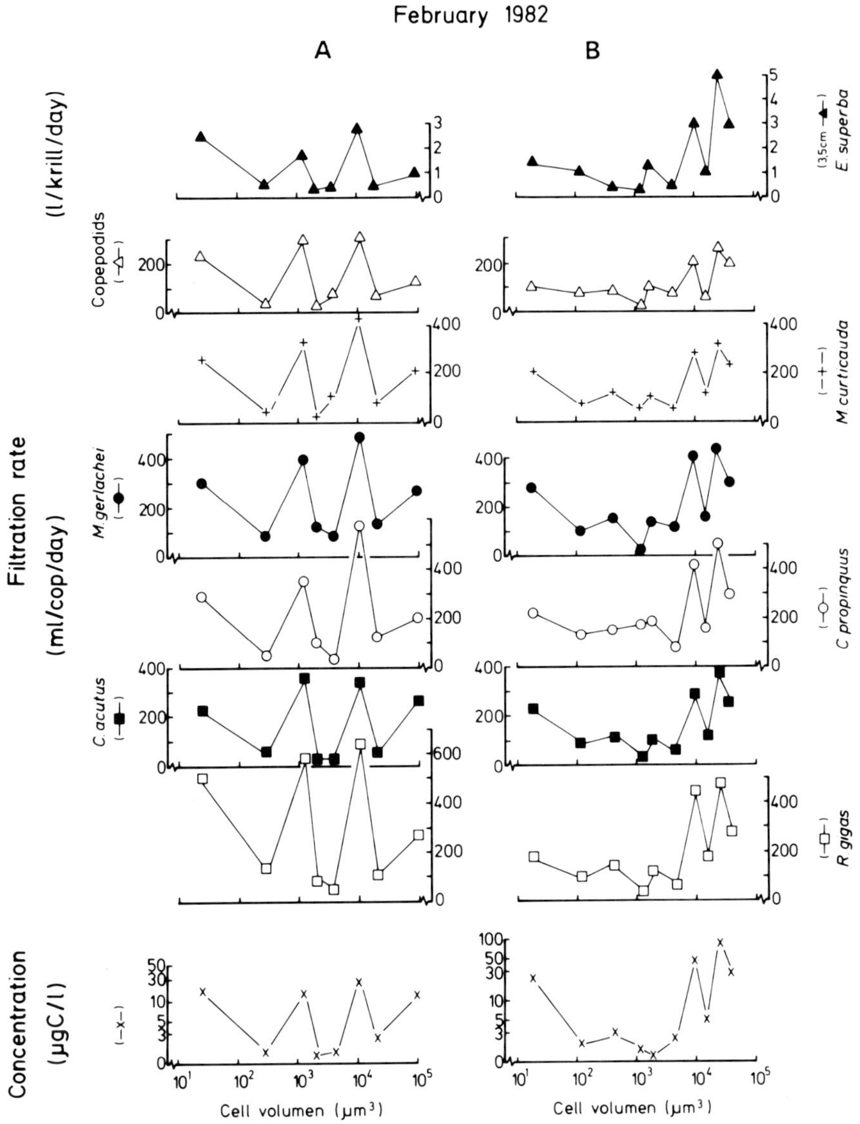

Fig. 5A, B. Filtration rates of *E. superba* and copepods in relation to two concentrations of phytoplankton (μ-flagellates, *Chaetoceros* spp., *Corethron criophilum, Biddulphia* spp.) food supply. Graphs on left-hand side **A** reflect an ambient food supply (61 μg C l^{-1}) and those on left-hand side **B** an enriched food supply (173 μg C l^{-1})

ary 1982. Total ingestion rate and assimilation efficiency are summarized for all experiments in Table 3. Apart from copepodids, *M. gerlachei* and (in February 1982) *M. curticauda* had higher efficiencies than the other copepod species.

4 Discussion

It has been stated often that there is a relationship between the body size of an animal and the size of its food; large animals feeding on large particles, small animals feeding on small particles (e.g., Fryer 1957; Parsons and LeBrasseur 1970). Size differences between similar animal species are supposed to be necessary for co-existence. However, according to Maly and Maly (1974), in certain freshwater species extreme size differences are probably unimportant compared to differences in morphology of feeding appendages and in behavioural patterns. For marine environments, Gamble (1978) and Harris (1982) described rather similar particle-size feeding by copepods differing considerably in size (ca. 10–20 times, in terms of body-weight). The large copepods fed more efficiently on large particles than the smaller copepods, but only by a factor of 1.2 times (in terms of the diameter of the food cells). Cowles (1979) found that four copepod species, which varied in body-weight by a factor of 3.5, consumed a wide range of particles, but he noted differences between the species in relation to food levels. At high food concentrations, all the copepods fed on those particles which were most abundant or selectively ingested the largest particles. This disproportionate feeding decreased with decreasing food abundance, which corresponds to optimal foraging theory (i.e. diet becomes less specialized with decreasing food abundance).

In my experiments, the differences in body weight between ecologically similar copepod species ranged between 89 and 505 μg C per individual in November–December 1980 and between 59 and 392 μg C per individual in Feb-

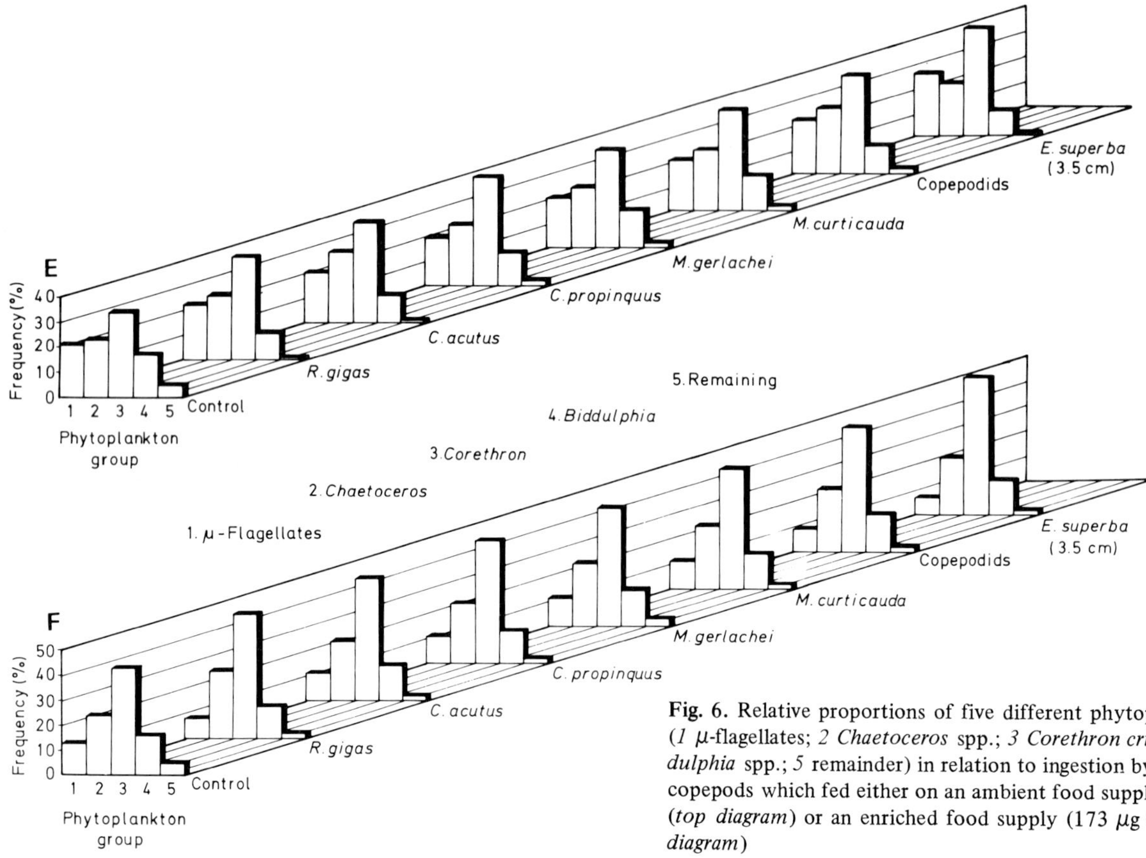

Fig. 6. Relative proportions of five different phytoplankton groups (*1* μ-flagellates; *2 Chaetoceros* spp.; *3 Corethron criophilum; 4 Biddulphia* spp.; *5* remainder) in relation to ingestion by *E. superba* and copepods which fed either on an ambient food supply (61 μg C l^{-1}) (*top diagram*) or an enriched food supply (173 μg C l^{-1}) (*bottom diagram*)

ruary 1982. These ratios (5.7 and 6.6) are in between the values given by Gamble (1978), Harris (1982) and Cowles (1979). I found no evidence for large copepod species feeding preferentially on relatively large particles. All animals fed most efficiently when food was most abundant, irrespective of size and shape of the food particles. This kind of feeding behaviour is described as opportunistic by Poulet and Chanut (1975), and my results are in agreement with those of Poulet (1978) who reported opportunistic feeding by five small copepod species. Similar behaviour has been described by Richman et al. (1977) for three copepod species which, however, showed a capacity for feeding selectively on large particles. Similar results were obtained by Schnack (1983) for *M. gerlachei* and *C. propinquus,* and it was shown that *M. gerlachei* handled large particles more efficiently than *C. propinquus.*

In the experiments reported here, *C. acutus* and *R. gigas* filtered food at higher rates when feeding on μ-flagellates than on the biomass peak made up of "centric diatoms" (experiment A). In the zone close to the ice-edge, where "centric diatoms" predominated (Schnack et al., to be publ.), *C. acutus* and *R. gigas* were still mainly in an overwintering condition (i.e. drifting in the water). However, only actively swimming animals were taken for the experimental work. Nevertheless, one explanation for the relatively low feeding activities of these two copepod species in experiment A (ambient food supply) may be that, although the animals were swimming, their feeding activities were still somewhat slower than those of the animals used in experiment B (enriched food supply). In experiment C, in which mainly small cells of *C. criophilum* dominated, all copepod species fed with highest rates on *C. criophilum,* followed by the largest food particles ("centric diatoms") which made up only 5% of the initial food concentration in terms of C. These copepods had a positive selectivity index for feeding on *C. criophilum* and "centric diatoms", which was not found in experiment D in which the concentration of food was about 2.5 times higher. In experiment D, the copepods had a positive selectivity index on *C. criophilum* but not for *Chaetoceros* spp. A great difference between these two experiments was in the mean size of the *C. criophilum* cells. In the unconcentrated experiment (C) the mean size of *C. criophilum* was about 10 × 90 μm, whereas in the concentrated experiment (D) it was 15 × 185 μm. No great difference was found for the lengths of the chains of "centric diatoms"; they varied between 130–170 μm in both experiments. This clearly demonstrated a length difference between *C. criophilum* and the "centric diatoms" in experiment C in which the copepods selectively fed on "centric diatoms". No such size difference occurred in experiment D, in which copepods fed most efficiently only on the most abundant particle (*C. criophilum*). No evidence was found for higher efficiency in feeding of larger particles with increasing food abundance. The most abundant, large, calanoid copepod species are trophically similar, and competition for food is possibly severe. *Rhincala-*

Table 2. Weight-specific filtration rates (ml μg C^{-1} d^{-1}) for *E. superba* of three size classes (1 = ~30 mm, 2 = ~40 mm, 3 = ~50 mm) and 5 copepod species (R.g. = *R. gigas*, C.a. = *C. acutus*, C.p. = *C. propinquus*, M.g. = *M. gerlachei*, M.c. = *M. curticauda*, Cop. = Copepodids) feeding on dominant phytoplankton groups in various concentrations (μg C l^{-1})

Main phytoplankton organisms	Cell size (µm)	Initial concentration	(%)	Filtration rate R.g.	C.a.	C.p.	M.g.	M.c.	Cop.	*E. superba* 1	2	3
"Centric diatoms"	37	61.9	(80.1)	0.19	0.47	3.30	4.54	–	1.99	1.00	0.23	0.25
µ-flagellates	3	6.9	(8.9)	0.14	0.49	0.90	0.93	–	0.81	0.45	0.09	0.09
Phaeocyst. swarmers	5	4.7	(6.1)	0.32	0.56	1.34	1.50	–	1.56	0.73	0.13	0.08
	Total:	77.3		0.19	0.47	2.56	3.30	–	1.76	0.87	0.20	0.21
"Centric diatoms"	37	146.8	(77.8)	0.84	2.43	2.86	3.73	–	2.52	1.88	0.56	0.61
µ-flagellates	3	13.0	(6.9)	0.44	0.61	1.92	2.29	–	2.64	1.19	0.26	0.22
Nitzschia	6x75	12.4	(6.6)	0.19	0.47	1.78	1.92	–	1.18	1.67	0.41	0.39
	Total:	188.7		0.68	1.82	2.35	3.00	–	2.22	1.66	0.45	0.48
Corethron	10x90	20.1	(27.8)	1.20	3.07	3.18	3.59	–	6.05	1.97	–	–
µ-flagellates	3	17.9	(24.8)	0.42	1.27	1.23	1.75	–	1.96	1.12	–	–
Dinoflagellates	30x50	11.2	(15.5)	0.28	0.54	1.11	1.60	–	1.20	0.75	–	–
Nitzschia	3x75	10.0	(13.8)	0.22	0.30	1.42	2.47	–	2.50	1.20	–	–
D. speculum	25	4.4	(6.1)	0.34	0.83	1.50	2.02	–	2.71	1.28	–	–
Chaetoceros	7–10	3.6	(5.0)	0.10	0.35	1.27	1.80	–	1.48	0.92	–	–
"centric diatoms"	37	3.6	(5.0)	0.63	1.96	2.69	3.15	–	4.49	0.76	–	–
	Total:	72.3		0.48	1.18	1.64	2.25	–	2.62	1.20	–	–
Corethron	15x185	119.2	(67.3)	1.08	2.38	3.47	4.76	–	6.02	2.66	–	–
D. speculum	25	13.0	(7.3)	0.30	0.62	1.25	2.72	–	3.08	1.30	–	–
Nitzschia	3x45	11.6	(6.5)	0.44	0.44	1.81	2.24	–	2.66	1.22	–	–
"centric diatoms"	37	8.9	(5.0)	0.52	1.32	2.53	3.10	–	3.75	1.55	–	–
	Total:	177.1		0.89	1.61	2.67	3.41	–	4.18	2.02	–	–
Corethron	15x150	20.8	(34.3)	1.63	3.16	3.05	4.81	4.39	5.06	0.23	–	–
Chaetoceros	16x4	14.0	(23.1)	1.48	3.29	1.87	3.95	3.36	5.11	0.15	–	–
µ-flagellates	3–7	12.8	(21.1)	1.26	2.04	1.53	2.97	2.61	3.87	0.21	–	–
Biddulphia	50x90	10.4	(17.2)	0.66	1.23	1.08	2.68	2.10	2.02	0.08	–	–
	Total:	60.6		1.23	2.33	1.78	3.46	2.97	3.82	0.16	–	–
Corethron	20x150	75.9	(43.8)	1.21	3.42	2.67	4.32	3.27	4.20	0.42	–	–
Chaetoceros	20–30	41.5	(23.9)	1.11	2.64	2.18	4.02	2.82	3.40	0.25	–	–
Biddulphia	27x75	27.2	(15.7)	0.69	2.32	1.55	2.94	2.24	3.01	0.24	–	–
µ-flagellates	3–7	22.3	(12.9)	0.42	2.09	1.69	2.79	1.99	1.78	0.11	–	–
	Total:	173.3		0.90	2.67	2.07	3.56	2.62	3.22	0.27	–	–

nus gigas, *C. acutus* and *C. propinquus* are filter-feeding herbivores, whereas *M. gerlachei* and *M. curticauda* are omnivorous feeders (Schnack, unpubl.). Niche separation based on feeding alone does not seem to apply, due to considerable overlapping in feeding habits. Possible mechanisms through which zooplankton populations may avoid or reduce interspecific competition include, besides differences in selective feeding, different spatial and temporal distributions connected with different over-wintering strategies. Differences in vertical distribution are obvious: *M. gerlachei* occurred mainly in the deeper layer, whereas *C. propinquus* and *R. gigas* were more abundant in the upper layers (Schnack, unpubl.).

I do not want to dwell on the quantitative aspects of the results obtained for *E. superba*. The constant-volume technique that I used has been described by Morris (to be publ.) as being inappropriate for an active, swimming animal such as *E. superba*. However, Morris (to be publ.) pointed out that "the size frequency distribution of available as opposed to ingested cells of the different phytoplankton are similar using both techniques, the constant volume one and the flowthrough one".

No great difference could be found between the feeding behaviour of *E. superba* and the copepods. However, there is a great difference between the distributional patterns of copepods and Krill. Adults and sub-adults of *E. superba* occur in great swarms. Virtually no copepods occurred in regions where large swarms of *E. superba* were found.

In conclusion, both *E. superba* and the dominant copepods have very similar patterns of feeding behaviour, and severe competition for phytoplankton food in the Southern Ocean might occur. Niche separation on the basis of differences in distribution during the life histories of Krill and copepods could be a mechanism for reducing potential

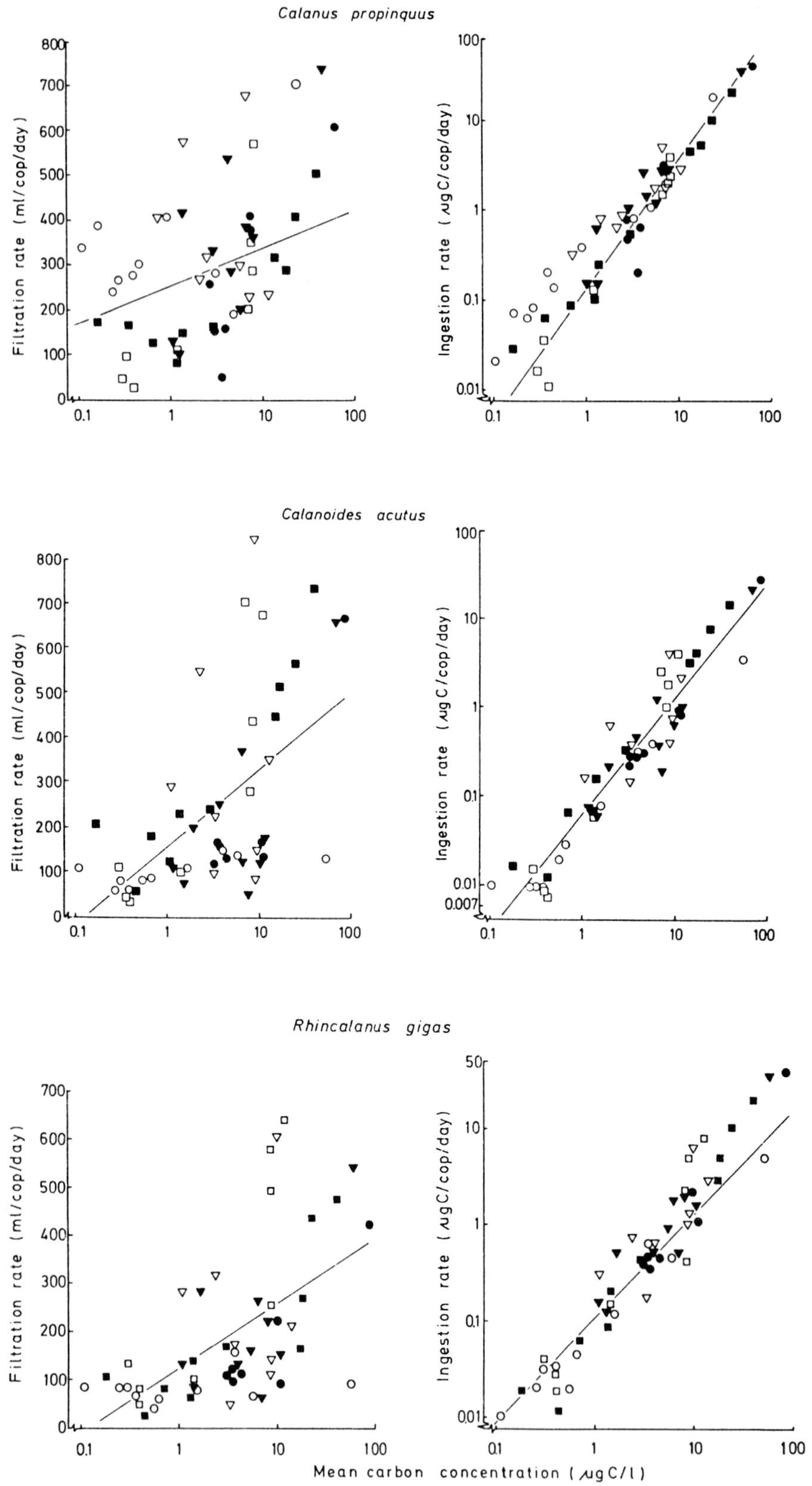

Calanus propinquus
Filtration rate (ml/cop/day)
Ingestion rate (µgC/cop/day)
Calanoides acutus
Filtration rate (ml/cop/day)
Ingestion rate (µgC/cop/day)
Rhincalanus gigas
Filtration rate (ml/cop/day)
Ingestion rate (µgC/cop/day)
Mean carbon concentration (µgC/l)

Fig. 7 (Part I)

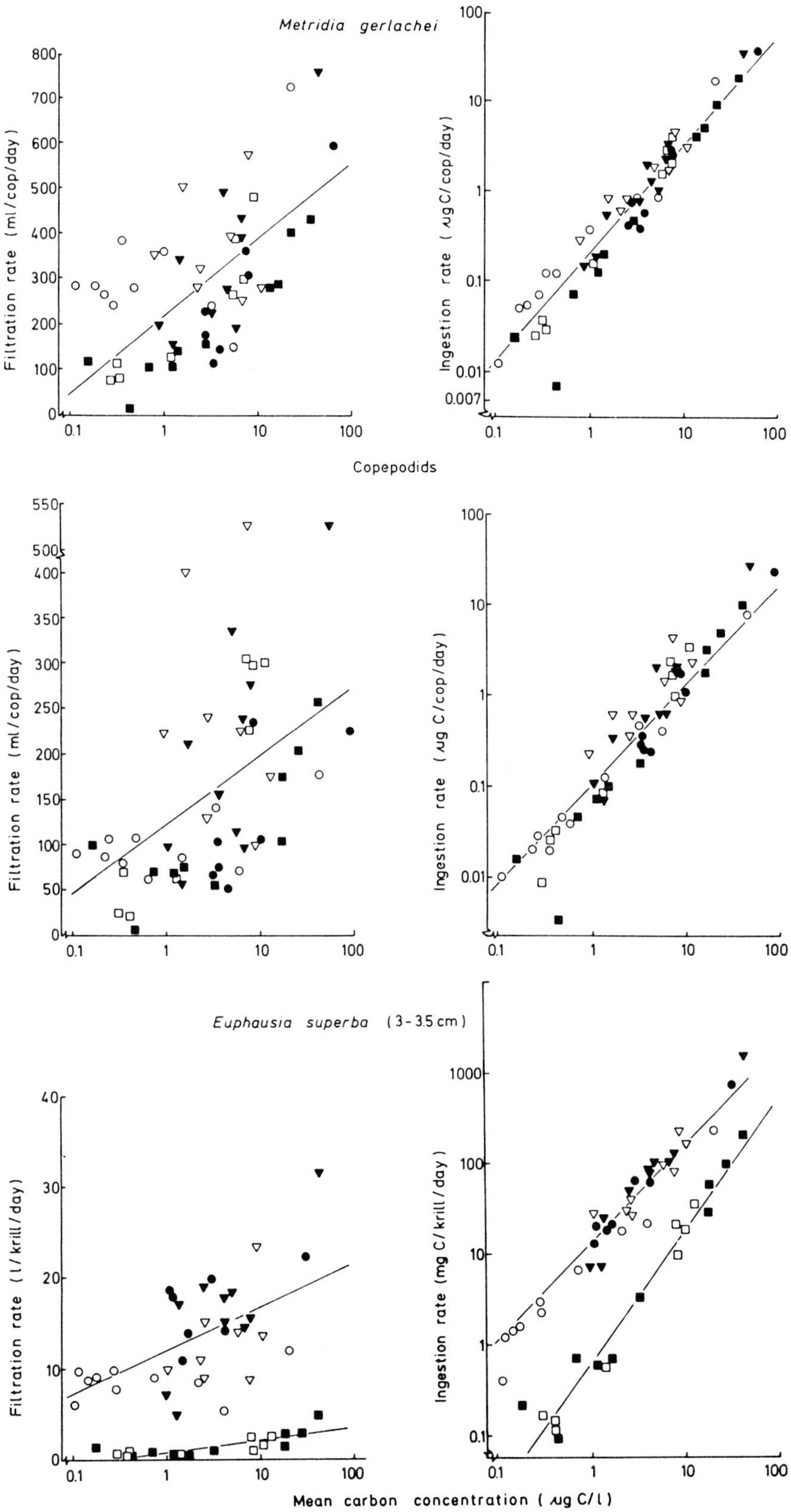

Fig. 7. Filtration and ingestion rates of *E. superba* and copepods in relation to the concentration of particles in the food supply. See Fig. 8 for explanation of symbols

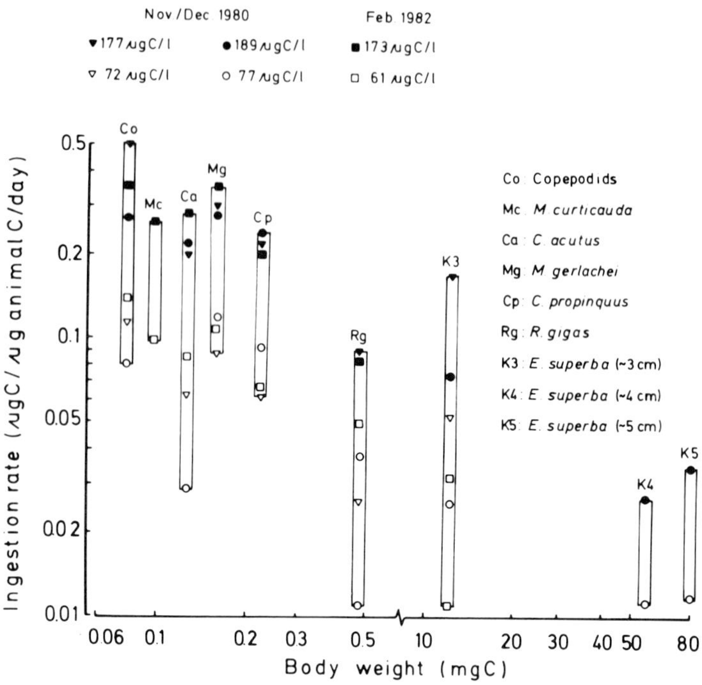

Fig. 8. Ingestion rates of food particles by *E. superba* and copepods in relation to body-weight

competition. Further behavioural studies are needed, which should be focused on comparisons between copepods (including small species) and juvenile Krill, rather than on swarming sub-adults and adults. In addition, work is required on the different over-wintering strategies of the animals in relation to their reproductive cycles and growth.

Acknowledgements. Thanks are due to Claudia Dieckmann, Sigrid Marschall and Elke Mizdalski for technical assistance, and to Dr. Victor Smetacek for reading the manuscript.

References

Antezana T; Ray K; Melo C (1982) Trophic behavior of *Euphausia superba* Dana in laboratory conditions. Polar Biol 1:77–82

Boyd CM; Heyraud M; Boyd CN (to be published) Feeding behavior of Antarctic krill, *Euphausia superba*. J Crust Biol

Conover RJ (1966) Assimilation of organic matter by zooplankton. Limnol Oceanogr 11:338–345

Cowles TJ (1979) The feeding response of copepods from the Peru upwelling system: Food size selection. J Mar Res 37:601–622

Frost BW (1972) Effects of size and concentration of food particles on the feeding behavior of the marine planktonic copepod *Calanus pacificus*. Limnol Oceanogr 17:805–815

Fryer G (1957) The food of some fresh-water cyclopoid copepods and its ecological significance. J Animal Ecol 26:263–286

Gamble JC (1978) Copepod grazing during a declining spring phytoplankton bloom in the northern North Sea. Mar Biol 49:303–315

Harris RP (1982) Comparison of the feeding behaviour of *Calanus* and *Pseudocalanus* in two experimentally manipulated enclosed ecosystems. J Mar Biol Assoc UK 62:71–91

Jacobs J (1974) Quantitative measurements of food selection. Oecologia (Berlin) 14:413–417

Kato M; Murano M; Segawa S (1979) Estimation of the filtering rate of the Antarctic krill under laboratory conditions. Trans Tokyo Univ Fish 3:107–112

Kato M; Segawa S; Tanoue E; Murano M (1982) Filtering and ingestion rates of the Antarctic krill *Euphausia superba* Dana. Trans Tokyo Univ Fish 5:167–175

Lenz J (1974) Untersuchungen zum Nahrungsgefüge im Pelagial der Kieler Bucht, der Gehalt an Phytoplankton, Zooplankton und or-

Table 3. Total ingestion rate (I, µg C ind. d^{-1}) and assimilation efficiency (A, %) for *E. superba* and five copepod species which fed on two food supplies ("centric diatoms" and *"Corethron"*) at different concentrations (77 and 189, 72 and 177, 61 and 173 µg C l^{-1}, respectively) in November–December 1980 and February 1982

	November–December 1980								February 1982			
	"Centric diatoms"				"Corethron"				"Corethron"			
	77		189		72		177		61		173	
	I	A	I	A	I	A	I	A	I	A	I	A
Rhincalanus gigas	6.4	68.5	43.6	77.9	13.0	67.9	46.2	49.2	19.8	73.7	37.7	61.6
Calanoides acutus	4.5	64.8	31.1	71.1	8.5	59.5	27.4	61.2	9.4	78.1	30.2	67.7
Calanus propinquus	20.2	79.1	49.0	76.0	13.9	71.8	49.3	66.2	11.8	80.6	40.3	79.0
Metridia gerlachei	19.1	88.0	45.8	82.7	14.0	78.4	45.8	70.2	10.9	86.5	35.6	80.0
Metridia curticauda	–	–	–	–	–	–	–	–	9.8	84.7	25.7	79.2
Copepodids	8.4	83.4	25.8	76.9	10.4	70.8	35.2	70.4	8.2	80.2	20.7	81.7
Euphausia superba (30 mm–35 mm)	311.1	75.8	926.9	73.3	638.0	64.9	2054.6	75.6	82.3	58.2	380.6	49.8
Euphausia superba (40 mm–45 mm)	418.1	80.1	1367.7	70.7	–	–	–	–	–	–	–	–
Euphausia superba (50 mm–55 mm)	736.1	51.9	2600.2	42.2	–	–	–	–	–	–	–	–

ganischem Detritus in Abhängigkeit von Wasserschichtung, Tiefe und Jahreszeit. Habilitationsschrift, Universität Kiel

Mackintosh NA (1934) Distribution of the macroplankton in the Atlantic Sector of the Antarctic. Discovery Rep 9:65–160

Maly EJ; Maly MP (1974) Dietary differences between two cooccurring calanoid copepod species. Oecologia (Berlin) 17:325–333

Meyer MA; El-Sayed SZ (1983) Grazing of *Euphausia superba* Dana on natural phytoplankton populations. Polar Biol 1:193–197

Morris DJ (to be published) Filtration rates of *Euphausia superba* Dana: Under or over estimates? J Crust Biol

Morris DJ; Ward P; Clarke A (1983) Some aspects of feeding in the Antartic krill, *Euphausia superba.* Polar Biol 2:21–26

Poulet SA (1978) Comparison between five coexisting species of marine copepods feeding on naturally occurring particulate matter. Limnol Oceanogr 23:1126–1143

Poulet SA; Chanut JP (1975) Nonselective feeding of *Pseudocalanus.* J Fsh Res Board Can 32:706–713

Richman S; Heinle DR; Huff R (1977) Grazing by adult estuarine calanoid copepods of the Chesapeake Bay. Mar Biol 42:69–84

Schnack SB (1983) Feeding of two Antarctic copepod species (*Calanus propinquus* and *Metridia gerlachei*) on a mixture of centric diatoms. Polar Biol 2:63–68

Schnack SB; Smetacek V; v. Bodungen B; Stegmann P (to be published) Utilization of phytoplankton by copepods in Antarctic waters during spring. In: Gray S, Christiansen ME (eds) 18th EMBS Marine Biology of Polar Regions and Effects of Stresseson Marine Organisms. Wiley, Chichester.

Strathmann RR (1967) Estimating the organic carbon content of plankton from cell volume and plasma volume. Limnol Oceanogr 12:411–418

Metabolism of Antarctic Krill, *Euphausia superba*, and Its Tropho-Dynamic Implications

R.Y. GEORGE[1]

Summary. The paper focuses on the nitrogenous excretion of various ontogenetic stages of the Antarctic Krill, *Euphausia superba*, with particular emphasis on contributions of ammonia to the ambient marine environment. As a reflection of a high feeding rate and elevated metabolism, Krill exhibit enhanced catabolic activity under favourable feeding conditions. Within various stages of Krill, ammonia excretion rate averages >260 μg NH_3 g^{-1} (dw) h^{-1}. Consequently, a very large quantity of ammonia is released to the ambient environment from Krill swarms in which density can attain 1000 individuals m^3. Swarms of significant magnitude, such as the great patches of Krill commonly encountered northwest of Elephant Island and the vast aggregations of Krill in the neritic waters of the Bransfield Strait in the Antarctic Peninsula region, may exert a profound impact on biological processes. The tropho-dynamic implications of the production of ammonia by Krill are discussed in terms of an hypothesis dealing with the positive feed-back of ammonia production by zooplankton to the Antarctic marine ecosystem.

1 Introduction

As a reflection of its ecological maturity, stability and age, the Antarctic marine ecosystem, cold since the Miocene glaciation, supports a considerable diversity of life both in the water column and over the sea floor (George 1977). A high biological productivity in the region is induced primarily by the upwelling of water (El-Sayed 1968) and, therefore, there have been several studies of primary production and the factors controlling phytoplankton growth in the Southern Ocean (see Holm-Hansen et al. 1977).

The bulk of the biomass in the Antarctic water column is often attributed to swarms of Krill, *Euphausia superba*, (Macaulay et al. 1984) which feed on diatoms to sustain high metabolic activity and growth during the Antarctic summer. Our knowledge of the life cycle of Krill, and the distribution of its larval and adult stages, is based primarily on studies made during the Discovery investigations (Marr 1962). In recent years the results of experimental studies, dealing with metabolic activities of Krill in laboratories, have shown that the animals are able to swim at considerable speeds, ranging from 100 to 400 mm s^{-1} (Kils 1979; Hamner 1984). Krill utilize 30% of their metabolic energy in maintaining a high feeding activity that calls for filtering 450 ml of water h^{-1} for an animal of 120 mm dry wt. (Boyd et al. 1984; Morris 1984), and a Krill swarm at a density of 270 mg dry wt. m^{-3} requires 2.4–5.4 mg C mg^{-3} d^{-1} for sustaining maintenance metabolism (Holm-Hansen and Huntley 1984). Recent data also suggest that Krill have relatively high respiratory and excretory rates (Ikeda 1984; George 1984).

Biggs (1982) emphasized the importance of ammonium dynamics, as a result of excretion by several Antarctic zooplankton species, including *E. superba*, in the Ross Sea. Recently, Hirsche (1983) reported on the excretion of ammonia by *E. superba* in the Scotia and Weddell seas, and George and Fields (1984) examined ammonia excretion by *E. superba* in the neritic waters of the Bransfield system of the Antarctic Peninsula. The results of these studies showed significant differences in flux rates between populations of Krill from different geographical regions. Johnson et al. (1984) monitored in situ ammonia excretion by Krill aggregations in the vicinity of Elephant Island, and arrived at some fundamental conclusions concerning the residence time of Krill schools in relation to in situ levels of oxygen and ammonia. In this report, I evaluate ammonia excretion for various ontogenetic stages of Krill, and extend measurements for computing the production of ammonia by Krill patches and swarms of known composition. The tropho-dynamic implications of this ammonia flux are discussed in terms of an hypothesis dealing with the positive feed-back of ammonia production by zooplankton to the Antarctic marine ecosystem.

2 Materials and Methods

Ammonia excretion measurements in different stages of Krill were made in the laboratory at Palmer Research Station during the austral summers of 1981 and 1982. The experimental Krill were captured during cruises of R.V. Hero in February and March in the following four regions: Admiralty Bay (east of King George Island); Charlotte Bay (on the eastern side of the Antarctic Peninsula); Bransfield Strait (between Deception Island and the Antarctic Peninsula); and, Dallman Bay (north of Anvers Island).

1 Institute for Marine Biomedical Research, University of North Carolina at Wilmington, Wilmington, NC, USA

Antarctic Nutrient Cycles and Food Webs (ed. by W.R. Siegfried, P.R. Condy, and R.M. Laws)

Krill patches and large swarms were surveyed with an acoustic recorder (50 kHz). All Krill patches were traced individually, and their dimensions were calculated. A standardization scheme was developed to estimate Krill biomass in each patch, based on speed of the ship and the 3-dimensional extent of the patch. The total Krill biomass of each patch was calculated for each of the four regions. Krill swarms in each patch were also sampled by IKMT trawl. The proportional representation of the various ontogenetic stages of Krill in the trawl samples for each of the four regions was determined and the data are used here to estimate the production of ammonia by the various Krill swarms.

In the laboratory, the ammonia excretion rate was determined for Krill of different stages in experiments lasting 4–6 h. In earlier studies, ammonia measurement was made mostly by the standard titration technique with phenol-hypochlorite (Biggs 1982). However, an Orion ammonia probe, in conjunction with a Corning pH meter, was used by Nelson et al. (1977) to obtain accurate measurements of ammonia production in palaemonid shrimps. Recently, George and Fields (1984) used an HNU probe, in conjunction with a Corning pH-ion meter 135, in studying ammonia production by Krill. This procedure facilitated continuous monitoring of ammonia excretion, with the same sensitivity as the titration method. The same procedure was used in the present study. Water samples of 10 ml were drawn from the experimental chamber, and the pH was adjusted to 13 by adding concentrated sodium hydroxide solution. The NH_4 (ammonium) in the water sample is converted to NH_3 (ammonia), as a consequence of the reverse reaction. Subsequently the ammonia content (in mol l^{-1}) was read directly on the monitor and then converted to μg ammonia.

3 Results and Discussion

3.1 Ammonia Excretion under Starvation and Satiation

The rate of excretion of ammonia in Krill was consistently high, particularly when well-fed animals, recently captured from the field, were used. In well-fed animals, the excretion rate ranged between 200 and 340 μg NH_3 g^{-1} (dry wt.) h^{-1} (Fig. 1). Post-spawned Krill exhibit the lowest level of ammonia excretion. This may reflect a depressed level of metabolism associated with the initiation of a "degrowth" process.

The high ammonia excretion rate reported here is almost an order of magnitude higher than the values given by Biggs (1982) and Hirsche (1983). This obvious difference between populations of *E. superba* calls for explanation. Biggs' data are based on specimens of *E. superba* taken in the Ross Sea, particularly in oceanic areas, and Hirsche's data originate from specimens captured in oceanic areas of the Scotia and Weddell seas. Neritic populations of *E. superba* were

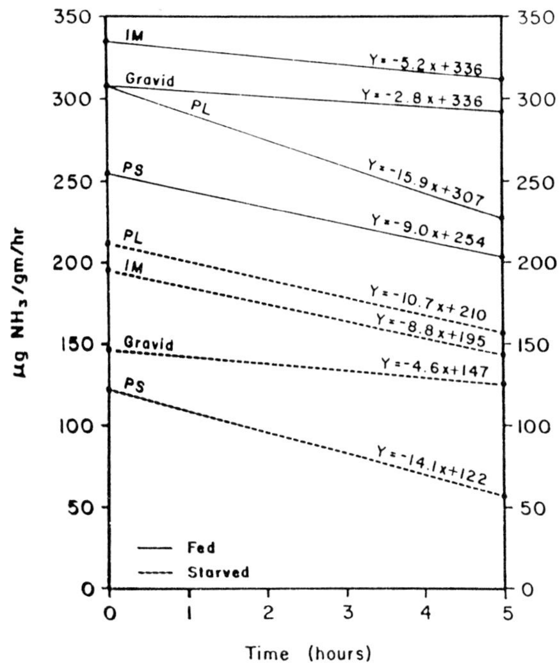

Fig. 1. Ammonia excretion rate in *Euphausia superba* at 2°C. *Broken line* represents Krill maintained without daily phytoplankton diet for 10 d; *solid line* represents freshly fed Krill. *PL* Post-larval Krill (15 mm–20 mm in length); *IM* Immature Krill (25 mm–30 mm in length; *PS* Post-spawned Krill. (> 40 mm in length)

used in my study, and the relatively high rate of ammonia excretion may also be attributed to a rich diatom community consisting of *Sysnedra spatula* and *Fragilariopsis antarctica* in the area from which the Krill were taken. Furthermore, a high respiration rate was found in these neritic Krill which had fed on fresh diatoms (George 1984). Hence, it appears that a high respiration rate is coupled with an increased rate of ammonia excretion. Under well-fed conditions, Krill seem to exhibit an enhanced metabolic rate. During swarming events, Krill tend to consume diatoms. The presence of *S. spatula* and *F. antarctica* in the guts of the experimental Krill suggests that these two species were consumed by the Krill while swarming in the Bransfield Strait. This observation agrees with the recent report of Antezana and Ray (1984) that Krill feed actively during swarming. Food rationing, for as long as 10 days substantially reduced the ammonia excretion rate of Krill (in all stages) in a through-flow tank by at least 1 or 2 times (Fig. 1). Table 1 shows that differences occurred in the amount of ammonia produced by the various ontogenetic stages (e.g. 1 mg post-larval Krill produces 3.66 mg NH_3^+ m^3 h^{-1} at density of ca. 1000 individuals m^3, and a gravid female produces 58.12 mg NH_3^+ m^3 h^{-1} at the same density).

Despite the fact that zooplankton in the Antarctic Ocean constitute an important element in the water column, with at least 180 species, our knowledge of the metabolic turnover rates of these species is very limited. Ikeda and Fay 1984 (in proofs) kept live representatives of 8 species of antarctic zooplankton in the laboratory, in tropical Austra-

Table 1. Production of ammonia by individual Krill of different ontogenetic stages and in swarms of 1000 individuals per m^3

	Weight of Krill		Individual production	Production in swarms
Krill stage	Wet (g)	Dry (g)	($\mu g\ NH_3\ g^{-1}$ dry wt.)	($\mu g\ NH_3\ m^3\ h^{-1}$)
Post-larva	0.1141	0.0183	200	3.66
Immature	0.7620	0.1335	225	30.04
Gravid	1.4627	0.2835	205	58.12
Post-spawned	1.2376	0.2425	175	42.44

lia, and their data on ammonia excretion rate are somewhat lower than mine for *E. superba;* the difference possibly being due to differences in culture techniques. The impact of temperature on ammonia excretion in *E. superba* revealed a Q_{10} value of 2.58 between 1.8° and 3.66°C (Ikeda and Fay 1984 (in proofs). George (1984) found that hydrostatic pressure at 25 atm doubled the ammonia excretion in juvenile Krill; adult Krill showing only a slight increase in excretion rate. This new information emphasizes that temperature and pressure cannot be ruled out in arriving at accurate estimates of ammonia flux from Krill excretion.

3.2. Ammonia Production by Krill Swarms

The proportional representation of various Krill stages in a patch was determined from net samples. The amount of ammonia released in 1 h from each patch was calculated from values obtained for the different ontogenetic stages in the laboratory, as reported in the preceding section.

The estimated amounts of ammonia flushed into the ocean from Krill patches are shown in Tables 2–5. The data suggest that, within a swarm, there is a constant release of ammonia into the medium. The maximum contribution in Admiralty Bay (Table 2) was made largely by sub-adult Krill, responsible for approximately 50% of the ammonia production. Post-larval Krill produce slightly more ammonia than immature animals. In Charlotte Bay (Table 3), however, the bulk of the ammonia was produced by immature Krill which constituted the main contributor in Dallman Bay (Table 4) as well. The Bransfield Strait is known as an area in which mixed populations of Krill from different spawning grounds, occur; 90% of the population sampled belonging to 1 yr and 2 yr classes of immature indivi-

Table 2. Production of ammonia (g NH_3) by three different stages of Krill in Admiralty Bay

	Dimensions of patch			Production of Krill stages		
Krill patch code	Length (m)	Depth (m)	Width (m)	Post-Larva	Immature	Sub-Adult
A_1	736	12	10	226.4	530.8	265.6
A_2	360	20	10	184.4	432.4	216.4
A_3	352	20	10	180.4	422.8	211.6
A_4	272	12	10	83.6	196.0	98.0
A_5	160	12	10	49.2	115.2	57.6
A_6	160	12	10	49.2	115.2	57.6
A_7	128	12	10	36.0	92.4	46.4
A_8	128	12	10	36.0	92.4	46.4
A_9	252	12	10	77.6	181.6	90.8
A_{10}	920	20	10	471.6	1105.6	552.8
A_{11}	420	20	10	214.4	504.8	252.4
A_{12}	184	12	10	56.4	132.8	66.4
Totals				1672.4	1392.2	1966.0

Table 3. Production of ammonia (g NH_3) by three different stages of Krill in patches in Charlotte Bay

	Dimensions of patch			Production of Krill stages		
Krill patch code	Length (m)	Depth (m)	Width (m)	Post-Larva	Immature	Sub-Adult
B_1	748	16	10	2.190	270.0	72.0
B_2	720	20	10	2.782	342.0	91.0
B_3	365	28	10	0.510	230.0	61.0
B_4	748	20	10	0.748	337.0	90.0
B_5	374	12	10	0.024	112.0	29.0
B_6	654	16	10	0.523	230.0	63.0
B_7	800	4	10	0.060	27.0	7.0
Totals				6.777	1548.0	413.0

Table 4. Production of ammonia (g NH_3) by three different stages of Krill in patches in Dallman Bay

	Dimensions of patch			Production of Krill stages		
Krill patch code	Length (m)	Height (m)	Width (m)	Post-Larva	Immature	Sub-Adult
C_1	1800	19.7	10	129.7	852.1	106.5
C_2	4300	31.5	10	495.7	3255.1	406.9
Totals				625.4	4107.2	513.4

duals. Post-larvae and sub-adults contributed only 10% of the ammonia flux in this area (Table 5).

If the total amount of ammonia produced is calculated on the basis of size of Krill patch and the contributions of the various stages of Krill, it emerges that in Dallman Bay the flux was 46 kg ammonia h^{-1}. The Admiralty Bay patch was larger, but the total production was only 7.5 kg ammonia h^{-1} (Table 6). The ammonia flux was 15.4 kg h^{-1} in the Bransfield Strait, and lowest (2.0 kg h^{-1}) in Charlotte Bay. The proportional flux of ammonia from the patches in the four regions, in relation to three ontogenetic stages of Krill, is given in Fig. 2. It is clear that the production of ammonia by Krill in Dallman Bay is potentially of considerable ecological consequence. However, we do not know the residence time of ammonia within a Krill swarm, and its tropho-dynamic and physiological implications.

Johnson et al. (1984) arrived at some generalizations concerning in situ ammonia profiles, and they believe that the residence time of a Krill swarm is governed largely by oxygen content and ammonia levels. I have data (unpubl.) to show that the production of ammonia is significant in this context, particularly if the Krill swarm is large. The large swarms reported off Elephant Island by Macaulay et al. (1984) could be responsible for a substantial part of the ammonia budget in the Antarctic ecosystem. From a tropho-dynamic point of view, the immediate consequence of ammonia regeneration through zooplankton is its availability to members of the diatom community on which Krill feeds. Ammonia can constitute a principal source of N if its concentration is sustained at μg at. l^{-1}, in regions of Krill swarms. Eppley et al. (1969) found a clear trend in the growth of large cells of *Coccolithus huxleyi,* and other neritic diatoms, in optimum conditions of ammonia. Maximum growth was about twice that of other species at a concentration of 0.5 μg at NH_4^+ l^{-1}. Such growth studies should be carried out in situ in Antarctic waters where Krill swarms are prevalent.

Dugdale and Goering (1967) have demonstrated that phytoplankton take up ammonia more rapidly than nitrate. It is also possible that conversion of ammonia to nitrite and nitrate may take place by a series of reversible reactions which may be aided by bacteria. In the light of these considerations, more carefully designed studies are needed to delineate the pathway of ammonia regeneration by Krill swarms in the N cycle of the Antarctic ecosystem. In a recent investigation, Biggs (1982) implied that ammonium (NH_4^+) is evidently a favoured N source for the growth of antarctic phytoplankton. However, he attributed only 2%

Table 5. Production of ammonia (g NH_3) by three different stages of Krill in patches in Bransfield Strait

	Dimensions of patch			Production of Krill stages		
Krill patch code	Length (m)	Height (m)	Width (m)	Post-Larva	Immature	Sub-Adult
D_1	720	19.0	10	225.3	2,055.0	206.0
D_2	720	19.0	10	225.3	2,055.0	206.0
D_3	700	19.0	10	219.1	1,998.0	200.0
D_4	700	19.0	10	219.1	1,998.0	200.0
D_5	820	12.5	10	116.1	1,539.0	154.0
D_6	430	12.5	10	88.5	807.0	81.0
D_7	250	25.0	10	102.9	939.0	94.0
D_8	280	19.0	10	87.6	799.0	80.0
D_9	340	12.5	10	70.0	638.0	64.0
Totals				1,353.9	12,828.0	1,285.0

Table 6. Estimated total production of ammonia (Kg NH_3) of Krill patches in four areas of the Antarctic Ocean

		Production of Krill stages			
Krill patch areas	Volume of computed Krill patches (m)	Post-Larva	Immature	Sub-Adult	Totals
Admiralty Bay	6,528,000	1.672	3.922	1.966	7.560
Charlotte Bay	64,480	0.067	1.548	0.413	2.028
Bransfield Strait	854,050	1.353	12.828	1.285	15.466
Dallman Bay	1,709,100	0.625	41.072	5.134	46.831

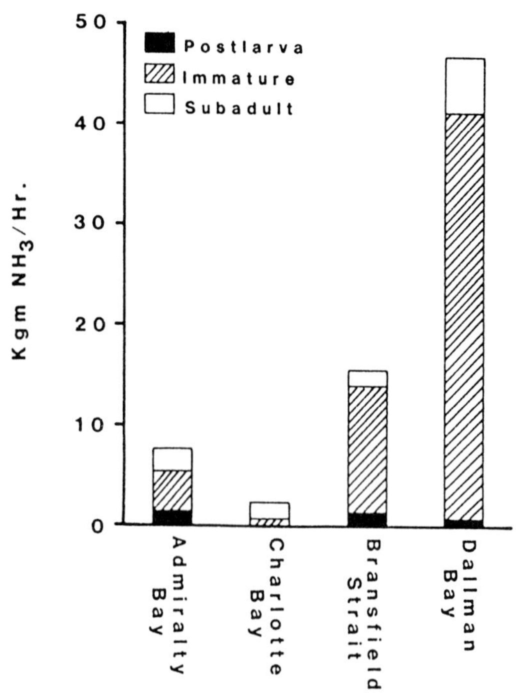

Fig. 2. Ammonia flux (kg h^{-1}) in Krill patches of known composition in four different areas of the Antarctic Ocean

of the daily NH_4^+ requirements of the phytoplankton in the Ross Sea to the ammonia regenerated by the average summer-time standing crop of zooplankton. Biggs (1982) hypothesized that the major sources of NH_4^+ may involve significant contributions from seasonal melting of pack-ice, non-biogenic flow from maritime air, and excretory ammonia products from seabirds in the Antarctic marine ecosystem. In view of the enormous biomass of Krill swarms (Macaulay et al. 1984) and on the basis of the estimated ammonia flux from Krill patches as reported in this paper, it is important to consider Krill swarms as a major source of ammonia flux in certain regions of the Antarctic marine ecosystem during the austral summer.

Accumulation of ammonia in the medium beyond a certain threshold level can lead to a sub-lethal toxic effect on the metabolic performance of Krill. Johnson et al. (1984) argued that, in the field, Krill aggregations exhibit a limited residence time in response to oxygen saturation levels. Kils (1979) also discussed the physiological performance of Krill in relation to oxygen tension. George (1984) showed that Krill maintain a "normal" metabolic rate, with oxygen consumption and ammonia excretion not showing any reduction even down to 40% oxygen saturation level.

As part of the present study, I also monitored dissolved oxygen and ammonium levels in the core of Krill patches, at a 24-h station in the Bransfield Strait. No drastic depletion in the oxygen saturation or ammonium content was encountered.

Acknowledgements. This research was supported by the Division of Polar Programs, National Science Foundation, Grant #DPP 8026535. I am grateful to Captain Peiter Lenie of the R.V. Hero for providing data on Krill acoustic signatures, which I have used for computing Krill biomass. I thank the sponsors of the Fourth SCAR Symposium on Antarctic Biology and NSF (international Travel grant) for providing me with the opportunity to participate in the Symposium.

References

Antezana TJ; Ray K (1984) Feeding of *Euphausia superba* in a swarm north of the Elephant Island. J Crust Biol 4 (Spec No. 1):142–155

Biggs DC (1982) Zooplankton Excretion and NH_4^+ Cycling in Near-Surface Waters of the Southern Ocean I. Ross Sea, Austral Summer 1977–1978. Polar Biol 1:55–67

Boyd CM; Heyraud M; Boyd CN (1984) Feeding of the Antarctic Krill *Euphausia superba.* J. Crust Biol 4 (Spec No. 1):123–141

Dugdale RC; Goering JJ (1967) Uptake of New and Regenerated Forms of Nitrogen in Primary Productivity. Limnon Oceanogr 12:196–206

El-Sayed SZ (1968) On the Productivity of the Southwest Atlantic Ocean and the Waters West of the Antarctic Peninsula: In: Llano GA, Schmitt WL (eds) Biology of the Antarctic Seas III, Antarctic Research Series II. American Geophysical Union, Washington, DC pp 15–47

Epply RW; Rogers JN; McCarthy JJ (1969) Half Saturation Constants for Uptake of Nitrate and Ammonium by Marine Phytoplankton. Limnol Oceanogr 14:912–920

George RY (1977) Dissimilar and Similar Trends in Antarctic and Arctic Marine Benthos: In: Dunbar MJ (ed) Polar Oceans. Arctic Institute of North America, Calgary, Alberta pp 391–407

George RY (1984) Ontogenetic Adaptions in Growth and Respiration of *Euphausia superba* in Relation to Temperature and Pressure. J Crust Biol 4 (Spec No. 1):252–262

George RY; Fields JR (1984) Ammonia Excretion in the Antarctic Krill *Euphausia superba* in Relation to Starvation and Ontogenetic Stages. J Crust Biol 4 (Spec No. 1):263–272

Hamner W (1984) Aspects of Schooling in *Euphausia superba.* J Crust Biol 4 (Spec No. 1):67–74

Hirsch HJ (1983) Excretion and Respiration of the Antarctic Krill *Euphausia Superba.* Polar Biol 1:205–209

Holm-Hansen O, El-Sayed SZ; Franceschini GA; Cuhel RL (1977) Primary Production and the Factors Controlling Phytoplankton Growth in the Southern Ocean. In: Llano GA (ed) Adaptations Within Antarctic Ecosystems. Gulf Publishing Co, Houston, Texas pp 11–50

Holm-Hansen O; Huntley M (1984) Feeding Requirements of Krill in Relation to Food Sources. J Crust Biol 4 (Spec No. 1):156–173

Ikeda I (1984) Sequence in Metablic Rates and Elemental Composition (C,N,P) During the Development of *Euphausia superba* and Estimated Food Requirements During its Life Span. J Crust Biol 4 (Spec No. 1):273–284

Johnson M; Macaulay M; Biggs D (1984) Respiration and Excretion Within a Mass Aggregation of *Euphausia superba:* Implications for Krill Distribution. J Crust Biol 4 (Spec No. 1):174–184

Kils U (1979) Performance of Antarctic Krill *Euphausia superba* at Different Levels of Oxygen Saturation. Meeresforschung 27:35–48

Macaulay MC; English TS; Mathiesen DA (1984) Acoustic Characterization of Antarctic Krill (*Euphausia superba*) Swarms from the Elephant Island and Bransfield Strait. J Crust Biol 4 (Spec No. 1): 16–44

Marr J (1962) The Natural History and Geography of the Antarctic Krill (*Eupausia superba* Dana). Discovery Re 32:33–464

Morris D (1984) Filtration Rates of *Euphausia superba:* Under or Over Estimates? J Crust Biol 4 (Spec No. 1):185–197

Nelson SG; Knight AW; Li WH (1977) The metabolic cost of food Utilization and Ammonia Production by Juvenile *Macrobrachium rosenbergii* (Crustacea: Paloemonidae). Comp Biochem Physiol 57A:67–72

Aspects of the Nutritional Metabolism of Two Antarctic Euphausiids: *Euphausia superba* and *Thysanoessa macrura*

P. MAYZAUD, J. FARBER-LORDA, and M.C. CORRE[1]

Summary. A survey of the nutritional characteristics of 2 euphausiids (*Euphausia superba* and *Thysanoessa macrura*) and of their trophic environment was made in an area encompassed by 60–64°S and 30–50°E. The euphausiids' potential food supply was defined by the chlorophyll, protein, carbohydrate and lipid contents of particulate matter collected at 15 m depth. Except for protein and to a minor extent lipid, the horizontal distribution of the particle standing stock was relatively homogeneous. For each of the 2 species, 5 digestive enzyme activities were identified and measured (amylase, laminarinase, cellulase, galactosidase and trypsin-like activities). As expected, *E. superba* displayed a spectrum of enzyme activities characteristic of an omnivorous organism with affinities for phytoplanktonic material (strong laminarinase). *Thysanoessa macrura,* generally considered to be more carnivorous, appears to be omnivorous and to include some phytoplankton in its diet (strong laminarinase and galactosidase). Changes in enzyme activities were measured in 15–20 different individuals at 16 stations, and related to the biochemical composition and the size characteristics of individuals of the same population. In *E. superba*, the western, smaller, lipid-rich individuals displayed lower enzyme activities than the eastern, larger individuals. Such variation is attributed to different physiology induced by differences in developmental stages. The subadult western population showed a negative correlation with the integrated content of chlorophyll-a, which is masked in the adult eastern population by specific energy needs related to reproductive metabolism.

1 Introduction

Krill (Euphausiacea) are successful in Antarctic waters despite several ecological paradoxes. Many studies on the primary production of this region have shown that the open waters comprise a high nutrient, low biomass and low productivity system (El-Sayed 1967, 1968a, 1968b, 1970; Holm-Hansen et al. 1977; Plancke 1977; Jacques et al. 1982). Phytoplankters are generally considered to be a major food source for *Euphausia superba*, and scientists usually have found it difficult to match the energy expenditure and abundance of krill with the available food supply (Holm-Hansen and Huntley 1983; Clarke and Morris 1983a). As pointed out by Clarke and Morris (1983) and Antezana and Ray (1983), these discrepancies arise from methodological problems related to the elaborate feeding behaviour of these animals; swarming as well as horizontal and vertical migration being associated with the optimization of food gathering. Thus, the field definition of the food supply is made difficult by spatial and temporal differences between the location of the animals at the time of capture and the particulate environment in which they were feeding. Such differences are likely to introduce bias in the measurement of all rapidly adapting physiological functions, such as filtration rate, or ingestion rate (Donaghay and Small 1979), but will have little influence on slowly adapting functions such as assimilation or growth. Therefore, to assess the spatial changes in euphausiid nutritional metabolism, we consider here the variations in digestive enzyme levels in relation to potential food supply.

Digestive enzyme activity of zooplankton is the functional link between ingestion and assimilation. It has been shown to be representative of the feeding adaptability to changes in quantity or quality of the food supply (Mayzaud and Conover 1976; Mayzaud and Poulet 1978; Cox 1981; Cox and Willason 1981; Tande and Slagstad 1982) for given stages of development. For *E. superba,* which encounters diversified food environments, such an index should be indicative of the recent nutritional history of the populations sampled, and of their metabolic status.

2 Material and Methods

Samples of *Euphausia superba* and *Thysanoessa macrura* were obtained from RMT 8 oblique tows made to a depth of 100 m. Details of the sampling stations and their positions are given in Fig. 1. Individuals were handled gently and frozen at –70°C before being analyzed. Water samples for chlorophyll-a determination were taken at each station from 6 depths (0, 10, 20, 30, 50, and 75 m) using 6-l Niskin bottles. Additional water samples for particulate protein, carbohydrate, polysaccharide and lipid analysis were collected at 15 m, using a 30-l Niskin bottle. All water samples were pre-filtered through Nitex (200 μm mesh) to eliminate large zooplankters and debris. Known volumes were then filtered on pre-baked (450°C for 30 min) Whatman GF/C filters for all chemical measurements. All filters were immediately frozen at –70°C before analysis.

1 Biochimie Marine, Station Zoologique, F-06230 Villefranche-sur-mer, France

Antarctic Nutrient Cycles and Food Webs
(ed. by W.R. Siegfried, P.R. Condy, and R.M. Laws)

Particulate protein, carbohydrates and lipids were analyzed according to Mayzaud and Martin (1975). Chlorophyll-a was measured by fluorometry according to the method of Yentsch and Menzel (1963). Additional data on chlorophyll-a were supplied by M. Panouze (Laboratoire ARAGO, Banyuls/mer). Particulate polysaccharides were estimated after two extractions of the material with 0.05 N HCl and evaporation to dryness of the supernatant. After dissolution in distilled water, the amount of polysaccharide was estimated using the colorimetric method of Dubois et al. (1956), standardized with commercial laminarin.

Euphausiids were thawed rapidly, measured and identified for developmental stages. Proteins were extracted from individual animals, using acetone and digested in 0.1 N NaOH at 70°C for 30 min. Three 1-ml fractions were withdrawn for protein determination using the Biuret method described by Raymont et al. (1964). Preparations of soluble krill protein were used to establish the calibration curves (Mayzaud, unpubl.). Lipid measurements were made using the solvent extraction method of Bligh and Dyer (1959), as described in Mayzaud and Martin (1975). After evaporation to dryness of the chloroform phase, the lipids were re-dissolved in a known volume of chloroform and aliquot fractions were analyzed according to the macro-method of Pande et al. (1963).

Enzyme analyses were conducted after extractions in an appropriate buffer. Each individual was measured, identified for growth stage and its cephalothorax homogenized. After centrifugation at 10,000 g and 2°C for 15 min, the supernatant was collected and used as total enzyme preparation. Amylase, laminarinase, cellulase, β-galactosidase and trypsin activities were estimated using commercial amylose, laminarin, carboxymethylcellose, β-lactose and BAPNA, respectively, as substrates. Details of the methods are given in Mayzaud (1980). Each method was adapted to be used in a kinetic mode on an automated Eppendorf 5040 analyzer.

3 Results

3.1 Definition of the Particulate Environment of Krill

Particulate matter was sampled at 21 stations located between 30°–50°E and 60°–64°S (Fig. 1). The values obtained for particulate chlorophyll at different depths were integrated over a 75 m column and the results presented as a horizontal profile. Although the stations are organized in Fig. 2a on an east-west basis, they corresponded to three transects along latitudes 64°S (Stations 3–11), 63°S (Stations 2, and 12–19) and 62°S (Stations 20–23). The levels of chlorophyll-a were generally low, with an overall mean value of 24.8 mg m^{-2}. If we except the level observed at Station 17, such a mean value reflects a rather uniform situation with no significant differences among transects $F^{1}_{15} = 2.23$).

Biochemical composition of the particulate matter was only considered at a single depth (15 m). Values for total proteins, carbohydrates and lipids are also presented as inter-station profiles (Figs. 2 and 3). The mean value for particulate protein for the entire zone was 97.4 μg l^{-1}, with maximum concentrations at Stations 5, 15, 22, and 23. Such means are indicative of a moderately contrasted situation with generally higher values along the 63°S transect and lower values along the 64°S transect ($F^{1}_{15} = 6.57$; $P < 0.025$) (Fig. 2b), which oscillated around a mean concentration of 39.7 μg l^{-1}. Contrary to the other constituents, the levels of particulate lipids were relatively high with a mean value of 51 μg l^{-1}. The profile of variation observed (Fig. 3a) was somewhat reversed from the one described for protein, since the higher values were recorded along 64°S and the lower ones along 63° and 62°S. Changes in particulate polysaccharides (Fig. 3b) were highly correlated with those of particulate proteins ($r = 0.87$; $P < 0.01$) and significantly correlated with chlorophyll at 15 m ($r = 0.477$;

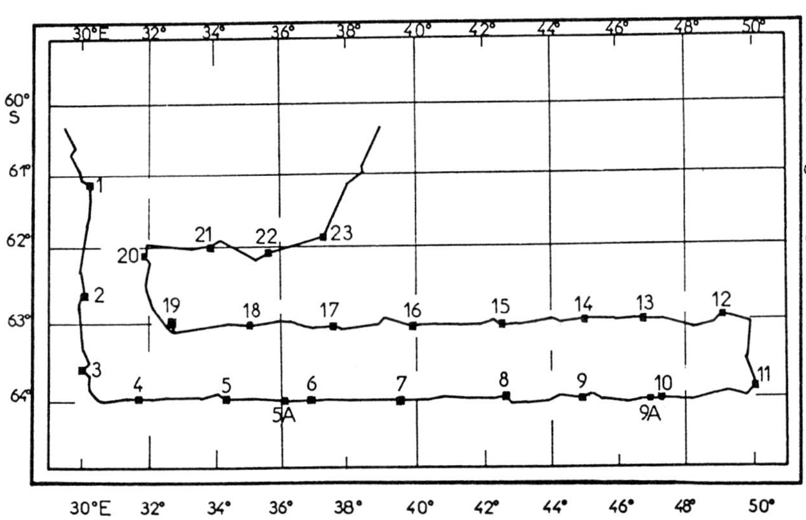

Fig. 1. Cruise track and stations operated during February, 1981 as part of FIBEX (First International BIOMASS Experiment)

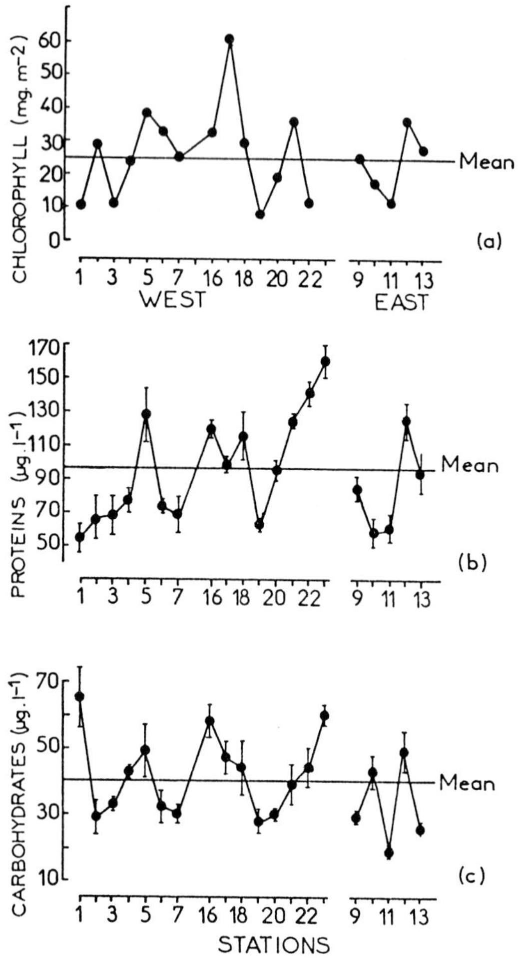

Fig. 2a–c. Horizontal profile of integrated chlorophyll concentration (**a**) and particulate protein (**b**) and carbohydrate (**c**) concentration at 15 m depth

$P < 0.05$) but not with total carbohydrates ($r = 0.42$). Although the average level was relatively low (9.9 µg l^{-1}), the percentage of polysaccharides varied between 2%, and 50% of the total sugars and most often lay in the vicinity of 25%.

The definition of the particulate environment of Krill would not be complete without some estimate of the changes in food quality. The carbohydrate to protein ratio has been used often to estimate the physiological state of diatom populations (Myklestad and Haug 1972; Haug et al. 1973; Myklestad 1977; Mayzaud and Taguchi 1979). If we compute the ratio with the concentrations of total carbohydrates (Fig. 4a) it appears that, except at Station 1, the populations displayed low values of the ratio, suggesting cells in an exponential growth phase (Myklestad 1977). When computed with the polysaccharide concentrations, the ratio appears more variable but, assuming that the phytoplankton population is dominated by diatoms, should also be indicative of an early stage of development. The very low levels of polysaccharides per unit protein at Stations 1 and 3 and the high levels at Stations 6 and 23 are in-

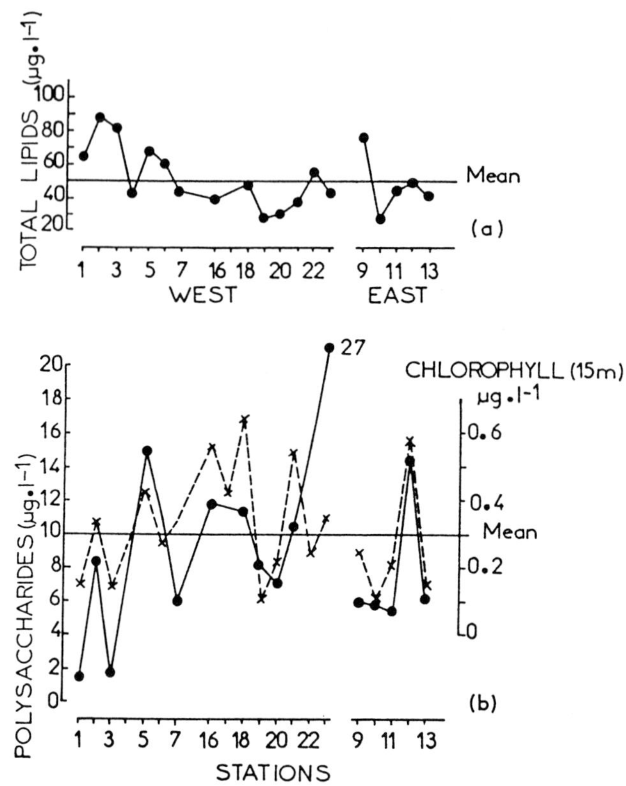

Fig. 3a, b. Horizontal profile of particulate lipid (**a**) and polysaccharides (**b**) at 15 m depth

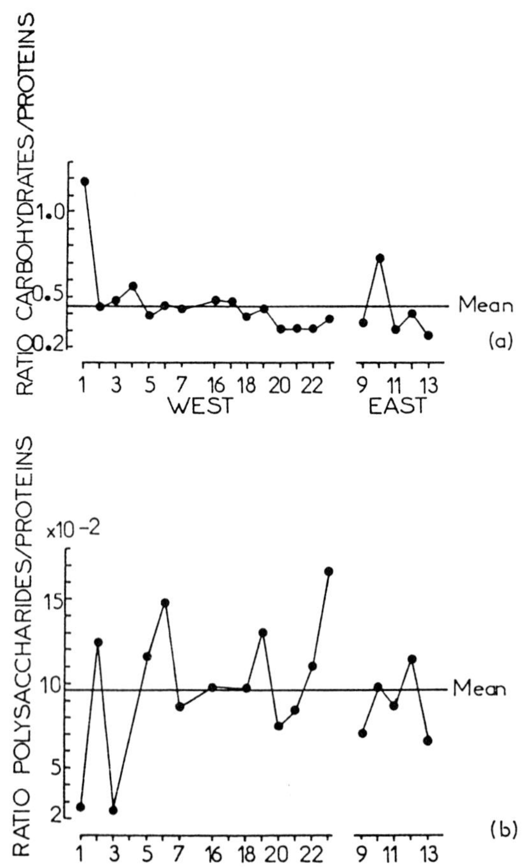

Fig. 4a, b. Values for the ratios of particulate total carbohydrate: protein (**a**) and polysaccharide: protein (**b**)

dicative of large variations in the proportion of structural vs. metabolic carbohydrates in the particules sampled and of their nutritive value.

3.2 Size and Digestive Enzyme Characteristics of Euphausiids

Enzyme analyses were carried out on sub-samples of the krill catches. To be meaningful, such sub-samples should be as homogneous as possible and representative of the original sample. For the 12 stations, where sufficient numbers of *E. superba* could be obtained, three criteria were used: size, maturity stage and soluble protein content. Size and protein content were estimated from measurements of the cephalothorax of each individual. Samples from Stations 2, 3, 5A, 17, and 19 comprised sub-adults (12–16 mm) and a few adults (16–18 mm). Samples from Stations 19 and 23 contained smaller sub-adults and juveniles. Samples from Stations 9, 9A, 10, 11, 13 were composed of sub-adults and larger adults (16-20 mm) with both males and ovigerous females. The distribution of the frequency of size in each samples (Fig. 5) shows that individuals were chosen to limit size variability, except at Stations 11 and 13 where too few animals were available. For the four stations where specimens of *T. macrura* were obtained, the relatively small size of the animals allowed pooled samples of five individuals of the same size.

Five enzymatic activities were used to define the hydrolytic capacities of the euphausiids: amylase, laminarinase, cellulase, ß-galactosidase and trypsin. All five activities were measured for *E. superba,* whereas only four could be detected for *T. macrura*. For *T. macrura*, amylase activity was most often negligible and only laminarinase, ß-galactosidase and trysin were measured at each station. pH kinetics for carbohydrases were assayed with 0.05 M citrate-phosphate buffer adjusted between 2.5. and 7.5. As shown in Fig. 6a, most pH optima lay between 4.0 (for ß-galactosidase) and 7.0 (for amylase). The assay for trypsin pH kinetic used 0.1 M tris-glycine or tris-Maleate buffers in the range 7–10. Both species showed very different optima (Fig. 6), since in *E. superba* trypsin is maximum at pH = 8 and in *T. macrura* it is maximum at pH = 10. Similar conclusions can be reached for the carbohydrases optima which also varied between the two species.

Size and enzyme measurements were made at each station on 12–17 individuals of the species present. For *E. superba*, the results, presented in Figs. 7–9, correspond to mean values per station. Over the entire zone sampled, the mean values for soluble protein and cephalothorax length were 21.48 mg and 14.5. mm per animal, respectively. Ex-

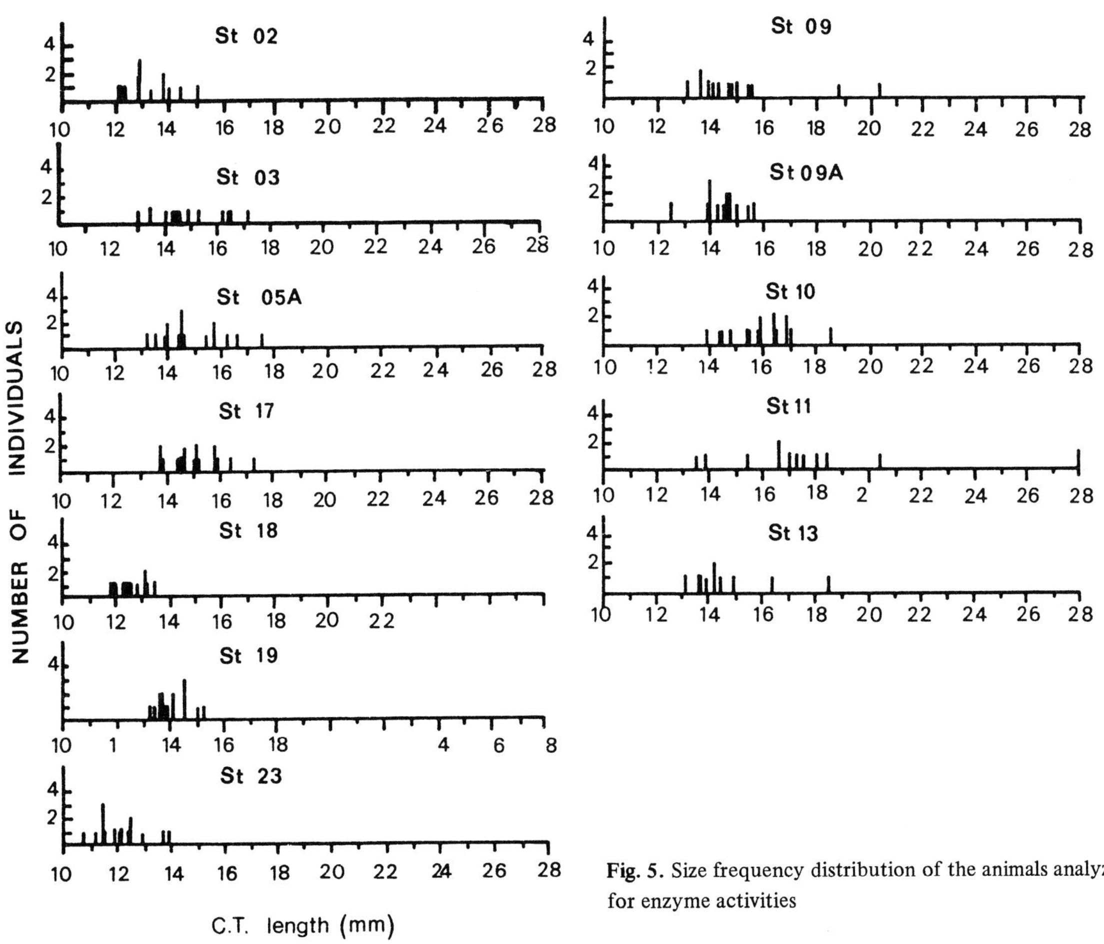

Fig. 5. Size frequency distribution of the animals analyzed for enzyme activities

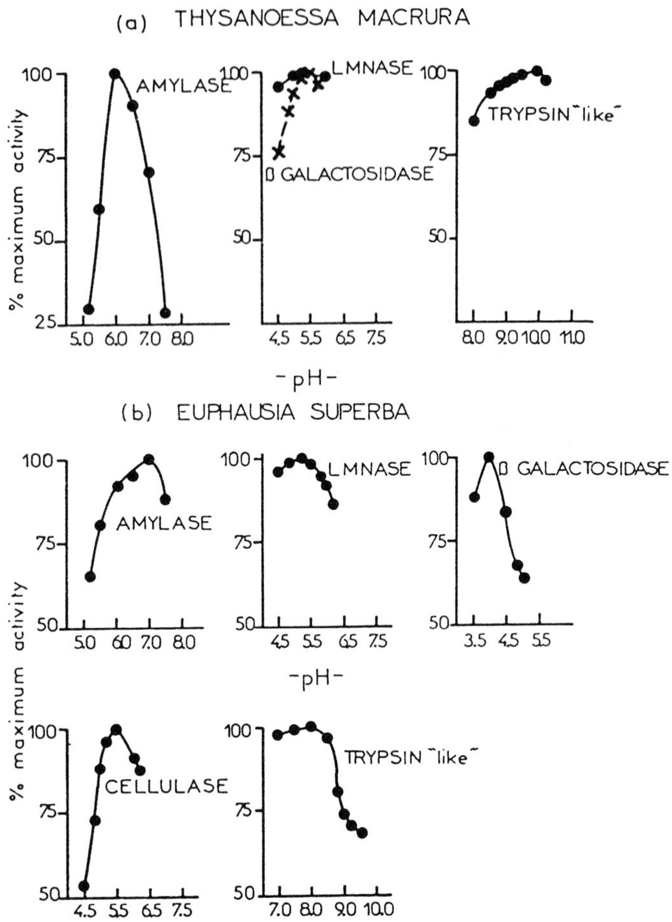

Fig. 6a, b. pH influence on the enzyme activities of *T. macrura* (**a**) and *E. superba* (**b**)

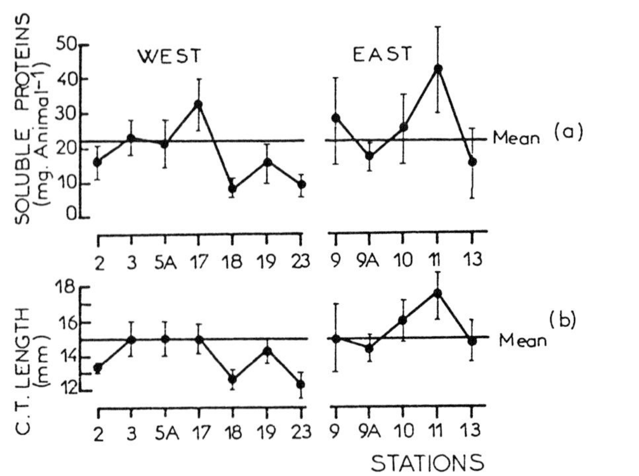

Fig. 7a, b. Soluble protein content (**a**) and cephalothorax (C.T.) length (**b**) of *E. superba* specimens analyzed for enzyme activities

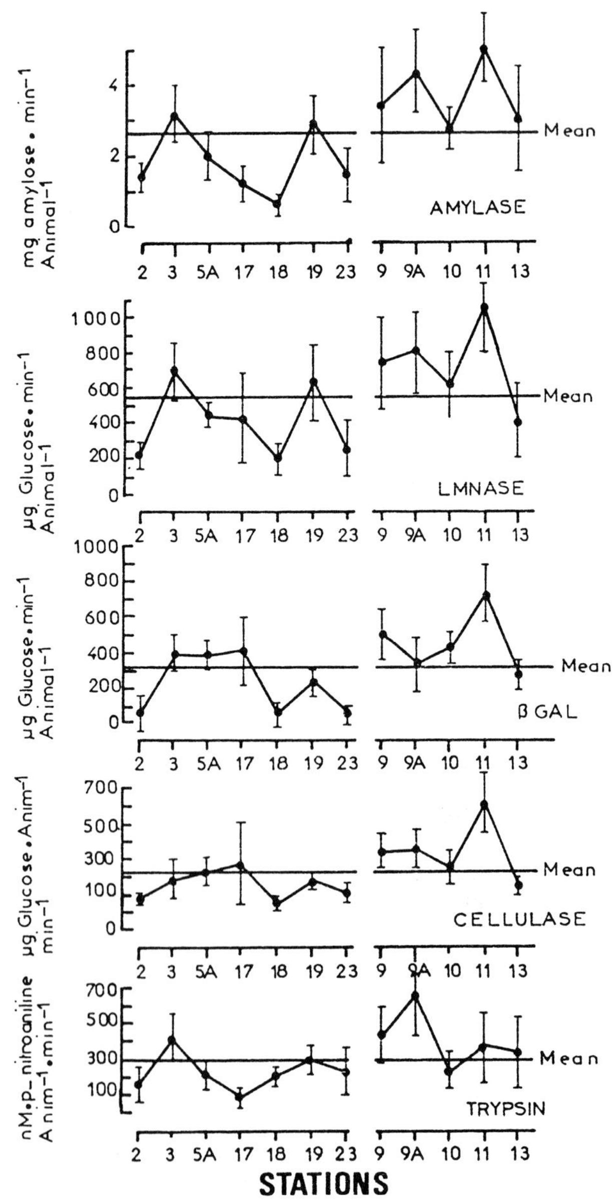

Fig. 8. Horizontal profiles of enzyme activities of *E. superba,* on individual animal basis

amination of the horizontal profile (Fig 7) showed that: the larger the animal, the greater the soluble protein content; and, despite a relatively high level of variability, most stations of the western zone were inhabited by average or smaller than average individuals with correspondingly low protein content. In general, the eastern-zone animals displayed equal or larger size and protein level. Similar differences occurred in enzyme activity (Figs. 8 and 9), regardless of the reference system used. On a per animal basis, the horizontal profiles of all carbohydrase enzymes showed strong similarities, with minimum activities of Stations 2 and 18 with maximum activities at Stations 11 and 9A (Fig. 8). Changes in trypsin activities are somewhat different with a minimum at Station 17 and a maximum at Station 9A (Fig. 8). Removal of the influence of size (dividing activity by the soluble protein content) did not reduce the high level of variability, but modified to some extent the per animal profiles (Fig. 9). Such changes were influential mainly in the eastern zone where the activities observed at Station 9A were enhanced, whereas those at Station 11 were reduced most often drastically. In the case of ß-galactosidase and cellulase, this led to a decrease in the overall amplitude of variation. In the western zone, the

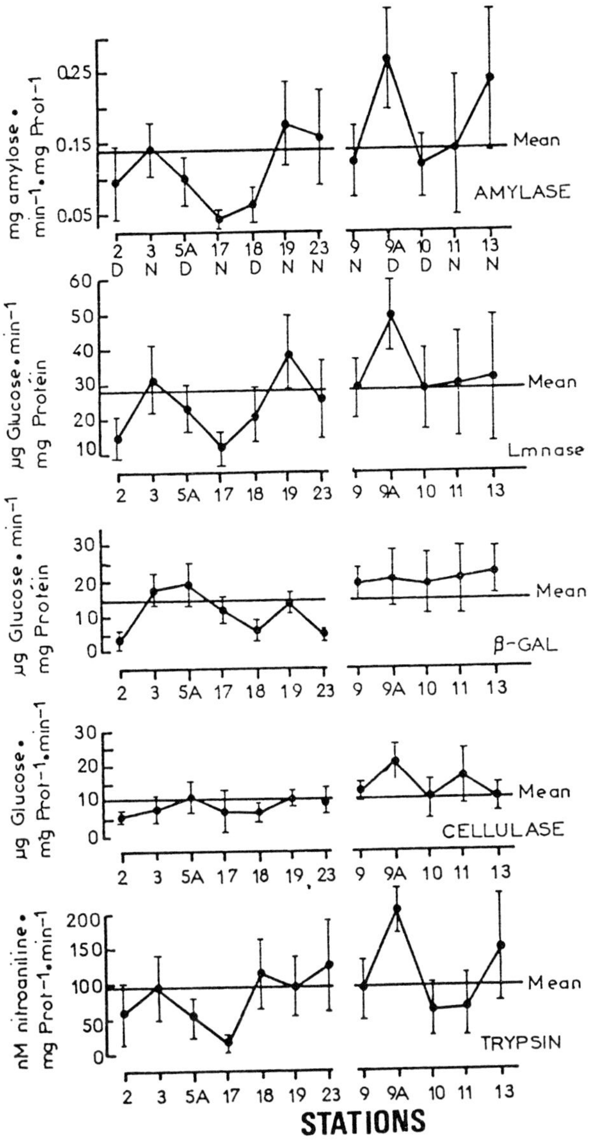

Fig. 9. Horizontal profiles of specific enzyme activities of *E. superba* D = day; N = night

major effect was to shift the minimum activity of amylase, laminarinase and, to a minor extent, cellulase from Station 18 to Station 17. None of these changes modified the differences between the two geographical zones. The highest activities remained associated with the eastern stations and the lowest with the western ones.

Horizontal profiles of enzyme variation are more difficult to establish for *T. macrura* since enough animals were captured at only four stations. Stations 7 and 8 were located at 64°S, Station 17 at 63° and Station 20 at 62°S. In terms of size or soluble proteins, the southern animals appeared to be smaller and contained less soluble proteins per individual (Fig. 10). On a per animal basis, ß-galactosidase and to a minor extent laminarinase activities displayed the same kind of north-south separation which was not revealed for trypsin. Expressed on a per mg of protein basis, all geographical distribution disappeared (Fig. 10), since laminarinase was maximum for Station 20, ß-galactosidase for Station 8 and trypsin for Station 16.

3.3. Biochemical Composition of Krill

At each station 10–15 individual *E. superba* were analyzed for protein and lipid content. To obtain an estimate for the total population, all the results were pooled, regardless of the differences in developmental stages. This procedure explains the very high levels of variability observed around each mean in Fig. 11. Nevertheless, it appears that the animals from the western zone had a higher lipid content than those from the eastern zone. Except for a high and a low value recorded for the animals caught at Stations 17 and 2, respectively, the protein levels were fairly stable around a mean value of 58% of dry wt. mass.

4 Discussion

The assimilation of ingested food requires the hydrolysis of macromolecules by digestive enzymes. Studies on mammals and invertebrates, including zooplankton, have shown that the types of enzymes and their levels of activity are related to the quality and the quantity of the food ingested (Grossman et al. 1944; Van Weel 1960; Desnuelle et al. 1961; Marchis-Mouren et al. 1963; Abdeljlil and Desnuelle 1964; Shimke and Doyle 1970; Ishaaya and Meisner 1973; Van

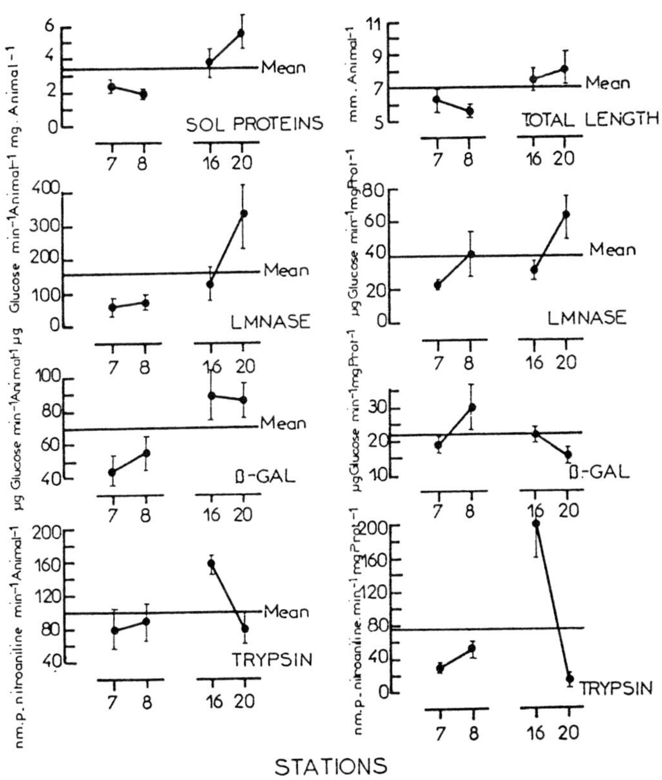

Fig. 10. Spatial changes of enzyme activities of *T. macrura*

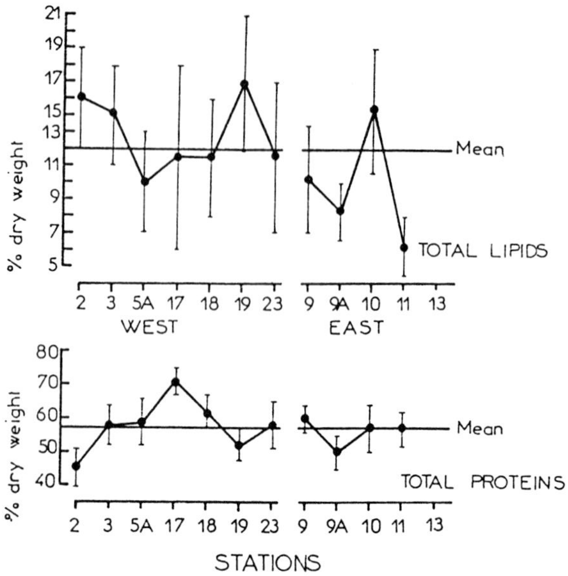

Fig. 11. Horizontal profiles of lipid and protein contents of *E. superba*

Wormoudt 1977, Mayzaud and Poulet 1978; Samain et al. 1980), although all digestive enzymes are not equally inducible. Based on our knowledge of the biochemical composition of natural phytoplankton, Mayzaud and Mayzaud (1981) postulated that a phytoplankton diet induced specifically certain carbohydrases such as laminarinase or cellulase. Later work by Cox (1981) confirmed experimentally that the degree of herbivory was associated with the activity of laminarinase. Most data obtained suggest that the ratio of carbohydrase to protease activity is a good index of the degree of carnivory (Vonk 1960; Boucher et Samain 1974, 1975; Van Wormoudt 1980). Thus, as expected, the high levels of activity of laminarinase, cellulase, ß-galactosidase and amylase define *E. superba* as a strong feeder on phytoplankton. The marked occurrence of laminarinase and ß-galactosidase in the digestive activities of *T. macrura,* implies the existence of a phytoplankton component in its diet, but the absence of amylase and cellulase suggests a more omnivorous feeding pattern.

In our survey, correlation analysis showed a good level of metabolic coherence between amylase, laminarinase, cellulase and to a lesser extent trypsin (Table 1). The lack of correlation between enzyme activities and size or soluble proteins confirms that size differences were not responsible for the changes observed. Despite the relatively large variability, the horizontal profiles of all five enzymes revealed that the western populations of *E. superba* had lower levels of activities than the eastern ones. If we consider that the western individuals seemed to have a relatively higher lipid content, such geographical variation could be related either to differences in metabolic strategy of the developmental stages or differences in food supply. No correlation could be established with any of the chemical descriptors of the particulate matter collected at 15 m, which suggests that such measure-nets were not representative of the diet of *E. superba.*

A better description of the trophic relationships seems possible by using the values of chlorophyll integrated over the 75m water column. Over the entire area suveyed, amylase and laminarinase showed negative correlations with integrated chlorophyll-a (Table 2). Examination of the same correlations for the western and the eastern zones separately revealed that although amylase, laminarinase and trypsin were correlated significantly and negatively in the western sector, no such relationship could be established in the eastern sector (Table 2). Why the western, smaller, lipid-rich individuals have a lower enzyme activity level than the eastern larger adults, is probably related to the dif-

Table 1. Correlation analysis between digestive enzyme activities per mg of mg of soluble protein and biochemical characteristics of *E. superba*

	Amylase	Lmnase	Cellulase	ß-Gal	Tryp	Sol Protein	C.T. Length	Tot. Protein
Amylase	1.000							
Lmnase	0.875	1.000						
Cellulase	0.667	0.773	1.000					
ß-Gal	–	–	0.627	1.000				
Tryp	0.884	0.773	–	–	1.000			
Sol. Protein	–	–	–	–	–	1.000		
C.T. Length	–	–	–	0.753	–	0.894	1.000	
Tot. Protein	0.612	–	–	–	–	–	–	1.000
Tot. Lipid	–	–	0.676	–	–	–	–	–

Abbreviations: Lmnase = laminarinase; ß-Fal. = ß-galactosidase; Tryp = Trypsin; Sol. prot. = Soluble protein; C.T. = cephalothorax; Tot. = Total. No. observations = 12 ($r_{0.01}$ = 0.708; $r_{0.05}$ = 0.576)

Table 2. Correlation analysis between enzyme activties and the integrated chlorophyll content of a 75-m water column

	All sector	Western sector	Eastern sector
Amylase	– 0.602[a]	–0.921[b]	0.617
Laminarinase	– 0.804[b]	–0.856[a]	0.362
Cellulase	– 0.410	–0.375	–0.639
ß-galactosidase	– 0.131	–0.014	0.402
Trypsin	– 0.510	–0.826[b]	0.851
n	12	7	4

a Significant at 5% level
b Significant at 1% level

ferent physiological states of the populations sampled. Results from the concurrent hydrological studies suggest that this geographical difference corresponded to parts of two gyres (Simon 1983) where separate *E. superba* populations were in different stages of their development (Gely 1983). Enzyme activities of younger sub-adults are probably lower than those of mature males or females and their profiles of variation are related to feeding activity. The negative relationships with chlorophyll is evidence that, for such fast-swimming zooplankters, the time required for "grazing down" locally available phytoplankton concentrations is of the same order of magnitude as the time required to establish a new equilibrium in enzyme rate. For the eastern adults, such a trophic relationship was masked by specific energy needs of reproductive metabolism, and probably of the intensity of lipid synthesis and/or by the smaller number of data points.

References

Abdeljlil B; Desnuelle P (1964) Sur l'adaptation des enzymes du pancréas à la composition du régime. Biochem Biophys Acta 81: 136–149

Antezana T, Ray K (1983) Swarming of *E. superba* as an adaptative group strategy to the antarctic ecosystem. Workshop on krill Biology – Bremerhaven 16–21 May

Bligh EG, Dyer WJ (1959) A rapid method for total lipid extraction and purification. Can J Biochem Physiol 37:911–917

Boucher J, Samain JF (1974) L'activité amylasique, indice de la nutrition du zooplancton: Mise en évidence d'un rythme quotidien en zone d'upwelling. Tethys 6:179–188

Boucher J, Samain JF (1975) L'étude de la nutrition du zooplancton en zone d'upwelling par la mesure des activités enzymatiques digestives, pp 329–341. In: Barnes H (ed) Proc 4th European Mar Biol Symp, Aberdeen University Press

Clarke A, Morris DJ (1983) Development of an energy budget for *Euphausia superba.* Workshop on krill Biology. Bremerhaven 16 –28 May

Clarke A, Morris DJ (1983a) Towards an energy budget for krill: The physiology and biochemistry of *Euphausia superba.* Polar Biol 2:69–86

Cox JL (1981) Laminarinase induction in marine zooplankton and its variability in zooplankton samples. J Plankton Res 3:345 –356

Cox JL, Willason SW (1981) Laminarinase induction in *Calanus pacificus.* Mar Biol Lett 2:307–311

Desnuelle P; Reboud JP; Ben Abdeljlil A (1961) Influence of the composition of the diet on the enzyme content of rat pancreas. p 90–120. In: Churchill (ed) CIBA Foundation Symposium on the exocrine pancreas. London

Donaghay PL, Small LF (1979) Long term food modification by *Acartia clausi:* A preliminary view. Mar Biol 52:129–136

Dubois M, Gilles KA; Hamilton JK; Rebers PA; Smith F (1956) Colorimetric method for the determination of sugars and related substances. Anal Chem 28:350–356

El-Sayed SZ (1967) Biological productivity investigation of the Pacific sector of Antarctica. Antarct J US 11:200–201

El-Sayed SZ (1968a) Prospects of primary productivity studies in Antarctic waters. In: Symp on Antarctic Oceanography pp 227 –239. Cambridge, Scott Polar Res Inst

El-Sayed SZ (1968b) On the productivity of the southwest Atlantic ocean and the waters west of the Antarctic peninsula. In: Llano GA, Schmitt WL (eds) Biology of the Antarctic Seas, Antarctic Res Ser N° 11 p 15–47. Am Geophys Union

El-Sayed SZ (1970) On the productivity of the southern ocean (Atlantic and Pacific sectors). In: Holdgate MW (ed) Antarctic Ecology vol I Academic Press, New York pp 119–135

Gely C (1983) Contribution à l'étude du krill *Euphausia superba.* Thèse Docteur 3ème cycle Université Pierre et Marie Curie 201 p

Grossman MI, Greengard H, Ivy AC (1944) On the mechanism of the adaptation of pancreatic enzymes to diet. Am J Physiol 141: 38–45

Haug A, Myklestad S, Sakshaug E (1973) Studies on the phytoplankton ecology of the Trondheimsfjord. I. The chemical composition of phytoplankton populations. J Exp Mar Biol Ecol 11: 15–26

Holm-Hansen O, Huntley M (1983) Feeding requirements of krill in relation to food sources. J Crust Biol (to be published)

Holm-Hansen O, El-Sayed SZ; Franceschini GA; Cuhel R (1977) Primary production and the factors controlling phytoplankton growth in the southern Ocean. In: Llano GA (ed) Adaptation within Antarctic Ecosystem, IIIrd SCAR Symposium Antarc Biol pp 11–50 Smithsonian Institution

Ishaaya I, Meisner J (1973) Physiological effects of sugars on various digestive enzymes of *Spodoptera littoralis* larvae. J Comp Physiol 86:117–124

Jacques G, De Billy G, Panouze M (1982) Biomasse et production primaire dans les secteurs antarctiques et subantarctiques de l'Ocean Indien. CNFRA 55:87–99

Marchis-Mouren G, Pasero L; Desnuelle P (1963) Further studies on amylase biosynthesis by pancreas of rats fed on a starch rich or casein rich diet. Biochem Biophys Res Commun 13:262–266

Mayzaud P (1980) Some sources of variability in determination of digestive enzyme activity in zooplankton. Can J Fish Aquat Sci 37:1426–1432

Mayzaud P, Conover RJ (1976) Influence of potential food supply on the activity of digestive enzymes of neritic zooplankton. In: Proc 10th European Mar Biol Symp, Ostend. Universa, Wetteren Vol 2:415–427

Mayzaud P, Martin JLM (1975) Some aspects of the iochemical and mineral composition of marine plankton. J Exp Mar Biol Ecol 17: 297–310

Mayzaud P, Mayzaud O (1981) Kinetic properties of digestive carbohydrases and proteases of zooplankton. Can J Fish Aquat Sci 38:535–543

Mayzaud P, Poulet SA (1978) The importance of the time factor in the response of zooplankton to varying concentrations of naturally occurring particulate matter. Limnol Oceanogr 23:1144–1154

Mayzaud P, Taguchi S (1979) Size spectral and biochemical characteristics of the particulate matter in Bedford Basin. J Fish Res Board Can 36:211–218

Myklestad S (1977) Production of carbohydrates by marine planktonic diatoms II. Influence of the N/P ratio in the growth medium on the assimilation ratio, growth rate and production of cellular and extracellular carbohydrates by *Chaetoceros affinis* and *Skeletonema costatum.* J Exp Mar Biol Ecol 29:161–179

Myklestad S, Haug A (1972) Production of carbohydrates by the marine diatoms *Chaetoceros affinis.* I. Effect of the concentration of nutrients in the culture medium. J Exp Mar Biol Ecol 9:125–136

Pande SV, Parvin-Khan R, Venkitasubramanian TR (1963) Microdetermination of lipids and serum total fatty acids. Anal Biochem 6:415–423

Plancke J (1977) Phytoplankton biomass and productivity in the subtropical convergence area and shelves in the western Indian Subantarctic islands. In: Llano GA (ed) Adaptation within Antarctic Ecosystems pp 51–73. IIIrd SCAR Symposium Antarct Biol Smithsonian Inst

Raymont JEG, Austin J; Linford E (1964) Biochemical studies in marine zooplankton. I. The biochemical composition of Neomysis integer- J Cons Perm Int Explor Mer 28:354–363

Samain JF, Moal J, Daniel JY, Lecoz JR Jezequel M (1980) The digestive enzyme amylase and trypsin during the development of *Artemia.* Effect of food conditions. In: Persoone G, Sorgeloos P, Roels O (eds) The Brine Shrimp Physiology, Biochemistry, Molecular Biology Universa, Wetteren Vol 2 pp 239–255

Shimke RT, Doyle D (1970) Control of enzyme levels in animals tissues. Annu Rev Biochem 39:929–976

Simon V (1983) Assimilation et regénération des sels nutritifs dans l'Océan austral: Caractérisation et relation avec la structure hydrologique, Thèse Doctorat 3ème cycle Université Aix-Marseille II. 99 pp

Tande KS, Slagstad D (1982) Ecological investigation on the zooplankton community of Balsjjorden. Northern Norway. Seasonal and short time variation in enzyme activity in copepodite stage V and VI males and females of *Calanus finmarchicus.* Sarsia 67: 63–68

Van Weel PB (1960) On the secretion of digestive enzymes by the marine crab *Thalamita crenata.* Z Vgl Physiol 43:567–577

Van Wormoudt A (1977) Activités enzymatiques digestives chez *Palaemon serratus:* variations annuelles de l'acrophase des rythmes circadiens. Biochem Syst Ecol 5:301–307

Van Wormoudt A (1980) Adaptation des activités digestives, de leurs cycles et de leur conrole aux facteurs du milieu chez *Palaemon serratus.* Thèse Doct es Sci Université Aix-Marseille II 351 pp

Vonk HJ (1960) Digestion and metabolism. In: Waterman TB (ed) The physiology of the Crustacea Academic, New York 296 pp

Yentsch CS, Menzel D (1963) A method for the determination of phytoplankton chlorophyll and phaeophytin by fluorescence. Deep-Sea Res 10:221–231

Moult and Growth in Euphausiids

F. BUCHHOLZ[1]

Summary. Laboratory experiments showed that growth rates in adolescent Antarctic Krill, *Euphausia superba*, were similar to calculated hypothetical values. Body shrinkage occurred in starved animals, but apparently was induced by laboratory effects. Reduction in body length during winter is not common to euphausiids; it was not found to occur in *Meganyctiphanes norvegica* (Euphausiacea) which continues to moult in winter. Euphausiid growth can be monitored by moult-staging, as tank experiments and field investigations have shown. Simultaneous moulting is a feature of *E. superba* in swarms, which can be observed either directly in the field or derived from the species' pattern of moult stages.

1 Introduction

The Antarctic Krill, *Euphausia superba*, occupies a central position in the Antarctic food web (Everson 1977). The international BIOMASS (Biogical Investigations of Marine Antarctic Systems and Stocks) research programme aims to assess the standing stock and production of Krill. Data on growth and longevity of Krill are required for this and, in particular, the frequency of moulting and growth increments per moult need to be determined. An obvious way to accumulate such data is to observe individual Krill maintained in aquaria for prolonged periods.

A number of studies report little growth and even body shrinkage in Krill maintained under laboratory conditions, and Ikeda and Dixon (1982) postulated that size reductions during winter serve as a hibernation strategy in Krill. This hypothesis could be tested by following growth and development in a given population of *E. superba* over an annual cycle in the field. However, for obvious reasons, this is difficult if not impossible. One approach to overcoming this problem is to determine whether other euphausiids exhibit negative growth in winter. Hence, a comparative study of annual growth and moult in an isolated natural population of a boreal euphausiid, *Meganyctiphanes norvegica*, was undertaken (Boysen and Buchholz 1984).

Here I present a preliminary report on my observations of growth and moult in captive Krill, maintained in aquaria, and relate the results to comparable data obtained for Krill and *M. norvegica* under natural conditions. More particularly, I examine the concept of synchronized moulting of Krill in swarms. A moult staging system, originally designed to be used in arthropod physiology is employed to monitor the moult development of Krill populations.

1 Institut für Meereskunde an der Universität Kiel, Düsternbrooker Weg 20, D-2300 Kiel, FRG

2 Materials and Methods

A 2 m^2 rectangular mid-water trawl (RMT) was used to capture Krill at various depths in Admiralty Bay, King George Island, South Shetland Islands (Rakusa-Suszczewski and Stepnik 1980). Open-ocean Krill were also caught by means of an RMT 1 + 8 net (Baker et al. 1973), operated from the R.S. Meteor. Ring-trawls (0.5 or 1.0 m^2) were used to capture pelagic *M. norvegica* in the Scandinavian Kattegat (Boysen and Buchholz 1984). The development of this population was monitored during the course of one year, when phytoplankton and zooplankton populations were monitored as well.

Two maintenance systems were used for keeping live Krill individually at the Polish Arctowski Research Station, at King George Island. Both systems minimized the need for handling the animals. In addition, 250 Krill were kept in a 700-1 through-flow tank (1000 l $\times$ h^{-1}) at 2.0 ± 0.5°C. The first system consisted of 3 through-flow aquaria, each containing 18 1.1.–1 perspex jars. The seawater in the aquaria (a total volume of 80 l and at 2°C) was exchanged three times $\times$ h^{-1}, and carried sufficient phytoplankton to feed the Krill. During starvation experiments, a reduced water supply was circulated by a pump; the water being exchanged every 4 weeks.

Four basins containing 96 1.0–1 plastic jars were used in the second system. Water was drawn off every 4 days and exchanged. The animals were fed intensively on a suspension of *Chlorella* algae. This (closed) system was also used for keeping *M. norvegica*. These animals were fed in excess on 3-day-old nauplii of *Artemia salina*.

Moult increments were determined by measuring, with a micrometer, linear increases of carapax and uropod length of the animals' exuviae. Mean values were calculated for the two sets of measurements. Standard length, as a control, was measured only occasionally to reduce handling the live animals. The original moult-staging method of Drach (1939)

Antarctic Nutrient Cycles and Food Webs
(ed. by W.R. Siegfried, P.R. Condy, and R.M. Laws)

Table 1. Characteristics of six stages of moult in *Euphausia superba*

Stage	Phase	Cuticle	Uropod tissues	% of moult-cycle duration
A_1	Early post-moult	Soft to flexible in all parts	Unstructured	7
BC	Inter-moult	Rigid	Characteristic stripe pattern	25
D_0	Early premoult	Rigid	Epidermis detaches from cuticle	18
D_1'	Premoult	Rigid	Differentiation of new setae starts	5
D_1''	Premoult	Rigid	Progressing development of setae	10
D_2	Late pre-moult (animals ready to moult)	Rigid to soft	All cuticle structures doubled	35

was modified for euphausiids; a detailed description is in Buchholz (1982). Moult stages can be identified microscopically by observing the state of differentiation of the uropod tissues. As many as 17 sub-stages can be described but, in analyzing animals caught in nets, a reduction to 6 main phases is necessary. These are characterized in Table 1. The duration of moult stages was determined for 152 animals, after defined post-moult intervals (Buchholz, unpubl.). A preliminary temporal correlation for the moult stages has been established by Morris and Keck (1984).

3 Results and Discussion

3.1. Moult and Growth in Aquaria

Mortality of Krill in both aquarium systems (through-flow and closed) did not exceed 6% per month and the animals moulted regularly. The mean moult interval of 14–17 days is within the range obtained by other workers, who used similar maintenance temperatures (Mackintosh 1967; Clarke 1976; Poleck and Denys 1982; Morris and Keck, to be publ.). The mean moult increment of fed animals at the first moult ranged between 3% and 5% of the pre-moult length, but decreased to 1% and 2% in two to four of the consecutive moults. The growth increment varied between +21% and –15%. In starved Krill, positive growth only occurred as the exception; it being negative (mean = –1% to –2%) over four consecutive moults separated by a considerable increase in the mean moult interval (20–33 days). The average growth rate was 0.132 mm d^{-1} for 25 animals, based exclusively on intervals with positive growth increments of first experimental moults. This agrees with the 0.141 mm d^{-1} reported by Morris and Keck (to be publ.) for 12 slightly smaller Krill (26 vs. 31 mm). These workers employed a prototype (Buchholz, unpubl.) of the through-flow system used in the present study. Both values also fit the theoretical growth curve of Mauchline (1980), from which a rate of 0.133 mm d^{-1} can be derived for the same size range of animals. The maximum growth rate found in other experiments is 0.068 mm d^{-1} (Poleck and Denys 1982).

In the context of the present paper, the following preliminary conclusions emerge. In aquarium experiments, the first moult-increase is always the greatest. Subsequently, consecutive moult-increases are smaller, but remain relatively constant. This means that the first moult in captivity probably still reflects natural growth in the wild, whereas subsequently the Krill adjust their growth to the experimental conditions. In comparison to the results of other workers (e.g., Murano et al. 1979; Ikeda and Dixon 1982 a,b), who usually employed simple closed-volume systems, the growth rates obtained with my experimental set-up seem more realistic. Nevertheless, a wide range of moult increments suggests that the flow-through system may still be sub-optimal in relation to natural conditions. A general reduction in growth increments per moult and extensions of intervals between moults have been demonstrated to occur in captive crustaceans (Hartnoll 1982). My experiments with starved Krill confirmed the findings of Ikeda and Dixon (1982): body shrinkage prevails under prolonged laboratory maintenance. Caution is, however, neccessary in using negative moult increments, observed in the laboratory, to calculate the growth of Krill in the field during winter. The available laboratory data are too strongly influenced by the synergistic effects of inadequate maintenance and experimental starvation, which are particularly difficult to quantify. Moreover, it cannot be excluded, on the basis of current knowledge, that moulting is delayed by winter conditions or that it may even cease during winter.

3.2. Growth under Natural Conditions

Field studies of growth in *E. superba* have usually involved analyzing annual size-frequency distributions. On this basis estimates of Krill longevity show a wide variation from 2 to > 4 yrs (Ivanov 1970; Mackintosh 1972). A major reason for this wide variation lies in difficulties encountered with continuously following a discrete population cohort over a prolonged period. Comparative studies with other euphausiid species, however, may facilitate the description of annual growth patterns which could be generalized to specific members of the group. Thus, Fig. 1 shows that in *M. norvegica*, under natural conditions, significant growth was evident only during spring and summer. Growth diminished substantially directly after the autumn plankton bloom, and almost ceased in November. Thereafter, growth re-commenced slowly over several months, apparently in response to an increasing food supply. No decrease in body size was observed. In contrast, in preliminary maintenance experiments, body shrinkage prevailed in captive specimens. Thus, apparently there is a tendency in both *M. norvegica* and *E. superba* to exhibit negative growth under unfavourable conditions, especially as manifest by inadequate maintenance in the laboratory. Reduction in size during winter was not observed for *M. norvegica* in the Kattegat. In a fjord-population of *M. norvegica,* growth in body length also stagnated during winter, but did not become negative

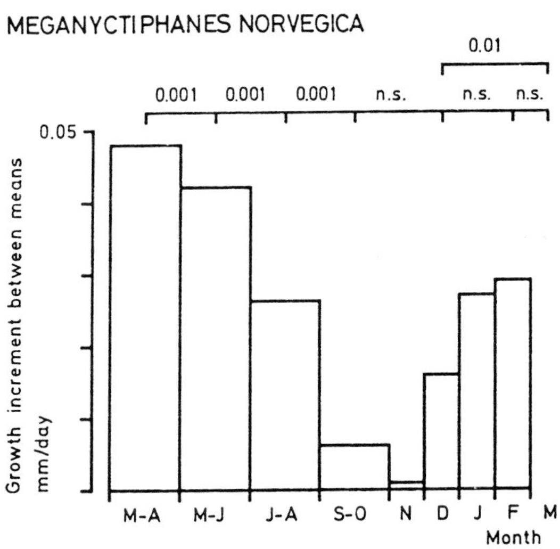

Fig. 1. Growth in a population of *Meganyctiphanes norvegica* in the Kattegat, based on monthly size-frequency distributions of age-group I (n = 72–241) and taken from Boysen and Buchholz (1984 and unpublished data). Values were compared by analysis of variance (Duncan-Test); levels of significance shown above histograms. (n.s. = not significant)

(Matthews 1973), whereas the weight of the animals of both populations (Kattegat and fjord) decreased slightly during December and November (Båmstedt 1976; Boysen and Buchholz 1984).

Fowler (1971) noted that starved *M. norvegica* and *Euphausia pacifica* moulted at the same rate as fed specimens, although the starved animals did not live for long. Moulting continues to occur rhythmically in captive *E. superba* but my results show that moult intervals tend to increase in starved animals. Since Hartnoll (1982) has demonstrated that the availability of food, as well as ambient temperature, markedly influence crustacean growth, I attempted to determine whether moulting ceases in *M. norvegica* under winter conditions when low temperatures and a low abundance of planktonic food prevail in the Kattegat. Moulting was monitored by examining specimens from net samples taken during summer and winter. It was expected that if animals did not moult, the majority would be found in stage B–C including the resting stage C_4. Normally, crustaceans remain in the latter phase of intermoult or "anecdysis" when the next ecdysis is delayed, either by adverse natural environmental conditions (Bückmann and Adelung 1964) or by artificial (experimental) factors (Passano 1960). This phenomenon is also evident in Krill in the laboratory (pers. obs.).

Figure 2 shows that a significant proportion of the *M. norvegica* population was in the "active" moult stage (i.e., the premoult condition D or the post-moult phase A_1) throughout the annual cycle. The inter-moult phase B–C was not over-represented in the winter samples, in comparison to the moult stage patterns found during spring and summer. Thus, the proportion of animals found in active moult remained relatively constant throughout the year. Even if no growth occurred, as in November (see Fig. 1), a high percentage of actively developing animals was evident. Also, from December to February a larger proportion of animals was observed to be preparing for the next ecdysis or to have moulted recently. At the same time (December–February) the levels of zoo- and phytoplankton are almost nil (Boysen and Buchholz 1984).

Two conclusions can be drawn from these observations. First, moulting does not cease in *M. norvegica* during winter, although it is not possible to derive the actual moult frequency from this qualitative account. Secondly, animals apparently moult even if no growth is achieved, as the contemporaneous size-frequency histograms show (Fig. 1; Boysen and Buchholz 1984). Associated body shrinkage did not occur.

3.3. Implications of Synchronous Moulting

Mackintosh (1967) and Clarke (1976) noted that, in their maintenance experiments, Krill tended to moult simultaneously even when individual animals were kept separate from each other. According to Clarke, such synchrony may

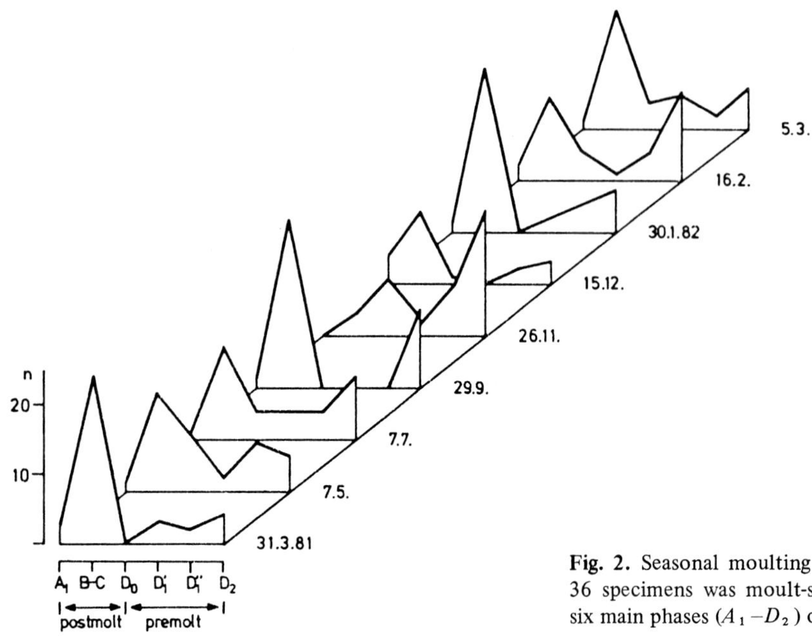

Fig. 2. Seasonal moulting activity in *Meganyctiphanes norvegica.* A sample of 36 specimens was moult-staged on the dates given and classified according to six main phases (A_1–D_2) of the moult cycle

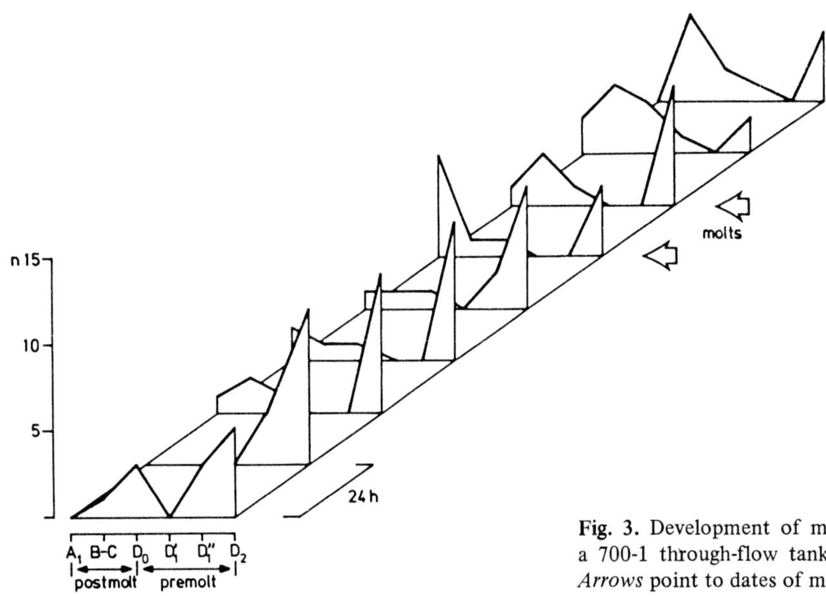

Fig. 3. Development of moult stages in *Euphausia superba* kept for 8 days in a 700-1 through-flow tank. A daily sample of 12 specimens was moult-staged. *Arrows* point to dates of mass occurrence of exuviae

be the result of physiological shock induced by trawl capture. I also observed moulting in groups during my maintenance experiments, and moult-staging indicates that post-trawl shock was not necessarily the cause (Buchholz, pers. obs.). Observations by SCUBA divers also indicate that synchronous moulting may be a feature of Krill behaviour in nature (Hamner et al. 1983).

Since a close temporal relationship exists between the progress of moult stages in individual Krill (Morris and Keck, 1984; Buchholz, pers. obs.), coincident development should also be present in large aggregations of moulting Krill. This could be easily tested by taking continuous random samples of Krill and moult-staging them. With the help of the known moult-time correlation it should be possible to extrapolate the complete moult interval, even if only a portion of the moult cycle was monitored during a given period of observation. Such an experiment was undertaken with ocean-caught Krill kept on-board ship in a 700 l through-flow system. At the start of the experiment a bimodal moult-stage distribution was found, which soon became unimodal with animals immediately preparing for ecdysis and getting ready to moult (D_2). After a delay of about 4 days which corresponds to the length of the phase D_2, moulting was observed, as represented by a number of freshly moulted animals and by the many moult exuviae found in the holding tank (Fig. 3). During the next 2 days

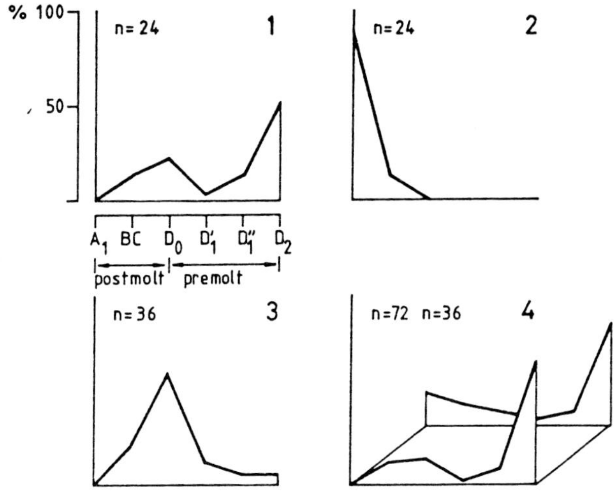

Fig. 4. Patterns of moult stages in five samples of *Euphausia superba* from four stations in the northern Weddell Sea and Scotia Sea during February 1981. Figure *4.4* represents samples taken from the same swarm of Krill on 2 consecutive days

mainly post-moult stages were evident and the moult-stage distribution again assumed the bimodal pattern observed initially. For technical reasons the experiment was terminated at this stage, but from the available data a moult interval of approximately 14 days could be extrapolated. This figure corresponds well with the data obtained from individual maintenance experiments.

Additional samples of oceanic Krill, taken by RMT 8 catches, from the northwestern Weddell Sea and north of Elephant Island, and immediately moult-staged, had an even distribution of moult stages (not shown here), but other samples from the same areas showed less uniform distributions pointing to a simultaneous moult development (Fig. 4). A similar approach was adopted for Krill taken in Admiralty Bay. It was hoped that a distinct and undisturbed Krill population could be sampled continuously in this relatively secluded bay. Figure 5 shows that uniform moult-stage distributions alternated with other patterns which clearly are biased towards specific moult stages. Thus, in haul 1 (26 Dec.), haul 2 (3 Jan.) and hauls 8–9 (16–17 Jan.), samples of intensively moulting swarms of Krill were encountered. Accordingly, most of the animals examined showed signs of moult-stages A_1 or D_2, being freshly moulted or in a stage just prior to ecdysis. In another freshly-moulted Krill swarm from Admiralty Bay, out of a sample of 100 moult-staged animals 73 were soft-shelled (Fig. 6) and 27 had rigid exoskeletons; the latter being in various stages of the moult cycle. Obviously, a size-disassociation occurred in this actively moulting swarm.

It can be concluded that synchronous moulting is a feature of Krill swarms. Regular development of moult stages, as well as mass moulting, were observed in animals kept in tanks. Mass moulting was observed directly or detected through moult-staging in isolated swarms of oceanic Krill. However, the distributions of moult stages in the samples analyzed in the field were often uniform, and gave few clues as to whether simultaneous development had in fact occurred. The growth of *E. superba* in Admiralty Bay has been measured by Suszczewski and Stepnik (1980, and pers. comm.) who showed frequent discontinuities in growth patterns. Since Admiralty Bay is an area of strong tidal currents, it is reasonable to assume that Krill swarms inside

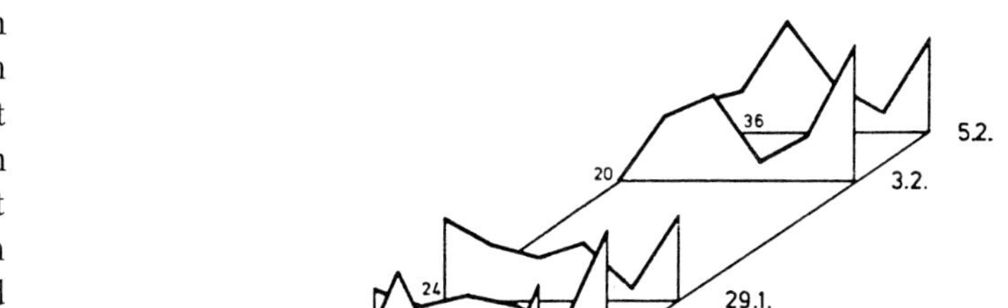

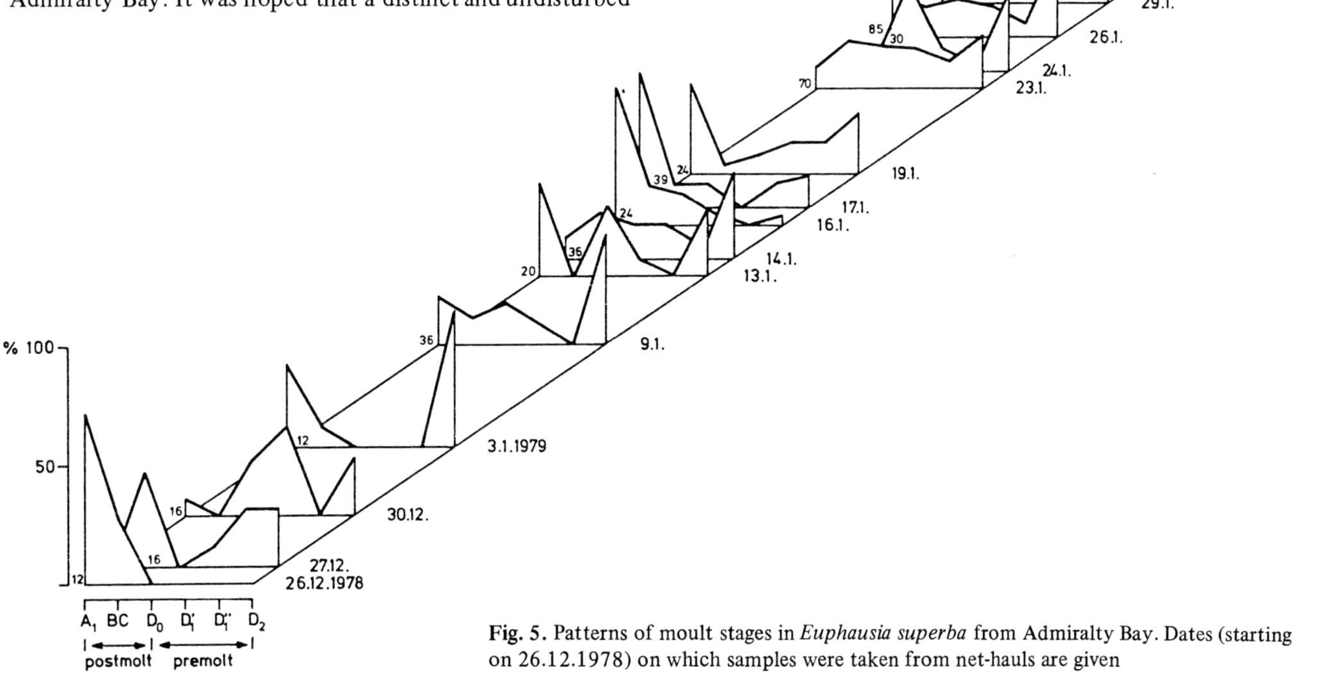

Fig. 5. Patterns of moult stages in *Euphausia superba* from Admiralty Bay. Dates (starting on 26.12.1978) on which samples were taken from net-hauls are given

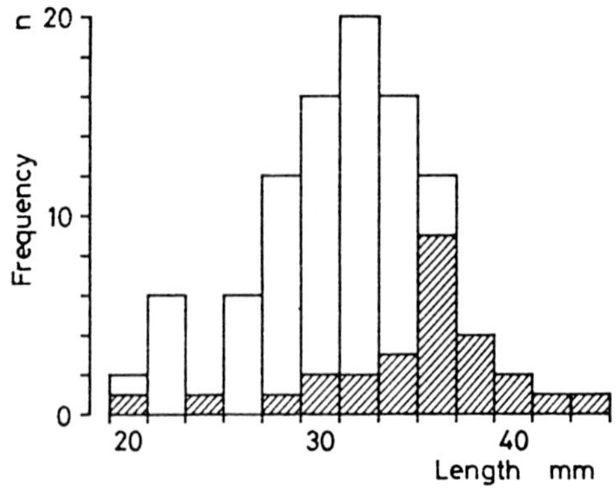

Fig. 6. Size-frequency distribution of an actively moulting Krill swarm taken from Admiralty Bay during December 1982. *Clear histograms* show specimens with soft or flexible exoskeletons; *hatched histograms* show specimens with rigid exoskeletons

the bay could become mixed with swarms of different size composition and moulting status from the nearby Bransfield Strait. Such frequent mixing would mask continuous moult development in field surveys, as well as in subsequent tank experiments with animals from Admiralty Bay.

Hamner et al. (1983) hypothesized that synchronous moulting can be triggered by external stimuli, such as the interference of a diver intruding into a Krill swarm. I believe that Hamner's divers met a patch of Krill just ready to moult. My observations in the laboratory show that ecdysis itself takes only a few seconds, with the animal literally shaking off its skin by flicking its tail. Accordingly, tail flicking could be induced as a fright response to the presence of a diver, resulting in the already loose skin being shed. It is doubtful whether this response could have occurred with animals in any stage of moulting other than that just prior to ecdysis. If so, one would expect to find Krill exuviae more regularly in net catches, since a similar fright reaction could presumably be induced by the net pressure wave. However, such behaviour has not been observed in Krill kept in tanks. Furthermore, histological evidence also militates against moulting occurring as a fright response, since both the new and the old cuticle stay in close contact until just prior to moulting (Buchholz, unpubl.).

Supportive evidence for moult synchrony is apparent from reports that Krill swarms often consist of individuals of a limited size range (Marr 1962). In addition, since development of the sexual appendages is coupled with the moulting process, individual animals would attain sexual maturity simultaneously. Coincident moulting is frequently observed in other crustacean taxa and is expressed, for example, as a regular succession of at least the first 5 to 10 moults after metamorphosis (Hartnoll 1983). In the benthic crustacean *Carcinus maenas* (Bückmann and Adelung 1964) moulting in the field is synchronized by an environmental stimulus: if the temperature after winter rises above 10°C, moulting is triggered and subsequently assumes a regular pattern. In Krill probably not rising temperatures, but the sudden onset of the phytoplankton bloom, following the post-winter break up of the ice-cover, may cause an analogous stimulus. Moreover, the commencement of moulting could be induced by the onset of high incident light levels characteristic of the austral spring and summer.

One more possible cause of moult synchrony is a consequence of an idea proposed by Kils (1982). He suggested that uni-sized Krill swarms are generated as a result of variations in sinking and swimming speed. Since ecdysis is an energy-intensive process, moulting Krill are likely to swim more slowly just before or after ecdysis and, therefore, are left behind by their faster moving swarm mates. These animals would then form synchronous aggregates of their own.

4 Conclusions

It has been demonstrated that it is possible for Krill maintained under artificial conditions to exhibit growth rates close to hypothesized "natural" values. Further consideration must be given to artificial maintenance systems which reduce direct handling of the animals, and which are of adequate size. Only then will it be possible to determine whether extensive reductions in body-size of Krill are the results of an artifact induced by inadequate maintenance conditions. The present data suggest that body shrinkage is not common to euphausiids.

The technique of moult-staging can be a valuable tool in determining whether moulting in *E. superba* continues during winter, as is the case in *M. norvegica*. Based on moult-staging, it seems highly probable that synchronous moulting occurs in *E. superba*. However, information is still lacking on the extent of synchrony in whole swarms, or whether large aggregations are structured in relation to their general moulting status. To answer these questions, the continuous monitoring of specific swarms and the use of discrete depth-sampling techniques are necessary. If swarms could be followed for several days, data on moult frequency and direct growth should be gained by moult-staging. Should a specific moulting pattern be detected within a swarm, this could then also be used to answer questions regarding the effects of feeding, reproduction and migration on growth in individual animals. This could help to elucidate the dynamics and coherence of Krill swarms more clearly.

Acknowledgements. My studies would not have been possible without the generous help of members of the 3rd and 6th Polish Antarctic expeditions. My sincere thanks are due to Professors D. Adelung and G. Hempel for their constant support. I acknowledge the financial assistance of the Deutsche Forschungsgemeinschaft, grants Ad 24/6 + 9 and Bu 548.

References

Båmstedt U (1976) Studies on the deep water pelagic community of Korsfjorden: Changes in size and biochemical composition of *Meganyctiphanes norvegica* in relation to its life cycle. Sarsia 61:15–30

Baker AC, Clarke MR, Harris MJ (1973) The N.I.O. Combination Net (RMT 1+8) and further developments of Rectangular Midwater Trawls, J M Biol Assos U K 53: 167-184

Boysen E, Buchholz F (1984) *Meganyctiphanes norvegica* in the Kattegat. Studies on the annual development of a pelagic population. Mar Biol 79: 195–207

Buchholz F (1982) Drach's moult staging system adapted for euphausiids. Mar Biol 66: 301–305

Bückmann D, Adelung D (1964) Der Einfluß der Umweltfaktoren auf das Wachstum und den Häutungscyklus der Strandkrabbe *Carcinus maenas.* Helg Wiss Meeresunters 10:91–103

Clarke A (1976) Some observations on Krill (*Euphausia superba* Dana) maintained alive in the laboratory. Bull Br Antarct Surv 43: 111–118

Drach P (1939) Mue et cycle d'intermue chez les crustacees decapodes. Ann Inst Oceanogr Monaco 19:103–391

Everson I (1977) The living resources of the Southern Ocean. Rome, FAO United Nations Development Programme, Southern Ocean Fisheries Survey Programme GLO/SO/77/1:156 pp

Fowler SW, Small LF, Kečkeš S (1971) Effects of temperature and size on molting of euphausiid crustaceans. Mar Biol 11:45–51

Hamner WM, Hamner PP; Stand SW; Gilmer RW (1983) Behaviour of Antarctic Krill, *Euphausia superba* : Chemoreception, feeding, schooling, and molting. Science 220: 433–435

Hartnoll RG (1982) Growth. In: Bliss D E (ed) Biology of crustacea. Academic, New York Vol 2, pp 111–195

Ikeda T, Dixon P (1982 a) Body shrinkage as a possible over-wintering mechanism of the Antarctic Krill, *Euphausia superba* Dana. J Exp Mar Biol Ecol 62: 143–152

Ikeda T, Dixon P (1982 b) Observations on moulting in Antarctic Krill (*Euphausia superba* Dana). Aust J Mar Freshwater Res 33: 71–76

Ivanov BG (1970) Biology of Antarctic krill (*Euphausis superba* Dana). Mar Biol 7: 340–351

Kils U (1982) Size dissociation in Krill swarms. Kiel Meeresforsch Sonderh 5: 262–263

Mackintosh NA (1967) Maintenance of living *Euphausia superba* and frequency of moults. Nor Hvalfangs-tid 56: 97–102

Mackintosh NA (1972) Life-cycle of antarctic krill in relation to the ice conditions. Discovery Rep 34: 1–94

Marr JWS (1962) The natural history and geography of antarctic krill (*Euphausia superba* Dana). Discovery Rep 32: 33–464

Matthews JB (1973) Ecological studies on the deep water pelagic community of Korsfjorden, Western Norway: Population dynamics of *Meganyctiphanes norvegica* in 1968 and 1969. Sarsia 54: 75–90

Mauchline J (1980) The biology of mysids and euphausiids. Adv Mar Biol, Academic, London Vol 18, 681 pp

Morris DJ, Keck A (1984) The time course of the moult cycle and growth of *Euphausia superba* in the laboratory. A preliminary study. Meeresforschung 30: 94–100

Murano M, Segawa S, Kato M (1979) Moult and growth of the Antarctic Krill in the laboratory. Trans Tokyo Univ Fish 3: 99–106

Passano LM (1960) Molting and its control. In: Waterman TH (ed) The physiology of crustacea. Academic Press, New York Vol 1, pp 473–536

Poleck TP, Denys CJ (1982) Effect of temperature on the moulting, growth and maturation of the Antarctic Krill *Euphausia superba* (Crustacea: Euphausiacea) under laboratory conditions. Mar Biol 70: 255–265

Rakusa-Suszczewski S, Stepnik SR (1980) Three Species of Krill from Admiralty Bay (King George, South Shetland) in summer 1978/79. Pol Arch Hydrobiol 27 (2): 273–284

Structural and Functional Analyses of the Feeding Appendages of Krill Larvae

H.-P. MARSCHALL[1]

Summary. This report describes the feeding appendages and the gut contents of 1st-calyptopis to 1st-furcilia stages of the Antarctic Krill, *Euphausia superba,* based mainly on scanning electron-microscopy. Distances between the second-degree setae of the second maxillae (which are the actual filtering appendages) indicate that larvae might be able to feed on particles with a minimum diameter of approximately 1 μm. There is some evidence for larvae cracking larger particles before the food reaches the mandibles. This might cause losses of diluted and particulate organic carbon. Mandibular structures are well suited for grinding food. Hence, only small fragments of ingested food are present in the gut. Analysis of gut contents revealed that diatoms are an important food source. The size of ingested diatoms covered a wide range, including *Thalassiosiara* sp. (ca. 5 μm in diam.), *Chaetoceros* sp., and *Corethron criophilum* (> 200 μm in length).

1 Introduction

In contrast to what is known about adult Antarctic Krill, *Euphausia superba,* there is very little information on the feeding behaviour of the larvae. Studies of the functional morphology of the feeding appendages of adult Krill have yielded valuable information on the species' feeding (Barkley 1940; Alberti and Kils 1980; Kils 1983; McClatchie and Boyd 1983) and, similarly so, for copepods (Anraku and Omory 1963; Marshall 1973; Nival and Nival 1976; Schnack 1982). This approach provides information only on the potential food sources, but analysis of gut contents can reveal the food ingested. A combination of both methods was used in this study in addressing the question: are morphological differences in the feeding appendages of Krill at different developmental stages correlated with differences in diet?

The first detailed description of the feeding appendages of euphausiid larvae was made by Sars (1885) for *Nyctiphanes australis.* Fraser (1936) described the development of *Euphausia superba* from the nauplius to the last furcilia stage, and also gives a description of the feeding appendages. Since they lack a filtering basket, as described for adult Krill (e.g., Barkley 1940; Kils 1983), the feeding of the early larval stages (calyptopes I–III and furcilia I) must differ from that of adult Krill. The filtering basket apparently only becomes functional from the second furcilia stage onwards.

1 Alfred-Wegener-Institut für Polarforschung, Columbus-Center, D-2850 Bremerhaven, FRG

2 Materials and Methods

Krill larvae were captured, using a RMT 1 + 8 net, from the R.V. Walther Herwig (Piatkowski and Klages 1983). They were preserved in 4% borax-buffered formaldehyde immediately after capture. Larvae examined by scanning electron-microscopy (SEM) were rinsed with distilled water and dehydrated by means of 30, 50, 70, and 100% alcohol before being transferred to DMP and subsequently critical point dried using liquid CO_2. After mounting on stubs, larvae were coated with approximately 10 nm gold-palladium and examined with an ISI-DS 130 SEM. Three-dimensional and fine structure analyses of the feeding appendages of a filter-feeding crustacean are important in evaluating the animals' potential food. The SEM is more suitable than the light-microscope for this and, therefore, it was used to examine feeding appendages and food particles lodged between mouth-parts and within the gut of Krill larvae.

3 Results

3.1 Mouth-Parts

The number of feeding appendages and their essential morphology remain relatively constant between 1st-calyptopis and 1st-furcilia stages. Hence, feeding appendages are described once for all these larval stages. Figure 1 shows the general arrangement of the feeding appendages.

3.1.1 Labrum and Labium

The labrum and also the paired labium bear a great number of flat setae. The ventral surface of the labrum shows discrete fields of protuberances (Fig. 2a, b). Two types of pores (possibly chemoreceptors) were found regularly on the labrum: a great number of very small pores (ca. 0.1 μm diam.) in two fields on each side of the ventral surface of

Antarctic Nutrient Cycles and Food Webs
(ed. by W.R. Siegfried, P.R. Condy, and R.M. Laws)

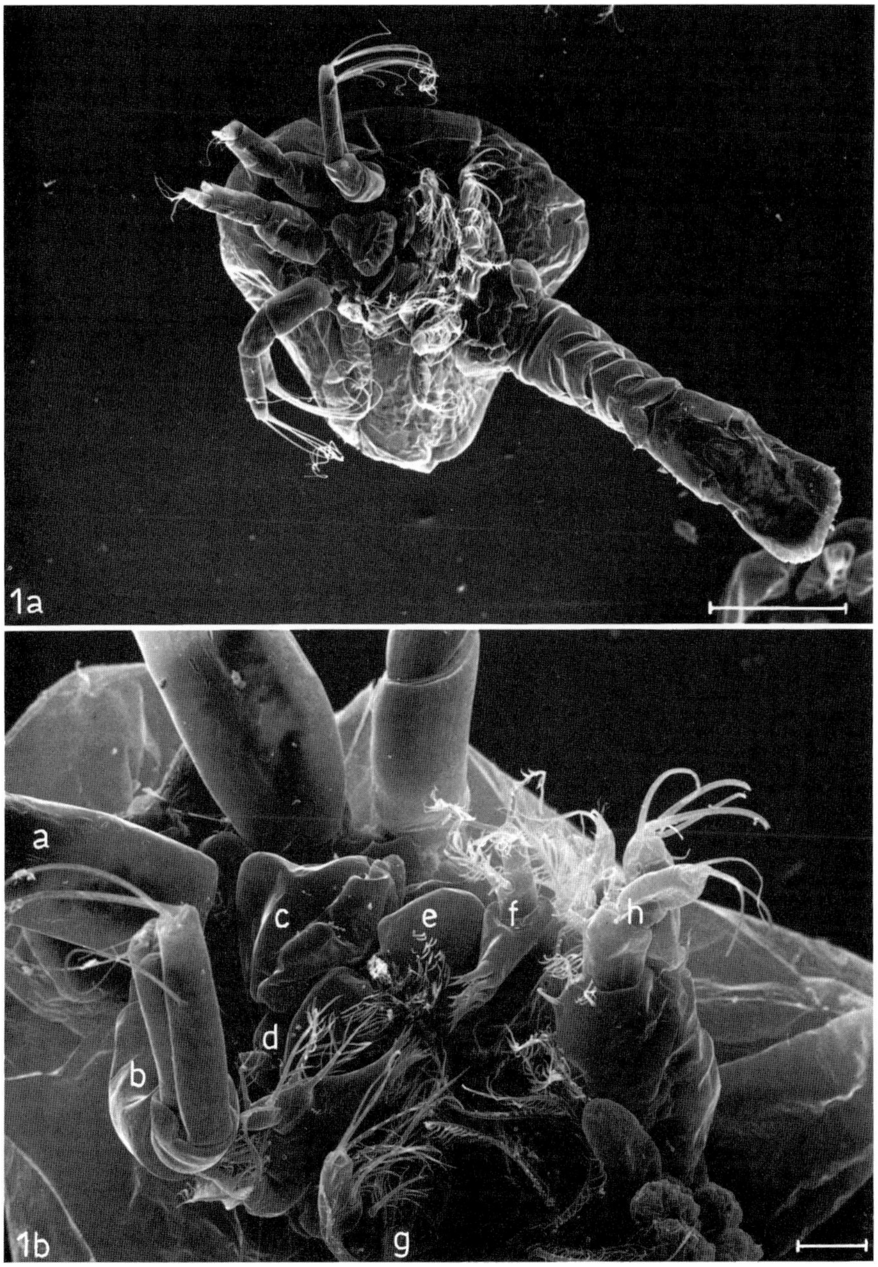

Fig. 1. **a** Calyptopis II (*bar* 500 μm). **b** Ventral view of calyptopis III. *a* 1st antenna; *b* 2nd antenna; *c* labrum; *d* mandible mostly covered by labium; *e* labium; *f* 1st maxillae; *g* 2nd maxillae; *h* 1st thoracopod (*bar* 100 μm)

the labrum (Fig. 2b); and, three pairs of larger pores (two pairs are shown in Fig. 2a, the other pair being on the anterior ventral surface of the labrum).

3.1.2 Mandibles

As in other eucarid and peracarid crustaceans (Gordon 1964) and in adult *E. superba,* the larval mandible consists of two principal components: a ventrally situated incissor process or cutting part, and a molar process located dorsally (Fig. 3a). In the gap between these two parts, the right mandible (viewed dorsally) of all investigated larval stages carries a *Lacinia mobilis* and a spine row (Fig. 3b), whereas the *Lacinia mobilis* does not occur on the left mandible (Fig. 3a, b). In adult and sub-adult Krill the *Lacinia mobilis* and the spine rows are absent from both mandibles (McClatchie and Boyd 1983). The molar process consists of a series of protuberances on each mandible, which correspond with notches on the other mandible. The end of each series bears a dentate spine, anteriorly on the right mandible and posteriorly on the left mandible (Figs. 3c, d, 7). This arrangement differs considerably from that of the adults (cf. McClatchie and Boyd 1983). Only the finally divided teeth of region five (see McClatchie and Boyd 1983) are somewhat similar to the dentate spines of the larvae. However, despite the morphological differences, the molar processes in both cases are very well suited for grinding food.

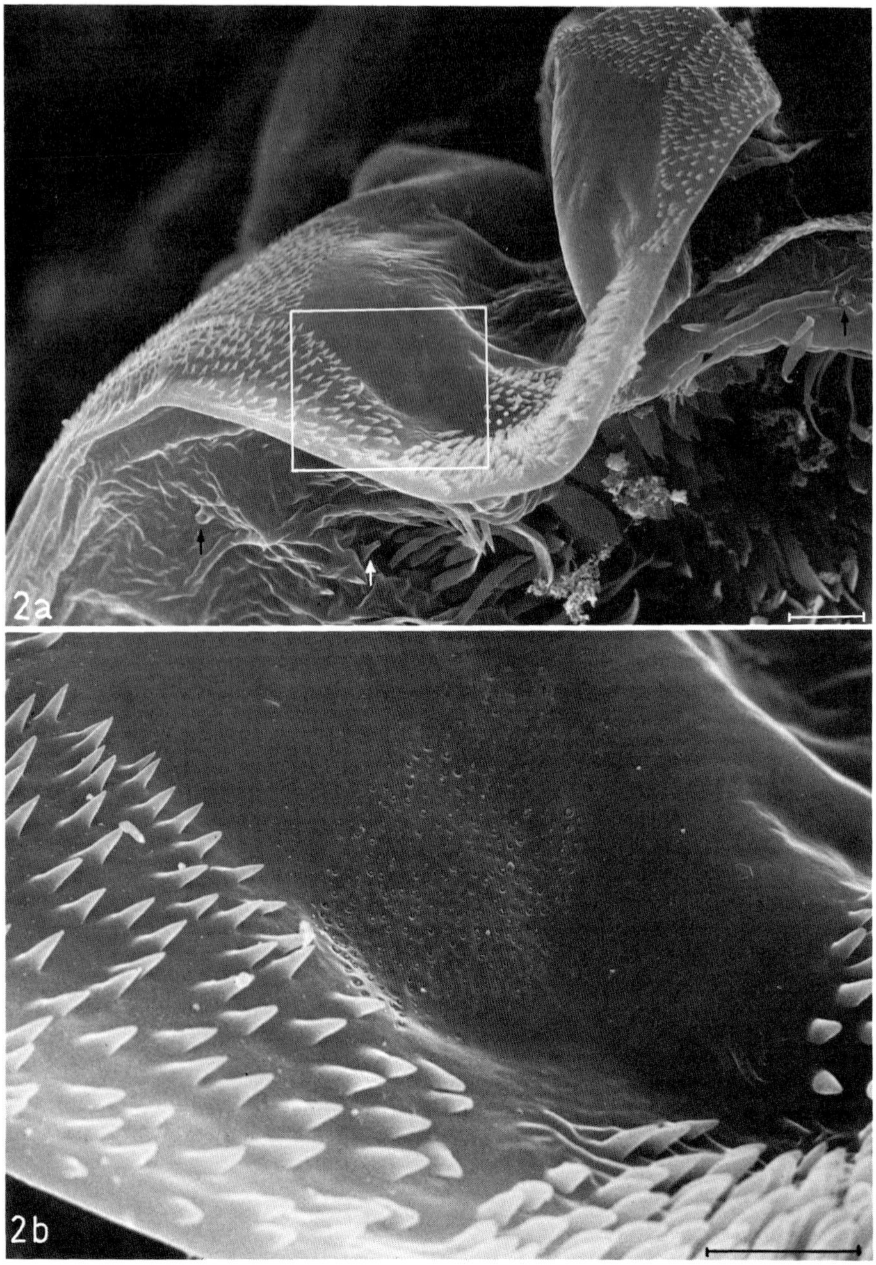

Fig. 2. a Ventral view of labrum of furcilia I seen from posterior. Large pores indicated by *arrows* (*bar* 10 μm). **b** Close-up of the area shown in *a.* A second field of small pores lies on the other side of the labrum (*bar* 1 μm)

The incissor process obviously functions in cutting up large food particles, and the *Lacinia mobilis* and the spine rows might facilitate the transfer of food particles.

3.1.3 First maxillae

The basal endite of the first maxilla normally bears three, four and five spines in the calyptopis I, II, and III stages, respectively, and seven spines in the 1st-furcilia stage. The spines are arranged in two rows, and there is no interruption between basal endite and spine in the anterior row (Fig. 4b, c). The bases of the posterior spines apparently consist of relatively weak material (Fig. 4a–c). This suggests that they are movable. The type of dentation is also different in each row, which supports this suggestion (Fig. 4b, c). Krill larvae may use these spines to handle and possibly also to crack food particles. The pores (Fig. 4a, c, d) are similar to the large type found on the labium. The number of these pores increases from 1 pair in the 1st-calyptopis to five pairs in the 1st-furcilia stage. The setae of the coxal endite are usually inserted in the gap between the two parts of the labium.

Fig. 3. a Left mandible (dorsal side up) with incissor process, spine row and molar process (*bar* 50 μm). **b** *Lacinia mobilis* and spine rows of right and left mandibles (dorsal side up, seen from posterior) fixed while handling food particles (*bar* 10 μm). c Incissor process (labrum up) fixed while cutting (*bar* 50 μm). **d** and **e** Molar processes of left and right mandible. Note the correspondence of protuberances of one mandible with notches on the other mandible (*bar* 10 μm)

Fig. 3a–e

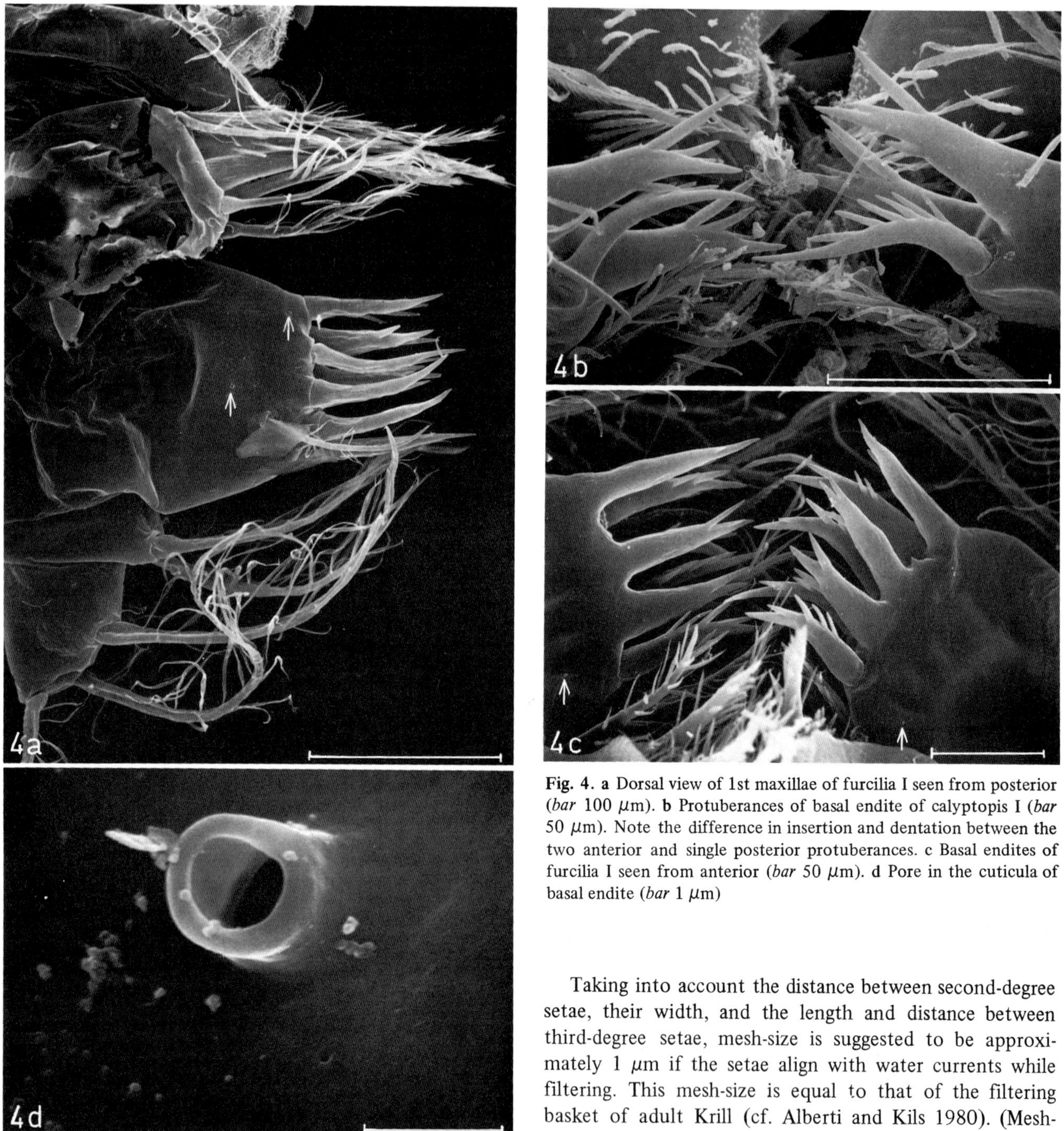

Fig. 4. **a** Dorsal view of 1st maxillae of furcilia I seen from posterior (*bar* 100 μm). **b** Protuberances of basal endite of calyptopis I (*bar* 50 μm). Note the difference in insertion and dentation between the two anterior and single posterior protuberances. **c** Basal endites of furcilia I seen from anterior (*bar* 50 μm). **d** Pore in the cuticula of basal endite (*bar* 1 μm)

3.1.4 Second maxillae

The second maxillae provide a very fine-mesh filter, in that they bear a series of second- and also third-degree setae (Fig. 5a). Gaps between the second-degree setae are within 2–5 μm wide. The distance between these setae increases from the coxopodite to the outer segments of the maxilla, and also between the base and the tip of each seta (Fig. 5a, c). Second-degree setae are flat and bear very fine third-degree setae, ca. 0.1 μm in diameter (Fig. 5c–e), which cannot be investigated by light-microscopy.

Taking into account the distance between second-degree setae, their width, and the length and distance between third-degree setae, mesh-size is suggested to be approximately 1 μm if the setae align with water currents while filtering. This mesh-size is equal to that of the filtering basket of adult Krill (cf. Alberti and Kils 1980). (Mesh-size of the filtering area can be over-estimated, using light-microscope measurements of distances between second-degree setae only.)

Small setae, as shown in Fig. 5b, were found on the cuticula of the second maxillae and also on all other appendages; they might be mechano-receptors.

3.2 Food

Food particles, mainly diatoms and the silicoflagellate *Distephanus* sp., were often found on the feeding appendages (Fig. 6a, b). All food species observed on and among

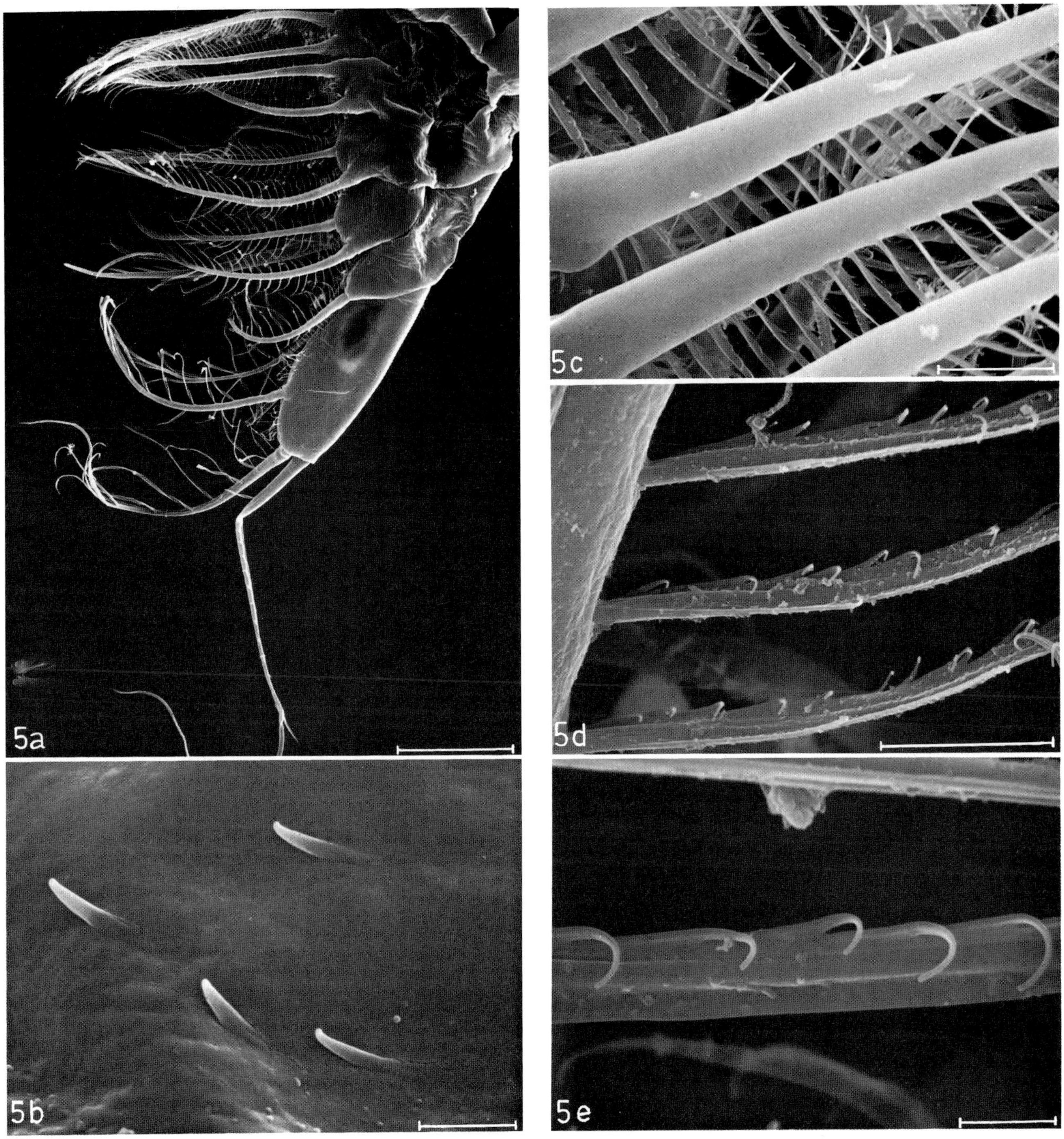

Fig. 5. **a** Dorsal view of 2nd maxillae seen from anterior (*bar* 100 μm). **b** Small setae, regularly found in discrete regions on the cuticula of the segments (*bar* 1 μm). **c** Part of the filtering area of 2nd maxillae (*bar* 10 μm). **d** First-degree setae bearing setae of second- and third-degree (*bar* 5 μm). **e** Setae of second and third degree (*bar* 1 μm)

mouth-parts were also found in the guts of Krill larvae, even in their posterior segments (Fig. 8a–d).

A wide size-range of food items was observed in the diet of Krill larvae. The smallest species which could be identified included *Thalassiosira* sp. (5 μm in diameter). The largest diatoms were *Chaetoceros* sp. and *Corethron criophilum* (often > 200 μm in length). Specimens of even the first feeding stage (calyptopis I) often ingested these large diatoms, including their spines (Fig. 8b).

It is obvious that Krill larvae cannot ingest relatively large food particles without first cutting them to bits. Food particles regularly found among the mouth-parts of larvae were already cracked (Fig. 6a, b), indicating that this is not only done by the mandibles but probably also by the spines

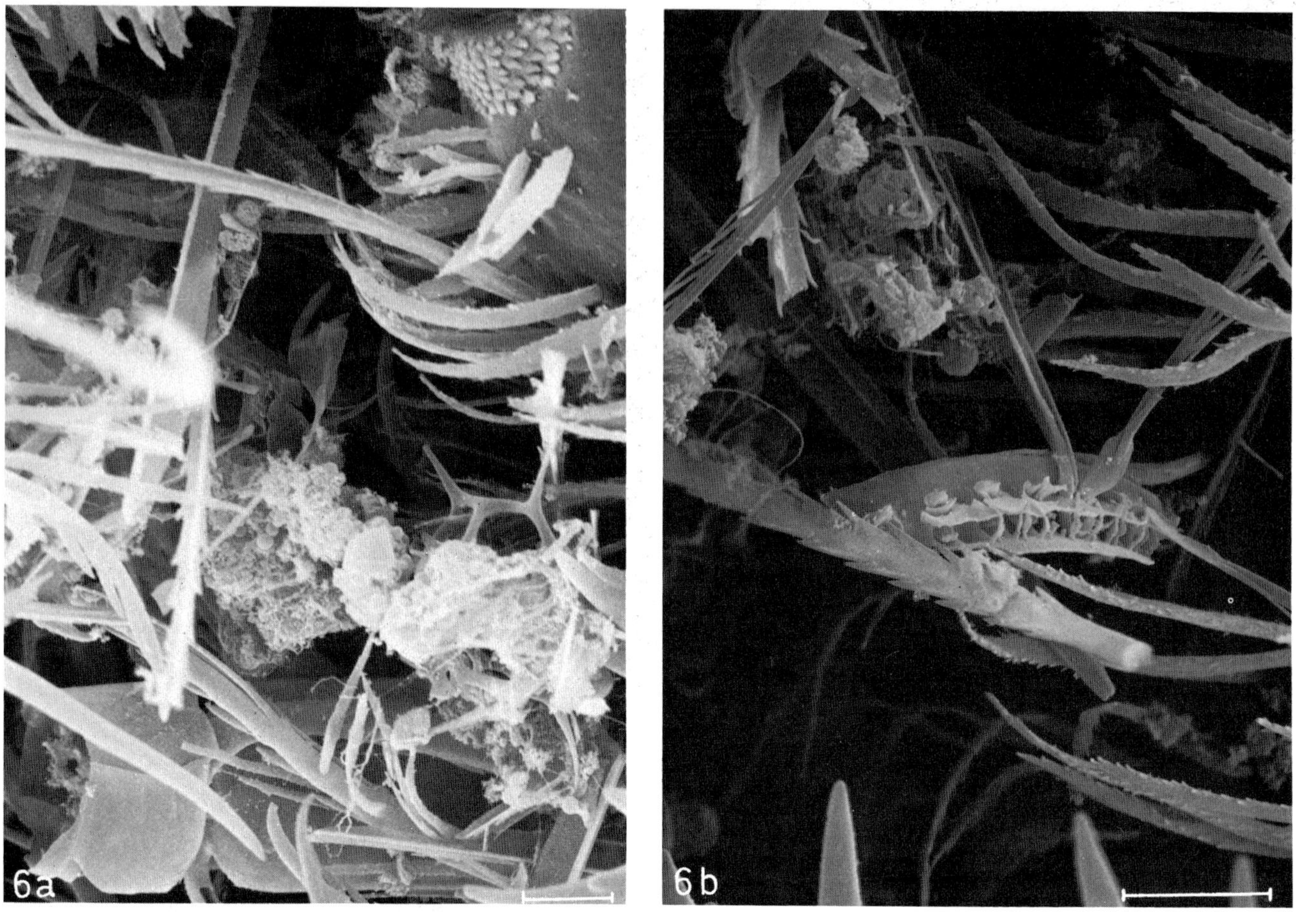

Fig. 6. a Labrum of calyptopis 1 with fragments of *Corethron criophilum, Distephanus* sp. and *Chaetoceros* sp. (*bar* 10 μm). **b** Labrum of calyptopis 1 with fragments of *Corethron criophilum* (*bar* 10 μm)

of the first maxillae. Small diatoms (e.g., *Nitzschia* sp.) were sometimes found intact on the outer part of the mandibles. Food in the gut was always macerated. Normally fragments in the gut did not exceed 10 μm in length. An anterior view of the molar processes (Fig. 7) of the 1st-calyptopis stage shows that most food particles will be ground up in passing the processes. (The spine of *Corethron criophilum* in the foreground (Fig. 7) gives an impression of the size relationship between food and grinding surface of the molar processes.)

4 Discussion

The nature of the diet of Krill larvae, as evaluated by gut content analysis, depends on the food supply available to the larvae just before fixation. Subsequently, delicate organisms, such as flagellates, can lose their identity due to fixation in formaldehyde. I have shown that Krill larvae ingest diatoms ranging from 5 to > 200 μm length. However, based on the suggested mesh-size of the filtering setae, larvae might be able to feed on particles as small as 1 μm

Fig. 7. Dorsal view of molar processes of calyptopis I seen from anterior. The spinous seta is from *Corethron criophilum* (*bar* 10 μm)

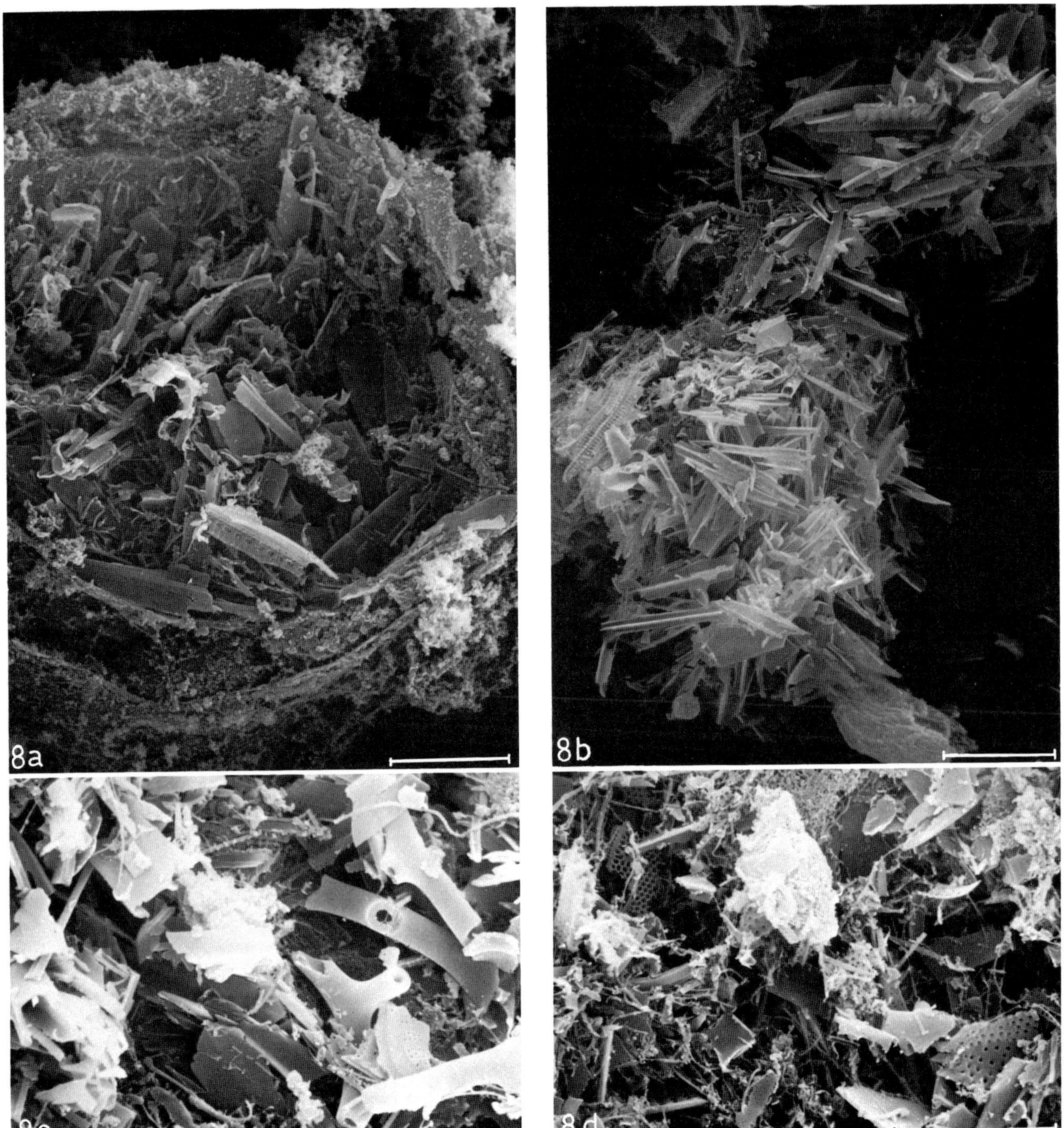

Fig. 8. a Gut and gut content of calyptopis I (*bar* 10 μm). **b** Gut content of calyptopis I. Fragments of *Nitzschia* sp. and *Corethron criophilum* can be identified (*bar* 10 μm). **c** Gut content of calyptopis I with fragments of *Nitzschia* sp. and *Distephanus* sp. (*bar* 1 μm). **d** Gut content of furcilia I. Fragments of *Nitzschia* sp. can be identified (*bar* 1 μm)

in diameter. Thus, they might be able to utilize food sources such as flagellates and bacteria.

I am well aware of the problems involved in predicting a minimum size of retained particles, from measurements of distances between filtering setae. Moreover, retention efficiency might not be equal for different size classes and/or different species of prey. Nevertheless, the guts examined in my study were filled almost entirely with diatom shells. Hence, diatoms apparently are an important food of Krill larvae, at least during the austral summer. From the structure of the feeding appendages (second maxillae and mandibles), I suggest that Krill larvae are herbivorous filter feeders, rather than omnivorous or carnivorous. However, it cannot be excluded that they by chance feed on zooplankton as well. During culturing of Krill larvae cannibalism was observed several times (H.-P. Marschall, unpublished). Thus, it seems as if there are no apparent differences between the food of larval and adult Krill.

Sullivan et al. (1975) examined the morphology of the mandibles of 11 herbivorous copepod species with SEM. They pointed out that copepod mandibles are built for cracking rather than grinding food. Krill larvae, which are of similar size, are well suited for grinding their food and obviously do so in spite of their feeding appendages differing morphologically from those of adult Krill.

Acknowledgements. I am grateful to Dr. R. Simonsen, Institut für Meeresforschung, Bremerhaven, for his support of the SEM studies.

References

Alberti G; Kils U (1980) The filtering basket of *Euphausia superba.* ICES 1980 L.54

Anraku M; Omori M (1963) Preliminary survey of the relationship between the feeding habit and the structure of the mouth parts of marine copepods. Limnol Oceanogr 8:116–126

Barkley E (1940) Nahrung und Filterapparat des Walkrebschens *Euphausia superba* Dana. Z Fisch 1:65–156

Fraser FC (1936) On the development and distribution of the young stages of krill (*Euphausia superba*). Discovery Rep 14:3–192

Gordon I (1964) On the mandible of the stygocaridae (Anaspidacea) and some other Eumalacostraca, with special reference to the lacinia mobilis. Crustaceana 7:150–158

Kils U (1983) Swimming and feeding of Antarctic krill, *Euphausia superba* – some outstanding energetics and dynamics, some unique morphological details. Ber Polarforsch Sonderh 4:130–155

Marshall SM (1973) Respiration and feeding in copepods. Adv Mar Biol 11:57–120

McClatchie S; Boyd CM (1983) Morphological study of sieve efficiencies and mandibular surfaces in the Antarctic krill, *Euphausia superba.* Can J Fish Aquat Sci 40:955–967

Nival P; Nival S (1976) Particle retention efficiencies of an herbivorous copepod, *Acartia clausii* (adult and copepodite stages): Effects of grazing. Limnol Oceanogr 21:300–308

Piatkowski U; Klages N (1983) German Antarctic Expedition 1980/81 with FRV "Walter Herwig" and RV "Meteor". First International BIOMASS Experiment. Data of micronekton and zooplankton hauls. Ber Polarforsch 15

Sars GO (1885) Report on the Scizopoda collected by H.M.S. "Challenger" during the years 1873–76. The Voyage of H.M.S. "Challenger", 13(37):1–228

Schnack SB (1982) The structure of the mouth parts of copepods in Kiel Bay. Meeresforschung 29:89–101

Sullivan BK; Miller CB; Peterson WT; Soeldner AH (1975) A scanning electron microscope study of the mandibular morphology of boreal copepods. Mar Biol 30:175–182

Marine Macro-Plankton of Two Sub-Antarctic Islands

D.G.M. MILLER[1]

Summary. Surveys of the marine macro-plankton and hydrography were made in the Gough Island (South Atlantic) and Prince Edward Island (South Indian) regions. There was a marked faunal dissimilarity between the 2 regions. This is attributed to hydrographic differences, resulting from the close proximity of Gough Island to the Sub-Tropical Front and the Prince Edward Islands to the Antarctic Polar Front, respectively. Information statistics were used in an attempt to isolate 'indicator species' for each region. Although perfect indicator species were not evident (i.e. occurring in all samples from 1 particular region), 12 species were sufficiently dissimilar in abundance to warrant separating the macro-plankton of the 2 regions. Three species characterized the Prince Edward Island region. Many of the indicator species were found in the stomach contents of island-based predators.

1 Introduction

Many Southern Ocean macro-plankton species occupy a characteristic circumpolar distributional zone, at least in the surface layers (Baker 1954; Mackintosh 1960). The latitudinal width of this zone may vary, but within it the distribution of populations tends to be continuous. Mackintosh (1960) hypothesized, however, that some regional differentiation does in fact occur and that the bulk of a single population is maintained within a limited biogeographical area or "faunal province". This means that, even within the open expanses of the Southern Ocean, a complex pattern of macro-plankton distribution is possible and considerable zonal segregation may occur, particularly between Antarctic and Sub-Antarctic organisms (Knox 1960; Mackintosh 1960; Boltovskoy 1975). Both Ekman (1953) and Deacon (1982) maintain that this segregation is influenced by prevailing hydrographic conditions, especially by the boundary effects of various frontal systems.

The Sub-Antarctic islands are regarded as small stenotopic environments which contribute little to the circumpolar zonation patterns of plankton distribution (Hedgepeth 1970). However, it has become increasingly apparent that certain Sub-Antarctic islands do in fact exert considerable influence on neighbouring hydrodynamic events (Plancke 1977; Grindley and Lane 1979; Allanson et al. 1985). This influence, when in close proximity to a major frontal system, may cause considerable interchange across the front and preclude the persistence of rigid biogeographical boundaries nearby (Deacon 1983).

Gough Island (40° 20′S, 09° 54′W) and the Prince Edward Island Group (46° 45′S, 37° 55′E) are situated close to two major frontal systems (Knox 1960). Gough Island lies close to the Sub-Tropical Front (STF), whereas the latter group is located near the Antarctic Polar Front (APF) (Deacon 1937). Consequently, it seems logical to expect that each region has its own particular macro-plankton community and that the differences between the communities reflect the prevailing hydrographic influences of the respective frontal systems. Recent surveys, using the same methods in each region (Miller 1982a, b; Miller et al. 1983), offer an unique opportunity to compare macro-plankton species diversity and abundance in the two areas.

2 Materials and Methods

Two hydro-acoustic and mid-water trawling surveys were made at the Prince Edward Islands and at Gough Island (Miller 1982a,b; Miller et al. 1983). A modified neuston net (Cheng 1975) and either an 8 m^2 or a 2 m^2 rectangular mid-water trawl (RMT) (Robertson et al. 1981) were used for collecting macro-plankton. The neuston net sampled surface waters and the RMTs were fished obliquely from the suraface to 100 m depth. The neuston net was fitted with 900 μ mesh, and the cod-ends of both RMTs were lined with the same mesh. Net towing speed was 3.6 km h^{-1}, and the volume of water filtered to obtain each sample was measured. Organisms were identified and counted; the number of individuals per 1000 m^3 of water filtered was calculated.

Catches were extracted where organism identification was successful to species. Of a total of 67 net hauls, 26 at Prince Edward and 30 at Gough fulfilled this criterion. Catch data were grouped, in the absence of demonstrable differences between catches by the various nets and between day and night ($P = < 0.01$). A search for biotic patterns in the macro-plankton samples was undertaken, using the "Bray-Curtis" dissimilarity analysis (Bray and Curtis 1957; Field et al. 1982). A normal (q-type) analysis was carried

1 Sea Fisheries Research Institute, Private Bag X2, Roggebaai 8012, South Africa

Antarctic Nutrient Cycles and Food Webs
(ed. by W.R. Siegfried, P.R. Condy, and R.M. Laws)

out (Field et al. 1982) on the lumped-haul data to identify sample groups. Data were logarithmically transformed to minimize the weighting of more abundant species (Clifford and Stephenson 1975). The application of the similarity measure gave a matrix whose entries compared each of n samples with every other sample. Group-average sorting (Clifford and Stephenson 1975) was then used to obtain a dendrogram in which sample groups were linked at an average level of similarity between all members of each and every group. A second, complementary, analysis grouped species. The total number of species collected at both localities was reduced from 69 to 51 by using only those which had > 4% dominance in any single haul. Species abundance was standardized for each haul as a percentage of the total abundance in all hauls (i.e. if a species was found in only one haul its abundance there was 100%). An inverse (r-type) analysis was then done on the standardized species data (Field et al. 1982).

Both the normal haul and standardized species data were ordinated using multi-dimensional scaling (MDS) (Kruskal and Wish 1978). A starting map of n stations or species was constructed in the required number of dimensions and the interpoint distances of this configuration were then regressed on the corresponding dissimilarity. The "goodness-of-fit" of this regression was measured by a criterion known as stress. Stress may be thought of as the distortion involved in "compressing" the data to a small number of dimensions. A low stress value indicates that a sample or species relationship is well represented by the original map in a specified number of dimensions.

3 Results

Figure 1 shows station affinities based on the log-transformed abundance of all 69 species encountered. A broken line drawn at the arbitrary similarity level of 10% delineates four major net-haul groups. Group 1 represents hauls at Gough only, and Group 2 consists essentially of hauls from the Prince Edward region. Three hauls (55, 42 and 44) in the latter group are from Gough; they are designated as a sub-division (Sub-Group 2/1). Groups 3 and 4 are from the Gough region only, and because they are minor groupings they are not discussed further. Three hauls (34, 7, and 39) remain which fall outside any major grouping and represent only single species' occurrences.

Figure 2 shows the result of MDS in 2-dimensions, using the similarity matrix above and delineating haul groups from the dendrogram (Fig. 1). Considering the resultant stress factor (0.19), it is clear that the map more than adequately portrays the relationship between hauls. Al-

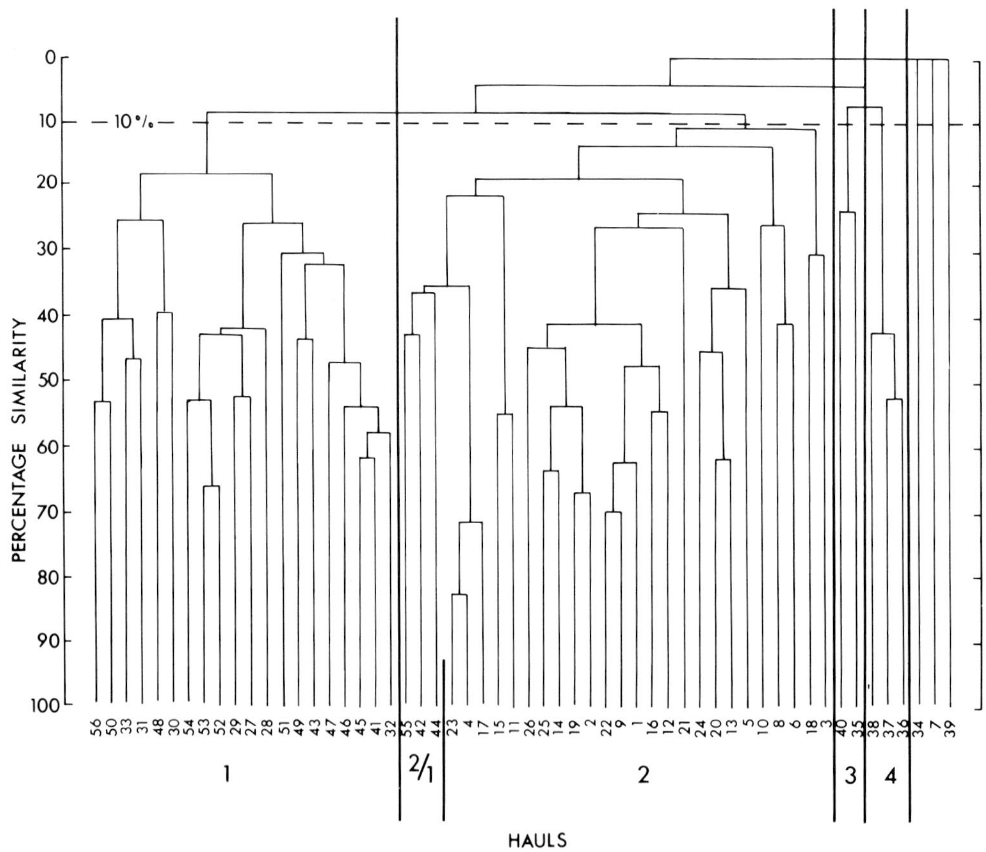

Fig. 1. Classification of 56 net-haul stations at Gough and the Prince Edward Islands. Abundances of species were log-transformed and the dendrogram formed by group average sorting. Main haul groups (*1–4*) are distinguished at an arbitrary similarity level of 10%

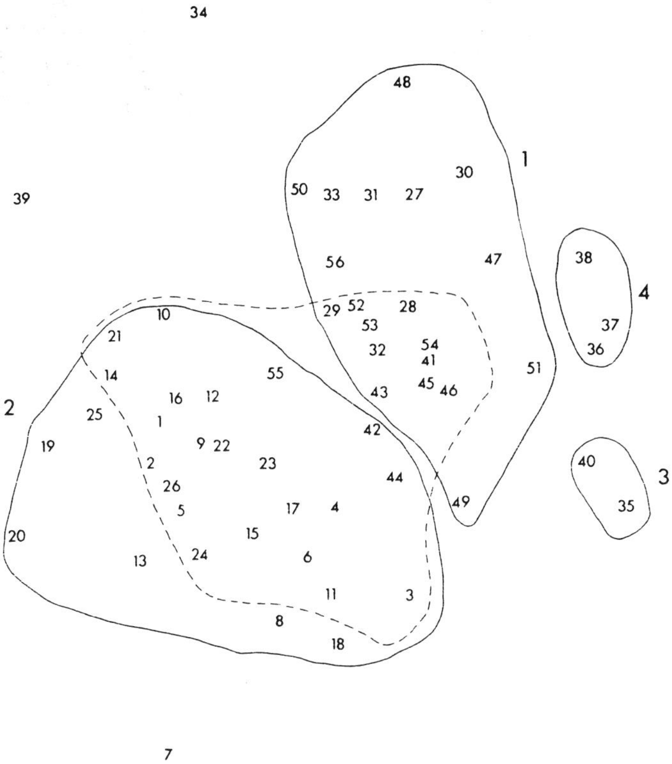

Fig. 2. Ordination of 56 net-hauls at Gough and the Prince Edward islands, using multi-dimensional scaling in the same similarity matrix as in Fig. 1. Stations with concurrent species are demarcated within the *broken lines*. Axis scales are relative and therefore are not given. Stress = 0.19

though this analysis gives essentially the same picture as the dendrogram (Fig. 1), it highlights the close proximity of the Gough hauls (55, 42 and 44), which are grouped in the Prince Edward cluster (Group 2), to Group 1 (the Gough haul cluster). This confirms the sub-group designation described above and indicates a weighting of these hauls by species coincident in both Groups 1 and 2. Hauls sharing species are identified and are demarcated by the broken boundary line in Fig. 2, indicating a close inter-group affinity which, however, does not withstand the specific differences between the two regions.

Figure 3 shows the dendrogram for the inverse analysis of relationships between species. To obtain the same number of species as station groups, an arbitrary line has been drawn at the 3% level of similarity. The four major groupings of species correspond to those defined in Fig. 1, and for convenience they have been given the same notation. A second short line has been drawn at the 5% level of similarity to further isolate the five species (two concurrent) which correspond with the Sub-Group 2/1. Lists of species in the various groups are presented in Appendix 1.

Figure 4 is a two-dimensional map of the species analysis obtained by MDS; groups of species being again designated from the dendrogram. Figure 5 shows the same configuration with the species' numbers replaced by the haul group/s in which they were found to represent > 4% of the specimens found in any one haul. A strong agreement between the species and haul clusters prevails. Concurrent species can be identified in both Figs. 4 and 5. Hence Fig. 5 pro-

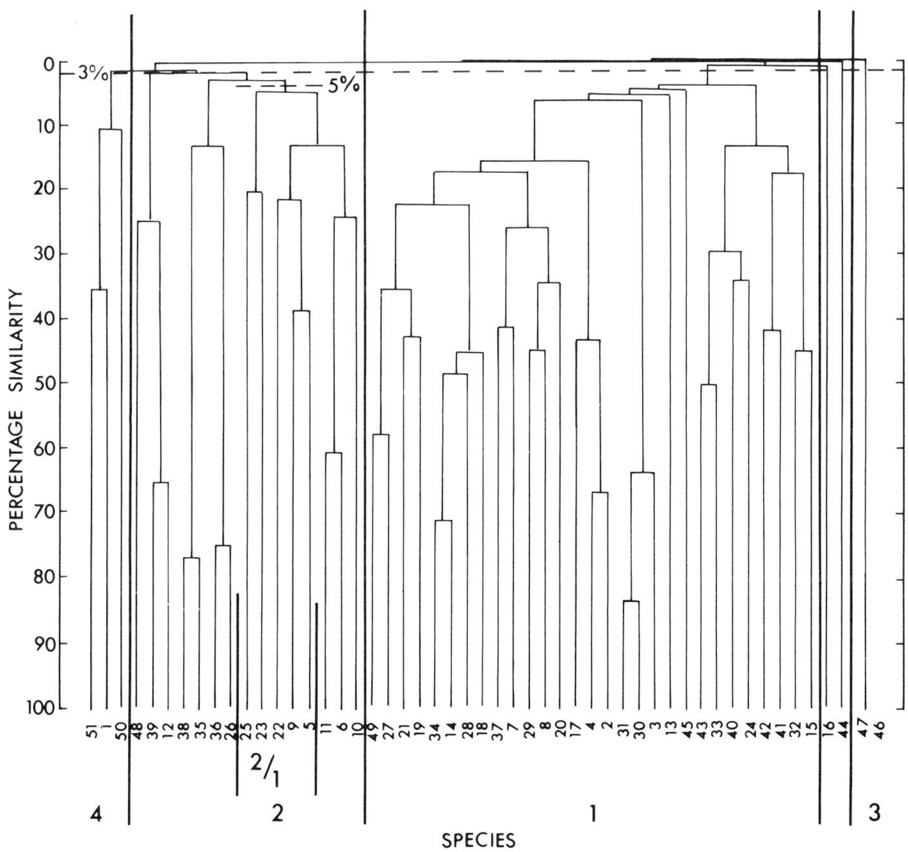

Fig. 3. Dendrogram of inverse analysis comparing 51 species occurring in > 4% dominance in any of the 56 net-hauls at Gough and the Prince Edward islands. Abundances of species were standardized and group-average sorting was used. The names and numbers of species are contained in Appendix 1

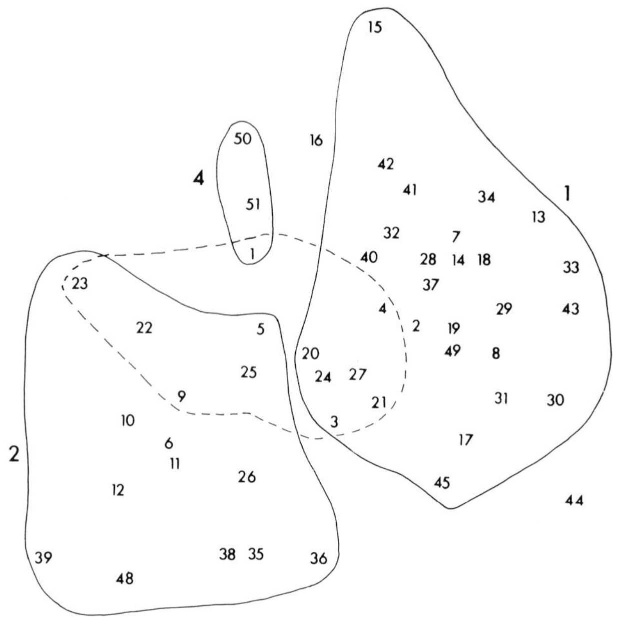

Fig. 4. MDS inverse ordination comparing 51 species at Gough and the Prince Edward islands, using standardized abundances. Groups of species delineated from the dendrogram in Fig. 3, whose numbers and names are contained in Appendix 1. Concurrent species are demarcated within the *broken line.* Stress = 0.11

Table 1. Frequencies of occurrence of indicator species, ranked by information statistics, which distinguish Group 1 hauls (Gough) from those of Group 2 (Prince Edward)

Species	Group 1	Group 2
Eukronia hamata	16	0
Dimophyes arctica	14	0
Thysanoessa vicina	14	0
Cavolina tridentata	12	0
Calanus simillimus	11	0
Electrona paucirastra	7	0
Electrona subaspera	6	0
Euphausia lucens	6	0
Euphausia similis var. *armata*	6	0
Palinurichthys antarcticus	5	0
Nemichthys scolopaceus	5	0
Serpula vermicularis	5	0

Species above the broken line have $2\Delta I\hat{i} > 6.63$ (P = 0.01); those below score > 3.84 (P = 0.05). Maximum possible occurrences are: Group 1 = 20 and Group 2 = 25

vides a good summary of differences in species composition between haul and subsequently regional groupings.

The above analyses separate the two regions by both haul and species' attributes. However, it is useful to identify those species which occur frequently enough and in sufficient abundance to characterize best one region and distinguish it from the other – the so-called "indicator species" (Field et al. 1982). Information statistic (I-) tests (Field 1969; Field et al. 1982) provide a means of assessing which species differed most between each island sample group. Comparing Group 1 (Gough) with Group 2 (Prince Edward) for any one species, i:

$$2\Delta I\hat{i} = 2\,(It\hat{i} - I1\hat{i} - I2\hat{i})$$

where $It\hat{i}$ = total information content of both groups combined; $It\hat{i} = N\hat{t}\log N\hat{t} - A\hat{t}\hat{i}\log A\hat{t}\hat{i} - (N\hat{t}-A\hat{t}\hat{i})\log(N\hat{t}-A\hat{t}\hat{i})$; $N\hat{t}$ =

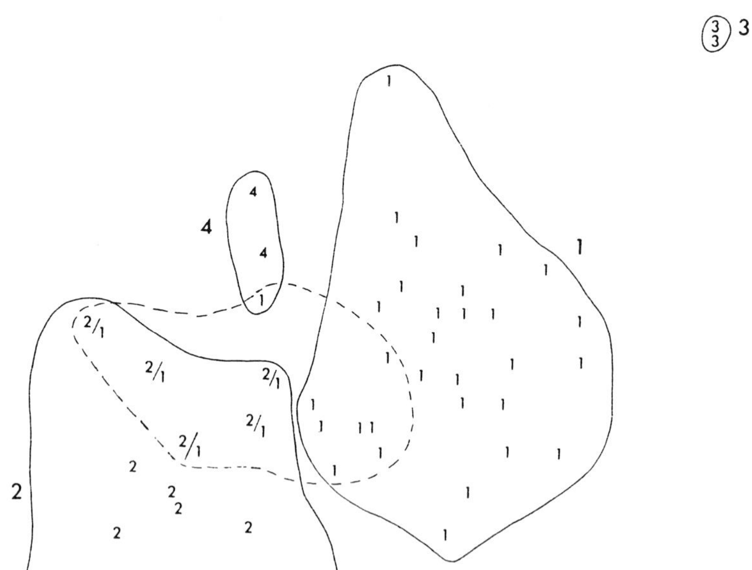

Fig. 5. Species numbers (from Fig. 4) replaced by haul group(s) where they represent > 4% of the specimens collected. Hauls containing concurrent species are demarcated within the *broken line*

Table 2. Frequencies of occurrence of indicator species, ranked by information statistics, which distinguish Group 2 hauls (Prince Edward) from those of Group 1 (Gough)

Species	Group 2	Group 1
Gaimardia trapesina	15	0
Notothenia marionensis	7	0
Sagitta maxima	8	0

Species above the broken line have $2\Delta I\hat{i} > 6.63$ (P = 0.01); those below score 3.84 (P = 0.05). Maximum possible occurrences are: Group 2 = 25 and Group 1 = 20

number of samples (45) in both clusters together ("potential presences"); $A\hat{t}i$ = number of samples in which species *i* is actually present; $(N\hat{t}\text{-}A\hat{t}i)$ = number of samples from which species *i* is absent. Similarly, the information contents $I\hat{1}i$ and $I\hat{2}i$ were obtained for Groups 1 and 2, respectively. A pairwise comparison of the groups gives the species which discriminate best between them. It is hypothesized that the probability of observing species is the same for samples in both cluster groups. In other words, $2\Delta I\hat{i}$ has an approximate chi-square distribution and the scores of each species can be compared from chi-square tables (for 1 df) to ascertain whether they differ significantly at either the 5% or 1% probability level.

Tables 1 and 2 list species which best characterize and separate the Gough and Prince Edward haul-species clusters. There are no "perfect" indicator species which occur in all samples for each group. Gough (Group 1) species having $2\Delta I\hat{i} > 6.63$ (P = 0.01) and > 3.84 (P = 0.05) are shown in Table 1. At the Prince Edward Islands, fewer species were found to be suitable as indicators (Table 2) and again no perfect indicators were evident.

4 Discussion

The analyses reported here not only quantify differences in the macro-plankton between the two island regions, but also take specific affinities into account. This provides positive and direct evidence for biogeographical separation, confirming more subjective assessments (Miller 1982a,b). The identification of indicator species highlights differences between the faunas of the two regions.

Gough and the Tristan da Cunha archipelago are grouped tentatively into a single biogeographical province located within "the Sub-Antarctic cold-temperature zone" (Knox 1960). Situated close to the Sub-Tropical Front (STF), these islands have a mixed fauna similar to that of the South American "Magellanic biotic province" and the Chatham Islands (Holdgate 1960; Knox 1960; Boltovskoy 1975). As such, the faunal composition reflects more of a warm sub-tropical than a cold Sub-Antarctic or Antarctic influence (Knox 1960). This is confirmed by the present results. Although, all the Gough indicator species (Table 1) can be considered as Sub-Antarctic in origin, at least eight have centres of distribution in the northernmost limits of the Sub-Antarctic zone. In fact, *Euphausia lucens, E. similis* var. *armata, Palinurichthys antarcticus* and *Nemichthys scolopaceus* are found concentrated north of the STF (Baker 1965; Penrith 1967; Nielsen and Smith 1978). *Eukronia hamata* was represented by the northern (small) racial variant. As an indicator species, its abundance also demonstrates the prevalent sub-tropical, as opposed to Sub-Antarctic, affinity of the Gough macro-plankton. Of the remaining indicator species, only *Thysanoessa vicina* is known to transgress the APF in significant numbers; although it is considered to be more of a Sub-Antarctic than an Antarctic organism (Nemoto 1966). Finally, *Dimophyes arctica* is a cosmopolitan species and exhibits a wide circumpolar distributional range (Mackintosh 1934).

The Prince Edward Islands are located farther south than Gough, and are close to the Antarctic Polar Front (APF). They are situated at the southern extremity of the Sub-Antarctic zone, within the "Kerguelenian biotic province" (Knox 1960). This area is more susceptible to the influences of cold temperate waters than the Gough region and, as such, is situated in a transition zone between the Sub-Antarctic and the Antarctic proper (Deacon 1983). Of the three indicator species (Table 2), *Sagitta maxima* and *Gaimardia trapesina* originate in the Sub-Antarctic, but exhibit a preference for temperate waters (David 1958; Dell 1964). The remaining indicator species, *Notothenia marionensis* is endemic to the Prince Edward region (Norman 1938). Of the more common (non-indicator) species, *Euphausia vallentini, Thysanoessa macrura* and *Clione limacina antarctica* have an affinity for cold Antarctic water (Baker 1965; Nemoto 1966; Van der Spöel 1976).

These biological observations are supported by hydrographical data. A southward instrustion of sub-tropical water has been shown to occur to the north of Gough Island (Miller and Tromp 1982). In combination with the island's close proximity to the STF (Deacon 1937), this could serve as a vehicle for the infusion of relatively warm sub-tropical water and species into the region. Similarly, it is proposed that sporadic advection of cold Antarctic surface water into the Prince Edward environs would promote violation of the APF boundary by species normally confined to the Antarctic zone (Miller et al. 1983). In the Prince Edward region, there also seems little doubt that topographical effects result in more interchange and less gradation between Antarctic and Sub-Antarctic waters than in other longitudes (Deacon 1983). In combination with the group's possible influence on neighbouring hydro-dynamic events (Allanson et al. 1984), this could account for the Antarctic character of much of the macro-plankton, which is probably not as unusual as originally thought by Grindley and Lane (1979).

The implication, therefore, is that macro-plankton composition at both localities is modified by hydrographical variability and hence rigid latitudinal zonation is pre-empted in favour of more localized differences between the two regions. On the other hand, a lack of rigid longitudinal separation is clearly illustrated by the appearance of concurrent species at both localities, particularly *Salpa fusiformis, Euphausia longirostris, S. gazellae* and *Themisto gaudichaudii.* Although endemic species were encountered, these were all coastal forms and their impact on widespread biogeographical affiliations is expected to be minimal.

In any study of this nature there is always an element of doubt concerning the validity of interpretations of community composition. This is directly attributable to latent bias caused by undefined sampling techniques and parameters (Angel 1977). A relatively small number of samples may compound such bias. For this reason, it is desirable to use alternative assessments of faunal composition to substantiate net-dependent data. Such measures are available from several analyses which confirm the dominance of indicator and other more common species in the stomach contents of island-based predators. At Prince Edward, *T. gaudichaudii, Nauticaris marionsis, E. vallentini, G. trapesina, T. macrura* and *Notothenia macrocephala* have all been found in the stomachs of breeding penguin and inshore fish species (Williams and Laycock 1981; Blankley 1982; Miller, unpubl. data). At Gough, *T. vicina* occurs in the stomachs of Rockhopper Penguins, *Eudyptes chrysocome,* (Williams and Laycock 1981), and *T. gaudichaudii* and *Cavolina tridentata* have been found in the faeces and regurgitations of the Cape Pigeon, *Daption capense,* and many prions, *Pachyptilla* spp., (Miller, unpubl. data). Previously, both *T. macrura* and *T. vicina* were thought to be absent. However, late-larval forms of these species, as opposed to only adult animals in the stomachs of predators, have only recently been identified in the net samples (Miller, unpubl. data).

In conclusion, Mackintosh's (1960) hypothesis is substantiated by my results in that the existence of some separation in the composition of the macro-plankton between two Sub-Antarctic regions is shown. It is also apparent that zonal differentiation is influenced directly by prevailing hydrographical conditions. Consequently, as Deacon (1982, 1983) suggests, rigid biogeographical boundaries are not necessarily persistent. Considerably more information is required in order to fully define the faunal and hydrographical boundaries which separate the macro-plankton of Gough Island from that of the Prince Edward region. The identification of indicator species in the dietary remains of predators holds promise for future integrated studies. Through their accessibility to both predators and net samplers, indicator species may well come to typify the biotic variability of the two regions.

Appendix 1

Groups of species distinguished by inverse (r-type) analysis. Groups are based on the dendrogram in Fig. 3 which shows the species' numbers.

Group 1.

49. *Nemichthys scolopaceous* 27. *Euphausia longirostris* 21. *Phoronima sedentaria* 19. *Serpula vermicularis* 34. *Nematoscelis atlantica* 14. *Clio pyrimidata sulcata* 28. *Euphausia similis* var *armata* 18. *Nereis callaoana* 37. *Thysanoessa vicina* 7. *Eukronia hamata* 29. *Euphausia lucens* 8. *Sagitta planktonis* 20. *Themisto gaudichaudii* 17. *Harmöthoe bervipalpa* 4. *Salpa thompsoni* 2. *Dimophyes arctica* 31. *Calanus simillimus* 30. *Jasus tristani* 3. *Salpa fusiformis* 13. *Cavolina tridentata* 45. *Acantholatris monodactylus* 43. *Bovichthys diacanthus* 33. *Idotea metallica* 40. *Electrona subaspera* 24. *Rhinocalanus gigas* 42. *Palinurichthys antarcticus* 41. *Electrona paucirastra* 32. *Cyllopus magellenicus* 15. *Cleodora sulcata*

Group 2/1.

25. *Calanus propinquus* 23. *Oraderea ocellata* 22. *Aytoella magellenica* 9. *Gaimardia trapesina* 5. *Sagitta gazellae*

Group 2.

48. *Harpagifer marionensis* 39. *Notothenia macrocephala* 12. *Clio limacina antarctica* 38. *Notothenia marionensis* 35. *Thysanoessa macrura* 36. *Nauticaris marionis* 26. *Euphausia vallentini* 11. *Clio pyrimidata antarctica* 6. *Sagitta maxima* 10. *Spongiobranchia australis*

Group 3.

47. *Decapterus longimanus* 46. *Thyrsites atun*

Group 4.

51. *Vellella vellella* 1. *Obelia geniculata* 50. *Notopogon notopogon*

Ungrouped.

16. *Clione antarctica* 44. *Trachipterus trachtipterus*

Acknowledgements. This study was sponsored by the Sea Fisheries Research Institute, the South African National Committee for Oceanographic Research, and the South African Scientific Committee for Antarctic Research. I also thank Prof JG Field for his assistance.

References

Allanson BR, Boden P, Parker L, Rae CD (1985) A contribution to the oceanology of the Prince Edward Islands. In: Siegfried WR, Condy PR, Laws RM (eds) Antarctic nutrient cylces and food webs (Proceedings of the fourth SCAR symposium on Antarctic biology). Springer, Berlin Heidelberg New York

Angel MV (1977) Windows into a sea of confusion: Sampling limitations to the measurement of ecological parameters in oceanic mid-water environments. In: Anderson NR, Zahuranec BJ (eds) Oceanic sound scattering prediction. Plenum, New York p 859

Baker A de C (1954) The circumpolar continuity of Antarctic plankton species. Discovery Rep 27:201–218

Baker A de C (1965) The latitudinal distribution of *Euphausia* species in the surface waters of the Indian Ocean. Discovery Rep 33:309–334

Blankley WO (1982) Feeding ecology of three inshore fish species at Marion Island (Southern Ocean). S Afr J Zool 17:164–170

Boltovskoy D (1975) Ecological aspects of zooplankton of the Southwestern Atlantic Ocean. Veliger 18:203–216

Bray JR, Curtis JT (1957) An ordination of the upland forest communities of Southern Wisconsin. Ecol Monogr 27:325–349

Cheng L (1975) Marine pleuston – Animals at the sea-air interface. Oceanogr Mar Biol Annu Rev 13:181–212

Clifford HT, Stephenson W (1975) An introduction to numerical classification. Academic Press, New York

David PM (1958) The distribution of the Chaeotognatha of the Southern Ocean. Discovery Rep 29:199–228

Deacon GER (1937) The hydrology of the Southen Ocean. Discovery Rep 15:1–124

Deacon GER (1982) Physical and biological zonation in the Southern Ocean. Deep-Sea Res 29:1–15

Deacon GER (1983) Kerguelen, Antarctic and subantarctic. Deep-Sea Res 30:77–81

Dell RK (1964) Antarctic and Subantarctic Mollusca: Amphineura, Scaphopoda and Bivalvia. Discovery Rep 33:93–250

Ekman S (1953) Zoogeography of the sea. Sidgwick and Johnson, London p 417

Field JG (1969) The use of the information statistic in the numerical classification of heterogeneous systems. J Ecol 57:565–569

Field JG, Clarke KR, Warwick RM (1982) A practical strategy for analysing multispecies distribution patterns. Mar Ecol Prog Ser 8:37–52

Grindley JR, Lane SB (1979) Zooplankton around Marion and Prince Edward Islands. Com Natnl Franc Rech Antarct 44:111–125

Hedgepeth JW (1970) Marine biogeography of the Antarctic regions. In: Holdgate MW (ed) Antarctic ecology, Vol I. Academic Press, London p 97

Holdgate MW (1960) The fauna of the mid-Atlantic islands. Proc R Soc Lond B Biol Sci 152:550–567

Knox GA (1960) Littoral ecology and biogeography of the southern oceans. Proc R Soc Lond Biol Sci 152:577–624

Kruskal JB, Wish M (1978) Multidimensional scaling. Sage Publications, Beverley Hills, California

Mackintosh NA (1934) Distribution of the macroplankton in the Atlantic sector of the Antarctic. Discovery Rep 9:65–160

Mackintosh NA (1934) Distribution of the macroplankton in the Atlantic sector of the Antarctic. Discovery Rep 9:65–160

Miller DGM (1982a) Results of a combined hydroacoustic and midwater trawling survey of the Prince Edward Island group. S Afr J Antarct Res 12:3–10

Miller DGM (1982b) Results of a combined hydroacoustic and midwater trawling survey of the Gough Island region. S Afr J Antarct Res 12:17–22

Miller DGM, Tromp BBS (1982) The hydrology of waters close to Gough Island. S Afr J Antarct Res 12:23–33

Miller DGM, Boden BP, Parker L (1983) Hydrology and bio-oceanography of the Prince Edward Islands (South-West Indian Ocean). S Afr J Antarct Res 14 (to be published)

Nemoto T (1966) *Thysanoessa* euphausiids, comparative morphology, allomorphosis and ecology. Sci Rep Whales Res Inst Tokyo 20:109–155

Nielsen JG, Smith DG (1978) The eel family Nemichthyidae. Dana Rept 88:3–71

Norman JR (1938) Coast fishes. Part III, The Antarctic Zone. Discovery Rep 18:1–104

Penrith MJ (1967) The fishes of Tristan da Cunha, Gough Island and Vema Seamount. Ann S Afr Mus 48:523–548

Plancke J (1977) Phytoplankton biomass and productivity in the Subtropical Convergence area and shelves of the western Indian Subantarctic islands. In: Llano GA (ed) Adaptation within Antarctic ecosystems. Smithsonian Institution, Washington DC, p 51

Robertson AA, Alexander DGW, Miller DGM (1981) Modified opening and closing midwater trawls (RMT-8 and RMT-2). Fish Bull S Afr 14:103–113

Van der Spoel S (1976) Pseudothecosomata, Gymnosomata and Heteropoda (Gastropoda). Bonn, Scheltma and Holkema, Utrecht p 484

Williams AJ, Laycock PA (1981) Euphausiids in the diet of some Antarctic *Eudyptes* penguins. S Afr J Antarct Res 10/11:27–31

The Relationship Between Krill Food Requirements and Phytoplankton Production in a Sector of the Southern Indian Ocean

D.G.M. MILLER[1], I. HAMPTON[1], J. HENRY[1],
R.W. ABRAMS[2], and J. COOPER[2]

Summary. Acoustic measurements of Antarctic Krill, *Euphausia superba*, standing stock and concurrent primary productivity measurements are used with recently published data on Krill energy requirements to assess the importance of the species as a herbivore in a sector of the southern Indian Ocean surveyed during FIBEX. It is estimated that on average Krill could satisfy daily energy requirements by consuming about 3.0% of the daily phytoplankton productivity available at the time of the survey. Assuming linear growth and exponential mortality rates, it is shown that energy requirements ments earlier mortality rates, it is shown that energy requirements earlier in the growing season were probably even lower than during the survey. Since phytoplankton growth rates were high (0.51–2.10 doublings d^{-1}), it is suggested that production in the area was being rapidly exploited by other consumers (particularly micro-organisms), with some contributory loss due to sinking and vertical mixing. Results from a bird census carried out during the survey, and from an IWC Minke Whale, *Balaenoptera acutorostrata*, survey the previous summer, are used to estimate the seasonal consumption of Krill by flying birds, squid and whales. Predation by seals and penguins is estimated crudely by extrapolation from circumpolar data, giving an estimated total predator consumption of 1.4×10^6 t. This is in reasonable agreement with seasonal Krill production estimated from the linear growth-exponential mortality model. It is suggested that similar analyses should be undertaken for other areas of the Antarctic for which data are available. Future tropho-dynamic studies should concentrate on alternative energy pathways, particularly those involving micro-zooplankton and other micro-organisms.

1 Introduction

Krill, *Euphausia superba,* has long been regarded as a key organism in the Antarctic marine food chain (Hart 1942; Marr 1962; Mauchline and Fisher 1969; Everson 1977). To date, postulated trophic relationships have been more qualitative than quantitative (Hart 1942; Holdgate 1967; Knox 1970), largely as a result of inadequate data on Krill biomass and production (Everson 1977; Nemoto and Harrison 1981).

In this paper we attempt to assess the importance of Krill as a primary herbovore in a sector of the southern Indian Ocean surveyed during FIBEX (First International BIOMASS Experiment) in late summer 1981. Use is made of acoustic estimates of Krill abundance and concurrent measurements of primary productivity. Recently published data on Krill metabolic processes are used to assess daily energy requirements as a function of body size. Published data on growth and mortality rates for the two predominant year classes present are used to assess Krill standing stock, production and food requirements in the area during a 5-month growing season. The seasonal Krill production is compared with crude predator estimates for the area based as far as possible on direct predator counts. It is hoped that the resulting tropho-dynamic relationships are more realistic than those previously derived by less direct methods (e.g., Gulland 1970; Everson 1977).

In an analysis of this kind, many factors have to be included which are of unknown accuracy. When presented with a choice of factors, and with not strong a priori reason for accepting any particular one, we have deliberately opted for those leading to the highest feeding rate so as to arrive at an upper limit for Krill food requirements.

2 Materials and Methods

An area (62°–69°S, 15°–30°E) of the Antarctic Ocean was surveyed hydro-acoustically by the M.V. SA Agulhas between 16 February and 10 March 1981 (Hampton 1985). The survey consisted of 9 zonal transects, extending from the pack-ice edge northward (Fig. 1). Water samples were collected from seven discrete depths at 32 stations, at 21 of which phytoplankton productivity was measured (Fig. 1). All samples were analyzed according to prescribed methods for chlorophyll measurement (Evans and O'Reilly 1982). Integrated primary productivity in the euphotic zone (above the 1% light level) was determined by simulated in situ incubation on-board the ship (Strickland and Parsons 1972; O'Reilly and Thomas 1982). It was assumed that the primary productivity measured was net production. Flying birds were counted during the survey according to prescribed methods for censusing birds at sea (Anon. 1982).

1 Sea Fisheries Research Institute, Private Bag X2, Roggebaai 8012, South Africa

2 FritzPatrick Institute, University of Cape Town, Rondebosch 7700, South Africa

Antarctic Nutrient Cycles and Food Webs
(ed. by W.R. Siegfried, P.R. Condy, and R.M. Laws)

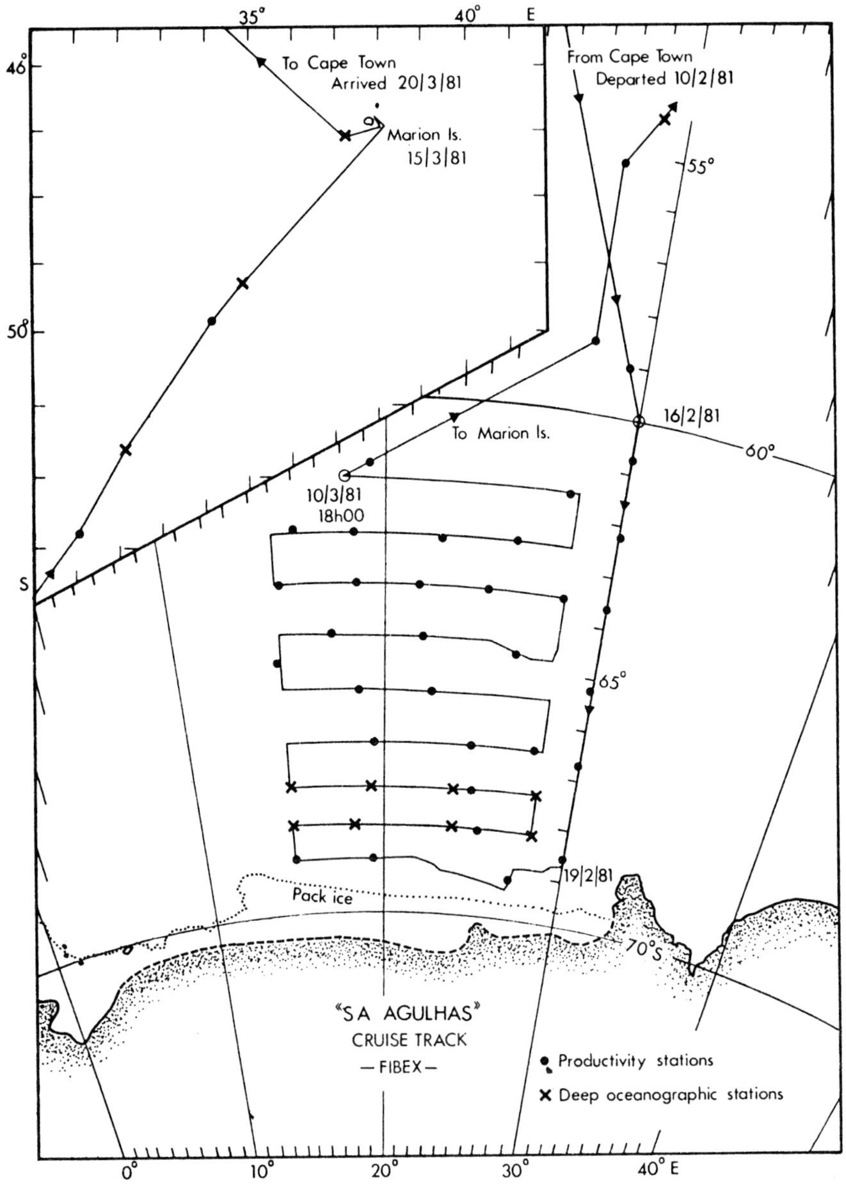

Fig. 1. Cruise track of M.V. SA Agulhas during FIBEX

3 Results

3.1 Relationship Between Krill Food Requirements and Phytoplankton Production

The mean phytoplankton production measured during the survey was 188 mg C m^{-2} d^{-1} (range = 80–364, S.D. = 79), or 1.88 g (wet wt.) phytoplankton m^{-2} d^{-1}, assuming a phytoplankton-C ratio of 10:1 (Holm-Hansen et al. 1977; Sorokin and Mikheev 1979). Expanding this over the survey area of 6.2×10^5 km^2 gives a daily phytoplankton production of approximately 1.2×10^6 t. The acoustically-derived estimate of mean Krill density was 1.46 g m^{-2} (Hampton 1984), giving a standing stock of 9.1×10^5 t at the time of survey.

From these values, the proportion of the daily phytoplankton production consumed by Krill each day can be estimated. P_i, the amount of phytoplankton consumed by Krill on day i, can be expressed by

$$P_i = \frac{100}{kK} \sum_{j=1}^{m} (N_i)_j \, (G_j + R_j) \qquad (1)$$

where $(N_i)_j$ is the number of animals in length class j on day i, G_j and R_j are the daily energy requirements of an animal in length class j for growth and respiration, respectively, and m is the number of length classes. k is a factor accounting for the calorific content of Antarctic phytoplankton and K is the assimilation efficiency in percent.

The $(N_i)_j$ values were calculated from N_s, the total number of Krill at the time of the survey, using an overall length promille derived by combining length distributions from individual net samples (Fig. 2). N_s was assumed to have been constant during the survey and was estimated from $B_s/\overline{W}_s$,

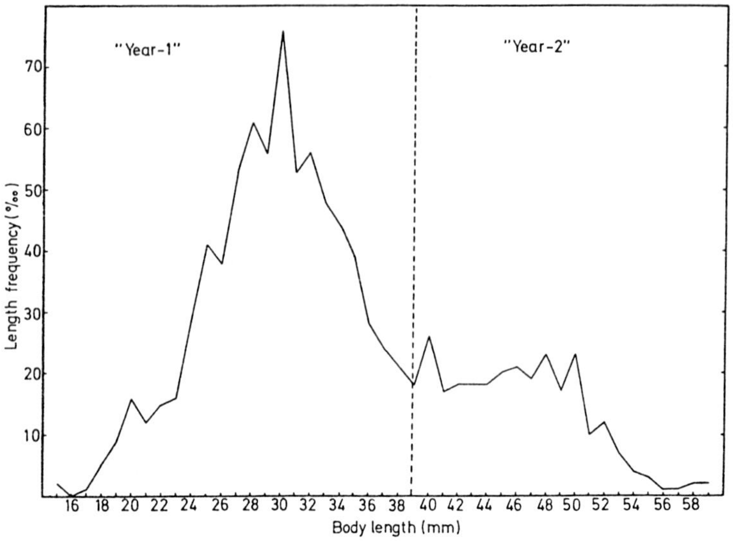

Fig. 2. Length frequency promille for all *E. superba* collected

where B_s is the total biomass estimated acoustically, and $\overline{W}_s$ is the mean Krill weight (wet wt.) at the time of the survey (348.2 mg).

To calculate G_j, we take the growth in length to be linear for the period under study. This appears reasonable from Mauchline's (1980) theoretical fit to Mackintosh's (1972) data. Using Jazdzewski et al's (1978) length-weight relationship for Krill,

$$W = 0.0018\, l^{3.38} \tag{2}$$

enables ΔW_j, the daily weight increase of animals in length class j, to be calculated from

$$\Delta W_j = 0.0018\,[(l_j + \Delta l)^{3.38} - l_j^{3.38}],$$

where Δl is the daily length increase. Weights and lengths are in mg (wet wt.) and mm, respectively. Expanding $(l_j + \Delta l)$ as a Taylor series and discarding higher order terms gives

$$\Delta W_j \cong 0.0061\, \Delta l\, l_j^{2.38} \tag{3}$$

Mauchline's Fig. 25 shows a daily length increment of about 0.13 mm in summer for animals between 10 and ca. 40 mm long (his so-called "year-1" group) and an increment of about 0.09 mm for animals between 40 and 53 mm ("year-2" group). Our catches were made up predominantly of these two groups (Fig. 2), so we have used the expression

$$\Delta W_j = 7.93 \times 10^{-4}\, l_j^{2.38}$$

for the yr-1 group, and

$$\Delta W_j = 5.49 \times 10^{-4}\, l_j^{2.38}$$

for the yr-2 group. We have ignored the 0-yr-old-group, since Allen (1971) demonstrated that animals in this group do not contribute significantly to the overall production and biomass. They are therefore unlikely to consume a significant proportion of the phytoplankton.

Taking the energetic value per unit mass of Krill tissue as 4.51 kJ g^{-1} (wet wt.) (Clarke and Prince 1980) and a tissue synthesis efficiency of 60%, the lower limit quoted by Clarke and Morris (1983), gives

$$G_j = 5.96 \times 10^{-3}\, l_j^{2.38} \tag{4}$$

and

$$G_j = 4.13 \times 10^{-3}\, l_j^{2.38} \tag{5}$$

for the yr-1 and yr-2 classes respectively.

R_j can be estimated from the rate of oxygen uptake by Krill as a function of length. Several workers (e.g., Chekunova and Rynkova 1974; Rakusa-Suszczewski and Opalinski 1978; Kils 1981; Ikeda and Mitchell 1982) have measured oxygen uptake of *E. superba* as a function of body weight under a variety of experimental conditions. We have chosen Ikeda and Mitchell's (1982) expression,

$$Q = 0.111\, W^{0.969}$$

(Q in $\mu l\ O_2$ individual^{-1} h^{-1}, W in mg wet wt.), since their work is the most recent and appears to have approached the natural state most closely, evidenced by a very low mortality during experimentation (Ikeda and Dixon 1982). Although their animals were not able to move freely, the expression should still apply to Krill under natural conditions in the open sea, since Kils (1981) showed that Krill can swim up to a speed of 0.13 m s^{-1} without exceeding their "standard" (i.e. hovering) metabolism. This speed is close to that reported by Kanda et al. (1982) for sustained swimming by large Krill swarms over many days. We have ignored the metabolic cost of ovarian development, considered to be important by Clarke and Morris (1983), as fewer than 1% of all female animals collected in our area exhibited advanced sexual maturity.

Taking W from Eq. 2, and applying an oxy-calorific factor of 2.03×10^{-2} J $(\mu l\ O_2)^{-1}$ gives the following expression for R_j;

$$R_j = 1.18 \times 10^{-4}\, l_j^{3.28} \quad \text{J individual}^{-1}\ d^{-1} \tag{6}$$

In Fig. 3. the proportion of energy being utilized for growth is shown as a function of length for the two age classes. It

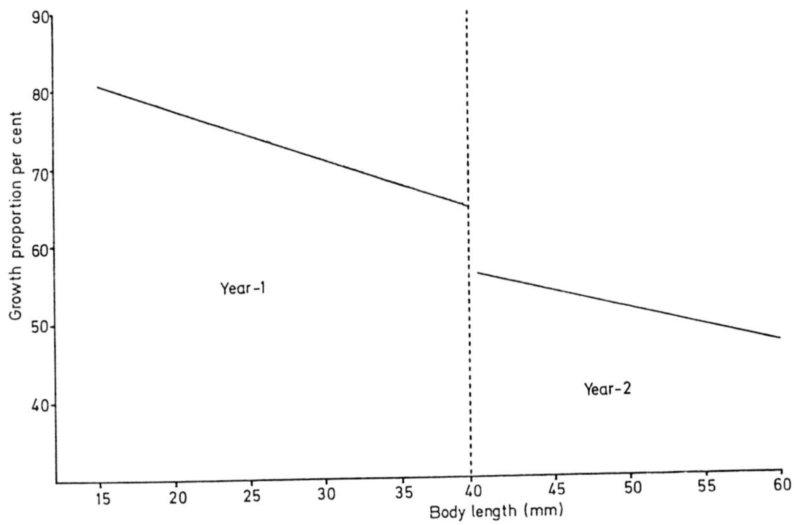

Fig. 3. Estimated proportion of the energy intake used by *E. superba* for growth, as a function of body length

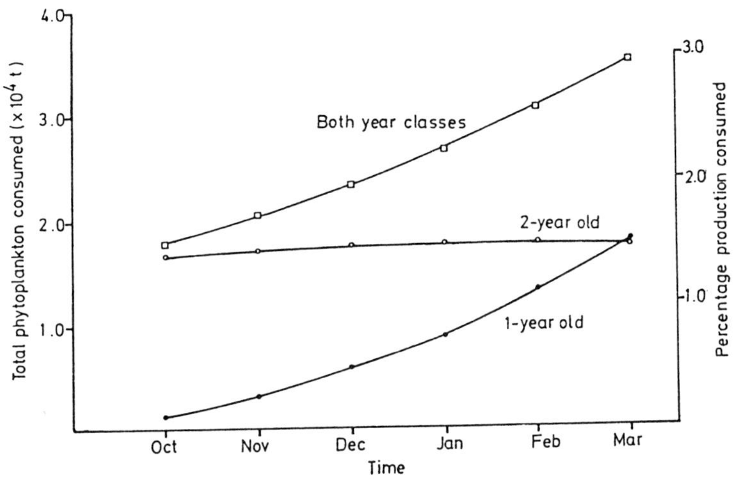

Fig. 4. Estimated daily phytoplankton consumption by 1-yr- and 2-yr-old Krill during an arbitrary 150-d feeding season (October–February) in the survey area

can be seen that the ratio is high even for a polar animal with a limited growing season (Remmert 1980). As expected, the proportion of energy required for growth is greater in smaller animals.

Since there are no published data on the assimilation efficiency of *E. superba,* we have taken a value of 71.9%, being the lowest limit of the only figures available for this species (Ikeda and Dixon, pers. comm. reported in Kato et al. 1982). Finally, k, the calorific value of Antarctic phytoplankton was taken as 4.0 J mg^{-1} phytoplankton (wet wt.) (Vinogradova 1967). This figure was originally quoted in dry weight. A factor of 5 has been applied to convert from dry to wet weight, after Kils (1981).

Substituting these constants and Eqs. 4, 5 and 6 into Eq. 1 gives

$$P = 0.347 \sum_{j=1}^{m} (N_i)_j \, (a \times 10^{-3}\, l_j^{2.38} + 1.18 \times 10^{-4}\, l_j^{3.28}),$$

where a = 5.96 for $l_j \lesssim 39$ mm and 4.13 for $l_j > 39$ mm. Using the length promille in Fig. 2, a daily phytoplankton consumption of 3.66 x 10^4 t is obtained. This amounts to 3.05% of the estimated daily phytoplankton production. Note that since the N_j values were taken from the accumulated length distributions for the whole survey, the phytoplankton consumption is effectively an average for the survey period.

Our analysis has indicated that an insignificant proportion of the phytoplankton production during the survey was consumed by Krill. Fig. 4, showing the estimated daily phytoplankton consumption throughout a 5-month season (October to February), indicates that consumption earlier in the growing season was probably even less significant, assuming similar phytoplankton production levels throughout the season. The curves in Fig. 4 were constructed for both 1-yr- and 2-yr-old animals assuming, after Allen (1971), an exponential mortality for both age classes with Z = 0.69 and 2.30 yr^{-1}, respectively. Daily length increments for the two classes were taken as 0.13 mm and 0.09 mm as before.

3.2 Krill Production

The total Krill production, P_r, and mean biomass during the growing season, $\bar{B}_t$, were estimated by solving the

Table 1. Estimated total production (P_r) and mean biomass ($\bar{B}_t$), both in 10^6 t, of 1- and 2-yr-old groups of Krill over a 150 d growing season

	1-yr-old group	2-yr-old group	Totals
P_r	0.36	0.56	0.92
$\bar{B}_t$	0.13	0.47	0.60
$\frac{P_r}{\bar{B}_t} = \frac{P}{B}$	2.77	1.19	1.53

following integrals numerically for both year classes:

$$P_r = \int_0^T N_t \Delta W_t \, dt$$

and

$$\bar{B}_t = 1/T \int_0^T N_t W_t \, dt$$

N_t, the number of Krill at time t, was calculated from $N_s e^{Zt/365}$. W_t and ΔW_t were calculated from Eqs. 2 and 3, respectively, with l_t taken as l_s-Δl (T-t). l_s is the mean length of animals at the time of the survey (29.5 mm and 45.9 mm, respectively, for the 2-yr classes). T is the length of the growing period up to the time of the survey (taken as 150 days). Using the same values for Z and Δl as before gave the values shown in Table 1.

3.3 Predation on Krill

3.3.1 Birds

From the FIBEX bird census the biomass of planktivorous birds was estimated at 400 t (4.0 birds km^{-2}) and that of mixed-feeders at 87 t (0.3 birds km^{-2}). Using Walsberg's (1980) energy-mass relationship, the mean daily food requirement per bird was calculated to be 396 kJ for the planktivores and 806 kJ for the mixed-feeders. Assuming that Krill comprised 100% and 50%, respectively, of the diets of birds in these two groups, a food assimilation figure of 80% (Furness 1978) and an energetic content of Krill of 4.51 kJ g^{-1} as before, the seasonal (150 d) consumption for both groups combined amounts to 4.5 × 10^4 t. This estimate excludes predation by penguins, which were not censused.

3.3.2 Cetaceans

The predominant cetacean consumer of Krill in recent times is the Minke Whale, *Balaenoptera acutorostrata*, (Anon. 1979). From the 1979–80 International Whaling Commission (IWC) survey of IWC Area III (Horwood 1981), we estimate by areal proportion that our survey area contained 2 × 10^4 Minke Whales. Taking the mean weight of each whale as 7 ton (Lockyer 1976) gives a biomass of 1.40 × 10^5 t. The daily food consumption is approximately 4% of the Whale's body weight (Ohsumi 1979), giving a total daily Krill consumption of 5.60 × 10^3 t, assuming that Minke Whales feed exclusively on *E. superba.* If the residence period is 120 d (Anon. 1979), total seasonal consumption amounts to 6.72 × 10^5 t. We note, however, that this is probably an over-estimate, because Minke Whales are known to concentrate close to the pack-ice edge (Butterworth 1982) which lay outside the present survey area.

3.3.3 Seals

Few data are available on the abundance of seals in the survey area specifically, but a rough estimate of seal predation on Krill can be made using Laws' (1977a) figure of 2.23 × 10^6 t for the circumpolar biomass of Crabeater Seals, *Lobodon carcinophagus,* which are by far the most abundant krill-eating seals in Antarctic waters (Laws 1977a, b). This figure gives a biomass of 9.6 × 10^4 t for the sector 15°E–30°E, assuming the species to be uniformly distributed around the Antarctic continent. Since Crabeater Seals are primarily residents of the pack-ice (Gilbert and Erickson 1977), which was essentially outside the survey area (see Fig. 1), consumption of Krill by the species would be limited to animals foraging away from the ice edge. Taking the effective foraging range to be 100 km (Gilbert and Erickson 1977), the width of the pack-ice zone as 100 km, and assuming Crabeater Seal density to be the same in the two zones, gives a biomass of 4.80 × 10^4 t in the survey area. If the daily food intake is 7% of the body weight (Øristland 1977), Crabeater Seal consumption of Krill would be 3.36 × 10^3 t of Krill d^{-1} or 4.03 × 10^5 t per 120-d season. It should be pointed out that these are likely to be severe over-estimates, since Crabeater Seal densities decrease progressively with distance from the pack-ice edge. Also, because of the comparatively sparse pack-ice cover adjacent to the Drönning Maud Coast, Crabeater Seals are probably substantially less abundant here than in the Ross Sea and Antarctic Penninsula areas where Laws' and Gilbert and Erickson's (1977) surveys were made, and on which estimates of circumpolar populations are based.

3.3.4 Other Predators (Penguins, Fish and Squid)

The most important Krill predators not yet considered are the penguins. Prévost (1981) maintains that in the Antarctic approximately 5.0 × 10^6 t of crustaceans (not specifically *E. superba*) are consumed per annum (365 d) by the Spheniscidae, of which the Adélie Penguin, *Pygoscelis adeliae,* consumes by far the highest proportion. Following the same approach as used for the Crabeater Seal, and assuming that the foraging range of the Adélie Penguin is similar to that reported by Williams and Siegfried (1980), results in an estimate of 1.03×10^5 t of Krill consumed by penguins per year. This is also likely to be an over-estimate, for the same reasons as given for Crabeater Seals. Furthermore, a significant proportion of the Adélie Penguin's diet is known to consist of *Euphausia crystallorophias* (Emison 1968), which tends to be more an inhabitant of the pack-ice (the favoured habitat of the Adélie Penguin) than *E. superba* (Mauchline and Fisher 1969).

Of the other Krill predators, we discount fish, since the survey area lay well away from the shallow waters of the continental shelf where most fish would be expected to concentrate (Everson 1977). We are unable to assess the impact of predation by squid due to a lack of information about this group in any area of the Antarctic, but attempt an estimate of consumption based on the amount of squid taken by squid-eating birds (mainly Diomedeidae) censused during the survey. Using census figures for squid-eaters and mixed-feeders (one-third of whose diet was assumed to be squid) and executing the same calculation used for the planktivores, we arrive at a seasonal squid consumption of 1.05×10^4 t. If the squid-bird system was in equilibrium and the ratio of biomass between the two trophic levels was 10:1, the seasonal consumption of Krill by squid is 1.05×10^5 t. This is seen as an under-estimate, because it excludes squid taken by other predators such as Sperm Whales, *Physeter catodon*, and penguins.

3.3.5 Total Consumption

The preceding estimates are summarized in Table 2 which shows that the total estimated seasonal consumption by all predators considered is about 1.4×10^6 t. This is consistent with the estimate of 9.2×10^5 t for seasonal Krill production (Table 1), especially bearing in mind that on balance predation estimates are probably too high. Such agreement increases confidence in the Krill biomass and production figures reported in the previous section.

4 Discussion

Our analysis has shown that in the area studied, Krill were consuming only a very small proportion of the phytoplankton production. This is surprising considering that Krill are generally regarded as the dominant planktonic herbivore in Antarctic waters (Marr 1962; Mauchline and Fisher 1969; Everson 1977). Certainly, phytoplankton consumption by Krill appears significantly lower than that reported for dominant herbivores in other oceanic systems (e.g., Paffenhöffer 1971; Frost 1972; Sorokin and Mikheev 1979) where consumption levels may match, and sometimes exceed, primary production. The present results suggest that either there was a general surplus of phytoplankton in the area, implying that the major proportion of the production was removed out of the system by sinking as detritus, or that

Table 2. Estimated seasonal consumption of Krill (10^3 t) by various predators in the survey area

Predators	Consumption	Feeding period days	Calculation method	Comment
Flying birds	45	150	FIBEX census	–
Penguins	103	365	Extrapolation ex global data	Probably overestimate
Minke Whales	672	120	IWC census 1979–80	overestimate
Crabeater Seals	403	150	Extrapolation ex global data	Overestimate
Squid	105	150	Indirectly-FIBEX bird census	Only squid taken by birds-underestimate?
Fish	insignificant	–	–	Survey off continental shelf
Total	1328	–	–	–

most of the production was being removed via some other route (e.g., consumption by herbivores other than Krill, bacterial degradation or by vertical mixing). We examine these possibilities.

Phytoplankton productivity to standing stock ratios measured during the survey indicated a high specific phytoplankton growth rate (μ). Estimates of μ at the 21 productivity stations varied between 0.51 and 2.10 doublings d^{-1} (mean – 1.23), assuming a plant C to chlorophyll-a ratio of 30:1 (El-Sayed and Taguchi 1981). A mean specific growth rate of 1.23 doublings d^{-1} is significantly higher than maximum values reported for Antarctic waters by both Holm-Hansen et al. (1977) (0.33 doublings d^{-1}) and El-Sayed and Taguchi (1981) (0.88 doublings d^{-1}). It also exceeds Eppley's (1972) theoretically predicted maximum growth rate for phytoplankton at 1°C (0.91 doublings d^{-1}). Exceptionally high phytoplankton growth rates are confirmed, however, by biomass standardized production (P_b) values which averaged 3.62 mg C (mg Chl-a $h)^{-1}$ calculated by Strickland's (1960) method. This is significantly higher than the range of P_b values [1.08–1.56 mg C (mg Chl-a $h)^{-1}$] obtained by El-Sayed and Mandelli (1965) in the southern Weddell Sea and exceeds the maximum value [3.30 mg C (mg Chl-a $h)^{-1}$] recorded by Jacques and Minas (1981) near Kerguelen Island. Assuming equilibrium between production and decay, the high growth rates imply that phytoplankton was being rapidly removed from the system. Neither Krill feeding nor phytoplankton sinking rates (typically 1–10 m d^{-1} for the particle sizes encountered; Simpson 1982) appear sufficient to have removed phytoplankton from the surface waters rapidly enough to have maintained equilibrium. This leads us to consider the role of other consumers and physical processes in the removal of phytoplankton.

Known Antarctic meso-planktonic herbivores other than Krill include copepods, pteropods and salps (Hart 1942; Mauchline and Fisher 1969), none of which were assessed during the survey. Although speculation that these herbivores are more abundant than Krill in the Southern Ocean largely has been dismissed (e.g., Marr 1962; Everson 1977; Nemoto and Harrison 1981), the fact that our mean Krill density was about 10 times lower than in other FIBEX areas (Hampton 1983) suggests that they may have been relatively important consumers of primary production in the survey area. More significantly, consumption by micro-zooplankton could have been important. There is evidence to suggest that micro-zooplankton may play an important role in Antarctic food chain processes (Hart 1942; Mauchline and Fisher 1969; von Bröckel 1981; Marchant 1984). For example, it has been shown elsewhere in the Antarctic that the biomass of proto-zooplankton may be as much as 2.5 times that of Krill (von Bröckel 1981). Unfortunately, micro-zooplankton standing stocks were not assessed during the survey but the high proportion (mean –70%) of nanoplankton in the phytoplankton collected (L. Weber, pers. comm.) would have constituted a suitable food resource for micro-organisms.

Micro-heterotrophic activity should also be considered. Little is known about the role of micro-heterotrophs in Antarctic waters (Knox 1970), although Morita et al. (1977) have established that in the Ross Sea the level of micro-heterotrophic activity is comparable to that reported for biologically productive temperate waters. In addition, micro-heterotroph abundance was sufficiently great to ensure substantial remineralization of organic matter. Nevertheless, as Joint and Morris (1982) emphasize, high bacterial activity does not necessarily imply competition with zooplankton for organic matter produced by phytoplankton, since a significant proportion of the primary production may become available to bacteria through excretion by pelagic herbivores, and ultimately pelagic carnivores as well. In the absence of more information, the role of bacterial and micro-heterotrophic pathways in the survey area cannot be examined further. However, in view of the importance of such pathways in other oceanic systems (Sorokin and Mikheev 1979; Williams 1981; Azam et al. 1983; Ducklow 1983), it is possible that they are also significant in Antarctic waters. It is essential that future studies attempt to assess microtroph feeding and turnover rates, their distribution and abundance, and their importance as a food source and/or competitor with other herbivores (especially Krill).

Removal of phytoplankton from the euphotic zone by deep mixing is regarded as an important factor limiting primary production in the open expanses of the Southern Ocean, particularly when high wind stress may mix phytoplankton to well below the critical depth for photosynthesis (Sverdrup 1953; Tranter 1982). Vertical temperature, salinity and density ($\sigma \hat{t}$) profiles (Fig. 5) indicate thorough mixing to between 25 and 50 m. Maximum chlorophyll-a concentrations were generally found below this, at the bottom or beneath the euphotic zone (mean depth 46.5 m) while peak primary productivity occurred between 5 and about 20 m. This does not appear as a "classical" vertical mixing effect (see review by Raymont 1980) and therefore requires explanation. The deep chlorophyll maximum cannot have been the result of high grazing pressure by Krill in the euphotic zone, since Krill were found over the entire depth range sampled (0–100 m), with a broad peak between 30 and 55 m (Hampton 1984). In addition, deep chlorophyll maxima were observed in areas of low Krill density (Hampton 1984). Finally, as shown above, it is unlikely that Krill were removing a significant proportion of the available phytoplankton. The only other explanations we can offer are that: (1) there was removal of phytoplankton in the euphotic zone by some or all the other organisms discussed; and, (2) there was accumulation of sinking particles beneath the pycnocline, which although unusual does occur (e.g., Uno 1983). In view of the imballance between sinking and production rates already noted, we tend to favour the former explanation.

In summary, we conclude that the most likely explanation for the rapid phytoplankton turnover in the absence of significant "grazing" by Krill is that production was

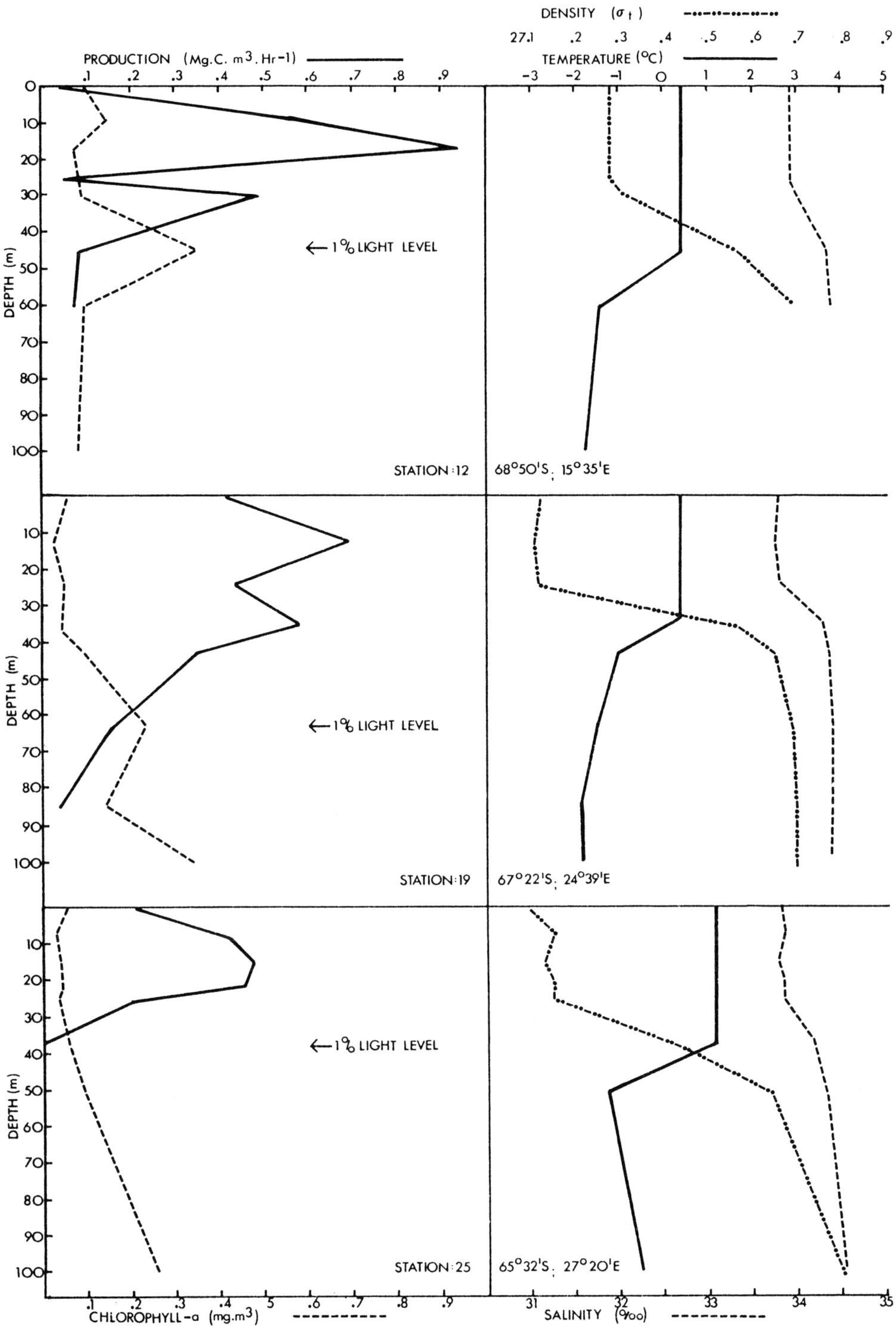

Fig. 5. Vertical profiles of primary production, chlorophyll-a temperature, salinity and density (σt) at three typical FIBEX stations

being taken up mainly by other consumers with a small contributory loss being due to sinking and vertical mixing. This is consistent with the low variability in integrated primary production values when compared to variability in Krill density (Hampton 1985). Removal of phytoplankton by a combination of partially independent mechanisms rather than by a dominant, highly patchy herbivore such as Krill, would tend to reduce variability in both phytoplankton abundance and productivity (Longhurst 1981).

5 Conclusions

Our analysis underscores the complexity of tropho-dynamic relationships in what appears superficially to be a simple system. This type of system analysis, ranging over many trophic levels and using many parameters of unknown accuracy, has obvious pitfalls. Nonetheless, the approach can be useful if, as in this case, the calculations indicate a major inconsistency in conventionally-assumed trophic relationships. We consider that similar analyses should be undertaken in other areas of the Antarctic for which concurrent acoustic data on Krill abundance and primary production measurements are available. In future trophodynamic work, more attention should be given to the impact of other herbivores and particularly to the role of micro-zooplankton and micro-trophic pathways. Sampling should extend to well below the euphotic zone to establish detrital concentrations (using sediment traps) and the full extent of vertical mixing. Only with the availability of such information can we hope to understand the nature and efficiency of the Southern Ocean food web.

Acknowledgements. We thank Profs RC Newell and JG Field for valuable assistance in the early stages of analysis, Miss DA Leach for preparing the figures, Miss P Brown for many useful discussions and Mr L Weber for the use of unpublished data. Financial support was provided by the South African National Committee for Oceanographic Research, the South African Scientific Committee for Antarctic Research and the Sea Fisheries Research Institute. The South African Department of Transport Affairs made M.V. SA Agulhas available for the work.

References

Allen KR (1971) Relation between production and biomass. J Fish Res Board Canada 28:1573–1581

Anon. (1979) Euphausiid consumption of baleen whales (excluding right whales) in the Antarctic (preliminary data). Rep Int Whal Comm 29:357–358

Anon. (1982) Recording observations of birds at sea. BIOMASS Handbook 18

Azam F, Fenchel T, Field JG, Gray JS, Meyer-Reil LA, Thingstad F (1983) The ecological role of water-column microbes in the sea. Mar Ecol Prog Ser 10:257–263

Bröckel K von (1981) The importance of nanoplankton within the pelagic Antarctic ecosystem. Kiel Meeresforsch 5:61–67

Butterworth DS (1982) An alternative sightings estimate of the minke whale population in Area III from data from the 1979/80 assessment cruise. Rep Int Whal Comm 32:875–882

Chekunova VI, Rynkova TI (1974) Energy requirements of the Antarctic crustacean *Euphausia superba* Dana. Oceanology (Berl) 14:426–432

Clarke A, Prince PA (1980) Chemical composition and calorific value of food web to mollymawk chicks *Diomedea melanophris* and *D. chrysostoma* at Bird Island, South Georgia. Ibis 122:488–494

Clarke A, Morris DJ (1983) Towards an energy budget for krill: The physiology and biochemistry of *Euphausia superba* Dana. Polar Biology 2:69–86

Ducklow HW (1983) Production and fate of bacteria in the oceans. BioSci 33:494–501

El-Sayed SZ, Mandelli EF (1965) Primary production and standing crop of phytoplankton in the Weddell Sea and Drake Passage. Antarct Res Ser 5:87–106

El-Sayed SZ, Taguchi S (1981) Primary production and standing crop of phytoplankton along the ice-edge in the Weddell Sea. Deep-Sea Res 28A:1017–1032

Emison WB (1968) Feeding preferences of the Adelie penguin at Cape Crozier, Ross Island. Antarct Res Ser 12:191–212

Eppley RW (1972) Temperature and phytoplankton growth in the sea. Fish Bull 70:1063–1085

Evans CA, O'Reilly SE (1982) A manual for measurement of chlorophyll A, net phytoplankton and nanoplankton. BIOMASS Handbook 9

Everson I (1977) The living resources of the Southern Ocean. Southern Oceans Survey Programme GLO/SO/77/1. FAO UNDP, Rome, 156 pp

Frost BW (1972) Effects of size and concentrations of food particles on the feeding behaviour of the marine planktonic copepod *Calanus pacificus.* Limnol Oceanogr 17:805–915

Furness RW (1978) Energy requirements of seabird communities: A bioenergetics model. J Anim Ecol 43:39–53

Gilbert JR, Erickson AW (1977) Distribution and abundance of seals in the pack ice of the Pacific sector of the Southern Ocean. In: Llano GA (ed) Adaptations within Antarctic ecosystems. Smithsonian Institution, Washington DC, pp 703–740

Gulland JA (1970) The development of the resources of the Antarctic seas. In: Holdgate MW (ed) Antarctic ecology, Vol I, Academic Press, London, pp 217–223

Hampton I (1983) Preliminary report on the FIBEX acoustic work to estimate the abundance of *Euphausia superba.* In: Proceedings of the BIOMASS colloquium in 1982, Mem Natnl Inst Polar Res Special Issue 27:165–175

Hampton I (1985) Abundance, distribution and behaviour of *Euphausia superba* in the Southern Ocean between 15° E and 30° E during FIBEX. In: Siegfried WR, Condy PR, Laws RM (eds) Antarctic nutrient cycles and food webs (Proceedings of the fourth SCAR symposium on Antarctic biology). Springer, Berlin Heidelberg New York

Hart TJ (1942) Phytoplankton periodicity in Antarctic surface waters. Discovery Rep 21:261–356

Holdgate MW (1967) The Antarctic ecosystem. Philos Trans R Soc Lond B Biol Sci 252:363–383

Holm-Hansen O, El-Sayed SZ, Franceschini GA, Cuhel RL (1977) Primary production and factors controlling phytoplankton growth in the Antarctic seas. In: Llano GA (ed) Adaptations within Antarctic ecosystems. Smithsonian Institution, Washington DC, pp 11–50

Horwood JW (1981) Results from the IWC/IDCR minke marking and sightings cruise. Rep Int Whal Comm 31:287–308

Ikeda T, Dixon P (1982) Body shrinkage as a possible overwintering mechanism of the Antarctic krill, *Euphausia superba* Dana. J Exp Mar Biol Ecol 62:143–151

Ikeda T, Mitchell AW (1982) Oxygen uptake, ammonia excretion and phosphate excretion by krill and other Antarctic zooplankton in relation to their body size and chemical composition. Mar Biol 71:283–298

Jacques G, Minas M (1981) Production primaire dans le secteur indien de l'océan Antarctique en fin d'été. Oceanol Acta 4:33–41

Jazdzewski K, Porebski J, Rakusa-Suszczewski S, Witek Z, Wolnomiejski N (1978) Biological and populational studies on krill near South Shetland Islands, Scotia Sea and South Georgia in the summer 1976. Pol Arch Hydrobiol 25:607–631

Joint IR, Morris RJ (1982) The role of bacteria in the turnover of organic matter in the sea. Oceanogr Mar Biol Annu Rev 20:65–118

Kanda K, Takagi K, Seki Y (1982) Movement of larger swarms of Antarctic krill *Euphausia superba* population off Enderby Land during 1976–1977 season. J Tokyo Univ Fish 68:25–42

Kato M, Segawa S, Tanoue E, Murano M (1982) Filtering and ingestion rates of the Antarctic krill, *Euphausia superba* Dana. Trans Tokyo Univ Fish 5:167–175

Kils U (1981) Swimming behaviour, swimming performances and energy balance of Anarctic krill, *Euphausia superba.* BIOMASS Sci Ser 3:1–122

Knox GA (1970) Antarctic marine ecosystems. In: Holdgate MW (ed) Antarctic ecology. Vol I, Academic Press, London, pp 69–96

Laws RM (1977a) The significance of vertebrates in the Antarctic marine ecosystem. In: Llano GA (ed) Adaptations within Antarctic ecosystems. Smithsonian Institution, Washington DC, pp 411–438

Laws RM (1977b) Seals and whales of the Southern Ocean. Philos Trans R Soc Lond B Biol 279:81–96

Lockyer C (1976) Body weights of some species of large whales J Cons pérm int Explor Mer 36:259–273

Longhurst AR (1981) Significance of spatial variability. In: Longhurst AR (ed) Analysis of marine ecosystems. Academic Press, London, pp 415–441

Mackintosh NA (1972) Life cycle of Antarctic krill in relation to ice and water conditions. Discovery Rep 36:1–94

Marchant HJ (1985) Choanoflagellates in the Antarctic marine food chain. In: Siegfried WR, Condy PR, Laws RM (eds) Antarctic nutrient cycles and food webs (Proceedings of the fourth symposium on Antarctic biology). Springer, Berlin Heidelberg New York

Marr JWS (1962) The natural history and geography of the Antarctic krill (*Euphausia superba* Dana). Discovery Rep 32:33–464

Mauchline J (1980) The biology of mysids and euphausiids. Adv Mar Biol 18:373–637

Mauchline J, Fisher LR (1969) The biology of euphausiids. Adv Mar Biol 7:1–454

Morita RY, Griffiths RP, Hoyosaka SS (1977) Heterotrophic activity of microorganisms in Antarctic waters. In: Llano GA (ed) Adaptations within Antarctic ecosystems. Smithsonian Institution, Washington DC, pp 99–113

Nemoto T, Harrison G (1981) High latitude ecosystems. In: Longhurst AR (ed) Analysis of marine ecosystems. Academic Press, London, pp 95–126

Ohsumi T (1979) Feeding habits of the minke whale in the Antarctic. Rep Int Whal Comm 29:473–476

Øristland T (1977) Food consumption of seals in the Antarctic pack ice. In: Llano GA (ed) Adaptations within Antarctic ecosystems. Smithsonian Institution, Washington DC, pp 703–740

O'Reilly JE, Thomas JP (1982) A manual for the measurement of total daily primary productivity. BIOMASS Handbook 10

Paffenhöffer GA (1971) Grazing and ingestion rates of nauplii, copepodites and adults of the marine planktonic copepod *Calanus helgolandicus*. Mar Biol 11:286–298

Prévost J (1981) Population, biomass and energy requirements of Antarctic birds. BIOMASS II:125–137

Rakusa-Suszczewski S, Opalinski W (1978) Oxygen consumption in *Euphausia superba*. Pol Arch Hydrobiol 3:633–641

Raymont JEG (1980) Plankton and productivity in the oceans – 2nd Edition, Volume 1: Phytoplankton. Pergamon, Oxford, pp 489

Remmert H (1980) Arctic animal ecology. Springer, Berlin Heidelberg New York, pp 123

Simpson WR (1982) Particulate matter in the oceans – Sampling methods, concentration, size distribution and particle dynamics. Oceanogr Mar Biol Annu Rev 20:19–172

Sorokin YuI, Mikheev VN (1979) On characteristics of the Peruvian upwelling ecosystem. Hydrobiologia 62:165–189

Strickland JDH (1960) Measuring the production of marine phytoplankton. Bull Fish Res Board Can 122:1–172

Strickland JDH, Parsons TR (1972) A practical handbook of seawater analysis. Bull Fish Res Board Can 167:1–311

Sverdrup HU (1953) On conditions for the vernal blooming of phytoplankton. J Cons pérm int Explor Mer 18:287–295

Tranter DJ (1982) Interlinking of physical and biological processes in the Antarctic ocean. Oceanogr Mar Biol Annu Rev 20:12–35

Uno S (1983) The relation between phytoplankton standing stock and water temperature in the Antarctic Ocean in summer, 1980–1981. In: Proceedings of the BIOMASS colloquium in 1982, Mem Natnl Inst Polar Res Special Issue 27:37–49

Vinogradova Z (1967) The biochemical composition of Antarctic plankton. In: Biochemistry of marine organisms. Ukrainian Academy of Sciences 7:17

Walsberg GE (1980) Energy expenditure in free-living birds; patterns and diversity. In: XVII Congresces Auternationales Ornithologicius. Nolsring, R. Deutsche Ornithologen-Gesellschaft, Berlin

Williams AJ, Siegfried WR (1980) Foraging rages of krill-eating penguins. Polar Rec 20:159–175

Williams PJ le B (1981) Incorporation of microheterotrophic processes into the classical paradigm of the planktonic food web. Kiel Meeresforsch 5:1–28

Feeding by Antarctic Krill, *Euphausia superba:* Does Size Matter?

L.B. QUETIN and R.M. ROSS[1]

Summary. Antarctic Krill, *Euphausia superba* (25–30 mm total length), were fed uni-algal cultures of a flagellate, *Isochrysis galbana* (= 4 μm sphere), a pennate diatom, *Phaeodactylum tricornutum* (equivalent spherical diameter (ESD) = 4 μm), and 2 centric diatoms, *Ditylum brightwellii* (ESD = 30 μm) and *Thalassiosira eccentrica* (ESD = 40 μm). Experiments were conducted at –0.5°C for 3–4 d, after an initial acclimation period of 4 d. The feeding response varied with the size of the phytoplankton cell. The maximum clearance rates for any individual feeding on the 4 phytoplankton species were: 103 ml animal^{-1} h^{-1} for *I. galbana,* 167 ml animal^{-1} h^{-1} for *P. tricornutum,* 281 ml animal^{-1} h^{-1} for *D. brightwellii,* and 357 ml animal^{-1} h^{-1} for *T. eccentrica.* The maximum clearance rates of all individuals were directly proportional to the calculated spherical radius squared $(CSR)^2$ of the cells, described by the equation $y = 61.75 + 0.233\ CSR^2$ where y is the clearance rate (N = 20, r = 0.694). No critical concentration was established for *I. galbana* at concentrations as great as 25000 cells ml^{-1}, indicating that *E. superba* does not feed very effectively on small round cells. Because the variation in the maximum clearance rate is closely linked to the size, but not the species, of phytoplankton being ingested, the physical dimensions, not the chemical composition, of the cell are probably the most important parameters affecting the maximum clearance rate. These feeding responses are important both to the energetics of *E. superba,* and to the modification of the phytoplankton community in which Krill feed.

1 Introduction

The feeding activities of the Antarctic Krill, *Euphausia superba,* have consequences of great importance to the structure and functioning and the management of the Antarctic marine ecosystem. Because of its abundance and pivotal position in the food web between primary producers and secondary consumers, *E. superba* assumes an importance in the Southern Ocean shared by few other species in the oceans of the world. The Peruvian Anchoveta, *Engraulis ringens,* is one of the few comparable examples of a single species whose abundance is crucial to the structure of a large marine ecosystem (Clark 1976).

Although little is known about its winter diet, the evidence is clear that during the time of high phytoplankton productivity in the Antarctic summer, *E. superba* feeds predominantly on phytoplankton (Hart 1934; Barkley 1940; Hustedt 1958; Marr 1962; Clarke 1980). The species has been observed to feed in a variety of ways. Krill are capable of scraping algae off a substratum with their pereiopods, raising speculation that they feed on algae from the benthos or sea-ice during the Antarctic winter when phytoplankton production is low (McWhinnie and Denys 1978; Antezana et al. 1982; Hamner et al. 1983). Krill are also capable of raptorially capturing and holding large particles, including moulted cuticles of crustaceans, in their feeding baskets (Hamner et al. 1983; Boyd et al. 1984). However, the dominant feeding behaviour observed by us and other investigators during the Antarctic summer appears to involve the ingestion of phytoplankton suspended in the water column.

Recent observations of Krill feeding on suspended particles (Hamner et al. 1983; Boyd et al. 1984) provide a more detailed description of the process than those reported previously (Kato et al. 1979; Berkes 1975; Antezana et al. 1982; Meyer and El-Sayed 1983). When feeding, the animal's pereiopods are extended to form a "packet" of water which is pressed through the mesh-basket formed by the pereiopods. Hence, the pereiopods act as a sieve in concentrating phytoplankton from discrete packets of water. We call this process compression feeding, to distinguish it from the more passive process of filter feeding.

The clearance rate (filtration rate) and apparent selectivity of food particles by *E. superba* are of equal importance in interactions between this species and certain members of the phytoplankton community. Previous investigators have determined clearance rate in a variety of ways: from the rate of chlorophyll-a depletion in closed vessels containing euphausiids and natural phytoplankton (Kato et al. 1979; Meyer and El-Sayed 1983); from the difference in chlorophyll-a in the seawater from the inlet and outlet of a large aquarium filled with thousands of animals (Antezana et al. 1982); and, most recently, from changes in cell numbers of natural phytoplankton, as counted by particle counters, consumed by Krill in flow-through (Morris 1984) or closed systems (Boyd et al. 1984). The clearance rates obtained with the flow-through system used by Morris are several orders of magnitude higher than those reported by other investigators. Morris (1984) concluded that *E. superba* feeds more effectively on certain sizes of phytoplankton than others and that particles 30 μm in equivalent spherical

1 Marine Science Institute, University of California, Santa Barbara, CA 93106, USA

Antarctic Nutrient Cycles and Food Webs
(ed. by W.R. Siegfried, P.R. Condy, and R.M. Laws)

diameter (ESD) are apparently rejected, whereas Boyd et al. (1984) found that all particles > 12 μm were retained with equal efficiency. The conflicting results from previous investigations, involving the use of both flow-through and closed aquarium systems, suggest that simpler, more controlled, experiments are needed to define and measure both clearance rate and selectivity.

Our investigation is the first to make use of uni-algal cultures in feeding experiments involving *E. superba,* thus avoiding the pitfalls associated with the use of natural phytoplankton assemblages (Harbison and McAlister 1980). Based on our experiments which employed four phytoplankton species, spanning wide ranges in the size and shape of organisms, we were able to determine both whether and how the size of the food particles affects the apparent feeding response of Krill to phytoplankton suspended in the water column.

2 Methods

Six live specimens of *E. superba,* collected in the Southern Ocean (in January 1980), were flown to the Australian Institute of Marine Science, in Townsville, Australia. These animals were kept at –0.5°C and fed on *Dunaliella tertolecta* and *Phaeodactylum tricornutum* until October 1980 when we began our experiments. The diet of the animals varied randomly, since they were acclimated to a new species of phytoplankton before each experiment. Each euphausiid was held in a glass beaker containing 2500–3000 ml of seawater in dim light in a cold-room at –0.5°C. The animals were kept in the laboratory for at least 15 months, with no mortality.

The length of each animal was calculated from measurements of the uropods of moulted exoskeletons. Poleck's (1981) equation $y = 5x + 0.48$ where x is the uropod length and y is the total length, was used to calculate a total length from the tip of the rostrum to the end of the sixth abdominal segment, the length of the uropod being added to obtain a Standard-1 length measurement (Mauchline 1980). We used the uropod lengths of exoskeletons cast during November 1980–February 1981 to calculate the growth rates of individuals.

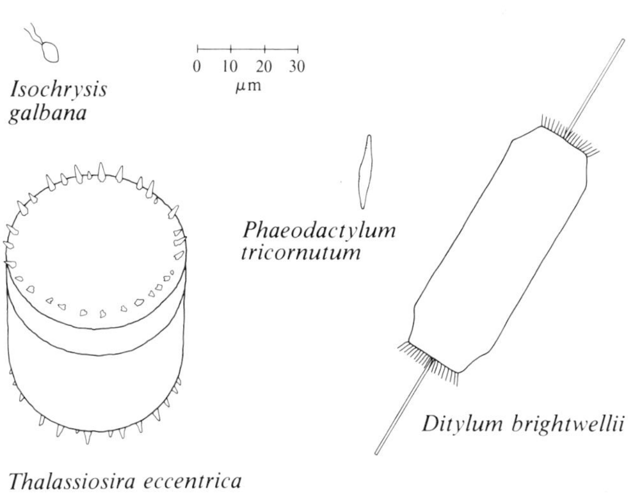

Fig. 1. Size and shape of the four phytoplankton species used in *E. superba* feeding experiments

Four species of phytoplankton were employed in the experiments: three diatoms, *Thalassiosira eccentrica, Ditylum brightwellii* and *Phaeodactylum tricornutum;* and, a flagellate, *Isochrysis galbana* (Fig. 1). The species were selected to represent a wide size range and, in the case of *P. tricornutum* and *I. galbana,* to compare the clearance rates of Krill feeding on organisms with similar ESD's but very different linear dimensions (Table 1). The dimensions of cells, represented by a sample of 50 individuals of each species of phytoplankton, were measured with a compound microscope. Phytoplankton sizes ranged from a calculated spherical diameter (CSD) of 5.3 μm for *I. galbana* to 50 μm for *T. eccentrica.* The ESD measured by the TA II Coulter Counter under-estimated the CSD by 20–46% (Table 1). A difference between measured and calculated estimates of the spherical diameter has been noted previously (Harbison and McAlister 1980). Here, we use the CSD or CSR (calculated spherical radius).

The phytoplankton species were grown in f/2 growth medium (Guillard and Ryther 1962), except *D. brightwellii* which was grown in G/2 growth medium (S. Jeffrey, pers. comm.), under constant fluorescent illumination at 15°C. We used a culture temperature of 15°C to obtain a growth rate faster than the one possible at the experimental temperature of –0.5°C. Only cells in the log phase of growth were used in the experiments. The phytoplankton cultures were transferred to the –0.5°C cold-room at least 5 h before starting an experiment. The phytoplankton was diluted to the desired concentration with filtered (Millipore) seawater at –0.5°C, and 2100 ml of the resulting seawater-phytoplankton mixture was added to each one of three to six experimental jars and two control jars. A single euphausiid in 120 ml of water was transferred gently to each experimental jar. The jars were then sealed, placed on the stirring machine and "rolled" at 2 rpm to keep the phytoplankton in suspension. The experiments were conducted in the dark to retard phytoplankton growth. We used only one species of algae in each experiment (i.e. mixtures of different species were not used).

An initial sample of 200 ml was taken from the centre of each jar 6–8 h after it was placed on the stirring machine. The concentration of cells in each jar was determined by averaging 8 replicate counts made with a TA II Coulter Counter on the 200 ml sample. After the initial sample, cell concentrations were determined at about 8 h intervals for up to 3 d. We used the equations of Frost (1972) to calculate the clearance rates of Krill feeding on a specific phytoplankton species at stated concentrations. At the end of an experiment, each euphausiid was returned to its beaker to acclimate to a new species of phytoplankton. The four species of phytoplankton were used in a random sequence.

Table 1. Comparison between calculated spherical diameter (CSD) and the estimated spherical diameter (ESD) for four species of phytoplankton. ESD was determined with a TA II Coulter Counter

Species	Date Measured	Width, diameter (μm)	Length, height (μm)	Volume equation used	CSD (μm)	ESD (μm)	% Error
Thalassiosira eccentrica	30 Dec. 80	57.5	25	Cylinder	49.7	40	20
Ditylum brightwellii	22 Jan. 81	25.7	73.8	Cylinder	40.9	30	27
Phaeodactylum tricornutum	28 Jan. 81	4.4	22.2	2 Cones	7.5	4	46
Isochrysis galbana	25 April 81	4.8	5.6	Oblate spheroid	5.3	4	25

3 Results

The individual Krill used in our experiments ranged from 23 to 27 mm in total length. The growth rate of the animals between November and February of the experimental period ranged from 0 to 0.039 mm d^{-1}. The inter-moult interval was about 30 d (Table 2).

Maximum clearance rates prevailed at phytoplankton concentrations less than the critical concentration and above a feeding threshold (Frost 1972; Mullin 1963) (Fig. 2). We estimated, by eye, the concentration of phytoplankton at which the clearance rate began to decrease (critical concentration) in Krill feeding on each of the four species. No ciritical concentration could be determined for *I. galbana;* the clearance rate remaining low and about the same for concentrations between 2000 and 24000 cells ml^{-1}. The critical concentrations for the other species of phytoplankton were 12000, 45 and 30 cells ml^{-1} for *P. tricornutum, D. brightwellii* and *T. eccentrica,* respectively.

The relationship between maximum clearance rate and food size was assessed by three methods (Fig. 3). First, for each species of phytoplankton we used the single highest clearance rate attained by any animal (Fig. 3, top line). Secondly, we used all the highest clearance rates of each individual; thus, including the variation in maximum clearance rate between individuals (Fig. 3, middle line). Thirdly, we

Table 2. Size, growth rate and moulting of six individual *E. superba,* used in feeding studies

Animal	Date	Uropod length (mm)	Calculated total length (mm)	Moult interval (d)	Growth (mm d^{-1})
1	15 Dec. '80	4.00	26.08		
	11 Jan. '81	4.06	26.46	27	0.014
	08 Feb. '81	4.10	26.72	28	0.009
2	15 Dec. '80	3.90	25.44		
	15 Jan. '80	4.00	26.08	31	0.021
3	20 Dec. '80	4.00	26.08		
	22 Jan. '81	4.00	26.08	33	0
4	06 Nov. '80	3.50	22.88		
	26 Dec. '80	3.60	23.04	50[a]	0.003
5	15 Jan. '81	3.80	24.80		
6	26 Dec. '80	3.80	24.80		
	21 Jan. '81	4.00	26.08	26	0.049
	21 Feb. '81	4.15	27.04	31	0.031

a Two inter-moult periods

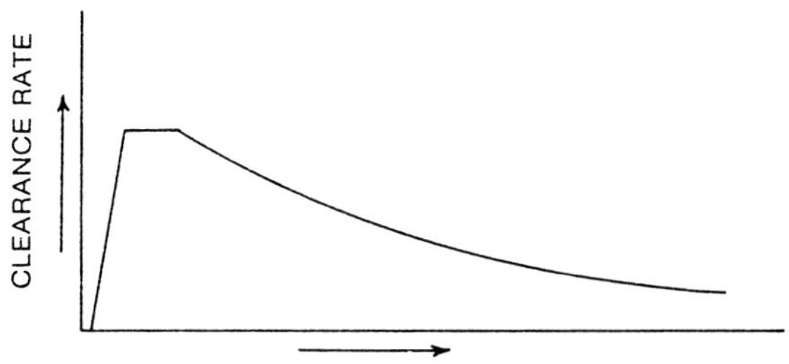

Fig. 2. Generalized relationship between clearance rate in *E. superba* and concentration of phytoplankton cells

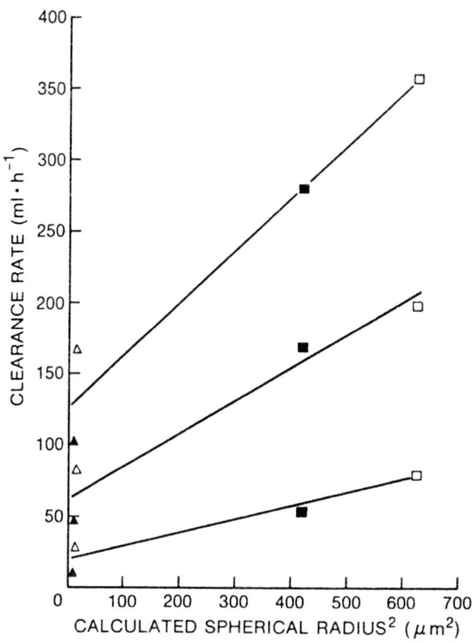

Fig. 3. The relationship between clearance rate in *E. superba* and CSR^2, calculated three ways (see text). *Top line* based on individual with highest clearance rate; *middle line* based on maximum clearance rates of all individuals; *bottom line* based on average of all clearance rates in all experiments determined below the critical concentration. △ = *Phaeodactylum tricornutum*, ▲ = *Isochrysis galbana*, ■ = *Ditylum brightwellii*, □ = *Thlassiosira eccentrica*

combined the clearance rates measured at concentrations of phytoplankton below the critical concentration for all experiments and all euphausiids and calculated an average clearance rate for each species of phytoplankton below the critical concentration (Fig. 3, bottom line). The first and third regression lines were fitted by eye. The equation for the relationship between maximum clearance rate (y) and CSR^2, as calculated by method 2, was $y = 61.75 + 0.233$ CSR^2 ($N = 20, r = 0.694$).

The maximum clearance rate increased linearly with size of the species of phytoplankton, when size was expressed as the CSR^2 for all methods (Fig. 3). This result suggests that the higher clearance rates of Krill feeding on larger cells reflects the larger cross-sectional area of these cells, rather than the larger volume. Given the results of Harbison and McAlister (1980) and Runge and Ohman (1982), we also feel that if two species of phytoplankton have the same ESD or CSD, the species with the longest linear dimensions should be retained more effectively if retention is primarily a function of size. Qualitatively, we can verify this prediction. The clearance rate of Krill feeding on the long, thin cells of *P. tricornutum* was always higher than for the oblate spheroid *I. galbana.* However, the difference between the two clearance rates is not significant (Mann-Whitney U test, $P > 0.05$) (Siegel 1956). The variation in clearance rates of Krill feeding on cells with similar ESD's but very different linear measurements needs to be examined further.

4 Discussion

The compression method of feeding employed by *E. superba* is analogous to that of other animals feeding in the water column. Baleen whales gulp huge volumes of water and then concentrate Krill by compressing the water through the baleen with their tongues. Copepods trap phytoplankton cells with the second maxillae which then press together to remove water and concentrate the cells. Compression feeding is especially important to copepods and *E. superba,* since the feeding appendages operate at very low Reynolds numbers. At low Reynolds numbers, water acts as a viscous fluid and would flow around and not through the feeding appendages if it was not actively compressed as a discrete volume (Jorgensen 1983; McClatchie and Boyd 1983). Since compression feeding by copepods and *E. superba* is similar, it is not surprising that both we and Boyd et al. (1984) regard the functional feeding response of *E. superba* to be analogous to that of copepods feeding on phytoplankton.

Several recent studies indicate that *E. superba* does not feed equally well on all sizes of phytoplankton (Meyer and El-Sayed 1983; Clarke and Morris 1983; Morris 1984; Boyd et al. 1984). This is not surprising, since even mechanical filters appear to behave differently when used to sieve different shapes and sizes of particles (Harbison and McAlister 1980; Runge and Ohman 1982). Our results support the general conclusion that particle size is important in the feeding process of Krill, but differ in detail from those of other investigators. Morris (1984) found that 30 μm particles were maximally retained, and that particles > 30 μm in diameter apparently were rejected. Boyd et al. (1984) found that phytoplankton with diameters > 12 μm were retained maximally, and that cells only 6 μm in diameter were retained with a 50% efficiency. In contrast, we found a linear relationship between areal size and maximum clearance rate for phytoplankton cells between 6 and 50 μm in spherical diameter, that the shape of the cell is of minor importance when compared to total area, and that particles > 30 μm were not rejected.

Table 3. Percentage cells retained by *E. superba* feeding on four species of phytoplankton

Species	CDS (μm)	A	B	McClatchie and Boyd (1983)
Thalassiosira eccentrica	49.7	48	45	78
Ditylum brightwelli	40.9	38	35	58
Phaeodactylum tricornutum	7.5	18 (22)	14 (18)	35
Isochrysis galbana	5.3	18 (14)	14 (10)	17

Estimates are based on a 95% retention efficiency for an 80 μm particle (McClatchie and Boyd 1983) and are computed using the CDS (μm) for each phytoplankton species. A = estimates using the regression on the single highest clearance rate; and, B = estimates using the regression on the average of the highest clearance rates for each animal. Retention efficiency (%) for each species not based on the regression given in parentheses

The retention efficiency of Krill feeding on the four types of cells was estimated by assuming that 95% of cells with an 80 μm spherical diameter are retained (McClatchie and Boyd 1983) (Table 3). Close agreement occurred between estimates of the percentage retention efficiency for cells of the same size as calculated from our maximum and average maximum clearance rates. Our estimates are uniformly lower than those derived from McClatchie and Boyd (1983) (Table 3). We found no maximum or plateau in retention efficiency for particles at least as large as 50 μm CSD (Table 3). Unfortunately, our experiments did not test the model of McClatchie and Boyd (1983), since the active component of filtering, associated with the rate at which the pereopods capture water, cannot be separated from the passive component of particles sieved through the feeding basket. This model cannot be tested until the feeding activity of Krill on a larger selection of phytoplankton species of different size and shape is determined.

Since the clearance rate of *E. superba* varies with particle concentration (Antezana et al. 1982; Boyd et al., 1984; this study), estimates of a maximum rate require that the rate be determined over a wide range of particle concentrations. Our maximum clearance rate of 357 ml h^{-1} may be an under-estimate, since maximum rate may also increase with particle size beyond the largest particle used. Despite this, however, our data agree generally with those of Boyd et al. (1984) who obtained a clearance rate of 500 ml h^{-1} for euphausiids 45 μm in total length, and the value of 210 ml h^{-1} measured by Antezana et al. (1982). Our clearance rates are lower than those of Boyd et al. (1984) in part because our experimental animals were much smaller. Within a species, smaller animals filter at lower rates than larger ones (Kato et al. 1979). The maximum clearance rate of Antezana et al. (1982) may also be low, since they had no measurements at chlorophyll-a concentrations between 6 and 10 μg l^{-1} at which Boyd et al. (1984) found maximum clearance rates.

Morris (1984) reported a maximum clearance rate for *E. superba* of 66816 ml h^{-1}. This is more than 130 times the maximum clearance rates found by other workers, and cannot be explained by differences in temperature or animal size. Although we do not understand why the clearance rates found by Morris (1984) are so much higher than previous estimates, indirect evidence suggests they are, indeed, much too high. Boyd et al. (1984) estimated the volume of water captured by one extension-contraction cycle of the feeding basket to be 0.20 ml. By dividing the maximum clearance rate found by Morris (1984) by the basket volume we calculated an expansion-contraction rate for the feeding basket to be 5568 cycles min^{-1} or 93 cycles s^{-1}. This, clearly, is too rapid, even if the volume of the feeding basket was under-estimated. Also, *E. superba* does not continually extend and contract its pereiopods at a constant rate (Antezana et al. 1982; Hamner et al. 1983; Boyd et al., 1984; pers. obs.) Hamner et al. (1983) found that extension-contraction rates varied between 1 and 5 cycles s^{-1}, and Antezana et al. (1982) and Boyd et al. (1984) estimate average rates of about 2–3 cycles s^{-1}, all being consistent with our observations. Another simple calculation serves to show that it would be virtually impossible to measure an instantaneous clearance rate of 66819 ml h^{-1} in a flow-through system with a volume of 1 l and a flow rate of 1 l h^{-1}. Such a clearance rate means that the euphausiid must process the entire volume of the chamber in less than 1 min, and sampling the flowing water twice in 1 min would be difficult. Thus, our experiments do not support the conclusion of Clarke and Morris (1983) that clearance rates of *E. superba* are so large that most of the particle removal occurs in the first 15–30 min of a closed-volume experiment.

The response curve of *E. superba* feeding on high concentrations of phytoplankton demonstrates that average ingestion rates are maximal at clearance rates far below those found by Morris (1984), especially at high concentrations of phytoplankton (Antezana et al. 1982; Boyd et al., to be publ.; this study). We conclude that unless *E. superba* was feeding differently to what was observed in these studies, the filtration rates found by Morris (1984) are too high. The flow rate of $1\ l\ h^{-1}$ through the 1 l experimental vessel is also probably too low to elicit feeding or swimming behaviour different to that of a euphausiid swimming in a closed, larger, experimental vessel.

What is the effect of *E. superba* on the phytoplankton community? Dense aggregations of Krill can drastically reduce the standing crop of phytoplankton in the water column (Antezana and Ray 1984). Furthermore, our results suggest that these schools could change the size composition of phytoplankton in the water column, shifting it toward smaller sizes. The effect of selectively reducing the abundance of certain species would be a profound restructuring of the phytoplankton community. Large aggregations of Krill have also been shown to increase the level of ammonium ions in the water column (Biggs et al. 1985). Whether the increased nutrient levels are utilized or needed by the phytoplankton are still open questions.

In conclusion, we have shown that *E. superba* feeds most effectively on relatively large species of phytoplankton, and that this apparently reflects the area and not the volume of the phytoplankter. These results are important in considerations of the feeding energetics of *E. superba* and the composition of phytoplankton communities in the Antarctic marine ecosystem.

Acknowledgements. We are grateful to T. Ikeda for providing the Krill, S. Jeffrey for sending us the initial innocula of the four phytoplankton species, P. Dixon for his help, and C. Boyd for a helpful review of an early draft of the manuscript. This research was carried out at the Australian Institute of Marine Science, while L. Quetin held a Queen Elizabeth II Fellowship in Marine Biology.

References

Antezana T, Ray K (1984) Feeding of *Euphausia superba* in a swarm north of Elephant Island. J Crust Biol 4 (Spec No 1):142–155

Antezana T, Ray K, Melo C (1982) Trophic behaviour of *Euphausia superba* Dana in laboratory conditions. Polar Biol 1:77–82

Barkley E (1940) Nahrung und Filterapparat des Walkrebschens *Euphausia superba* Dana. Z Fisch 1:65–156

Berkes F (1975) Some aspects of feeding mechanisms of euphausiid crustaceans. Crustaceana 29:266–269

Bidigara RR, Johnson MA, Guffy JD, Biggs DC (1981) Nutrient chemistry of ammonium in antarctic surface waters. Ant J US 26:168–170

Biggs DC, Amos AF, Holm-Hansen O (1985) Oceanographic studies of epipelagic ammonium distributions: the Ross Sea NH_4^+ flux experiment. In: Siegfried WR, Condy PR, Laws RM (eds) Antarctic nutrient cycles and food webs (Proceedings of the 4th SCAR symposium on Antarctic biology). Springer, Berlin Heidelberg New York

Boyd CM, Heyraud M, Boyd CN (1984) Feeding behaviour of krill. J Crust Biol 4 (Spec No 1):123–141

Clark WG (1976) The lessons of the Peruvian anchovet a fishery. CALCOFI Rep 19:57–63

Clarke A (1980) The biochemical composition of krill, *Euphausia superba* Dana, from South Georgia. J Exp Mar Biol Ecol 43: 221–236

Clarke A, Morris DJ (1983) Towards an energy budget for krill: the physiology and biochemistry of *Euphausia superba* Dana. Polar Biol 2:69–86

Frost BW (1972) Effects of size and concentration of food particles on the feeding behaviour of the marine planktonic copepod Calanus pacificus. Limnol Oceanogr 17:805–815

Guillard RRL, Ryther JH (1962) Studies of marine planktonic diatoms. 1 *Cyclotella nana* Justedt and *Detonula confervacea* (Cleve) Gran. Can J Microbiol 8:229–233

Hamner WM, Hamner PP, Strand SW, Gilmer RW (1983) Behaviour of Antarctic krill, *Euphausia superba:* chemoreception, feeding, schooling, molting. Science 220:433–434

Harbison GR, McAlister VL (1980) Fact and artifact in copepod feeding experiments. Limnol Oceanogr 25:971–981

Hart TJ (1934) On the phytoplankton of the southwest Atlantic and the Bellingshausen Sea, 1929–31. Discovery Rep 8:1–268

Hustedt F (1958) Diatomeen aus der antarktis und dem Südatlanktik Deutsche Antarktische Expedition 1938–39. Wiss Erg dt antarkt Exped 2:103–191

Jorgensen CB (1983) Fluid mechanical aspects of suspension feeding. Mar Ecol Prog Ser 11:89–103

Kato M, Murano M, Segawa S (1979) Estimation of the filtering rate of the Antarctic krill under laboratory conditions. Trans Tokyo Univ Fish 3:107–112

McClatchie S, Boyd CM (1983) Morphological study of sieve efficiencies and mandibular surfaces in the Antarctic krill *Euphausia superba.* Can J Fish Aquat Sci 40:955–967

McWhinnie MA, Denys CJ (1978) Antarctic marine living resources with special reference to krill, *Euphausia superba:* assessment of adequacy of present knowledge. Report to USNSF 1978, 209 pp

Marr JWS (1962) The natural history and geography of Antarctic krill (*Euphausia superba* Dana). Discovery Rep 32:33–464

Mauchline J (1980) Measurement of body length of *Euphausia superba* Dana. BIOMASS Handbook 4 p 9

Meyer MA, El-Sayed SZ (1983) Grazing of *Euphausia superba* Dana on natural phytoplankton populations. Polar Biol 1:193–197

Morris DJ (1984) Filtration rates in *Euphausia superba:* under or over estimates? J Crust Biol 4 (Spec No 1):185–197

Mullin MM (1963) Some factors affecting the feeding of marine copepods of the genus *Calanus.* Limnol Oceanogr 8:239–250

Poleck TP (1981) The effect of temperature on the molting, growth, and maturation of the Antarctic krill, *Euphausia superba* Dana. M.S. De Paul University Chicago p 122

Runge JA, Ohman MD (1982) Size fractionation of phytoplankton as an estimate of food available to herbivores. Limnol Oceanogr 27:570–575

Siegel S (1956) Nonparametric statistics for the behavioural sciences. McGraw Hill, New York

Nutrition of *Allotanais hirsutus* (Crustacea: Tanaidacea) at Kerguelen Island

D. DELILLE, L.D. GUIDI, and J. SOYER[1]

Summary. High densities (56000–146000 individuals m^2) of *Allotanais hirsutus* were observed in the intertidal zone of Morbihand Bay at Kerguelen Island. A new method, involving ^{14}C-labelled tracers, was used in a preliminary in situ investigations of the ingestion of bacteria by *A. hirsutus*. Feeding activity appeared to be influenced by the tide, being higher during retreating and rising tides than during the slack. The average rate of total bacterial carbon incorporated by *A. hirsutus* was 12.8×10^{-3} μg C μg individual (dw^{-1}) h^{-1}, indicating that bacteria could contribute significantly to the diet of the dense population of *A. hirsutus*.

1 Introduction

A remarkably dense population (56000–146000 individuals m^{-2}) of the tubiculous tanaid, *Allotanais hirsutus*, was found in the muddy sand of the intertidal zone in a small cove in Morbihan Bay at Kerguelen Island. Here we report on a preliminary study designed to determine the energy requirements of *A. hirsutus* and to investigate the species' utilization of bacteria as a source of food. Our study used a new in situ method, involving ^{14}C-labelled bacteria.

2 Materials and Methods

2.1 Preparation of Labelled Bacteria ("Tracers")

Bacterial populations at the experimental site were enumerated by means of epifluorescence, making direct counts under an Olympus BHA microscope, after acridine orange staining. This technique was modified slightly from that described by Hobbie et al. 1977.

Three bacterial strains (nos 2, 3, and 5) were isolated at random from a 2216 solid-medium (Oppenheimer and Zobell 1952). After purification, these strains appeared to be gram-negative rods with a mean carbon content of 1.6 times 10^{-13} g C $cell^{-1}$ (Leco Carbon Analyzer WR-12). Pure cultures were grown in the dark at 20°C in liquid 2216 medium, and labelled with 1000 μ Ci l^{-1} of ^{14}C-glutamic acid 24 h after inoculation. The next day, the cultures were centrifuged, washed in sterile 0.2 μm-filtered seawater, and adjusted to the required concentrations (Fig. 1) in 10 ml portions.

2.2 Feeding Experiments

In a first series of experiments, plastic cores (25 mm diameter) containing samples of the top 100 mm of sediment were obtained at the field site, sealed with plastic stoppers and taken to the laboratory where they were kept in water at the in situ temperature (5°C). Experiments started by injecting through the bottom stopper 10 ml of labelled bacteria per core into the sediment; they were stopped with neutralized formalin (conc. approximately 4%) after incubation times ranging from 0.5 to 3.0 h.

In a second series of experiments, the cores were left in place, without stoppers, along a transect in the intertidal zone. At different periods in the tidal cycle, experiments started by adding 10 ml of labelled bacteria into each core; they were stopped after different incubation times (20–60 min) by addition of neutralized formalin (conc. approximately 4%), and taken to the laboratory. The plastic tubes were sufficiently high above ground and incubation times sufficiently short to avoid penetration of water through the top opening.

2.3 Treatment of Labelled Samples

At the end of all incubations, the tanaids were removed from the sediment, and rinsed in seawater and distilled water. They were measured under a dissecting microscope (the length of individuals found in the cores ranged between 3 and 8 mm) and dried overnight at 60°C. Each individual was then combusted in a Packard Tri-Carb Oxidizer to oxydize all ^{14}C into $^{14}CO_2$ which was retained by Carbo-Sorb II (Packard) and automatically mixed to a scintillation cocktail (Permafluor V, Packard). Liquid samples of bacterial cultures were mixed directly with the scintillation cocktail (Instagel, Packard). All samples were counted for at least 10 min or 10,000 counts in a Beckman LS 7000 scintillation counter. Values were corrected for quench, background, and machine efficiency.

1 Laboratoire Arago, F-66650 Banyuls-sur-Mer, France

Antarctic Nutrient Cycles and Food Webs
(ed. by W.R. Siegfried, P.R. Condy, and R.M. Laws)

2.4 Calculations

Length (L, mm) of tanaids was converted to dw μg by means of the equation dry wt. = 8.14 $L^{2.830}$ (r = 0.97, P = 0.99), established from 114 specimens (size range = 1.6–8.5 mm) measured under a dissecting microscope equipped with a *camera lucida,* dried at 60°C for 24 h, and weighed on a Cahn 21 Automatic Electrobalance. Weight-specific uptake of natural bacteria (Bact. μg individual dw^{-1}) by *A. hirsutus* was calculated from the weight-specific activity of the tanaids (DPM μg dry $wt.^{-1}$) the specific activity of the "tracer" (DPM $cell^{-1}$) and the ratio of labelled bacteria to natural bacteria in the cores. Results are averages of 19–54 independent determinations.

3 Results and Discussion

The use of radio-isotopes in studies of the nutrition of aquatic animals has been described: usually experiments are conducted in the laboratory, and labelled food sources, algal or bacterial cultures (e.g., Hargrave 1970), or surface sediment containing natural populations of benthic micro-organisms (e.g., Brown and Sibert 1977), are offered alone in controlled conditions to the animals. Fewer studies have been performed in close-to-natural conditions. For example, Meyer-Reil and Faubel (1980) injected labelled glucose directly into sediment cores to assess the different participation of individual groups of meiofauna in the incorporation of labelled material. In all these studies the tracer was

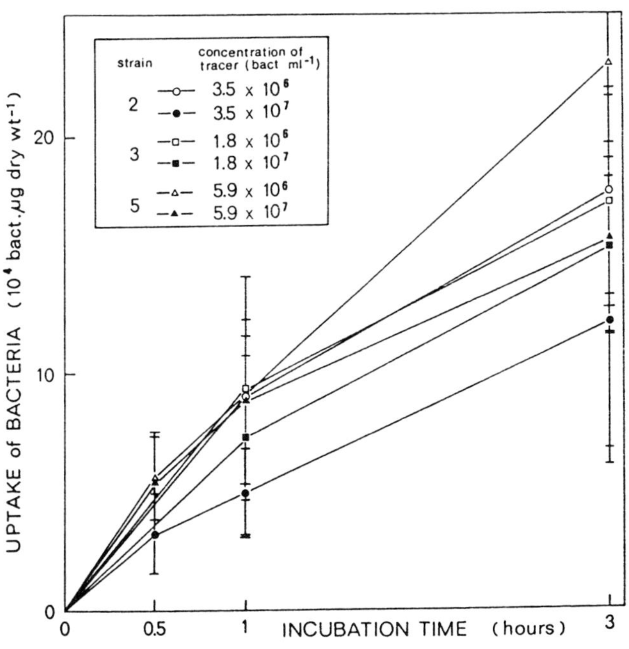

Fig. 1. Mean (± S.E.) weight-specific uptake of total bacteria by *A. hirsutus* after different incubation times in cores brought back to the laboratory, using different strains and concentrations of labelled bacteria as "tracers". Concentrations stated are those in the 10 ml added to the cores

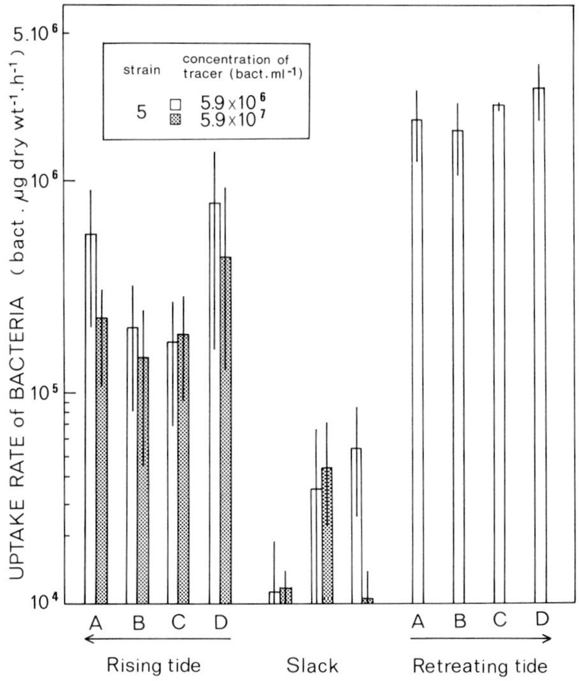

Fig. 2. Mean (± S.E.) weight-specific uptake rates of total bacteria by *A. hirsutus* during rising and retreating tides, in cores placed along a transect from shore (*A*) to sea (*D*) in the intertidal zone. At the slack, the cores were placed in the middle to the intertidal zone. Bacterial strain *5* was used as "tracer". Concentrations stated are those in the 10 ml added to the cores

the radio-isotope, whereas in our study the labelled bacteria themselves served as tracers of natural bacterial populations.

Depending on the experiment, the ratio of labelled bacteria to natural bacteria in the cores ranged from 1:50 to 11560. Fig. 1 suggests that under laboratory conditions, uptake was not affected by the concentration and/or the nature (strain nos. 2, 3, and 5) of the "tracer" used. Indeed, for each incubation time, the Kruskal-Wallis one-way analysis of variance (Sieger 1956) showed that uptake values, calculated with the different strains and concentrations, did not differ significantly at the 5% level. These results indicate that the labelled cells truly played the role of "tracers" of natural bacterial populations.

The slopes of the feeding curves (Fig. 1) decreased as experimental duration increased. This increase in uptake rate with time may reflect egestion of ingested bacteria as well as respiration of some ^{14}C from the metabolic pool, indicating that bacteria were not merely ingested, but also assimilated. In order to minimize errors, ingestion rate was calculated after the shortest experimental duration (i.e. 30 min): mean ingestion rate of bacteria by *A. hirsutus* was approximately 8 × 10^4 bact. μg individual dry $wt.^{-1}$ h^{-1}, or 12.8 × 10^{-3} μg C μg individual dry $wt.^{-1}$ h^{-1} (given that the average bacterial cell C content is 1.6 × 10^{-13} g). This value compares with those measured under laboratory conditions for other crustaceans of similar size, such as amphipods (e.g., *siphonoecetes dellavallei* fed at a rate of 8.5 × 10^{-3} μg C μg individual dry $wt.^{-1}$ h^{-1} on the diatom *Nitzschia acicularis* (Guidi 1981), and *Hyalella azteca* ingested surface sediment at a rate of 8.1×10^{-3} μg C μg individual dry wt.$^{-1}$ h^{-1} (Hargrave 1972).

Figure 2 shows the uptake rate of bacteria by *A. hirsutus* under different tidal conditions. Uptake was highest during retreating tide, and reached a maximum value of 3×10^6 bact. μg individual dry wt.$^{-1}$ h^{-1}. Slightly lower rates were recorded during rising tide (maximum = 8×10^5 bact. μg individual dry wt.$^{-1}$ h^{-1}). Uptake, which was relatively low during the slack (maximum = 5×10^4 bact. μg individual dry wt.$^{-1}$ h^{-1}), was comparable to that measured under laboratory conditions. It is likely that currents, which are generated through the vertical tubes inhabited by the tanaids during the period of tidal surge, increase the effect of the pumping motion of the pleopods for food conveyance. However, the apparent tidal influence on uptake rate could also be an artefact of the experimental design; during the period of tidal surge, the tracer would be diluted much more than during the slack tide, and this would tend to accentuate the "apparent" uptake rates. Further investigation is necessary. The problem can be resolved only by determining the initial and final ratios of labelled to unlabelled bacteria in the cores. The position of the cores in the intertidal zone had little influence on uptake rate.

4 Conclusion

The new method described and tested here appears to be suitable for in situ feeding investigations. Our results indicate that bacteria probably contribute significantly to the diet of *A. hirsutus* (average uptake rate of total bacterial C = 12.8×10^{-3} μg C μg dry wt.$^{-1}$ h^{-1}). The high densities of natural bacteria (2×10^9 ml^{-1}), together with the turnover of overlying water-masses by tidal currents, might be the clue for a better understanding of the remarkably high densities of organisms observed in the intertidal sediment of Morbihan Bay at Kerguelen Island.

Acknowledgements. Our research was supported by the *Territoire des Terres Australes et Antarctiques Francaises.*

References

Brown TJ, Sibert JR (1977) The food of some benthic harpacticoid copepods. J Fish Res Board Can 34:1028–1031

Guidi L (1981) Etude expérimentale de la nutrition d'un amphipode épibenthique *Siphonoecetes dellavallei:* relation qualité–qualité de la nourriture et ingestion. Thèse 3e cycle, Univ, Pierre et Marie Curie, Paris, 90 p

Hargrave BT (1970) The utilization of benthic microflora by *Hyalella azteca* (Amphipoda). J Anim Ecol 39:427–437

Hargrave BT (1972) Prediction of egestion by the deposit-feeding amphipod *Hyalella azteca.* Oikos 23:116–124

Hobbie JE, Daley RJ, Jasper S (1977) Use of nuclepore filters for counting bacteria by fluorescence microscopy. Appl Environ Microbiol 33:1225–1228

Meyer-Reil LA, Faubel A (1980) Uptake of organic matter by meiofauna organisms and interrelationships with bacteria. Mar Ecol Prog Ser 3:251–256

Oppenheimer CH, Zobell CE (1952) The growth and viability of sixty three species of marine bacteria as influenced by hydrostatic pressure. J Mar Res 11:10–18

Siegel S (1956) Non parametric statistics for behavioral sciences. McGraw-Hill Kogakusha Ltd., Tokyo, 312 pp

The Trophic Role of the Stone Crab, *Lithodes murrayi*, in the Benthic Ecosystem of the Crozet Islands

P.M. ARNAUD[1] and J.C. MIQUEL[2]

Summary. The contents of some 500 stomachs of the Stone Crab, *Lithodes murrayi*, from the benthos around the Crozet Islands were analyzed. The species has a very diverse diet (including sessile and vagile organisms of various groups), and is largely opportunistic (including necrophagy) in its foraging. The relative frequency of different prey groups varies in relation to the size class of Crab and to the migrations made by the species around the islands. *Lithodes murrayi* apparently occupies a trophic niche similar to those of Brachyura and Pagurida in temperate and tropical regions, and to that of isopods and amphipods in Antarctic waters. The paucity of the Crozet area in Brachyura and Pagurida favours *L. murrayi* there, but the species apparently is competing with isopods and amphipods which are very abundant.

1 Introduction

The Stone Crab, *Lithodes murrayi*, was studied at the Crozet Islands, in the southern Indian Ocean, during visits by the M.V. Marion-Dufresne in 1976 and 1982 (Miquel et al., to be publ.). The species is found in shallow water and down to 1050 m. It migrates towards the coast at the end of the austral summer, as indicated by variations in the sex ratio and size classes of a total of 1178 specimens taken by a beam trawl, or by traps similar to those used for capturing King Crabs, *Paralithodes camtschatica*, in the north Pacific (Miquel et al., to be publ.). We analyzed the stomach contents of Stone Crabs to determine the diet of the species, which was largely unknown (Arnaud and Do-Chi 1977), and to assess the species' trophic role in the benthos at the Crozet Islands.

2 Materials and Methods

The stomachs of Stone Crabs, separated from oesophagus and proximal part of the intestine, were obtained from 568 specimens (457 males, 111 females) caught by a beam trawl (n = 254) and in traps (n = 314). Catch locations, and the distributions of specimens by size classes and bathymetric zones, are shown in Figs. 1 and 2, respectively. The size classes are numbered 1–13 (class 1 containing specimens 10–19 mm in standard length, class 2 = 20–29 mm, class 3 = 30–39 mm, etc.). The bathymetric zones used in the analysis were dictated by trawling depths. Each stomach was fixed immediately in 10% neutralized formalin, and later opened and washed with formalin.

A vacuity index (V) was calculated as

$$V = \frac{Ev \times 100}{N}$$

where Ev is the number of empty stomachs and N is the number of stomachs examined. A repletion index (R) was evaluated visually, and scored from 0 (empty) to 5 (full). A frequency index (F) for prey taxa identified in stomachs was calculated as

$$F = \frac{n \times 100}{N'}$$

where n is the number of stomachs containing a particular taxon and N′ is the number of stomachs containing prey. For these stomachs the approximate dominance (d) of each prey taxon was assessed visually and scored (starting with 1 = 10% of total volume of prey in stomach). Since the total volume varies among stomachs, each value of d was corrected to R to obtain a corrected dominance (d′) for calculating the volumetric dominance (D) of the various groups of prey. Thus,

$$D = \frac{(d' \times R) \text{ of the prey taxon}}{(d' \times R) \text{ of all prey taxa}}$$

The presence of bait (fish, meat), as well as amphipods and isopods likely to have been attracted to the bait in our traps, which occurred in the stomachs of specimens of Stone Crabs were ignored in our analysis, in order to facilitate comparisons between the results obtained from the trawls and traps.

3 Results

3.1 Vacuity Index (V)

We anticipated high values of V from specimens caught in traps, since traps are likely to sample relatively more

1 Station marine d'Endoume, F-13007 Marseille, France
2 Laboratoire de Biologie et Ecologie marines, Faculte des Sciences, 2 avenue de Valrose, F-06034 Nice, France

Antarctic Nutrient Cycles and Food Webs
(ed. by W.R. Siegfried, P.R. Condy, and R.M. Laws)

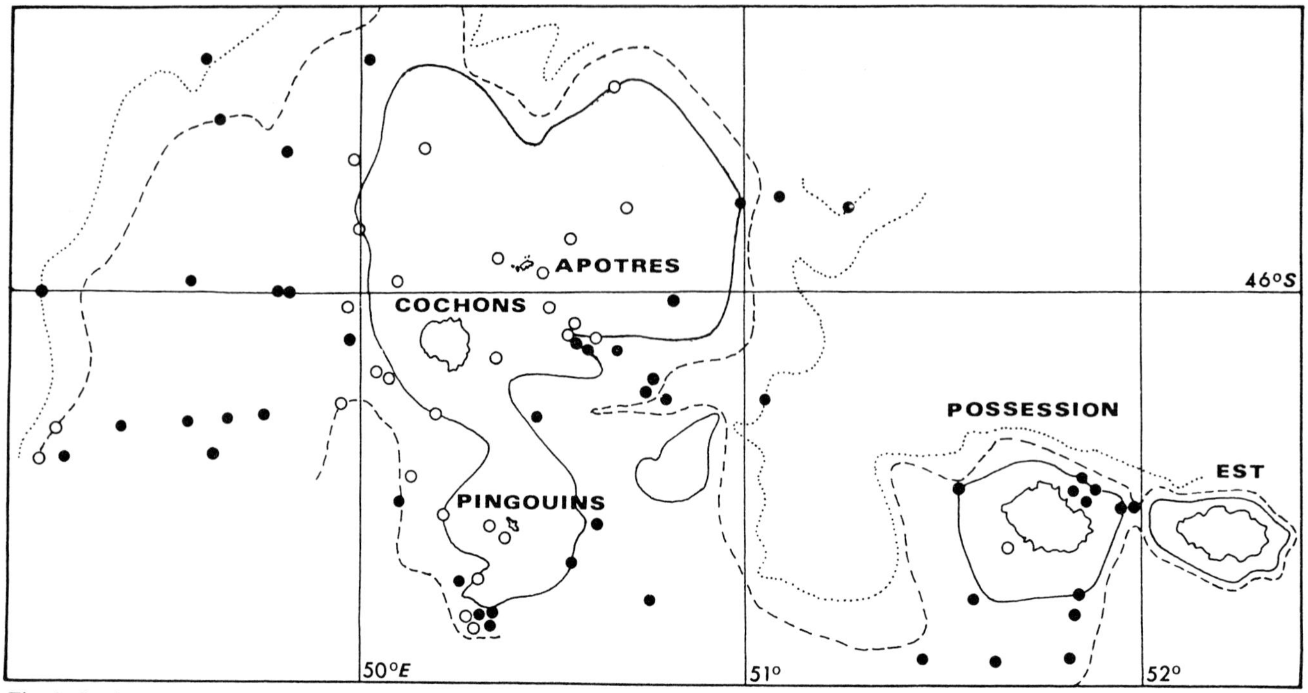

Fig. 1. Stations at which either trawling or trapping for *Lithodes murrayi* was carried out at Crozet Islands in April 1976 and February 1982. Stations at which Stone Crabs were captured are represented by *dots* and those at which crabs were not caught are represented by *circles*. The 100, 500 and 1000 m isobaths are indicated by *solid, broken* and *dotted lines*, respectively

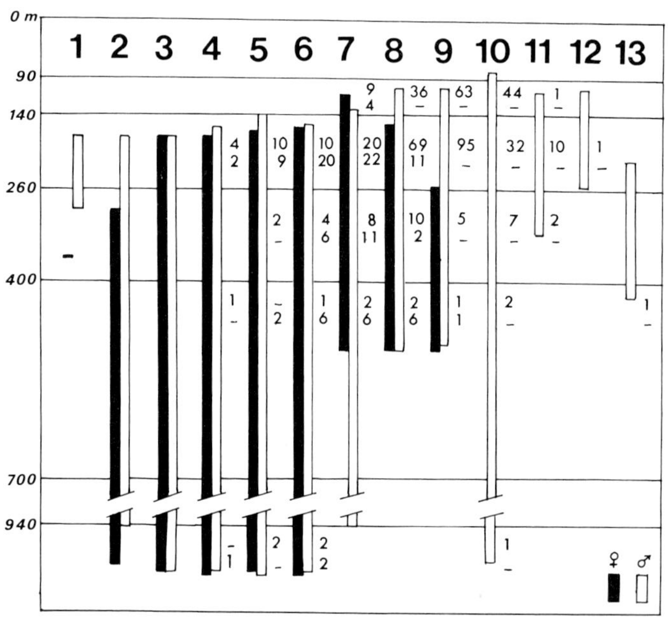

Fig. 2. Depth distribution (0 m–1000 m) of 13 size classes (10 mm standard-length intervals) of *Lithodes murrayi* according to sex at the Crozet Islands. The numbers of specimens whose stomach contents were analyzed are shown for each size class (*upper* and *lower* nos. for males and females, respectively)

hungry Stone Crabs than trawls. The data in Table 1 confirm this assumption; the mean value for V being 74.4% for trapped vs. 42.1% for trawled specimens. Considering only the 90–400 m bathymetric zone (the only zone in which both trapping and trawling occurred), the means for V are 74.4% and 39.7% for trapped and trawled specimens, respectively.

Based on trawled specimens exclusively, V is lower for males than for "berried" females for which, in turn, V averaged lower than for non-berried females (Table 1), suggesting that males were relatively more active and more successful at foraging.

The total catch by trawls comprised 56.5%, 22.8% berried females and 20.6% non-berried females. In traps, the comparable percentage were 98.6%, 1.4%, and 0% (berried females were not taken by traps). This suggests that males exclude both berried and non-berried femals from traps, being more active and larger than females. Such a behavioural dominance seems to be correlated with a size hierarchy, similar to that in the Rock Lobster, *Jasus lalandei* (see Fielder 1965). Table 1 also indicates that V is especially high for specimens taken between 146 and 260 m and between 401 and 700 m. The reason for this is not clear, but it is possible that food is less abundant at these depths.

The values of V vary with the different size classes (Table 2). In males measuring < 100 mm in standard length V ranges from 20–33%, but suddenly increases after this size. Apparently, relatively large males eat less during their summer migration. A similar increase of V is observed for berried females. Non-berried females tend to have high values of V for small sizes (classes 4 and 5), as well as for the large ones (classes 7 to 9).

3.2 Repletion Index (R)

This index was applied to trawled specimens only, because in the specimens caught by traps digestion may affect R. In

Table 1. Vacuity index V (%) for stomachs of *Lithodes murrayi* taken by trawl or trap in relation to sex and depth

	Trawl			Trap		
Depth (m)	Males	Females	B. females[a]	Males	Females	B. Females[a]
90– 145	22.5 (31)	–	–	72.9 (122)	–	100 (4)
146– 260	45.0 (72)	56.2 (32)	53.5 (28)	74.8 (179)	–	33.3 (3)
261– 400	28.1 (32)	42.8 (7)	16.6 (12)	100 (6)	–	–
401– 700	40.0 (10)	75.0 (16)	66.6 (6)	–	–	–
940–1015	20.0 (5)	33.3 (3)	–	–	–	–

Numbers of specimens in parenthesis
Females[a] in "berry"

Table 2. Vacuity index V (%) for stomachs of *Lithodes murrayi* taken by trawl in relation to sex and size class

Size class	Males	Females	B. Females[a]
4	20.0 (5)	80 (5)	–
5	25.0 (12)	70 (10)	100 (1)
6	31.2 (16)	40.9 (22)	33.4 (11)
7	33.3 (27)	63.6 (11)	37.5 (24)
8	27.2 (33)	66.6 (9)	70.0 (10)
9	31.2 (32)	100 (1)	–
10	55.0 (20)	–	–
11	75.0 (4)	–	–
12–13	100 (1)	–	–

Numbers of specimens in parentheses
Females[a] in "berry"

males and berried females, the variations of R with depth are inversely related to those of V (Tables 1–3), which tends to confirm a lesser abundance of food between 146 and 260 m and below 400 m. But, curiously, the decrease in R for specimens from between 146 and 260 m is replaced in non-berried females by a noticeable increase. The variation of R with size class (Table 4) is low in both sexes, except in the R of males of classes 6 and 7 which shows a marked decrease. The mean values of R, according to size classes, are such that they are lower for females than for males for which R is lower than for berried females (Table 3). This means that among stomachs containing food, those of non-berried females are less full than those of males (confirming the females' dominance by males). Apparently, berried females are less dominated than the non-berried ones.

3.3 Frequency Index (F) and Dominance (D)

Figure 3 compares the frequency of each food item in the stomachs of Stone Crabs from trawls (n = 92) and from traps (n = 80), obtained at depths < 260 m (the zone in which specimens were taken by both types of gear). Crabs from trawls included polychaetes and bivalves and, to a lesser extent bryozoa, asteroids and brachiopods, more frequently than crabs from traps, which contained algae and nematodes more frequently. These differences indicate that prey from soft substrata tend to be relatively more frequent in trawled crabs; prey from hard-bottom substrata dominating generally in the specimens taken in traps. This difference may be tentatively explained as follows: on the very mixed bottom deposits surrounding the Crozet Islands, trawls tend to collect Stone Crabs on the less rough sub-

Table 3. Mean (and 1 ± S.D.) repletion index R for stomachs of *Lithodes murrayi* taken by trawl in relation to sex and depth

Depth (m)	Males	Females	B. Females[a]	All sexes combined
90– 145	3.63 ± 1.13 (24)	–	–	3.63 ± 1.13 (24)
146– 260	2.73 ± 1.24 (41)	2.83 ± 1.03 (12)	2.87 ± 1.19 (15)	2.78 ± 1.18 (68)
261– 400	3.09 ± 1.50 (23)	2.25 ± 0.96 (4)	3.90 ± 1.20 (10)	3.22 ± 1.44 (37)
401– 700	2.17 ± 1.83 (6)	1.25 ± 0.50 (4)	3.50 ± 0.71 (2)	2.08 ± 1.50 (12)
940–1015	2.50 ± 1.29 (4)	1.00 ± 0.00 (2)	–	2.00 ± 1.26 (6)
90–1015	2.99 ± 1.37 (98)	2.27 ± 1.12 (22)	3.30 ± 1.23 (27)	

Numbers of specimens in parentheses
Females[a] in "berry"

Table 4. Mean (and 1 ± S.D.) repletion index R for stomachs of *Lithodes murrayi* taken by trawl in relation to sex and size class

Size class	Males	Females	B. Females[a]
4	3.25 ± 2.06 (4)	–	–
5	3.11 ± 1.36 (9)	2.33 ± 0.58 (3)	–
6	2.64 ± 1.43 (11)	2.31 ± 1.32 (13)	3.63 ± 1.19 (8)
7	2.67 ± 1.24 (18)	2.00 ± 1.00 (3)	3.06 ± 1.34 (16)
8	3.08 ± 1.41 (24)	2.33 ± 1.15 (3)	3.67 ± 0.58 (3)
9	3.05 ± 1.33 (22)	–	–
10	3.22 ± 1.56 (9)	–	–
11	4 (1)	–	–

Numbers of specimens in parentheses
Females[a] in "berry"

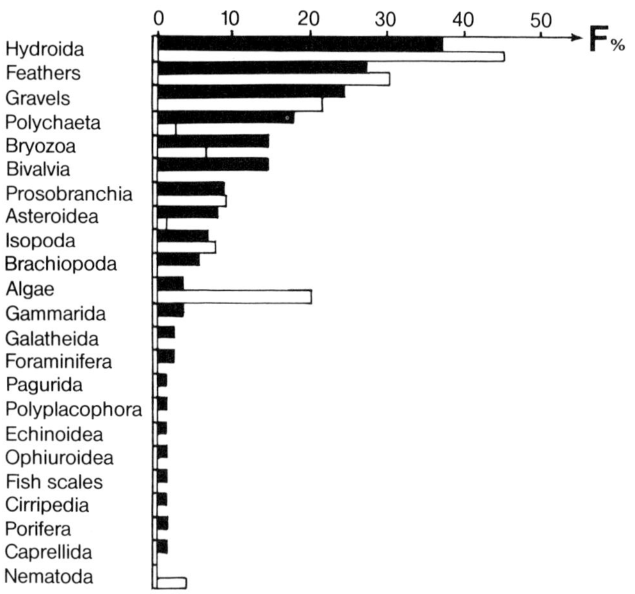

Fig. 3. Frequency of prey groups in the stomachs of *Lithodes murrayi* caught by beam trawl (*dark boxes*) and traps (*open boxes*) at depths < 260 m

strata, whereas the traps immersed at random (on soft, hard or mixed bottoms) catch relatively more specimens which forage over hard substrata. However, these differences are small, and so the results from both types of gear are taken together in considering variations in frequency and dominance of prey items according to depth and according to size classes of Stone Crabs.

3.4 Variations of F and D According to Depth

Table 5 shows that a high diversity of prey occurred in Stone Crabs taken from 146–260 m (perhaps because this is the range for which the greatest number of stomachs was examined). The vacuity index (V) and R here reach their highest and lowest values, respectively, in specimens from the first 400 m. Hence, we conclude that in this area the increase of prey diversity is not linked to an increase in prey abundance. Certain food items (e.g., feathers, isopods, hydroids) decreased in Stone Crabs with increasing depth, whereas polychaetes and amphipods increased. Other groups occurred more irregularly in relation to depth, or occurred only in the deepest part of the first 400 m zone (e.g., ascidians, tanaidaceans). Hydroids were dominant in specimens down to 400 m, and were present in about 50% of the Stone Crabs sampled. Curiously, after hydroids, penguin feathers, followed by gravel and various non-sessile prey (e.g., isopods, prosobranchs, bivalves, polychaetes) and sessile items (e.g., algae, bryozoans and asteroids down to 260 m) featured prominently in these specimens.

The number of prey groups apparently decreased in specimens below 400 m but this may be a reflection of the few stomachs studied. It is, nevertheless, interesting to record that a very sharp decrease occurred in hydroids (more or less replaced by ascidians and isopods), and an absence of brachiopods, cirripedes, molluscs and penguin feathers. Tanaidaceans first appeared and, paradoxically, algae reached their highest dominance in specimens from deep waters. This points to the unsuspected trophic importance of algae (in wrecked condition) at these great depths.

3.5 Variations of F and D Accorrding to Size Class

Since there is no significant difference for F and D in males, as well as in berried and non-berried females, between size classes where such a comparison is possible, values of F and D are given here regardless of sex. Table 6 shows that the diet of *L. murrayi* varies with size class. Diet diversity is greatest in class 7 (18 prey groups), followed by classes 8 and 9 (15 and 16 prey groups, respectively). The extreme classes (4–5 and 10–11) of specimens contain only 9 and 10 prey groups, respectively. However, these variations in prey groups are mostly irregular; molluscs showing the only clear gradient in diversity. With increasing size (= age) of Stone Crabs, predation on prosobranchs apparently increases, whereas it decreases on bivalves. Hydroids dominated in specimens of all size classes and down to 400 m, followed by feathers and gravel, and then algae, bryozoa and polychaetes. These five groups were found also in the stomachs of specimens of all size classes. Five other groups (brachiopods, isopods, prosobranchs, bivalves and asteroids) occurred in specimens of four or five size classes, one (amphipods) in three classes, seven (ascidians, amphipods, galatheids, ophiuroids, nematodes, foraminiferans) in only two classes, and the last six groups in only one size class, and always in low numbers.

3.6 Variations of F and D According to Geographical Sectors

In view of the peculiarities (gentle slope, dominance of hard substrata) of American Bay, north east of ile de la Posses-

Table 5. Frequency index *F* (%) and volumetric dominance *D* (%) of groups of prey in the stomachs of *Lithodes murrayi* according to 4 bathymetric zones

Prey group	Zone 1[a]		Zone 2[b]		Zone 3[c]		Zone 4[d]	
	F	D	F	D	F	D	F	D
Attached								
Hydroida	50.9	30.54	47.0	22.26	56.8	45.64	16.7	7.03
Algae	8.8	4.77	13.0	5.65	5.4	1.88	5.5	7.57
Brachiopoda	5.3	2.55	4.3	0.59	2.7	3.08	–	–
Bryozoa	10.5	5.77	13.9	5.39	–	–	–	–
Porifera	–	–	0.9	0.22	–	–	–	–
Actiniaria	–	–	–	–	2.7	1.28	–	–
Ascidiacea	–	–	–	–	–	–	11.1	17.30
Cirripedia	–	–	0.9	0.66	2.7	0.85	–	–
Free								
Isopoda	10.5	9.19	7.8	3.77	8.1	5.47	5.5	0.54
Amphipoda	1.8	0.27	3.5	1.33	10.8	4.44	11.1	4.86
Tanaidacea	–	–	–	–	–	–	5.5	6.49
Galatheida	–	–	1.7	2.58	–	–	–	–
Pagurida	–	–	0.9	0.18	–	–	–	–
Prosobranchia	14.0	3.62	10.4	3.54	18.9	4.44	–	–
Bivalvia	5.3	2.48	11.3	3.99	2.7	0.85	–	–
Polyplacophora	–	–	0.9	0.44	–	–	–	–
Asteroidea	1.8	3.56	7.0	6.57	–	–	5.5	4.86
Echinoidea	1.8	2.68	–	–	–	–	–	–
Ophiuroidea	–	–	0.9	0.37	–	–	5.5	5.41
Polychaeta	1.8	0.34	16.5	9.82	18.9	10.26	16.7	18.92
Nematoda	–	–	2.6	0.59	–	–	–	–
Foraminifera	–	–	1.7	0.89	–	–	–	–
Fish scales	–	–	0.9	0.07	–	–	–	–
Indet		3.62		6.24		4.27		1.86
Non-living								
Feathers	29.8	18.39	32.2	14.88	24.3	12.65	–	–
Gravel	29.8	12.42	27.8	9.97	21.6	4.87	27.8	12.16
No. prey groups	(13)		(21)		(12)		(10)	
No. specimens	(57)		(115)		(37)		(18)	

a 90– 145 m b 146– 260 m c 261– 400 m d 401–1015 m

sion (Fig. 1), and an exceptionally high concentration of lithodid crabs there (Miquel et al., to be publ.), it is of interest to compare the stomach contents of Stone Crabs from this bay with those of specimens from other parts of the Crozet archipelago (Table 7). Significant differences are obvious. The numbers of groups of prey are analogous, but fewer sessile groups (5 vs. 7) and more sedentary or vagile groups (13 vs. 8) were found in specimens from American Bay. This could reflect the predominance of soft substrata in American Bay where relatively more bryozoans, algae, bivalves, isopods, asteroids and penguin feathers, but fewer amphipods and polychaetes, occur.

4 Discussion

Generally speaking, *L. murrayi* feeds on all available prey with no noticeable qualitative selection (except for octocorallia which are always avoided, even where they are abundant) and probably without quantitative selection, since the various groups seem to be preyed on more or less in proportion to their natural abundance (Fig. 4). Both sessile and vagile species are preyed upon, as well as organic detritus (penguin feathers, carrion) and gravel (possibly to aid digestion). The dominant prey groups are hydroids (or ascidians in deep waters) and polychaetes. Molluscs (prosobranchs and bivalves), small crustaceans (isopods and amphipods) and bryozoans are also well represented in the diet. Algae are a very important dietary item. In spite of an abundance of feathers and gravels, these cannot be considered as proper food (unless there is some feeding on their bacterial or microbial coating).

For any given size of Stone Crabs, the diet is very similar for both sexes. The composition of the diet, however, changes in relation to size of the predator, depth and even

Table 6. Frequency index F (%) and volumetric dominance D (%) of groups of prey in the stomachs of *Lithodes murrayi* according to size classes of specimens (4–5, 6, 7, 8, 9 and 10–11)

Prey group	4–5		6		7		8		9		10–11	
	F	D	F	D	F	D	F	D	F	D	F	D
Attached												
Hydroida	58.8	29.60	34.4	28.63	39.5	30.18	36.7	23.39	43.5	32.43	41.7	24.10
Algae	11.8	8.00	3.1	0.45	7.0	4.37	8.2	3.88	17.7	7.71	8.3	3.77
Brachiopoda	5.9	2.00	–	–	7.0	4.78	2.0	1.32	–	–	4.2	1.64
Bryozoa	11.8	8.20	9.4	3.86	7.0	2.39	4.1	1.65	9.7	6.39	8.3	2.95
Porifera	–	–	–	–	2.3	0.53	–	–	–	–	–	–
Actiniaria	–	–	–	–	–	–	2.0	1.24	–	–	–	–
Ascidiacea	–	–	3.1	4.09	2.3	2.48	–	–	–	–	–	–
Cirripedia	–	–	–	–	2.3	0.88	–	–	1.6	1.25	–	–
Free												
Isopoda	–	–	9.4	11.70	7.0	2.57	2.0	1.65	11.3	8.40	8.3	5.25
Amphipoda	–	–	–	–	11.6	7.43	6.1	1.82	1.6	0.28	–	–
Tanaidacea	–	–	–	–	–	–	2.0	1.98	–	–	–	–
Galatheida	5.9	8.00	3.1	3.41	–	–	–	–	–	–	–	–
Pagurida	–	–	–	–	2.3	0.44	–	–	–	–	–	–
Prosobranchia	–	–	3.1	0.68	9.3	3.54	14.3	5.62	9.7	3.13	16.7	7.05
Bivalvia	23.5	6.40	9.4	3.86	4.7	3.72	6.1	2.64	3.2	1.01	–	–
Polyplacophora	–	–	–	–	2.3	1.06	–	–	–	–	–	–
Asteroidea	–	–	6.3	6.59	4.7	3.19	8.2	8.76	1.6	3.47	–	–
Echinoidea	–	–	–	–	–	–	–	–	1.6	2.78	–	–
Ophiuroidea	5.9	4.00	–	–	–	–	–	–	1.6	0.69	–	–
Polychaeta	11.8	2.00	12.5	3.18	9.3	3.01	20.4	16.12	6.5	2.92	16.7	24.92
Nematoda	–	–	–	–	–	–	2.0	0.83	3.2	0.42	–	–
Foraminifera	–	–	3.1	0.45	2.3	1.77	–	–	–	–	–	–
Fish scales	–	–	–	–	–	–	–	–	1.6	0.14	–	–
Indet		12.00		10.34		7.43		4.13		2.85		1.64
Non-living												
Feathers	5.9	2.00	31.3	16.59	25.6	12.57	28.6	18.18	22.6	13.40	33.3	19.18
Gravel	29.4	17.80	18.8	6.14	23.3	7.70	18.4	6.78	27.4	12.71	20.8	9.51
No. prey groups	(10)		(13)		(18)		(15)		(16)		(9)	
No. specimens	(17)		(32)		(43)		(49)		(62)		(24)	

sector of the archipelago. This points to broad opportunistic foraging by Stone Crabs during their seasonal migrations. Cannibalism is rare (Arnaud and Do-Chi 1977), which probably is correlated with the existence of a remarkable size hierarchy (behavioural dominance) tending to minimize chance meetings between individuals of different sizes, both between and within sexes. The hierarchy acts to prevent non-berried females, and to a great extent berried females, from entering traps. This may also apply to natural carrion and, as a result, females are always more or less under-fed in comparison with males.

There are very few studies with which ours can be compared. Stuardo and Solis (1963) mention the dominance of bryozoans and crustaceans in the diet of *Lithodes antarctica* from the Magellan Strait, based on 25 stomachs of which 20 were empty. In contrast, the diet of the North Pacific King Crab, *Paralithodes camtschatica,* has been much studied (see the bibliography in Jewett and Feder (1982)) and it shows great variation in composition and in the dominance of various prey types. In summary, bivalves, gastropods, cirripedes and echinoderms dominate the species' diet. These items are mostly sedentary, and live on soft substrata (the exception being cirripedes). The same groups (except prosobranch gastropods) are less well represented in the diet of *L. murrayi* which feeds mainly on members of the sessile epifauna and epiflora (hydroids, ascidians, algae) which are particularly abundant on the mixed substrata at the Crozet Islands. The diets of both species are good expressions of the composition of the available benthic assemblages, as suggested by Takeuchi (1967).

At the Crozet Islands, *L. murrayi* apparently benefits through the absence of brachyurans (only one small species, *Halicarcinus planatus*) and pagurids (only one species) which occupy a similar trophic niche in temperate and tropi-

Table 7. Frequency index *F* (%) and volumetric dominance *D* (%) of groups of prey in the stomachs of *Lithodes murrayi* caught in American Bay and other areas of the Crozet Islands

Prey group		American Bay		Other areas	
		F	D	F	D
Attached	Hydroida	43.6	26.96	36.6	31.25
	Algae	12.2	5.82	5.6	2.60
	Brachiopoda	3.2	1.40	1.4	1.88
	Bryozoa	10.9	5.77	1.4	0.52
	Porifera	0.6	0.16	–	–
	Actiniaria	–	–	1.4	0.78
	Ascidiacea	–	–	2.8	3.34
	Cirripedia	–	–	2.8	1.46
Free	Isopoda	6.4	6.21	5.6	3.44
	Amphipoda	1.3	0.94	11.3	3.85
	Tanaidacea	–	–	1.4	1.25
	Galatheida	1.3	1.82	–	–
	Pagurida	0.6	0.13	–	–
	Prosobranchia	9.6	3.64	9.9	3.23
	Bivalvia	7.7	3.25	2.8	1.56
	Polyplacophora	0.6	0.31	–	–
	Asteroidea	5.1	6.03	1.4	0.94
	Echinoidea	0.6	1.04	–	–
	Ophiuroidea	–	–	2.8	1.56
	Polychaeta	5.8	1.32	28.2	21.04
	Nematoda	1.9	0.42	–	–
	Foraminifera	1.3	0.62	–	–
	Fish scales	0.6	0.05	–	–
Indet.			5.64		6.20
Non-living	Feathers	32.0	17.40	12.7	7.41
	Gravel	22.4	10.83	23.9	7.08
No. prey groups		(20)		(17)	
No. specimens		(156)		(71)	

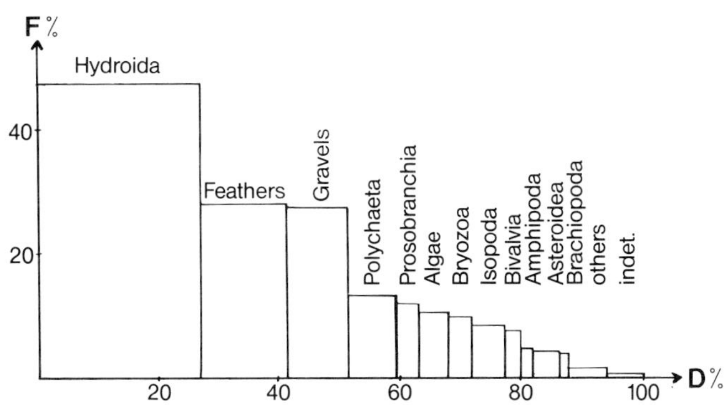

Fig. 4. Frequency (*F*) and dominance (*D*) of prey groups in all stomachs of *Lithodes murrayi* from Crozet Islands. Relative importance of a group is expressed by the area of the box

cal regions. The only potential invertebrate competitor for live prey is another, broadly sympatric, species of lithodid crab, *Paralomis aculeata,* although it is far less abundant than *L. murrayi.* Among competitors for dead prey (naturally occurring carrion, bait in the traps) are gammarid amphipods (mostly Lysianassidae) and cirolanid isopods. Thousands of these animals, as in Antarctic waters (Arnaud 1977), are caught in traps at the Crozet Islands. Among vertebrates, fish are potential competitors for live as well as dead prey. Thus, in the benthic marine ecosystem of the Crozet Islands, the usual trophic niche of pagurid and true crabs is occupied by *L. murrayi,* and shared with other crustaceans (gammarid amphipods and *Paralomis aculeata*) and also with fish. In Antarctic waters where there is a complete lack of pagurids, true crabs and lithodid crabs, this niche is again shared by gammaridean amphipods, isopods and fish, and also echinoderms and nemerteans.

Acknowledgements. The financial and logistic support of Terres Australes et Antarctiques Françaises (Paris) is gratefully acknowledged.

References

Arnaud PM (1977) Adaptation within the Antarctic benthic ecosystem. In: Llano G (ed) Adaptations within Antarctic ecosystem, Proceedings of the 3rd SCAR symposium on Antarctic Biology. Gulf Publ Co Houston pp 135–157

Arnaud PM, Do-Chi T (1977) Données biologiques et biométriques sur les lithodes *Lithodes murrayi* (Crustacea : Decapoda : Anomura) des iles Crozet (SW Océan indien). Mar Biol 39:147–159

Fielder DR (1965) A dominance order for shelter in the spiny lobster *Jasus lalandei* (H Milne-Edwards). Behaviour 24:236–245

Jewett SC, Feder HM (1982) Food and feeding habits of the king crab *Paralithodes camtschatica* near Kodiak island, Alaska. Mar Biol 66:243–250

Miquel JC, Arnaud PM, Do-Chi T (to be published) Structure of populations and migrations in *Lithodes murrayi* (Hend. 1888) at Crozet islands, SW Indian ocean (Crustacea, Decapoda, Anomura). Mar Biol

Stuardo J, Solis I (1963) Biometrica y observaciones sobre la biologia de *Lithodes antarcticus* Jacquinot. Gayana (Zool) 11:1–57

Takeuchi I (1967) Food of king crab *Paralithodes camtschatica* off the west coast of the Kamchatka peninsula 1958–1964 (in Jap Engl Abstr). Bull Hokkaido Reg Fish Res Lab 33:32–44

The Energy Budget of a Population of the Antarctic Isopod *Serolis polita*

R.A. LUXMOORE[1]

Summary. A population of the predatory benthic isopod *Serolis polita* was studied at Signy Island, South Orkney Islands, for 2 consecutive years. Field observations of population structure, growth rate and reproductive biology were complemented by laboratory measurements of respiratory rate and assimilation efficiency, and used to construct a computer model of the energy expenditure of all sections of the population. The complete annual energy budget (in kJ m^{-2} yr^{-1}) was calculated as Consumption (289) = Production (115) + Respiration (108) + Faeces (58) + Excretion (9), with a mean biomass of 113 kJ m^{-2}. Energy consumption was found to be markedly seasonal with the majority of growth and reproductive output occurring in the summer. Total energy consumption was lower in winter with respiratory costs predominating. The relatively high assimilation efficiency (80%) on a diet of amphipods and polychaetes is characteristic of other isopods, but the gross growth efficiency (40%) and the net growth efficiency (52%) are surprisingly high in view of the organism's 4-yr life cycle. The ecological implications of this are discussed. The turnover rate ($P/\bar{B}$) is low (1.02) in comparison with other marine isopods studied, and is considered to be related to the large body size. This is discussed as a general feature of the Antarctic marine ecosystem and may help to explain the observed high biomasses without the need to infer unusually high productivity.

1 Introduction

The study of the energy budget of an organism is one of the most comprehensive ways of examining its ecology. It expresses what proportion of the energy intake is allocated to the various life processes, and shows how this can fluctuate throughout the year or the life-cycle. For the individual animal it can identify critical periods of energy demand, and at the population level it can be used to study trophic relationships.

All calculations of energy budgets entail some estimates and assumptions, the magnitude of which modifies the validity of the conclusions drawn. The present study is no exception, but it was attempted to measure all the major components with some care and to make reasonable assumptions for the minor components. Wherever possible the magnitude of these errors was evaluated.

There has been very little work carried out on Antarctic ecological energetics and this investigation was planned to enable the compilation of an energy budget for *Serolis polita,* a common benthic isopod at Signy Island, South Orkney Islands.

Peracarid Crustacea (especially amphipods and isopods) are very prominent in the Antarctic benthic marine fauna, but although the amphipods have been extensively studied in the past (Bregazzi 1972; Thurston 1972; Rakuska-Suszczewski and Klekowski 1973) less is known of the isopods. The genus *Serolis* is characteristic of the Antarctic, about 20 of the 64 known species being endemic to the region. All species are similar in appearance, being highly dorso-ventrally flattened, and from the little that is known they mostly seem to have a similar ecology, being typically predacious and living in areas of soft sediment. Since the species range from near the equator (Glynn 1976) to the Ross Sea (Lipps et al. 1978), the genus would make an interesting model for testing theories of Antarctic adaptations if comparative data could be collected from temperate or tropical regions.

Serolis polita is confined to the Scotia Arc and is common in shallow water, the deepest record being from 172 m (Kusakin 1967). It is a predator with an unspecialized diet in which amphipods and polychaetes predominate. The growth rate is slow and markedly seasonal, taking 28 months to reach maturity at a length of about 20 mm (Luxmoore 1982a). Eggs are brooded in a marsupium, the mean clutch size being 40.2, and incubation lasts for 20 months. The welldeveloped juveniles are released in the spring, following which females may lay a second brood of eggs in the autumn. Females can thus live for more than 6 years (Luxmoore 1982b).

To standardize this study, the recommended I.B.P. approach (Phillipson 1975) was adopted and the energy budget was divided into its components:

$$C = P + R + F + U$$

where

consumption (C) = total intake of food energy;
production (P) = energy content of biological tissues produced;

1 British Antarctic Survey, NERC, High Cross, Madingley Road, Cambridge, CB3 OET, United Kingdom
Present address: International Union for Conservation of Nature and Natural Resources, 219c Huntingdon Road, Cambridge, CB3 ODL, United Kingdom

Antarctic Nutrient Cycles and Food Webs
(ed. by W.R. Siegfried, P.R. Condy, and R.M. Laws)

respiration (R) = heat lost by respiratory processes including contributions associated with growth and vitellogenesis;

egesta (F) = that portion of consumption which is not assimilated through the gut lining (faeces and rejecta); and,

excreta (U) = the energy content of digested material which is passed from the body (mostly in urine).

Production can further be broken down into reproductive production (Pr) = the energy devoted to gonads and gametes; somatic production (Pg) = the energy increment in other body tissues; and, exuvia (Px) = the energy content of cast cuticles.

In order to estimate the population energy turnover it was first necessary to find the energy requirement of an individual. Individual energy consumption is very variable: young animals may devote a greater proportion of energy to growth, whereas older animals may expend more on reproduction. In addition, energy consumption may fluctuate throughout the year, and so it is not reasonable to calculate the budget for an "average animal" and to extrapolate the population energy consumption from this. Instead, the approach adopted was to divide the population into cohorts and to calculate the energy expended by a representative animal in each cohort in every month of the sampling period. This was achieved by estimating R, Pg, Pr, Px, U, and the assimilation efficiency, K, defined as (C–F)/C. Total food consumption was then calculated from the formula.

$$C = (R + Pg + Pr + Px + U)/K$$

This generated a large amount of data, and necessitated the use of a computer model for calculations.

2 Materials and Methods

The majority of the methods used in this study have been described elsewhere (Luxmoore 1982a, 1982b, 1982c, 1984), and so only brief details are given here.

2.1 Field Sampling

Samples were taken at monthly intervals for 2 years from an area of flat sand at 14 m depth in Borge Bay, using an under-water suction sampler. Each month 9 quadrats of 0.25 m^2 were collected to permit the estimation of population size, and all animals were measured. Size frequency analysis permitted the recognition of distinct cohorts and the determination of growth rate.

Sea temperatures have been measured at 6 m depth in Borge Bay at monthly intervals, as part of long-term routine monitoring. Little thermal stratification was found, and so these values were taken as representative of the sampling site for *S. polita* at 14 m.

2.2 Laboratory Measurements

Dry wt. determinations were made after rinsing in isotonic ammonium formate and drying at 60°–65°C for 1–3 days. Calorific values were measured in a micro-bomb calorimeter calibrated with benzoic acid. Pellet weights ranged from 10–40 mg. Oxygen consumption of individual animals was measured in a modified sealed-chamber apparatus at temperatures from –1.5° to +4.5°C, and a multiple linear regression equation was calculated relating ln oxygen consumption to ln dry wt. and temperature (Luxmoore, 1984).

The timing of reproduction and the fecundity were investigated by studying the embryos in marsupia of females collected in the monthly samples. The development of ovaries was quantified by dissecting them out of immature females and measuring the diameter of the oocytes (Luxmoore 1982b).

Analysis of stomach contents showed that the diet comprised mostly amphipods and polychaetes. Accordingly, assimilation efficiency was estimated gravimetrically using a diet of live amphipods. Groups of 3–4 *S. polita* were starved for 36 h to allow clearance of the digestive tract. They were then placed in containers of filtered seawater and presented with a pre-weighed quantity of live *Cheirimedon femoratus*. During the course of the experiments, which lasted for 4–10 days, additional amphipods were provided as necessary, and at the end any uneaten food (live and partially eaten amphipods) was removed; dried and weighed. The isopods were left for a further 36 h for gut clearance, and the faeces were then separated by filtration. An isotonic solution of ammonium formate was passed through the filter to remove thermo-stable salts and the filter and faeces were then dried and weighed. The dry wt of amphipods fed was calculated from the fresh weight using their dry matter content determined from a replicate batch of amphipods. The weight of uneaten food was subtracted from this to determine the total amount of food eaten. In each experiment there was a control chamber containing only amphipods which was subsequently filtered as before. The weight of extracted solids was subtracted from the weight of faeces to give a corrected faecal weight. The calorific contents of both food and faeces were determined and all dry wts. were converted to caloric equivalents. Assimilation efficiency was then calculated as

$$100\,(C-F)/C$$

Twenty-four separate determinations of assimilation efficiency were made in this way, and the results ranged from 49.6%–84.1% with a mean of 73.7% (S.D. = 7.41). It was found (Luxmoore 1981) that the stomachs of only 43% of *S. polita* contained crustacean material, whereas 23% contained polychaete chaetae, although this probably does not reflect the true prominence of polychaetes in the diet, because of the inconspicuous nature of the chaetae. If it is assumed that the remaining soft parts are digested with the same efficiency as polychaete material, the overall assimilation efficiency for the whole diet can be calculated as 80%. This is similar to values obtained for other carnivorous isopods (Johnson 1976; Shafir and Field 1980).

2.3 Calculations

The sampling period extended over 24 months, and during this period the population was found to contain members of 12 cohorts (7 female and 5 male). Data were therefore arranged in matrices of dimensions 24 × 12. An outline of the calculations used is given below; full details, together with the complete computer programme, may be consulted in Luxmoore (1981) where tables containing both the original and derived data are also available.

2.3.1 Growth (Pg)

The determination of growth rate was described in Luxmoore (1982a). The mean weight of an animal in each cohort was calculated each month from the regression equations relating ln weight to time. The weight gain during the month was then multiplied by the mean caloric content of body tissues, 11.99 J mg^{-1}.

2.3.2 Respiration (R)

Respiratory rate (R, in $\mu l\ O_2\ h^{-1}$) was found (Luxmoore, 1984) to be a function of dry wt. (W, in mg) and environmental temperature (T) according to the formula

$$\ln R = 0.6449 \ln W + 0.0692\,T - 1.1619.$$

In order to convert respiratory rate from units of oxygen consumption to energy it is necessary to know the oxycalorific coefficient. This varies with the biochemical composition of the food, but the total variation over the full range of food types is less than 5% (Petrusewicz and McFadyen 1970). These authors suggested a value of 19.68 J ml^{-1} O_2 for the average human mixed diet, but because *S. polita* is carnivorous it has a lower percentage of carbohydrate in its diet. The oxycalorific coefficient for *S. polita* was therefore taken as 19.38 J ml^{-1} O_2 (on the basis of a diet of 50% protein, 20% fat, 5% carbohydrate and 25% ash).

Monthly values of sea temperatures, which ranged from −1.86° to +0.48°C, were used together with the mean dry wt. to find the respiratory rate from the equation.

2.3.3 Excretion (U)

In aquatic organisms the majority of the nitrogenous excretion is in the form of ammonia, and Parry (1960) states that in the isopod *Ligia oceanica* 83% is ammonia. Brafield and Solomon (1972) have shown that when an organism is metabolizing protein, 3.05 J are lost as ammonia per ml of oxygen respired (0.51 cal mg^{-1}). Using the same biochemical composition of the diet, 54% of the oxygen consumption can be attributed to metabolizing protein and, therefore, the urinary loss amounts to 1.65 J ml^{-1} O_2.

2.3.4 Reproductive Production (Pr)

The majority of the energy cost of reproduction is involved in oogenesis. Except in species which employ external fertilization and broadcast large quantities of sperm into the environment, the cost of sperm production is usually low. *Serolis polita* uses direct fertilization and has relatively small testes. Since the energy content of the testes was included in the measurement of the somatic energy content, spermatogenesis was not considered separately. It was shown (Luxmoore 1982b) that the weight of material accumulated in the ovaries (WO) could be approximated by an equation relating to the age (A, in months) of the female ($\ln WO = 0.281\,A - 4.568$). This was converted to caloric units assuming a constant energy content of 25.18 J mg^{-1} (Luxmoore 1982b). Female *S. polita* can breed more than once; as soon as 1 brood of eggs has been spawned the next batch of eggs starts developing and becomes ready for spawning 24 months later. Thus, spent and ovigerous females were considered to be sustaining the same cost of oogenesis as that of immature females.

2.3.5 Moulting (Px)

The determination of the dry wt. of cast cuticles together with the mean dates of moulting were described in Luxmoore (1982a). The energy content of cast cuticular material was determined in a micro-bomb calorimeter after mixing with benzoic acid to improve combustion, and was found to be 6.01 J mg^{-1}.

2.3.6 Biomass (B)

The caloric biomass each month was calculated as the sum of the energy contents of the body tissues, the ovaries and the eggs in the marsupium.

2.3.7 Parameters of the Population Energy Budget

All of the parameters determined so far have been for individual animals in each month of the year. In order to convert these to their respective population totals it is necessary to multiply each by the corresponding cohort density (Luxmoore 1982c). These products were then summed to give a total over the 23-month sampling period, and an average annual total was obtained by multiplying by 12/23. It is unfortunate that the sampling period did not extend over an extra month to give a full 2-year data set, but it is likely that the 23-month average will not be greatly different.

2.4 Discussion of Methodology

This technique of constructing an energy budget by summing P, R, and U and dividing by the assimilation efficiency to find C, has been criticized (Hargrave 1971; Shafir and Field 1980), since it allows no cross-checking of the individual components. While it is obviously preferable to measure C as well, this is rarely possible in field conditions, and it is seldom justified to predict food intake in the wild from laboratory experiments. Shafir and Field (1980) used this approach in the study of *Cirolana imposita,* and the observed imbalance in the budget was almost certainly the result of a gross over-estimation of food consumption. Wright and Hartnoll (1981) similarly concluded that their estimate of consumption for *Patella vulgata,* extrapolated from feeding experiments, was excessive. Hence, it was decided that, provided the other parameters were accurately determined, the technique in the present study was more reliable, since a laboratory estimate of food consumption is of little value in cross-checking.

The rate of weight gain was determined for a wild population and, as discussed in Luxmoore (1982a), it is therefore reasonably accurate. All weights and calorific values were determined for males and immature or spent females, and consequently the term Pg can be assumed to exclude the great majority of female reproductive tissue. No allowance was made for any seasonal variation in caloric content, and this may introduce a small inaccuracy. However, Clarke (1983) suggested that there may be little variation in lipid content of polar benthic invertebrates other than that associated with reproductive tissue. The rate of accumulation of reproductive tissue was discussed in Luxmoore (1982b) and, although the mathematical equation describing it is only an approximation, the total amount of reproductive tissue produced is a good estimate. Small variations in the rate of accumulation will make little difference to the overall budget.

The respiratory rate is clearly an under-estimate of the true value. Oxygen consumption was determined for resting animals and no allowance was made for activity. *Serolis polita* is not a highly active organism, and spends 75–78% of its time buried under the surface of the sediment (Luxmoore 1981). Even when the animals are on the surface, observation showed that only about 20% of the population was active at any one time, and thus the time spent in activity amounted to about 4.7% of the total time. Activity was shown to entail an increased respiratory rate, an average of 2.4 times the resting rate (Luxmoore, 1984), and therefore the cost of respiration ought to be increased by about 6.6% to allow for activity. Other physiological parameters also increase the metabolic rate, the most important of these being growth. It has recently become apparent that many of the seasonal changes in metabolic rate may, in fact, be due to changes in the growth rate (Parry 1983). Parry (1978) showed that most of the increased metabolism of the limpet *Cellana tramoserica* in summer was attributable to an increased growth rate. The experiments with *S. polita* were conducted in the winter and spring when growth is low (Luxmoore 1982a) and, therefore, probably give a reasonable estimate of basal metabolic rate (Luxmoore, 1984). However, somatic and reproductive growth both increase markedly in summer and it is likely that this would induce an increase in the metabolic rate, possibly in the order of 40–80% (Parry 1983). Averaged over the year this would mean that the cost of respiration might have to be increased by a factor of 27–47%, including the allowance for activity.

The cost of moulting is not solely confined to the energy content of the cuticle, as was assumed in this study. Dissolved organic compounds are also lost at the same time and the tissues comprising the cuticle must have been laid down at an earlier date. Ecdysis also involves increased respiratory rates (Bulnheim 1974). However, it was felt that the approximations made were preferable to leaving the term Px out altogether. A similar argument applies to the urinary excretion, U, which is an approximation, since it assumes a constant composition of food and only considers the excretion of ammonia, but its introduction involves less error than excluding it.

The estimate of population biomass is reasonably accurate, since it takes into account all the different components of the population but it is obviously controlled by the accuracy of the population density estimate. This, however, also affects the other parameters equally and so it will only alter the absolute values and not their relative proportions.

Because of the many complex factors involved in the construction of the energy budget, it is not possible to assign confidence limits. However, from the assumptions made, the overall food consumption is clearly an underestimate, the major error being the value of respiration.

3 Results and Discussion

3.1 Individual Energy Budget

The lifespan of *Serolis polita* is in excess of 4 years and, since the data only span 2 years it was impossible to follow the fate of any animal over its entire lifetime. Instead, a composite energy budget was compiled from cohorts of different ages, averaging the data for the 2 successive years. The composite energy expenditures are given in Fig. 1.

There was an overall increase in the energy used during the life-time, reaching a plateau at maturity, but superimposed on this were large peaks confined to the summer period. In the male these peaks were mainly caused by the higher rate of somatic growth, combined with increased moulting frequency. In the female the peaks were further

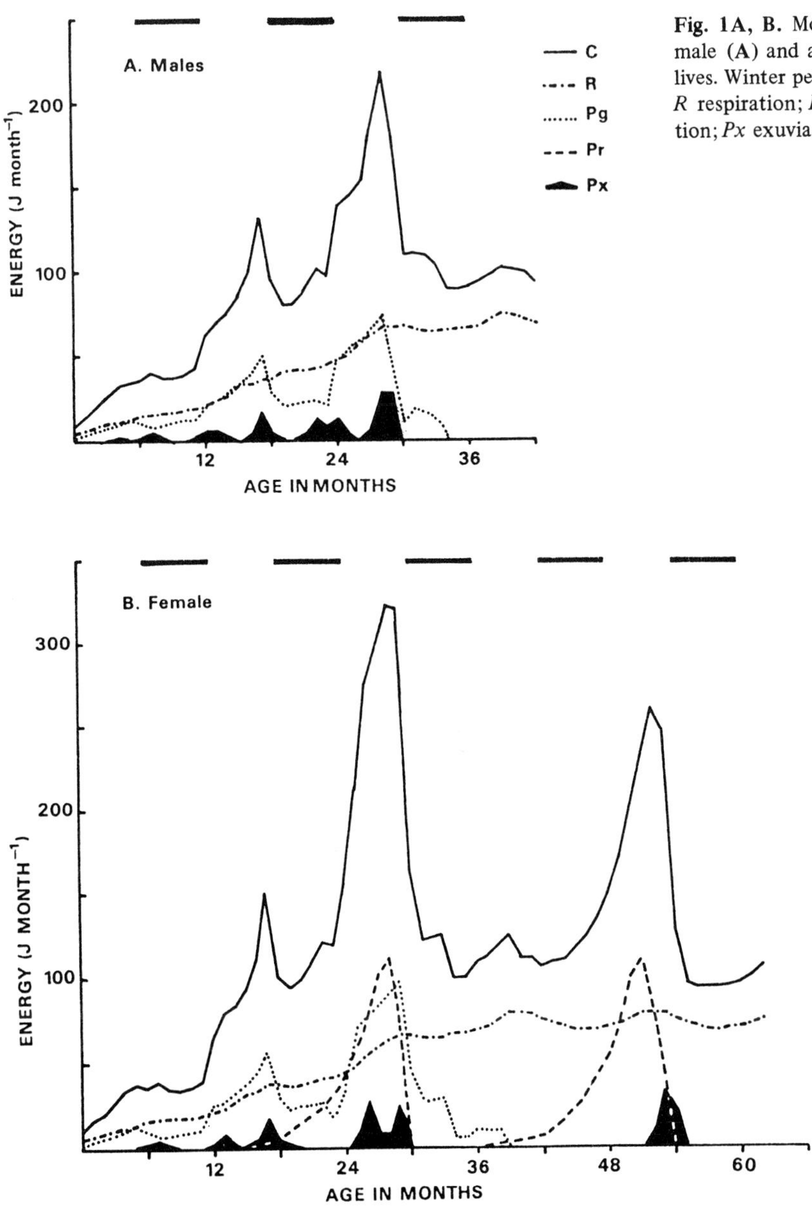

Fig. 1A, B. Monthly variation in the energy budget of an average male (**A**) and an average female (**B**) *Serolis polita* throughout their lives. Winter periods are shown by *black bars*. *C* energy consumption; *R* respiration; *Pg* somatic production; *Pr* gonad and gamete production; *Px* exuvia. (see text)

enhanced by the large energy drain imposed by reproduction. Respiration followed a much smoother curve with only minor increases in the summer, due to the small amplitude of the temperature fluctuations. Consequently, during the summer production comprised the major energy drain, whereas during the winter respiratory coasts tended to predominate. It should, however, be noted that if growth entails an increased respiratory rate this pattern might change, although the seasonality of energy consumption would become even more pronounced.

3.2 Population Energy Budget

The total energy budget for the population of *Serolis polita* over the whole sampling period can be represented by the equation (in kJ m^{-2} yr^{-1}):

$$C = Pg + Pr + Px + R + U + F$$

$$289 = 78 + 24 + 13 + 108 + 9 + 58$$

The small imbalance is due to a net change in biomass.

Whitaker (1982) estimated the primary production by phytoplankton in Borge Bay to be 89–289 g C m^{-2} yr^{-1}. Using the calorific content of a representative diatom, *Navicula glaciei* (Whitaker and Richardson 1980), this gives a value of 4.0–13.1 MJ m^{-2} yr^{-1}. The total energy consumed by *S. polita*, therefore, amounts to 2–7% of the annual primary production.

The only other full energy budget so far compiled for an Antarctic marine ectotherm is for the fish *Notothenia neglecta*, which was also studied in Borge Bay (Everson 1970). This species is the dominant member of the shallow-

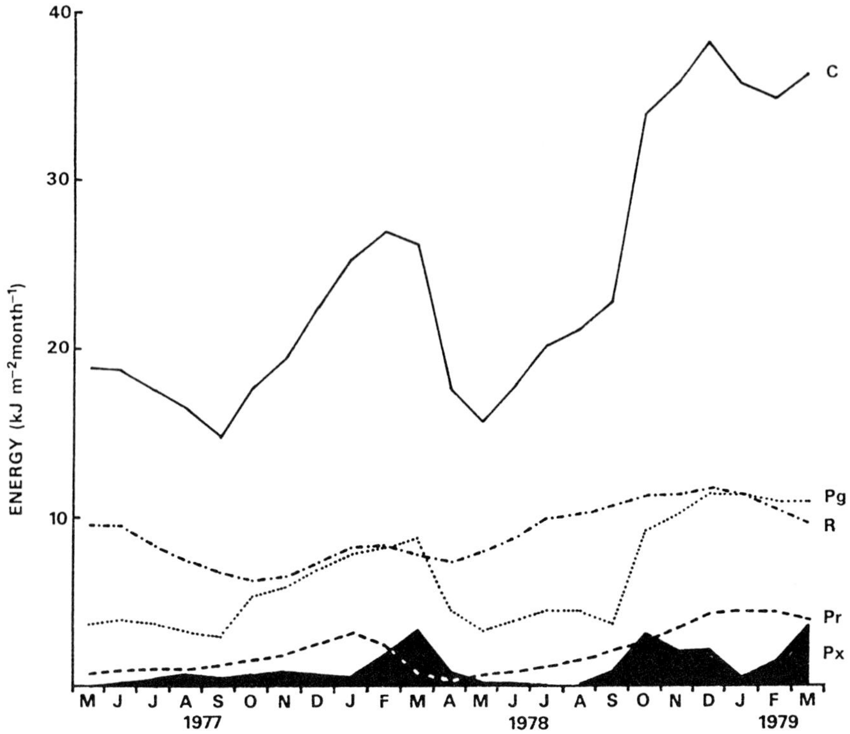

Fig. 2. Monthly variation in the energy budget of a population of *Serolis polita*. See Fig. 1 for explanation of letters identifying graphs

water fish fauna and its energy budget was calculated as:

$$C = Pg + Pr + R + FU$$

$$454 = 23 + 3 + 338 + 91$$

It, therefore, has a much higher energy consumption than *S. polita,* but its production is lower. *Notothenia neglecta* is a large fish with a lifespan of about 17 years, and because of this totally different life-style further comparison with *S. polita* is hardly justified.

The energy budget as presented in an equation represents an average for 2 years, but it is perhaps more instructive to examine how the various components fluctuated throughout the year. An analysis of this is given in Fig. 2, and a marked seasonality is again apparent. As with the individual energy budgets, production showed the greatest variation and over-shadowed respiration in the summer, falling back to a low level in the winter. The excretory energy (U) is not shown in this graph, since it was relatively small and paralleled respiration because of the method of calculation. The total energy lost in exuvia (Px) was of the same order of magnitude but, because it was concentrated in the summer months, it made a significant contribution to the seasonality of energy consumption. The two parameters Px and U are often omitted from energy budget calculations, because they are assumed to be small. However, this analysis shows that this omission may not be justified because their impact may be significant at certain times of the year.

While the overall population fluctuations did not show any seasonal regularity (Luxmoore 1982c), the total energy consumption seemed to be more closely controlled. In winter the level of consumption was low and fairly constant, whereas in summer it increased to higher values. Thus, it is possible that the population was regulated in a density-dependent way to match the energy demand to a fluctuating energy supply. The numerical population density is not in this case a good measure of the energy turnover.

3.3 Comparison with Other Studies

The absolute values of energy budget parameters for different organisms vary greatly, due to differences in population density and techniques. Consequently, when comparing the results of different studies it is preferable to consider the relative proportions of the components. McNeil and Lawton (1970) investigated the relationship of production to respiration for a large number of species and were able to demonstrate a straight-line logarithmic relationship between the two parameters. This analysis was later expanded by Humphreys (1979). The relationship is conveniently expressed by the term "net growth efficiency" or "production efficiency", defined as P/(P + R). Humphreys arranged species according to their life-styles, and showed that the mean net growth efficiency for carnivorous invertebrates (excluding insects) was 27.6% (S.E. = 5.09). The corresponding figure for *S. polita* was 51.8% which is clearly higher than expected, although it is similar to Humphreys' (1979) value for carnivorous insects (55.6% S.E. = 0.64). McNeil and Lawton (1970) and Humphreys (1979) also showed that long-lived anthropods, defined as having life-cycles of more than 2 yrs, tended to have lower net growth

Table 1. The energy budget parameters of various populations of marine isopods

Species	Assimilation efficiency 100 (C–F)/C	Net growth efficiency 100 P/P + R	Gross growth efficiency 100 P/C	Turnover rate P/B	Source
Serolis polita	80	52	40	1.02	This study
Cirolana imposita	88–98	55	15–26	5.56	Shafir and Field (1980)
Cirolana hardfordi	88	40	35	2.39	Johnson (1976)
Idotea baltica[a]	56	42	23	–	Strong and Daborn (1979)
Ligia dilatata[b]	5	17	0.8	2.50	Koop (1979) (vide, Shafir and Field)

a Herbivore.
b Detritivore.

efficiencies, which they attributed to the high energetic cost of sustaining unproductive over-wintering stages.

The energetics of several other species of marine isopods have been studied in detail, and the results are summarized in Table 1. Ignoring for the moment the herbivore and the detritivore, which tend to have lower efficiencies (Humphreys 1979), the net growth efficiencies of the carnivores are all broadly comparable and are all higher than average for carnivorous invertebrates (apart from insects). Isopods tend to have high assimilation efficiencies (Strong and Daborn 1979), but although this can explain the high gross growth efficiency (P/C) it should not affect the net growth efficiency (P/(P + R)) which expresses the relative proportions of energy allocated to production and respiration. This depends largely on the life-style of the organism, whether it expends much energy in food acquisition, and whether there are any lengthy unproductive phases in the life-cycle with high respiratory costs. Shafir and Field (1980) suggested that the high net growth efficiency of *Cirolana imposita* was due to its low metabolic requirements and an abundance of food, presumably reducing feeding costs. *Serolis polita*, although well fed in the summer, almost certainly experiences food shortage in the winter, and its protracted life-cycle, extending over several winters, should result in high respiratory costs; consequently, its high net growth efficiency seems all the more surprising. It is true that the respiration is probably an under-estimate but this makes little difference: if R were increased by 30% the net growth efficiency would only fall to 45%. It is difficult to see what could account for this unexpectedly high efficiency. Respiratory rates are low due to the low temperatures, but production, being another metabolic process, should also be linked to this. Clarke (1980, 1983) has shown that many metabolic processes are not absolutely linked to temperature but are capable of some thermal compensation, and this may allow a small increase in growth relative to respiration.

It appears that *S. polita* is very well adapted to life in a seasonally fluctuating food supply. The protracted life-cycle, which results from the large body size and slow growth, has somehow been achieved without any loss of efficiency. The energy turnover rate can be represented by the proportion of annual production to mean biomass ($P/\bar{B}$). The value for *S. polita* (1.02) is below that of the other isopods in Table 1, but this result must be interpreted with care. Production is a function of respiration (McNeil and Lawton 1970), and therefore $P/\bar{B}$ is a function of weight-specific respiration (R/W). This in turn is a function of temperature and an inverse function of weight, governed by the equation:

$$\ln R = k \ln W + qT + c,$$

and hence

$$P/\bar{B} = w^{k-1} \times e^{qT} + d,$$

where k, q, c, and d are constants. Thus, if two species (1 and 2) have the same net growth efficiency, the ratio of their turnover rates is given by the equation:

$$(P_1/\bar{B}_1) / (P_2/\bar{B}_2) = (W_1/W_2)^{k-1} \times e^{q\,(T_1 - T_2)}.$$

Cirolana imposita and *S. polita* can now be compared in this way. It was shown (Luxmoore, 1984) that for isopods k = 0.75 and q = 0.109. If representative weights are taken as 5 mg and 50 mg, at environmental temperatures of 10°C and 0°C for the two species, respectively, these values can be substituted into the equation to show that *C. imposita* should have a turnover rate 5.3 times higher than *S. polita.* This is remarkably close to the observed value, and it shows that the lower turnover rate of *S. polita* is entirely explicable in terms of its larger body size and lower environmental temperature as a result of these empirical biophysical laws.

Large body size has been reported as a widespread feature of Antarctic invertebrates (Dell 1972; Arnaud 1974) and this, together with the low temperatures, should result in a low $P/\bar{B}$ ratio. Thus, for a given amount of primary production, an Antarctic environment can support a higher biomass than a comparable temperate environment. High biomasses of benthic invertebrates have frequently been reported from the Antarctic (Hardy 1972), and this in the past has been construed as an indication of high productivity (Holdgate 1967). El-Sayed (1978) and Whitaker (1982) have recently shown that Antarctic seas are not outstanding in their levels of primary productivity but this can now be reconciled with the high biomass by proposing a low turnover rate.

Acknowledgements. The initial computer program on which this study relied was written by G.L.M. Hughes, although no blame for my subsequent modifications can be attributed to him. I also thank many members of the British Antarctic Survey, notably Dr. A. Clarke and M.G. White, for much helpful advice and discussion throughout the planning and final analysis of this work. Funds to attend the Fourth SCAR Symposium on Antarctic Biology were generously provided by the Royal Society and the British Antarctic Survey. Finally, I express my appreciation for the invaluable companionship and assistance given by all the staff at Signy Island from 1976 until 1979.

References

Arnaud PM (1974) Contribution à la bionomie marine benthique des régions antarctiques et subantarctiques. Téthys 6:465–656

Brafield AE, Solomon DJ (1972) Oxy-calorific coefficients for animals respiring nitrogenous substrates. Comp Biochem Physiol 43A:837–841

Bregazzi PK (1972) Life cycles and seasonal movements of *Cheirimedon femoratus* Pfeffer and *Tryphosa kergueleni* Miers (Crustacea; Amphipoda). Br Antarct Surv Bull 30:1–34

Bulnheim H-P (1974) Respiratory metabolism of *Idotea baltica* (Crustacea, Isopoda) in relation to environmental variables, acclimation processes and moulting. Helgol Wiss Meeresunters 26:464–480

Clarke A (1980) A reappraisal of the concept of metabolic cold adaptation in polar marine invertebrates. Biol J Linn Soc 14:77–92

Clarke A (1983) Life in cold water: the physiological ecology of polar marine ectotherms. Oceanogr Mar Biol Annu Rev 21:341–453

Dell RK (1972) Antarctic benthos. Adv Mar Biol 10:1–216

El-Sayed SZ (1978) Primary productivity and estimates of potential yields of the Southern Ocean. In: McWinnie MA (ed) Polar Research. Westview, Boulder, Colorado pp 141–160

Everson I (1970) The population dynamics and energy budget of *Notothenia neglecta* Nybelin at Signy Island, South Orkney Islands. Br Antarct Surv Bull 23:25–50

Glynn PW (1976) A new shallow-water serolid (Isopoda, Flabellifera) from the pacific coast of Panama. J Nat Hist 10:7–16

Hardy P (1972) Biomass estimates for some shallow-water infaunal communities at Signy Island, South Orkney Islands. Br Antarct Surv Bull 31:93–106

Hargrave BT (1971) An energy budget for a deposit-feeding amphipod. Limnol Oceanogr 16:99–103

Holdgate MW (1967) The Antarctic ecosystem. Philos Trans R Soc Lond B Biol Sci 252:363–383

Humphreys WF (1979) Production and respiration in animal populations. J Anim Ecol 48:427–453

Johnson WS (1976) Population energetics of the intertidal isopod *Cirolana hardfordi.* Mar Biol 36:351–357

Kusakin O (1967) Fauna of Isopoda and Tanaidacea in the coastal zones of the Antarctic and Subantarctic waters. Biological Reports of the Soviet Antarctic Expedition (1955–1958) 3:220–389

Lipps JH, Ronan TE, De Laca TE (1978) Life below the Ross Ice Shelf. Science (NY)203:447–449

Luxmoore RA (1981) The ecology of Antarctic serolid isopods. PhD Thesis Council for National Academic Awards, UK

Luxmoore RA (1982a) Moulting and growth in serolid isopods. J Exp Mar Biol Ecol 56:63–85

Luxmoore RA (1982b) The reproductive biology of some serolid isopods from the Antarctic. Polar Biol 1:3–11

Luxmoore RA (1982c) Ecological and behavioural adaptations to the Antarctic environment by the isopod *Serolis polita.* In: Blanchard J, Mair J, Morrison I (eds) Proceedings of the 6th international scientific symposium of the World Underwater Federation (CMAS) Sept 14–18 1980. Natural Environment Research Council, UK, pp 140–151

Luxmoore RA (1984) A comparison of the respiration rate of some Antarctic isopods with species from lower latitudes. Br Antarct Surv Bull 62:53–65

McNeil S, Lawton JH (1970) Annual production and respiration in animal populations. Nature (Lond) 255:472–474

Parry GD (1960) Excretion. In: Waterman TH (ed) The physiology of Crustacea, Vol 1. Academic, New York, pp 341–366

Parry GD (1978) Effects of growth and temperature acclimation on metabolic rate in the limpet, *Cellana tramoserica* (Gastropoda: Patellidae). J Anim Ecol 47:351–368

Parry GD (1983) The influence of the cost of growth on ectotherm metabolism. J Theor Biol 101:453–477

Petrusewicz K, McFadyen A (1970) Productivity of animals, IBP Handbook No. 13 Blackwell Scientific Publications, Oxford, UK, 190 pp

Phillipson J (1975) Introduction to ecological energetics. In: Grodzinski W, Klekowski RZ, Duncan A (eds) Methods for ecological bioenergetics, IBP Handbook No. 24. Blackwell Scientific Publications, Oxford, UK pp 3–14

Rakuska-Susczewski S, Klekowski RZ (1973) Biology and respiration of the Antarctic amphipod *Paramoera walkeri* Stebbing in the summer. Pol Arch Hydrobiol 20:475–488

Shafir A, Field JG (1980) Importance of a small carnivorous isopod in energy transfer. Mar Ecol Prog Ser 3:203–215

Strong KW, Daborn GR (1979) Growth and energy utilization of the intertidal isopod *Idotea baltica* (Pallas). J Exp Mar Biol Ecol 41:101–123

Thurston MH (1972) The Crustacea Amphipoda of Signy Island, South Orkney Islands. Br Antarct Surv Sci Rep 71:1–127

Whitaker TM (1982) Primary production of phytoplankton off Signy Island, South Orkneys, the Antarctic. Proc R Soc Lond B Biol Sci 214:169–189

Whitaker TM, Richardson MG (1980) Morphology and chemical composition of a population of an ice-associated Antarctic diatom *Navicula glaciei.* J Phycol 61:250–257

Wright JR, Hartnoll RG (1981) An energy budget for a population of the limpet *Patella vulgata.* J Mar Biol Assoc UK 61:627–646

Non-Cold-Adapted Metabolism in the Decapod *Chorismus antarcticus* and Other Sub-Antarctic Marine Crustaceans

J.G.H.MAXWELL[1,2] and R. RALPH[1]

Summary. A study on the metabolic rate of the Antarctic decapod *Chorismus antarcticus*, determined by oxygen consumption rate, indicated the absence of "cold-adapted" metabolism. Experiments were carried out under a number of temperature regimes, and the influence that experimental technique could have on the results was examined. The oxygen consumption rate was compared with that of temperate decapods, and the low level of metabolic activity of *C. antarcticus* and other Antarctic benthic invertebrates, at their normal environmental temperature was confirmed. This finding supports the observation that metabolism at low ambient temperatures is commonly at a level expected empirically from the temperature dependent behaviour of physico-chemical reaction rates. This physiological behaviour of *C. antarcticus* is coupled with a life-history indicative of a K-selection strategy which is an evolutionary trend observed among a number of Antarctic marine species.

1 Introduction

The suggestion of "cold-adapted" metabolism at low temperatures was proposed by Krogh (1916) who suggested that the normal van't Hoff rule, which implies that over a 10°C range chemical reaction rates should alter by a factor of 2 should somehow be overriden in the case of animals living at low temperatures. Subsequently, Sparck (1926) and Bruce (1926) suggested that a special metabolic regime might exist for cold-water poikilotherms. Sparck (1936) and Thorson (1936) concluded that such a mechanism did exist and Fox (1936), Fox and Wingfield (1937) and Wingfield (1939), while concluding that aquatic poikilotherms living in cold water had a lower metabolic rate than similar species in warmer water, suggested that the proportion of the oxygen consumption required to "fuel" the metabolism was the same in both cases. Berg (1953) determined that there was a degree of cold adaptation in the cold-water species used in the work of Fox and Wingfield. Subsequent studies (Edwards and Irving 1943a; Thorson 1952; Scholander et al. 1953a; Bullock 1955; Prosser 1955; Demeusey 1957) established the concept of "cold adaptation" as a generally accepted physiological principle (Prosser 1958; Wohlschlag 1960, 1964; Brett 1971; Hochachka and Somero 1973; Belman 1975; Umminger 1977; Remmert 1980). However, a number of studies indicating an absence of "cold-adapted" metabolism have been published (Edwards and Irving 1943b; Wohlschlag 1963; Holeton 1974; White 1975; Ralph and Maxwell 1977a,b).

In this study, four marine benthic crustacean species, three from South Georgia (*Chorismus antarcticus, Notocrangon antarcticus* and *Serolis pagenstecheri*) and one from the Falkland Islands (*Acanthocyclus albatrossis*) were examined for their oxygen consumption rates in conjunction with previously reported findings for an Antarctic limpet, *Nacella concinna,* and a bivalve mollusc, *Gaimardia trapesina.* The results suggest that the concept of "cold adaptation" as a metabolic adaptation for cold-water poikilotherms can no longer be supported.

2 Materials and Methods

All the oxygen consumption measurements were made at South Georgia, during the austral summer. The animals were collected by a 0.5 m beam-trawl or by SCUBA diving, and maintained in a thermostatically controlled closed marine aquarium system accurate to ± 0.01°C. Specimens of the brachyuran crab, *A. albàtrossis,* were collected from the littoral zone in the Falkland Islands and transported to South Georgia.

The rates of oxygen consumption were measured by the closed-bottle Winkler technique, as described by Maxwell (1976) and Ralph and Maxwell (1977a). The animals, depending on their size, were placed individually in 135 ml or 425 ml bottles at the experimental temperature, after being taken directly from the holding tank at 2°C. The duration of the experiment also depended on the animal's size and in no case was the oxygen level at the end of the experiment $< 75\%$ of the initial value. The experiments were carried out in batches of 20 with appropriate controls and blanks. With care, the Winkler technique can be very accurate (Carpenter 1965; Carritt and Carpenter 1966) and the results are known to be comparable with other methods commonly employed to measure respiration rates (Lawton and Richards 1970).

1 Aberdeen University Marine Studies Ltd., Department of Zoology, University of Aberdeen, Aberdeen AB9 2TN, Scotland
2 British Antarctic Survey, High Cross, Madingly Road, Cambridge CB3 OET, Great Britain
Present address: Marine Science and Resource Investigations, PO Box 134, Engadine, New South Wales 2233, Australia

Antarctic Nutrient Cycles and Food Webs
(ed. by W.R. Siegfried, P.R. Condy, and R.M. Laws)

3 Results

3.1 Chorismus antarcticus

3.1.1 Oxygen Consumption

A set of observations was made at 2°C which approximated the normal ambient water temperature from which the animals were collected. There was a high degree of replicability of the experimental data. Most of the observations were made during an intensive 3-week period (Fig. 1) with a concluding set of observations made some 12 weeks later. The regressions of oxygen consumption (y) to dw animal body-mass (x), on logarithmically transformed data, were not significantly different ($P < 0.05$) with respect to the regression coefficient or elevation. Similarly, the slopes or elevations within the initial set of experiments were not significantly different.

The influence of temperature on the relationship between oxygen consumption and dw mass was measured after the animals had been transferred from the standard 2°C holding temperature. Whereas there was no significant difference between the regression coefficients ($P < 0.05$), there was a significant difference in the elevations of the relationship between oxygen consumption and dw mass for the experiments carried out at 7° and 12°C but, since a significant difference was not found between the regression coefficients or elevations for the 0°C and 2°C data, a common equation was calculated (Fig. 2). The relationships at different temperatures are given in Table 1. The common regression coefficient was 0.7845, and the corresponding set of equations is given in Table 1. Table 2 gives a set of Q_{10} values, which standardize the rate of change of oxygen consumption to that experienced over a 10°C range, for oxygen consumption rates between the different experimental temperatures.

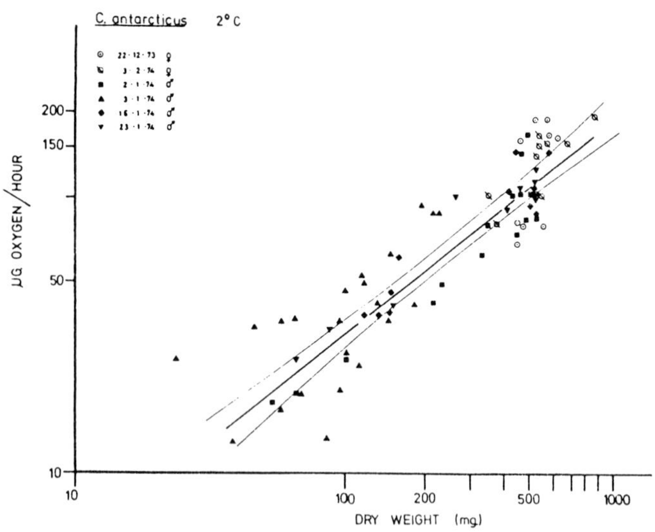

Fig. 1. Rate of oxygen consumption as a function of animal dry wt. mass for *Chorismus antarcticus* at 2°C. The line of best fit (calculated by least squares regression) is shown with the associated 95% confidence interval boundary

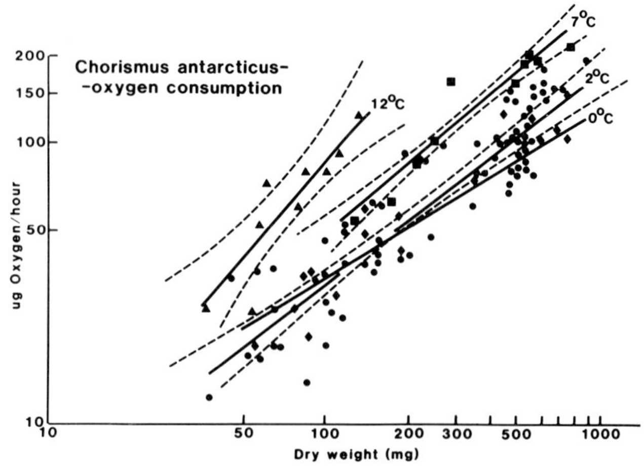

Fig. 2. Rate of oxygen consumption as a function of animal dw mass for *Chorismus antarcticus* at 0°, 2°, 7° and 12°C, respectively. Fitted lines and boundaries as in Fig. 1

3.1.2 Acclimation to 7°C

It is unlikely that *C. antarcticus* encounters ambient temperatures of 12°C, but the population in Terra del Fuego, north of the Antarctic Convergence, can be expected to experience water at 7°C. The ability of the species to acclimate to temperatures higher than normal was demonstrated by a number of experiments in which the animals were exposed to 7°C for different periods. There was no significant difference between the regression coefficients of the relationship of oxygen consumption to dw mass at periods of zero, 1, 6 and 24 h (Table 3). The effect of temperature with respect to time and the ability of *C. antarcticus* to acclimate is shown in Fig. 3 for a 500 mg animal; oxygen consumption rate rising rapidly after the initial increase in temperature but returning after 6 h to a level not significantly different from the initial rate at 2°C.

3.2 Acanthocyclus albatrossis

3.2.1 Oxygen Consumption

The measurements were made at 2° and 7°C and whereas the relationship of oxygen consumption to dw mass (Table 4) indicated that there was no significant difference in the values of the regression coefficient, from which a common value of 0.7002 was derived (Table 4), there was a difference (significant at the 5% level) in the elevations of the 2 regressions. The Q_{10} value over the 5°C temperature range was 2.2.

3.3 Notocrangon antarcticus

3.3.1 Oxygen consumption

The relationship of oxygen consumption to dry wt. mass for this species was obtained at 2°C, as $\log y = 0.8430 \log x - \log 0.3629$ ($n = 8$, $r = 0.8702$).

Table 1. Relationships between rate of oxygen consumption (y in $\mu g\ O_2\ h^{-1}$) and dry wt. animal mass (x in mg) for *Chorismus antarcticus* at 0°C, 2°C, 7°C, and 12°C, respectively

Temperature (°C)	First equations	Second equations	
0	log y = 0.7175 log x+log 0.0839	log y = 0.7845 log x–log 0.0730	n = 14, r = 0.9406
2	log y = 0.7825 log x–log 0.0552	log y = 0.7845 log x–log 0.0730	n = 76, r = 0.8779
7	log y = 0.8430 log x–log 0.0104	log y = 0.7845 log x+log 0.1356	n = 10, r = 0.9848
12	log y = 1.2176 log x–log 0.4821	log y = 0.7845 log x+log 0.3272	n = 9, r = 0.8841

The second equations are based on the assumption of a common regression coefficient of 0.7845.

Table 2. Q_{10} values for rate of oxygen consumption of *Chorismus antarcticus* at different temperature ranges, based on second set of equations in Table 1

Temp. range (°C)	Q_{10} value
0– 7	1.7
0–12	1.8
2– 7	2.1
2–12	2.1
7–12	2.0

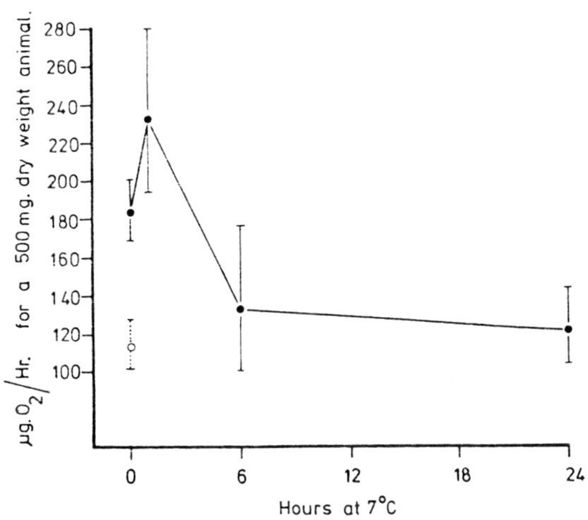

Fig. 3. Acclimation of *Chorismus antarcticus* for oxygen consumption rate in relation to time, measured at 7°C after direct transfer from 2°C. *Dots* indicate animals at 7°C and circle shows control at 2°C. Upper and lower 95% confidence intervals are shown

3.4 Serolis pagenstecheri

3.4.1 Oxygen Consumption

The relationship of oxygen consumption to dry wt. mass was determined at 2°C (Fig. 4), and the equation is log y = 0.9860 log x–log 0.9456 (n = 18, r = 0.8410). One specimen which moulted during the experiment (Fig. 4) was excluded from the overall regression, since the activity of moulting elevated the oxygen consumption above the standard level. This animal, with a 554.5 mg dry wt. body-mass, had an oxygen consumption rate of 183.4 $\mu g\ O_2\ h^{-1}$ which is 3.2 times the predicted value for an animal of this weight as calculated from the regression for the remainder of the sample.

Table 3. Relationships between rate of oxygen consumption after different time intervals of *Chorismus antarcticus* at 7°C after direct transfer from 2°C (see Table 1 for fuller particulars)

Temperature (°C)	Treatment	Equation	
7	Direct	log y = 0.8430 log x–log 0.0104	n = 10, r = 0.9848
7	1 h	log y = 0.8354 log x+log 0.1130	n = 15, r = 0.9614
7	6 h	log y = 0.7760 log x+log 0.0291	n = 13, r = 0.9369
7	24 h	log y = 0.8358 log x–log 0.1711	n = 11, r = 0.9752
2	Control	log y = 0.7825 log x–log 0.0552	n = 76, r = 0.8779

Table 4. Relationships between rate of oxygen consumption and dw animal mass for *Acanthocyclus albatrossis* at 2° and 7°C, respectively. (see Table 1 for fuller particulars)

Temperature (°C)	First equation	Second equation	
2	log y = 0.7104 log x+log 0.0226	log y = 0.7002 log x+log 0.0538	n = 9, r = 0.9448
7	log y = 0.6895 log x +log 0.2612	log y = 0.7002 log x+log 0.2273	n = 11, r = 0.2612

The second equations are based on the assumption of a common regression coefficient of 0.7002

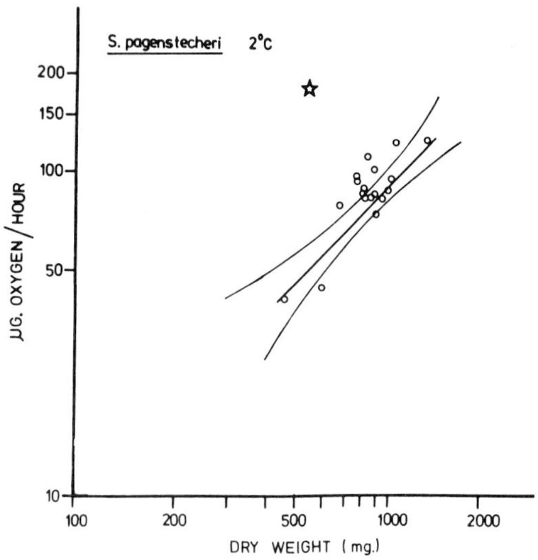

Fig. 4. Rate of oxygen consumption as a function of animal dry wt. mass for *Serolis pagenstecheri* at 2°C. The *star* represents an animal which moulted during the experiment and whose metabolic rate was excluded from the calculation of the line of best fit (derived by least squares regression) with the associated 95% confidence interval boundary

4 Discussion

4.1 Comparison of the Regression Coefficient 'b'

There are certain values for the regression coefficient relating metabolism to body weight that appear characteristic for certain groups of animals. While the subject has been a focus for much review and debate (Brody 1945; Kleiber 1947, 1961) there is still no completely satisfactory explanation of the parameters that might control and govern the value for any particular species. There are two values for b that appear attractive from the theoretical standpoint. It is possible that respiration is a surface-related process which would lead to a b value of 0.66 (surface area = K.mass$^{2/3}$) (Rubner 1883). Alternatively, if it is directly proportional to weight the expected value would be 1, a value that holds true for some unicellular organisms (Hemmingsen 1960). In practice most multi-cellular animals are found to have b values between these two theoretical values, in the range 0.7–0.8. Winberg (1956) obtained a mean value of 0.8 for many species of fish and Hemmingsen (1960) proposed 0.75 as a general value for crustaceans and other poikilotherms. Ansell (1973) refers to a value of 0.703 for 16 species of bivalves from the west coast of Scotland.

For *C. antarcticus* in this study, and also for *G. trapesina* (Ralph and Maxwell 1977b), there was a close agreement with Hemmingsen's predicted value of 0.75, with the calculated coefficients being statistically different from 0.66 and 1.0. *Acanthocyclus albatrossis* and *N. concinna* (Ralph and Maxwell 1977a) had b values significantly different (at the 5% level) from 1.0 but not from either 0.66 or 0.75. The regression coefficient of 0.8430 for *N. antarcticus* was not significant from any of the three predicted values. The reason for this is taken to be the small sample size which resulted in a narrow weight range of the animals used. For a meaningful value of b, the weight range of the animal used should extend over most of the known size – weight range of the species (Hemmingsen 1960; Laird 1965; Ralph and Everson 1968).

The calculated regression coefficients for the species examined were < 1 in each case, which means that the oxygen consumption per unit weight vs total dry wt. has a negative slope, indicating that metabolism per unit body weight decreases with size in the expected fashion (Brody 1945; Hemmingson 1950; Kleiber 1961; Zeuthen 1947, 1953).

4.2 Interspecific Comparison of Results

A combined plot of the results (Fig. 5) from this study and the species examined by Ralph and Maxwell (1977a,b) indicates the good agreement for the relationship between these species for the regression of oxygen consumption against body weight at their normal environmental temperatures. Further, when an interspecific comparison is made between the Antarctic species and those temperate-water species for which it is possible to make a comparison, the values for the oxygen consumption of the cold-water species fall on the general line (Fig. 6) relating oxygen consumption rate to temperature. There does not appear to be any special modification of their oxygen consumption rate that might be related to their evolutionary history or zoogeographical position, as is suggested by the concept of "cold adaptation" reported by Scholander et al. (1953a) or Prosser (1955).

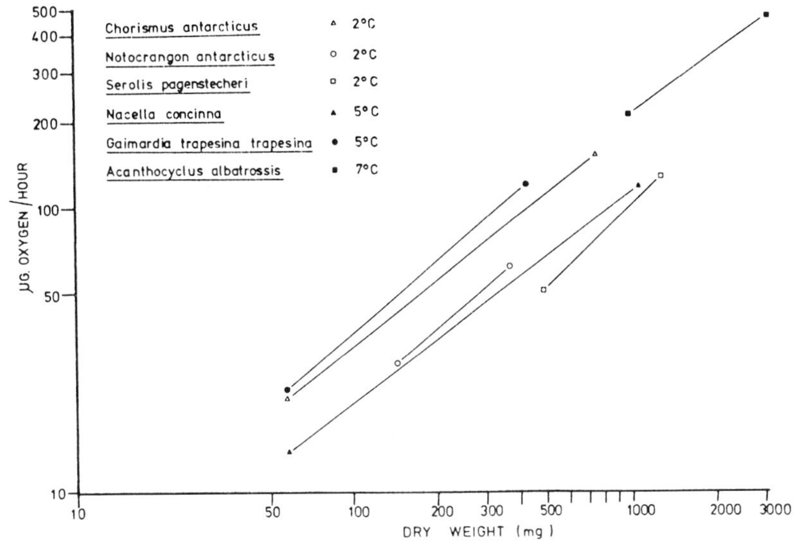

Fig. 5. Composite rate of oxygen consumption as a function of animal dry wt. mass for five cold-water benthic crustaceans from South Georgia and one (*A. albatrossis*) from the Falkland Islands

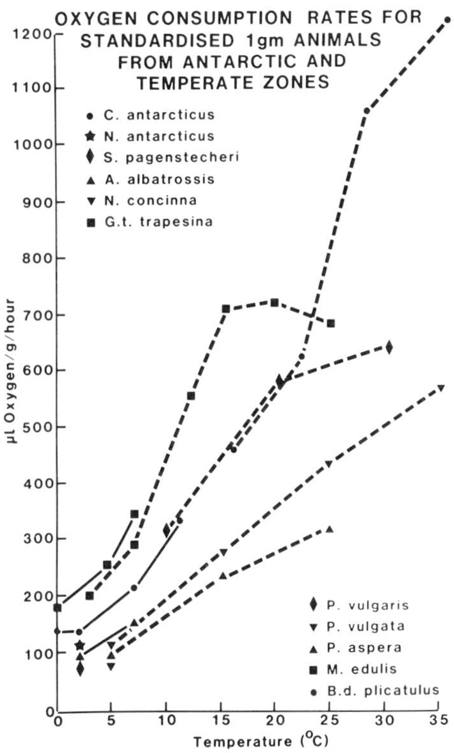

Fig. 6. A comparison of the rate of oxygen consumption for standard 1 g dry wt. animals from the Antarctic (*clear symbols*), based on this study and Ralph and Maxwell (1977a, b), and allied temperate-water species: *Palaemonetes vulgaris* (McFarland and Pickens 1965); *Patella vulgata* and *P. aspersa* (Davies 1966); *Mytilus edulis* and *Brachidontes demissus* (Read 1962)

4.3 Influence of Experimental Procedure on Oxygen Consumption Measurements

The physical effect that experimental procedures may have on the results of physiological work is well documented (Fry 1958; Brett 1962; Beamish 1964; Beamish and Mookherjii 1964; Saunders 1963), and it is known that the activity or "stressed" condition of an experimental animal can mask the true relationship of oxygen consumption to body weight (Holeton 1974), especially in cold-water or polar species. Halcrow and Boyd (1967) showed that spontaneous activity in *Gammerus oceanicus* elevated its respiratory rate 2.5 times the normal standard level at temperatures between 10° and 15°C. Thomson and Bayne (1972) showed an increase of active over standard metabolism by a factor of 2.67 for the mussel, *Mytilus edulis,* which is comparable with the change in rate shown by Newell and Bayne (1973) and Widdows (1973) of 2.73 for the same species. In the case of the isopod *Serolis pagenstecheri,* the elevated oxygen consumption rate by the moulting animal (Fig. 4) was due to the high locomotory activity associated with this process. The increase by a factor of 3.2 is of the correct order expected in the level of active metabolism over the standard rate.

In our study, the level of oxygen consumption measured was the standard rate and not some elevated value. The results from the experiment with *C. antarcticus* can be interpreted to show that the experimental procedure has no effect on the results. A comparison of the data for the oxygen consumption rate measured at 2°C and 7°C showed that there was s significant difference in the rate at 2°C and after a direct change to 7°C after 1 h. After 6 h the rates at the two temperatures are comparable (Fig. 3) and no different from the control at 2°C. The evidence suggests that the initial elevation on transfer to the higher temperature is due to the increase in temperature, and not the effect of active metabolism and experimental manipulation, since the animals in the control were treated in the same experimental manner and did not exhibit an elevated oxygen consumption rate.

4.4 Non-Cold-Adapted Metabolism in Antarctic Invertebrates

Central to the problem of determining whether or not animals are "cold adapted" is the assumption that is made as to what the expected metabolism might be by extrapolation of the metabolic rate of tropical and temperate species to lower temperatures. Figure 7 gives the general relationship between cold-adapted metabolism in Arctic and Antarctic poikilotherms which shows elevated oxygen consumption rates over the expected rate predicted from the downward trend of tropical and temperate metabolism from the empirical relationship between temperature and chemical reaction rates and the trend from "Krogh's Curve". The Antarctic species studied by us did not exhibit the high level of oxygen consumption predicted from this relationship and, thus, are described as "non-cold adapted". Although *G. trapesina* has been shown previously to be conclusively "non-cold adapted" (Ralph and Maxwell 1977b), it appears to have an elevated metabolism as represented in Fig. 7. This results from the extrapolation of the small tissue weight of the species to a much larger weight base which introduces a considerable error. The fish, *Rhizophila dearboni,* described by Wohlschlag (1963) as having an "unusually" low metabolism can be seen (Fig. 7) to have had its oxygen consumption measured accurately but, nevertheless, because it did not conform to the concept of "cold adaptation" it was regarded as "unusual". Much of the assumption regarding "cold adaptation" lies in the relationship derived by Ege and Krogh (1914) in a study on a single narcotized goldfish, *Carassius* sp. Ege and Krogh admitted themselves that the high Q_{10} value of 9.8 for the lowest temperature range of 0°–5°C was obviously erroneous and Krogh (1916), in a further discussion of the fit of biological reaction rates to the van't Hoff rule, which

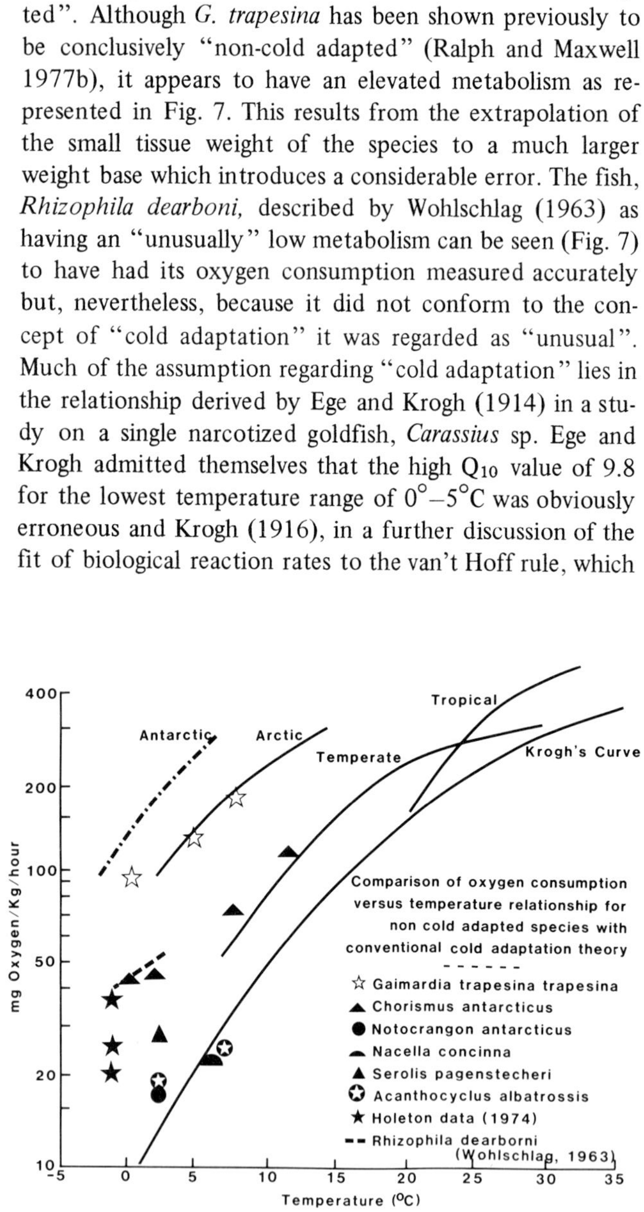

Fig. 7. Diagrammatic representation of the relationship between oxygen consumption and temperature showing "Krogh's Curve" and the expected range of metabolic activity for tropical, temperate and cold-water poikilotherms, following the concept of "cold adaptation" (redrawn from Holeton 1974). Values for "non-cold-adapted" Antarctic marine species obtained from this study are given for comparison

for most biological reactions gives a Q_{10} between 2 and 3, noted the high value in the previous study and pointed out that the low temperature end of his extrapolated standard line of metabolism was based on very little experimental work and that interpretation of the results at the low end of the temperature scale as representing the real situation could not be justified. This fact was overlooked for some time, a typical example being Fry and Hart (1948). Rao and Bullock (1954) suggested that there was a general increase of the Q_{10} value at lower temperatures, although they were aware of the problem of relating results they had obtained from animals outside their normal temperature ranges to normal oxygen consumption values and also the difficulty of making interspecific comparisons. Belehradek (1935) gave several explanations for the variation of the Q_{10} value with temperature and, in developing the theme, went so far as to derive a mathematical expression relating rate (velocity) to temperature (Belehradek 1930). This has been used in several studies and especially by McLaren (1963) who discussed a mathematical fit between Belehradek's temperature function and "Krogh's Normal Curve" when applied to data from Winberg (1956). Winberg (1956) attempted to establish a standard set of constants to convert metabolism to a standard value at 2°C. Precht (1958) made a similar attempt to derive a comprehensive expression to describe the inter-relationship of metabolic types. In the intervening time several authors (Kanwisher 1959; Scholander et al. 1953b) had reported high experimental Q_{10} values for cold-water poikilotherms. In one extreme case (Klekowski et al. 1973) a Q_{10} of 71.4 was reported over the temperature range −1.9° to 0°C. Re-calculating their data indicates that the value from their results was 94.9, still unreasonably high and unlikely to contribute any real understanding about the mechanics of the physiological system in the study. These high Q_{10} values became acceptable on the grounds that at low temperatures the normal Q_{10} relationship could not be expected to hold (Rao and Bullock 1954; Dunbar 1968, 1970). Results such as these, based as they were on the acceptance of "Krogh's Normal Curve", are therefore placed in considerable doubt if it be recalled that any failure of the relationship suggested by Krogh (1916), and later re-emphasized by Holeton (1947), makes the basis for the concept of "cold adaptation" as a general physiological adaptation unlikely. The results of our study indicate that a low oxygen consumption rate is found in a number of species and that the behaviour of the Q_{10} value for *C. antarcticus* and *A. albatrossis,* and a number of other cold-water species (McWhinne 1964; Paranjape 1967; Holeton 1973; White 1975; Ralph and Maxwell 1977b), fall in the range one would empirically expect for organisms adapted for life at low temperatures. Beamish (1964) and Beamish and Mookherjii (1964) established a semi-logarithmic relationship between oxygen consumption and temperature, in accord with the relationship empirically derived from the van't Hoff rule (Jones 1976). On the evidence now available it seems unnecessary to invoke a concept of "cold adaptation" for life at low environmental temperatures for marine organisms.

4.5 Cold Adaptation at the Biochemical Level

In physico-chemical terms the viable biological temperature range is narrow, 0°–40°C. There is, therefore, the opportunity for organisms to adapt to low (and high) environmental temperatures by molecular adaptation at the biochemical level, allowing their physiological systems to function as one might empirically expect of the particular environmental regime which they inhabit. Thus, there would be no need to invoke special mechanisms which release them from the expected constraints imposed by normal physico-chemical relationships. Lowering of the activation energy makes it possible for a greater proportion of molecular species to take part in a reaction, and it is known that enzymatic adaptation has taken account of this fact in some low-temperature systems (Somero 1969; Somero et al. 1968; Hochachka and Somero 1973; Johnston et al. 1973), whereas the opposite tendency has been noted in high-temperature systems (Kwon and Olcott 1965). Further, biochemical adaptations now recorded include the evolutionary adaptations within species for enzymes with low K values (Hochachka and Somero 1973; Somero and Hochachka 1968; Somero and Hochachka 1971) at low temperature and the relationship of molecular stability related to temperature (Smith 1973a,b; Johnston et al. 1975). The acknowledgement of the importance of such biochemically adaptive solutions over the "cold adapted" physiological solution is evident in the review by Hazel and Prosser (1974). If the type of "cold adaptation" proposed by Scholander et al. (1953a) was a universal answer to metabolism at low temperatures then it would have had important implications for the energy flow through much of the world's marine ecosystems, since the majority of the world's oceans, apart from the surface and coastal regions, are typified by waters of < 5°C. The elevation of metabolism in these cold and usually poorly productive waters would be difficult to justify energetically. The work that has been done on the metabolism of deep-water species suggests that they, too, are biochemically adapted at the enzymatic level, which results in a lowering of the metabolic rate at low temperature in the manner that would be expected from physico-chemical relationships (McLaren 1963; Pearcy and Small 1968; Hochachka et al. 1972; Sleigh and MacDonald 1972; Childress 1977; MacDonald 1975). The situation is further complicated by the fact that whereas the respiration rates of certain mid-water species have been found to be determined directly by temperature effect in others respiration was depressed by temperature but elevated by pressure which resulted in no net change in respiration rate with depth (Teal 1971). In other species sophisticated mechanisms have evolved for the reduction of metabolism with depth, thus minimizing the temperature effect almost totally (Torres et al. 1979). The oxygen consumption characteristics of the species inhabiting the cold waters of the majority of the world's oceans does little to support a concept of "cold adaptation".

4.6. Life-History Strategies of Antarctic Marine Fauna

It has been observed that many polar species exhibit a life-history characterized by large size, slow growth and low metabolism (Dunbar 1960). To these attributes can be added the faculty for reduced fecundity, parental care, truncated larval life and other reproductive specializations (Thorson 1936, 1946, 1950, 1961; Dell 1972; Wyatt 1976; Maxwell 1977). These life-history parameters are indicative of what has become generally known as the K-selective strategy, on the r-K continuum (MacArthur and Wilson 1967; Pianka 1970, 1972; Southwood et al. 1974). While the use of the r-K terminology has acquired a number of meanings (Parry 1981) and the debate between the selective value of either strategy tends to be tedious (Stearns 1977), it is generally accepted that K-selected species are found in stable environments where density dependence is important, and exhibit late sexual maturity with parental care of a few large young, whereas r-selected organisms tend to occur in fluctuating environments where density independence is important, and exhibit early maturity and produce many small young which are invested with little or no parental care (Stearns 1976; Stubbs 1977). The assertion that the tropics favour K-selected species (Dunbar 1960) can be seen to be an over-simplification with the case of many warm-water prawn species which have an annual life cycle, rapid growth and high fecundity. The banana prawn, *Penaeus merguiensius*, for example, is a classic r-selected species with high fecundity, rapid growth and development, and the capacity for sustaining an annual mortality of over 80% due to natural and fishing mortality without affecting the following year's recruitment (Lucas et al. 1979). The relatively large size of many Antarctic marine poikilotherms (e.g. *Chorismus antarcticus* (Maxwell 1976, 1977), *Euphausia superba* (Mauchline and Fisher 1969) can be interpreted as a response to the development of K-selective traits, such as slow growth and delayed maturity (Bonner 1965), and increased life span (Rensch 1959) and, important to the theme of this study, the reduced metabolic activity which results from the inverse relationship of metabolic activity to life span (Bourliere 1957). There is a further selective advantage at the ecosystem level in terms of adopting the K-strategy with its associated large size and low metabolism among the other attributes. This is a result of the fact that the larger the organism the lower the respiration rate per unit weight of tissue (Zeuthen 1947, 1953; Hemmingsen 1950, 1960; Von Bertalanffy 1951) which implies that for the same food resource availability a larger population of slow-growing animals with a low metabolic requirement can be sustained per unit of time in comparison to a population of small, fast-growing organisms with a high rate of metabolism (Harvey 1955). It is, therefore, suggested that there exists an inter-relationship between life-history parameters and the physiological mechanism that characterize the evolution of species at the ecosystem level. When the constraints imposed by the Antarctic marine ecosystem are viewed in this fashion, the value of a low metabolic activity (i.e. "non-cold-adapted" metabolism) at low environmental temperatures, suggested from the empirial evidence, would be the most likely selective response. The data presented in our paper support this suggestion and allow Antarctic researchers to move close to the type of predictive ecology suggested by Blueweiss et al. (1978).

5 Conclusion

The oxygen consumption rates of the species examined by us indicated the absence of "cold adaptation". A low oxygen consumption rate is more common among Antarctic marine poikilotherms than was thought previously, and the concept of "cold adaptation" as a physiological adaptation for life at low environmental temperatures is doubtful, since the basis of the original concept is accepted as having been founded on insufficient experimental evidence. The results of most of the experimental studies which supported "cold adaptation" are questionable, since the techniques employed may have been responsible for the high oxygen consumption rates observed. Many biochemical adaptations have been demonstrated to exist in cold-water poikolotherms, making it likely that such features have evolved for life at low environmental temperatures and which can function at the low metabolic levels that are implied from the low levels of oxygen consumption in this study. From an eco-physiological standpoint, "cold adaptation" provides no clear selective value. The adjustment of the metabolism to a low level at low environmental temperatures, which would be expected empirically, would confer considerable advantage by allowing organisms to conform to fundamentally inter-related strategies of growth, metabolism and life history.

Acknowledgements. We wish to thank all those at South Georgia and the officers and crew of the RRS John Biscoe for assistance during the course of the field work and Drs RM Laws, I. Everson and MG White for ancouragement and advice. The senior author wishes to thank Professor F Holliday, Department of Zoology, University of Aberdeen, Scotland for making available facilities during the preparation of his PhD thesis of which this work formed a part, and the second author wishes to thank the Royal Society for a travel grant, the University of Aberdeen for financial support and the British Antarctic Survey for the invitation to work at South Georgia.

References

Ansell JA (1973) Oxygen consumption by the bivalve *Donax vittatus* (da Costa). J Exp Mar Biol Ecol 11:311–328

Beamish FWH (1964) Respiration in fishes, with special emphasis on standard oxygen consumption. II. Influence of weight and temperature on respiration of several species. Can J Zool 42:177–188

Beamish FWH, Mookherjii PS (1964) Respiration of fishes with special emphasis on standard oxygen consumption. I. Influence of weight and temperature on respiration of goldfish, *Carassius auratus* L. Can J Zool 42:161–175

Belehradek J (1930) Temperature coefficients and biology. Biol Rev 5:30–58

Belehradek J (1935) Temperature and living matter. Protoplasma Monogr 8 Berlin

Belman BW (1975) Oxygen consumption and ventilation of the Antarctic isopod *Glyptonotus*. Comp Biochem Physiol 50A: 149–151

Berg K (1953) The problem of respiratory acclimatization. Hydrobiologia 5:311–350

Von Bertalanffy L (1951) Metabolic types and growth rates. Am Nat 85:11–117

Blueweiss L, Fox H. Kudzma V, Nakashima D, Peters R, Simms S (1978) Relationships between body size and some life history parameters. Oecologia (Berl) 37:257–272

Bonner JT (1965) Size and cycle. Princeton University, Princeton

Bourliere F (1957) Ageing and metabolism. In: Yapp WB, Bourne GH (eds) The biology of ageing. Symposium of the Institute of Biology 6 pp 27–34

Brett JR (1962) Some considerations in the study of respiratory metabolism in fish, particularly salmon. J Rish Res Board Can 19:1025–1038

Brett JR (1971) Temperature - animals - fishes. In: Kinne 0 (ed) Marine Ecology I. Wiley, London pp 513-560

Brody S (1945) Bioenergetics and growth. Reinhold Pub Corp, New York

Bruce JR (1926) The respiratory exchange of the mussel, *Mytilus edulis*, L Biochem J 20:829

Bullock TN (1955) Compensation for temperature in the metabolism and activity of poikilotherms. Biol Rev 30:311–342

Carpenter JH (1965) The accuracy of the Winkler method for dissolved oxygen analyses. Limnol Oceanogr 10:135–140

Carritt DE, Carpenter JH (1966) Comparison and evaluation of currently employed modifications of the Winkler method for determining dessolved oxygen in seawater, a NASCO Report. J Mar Res 24:286–318

Childress JJ (1977) The respiratory rates of midwater crustaceans as a function of depth of occurrence and relation to the oxygen minimum layer off southern California. Comp Biochem Physiol 50A:787–799

Davies PS (1966) Physiological ecology of *Patella*. I. The effect of body size and temperature on the metabolic rate. J Mar Biol Ass UK 46:647–658

Dell RK (1972) Antarctic benthos. In: Russel FS, Yonge M (eds) Advances in marine biology, 10 pp 1–216

Demeusy N (1957) Respiratory metabolism of the fiddler crab *Uca pugilator* from two different populations. Biol Bull (Woods Hole) 113:245–253

Dunbar MJ (1960) The evolution of stability in marine environments; natural selection at the level of the ecosystem. Am Nat 94:129–136

Dunbar MJ (1968) Ecological development in Polar regions. Prentice-Hall, New Jersey

Dunbar MJ (1970) Marine ecosystem development in Polar regions. In: Steele JH (ed) Marine Food Chains. Oliver and Boyd, Edinburgh pp 528–534

Edwards GA, Irving L (1943a) The influence of temperature and season upon the oxygen consumption of the sand crab, *Emerita telpoida* Say. J. Cell Comp Physiol 21:169–182

Edwards GA, Irving L (1943b) The influence of the season and temperature upon the oxygen consumption of the beach flea, *Talorchestia megalopthalma*, J Cell Comp Physiol 21:183–189

Ege R, Krogh A (1914) On the relation between the temperature and the respiratory exchange in fishes. Int Rev Gesamten Hydrobiol Hydrog 7:48–55

Fox HM (1936) The activity and metabolism of poikilothermal animals in different latitudes. I. Proc Zool Soc Lond 106A:945–955

Fox HM, Wingfield CA (1937) The activity and metabolism of poikilothermal animals in different latitudes. II Proc Zool Soc Lond 107A:275–282

Fry FEJ (1958) Temperature compensation. Ann Rev Physiol 20: 207–224

Fry FEJ, Hart JS (1948) The relation of temperature to oxygen consumption in goldfish. Biol Bull (Woods Hole) 94:66–77

Halcrow K, Boyd CH (1967) The oxygen consumption and swimming activity of the amphipod *Gammerus oceanicus* at different temperatures. Comp Biochem Physiol 23:233–242

Harvey HW (1955) The chemistry and fertility of sea waters. Cambridge University Press, Cambridge

Hazel JR, Prosser CL (1974) Molecular mechanisms of temperature adaptation in poikilotherms. Physiol Rev 54:620–677

Hemmingsen AM (1950) The realationship of standard (basal) energy metabolism to total fresh weight of living organisms. Rep Steno Mem Hosp Nord Insulin lab 4:7–58

Hemmingsen AM (1960) Energy metabolism as related to body size and respiratory surfaces, and its evolution. Rep Steno Mem Hosp Copenhagen 9:7–10

Hochachka PW, Moon TW, Mustafa T (1972) The adaptation of enzymes to pressure in abyssal midwater fishes. In: Sleigh MA, Mac Donald AG (eds) Symposium of the Society for Experimental Biology, 26, Cambridge University Press, Cambridge

Hochachka PW, Somero GN (1973) Strategies of biochemical adaptation. WB Saunders, Philadelphia

Holeton GF (1973) Respiration of Arctic char (*Salvelinus alpinus*) from a high lake. J. Fish Res Board can 30:717–723

Holeton GF (1974) Metabolic cold adaption of polar fish; fact or artefact? Physiol Zool 46:137–152

Johnston IA, Frearson N, Goldspink G (1973) The effects of environmental temperature on properties of myofibrillar adenosine triophosphatase from various species of fish. Biochem J. 133: 735–738

Johnston IA, Walesby NJ, Davidson W, Goldspink G (1975) Temperature adaption in myosin of Antarctic fish. Nature (Lond) 254: 74–75

Jones R (1976) Growth of fishes. In: Cushing DH, Walsh JJ (eds) The ecology of the seas, Blackwell, Oxford pp 251-279

Kanwisher J (1959) Histology and metabolism of frozen intertidal animals. Biol Bull (Woods Hole) 116:2580264

Kleiber M (1947) Body size and metabolic rate. Physiol Rev 27: 511–541

Kleiber M (1961) The fire of life. J Wiley & Sons, New York

Klekowski RZ, Opalinski KW, Rakussa-Suszczewski S (1973) Respiration of Antarctic amphipoda *Paramoera walkeri* Stebbing during the winter season. Pol Arch Hydrobiol 20:301–308

Krogh A (1916) The respiratory exchange of animals and man. Longmans Green, London

Kwon TW, Olcott HS (1965) Tuna muscle alsolase I. Purification and properties. Comp Biochem Physiol 15:7–16

Laird AK (1965) Dynamics of relative growth. Growth 29:249–263

Lawton JH, Richards J (1970) Comparability of Cartesian Diver, Gilson, Warburg and Winkler methods of measuring respiratory rates of aquatic invertebrates in ecological studies. Oecologia (Berl) 4:319–324

Lucas C, Kirkwood G, Somers I (1979) An assessment of the stocks of the banana prawn *Penaeus merguinsis* in the Gulf of Carpentaria. Aus J. Mar Freshwater Res 30:639–652

MacDonald AG (1975) Physiological aspects of deep sea biology. Physiological Society Monographs, 31, Cambridge University Press, Cambridge

Mauchline J, Fisher LR (1969) The biology of euphausiids. In: Russell FS, Yonge M (eds) Advances in Marine Biology. 7 pp 1–94

Maxwell JGH (1976) Aspects of the biology and ecology of selected Antarctic invertebrates. PhD Thesis, University of Aberdeen, Aberdeen

Maxwell JGH (1977) The breeding biology of *Chorismus antarcticus* (Pfeffer) and *Notocrangon antarcticus* (Pfeffer) (Crustacea, Decapod) and its bearing on the problems of the impoverished Antarctic decapod fauna. In: Llano G (ed) Adaptations within Antarctic Ecosystems. Gulf Publishing Co. Houston, pp 335–342

McArthur RH, Wilson EO (1967) The theory of island biogeography. Monographs in population biology. I. Princeton University Press, Princeton

McFarland WN, Pickens PE (1965) The effects of season, temperature and salinity on standard and active oxygen consumption of the grass shrimp, *Palaemonetes vulgaris* (Say). Can J Zool 43: 571–585

McLaren LA (1963) Effects of temperature on growth of zooplankton, and the adaptive value of vertical migration. J. Fish Res Board Ca 20:685–727

McWhinnie MA (1964) Temperature responses and tissue respiration in Antarctic crustacea with particular reference to the krill *Euphausia superba*. Antarct Ser (Am Geophys Un) 1:63–72

Mortensen T (1936) Echinoidea and Ophiuroidea. Discovery Rep 12:199–348

Newell RC, Bayne BL (1973) A review on temperature and metabolic acclimation in intertidal marine invertebrates. Neth J Sea Res 7:421–433

Paranjape MA (1967) Moulting and respiration of euphausiids. J Fish Res Board Can 24:1229–1240

Parry GD (1981) The meanings of r- and K-selection. Oecologia (Berl) 48:260–264

Pearcy WG, Small LF (1968) Effects of pressure on the respiration of vertically migrating crustaceans. J Fish Res Board Can 25:1311–1316

Pianka ER (1970) On r and K selection. Am Nat 104:592–597

Pianka ER (1972) r and K selection or b and d selection? Am Nat 106:581–587

Precht H (1958) Concepts of temperature adaptation of unchanging reaction systems of cold blooded animals. In: Prosser CL (ed) Physiological adaptation. Am Physiol Soc Wash

Prosser CL (1955) Physiological variation in animals. Biol Rev 30: 229–262

Prosser CL (1958) General summary: the nature of physiological adaptation. In: Prosser CL (ed) Physiological adaptation. Am Physiol Soc Wash

Ralph R, Everson Inigo (1968) The respiratory metabolism of some Antarctic fish. Comp Biochem Physiol 27:299–307

Ralph R, Maxwell JGH (1977a) The oxygen consumption of the Antarctic limpet *Nacella (Patinigera) concinna*. Br Antarct Surv Bull 45:19–23

Ralph R, Maxwell JGH (1977b) The oxygen consumption of the Antarctic lamellibranch *Gaimardia trapesina trapesina* in relation to cold adaptation in polar invertebrates. Br Antarct Surv Bull 45:41–46

Rao KP, Bullock TH (1954) Q_{10} as a function of size and habitat temperature in poikilotherms. Am Nat 88:33–44

Read KRH (1962) Respiration of the bivalve mollusts *Mytilius edulis* L. and *Brachiontes demissus plicatus* Lamarch a s a function of size and temperature. Comp Biochem Physiol 7:89–101

Remmert H (1980) Arctic animal ecology. Springer, Berlin Heidelberg New York

Rensch B (1959) Evolution above the species level. J Wiley & Sons, New York

Rubner M (1883) Über den Einfluß der Körpergröße auf Stoff- und Kraftwechsel. Z Biol 19:535–562

Saunders RL (1963) Respiration of the Atlantic cod. J Fish Res Board Can 20:373–386

Scholander PF, Flagg W, Walters V, Irving L (1953a) Climatic adaptation in the arctic and tropical poikilotherms. Physiol Zool. 26: 67–92

Scholander PF, Flagg W, Hogh RJ, Irving L (1953b) Studies on the physiology of frozen plants and animals in the Arctic. J Cell Comp Physiol 42:suppl 1:1–56

Sleigh MA, MacDonald AG (1972) The effects of pressure on organisms. Symp of the Soc for Experimental Biol 26, Cambridge University Press, Cambridge

Smith CL (1973a) Thermostability of some mitochondrial enzymes of lower vertebrates. I. general survey. Comp Biochem Physiol 44b:779–788

Smith CL (1973b) Thermostability of mitochondrial enzymes of lower vertebrates. II. Fresh water teleosts. Comp Biochem Physiol 44B:789–801

Somero GN (1969) Enzymatic mechanisms of temperature compensation: immediate and evolutionaly effects of temperature on enzymes of aquatic poikilotherms. Am Nat 103:517–530

Somero GN, Geise AC, Wholschlag DE (1968) Cold adaptation of the Antarctic fish *Trematomus bernacchii.* Comp Biochem Physiol 26:223–233

Somero GN, Hochachka PW (1968) The effects of temperature on catalytic and regulatory properties of pyruvate kinases of the rainbow trout and the Antarctic fish *Trematomus bernachii.* Biochem J 110:395–400

Somero GN, Hochachka PW (1971) Biochemical adaptation to the environment. Am,Zool 11:159–167

Southwood TRE, May RM, Hassell MP, Conway GR (1974) Ecological strategies and population parameters. Am Nat 108:791–804

Sparck R (1926) On the food problem in relation to marine zoogeography. In: Physiological papers: Dedicated to Professor August Krogh, Heineman, London, pp 268–283

Sparck R (1936) On the relation between metabolism and temperature in some marine lamellibranchs and its zoogeographic significance. K Danske Vidensk Selsk Skr, Biol Med 13 No 5, Copenhagen

Stearns SC (1976) Life-history tactics; a review of ideas. Q Rev Biol 51:3–47

Stearns SC (1977) The evolution of life history traits: a critique of the theory and a review of the data. Annu Rev Ecol Syst 8:145–171

Stubbs M (1977) Density dependence in the life-cycle of animals and its importance in K and r strategies. J Anim Ecol 46:677–688

Teal JM (1971) Pressure effects on the respiration and temperature on the respiration of Euphausiids. Deep-Sea Res 14:725–733

Teal TM, Carey FG (1967) Effects of pressure and temperature on the respiration of Euphausiids. Deep-Sea Res 14:725–733

Thompson RJ, Bayne BL (1972) Active metabolism associated with feeding in the mussel *Mytilus edulis* L. J Exp Mar Biol Ecol 9: 111–124

Thorson G (1936) The larval development, growth, and metabolism of Arctic marine bottom invertebrates. Medd Gronl 100:1–155

Thorson G (1946) Reproduction and larval development of Danish marine bottom invertebrates. Medd Komm Danm Fiskeri-og Havunders Ser Plankton 4:1–523

Thorson G (1950) Reproductive and larval ecology of marine bottom invertebrates. Biol Rev 25:1–45

Thorson G (1952) Zur jetzigen Lage der marinen Bodentierökologie. Verh Zool Ges 34:276–327

Thorson G (1961) Length of pelagic larval life in marine bottom invertebrates as related to larval transport by ocean currents. Am Assoc Adv Sci Publ 67:455–474

Torres JJ, Belman BW, Childress JJ (1979) Oxygen consumption rates of midwater fishes as a function of depth of occurrence. Deep-Sea Res 26A:185–197

Umminger BL (1977) Mechanisms of cold adaptation in polar marine animals. In: Llano G (ed) Adaptation within Antarctic ecosystems. Gulf Publishing Co, Houston, pp 397–409

White MG (1975) Oxygen consumption and nitrogen excretion by the giant isopod *Glyptonotus antarcticus* Eights in relation to cold-adapted metabolism in marine polar poikilotherms. In: Barnes H (ed) Proc 9th Europ Mar Biol Symp, Aberdeen University Press, pp 707–724

Widdows J (1973) The effects of temperature on the metabolism and activity of *Mytilus edulis.* Neth J Sea Res 7:387–398

Winberg GG (1956) Rate of metabolism and food requirements of fishes (original in Russian). J Fish Res Board Can Translation Series No 194

Wingfield CA (1939) The activity and metabolism of poikilothermal animals in different latitudes IV. Proc Zool Soc Lond 109A:103–108

Wohlschlag DE (1960) Metabolism of an Antarctic fish and the phenomenon of cold adaptation. Ecology 41:287–292

Wohlschlag DE (1963) An arctic fish with an unusually low metabolism. Ecology 44:557–564

Wohlschlag DE (1964) Respiratory metabolism and ecological characteristics of some fishes in McMurdo Sound, Antarctica. In: Lee MO (ed) Biology of the Antarctic Seas, Am Geophys Un, Washington, DC, Antarct Res Ser 1:33–62

Wyatt T (1976) Plants and animals of the sea. In: Cushing DH, Walsh JJ (eds) Ecology of the seas. Blackwell, Oxford, pp 59–80

Zeuthen E (1947) Body size and metabolic rate in the animal kingdom. CR Trav Lab Carlsberg Ser Chim 26:17–161

Zeuthen E (1953) Oxygen uptake as related to body size in organisms. Q Rev Biol 28:1–12

Food Webs and Functional Aspects of the Kelp, *Macrocystis pyrifera*, Community in the Beagle Channel, Chile

J.C. CASTILLA[1]

Summary. Food webs were studied in coastal belts of *Macrocystis pyrifera* in the Beagle Channel, mainly at Puerto Toro (55° 05′S, 67° 04′W), by SCUBA diving during 6 field trips between August 1979 and December 1980. I report a general connectedness web, showing trophic relationships among macro invertebrates and algae. The web is capped by the trophic generalist starfish, *Cosmasterias lurida.* Four species of sea-urchins are present in the community: *Loxechinus albus, Pseudechinus magellanicus, Arbacia dufresnei* and *Austrocidaris canaliculata.* Experimental manipulation of the 2 most abundant species, *L. albus* and *P. magellanicus,* demonstrated that they did not over-exploit live plants of *M. pyrifera;* instead they used as their main food item the abundant subtidal algal material drifting in the area. Since I found no high-trophic-level keystone species in the community, I discuss certain density independent factors that could control sea-urchin populations. I predict that, even within a restricted geographical area, different degrees of grazer-plant (sea-urchins – *M. pyrifera*) interactions can be found within the same species assemblage. I suggest that in kelp communities the sea-urchin – live *Macrocystis* ecological interactions are changeable and cannot be determined *a priori.*

1 Introduction

Macrocystis pyrifera kelp communities have attracted the attention of numerous investigators in different parts of the world. Such communities are present both in the northern and southern hemispheres (Neushul 1971; North 1971; Barrales and Lobban 1975) and embody abundant and diverse assemblages of species. Dayton et al. (1973), after visiting southern Chilean *Macrocystis* beds (Corcovado Gulf) and making preliminary and qualitative observations, stated that food webs in these beds were marked by an extreme simplicity unobserved in other kelp communities. Furthermore, Dayton (1974) suggested that in southern Chile the *Macrocystis* kelp community has apparently evolved in the absence of key predators which control population increases of herbivorous invertebrates such as sea urchins. It is puzzling how this community could have evolved under such conditions, particularly when it is recalled that Dayton et al. (1973) and Dayton (1974) mentioned that sea-urchins, mainly *Loxechinus albus,* over-exploit Chilean *Macrocystis* beds.

I have analyzed trophic webs in coastal belts of *M. pyrifera* in the most southern region of Chile. Based on the results of these analyses, and previous studies in the same localities (Castilla and Moreno 1982), I present here information on aspects of the functioning of these communities. Unless stated otherwise, the trophic study refers exclusively to *M. pyrifera* non-holdfast assemblages of species. The species associated with holdfasts have been described by Ojeda and Santelices (1980). Information on the fish community is given in Moreno and Jara (1981).

2 Materials and Methods

The study area was the Beagle Channel, south of Tierra del Fuego Island. The main site (Fig. 1) was Puerto Toro (Picton Channel) on Navarino Island (55° 05′S, 67° 04′W). This site was visited six times during November 1979 to December 1980. Figure 1 also shows three other sites visited during April 1980. The study was conducted in and around coastal belts of *M. pyrifera* (Fig. 2), defined by Castilla and Moreno (1982). A total of 13 submarine transects (each 100 m^2) was surveyed, using methods described by Castilla and Moreno (1982). Food items of the major species were identified by means of: direct underwater observation (SCUBA); individual collections of sea-urchins, and analysis of their stomach contents in the laboratory (see Castilla and Moreno 1982); and information in the literature. The diving was carried out by a team of four divers, completing about 170 h of diurnal and 6 h of nocturnal work at depths between 2 m and 25 m. Surface-water temperatures ranged between 4° and 9°C.

3 Results

3.1 Generalized Trophic Web

Figure 3 shows a generalized connectedness trophic web which includes 34 species or taxonomic groups of marine invertebrates and algae. In terms of the number of connections, the starfish, *Cosmasterias lurida,* was the most not-

1 Ecología Marina, Facultad de Ciencias, Biológicas, Pontificia Universidad Católica de Chile, Casilla 114-D, Santiago, Chile

Antarctic Nutrient Cycles and Food Webs
(ed. by W.R. Siegfried, P.R. Condy, and R.M. Laws)

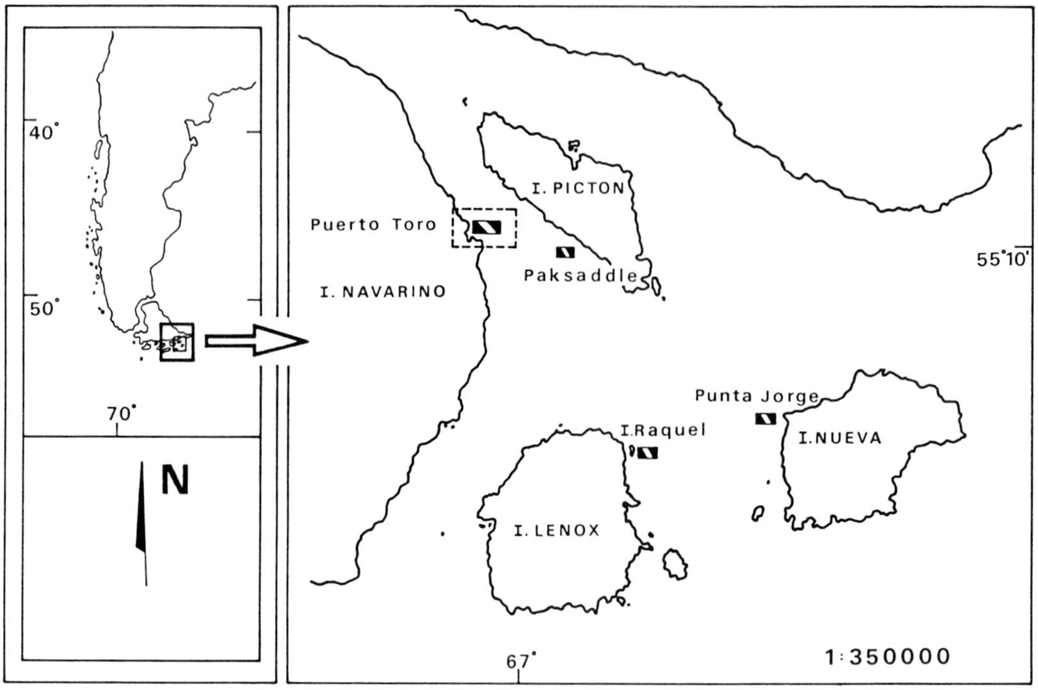

Fig. 1. Study sites in southern Chile

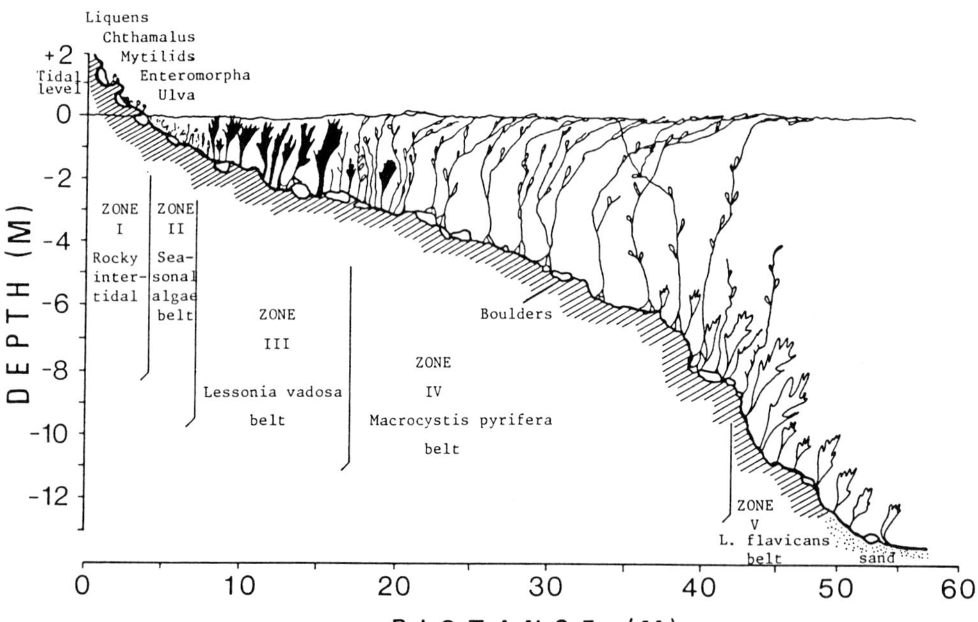

Fig. 2. Schematic representation of coastal belts of *Macrocystis pyrifera* and adjacent zones at Puerto Toro, Picton Channel, Chile

able consumer in the 3rd trophic level. Table 1 shows the number and percentage frequency of 25 food items consumed by *C. lurida*. It can be seen that, with the exception of *Balanus* sp. (27.3%) and *Crepipatella dilatata* (19.7%), most of the food items do not reach the 5% frequency level in the diet of *C. lurida* which suggests that this high-level predator is a trophic generalist. Indeed, a parallel study (Vàsquez and Castilla 1980) showed that prey consumption in *C. lurida* was correlated directly with the frequency of prey in the environment. During the study, 249 specimens of *C. lurida* were collected. Mean density was 0.097 starfish m^2. Only two starfish showed evidence of arm regeneration, and no predator of the species was found. The size of starfish within the kelp belts ranged between 33 and 208 mm.

Table 2 shows a group of 15 species of invertebrates and vertebrates present in or around the kelp belts, which can also be considered as predators in the 3rd trophic level

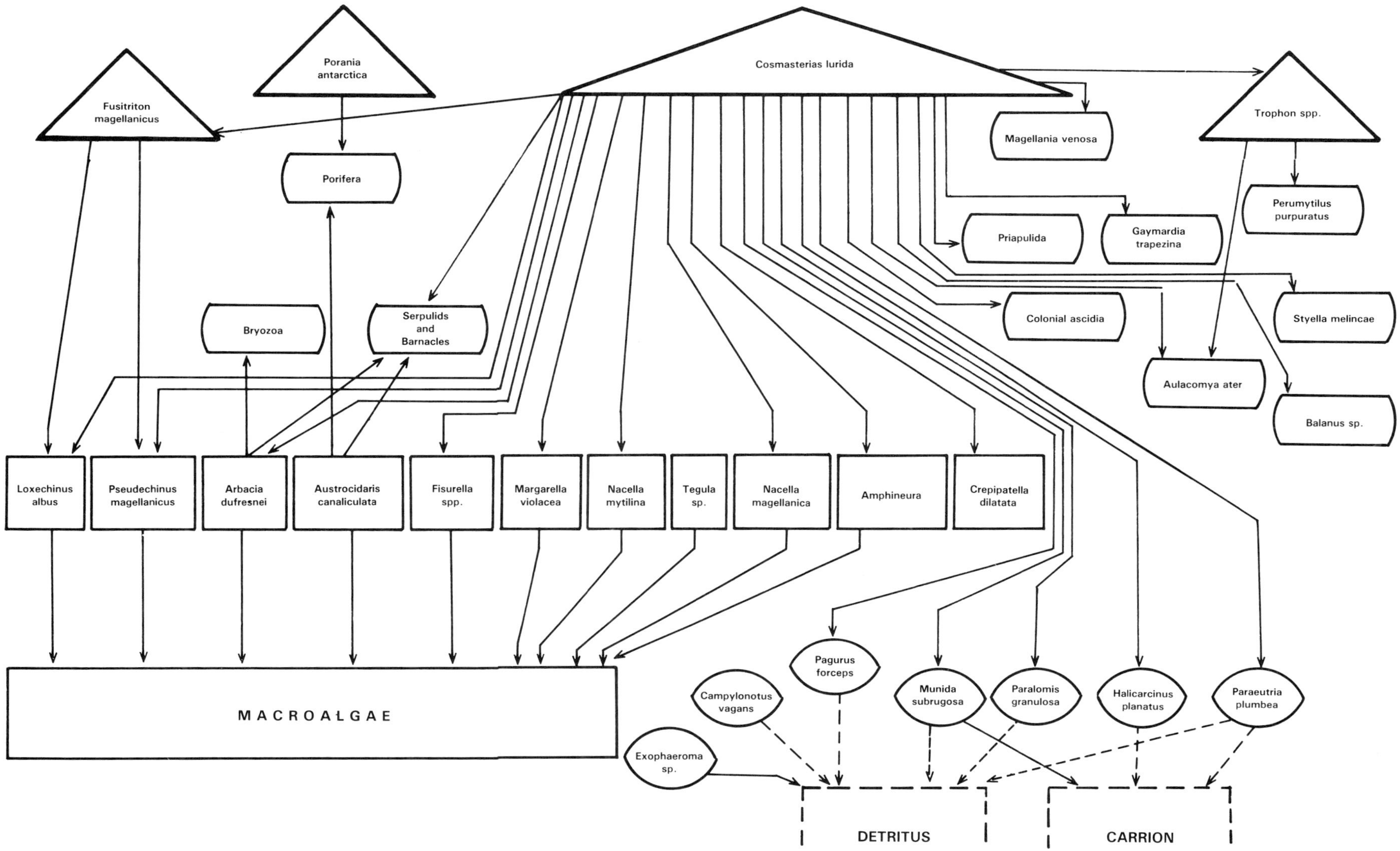

Fig. 3. A generalized trophic web showing connectedness in southern Chile *Macrocystis pyrifera* coastal belts. *Arrows* point towards prey

Table 1. Frequency of occurrence of prey of *Cosmasterias lurida* observed feeding in a *Macrocystis pyrifera* kelp community in the Beagle Channel, Chile

Prey items	Frequency	
	No.	%
Fusitriton magellanicus	1	0.76
Fissurella spp.	5	3.79
Nacella mytilina	2	1.52
Crepipatella dilatata	26	19.70
Margarella violacea	6	4.54
Thophon spp.	4	3.03
Paraeuthria plumbea	3	2.27
Aulacomya ater	10	7.57
Gaymardia trapezina	2	1.52
Amphineura	7	5.30
Nacella magellanica	2	1.52
Halicarcinus planatus	1	0.76
Munida subrugosa	5	3.79
Balanus sp.	36	27.27
Paralomis granulosa	2	1.52
Pagarus forceps	3	2.27
Styela melicae	3	2.27
"White Colonial ascidia"	1	0.76
Magellanica (Neothyris) venosa	2	1.52
Priapulido	1	0.76
Loxechinus albus	2	1.52
Pseudechinus magellanicus	1	0.76
Arbacia dufresnei	1	0.76
Notothenia sp.	2	1.52
Carrion (rest of *L. albus* and *Lithodes antarctica*)	4	3.03

(only the starfish *Porania antarctica* is included in the web shown in Fig. 3). According to Table 2, the density of these predators in the study area was extremely low, in comparison with *C. lurida.*

The fish fauna calls for a special comment. There were nine species present: *Cottoperca gobio, Champsocephalus esox* and seven species of *Notothenia.* The first two prey on *Notothenia* spp., and the Notothenidae are mostly bottom detritivore and small crustacean feeders (Moreno and Jara 1982). None of them plays the carnivorous role of, for instance, *Semicossiphus pulchrum* in Californian kelp beds, or shows direct interactions with live *Macrocystis* plants.

Cosmasterias lurida preys on a group of 2nd-trophic-level carnivorous snails, *Trophon* spp. and *Fusitriton magellanicus.* The former species are characteristic of intertidal and shallow subtidal zones (II and III, see Fig. 2) and prey on the mussels, *Perumytilus purpuratus* and *Aulacomya ater,* and the barnacle, *Chthamalus* sp., *Fusitriton magellanicus* is typical of the landward border of *Macrocystis* belts. Table 3 shows the frequency of prey items consumed by these species; out of 79 snails observed (daytime dives), only 19 were seen eating. The sea-urchins, *Loxechinus albus* and *Pseudechinus magellanicus,* were the only prey items taken by *F. magellanicus.* This predator attained a maximum size of about 110 mm, and mean densities (4 transects) ranged between zero and 0.10 snails m^2. Individuals gather in crevices and prey on sea-urchins by boring a hole through the peristomial membrane, using the proboscis. During the study, four attacks of *F. magellanicus* on *L. albus* and three on *P. magellanicus* were monitored. In each case, the size (diameter) of the prey was similar to the length of the snail, and the snails spent between 23 and 36 h to consume their prey.

Herbivorous 1st-trophic-level consumers are diverse in southern Chile *Macrocystis* coastal belts. The web (Fig. 3) shows four species of sea urchins: *L. albus, P. magellanicus, Austrocidaris canaliculata, Arbacia dufresnei,* and several species of gastropod molluscs and Amphineura. They graze mainly, but not exclusively, on several species of macro-algae, among which *M. pyrifera, Lessonia vadosa* and *L. flavicans* should be noted.

The trophic web (Fig. 3) also contains a number of filter-feeder species belonging to diverse phyla, from Porifera to Chordata. All of them play important roles in the food web through trophic connections with representatives of all three trophic levels. Particularly important in these kelp belts appears to be the role of species which eat drifting benthic algae, and detritus feeders, such as the isopods *Exosphaeroma gigas* and *Cassidinopsis emarginata,* several species of amphipods, crabs, such as *Halicarcinus planatus,* and other groups. Only a few species are represented in Fig. 3, since knowledge about their trophic roles is rather poor. The amount of drifting algae is extremely high on the bottoms of the kelp belts, and the recycling processes assumed for the organisms mentioned above could be highly significant for energy flow in the system.

3.2 Experimental Manipulations

Castilla and Moreno (1982) performed a set of manipulations whereby a series of exclusion and addition experiments involving sea-urchins demonstrate no significant differences in recruitment or persistence of adult plants of *M. pyrifera* and *Lessonia* spp., when compared with control sites. *Loxechinus albus* and *P. magellanicus* actively use drifting algae, mainly *M. pyrifera* and *Lessonia* spp., for both cover and food. For instance, 69.9% of all individuals of *L. albus* fed during the daytime on drifting algae (mostly stipes and fronds of *Macrocystis*), 7.1% on plants of *Lessonia* spp. and 18.5% on other species of benthic macro-algae. Similar results were obtained for populations of *P. magellanicus* not associated with kelp holdfasts.

Based on the above results, I have introduced functional aspects in the 1st consumer – primary producer level of the trophic web. They are shown in Fig. 4. by linking *L. albus* and *P. magellanicus* to drifting algae. Furthermore, based on the number of grazer attacks on algae and detachment of *M. pyrifera* plants due to holdfast-boring organisms, I have represented in Fig. 4 two additional categories of func-

Table 2. Mean density (no. animals 100 m^{-2}) of high-trophic-level predators in a *Macrocystis pyrifera* kelp community in the Beagle Channel, Chile

Species	Density	Main prey items	Source
Nimertea			
Linneus sp.	?	?	
Echinodermata			
Porania antarctica	0.015	Sponges	This study
Odontaster validus	0.014	(?) omnivorous	Dayton et al. (1974)
Asterodon singularis	0.007	?	
Patiriella fimbriata	0.007	?	
Cycetra verrucosa	0.015	?	
Labidiaster annulatus	0.007	(?) carnivorous	Dearborn (1977)
Mollusca			
Octopus sp.	0.054	?	This study
Arthropoda			
Pycnogonida	?	hydroids	This study
Lithodes antarctica	0.054	*Magellanica venosa*	Stuardo and Solis (1963)
Osteichthyes			
Cottoperca gobio	?	Nototheniidae and *Munida*	Moreno and Jara (1981)
Champsocephalus esox	?	Nototheniidae and *Munida*	Moreno and Jara (1981)
Aves			
Tachyeres pleneres	0.001	*Fisurella* spp. Amphineura *Nacella magellanica* *Aulacomya ater* *Mytilus chilensis* *Munida subrugosa* *Paralomis granulosa*	This study; Humphrey et al. (1970)
Mammalia			
Lutra felina	0.007	Fishes and invertebrates	Castilla and Bahamondes (1979)
Arctocephalus australis	0.031	?	This study
Otaria flavescens	0.060	fishes	Aguayo and Maturana (1973)

Table 3. Prey taken by two second-leveltrophic consumers in a *Macrocystis pyrifera* kelp community in the Beagle Channel, Chile

Consumers	Prey species	No. prey animals eaten	$\bar{x}$ size ± S.E. *Fusitriton* (mm)	Prey (mm)
Fusitriton magellanicus	*Loxechinus albus*	13	79.0 ± 16.0	68 ± 10.0
	Pseudechinus magellanicus	6	62.0 ± 26.0	17 ± 3.0
Trophon spp.	*Perumytilus purpuratus*	5		
	Aulacomya ater	4		
	Chthamalus sp.	11		

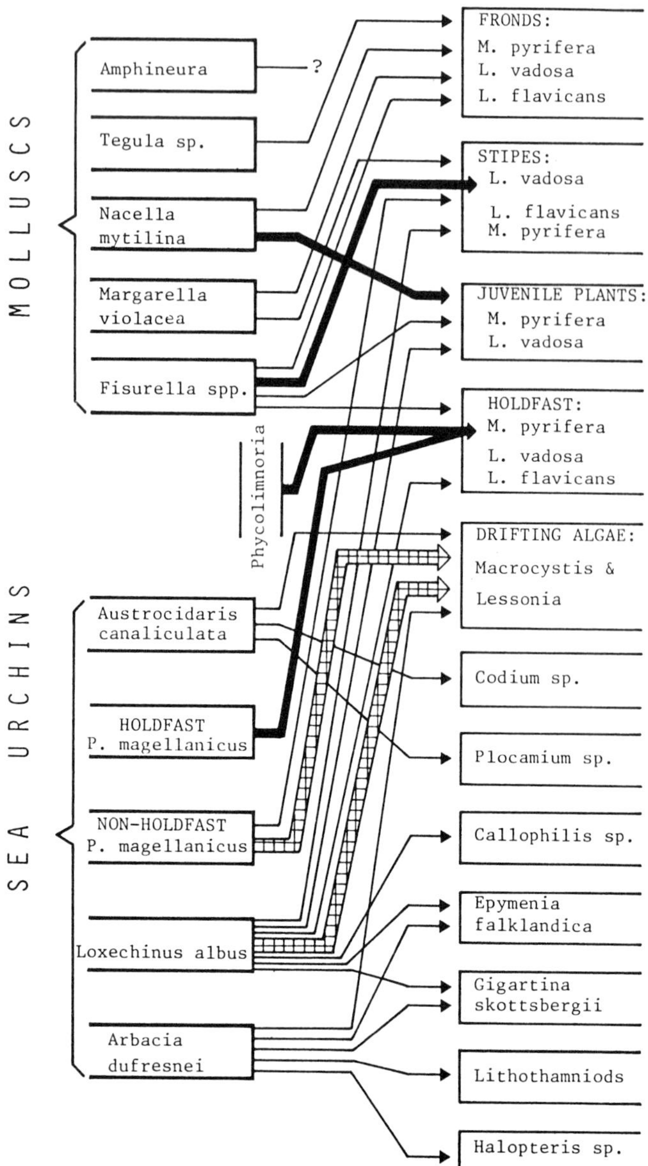

Fig. 4. Trophic links between first-level consumers (grazers) and macro-algae. *Arrows* point towards prey; moderate and weak interactions are represented by *broad solid lines* and *thinner solid lines*, respectively. *Hatched lines* link *Loxechinus albus* and *Pseudechinus magellanicus* with drifting algae (see text)

tional relationships (i.e., weak and moderate interactions, respectively). I suggest that *Nacella mytilina, Fissurella* spp. and particularly *Phycolimnoria* sp., which bores in *M. pyrifera* holdfasts, could show such functional interactions. I found various degrees of *Phycolimnoria* infestation, the most severe being in detached "floating" plants of *M. pyrifera.*

4 Discussion

Castilla and Moreno (1982) suggested that there is no single, efficient sea-urchin predator in the *M. pyrifera* kelp communities of southern Chile. The present study confirms this and shows that the community lacks a high-trophic-level keystone predator. Keystone species are strongly interacting species (Paine 1980) which play major roles in community organization (see Paine 1969, 1971, 1974; Simenstead et al. 1978; Estes and Palmisano 1974; Power and Gregoire 1978; Castilla 1980). These roles affect control of the biological pattern and the structure of a community, via, for instance, preferential consumption of a competitive, dominant species (Paine 1971) or by predation on herbivorous species, such as sea-urchins in *Macrocystis* communities, which in turn show strong plant-herbivore links (Palmisano and Estes 1977; Estes et al. 1978). The latter process has been reported for *M. pyrifera* beds in California (North 1965; Leighton et al. 1966; Pearse and Hines 1979).

The carnivorous starfish *C. lurida* (the predator exploiting the greatest trophic spectrum in Chilean kelp communities) cannot be regarded as a keystone species. Nor do any of the other high-trophic-level carnivores in the community play such an ecological role. It is not yet known whether the total constellation of high-trophic-level carnivores could be responsible for such a role, as was suggested by Castilla (1980). However, I will assume here that this is not the case, since only one member (*F. magellanicus*) of this category is a specialized feeder, although not a very efficient one. The density of the remaining carnivores was extremely reduced in the study area, and their diets were very diverse.

On the other hand, Castilla and Moreno (1982) have shown that sea-urchins (mainly *L. albus*) do not display a strong "live *Macrocystis*-herbivore" interaction in kelp communities in the Beagle Channel. Moreover, I propose here that this holds for all other single herbivorous species present in these communities. although confirming experiments are still lacking. This contradicts the opinion of Dayton et al. (1973) and Dayton (1974) who reported that *L. albus* is the species that most efficiently over-exploits *M. pyrifera* beds in southern Chile. In effect, my proposal is in agreement with the findings of Moreno and Sutherland (1982). Indeed, the trophic web presented here and the findings of Castilla and Moreno (1982) are in tune with prevailing ecological theory. In the absence of keystone carnivores in *Macrocystis* communities, one should predict a lack of, or a significant reduction in, interactions between live *Macrocystis* plants and sea-urchins, if the *Macrocystis* community is to persist in time.

Loxechinus albus and the non-holdfast populations of *P. magellanicus* have morphological (i.e. long tube-feet) and behavioural (i.e. covering feeding behaviour) adaptations, allowing them to utilize an abundant, drifting, macroalgal resource (Castilla and Moreno 1982). This reduces the attacks on live *M. pyrifera* plants. According to Yelling et al. (1977), *M. pyrifera* canopies ("forests") along sheltered of the Californian coast support very low densities (0–0.3 animals m^2) of the sea-urchin *Stronylocentrotus franciscanus* within them, but dense populations (29.9–49.5 animals m^2) of this species occur along the seaward borders of the kelp forests. This was correlated with large amounts of drifting kelp found along the seaward edge of the forests (Mattison et al. 1977), and it was suggested that perhaps the stimulus for sea-urchins to attack whole live plants is removed by large amounts of drift-kelp.

A crucial question is: how are sea-urchin populations controlled either in southern Chilean kelp belts, where keystone species are absent, or in Californian beds where keystone predators, such as *Enhidra lutris,* are rare or nonexistent (e.g., at Santa Catalina Island)? Apart from density-dependent controlling factors acting on sea-urchins, the following factors should be considered. (1) Habitat availability: for instance, the type or area of rock which can be bored by Californian strongylocentrotid sea-urchins, or the number of rocky crevices (see Lowry and Pearse 1973). No southern Chilean species of sea-urchin bores into rocks, but they do actively seek crevices. (2) The highly variable size structure of sea-urchin populations in space and time due to unpredictable larval recruitment. (3) Bottom anoxia conditions in the kelp belts, due to excessive production of drifting *Macrocystis* under low temperature conditions, could prevent or reduce the success of invertebrate recruitment.

Finally, the findings reported here permit at least two important predictions. (1) *Macrocystis pyrifera* communities can persist and evolve in the absence of keystone carnivorous species. Changes in the foraging behaviour of grazers (i.e., sea-urchins) can lead to the use of drifting algal resources, instead of live *Macrocystis* plants. This is perhaps triggered by a high production of drifting kelp material, which appears to be a prerequisite for the process. (2) Given elastic foraging behaviour (as in sea-urchins), it is natural to find, even within a restricted geographical area, different degrees of grazer-plant interactions within the same species assemblage in the *Macrocystis* community. Hence, it could be said that the quantity and even the quality of the ecological interaction (*sensu* Paine 1980) between sea-urchins and *Macrocystis* is changeable and cannot be determined a priori. The evidence available shows that both the ecological role of key-predators and the mechanism proposed here are present in *Macrocystis* communities in the northern hemisphere. In Navarino, Chile, however, the former is greatly diminished and the latter enhanced.

Acknowledgements. This paper is part of a comprehensive study of the biology and utilization of *Macrocystis pyrifera* in southern Chile. The work was financed by a research grant negotiated between the Armada de Chile and the Pontificia Universidad Catolica de Chile. I sincerely thank my colleages: J. Vàsquez, A. Larrea, A. Jullian, P. Ojeda, C. Moreno, F. Jara and B. Santelices. Linda Nelson and Robert Paine made useful comments on the original text. I appreciate suggestions made by an anonymous referee. I thank the Instituto Antàrctico de Chile for making possible my participation in the Fourth SCAR Antarctic Biology Symposium in South Africa.

References

Aguayo A, Maturana R (1973) Presencia del lobo marino comùn (*Otaria flavescens*) en el litoral chileno. Biol Pesq, Chile 6:45–75

Barrales HL, Lobban CS (1975) The comparative ecology of *Macrocystis pyrifera,* with emphasis on the forests of Chubut, Argentina. J Ecol 63:657–677

Castilla JC (1980) Perspectivas de investigacion en estructura y dinàmica de comunidades intermareales rocosas de Chile Central. Il Depredadores de alto nivel trófico. Medio Ambiente 5(1–2):190–215

Castilla JC, Bahamondes I (1979) Observactiones conductuales y ecológicas sobre *Lutra felina* (Molina) 1782 (Carnivora: Mustelidae) en las zonas Central y Centro-Norte de Chile. Arch Biol Med Ex 12:119–132

Castilla JC, Moreno C (1982) Sea urchins and *Macrocystis pyrifera:* Experimental test of their ecological relations in southern Chile. International Echinoderm Conference, Tampa Bay. Lawrence JM (ed) (AA Balkema, Rotterdam) pp 257–263

Dayton PK (1974) Kelp communities of southern South America. Antarct J US 9(1):22–23

Dayton PK, Rosenthal RJ, Mahan LC (1973) Kelp community in the Chilean archipelago: R/V Hero Cruise 72-5. Antarct J US 8(2):34–35

Dayton PK, Robilliand GA, Paine RT, Dayton LB (1974) Biological accomodation in the benthic community at McMurdo Sound, Antarctica. Ecol Monogr 44(1):105–128

Dayton PK, Rosenthal RJ, Mahan LC, Antezana T (1977) Population structure and foraging biology of the predaceous Chilean asteroid *Meyenaster gelatinosus* and the escape biology of its prey. Mar Biol 39:361–370

Deaborn JH (1977) Food and feeding characteristics of antarctic asteroids and ophiuroids. In: Llanos GA (ed) "Proceedings of the 3rd SCAR Symposium on Antarctic Biology: Adaptations within Antarctic Ecosystems" pp 293–326

Estes JA, Palmisano JF (1974) Sea otters: their role in structuring nearshore communities. Science (Wash DC) 185:1058–1060

Estes JA, Smith NS, Palmisano JF (1978) Sea otter predation and community organization in the western Aleutian Islands, Alaska. Ecology 59(4):822–833

Humprey PS, Bridge D, Reynolds PW, Peterson RT (1970) Birds of Isla Grande (Tierra de Fuego). Preliminary Smithsonian Manual University of Kansas, Museum of Natural History 411 pp

Leighton DL, Jones LG, North WJ (1966) Ecological relationship between giant kelp and sea urchins in Southern California. 5th International Symposium on Seaweeds pp 141–153

Lowry LF, Pearse JS (1973) Abalones and sea urchins in an area inhabited by sea otters. Mar Biol 23:213–219

Mattison JE, Trent JD, Shanks AL, Akin TB, Pearse JS (1977) Movement and feeding activity of the red sea urchins (*Strongylocentrotus franciscanus* adjacent to a kelp forest. Mar Biol 39:25–30

Moreno C, Jara F (1981) Relaciones entre los peces y *Macrocystis pyrifera* en cinturones costeros de islas Sur-Fueginas, Chile. In: Informe Final, Biología y Factibilidad de Utilización de Praderas de *Macrocystis pyrifera* en Puerto Toro, Isla Navarino, Chile. Prep by Santelices B pp 440–482

Moreno C, Sutherland J (1982) Physical and biological processes in a *Macrocystis pyrifera* community near Valdivia, Chile. Oecologia 55:1–6

Neushul M (1971) The species of *Macrocystis.* In: North WJ (ed) The Biology of giant kelp beds (*Macrocystis*) in California. Nova Hedwigia 32:211–222

North WJ (1965) Urchin predation. In: North WJ (ed) Kelp habitats improvement project, Annual Report. 1964–1965, California Institute of Technology, Pasadena California pp 95–108

North WJ (1971) Introduction and background. In: North WJ (ed) The biology of giant kelp beds (*Macrocystis*) in California. Nova Hedwigia 32:1–97

Ojeda P, Santelices B (1980) Comunidades de macro-invertebrados en discos de *Macrocystis* en Puerto Toro, Provincia Antàrctica,

Chile. In: Informe Final, Biología y Factibilidad de Utilización de Praderas de *Macrocystis pyrifera* en Puerto Toro, Isla Navarino, Chile. Prep by Santelices B, pp 193–236

Paine RT (1969) The *Pisaster–Tegula* interaction: prey patches predator food preference and ecology intertidal community structure. Ecology 50:950–961

Paine RT (1971) A short-term experimental investigation of resource partitioning in a New Zealand rocky intertidal habitat. Ecology 52:1096–1106

Paine RT (1974) Intertidal community structure. Experimental studies on the relationship between a dominant competitor and its principal predator. Oecologia 15:93–120

Paine RT (1980) Food Webs: Linkage, interaction strength and community infrastructure. J Anim Ecol 49:667–685

Palmisano JF, Estes JE (1977) Sea otters: pillars of the nearshore community. Northwest and Alaska, Fisheries Center Processed Report US Department of Commerce NOAA pp 1–7

Pearse JS, Hines AH (1979) Expansion of a Central California Kelp forest following the mass mortality of Sea Urchins. Mar Biol 51: 83–91

Power G, Georgoire J (1978) Predation by freshwater seals on the fish community of Lower Seal Lake, Quebec. J Fish Res Board Can 35:844–850

Simenstad CA, Estes JA, Kenyon KW (1978) Aleuts, Sea Otters, and alternate stable-state communities. Science (Wash DC) 200:403 –411

Stuardo J, Solis I (1963) Biometría y observationes generales sobre la biología de *Lithodes antarctica* Jacquinot. Gayana (11):1–49

Vàsquez J, Castilla JC (1980) Estructura poblacional, dieta y preferencia alimentaria de *Cosmasterias lurida* (Asteroidea, Asteriinae) en Puerto Toro, Isla Navarino. In: Informe Final. Biología y Factibilidad de Utilización de Praderas de *Macrocystis pyrifera* en Puerto Toro, Isla Navarino, Chile. Prep by Santelices B pp 414 –439

Yelling MB, Agegian CR, Pearse JS (1977) Ecological benchmarks in the Santa Cruz County Kelp forests the re-establishment of sea otters. Center for Coastal Marine Studies. University of California, Santa Cruz. Techn Rep (6):1–125

Aspects of the Roles of Squid in Food Chains of Marine Antarctic Ecosystems

T. NEMOTO[1], M. OKIYAMA[1], and M. TAKAHASHI[2]

Summary. Squid (Cephalopoda) are important organisms in marine food chains. However, there are very few studies of their ecological roles in marine ecosystems of the Antarctic. Seventy-six specimens of squid, both large (*ca* 355 mm in mantle length) and small (*ca* 40 mm), have been collected by large mid-water trawl operations of the Japanese pilot fishery for Antarctic Krill, *Euphausia superba,* in the Antarctic during 1980–81 and 1981–82. The squid collected are all considered to be young juveniles distributed at shallower depths than adults. Two species, *Kondakovia longimana* and *Moroteuthis knipovitchi,* are dominant. *Alluroteuthis antarcticus, Pholidoteuthis boschmai, Brachioteuthis picta* and *Galiteuthis glacialis* also occurred in the collections. These squid consume various kinds of food, but they feed mainly on macro-zooplankton and micro-nekton. Krill is the main food item of squid in the Antarctic. An amphipod, *Themisto gaudichaudii,* is also a major component of the food of *Kondakovia longimana.* Small euphausiids, *Thysanoessa macrura,* and large chaetognaths, *Sagitta gazellae,* are also found in the stomachs of *K. longimana,* which indicates that the species is a plankton feeder. Fish are considered to be a major food item for *Moroteuthis knipovitchi,* together with *E. superba.* Cannibalism appears to be rather common in these squid. Predation by deep-diving Sperm Whales, *Physeter catodon,* on squid larger than 400 mm in mantle length was observed, and this, associated with the fact that such large squid were not caught in mid-water trawls, suggests that squid are segregated into different water depths according to size and maturity in the Antarctic.

1 Introduction

It is clear that squid (Cephalopoda) play a key role in the marine ecosystems of the Antarctic. They are significant predators of smaller organisms, including Antarctic Krill, *Euphausia superba,* and in turn act as prey for large Sperm Whales, *Physeter catodon,* and seals (Laws 1977). However, their ecological characteristics rarely have been examined, except for the recent analysis of "beaks" found in the stomachs of Sperm Whales (Clarke 1980b) and descriptions of specimens found in large trawls and nets (Filippova 1972; McSweeny 1978). The main reason for the paucity of squid studies probably stems from the animals' strong swimming ability which enables them to avoid being caught by ordinary sampling nets. According to Clarke (1977), no net currently available can effectively sample any of the species of squid forming the main food of Sperm Whales. The SCOR (Scientific Committee on Oceanic Research) Working Group 52, which reviewed the techniques used to assess micro-nekton abundance, reported in 1980, that the methods of sampling micro-nektonic squid are still inadequate and require further study.

In the Japanese BIOMASS (Biological Investigations of Marine Antarctic Systems and Stocks) programme, relatively large micro-nektonic squid have been collected using large trawling nets. In this paper, we describe the species composition, and the food and feeding of these squid, and compare the results with those obtained through an analysis of squid in the stomachs of Sperm Whales in the Antarctic.

2 Materials and Methods

During the summers of 1980–81 and 1981–82, the R.V. Yoshino-Maru, of the Japan Marine Fisheries Resources Research Centre, carried out mid-water trawling for Krill swarms in Antarctic waters between 90°W and 50°W north of the South Shetland Islands. The trawl employed had a large mouth with an opening of 615.4 m^2, a length of 62 m and a 26 m cod-end. The inner mesh of the anterior part of the net was 20 mm, and that of the posterior part 13 mm. The sampling depth was between 10 m and 200 m. Specimens of squid were frozen on board at once, thawed in the laboratory and fixed with 10% formalin in seawater. All examinations were made after fixation with formalin. The data on squid in the stomachs of Sperm Whales were obtained by Japanese whaling operations during the 1951–52 whaling season in the Antarctic.

3 Results

Six species (one is unknown) of squid were identified from 76 specimens (Table 1). *Kondakovia longimana* and *Moroteuthis knipovitchi* accounted for more than 80% of the specimens examined. *Galiteuthis glacialis, Brachioteuthis*

1 Ocean Research Institute, University of Tokyo, Nakano, Tokyo 164, Japan
2 Japan Marine Fisheries Resources Research Centre, Chiyoda, Tokyo 102, Japan

Antarctic Nutrient Cycles and Food Webs
(ed. by W.R. Siegfried, P.R. Condy, and R.M. Laws)

Table 1. Species and number of specimens of cephalopods collected in the Antarctic by trawling operations during the summers of 1980–81 and 1981–82

Species	1980–81	1981–82	Total
Alluroteuthis antarcticus	–	3	3
Moroteuthis knipovitchi	5	8	13
Kondakovia longimana	20	31	51
Pholidoteuthis boschmai	–	1	1
Brachioteuthis picta	–	3	3
Galiteuthis glacialis	–	4	4
Unknown	–	1	1
Totals	25	51	76

picta, Pholidoteuthis boschmai and *Alluroteuthis antarcticus* also occurred, but in smaller numbers.

The relationship between body mantle length and body weight is shown in Fig. 1. *Kondakovia longimana* and *M. knipovitchi,* belonging to the family Onychoteuthidae, form related allometric lines but *A. antarcticus* and *G. glacialis,* which belong to different families, form quite separate lines. The maximum mantle length of *K. longimana* and *M. knipovitchi* is 355 mm (Table 2); a value far greater than those of squid generally caught by smaller nets (Roper 1977). The mode for the mantle length of *K. longimana* occurs at 200 mm, which is much greater than comparable values (generally < 100 mm) given by Roper (1977) for squid caught with RMT and IKMT nets.

Clarke (1980b) reported the size distribution of the lower rostral length of *K. longimana* consumed by Sperm Whales in the Antarctic. This size distribution can be converted into mantle length distribution, using the regression line given by Clarke (1980b, Fig. 55). The relationship between lower rostral length and mantle length of *M. knipovitchi* can also be calculated as: y (mantle length) = 87.35 × (lower rostral length) –27.81, where the correlation coefficient is r = 0.880, using the data in Clarke (1980b, Table 24). The mantle length distribution of the squid caught in trawls is clearly different from that estimated from beaks obtained from the stomach contents of Sperm Whales (Fig. 2). Since the trawl was operated in the upper water column, from 10 m down to about 200 m, it is concluded that the larger *K. longimana* eaten by Sperm Whales probably occur in deeper waters.

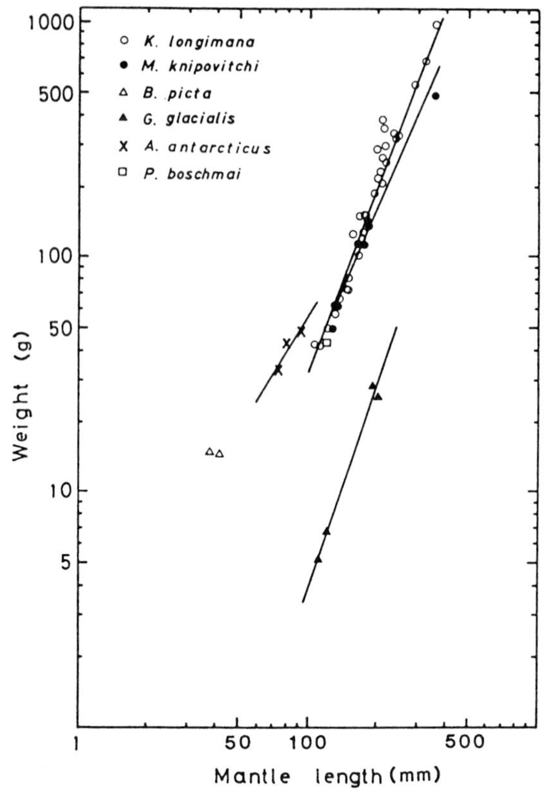

Fig. 1. Relationship between body length and weight of squid

The largest specimens of *M. knipovitchi* observed in the stomachs of Sperm Whales taken near the South Shetland Islands were > 400 mm in mantle length (Clarke 1980b), but the largest specimen of *M. knipovitchi* in the trawl collections is 355 mm and the other 12 specimens are <300 mm in mantle length.

Table 2. Size of squid collected in the Antarctic by trawling operations during the summers of 1980–81 and 1981–82

Species	No. of specimens measured	Range of mantle length (mm)	Average mantle length (mm)
K. longimana	49	105–355	253
M. knipovitchi	13	135–355	213
G. glacialis	4	110–200	167
A. antarcticus	3	73– 92	81
B. picta	2	41 and 56	49
P. boschmai	1	114	–

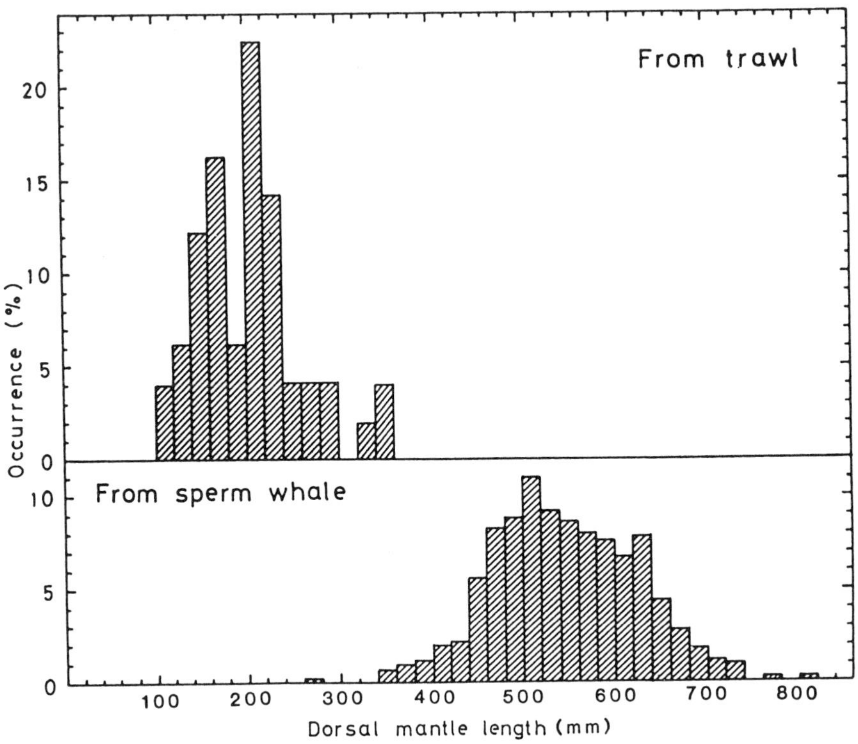

Fig. 2. Size distribution of *Kondakovia longimana* collected by trawling operations and found in the stomach contents of Sperm Whales

Food items in the stomach contents of squid caught by trawl are shown in Table 3. *Euphausia superba* was the main food. Among other dominant zooplankters in the Antarctic, the euphausiid *Thysanoessa macrura* and the amphipod *Themisto gaudichaudii* were also common in the stomach contents of *K. longimana.* One species of chaetognath, *Sagitta gazellae,* occurred in the stomachs of *K. longimana.* We found fish to be common in the stomachs of *M. knipovitchi,* but only on a few occasions were fish observed both in *K. longimana* and *A. antarcticus.* Cannibalism was observed in *A. antarcticus, K. longimana, M. knipovitchi* and *B. picta.*

The eyes of food items are very clearly detected in the stomach contents of squid, although other body parts are broken and digested. The counting and measuring of the eyes of euphausiids allows determination of the species, size and numbers taken by squid. The numbers of Krill found in the stomachs of *K. longimana* ranged from 2 to 54. The probable size distribution and biomass of euphausiids were also calculated from an equation relating eye size to body size or weight. In fresh stomach contents, reasonable agreement was observed between stomach content weight and calculated weight of total Krill fed on by the squid. On the other hand, in stomachs in which digestion had occurred, the stomach content weight was far less than the weight calculated using the equation.

Five examples of size distribution of the body length of Krill collected from the stomach contents of individual squid are given in Fig. 3, in which the size of *E. superba* centred around a particular length. This might reflect feed-

Table 3. Food items of 56 specimens of squid collected in the Antarctic by trawling operations during the summers of 1980–81 and 1981–82

	Food items						
Squid species	*E. superba*	*T. macrura*	*T. gaudichaudii*	*S. gazellae*	Fish	Squid	Examined specimens
K. longimana	40	3	10	3	3	2	40
M. knipovitchi	4	–	–	–	8	1	11
A. antarcticus	1	–	1	–	1	1	2
G. glacialis	1	–	1	–	–	–	1
B. picta	–	–	–	–	–	1	1
Unknown	–	–	–	–	1	–	1

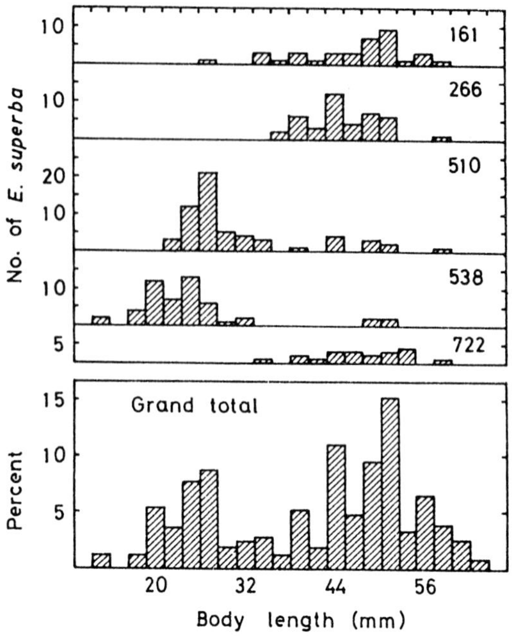

Fig. 3. Five examples (nos. 161–722) of the size distribution of *Euphausia superba* in the stomachs of individual *Kondakovia longimana*

ing by the squid on Krill swarms in which the body length of the animals is rather similar in each swarm. The size distribution of all Krill collected from the individual squid shows peaks at 25–26 mm and 44–52 mm (Fig. 3). Large-sized Krill (44–52 mm) were more common in the food of *K. longimana.* It should be noted that these two size ranges of Krill also occurred in samples of the food eaten by baleen whales in the Antarctic (Marr 1962). The eyes of *Themisto gaudichaudii* found in the stomachs of *K. longimana* were usually broken, but the body length of these amphipods ranged from 13 to 15 mm from the head to the tip of the telson.

Several eye lenses of fish and squid were detected in the remains of food in the stomachs of squid. The size of eye lenses of fish ranged from 2.1 mm to 2.5 mm, and there apparently is a tendency for *M. knipovitchi* to feed on somewhat larger fish than *K. longimana.* The species to which these juvenile fish belong is not certain but their size (as estimated from their eyes) is smaller than that of their squid predators.

Cannibalism was observed in 5 out of 56 specimens examined for food items. The fragments of cephalopods found in the stomachs of these squid were as follows. Two *Kondakovia* (200 mm in mantle length) consumed two cephalopods, indicated by the presence of two different pairs of eyeballs ranging from 2.7 to 4.0 mm in diameter. If these eyes are of common species, such as *K. longimana* and *M. knipovitchi,* then the prey are estimated to be smaller than their predators. One broken beak (upper) and a cusp occurred in the stomach of a *Brachioteuthis picta* specimen having a mantle length of 56 mm. The size of the beak suggests that this prey object was almost the same size or slightly smaller than its predator.

The feeding activity of squid during the day was examined, using the ratio of food-weight: body-weight in relation to the time of day when the squid were captured. Figure 4 shows that feeding activity peaked in the morning and again in the evening. Thus, the maximum daily ration of a squid comprises up to 30% of the squid's body weight if morning and evening stomach contents are combined.

4 Discussion

There have been relatively few studies of the roles of squid in trophic systems of Antarctic waters. However, cephalopods are the main food of Sperm Whales in the Antarctic (Clarke 1980b); the most important species being *K. longimana, Mesonychoteuthis hamiltoni* and *Moroteuthis knipovitchi. Kondakovia longimana* and *M. knipovitchi* were also common in the trawl catches reported in our study but no specimens of *Mesonychoteuthis* were found in the shallow-water trawl catches, suggesting that the larger species of this genus are distributed in deeper layers. Another genus belonging to the Cranchiidae is *Galiteuthis* which is commonly caught by net sampling in the Antarctic. However,

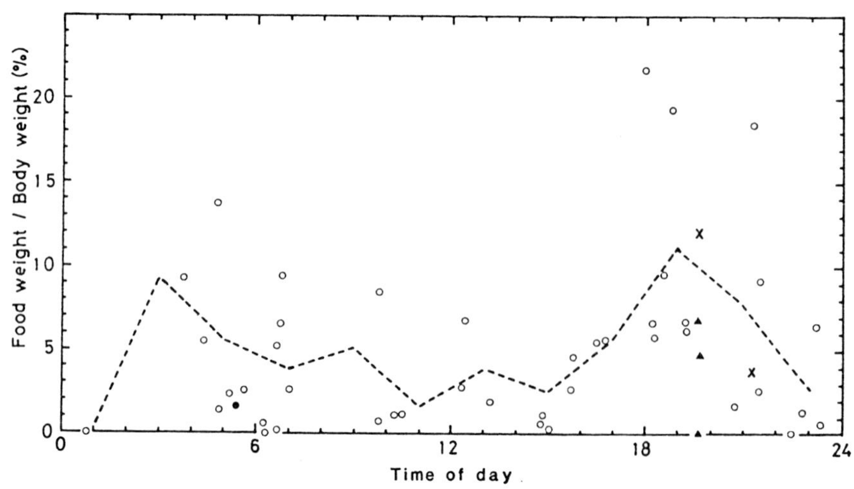

Fig. 4. Daily variation of food weight/body weight ratio in squid collected by trawling operations. The *broken line* traces mean values for 2-h intervals over a 24-h day. Symbols as in Fig. 1

large-sized specimens of *Galiteuthis glacialis* (up to 1.0 m) were not recorded by us which suggests that they also live in deeper waters.

The size distribution of cephalopods obtained from the stomach contents of Sperm Whales caught during the Japanese whaling operations in the 1951–52 season in the Antarctic ranged from 260 mm to 2.4 m in mantle length. Most of these animals were between 400 and 700 mm in mantle length; the peak being between 500 and 600 mm. When we compared this with the size distribution of *K. longimana* and *M. knipovitchi,* estimated from the beaks collected in the stomachs of Sperm Whales, the maximum peak at 500 –600 mm is considered to be due to *K. longimana* which is the dominant food of Sperm Whales in the Antarctic. Thus, the lower size limit of squid fed on by Sperm Whales coincides with the upper size limit of squid caught by the midwater trawls of our study. Clarke (1977, 1980b) also described the size of squid samples obtained from nets and Sperm Whales, and showed clearly that even the smallest squid taken by Sperm Whales are bigger than the largest squid taken by ordinary nets with a mouth opening of 9 m^2. The clear separation in the size distribution of *K. longimana* between specimens from trawl catches and those from Sperm Whale stomachs (Fig. 3) is due to two factors. One is the selective feeding by Sperm Whales which may take only relatively large specimens. The mean weight of *K. longimana* specimens in Sperm Whale stomachs, calculated by Clarke (1980b), is about 5 kg, corresponding to a mantle length of about 0.6–0.7 m, which is far larger than the specimens caught by trawling. Another reason for the discrepancy is that Sperm Whales are generally deep-sea feeders which do not feed on squid in shallow waters, and large squid are generally also not found at shallower depths. Tarasevich (1968) observed that Sperm Whales larger than 13 m in length fed on larger squid than did smaller individuals in the Gulf of Alaska. He suggested that younger squid live in shallow water, and that small Sperm Whales dive less deeply than large ones. In the Antarctic, most Sperm Whales are comparatively large males which probably go down to deep waters to feed.

Clarke and Lu (1975) confirmed the clear vertical descent of the squid *Japetella diaphana* with ontogenetic development in the North Atlantic. Their results indicated that there might be shallow and deep populations of squid, since large specimens of *J. diaphana* (20 mm or more in mantle length) were found below 800 m, whereas specimens smaller than 5 mm were concentrated in shallower waters especially at night.

Although the food of the six species of squid in this study varied, two main conclusions are evident. *Kondakovia longimana* is a plankton feeder, taking not only euphausiids, *Euphausia superba* and *Thysanoessa macrura,* but also amphipods and chaetognaths. On the other hand, *M. knipovitchi* is a fish feeder which took myctophid and other Antarctic fish in the mesopelagic layer. Several differences in feeding characteristics between these two species, such as differences in tentacles, arm-sucker, tentacular club, ring, etc., have been described by Filippova (1972), and her observations indicate that *K. longimana* is adapted to feed on macro-zooplankton, mainly euphausiids. According to Filippova (1972), *K. longimana* is confined to the limits of the distribution of *E. superba.*

The stomachs and caecae of *K. longimana* were examined by Clarke (1980b). He found remains of teleost fish, crustaceans, squid and sponges; teleost bones being most common. The reason why fish were common in Clarke's specimens of *K. longimana* could be that the specimens came from Sperm Whales which may have fed at greater depths than those sampled in our study. McSweeny (1978) reported that the most common remains in the stomachs of *Galiteuthis glacialis* were crustacean exoskeletons, copepods, amphipods and euphausiids. Chaetognath spines were the next most common item. Fish remains were found occasionally but by no means commonly.

Thus, it is believed that there are two trophic groups of squid in the Antarctic, those that feed in the euphotic zone chiefly on Krill and zooplankton, and meso- and bathy-pelagic dwellers that feed basically on fish, squid and other benthic organisms, as first suggested by Filippova (1972). These two groups are segregated according to species and ontogenetically within the same species.

The diets of pelagic squid have been examined to some extent in the North Pacific (Okiyama 1965) and in the Atlantic (Clarke 1980b). Both of these workers described some local differences in the food of squid in these areas. Okiyama (1965) observed that the stomach contents of the Common Squid, *Todarodes pacificus,* in the inshore waters in the Sea of Japan contained more than 50% of fish by number, and that *Eupausia pacifica* and *Themisto japonica* were dominant in specimens from offshore waters. Clarke (1980b) described the food of squid in the Antarctic, as well as in areas of the Atlantic, and noted that benthic organisms, including fish, were the dominant prey especially in coastal waters. There are no such clear differences in the results of our study, probably because of our relatively small survey area.

The weight of stomach contents was examined in specimens of *Todarodes pacificus,* taken in the Sea of Japan, by Okiyama (1965) who found that the maximum amount of food accounted for 8% of body weight in small specimens and about 4% in larger specimens. Okutani (1962) also reported that *T. pacificus* eats the equivalent of only a very small proportion of its body weight in Pacific waters. In contrast, the daily ration and weight of stomach contents of squid revealed in our study are very high.

Japanese fishermen report that *T. pacificus* feeds actively in the morning and again in the evening. Okiyama (1965) observed active feeding in the evening, peaking between 1800 h and 1900 h local time. However, he did not examine morning and daytime feeding activity, and, thus, our finding of a peak in morning feeding activity is not well understood.

The length of the nidamental gland is used to indicate the developmental stage of squid in general. Clarke (1980b) observed rapid growth in the nidamental gland at mantle lengths of > 400 mm in *K. longimana,* and > 250 mm in *M. knipovitchi.* In all our specimens belonging to these two genera the glands were shorter than the values given by Clarke. Thus, the squid collected by our trawls in shallow waters apparently were immature.

The depth at which most specimens of *K. longimana* were taken became shallower with the advance of the summer season in the Antarctic, since the main concentration of swarms and layers of *E. superba* also became shallower. In November and December, swarms and layers of Krill were dispersed from the surface down to about 100 m and most specimens of *K. longimana* were caught at 60 m. In January and February, both layers and swarms of Krill moved up to between 30 and 40 m and we caught most Krill and squid at this level.

References

Clarke MR (1977) Beaks, nets and numbers. Symp Zool Soc Lond 38:89–126

Clarke MR (1980a) Squids and whales. NERC News J 2(9):4–6

Clarke MR (1980a) Cephalopoda in the diet of sperm whales of the Southern hemisphere and their bearing on sperm whale biology. Discovery Rep 27:1–324

Clarke MR, Lu CC (1975) Vertical distribution of cephalopods at 18°N 25°W in the North Atlantic. J Mar Biol Assoc UK 55:165–182

Filippova JA (1972) New data on the squids (Cephalopoda: Oegopsida) from the Scotia Sea (Antarctica). Malacologia 11(2):391–406

Laws RM (1977) Seals and whales of the Southern Ocean. Phil Trans R Soc Lond B 279:81–96

Marr JWS (1962) The natural history and geography of the Antarctic krill (*Euphausia superba*). Discovery Rep 32:33–464

McSweeny ES (1978) Systematics and morphology of the Antarctic cranchiid squid *Galiteuthis glacialis* (Chun). Antarc Res Ser 27: 1–39

Okiyama M (1965) On the feeding habits of the common squid, *Todarodes pacificus* Steenstrup in the off-shore region of the Japan Sea. Bull Jpn Sea Reg Fish Res Lab 14:31–41

Okutani T (1962) Diet of the common squid, *Ommatrephes sloani pacificus* landed around Ito Port, Shizuoka Prefecture. Bull Tokai Reg Fish Res Lab 32:4–47

Roper CFE (1977) Comparative capture of pelagic cephalopods by midwater trawls. Symp Zool Soc Lond (1977) 38:61–87

Tarasevich MN (1968) Dependence of distribution of the sperm whale males upon the character of feeding. Zool Z Ukrayiny 47(11):1683–1688

The Role of Zooplankton in the Diets of Certain Sub-Antarctic Marine Fish

G. DUHAMEL and J.C. HUREAU[1]

Summary. The so-called benthic fish of the Southern Ocean are classically known as benthophagous species. However, recent studies have shown that in some cases zooplankton predominates in their diets. Nototheniids and channichthyids were studied at Kerguelen and Crozet islands, where it was found that the young of *Notothenia rossii* are coastal and feed on benthic invertebrates but the adults, which live offshore at depths of 100–400 m, have diets mainly composed of planktonic organisms such as euphausiids, medusae and salps. *Notothenia squamifrons* is widely distributed in the area, but rarely occurs in shallow waters. It feeds on salps, medusae and amphipods. *Dissostichus eleginoides* is a predator on other fish, its young feeding on euphausiids and other zooplankters. The diet of *Champsocephalus gunnari* is based on euphausiids, pelagic amphipods and pelagic fish, such as myctophids. *Channichthys rhinoceratus* is mainly a predator of young fish. These plankton-based diets suggest that the fishes' foraging behaviour is attuned to the movements of their prey. For example, *Champsocephalus gunnari* migrates upwards during the night to catch its prey. The other species probably behave similarly, but their time of feeding is not known. This type of diet may also reflect the lack of rich benthic communities at the depths where the fish are most abundant.

Résumé. Les poissons dits benthiques sont connus pour être classiquement benthophages. De récentes études ont cependant montré que dans quelques cas, le zooplancton prédomine dans leur régime alimentaire. Les Nototheniidae et les Channichthyidae on été étudiés au large des iles Kerguelen et Crozet où il a été démontré que les juvéniles de *Notothenia rossii* sont côtiers et qu'ils tirent leur nourriture d'invertébrés benthiques mais les adultes qui vivent au large, à des profondeurs comprises entre 100 et 400 mètres, possèdent un régime alimentaire à base d'organismes planctoniques tels que les euphausiacés, les méduses et les salpes. *Notothenia squamifrons* possède une vaste distribution géographique mais est rarement trouvé dans les eaux côtières. Sa nourriture se compose de salpes, méduses et amphipodes. *Dissostichus eleginoides* connu à l'état adulte comme prédateur d'autres poissons se nourrit pendant sa phase juvénile d'euphausiacés et d'autres organismes zooplanctoniques. *Champsocephalus gunnari* présente un régime alimentaire composé d'euphausiacés, d'amphipodes pélagiques et de poissons pélagiques tels que les Myctophidae. Enfin *Channichthys rhinoceratus* est principalement un prédateur de jeunes poissons. Un tel régime alimentaire de type planctonique suggère que ces espèces ont un comportement lié aux mouvements de leurs proies. Par exemple: *Champsocephalus gunnari* effectue des migrations verticales nycthémérales pour capturer ses proies. Les autres espèces possèdent probablement un comportement similaire, mais leur période d'alimentation est encore difficile à préciser. Ce type de régime alimentaire peut aussi refléter l'absence de riches communautés benthiques aux profondeurs où ces poissons sont les plus abondants.

1 Laboratoire d'Ichtyologie générale et appliquée, Muséum National d'Histoire naturelle, 43 rue Cuvier, F-75231 Paris, Cedex 05, France

1 Introduction

Apart from the typically pelagic fish, such as the myctophids whose diet is based on zooplankton (Rembiszewski et al. 1978) but whose abundance is still virtually unknown, two families of fish may be predators on zooplankton in the Southern Ocean. In this region, nototheniids and channichthyids, which include several zooplankton species in their diets, are abundant enough to justify commercial exploitation particularly in the southwestern Atlantic and the Indian sectors. In these sectors, the shelf areas around the main islands (South Georgia and Kerguelen) provide certain species with feeding or spawning grounds which are large enough to support very large populations of fish. Here we report on the diets of certain species from the Indian sector of the Southern Ocean, and evaluate the possible role of zooplankton in the food of these fish.

2 Materials and Methods

Collections of fish were made around Kerguelen and Crozet islands, using research ships and commercial trawlers. Because only species abundant in the catches were retained, this study is limited to four species of nototheniids (*Notothenia rossii rossii, Notothenia squamifrons, Notothenia larseni*, and *Dissostichus eleginoides*) and two species of channichthyids (*Champsocephalus gunnari* and *Channichthys rhinoceratus*). The diets of these species were analyzed qualitatively and quantitatively. The prey were identified to the order, family or, when possible, species level. The frequency of each group of prey was assessed against the total number of prey included in non-empty stomachs. The "emptiness coefficient" and the "repletion index" (Hureau 1970) were evaluated; for the latter we used a scale with six stages (Ball 1961). *Champsocephalus gunnari*, for which stomach contents remain unchanged (i.e., no devagination under pressure effect when the trawl is brought up), was

Antarctic Nutrient Cycles and Food Webs
(ed. by W.R. Siegfried, P.R. Condy, and R.M. Laws)

Table 1. Percentage occurrence of prey items in stomachs of *Notothenia rossii* at Kerguelen

	Oct. 1979–Feb. 1980	Aug. 1981	June 1983
Cnidaria	13.7	42.4	4.5
Ctenaria	6.6	–	13.4
Euphausiids	8.2	1.5	12.5
Hyperiid amphipods	7.1	–	9.2
Isopods	5.5	–	3.0
Salps	4.4	–	18.4
C. gunnari	7.1	47.0	15.8
Other fish	22.4	9.1	17.3
Other prey	25.0	–	5.9
No. stomachs examined	254	71	1137
No. empty stomachs	14	5	471
Emptiness coefficient	5.5	7.0	41.4
Source	Duhamel (1981)	This study	This study

studied in order to measure precisely the effect of predation on zooplankton; it was possible to determine the daily feeding rhythm of this species.

3 Results and Discussion

The first information on the diets of fish in the Indian sector of the Southern Ocean was obtained for certain species caught with a gillnet (Hureau 1966, 1970) in the coastal waters of the Kerguelen Islands. Three nototheniids (*Notothenia rossii rossii, N. cyanobrancha* and *N. magellanica*) and one channichthyid (*Channichthys rhinoceratus*) were included in these catches. The diet of the nototheniids was made up mainly of isopods (*Exasphaeroma gigas* and *Serolis* sp.) which represented more than 50% of the prey. The balance varied according to the species of fish. *Notothenia cyanobrancha* took advantage of sedentary or relatively immobile prey living close to the bottom, such as *Halicarcinus planatus* and *Patigera kerguelensis. Notothenia magellanica* fed on small molluscs *Gaimardia trapesina* living on kelp, and *N. rossii rossii* showed a tendency to feed on zooplankters such as young pelagic isopods, amphipods, euphausiids or young fish. *Channichthys rhinoceratus* was classified as piscivorous, because *N. cyanobrancha* represented 90% of its prey. However, these results, obtained in coastal or inshore waters, are not representative of the possible impact of fish on zooplankton. In fact, large concentrations of fish are rarely found in these waters and certain species are never observed there, whereas they are abundant over the shelf extending to 18 km or more offshore. Moreover, certain sea-mounts or banks can be attractive for a few particular species (Duhamel 1981).

The species found offshore differ from those in the coastal waters, except for *N. rossii rossii* which uses coastal waters as a nursery for juvenile fish (Duhamel 1982) and for *C. rhinoceratus* which spawns inshore but feeds offshore (Hureau 1966). Before dealing with the diets of the fish living offshore, or on the sea-mounts, we point out that, in the Indian sector of the Southern Ocean, channichthyids are only present around Kerguelen and Heard islands, and *N. larseni* occurs only around the Crozet and Marion archipelagoes and in association with nearby sea-mounts: the other species being present elsewhere as well, according to their vertical distribution (Duhamel et al. 1983).

3.1 Notothenia rossii rossii

This relatively large species (900 mm) was originally very abundant around the Kerguelen Islands, but its population has been significantly reduced by commercial fishing. After a pelagic stage, during which its diet is mainly euphausiids and copepods (Rembiszewski et al. 1978, based on *N. rossii marmorata* which is a South-Atlantic sub-species), the young immature fish live in the coastal bottom waters to a depth of about 150 m. They stay there until they are sexually mature which occurs at an age of about 5 or 6 years. Then the adults disperse offshore, over the shelf and the banks.

The diet, already based on plankton in the inshore waters (Hureau 1966), becomes typically zooplanktonic over the shelf (Table 1), being based mainly on Cnidaria, Ctenaria, euphausiids, pelagic amphipods and isopods, salps and fish (many being young and adults of *Champsocephalus gunnari*). It is noticeable that the emptiness coefficient is

Table 2. Repletion index for *Notothenia rossii rossii* in different seasons at Kerguelen. The numbers are percentages of stomachs at each stage

Season	Stage						No. Stomachs
	0	I	II	III	IV	V	
Oct. 1979	2.5	10.1	12.6	13.6	27.6	33.7	199
Nov. 1979–Jan. 1980	16.4	14.5	32.7	9.1	5.5	21.8	55
Aug. 1981	7.0	4.2	21.1	12.7	4.2	50.7	71
June 1983	41.4	20.7	15.3	17.6	3.7	1.2	1137

rather low outside of the spawning season (June), when the gonads filled the whole of the abdominal cavity. Analysis of the variation in the repletion index (Table 2) showed a very active feeding period after spawning; more than 50% of the stomachs being at stage V, whereas there was a predominance of empty or little-filled stomachs during spawning.

The composition of the stomach contents of *N. rossii rossii* varied according to the available food in the areas in which the adult population was encountered, because of the species' migrations (Duhamel 1982). At Kerguelen, apart from invertebrate members of the zooplankton which form the bulk of the diet of adults and juveniles, *N. rossii rossii* preys extensively on post-larvae, juveniles and adults of *C. gunnari;* certain stomachs containing as many as 100 post-larvae (60 mm long) or 10–12 adults (260–300 mm long). Thus, we assume that the overfishing of the population of *N. rossii rossii* will benefit *C. gunnari* by the elimination of its main predator.

The diet of *N. rossii rossii* is similar to that of *N. rossii marmorata* in the Scotia Sea (Tarverdiyeva 1972; Permitin and Tarverdiyeva 1972; Tarverdiyeva and Pinskaya 1981), but with the absence in the former species of Antarctic Krill, *Euphausia superba,* which does not occur around the Kerguelen Islands. An evaluation of the daily food intake was made by Tarverdiyeva (1972) who estimated the Krill consumption of *N. rossii marmorata* to be 15% of the body mass of an individual adult fish, which means between 700 and 750 g. Feeding sometimes occurs near the surface (Shust and Sil'yanova 1971), which explains why the species cannot be caught by bottom gear during part of the year. Moreover, the fish shoals move in relation to the areas in which food is abundant.

3.2 Notothenia squamifrons

This species is very abundant all over the study area including isolated sea-mounts (e.g., Ob and Lena). The species does not require coastal waters for its young stages, unlike *N. rossii.* Its diet has been studied around Kerguelen (Duhamel 1981) and Crozet islands (Duhamel and Pletikosic 1983) in summer and recently (1983) in winter (May–June) by us (Table 3). All the samples taken at Kerguelen or at Crozet in May 1983 included mainly adult and subadult fish (250–540 mm), but those taken at Crozet in February 1982 included only juveniles. This may explain the differences observed in the stomach contents of fish from the Crozet Islands: the juveniles' diet was based on benthic organisms, mainly amphipods (gammarids), whereas the adults and subadults, which live on the same type of bottom, fed on zooplankters such as salps, amphipods (hyperiids), isopods and euphausiids. Salps were dominant in the diet of *N. squamifrons* from Kerguelen, since they always represented at least one-third of all the prey. The other prey were also zooplanktonic: Cnidaria, Ctenaria and amphipods (hyperiids); eggs and post-larvae of fish always being observed. The composition of the diet remained almost unchanged throughout the year.

The emptiness coefficients for fish populations from Crozet and Kerguelen islands at various seasons were remarkably stable, varying between 21.3 and 29.3%. The repletion indexes (Table 4) confirm the rather large proportion of empty stomachs but also show that feeding was rather constant, since their values are distributed regularly between the six stages, even during the spawning in late October. Because *N. squamifrons* is abundant, it has a strong impact on zooplankton, mainly on salp but also on Cnidaria and Ctenaria and, secondarily, on amphipods (hyperiids) and euphausiids.

3.3 Notothenia larseni

This small nototheniid (170 mm max. length at Crozet Islands) has been recorded only on the western side of a line between Kerguelen and Heard islands, around the islands themselves, and on the sea-mounts. Its diet is known only from a small sample (49 specimens) obtained during the spawning season in February 1982 (Duhamel and Pletikosic 1983). Ctenaria, euphausiids, amphipods (gammarids) and fish larvae have been recorded as prey. This diet is similar to that of populations in the Scotia Sea (Permitin and Tarverdiyeva 1972; Tarverdiyeva and Pinskaya 1981) and shows that this species' life is pelagic in part. The impact by

Table 3. Percentage occurrence of prey items in stomachs of *Notothenia squamifrons* at Crozet and Kerguelen

	Kerguelen Oct. 79–Feb. 80	Crozet Feb. 82 (immatures)	Crozet May 1983	Kerguelen June 1983
Cnidaria	23.5	0.5	3.0	9.4
Ctenaria	13.8	6.5	1.0	12.9
Annelida	0.8	5.4	2.0	0.4
Mollusca				
Cephalopods	–	–	–	1.1
Others	0.1	3.8	1.0	–
Crustacea				
Mysids	1.0	–	–	–
Euphausiids	2.7	8.4	9.4	2.3
Hyperiids	11.9	–	16.7	14.0
Gammariids	–	40.5	–	–
Caprellids	–	2.7	3.1	–
Isopods	1.0	8.7	18.7	0.2
Decapods	0.1	–	1.0	–
Undetermined	–	1.4	–	–
Echinodermata				
Crinoids	1.0	–	–	0.5
Ophiurids	–	0.3	–	–
Tunicata				
Salps	35.8	2.7	34.4	39.5
Fish	6.1	9.2	3.0	5.0
Fish Eggs	1.2	0.3	–	0.2
Undetermined	0.9	8.4	6.2	14.6
Brachiopoda	–	0.3	–	–
Chaetognatha	–	0.8	–	–
No. stomachs examined	2880	441	75	417
No. empty stomachs	780	129	16	117
Emptiness coefficient	27.0	29.3	21.3	28.1
Source	Duhamel (1981)	Duhamel and Pletikosic (1983)	This study	This study

N. larseni on zooplankton is probably negligible, because of the small size of the fish which makes for a relatively small total consumption of the populations. Moreover, the species is not as abundant as *Dissostichus eleginoides* or *Notothenia squamifrons* in the Crozet Islands.

3.4 Dissostichus eleginoides

This nototheniid is abundant in the Indian sector of the Southern Ocean, attaining a length of 1.70 m at Crozet and Kerguelen islands. Its biology is still largely unknown.

After hatching, the larvae have a pelagic life and then move to shallow waters around the islands. They feed on small zooplankton, mainly euphausiids (Rembiszewski et al. 1978; Duhamel and Pletikosic 1983). Members of the first age-class (80–110 mm, in February) and the second (200–260 mm, in June) feed on zooplanktonic organisms (euphausiids and hyperiids). Subsequently, as soon as the fish reach their third year (> 300 mm, in June), their diet is based on fish. This tendency is clearly shown by an analysis of stomach contents in relation to the size of the fish at the Crozet Islands (Fig. 1). The diets of sub-adult and adult

Table 4. Repletion index for *Notothenia squamifrons* in different seasons at Kerguelen. The numbers are percentages of stomachs at each stage

Season	Stage						No. Stomachs
	0	I	II	III	IV	V	
Oct. 1979	26.1	10.6	20.1	10.6	12.1	20.6	199
Nov. 1979	54.6	17.9	20.1	9.5	2.5	1.0	240
Dec. 1979	31.5	8.4	16.4	17.2	8.1	18.3	847
Jan. 1980	32.4	13.9	19.8	15.9	7.0	11.0	555
June 1983	28.1	19.2	25.2	16.8	7.9	2.9	417

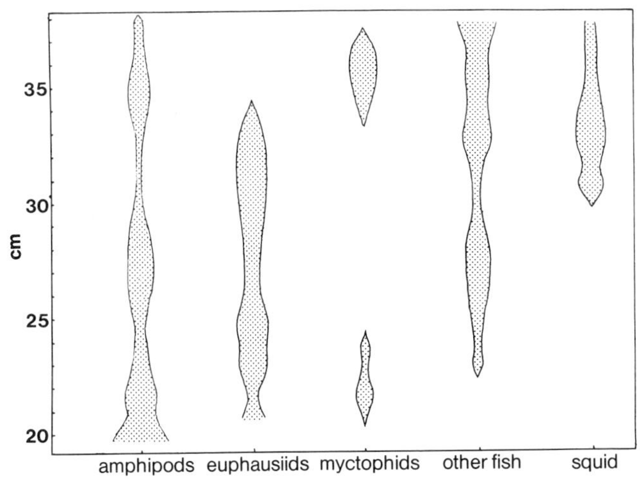

Fig. 1. Relative change in composition of the diet of *Dissostichus eleginoides* according to size of fish (Crozet Islands, May 1983)

members of *D. eleginoides* were based almost entirely on fish (Table 5), which is noteworthy because it represents more than 50% of the diet in any season in both archipelagoes. Nevertheless, the percentage is lower at Crozet (50–52%) than at Kerguelen (82–90%). Apart from the fish, squids formed a large part of the diet of the adult specimens living in the deeper water layers. Mesopelagic fish, such as myctophids, always accounted for a large proportion of the food in all the stomach contents.

The emptiness coefficient for this species was rather high, varying between 32 and 58%. The repletion index was generally not very high, since the proportion of stages IV and V was low (Table 6). However, in the austral summer, in both study areas, there were more full stomachs (stage V) than in the winter. This can be explained by a greater activity of the fish in summer or by a greater abundance of food in that season.

In short, *D. eleginoides* is a zooplankton predator during its first 2 or 3 years of life. Thereafter this form of predation decreases, and subsequently the species becomes more and more piscivorous. Juvenile fish are very abundant in the shallow waters of the shelf, so their predation on euphausiids and amphipods (hyperiids) is not negligible. A similar diet was described for the species in the South Atlantic sector of the Southern Ocean (Tarverdiyeva 1972; Permitin and Tarverdiyeva 1972).

3.5 Channichthys rhinoceratus

This demersal species is endemic to the Kerguelen-Heard plateau and plays a predatory role similar to that of *Dissostichus eleginoides.* It occurs regularly in trawls, but it is never very abundant. Its predatory impact on zooplankton can be considered as very low, apart from the juvenile period of life. During that period, the main prey are young fish or euphausiids (Duhamel 1981). The adults, however, are exclusively piscivorous (Hureau 1966). The emptiness coefficient was always high (73%), and greater than that for *D. eleginoides.* In fact, *C. rhinoceratus* rarely moves into open waters and is much more dependent on the abundance of fish shoals (nototheniids or channichthyids) on the bottom of the shelf.

3.6 Champsocephalus gunnari

Together with *N. rossii* and *N. squamifrons, C. gunnari* is one of the most abundant fish of the Indian sector of the Southern Ocean, although it is only found in the vicinities of Kerguelen and Heard islands. It forms very dense shoals. Some 26,000 t were fished commercially from a small area (1° long. × 0.5° lat.) in the northeastern shelf of Kerguelen from October 1982 until April 1983. This mass (26,000 t) represents more than 187 × 10^6 individual fish, but does not reflect the real abundance of the species which was still great when the commercial fishing ceased.

The diet of this bentho-pelagic species at Kerguelen is now well known (Komposki 1980; Duhamel 1981). The diversity of the prey is low, since the diet is mainly composed of euphausiids (*Euphausia similis*), hyperiid amphi-

Table 5. Percentage occurrence of prey items in stomachs of *Dissostichus eleginoides* at Crozet and Kerguelen

	Kerguelen Oct. 79–Feb. 80	Crozet Feb. 82	Crozet May 1983	Kerguelen June 1983
Cnidaria	–	–	0.8	0.1
Annelida	0.1	–	–	–
Cephalopoda				
Octopus	–	–	2.9	0.3
Squids	0.9	7.7	24.9	6.9
Crustacea				
Mysids	0.1	3.8	–	–
Euphausiids	3.2	7.7	7.3	9.6
Amphipods	1.4	23.1	7.6	1.0
Isopods	–	1.9	0.5	–
Lithodes	–	–	0.5	–
Salps	–	–	0.3	0.1
Fish	89.6	49.9	52.3	81.5
(Myctophidae included)	(27.2)	(11.5)	(14.3)	(17.2)
Undetermined	4.8	5.8	2.7	0.4
No. stomachs examined	1514	74	929	1481
No. empty stomachs	748	31	559	477
Emptiness coefficient	50.8	41.9	58.3	32.2
Source	Duhamel (1981)	Duhamel and Pletikosic (1983)	This study	This study

Table 6. Repletion index for *Dissostichus eleginoides* in different seasons at Crozet and Kerguelen. The numbers are percentages of stomachs at each stage

Season	Stage						No. Stomachs
	0	I	II	III	IV	V	
Kerguelen							
Oct. 1979–Feb. 1980	50.8	13.0	11.3	7.6	4.8	12.5	1514
Kerguelen							
June 1983	32.2	17.8	21.7	17.1	9.8	1.4	1481
Crozet							
Feb. 1982	41.9	13.5	16.2	9.5	8.1	10.8	74
Crozet							
May 1983	58.3	14.3	11.6	8.1	3.2	1.3	929

Table 7. Percentage occurrence of prey items in stomachs of *Champsocephalus gunnari* in summer and winter at Crozet and Kerguelen

	Oct. 1979–Jan. 1980	May–June 1983
Cephalopoda	0.1	–
Crustacea		
Mysids	0.8	–
Euphausiids	75.0	71.4
Hyperiid amphipods	13.8	8.4
Fish		
Undetermined	0.1	2.2
Myctophids	9.5	1.9
C. gunnari fingerlings	–	15.9
Undetermined	0.6	0.7
No. stomachs examined	1305	1210
No. empty stomachs	516	842
Emptiness coefficient	39.5	69.6
Source	Duhamel (1981)	This study

Table 8. Repletion index for *Champsocephalus gunnari* in different seasons at Kerguelen. The numbers are percentages of stomachs at each stage

Season	Stage						No. Stomachs
	0	I	II	III	IV	V	
Nov. 1979	35.3	30.7	16.1	5.8	4.6	7.5	411
Dec. 1979	38.0	19.3	19.1	8.4	5.3	9.8	692
May–June 1983	69.6	15.2	6.4	5.0	2.4	1.4	1210

pods (*Themisto gaudichaudii*), myctophids and young fish of its own species (*C. gunnari*) (Table 7). The percentage of the various items in the stomach contents varied according to the study area or season (Tables 7 and 9) with a consistent dominance by euphausiids and hyperiids. Cannibalism occurred as soon as the post-larvae or young appeared in the zooplankton. This has been observed from the end of the austral summer but it is particularly great in winter when food is less abundantly available, as indicated by the increase of the emptiness coefficient (69.6% in winter against 39.5% in summer). The repletion indexes also varied according to season, indicating that the fish feed actively in summer. In winter, the empty or little-filled stomachs (stages 0 and I) represented 84.8% of the stomachs examined (Table 8). Thus, *C. gunnari* has a typical zooplanktonic diet and, as a result, it is a species which is strongly dependent on the seasonal abundance of the plankton. In the Scotia

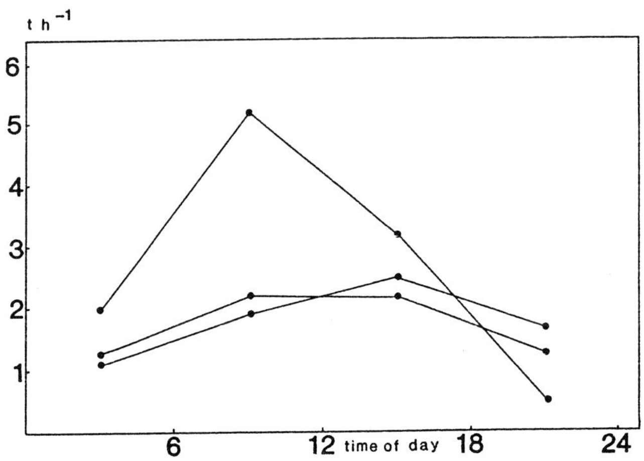

Fig. 2. Mean values of CPUE (t fishing h^{-1}) for three commercial trawlers in November–December 1979 at different times of the day for *Champsocephalus gunnari*

Table 9. Qualitative and quantative analyses of the stomach contents of *Champsocephalus gunnari*. (Repletion index at stage V)

	North-East Kerguelen	Kerguelen-Heard Ridge	North-East Kerguelen
% Euphausiids	75.1	18.7	39.1
% Hyperiid amphipods	24.9	81.3	59.1
% Undet. fish	< 1		< 1
% *C. gunnari* fingerlins			1.8
Total no. prey objects	2623	604	2102
Mean no. prey objects per stomach	145.7	120.8	110.6
Total weight (g) of prey objects	249.8	45.4	148.2
Mean weight (g) of prey per stomach	13.9	9.1	7.8
No. stomachs	18	5	19
Time samples taken	December 1977	February 1978	Nov.–Feb. 1983
Source	Komposki (1980)	Komposki (1980)	This study

Sea, the same species has a very similar diet based on *Euphausia superba, Themisto gaudichaudii* and mysids (Permitin and Tarverdiyeva 1972, 1978; Kock 1981).

Champsocephalus gunnari is one of the few Antarctic fish which can be detected with acoustic methods, and analyses of the records (mainly made from a colour-chromascope KODEN echo-sounder) show daily fluctuations in the depth and density of the shoals. At twiglight, in the evening, the fish rise up to tens of metres above the bottom and then spread out in the water column. At the end of the night, before day-break, the fish concentrate close to the bottom where they spend a large part of the day. This daily migration is also shown by variation in the catch-per-unit-effort (CPUE) of bottom gear (Fig. 2): the CPUE is highest during the daytime when the fish are close to the bottom, and then becomes lower and lower until twilight when the fish spread out. From an examination of the state of digestion of the prey, it is possible to correlate the species' daily migration with its feeding rhythm. The fish need during the night, and stay quiet on the bottom during daytime.

The evaluation of the predatory impact of the population of *C. gunnari* on the zooplankton (mainly euphausiids and hyperiids) requires knowledge of several parameters, such as the digestive transit duration of the prey, the monthly variation in the repletion index of the stomachs, the total biomass of the population, and the mean size and weight of the fish. A first preliminary evaluation of this predation of *C. gunnari* can, however, be made, with the following information. An analysis of full stomachs gives 10.25 g as the mean weight of a stomach at stage V (Table 9). If we consider the repletion indexes in December 1979 (Table 8), 9.8% of the fish have a stomach content of 10.25 g, 5.3% of 8.97 g, 8.4% of 6.40 g, 19.1% of 3.84 g and 19.3% of 1.28 g; the rest of the fish having empty stomachs. The mean individual size and weight of the 26,000 t of fish caught in 1982–83 are 290 mm and 139 g, respectively. Thus, we can estimate the daily consumption of that population of *C. gunnari* at about 560 t of zooplankton. This is considerable, considering the small area concerned. Using the calorific values given by Clarke and Prince (1980) for crustaceans and fish (1.04 and 0.95 kcal g^{-1} ((fr. wt.), respectively) and the data in Table 9, the daily consumption of the *C. gunnari* population represents an amount of energy equal to 582×10^6 kcal.

This preliminary evaluation shows very clearly that *C. gunnari* can be found abundantly only in areas where zooplankton is abundant. If environmental variations contribute to the disappearance of this source of food, a high mortality of the fish or a diminution of their fecundity might occur.

4 Conclusion

The six species of fish studied in the waters of Crozet and Kerguelen can be classified in three groups, according to their type of predation on zooplankton: (1) Those whose juveniles feed on benthic organisms, and whose adults feed on zooplankton (*Notothenia squamifrons* and, less typically, *N. rossii*); (2) Those whose juveniles feed on zooplankton and whose adults feed on fish (mainly on *Champsocephalus gunnari*) or also on squid and octopus (*Dissostichus eleginoides* and *Channichthys rhinoceratus*); and, (3) those which feed on zooplankton throughout their lives (*Notothenia larseni* and *Champsocephalus gunnari*).

The importance of certain categories of prey (euphausiids, salps, hyperiid amphipods, coelenterates and pelagic fishes (mainly young *C. gunnari*)) is noteworthy. The predation on zooplankton varies according to the six species. Those which have the greatest impact on pelagic organisms are *Notothenia squamifrons, N. rossii, Dissostichus eleginoides* (because of the abundance of its juveniles) and *Champsocephalus gunnari* (because of the abundance of its population). A preliminary evaluation of the impact of *C. gunnari* on zooplankton shows clearly the great importance of zooplankters in the diets of the so-called benthic fishes. This points to a need for hydrological studies, so that a better understanding might emerge of the factors that influence the distribution of the zooplankton and consequently the fishes.

References

Ball JN (1961) On the food of the brown trout of Llyn Tegid. Proc Zool Soc Lond 137:599–622

Clarke A, Prince PA (1980) Chemical composition and calorific value of food fed to mollymauk chicks *Diomedea melanophris* and *D. chrysostoma* at Bird Island, South Georgia. Ibis 122: 488–494

Duhamel G (1981) Caractéristiques biologiques des principales espèces de poissons du plateau continental des îles Kerguelen. Cybium 3e série, 5(1):19–33

Duhamel G (1982) Biology and population dynamics of *Notothenia rossii rossii* from the Kerguelen Islands (Indian sector of Southern Ocean). Polar Biol 1:141–151

Duhamel G, Hureau JC, Ozouf-Costaz C (1983) Ecological survey of the Notothenioid fishes in the Southern Ocean from Bouvet to Kerguelen Islands. Mem Natl Inst Polar Res (Tokyo), Special issue 27:176–182

Duhamel G, Pletikosic M (1983) Données biologiques sur les Nototheniidae des îles Crozet. Cybium, 3e série 7(3):43–57

Hureau JC (1966) Biologie de *Chaenichthys rhinoceratus* Richardson et problème du sang incolore des Chaenichthyidae, poissons des mers australes. Bull Soc Zool Fr 91(4):735–751

Hureau JC (1970) Biologie comparée de quelques poissons antarctiques (Nototheniidae). Bull Inst Oceanogr (Monaco) 68(1391): 1–244

Kock KH (1981) Fischereibiologische Untersuchungen an drei antarktischen Fischarten: *Champsocephalus gunnari* Lönnberg, 1905 *Chaenocephalus aceratus* (Lönnberg, 1906) und *Pseudochaenichthys georgianus* Norman, 1937 (Nototheniidae, Channichthyidae). Mitt Inst Seefisch (Hamburg) 32:1–226

Komposki A (1980) On feeding of *Champsocephalus gunnari* Lönnberg 1905 (Pisces, Chaenichthyidae) off South Georgia and Kerguelen Islands. Acta Ichthyol Piscat 10(1):25–43

Permitin YE, Tarverdiyeva MI (1972) Feeding of some species of Antarctic fishes in South Georgia area. J Ichthyol 12(1):104 –114

Permitin YE, Tarverdiyeva MI (1978) Feeding of antarctic cods and ice fishes near the South Orkney Islands (in Russian). Biol Morya (Vladivost) 2:75–81

Rembiszewski JM, Krzeptowski M, & Linkowski TB (1978) Fishes (Pisces) as by catch in fisheries of krill *Euphausia superba* Dana (Euphausiacea, Crustacea). Pol Arch Hydrobiol 25(3):677–695

Shust KV, Silyanova ZS (1971) New data on the biology of some fish in the South-West Atlantic (in Russian). Tr VNIRO 86:70 –81

Tarverdiyeva MI (1972) Daily food consumption and feeding pattern of the South Georgia cod (*Notothenia rossii marmorata* Fischer) and the Patagonian toothfish (*Dissostichus eleginoides* Smitt) (family Nototheniidae) in the South Georgia area. J Ichthyol 12:684–692

Tarverdiyeva MI, Pinskaya LA (1981) The feeding of fishes of the families Nototheniidae and Channichthyidae on the shelves of the Antarctic Peninsula and the South Shetlands. J Ichthyol 20(4):50–60

The Evolution of Neutrally Buoyant Notothenioid Fishes: Their Specializations and Potential Interactions in the Antarctic Marine Food Web

J.T. EASTMAN[1]

Summary. Since all notothenioid fishes lack swim bladders, one would expect them to be confined to predominantly benthic habitats. There is, however, a trend towards the evolution of neutrally buoyant species which have become permanent members of the mid-water community. The absence of competition from other mid-water fishes, and trophic factors, have probably been responsible for this trend. *Pleuragramma antarcticum, Aethotaxis mitopteryx* and *Dissostichus mawsoni* are neutrally buoyant. *Dissostichus mawsoni* is the largest piscine predator in the mid-waters of the Southern Ocean. In McMurdo Sound, *D. mawsoni* eats predominantly fish and mysid shrimps. The recently described nototheniid *Cryothenia peninsulae,* a krill-eating species from West Antarctica, is not neutrally buoyant. Evolutionary alterations in buoyancy are reflected as morphological specializations in the following body systems: skeleton, integument, lipid storage, muscles and viscera.

1 Introduction

The notothenioids, a primarily benthic group, are the dominant element of the Antarctic fish fauna. These fishes have undergone evolutionary diversification, including specialization of that part of the niche representing position in the food web (Dunbar 1970). I have studied several species that, in the absence of swim bladders, have experienced significant evolutionary modification to become neutrally buoyant members of the mid-water community.

In this paper I discuss the evolution of neutrally buoyant notothenioids, comment on the interactions of these species, especially *Dissostichus mawsoni,* in the mid-water food web of McMurdo Sound, review morphological specializations promoting neutral buoyancy, and suggest priorities for future research.

2 The Antarctic Marine Environment as a Setting for the Adaptive Radiation of Notothenioid Fishes

About 40 million years ago substantial Antarctic sea-ice began to form, resulting in a rapid drop in water temperature (Kennett 1978). It has been hypothesized that at about this time a bottom-dwelling stock of perciform fishes became associated with the water surrounding the continent (Regan 1914; Norman 1938; DeWitt 1971). With 90–100 species, these members of the suborder Notothenioidei became the dominant group of fishes in Antarctic waters. A long period of evolution in isolated, cold waters produced a highly endemic assemblage with 86% of the genera and 95% of the species confined to Antarctic waters (DeWitt 1971). As DeWitt (1971) has indicated, the fossil record is too meagre to provide significant information concerning the relationships and prior distribution of this group, although circumstantial evidence, including the uncertainty of their relationships and their isolated southern distribution, suggests an independent evolution throughout most of the Tertiary.

Low water temperatures, limited continental shelf habitat, extreme geographical isolation and lack of south-flowing surface currents have prevented most non-notothenioid fish groups from becoming established in Antarctic waters. The Antarctic fish fauna is therefore much less diverse than that of the Arctic (Llano 1978). Cods, herrings, salmons, smelts, sculpins and flatfishes are either absent or poorly represented in the Antarctic. There are, for example, no obligatory mass consumers of planktonic crustaceans equivalent to the vast schools of Pacific Herring, *Clupea harengus pallasi,* in the Arctic. Nor are there phytoplankton feeders like the Arctic Cod, *Boreogadus saida* (Andriashev 1954). Since seawater has a high kinematic viscosity at low temperatures (Vogel 1981), it may be that the evolution of continuously swimming, filter-feeding notothenioids was constrained by the great energetic cost of these activities in sub-zero sea water.

The role of notothenioid fishes in the Antarctic marine ecosystem is poorly understood (Knox 1970, 1981; Laws 1977). It does appear, however, that the mid-waters of the Southern Ocean are, in an evolutionary sense, under-utilized by fishes and could support more species. Although notothenioids lack swim bladders, there is a trend towards the evolution of pelagic species (Nybelin 1947), and this trend may be influenced by trophic factors, especially Antarctic Krill, *Euphausia superba,* (Andriashev 1964). The absence of competition from mid-water fish groups and the isolation of notothenioids have provided the opportunity for separation, adaptation and speciation within this group.

1 Department of Zoology, Irvine Hall, Ohio University, Athens, OH 45701, USA

Antarctic Nutrient Cycles and Food Webs
(ed. by W.R. Siegfried, P.R. Condy, and R.M. Laws)

Notothenioids have evolved to fill roles normally occupied by other fishes.

3 The Evolution of Neutrally Buoyant Notothenioids and Their Role in the Mid-Water Food Web

Knowledge of the composition of the mid-water food web in the Southern Ocean has been expanded by the discovery that several species of notothenioids are neutrally buoyant as adults and are therefore permanent members of the mid-water community.

3.1. Pleuragramma

Pleuragramma antarcticum is within 0.6% of neutral buoyancy (DeVries and Eastman 1978; Eastman and DeVries 1982). This species is the dominant fish in both the Ross (DeWitt 1970) and Weddell seas (Hubold 1985). It feeds on euphausiids and copepods in the Ross Sea (DeWitt and Hopkins 1977). In McMurdo Sound it eats amphipods and is cannibalistic (Eastman, pers. obs.). The species is the major item in the diet of *Dissostichus mawsoni* (see below) and Weddell Seals, *Leptonychotes weddelli,* (Testa et al. 1985) in McMurdo Sound. It is, therefore, a pivotal component in the food web and knowledge of its biology is vital to an understanding of the mid-water ecosystem of McMurdo Sound and the Southern Ocean in general.

3.2. Aethotaxis

Although fresh specimens were not available for study, Eastman and DeVries (1982) decided that, on the basis of specializations mentioned below, *Aethotaxis mitopteryx* is probably neutrally buoyant. Although its diet is unknown, the numerous (51), long, closely set gill rakers indicate that it eats smaller food items than *Pleuragramma* and other nototheniids. The species also has more extensive black pigmentation in the gut and body walls than other nototheniids (Eastman 1981). This indicates that the gut contents are probably bio-luminescent and that the pigmentation serves to dampen the bio-luminescence, so that the location of the fish will not be revealed to potential predators during the long periods of darkness in the Southern Ocean.

3.3 Dissostichus

Eastman and DeVries (1981a) discovered that the largest Antarctic fish, *Dissostichus mawsoni*, is neutrally buoyant. This species is a mid-water predator exhibiting low diet diversity; fishes and mysid shrimps being the most frequently occurring dietary items at McMurdo Sound (Table 1). These two items comprised 89.2 and 10.8%, respectively, of

Table 1. Stomach contents of 58 *Dissostichus mawsoni* collected at McMurdo Sound, Antarctica, October–December, 1978, 1979, 1981

Stomach contents	No. of individuals containing item	% frequency of occurrence[a]
Fishes	44	84.6
Pleuragramma antarcticum	37	71.2
Pagothenia borchgrevinki	4	7.7
Pagetopsis macropterus	4	7.7
Unidentifiable channichthyids	5	9.6
Trematomus loennbergii	2	3.8
Paraliparis devriesi	1	1.9
Gymnodraco acuticeps	1	1.9
Unidentifiable fish remains	4	1.7
Mysidacea	31	59.6
Antarctomysis sp.	31	59.6
Cephalopoda	7	13.5
Unidentifiable squid	2	3.8
Squid beaks (mandibles)	5	9.6
Amphipoda	2	3.8
Orchomene (Orchomenella) sp.	2	3.8
Empty stomach	6	10.3

[a] Only the 52 stomachs containing food were taken into consideration when calculating the frequency of occurrence of food items.

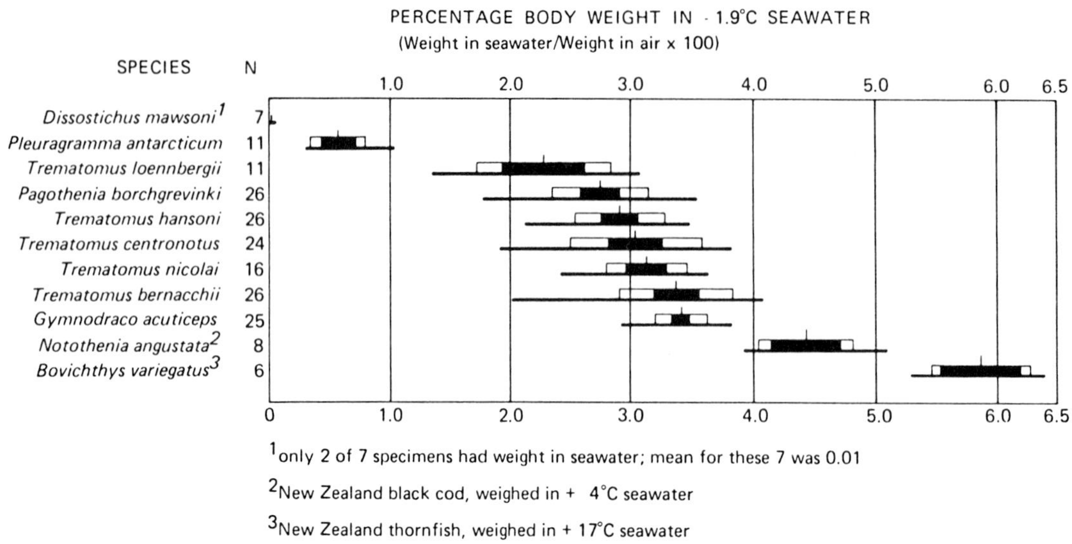

Fig. 1. Dice-Leraas diagram (modification of Hubbs and Hubbs 1953) comparing measurements of buoyancy among notothenioids from McMurdo Sound and New Zealand. Original data from Eastman and DeVries (1981a, 1982) and DeVries and Eastman (1981). Measurements for gravid females are not included. *Heavy horizontal* line is range; *short vertical line* is mean; *black rectangle* is 2 S.E.M. on each side of the mean; one-half of each *black rectangle* plus *white rectangle* is 1 S.D. on each side of the mean

the dry weight stomach contents in a sample of 12 *D. mawsoni* captured in 1981.

The presence of both cryo-pelagic and bentho-pelagic fishes in the diet indicates that *D. mawsoni* may occasionally feed throughout the water column. However, the mid-water inhabitant *Pleuragramma antarcticum* is the primary prey item both by occurrence (71.2%) and by dry wt. (89.2%). Channichthyids and squid are probably also captured in the mid-waters. Amphipods are so abundant in McMurdo Sound that they may be angulfed incidentally with other prey.

Dissostichus mawsoni is unlike many other notothenioids in that Krill, polychaetes, amphipods and isopods are not important dietary items (Arnaud and Hureau 1966; Hureau 1970; Richardson 1975; Permitin and Tarverdiyeva 1972, 1978; Duarte and Moreno 1981; Targett 1981; Daniels 1982). It is possible, however, that *D. mawsoni* may consume Krill at locations other than McMurdo Sound. It eats primarily squid near the Antarctic Covergence (Yukhov 1971). Thus, the species probably can be considered an opportunistic feeder in the mid-waters.

In McMurdo Sound, *D. mawsoni* is subject to minor depredation by Weddell Seals (Calhaem and Christoffel 1969; Testa et al. 1985) and Killer Whales, *Orcinus Orca,* (Thomas et al. 1981). Similarly to *Pleuragramma,* the species is an important component in the mid-water food web in McMurdo Sound, and it would be desirable to know more about its biology.

The evolution of neutral buoyancy in *Pleuragramma, Aethotaxis* and *Dissostichus*, species with a circum-Antarctic distribution (Permitin 1977), indicates that the mid-waters of the Southern Océan, sometimes regarded as nearly devoid of notothenioids, are inhabited by some specialized species, including the largest piscine predator in the region. By virtue of their neutral buoyancy these fishes are members of the mid-water food web. More research is needed, however, to assess the significance of their presence in the web.

3.4 Cryothenia

Daniels (1981) suggested that the recently described nototheniid *Cryothenia peninsulae*, from the Antarctic Peninsula, may be pelagic or bentho-pelagic. Based on a study of 13 museum specimens, I found that this species is, however, not neutrally buoyant. After re-hydrating eviscerated, alcohol-preserved specimens in seawater for 2 h at 2.6°–2.9°C, *C. peninsulae* had a mean percentage weight of 4.17 (S.D. = 1.72; S.E.M. = 0.478). This value is towards the heavy end of the spectrum for notothenioids (Fig. 1). However, it is possible that the presence of the viscera would have resulted in a lower percentage weight. Morphological examination of specimens revealed no obvious lipid deposits, and no skeletal reduction was evident in radiographs. The anal and pelvic fins were covered by thick skin. In addition, the rays of the anterior three-fourths of the anal fin were slightly club-shaped with a rudimentary branching pattern. These are probably substratum contact adaptations.

3.5. Buoyancy and Krill Feeding

Although Krill is the major dietary item of *C. peninsulae* (Daniels 1981), the species is neither neutrally buoyant nor morphologically specialized for permanent life in the mid-waters. It probably is capable of feeding on benthic organisms, as well as on seasonally abundant items in the mid-waters. Certain typical benthic notothenioids show similar feeding behaviour. For example, *Trematomus bernacchii* consumes mostly benthic organisms in Adélie Land (Hureau 1970), although Krill is taken near the Antarctic Peninsula (Moreno 1980). It should be emphasized that Krill is available as food to a variety of organisms living in the water column and on the bottom. It is not neccessary for an organism to be neutrally buoyant to feed on Krill. As Dearborn (1977) has indicated, Krill is an important dietary item for many benthic invertebrates.

While there has been a trend towards the evolution of neutrally buoyant species, these species are not as numerous or as diverse as the benthic species. Roberts (1982) noted that there is a greater variety of food and micro-habitats available to bottom-dwelling fishes than to mid-water fishes. Bottom-dwelling fish communities are consequently more diverse in higher taxa and richer in species than mid-water communities. Furthermore, for Antarctic species, there are disadvantages in a permanent pelagic existence, including severe environmental oscillations in ice cover, light and productivity. Nevertheless, some pelagic notothenioids have evolved and many other benthic and bentho-pelagic notothenioids are probably capable of exploiting the high secondary productivity during the summer months, since they are lighter than other teleosts without swim bladders (Eastman and DeVries 1982).

4 Morphological Specializations in Neutrally Buoyant Notothenioids

4.1. General Matters Pertaining to Buoyancy

Neutral buoyancy conserves muscular energy and, as Alexander (1967) has indicated, evolution by natural selection offers a plausible explanation for adaptations resulting in small savings in energy. A neutrally buoyant fish has no weight to support in water and no component of the forward locomotion must be diverted to provide hydro-dynamic lift.

Many marine fishes are made neutrally buoyant by a swim bladder with a capacity of 5% of the body volume (Marshall 1966). Assuming volume is roughly equal to weight, a marine fish without a swim bladder and with no special buoyancy adaptations must support about 5% of its weight in air when in the water column. Since notothenioids do not have swim bladders, evolutionary alterations in buoyancy are associated with modifications in the structure and functioning of other body systems. The New Zealand Black Cod, *Notothenia angustata,* and the New Zealand Thornfish, *Bovichthys variegatus,* are non-Antarctic notothenioids and relatively heavy, bottom dwellers. They serve as a standard of comparison in determining the degree of radiation in buoyancy exhibited by Antarctic notothenioids.

Eastman and DeVries (1981a, 1982) have described the method of determining buoyancy by weighing heavily anaesthetized fishes in seawater. The measurement is expressed as a percentage weight in seawater. Species supporting a mean of $< 0.6\%$ of their weight in air when in the water are considered neutrally buoyant. Most benthic, bentho-pelagic and cryo-pelagic species from McMurdo Sound are clustered in the 2.0–3.5% range (Fig. 1). Values for neutrally buoyant species and heavy, benthic non-Antarctic notothenioids lie at the extremes of this diagram. Fig. 1 also demonstrates that, for sample sizes greater than 20, the measurements of buoyancy approach statistical adequacy in terms of the ratio of dispersion (SD) to reliability (SEM) (Hubbs and Hubbs 1953).

Having established that there are neutrally buoyant notothenioids, we are confronted with investigating the modern results of an historical process: the evolution of buoyancy alterations in fishes without swim bladders whose ancestors were bottom dwellers. How is neutral buoyancy attained in the absence of a swim-bladder? Presumably there have been major evolutionary modifications in a variety of body systems. Among the most obvious buoyancy adaptations are a reduction in the quantity of heavy body components and an increase in the proportion of lighter components. My research (Eastman and DeVries 1981a, 1982; DeVries and Eastman 1978, 1981) indicates that evolutionary alterations in buoyancy are reflected as morphological specializations in a variety of body systems, especially those considered below.

4.2. Skeleton

The skeleton is the densest body component. The ash content provides a quantitative measure of the degree of mineralization of the skeleton. In neutrally buoyant species the ash content is less than 0.6% of the body weight. In notothenioids which are not neutrally buoyant this figure is 0.7%–3.8%, whereas in other teleosts it is 2.0–3.6% (Vinogradov 1952; Childress and Nygaard 1973). With a density of 1.1, cartilage is substituted for bone (density = 2.0) in some parts of the skeleton of neutrally buoyant species. This is especially marked in the skull, pectoral girdle and caudal skelton of *D. mawsoni*. Since the vertebral column represents the largest component of the skeleton by weight, any reduction in the mineralization of this structure yields a significant saving in weight. The vertebrae are unconstricted, vertebral processes are reduced and there is a persistent notochord in *P. antarcticum.*

4.3. Integument

Most notothenioids have well developed ctenoid scales. Since scales contain heavy bone salts, it is not surprising that scale mineralization is reduced in neutrally buoyant species. In *P. antarcticum* and *A. mitopteryx* the scales are weakly mineralized and deciduous. In *D. mawsoni* the scales are nonctenoid and the posterior margin of each scale is unmineralized.

4.4. Lipid Storage

Lipid deposits provide static lift in neutrally buoyant species. Lipid deposits in fishes may be subcutaneous, muscular, hepatic or mesenteric. Only the first two of these locations are extensively utilized in the neutrally buoyant species. *Dissostichus mawsoni* has a 2–8 mm thick subcutaneous layer of adipose tissue that accounts for 4.7% of the body weight. In *A. mitopteryx* the fishes are also almost completely encircled by a subcutaneous adipose layer that is 1.1–1.5 mm thick in a 151 mm specimen. In *P. antarcticum* there are subcutaneous lipid sacs (0.2–1.5 mm diam.) along the sides of the body, with greatest concentration near the centre of mass of the body in the pectoral region. *Pleuragramma* also has larger (0.5–3.0 mm diameter) intermuscular lipid sacs proximal to the bases of the dorsal and anal fins and adjacent to the dorsal and ventral median septa. *Aethotaxis mitopteryx* has typical adipose tissue in these same locations. The implications of lipid storage in connective tissue sacs compared to storage in adipose tissue have not been investigated. It is possible that lipid stored in sacs is not available for metabolism, since there is cellular mechanism to effect its removal.

Neutrally buoyant nototheniolds store lipid primarily in the form of triglycerides having a density of 0.92–0.93. Unlike some other teleosts, notothenioids studied to date do not store significant amounts of wax esters or other less dense lipids that would provide more static lift than triglycerides. It may be that wax esters would be in the solid state at sub-zero water temperatures, since they have higher melting points than triglycerides. This would impair membrane fluidity and body flexibility. Yayanos et al. (1978) have also noted that the buoyant force of wax esters decreases with decreasing temperatures.

4.5 Muscles

As in most other teleosts, the axial white musculature comprises about 50% of the body weight in neutrally buoyant notothenioids. Although muscles high in water content are used for buoyancy in some fishes without swim bladders, this is not the case in notothenioids. In *D. mawsoni*, for example, water content is 59% compared to 80% for most teleosts. *Dissostichus mawsoni* is also unusual among the neutrally buoyant notothenioids in having a large muscular lipid store that contributes to static lift. This deposit represents 4.8% of the body weight and is contained in adipose tissue among the fasciculi of the white axial musculature. The low water content of the muscle is attributable to the high lipid content.

4.6 Viscera

Neutrally buoyant notothenioids do not have extensive lipid deposits located among the viscera or in the mesenteries. The liver is not an organ of buoyancy in the sense that its large size and high lipid content provide static lift. However, the liver of *D. mawsoni* does contain significant amounts of intracellular lipid and is unique among vertebrates in possessing large numbers of perisinusoidal cells (Eastman and DeVries 1981b). Perisinusoidal cells are specialized for protein synthesis and it is possible that the protein may be involved in transporting lipid in the blood.

5 Priorities for Future Research

In addition to displaying the results of evolution within the Notothenioidei, buoyancy studies provide important information about a fish's position in the water column, mode of feeding, type of swimming and potential interactions with other members of the ecosystem. In this regard it would be interesting to study the diverse notothenioid fauna of West Antarctica, which is suspected to contain several species of pelagic juveniles and adults (Regan 1913, 1916; Nybelin 1947; Olsen 1955; Andriashev 1965; DeWitt 1971). Possible members of this group are the channichthyids *Champsocephalus gunnari, Cryodraco antarcticus, Neopagetopsis ionah* and *Pseudochaenichthys georgianus,* and some bathydraconids. There is a good chance that most of these species are also neutrally buoyant and would therefore be good subjects for investigating the evolution of buoyancy mechanisms in the most specialized notothenioid families.

Four of the five species in the Coriiceps group of the genus *Notothenia* have silvery coloured juveniles with forked caudal fins and long pectoral fins (DeWitt 1966). These adaptations for pelagic life may assist in widely distributing these species in the Southern Ocean. As recently described by Burchett (1983), one member of this group, *Notothenia rossii marmorata,* experiences habitat shifts during its life cycle. Fingerlings are pelagic, nearshore juveniles are demersal and offshore adults are both demersal and pelagic. These changes in habitat are accompanied by changes in colour, caudal fin shape and body shape, but nothing is known about buoyancy of a species with such varied habitats. How are buoyancy shifts accomplished in species experiencing ontogenetic habitat shifts during their lives? What is the buoyancy of an adult that lives both on the bottom and in the mid-waters? Is there a complex of morphological and physiological specializations that facilitate this dual existence? Does feeding on lipid-rich Krill increase the fat content (and buoyancy) of adult *N. r. marmorata*?

Acknowledgements. This research was supported by NSF grants DPP 77-15612 and DPP 79-19070 and by funds from the Ohio University College of Osteopathic Medicine. For help in the field I thank Art DeVries, Wayne Van Voorhies, Bob Boyd, Fred

Whoriskey, Joe Schrag, Scott O'Grady, Jeff Turner, Martha Wolfe, and Beth Marks. For the loan of specimens, I am grateful to Art DeVries and Hugh DeWitt. I also thank Diana Guthrie, Alberta Montgomery and Kathy Roberson for their expertise in word processing.

References

Alexander RMcN (1967) Functional design in fishes. Hutchinson, London

Andriashev AP (1954) Fishes of the northern seas of the USSR. Israel Program for Scientific Translations, Jerusalem

Andriashev AP (1964) On the composition and origin of the Antarctic pelagic fish fauna. In: Carrick R, Holdgate M, Prévost J (eds) Biologie Antarctique (Prem symp SCAR). Hermann, Paris pp 271–272

Andriashev AP (1965) A general review of the Antarctic fish fauna. In: van Oye P, van Mieghem J (eds) Biogeography and ecology in Antarctica (Monogr Biol Vol XV). Junk, The Hague pp 491–550

Arnaud P; Hureau J-C (1966) Régime alimentaire de trois Téléostéens Nototheniidae antarctiques (Terre Adélie). Bull Inst Océanogr (Monaco) 66:1–24

Burchett MS (1983) Morphology and morphometry of the Antarctic nothotheniid *Notothenia rossii marmorata.* Br Antarct Surv Bull 58:71–81

Calhaem I; Christoffel DA (1969) Some observations on the feeding habits of a Weddell seal, and measurements of its prey, *Dissostichus mawsoni,* at McMurdo Sound, Antarctica. NZ J Mar Freshwater Res 3:181–190

Childress JJ; Nygaard MH (1973) The chemical composition of midwater fishes as a function of depth of occurrence off southern California. Deep-Sea Res 20:1093–1109

Daniels RA *Cryothenia peninsulae,* a new genus and species of nototheniid fish from the Antarctic Peninsula. Copeia 1981:558–562

Daniels RA (1982) Feeding ecology of some fishes of the Antarctic Peninsula Fish Bull 80:575–588

Dearborn JH (1977) Foods and feeding characteristics of Antarctic asteroids and ophiuroids. In: Llano GA (ed) Adaptations within Antarctic ecosystems (Proceedings of the 3rd SCAR symposium on Antarctic biology). Smithsonian Institution, Washington pp 293–326

DeVries AL; Eastman JT (1978) Lipid sacs as a buoyancy adaptation in an Antarctic fish. Nature 271:352–353

DeVries AL; Eastman JT (1981) Physiology and ecology of notothenioid fishes of the Ross Sea. J R Soc NZ 11:329–340

DeWitt HH (1966) A revision of the Antarctic and southern genus *Notothenia* (Pisces, Nototheniidae). PhD Dissertation, Stanford University

DeWitt HH (1970) The character of the midwater fish fauna of the Ross Sea, Antarctica. In: Holdgate MW (ed) Antarctic ecology. Academic, London Vol 1 pp 305–314

DeWitt HH (1971) Coastal and deep-water benthic fishes of the Antarctic. Antarctic map folio series. American Geographical Society, New York, Folio 15 pp 1–10

DeWitt HH; Hopkins TL (1977) Aspects of the diet of the Antarctic silverfish, *Pleuragramma antarcticum.* In: Llano GA (ed) Adaptations within Antarctic ecosystems (Proceedings of the 3rd SCAR symposium on Antarct biology). Smithsonian Institution, Washington pp 557–567

Duarte WE; Moreno CA (1981) The specialized diet of *Harpagifer bispinis:* Its effect on the diversity of Antarctic intertidal amphipods. Hydrobiologia 80:241–250

Dunbar MJ (1970) Ecosystem adaptation in marine polar environments. In: Holdgate MW (ed) Antarctic ecology. Academic Press, London, Vol 1 pp 105–111

Eastman JT (1981) Morphological specializations in Antarctic fishes. Antarct J US 16:146–147

Eastman JT; DeVries AL (1981a) Buoyancy adaptations in a swim-bladderless Antarctic fish. J Morphol 167:91–102

Eastman JT; DeVries AL (1981b) Hepatic ultrastructural specialization in Antarctic fishes. Cell Tissue Res 219:489–496

Eastman JT; DeVries AL (1982) Buoyancy studies of notothenioid fishes in McMurdo Sound, Antarctica. Copeia 1982:385–393

Hubbs CL; Hubbs C (1953) An improved graphical analysis and comparison of series of samples. Syst Zool 2:49–57

Hubold G (1985) On the early life history of the high Antarctic fish *Pleuragramma antarcticum.* In: Siegfried WR, Condy PR, Laws RM (eds) Antarctic nutrient cycles and food webs (Proceedings of the 4th SCAR symposium on Antarctic biology). Springer, Berlin Heidelberg New York

Hureau J-C (1970) Biologie comparée de quelques Poissons antarctiques (Nototheniidae). Bull Inst Océanogr (Monaco) 68:1–244

Kennett JP (1978) The development of planktonic biogeography in the Southern Ocean during the Cenozoic. Mar Micropaleontol 3:301–345

Knox GA (1970) Antarctic marine ecosystems. In: Holdgate MW (ed) Antarctic ecology. Academic Press, London, Vol 1 pp 69–96

Knox GA (1981) Biological oceanography of the Ross Sea. J R Soc NZ 11:341–347

Laws RM (1977) The significance of vertebrates in the Antarctic marine ecosystem. In: Llano GA (ed) Adaptations within Antarctic ecosystems (Proceedings of the 3rd SCAR symposium on Antarctic biology). Smithsonian Institution, Washington pp 411–438

Llano GA (1978) Polar research: A synthesis with special reference to biology. In: McWinnie MA (ed) Polar research: To the present, and the future (AAAS selected symp ser No 7). Westview, Boulder pp 27–61

Marshall NB (1966) The life of fishes. Universe Books, New York

Moreno CA (1980) Observations on food and reproduction in *Trematomus bernacchii* (Pisces: Nototheniidae) from the Palmer Archipelago, Antarctica. Copeia 1980:171–173

Norman JR (1938) Coast fishes. Part III. The Antarctic zone. Discovery Rep 18:1–104

Nybelin O (1947) Antarctic fishes. Sci Results Norweg Antarct Exped 1927–1928 26:1–76

Olsen S (1955) A contribution to the systematics and biology of chaenichthyid fishes from South Georgia. Nytt Mag Zool 3:79–93

Permitin YY (1977) Species composition and zoogeographical analysis of the bottom fish fauna of the Scotia Sea. J Ichthyol 17: 710–726

Permitin YY; Tarverdiyeva MI (1972) The food of some Antarctic fish in the South Georgia area. J Ichthyol 12:104–114

Permitin YY; Tarverdiyeva MI (1978) Feeding of fishes of the families Nototheniidae and Chaenichthyidae in the South Orkney Islands. Sov J Mar Biol 4:619–622

Regan CT (1913) The Antarctic fishes of the Scottish National Antarctic Expedition. Trans R Soc Edin 49:229–292

Regan CT (1914) Fishes. Br Antarct ("Terra Nova") Exped 1910 Nat Hist Rep Zool 1:1–54

Regan CT (1916) Larval and post-larval fishes. Br Antarct ("Terra Nova") Exped 1910 Nat Hist Rep Zool 1:125–156

Richardson MG (1975) The dietary composition of some Antarctic fish. Br Antarct Surv Bull 41/42:113–120

Roberts TR (1982) Unculi (horny projections arising from single cells), an adaptive feature of the epidermis of ostariophysan fishes. Zool Scr 11:55–76

Targett TE (1981) Trophic ecology and structure of coastal Antarctic fish communities. Mar Ecol Prog Ser 4:243–263

Testa JW, Siniff DB, Ross MJ, Winter JD (1985) Weddell Seal – Antarctic Cod interactions in McMurdo Sound, Antarctica. In: Siegfried WR, Condy PR, Laws RM (eds) Antarctic nutrient cylces and food webs (Proceedings of the 4th SCAR symposium on Antarctic biology). Springer, Berlin Heidelberg New York

Thomas JA, Leatherwood S, Evans WE, Jehl JR, Awbrey FT (1981) Ross Sea killer whale distribution, behavior, color pattern and vocalizations. Antarct J US 16:157–158

Vinogradov AP (1953) The elementary chemical composition of marine organisms. Sears Foundation for Marine Research, Yale University, Memoir No II, New Haven

Vogel S (1981) Life in moving fluids: The physical biology of flow. Willard Grant, Boston

Yayanos AA, Benson AA, Nevenzel JC (1978) The pressure-volume-temperature (PVT) properties of a lipid mixture from a marine copepod, *Calanus plumchrus:* Implications for buoyancy and sound scattering. Deep-Sea Res 25:257–268

Yukhov VL (1971) The range of *Dissostichus mawsoni* Norman and some features of its biology. J Ichthyol 11:8–18

Krill Consumption by Antarctic Notothenioid Fish

K-H. KOCK[1]

Summary. The feeding ecology and the food intake of 4 species of channichthyid fishes (*Champsocephalus gunnari, Chaenocephalus aceratus, Pseudochaenichthys georgianus, Chionodraco rastrospinosus*) and 3 species of nototheniids (*Notothenia rossii marmorata, N. gibberifrons, N. larseni*) were studied in western Antarctic waters and off South Georgia. Based on the biomasses of these species, assessed by means of a 'swept area' method, estimates of their consumption of Krill, *Euphausia superba*, were made. Krill consumption by *C. gunnari*, for example, was about 800,000 t yr^{-1} off South Georgia, and 2400,000 t yr^{-1} in western Antarctic waters. The decrease in stocks of *C. gunnari*, due to heavy fishing, has reduced overall consumption of Krill to < 1 times 10^6 t yr^{-1}.

1 Introduction

Quantitative accounts of the trophic relationships and the food consumption of major groups of predators, such as fish, are needed in order to obtain the knowledge that is necessary for an understanding of the functioning of the Antarctic marine ecosystem and the management of its renewable resources.

In the Sub-Antarctic and Antarctic regions, more than 60% of the coastal fish species, and probably more than 90% of the individuals, belong to the endemic sub-order Notothenioidei (DeWitt 1971). Notothenioids are primarily bottom-dwelling fish which lack swim bladders. However, many of them especially nototheniids and channichthyids, are adapted for temporary or even permanent pelagic life and, thus, utilize the abundant food resources (mainly Krill, *Euphausia superba*) of the mid-water zone (Kompowski 1980b; Tarverdiyeva and Pinskaya 1980; Sosinski 1981; Kock 1981; Targett 1981; Daniels 1982; Takahashi 1983). Data on daily food consumption (Tarverdiyeva 1972, 1982; Gorelova and Gerasimchuk 1981), mean stomach content weights (Kompowski 1980a; Kock 1981; Rowedder 1984; Takahashi 1983), energy requirements (Chekunova and Naumov 1982a, b), gastric evacuation rates (Crawford 1978), standing stock biomass (Kock 1981; Mucha 1982) and productivity of fish populations (Everson 1970) are, however, scarce and confined to a few species. Consumption of euphausiids (mainly *E. superba* and *E. crystallorophias*) by fish inhabiting Antarctic coastal (= shelf) waters was estimated as 24–28 times 10^6 t yr^{-1} (Lubimova and Shust 1980). Consumption of Krill by the Icefish, *Champsocephalus gunnari,* in western Antarctic waters was assessed as 3–4 times 10^{-6} t in 1975–76 (Kock 1981).

In this paper I report a preliminary estimate of the amount of Krill consumed by seven species of notothenioid fish in western Antarctic waters. The species are: *Notothenia rossii, N. gibberifrons, N. larseni* (probably identical with *N. nybelini,* Balushkin 1976), *Champsocephalus gunnari, Chaenocephalus aceratus, Pseudochaenichthys georgianus* and *Chionodraco rastrospinosus* (probably identical with *Chionodraco hamatus* of other authors except Lönnberg (1905)). My estimates are based on data on fish-stock biomass, daily food consumption and the proportionate incidence of Krill in the stomach contents of fish.

2 Materials and Methods

Based on data from bottom-trawl surveys carried out in the Scotia Arc area during 1975–76 and 1977–78 (Table 1), fish-stock biomass was estimated by means of the "swept-area" method (Saville 1977). In general, sampling stations were chosen randomly, although some directed fishing was carried out on concentrations of *N. rossii* and *C. gunnari* detected in the course of the surveys (Sahrhage et al. 1978; Hempel et al. 1979). The application of the "swept-area" method, the stratification of the survey area by depth zones, and density of fish (for *N. rossii* and *C. gunnari*), as well

Table 1. Number of bottom-trawls made in the Scotia Arc area. After Kock (1981) and Mucha (1982)

	1975–76	1977–78	1980–81
South Georgia	94	63	507
South Orkney Islands	22	12	–
Elephant Island	17	30	–
South Shetland Islands	23	14	–
Totals	156	119	507

1 Institut für Seefischerei, Palmaille 9, D-2000 Hamburg 50, FRG

Antarctic Nutrient Cycles and Food Webs
(ed. by W.R. Siegfried, P.R. Condy, and R.M. Laws)

Table 2. Daily consumption of food expressed as a proportion (R) of the body weights of eight Antarctic fish species and the Atlantic Cod, *Gadus morhua*

Species[a]		Season	R (%)	Source
Notothenia rossii	Adult	Autumn (Ctenophores)	4.24–5.7	Tarverdiyeva (1972)
	Adult	Autumn (Krill)	0.94–1.26	Tarverdiyeva (1972)
	Juvenile	Summer–autumn	2.8 –3.6	Chekunova and Naumov (1982a)
	Juvenile	Winter–spring	1.8 –2.4	Chekunova and Naumov (1982a)
	Adult	Summer–autumn	7.4 –8.9	Chekunova and Naumov (1982a)
	Adult	Summer (Krill)	4.7 –5.7	Chekunova and Naumov (1982a)
	Adult	Winter–spring	1.1 –1.3	Chekunova and Naumov (1982a)
	Adult	Winter (Krill)	0.24–0.29	Chekunova and Naumov (1982a)
N. neglecta		Annual	1.4 –2.2	Everson (1970)
		Annual	2	Crawford (1978)
N. larseni		Annual	2	Crawford (1978)
		Summer–autumn	1.2 –2.0	Targett (1981)
N. gibberifrons		Annual	< 2.0	Crawford (1978)
		Summer–autumn	0.24–0.60	Targett (1981)
		Summer–autumn	1.51–2.76	Chekunova and Naumov (1982b)
		Winter–spring	0.85	Chekunova and Naumov (1982b)
N. angustifrons		Autumn	< 2.0	Crawford (1978)
		Autumn	1.01	Targett (1981)
Dissostichus eleginoides		Autumn	5.1	Tarverdiyeva (1982)
Pleuragramma antarcticum		Summer	0.5 –1.2	Gorelova and Gerasimchuk (1981)
Champsocephalus gunnari		Summer–autumn	10.1	Tarverdiyeva (1982)
Gadus morhua	Juvenile		0.85	Tyler (1970)

a If not otherwise stated juveniles + adults

as the operational characteristics of the trawls used (e.g., wing spread) have been described elsewhere (Kock 1980; Kroeger and Kock 1982). Biomass estimates for the 1980–81 season off South Georgia were derived from data in Mucha (1982).

Estimates of the daily food consumption of Antarctic fish species were taken from the literature (Table 2). These estimates had either been obtained from stomach contents analysis and observed or estimated gastric evacuation rates (Everson 1970; Tarverdiyeva 1972, 1982; Crawford 1978; Gorelova and Gerasimchuk 1981; Targett 1981) or by extrapolations from energy requirements (Chekunova and Naumov 1982a, b). Weight percentages of Krill in the diets were based either on empirical data (Kompowski 1980a; Kock 1981; Targett 1981; Chekunova and Naumov 1982a; Rowedder 1984; Takahashi 1983) or rough extrapolations from a number of qualitative studies (Latogurskij 1972; Naumov and Permitin 1973, Linkowski and Rembiszewski 1978; Kompowski 1980b; Tarverdiyeva and Pinskaya 1980; Daniels 1982).

3 Results

3.1 Biomass Estimates

In general, calculations were carried out separately for each sub-area and split-year. Due to the low number of hauls, catch data from both surveys of *Ch. aceratus* and *C. rastrospinosus* off the South Orkney Islands as well as those for all species off the South Shetland Islands – Antarctic Peninsula have been lumped together (Table 3), since neither substancial catches of the species (South Orkney Islands) nor any fishing on a commercial scale (South Shetland Islands) were reported until 1977–78 (FAO 1982). Most of the differences between the biomass estimates, as well as the somewhat high standard deviation (especially in areas other than South Georgia), are due to the high variability between individual hauls and the comparatively low number of hauls. Comparison with length-frequency distribution from the catches (Kock 1981, unpubl.) demonstrate that except for South Georgia sampling covered only parts of the populations of most of the species. This further lowers the accuracy of the estimates. Since part of the fishing effort around the South Orkney Islands and off Elephant Island concentrated on previously detected concentrations, no significant estimate for *C. gunnari* could be obtained for both seasons. Hence, the figures in Table 3 are extrapolations based on biomass estimates for South Georgia in 1975–76 (Kock 1981). The figure for the biomass of *C. gunnari* in 1980–81 is probably an over-estimate, whereas those for the other species are probably under-estimates, since fishing was directed mainly on concentrations of *C. gunnari* (Mucha 1982).

Table 3. Trawlable biomass and standard deviation (S_D) in % (95% confidence limits) of seven Antarctic notothenioids in west Antarctic waters (Shag Rocks to 69°S, 70°W) in the (split) years 1975–76, 1977–78 and 1980–81

	Year	South Georgia		S. Orkney Islands		Elephant Island		S. Shetland Islands Ant. Peninsula	
		Ton	S_D	Ton	S_D	Ton	S_D	Ton	S_D
Notothenia rossii	1975–76	35,682[a]	43.6	133	100.5	9,370	74.0	1,626	88.7
	1977–78	9,326	74.3	284	43.6	15,663	76.7	–	–
	1980–81	2,071	31.0	–	–	–	–	–	–
Notothenia gibberifrons	1975–76	40,094	64.7	68,430	80.4	16,471	192.0	53,680	43.9
	1977–78	20,100	35.2	29,187	77.6	17,824	145.6	–	–
	1980–81	12,186	< 15	–	–	–	–	–	–
Notothenia larseni	1975–76	449	63.3	562	57.9	100[b]	–	4,065	81.9
	1977–78	422	78.6	505	62.5	90	113.7	–	–
Champsocephalus gunnari	1975–76	141,469	125.9	140,000[b]	–	20,000[b]	–	22,162[c]	82.6
	1977–78	34,713	141.2	40,000[b]	–	20,000[b]	–	–	–
	1980–81	78,681	< 15	–	–	–	–	–	–
Chaenocephalus aceratus	1975–76	18,719	41.4	–	–	–	–	25,084	35.6
	1977–78	18,399	76.5	9,854	52.4	1,249	57.9	–	–
	1980–81	6,870	< 15	–	–	–	–	–	–
Pseudochaenichthys georgianus	1975–76	36,401	47.5	–	–	–	–	2,282	52.9
	1977–78	31,057	79.2	8,270	139.3	–	–	–	–
	1980–81	7,757	< 15	–	–	–	–	–	–
Chionodraco rastrospinosus	1975–76	–	–	8,759	64.0	–	–	22,144	44.9
	1977–78	–	–	–	–	1,015	59.6	–	–

a Different from Kock (1980) due to stratification by depth and fish density, whereas Kock (1980) stratified only by depth
b Rough estimate (see Kock 1981 p 75)
c Overestimated by Kock (1981)

3.2 Analysis of Stomach Contents

3.2.1 Empty Stomachs

Studies of the diets of Antarctic notothenioid fish (especially channichthyids) report substantial proportions of stomachs lacking food. Although this may influence assessments of food consumption rates to a large extent, most authors tend to confine their considerations to speculations relating this either to low overall feeding intensities (e.g., Holeton 1970; Linkowski and Rembiszewski 1978) or frequent regurgitation of the prey during sampling operations (e.g., Permitin and Tarverdiyeva 1972, 1978; Twelves 1972). Kock (1981), however, demonstrated that in *Ch. aceratus* 35.7–57.8% and in *Ps. georgianus* 25.3–58.2% of the empty stomachs collected in the spring-summer period (feeding season) were lacking food due to regurgitation. The proportion of non-feeding specimens in icefish is therefore probably much lower than previous studies have indicated. The proportion of empty stomachs only increases from autumn onwards, due to the fish feeding less intensively in the pre-spawning periods (Tarverdiyeva 1972; Linkowski and Rembiszewski 1978; Kompowski 1980a; Kock 1981), which is well known for many temperate-water species.

The proportions of empty stomachs for *N. rossii, N. gibberifrons* and *N. larseni* (Table 4) agree with the published data for the summer period (Tarverdiyeva 1972; Permitin and Tarverdiyeva 1972, 1978; Linkowski and Rembiszewski 1978). Although already corrected for regurgitation, the incidence of empty stomachs for *Ch. aceratus* and *Ps. georgianus* may be slightly overestimated for the feeding period,

Table 4. The proportion (%) of empty stomachs for seven species of Antarctic fish, based on data in Kock (1981) and Rowedder (1984)

Species	No. stomachs studied	% stomachs empty
Notothenia rossii	751	6.0
N. gibberifrons	1,671	7.1
N. larseni	256	6.2
Champsocephalus gunnari	3,052	10
Chaenocephalus aceratus	1,308	27
Pseudochaenichthys georgianus	541	25.7
Chionodraco rastrospinosus	1,344	17.1

Table 5. Relationship between mean (wet wt.) mass $\bar{w}$ (g) of stomach contents and total fish length (cm) and body-mass (g fr. wt.) of seven species of Antarctic notothenioid fish

Species	$\log \bar{w} = \log a + b \cdot \log L$ (W)	r	Source
Notothenia rossii	$\log \bar{w} = -3.43 + 2.85 \cdot \log L$	0.93	Rowedder (1984)[a]
	$\log \bar{w} = -1.63 + 0.94 \cdot \log W$		Rowedder (1984)[a]
N. gibberifrons	$\log \bar{w} = -3.93 + 2.85 \cdot \log L$	0.91	Rowedder (1984)[a]
	$\log \bar{w} = -1.86 + 0.89 \cdot \log W$		Rowedde (1984)[a]
N. larseni	$\log \bar{w} = -3.89 + 2.79 \cdot \log L$	0.94	Rowedder (1984)[a]
	$\log \bar{w} = -1.66 + 0.82 \cdot \log W$		Rowedder (1984)[a]
Champsocephalus gunnari	$\log \bar{w} = -4.51 + 3.64 \cdot \log L$	0.96	Kock (1981)[b]
	$\log \bar{w} = -1.52 + 1.08 \cdot \log W$		Kock (1981)[b]
Chaenocephalus aceratus	$\log \bar{w} = -6.15 + 4.74 \cdot \log L$	0.90	Kock (1981)[b]
	$\log \bar{w} = -2.05 + 1.34 \cdot \log W$		Kock (unpubl.)[b]
Pseudochaenichthys georgianus	$\log \bar{w} = -3.14 + 2.90 \cdot \log L$	0.94	Kock (unpubl.)[b]
	$\log \bar{w} = -0.75 + 0.82 \cdot \log W$		Kock (unpubl.)[b]
Chionodraco rastrospinosus	$\log \bar{w} = -2.85 + 2.4 \cdot \log L$	0.93	Rowedder (1984)[a]
	$\log \bar{w} = -0.83 + 0.65 \cdot \log W$		Rowedder (1984)[a]

a Based on stomach contents preserved in formaldehyde
b Based on stomach contents preserved by deep freezing

since some of the samples included specimens in pre-spawning condition.

3.2.2 Weight of Stomach Contents

As for example, in the North Sea Cod, *Gadus morhua* (Daan 1973) and the Antarctic myctophid *Electrona antarctica* (Rowedder 1979), changes in mass (wet wt.) of stomach contents could be closely related to fish length. A double logarithmic regression analysis of mean stomach-content weights (empty stomachs included) against length is set out in Table 5; length was easily transformed into weight, by means of length-weight relationships (Kock 1981; Rowedder 1984).

4 Proportion of Krill in Diets

In spite of annual fluctuations in abundance, Krill is continuously present in large quantities around the South Orkney Islands, South Shetland Islands and west of the Antarctic Peninsula. This permanent availability is clearly reflected in the food of the primarily pelagic feeding *N. rossii* and *C. gunnari;* Krill making up 90–100% in frequency of occurrence in a number of qualitative studies (Freytag 1977; Permitin and Tarverdiyeva 1978; Tarverdiyeva and Pinskaya 1980; Kock et al. 1980; Kock 1981, 1982) and at least 95% of the stomach-content mass (Kock 1981; Tarverdiyeva 1982; Takahashi 1983). Off South Georgia, the significance of Krill in the diets was much lower probably due to discontinuity in its abundance around the island. Its frequency of occurrence in stomachs of *N. rossii* and *C. gunnari* varied considerably between 15.2 and 90.4% and 18.1 and 94.1%, respectively, for studies made in different years as well as within one season (e.g., 52.9–90.4% for *N. rossii* in 1969–70 and 18.0–76.4% for *C. gunnari* in 1975–76) (Latogurskij 1972; Permitin and Tarverdiyeva 1972; Naumov and Permitin 1973; Linkowski and Rembiszewski 1978; Kock 1981). The proportion by weight of Krill in the diet of *N. rossii* has been estimated to be 43% (Chekunova and Naumov 1982a), and 56.2% (1975–76) and 40.2% (1977–78) (Rowedder 1984), and in stomachs of *C. gunnari* Krill made up 90% (1975–76), 56.5–72.5% (March–May 1977) and 42% (1977–78) (Kompowski 1980a; Kock 1981). Therefore, an overall mean weight proportion of 50% for both species off South Georgia seems to be a realistic value for further calculations.

The frequency of occurrence of Krill in the pelagic feeding *N. larseni* varied from 20.6 to 95% in the Scotia Arc region (Permitin and Tarverdiyeva 1972, 1978; Tarverdiyeva and Pinskaya 1980; Daniels 1982; Takahashi 1983). Weight percentages of Krill have been found to be 74.9% (May 1975, South Georgia), 41% (February 1976, South Orkney Islands) (Targett 1981), 38.5% (1975–76, 1977–78, South Orkney and South Shetland Islands) (Rowedder 1984) and 92% (Jan. 1981, South Shetland Islands) (Takahashi 1983). The data of Targett (1981) for one community off South Georgia and those of Rowedder (1984), however, indicate a decrease in the proportion of Krill with increasing length of fish. Thus, an overall mean weight proportion of 50% for all areas was chosen for further consideration.

Notothenia gibberifrons is a benthic feeder preying primarily on in-faunal organisms. Krill, nevertheless, may sometimes make up a considerable part of the diet. The frequency of its occurrence was in general low: 22.2–29.7% (Feb.–April 1965, 1967, 1969, South Georgia; Feb.–March 1967, 1970, South Orkney Islands) (Permitin and Tarverdiyeva 1972, 1978) but increased occasionally up to 60–97.6% (Feb. 1965, Feb. 1976, March 1976, Jan. 1981) (Linkowski and Rembiszewski 1978; Permitin and Tarverdiyeva 1978; Tarverdiyeva 1980; Takahashi 1983). Krill proportion in the prey of *N. gibberifrons* off South Georgia increased in winter when it is more often found close to the bottom (Chekunova and Naumov 1982b). Weight proportion of Krill was 3.7 and 5.7% in the Scotia Arc region in 1975–76 and 1977–78, respectively (Rowedder 1984) and 18% around South Georgia in winter (Chekunova and Naumov 1982b). Hence, a mean weight proportion of 10% was used in further calculations.

Several food studies have revealed that in *Ch. aceratus* (> 250 mm) and *C. rastrospinosus* (> 350 mm) Krill consumption steadily diminishes with increasing length of fish (Kock 1981; Takahashi 1983; Rowedder 1984). This was not apparent in the rather limited material (of similar length range) of Tarverdiyeva and Pinskaya (1980) in which Krill was present in 98.1% of the stomachs of *C. rastrospinosus.* Weight proportion of Krill in the diet of *Ch. aceratus* decreased from 90% in juveniles of 140–210 mm to < 10% in adults > 600 mm (Kock, unpubl.), although adults may also occasionally feed exclusively on Krill (Takahashi 1983). Furthermore, regional differences have been demonstrated by Gubsch (1982). The proportion of Krill in the stomachs of *C. rastrospinosus* decreased from 97.2% in specimens of 260 mm to 6.8% in those of 420 mm, with larger variations between the size groups (Rowedder 1984) and 100% in individuals of 100–200 mm to 72% in those of 400–500 mm in Takahashi's (1983) limited study. A mean weight proportion of Krill of 30% in *Ch. aceratus* and 60% in *C. rastrospinosus* was chosen for further calculations.

Pseudochaenichthys georgianus feeds mainly on Krill and fish. The frequency of Krill occurrence in different years varied from 25.0 to 87.4% off South Georgia and 38.5 to 100% in other areas of the Scotia Arc, but was mainly more than 60% (Permitin and Tarverdiyeva 1972, 1978; Linkowski and Rembiszewski 1978; Tarverdiyeva and Pinskaya 1980; Kock 1981). The rather limited material in quantative studies (Kock 1981; Takahashi 1983) allows no reliable estimate to be made for the weight proportion of Krill in the species' diet. A value of 60% for all areas was assumed for further consideration.

4.1 Daily Food Consumption

The validity of the data on daily food consumption (Table 4) is difficult to assess, since different methods and often different amounts of material were involved in the analyses. Results of studies on energy requirements, which indicate that daily food demands in *N. rossii* and *N. gibberifrons* slow down by a factor of 2–3 between summer and winter (Chekunova and Naumov 1982a, b), are somewhat contradictory to information given by Tarverdiyeva (1972; citing observations by Permitin and Shust) that *N. rossii* feeds intensively throughout the year except for a 2-month spawning season.

The daily food demands of *N. rossii, C. gunnari* and *D. eleginoides,* which were estimated from intervals between periods of intensive feeding (Tarverdiyeva 1972, 1982), are considerably greater than those for the other species set out in Table 2, as well as those derived from equations in Table 5, which are all assessments from mean stomach-content weights and estimated gastric evacuation rates (based on gastric evacuation rates of 2 d for Krill and 3 d for fish). Providing that gastric enzymes of Antarctic fish are not cold-adapted (Crawford 1978), the latter data are similar to those obtained from juvenile Atlantic Cod at similar temperatures feeding on shrimp tails (Tyler 1970). Although the data vary up to a factor of 6–8 within one species, they fall within anticipated limits for other cold-water species (Crawford 1978).

Following a new concept of cold adaptation (Clarke 1980, Kock 1985), the high daily food consumption for *N. rossii, C. gunnari* and *D. eleginoides* cannot be explained only by higher maintenance costs but may be due either to methodological differences or to extreme values for periods of maximum feeding intensity. In order to overcome these difficulties in the interpretation of the variability of the data in Table 2, daily rations of 5% in *N. rossii* and *C. gunnari,* 3% in *Ch. aceratus* and *Ps. georgianus* (due to the higher percentage of empty stomachs) and 2% in the other species in the feeding period (summer–autumn), and 1% in all species in the post-spawning period (winter–spring) have been used in my calculation of overall Krill consumption.

4.2 Krill Consumption

Evaluations of Krill consumption (Table 6) are based on biomass estimates (Table 3), an extrapolated initial stock size of 500,000 t for *N. rossii* around South Georgia before the onset of exploitation (Everson 1977), estimated daily food requirements and the proportions of Krill in the species' diet. Feeding during the spawning season (about 2 months) was ignored, since in *C. gunnari* off South Georgia more than 90% of pre-spawning fish had empty stomachs (Kock 1981).

Before the onset of commercial exploitation, total consumption of Krill by the seven species was probably about 5 times 10^6 t yr^{-1}. *Notothenia rossii* and *C. gunnari* were by far the most important species, sharing more than 80% of the annual consumption. Due to a substantial reduction in stock size, the significance of *N. rossii* as a predator on Krill around South Georgia may have become almost negli-

Table 6. Consumption of Krill (ton times 1,000) by seven Antarctic notothenoids in different localities and seasons

		South Georgia			S. Orkney Islands		Elephant Island		S. Shetland Islands	
Species	Before exploitation	1975–76	1977–78	1980–81	1975–76	1977–78	1975–76	1977–78	1975–76	1977–78
Notothenia rossii	2,250	160.8	42.0	9.3	1.1	2.4	80.1	133.9	13.9	
N. gibberifrons	–	18.0	9.0	5.5	30.7	13.1	7.4	8.0	24.1	
N. larseni	–	1.0	0.9	–	1.3	1.1	0.2	0.2	9.1	
Champsocephalus gunnari	–	630.0	156.2	354.0	119.7	342.0	171.0	171.0	189.5	
Chaenocephalus aceratus	–	33.7	33.1	12.4	17.7		–	2.2	45.2	
Pseudochaenichthys georgianus	–	131.0	11.8	27.9	–	29.8	–	–	8.2	
Chionodraco rastrospinosus	–	–	–	–	31.5		–	3.7	79.7	
Totals		974.5	353.0	409.1	1,300	437	258	319	369.7	

gible, whereas *C. gunnari* is still the most important consumer in spite of its consumption probably having decreased by a factor of 4–5. Consumption estimates for *C. gunnari* off South Georgia agree with those of Kock (1981) which were based on mean stomach-content weights and observed weight proportions of Krill. The estimate was about 800,000 in 1975–76 when Krill was very abundant around South Georgia, resulting in weight proportions of 80–90% in the diet (i.e., above the mean value of 50% of used here) and about 70,000 t in 1977–78 when Krill abundance was extremely low and Krill made up only about 30–40% of the diet.

5 Discussion

Experience has shown that two major sets of factors influence the accuracy of fish biomass estimates based on bottom-trawl surveys: factors contributing to the systematic error (e.g., vessel type, type of net, officers and crew), which may become important when more than one vessel is used in surveys; and, factors contributing to the variance. A significant fraction of the variance can be related to the (present) lack of control over the operation of the trawl (e.g., the speed of the trawl through the water), which expose the trawl to continuously changing conditions during the tow (see Carrothers 1981). The uneven spatial distribution of certain species due to differences in environmental conditions should increase the variance quite substantially, especially for the highly gregarious *N. rossii* and *C. gunnari* which tend to swarm when Krill is abundant and may disperse when the availability of Krill is relatively low (Kock et al. 1980; Kock 1981). This can be compensated only partially by a re-stratification of the survey area according to the observed differences in abundance during the survey. The efficiency of the trawl (the proportion of fish in the area sampled which is actually caught) was not constant from species to species. The efficiency was most probably higher in the more bottom-dwelling "slurp" and "ambush" feeders (*N. gibberifrons* and *Ch. aceratus*) than in the more actively hunting *N. rossii, C. gunnari* and *Ps. georgianus* which exhibit regular vertical migrations. Due to the lack of data for estimating the "catchability coefficient" (c) of any species, it was assumed to be equal to 1 (i.e., 100% efficiency) although it actually is most probably lower. Thus, the estimate of trawable biomass is most surely an under-estimate of the actual standing stock biomass, if not balanced by a "shepherding" effect of the trawl otter-boards and bridles. The magnitude of these factors and the contribution of each of them to the variance is impossible to assess. Standard deviations of biomass estimates for at least certain species, however, are in good accordance with those estimated for species occurring on Georges Bank, which were in general ± 50% of the mean (Byrne et al. 1981).

Part of the decrease in the estimated fish biomass off South Georgia and the South Orkney Islands obviously reflects the substantial depletion of certain stocks from 1975–76 to 1977–78. Around South Georgia, catches of *N. rossii* and *C. gunnari* were reported to be 19,000 t and 132,000 t in 1975–76 and 1976–77, respectively. Landings of *C. gunnari* in 1977–78 amounted to 154,000 t (FAO, 1982), most of them probably taken around the South Orkney Islands.

The seven species studied here made up most of the biomass of the nototheniids and channichthyids in the Scotia Arc region before the onset of commercial exploitation. The biomass of all nototheniid and channichthyid species from 50–500 m depth then would have been about 1.1–1.2 times 10^6 t. This is lower than the 1.8–2.2 times 10^6 t given by Lubimova and Shust (1980) which is based on a somewhat larger dimension of the shelf area and the production estimates of Everson (1977).

Krill is permanently abundant in sub-surface waters of the southern Scotia Arc region and in some years also around South Georgia. Regular sampling during 5 months around Elephant Island, nevertheless, indicated that considerable temporal and spatial differences may exist (Kock et al. 1980). In shelf-waters, where no stratification of the water column is present in summer, Krill irregularly descends to the bottom. Many Antarctic fish species have opportunistic feeding habits, taking advantage of Krill when it is abundant. This most probably leads to considerable fluctuations in the intake of Krill by different species, especially in those which tend to be bottom-dwelling forms (e.g., *N. gibberifrons* and adult *Ch. aceratus*). Most food studies, however, have been limited to relatively short periods (often a few days) and small areas of sampling. Their results may, therefore, reflect either, at most, the actual temporal or spatial availability of Krill in the summer–autumn or are merely fortuitous rather than representative. Furthermore, certain authors have not taken into account the fact that several species change their diet with size. This may have biased, to an unknown extent, the estimation of the mean weight proportion of Krill in the diet of the fish.

Information on winter feeding is still scarce and contradictory. Extrapolations from energy requirements suggest that *N. rossii* and *N. gibberifrons* around South Georgia reduce their daily summer–autumn and winter–spring food intake by a factor of 2–3 (Chekunova and Naumov 1982a, b). Tarverdiyeva (1972) for adult *N. rossii* off South Georgia and Everson (1970), Marshall (1964) and Daniels (1982) for *N. neglecta* off Signy Island (South Orkney Islands) and off the Antarctic Peninsula, however, report intensive feeding in the winter–spring period. Burchett (1983) found that condition in juvenile *N. rossii* in Cumberland Bay (South Georgia) was poorest in late winter–early spring, but this was attributed to a lower energy content of the prey and not to a lower feeding intensity by the fish. Except for South Georgia, virtually no data exist on Krill abundance and availability as food in winter. Off South Georgia, Krill often sinks to the bottom in winter; thus increasing the proportion of Krill in the diet of the benthic feeding *N. gibberifrons* in comparison to the summer (Chekunova and Naumov 1982b). If Krill is available in similar quantities in both winter and summer, and feeding intensity of fish is not much reduced in winter, except for the spawning season, then annual Krill consumption would be much higher than is given in Table 6.

Although digestion time primarily influences the rate of daily food intake, very few studies have been dedicated to the determination of gastric evacuation rates in Antarctic fish. Different methods here revealed digestion times ranging from 28 to about 120 h, depending on the type, size, number and weight of the prey (Tarverdiyeva 1972; Crawford 1978; Rowedder 1979). Gastric evacuation rates of fish feeding on Krill and comparable crustaceans range from 1.5 to 2.5 d; similar to what has been found for Atlantic Cod kept at 2°C and fed on shrimp tails (Tyler 1970). More work on gastric evacuation rates in fish is needed.

Consumption of Krill by Antarctic nototheniid and channichthyid fish has been estimated to be 5.4–6.6 times 10^6 t yr^{-1} in the Scotia Arc region (Lubimova and Shust 1980). If my biomass estimate is an under-estimate of the true standing-stock biomass, it suggests that the total annual consumption of Krill would have been about 9–10 times 10^6 t before the onset of commercial exploitation of the fish. This figure may be doubled, if the species do feed intensively in winter. Although my figures are still very preliminary and provisional, it is obvious that the consumption of Krill by Antarctic notothenioid fish species in the Scotia Arc region has been substantially reduced due to their heavy depletion in the course of large-scale exploitation in recent years.

Acknowledgements. This paper is based on work carried out during the FRG Antarctic expeditions of 1975–76 and 1977–78, and was supported financially and logistically by the Ministry of Research and Technology and the Ministry of Food, Agriculture and Forestry.

References

Burchett MS (1983) Age and growth of the Antarctic fish *Notothenia rossii* from South Georgia. Br Antarct Surv Bull 60:45–61

Byrne CJ; Azarovits TR; Sissenwine MP (1981) Factors affecting variability of research vessel trawl surveys. In: Doubleday WG, Rivard D (eds) Bottom Trawl Surveys. Can Spec Publ Fish Aquat Sci 58:258–273

Carrothers PJG (1981) Catch variability due to variations in groundfish otter trawl behaviour and possibilities to reduce it through instrumented fishing gear studies and improved fishing procedures. In: Doubleday WG, Rivard D (eds). Bottom Trawl Surveys. Can Spec Publ Fish Aquat Sci 58:247–257

Chekunova VJ; Naumov AG (1982a) Energy metabolism and food requirements of the marbled notothenia *Notothenia rossii marmorata* (Nototheniidae). Vopr Ikthiol 22(2):294–302

Chekunova VJ; Naumov AG (1982b) Energy metabolism of *Notothenia gibberifrons* Lönnberg off South Georgia. Trudy VNIRO, Characteristics of the Pelagic Community from the Sea of Scotia and Adjacent Waters, pp 76–85

Clarke A (1980) A reappraisal of the concept of metabolic cold adaptation in polar marine invertebrates. Biol J Linn Soc 14:77–92

Crawford RE (1978) Digestive system morphology, gastric evacuation rates and energetics in Antarctic notothenia. PhD Thesis. University of Maine, Orono 114 p

Daan N (1973) A quantitative analysis of the food intake of North Sea cod *Gadus morhua*. Neth J Mar Res 6:479–517

Daniels RA (1982) Feeding ecology of some fishes of the Antarctic Peninsula. Fish Bull 80(3):575–588

DeWitt HH (1971) Coastal and deep-water benthic fishes of the Antarctic. Antarct Map Folio Series, AGU, 15:1–10

Everson I (1970) The population dynamics and energy budget of *Notothenia neglecta* Nybelin at Signy Island, South Orkney Islands. Br Antarct Surv Bull 23:25–50

Everson I (1977) The living resources of the Southern Ocean. FAO, GLO/SO/77/1, Rome, 156 p

FAO (1982) Yb Fish. Statistics Vol 53. Fishery commodities, 1981, FAO, Rome

Freytag GE (1977) Beiträge zur Biologie von *Notothenia rossii marmorata* Fischer. MS Thesis University of Kiel, 72 p

Gorelova TA; Gerasimchuk VV (1981) Data on nutrition and daily consumption of juvenile *Pleuragramma antarcticum* Boul. Fishes of the Open Ocean, Shirsov Institute of Oceanology Moscow, pp 103–109

Gubsch G (1982) Zur Verbreitung und Biologie der Eisfische (Chaenichtyidae) im atlantischen Sektor der Antarktis. Fischereiforsch 20(2):39–47

Hempel G; Sahrhage D; Schreiber W; Steinberg R (1979) Antarktis-Expedition 1977/78 der Bundesrepublik Deutschland. Arch Fisch Wiss 30(Beih 1):1–119

Holeton GF (1970) Oxygen uptake and circulation by a haemoglobinless Antarctic fish (*Chaenocephalus aceratus* Lönnberg) compared with 3 red blooded Antarctic fish. Comp Biochem Physiol 34(2):457–471

Kock KH (1980) On the estimation of fish stock sizes off South Georgia using "swept area" method. Paper presented at 2nd Meeting of the Working Party of Antarctic Fish Biology, Dammarie-les-Lys, France 1980 (mimeo)

Kock KH (1981) Fischereibiologische Untersuchungen an drei antarktischen Fischarten: *Champsocephalus gunnari* Lönnberg 1905, *Chaenocephalus aceratus* (Lönnberg 1906) und *Pseudochaenichthys georgianus* Norman 1937 (Notothenioidei, Channichthyidae). Mitt Inst Seefisch Hamb 32:1–226

Kock KH (1982) Fischereibiologische Untersuchungen bei Elephant Island im März 1981. Arch Fischereiwiss 33 (Beih. 1):127–142

Kock KH (1985) Antarctic fish. In: Bonner WN, Walton DJ (eds) Key Environments – Antarctica. Pergamon, London pp. 173–192

Kock KH; Nast F; Stein M (1980) Some interactions between hydrography, krill and fish near Elephant Island. ICES CM 1980/L:3 (mimeo)

Kompowski A (1980a) On feeding of *Champsocephalus gunnari* Lönnberg 1905 (Pisces, Chaenichthyidae) off South Georgia and Kerguelen Islands, Acta Ichthyol Piscat 10:25–43

Kompowski A (1980b) Studies on juvenile *Chaenocephalus aceratus* (Lönnberg 1906, Pisces, Chaenichthyidae) from off South Georgia. Acta Ichthyol Piscat 10:45–53

Kroeger M; Kock KH (1982) First attempts to quantify variations in behaviour of groundfish otter trawls used by the Federal Republic of Germany in North Atlantic surveys. NAFO Scr Doc 82/VI/51, 15 p

Latogurskij VL (1972) *Euphausia superba* Dana as a food item of *Notothenia rossii marmorata* (russ.) Trudy Atlant NIRO 42:167–169

Linkowski TB; Rembiszewski JM (1978) Ichthyological observations off the South Georgia coasts. Pol Arch Hydrobiol 25(3): 697–704

Lönnberg AJE (1905) The fishes of the Swedish South Polar Expedition. Wiss Erg Schwed Südpol Exped 5(6):1–69

Lubimova TG; Shust KV (1980) Estimation of the Antarctic krill consumption by the main consumer groups (in russian). In: Biologicheskie Resursy Antarkticheskogo Krilya. ONTI VNIRO, Moscow:203–224

Marshall NB (1964) Fish. In: Priestley R, Adie RY, Robin G de Q (eds) Antarctic Research. A review of British scientific achievement in Antarctica. Butterworths, London, pp 206–218

Mucha M (1982) Biomass estimates of commericial fishes in the South Georgia region (by "swept area" method) Paper presented at the Meeting of the Working Party on Antarctic Fish Ecology, Hamburg, Federal Republic of Germany, 1982 (mimeo)

Naumov AG; Permitin YuYe (1973) Trophic relationships of *Euphausia superba* Dana and fish of the Southern Ocean (Scotia Sea) (russ.). Trudy VNIRO 93:216–229

Permitin YuYe (1970) The consumption of krill by Antarctic fishes. In: Holdgate MW (ed) Antarctic Ecology Vol 1, Academic, London, pp 177–182

Permitin YuYe; Tarverdiyeva MI (1972) Feeding of some species of Antarctic fishes in the South Georgia Island area. J Ichthyol 12 (1):104–114

Permitin YuYe; Tarverdiyeva MI (1978) Feeding of Antarctic cods and ice fishes near the South Orkney Islands. Biol Morya 2:75–81

Rowedder U (1979) Feeding ecology of the myctophid *Electrona antarctica* (Günther 1878) (Teleostei). Meeresforschung 27:252–263

Rowedder U (1984) Nahrungsökologie antarktischer Fische. PhD Thesis, University of Kiel, 163 p

Sahrhage D; Schreiber W; Steinberg R; Hempel G (1978) Antarktis-Expedition 1975/76 der Bundesrepublik Deutschland Arch Fisch Wiss 29(Beih 1):1–96

Saville A (1977) Survey methods of appraising fishery resources. FAO Fish Techn Rap 171:1–76

Sosinski J (1981) Biologia porownawcza Kergulen (*Champsocephalus gunnari*) z rejonow Antarktyki. Studia i Materialy Seria B, number 48, Morskiego Instytutu Rybackiego Gdynia, 91 p

Takahashi M (1983) Trophic ecology of demersal fish community north of the South Shetland Islands with some note on the ecological role of the krill. Mem Natl Inst Polar Res Spec Issue 27: 183–192

Targett TE (1981) Trophic ecology and structure of coastal Antarctic fish communities. Mar Ecol Progr Ser 4:243–263

Tarverdiyeva MI (1972) Daily food consumption and feeding pattern of the South Georgian cod (*Notothenia rossii marmorata*) and the Patagonian toothfish (*Dissostichus eleginoides* Smitt) (fam Nototheniidae) in the South Georgia area. J Ichthyol 12(4):684–692

Tarverdiyeva MI (1982) The food composition, daily ration and feeding habits of the icefish *Champsocephalus gunnari* Lönnberg off the South Orkneys. Trudy VNIRO, Characteristics of the Pelagic Community from the Sea of Scotia and Adjacent Waters, pp 69–76

Tarverdiyeva MI; Pinskaya IA (1980) The feeding of fishes of the families Nototheniidae and Chaenichthyidae on the shelves of the Antarctic Peninsula and the South Shetlands. J Ichthyol 20 (4):50–60

Twelves EL (1972) Blood volume of two Antarctic fishes. Br Antarct Surv Bull 31:85–92

Tyler AV (1970) Rates of gastric emptying in young cod. J Fish Res Bd Can 27(7):1177–1189

The Early Life-History of the High-Antarctic Silverfish, *Pleuragramma antarcticum*

G. HUBOLD[1]

Summary. During the German Antarctic expeditions of 1979–1980 and 1980–1981 plankton and fish fry were collected quantitatively in the eastern and southern Weddell Sea. The samples were obtained mainly by double-oblique bongo with 300 or 500 μ nylon gauze; they carried digital flow-meters and time-depth recorders. The 1979–80 cruise yielded more than 28,000 fish larvae in 102 hauls (275 per haul), and 939 larvae were caught in 55 hauls (17 per haul) in 1980–81. More than 95% of the fish larvae belonged to *Pleuragramma antarcticum*. Maximum densities of 774 larvae m^2 and 59 larvae m^2 were found in 1979–80 and 1980–81, respectively. Length of age-group 0 fish ranged from 8–16 mm in January of both years, and age-group 1 fish were 25–47 mm at the same time. By the end of February, lengths of age-group 0 fish were 18–22 mm. Larval growth of *P. antarcticum* during the 6-week observation period was 0.24 mm d^{-1} for 0 age-group fish. These larvae were concentrated in the 25–50 m water layer. Changes in relative eye-size, from 5.5% in 12 mm larvae to > 8% in adults, are related to life at increasing depths. *Pleuragramma antarcticum* larvae co-occurred frequently with larval channichthyids which preyed on them. A hypothetical account of the life-cycle of the species in the Weddell Sea is given.

1 Introduction

The Antarctic Silverfish, *Pleuragramma antarcticum*, is the only true pelagic species among notothenioid fish (Andriashev 1965). In the Ross Sea, the pelagic fish fauna is dominated by this species (92% in numbers and 96.7% in biomass), and northern mid-water fish, such as myctophids, are excluded from the system by the steep temperature gradient over the edge of the outer shelf (DeWitt 1970). The description of *P. antarcticum* by Boulenger (1902) was based on a few specimens obatined at the ice barrier (at 78° 35′S) during the expedition of the *Southern Cross*. Pappenheim (1912) reported information, based on some 30 adults and more than 1000 post-larvae from Wilhelm II Land (92°E). Regan (1916) described the species' larval development in the Ross Sea, and gave sizes of larvae and young fish at different times of the year. Norman (1938) reviewed this early material. Recent descriptions of the post-larvae and juveniles of *P. antarcticum* were provided by Efremenko (1983).

1 Alfred-Wegener-Institut für Polarforschung, Columbus Center, D-2850 Bremerhaven, FRG

The importance of *P. antarcticum*, as the most abundant pelagic fish in high-Antarctic waters, is reflected by its occurrence in the stomachs of predators. The species is the main food of the Antarctic Cod, *Dissostichus mawsoni*, (Eastman and DeVries 1981) and it has also been found in Weddell Seals, *Leptonychotes weddelli*, (DeWitt and Tyler 1960), in Whales (Andriashev 1965), in Gentoo, *Pygoscelis papua*, and Adélie, *P. adeliae*, Penguins (Volkman et al. 1980), in Emperor Penguins, *Aptenodytes forsteri*, (Norman 1937), in South Polar Skuas, *Catharacta maccormicki*, (P. Pietz, pers. comm.), and in Antarctic Petrels, *Thalassoica antarctica*, (H-E. Drescher, pers.comm.). Lubimova et al. (1973), in a much cited paper, described *P. antarcticum* as a "widely distributed species which lives close to the edge of the pack ice . . . and can be used for human consumption". Consequently, in 1981, some 1500 t of "Antarctic Side Stripe" was caught by the U.S.S.R fishing fleet in the Pacific sector of the Antarctic Ocean (Anon 1983).

The present paper deals with the distribution, abundance and growth of *P. antarcticum* in the southern and eastern Weddell Sea, based on larval specimens obtained by means of plankton nets during cruises in 1979–80 and 1980–81 Hempel et al. (1983) and Keller (1983) provide preliminary accounts of the ichthyoplankton collected during these cruises.

2 Materials and Methods

Plankton samples were collected during cruises of the Norwegian ice-breaker R.V. Polarsirkel, chartered for German expeditions to the Weddell Sea in the austral summers of 1979–80 and 1980–81.

In 1979–80, 102 bongo net-hauls were made along a cruise track from South Georgia via Kap Norwegia to the southern Weddell Sea, as far west as 61° 19′W, and then to Atka Ice Port and Bouvet Island (Schneppenheim 1982). A total of 165 vertical hauls was made with a 700 mm diameter Nansen closing net, mainly at a time-series station and in a "box" of 75 × 75 km of the Ronne-Filchner Ice Shelf. The bongo nets were of 500–500 μ gauze, except at the first 17 stations where nets of 300–500 μ mesh were used. They were fitted with digital flow-meters and time-

Antarctic Nutrient Cycles and Food Webs
(ed. by W.R. Siegfried, P.R. Condy, and R.M. Laws)

depth recorders (Benthos Corp.). The hauls were made through the upper 140 m water layer, with the ship moving at 4.6 km h^{-1}. The Nansen nets of 335 μ mesh were hauled vertically at low speed through discrete water layers between 700 m and the surface.

In 1980–81, 56 bongo hauls with a 300–300 μ mesh net and 42 vertical hauls with a Nansen net of 335 μ mesh were made along the cruise track from South Georgia via Kap Norwegia to the southern Weddell Sea (Gould Bay), and then to Atka Ice Port where a 111 × 111 km "box" was sampled. The standard hauling depth for the bongo nets was 200 m, at a ship's speed of 3.7–5.5 km h^{-1} (Hubold and Drescher 1982). The vertical net was used as a nonclosing net only, between 2400 m and the surface.

All plankton samples were fixed and stored in 4% borax-buffered formaldehyde. Fish larvae usually were sorted from unsplit samples.

3 Results

3.1 Abundance

More than 28000 fish larvae were caught in the 102 bongo hauls made during 1979–80. Out of a total 14466 larvae in the bongo I (500 μ) samples, 13707 (95%) were *P. antarcticum* (Keller 1983). The species was most abundant at Station 27 (73° 30′S, 22° 22′W), between Camp Norway and Halley Bay, where it accounted for 99.8% of all (3920) fish larvae caught in the bongo I. At Station 33 (77° 43′S, 36° 13′W), in Vahsel Bay, all 2900 larvae were *P. antarcticum*. At Station 87 (77° 14′S, 41° 43′W), in Gould Bay, the species contributed 99.6 % of all (3920) fish larvae.

Considerably fewer fish larvae were caught during 1980–81. Only 939 larvae were obtained in 55 bongo (335 μ) hauls, of which 798 (85%) were *P. antarcticum*. Maximum abundance was at Station 97 (77° 42′S, 37° 00′W), in Vahsel Bay, where all 427 larvae were *P. antarcticum*. Similarly, only *P. antarcticum* was represented at Station 99 (77° 31′S, 39° 12′W), also in Vahsel Bay where 148 larvae were caught.

Figure 1 shows the standardized abundance of larvae of *P. antarcticum* at stations in the Weddell Sea in 1979–81.

3.2 Length

The size of post-larvae ranged from 8–24 mm, based on a sub-sample of 2810 *P. antarcticum* taken from the 1979–80 bongo I nets (Keller 1983). Only 20 specimens in the entire collection were between 25 and 84 mm SL (Standard Length). These fish were found at Station 29 (74° 58′S, 25° 19′W; n = 14) and off the Filcher and Ronne ice shelves. The size frequency distribution of post-larvae from Stations 27 and 29 is shown in Fig. 2. In 1980–81, the post-larvae ranged between 9 and 45 mm, and a bimodal SL distribution, with maxima at 10–31 mm, was found at Station 97 (Fig. 2).

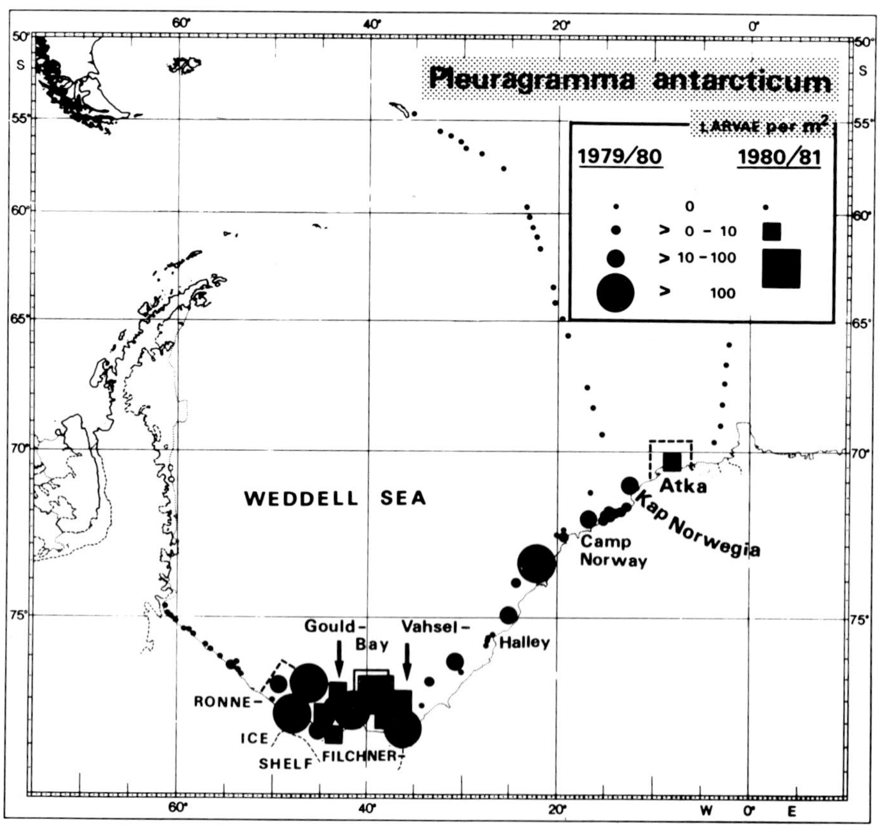

Fig. 1. Abundance of post-larvae of *P. antarcticum* in the southern and eastern Weddell Sea in 1979–80 and 1980–81, based exclusively on samples taken by bongo nets

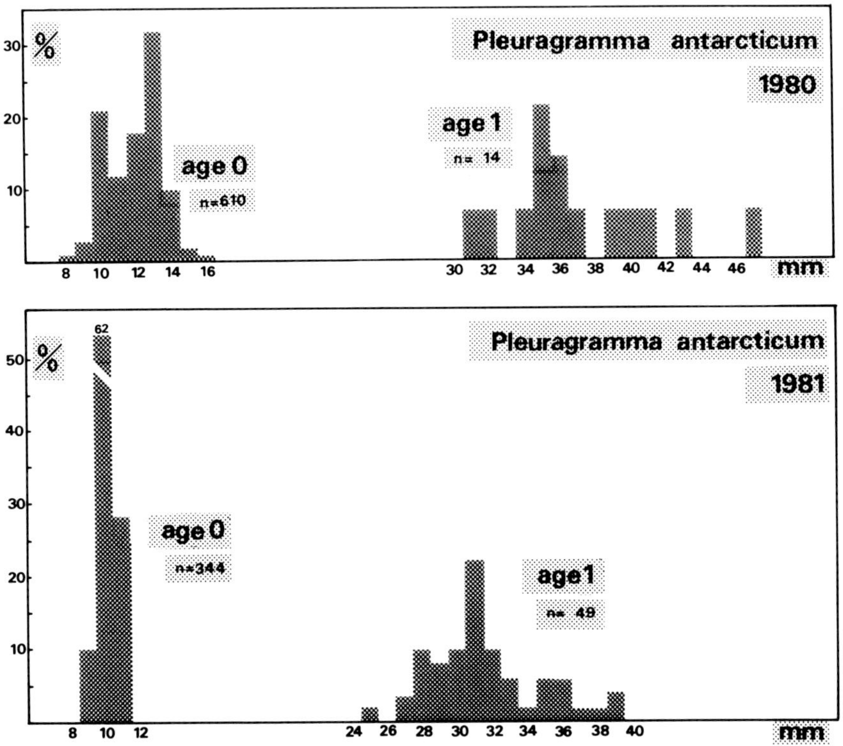

Fig. 2. Size-frequency distribution of post-larvae of *P. antarcticum* at Stations 27 and 29 (01 Jan. 1980) and Station 97 (04 Jan. 1981). Data for 1980 taken from Keller (1982)

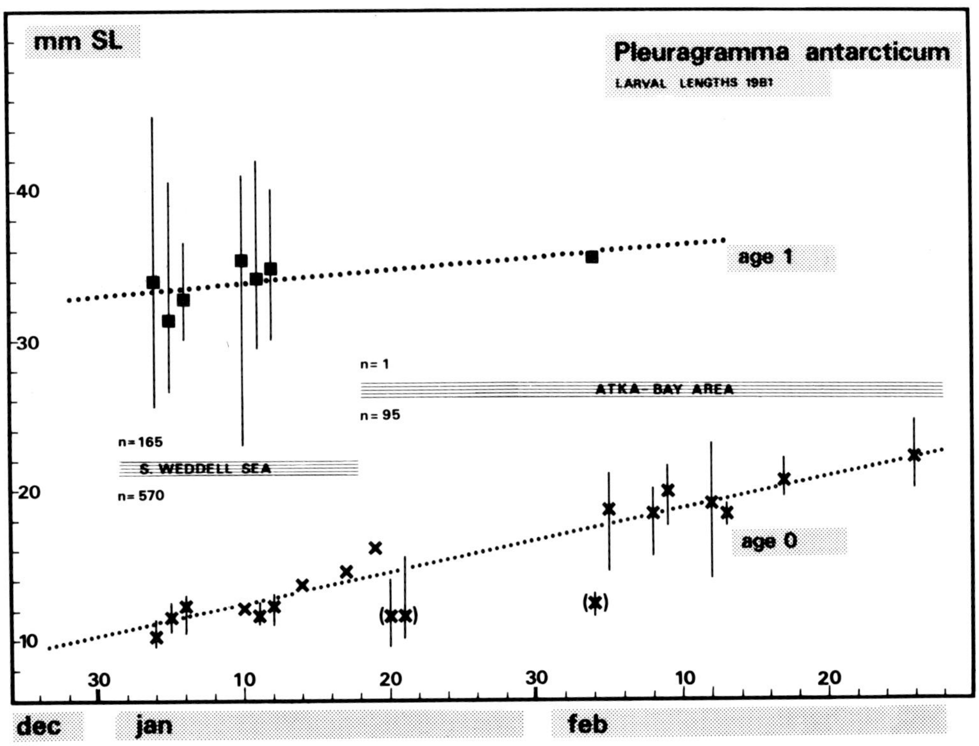

Fig. 3. Mean (and range) standard length of post-larvae of *P. antarcticum* in 1981. Values in brackets omitted from linear-curve fitting (AC 0:r^2 = 0.95; n = 16. AC 1:r^2 = 0.38; n = 7)

Based on the observed size distribution (Fig. 2), post-larvae, between 8 and 16 mm, and specimens of 25–39 mm were regarded as belonging to age classes (AC) 0 and 1, respectively. The proportion of AC 0 to AC 1 in 1980–81 in the southern Weddell Sea was 570:165. In the Atka Bay area, in spite of intense sampling in the "box", only one AC 1 fish was caught.

3.3 Growth

Length increments during the summer of 1979–1980, as calculated from the mean and modal length distribution of AC 0 post-larvae caught during the 6-week investigation period, were between 0.19 and 0.25 mm d^{-1} (Keller 1983). The mean length of post-larvae in AC 0 increased from

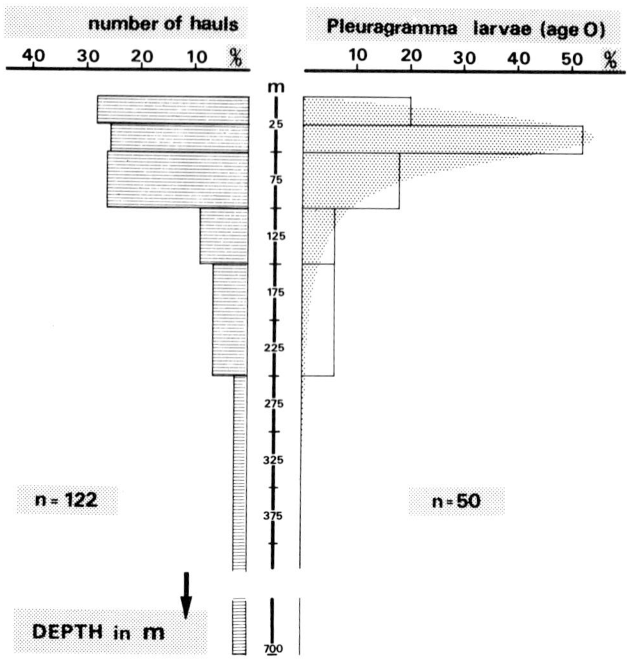

Fig. 4. Vertical distribution of post-larvae of *P. antarcticum* (AC 0) off the Ronne Ice Shelf in 1980 as obtained by Nansen closing net samples

12.5–20 mm in the period 1 January–15 February 1980. In 1981, mean size increased from 10 mm in January to 22 mm by the end of February (50 d); a mean daily increment of 0.24 mm was derived by fitting a linear curve (Fig. 3). The mean length of AC 1 post-larvae increased only slightly during the same period, from 32–35 mm.

Data in SL of larval *P. antarcticum* at other times of the year are given by Regan (1916) who reported lengths of 6–7 mm in December, 15–25 mm in April, 24 mm in May, 15–25 mm in August, and 20–25 mm in September.

3.4 Vertical Distribution

Fifty post-larvae of AC 0 were obtained by 122 vertical hauls between 700 m and the surface, mainly in the "box" off the Ronne Ice Shelf in 1979–80. The numbers of specimens per haul were very low, thus, provide no more than a preliminary, qualitative, indication of the species' vertical distribution (Fig. 4). Larvae were most abundant in the 25–50 m layer. Larvae were not obtained in hauls deeper than 250 m.

3.5 Change in Eye-Size

Regan (1916) suggested that the adults of *P. antarcticum* live at great depths, and that this is indicated by their relatively large eyes. A change in eye-size (horizontal diameter of eye) in relation to SL of larvae, during development, is shown in Fig. 5, based on data from Keller (1982). The eye diameter is 5.5–6.0% of the 12 mm SL of post-larvae and then decreases to < 4% in larvae of 25 mm. Young fish of 30 mm and longer (AC 1) had relatively large eyes. (6.5–7.0% of SL); the proportion decreasing slightly to about 6% in fish of 50 mm SL. Fish > 55 mm had bigger eyes (7% of SL). The proportion was about 8% in adults, which is larger than that given by Norman (1938).

3.6 Co-occurrence with Other Fish Larvae

In 1979–80, *P. antarcticum*, represented by post-larvae (AC 0), was the only fish species in 56 samples taken by bongo nets I and II (examined separately). The species occurred in 53 samples together with nototheniids (undescribed larvae of *Trematomus*). The former were post-larval *Pagetopsis* (22 samples), followed by *Cryodraco* (14 samples) and *Chionodraco* (13 samples) species, and unidentified channichthyids (1 sample). Post-larvae of *Page-*

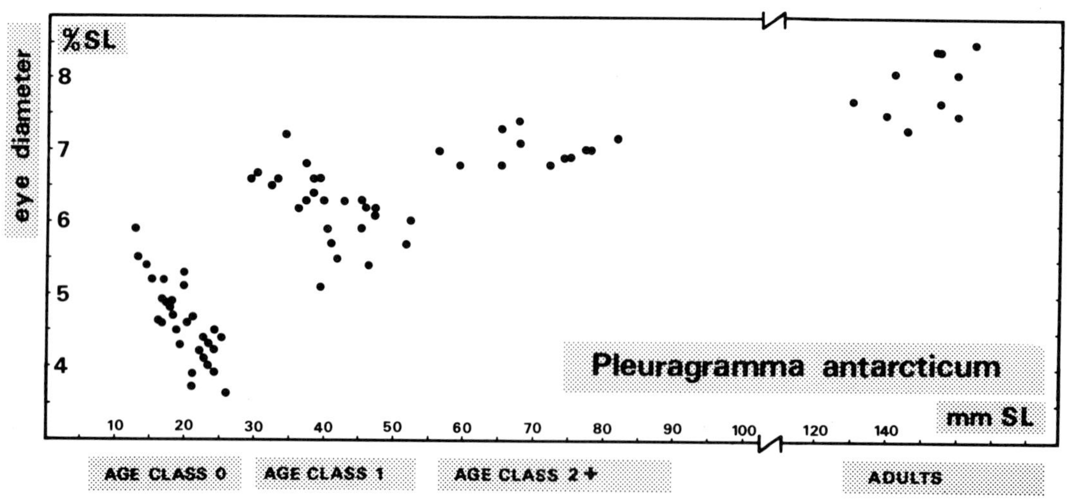

Fig. 5. Relationship between eye-size and standard length of *P. antarcticum*

topsis were also found together with other nototheniid larvae (14 samples) but never alone. A similar pattern prevailed in 1980–81.

4 Discussion

The overwhelming dominance of post-larval *P. antarcticum* in the plankton points to the importance of the species in the high-Antarctic shelf waters of the Weddell Sea. The relative abundance of the species' larvae in the total ichthyoplankton corresponds well with the high proportion (92%) of adults found in the pelagic system of the Ross Sea (DeWitt 1970). This is comparable with the high dominance of certain single species in sub-tropical upwelling areas and estuaries. For example, the larvae of the South-West Atlantic Anchovy, *Engraulis anchoita*, account for approximately 60% of the ichthyoplankton in the La Plata Estuary, Argentina, and > 90% in its main centres of distribution; the maximum densities (no. larvae m^{-3}) of the two species (*P. antarcticum* and *E. anchoita*) being similar (Hubold 1982). In both species, adaptation to particular hydrographic features apparently result in the occupancy of similar "low-competition" niches. The temperature gradient over the outer shelf does not allow "northern" fish, such as myctophids, to penetrate into the pelagic system of the high-Antarctic shelves. This leaves the system open to *P. antarcticum*, the only cold-adapted pelagic species (DeWitt 1970). The dominance of the species is not maintained through its predation on other species of fish (DeWitt 1970; DeWitt and Hopkins 1977).

Post-larvae of *P. antarcticum* were found all along the coast of the Weddell Sea, as far west as 60°W. The overall pattern of distribution appears to be correlated with the cold water (−1.4 to < −2.0°C) of the ice shelf. The species' occurrence was not confined to shallow areas; the larvae being also abundant over the deep trenches (Filchner depression and the trench off Camp Norway). These trenches or "inner-shelf depressions" were defined as a "pseudo-abyssal zone" by Andriashev (1965) and belong to the cold, ice-shelf ecosytem. Larvae of *P. antarcticum* were scarce, and occurred together with myctophids, where the warmer water of the East Wind Drift comes closer to the continent (i.e. near Atka Bay). Co-occurrence of the species and myctophids was observed only in samples taken in the Atka "box" during 1980–81.

Concentrations of post-larvae in 1980–81 were lower by a factor of 10 than in 1979–80. Despite possible sources of error in sampling, the difference in larval abundance between the two summers appears to be real. In both years, high concentrations of fish larvae, almost exclusively of *P. antarcticum*, were found in the southern Weddell Sea (in Vahsel Bay and Gould Bay). The larvae were preyed on by channichthyid larvae (especially of *Pagetopsis* sp.) which occurred in several samples. These concentrations of fish larvae in the southeastern Weddell Sea are associated with a stationary cyclonic gyre over the Filchner depression (Carmack and Foster 1975).

The relatively high numbers of post-larvae in 1979–80 stemmed from AC 0 fish hatched in 1979. Only 20 specimens of AC 1 (0.2%) were collected in that year. The proportion of one-yr-old fish was significantly higher (25%) in 1980–81, despite the use of nets with finer mesh (300 μ). This may indicate an exceptionally strong 1979 year-class. Larvae caught in 1980 were relatively large (Fig. 2). The difference between years in mean length of larvae at the beginning of January could be partly due to larvae avoiding the front of the narrower mesh (300 μ) net used in 1981. The 300 μ net did, however, catch a high proportion of AC 1 fish, indicating that net-avoidance was low. Larval abundance, growth, and year-class strength seem to be related to the same environmental factor. An extraordinary feature of the summer of 1979–80, which could account for improved survival of larvae, was the early and extremely wide opening of the coastal polynya which, for the first time, allowed a ship to penetrate as far west as the base of the Antarctic Peninsula (Kohnen and Schwarz 1981).

Young *P. antarcticum* of 38–76 mm SL were found in 1980–81 near the northern part of the Antarctic Peninsula (Kellermann and Kock 1984). These fish were caught by RMT nets which might have operated selectively to catch relatively large specimens, since fish in AC 1 in the Peninsula area may attain 50 mm SL by end of their second summer. However, drifting of post-larvae, with the Weddell Gyre, from spawning centres in the southwest, to the northern Weddell Sea (Weddell-Scotia confluence), could also account for the presence of relatively large young fish in the Peninsula area. Records of young *P. antarcticum* from the northern Weddell Sea are common, and can be partly a reflection of intensive sampling with large nets in waters near the Antarctic Peninsula. However, in spite of this intensive sampling, involving the use of a 350 μ mesh RMT-1 net, few small (AC 0) larvae of *P. antarcticum* have been taken in the area (A. Kellermann, pers. comm.).

The daily growth increments (0.20–0.24 mm d^{-1}) of *P. antarcticum* AC 0 post-larvae, as calculated from data on length frequencies, are similar to those of larvae of the Herring, *Clupea harengus*, in the North Sea and the Schlei Estuary (Schnack 1972). The growth rate is, however, considerably lower than that (0.34–0.55 mm d^{-1}) of sub-tropical engraulids (Method and Kramer 1979) or the 0.5 –1.0 mm d^{-1} by tropical Scaled Sardine, *Harengula jaguana*, (Houde 1977). Considering the probably unlimited food and the 24 h daylight for foraging during summer in the Antarctic, the observed growth rate of *P. antarcticum* is comparatively low. Furthermore, growth seems to be restricted to the period November–March, judging from the lengths of larvae reported by Regan (1916). Otoliths of four adult *P. antarcticum* measuring 253/225 mm, 225/210 mm, 195/175 mm and 185/165 mm (total length/standard length) were evaluated for growth zones, and ages of 12, 11, 9 and 7 years, respectively, were estimated.

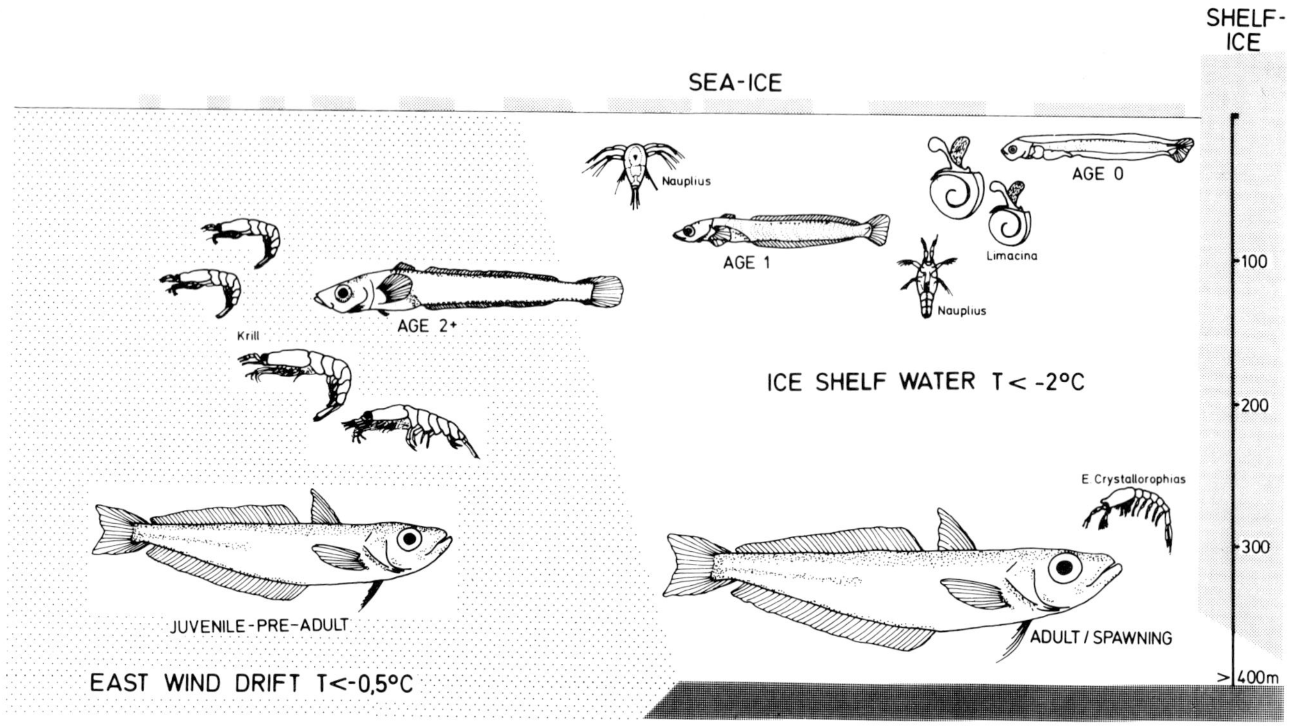

Fig. 6. Schematic life-cycle of *P. antarcticum* in the Weddell Sea

Specimens of 300–350 mm, the maximum size given by Lubimova et al. (1975), if they exist, must be of very old age. The largest specimen obtained by me, by means of a 42 m bottom-trawl in Gould Bay in 1983, had dimensions of 250/225 mm (Hubold, unpubl.).

Pleuragramma antarcticum specimens larger than 60 mm SL appear to occupy a habitat different to that of the post-larvae and young of AC 0 and 1, judging by the size of their eyes. Fingerlings have been reported as a by-catch in hauls of Antarctic Krill, *Euphausia superba,* they had juvenile euphausiids in their stomachs (Rembiszewski et al. 1978). Many juvenile specimens of *P. antarcticum* have been caught while associating with Krill swarms in the East wind Drift near Prydz Bay (Sloszarcyk 1982). Hence, it is likely that young fish leave the cold waters of the shelf to feed on Krill. The food of the youngest post-larvae in the southern Weddell Sea consists mainly of small molluscs (*Limacina* spp.). These molluscs are a typical and very abundant member of the plankton in the cold waters of the shelf (U. Piatkowski, pers. comm.). One-year-old fish (AC 1) near the Antarctic Peninsula had fed mainly on copepods, whereas older fish had eaten copepods, polychaetes and chaetognaths (A. Kellermann, pers. comm.). Adult *P. antarcticum* prey on Krill exclusively (Takahashi 1982), on Krill and copepods, over the outer shelf of the Ross Sea, or on other euphausiids in the Ross Sea, reflecting the abundance of the prey (DeWitt 1970). The stomachs of adult fish in the southern Weddell Sea contained *Euphausia crystallorophias* (Hubold, unpubl.).

Pleuragramma antarcticum is the most abundant fish species, and possibly the species with the highest biomass in the pelagic shelf ecosystem of the high-Antarctic. Early post-larvae and maturing adults are abundant near the ice shelf in the southern Weddell Sea. Juvenile and sub-adult fish apparently join Krill swarms to feed, in the East Wind Drift. Thus, the species represents a link between the East Wind Drift system, dominated by Krill, and the shelf biotope where the main euphausiid is *E. crystallorophias* (Fig. 6). Here, to a certain degree, *P. antarcticum* substitutes Krill in the diets of top predators, being the main food of the Antarctic Cod, Weddell Seal and, possibly, the Killer Whale, *Orcinus orca.*

References

Andriashev AP (1965) A general review of the Antarctic fish fauna. Monogr Biol 25:491–550

Anon. (1983) FAO Statistical yearbook N′53. Rome

Boulenger GA (1902) Pisces. In: Report on the collection of natural history made in the Antarctic regions during the voyage of the 'Southern Cross'. London, pp 174–189

Carmack EC, Foster TD (1975) Circulation and distribution of oceanographic properties near the Filchner Ice Shelf. Deep-Sea Res 22:77–90

DeWitt HH (1970) The character of the midwater fish fauna of the Ross Sea, Antarctica. In: Holdgate (ed) Antarctic Ecology, vol I, pp 305–314

DeWitt HH, Tyler JC (1960) Fishes of the Stanford Antarctic Program 1958–59. Stanford Ichthyol Bull 7(4):162–199

DeWitt HH, Hopkins TL (1977) Aspects of the diet of the Antarctic Silverfish *Pleuragramma antarcticum*. In: Llano, BA (ed) Adaptations within Antarctic Ecosystems. Gulf, Houston, pp 557–567

Eastman JT, DeVries AL (1981) Buoyancy adaptations in a swimbladderless Antarctic fish. J Morphol 167:91–102

Efremenko FN (1983) Atlas of fish larvae of the Southern Ocean. BIOMASS Handbook No 22, 74 pp

Hempel I, Hubold G, Kaczmaruk B, Keller R, Weigmann-Haass R (1983) Distribution of some groups of zooplankton in the inner Weddell Sea in summer 1979/80. Ber Polarforsch 9:1–55

Houde ED (1977) Abundance and potential yield of the scaled sardine, *Harengula jaguana,* and aspects of its early life history in the eastern Gulf of Mexico. Fish Bull 75:613–628

Hubold G (1982) Eggs and larvae of *Engraulis anchoita* Hubbs and Marini, 1935 in the southwest Atlantic between 25°S and 40°S. Meeresforschung 29(4):208–218

Hubold G, Drescher HE (1982) Filchner-Schelfeis-Expedition 1980/81 mit MS "Polarsirkel". Liste der Planktonfänge und Lichtstärkemessungen. Ber Polarforsch 4:1–30

Keller R (1982) Untersuchungen am Ichthyoplankton der Weddell-See, insbesondere an *Pleuragramma antarcticum* (Boul 1902). Masters thesis, Kiel

Keller R (1983) Contributions to the early life history of *Pleuragramma antarcticum.* Meeresforschung, 30:10–24

Kellermann A, Kock KH (1984) The distribution of postlarval and juvenile notothenioids (Pisces, Perciformes) in the southern Scotia Sea and northern Weddell Sea (57–64°S/48–56°W) during FIBEX 1981. Meeresforschung, (to be published)

Kohnen H, Schwarz J (1981) The average pack ice conditions in the southern Weddell Sea and the actual situation during the summer of 1979/80. Polarforsch 51(1):9–15

Lubimova TG, Naumov AG, Lagunov LL (1975) Prospects of the utilization of krill and other nonconventional resources of the world ocean. J Fish Res Board Can 30:2196–2201

Method RD jr, Kramer D (1979) Growth of Northern Anchovy, *Engraulis mordax,* larvae in the sea. Fish Bull 77(2):913–423

Norman JR (1937) B.A.N.Z. Antarctic research expedition 1929–1931, under the command of Douglas Mawson, Kt OBE, BE, DSc, FRS Fishes. Rep B.A.N.Z. Antarct Res Exped Ser B 1(2):49–88

Norman JR (1938) Coast Fishes. P III: The Antarctic Zone. Discovery Rep, vol XVIII, pp 1–105

Pappenheim P (1912) Die Fische der deutschen Südpolarexpedition 1901–1903. I: Die Fische der Antarktis und Subantarktis. Dt Südpol Exped 13, 160–183

Regan CT (1916) Larval and post-larval fishes. British Antarctic ("Terra Nova") Exped 1910. Zoology 1(4):125–156

Rembizewski JM, Krzeptowski M, Linkowski TB (1978) Fishes as by-catch in fisheries of krill *Euphausia superba* Dana (Euphausiacea-Crustacea). Pol Arch Hydrbiol 25(3):677–695

Schnack D (1972) Nahrungsökologische Untersuchungen an Heringslarven. Ber Dtsch Wiss Komm Meeresforsch 22(3):213–343

Schneppenheim R (1982) Marin-biologisches Begleitprogramm zur Standorterkundung 1979/80 mit MS "Polarsirkel" (Pre-Site-Survey) – Stationsliste der Micronekton- und Zooplanktonfänge sowie der Bodenfischerei. Ber Polarforsch 7:1–32

Slósarczyk W (1982) Preliminary estimation of distribution and abundance of juvenile Notothenioidei within krill concentrations in the east Antarctic region. Mimeo, BIOMASS Working Party on Antarctic Fish Ecology, Hamburg Sept 1982

Takahashi M (1982) Fishes caught by bottom trawl in north of South Shetland Islands. Mimeo, submitted by Dr Naito at Meeting of Working Party on Antarctic Fish Ecology, Hamburg Sept 1982

Volkmann NJ, Presler P, Trivelpiece W (1980) Diets of pygoscelid penguins at King-George Island, Antarctica. Condor 82(4):373–378

Trophic Relationships Between Pelagic Fish and Euphausiids in Antarctic Waters

R. WILLIAMS[1]

Summary. This paper reports on work conducted in the Prydz Bay region on the role of *Euphausia superba* and other euphausiids in the diet of Antarctic ocean fishes (mainly *Electrona antarctica, Notolepis coatsi, Pleuragramma antarcticum* and *Gymnoscopelus* spp.). The stomach contents of fish collected both from the vicinity of Krill swarms and from areas of low abundance of Krill were examined.

1 Introduction

The importance of Antarctic Krill, *Euphausia superba,* in the diet of fish has been recognized for a long time (Permitin 1970), and for the management of any commercial exploitation of Krill it is obviously necessary to have reliable estimates of its natural mortality due to predation by fish. Despite this, there is relatively little information on the diets of fish in the Southern Ocean, compared to that available for whales, seals and seabirds, and such data do not in general concern pelagic fish.

Most studies have concentrated on adult benthic or bentho-pelagic fish in the Atlantic sector (e.g., the South Shetland Islands (Tarverdiyeva and Pinskaya 1980; Daniels 1982; Takahashi 1983), Signy Island (Richardson 1975) and South Georgia (Permitin and Tarverdiyeva 1972; Targett 1981). The only detailed work in eastern Antarctic waters is that of Hureau (1970). Data on pelagic fish are limited to general information on the diet of *Notolepis coatsi* (Solyanik 1965), *Electrona antarctica* (Solyanik 1967), and various species caught in the Scotia Sea area (Rembiszewski et al. 1978). The diets of *Pleuragramma antarcticum* in Prydz Bay (Gorelova and Gerasimchuk 1981) and in the Ross Sea (DeWitt and Hopkins 1977) and of *E. antarctica* in the Scotia Sea (Rowedder 1979) have been studied in more detail.

This study presents information on the stomach contents of fish caught during Krill surveys by the Australian National Antarctic Research Expeditions (ANARE) in the region of Prydz Bay, East Antarctica. Although the samples for most species are small, the data they provide give valuable information on the diets of adult pelagic fish, of which only *P. antarcticum* and *E. antarctica* have been reported on before. A distinctive feature of the Prydz Bay region is the relatively large area of continental shelf, which is ice-free in the austral summer. Plankton hauls over the shelf contain juvenile stages of normally demersal fish, and data on the stomach contents of these specimens complement information already published on the adults of the same or closely related species.

1 Antarctic Division, Department of Science and Technology, Channel Highway, Kingston 7150, Tasmania, Australia

2 Materials and Methods

All specimens were caught with an RMT-8 net towed at 5.5 km h^{-1} in depths of 75 m or less. Most material was obtained from the FIBEX (First International BIOMASS Experiment) cruise from January to March 1981, but some additional specimens of the juvenile channichthyids and *Trematomus* spp. are included from a subsequent cruise in November and December 1982. Both cruises operated in the area between 62°S and the Antarctic coast and between 60°E and 88°E. Full details for stations and catches during the FIBEX cruise are available in Williams et al. (1983).

The fish were sorted from the catch and preserved as soon as possible in preserving solution (Steedman 1974). In the laboratory, the standard length (SL) of each fish was measured to the nearest mm below, and its weight determined to the nearest 0.1 g. The stomach was removed and the contents analyzed according to the mixed method of Hureau (1970), which is the method recommended for the BIOMASS programme (Anon. 1981). This method involves obtaining the product of the percentage by weight of each prey type and its percentage by number to give a coefficient (Q). This reduces anomalies caused by the presence of large numbers of very small prey items or few very large prey items. The euphausiid components of the stomach contents, when sufficiently intact, were measured to the nearest mm below, from the posterior margin of the eye socket to the tip of the telson.

3 Results

3.1 Gross Diet

Tables 1–8 show that 10 of the 11 fish species had euphausiids as a significant part of their diet. Only *Trematomus*

Antarctic Nutrient Cycles and Food Webs
(ed. by W.R. Siegfried, P.R. Condy, and R.M. Laws)

Table 1. Stomach contents of *Gymnoscopelus nicholsi* and *G. braueri* (41 mm–150 mm standard length)

Food item	No. items	No. stomachs	Frequency of occurrence (%)	Weight of items	% by no.	% by weight	Q
Copepods							
Euchaeta antarctica	6	1	12.5	0.092	27.27	2.54	69.3
Rhincalanus gigas	1	1	12.5	0.012	13.64	0.33	4.5
Unidentified	2	1	12.5				
Amphipods							
Themisto gaudichaudii	3	1	12.5	0.151	13.64	4.17	55.9
Primno macropa	1	1	12.5	0.021	4.55	0.58	2.7
Euphausiids							
Euphausia superba	6	3	37.5	3.26	27.27	89.98	2,453.8
Thysanoessa macrura	3	3	37.5	0.087	13.64	2.40	32.7
(10 stomachs examined, 2 empty)							

Table 2. Stomach contents of *Notolepis coatsi* (52 mm–280 mm standard length)

Food item	No. items	No. stomachs	Freqency of occurrence (%)	Weight of items	% by no.	% by weight	Q
Polychaetes							
Pelagobia longicirrata	13	2	25.0	0.021	18.31	0.22	4.0
Chaetognaths	26	3	37.5	0.047	36.62	0.50	18.3
Copepods	2	1	12.5	0.068	12.68	0.72	9.1
Euphausiids							
Euphausia superba	20	2	25.0	9.02	28.17	96.24	271.1
Thysanoessa macrura	9	1	12.5	0.068	12.68	0.72	9.1
Decapods							
Nematocarcinus sp. larvae	1	1	12.5	0.31	1.41	3.27	4.6
(10 stomachs examined, 2 empty)							

Table 3. Stomach contents of Channichthyidae species

Food item	No. items	No. stomachs	Frequency of occurrence (%)	Weight of items	% by no.	% by weight	Q
		Cryodraco antarcticus	(45 mm–62 mm standard length)				
Euphausia superba	1	1	50.0	0.047	50.0	97.92	4,896.0
Crustacean unidentified	1	1	50.0	0.001	50.0	2.08	104.2
(3 stomachs examined, 1 empty)							
		Pagetopsis macropterus	(33 mm–47 mm standard length)				
Euphausia crystallorophias	3	1	50.0	0.028	60.0	22.53	1,351.8
Euphausiid unidentified	1	1	50.0	0.085	20.0	68.0	1,360.0
Chionodraco sp.	1	1	50.0	0.012	20.0	9.60	192.0
(3 stomachs examined, 1 empty)							
		Chaenodraco wilsoni	(37 mm–62 mm standard length)				
Euphausia crystallorophias	3	3	50.0	0.126	37.5	73.68	2,763.2
Euphausiid unidentified	5	3	50.0	0.045	62.5	26.32	1,645.0
(9 stomachs examined, 3 empty)							

Table 4. Stomach contents of *Electrona antarctica* (20 mm–95 mm standard length)

Food item	No. items	No. stomachs	Frequency of occurrence (%)	Weight of items	% by no.	% by weight	Q
Polychaetes							
Alciopidae unidentified	9	7	2.7	0.010	1.49	0.14	0.2
Chaetognaths	5	5	1.9	0.011	0.83	0.15	0.1
Ostracods	17	9	3.4	0.024	2.81	0.33	0.9
Copepods							
Metridia gerlachei	89	113	43.1	0.872	47.85	12.02	575.2
Euchaeta antarctica	15						
Calanus propinquus	69						
Calanoides acutus	13						
Unidentified	102						
Amphipods							
Themisto gaudichaudii	9	35	13.4	1.977	7.45	27.24	202.9
Cyllopus sp.	8						
Other hyperiids	28						
Euphausiids							
Euphausia superba	6	6	2.3	0.890	0.99	12.26	12.1
E. frigida	6	4	1.5	0.203	0.99	2.80	2.8
Unidentified	29	19	7.3	0.414	4.80	5.66	27.2
Thysanoessa macrura	31	23	8.8	0.711	5.13	7.80	50.3
Decapods							
Nematocarcinus sp. larvae	33	11	4.2	0.073	5.46	1.01	5.5
Crustaceans unidentified	49	49	18.7	0.493	8.11	6.79	55.1
Molluscs							
Clio pyramidata	36	25	9.5	1.32	6.29	18.19	114.4
Limacina helicina	2	2	0.8				
Fish							
Myctophid juvenile	1	1	0.4	0.34	0.33	0.47	0.16
Unidentified	1	1	0.4				
Unidentifiable remains	45	45	17.2	0.225	7.45	3.10	23.1
(347 stomachs examined, 85 empty)							

Table 5. Stomach contents of *Krefftichthys anderssoni* (27 mm–66 mm standard length)

Food item	No. items	No. stomachs	Frequency of occurrence (%)	Weight of items	% by no.	% by weight	Q
Copepods							
Calanoides acutus	1,184	15	68.2	1.411	76.22	64.58	4,922.3
Calanus propinquus	79						
Unidentified	320						
Amphipods							
Primno macropa	22	4	18.2	0.100	1.11	4.58	5.1
Hyperia sp. larva	1	1	4.5				
Euphausiids							
Thysanoessa macrura furciliae	459	11	50.0	0.475	22.10	21.74	480.5
adult	2	2	9.1	0.031	0.10	1.42	0.1
Crustacean unidentified	1	1	4.5	0.001	0.048	0.05	0.002
Others							
Chaetognaths	1	1	4.5	0.167	0.436	7.64	3.3
Fish	1	1	4.5				
Unidentified	7	7	31.8				
(26 stomachs examined, 4 empty)							

Table 6. Stomach contents of *Pleuragramma antarcticum* (56 mm–160 mm standard length)

Food item	No. items	No. stomachs	Frequency of occurrence (%)	Weight of items	% by no.	% by weight	Q
Copepods							
Calanoides acutus	487						
Metridia gerlachei	91						
Euchaeta antarctica	14	42	93.3	1.378	64.21	48.28	3,100.1
Copepodites unidentified	70						
Unidentified	88						
Euphausiids							
Euphausia superba							
nauplii & calyptopes	45			0.127	9.16	4.45	40.8
furciliae	62						
E. crystallorophias calyptopes	4			0.469	20.89	16.43	343.2
furciliae	239	35	77.8				
adult	1			0.04	0.08	1.40	0.1
Unidentified furciliae	54			0.119	4.71	4.17	19.6
adult	1						
Crustacea unidentified	2	2	4.4	0.02	0.17	0.70	0.2
Fish							
Notolepis coatsi	1	1	2.2				
myctophid juvenile	1	1	2.2	0.596	0.60	20.88	12.5
Unidentified pieces	5	4	8.8				
Others							
Polychaetes	1	1	2.2	0.106	0.26	3.69	1.0
Molluscs	2	2	4.4				
(45 stomachs examined, 0 empty)							

Table 7. Stomach contents of *Trematomus* spp.

Food items	No. items	No. stomachs	Frequency of occurrence (%)	Weight of items	% by no.	% by weight	Q
		Trematomus scotti	(32 mm–38 mm standard length)				
Copepods							
Metridia gerlachei	8						
Calanus propinquus	3	3	37.5	0.030	1.29	53.57	765.5
Calanoides acutus	1						
Copepodites unidentified	29	3	37.5	0.013	34.52	23.21	801.2
Euphausiids unidentified furciliae	43	1	12,5	0.013	51.19	23.21	1,188.1
(9 stomachs examined, 1 empty)							
		Trematomus sp.	(32 mm–57 mm standard length)				
Copepods							
Metridia gerlachei	3						
Calanus propinquus	2	3	60.0	0.018	98.11	90.08	8,837.7
Copepodites unidentified	151						
Molluscs							
Limacina helicina	2	1	20.0	0.008	1.26	6.11	7.7
Unidentified	1	1	20.0	0.005	0.63	3.82	2.4
(7 stomachs examined, 2 empty)							

sp. juveniles had no euphausiids whatsoever. The 10 species fall into three groups on the basis of the contribution of euphausiids to their diet, as shown by the values of Q. In the first group which includes *N. coatsi*, *Gymnoscopelus* spp. and the juveniles of *Cryodraco antarcticus*, *Pagetopsis macropterus*, and *Chaenodraco wilsoni* euphausiids are the dominant part of the diet, with other components definitely secondary. The second group contains general plankton feeders in which euphausiids form a major part, but are secondary to copepods (*P. antarcticum*, *Krefftichthys an-*

Table 8. Stomach contents of *Chionodraco* sp. (21 mm–78 mm standard length)

Food item	No. items	No. stomachs	Frequency of occurrence (%)	Weight of items	% by no.	% by weight	Q
Euphausiids							
Euphausia superba	4	4	7.14	0.591	7.27	31.94	232.2
E. crystallorophias	1	1	1.79	0.027	1.81	1.44	2.6
Unidentified furciliae	3	2	3.57	0.01	5.45		
adults	3	3	5.36	0.136	5.45	7.81	85.1
Decapods							
Unidentified larva	1	1	1.79	0.060	1.82	3.21	5.8
Crustacean unidentified	2	2	2.57	0.01	3.64	0.54	2.0
Fish							
Trematomus scotti	1	1	1.79	0.244	1.82	13.06	23.8
Pleuragramma antarcticum	18	18	32.14	0.410	32.73	21.94	718.1
Chionodraco sp.	6	6	10.71	0.156	10.91	8.35	91.1
Unidentified	15	15	26.79	0.222	27.27	11.88	324.0
(124 stomachs examined, 68 empty)							

derssoni, and *Trematomus scotti*) or co-dominant with copepods and amphipods (*E. antarctica*). In the last group, euphausiids are a relatively minor component in *Chionodraco* sp. whose major diet component is fish.

3.2 Change of Diet with Size of Predator

Figure 1 shows how the composition of the diet changes with size of fish in the three species with sufficient available to allow th data to be subdivided. Results using frequency of occurrence, rather than percentage by number of prey items as in Fig. 1, were very similar. In *E. antarctica* there is a clear trend from copepod dominance in individuals < 60 mm long to a more mixed diet consisting of the larger amphipods, euphausiids and pteropods in the larger specimens. The more important subsidiary items in the smaller fish are small euphausiids (mostly *Thysanoessa macrura*) and decapod larvae. *Chionodraco* sp. also shows a clear

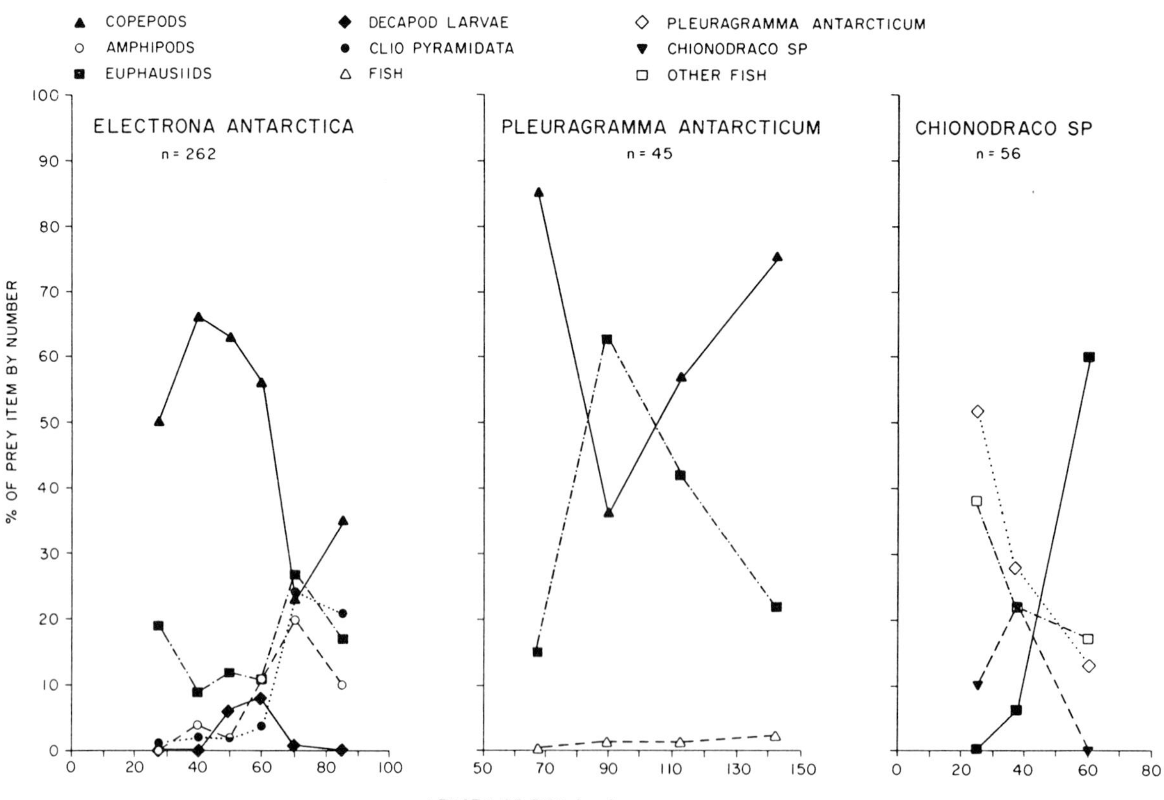

Fig. 1. Variation in diet in relation to size of fish for *E. antarctica, P. antarcticum* and *Chionodraco* sp.

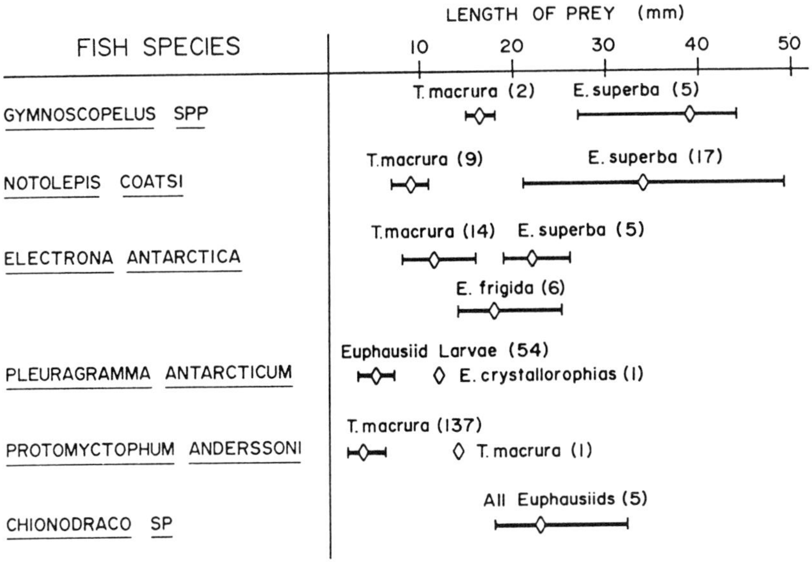

Fig. 2. Length range (*horizontal bars*) and mean length (*diamonds*) of euphausiid prey from the stomachs of six species of fish. Figures in parentheses are numbers of specimens measured

trend from the dominance of small fish in the diet of specimens < 40 mm long to an equally clear dominance by euphausiids in the larger individuals. *Pleuragramma antarcticum* shows no consistent trend, but throughout the size range the diet is dominated by copepods and euphausiids in various proportions. In such a small sample, too much emphasis should not be placed on the significance of single points in the graph, especially as some net-feeding is suspected in this species (see Discussion). Therefore, the diet of *P. antarcticum* (within the size range reported here) should be regarded as generally dominated by copepods, with varying proportions of euphausiids.

3.3 Euphausiid Species in the Diets

Small sample sizes and the high incidence in some cases of unidentifiable euphausiids make it difficult to provide from information on the species of euphausiids taken by the various fish. It is clear from Tables 1 and 2, however, that among fish which feed predominantly on euphausiids, both *Gymnoscopelus* spp. and *N. coatsi* are taking the majority of that component as *Euphausia superba.* The channichthyid species, as far as can be judged from such small samples, take mostly *E. crystallorophias* (Table 3).

Among the more generalized plankton feeders, there is a more even spread of euphausiid species in the diet, except in *K. anderssoni,* in which only *Thysanoessa macrura* occurred. Whereas *E. antarctica* took only post-larval euphausiids, *K. anderssoni* and *T. scotti* took mostly larval stages (Tables 4–7).

The *Chionodraco* sp. juveniles only had euphausiids as a significant part of the diet in the upper end of the size range sampled here. *E. superba* occurred in much greater quantities than other euphausiids.

3.4 Size of Euphausiids

Figure 2 shows the size ranges of those euphausiids from stomachs of various fish species that were sufficiently intact to measure. *Gynmnoscopelus* spp. and *N. coatsi* were the only species taking adult *E. superba. Electrona antarctica* and *Chionodraco* sp. took sub-adult Krill. The euphausiid diet of *P. antarcticum* and *K. anderssoni* is almost entirely restricted to larval stages < 5 mm long.

4 Discussion

The range of fish examined in this study covers a wide spectrum of species, from juvenile pelagic phases of fish that are demersal when adult (the channichthyids and nototheniids caught over the continental shelf) to the myctophids which are pelagic for their entire life history and are caught exclusively offshore from the continental shelf. Occupying an intermediate position is *P. antarcticum,* a holopelagic nototheniid, but encountered mostly over the continental shelf. This study shows that euphausiids form a significant and sometimes dominant part of the diet of fish from all habitats, and that *E. superba* forms a varying but generally significant percentage of this group. *Euphausia crystallorophias* replaces *E. superba* in those channichthyid juveniles and *P. antarcticum* taken from the southernmost stations. The present results agree in general with the limited data from previous studies. The two best known fish, *E. antarctica* and *P. antarcticum,* are both shown in all studies to be generalized plankton feeders, with copepods and small euphausiids forming the bulk of their diet. However, most previous studies have shown that euphausiids, and specificall *E. superba,* form a larger part of the diet of *E. antarctica* than is indicated here. Both Rowedder (1979) and Permitin (1970) give the frequency of occurrence of *E. superba* in the range 21–27%, whereas Table 4 shows that all euphausiids combined occur in only 14.1% of the stomachs in this study. Even in the 65–74 mm length group, which had the highest incidence of euphausiids (Fig. 1), the frequency of occurrence of all *Euphausia* species was only 18.9%. Rowedder (1979), working in the

South Atlantic area, found differences in the stomach contents of *E. antarctica* between South Georgia, Elephant Island the South Sandwich Islands area, but in no case did *E. superba* appear in < 20% of the stomachs or comprise < 50% of the stomach contents by weight. Rembiszewski et al. (1978) found *E. antarctica* from the Scotia Sea to contain mainly *E. superba* and *Thysanoessa* sp. The data of Solyanik (1967), while very limited, appear to be more similar to those in this study, with a maximum frequency of occurrence of *E. superba* of only 13.3%, and high percentages of copepods and amphipods. It is perhaps significant that the reports of high incidences of Krill in diets of *E. antarctica* are from the Scotia Sea areas, whereas the studies showing lower incidences of Krill are from the Indian Ocean (this study) and the "Indian Ocean and Pacific sectors of Antarctica" (Solyanik 1967). All studies agree that the mean size of *E. superba* taken is in the range 22–27 mm.

Whereas the classes of animals found in the diet of *P. antarcticum* agree well with those found by DeWitt and Hopkins (1977), the relative importance of the two major groups (copepods and euphausiids) differs markedly. DeWitt and Hopkins (1977) found copepods to be numerically dominant and euphausiids to account for the greatest proportion by weight. This was reflected in the size frequency distribution of prey items, which was bimodal with the < 10 mm peak representing copepods and the > 20 mm group comprising euphausiids. Table 6 shows, however, that while copepods and euphausiids account for about equal percentages by number in this study, euphausiids comprised only 50% of the weight of copepods. This is explained in Fig. 2 where euphausiids in the diet were almost entirely 5 mm–8 mm long larvae. It is apparent from this study that *P. antarcticum* is primarily a copepod feeder and takes euphausiid larvae as 'alternative copepods', a pattern generally maintained throughout the size range covered (Fig. 1). Gorelova and Gerasimchuk (1981), also working in Prydz Bay, found similar proportions of copepods and larval euphausiids in young fish (< 125 mm length), but observed a switch to adult euphausiids in larger fish. A problem with the diet of *P. antarcticum* that DeWitt and Hopkins (1977) recognized is that of net-feeding, as evidenced by the presence in the stomachs of fish scales and a variety of food closely similar to that found in the net. In this study, evidence of net feeding (scales and entirely recently ingested food items) was found in *P. antarcticum* from one haul only. To this extent, the results should be treated cautiously, but those for other species, except some *Chionodraco* sp. to be discussed later, are reliable in this respect.

Krefftichthys anderssoni has a very similar diet to *P. antarcticum,* taking mostly copepods with small euphausiids as alternatives. As with *E. antarctica,* this differs from the diet of the same species in the South Atlantic area, where Rembiszewski et al. (1978) found *E. superba* in 63% of stomachs examined.

Notolepis coatsi and the two *Gymnoscopelus* species are undoubtedly the specialist feeders on *E. superba.* All three species were caught exclusively in the deep ocean zone offshore from the continental shelf and only in association with high Krill catches. These observations agree well with those of Solyanik (1965) and Kawamura (1981), who both reported *N. coatsi* from the Indian Ocean sector to be feeding heavily on large *E. superba.* Permitin (1970) also reported on the heavy predation of Krill by *Gymnoscopelus nicholsi.*

Several studies have shown the importance of *E. superba* in the diets of adult members of the family Channichthyidae (Permitin 1970; Permitin and Tarverdiyeva 1972), even those normally with a benthic habit (Targett 1981). Takahashi (1983) has also shown that in sub-adult channichthyids (> 200 mm long), *E. superba* is most important in the diet of smaller individuals, and that adult Krill (>40 mm long) were being taken. This was related to the more pelagic habit of the younger channichthyids. Kompowski (1980) also stressed the importance of *E. superba* in the diet of the smaller juvenile stages (72 mm–98 mm SL) of *Chaenocephalus aceratus.* This study extends the size range of pelagic juvenile channichthyids examined down to the smallest post yolk-sac stage, and, despite the small sample sizes, it is apparent from Table 3 that all these juveniles are heavily dependent on euphausiids, either *E. superba* or *E. crystallorophias,* depending on where they were caught.

Although *Chionodraco* sp. has an apparently lower euphausiid component in its diet than the other channichthyids, I suspect that some net-feeding in one haul enhanced the content of fish, especially of *P. antarcticum* and *Chionodraco* sp. Even based on the figures in Table 8, however, the diet of juveniles of *Chionodraco* sp. still has a significant euphausiid component, especially in individuals > 40 mm long (Fig. 1). The euphausiids (mostly *E. superba*) are in the size range 18 mm–32 mm (Fig. 2), which is also the size of euphausiids taken by the other channichthyid species.

This study shows that the most commonly encountered pelagic fish in the Prydz Bay region depend heavily on euphausiids, and in some cases specifically on *E. superba* for food. In addition to the well-known predation by large fishes on adult Krill, there is significant predation by juvenile fish on the younger stages of Krill. Thus, the entire size range of Krill is a source of food or fish, and some fish species apparently prey on euphausiids throughout their entire post-larval life history.

Acknowledgements. I thank Mr. J. Kirkwood for identifying the crustacean remains in the stomach contents, and the anonymous referee for some most helpful comments.

References

Anon (1981) Recommended methods for standardisation of measurements of fish. BIOMASS Handbook No. 13

Daniels RA (1982) Feeding ecology of fishes in the Antarctic Peninsula. US Nat Mar Fish Serv Fish Bull 80(3):575–588

DeWitt HH; Hopkins TL (1977) Aspects of the diet of the Antarctic Silverfish, *Pleuragramma antarcticum.* In: Llano GA (ed) Adaptations within Antarctic ecosystems. Smithsonian Institution, Washington, pp 557–567

Gorelova TA; Gerasimchuk VV (1981) Data on nutrition and daily consumption of juvenile *Pleuragramma antarcticum* Boulenger. In: Parin NV (ed) Fishes of the open ocean. Academy of Sciences of the USSR, Moscow, pp 103–109

Hureau JC (1970) Biologie comparée de quelques poissons antarctiques (Nototheniidae). Bull Inst Oceanogr (Monaco) 68:1–244

Kawamura A (1981) Stomach contents of some fish caught during exploratory fishing for krill. In: El-Sayed SZ (ed) Biological Investigations of Marine Antarctic Systems and Stocks (BIOMASS) vol II. Scott Polar Research Institute, Cambridge pp 69–71

Kompowski A (1980) Studies on juvenile *Chaenocephalus aceratus* (Lonnberg, 1906) (Pisces, Chaenichthyidae) from off South Georgia. Acta Ichthyol Piscat 10(1):45–53

Permitin YE (1970) The consumption of krill by Antarctic fishes. In: Holdgate MW (ed) Antarctic Ecology vol I. Academic Press, London, pp 177–182

Permitin YE; Tarverdiyeva MI (1972) The food of some Antarctic fish in the South Georgia area. J Ichthyol 12:104–114

Rembiszewski JM; Krzeptowski M; Linkowski T (1978) Fishes (Pisces) as by-catch in fisheries of krill *Euphausia superba* Dana (Euphausiacea, Crustacea). Pol Arch Hydrobiol 25:677–695

Richardson MG (1975) The dietary composition of some Antarctic fish. Br Antarct Surv Bull 41:113–120

Rowedder U (1979) Feeding ecology of the myctophid *Electrona antarctica* (Gunther, 1878) (Teleostei). Meeresforschung 27: 252–263

Solyanik GA (1965) Distribution and nutrition of bathypelagic fish *Notolepis coatsi* (Paralepididae Family). Inf Bull Sov Antarct Exped 5:365–6

Solyanik GA (1967) The food of the Antarctic (*Electrona antarctica* Gunth.). Inf Bull Sov Antarct Exped 6:443

Steedman HF (1974) Laboratory methods in the study of zooplankton. J Cons Int Explor Mer 35:351–358

Takahashi M (1983) Trophic ecology of demersal fish community north of the South Shetland Islands, with notes on the ecological role of krill. Mem Natl Inst Polar Res Spec Issue 27:183–192

Targett TE (1981) Trophic ecology and structure of coastal Antarctic fish communities. Mar Ecol Prog Ser 4:243–263

Tarverdiyeva MI; Pinskaya IA (1980) The feeding of fishes of the Families Nototheniidae and Channichthyidae on the shelves of the Antarctic Peninsula and the South Shetlands. J Ichthyol 20(4):50–60

Williams R; Kirkwood JM; O'Sullivan D (1983) FIBEX cruise zooplankton data. ANARE Res Notes 7:1–163

Chick Energy Balance in the Whitechinned Petrel, *Procellaria aequinoctialis*

A. BERRUTI[1,2], N.J. ADAMS[1], and C.R. BROWN[1]

Summary. We investigated the energy delivery to, and energy expenditure of, Whitechinned Petrel (*Procellaria aequinoctialis*) chicks at Sub-Antarctic Marion Island. Meal mass and energy content of the food fed to chicks were variable. Feeding frequency was generally low and irregular. We calculated an energy budget based on measurements of energy expenditure and rates of mass accumulation. Energy received soon after hatching was in excess of maintenance requirements and was presumably laid down as an energy reserve. The chick experienced periods of relative fasting during growth when energy expenditure for maintenance and biosynthesis alone exceeded total energy delivery. Energy reserves sustained the chick during these periods, including that immediately prior to departure from the nest, allowing the adults to extend foraging periods when food was in short supply. Total energy ingested by a chick from 8 to 96 days of age was estimated to be ca. 67,000 kJ compared to an energy expenditure over the same period of ca. 54,000 kJ. Energy accumulated as chick biomass was estimated to reach an maximum of ca. 9,000 kJ.

1 Introduction

The growth of procellariiform chicks is characteristically slow, with chicks accumulating large lipid reserves (Ricklefs et al. 1980a). Lack (1968) suggested that this slow growth is an adaptation to reduce chick energy requirements in response to a scarce and unpredictable food supply. Recent studies on the growth and energetics of pelagic seabird chicks have concentrated on the measurement of energy expenditure and tissue accumulation through growth in the light of this hypothesis (Ricklefs et al. 1980b; Ricklefs and White 1981).

We investigated growth, food supply, and metabolism of nestling Whitechinned Petrels, *Procellaria aequinoctialis,* at Sub-Antarctic Marion Island (46° 52′S, 37° 51′E). The long interval between feeds in this species, characteristic of procellariiforms, allowed determination of meal mass and feeding frequency, and hence estimation of energy delivery to the chick, by repeated weighings of chicks. In addition, energy expenditure was measured using rates of oxygen consumption. Such combined measurements of both energy delivery to, and energy expenditure of, chicks of pelagic seabirds through the growth period have not previously been reported.

1 FitzPatrick Institute, University of Cape Town, Rondebosch, 7700 South Africa

2 Present address: Sea Fisheries Research Institute, Private Bag X2, Roggebaai, 8012, South Africa

2 Materials and Methods

During December 1980–April 1981 we investigated feeding rates, meal mass and chick growth. Further measurements of chick growth, and the study of energy expenditure based on oxygen consumption, were made during January 1982 –April 1982.

We examined Whitechinned Petrel burrows by removing earth plugs cut into the side of the nest chamber (Sinclair 1981). Growth was measured by weighing 22 chicks at 1–5-day intervals in 1980–1981, and three chicks at 7-day intervals in 1982. We estimated feeding rates and meal mass by weighing 22 known-age chicks every 6 h over 30–48 h periods on nine occasions during growth. Chicks were weighed in bags using handheld Pesola balances. Weighing was possible at average windspeeds of up to 70 km h^{-1}, since burrows are sheltered, and portable windbreaks were used when chicks were less than 1,000 g. We regarded a mass increase of 30 g or more between weighings as a meal but ignored mass increases < 30 g, regarding these as possible errors in field weighings. To compensate for mass lost from meals as a result of digestion and excretion occurring between the meal delivery and subsequent weighing, a correction factor of 19 g up to age 60 days or 26 g from 61 days onward was added to the measured meal mass. This was calculated as half the mass lost during a 6 h period without feeding and following a detected mass increase. High winds prevented weighing on four occasions but subsided sufficiently to allow weighings to continue 12 h later. A correction factor of 38 g up to age 60 days and 52 g thereafter was added to 22 meals delivered during these periods.

We dried whole regurgitated food samples from chicks, and stomach contents of culled adults and chicks, to constant mass in a forced-draught oven at 60°C. Lipid content was estimated by diethyl-ether extraction over 16 h and the remainder of the sample was assumed to be protein (Clarke and Prince 1980). We calculated energy content using

Antarctic Nutrient Cycles and Food Webs
(ed. by W.R. Siegfried, P.R. Condy, and R.M. Laws)

values of 39.7 and 16.7 kJ g^{-1} for lipid and protein, respectively (Petrusewicz and Macfadyen 1970). Growth, feeding rates, meal mass and total energy delivery were calculated for the same 5–8-day age classes used for the oxygen consumption determinations in 1982 (see below).

At 5–8-day intervals from hatching we measured oxygen consumption (VO_2), in a translucent, airtight chamber (volume 28 l), using an open flow-through system coupled to a paramagnetic oxygen analyzer as described in Adams and Brown (1984). We adjusted the flow rate to between 560 and 3,075 ml min^{-1}, depending on the mass of the chick, in order to produce a drop in O_2 concentration in the expired air of 1–2% below that of ambient air. The chamber temperature could not be controlled and ranged from 8.5°–21.0°C (mean 13.7°C). Mougin (1970) measured temperatures of about 4°–10.5°C in occupied Whitechinned Petrel burrows at the Crozet Island, with mean temperatures of 6°–8°C. To minimize disturbance, we made no attempt to measure body temperature. However, Whitechinned Petrel chicks rapidly acquire the ability to thermoregulate (Mougin 1970), being brooded for only 2 to 6 days after hatching (pers. obs.). We allowed an initial period of 45 min for chicks to settle and for air in the chamber to equilibrate. Thereafter, readings of percentage oxygen, flow rate and chamber temperature were taken at 15 min intervals over 2 h and corrected to STPD. We made all measurements during daylight hours, the non-active phase of the adult Whitechinned Petrel's circadian cycle (Adams and Brown 1984), and chicks were generally passive during measurement. We converted oxygen consumption measurements to an energy equivalent using 1 l O_2 = 20.084 kJ. We considered this to be a good estimate of the energy expenditure of an inactive chick in a burrow. Additional expenditure due to activity as a result of wing flapping may, however, represent a significant proportion of the energy budget close to departure from the nest. We weighed and returned chicks to the burrow after each measurement. The three chicks fledged successfully.

3 Results

Adult Whitechinned Petrels delivered 133 meals during the total nest observation time of 6,556 h. The mean rate of meal delivery was 0.53 meals d^{-1} (range = 0.31–0.78 meals d^{-1}) per 5–8-day age class (Table 1). The interval between feeds varied from 1–3 days between 8 and 21 days age, but up to 8 days in the period immediately prior to fledging (mean 3.5± 2.4). The last meal of 10 chicks was delivered a maximum of 9 days prior to nest departure (mean 4.2 ± 3.1 days, range 1–9 days). Individual meal masses received by Whitechinned Petrel chicks were highly variable. However, the mean meal mass was 190.3 (range = 127–217 g, n = 133), remaining relatively constant between 8 and 91 d, but being significantly lower ($t = 3.252$, $P < 0.005$) immediately prior to fledging when compared to meal mass from the 84–91-day age class (Table 1).

The lipid, protein and water composition of 14 food samples varied greatly (Table 2). Energy content of food samples from adults and chicks did not differ significantly ($t = 0.65$ $P > 0.1$), so we used the overall mean energy content of 9.3 kJ g (wet wt.) mass from pooled adult and chick samples to calculate energy delivery to chicks. The assimilation efficiency of Whitechinned Petrel chicks is not known. Measurements of assimilation efficiency in other seabird vary from 75–85% (Dunn 1975; Cooper 1978; Kooyman et al. 1982), so an assimilation efficiency of 80% was assumed for the chicks in this study.

Growth of the three chicks in 1982 was significantly different from growth in the 1981–1982 season only at

Table 1. Feeding rate and mass of meals delivered to Whitechinned Petrel, *Procellaria aequinoctialis*, chicks at Marion Island

Chick age (d)	Sample size	Feeding rate (meals d^{-1})	Sample size	Meal mass ± S.D. (g)
8–14	9	0.64	18	151 ± 38
15–21	16	0.44	9	183 ± 67
22–28	13	0.48	13	198 ± 58
29–34	4	0.34	6	208 ± 75
35–40	10	0.47	7	198 ± 71
41–48	15	0.76	13	193 ± 58
49–55	16	0.52	12	203 ± 56
56–62	13	0.44	12	206 ± 41
63–69	7	0.48	6	175 ± 65
70–76	10	0.77	14	217 ± 65
77–83	9	0.78	8	183 ± 68
84–91	9	0.31	10	212 ± 62
92–96	5	0.44	5	127 ± 29[a]

a Significantly different from meal mass for 84–91-day age class $t = 3.252$, $P < 0.005$.

Table 2. Percentage composition and energy content of food from Whitechinned Petrel, *Procellaria aequinoctialis,* chicks and adults at Marion Island

		% Water		% Lipid		% Protein		Energy content kJ g^{-1} (wet wt.)	
	n	Mean	Range	Mean	Range	Mean	Range	Mean	Range
Adult	4	72.4	69.5–75.7	14.8	12.9–17.8	12.9	11.3–16.0	8.0 ± 0.9	7.0– 9.0
Chick	10	66.1	20.1–81.3	17.9	8.3–43.9	16.0	5.7–47.9	9.8 ± 5.2	5.3–20.7
Overall	14	67.8		17.0		15.2		9.3 ± 4.7	

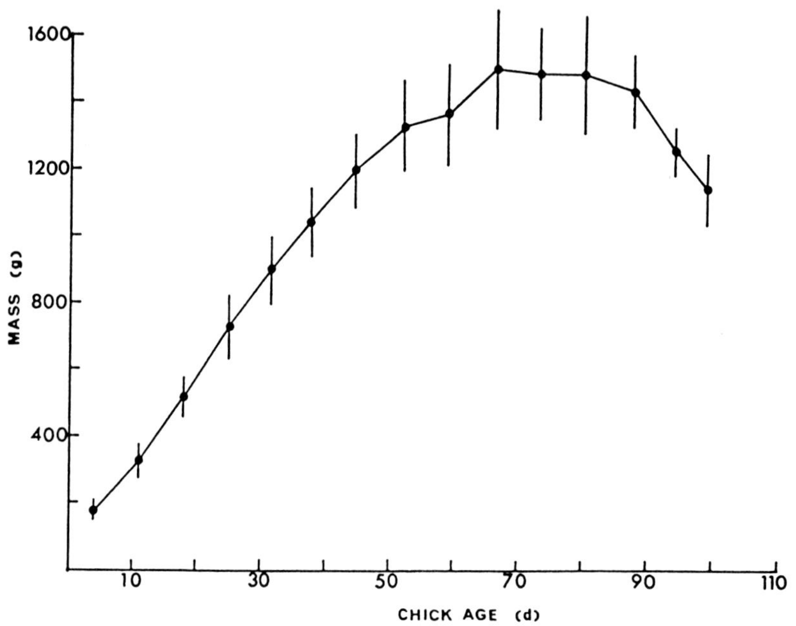

Fig. 1. Growth of Whitechinned Petrel, *Procellaria aequinoctialis,* chicks at Marion Island by 5–8-day age class. Mean ± 1 S.D

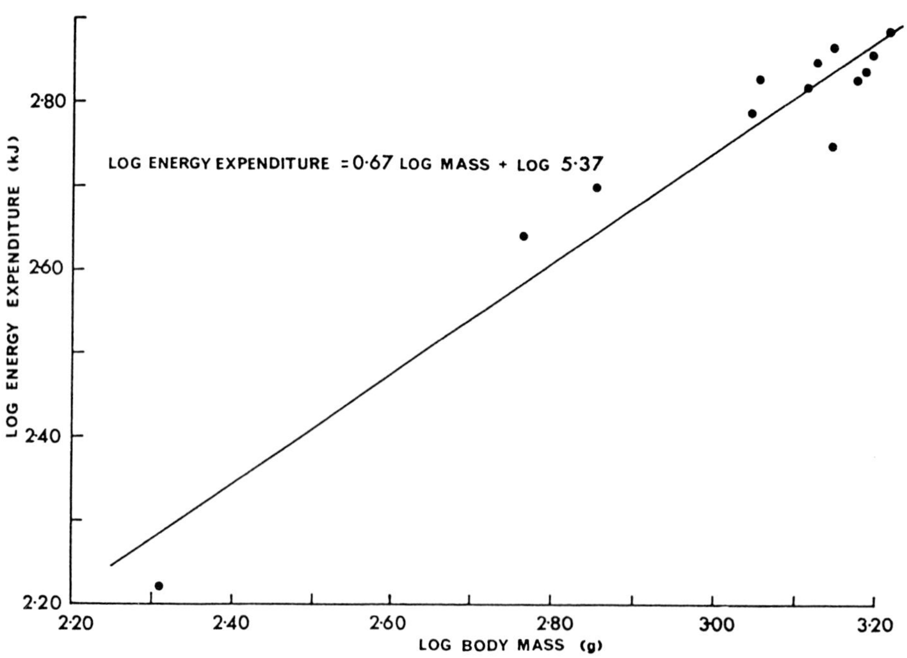

Fig. 2. Regression of log body mass aginst log energy expenditure of three Whitechinned Petrels, *Procellaria aequinoctialis,* chicks throughout the growth period in the 1982 season

27 days of age ($t = -2.97$, $P < 0.005$). Thirteen t-tests were performed, thus one significant result might be expected. The growth measurements for the 2 seasons were therefore combined (Fig. 1). Total energy expenditure of growing chicks in the two seasons was therefore comparable. Log energy expenditure correlated highly with log chick mass in 1982 ($r = 0.96$, $P < 0.001$, $n = 13$) (Fig. 2) and maximum energy expenditure therefore coincided with maximum chick mass at 62 days of age. Mass specific energy expenditure decreased from a maximum of 776 kJ kg^{-1} d^{-1} at 7 days to a minimum of 393 kJ kg^{-1} at 91 d, lower than the 423 kJ kg^{-1} d^{-1} measured for Whitechinned Petrel adults (Adams and Brown 1984).

4 Discussion

The Whitechinned Petrel chick receives on average one meal every 2 days at both South Georgia (Croxall and Prince 1980) and at Marion Island (this study), although individual chicks are fed irregularly. Variation in feeding intervals was probably underestimated in the present study, since double feeds in periods of < 6 h would have been recorded as a single meal and weighings often ceased before lengthy fasts were broken. However, despite this, the irregularity in meal delivery is apparent, supporting the view that food is sparse and patchily distributed (Croxall and Prince 1980). Nevertheless, some of the variability in feeding frequency may be attributable to weather changes. Adverse weather conditions associated with the movement of intense low pressure systems across the Sub-Antarctic region may markedly affect the feeding of pelagic seabirds, as postulated for the Broadbilled Prion, *Pachyptila vittata,* at Marion Island (Mendelsohn 1981).

There was a considerable variation in the water, lipid and energy content of food obtained from stomachs of White-chinned Petrel adults and chicks (Table 2). This may reflect differences in meal composition or, more likely, differing stages of digestion. Procellariiforms are well known for their ability to accumulate stomach oils of dietary origin (Clarke and Prince 1976; Imber 1976; Warham et al. 1976). Cheah and Hansen (1970) and Ashmole (1971) suggested that procellariiforms could selectively assimilate nitrogenous components of the diet and excrete much of the water. The food samples from Whitechinned Petrel chicks are considerably more variable in lipid and water than are those from adults (Table 2), consistent with lipid concentration in the stomach through differential digestion.

Estimates of the total energy assimilated (feeding rate × meal mass × energy content × assumed assimilation efficiency) are given in Fig. 3. Total energy assimilated for chick growth period of 96 days was calculated to be 66,938 kJ. Correcting for assimilation efficiency, this represents an intake of 8,637 g of food.

We propose a chick energy budget per 5–8-day age class based on combined data for the two seasons of chick growth and energy expenditure determined from Fig. 2. The budget is broken down into measured energy expenditure (cost of maintenance plus cost of biomass synthesis) and energy contained in accumulated chick biomass (Table 3). Energy contained in accumulated biomass was calculated using an assumed mean energy value of 25.4 kJ g^{-1} (dry wt.) mass of chicks, calculated from the carcass analysis of 10 species of petrel (Williams 1978) and an

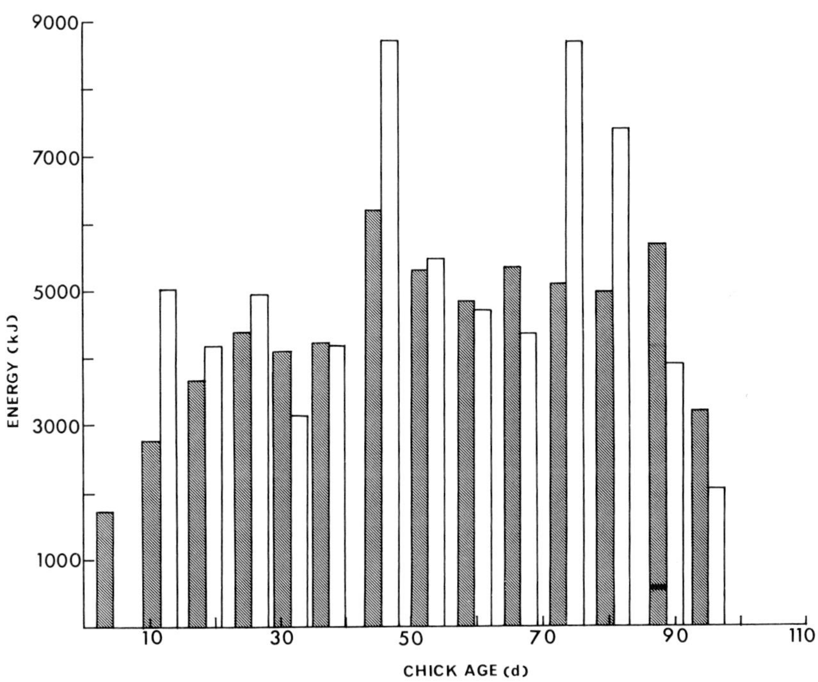

Fig. 3. Energy assimilated and utilized by Whitechinned Petrels, *Procellaria aequinoctialis,* chicks by 5–8-day age class □ assimilated energy ▧ utilized energy

Table 3. Energy budget of Whitechinned Petrel, *Procellaria aequinoctialis*, chicks by 5–8-day age class

Chick age class (d)	Energy expenditure kJ d^{-1}	Energy expenditure per age class (kJ) (1)	Energy accumulated in biomass (kJ) (2)	(1) + (2) (kJ)
1– 7	175.8	1,231	478.3	1,709
8–14	262.1	1,835	923.3	2,758
15–21	356.9	2,498	1,202.0	3,700
22–28	451.0	3,157	1,360.0	4,517
29–34	516.5	3,099	1,033.1	4,132
35–40	569.3	3,416	980.7	4,297
41–48	630.3	5,042	1,068.7	6,111
49–55	676.0	4,732	836.3	5,568
56–62	686.2	4,803	190.5	4,994
63–69	731.8	5,123	866.7	5,989
70–76	727.7	5,094		5,094
77–83	727.2	5,090		5,090
84–91	712.7	5,702		5,702
92–96	649.6	3,248		3,248
Totals		54,070	8,840	62,909

assumed body water content of 75% (= water content of a newly hatched semi-precocial tern; Ricklefs and White 1981). Body water content of chicks generally decreases from around 85% at hatching to about 60% as adult mass is attained (Skadhauge 1981). Biomass accumulation from hatching to peak mass at 63–69 days was calculated for 5–8 day age from masses presented in Fig. 1. This was calculated by dividing the differences in mass between age classes by the number of days between each age class. This gives a total energy value of accumulated biomass of 8,840 kJ for a bird of mean peak mass of 1,496 g. Mass recession prior to departure from the nest will release energy. Measurements of energy expenditure will take into account oxidation of these energy reserves.

The semi-precocial development of the chick with rapid attainment of thermal independence allows both parents to forage within 3 to 7 days of the chick hatching. Initially the rate of energy delivery by the adults is high in relation to the maintenance energy requirements of the chick. Average meal mass is 46% of the body mass of chicks at 10 days of age, decreasing to 10% at 92 days. Thus, the parents can easily provide an excess of energy which the chick presumably stores as lipid. A large proportion of the energy input to the chick is, thus, allocated to accumulation of biomass (30%). This decreases to 15% between 49 and 55 days of age and to 4% between 56 and 62 days, at which stage most of the energy is allocated to maintenance. The pattern during the initial part of growth is consistent with that described for Leach's Stormpetrel, *Oceanodroma leucorhoa*. Ricklefs et al. (1980b) estimated that energy requirements for growth in the smaller Leach's Stormpetrel decrease from 60% of the total energy budget just after hatching to 25% before departure from the nest. Summation of the energy partitioning for 5–8-days age classes (Table 3) gives a total energy utilization by the chick of 62,909 kJ.

There is a surplus of 4,028 kJ when energy delivery is compared to energy utilization. Energy delivery to the chick does not, however, include that delivered to the chick in the first 7 days after hatching, and energy expenditure does not take into account activity such as wing flapping which occurs prior to departure to sea (pers. obs). Moreover, temperatures in natural burrows are generally cooler than the temperatures at which measurements of oxygen consumption were made. Energy expenditure required for thermoregulation will therefore be under-estimated.

Parental investment in the chick must be balanced against risk incurred by the adults in raising the chick. In a longlived species with low fecundity, typical of the procellariiforms (Lack 1968), selection will tend to favour reduction in parental risk in any particular year to protect adult survival and future productivity. The additional food necessary for the deposition of energy reserves by the chick early in the growth period presents little risk to the adult. Parents can thus extend their foraging trips if necessary and maintain body condition later in the breeding cycle despite subsequent increases in chick energy requirements.

Acknowledgements. Research at Marion Island is carried out under the auspices of the South African Scientific Committee for Antarctic Research. Financial and logistical support was provided by the South African Department of Transport. We thank Jon Maltby and Pieter van Litsenborgh for help in the field.

References

Adams NJ, Brown CR (1984) Metabolic rates of sub-Antarctic Procellariiformes: a comparative study. Comp Biochem Physiol 77A:169–173

Ashmole NP (1971) Seabird ecology and the marine environment. In: Farner DS, King JR (eds) Avian biology. Academic, New York, vol 1, pp 223–286

Cheah CC, Hansen IA (1970) Stomach oil and tissue lipids of the petrels *Puffinus pacificus* and *Pterodroma macroptera.* Int J Biochem 1:203–208

Clarke A, Prince PA (1976) The origin of stomach oil in marine birds: the analyses of stomach oil from six species of Sub-Antarctic procellariiform birds. J Exp Mar Biol Ecol 23:15–30

Clarke A, Prince PA (1980) Chemical composition and calorific value of food fed to mollymauk chicks *Diomedea melanophris* and *D. chrysostoma* at Bird Island, South Georgia. Ibis 122:488–494

Cooper J (1978) Energetic requirements for growth and maintenance of the Cape Gannet (Aves: Sulidae). Zool Afr 13:305–317

Croxall JP, Prince PA (1980) Food, feeding ecology and ecological segregation of seabirds at South Georgia. Biol J Linn Soc 14:103–131

Dunn EH (1975) Caloric intake of nestling double-crested cormorants. Auk 92:553–565

Imber MJ (1976) The origin of petrel stomach oils – a review. Condor 78:366–369

Kooyman GL, Davis RW, Croxall JP, Costa DP (1982) Diving depths and energy requirements of King Penguins. Science (Wash DC) 217:727

Lack D (1968) Ecological adaptations for breeding in birds. Methuen, London

Mendelsohn J (1981) Movements of *Pachyptila* spp. and low pressure systems at Marion Island. In: Cooper J (ed) Proceedings of the Symposium on Birds of the Sea and Shore, 1979. African Seabird Group; Cape Town pp 223–231

Mougin JL (1970) Le Pétrel à menton blanc *Procellaria aequinoctialis* de l'île de la Possession (Archipel Crozet). L'Oiseau et RFO 40, no. spécial 62–90

Petrusewicz K, Macfadyen A (1970). Productivity of terrestrial animals: principles and method. IBP Handbook 13. Blackwell Scientific Publications, Oxford

Ricklefs RE, White SC, Cullen J (1980a) Postnatal development of Leach's Storm-Petrel. Auk 97:768–781

Ricklefs RE, White SC, Cullen J (1980b) Energetics of postnatal growth in Leach's Storm-Petrel. Auk 97:566–575

Ricklefs RE, White SC (1981) Growth and energetics of chicks of the Sooty Tern (*Sterna fuscata*) and Common Tern (*S. hirundo*). Auk 98:361–378

Sinclair JC (1981) Techniques for observing Sub-Antarctic burrowing petrels at the nest. Cormorant 9:67–72

Skadhauge E (1981) Osmoregulation in bird. Springer, Berlin Heidelberg New York

Warham J, Watts R, Dainty RJ (1976) The composition, energy content and function of the stomach oils of petrels (Order, Procellariiformes) J Exp Mar Biol Ecol 23:1–13

Williams AJ (1978) Mineral and energy contribution of petrels (Procellariiformes) killed by cats, to the Marion Island terrestrial ecosystem. S Afr J Antarct Res 8:48–52

Energy and Food Requirements of Pelagic Aerial Seabirds in Different Regions of the African Sector of the Southern Ocean

R.W. ABRAMS[1]

Summary. The African sector of the Southern Ocean includes 3 regions separated by oceanic fronts. The frontal areas between the sub-tropical, sub-Antarctic and Antarctic regions are characterized by relatively high nutrient and biotic production levels. In addition, the Benguela upwelling system over the southwestern African continental shelf is especially productive. Daily Energy Expenditures (DEE) of seabirds capable of flight were compared to see whether differences in primary productivity between different areas at sea are reflected at higher trophic levels. Theoretical carbon flux to seabirds at sea during 1979–1981 was calculated to be between 0.040 g C m^2 yr^{-1} and 0.224 g C m^2 yr^{-1}. The greatest C flux to seabirds occurred in the sub-tropical, and in certain frontal, areas.

1 Introduction

It has been suggested that seabirds could serve as indicators of the distribution of prey populations in the Southem Ocean, because they are the most visible carnivores at sea (Anon. 1977). High seabird abundance has been related to enhanced prey availability at oceanic fronts in the Bering Sea (Hunt et al. 1981; Schneider 1982), and near Antarctica (Ainley and Jacobs 1981). Aerial seabirds (species capable of flight) in the Southern Ocean are concentrated in convergence, divergence and frontal zones (Abrams and Griffiths 1981; Ainley and Jacobs 1981; Abrams 1982; Griffiths et al. 1982). These zones are characterized by steep thermohaline gradients and high relative abundance of nutrients, plankton, squid and fish (El-Sayed 1970; Holm-Hansen et al. 1977; Iverson 1977; Hulley 1981; Tranter 1982).

In this paper I report the minimum amount of prey biomass required by aerial seabirds, based on their energy needs, in the African sector of the Southern Ocean during 1979–81. More particularly, the amount of carbon theoretically transferred through plankton (principally Krill, Euphausiacea), squid and fish to aerial seabirds is compared between nine oceanographic zones. The distribution of seabirds is discussed in the context of the physical processes which could influence the availability of prey. Comparisons are made of the C flux and food required by seabirds in the African sector of the Southern Ocean and other oceanic regions.

2 Methods

Seabirds (excluding penguins) were recorded during 12 separate Cruises of the M.V. S.A. Agulhas between April 1979 and June 1981 (Fig. 1). Similar areas were covered during the same seasons in 1979–80 and 1980–81. Attempts were made to count all birds observed flying past the moving ship (mean speed = 23.4 km h^{-1}) or sitting on the sea in a 1 km-wide transect (abeam of the ship) during 5923 10-min (in 1-h periods) observation sessions (Griffiths 1981). No counts were made within 100 km of breeding sites of any species.

Seabird density estimates (no. birds km^{-2}) were calculated for four frontal zones and five zones between the fronts and continents (Fig. 1). The locations of fronts were taken from Deacon (1982) and Lutjeharms and Valentine (to be publ.). The inter-frontal zones are named according to the fronts bordering the zones (i.e. Sub-Tropical Zone (STZ) and Sub-Tropical Convergence (STC); Sub-Antarctic Zone (SAZ) and Sub-Antarctic Front (SAF); Polar Frontal Zone (PFZ) and Antarctic Polar Front (APF); Antarctic Water Zone (AWZ) and Continental Water Boundary (CWB); Continental Water Zone (CWZ) and Antarctica). Mean mass per bird in plankton-, squid-, fish-, and mixed-diet classes was based on the proportionate abundance of each species which occurred in each of the nine zones (Appendix 1, see page 471).

Theoretical energy requirements (Daily Energy Expenditure) (DEE in kJ per bird of mean mass per day: Walsberg 1980) were calculated for each diet class in each zone. The food (wet wt. mass) of plankton- (including krill), and squid-eating birds was assumed to have an energy value of 4.51 kJ g^{-1} (Mauchline 1980), and that of fish-eaters to be 21 kJ g^{-1} (Schneider and Hunt 1982). Equal proportions of plankton, squid and fish were assigned to the mixed-diet species. All birds were assumed to have an assimilation efficiency of 80% (Clark and Prince 1981). Carbon flux was based on a value of 0.4 g C g^{-1} dry wt. prey (Curl 1962). The seabirds' requirements for the area north of the Ant-

1 FitzPatrick Institute, University of Cape Town, Rondebosch 7700, South Africa

Antarctic Nutrient Cycles and Food Webs (ed. by W.R. Siegfried, P.R. Condy, and R.M. Laws)

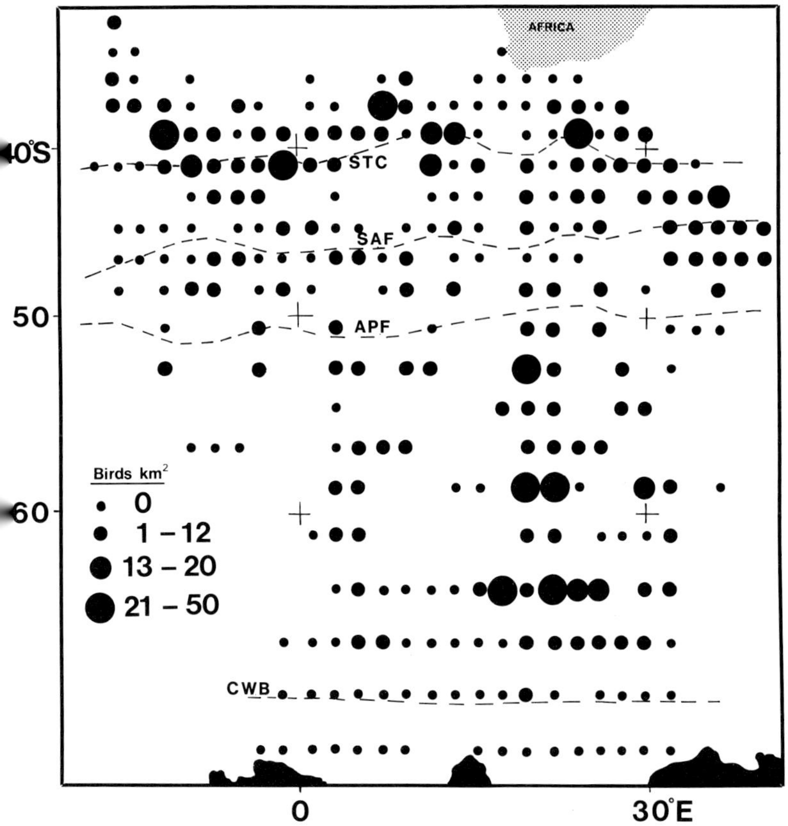

Fig. 1. Pelagic distribution (no. birds km²) of planktivorous seabirds in the African sector of the Southern Ocean, 1979–1981. Approximate positions are given for the Sub-Tropical Convergence (*STC*); Sub-Antarctic Front (*SAF*); Antarctic Polar Front (*APF*); Continental Water Boundary (*CWB*)

arctic Polar Front (maximum of 53°S) (Lutjeharms and Valentine, to be publ.) were full-year composites, and south of the winter pack-ice limit half-year composites (Nov.–March), because aerial seabirds cannot forage in ice-covered seas.

The Daily Energy Expenditure (Walsberg 1980) equations used here produced generally higher values than any of the formulae reviewed by Schneider and Hunt (1982). However, Weathers and Nagy (1980) found Walsberg's (1980) formula to be conservative, based on determinationas of metabolic rate using labelled water turnover rates in free-living animals. Weathers and Nagy (1980) made the methodological comparison using a bird with an inordinately high water turnover rate, so that their D_2O^{18} method would perhaps slightly under-estimate actual metabolism. This reinforces the conservative bias of Walsberg's (1980) formula.

3 Results

The sub-tropical areas supported the highest avian density (Fig. 1, Table 1); the birds being concentrated in the Sub-Tropical Zone. In abundance, the avifauna was dominated by planktivores at the Sub-Tropical Convergence and large-bodied squid-eaters nearer the African continent (Figs. 1–2).

High abundance of large-bodied species in the north of the survey area was reflected as a maximum flux of C (0.181–0.224 g C m^{-2} yr^{-1}, Table 2, Fig. 3) to seabirds in the sub-tropical areas. The relative requirements of plankton-eaters increased in high latitudes (Fig. 3); the Antarctic Water Zone theoretically supporting, per unit time, a C flux (0.086 g C m^{-2} 0.5 yr^{-1}) similar to those of the sub-tropical areas (Table 2). The avifauna of high latitudes contained a high abundance of relatively small-bodied birds adapted to utilizing patchily but locally rich prey concentrations typical of areas south of 50°S (Mauchline 1980; Tranter 1982; Abrams 1982).

Large squid-eating albatrosses and petrels require 10–12% of their DEE for flight (Table 3). To sustain continual flight, albatrosses require between 85 and 160 g squid daily. Thus, based on a meal of 650 g (Prince 1980), a relatively small albatross could forage for six days and a Wandering Albatross for 3–4 days before losing body mass. Prions, on the other hand, use daily 67 g food for flight alone, which is over 50% of their DEE (Table 3). At 12 g per meal (Prince 1980), they require 5–6 meals d^{-1}, or a meal every 4–5 h, just for flight.

4 Discussion

4.1 Regional Carbon Flux to Seabirds in the African Sector

In the African sector of the Southern Ocean, the distribution of the different diet classes of seabirds apparently reflects the gross structure of the habitat and availability of

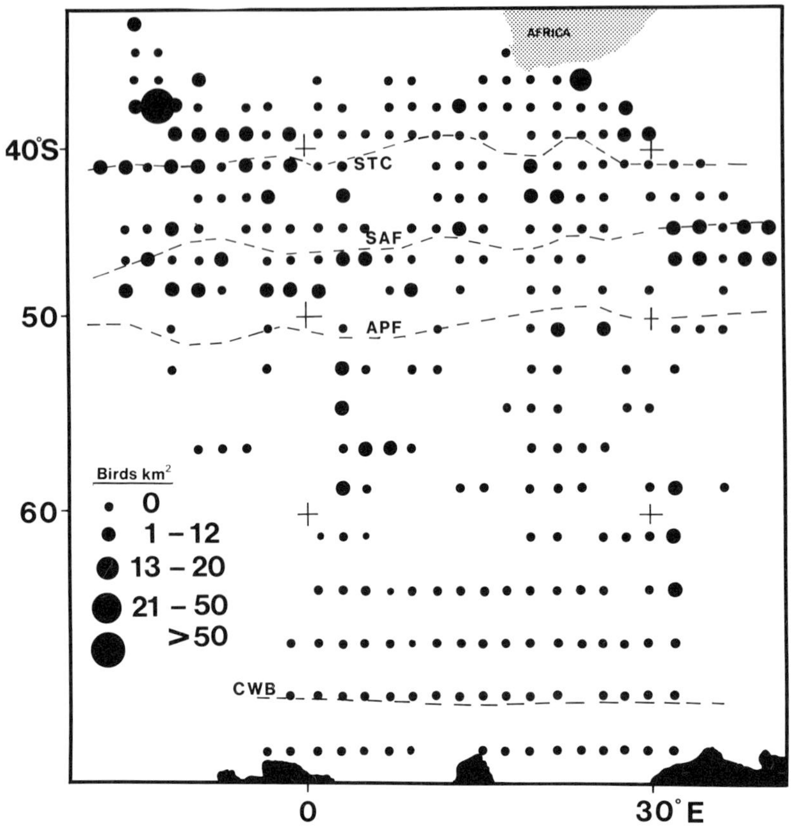

Fig. 2. Pelagic distribution (no. birds km^{-2}) of squid-eating seabirds in the African sector of the Southern Ocean, 1979–1981. See Fig. 1 for explanation of abbreviations

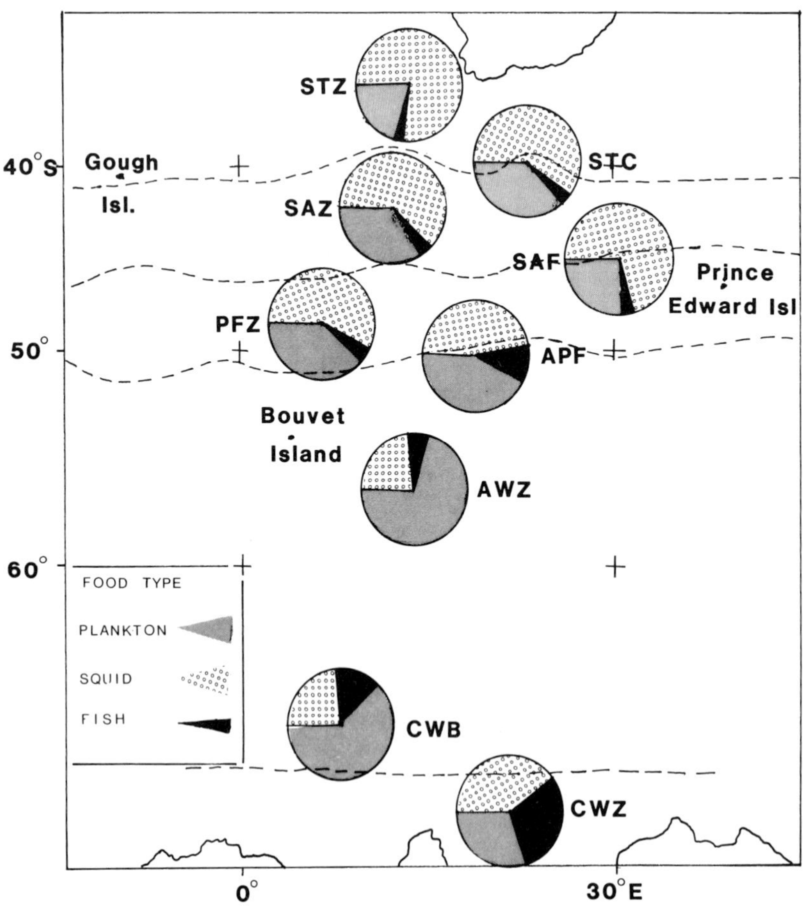

Fig. 3. Proportionate distribution of the amounts of seabird food-types based on diets and carbon requirements (g C m^{-2} yr^{-1}) of seabird populations in different zones of the African sector of the Southern Ocean, 1979–1981. See text for explanations of abbreviations of names of oceanic fronts and zones

Table 1. Mean densities of seabirds (no. birds km^{-2}) and their energy requirements (Daily Energy expenditure, kJ km^{-2}) in the African sector of the Southern Ocean, 1979–1981. See text for names of zones

Zone	Diet class								No. of seabird counts
	Planktivores		Squid-Eaters		Piscivores		Mixed		
	Density	Energy	Density	Energy	Density	Energy	Density	Energy	
STZ	2.6	983	2.7	4,166	0.4	381	0.5	396	1,363
STC	4.0	1,503	1.3	2,575	0.3	209	1.1	864	865
SAZ	1.9	718	0.7	1,409	0.1	73	0.8	523	520
SAF	1.2	482	0.7	1,426	0.1	54	0.6	460	497
PFZ	1.9	721	0.5	1,158	0.1	30	1.1	730	255
APF	2.3	866	0.6	960	0.5	516	1.6	1,041	115
AWZ	7.0	2,838	0.5	852	0.4	449	1.8	1,315	661
CWB	2.6	1,113	0.2	348	0.7	718	1.0	944	1,224
CWZ	0.1	54	0.1	154	0.8	281	2.3	2,175	423

Table 2. Annual prey biomass (g (fr. wt.) m^{2}) and carbon flux (g C m^{-2}) required by seabirds in the African sector of the Southern Ocean, 1979–81

Zone	Diet class									
	Planktivores		Squid-Eaters		Piscivores		Mixed		Total	
	Biomass	C	Biomass	C	Biomass	C	Biomass	C	Biomass	C
STZ	0.39	0.040	1.57	0.175	0.03	0.003	0.06	0.006	2.05	0.224
STC	0.57	0.061	0.98	0.104	0.02	0.002	0.13	0.014	1.63	0.181
SAZ	0.27	0.029	0.53	0.056	0.01	0.001	0.08	0.009	0.89	0.095
SAF	0.18	0.020	0.51	0.058	0.01	0.001	0.07	0.007	0.77	0.085
PFZ	0.27	0.029	0.43	0.047	0.01	0.001	0.11	0.011	0.82	0.088
APF	0.33	0.035	0.36	0.039	0.04	0.005	0.15	0.016	0.88	0.095
AWZ	0.53	0.058	0.16	0.017	0.02	0.002	0.10	0.010	0.81	0.087
CWB	0.21	0.023	0.07	0.007	0.03	0.003	0.07	0.007	0.38	0.040
CWZ	0.01	0.001	0.03	0.003	0.01	0.001	0.16	0.017	0.21	0.020

prey at the sea surface. Primary production at the Sub-Tropical Convergence is higher than at the Antarctic Polar Front (El-Sayed 1970); Deacon 1982). Allanson et al. (1981) found relatively high chlorophyll levels during two crossings of the Sub-Tropical Convergence (40°S). Aerial seabirds required 0.2% of the primary production (93 g C m^{-2} yr^{-1}) reported for this frontal zone by Holm-Hansen et al. (1977). Seabird density was relatively low in the zones associated with the Sub-Antarctic and Antarctic Polar fronts, where Allanson et al. (1981) found high chlorophyll levels during one of two crossings. Aerial seabirds required 1.2% of the primary production (8 g C m^{-2} yr^{-1}) reported for the APF by Holm-Hansen et al. (1977). The seabirds in the Antarctic Water Zone required 0.6% of the C flux, based on productivity (16 g C m^{-2} yr^{-1}) for these latitudes given by Holm-Hansen et al. (1977) and El-Sayed (1978). Seabird energy requirements were lower at the Sub-Antarctic and Polar fronts than in the Sub-Tropical Convergence and the Antarctic Water zones (Table 2). Yet, the percentage of C flux to the birds was higher at the Antarctic Polar Front than in the Sub-Tropical Convergence or Antarctic Water zones.

Table 3. Daily Flight Energy (DFE, Pennycuick 1982) as a percentage of Daily Energy Expenditure (Walsberg 1980) for selected species of seabirds

Species	DEE ($kJ\ d^{-1}$)	DFE ($kJ\ d^{-1}$)	DFE / DEE × 100
Diomedea exulans	5,538	569	10.3
D. melanophris	3,052	354	11.6
D. chrysostoma	3,109	346	11.1
Phoebetria fusca	2,441	300	12.3
Procellaria aequinoctialis	1,509	179	11.9
Daption capensis	783	665	84.9
Pachyptila spp.	378	240	63.5
Oceanites oceanicus	157	45	28.8
Macronectes spp.	3,710	440	11.9

.4.2. The African Sector in a Global Context

The biomass of seabirds (excluding penguins) in the Southern Ocean has been estimated at 48000 t (Iverson 1977) and 56500 t (Prevost 1981). Based on counts of seabirds at sea, the African sector of the Southern Ocean contains 31 t yr^{-1} of aerial seabirds, or 0.07% and 0.06% of Iverson's and Prevost's estimates, respectively. Iverson (1977) estimated total annual food requirements for aerial seabirds in the Southern Ocean to be 6.0–8.5 times 10^6 t yr^{-1} of krill, squid and fish. Prevost's (1981) estimate was slightly lower (5.5×10^6 t). The minimum amount of food required by the seabird population in the African sector, based on the data presented here, is 1.4 times 10^4 t, or about 2.0–3.0% of Iverson's and Prevost's totals for the entire Southern Ocean population of aerial species. The area covered by my study is 0.4% of Iverson's area; meaning that my estimates of food consumption are comparable with his.

The sub-tropical region of southwestern Africa (Benguela upwelling system) supported the greatest avian density, requiring between 1.8 g and 2.1 g food m^{-2} yr^{-1}. Estimates of food required by breeding birds in the sub-tropical Peruvian upwelling area (11–45 g food m^{-2} yr^{-1}, Schaeffer 1970) constitute the only example of a higher seabird food requirement. The deep-ocean avifauna of the eastern Pacific Sub-Arctic zone require less food (0.05–0.10 g food m^{-2} yr^{-1}, Sanger 1972) than the birds of the sub-Antarctic zones of the Southern Ocean (0.77 g–0.89 g food m^{-2} yr^{-1}). A higher northern sub-polar seabird food requirement (1.9 g food m^{-2} yr^{-1} by seabirds in the vicinity of an Atlantic Sub-Arctic breeding colony) was reported by Furness (1978).

In polar latitudes (50°–65°) C flux to seabirds in the Southern Ocean (0.087 g C m^{-2} 0.5 yr^{-1}) was slightly higher than in the polar Bering Sea (0.050 g C m^{-2} 0.5 yr^{-1}, Schneider and Hunt 1982). Using a formula more like that of Walsberg (1980) on the same data set, Schneider and Hunt (1982) re-calculated DEE (after King 1974) which gave a C flux of 0.075 g C m^{-2} 0.5 yr^{-1} to seabirds in the Bering Sea. This is very close to the values for the Southern Ocean. Whereas Southern Ocean aerial seabirds required 0.1.–1.2% of the estimated primary production, the Bering Sea population required 0.02–0.03%, based on primary production estimates by McRoy and Goering (1976). With the inclusion of penguins at the same trophic level as aerial birds (Prevost 1981), close to 10% of the primary production may be required by the Southern Ocean seabirds.

A composite of Antarctic food-chain models by Nemoto and Harrison (1981) suggests only two or three trophic levels below seabirds. Sanger (1972) postulates five trophic levels below seabirds in the North Pacific. The food chain to Antarctic baleen whales is shorter than that to baleen whales in the North Pacific (Nemoto and Harrison 1981). The apparently shorter food chain to seabirds in the Antarctic is probably linked with the birds' requirements for a greater proportion of the primary production than in the North Pacific.

Caution is necessary in making generalizations about oceanic primary production and food chains, because the widest variation in productivity values and the greatest degree of plankton patchiness occur in the richest areas (Ryther et al. 1973; Longhurst 1981). In the past, oceanographic stations were operated on broad spatio-temporal scales (i.e. at 4–6 h intervals). Recently, oceanographers have shown the existence and complexity of small-scale events (i.e. 50–150 km diameter) (Longhurst 1981; Lutjeharms 1981; Peterson et al. 1982; Hofmann et al. to be publ.) which probably affect plankton productivity significantly.

Patchy concentrations of nutrients and krill are associated with the Antarctic Water Zone (Makarov et al. 1970; Tranter 1982). Prey patches at the sea surface must be frequent enough in the zone to support dense patches of plankton-eating birds, since they require frequent meals. Moreover, seabird mobility provides a "constancy in their habitat preferences" on an oceanic scale (Brown 1980), so that the birds adapt rapidly to different prey distributions. Plankton-eating birds occur more uniformly in high densities over areas where primary production is more uniformly distributed (i.e. the sub-tropical zones) than in high latitudes (i.e. Antarctic Water Zone) where they occur in dense patches in response to patchy prey distribution. Data on prey abundance and distribution, and seabird associations with small-scale oceanographic events, are needed to investigate further the extent to which seabird distribution is influenced by small-scale oceanic events which cause plankton patchiness.

Appendix 1. Principal food-type classes, body mass (live weight) and percentage abundance (numbers of individuals) of species observed in the African sector of the Southern Ocean, 1979–1981. Percentage abundance is presented for each species in each of the habitat zones as described in text. Food-type classes based on Ashmole (1971) and unpublished records taken from the FitzPatrick Institute which also maintains records of bird weights

		Mass (kg)	STZ	STC	SAZ	SAF	PFZ	APF	AWZ	CWB	CWZ
Plankton											
Pachyptila spp	prions	0.15	40.84	57.52	48.07	29.48	48.66	41.01	50.83	27.20	1.06
Haloboena caerulea	Blue Petrel	0.21	0	0.08	3.63	13.65	1.68	1.57	20.09	30.33	1.87
Oceanites oceanicus	Wilson's Stormpetrel	0.04	0.36	0.25	0.44	0.21	0.17	0.17	0.24	0.43	0.86
Oceanodroma leucorhoa	Leach's Stormpetrel	0.05	0.10	0.02	0.03	0	0	0	0	0.24	0
Fregatta tropica	Blackbellied Stormpetrel	0.06	0.13	0.13	0.73	0.59	0.33	1.97	0.83	< 0.01	0
Fregatta grallaria	Whitebellied Stormpetrel	0.05	0.03	0.15	0.14	0.16	0.06	0.17	0.02	0	0
Pelagadroma marina	Whitefaced Stormpetrel	0.03	0.01	0.03	0	0	0	0	0	0	0
Hydrobates pelagicus	European Stormpetrel	0.04	0.06	0.01	0.01	0	0.25	0.39	0.19	0.17	0
Pelecanoides spp	diving petrels	0.12	0.42	0.80	0.48	0.86	1.51	1.49	0.15	< 0.01	0
Phalaropus fulicarus	Grey Phalarope	0.03	0.03	0.09	0	0.54	1.65	0	0	0	0
Garrodia nereis	Greyrumped Stormpetrel	0.05	0	0.09	0	0.02	0	0	0	0	0
			41.56	59.17	53.53	45.51	54.31	46.77	71.53	57.95	3.79
Squid											
Diomedea exulans	Wandering Albatross	8.60	0.93	1.46	1.03	1.33	0.58	0.48	0.20	0.27	0.11
Diomedea melanophris	Blackbrowed Albatross	3.50	1.52	1.22	2.50	2.37	3.00	0.39	0.18	0.01	0
Diomedea chrysostoma	Greyheaded Albatross	3.60	0.03	0.15	0.90	2.50	1.21	0.13	0.18	0.01	0
Diomedea chlororhynchos	Yellownosed Albatross	2.00	1.04	0.44	0.16	0.06	0.03	0	0	0	0
Diomedea cauta	Shy Albatross	4.10	0.36	0.15	0.52	0.21	0.05	0	0.02	0	0
Phoebetria fusca	Sooty Albatross	2.50	0.31	1.57	0.59	1.05	0.41	0.12	0.18	0.06	0
Phoebetria palpebrata	Lightmantled Sooty Albatross	2.70	0	0.05	0.59	1.45	0.61	0.05	0.22	0.86	0.52
Fulmarus glacialoides	Antarctic Fulmar	1.00	< 0.01	0.02	0	0.66	2.64	0.17	1.42	0.37	0.58
Daption capensis	Pintado Petrel	0.45	0.87	1.01	0.65	1.15	1.24	0	0.62	0.39	0.40
Pterodroma macroptera	Greatwinged Petrel	0.58	1.65	1.65	4.09	3.57	1.45	2.80	0.43	0	0
Pterodroma lessonii	Whiteheaded Petrel	0.75	0.21	0.43	3.08	5.65	0.86	1.05	1.26	0.64	0
Pterodroma incerta	Atlantic Petrel	0.52	0.44	1.58	0.63	0.84	0.30	0	0.04	0	0
Procellaria aequinoctialis	Whitechinned Petrel	1.21	2.00	1.50	3.89	3.97	1.38	3.98	0.42	1.27	0.58
Procellaria cinerea	Grey Petrel	1.03	0.04	0.23	0.73	1.41	0.14	0.09	0.01	0	0
Puffinus gravis	Great Shearwater	0.95	34.06	7.61	0.87	0.63	0.80	2.32	0.06	0	0
			43.47	19.07	20.23	26.85	14.70	11.58	5.24	3.88	2.19
Fish											
Calonectris diomedea	Cory's Shearwater	0.96	1.81	0.03	0	0	0	0	0	0	0
Puffinus griseus	Sooty Shearwater	0.79	1.50	0.63	0.05	0.28	0.39	9.23	4.07	14.37	0.18
Puffinus puffinus	Manx Shearwater	0.48	0.10	0	0.01	0	0	0	0	0	0
Puffinus assimilis	Little Shearwater	0.23	1.97	4.18	3.55	3.09	0.69	0.22	0.01	0	0
Sterna vittata	Antarctic Tern	0.14	0.04	0.02	0	0	0	0	0	0.01	0.20
Sterna paradisaea	Arctic Tern	0.13	0.27	0.04	0.03	0	0.08	0	0.09	0.52	23.70
Sterna hirundu	Common Tern	0.12	0.37	0.01	0	0	0	0	0	0	0.75
Morus capensis	Cape Gannet	2.90	0.06	0	0	0	0	0	0	0	0
			6.12	4.91	3.64	3.37	1.16	9.45	4.17	14.90	24.83
mixed											
Macronectes giganteus	Southern Giant Petrel	4.10	0.05	0.26	0.27	0.70	0.44	0.08	0.27	0.20	0.42
Macronectes halli	Northern Giant Petrel	5.20	0.16	0.20	0.05	0.02	0	0	0	0.10	0.03
Thalassoica antarctica	Antarctic Petrel	0.70	0	0	0	0	0.03	0.52	1.83	12.29	48.37
Pagodroma nivea	Snow Petrel	0.30	0	0	0	0	0	0.09	0.65	2.29	14.96
Pterodroma brevirostris	Kerguelen Petrel	0.33	0.42	4.06	3.26	11.17	12.42	11.94	11.93	8.05	4.71
Pterodroma mollis	Softplumaged Petrel	0.31	7.08	11.86	17.81	10.90	16.52	19.24	3.56	0	0.11
Catharacta antarctica	Subantarctic Skua	1.63	0.09	0.05	0.08	0.08	0	0.13	0	0	0
Catharacta maccormicki	Maccormicks Skua	1.26	0.01	0	0	0	0	0	0.03	0	0
Stercorarius pomarinus	Pomarine Skua	0.67	0.13	0.05	0	0	0	0	0	0	0
Stercorarius parasiticus	Arctic Skua	0.53	0.05	0.12	0.03	0.39	0.03	0.17	0	0	0
Stercorarius longicaudus	Longtailed Skua	0.29	0.07	0	0	0	0	0	0	0	0
			8.06	16.60	21.50	23.26	29.44	32.17	18.27	22.93	68.60

Acknowledgements. This research was supported by the South African Scientific Committee for Antarctic Research. The data were collected by members of the FitzPatrick Institute, with the cooperation of the Department of Transport and the Captain and crew of the M.V. S.A. Agulhas.

References

Abrams RW (1982). Pelagic seabird distribution in relation to environmental features in the African sector of the Southern Ocean. In: Ilyichev VD, Gavrilov VM (eds) XVIII Congressus Inernationalis Ornithologicus. Abstracts of symposia and poster presentations. Moscow; Nauka pp 87–88

Abrams RW, Griffiths AM (1981) Ecological structure of the pelagic seabird community in the Benguela Current region. Mar Ecol Prog Ser 5: 269-277

Ainley DG, Jacobs SS (1981) Seabird affinities for ocean and ice boundaries in the Antarctic. Deep-Sea Res 28A: 1173–1185

Allanson BR, Hart RC, Lutjeharms JRE (1981) Observations on the nutrients, cholorophyll and primary production of the Southern Ocean south of Africa. S Afr J Antarct Res 10/11: 3–14

Anon. (1977) Biological Investigations of Marine Antarctic Systems and Stocks. BIOMASS Sci Ser 1 SCAR/SCOR, Scott Polar Research Institute, Cambridge, p 31

Brown RGB (1980) Seabirds as marine animals. In: Burger J, Olla BL, Winn HE (eds) Behavior of marine animals Vol 4, Plenum, New York, pp 1–39

Clarke MR, Prince PA (1981) Cephalopod remains and regurgitations of Black-browed and Grey-headed Albatrosses at South Georgia. Brit Antarct Surv Bull 54:1–7

Curl HS (1962) Analyses of carbon in marine plankton organisms. J Mar Res 20: 181–188

Deacon GER (1982) Physical and biological zonation in the Southern Ocean. Deep-Sea Res 29A: 1–15

El-Sayed SZ (1970) On the productivity of the southern ocean. In: Holdgate MW (ed) Antarctic ecology. Academic Press New York, pp 119–135

El-Sayed SZ (1978) Primary productivity and estimates of potential yields of the Southern Ocean. In: McWhinnie MA (ed) Polar research. AAAS selected symposium 7. Westview, Boulder

Emery KO, Milliman JD, Uchupi E (1973) Physical properties and suspended matter of surface waters in the southeastern Atlantic Ocean. J. Sediment Petrol 43(3):822–837

Furness RW (1978) Energy requirements of seabird communities: a bioenergetic model. J Anim Ecol 47:39–53

Griffiths AM (1981) Biases in censuses of pelagic seabirds at sea in the Southern Ocean. In: Cooper J (ed) Proc sym birds of sea and shore, 1979. African Seabird Group, Cape Town

Griffiths AM, Siegfried WR, Abrams RW (1982) Ecological structure of a pelagic seabird community in the Southern Ocean. Polar Biol 1: 39–46

Hofmann EE, Whitworth T, Nowlin WD (in press) Mesoscale flow variability at Drake Passage, J Geophys Res

Holm-Hansen O, El-Sayed SZ, Francerschini GA, Cuhel RL (1977) Primary production and the factors controlling phytoplankton growth in the Southern Ocean. In: Llano GA (ed) Adaptations within Antarctic ecosystems; Proc 3rd SCAR symposium on Antarctic Biology. Smithsonian Inst, Washington, DC pp 11–50

Hulley PA (1981) Results of the research cruises of FRV 'Walher Herwig' to South America. LVII Family Myctophidae. Arch Fischereiwiss 31(1):1–300

Hunt GL Jr, Gould PJ, Forsell DJ, Peterson FR (1981) Pelagic distribution of Marine Birds in the Eastern Bering Sea. In: Hood DW, Calder JA (eds) The eastern Bering Sea shelf: oceanography and resources. Vol 2. University Washington Press. Seattle, pp 689–717

Iverson I (1977) The Living Resources of the Southern Ocean, UNDP, FAO, Rome, p 156

King JR (1974) Seasonal allocation of time and energy resources in birds. In: Paynter RA (ed) Avian energetics. Nuttal Ornithological Club, Cambridge Mass, pp 4–85

Longhurst AR (1981) Significance of spatial variability. In: Longhurst AR (ed) Analysis of marine ecosystems. Academic Press, London, pp 415–442

Lutjeharms JRE (1981) Spatial scales and intensities of circulation in the ocean areas adjacent to South Africa. Deep-Sea Res 28A (11): 1289-1302

Lutjeharms JRE, Valentine HR (to be published) Southern Ocean thermal fronts south of Africa. Deep-Sea Res

Makarov RR, Naumov AG, Sherster VV (1970) The biology and distribution of the Antarctic krill. In: Holdgate MV (ed) Antarctic ecology. Academic Press, London, pp 173–176

Mauchline J (1980) Studies on patches of krill, *Euphausia superba* Dana. BIOMASS Handbook 6. SCAR/SCOR, Duntsaffnage Marine Research Laboratory, United Kingdom.

McRoy CP and Goering JT (1976) The influence of ice on the primary productivity of the Bering Sea. In: Hood D, Kelly E (eds) The oceanography of the Bering Sea. Univ Alaska Inst of Marine Sciences, pp 403–421

Nemoto T, Harrison G (1981) High latitude ecosystems. In: Longhurst AR (ed) Analysis of marine ecosystems. Academic Press, London, pp 95–126

Owen RW (1981) Fronts and eddies in the sea: mechanisms, interactions and biological effects. In: Longhurst AR (ed) Analysis of marine ecosystems. Academic, London, pp 197–234

Pennycuick CJ (1982) The flight of petrels and albatrosses (Procellariiformes), observed in South Georgia and its vicinity. Phils Trans R Soc Lond B 300:75–106

Peterson RG, Nowlin WD Jr, Whitworth T (1982) Generation and evolution of a cyclonic ring at Drake Passage in early 1979. J Phys Ocean 12: 712–719

Prevost J (1981) Population and energy requirements of Antarctic birds. BIOMASS Sci Ser II SCAR/SCOR, Scott Polar Research Institute, Cambridge, pp 125–137

Prince PA (1980) The food and feeding ecology of Grey-headed Albatross (*Diomedea chrysostoma*) and Black-browed Albatross (*D. melanophris*). Ibis 122:476–488

Ryther JH, Menzel DW, Barber RT (1973) Primary productivity and sizes of pools of organic carbon. Mar Biol 19(4): 298–306

Sanger GA (1972) Preliminary standing stock and biomass estimates of seabirds in the Subarctic Pacific Region. In: Takenouti A Y (ed) Biological oceanography of the northern Pacific Ocean. Idenitsu Shoten, Tokyo, pp 589–611

Schaeffer MB (1970) Men, birds and anchovies in the Peru current-dynamic interactions. Trans Am Fish Soc 99: 461–67

Schneider D (1982) Fronts and seabird aggregation in the southeastern Bering Sea. Mar Ecol Prog Ser 10:101–103

Schneider D, Hunt GL (1982) Carbon flux to seabirds in waters with different mixing regimes in the southeastern Bering Sea. Mar Biol 67:337–344

Tranter DJ (1982) Interliking of physical and biological processes in the Antarctic Ocean. Oceanogr Mar Biol Annu Rev 20:11–35

Walsberg GE (1980) Energy expenditure in free-living birds: patterns and diversity. In: Nohring R (ed) Acta XVII Congressus Internationalis Ornithologicus: 300–305 Deutsche Ornithologen Gesellschaft, Berlin

Weathers WW, Nagy KA (1980) Simultaneous doubly-labelled water (HHO) and time-budget estimates of daily energy expenditure in *Phainopepla nitens*. Auk 97: 861–867

Population and Energetic Aspects of the Relationship Between Blackbrowed and Greyheaded Albatrosses and the Southern Ocean Marine Environment

P.A. PRINCE[1]

Summary. The methods and results of the study of the tropho-dynamic relationships between 2 *Diomedea* albatrosses and the marine environment at South Georgia are described. They illustrate the technical and theoretical developments necessary to obtain certain empirical data essential for accurate assessments of the role of seabirds in marine ecosystems. Differences in breeding success during eight yr (consistent in *D. chrysostoma,* more variable in *D. melanophris*) are linked with important differences in breeding frequencies which affect the size and activities of populations at the breeding sites. Extensive dietary studies, based on sampling adults about to feed chicks, showed major inter-specific differences, resulting in chicks receiving meals of similar size and frequency but of different energy content. The frequency of chick feeding was determined initially by daily and 3-h weighing. Recently automatic equipment has recorded weights every 10 min, giving the frequency and size of meals and resulting digestive performances of the chicks. Experiments involving exchanging chicks between the two species were combined with new methods for analyzing growth curves. They showed that, while there was a species-specific genetic component to growth, the overall rate could be significantly modified by the nature of the diet. The slower growth rate of *D. chrysostoma* chicks, and the species' diet, are probably important factors affecting breeding frequency. Adult feeding performance is being studied by devices recording simple activity budgets at sea. Preliminary results are described and projected work linking this with the automatic weighing equipment and with assessment of foraging energy costs is outlined.

1 Introduction

Greyheaded and Blackbrowed albatrosses, *Diomedea chrysostoma* and *D. melanophris,* are very similar in size and weight. Their timing and duration of breeding is similar, although *D. chrysostoma* lays slightly earlier and takes about 25 d longer to rear its chick (Tickell and Pinder 1975). They have been studied in each season since 1975–76 at Bird Island, South Georgia, where they breed in large, usually separate, colonies.

Following Tickell and Pinder's (1975) basic study in 1962–64, these two mollymauks have been studied in regard to food and feeding ecology (Prince 1980), chick growth (Ricketts and Prince 1981; Prince and Ricketts 1981), population assessment and monitoring (Prince and Payne 1979; Croxall and Prince 1979), adult and chick energetics (Prince et al. 1981; Ricketts and Prince 1984) and adult activity patterns while foraging at sea (Prince and Francis 1984). In addition, data have been accumulating on population structure and dynamics and on breeding periodicity. Few seabirds, and no procellariiform species, have been subjected to such diverse and intensive study. This paper brings together for the first time many of the results from the study in order to review the main features of the comparative ecology and biology of these two species. The study is viewed as a case history, discussing the data that need to be acquired for any species of seabird in order to investigate adaptations related to both seasonal and life-time reproductive success and to predator-prey interactions.

2 Breeding Success and Periodicity

One typical colony of each species was selected for the main study. Each season, nests were checked daily throughout the laying period; both partners at all nests where eggs were laid were identified individually using both monel and 40-mm tall coloured plastic (Darvic) leg-bands. More than 95% of the total breeding population was re-trapped each season. In 1979–80, however, laying was well advanced when the observers arrived and birds which had failed earlier would not have been re-trapped and thus could not be classified correctly. Once laying was complete, nests were checked weekly to monitor breeding success. Although about 10% of fledglings breed in a colony other than their natal one, once an individual has bred it is entirely faithful to that colony (based on checks of ca. 10,000 birds of each species in nearby colonies during five yrs).

The breeding success of *D. melanophris* is very variable, compared with *D. chrysostoma,* ranging from 67% in 1976–77 to a complete failure in 1979–80 (Table 1). Overall, *D. chrysostoma* has had a slightly higher success, averaging 9% more chicks fledged during an 8-yr period. For *D. melanophris,* 1977–78 and 1979–80 were exceptionally poor breeding seasons and in both seasons swarms of Antarctic Krill *Euphausia superba* near the sea surface, were very scarce around South Georgia (Croxall and Prince 1979; I. Everson, pers. comm.). These events apparently had little

1 British Antarctic Survey, NERC, High Cross, Madingley Road, Cambridge CB3 OET, Great Britain

Antarctic Nutrient Cycles and Food Webs
(ed. by W.R. Siegfried, P.R. Condy, and R.M. Laws)

Table 1. Breeding success (chicks fledged from eggs laid) of Greyheaded (GHA) and Blackbrowed (BBA) albatrosses at Bird Island study colonies

Season	Total number of breeding pairs		Overall breeding success (%)	
	GHA	BBA	GHA	BBA
1975–76	406	230	46.5	46.9
1976–77	422	229	48.5	66.8
1977–78	420	253	64.0	11.4
1978–79	269	210	56.9	43.3
[a]1979–80	261	129	37.9	0.0
1980–81	250	118	21.6	30.5
1981–82	444	147	53.3	57.8
1982–83	333	183	42.6	45.9
Mean (± S.D.)			46.4 ± 12.9	37.8 ± 22.7

a Refers to number of pairs recovered breeding in late November and early December and therefore does not take into account breeding attempts which would have failed prior to recovery operations

effect on *D. chrysostoma.* Thus, in 1977–78, whereas *D. melanophris* reared chicks from only 11% of eggs laid, *D. chrysostoma* had its highest breeding success (64%). Similarly in 1979–80, when *D. melanophris* failed to rear any chicks, *D. chrysostoma* still achieved a 38% overall breeding success (Table 1).

Tables 2 and 3 show breeding periodicity for both species on the basis of the proportion of the population breeding in the original year, which has been recovered breeding in subsequent years. From the 1975–76 to 1977–78 breeding populations, a high proportion (76–86%) of *D. melanophris* returned to breed next year. However, in 1979–80 only 58% of the previous season's breeding population returned, with a further 11% and 10% returning 2 and 3 yr later, respectively. A similar pattern is suggested for the returns from 1980–81, but the pattern has apparently returned to "normal" for the 1981–82 population. The reduction in breeding frequency may relate to the poor breeding success of *D. melanophris* in 1977–78 and 1979–80. That the apparent reduction in the size of the breeding population reflects a change in breeding frequency, rather than permanent loss of birds from the population, is strongly suggested by the similarity (allowing for re-traps still to come) of the total of re-traps each season (Table 2 last column). There is also no indication of differential survivorship of birds for any cohort.

Although *D. chrysostoma* individuals successful in rearing a chick have a basically biennial breeding periodicity, unsuccessful birds usually attempt to breed the following year. Thus, fewer than 2% of successful but up to 83% of unsuccessful breeders return to breed 1 yr after they last bred. Similarly to *D. melanophris, D. chrysostoma* also shows population changes that may reflect adverse environmental conditions, perhaps especially food availability. In

Table 2. Breeding frequency of individually marked Blackbrowed Albatrosses

Original season	Population	Proportion (%) of birds next breeding in a subsequent season 1	2	3	4	5	6	7 years later	Total retrapped to date (%)
1975–76	332	86.4	6.3	1.8	0	0.3	0	0	95
1976–77	427	76.1	9.1	1.4	0.2	0.7	0	–	87.5
1977–78	403	79.9	4.2	1.7	1.7	1.7	–	–	89
1978–79	393	58.0	10.6	9.9	5.0	–	–	–	88
1979–80	258	52.7	25.1	10.8	–	–	–	–	89
1980–81	193	65.8	16.5	–	–	–	–	–	82
1981–82	265	84.1	–	–	–	–	–	–	84

Table 3. Breeding frequency of individually marked Greyheaded Albatrosses. (a) Unsuccessful birds and (b) successful birds

Original season	Population	Proportion (%) of birds next breeding in subsequent season 1	2	3	4	5	6	7 years later	Total retrapped to date (%)
a) 1975–76	106	76.4	18.8	9	0.9	0	0	0	96
1976–77	238	72.2	11.3	1.6	0.42	1.2	1.2	–	88
1977–78	262	34.3	24.4	4.6	6.4	3.0	–	–	73
1978–79	218	43.5	14.6	12.3	8.2	–	–	–	79
1979–80	324	26.2	45.6	14.5	–	–	–	–	86
1980–81	212	72.6	17.9	–	–	–	–	–	90.5
1981–82	240	83.0	–	–	–	–	–	–	83
b) 1975–76	366	1.9	74.5	4.9	2.7	0.5	1.6	1.1	87
1976–77	409	1.2	68.7	6.3	1.7	4.8	1.2	–	84
1977–78	528	0.8	50.0	11.4	10.4	4.9	–	–	77.5
1978–79	297	0.6	30.3	31.9	14.1	–	–	–	77
1979–80	198	0	62.1	17.1	–	–	–	–	79
1980–81	99	1.0	77.7	–	–	–	–	–	79
1981–82	474	1.6	–	–	–	–	–	–	

1979–80, *D. chrysostoma* had its second lowest breeding success (38%). Normally one would expect an especially large breeding population in the following year (1980–81), comprising many failed birds from 1979–80 and those successful in 1978–79. In fact, the breeding population (of 250 pairs: Table 1) was the lowest ever, combining the lowest recorded population of both successful breeders (30%) and unsuccessful breeders (26%) of any season. Not surprisingly it was the only season in which no successful birds from the previous year attempted to breed. This also suggests that 1979–80 was a particularly poor year for breeders and non-breeders alike, and that far fewer birds than usual were able to attain breeding condition by the following year.

These population fluctuations have significance beyond their intrinsic interest. First, population and therefore biomass estimates made during the years of lowest breeding populations would give totals up to 45% lower than the best years. Secondly, a series of counts up to 1980–81 would have suggested that a significant decrease in the breeding population of *D. melanophris* was in progress. However, not only did the breeding population increase in 1981–82 and 1982–83 but the total population (breeders and mature non-breeders) has remained approximately constant since 1975–76. Indeed, there is no evidence that even such a substantial change in breeding numbers is abnormal or will have deleterious long-term effects. For most Antarctic seabirds there has been insufficient monitoring of breeding populations to assess the extent and implications of short-term numerical changes.

3 Food and Energetics

To model chick bio-energetics from hatching to fledging, we need data on "input" (meal size, composition and energy content and frequency of feeding). We can then study "output" in terms of growth rates and assimilation, excretion and metabolic costs and even the partitioning of assimilated resources at various stages of growth.

Quantitative analysis of food from adults about to feed chicks (Prince 1980) showed that *D. chrysostoma* took, by weight, mainly squid (49%), followed by fish (35%; one-third of which was the Southern Lamprey, *Geotria australis*) and Krill (17%). *Diomedea melanophris* took mainly Krill (41%) and fish (37%; < 1% Lamprey), squid accounting for only 21%. Thus, while the three main classes of prey were common to both species there were important differences in their relative importance.

The most significant difference is probably the very different proportions of Krill and squid taken. However, the importance of Lampreys to *D. chrysostoma* (equivalent to 1.6 million individuals each breeding season (Potter et al. 1979) suggests further differences in feeding ecology. Although *Todarodes ?sagittatus* is the single most important squid eaten by both mollymauks, *D. melanophris* takes a greater diversity of squid than does *D. chrysostoma* (Clarke and Prince 1981). When the energy contents of these prey are taken into account, the dietary differences have enhanced significance in terms of the energy content of meals delivered to the chick. Thus, Krill and fish meals have a higher energy content than squid meals (Clarke and Prince

Table 4. Estimated energy content of meals delivered to mollymauk chicks

Species	Meal	Meal composition (%)			Meal energy content (kJ g^{-1})[a]			
	size (g)	Squid	Fish	Krill	Squid	Fish	Krill	Total
D. chrysostoma	ca. 600	49	35	16	1,020	834	418	2,272
D. melanophris	ca. 600	21	39	40	437	929	1,044	1,410

a Energy contents as in Croxall et al. (1985)

1980), which probably reflects the low energy content of squid, although data for Antarctic species are still lacking (Croxall and Prince 1982). Because, between the two species, there are neither significant differences in the sizes of meals fed to chicks nor, in a "normal" season, in the frequency with which they are received (Prince 1980) the differences in composition and energy content of the food should be linked to differences in chick bio-energetics and/or growth rates. To study the latter, eggs were exchanged between the species and the growth of fostered offspring compared with control populations (Prince and Ricketts 1981). The chicks of both species altered their growth patterns in response to dietary differences within the limits of growth constraints specific to each species. The influence of the energy content of meals fed to chicks of the two species and its potential influence on their growth can be shown in a simplified way by considering the respective contributions (Table 4): for meals of the same size and frequency, *D. chrysostoma* chicks receive 6% less energy per day than do *D. melanophris* chicks. The difference in time to attain peak weight in chicks of the two species (92–95 d in *D. chrysostoma*; 85–88 d in *D. melanophris*) is similar. However, the difference in overall fledging period (116 d in *D. melanophris*; 141 d in *D. chrysostoma*) is 18%, not 6%. Thus, differences in energy content of meals alone will not explain this discrepancy.

After reaching peak weight albatross chicks steadily decrease in weight until fledging, but there is still very active growth of body tissues, especially associated with plumage and bone development (e.g., as recorded by measurements of wing length). The differences in growth rates (slower in *D. chrysostoma*) of these structures (which presumably have a crucial bearing on fledging ability) might relate to differences in the availability of certain essential elements (e.g., calcium, sulphur) in the diet. The complex inter-relationships between meal size, composition, frequency and chick energy demands and growth show the need to collect such information throughout the chick-rearing period, particularly if variations in breeding success between seasons are to be explained.

The size and frequency of delivery of meals are best measured by using some form of automatic weighing equipment. Such a device, fitted inside an artificial albatross nest, has recently been developed (Prince and Walton, to be publ.). The experimental system consists of a number of fibre-glass nests each attached to its own weighing assembly. The weight of the chick is measured in terms of pressure exerted on an air-filled diaphragm within the assembly. The diaphragm is connected to an electronic low pressure transducer located away from the nest. The nests have an operating range of 1 kg–8 kg. Data collected are automatically stored on magnetic tape. The data-recording system can be set to scan each nest at time intervals as short as every 2 min, but this can be modified as dictated by the frequency of chick feeding and duration of each feed.

Albatross chicks, in common with other Procellariiformes, are fed at irregular intervals and frequently have to fast for up to several days. After a meal the total weight of a chick comprises two components: the chick's body weight and the weight of the meal. Thereafter, three processes occur simultaneously: metabolism, assimilation and elimination of faecal matter. Initially weight loss is rapid but eventually it decreases, producing a complex overall curve. The initial rapid decrease in weight is a result of both assimilation and excretion, especially of water. The flatter part of the curve represents weight loss as a result of metabolism. Analysis of such curves can assess the relative contribution of these three activities (Ricketts and Prince 1984).

For adults fasting during incubation, weight loss data allow estimation of metabolic costs (Prince et al. 1981). These, and the chick data, would be improved by information on the composition of the material lost. The conventional method of obtaining metabolic rates is by measuring oxygen consumption. The problem with present techniques, particularly with large birds, is the stress caused by experimental procedures, leading to artificially elevated results; certain values included in a recent study of Antarctic seabirds (Ricklefs and Matthew 1983) perhaps being cases in point. The use of radio-isotope methods to estimate metabolic rate from rates of water turnover and CO_2 production has been applied only very recently to Antarctic seabirds (Kooyman et al. 1982; Davis et al. 1983) and it is likely that for many purposes the use of expensive double-labelled water (D_2O^{18}) will be necessary (Nagy 1980).

The ability to measure automatically and accurately meal size and feeding frequency as well as to identify (e.g., by using radio transmitters) the individual parents responsible creates considerable scope for assessing the feeding performance of foraging adults. To do this meaningfully requires two further items of data. First, an estimate of the energy cost of individual trips to sea (this can presently only be obtained by using radio-isotope techniques); and, second, information on activity patterns of birds at sea.

Prince and Francis (1984) have developed an electrolytic device that measures the amount of time albatrosses sit on the surface of the sea during daylight and at night. From devices deployed on 13 birds for a total of 284 d during the chick rearing period, in *D. chrysostoma* 75% of the birds' time was spent in flight. During the hours of darkness 50% of the time was spent on the surface of the sea, compared with only 15% of daylight hours. These are average values and, not surprisingly, there was appreciable individual variation (e.g., total time on water ranging from 15–38%). Interpreting this in terms of variations in feeding success (e.g., frequency and size of meals delivered to the chick) and energy costs of feeding trips is the aim of current research. Other research will be investigating relationships between sex, age, breeding experience and breeding success.

4 Conclusions

There has been insufficient monitoring of Antarctic breeding seabird populations to assess the extent and implications of short-term numerical changes. Current evidence for *D. melanophris* suggests that changes in the breeding population reflect differences in the proportion of birds with previous breeding experience that actually breed, rather than changes in adult mortality rates.

Differences in diet composition and energy content may be significant factors in the evolution of basic differences in breeding frequency between *D. melanophris* and *D. chrysostoma.* Certain intra-specific and between-year variations in breeding frequency may also reflect the effect of, for instance, body condition which may be influenced by diet, feeding performance and recent reproductive experience. The relative importance of these, and other, factors is unknown at present.

Further understanding of the ecological and demographic differences between these two mollymauks requires both long-term studies of populations and detailed studies of individuals and pairs. Productive areas for research have been outlined in the review.

Acknowledgements. I thank the many members of past Bird Island field parties, especially L. Kearsley, B. Pearson and P.G. Copestake. For constructive help with this manuscript I am particularly grateful to J.P., Croxall and C. Ricketts and I thank F. Prince and E. Bailey for typing it.

References

Clarke A; Prince PA (1980) Chemical composition and calorific value of food fed to mollymauk chicks *Diomedea melanophris* and *D. chrysostoma* at Bird Island, South Georgia. Ibis 122:488–494

Clarke MR; Prince PA (1981) Cephalopod remains in regurgitations of black-browed and grey-headed albatrosses at South Georgia. Bull Br Antarct Surv 54:1–8

Croxall JP; Prince PA (1979) Antarctic seabird and seal monitoring studies. Polar Rec 19:573–595

Croxall JP; Prince PA (1982) Calorific content of squid (Mollusca: Cephalopoda). Bull Br Antarct Surv 55:27–31

Croxall JP; Prince PA; Ricketts C (1985) Relationships between prey life-cycles and the extent, nature and timing of seal and seabird predation in the Scotia Sea. In: Siegfried WR, Condy PR, Laws RM (eds) Antarctic nutrient cycles and food webs (Proceedings of the 4th SCAR symposium on Antarctic biology). Springer, Berlin Heidelberg New York

Davis RW; Kooyman GL; Croxall JP (1983) Water flux and estimated metabolism of free-ranging gentoo and macaroni penguins at South Georgia. Polar Biol 2:41–46

Kooyman GL; Davis RW; Croxall JP; Costa DP (1982) Diving depths and energy requirements of king penguins. Science 217: 726–727

Nagy KA (1980) CO_2 production in animals: analysis of potential errors in the doubly labelled water method. Am J Physiol 238: 454–465

Potter IC; Prince PA; Croxall JP (1979) Data on the adult marine and migratory phases in the life cycle of the southern hemisphere lamprey *Geotria australis* Gray. Environ Biol Fishes 41:65–69

Prince PA (1980) The food and feeding ecology of grey-headed albatross *Diomedea chrysostoma* and black-browed albatross *D. melanophris.* Ibis 122:746–488

Prince PA; Francis MD (1984) Activity budgets of foraging grey-headed albatrosses. Condor 96:297–300

Prince PA; Payne MR (1979) Current status of birds at South Georgia. Bull Br Antarct Surv 48:103–118

Prince PA; Ricketts C (1981) Relationships between food supply and growth in albatrosses: an interspecies chick fostering experiment. Ornis Scand 12:207–210

Prince PA; Ricketts C; Thomas G (1981) Weight loss in incubating albatrosses and its implications for their energy and food requirements. Condor 83:238–242

Prince PA; Walton DWH (1984) Automated measurements of feed size and feeding frequency in albatrosses. J Appl Ecol 21

Ricketts C; Prince PA (1981) Comparison of growth of albatrosses. Ornis Scand 12:120–124

Ricketts C; Prince PA (1984) Estimates of metabolic costs and food conversion efficiency in albatross chicks using field weighings. Auk 101:790–795

Ricklefs RE; Matthew KK (1983) Rates of oxygen consumption in four species of seabird at Palmer Station, Antarctic Peninsula. Comp Biochem Physiol 74A:885–888

Tickell WLN; Pinder R (1975) Breeding biology of the black-browed albatross *Diomedea melanophris* and grey-headed albatross *D. chrysostoma* at Bird Island, South Georgia. Ibis 117:433–450

The Foraging Zones of Seabirds in the Crozet Islands Sector of the Southern Ocean

J.C. STAHL[1], P. JOUVENTIN[1], J.L. MOUGIN[2],
J.P. ROUX[1], and H. WEIMERSKIRCH[1]

Summary. This paper deals with the foraging zones of seabirds in the Crozet Islands sector of the southwestern Indian Ocean. Data were collected at sea by means of 10-min counts. Habitat partitioning is described among foraging seabirds in February, when almost 80% of the species are feeding their chicks. Seasonal change in habitat utilization and partitioning is analyzed. The influence of hydrography, distance from breeding locality, marine productivity and seasonal vertical migration of zooplankton on the distribution of seabirds is discussed.

1 Introduction

A knowledge of the foraging zones of seabirds is important for two main reasons: to locate the impact of a given species on the marine ecosystem; and, to show the importance of the habitat axis in seabird niche differentiation. Specific habitat preferences are interpreted generally as being linked to prey species composition, abundance or availability (Ashmole 1971; Brown 1980). Furthermore, the foraging ranges of breeding birds in the vicinity of their nesting sites are limited by specific flight and fasting abilities (Jouventin and Mougin 1981). Our report aims at presenting a first synthesis of the habitat relationships of seabird species in the Crozet Islands sector of the southwestern Indian Ocean.

2 Study Area

The Crozet Islands sector is that part of the Southern Ocean which is limited westwards and eastwards by lines of equidistance between the Crozet Islands and the neighbouring Sub-Antarctic island groups (Prince Edward and Kerguelen). The hydrography of this oceanic area is complex, and is subject to debate (Deacon 1983). According to the most recent synthesis (Gamberoni et al. 1982), a number of hydrographical zones can be recognized. In the north, (25°–37°S) warm and highly saline sub-tropical waters are separated from the Sub-Tropical Convergence by a zone of large cyclonic eddies (37°–41°S). The Sub-Tropical and Antarctic convergences, situated at 41° and 43°S, respectively, are very close together in this area; the Sub-Antarctic zone being restricted to a narrow band or even replaced by a single broad convergence zone. On the other hand, the Antarctic Convergence (sinking of Antarctic surface waters) is here largely disassociated from the Polar Front (northern limit of the sub-surface temperature minimum, at 49°S). Thus, the surface waters south of the Antarctic Convergence can be divided into modified Antarctic waters (43°–49°S, i.e. between the Antarctic Convergence and the Polar Front), Antarctic surface waters (49°–63°S), and the Antarctic Divergence zone (south of 63°S).

The Crozet archipelago (46°S, 51°E) is situated in the zone of the modified Antarctic waters. Most of the five islands of the group are surrounded by shelves. Thirty-four seabird species breed on these islands (Jouventin et al., to be publ. a).

3 Methods

Data were collected in the austral winter, spring and summer months during several years, by means of 10-min counts of seabirds made from the M.V. Marion Dufresne, at a height of 11 m a.s.l. All birds observed within 360° of the ship were counted during each 10-min record. Ship-following birds were included, since a rapid turnover of such birds in the Southern Ocean (Griffiths 1982; La Cock and Schneider 1982; Stahl et al., to be publ.) reduces over-estimates of the abundance of species most attracted to vessels and maintains independence between consecutive censuses. An arbitrary exclusion of these ship-following birds might result in underestimates of the abundance of those species which are attracted strongly to ships.

Normally, two 10-min counts were made every hour while the ship was cruising. Additional counts were made in frontal zones or in the offshore zone (defined as the zone between 1 and 100 km of land; the pelagic zone being the area beyond 100 km of the Crozet Islands). These additional counts were excluded from general analyses reported here, in order to avoid biases due to enhanced sampling effort over these zones. Sampling biases have been further reduced

1 Institut des Sciences de l'Evolution, Place Bataillon, F-34060 Montpellier Cedex, France
2 Muséum National d'Histoire Naturelle, 55 rue Buffon, F-75005 Paris, France

Antarctic Nutrient Cycles and Food Webs
(ed. by W.R. Siegfried, P.R. Condy, and R.M. Laws)

by the exclusion of data collected when the ship's speed was $< 11\ km\ h^{-1}$, when visibility was < 5 km, and during gales (Stahl et al., to be publ.).

Only the foraging zones of those seabird species which nest on the Sub-Antarctic islands of the southern Indian Ocean are considered in this paper, since the ranges of Antarctic or tropical species were crossed only marginally during our surveys. Nevertheless, records of these species were taken into account in our biomass calculation. Calculations of the theoretical maximum foraging ranges of breeding birds were based on the rate of turnover of adults at nests (i.e. incubation shifts or chick-feeding trips). Flying or swimming speeds were determined by timing birds moving in a straight line over a known distance between two landmarks along the coast or inland (Stahl et al., to be publ.). The bird associations in the offshore zone were obtained by means of a cluster analysis (Diday 1970).

4 Results and Discussions

Since the foraging zones of most of the seabirds considered here are subject to significant and sudden seasonal changes, the first part of this section describes habitat relationships during a fixed period (February), whereas the second section deals with an analysis of seasonal changes in habitat utilization and partitioning.

4.1 Habitat Partitioning in February

4.1.1. General Habitat Relationships

In February, almost 80% of the species nesting in the southern Indian Ocean are feeding chicks (Appendix, see page 485). Figure 1 shows a general ordination (from data in Jouventin et al. 1982; Jouventin et al., to be publ. b) of habitat relationships among the different seabird species, based on three transects starting from the Crozet Islands and proceeding toward the north, southeast and south, respectively. The species were ordinated by their weighted means according to three variables: water temperature; depth (indicating the importance of shallow offshore waters in the foraging zone of a species); and, distance from the nearest breeding locality (reflecting, to a certain extent, the dispersion of breeding populations around the islands).

Four species are restricted to the immediate vicinity of the islands: *Phalacrocorax atriceps, Larus dominicanus, Sterna virgata,* and *S. vittata* (the latter two species have only once been observed from the ship, which explains why they are not represented in Fig. 1). A second group includes species which forage principally over the shelf and slope areas, but for which some individuals were recorded several 100s of km from the Crozet Islands. *Aptenodytes patagonicus, Eudyptes chrysolophus* and *Diomedea melanophris* are typical examples of this group. The remaining species all forage regularly over the pelagic waters of the study area. Several distribution profiles can, however, be distinguished

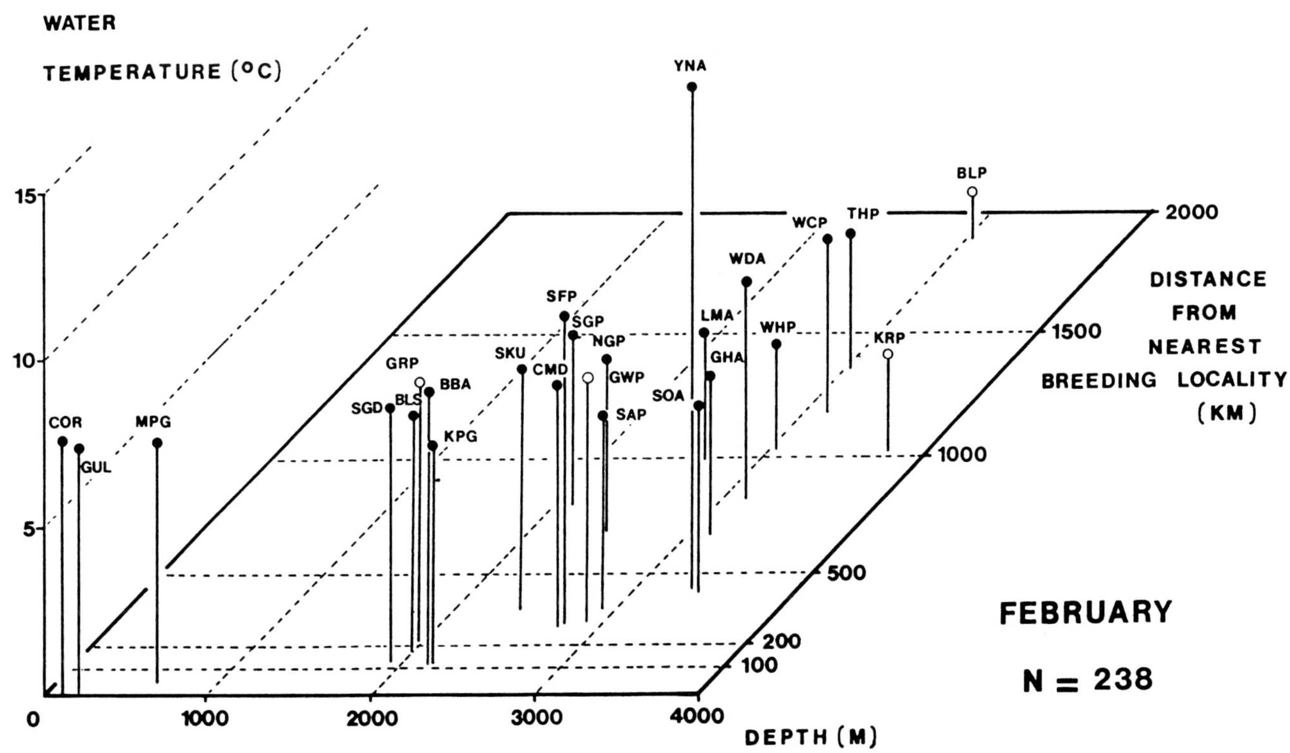

Fig. 1. Ordination of habitat relationships among seabirds in February. Breeding species are represented by *dots,* non-breeding species by *circles.* N = no. 10-min counts. See Appendix 1 for abbreviations of names

among them. In certain species only non-breeding birds are able to forage far away from land (e.g., *Pelecanoides urinatrix, Catharacta skua*), whereas for other groups, such as albatrosses and prions, distant sightings may apply to non-breeding as well as to breeding individuals. These species are well segregated on the distance axis, but much less so on the temperature axis, since very few species have narrow temperature associations in February. Only *Diomedea chlororhynchos,* is then mostly restricted to sub-tropical waters, *Halobaena caerulea* to the Antarctic Divergence zone, and the other species foraging over largely over-lapping areas south of the Sub-Tropical Convergence (Jouventin et al., 1982; Jouventin et al., to be publ. b).

Another feature shown in Fig. 1 is that, among the species which do not breed in February, species which are in a pre-breeding stage (e.g., *Procellaria cinerea* and *Pterodroma macroptera*) are then moreorless concentrated around the Crozet Islands, whereas species in a post-breeding stage (e.g., *H. caerulea* and *Pterodroma brevirostris*) disperse from the islands; the distribution of the latter species being quite different during their breeding period (Jouventin et al., to be publ. b).

4.1.2. Foraging Ranges of Breeding Birds

The maximum foraging ranges of breeding birds can be calculated from the duration of feeding trips and flying or swimming speeds. Although it is obvious that such theoretical foraging ranges are over-estimates, since the birds are assumed to move continuously and in a straight line, it is difficult to estimate the magnitude of over-estimation. Nevertheless, further information on the foraging ranges can be deduced for species which show gaps in their distribution at sea between different populations. In the Crozet Island sector, this applies to species with large distribution gaps between Antarctic and Sub-Antarctic populations (e.g., *Macronectes giganteus, Daption capense* and *Oceanites oceanicus*). On the other hand, several Sub-Antarctic species present large distribution gaps between island-bound population (breeding and non-breeding birds returning regularly to the colonies) and non-breeding populations encountered beyond the theoretical maximum foraging ranges of breeding birds. These non-breeding populations seem to be mostly restricted to particular oceanic habitats, such as the Polar Front zone or the Lena Bank south of the Crozet Islands (e.g., *E. chrysolophus* and *E. chrysocome*) or the Antarctic Divergence zone (e.g., *Phoebetria fusca*). As an example, the abundance of the Sooty Albatross, *P. fusca,* is shown in relation to distance from the nearest breeding locality (Fig. 2). If one considers these species, the farthest recorded distance from the breeding localities for potential breeding birds appears to be strongly correlated ($r = 0.96, P < 0.01$) with the theoretical maximum foraging range of breeding birds (Fig. 3). The equation of the regression line ($y = 0.53 \times + 11$) indicates that the theoretical foraging range is about twice as large as the probably more realistic observed foraging ranges for these species.

4.1.3 Habitat Relationships Within the Offshore Zone

In the offshore zone of the Crozet sector, where the extension of shelf areas is highly variable around the different islands (Fig. 4), and, as a consequence, depth is poorly correlated to the distance offshore, depth variations represent

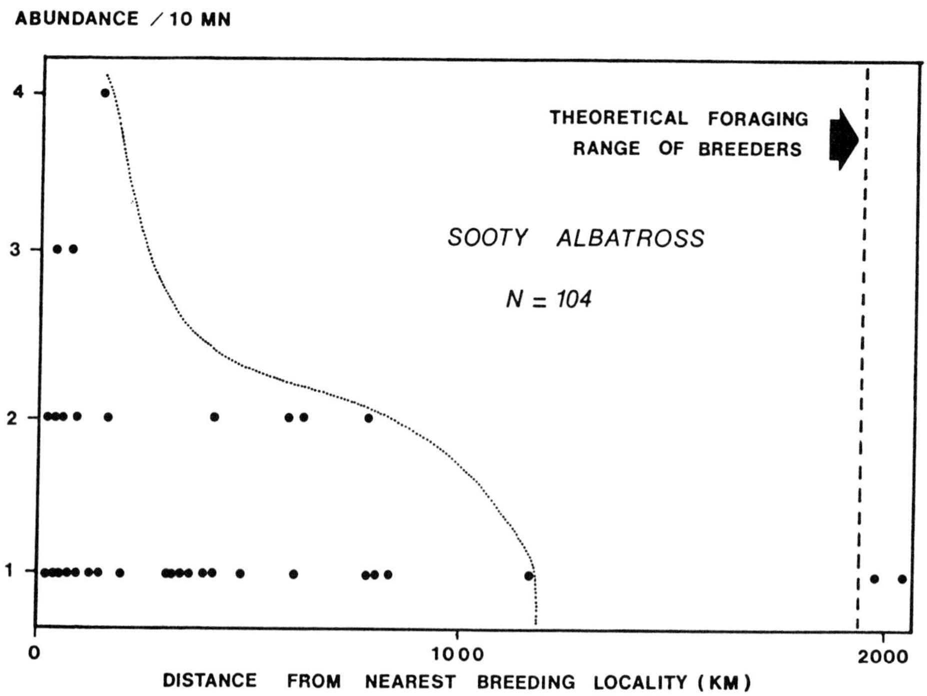

Fig. 2. Abundance of the Sooty Albatross in relation to distance from the nearest breeding locality. N = no. 10-min counts in which the species was present

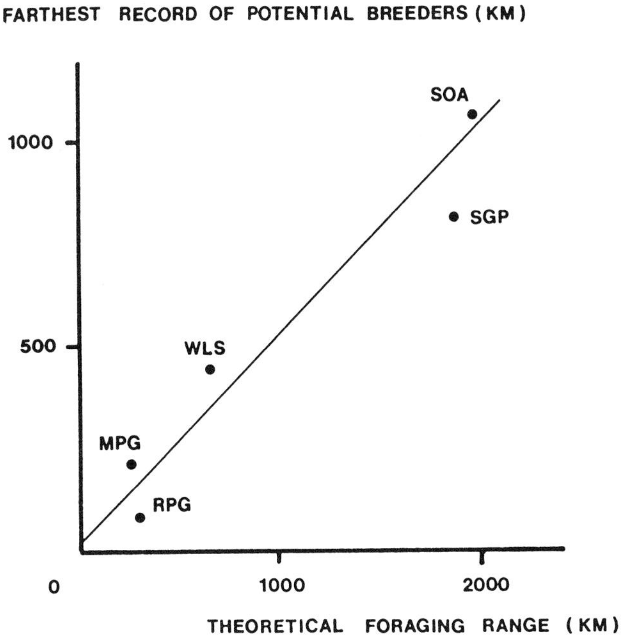

Fig. 3. Relationship between the observed and the theoretical foraging ranges for breeding seabirds of five species. See Appendix 1 for abbreviations of names

the major factor in seabird habitat diversification. Bird biomass and abundance are, in this zone, independent of distance from land, being correlated with depth ($r = 0.19$ and $r = -0.20$, respectively, for $n = 288$, $P < 0.01$ for both sets of results), which may be related to a higher amount of prey available over the shelf areas. Although data in prey are not yet available for the Crozet area, this idea is supported by a zooplankton study in the Kerguelen region, which showed that within the 0–100 m layer, the zooplankton biomass is on average 2.5 times higher over the shelf than in the surrounding pelagic waters (Razouls and Razouls 1982).

Depth not only influences bird biomass or abundance, but also bird species composition. The distribution of two of the six main bird associations recognized in the offshore zone (Stahl et al., to be publ.) illustrates this feature. Both associations, namely the *D. melanophris* – *D. chrysostoma* – *Procellaria aequinoctialis* association, and the *Fregetta tropica* – *Pterodroma mollis* – *P. Macroptera* association, show a strong spatial segregation (Fig. 4). Whereas the first association is encountered mainly over the large shelf area, the second association occurs principally beyond the shelves. Such a spatial segregation could be promoted by differences in prey species composition (i.e. neritic species vs. pelagic species over the deeper waters). However, the fact that only two species (*D. melanophris* and *D. capense*) characterizing the shelf association are specialized shelf-foragers in our study area, and the fact that the other species utilize a broad range of food resources, at least in the Indian Ocean (Voisin 1968; Mougin 1970, 1975, unpubl.) apparently do not support this hypothesis. Another hypothesis is that the high density of birds occurring over the shelf area enhances the frequency of competitive feeding interactions (Stahl et al., to be publ.), to such an extent, that if smaller and less aggressive species locate prey there, they will not have enough time to feed before being harassed by dominant species. This could explain why, in the medium-sized petrels for example, species of the genus *Pterodroma* avoid shelf areas where the dominant *P. aequinoctialis* occurs in great numbers.

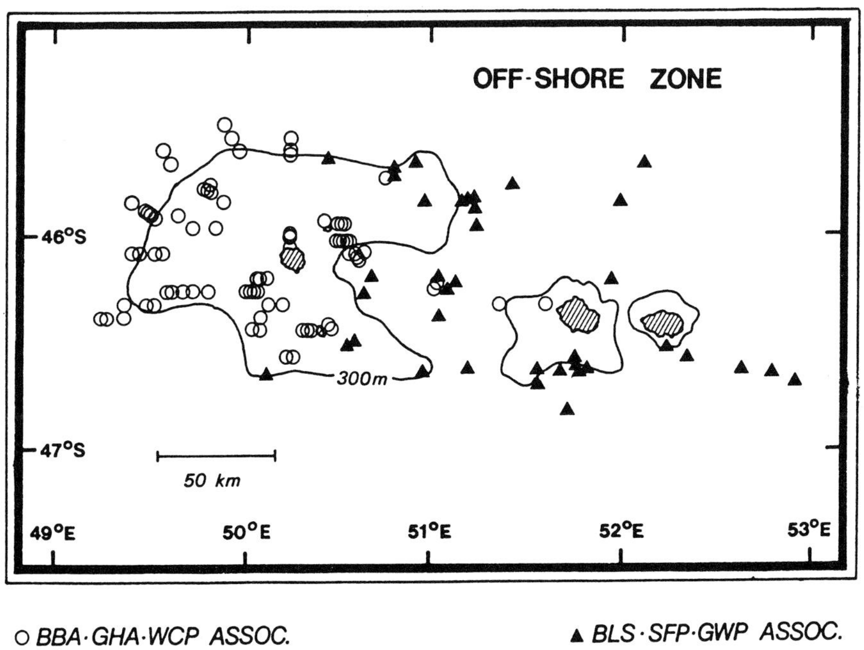

Fig. 4. Distribution of two associations of seabird species in the offshore zone of the Crozet Islands. See Appendix 1 for abbreviations of names

4.1.4. Relationships Between Marine Productivity and Bird Distribution in the Pelagic Zone

Observations in the pelagic zone support the hypothesis that interspecific competition is potentially an important factor affecting the distribution of medium-sized petrels. If the bird biomass is taken as an index of available prey abundance, an analysis of the relative mean abundance of different medium-sized petrels in four biomass classes (Fig. 5) shows that, as in the offshore zone, *P. aequinoctialis* reacts differently to the gadfly petrels (*Pterodroma* spp.). Indeed, whereas the abundance of *P. aequinoctialis* increases with increasing bird biomass, the abundance of gadfly petrels decreases beyond a given threshold of bird biomass, which can be variable from one species to another. This seems to confirm that within this group, only *P. aequinoctialis* is well adapted to feed in areas of high food and bird concentrations.

Hence, *P. aequinoctialis* can be expected to be concentrated over frontal zones which are known for enhanced primary productivity (Plancke 1977; Lutjeharms et al. 1985) and zooplankton biomass (Foxton 1956). This indeed is the case, as is shown by the results of a continuous series of 10-min counts made in the convergence zone north of the Crozet Islands. Variations in the abundance of *P. aequinoctialis*, and the distribution of isotherms (XBT) analyzed along the same transect, are shown in Fig. 6; the species' mean abundance in this convergence zone and adjacent pelagic waters being given in Table 1. The abundance of *P. aequinoctialis* increased over the convergence zone, decreased north of it, and increased again in an area of upwelling (shown in Fig. 6 by a sharp rişe in isotherms at 40° 20'S) due to a large cyclonic eddy (Gamberoni et al. 1982). The other species concentrated along this convergence were *Pachyptila salvini* and/or *P. desolata* (which cannot be separated at sea).

4.2 Seasonal Change in Habitat Utilization and Partitioning

During the course of a year, the distribution at sea of almost all the species considered here changed significantly. A detailed analysis of this change in relation to season and habitat for each species is beyond this paper (for detailed information, see Jouventin et al., to be publ. b). However, two aspects of seasonal change reveal the general trends observed in the study area.

Figure 7 shows monthly variation in bird biomass (mean biomass per 10 min count) in relation to different water-masses. The first feature to note is the relatively low biomass encountered during August over almost all the water-masses.

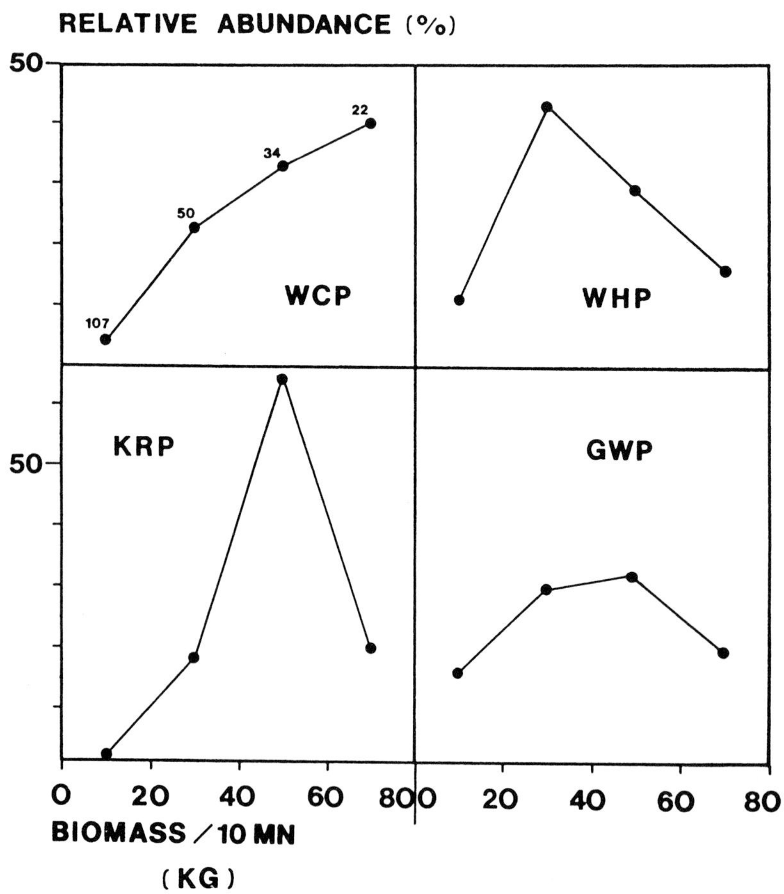

Fig. 5. Relative abundance of four medium-sized petrels in four biomass classes within the pelagic zone, based on 213 10-min counts. See Appendix 1 for abbreviations of names

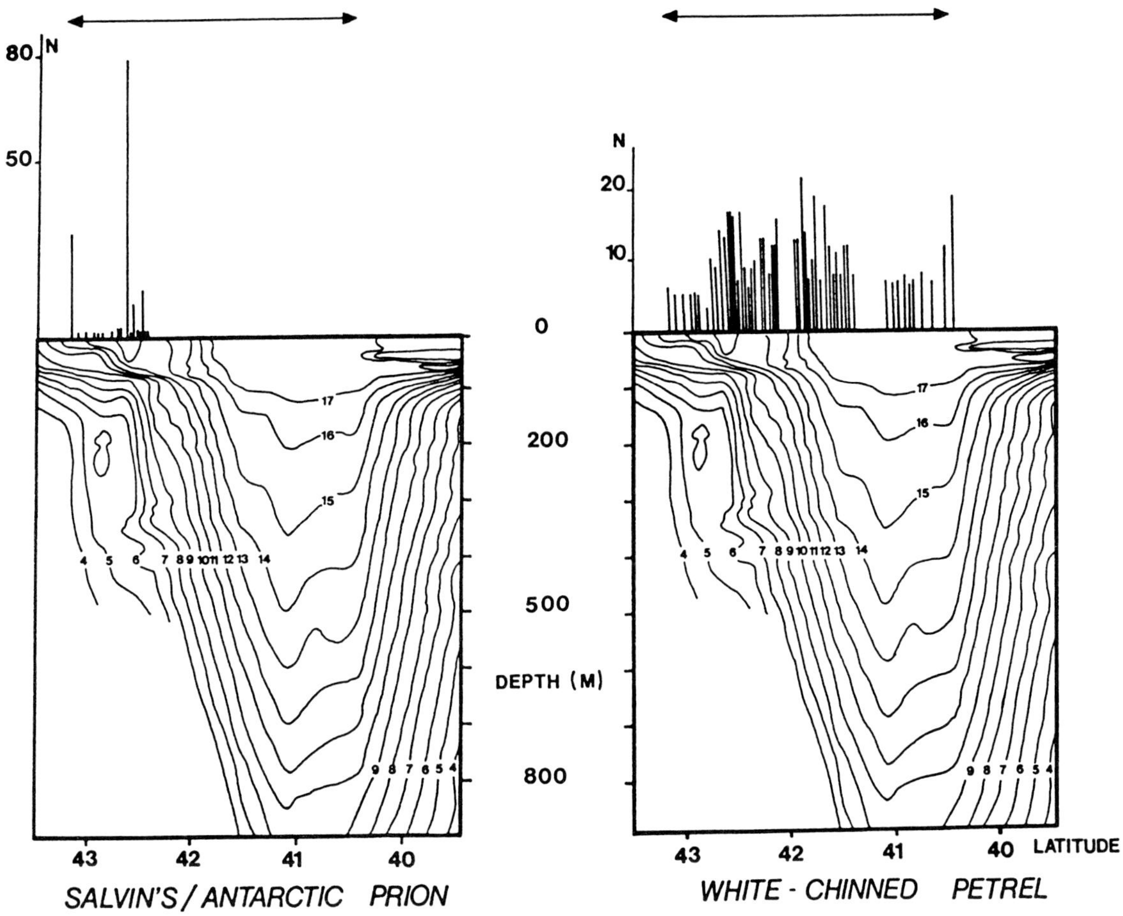

Fig. 6. Variation in abundance of Salvin's and/or Antarctic prions and Whitechinned Petrels over the convergence zone north of the Crozet Islands. *Arrows* indicate the ranges of continuous sampling. The distribution of isotherms is based on XBT soundings (L. Gamberoni and F. Murail, unpubl. data)

Several species (*P. aequinoctialis, P. salvini, P. desolata, O. oceanicus, F. tropica* and *Garrodia nereis*) had by then almost completely disappeared from the area, others (*D. exulans, D. chlororhynchos* and *P. fusca*) were still present, but much less abundant than during spring and summer. This winter situation changes rapidly after the beginning of September, with the return of numerous birds (probably also combined with passing migratory individuals). As in August, the biomass in September had a relatively homogeneous distribution over the water-masses sampled. The seabird biomass was virtually identical over sub-tropical waters (where the zooplankton standing crop is relatively low but close to the surface), as in Antarctic waters (where the standing crop is higher, but the zooplankton is concentrated mainly in the deep-water layer in winter) (Foxton 1956; Voronina 1968). As soon as the zooplankton returns to the surface layers south of the Sub-Tropical Convergence (November–December), the bird biomass over northern sub-tropical waters suddenly becomes much lower than over colder waters. It appears that the biomass peak moves to the south between December and February, which may be related to the delayed vertical migration of the zooplankton in

Table 1. Mean (± S.D. and range) abundance per 10-min count of the Whitechinned Petrel and Salvin's/Antarctic prions over four water-masses in February

	Northern sub-tropical waters (n = 41)	Southern sub-tropical eddy (n = 15)	Convergence Zone (n = 14)	Modified Antarctic waters (n = 59)
Whitechinned Petrel	2.6 ± 4.9 (0–21)	10.4 ± 5.4 (1–21)	10.3 ± 5.8 (3–22)	6.2 ± 5.4 (0–25)
Salvin's/Antarctic prions	0	0	7.2 ± 21.0 (0–78)	2.8 ± 3.7 (0–23)

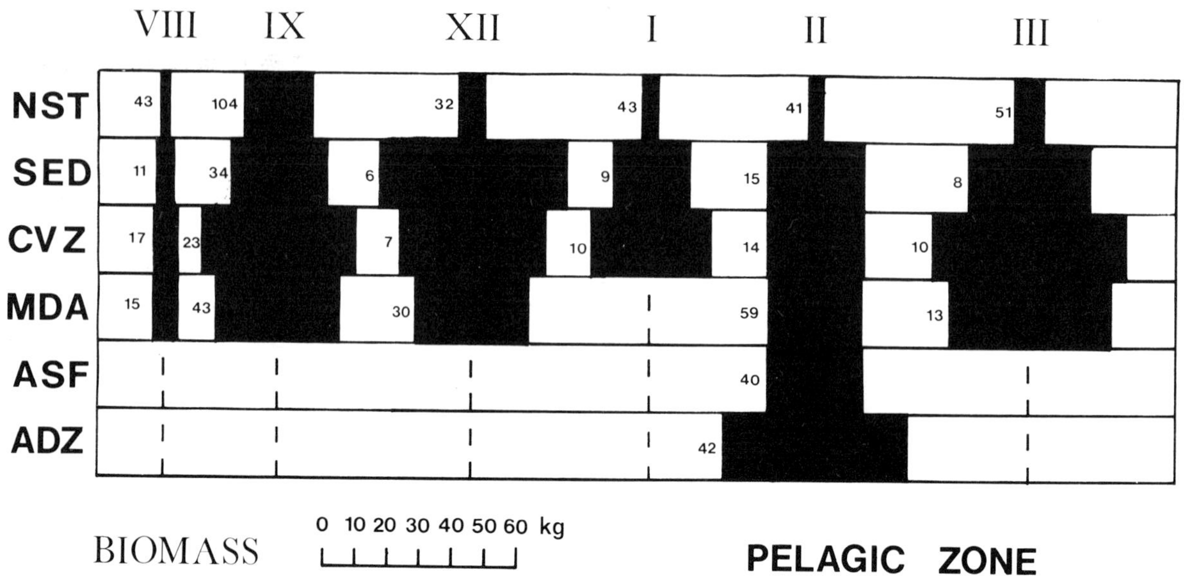

Fig. 7. Change in mean brid biomass according to month in six zones: *NST* northern sub-tropical waters; *SED* southern sub-tropical eddy; *CVZ* Convergence zone; *MDA* modified Antarctic waters; *ASF* Antarctic surface waters; *ADZ* Antarctic Divergence zone

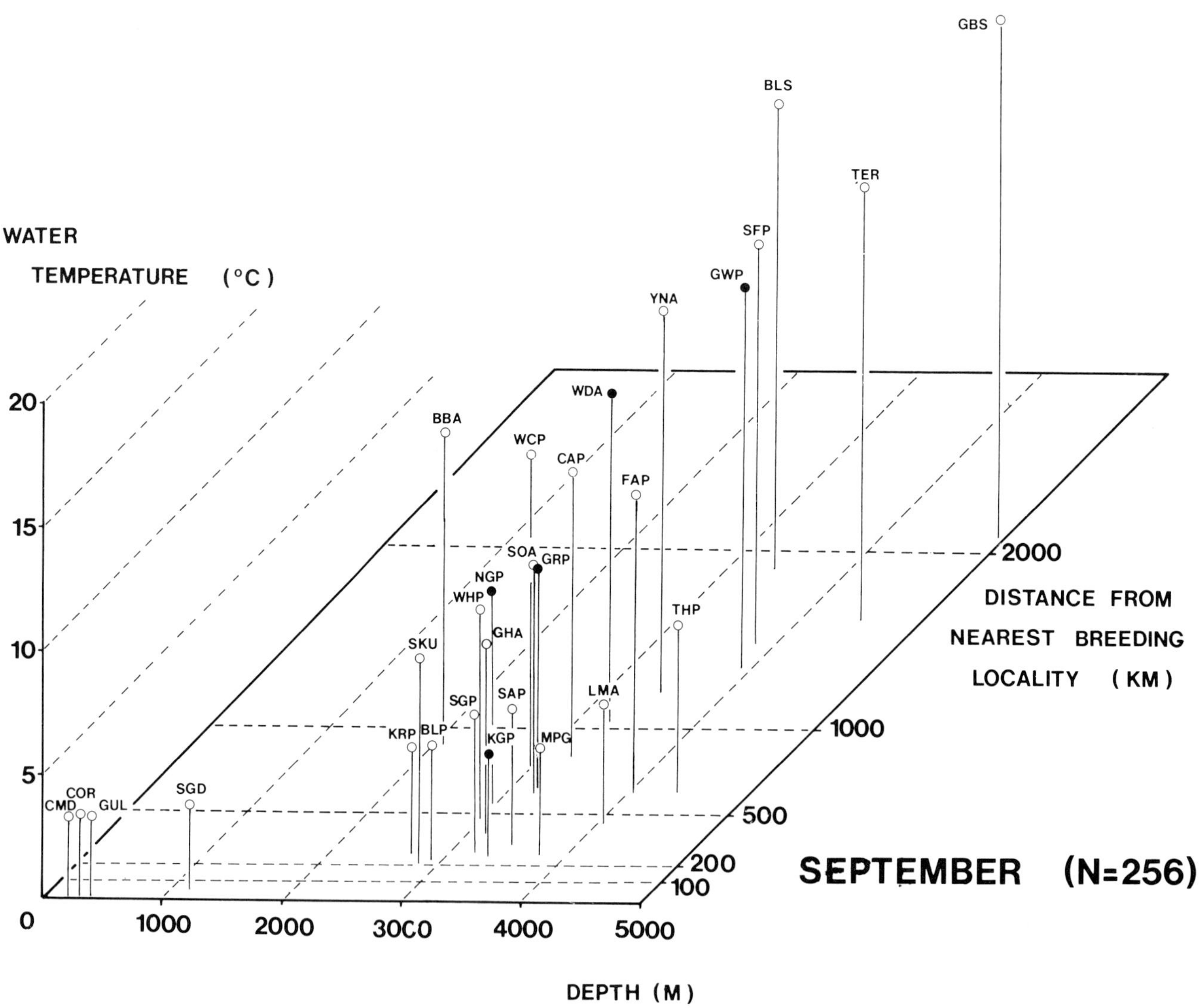

Fig. 8. Ordination of habitat relationships among seabirds in September. Breeding species are represented by *dots*, non-breeding species by *circles*. *N* no. 10-min counts. See Appendix 1 for abbreviations of names

higher latitudes. The general increase of bird biomass in March is more difficult to interpret, since it could be due to the beginning of a northward movement of seabirds, as well as to an immigration of non-breeding birds.

A different analysis is necessary to reveal the seasonal change in the habitat relationships of the different species. Figure 8 presents a general ordination of habitat relationships in September. It appears that the foraging zones of the different seabird species are segregated better than in February (Fig. 1); water temperature being an important factor. Sub-tropical waters are utilized by a larger number of species in September, while another group of species remains south of the Antarctic Convergence. A further change from the pattern in February is the reduced number of species foraging over shelf areas.

5 Conclusions

In the Crozet Islands sector of the Southern Ocean, the distribution of seabirds taken as a whole probably reflects variation in the abundance of marine organisms on which the birds feed. Thus, seabirds are concentrated over the productive shelf areas and frontal zones. However, the distribution of bird biomass over pelagic waters of this sector seems to be influenced significantly by the seasonal vertical migration of the zooplankton south of the Sub-Tropical Convergence. It appears that the foraging zones of the avian species are segregated better in winter and spring than in summer. Increased spatial segregation during the period of food shortage is realized mainly by emigration of several species out of the Crozet Islands sector during winter, or by foraging over sub-tropical waters during spring. The productivity of the marine environment apparently affects the distribution at sea of several seabird species. Thus, certain species (e.g., *D. melanophris*) are linked to productive waters; *D. melanophris* being almost restricted to shelf areas of the study area in summer. Elsewhere, notably between New Zealand and the Ross Sea, this species forages regularly over rich pelagic waters (Knox 1970; Marie Darby 1970; Hicks 1973).

Appendix. Names, codes and breeding status of seabird species in February and September at the Crozet Islands

			February	*September*
KPG	King Penguin	*Aptenodytes patagonicus*	Inc/Chf	Chf
GPG	Gentoo Penguin	*Pygoscelis papua*	Nbr	Inc/Chf
RPG	Rockhopper Penguin	*Eudyptes chrysocome*	Chf	Nbr
MPG	Macaroni Penguin	*Eudyptes chrysolophus*	Chf	Nbr
WDA	Wandering Albatross	*Diomedea exulans*	Inc	Chf
BBA	Blackbrowed Albatross	*Diomedea melanophris*	Chf	Nbr
GHA	Greyheaded Albatross	*Diomedea chrysostoma*	Chf	Nbr
YNA	Yellownosed Albatross	*Diomedea chlorohynchos*	Chf	Nbr
SOA	Sooty Albatross	*Phoebetria fusca*	Chf	Nbr
LMA	Lightmantled Sooty Albatross	*Phoebetria palpebrata*	Chf	Nbr
SGP	Southern Giant Petrel	*Macronectes giganteus*	Chf	Nbr
NGP	Northern Giat Petrel	*Macronectes halli*	Chf	Inc
CAP	Cape Pigeon	*Daption capense*	Chf	Nbr
THP	Thinbilled Prion	*Pachyptila belcheri*	Chf	Nbr
SAP	Antarctic Prion	*Pachyptila desolata*	Chf	Nbr
SAP	Salvin's Prion	*Pachyptila salvini*	Chf	Nbr
FAP	Fairy Prion	*Pachyptila turtur*	Chf	Nbr
BLP	Blue Petrel	*Halobaena caerulea*	Nbr	Nbr
GWP	Greatwinged Petrel	*Pterodroma macroptera*	Nbr	Chf
WHP	Whiteheaded Petrel	*Pterodroma lessonii*	Chf	Nbr
KRP	Kerguelen Petrel	*Pterodroma brevirostris*	Nbr	Nbr
SFP	Softplumaged Petrel	*Pterodroma mollis*	Chf	Nbr
WCP	Whitechinned Petrel	*Procellaria aequinoctialis*	Chf	Nbr
GRP	Grey Petrel	*Procellaria cinerea*	Nbr	Chf
WLS	Wilson's Stormpetrel	*Oceanites oceanicus*	Chf	Nbr
BLS	Blackbellied Stormpetrel	*Fregetta tropica*	Chf	Nbr
GBS	Greybacked Stormpetrel	*Garrodia nereis*	Chf	Nbr
CMD	Common Divingpetrel	*Pelecanoides urinatrix*	Chf	Nbr
SGD	South Georgia Divingpetrel	*Pelecanoides georgicus*	Chf	Nbr
COR	Imperial Cormorant	*Phalacrocorax atriceps*	Chf	Nbr
SKU	Antarctic Skua	*Catharacta skua*	Nbr	Nbr
GUL	Kelp Gull	*Larus dominicanus*	Nbr	Chf
TER	Antarctic Tern	*Sterna vittata*	Chf	Nbr

Abbreviations: Nbr = non-breeding; Chf = feeding chicks; and Inc = incubating

Acknowledgements. We thank the Mission de Recherche des Terres Australes et Antarctiques Francaises for financial and logistical support, J.A. Bartle, J.L. Martin, V. Ridoux, C.R. Viot, and L. Gamberoni for contributing data, and M. Van Beveren who formatted part of the data. We are especially grateful to H. Bigot statistician at the CNRS analysis centre of Strasbourg, for her help during the data analysis Finally, thanks are due to C. Thomas and S. Viljoen for their help.

References

Ashmole NP (1971) Seabird ecology and the marine environment. In: Farner DS, King JR (eds) Avian Biology, Vol. I. Academic Press, New York, pp 223–286

Brown RGB (1980) Seabirds as marine animals. In: Burger J, Olla BL, Winn HE (eds) Marine Birds. Plenum Publ. Corp, New York, pp. 1–33

Deacon GER (1983) Kerguelen, Antarctic and subantarctic. Deep-Sea Res 30:77–81

Diday E (1970) La méthode des nuées dynamiques et la reconnaissance des formes. IRIA

Foxton P (1956) The distribution and standing crop of zooplankton in the Southern Ocean. Discovery Rep 28:1–191

Gamberoni L, Geronimi G, Jeannin PF, Murail JF (1982) Study of frontal zones in the Crozet-Kerguelen region. Oceanol Acta 5(3): 289–299

Griffiths AM (1982) Reactions of some seabirds to a ship in the Southern Ocean. Ostrich 53:228–235

Hicks GRF (1973) Latitudinal distribution of seabirds between New Zealand and the Ross Sea, December 1970. Notornis 20:231–247

Jouventin P, Mougin JL (1981) Les stratégies adaptatives des oiseaux de mer. Terre Vie 35:217–272

Jouventin P, Mougin JL, Stahl JC, Weimerskirch H (1982) Données preliminaires sur la distribution pélagique des oiseaux des TAAF. In: Jouventin P, Massé L, Tréhen P (eds) Colloque sur les écosystèmes subantarctiques. CNFRA 51:427–436

Jouventin P, Mougin JL, Stahl JC, Weimerskirch H (in press a) The seabirds of the French Subantarctic Islands and Adélie Land, their status and conservation. ICBP Seabird Workshop, Cambridge 1982

Jouventin P, Bartle JA, Mougin JL, Roux JP, Stahl JC, Weimerskirch H (in press b) Atlas de distribution des Oiseaux et Mammiferes marins dans le secteur des iles Crozet, Kerguelen et Amsterdam (Sud-Ouest de l'Océan Indien). CNFRA.

Knox GA (1970) Biological oceanography of the South Pacific. In: Wooster WS (ed) Scientific Exploration of the South Pacific. Ac. of Sciences, Washington DC, pp 155–182

La Cock G, Schneider DC (1982) Duration of ship following by Wandering Albatrosses *Diomedea exulans.* Cormorant 10:105–109

Lutjeharms JRE, Walters NM, Allanson BR (1984) Oceanic frontal systems and biological enhancement. In: Siegfried WR, Condy PR, Laws RM (eds) Antarctic nutrient cycles and food webs (Proceedings of the 4th SCAR symposium on Antarctic biology). Springer Berlin Heidelberg New York

Marie Darby H (1970) Summer seabirds between New Zealand and MacMurdo Sound. Notornis 17:28–55

Mougin JL (1970) Observations écologiques sur les Grands Albatros (*Diomedea exulans*) de l'ile de la Possession (archipel Crozet) en 1968. Oiseau Rev Fr Ornithol 40:1–36

Mougin JL (1975) Ecologie comparée des Procellariidés antarctiques et subantarctiques. CNFRA 36:195 pp

Plancke J (1977) Phytoplankton biomass and productivity in the subtropical convergence area and shelves of the western Indian subantarctic islands. In: Llano GA (ed) Adaptations within Antarctic ecosystems. Smithsonian Inst., Washington, pp 51–73

Razouls C, Razouls S (1982) Element du bilan energetique du Mesoplanton antarctique. CNFRA 53:131–141

Stahl JC, Jouventin P, Weimerskirch H (in press) Habitats d'alimentation et stratégies d'exploitation des ressources des oiseaux marins dans le secteur des iles Crozet (Sud-Ouest de l'Océan Indien). Terre Vie.

Voisin JF (1968) Les Pétrels géants (*Macronectes halli* et *Macronectes giganteus*) de l'île de la Possession. Oiseau Rev Fr Ornithol) 39:95–121

Voronina NM (1968) The distribution of zooplankton in the Southern Ocean and its dependence on the circulation of Water. Sarsia 34:277–283

A Preliminary Comparison of Marine Bird Biomass and Food Consumption Between the Southeastern Bering Sea and Parts of the Southern Ocean

G.L. HUNT, Jr[1]

Summary. Seabird density, biomass and C demand were compared in open waters of the southeastern Bering Sea shelf and the Argentine shelf at comparable latitudes and seasons; comparisons were also made for waters near Bering Sea colonies and near the South Orkney Islands. Away from colonies, seabird densities were generally higher in the Bering Sea than off southeastern Argentina, but the biomass of seabirds was similar. However, since the mean weight of birds in the Bering Sea was less than that off Argentina, C demand was higher in the Bering Sea as a consequence of the allometric nature of the metabolic equations. Similarly, limited sampling near the South Orkney Islands encountered a lower density of birds than sampling near islands at comparable latitudes in the Bering Sea. Again, the biomass of birds near the South Orkney Islands was similar to that found near the Bering Sea colonies due to the larger mean size of the southern hemisphere birds, but overall C demand of seabirds was similar or somewhat greater in the Bering Sea. Seabird use of the waters near the South Orkney Islands may have been under-estimated due to the departure of birds for moulting.

1 Introduction

In recent years there has been a growing interest in the role of seabirds as consumers in marine ecosystems. Estimates of the trophic demands of seabird assemblages have been based either on the energy consumption at a colony (Furness 1978; Wiens et al. 1979; Croxall and Prince 1981; Furness and Cooper 1982) or on the energy requirements of the avian biomass of an oceanic region (Sanger 1972; Idyll 1973; Wiens and Scott 1975; Everson 1977; Mougin and Prevost 1980; Hunt et al. 1981; Jouventin and Mougin 1981; Schneider and Hunt 1982). Although most authors have used the allometric equations of Kendeigh (1970) or Lasiewski and Dawson (1967) to convert avian biomass to energy required, others have used a fixed percentage of body weight as a crude estimate of energy needs (Hunt et al. 1981) or weight loss during starvation (Jouventin and Mougin 1981). Schneider and Hunt (1982) provide a comparison of these methods.

It is of interest to compare estimated C flux to birds in pelagic and near-shore (colony) situations. In the northern hemisphere, attempts at modelling C budgets in offshore marine ecosystems have assumed that seabirds play a distinctly minor role in C cycling. However, colony-based studies provide evidence that the energy requirements of breeding seabirds in the vicinity of colonies may be considerable (Furness 1978; Wiens et al. 1979; Furness and Cooper 1982), and may be influenced by competition with man for resources (Furness 1983).

These comparisons of C flux to seabirds between pelagic and near-shore areas gain added interest when made between the northern and southern hemispheres. The taxa and size classes of seabirds found in the two hemispheres vary greatly. When compared with estimates of local productivity, comparisons between the hemispheres provide a means of identifying differences in the trophic roles of seabirds, and a clearer understanding of the ability of seabirds to exploit marine ecosystems. Unfortunately, the data required for such comparisons are widely scattered, meagre and in many cases missing. In this paper I present a beginning for such comparisons, based on recently completed pelagic counts of seabirds in the Bering Sea and in the Southern Ocean at comparable latitudes and seasons.

2 Methods

If useful comparisons are to be made between the northern and southern hemispheres, data from similar latitudes and seasons should be used since there are marked differences in seasonal and latitudinal distributions of seabirds (Shuntov 1972). Hence, the densities of seabirds within 60 km of four colonies (islands) in the Bering Sea (St. Lawrence Island (63°N, 169°–171°W), St. Matthew Island (60°N, 173°W) and the two main Pribilof Islands together (St. Paul 57°N, 170°W and St. George 55°N, 169°W)) were compared with the density of seabirds within 60 km of the South Orkney Islands (Fig. 1), the only high-latitude southern hemisphere colony for which comparable data are available. The distance of 60 km was chosen arbitrarily as the distance within which there was a reasonably good coverage of seabird distributions, although it is recognized that birds from colonies in both hemispheres fly farther than 60 km in search of food (Croxall and Prince 1980; Hunt et al. 1982). Data for similar seasons were taken (Bering Sea – August,

1 Department of Ecology and Evolutionary Biology, University of California, Irvine, CA 92717, USA

Antarctic Nutrient Cycles and Food Webs (ed. by W.R. Siegfried, P.R. Condy, and R.M. Laws)

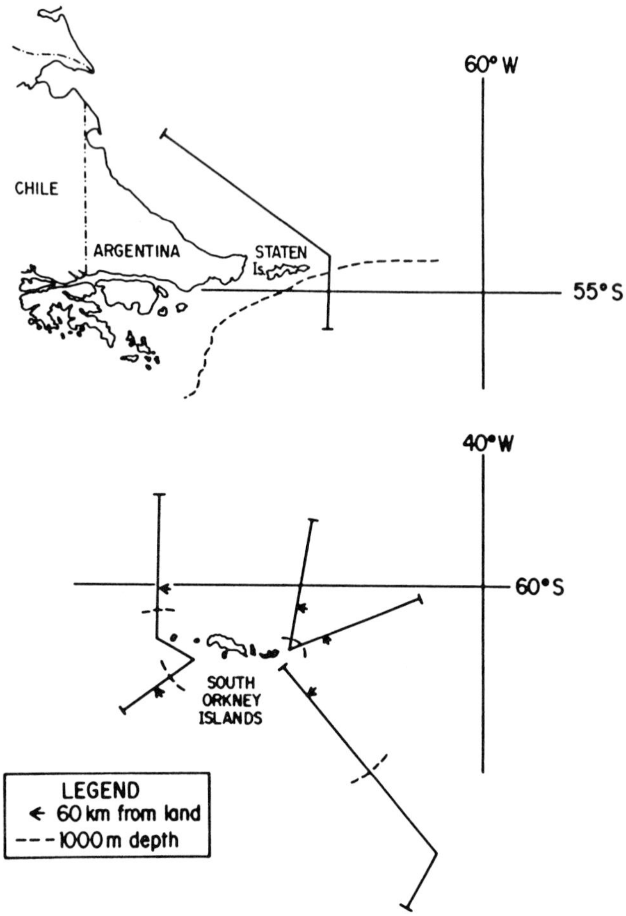

Fig. 1. Cruise tracks over the Argentine shelf and near the South Orkney Islands. Observations were made on a continuous basis along the lines indicated

September; Southern Ocean – February, March), although a broader range of seasonal coverage is also presented for the Bering Sea.

Pelagic distributions of seabirds in August–September in four domains (in Bristol Bay) of the shelf of the southeastern Bering Sea were selected for comparison with the distributions of seabirds in early March over the southeast Argentine continental shelf (Fig. 1). Four distinct oceanographic domains have been identified in the Bering Sea and differences in C flux to seabirds between these domains reflects inter-domain differences in C flow (Iverson et al. 1979; Schneider and Hunt 1982). Data are unavailable to make similar oceanographic distinctions on the Argentine shelf. Data from a survey made in the southern hemisphere on 9 March 1983 are compared with those from a cruise during 16 August–5 September 1980 in the Bering Sea.

Seabird densities were measured by counting birds within a 90° sector from abeam of the ship to directly ahead and extending out 300 m to the side with the best visibility while proceeding at a known speed. Bird numbers were then divided by the area surveyed in each 10 min period of observation to obtain densities (Hunt et al. 1982). Ship-following seabirds were excluded from density calculations and no attempt was made to correct for ship-attracted seabirds (Schneider and Hunt 1982), because independent assessments of densities away from the ship, using a helicopter, were not possible in the southern hemisphere.

Energy requirements of seabirds were estimated by calculating standard metabolic rate (SMR), based on the equation for non-passerine species (Lasiewski and Dawson 1967):

$$\mathrm{SMR} = 78.3\ \mathrm{M}^{0.723}$$

where SMR = kcal d^{-1} and M = kg. Weights of seabirds in the Bering Sea were obtained from specimens collected in the study area. Data for species in the southern hemisphere were taken from Jouventin and Mougin (1981). Food required by active seabirds was estimated by multiplying SMR by 2.5 to obtain existence energy needs (MacMillen and Carpenter 1977; Weathers and Nagy 1980; Schneider and Hunt 1982), allowing an assimilation ratio of 1.4 (Kendeigh et al. 1977), and setting energy values of 5 kcal g^{-1} (dry wt.) for fish (Nishiyama 1977) and 0.27 g (dry wt.) g^{-1} wet wt. for fish (Cooper 1979). A conversion of 0.4 g C g^{-1} (dry wt.) fish, squid and arthropods (Curl 1962) was used.

3 Results

3.1 Shelf-Waters away from Colonies

The C demand of seabirds in the Bering Sea was 1.59 times that found over the Argentine shelf for a comparable time of the year, a difference less than that found between domains in the Bering Sea (Table 1). Although a lack of oceanographic data precluded assigning the Argentine shelf observations to domains comparable to those of the Bering Sea shelf, the cruise track crossed from deep water into the centre of the shelf (Fig. 1). The similarity in C demand between hemispheres is striking, given the 2.4 times higher density of birds in Bristol Bay. However, the biomass of birds per unit area in Bristol Bay was only 1.25 times that in Argentine waters due to the smaller average size of birds encountered in Bristol Bay. This smaller average size contributed to the proportionately greater C demand in Bristol Bay, in accord with the allometric nature of the metabolic equations.

The rate C flux to surface vs. sub-surface feeding birds was also remarkably similar in Bristol Bay and the Argentine shelf (Table 1). Analysis of C flux pathways in different domains of the Bering Sea shelf show striking changes between domains, with most C going to surface foragers in the shelf-edge domain and to sub-surface foragers in the inner domain (Table 1). In the shelf-edge and outer domain, Northern Fulmars, *Fulmarus glacialis,* were the predominant consumers, whereas in the middle and inner domains Short-

Table 1. Density (no. birds km^{-2}), biomass (g birds km^{-2} × 10^4), mean mass per bird (g), carbon demand (g C m^{-2} d^{-1} × 10^4) and ratio of surface to sub-surface carbon flux for seabirds distant from their breeding colonies

Location	Density	Biomass	Mass	Carbon demand	Ratio surface: sub-surface carbon flux
Argentine shelf[a]	23.4	2.70	1,152	7.1	0.30
Bristol Bay[b]	55.2	3.37	610	11.31	0.28
Slope	19.6	1.16	591	2.66	1.60
Outer	57.9	3.60	621	9.09	0.23
Middle	59.0	3.80	645	9.61	0.42
Inner	77.0	4.95	642	12.42	0.09

a March 1983, 57 counts, see Fig. 1 for area surveyed
b Cruise 580; 16 Aug.–5 Sept. 1980, slope – 25, outer – 26, middle – 30, inner – 18 counts

tailed Shearwaters, *Puffinus tenuirostris,* which took euphausiids (*Thysanoessa* sp.), were dominant.

3.2 Waters near Colonies

St. Lawrence Island supports seabird colonies in the Bering Sea at a latitude comparable to that occupied by the South Orkney Islands in the Southern Ocean, but there was an inadequate sample for August–September (the period most comparable to February–March in the south). Thus, samples from throughout the breeding season at St. Lawrence were compared with those from near the end of the season at the South Orkney Islands.

The biomass of birds km^{-2} within 60 km of St. Lawrence Island was 1.11 times that near the South Orkney Islands, a remarkable similarity given the relatively meagre sampling effort available for the two islands (Table 2). However, energy consumption by seabirds near St. Lawrence Island was 1.74 times that near the South Orkney Islands due to the smaller size of the seabirds near St. Lawrence (mean mass = 305 vs. 1,581 g $bird^{-1}$).

Near the South Orkney Islands, major differences in avian density, standing crop biomass and C flux of seabirds were found between areas with waters < and > 1,000 m in depth. Density in an area of deep water was 50% lower than that found in water over the shelf (24.5 birds km^{-2}, n = 169 counts vs. 50.9 birds km^{-2}, n = 69 counts). Avian

Table 2. Density (birds km^{-2}), biomass (g birds km^{-2} × 10^3), mean mass per bird (g) and carbon demand (g C m^{-2} d^{-1} × 10^4) for seabirds within 60 km of colonies

Location and time of survey	No. 10-min counts	Density	Biomass	Mass	Carbon demand
Bering Sea (August–September)					
Pribilof Islands	746	72.3	54.68	757	17
St. Matthew Island	25	186.3	116.04	623	37.1
St. Lawrence Island	17[a]	–	–	–	–
Scotia Sea (February–March)					
South Orkney Islands	92	62.3	98.52	1,581	23.1
Bering Sea (May–October)					
Pribilof Islands	1,360	129.2	104.87	811	32.2
St. Matthew Island	140	93.3	71.40	774	22.1
St. Lawrence Island	107	360.2	109.84	305	40.1

a Sample too small to trust

Table 3. Species contributing significantly to avian biomass (10^3 g km^{-2}) near colonies. Numbers in parentheses are percentages

	Pribilofs May–October	Pribilofs August–September	St. Matthew May–October	St. Matthew August–September	St. Lawrence May–October	S. Orkney February–March
Bering Sea						
Fulmarus glacialis	3.28 (3)	2.39 (4)	3.19 (4)	0.56 (1)	0.85 (1)	
Puffinus spp.	3.68 (4)	6.36 (12)	0.39 (1)	0.91 (1)	0.00 (0)	
Rissa spp.	3.29 (3)	2.95 (5)	0.97 (1)	1.07 (1)	3.95 (4)	
Uria spp.	91.34 (87)	40.40 (74)	63.03 (88)	98.21 (85)	55.86 (51)	
Aethia pusilla	0.55 (1)	0.03 (< 1)	0.80 (1)	4.05 (3)	13.71 (12)	
other small auklets	0.63 (1)	0.32 (1)	2.08 (3)	7.61 (7)	33.73 (31)	
Fratercula corniculata	0.37 (< 1)	0.45 (1)	1.00 (1)	2.64 (2)	0.34 (< 1)	
South Orkney Islands						
Pygoscelis antarctica						60.42 (61)
Diomedea melanophris						8.96 (9)
Diomedea chrysostoma						2.55 (3)
Macronectes giganteus						1.74 (2)
Fulmarus glacialoides						21.08 (21)
Daption capensis						1.31 (1)
Pachyptila desolata						1.27 (1)

biomass was 3 times greater over the shelf (9.0 times 10^4 vs. 2.9 times 10^4 g km^{-2}), as was C flux (15.4 times 10^4 vs. 4.7 times 10^4 g C m^{-2} d^{-1}). Seabirds over the shelf were of greater mass on average (1,767 vs. 1,166 g bird^{-1}), and the most important contributors to avian biomass were the Chinstrap Penguin, *Pygoscelis antarctica,* Southern Fulmar, *Fulmarus glacialoides* and Blackbrowed Albatross, *Diomedea melanophris.* These same three species were also the most important over deeper water but, in addition, there was a dramatic increase in the Dove Prion, *Pachyptila desolata,* (3.1 to 8.7 birds km^{-2}).

For waters near all colonies examined, the major biomass of seabirds was concentrated in sub-surface foragers (Table 3). In the Bering Sea, these were represented by murres (*Uria* spp.) and small auklets (primarily *Aethia* spp.). Chinstrap Penguins were dominant near the South Orkney Islands. At South Georgia Island, where surface-foraging seabirds predominate in the colonies, the biomass of foraging seabirds close to the island mostly consists of sub-surface feeding penguins (J.P. Croxall, pers. comm.). At the South Orkney Islands, a substantial portion of the avian biomass was represented by Blackbrowed Albatrosses, whose closest large breeding colony is on South Georgia Island. In contrast, at St. Lawrence and St. Mattthew islands in the Bering Sea, a relatively small amount of the nearby avian biomass was represented by seabirds other than those associated with these colonies. However, near the Pribilof Islands and to the south in Bristol Bay, mirgrant shearwaters (*Puffinus* spp.) from the southern hemisphere made a significant contribution to avian biomass.

The values of avian biomass and C flux to seabirds near colonies in either region must be treated with great caution. Comparison of data from the Pribilof Islands for the entire breeding season with those from the end of the season show a 2 times decrease in C demand (Table 2). This drop reflects the departure of murres and Least Auklets, *Aethia pusilla,* during the second half of August (Hunt et al. 1982). Similar seasonal changes may influence seabird counts around St. Lawrence Island.

The seabird biomass supported by the near-shore waters of the South Orkney Islands may have been underestimated for two reasons (W.Z. Trivelpiece, pers. comm.). Adelie Penguin, *Pygoscelis adeliae,* populations there are large, but the birds had departed for moulting grounds on the Weddell Sea ice-pack prior to my visit. Thus, due to the time of my census, approximately 50% of the penguin biomass was absent. Additionally, the Chinstrap Penguin population may have been underestimated. Failed breeders and prebreeding younger birds are ashore moulting in mid-February to mid-March, and successful breeders may concentrate foraging closer to the islands than I was able to survey adequately. It is hard to estimate the effect of these sampling problems in the comparison between hemispheres, especially since similar arguments in modified form may apply to the waters near the Bering Sea colonies. Moreover, for large islands where seabird colonies are restricted to limited areas, such as at St. Lawrence and St. Matthew islands, the distribution of survey effort can affect density estimates greatly. Cruise tracks were not identical during the two surveys of St. Matthew Island, and this may partly explain the 1.7 times difference in C demand. Local patchiness in seabird distribution may also contribute to variance, if sampling effort is small (< 50 km^2 sampled).

4 Discussion

This paper reports a preliminary attempt to compare seabird densities, biomass, and C demand between areas of roughly equivalent latitudes and at roughly equivalent seasons in the northern and southern hemispheres. Densities were generally lower in the southern hemisphere for similar situations (close to breeding colony, removed from colony) but, because southern hemisphere birds were on the average larger, avian biomass per unit area was remarkably similar in the two hemispheres. Carbon flux to seabirds was, if anything, lower in the southern hemisphere, due to allometric considerations. This result suggests that, in relation to their ability to support marine birds, the waters studied in the Scotia Sea and in Bristol Bay of the Bering Sea were roughly equivalent. There was certainly no support for a relatively high productivity in the Southern Ocean with respect to birds occurring over shelf waters away from colonies (Bristol Bay vs. Argentine Shelf) where sampling problems were less severe than near colonies.

A notion that seabirds may be abundant in the Southern Ocean may result from the very high numbers of birds found breeding on the larger Sub-Antarctic islands such as South Georgia. Seabirds on South Georgia require 4.6 times 10^6 t of food each breeding season (Croxall et al. 1985), compared to 5.4 times 10^5 t per breeding season on the Pribilof Islands (Wiens et al. 1979). While at present we lack sufficient data to characterize the area of ocean required to support the bird populations on South Georgia, albatrosses breeding there forage over vast areas, frequently 900–2,650 km from their colonies, based on the calculations of Croxall and Prince (1980) and observations (Hunt and Veit, pers. obs.) of marked birds from South Georgia in the vicinity of the South Orkney Islands. In contrast, for the Pribilof Islands, densities of birds observed at sea within 110 km of the islands are sufficient to account for all birds from this colony, although it is reasonable to assume that fulmars and kittiwakes (*Rissa* spp.) fly greater distances in search of food. The southern hemisphere colonies may be larger because the Procellariiformes that nest there are adapted to search for food over greater distances than is the case for most northern hemisphere species. This ability is made possible not only by their lower wing loading (Ainley et al. 1974; Croxall and Prince 1980; Jouventin and Mougin 1981) but also by the adaptation of their chicks to survive periods of fasting while parents search for food (Croxall and Prince 1980).

It is also interesting to compare pathways of C flux to seabirds in different ocean regions or domains. Schneider and Hunt (1982) found significant differences in the extent of C flux to surface and sub-surface feeding birds in the domains of the shelf waters of the Bering Sea. These results have been confirmed and extended by recent work (Schneider, pers. comm.). The results presented here, based on a single cruise chosen because of its seasonal correspondence with the survey over the Argentine shelf, reflect, in part, the pattern observed in the more comprehensive studies. The similarities in seabird biomass and in the ratio of surface to sub-surface C flux to birds suggest that processes similar to those found in the Bering Sea shelf may be found in the waters off Argentina.

The predominance of sub-surface foragers in waters near colonies is undoubtedly the result of a combination of factors (Croxall and Prince 1980). In the northern hemisphere, wing-loading of alcids is greater than that of the surface-foraging Procellariiformes and gulls (Laridae) and, thus, the greater cost of flying would restrict the alcids' use of distant foraging grounds (Cody 1973). Similar energetic arguments would apply to penguins in the southern hemisphere. Additionally, sub-surface foragers can pursue prey through a large portion of the inshore water column, whereas surface foragers are restricted to catching prey when they are near the surface. One might expect that surface foragers would be at a competitive disadvantage to sub-surface foragers when foraging in inshore waters.

My data sets for regions other than Bristol Bay and the Pribilof Islands are small, and the information presented here must be considered provisional. However, the prospect of making quantitative comparisons between marine bird faunas of the northern and southern hemisphere is exciting. These comparisons provide a new tool for examining the avian communities of the two regions and their adaptations. The comparisons may also lead to a new insights regarding the dynamics of marine ecosystems in the two regions.

Acknowledgements. I thank the many field observers who aided in obtaining data on bird distributions around the Pribilof Islands and particularly N. Butowski who made the observations on UCI cruise 580 in Bristol Bay. R. Veit shared in taking observations in the southern hemisphere reported here. Earlier versions of the manuscript were commented on by Z. Eppley, N. Harrison, B. Obst, D. Schneider and W. Trivelpiece. This work was partially supported by the National Oceanic and Atmospheric Administration (NOAA) (through interagency agreement with the Bureau of Land Management under which a multiyear program responding to the needs of petroleum development of the Alaskan continental shelf is managed by the Outer Continental Shelf Environmental Assessment Program (OCSEAP)) and by NSF grants DPP 79-10386, DPP 82-06036 and DPP 82-06052 to G.L. Hunt, Jr.

References

Ainley DG; Morrell S; Lewis TJ (1974) Patterns in the life histories of storm petrels on the Farallon Islands. Living Bird 18:295–312

Cody M (1973) Coexistence, coevolution and convergent evolution in seabird communities. Ecology 54:31–44

Cooper J (1979) Energetic requirements for growth and maintenance of the Cape Gannet (AVES: Sulidae). Zool Afr 13:305–317

Croxall JP; Prince PA (1980) Food, feeding ecology and ecological segregation of seabirds at South Georgia. Biol J Lim Soc 14:103–131

Croxall JP; Prince PA (1981) A preliminary assessment of the impact of seabrids on marine resources at South Georgia. CNFRA 51:501–509

Croxall JP; Prince PA; Ricketts C (1985) Relationships between prey life-cycles and the extent, nature and timing of seal and seabird predation in the Scotia Sea. In: Siegfried WR, Condy PR, Laws RM (eds) Antarctic nutrient cycles and food webs (Proceedings of the 4th SCAR symposium on Antarctic biology). Springer, Berlin Heidelberg New York

Curl HS (1962) Analysis of carbon in marine organisms. J Mar Res 20:181–188

Everson I (1977) The living resources of the Southern Ocean. Food and Agriculture Organisation Southern Ocean Fisheries Survey Programme, Rome

Furness RW (1978) Energy requirements of seabird communities: a bioenergetics model. J Anim Ecol 47:39–53

Furness RW (1983) Competition between fisheries and seabird communities. Adv Mar Biol 20:225–307

Furness RW; Cooper J (1982) Interactions between breeding seabird and pelagic fish populations in the southern Benguela region. Mar Ecol Prog Ser 8:243–250

Hunt GL; Burgeson B; Sanger GA (1981) Feeding ecology of seabirds of the eastern Bering Sea. In: Hood DW, Calder JA (eds) The eastern Bering Sea shelf: Oceanography and resources, vol 2. US Govt Printing Office Washington DC pp 629

Hunt GL; Eppley Z; Burgeson B; Squibb R (1982) Reproductive ecology, foods and foraging areas of seabirds nesting on the Pribilof Islands, 1975–1979. Environmental assessment of the Alaskan continental shelf. Biol Stud 12:1–258

Idyll CP (1973) The Anchovy crisis. Sci Am 228:22–29

Iverson RL; Coachman LK; Cooney RT; English TS; Goering JJ; Hunt GL; Maccauley MC; McRoy CP; Reeburgh WR; Whitledge TE (1979) Ecological significance of fronts in the southeastern Bering Sea. In: Livingston RJ (ed) Ecological processes in coastal marine ecosystems. Plenum, New York pp 437

Jouventin P; Mougin JL (1981) Les strategies adaptives des oiseaux de mer. Rep Ecol (Tierre et Vie) 35:217–272

Kendeigh SC (1970) Energy requirements for existence in relation to size of bird. Condor 72:60–65

Kendeigh SC; Dol'nik VR; Garrilov VM (1977) Avian energetics. In: Pinowski J, Kendeigh SC (eds) Granivorous birds in ecosystems. International Biological Programme 12:127–402. Cambridge University Press, Cambridge

Lasiewski RC; Dawson WR (1967) A re-examination of the relation between standard metabolic rate and body weight in birds. Condor 69:13–23

MacMillen RE; Carpenter FL (1977) Daily energy costs and body weight in nectivorous birds. Comp Biochem Physiol 56A:439–441

Mougin JL; Prevost J (1980) Evolution annuelle des effectifs et des biomasses des oiseaux antarctiques. Rev Ecol (Tierre et Vie) 34:101–133

Nishiyama T (1977) Food-energy requirements of Bristol Bay Sockeye Salmon *Oncorhynchus nerka* (Walbaum) during the last marine stage. Spec Vol Res Inst N Pacific Fish Hokkaido Univ Fac Fish pp 289–320

Sanger GA (1972) Preliminary standing stock and biomass estimates of seabirds in the subarctic Pacific region. In: Takenouti AY et al. (eds) Biological oceanography of the northern North Pacific Ocean. Idemitsu Shoten, Tokyo p 626

Schneider D; Hunt GL (1982) Carbon flux to seabirds in waters with different mixing regimes in the southeastern Bering Sea. Mar Biol 67:337–344

Shuntov VP (1972) Marine birds and the biological structure of the Ocean (In Russian Dalnevastochnoe Knizhnoe Izdat, Vladivostok (Trans. 1974) Nat Tech Inf Service, Washington DC

Weathers WW; Nagy KA (1980) Simultaneously labelled water ($^3HH^{18}O$) and time-budget estimates of energy expenditure in *Phainopepla nitens*. Auk 97:861–867

Wiens JA; Scott JM (1975) Model estimation of energy flow in Oregon coastal seabird populations. Condor 77:430–452

Wiens JA; Ford RG; Heinemann D; Fieber C (1979) Simulation modelling of marine bird population energetics, food consumption and sensitivity to perturbation. Environmental assessment of the Alaskan continental shelf. Annu Rep Prinicp Invest 1:217–270

The Use of Lipid Profiles in Comparing the Diet of Seabirds

I.E. HORGAN and J.A. BARRETT[1]

Summary. Comparisons of lipid classes and fatty acid ratios in the stomach contents and stomach oils of seabirds can be useful for inferring the birds' diets. Background studies of phyto- and zooplankton, fish and squid suggest that certain biochemical compounds are particularly important in interpreting seabird diets. These are wax esters, alkylglycerolethers, phospholipids, sterols and triacylglycerols. The contents of stomachs of seabirds collected in the Benguela Current region are high in triacylglycerols, suggesting a diet dominated by fish. The higher proportions of sterols, wax esters and alkylglycerolethers in the stomach contents of seabirds collected in the Southern Ocean suggest a diet dominated by squid and euphausiids.

1 Introduction

Lipids are likely to be particularly important as energy resources in marine ecosystems with fluctuating food supplies. They also have specific physiological functions, and their patterns of distribution are influenced by environmental conditions (Malins and Wekell 1971; Morris and Culkin 1976).

In this paper, we discuss the results of analyses of seabird stomach contents and oils from two areas, the Benguela Current region and the Southern Ocean. This is part of an extended survey of lipid composition at different trophic levels in the Benguela ecosystem. We have also analyzed different organs (e.g., liver, muscle, fat deposits and stomach wall) of seabirds, and determined their lipid and fatty acid profiles (Horgan and Barrett, unpubl.), to ascertain how the lipids acquired in the diet are deposited in the birds' tissues.

2 Materials and Methods

Table 1 contains a summary of the species and samples analyzed for this study. Each sample was homogenized in a mixture containing chloroform:methanol (1:2) and subsequently sonicated in a Brandon cell disruptor model B-30 for 5 min, to improve the extraction. The procedure of Bligh and Dyer (1959) was followed to obtain a total lipid extract. The concentration of lipids in each sample was determined by weight from an aliquot of the extract (which was then discarded) and expressed as g % of fresh weight (wet wt.). Water content was the difference between wet wt. and dry wt. (freeze-dried). The remaining total lipid extract was stored at $-20^{\circ}C$ until used under N_2 to prevent lipid degradation and oxidation.

Another aliquot of the total extract was digested with 70% perchloric acid, and inorganic phosphorus determined as described in Dittmer and Wells (1969). Molecular weight relationships were used to calculate the phospholipid (PL) percentage contribution to the total lipids. Neutral lipids were separated from PL by adsorption on silicic acid and elution with chloroform and methanol, respectively (Wren 1960).

Phosphorus was determined (as above) in an aliquot of each fraction to calculate the recovery of PL and contamination of neutral lipids with PL. The neutral lipid fraction (free of PL) was used subsequently for further fractionation, identification and quantification of individual lipid classes by thin-layer chromatography (TLC), according to Walker (1971) and modified as follows. The fractionation was carried out on pre-coated 0.2 mm silica-gel TLC plates (Merck), pre-washed with petroleum ether:diethyl ether (95:5), sprayed with 10% $(NH_4)2SO_4$ and activated at $110^{\circ}C$ for 30 min. The plates were developed twice with petroleum ether:diethyl ether:acetic acid (90:10:1) and then placed in an oven at $180^{\circ}C$ for 30 min for visualization. Lipid classes were identified by comparing their relative mobilities with those of authentic standards and by differential staining. The plates were scanned quantitatively in a Vitatron TLD-100 densitometer, using different amounts of given sample and comparison with a standard mixture (Supelco Co). The identity of wax and sterol esters was confirmed by alkaline hydrolysis (0.3 N NaOH in methanol – 2 h reflux) of the respective components, previously separated by TLC and eluted. Re-chromatography of the hydrolysis products in the presence of standards was then performed. The same procedure was applied simultaneously to authentic standards of wax esters and cholesterol esters.

1 Department of Zoology, University of Cape Town, Rondebosch 7700, South Africa

Antarctic Nutrient Cycles and Food Webs
(ed. by W.R. Siegfried, P.R. Condy, and R.M. Laws)

Table 1. List of seabirds whose stomach contents (SC) and/or oils (SO) were analyzed

Species	Code	Age class	No. specimens	Sample	Date and place collected
Jackass Penguin	JPI	Adult	1	SC	August 1982 Benguela Region
Spheniscus demersus	JPII	Adult	1	SC	February 1983 Benguela Region
	JPIII	Adult	1	SC	May 1983 Benguela Region
	JPIV	Chick	3 (pooled)	SC	February 1983 Benguela Region
Blackbrowed Albatross *Diomedea melanophris*	BBA	Adult	3 (pooled)	SC	December 1982 Benguela Region
Whitechinned Petrel	WCPI	Adult	6 (pooled)	SC	December 1982 Benguela Region
Procellaria aequinoctialis	WCPII	Chick	1	SC + SO	January 1981 Southern Ocean
	WCPIII	Chick	1	SO	January 1981 Southern Ocean
	WCPIV	Chick	1	SO	January 1981 Southern Ocean
	WCPV	Chick	1	SO	January 1981 Southern Ocean
Wandering Albatross *Diomedea exulans*	WA	Chick	1	SO	September 1982 Southern Ocean
King Penguin	KPI	Chick	1	SC	December 1981 Southern Ocean
Aptenodytes patagonicus	KPII	Adult	1	SC	March 1982 Southern Ocean
Greatwinged Petrel *Pterodroma macroptera*	GWPI	Chick	2	SC + SO	September 1982 Southern Ocean

3 Results

The lipid compositions of seabird stomach contents from the Benguela Current region and the Southern Ocean are shown in Table 2. Levels of phospholipids in the stomach contents of seabirds from the Benguela region were approximately 10 times higher than those for seabirds in the Southern Ocean. In the neutral lipid fraction, fatty alcohols from Benguela seabird stomach contents were 2 times as concentrated as those from Southern Ocean seabirds. Fatty alcohols featured prominently in other pelagic organisms in the Benguela region, notably phyto- and zooplankters, squid (*Loligo* spp.) and Cape Anchovy, *Engraulis capensis* (Horgan and Barrett, unpubl.), and the overall pattern of the distribution of the neutral lipids of these organisms appears similar to that found for the stomach contents of seabirds from the same region.

Free fatty acids were present in the stomach contents of all seabirds taken in both regions. However, the proportions of free fatty acids were lower than expected, since, as digestion proceeds, lipids are hydrolyzed and liberate free fatty acids. Penguins, from both regions, had the highest proportions of free fatty acids in their stomach contents (Table 2), with esters accounting for 60% of the total lipids and mono- and diacylglycerides present in low proportions. Jackass Penguins, *Spheniscus demersus* (II and IV, Table 1), with a high level of free fatty acids, phospholipids, sterol esters and wax esters, showed only traces of diacylglycerides. With the exception of the Whitechinned Petrel, *Procellaria aequinoctialis* (II, Table 1) and Jackass Penguin (III, Table 1), triacylglycerides apparently occurred at similar levels in seabird stomach contents from both regions (Table 2). However, the ratios triacylglycerides:wax esters and triacylglycerides:alkylglycerolethers may indicate that Benguela seabirds have a diet richer in triacylglycerides, whereas wax esters and alkylglycerolethers constituted a relatively greater proportion of the dietary lipid of seabirds in the Southern Ocean.

An unknown compound "X" was detected in the stomach contents of three species of seabirds from the Benguela region. This compound had a TLC mobility between alkylglycerolethers and wax esters. A similarly behaving compound has been detected in zoo- and phytoplankton, squid, and Hake, *Merluccius capensis* (Horgan and Barrett, unpubl.), but not in other fish, or in the stomach contents and stomach oils of seabirds from the Southern Ocean. Wax esters were found in the stomach contents of Benguela seabirds in proportions approximately 2 times those of fish, and 50% those of squid, from the same region. These proportions were, on average, 50% of the comparable values for seabirds of the Southern Ocean.

Stomach oils were collected from three species of seabirds in the Southern Ocean: Wandering Albatross, *Diomedea exulans*, Greatwinged Petrel, *Pterodroma macroptera*, and Whitechinned Petrel (Table 3). Phospholipids were present in oils taken from the Wandering Albatross and Greatwinged Petrel samples, but were not found in the Whitechinned Petrel samples. Considering the possibility of predigestion occurring in the parent birds' stomachs, remarkable amounts of unhydrolyzed esters were present in oils taken from chicks, with low concentrations of fatty alcohols and free fatty acids which presumably had already been

Table 2. Percentage lipid composition in seabird stomach contents

Birds[a]	Total lipids[b]	PL	Glycerides MAG[c]	DAG 1,2	DAG 1,3	TAG	Sterols	Alcohols	FFA	AGE	X	WE	SE	HC
JPI	1.86	25.3	2.7	1.6	1.1	28.0	4.8	19.9	4.3	1.6	2.2	7.0	1.5	–
JPII	1.95	30.8	2.0	–	–	18.0	10.7	13.3	18.5	–	–	6.7	TR	–
JPIII	2.80	5.0	2.2	1.0	1.4	48.8	3.0	14.1	18.1	1.6	–	1.9	1.1	1.8
JPIV	2.23	48.0	6.3	–	–	4.9	4.9	8.5	6.3	–	–	8.1	13.0	–
BBA	2.13	23.5	2.8	0.9	0.5	14.6	6.1	35.6	3.8	–	4.7	5.6	1.9	–
WCPI	7.47	36.3	3.5	1.5	0.9	18.3	3.6	11.9	2.7	–	14.1	4.8	2.4	–
WCPII	8.90	1.8	2.4	1.7	0.9	57.4	3.8	7.5	1.9	9.6	–	8.7	4.3	TR
GWPI	7.53	2.3	3.7	2.0	1.3	21.2	5.8	6.8	3.9	27.0	–	12.4	8.5	5.1
KPI	3.27	6.7	1.2	2.4	1.8	21.0	6.5	11.5	6.3	12.6	–	20.2	8.6	1.2
KPII	4.57	8.8	3.2	2.8	1.7	11.7	5.0	15.5	3.5	21.4	–	18.4	4.0	4.0

Abbreviations: TR = traces, PL = phospholipids, MAG = monoacylglycerides, DAG = diacylglycerides, TAG = triacylglycerides, FFA = free fatty acids, AGE = alkylglycerolethers, WE = wax esters, SE = sterol esters, HC = hydrocarbons, X = unknown

Double line separates data for the Benguela region (top) and the Southern Ocean (bottom)

a Code in accordance with Table 1

b % wet wt.

c Includes pigments

Table 3. Percentage lipid composition of seabird stomach oils (SO) and relevant stomach contents (SC)

Birds[a]	Total lipids	PL[b]	Glycerides MAG[c]	DAG 1,2	DAG 1,3	TAG	Sterols	Alcohols	FFA	AGE	WE	SE	HC	Source[e]
WCPII (SC)	8.9	1.8	2.4	1.7	0.9	57.4	3.8	7.5	1.9	9.6	8.7	4.3	TR	1
WCPII (SO)	46.14	–	2.0	2.5	1.0	70.5	2.3	3.3	1.2	9.3	4.3	3.6	–	1
WCPIII (SO)	21.15	–	4.0	4.0	3.6	28.0	5.3	3.0	2.8	11.9	22.3	8.5	6.6	1
WCPIV (SO)	51.26	–	6.4	2.4	1.7	41.0	4.8	5.9	7.6	10.0	13.8	6.4	–	1
WCPV (SO)	82.29	–	12.0	2.2	1.7	28.7	3.6	9.9	5.0	19.9	1.7	13.1	2.2	1
WCP chick	–		1.8	4.3	....	9.2	5.6	42.5	10.3	9.1	3.8	13.4	–	2
WCI (SC)	7.47	36.3	3.5	1.5	0.9	18.3	3.6	11.9	2.7	–	4.8	2.4	–	1
GWPI (SC)	7.53	2.3	3.7	2.0	1.3	21.2	5.8	6.8	3.9	27.0	12.4	8.5	5.1	1
GWPI (SO)	1.65	29.7	0.6	0.6	0.6	22.4	6.1	4.2	3.0	21.3	7.3	4.2	–	1
GWP (SO)	–	2.0	–	–	–	60.9	2.4	–	4.5	–	30.2[d]	–	–	3
GWP (SO)	–	6.0	3.4	–	–	78.8	3.2	–	2.7	–	5.9	–	–	3
WAI (SO)	40.05	7.0	2.5	1.4	1.0	44.8	2.5	4.0	1.0	8.1	15.9	4.7	7.1	1
WA (SO)	–	TR	–	–	–	4.1	–	–	–	2.3	77.0[d]	–	8.6	4

a, b, c As in Table 2

d Probably includes sterol esters

e Data from this study, as well as results from other authors, included for comparison purposes
1 = this study, 2 = Warham et al. (1976), 3 = Cheah and Hansen (1970), 4 = Lewis (1969)

absorbed. Triacylglycerides predominated in the oils of all the birds samples, whereas the second largest neutral lipid class was either alkylglycerolethers or wax esters which occurred in inverse proportions. Thus, for example, in the Whitechinned Petrel (II and V, Table 3) and Greatwinged Petrel alkylglycerolethers offset wax esters, whereas the reverse occurred for Whitechinned Petrel (III and IV, Table 3) and Wandering Albatross.

4 Discussion

Considerable differences were found in the lipid profiles of stomach contents of seabirds collected from the Benguela region and the Southern Ocean. Phospholipids, alcohols and triacyclglycerides predominated on a ratio basis in the Benguela region, whereas alkylglycerolethers and wax esters prevailed in the Southern Ocean. Wax esters are present in substantial amounts in certain species of krill (Euphaussiacea), but not *Euphausia superba* (Clarke and Prince 1980a), fish, copepods and various other marine organisms (Malins and Wekell 1971; Morris and Culkin 1976). Since wax esters are associated with low temperatures, seasonal food supplies, vertical migrations, and deep-living organisms, they may be fairly abundant in the Southern Ocean at relatively low trophic levels. In the Benguela region, however, organisms at comparable trophic levels contain relatively small amounts (10%) of wax esters (Horgan and Barrett, unpubl.) which are concentrated at levels approximately 4 times higher in squid than in fish. Without enzymatic studies or chemical characterization, it is difficult to infer whether wax esters originate in fish and are accumulated in the stomachs of Benguela seabirds, or are hydrolyzed after originating in squid. The high concentrations of alcohols found in the stomach contents of Benguela seabirds may originate in squid, and could be indicative of wax ester hydrolysis. Thus, the composition and distribution of fatty acids and fatty alcohols may indicate from which trophic level the wax esters are derived.

Alkylglycerolethers have been associated with squid and anchovy (Warham 1977), but were not found in related organisms taken from the Benguela region (Horgan and Barrett, unpubl.). However, they were detected in the stomach contents of all Southern Ocean seabirds sampled, and their presence may be governed by environmental conditions.

The "X" compound was found in zoo- and phytoplankton, Hake and squid samples collected from the Benguela region, and when characterized may serve as a dietary marker of squid. Direct observations of the stomach contents of Blackbrowed Albatrosses, *Diomedea melanophris,* and Whitechinned Petrels from the Benguela region suggest that Blackbrowed Albatrosses favour fish, whereas Whitechinned Petrels favour squid (FitzPatrick Institute, unpubl. data). The "X" compound was found to be approximately 4 times higher in Whitechinned Petrel than in Blackbrowed Albatross stomach samples.

A dietary origin for procellariiform stomach oils appears to be well established (Warham et al. 1976; Clarke and Prince 1976; Watts and Warham 1976; Warham 1977). The functions of the stomach oils include: stable energy and water reserves (Warham 1977); preservation properties which may retard digestion of food ingested by adults to be fed to chicks (Warham 1977); defence (Warham et al. 1976; Warham 1977); conversion of wax esters to triacylglycerides as a means of depositing energy (Clarke and Prince 1976); and, the facilitation of certain breeding strategies of adults and the survival of their chicks, (Croxall and Prince 1980).

The prevalence of diets rich in wax esters, alkylglycerolethers and triacylglycerides is governed largely by environmental conditions. For example, wax esters are linked to low temperatures, depth, pulsating food availability, and energy production in the absence of oxygen (Malins and Wekell 1971; Morris and Culkin 1976; Spark 1982). Consequently, the inter- and intra-specific variability of lipids and the random distribution of fatty acids and fatty alcohols in the oils of seabirds (Watts and Warham 1976; Clarke and Prince 1976) are linked to diets containing random distributions of these compounds. Warham et al. (1976) found high concentrations of alkylglycerolethers in the stomach oils of Whitechinned Petrels, whereas our samples had higher concentrations of triacylglycerides, with alkylglycerolethers and wax esters equal or higher, sterol esters equal or lower, and fatty acids and fatty alcohols much lower. Alcohols in stomach oils are likely to be degradation products of wax esters. Alcohols are ubiquitous, since they can be metabolically transformed to fatty acids and aldehydes which in turn are precursors for different lipid classes. In addition, alcohols inhibit hydrolysis of triacylglycerides. Warham et al. (1976) found high concentrations of alcohols in the stomach oils of the Whitechinned Petrel and Pintado Petrel, *Daption capense.*

The generally low concentration of phospholipids in the stomach oils of Southern Ocean seabirds agrees with Warham's (1977) results. Warham reported low levels of phospholipids in the oils of Southern Giant Petrels, *Macronectes giganteus,* in spite of high concentrations of phospholipids in their potential prey items. In the Southern Ocean, phospholipids may be absorbed immediately, and this could be a form of adaptation in order to preserve phospholipids, since they are important constituents of plasma membrane (Malins and Wekell 1971). Warham et al. (1976) suggested that either a highly active acid phospholipase is present, or that phospholipids are absorbed pynocytotically. If chicks were to lack food for lengthy periods, as apparently is the case (Warham 1977; Croxall and Prince 1980), fundamental phospholipids may be absorbed and resynthesized to suit the chicks' requirements. When stomach contents and oils of Whitechinned Petrel II and Greatwinged Petrel (both chicks) were compared (Table 3) triacylglycerides were found to be higher in the oils, wheras alkylglycerolethers were approximately in the same proportion and wax esters reduced by half. A comparison of the stomach contents of King Penguins, *Aptenodytes patagonicus,* showed that chicks had slightly higher concentrations of triacylglycerides than adults which had an apparent lower percentage of wax esters (since the amount of total lipids was higher). According to Clarke and Prince (1976), chicks may split wax esters and accumulate triacylglycerides more rapidly, and Warham (1977) reported that Blackbrowed Albatross chicks accumulated 70.5% of triacylglycerides in their oils.

In conclusion, in the Benguela region the presence of phospholipids and fatty alcohols, as well as increased ratios of triacylglycerides vs. alkylglycerolethers and wax esters, in the stomach contents of seabirds, suggests a diet of predominantly fish with squid taken in small proportions. In the Southern Ocean, however, the higher proportions of sterols, wax esters and alkylglycerolethers may indicate a diet largely composed of squid and euphausiids. Thus, it would seem that gross comparisons of the qualitative and quantitative lipid compositions of stomach contents and oils with lipid profiles of putative prey may determine the sources of the specific lipids. Furthermore, the use of lipids as tracers of marine food webs may be feasible provided that supplementary data on environmental (temperature, depth, nutrient availability) and physiological (age, sex, gonadal development, seasonal change) features are collected and detailed chemical and biochemical analyses of lipids and fatty acids are performed. Ultimately, however, the study of lipids may be important for their role in affecting certain physiological processes such as reproduction and growth.

Acknowledgements. We thank the staff of the FitzPatrick Institute, University of Cape Town, for collecting the samples. Financial support by the South African National Committee on Oceanographic Research (SANCOR), through the Benguela Ecology Programme, is gratefully acknowledged. We thank Mrs. J Goode for typing assistance.

References

Bligh EG, Dyer WJ (1959) A rapid method of total lipid extraction and purification, Can J Biochem Physiol 37:911–917

Cheah CC, Hansen IA (1970) Stomach oil and tissue lipids of the petrels *Puffinus pacificus* and *Pterodroma macroptera.* Int J Biochem 1:203–208

Clarke A, Prince PA (1976) The origin of stomach oil in marine birds: Analysis of stomach oils from six species of subantarctic Procellariiform birds. J Exp Mar Biol Ecol 23:15–30

Clarke A, Prince PA (1980a) The biochemical composition of Krill, *Euphausia superba* Dana, from South Georgia. J Exp Mar Biol Ecol 43:221–236

Clarke A, Prince PA (1980b) Chemical composition and calorific value of food fed to mollymauk chicks *Diomedea melanophris* and *D. chrysostoma* at Bird Island, South Georgia. Ibis 122:488–494

Croxall JP, Prince PA (1980) Food, feeding ecology and ecological segregation of seabirds at South Georgia. Biol J Linn Soc 14:103–131

Dittmer JC, Wells MA (1969) Qualitative and quantitative analysis of lipids and lipid components. Methods Enzymol 14:482–530

Lewis RW (1969) Studies on the stomach oils of marine animals – II Oils of some Procellariiform birds. Comp Biochem Physiol 31: 725–731

Malins DC, Wekell JC (1971) The lipid biochemistry of marine organisms. Progr Chem Fats Other Lipids. 10:337–363

Morris RJ, Culkin F (1976) Marine lipids. Analytical techniques and fatty acid ester analysis. Oceanogr Mar Biol Annu Rev 14:391–433

Spark A (1982) New light in marine waxes, fats and oils S Afr J Sci 78:303–305

Walter BL (1971) A novel charring technique for detection of lipids on thin-layer chromatography. J Chromatogr 56:320–323

Warham J, Watts R, Dainty RJ (1976) The composition, energy content and function of the stomach oils of petrels (order, Procellariiforms). J Exp Mar Biol Ecol 23:1–13

Warham J (1977) The incidence, function and ecological significance of petrel stomach oils. Proc N Z Ecol Soc 24:84–93

Watts R, Warham J (1976) The structure of some intact lipids of petrel stomach oils. Lipids 11:423–429

Wren JJ (1960) Chromatography of lipids in silicic acid. J Chromatogr 4:173–195

Biomass of Birds and Mammals in the Ross Sea

D.G. AINLEY[1]

Summary. The distribution and biomass of birds and mammals were determined for the Ross Sea, Antarctica. A total of 548 30-min strip (4–10 km long) censuses were made during 5 cruises south of 65°S between 15 December and 18 January in 1976–1980. Within the boundaries of the Ross Sea, the biomass of birds, seals and whales was estimated to be 0.044–0.070, 0.068–0.089 and 0.182–0.394 g m^{-2}, respectively. The difference between bird and seal biomass was not statistically significant. The bird community was dominated in abundance (no. individuals) by Antarctic Petrels, *Thalassoica antarctica,* and in biomass by penguins. The mammal community was dominted in abundance by Crabeater Seals, *Lobodon carcinophagus,* and in biomass by Crabeater Seals and Minke Whales, *Balaenoptera acutorostrata.* Birds and mammals were concentrated in the same areas, in conjunction with oceanographic fronts, while large areas were devoid of these upper-trophic-level predators. Characteristics of pack-ice also affected the animals' distributions, especially those of mammals.

1 Introduction

Laws (1977) summarized available information on the abundance of vertebrates in the Antarctic, and attempted to estimate the biomass of each major taxonomic group on an ecosystem-wide basis. The data for certain groups (e.g., seals and whales) were fairly extensive, but for others Laws had to assume that limited data from particular regions represented the entire Antarctic. In the case of birds, Laws (1977) assumed that the biomass of penguins in the Ross Sea was typical of bird biomass throughout the remaining waters in the Southern Ocean. This, of course, was an oversimplification, but it was the best that could be done at the time.

I made censuses of birds, seals and whales in the Ross Sea region, in order to compare their biomass. I was also interested in the distribution of biomass among these upper-trophic-level vertebrates and, if possible, I wanted to explain any discontinuities in the abundance of the three groups. Findings on the abundance and distribution of birds in the Ross Sea have been reported recently (Ainley et al. 1983). In this paper, I report census results for seals and whales, and compare their relative abundance with that of birds.

1 Point Reyes Bird Observatory, Stinson Beach, CA 94970, USA

2 Methods

Censuses were conducted from U.S. Coast Guard ice-breakers during the early austral summer between 15 December and 18 January. This is slightly different from the "early summer" period in Ainley et al. (1983) during which a 4-day cruise in the far western Ross Sea (16–19 January), together with several other (later) cruises not reported on here, were placed in a "late summer" category. Switching this short cruise to the early summer increased the completeness of information on the occurrence of birds and mammals in the pack-ice of the western Ross Sea. As will become obvious later, this particular coverage of an important pack-ice area is crucial for understanding the distribution of marine mammals in the region. Nevertheless, the results of this report and those of Ainley et al. (1983) are comparable, because by 19 January birds are still tied to rookeries by breeding duties.

Ships and dates of cruises were: USCGC Northwind (15 December 1976–4 January 1977, 19 December 1979 –2 January 1980, and 5–9 January 1980); and, USCGC Burton Island (23–29 December 1977, and 16–19 January 1977). Census platforms were identical because the two vessels were sister ships. Counts were made from the bridge wings where eye-level was about 16 m above the sea surface. A count was made for 30 min of every hour that the ship travelled at speeds exceeding 11 km h^{-1} during daylight (which was continuous). The ships normally cruised at 19–22 km h^{-1} in open water. Each 30-min census was equivalent to one "transect". Ideal census conditions were experienced during all cruises. The censuses included only those birds and seals which passed within 300 m, and whales within 800 m, of the side (forequarter) of the ship on which observers were positioned to experience the least glare. Transect width was determined by means of a sighting board described by Cline et al. (1969). Ship's position, updated hourly, was determined by satellite navigation. The distance travelled during each 30-min transect, multiplied by the transect width, provided the area of the strip samples; dividing animal numbers by this area gave a estimate of density that was adjusted later according to various factors. Within the confines of the Ross Sea (see below for boundary definitions), 458 transects were carried out

Antarctic Nutrient Cycles and Food Webs
(ed. by W.R. Siegfried, P.R. Condy, and R.M. Laws)

Table 1. Summary of survey and census data for the Ross Sea

Cruise	Transects	Area surveyed (km²)		Number of individuals						
		Whales	Seals	SECR	SELE	SERO	SEWE	SESE	WMIN	KILL
NW 76	240	2,394.1	900.0	198	7	2	30	1	46	30
BI 77	58	434.5	163.4	4	1		9		12	
GL 77	79	446.8	168.0	19	4		18		12	
NW 79 A	54	477.9	179.7	1	1				17	23
NW 79 B	27	204.9	77.0	19	2		1		26	
Totals	458	3,958.2	1,488.1	241	15	2	58	1	113	53

Abbreviations for species' names are: SECR = Crabeater Seal; SELE = Leopard Seal; SERO = Ross Seal; SEWE = Weddell Seal; SESE = Southern Elephant Seal; WMIN = Minke Whale; KILL = Killer Whale

during the Dec.–Jan. study period, accounting for a survey of 3,958 km² for whales and 1,488 km² for birds and seals (Table 1). Most of the transects were logged in December.

Ice conditions during each transect were recorded according to World Meteorological Organization format. Because the ice-breaker generally followed the path of least resistance through heavy pack-ice ($\geq$ 6 oktas), observers always estimated ice concentration just outside the transect (i.e. from 300 m to 800 m) which gave the best approximation of overall ice conditions. Immediately after each 30-min transect, sea-surface temperature was measured, using a bucket thermometer. Other researchers periodically measured the vertical temperature-salinity profile of the water column (see Ainley and Jacobs 1981). These profiles were useful for locating particular features such as the Antarctic Slope Front. As a measure of phytoplankton density, whenever the ship stopped, water clarity was measured, using a secchi disk. When in transit, water colour (green vs. blue) was noted, as an indirect indication of gross changes in relative clarity.

I estimated faunal density in two ways. First, density estimates (no correction factors applied) for each species, in each 30-min transect were plotted on separate charts of the Ross Sea. Cruise tracks had been plotted on these charts which were drawn from Hayes and Davey (1974, polar stereographic projection). From visual inspection, lines were drawn for each species to define broad zones having densities of consistently similar orders of magnitude. The data themselves determined the levels of magnitude of density estimates used to define each zone: as density increased so did the disparity in the levels of magnitude among different zones and, ultimately, in the number of zones. The latter ranged from 1 to 5 depending on species. For seals and whales, my drawing of zone boundaries took into account species' preferences for pack-ice concentrations and floe size (see below; Gilbert and Erickson 1977). The zones of different densities were then plotted on charts to show species' distributions (Fig. 5–8, 10–11).

I estimated the total population of each species in the Ross Sea during the study period, using the distribution charts. This was done by determining the area of each abundance zone and multiplying by its respective overall average corrected density (see below), and then by adding the results for all zones. The area of zones was determined with a polar planimeter. The procedure for Adélie Penguins, *Pygoscelis adeliae,* was different, because direct counts of rookery populations were available and because about 60% of the Adélie Penguins were actually on land at any one time during the census period. For birds, Ainley et al. (1983) describe the estimation procedure and correction factors in much greater detail. For seals, crude density estimates were adjusted upward to account for their diurnal haul-out cycle and the effect this had on the probability of sighting (see below); corrections for pack-ice preferences were also made, as explained above. For Minke Whales, my abundance estimates were done roughly according to the method outlined by Butterworth and Best (1982) where I used mean school size, school density and factors correcting for probability of sighting, bias in distance and speed estimation and ship/whale movements. Because census platforms and census conditions were similar between the latter study and mine, I used the ("unsmeared") correction factors developed by these authors. Because my search area was much smaller than theirs (only one side of the ship instead of two, and outward 800 m, instead of to the limits of visibility), and because Minke Whales are apparently unaffected by the ship (Leatherwood et al. 1982a), and because I had been on a census cruise (in the North Pacific, 1979) with one of the observers who collected data on the cruises used by Butterworth and Best (1982), and thus had some gauge of my relative sighting abilities, I assumed that the number of whales missed was comparable between their data and mine. Instead of the contouring employed by Butterworth and Best, I used broader zones of abundance; in both studies, lines were drawn "by eye".

I also estimated density another way. The total bird, seal, and whale biomasses were plotted for 2.5° latitude times longitude blocks by averaging transect results within each block (Figs. 4, 9, and 12). I then averaged the figures for all blocks. As a function of cruise tracks, census information was available for 96 of the 160 blocks (60%) contained entirely or in part within Ross Sea boundaries. An additional 26 blocks not censused were bordered by at least 2 other blocks that had been censused. Densities in the bordering blocks were averaged to estimate the density of those 26 blocks. This gave 76% coverage of the Ross Sea in order to assess to overall distributions of birds and mammals.

My density estimations were confined to the boundaries of the Ross Sea (as defined below) because cruise coverage north of the continental slope was too sparse. Population and density estimates, and information on body weights (from field and literature) were used to estimate biomass; all weight data for marine mammals were taken from Laws (1977).

3 The Environment

The Ross Sea (Fig. 1) lies between Victoria Land and the King Edward VII Peninsula (Marie Byrd Land) on the west and east, respectively, and between the Ross Ice Shelf and the 3,000 m depth contour on the south and north, res-

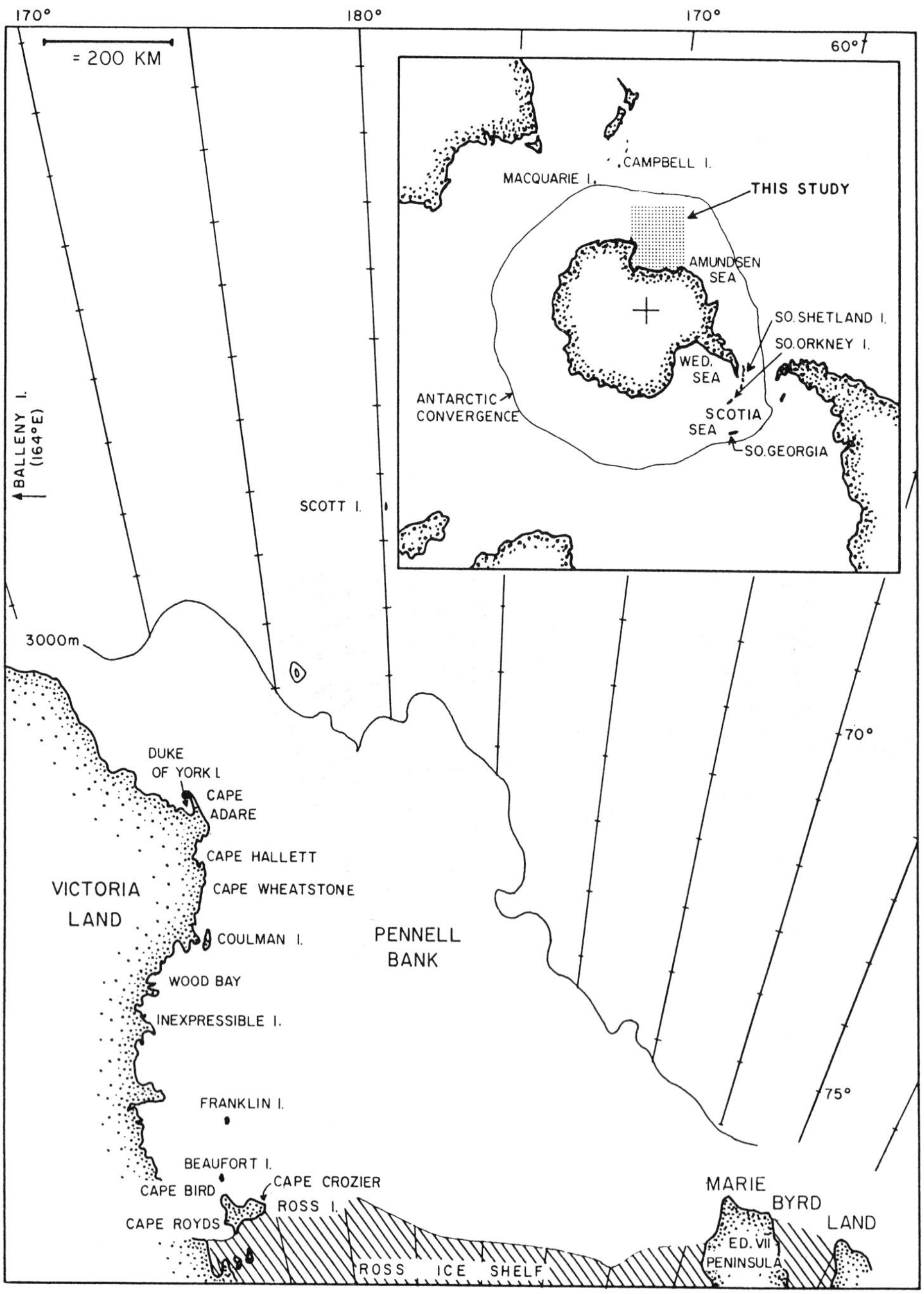

Fig. 1. The Ross Sea study area

pectively, The 3,000 contour is at the base of the continental slope. Shaped approximately like an equilateral triangle with a base and height of about 880 km each, the Ross Sea is about 598,000 km^2 in area (determined with a planimeter).

Circulation in the Ross Sea is cyclonic, with a westerly flow along the ice-shelf, a northerly flow along Victoria Land, and a southeasterly flow near the continental shelf break. An opposing current sets northwesterly over the continental slope (Sverdrup et al. 1942; Ainley and Jacobs 1981).

The Antarctic Slope Front described by Ainley and Jacobs (1981) lies over the Ross Sea continental slope. The front generally lies 10–55 km seaward of the shelf break, which places it just to the south of the 1,000 m contour (Fig. 2). Steep horizontal gradients in physical properties of the ocean in the western Ross Sea, parallel to the Victoria Land coast and adjacent to the edge of the persistent pack-ice there, indicate frontal activity in that region as well (Fig. 2). Biggs (1982) reported peaks in zooplankton biomass both over the slope front and in the western Ross Sea.

During the winter, pack-ice extends outward from the coast to beyond Scott Island. In late October a large open-water area appears in the south-western Ross Sea, and it continues to widen toward the north and the east as late summer approaches (Fig. 3). By late January, pack-ice

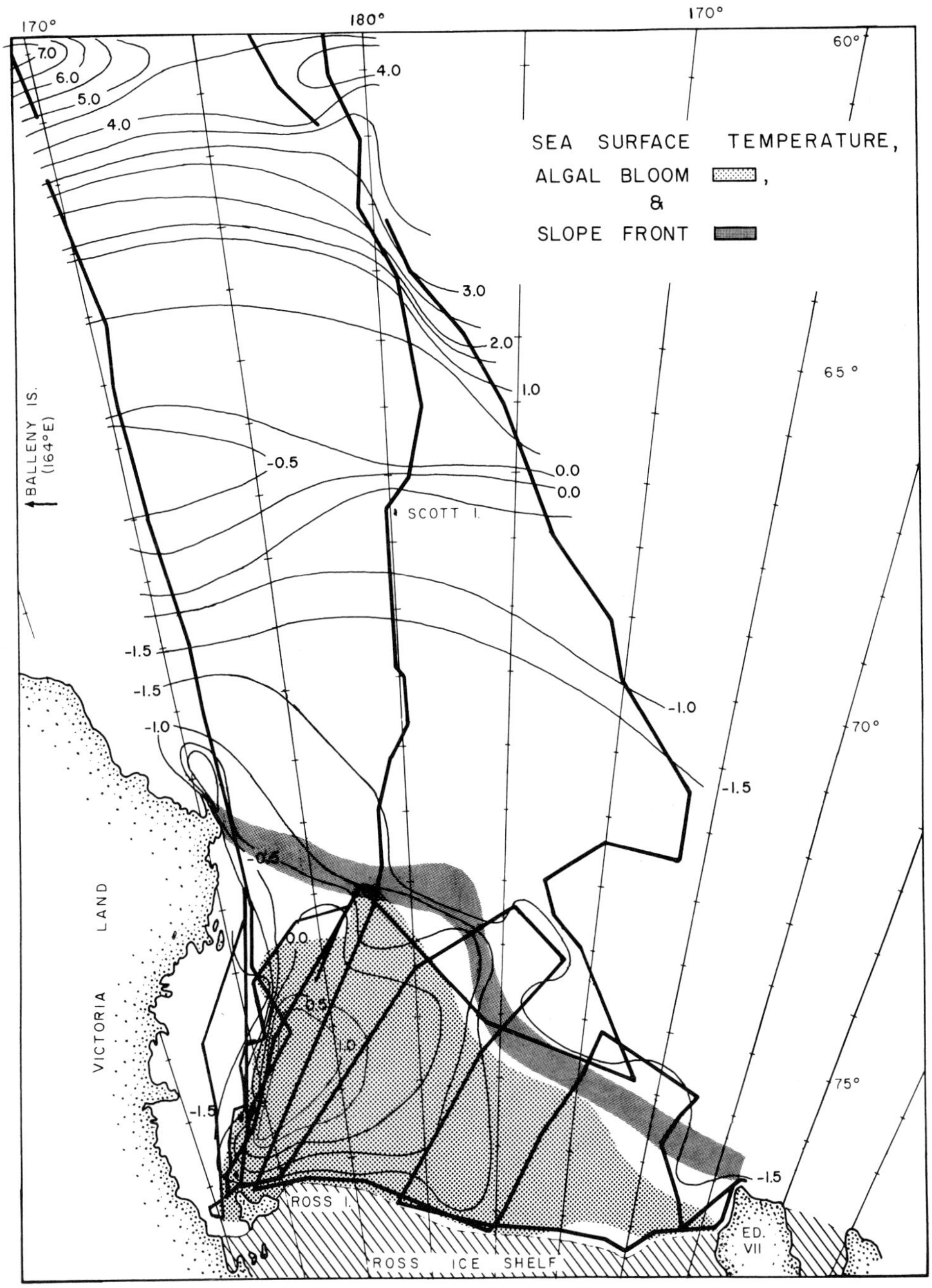

Fig. 2. Sea-surface isotherms and locations of an algal bloom and the Antarctic Slope Front

usually remains only along the Victoria Land coast and in a large tongue that extends northwestward from the King Edward VII Peninsula (see Gilbert and Erickson 1977; Fig. 7). In both areas the pack-ice is rather concentrated (6–8 oktas) with sparser ice cover around the edges. Fifty-seven per cent of the transects were within pack-ice and 43% were in open water.

An algal bloom, which persists throughout the summer and autumn, coloured surface waters over the continental shelf of the Ross Sea emerald green (El-Sayed et al. 1978; Ainley and Jacobs 1981, Fig. 2). A dramatic change in water colour from green to blue and an equally dramatic decrease in turbidity over a distance of just a few kilometers indicated the abrupt boundary of the bloom at the slope front and at the edge of the persistent pack-ice in the western Ross Sea. The bloom decreased in intensity from west to east in the same direction that pack-ice recedes during summer.

4 Distribution and Numbers of Species

4.1 Birds

The distribution, number and biomass of individual bird species are presented in Ainley et al. (1983). In summary,

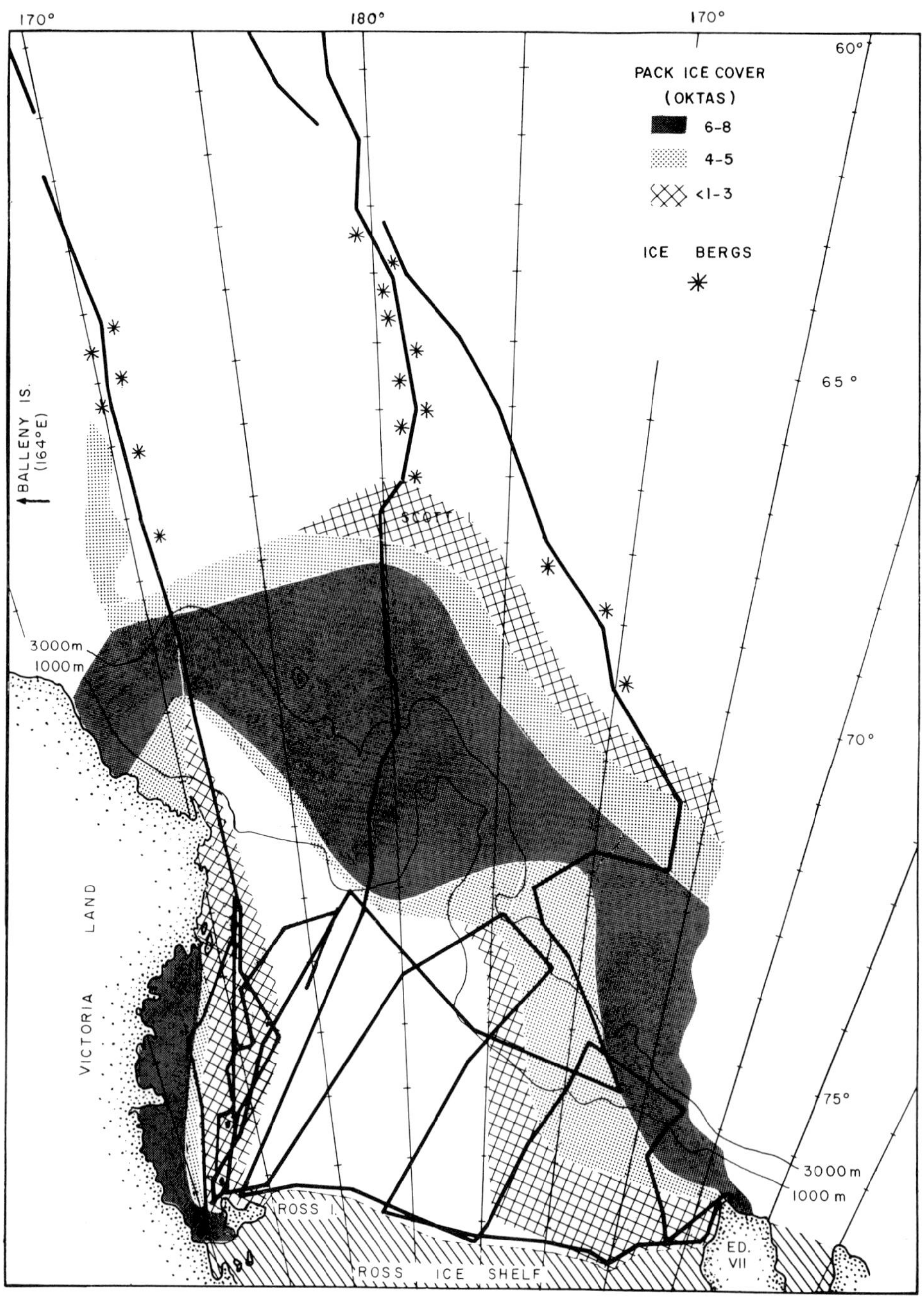

Fig. 3. The distribution of pack-ice. Data were integrated from direct observations and satellite imagery

a total of 9,733,000 individuals at an average density of 16.3 birds per km^2 and a biomass of 0.040 g m^{-2} occurred in the Ross Sea during December and early January. Including chicks dependent on parents for food, the biomass was 0.044 g m^{-2}. Bird biomass was concentrated over the slope front and in the pack-ice of the southwestern Ross Sea; few birds occurred in the centre of the algal bloom nor seaward of the slope front (Fig. 4). Avian biomass, determined by averaging totals for each of the blocks for which data were available (Fig. 4), was 0.070 ± 0.010 g m^{-2} ($\bar{x}$ ± S.E.). Nine species comprised the avifauna during the census period, which was dominated in numbers by the Antarctic Petrel, *Thalassoica antarctica,* (55%) and in biomass by Emperor, *Apenodytes forsteri,* and Adélie penguins, (42.8 and 39.3%, respectively).

4.2 Seals

Erickson, Siniff, and co-workers (e.g., Gilbert and Erickson (1977) and reports summarized in Brown et al. (1974) and Laws (1977)), adjusted counts of seals according to the ice-type preferences of each species (and the amount of that habitat available) and to diurnal cycles in probability of sighting dependent on the fact that more seals are hauled out, and, thus, are easier to see during certain periods of the

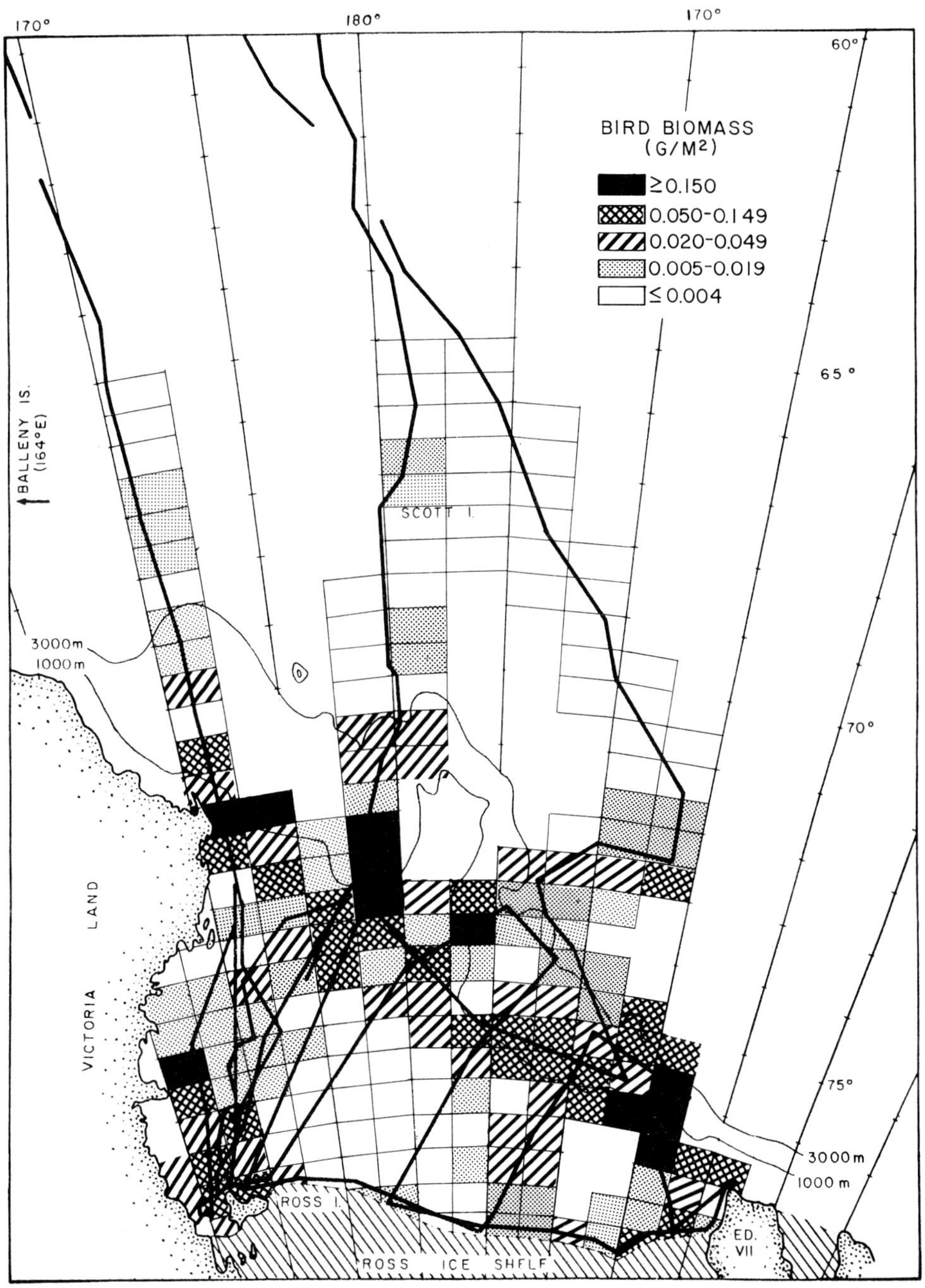

Fig. 4. The distribution of bird biomass in the Ross Sea region by 2.5° lat. times long. *blocks*

Table 2. Percentage of animals observed in relation to ice concentration

Ice concentration (oktas)	Transects	Species				
		WMIN	SEWE	SELE	SECR	SERO
None	244	5.8 (22.4)	0.5 (3.7)	0.0 (0.0)	0.2 (1.4)	0.0 (0.0)
0–1	104	26.3 (43.1)	8.1 (24.1)	9.7 (25.0)	7.5 (20.8)	0.0 (0.0)
2	30	9.1 (4.3)	8.7 (7.4)	11.5 (8.3)	4.0 (3.2)	12.7 (12.5)
3	36	32.0 (18.1)	23.5 (24.1)	28.5 (25.0)	12.8 (12.4)	0.0 (0.0)
4	33	13.3 (6.9)	5.9 (5.6)	30.9 (25.0)	7.6 (6.7)	23.2 (25.0)
5	36	7.6 (2.6)	12.7 (13.0)	9.7 (8.3)	22.0 (21.2)	0.0 (0.0)
6	41	0.0 (0.0)	6.4 (7.4)	0.0 (0.0)	17.4 (19.1)	23.2 (25.0)
7	28	5.8 (2.6)	7.0 (5.6)	9.7 (0.0)	14.1 (10.6)	40.0 (37.3)
8	12	0.0 (0.0)	27.2 (9.3)	0.0 (0.0)	14.3 (4.6)	0.0 (0.0)
Animals		116	54	12	283	8

The proportion of animals seen at a particular ice concentration level (numbers in parentheses) was multiplied by the total number of transects (564); that product was divided by the number of transects where that respective ice concentration occurred. The resulting numbers were then used to calculate the adjusted proportion of animals at each ice concentration. See Table 1 for explanation of species' abbreviations

day. I included corrections for these factors in my analysis, as well as for the fact that mammal distribution could be affected not just by ice characteristics but also by food availability. Gilbert and Erickson (1977) hoped that chlorophyll-a concentration could be used to indirectly indicate food availability (assuming a relationship with productivity in the water column), but probably because too many trophic levels lie between algae and seals in the food web, they could find no correlations to explain seal distribution. While I did not measure food availability directly, I assumed that it is increased for upper-trophic-level predators in frontal areas (see Brown 1980). Biggs' (1982) data indicate that this assumption is valid in the Ross Sea frontal systems.

I took the ice-type preferences of various mammals into consideration (as shown in Table 2) when drawing occurrence zones for later calculation of abundance. Ice-type preferences were thus automatically incorporated into 'crude estimates' of abundance (Table 4). Adjustment for haul-out pattern was basically similar to the procedure of Erickson et al. (1971), although they used slightly different correction factors. In my data (Table 3), the increased frequency of animals hauled out as a function of time was

Table 3. Percentage of seals hauled out by local time of day in 3- h intervals

Local time	Species[a]			
	SECR	SELE	SERO	SEWE
0000–0259 hours	9.0	16.7		22.1
0300–0550 hours	20.7	8.3		19.1
0600–0859 hours	18.0	16.7		2.9
0900–1159 hours	13.7	8.3	14.3	13.2
1200–1459 hours	20.7	25.0	42.9	19.1
1500–1759 hours	9.8		28.6	16.2
1800–2059 hours	4.7	8.3	14.3	5.9
2100–2359 hours	3.5	16.7		1.5

a See Table 1 for explanation of abbreviations. Census results are affected by time of day: SECR (χ^2 = 28.21, P < 0.01), SELE (χ^2 = 6.50, P < 0.50), SERO (χ^2 = 9.00, P > 0.50), SEWE (χ^2 = 22.33, P < 0.01); Sokal and Rohlf (1969)

Table 4. Population estimates of marine mammals in the Ross Sea

Species	Crude estimate	Adjusted estimate	Population weight (kg) (times 100)	Biomass (g m^{-2})
SEWE	25,594	31,992	6,398	0.011
SELE	6,390	7,988	2,396	0.004
SECR	153,157	203,699	30,555	0.051
SERO	3,368	5,052	758	0.001
SESE	40	40	32	<0.001
WMIN	5,721	14,277	999,390	0.167
KILL	3,442	3,442	88,290	0.015

See Table 1 for explanation of abbreviations.
Determine mean density of transects in zones of equivalent abundance, then determine area (km^2) of each zone; add the number of animals among all zones.
Population estimate for seals adjusted upward to compensate for periods when a large proportion of seals are feeding (see Table 3) and thus more difficult to detect: SECR increased by 33%, SELE by 25%, SERO by 50%, and SEWE by 25%. Adjustment to population estimates for whales is explained in the text.

statistically significant only for Crabeater Seals, *Lobodon carcinophagus*, and Weddell Seals, *Leptonychotes weddellii*, the two species for which I had a sufficiently large number of sightings. However, because the peak haul-out periods that I observed were identical to those described for each species by Gilbert and Erickson (1977), I made corrections for all species.

4.2.1 Weddell Seal, *Leptonychotes weddellii* (SEWE in Tables).

A total of 58 individuals of this species was observed, only in the pack-ice and only in ice overlying the continental shelf and slope (Fig. 5). Furthermore, the species occurred only in pack-ice within about 200 km of the fast-ice areas where it breeds (i.e., along the coast of Victoria Land and Ross Island, the Bay of Whales (ca. 78° 30′S, 165°E), and probably along the coast of the Edward VII Peninsula (see Brown et al. 1974). Only 17.3% of all individuals occurred where the ice was < 3 oktas (Table 2). Fewest Weddell Seals were hauled out during the late evening (Table 3).

I estimated that about 32,000 Weddell Seals occurred in the Ross Sea. This is equivalent to 0.054 individuals per km^2, and a biomass of 0.011 g m^{-2} (Table 4). This may be an under-estimate, because an appreciable number of individuals should frequent some fast-ice areas that still exist in early summer. Actually, however, a significant portion of fast-ice in the Ross Sea disintegrates into pack-ice by early January. Indeed, Stirling (1969) estimated a population of 50,000 Weddell Seals in the Ross Sea, not including any in the vicinity of the King Edward VII Peninsula. On the other hand, these estimates are much higher than what would be expected relative to the data of Gilbert and Erickson (1977). They estimated a population of only 111,000 Weddell Seals in the entire Pacific sector of the Antarctic, but did not consider that appropriate habitat for this species exists along the Victoria Land coast (cf. their Fig. 7, p. 714: 1977).

4.2.2 Leopard Seal, *Hydrurga leptonyx* (SELE in tables).

A total of 15 Leopard Seals was sighted (Table 1), all in the pack-ice. Although most sightings were in frontal areas, it is likely that the species occurred in low numbers throughout the pack-ice region (Fig. 6). Most sightings were in the pack-ice of the western Ross Sea. This area supports one of the densest concentrations of Adélie Penguins, a frequent prey species of this seal. The Leopard Seal, however, tended to prefer pack-ice of $\leq$ 4 oktas (Table 2), whereas the Adélie Penguin showed no such preference (Ainley et al. 1983). Fewest Leopard Seals were hauled out during the early evening (Table 3), a period when Adélie Penguins are also in the water feeding (Ainley et al. 1983).

I estimate that about 8,000 Leopard Seals occurred in the Ross Sea, and that the species' density was about 0.013 individuals per km^2. This translates to a biomass of 0.004 g m^{-2}. My estimate is at an order of magnitude compatible with the results of Gilbert and Erickson (1977), who estimated that there were 72,000 Leopard Seals in the Pacific sector.

4.2.3 Crabeater Seal, *Lobodon carcinophagus* (SECR in tables).

A total of 241 individuals of this species was sighted (Table 1). All except 4 individuals were in the pack-ice, but even they were in open water within 5 km of the pack. The Crabeater Seal showed a marked preference for pack-ice, and within the pack tended to occur mostly where the ice was ≥ 3 oktas (Table 2). Such a preference for relatively heavy ice concentrations is interesting, given the species' tendency to occur in the vicinity of the pack-ice edge (where ice-floe size is relatively small). Perhaps its preference for relatively heavy ice is a response to the preference for lighter ice of two important predators, the Killer Whale, *Orcinus orca,* and the Leopard Seal (Siniff 1980; Smith et al. 1981). Crabeater Seals were also markedly concentrated over the Antarctic Slope Front (Fig. 7). Fewest animals were hauled out during the morning (Table 3).

I estimated that about 204,000 Crabeater Seals occurred within the confines of the Ross Sea, and that the species' density was about 0.341 animals per km^2. This translates to a biomass of 0.051 g m^{-2} which is second only to that of the Minke Whale, *Balaenoptera acutorostrata* (Table 2). Gilbert and Erickson (1977) estimated 1.67 times 10^6 Crabeater Seals in the entire Pacific sector of the Antarctic. My estimate is compatible with theirs, given the relative size of the Ross Sea (cf. their Fig. 7, p. 714:1977).

4.2.4 Ross Seal, *Ommatophoca rossi* (SERO in tables).

Only two Ross Seals were seen in the Ross Sea. However, a concentration of animals occurred just to the north of the continental slope in the vicinity of 70°S, 180°W (Fig. 8). This was an area where a concentration of Crabeater Seals

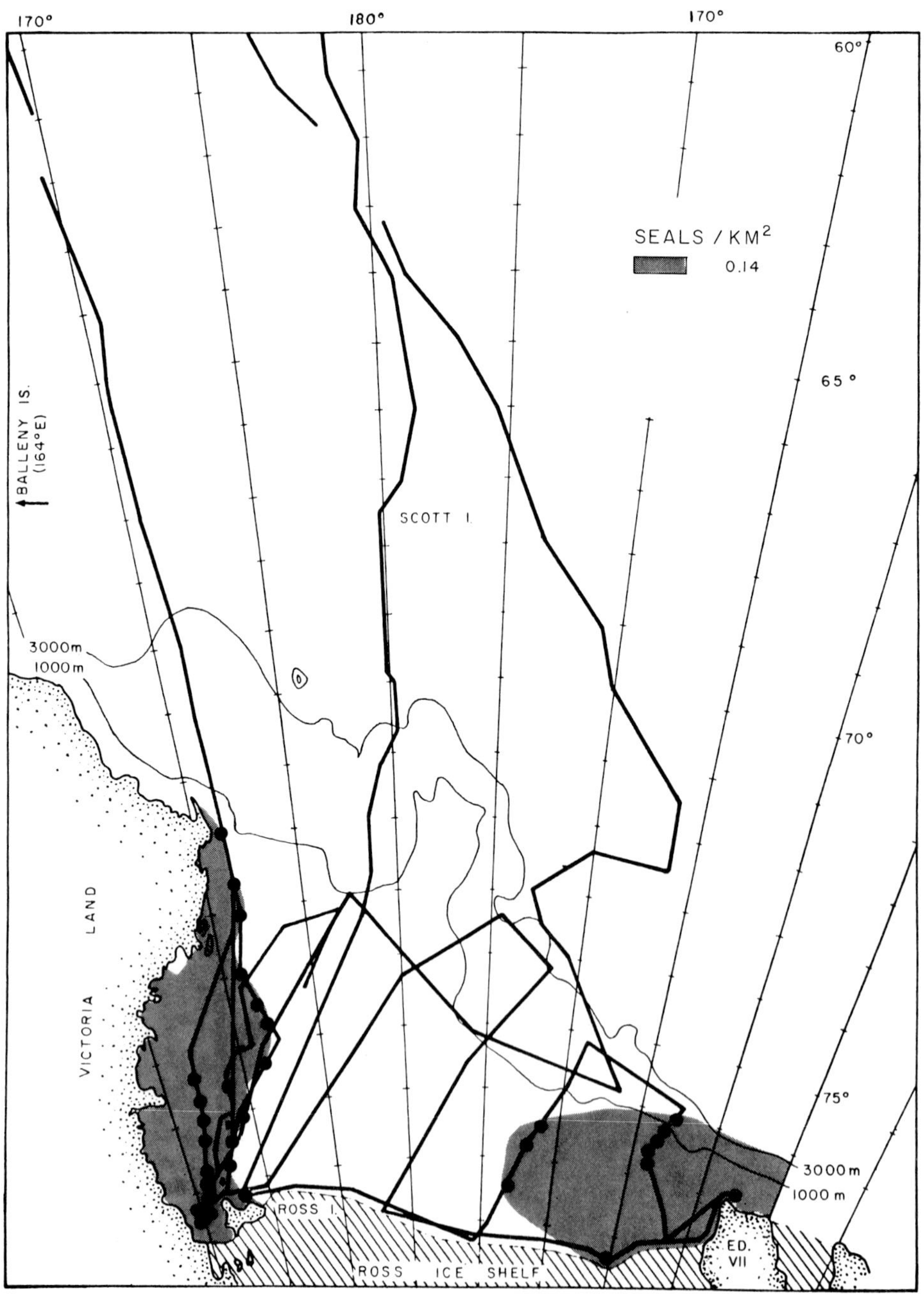

Fig. 5. The distribution of Weddell Seals. *Dots* indicate positions of sightings

occurred as well (Fig. 7), and, thus, I suspect that the area may have been locally productive. I projected the species to occur at very low densities throughout the pack-ice of the outer Ross Sea and adjacent parts of the South Pacific. Ross Seals were not observed where the ice was ≥ 4 oktas (Table 2), and none was seen between 2100 hours and 0800 hours (Table 3).

I estimated a population of about 5,000 Ross Seals in the Ross Sea. This is equivalent to a density of 0.008 individuals per km^2 and a biomass of 0.001 g m^{-2}. My population estimate is slightly lower than what would be expected. based on Gilbert and Erickson's (1977) estimate of 101,000 animals for the entire Pacific sector. It appears, though, that Ross Seals are rather irregularly distributed with high concentrations in localized areas (Hall-Martin 1974), and that the Ross Sea is apparently not an area where this species concentrates.

4.2.5 Southern Elephant Seal, *Mirounga Leonina* (SESE in tables).

One Elephant Seal was seen in ice-free waters over the continental slope at about 75°S, 175°W on 31 December 1976. The species has been seen increasingly, but in small numbers, at various sites in the Ross Sea (e.g. Brownell and Ainley 1976). I estimated that perhaps 40 animals occurred in the Ross Sea at any one time during the early summer (Tables 1 and 4).

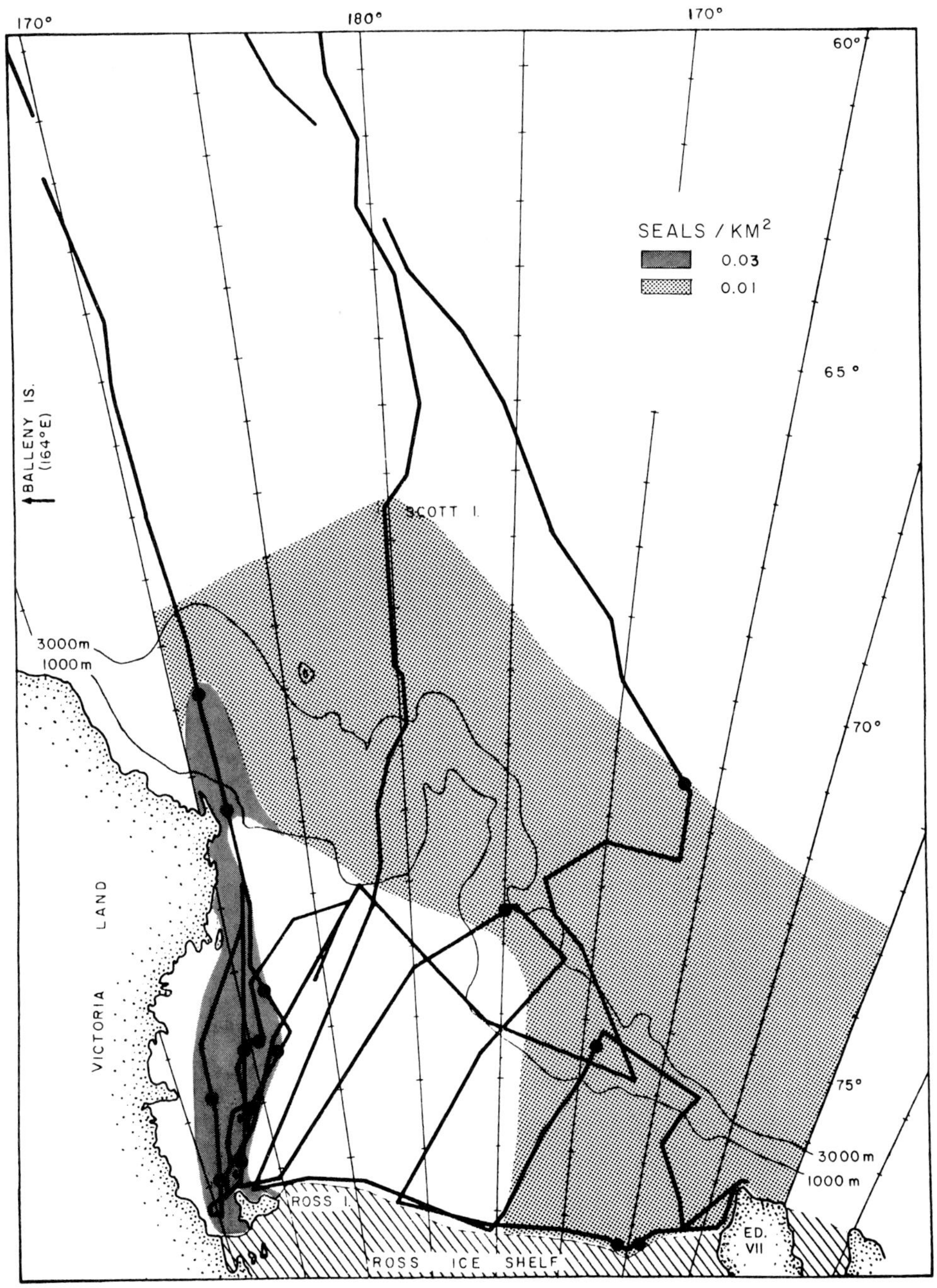

Fig. 6. The distribution of Leopard Seals. *Dots* indicate positions of sightings

4.2.6 Total Biomass of Seals

The total biomass of seals in the Ross Sea, a figure derived by summing the contributions of all species, was 0.068 g m^{-2} (Table 4). The estimate derived by summing the biomass of 2.5° lat. times long. blocks was 0.089 ± 0.190 g m^{-2}. Thus, as with the birds, agreement was relatively close for the two methods. Using figures derived by the second method, there was not statistical difference between the biomass of birds and seals (t = 1.761, 144 df). As in the avian community, seal biomass was concentrated in frontal areas, but unlike bird biomass, it was restricted to areas of pack-ice or fast-ice (Fig. 9). Seals, as in the case of birds, were especially concentrated where pack-ice coincided with frontal zones. In fact, considering only the 2.5° blocks that included ice, a significant correlation existed between bird and seal biomass (r = 0.6705, 40 df).

4.3 Whales

4.3.1 Sei Whale, *Balaenoptera borealis.*

Two individuals were seen at the edge of the pack-ice in the vicinity of Scott Island on 15 December 1976. None was seen elsewhere.

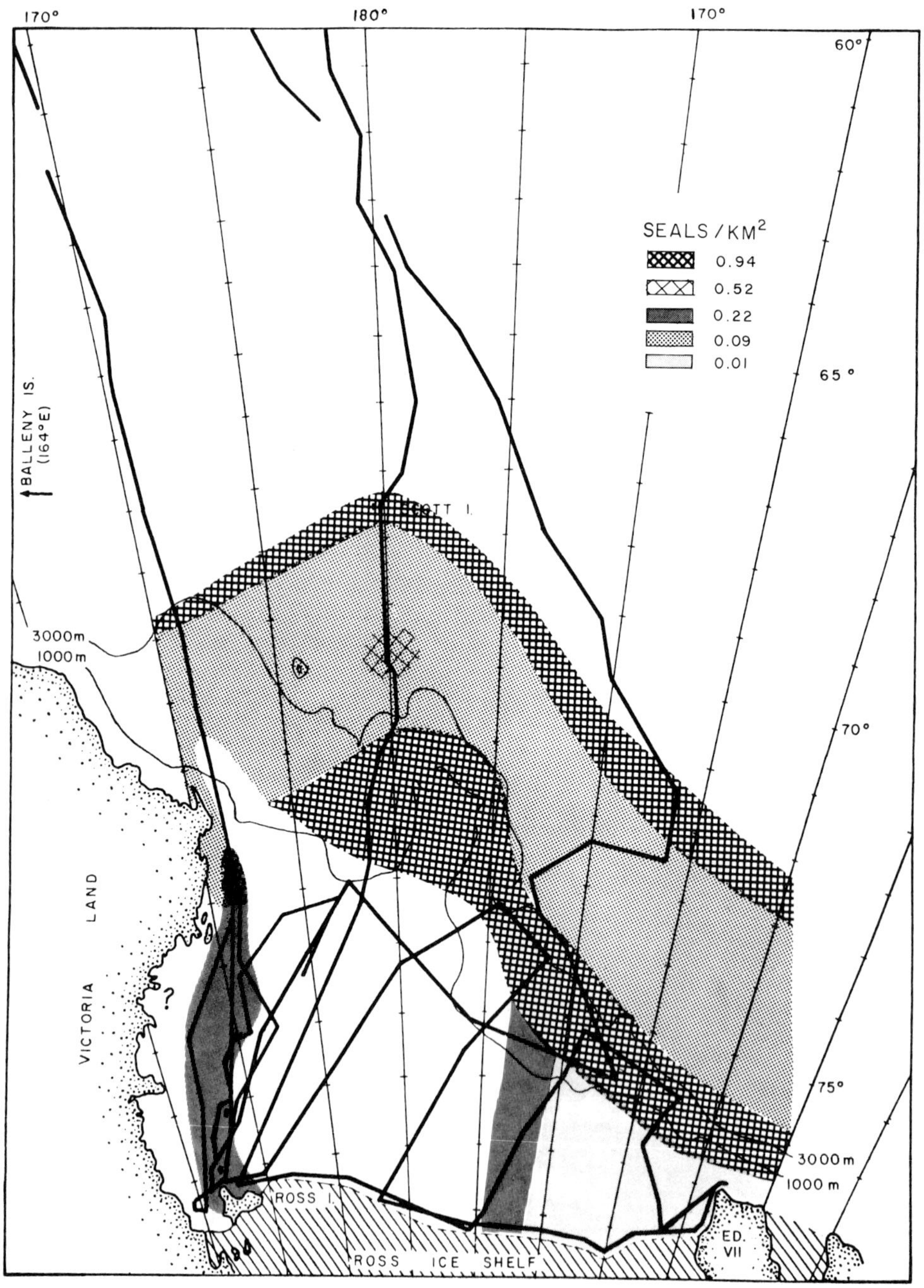

Fig. 7. The distribution of Crabeater Seals. Sightings were too numerous to show individually

4.3.2 Minke Whale, *Balaenoptera acutorostrata* (abbreviated WMIN in the tables).

Ice concentration was an important factor in the pattern of occurrence of this species, because 65.5% of individuals seen were in ice of ≤ 1 okta or in open water within 5 km of the ice. Adjusting these percentages by the number of censuses in various ice concentrations, indicated that only 5.8% of the animals were in open water and 67.4% were in ice of ≤ 3 oktas (Table 2). Only 13.4% were seen in ice of ≥ 5 oktas. Light ice concentration was characteristic of the marginal ice zone (Fig. 3) and, thus, essentially, the Minke Whale was a species of the ice-edge habitat. The sightings data also support this fact.

Another important factor governing the occurrence of this species was the Antarctic Slope Front and the frontal activity in the western Ross Sea associated with the ice edge and the northward flowing current there. Virtually no whales were observed within the centre of the plankton bloom (Figs. 2 and 10).

A total of 56 Minke Whale schools was observed: 44 in the western Ross Sea and 12 elsewhere in the Ross Sea (Fig. 10). Mean school size ($\bar{s}$) was 1.97 ± 2.50 (S.D.) which was not statistically different from the $\bar{s}$ of 2.19 ± 0.32 calculated by Butterworth and Best (1982) for approximately the same region in 1980–81 (t = 0.0324). The largest school I observed contained 29 animals, in the vicinity of Franklin Island, but most sightings were of 1 or 2 animals.

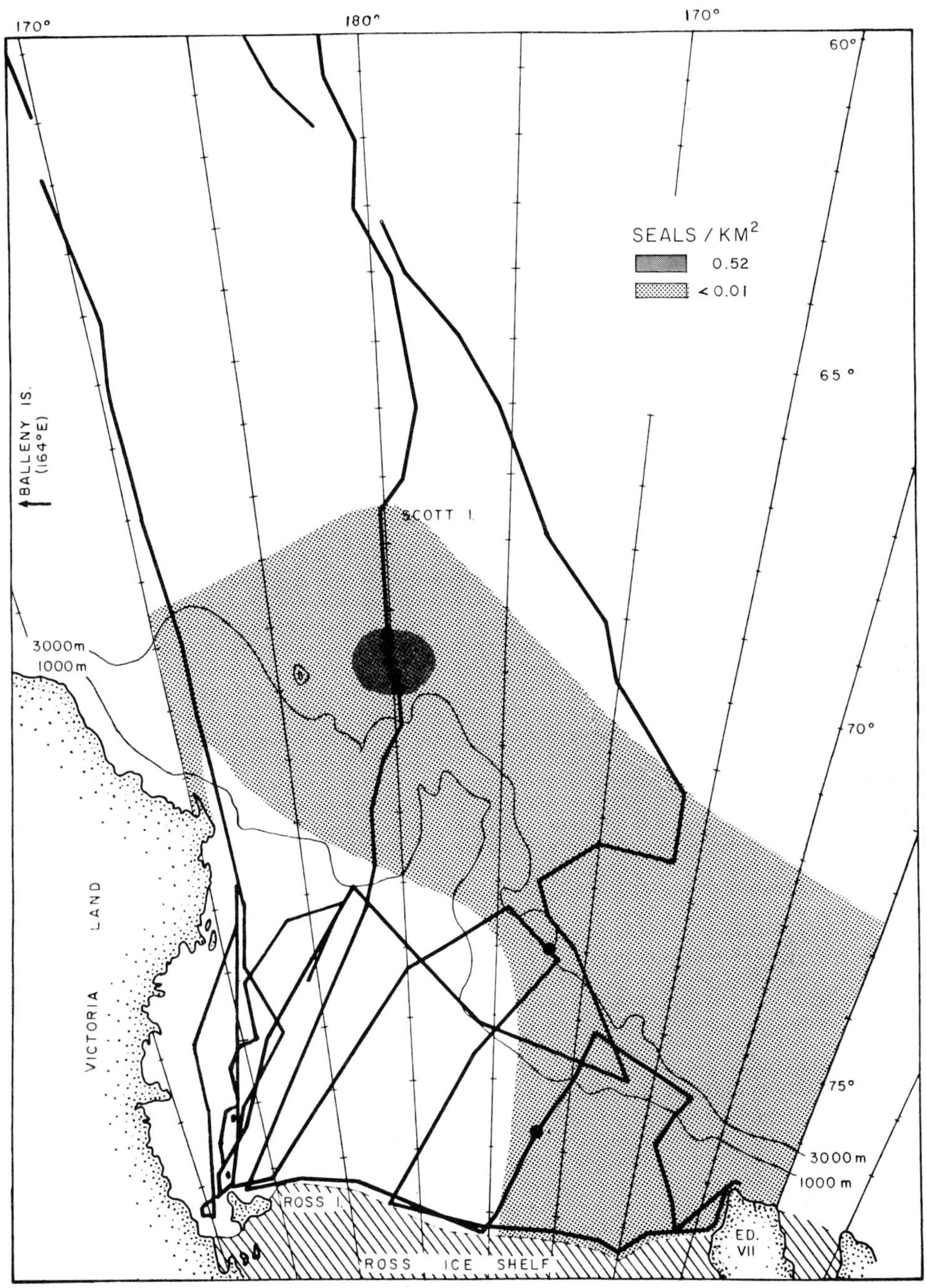

Fig. 8. The distribution of Ross Seals. *Dots* indicate positions of sightings

School density in the western Ross Sea was 0.0231 schools per km^2 of area surveyed (1,907.3 km^2), and in the remainder of the Ross Sea area where the species occurred (1,281.8 km^2 surveyed), it was 0.0094. Given that Minke Whales occurred in an area of 88,746 km^2 in the west and 91,534 km^2 elsewhere in the Ross Sea, the number of schools (N′) was 2,904. Butterworth and Best (1982) used the following formula to estimate Minke Whale abundance:

$$Pop = (e)\,(m)\,(h)\,(Pop'),$$

where

e = 1.87, a correction for biases in estimates of direction and distances of sightings;
m = 0.96, a correction for whale movement;
h = 1.39, a correction for whales missed; and
$Pop' = \bar{s}\,N'$.

They estimated a standard deviation (S.D.) for their population estimate by summing the S.D. estimates for $\bar{s}$ and N′; as they pointed out, however, the S.D. of N′, which I was not able to estimate with my analysis technique, added very little to the total S.D. of their population estimate. Using the above formula and correction factors (as explained in the Methods), I estimated a Ross Sea population of 14,277 ± 18,118 whales (Table 4), virtually all of which (97%) were adults. This was equivalent of an overall density in the Ross Sea of 0.0239 whales per km^2 and a biomass of 0.167 g

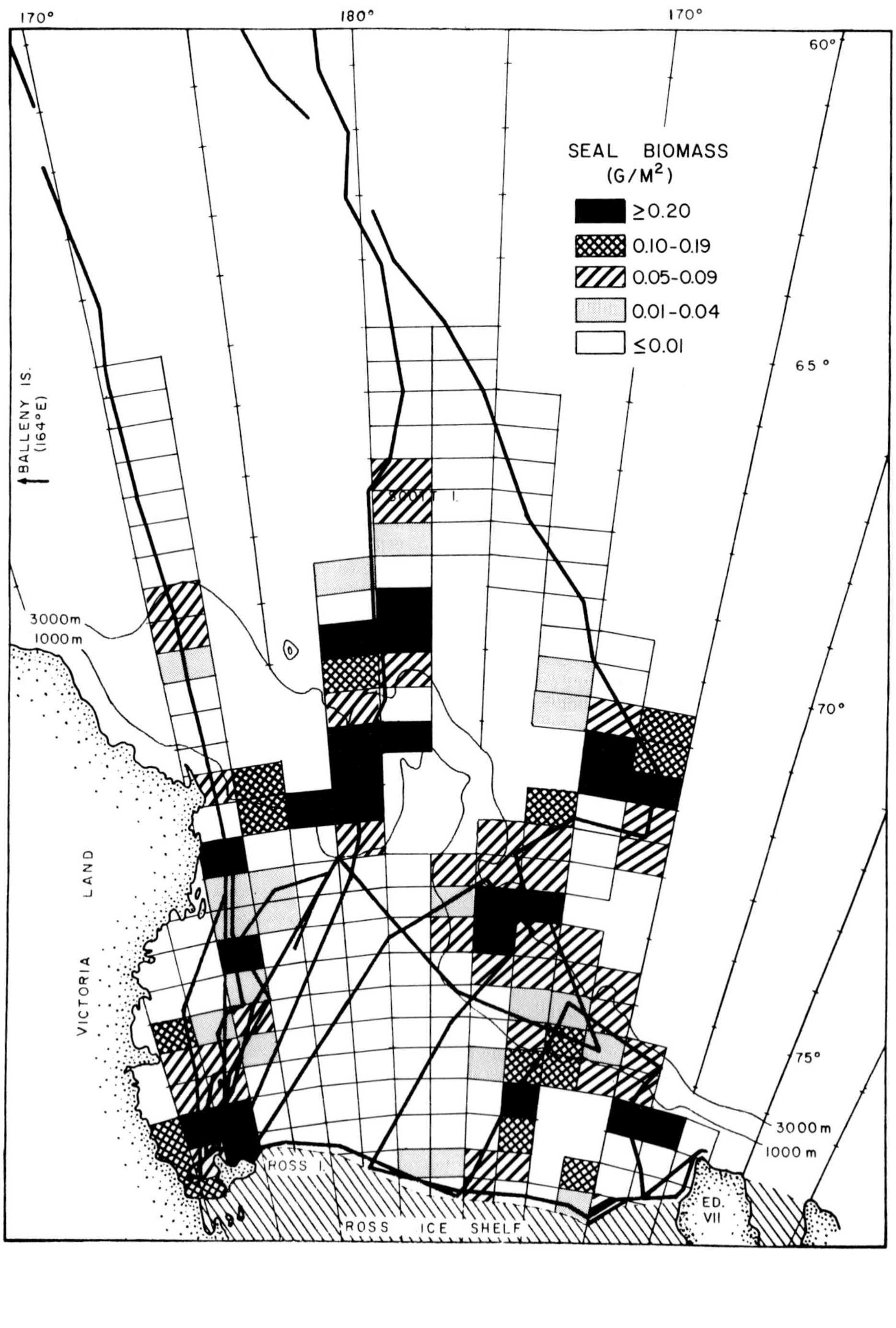

Fig. 9. The distribution of seal biomass in the Ross Sea region by 2.5° lat. times long. blocks

m^{-2}. In addition, Minke Whales occurred at the highest observed densities in a narrow strip of sparse ice concentration at the outer edge of the pack-ice in the South Pacific (Fig. 10), but, of course, these individuals were not included in the Ross Sea biomass estimates.

Butterworth and Best (1982) estimated a stock of 104,911 ± 16,600 adult Minke Whales in the entire region between 160°E and about 170°W and between 62° and 78°S, an area including, but about 2.5 times larger than, the Ross Sea. One reason for the discrepancy between their abundance estimate and mine might be that few whales had moved into the Ross Sea during December when most of my censuses were made: the majority of Butterworth and Best's (1982) censuses were more than a month later, and the significant band of heavy ice that earlier had closed the Ross Sea from the Pacific was gone. Another possible reason is that Butterworth and Best's (1982) estimate might be high, because their determination of school distribution (and numbers), by "contouring", extrapolated their sightings along the ice edge and the Antarctic Slope Front to areas where the whales do not generally occur and where their cruise track did not allow them to sample. Indeed, most of their cruise track was concentrated near to, or within, frontal areas and the pack-ice edge (cf. Figs. 2 and 3 and their Fig. 1a). A third possibility is that the greater statistical sophistication of their technique leads to a higher (and presumably more accurate) estimate, or that the difference between the two estimates is an artifact of their

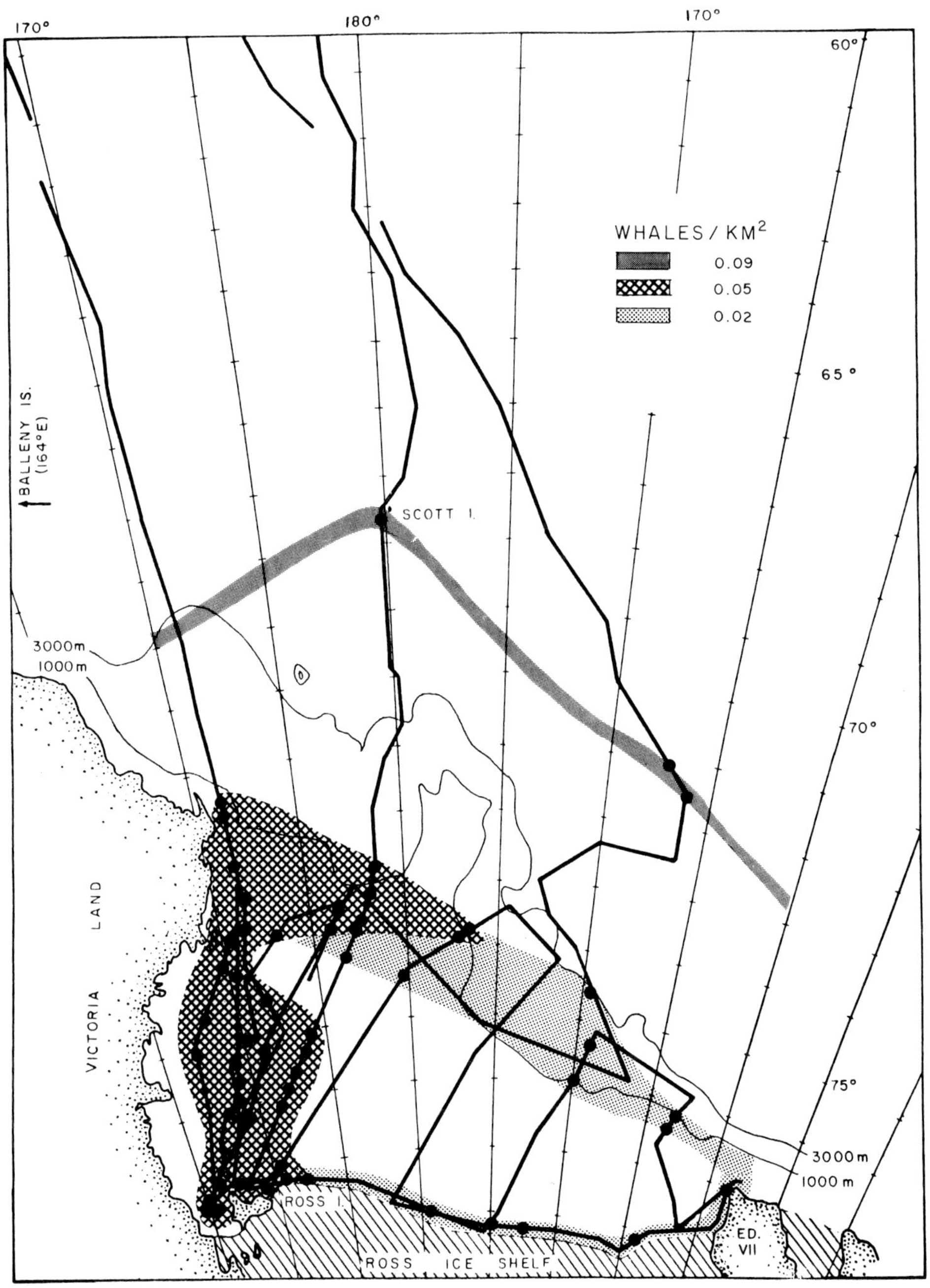

Fig. 10. The distribution of Minke Whales. *Dots* indicate positions of sightings

"contours" and my broader "zones". I do not believe I missed any more whales than did the observers contributing to the data used by Butterworth and Best (1982; see Methods).

4.3.3 Killer Whale, *Orcinus orca* (abbreviated KILL in Tables).

Except for 1 sighting of a pod of Killer Whales in the South Pacific, well north of the pack-ice at 65°S, 176° 30′W, and outside the census strip, all Killer Whales were observed over the continental shelf and at or within the pack-ice edge (Fig. 11, Table 2). Six schools were in ice ≤ 1 okta and 1 was in 6 oktas of pack-ice. Three of the seven schools occurred within the region of the Antarctic Slope Front, and another three were seen in the vicinity of the ice edge in the western Ros Sea near Ross Island. The animals were probably responding to both ice edge and the fronts, because none was encountered at ice edges away from the frontal areas (e.g., at the ice edge over the bloom in the southeastern Ross Sea). In the vicinity of Ross Island, Killer Whales approached the ship on several occasions when it was stationary and censuses were not being conducted. On some other instances, it was also obvious that schools approached the ship even when steaming. Leatherwood et al. (1982b), who reported cetacean sightings from three cruises spanning the breadth of South Pacific polar waters,

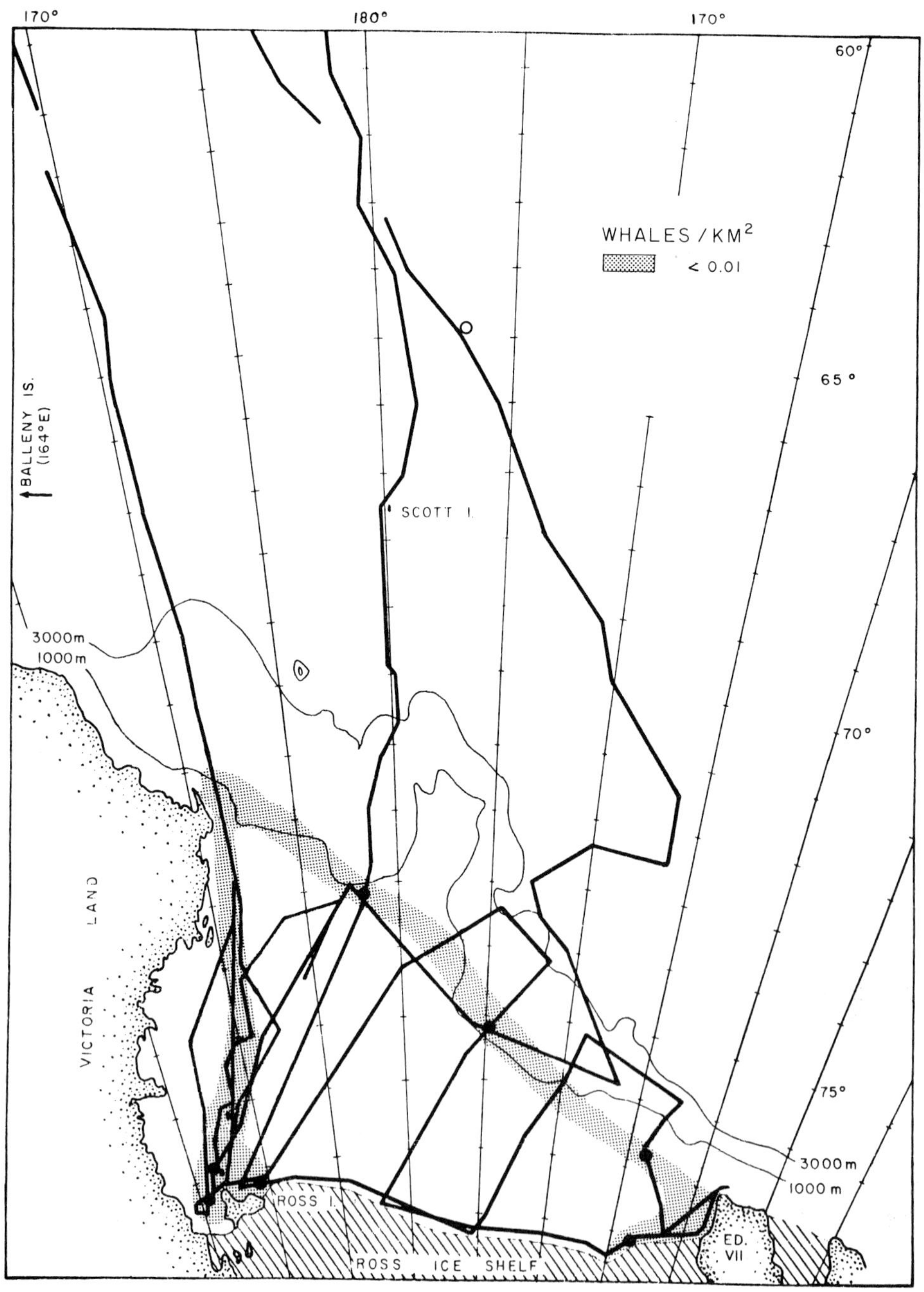

Fig. 11. The distribution of Killer Whales. *Dots* indicate positions of sightings during censuses. *Circle* indicates a sighting made outside the transect boundaries

logged seven of their nine sightings of this species in the vicinity of Ross Island.

A total of 54 individuals was observed, including 7 males, 38 female-sized individuals, and 17 sub-adults. Mean school size was 10.29 ± 8.52 whales. In the area of occurrence (74,342 km^2; Fig. 11), the area surveyed was 1,551 km^2, school density was 0.0045 per km^2, and the number of schools was estimated to be 335. Using the formula, Pop $= \bar{s}\ N'$, an estimated 3,442 ± 2,850 Killer Whales occurred in the Ross Sea. If anything, this is an over-estimate because of the observed tendency of Killer Whales to approach the ship. This population estimate, partitioned by size (sex), translated to a biomass of 0.015 g m^{-2} in the Ross Sea during the early summer (Table 4). In the entire Pacific sector of the Antarctic (130°E–60°W) south of 65°S, Ohsumi (1981) estimated a seasonal peak number of 7,340 Killer Whales during January.

4.3.4 Total Biomass of Whales

In total, whale biomass was concentrated over the slope front, in the frontal area of the western Ross Sea, and, especially, in the vicinity of the ice edge within the latter two areas (Fig. 12). Few whales were observed within the heavy pack-ice northeast of the slope front and along the Victoria Land coast, and none was observed in the open water of the central Ross Sea except over the Pennell Bank.

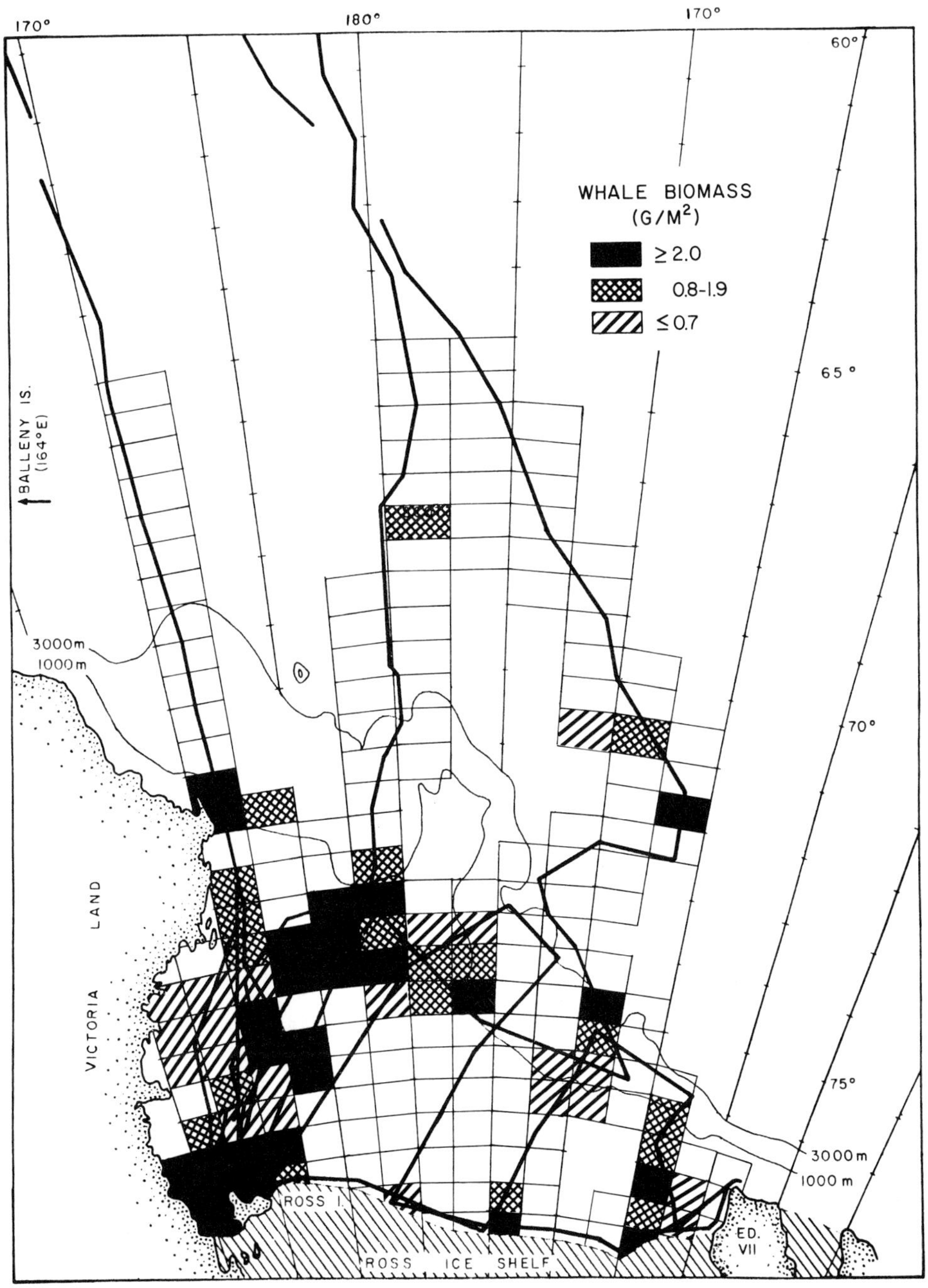

Fig. 12. The distribution of whale biomass in the Ross Sea region by 2.5° lat. times long. blocks

The average total biomass by blocks was 0.762 ± 1.377 g m^{-2}. This figure is much higher than total biomass estimated by the method above for at least two reasons, both of which are artifacts of sampling/analysis. (1) The large block size and the large weight of a single whale could completely swamp results for a single block; and, (2) the area of a block decreases with increasing latitude and, thus, its relative contribution to the total average increases. These factors, and the many blocks having 0 whale biomass, also produced the high standard deviation. The relative rarity of whales made the use of 2.5° blocks less satisfactory than for the more abundant and more evenly distributed seals and birds.

5 Discussion

The biomass of birds, seals and whales was each concentrated over the western Ross Sea and over the continental slope (Figs. 4, 9, and 12). Mammals, like birds, were largely absent from the central and south-central Ross Sea, and occurred in reduced numbers in the pack-ice north of the slope front. Thus, the distributions of all higher vertebrates in the Ross Sea region were strongly affected by the location of frontal areas. Within these areas, mammals and, to a lesser extent, birds sorted themselves by ice concentrations (see Ainley et al. 1983, for birds). The edge of the ice-pack was also a habitat where mammals and birds were concentrated, but whether this was a response to increased productivity, remains to be determined. The western and northern edges of an extensive, persistant algal bloom coincided with both the frontal areas and with the pack-ice edge. The factors generating and confining this bloom to shelf waters of the Ross Sea require study as well.

Compared to Laws' (1977) estimates, my estimates of biomass were significantly higher for birds (0.025 vs. 0.044–0.070 g m^{-2}), the same for seals (0.068 vs. 0.068 –0.089 g m^{-2}), and significantly lower for whales (0.399+ vs. 0.167+ g m^{-2}). As explained in Ainley et al. (1983), the discrepancy in bird biomass is due to the fact that aerial species in the Antarctic are much more abundant and are much less restricted to waters immediately adjacent to breeding sites than has generally been appreciated. The similarity in biomass estimates for seals is not surprising, given that Laws relied principally on the studies of Erickson and his co-workers and that the results for seals in the present study were by and large compatible with those of Gilbert and Erickson (1977) for the South Pacific sector. As for whales, explaining the discrepancy between Laws' estimate and mine is problematic because both are significantly lower than what would be expected from the currently acceptable population figures for the Ross Sea (i.e., Butterworth and Best 1982). The population equivalent of Laws' biomass estimate of 0.399 g m^{-2} (at 7,000 kg per whale) is 34,086 whales. This, interestingly, is similar to my estimate plus 1 S.D., or about 32,400 whales (a biomass of 0.379 g m^{-2}, or 0.394 g m^{-2} including Killer Whales). The similarity, however, is superficial because Laws intended his estimate to represent the annual peak in the whale population, which is reached during late January and February in the Antarctic; the estimate of Butterworth and Best (1982) also represents the population peak. My estimate is essentially for mid-December to early January. During the earlier season, when I conducted the large majority of my counts, a large number of whales was obviously blocked from entering the Ross Sea by a significant band of heavy pack-ice (Figs. 3, 10, and 12). Accounting for those whales might increase my estimate. In the Minke Whale section earlier in this paper, I discussed other possible reasons why my whale population estimates are lower than those of Butterworth and Best (1982).

Birds, unlike Antarctic marine mammals, are constrained by breeding duties to maintain a tie to land. Except for the Weddell Seal, and the general constraint of Antarctic seals to associate with ice, mammals should be free to move almost anywhere over the ocean. Nevertheless, within the Ross Sea region, both birds and mammals concentrated in the same areas, no doubt because of mutual attraction to highly productive waters. This extreme discontinuity in the distribution of higher vertebrates indicates that, within the Antarctic marine ecosystem, prey are probably not everywhere equally available, and that the size of habitat patches where prey are available can be rather large. It further indicates that, in order to discuss competition or relative ecological roles among upper-trophic-level predators (Laws 1977), it may be more valid to consider discrete areas rather than trying to extrapolate to the entire Antarctic marine ecosystem. On the local level, results presented here indicate that studies of the abundance and distribution of higher vertebrates at sea should define species' habitats in terms of biological as well as physical oceanographic factors. To do this effectively, more expeditions that integrate multiple facets of biological oceanography are needed.

Acknowledgements. I thank the officers and crews of the USCG ice-breakers NORTHWIND and BURTON ISLAND. United States Antarctic Research Program personnel from the National Science Foundation (NSF) and Holmes and Narver, Inc., particularly A.F. Betzel and M.A. Lanyon, helped to co-ordinate logistical needs. A. Amos, R. Boekelheide, G. Divoky, E. O'Connor, and S. Jacobs provided assistance and guidance on cruises, and L. Karl, P. Geis and E. Tuomi assisted in data analysis and manuscript preparation. I also appreciate the assistance of D. Butterworth and D. Miller in improving an early draft of the manuscript. S. Jacobs provided unpublished data on temperature/depth profiles along various cruise tracks. This project was funded by NSF Division of Polar Programs, Grants DPP-7615358, DPP-7616358 A01, DPP-7820755, DPP-780755 A01, and DPP-7920796. This is contribution number 260 of the Point Reyes Bird Observatory.

References

Ainley DG, Jacobs SS (1981) Seabird affinities for ocean and ice boundaries in the Antarctic. Deep-Sea Res 28:1173–1185

Ainley DG, O'Connor EF, Boekelheide RJ (1983) The marine ecology of birds in the Ross Sea, Antarctica. Am Ornithol Union Monogr No. 32

Biggs DC (1982) Zooplankton excretion and NH cycling in near-surface waters of the Southern Ocean. I. Ross Sea, austral summer 1977–1978. Polar Biol 1:55–67

Brown RGB (1980) Seabirds as marine animals, pp 1–39. In: Burger J, Olla BL, Winn HE (eds) Behaviour of marine animals, vol 4: Marine Birds, Plenum, New York

Brown SG, Brownell RL Jr, Eickson AW, Hofman RJ, Llano GA, Mackintosh NA (1974) Antarctic mammals. Ant Map Folio Ser, Folio 18. Am Geogr Soc New York

Brownell RL Jr, Ainley DG (1976) Southern elephant seals in the Ross Sea. Ant J US 11:189–190

Butterworth DS, Best PB (1982) Report of the Southern Hemisphere mine whale assessment cruise, 1980/81. Rep Int Whal Commn 32:835–873

Cline DR, Siniff DB, Erickson AW (1969) Summer birds of the pack ice in the Weddell Sea, Antarctica. Auk 86:701–716

El-Sayed SZ, Biggs DC, Stockwell D, Warner R, Meyer M (1978) Biogeography and metabolism of phytoplankton and zooplankton in the Ross Sea, Antarctica. Ant J US 13:131–134

Erickson AW, Siniff DB, Cline DR, Hofman RJ (1971) Distributional ecology of antarctic seals, pp 55–76. In: Deacon G (ed) Symposium on Antarctic ice and water masses. SCAR, Cambridge

Gilbert JR, Erickson AW (1977) Distribution and abundance of seals in the pack ice of the Pacific sector of the Southern Ocean, pp 703–740. In: Llano GA (ed) Adaptations within Antarctic ecosystems. Gulf Publ, Houston, Texas

Hall-Martin AJ (1974) Observations on population density and species composition of seals in the King Haakon VII Sea, Antarctica. Suid-Afrikaanse Tydskrif Ant. Navors Nas Mus 4:34–39

Hayes DE, Davey FJ (1974) Geophysical study of the Ross Sea, Antarctica, pp 887–907. In: Initial reports of the deep-sea drilling project, No. 28

Laws RM (1977) The significance of vertebrates in the Antarctic marine ecosystem, pp 411–438. In: Llano GA (ed) Adaptations within Antarctic ecosystems. Gulf Publ, Houston, Texas

Leatherwood S, Awbrey FT, Thomas JA (1982a) Minke whale response to a transiting survey vessel. Rep Int Whal Commn 32: 795–802

Leatherwood S, Todd FS, Thomas JA, Awbrey FT (1982b) Incidental records of cetaceans in southern seas January and February 1981. Rep Int Whal Commn 32:515–520

Mackintosh NA (1973) Distribution of postlarval krill in the Antarctic. Discovery Rep 36:95–156

Ohsumi S (1981) Distribution and abundance of killer whales in the Southern Hemisphere. Rep Int Whal Commn 31:1-9

Siniff DB (1980) Seal population dynamics and ecology. J R Soc NZ 11(4):317–327

Smith TG, Siniff DB, Reichle R, Stone S (1981) Coordinated behaviour of killer whales (*Orcinus orca*) hunting a crabeater seal (*Lobodon carcinophagus*). Can J Zool 59:1185–1189

Sokal RR, Rohlf FJ (1969) Biometry. WH Feeman, San Francisco

Stirling I (1969) Distribution and abundance of the Weddell seal in the western Ross Sea, Antarctica. NZ J Mar Freshwater Res 3: 191–200

Sverdrup HE, Johnson MW, Fleming RH (1942) The oceans: their physics, chemistry and general biology. Prentice-Hall, Englewood Cliffs, New Jersey

Relationships Between Prey Life-Cycles and the Extent, Nature and Timing of Seal and Seabird Predation in the Scotia Sea

J.P. CROXALL, P.A. PRINCE, and C. RICKETTS[1]

Summary. The impacts of seals and breeding seabirds on prey resources in the Scotia Sea are assessed, and for seabirds seasonal changes are modelled. Although the data available are the best for any part of the Southern Ocean important deficiencies exist, which are summarized. Total consumption is estimated as 23×10^6 t, fairly evenly divided between seabirds and seals and between South Georgia and the rest of the area. Krill, *Euphausia superba,* (70%) is much more important than squid (16%) or fish (8%) but there are few data on winter diets. The main consumers of Krill are Crabeater Seals, *Lobodon carcinophagus,* Macaroni, *Eudyptes chrysolophus,* and Chinstrap, *Pygoscelis antarctica,* Penguins, with penguins accounding for 76% of the intake by birds. Peak demand is in February; consumption estimates are high compared with stock assessments and reasons for this are discussed. Despite variations between seasons, species' and sites, Krill (mostly sexually mature individuals, 35–45 mm long) taken by predators and by net-hauls in the southern Scotia Sea are generally similar, although the largest Krill are absent from predator samples. At South Georgia, however, most predators take 45–55 mm long Krill which are rare in net-hauls; possible causes and implications of this important discrepancy are discussed. The significance of differential predation on male and female Krill (bias favouring males in the south and females at South Georgia) is evaluated in terms of seasonal changes in energy content (due to lipogenesis), and it is suggested that the latter might be a factor influencing timing of breeding in certain predators. Other aspects of krill-predator interactions, including diving patterns and catching rates, are reviewed; more data on Krill demography and on seasonal variations in the age, sex and status of Krill taken by predators are required to assess critically predator-prey interactions. Elephant Seals, *Mirounga leonina,* take perhaps 75% of squid eaten by predators. A comparison of quantitative data on squid diets of Sperm Whales, *Physeter catodon,* albatrosses and certain seals shows the importance of *Mesonychoteuthis, Moroteuthis* and *Kondakovia,* although other taxa are as important for certain species (e.g. octopods for seals). All these squid are rarely caught by net-hauls and data from predators presently provide the best conspectus of the abundance of the Southern Ocean epi- and meso-pelagic squid of potential commercial importance. Fish are less important to predators than Krill and squid, although the total consumption (77% by seals, especially Elephant, Fur, *Arctocephalus gazella,* and Weddell, *Leptonychotes weddelli,* Seals) is a substantial proportion (e.g. at least 38% for *Notothenia rossii* at South Georgia) of estimated standing stocks, which may therefore be appreciably under-estimated. Seabirds (cormorants, penguins) mainly take immature nototheniids; Fur Seals mostly eat adult *Champsocephalus* and are thus now in direct potential competition with commercial operations.

1 British Antarctic Survey, NERC, High Cross, Madingley Road, Cambridge CB3 OET, United Kingdom

1 Introduction

Knowledge of the nature of predator-prey relationships is fundamental for understanding the functioning of any ecosystem. In marine systems there are major difficulties in determining quantitatively the identity and status (age, sex) of the main prey and in assessing the significance of the predation in terms of prey population dynamics. Without such information the impact of predators cannot be adequately determined nor can valid comparisons be made with the nature of existing or potential commercial exploitation.

In this paper we make the first, and very preliminary, attempt at such an evalutation for any part of the Southern Ocean system, by reviewing details of predator-prey relationships in comparison with information on prey populations derived largely from conventional biological sampling. We confine our attention to interactions between breeding populations of seabirds and seals and Antarctic Krill, *Euphausia superba,* cephalopods and fish as these are substantially better documented than any other interrelationships. We consider only the Scotia Sea (as defined by Griffiths and Barker (1971) but extending to the base of the Antarctic Peninsula on its western side and including the satellite islands near its northeastern tip) and in particular the South Georgia area.

The Scotia Sea is a highly productive region supporting large concentrations of Krill, fish, seabirds, and seals, many of which are especially abundant around South Georgia. It has also been the site of most of the detailed quantitative studies of predator diets (e.g., Croxall and Prince 1980a) and bio-energetics (e.g., Croxall 1982, 1984), including the only published studies of the energy costs of swimming in penguins (Kooyman et al. 1982; Davis et al. 1983). In addition, there has been substantial research by many nations on the distribution, abundance and dynamics of Krill and fish resources, particularly around the Bransfield Strait, Elephant Island, the South Orkney Islands and South Georgia, and the region is a focus for commercial exploitation of Krill and fish. Together, this provides the most extensive source of information on predators and prey available for any area of the Southern Ocean.

Our approach is: firstly, to provide and briefly review estimates of food consumption by seabirds and seals; secondly,

Antarctic Nutrient Cycles and Food Webs
(ed. by W.R. Siegfried, P.R. Condy, and R.M. Laws)

to compare, for each of the main resources, the nature of the prey taken with data on the abundance and dynamics of the available stock; and, thirdly, to suggest profitable lines of future research.

2 Data Base and Methods

2.1 Data Base

A preliminary attempt to assess the consumption of breeding seabird populations (Croxall and Prince 1982a) included extensive details of the data used and we mainly note here differences from this. Population estimates are taken from Croxall et al. (1984a) and sex-specific weights are used (Croxall et al. 1984b). All sources of dietary information (mainly composition by weight) are from Scotia Sea studies and were reviewed by Croxall and Prince (1980a); we are almost entirely ignorant of winter diets. Empirical bio-energetic data are from Kooyman et al. (1982), Davis et al. (1983) and studies reviewed in Croxall (1982); other estimates, including flight costs, rely on theoretical relationships, chiefly from Kendeigh et al. (1977), as described in Croxall and Prince (1982a). The present model also incorporates information on the nature and timing of activities for each sex throughout the annual cycle. These data on arrival date, duration of prelaying departure to sea, laying date, share of incubation duties, hatching date, duration of brooding, fledging date, and for penguins also duration of post-fledging at sea and moult, will be summarized elsewhere (Croxall et al. 1984b). Mortality rates of eggs and chicks are taken from published (Pinder 1966; Beck and Brown 1972; Payne and Prince 1979; Croxall and Prince 1980a; Prince 1980a; Croxall and Hunter 1982; Thomas et al. 1983) and BAS unpublished studies for the region. Mean calorific contents of prey are: Krill 4.35 kJ g^{-1}, fish 3.97 kJ g^{-1}, squid 3.47 kJ g^{-1} (Clarke and Prince 1980; Croxall and Prince 1982b). Biennially breeding species are treated as if two entirely separate demi-populations exist. King Penguins, *Aptenodytes patagonicus,* with a still more complex breeding pattern, are modelled using three separate populations and annual cycles.

Sources of information for the Antarctic Fur Seal, *Arctocephalus gazella,* are summarized in Doidge and Croxall (1985). Much less detail is available for other seals. For Elephant Seals, *Mirounga leonina,* McCann (1985) is followed; Crabeater Seal, *Lobodon carcinophagus,* populations are calculated on the assumption that the average density is 0.75 individuals per km^2 (15×10^6 Seals (Gilbert and Erickson 1977); 20×10^6 km^2 of habitat (Laws 1977a)) and that there is 1.2×10^6 km^2 of suitable habitat in the Scotia Sea (unpubl. measurements from ice-charts). Weddell, *Leptonychotes weddelli,* and Leopard, *Hydrurga leptonyx,* Seal populations are calculated in proportion to Crabeater Seals, based on the figures for world populations given in Laws (1977b). The mean weights and food consumption requirements calculated by Laws (1977b) are used, together with the diet compositions documented by Øritsland (1977) as interpreted by Laws (1977b).

2.2 Methods

We allocate seabird activities to one of six categories standing around (including courtship), incubating, brooding, foraging for self, foraging for chick (i.e., with an extra daily food consumption equivalent to the meal size and frequency of meal delivery to the chick) and moult (penguins only), each with its own sex- and species-specific energy cost. Since the timing and duration of activities are determined by the "input specification" of the annual cycle, the energy consumption by birds of each sex on each day of the year can be calculated. This is converted to food consumption using the diet composition and energy content data and assuming assimilation efficiency is 75% (Wiens, 1984). The patterns of egg and chick losses are modelled by fitting the field data for each to separate exponential curves. Seabirds which have lost eggs or chicks are assumed to stay in the area (foraging for self) until the end of the breeding season.

At present we consider only breeding seabirds, prebreeding and non-breeding birds being excluded because information on the proportions of the populations present in the area and on their activity patterns is inadequate for modelling purposes. Species known to be migratory (e.g., Wilson's Stormpetrel, *Oceanites oceanicus,* Blackbrowed Albatross, *Diomedea melanophrys*) are assumed to leave the region immediately their chicks fledge.

Daily food consumption is calculated in a similar way for Antarctic Fur Seals (Doidge and Croxall 1985), but data for other seals are inadequate for this approach and only annual food consumption of the total population is estimated.

3 Results and Discussions

3.1 Overall Food Consumption

The food consumption of seabirds and seals in the area is summarized in Table 1. Krill comprises 82% of the total consumption (13×10^6 t) of seabirds, followed by copepods (8%), squid (4%), fish (3%), and amphipods (2%). Seabirds breeding on South Georgia account for 59% of the total. The portion taken by seals comprises 54% Krill, 31% squid, and 15% fish. Of the three main resources, Krill (70%) easily predominates and squid (16%) is twice as important as fish (8%). We lack any quantitative information on the occurrence of copepods and amphipods in seal diets. Such prey are probably present as a minor component in the diets of all predators but, although they are particularly important to Dove Prions, *Pachyptila desolata,* (Prince 1980b) and Common Diving Petrels, *Pelecanoides urinatrix,*

Table 1. Food consumption (t × 10^3) of seabirds and seals in the Scotia Sea

Species	Population (× 10^3)[a] South Georgia	Rest	South Georgia	Rest	Total	Krill	Squid	Fish	Copepod	Amphipod	Others[b]
King Penguin *Aptenodytes patagonicus*	34.3		63.9		63.9		44.8	19.1			
Adélie Penguin *Pygoscelis adeliae*		750		464.6	464.6	464.6					
Chinstrap Penguin *P. antarctica*	4	6,420	2.5	3,847.7	3,850.2	3,850.2					
Gentoo Penguin *P. papua*	100	42	90.1	35.5	125.5	85.3		40.3			
Macaroni Penguin *Eudyptes chrysolophus*	5,000	10	3,961.7	7.8	3,969.5	3,890.4		79.1			
Wandering Albatross *Diomedea exulans*	4.7		3.0		3.0	0.3	2.4	0.3			
Blackbrowed Albatross *D. melanophrys*	60		21.7		21.7	8.3	4.6	8.5	0.2	0.2	
Greyheaded Albatross *D. chrysostoma*	60		42.0		42.0	6.3	20.6	14.7		0.4	
Lightmantled Sooty Albatross *Phoebetria palpebrata*	8		4.9		4.9	1.8	2.3	0.6			0.2
Southern Giant Petrel *Macronectes giganteus*	5	17	3.3	11.4	14.7	11.7	2.0	1.0			
Northern Giant Petrel *M. halli*	3.5		2.3		2.3	1.5	0.6	0.2			
Antarctic Fulmar *Fulmarus glacialoides*		2,500		640.3	640.3	544.2	96.0				
Cape Pigeon *Daption capense*	20	1,000	3.6	179.6	183.2	155.7	27.5				
Snow Petrel *Pagodroma nivea*	3	500	0.5	77.8	78.3	62.6	7.8	7.8			
Dove Prion *Pachyptila desolata*	22,000	250	2,319.7	26.4	2,346.1	1,360.7	23.5	46.9	727.3	187.7	
Blue Petrel *Halobaena caerulea*	70		8.0		8.0	6.6	0.1	0.6	0.3	0.4	

Whitechinned Petrel *Procellaria aequinoctialis*	2,000		770.5		770.5	210.1	365.8	186.8		7.8	
Wilson's Stormpetrel *Oceanites oceanicus*	600	5,000	9.5	79.3	88.8	40.0	4.4	35.5	8.9		
Blackbellied Stormpetrel *Fregetta tropica*	10	40	0.5	2.2	2.7	1.3	0.1	1.1	0.3		
Common Divingpetrel *Pelecanoides u. exsul*	3,800		351.8		351.8	52.8			239.2	59.8	
South Georgia Divingpetrel *P. georgicus*	2,000		165.2		165.2	125.6			33.0	6.6	
Imperial Cormorant *Phalacrocorax atriceps*	7,500	13,500	4.7	8.4	13.1		1.0	11.3	0.3		0.5
Antarctic Tern *Sterna vittata*	2.6	47	0.3	5.2	5.5	0.8	2.7	0.8	1.1		
Crabeater Seal *Lobodon carcinophagus*	1,000			4,526	4,526	4,300	90	136			
Weddell Seal *Leptonychotes weddelli*	44	0.5	3.0	254	257		29.5	135.5			92
Leopard Seal *Hydrurga leptonyx*	13			49	49	31	7	11			
Elephant Seal *Mirounga leonina*	350	1.5	3,682	16	3,698		2,773	925			
Antarctic Fur Seal *Arctocephalus gazella*	864	50	1,560	17	1,577	1,095.4	182.2	299.4			
Total			13,075	10,248	23,323	16,307	3,690	1,961	1,010	263	93

a Seabirds: breeding population in pairs; seals: total population

b Does not include carrion and other non-marine derived prey

(Payne and Prince 1979) they will not be discussed further here.

It is unlikely that all breeding individuals of non-migratory species stay in the region all year and therefore we have probably overestimated their annual food consumption. As an alternative, we can assume that breeding individuals of all species are only present in the region from the date of first arrival ashore to the last fledging date of any species. This overall attendance period is 233 days at South Georgia and 217 days for the rest of the Scotia Sea. This assumption would reduce consumption by nearly 40% (i.e., to 10×10^6 t of Krill overall). However, as we have taken no account of the substantial population of prebreeding seabirds, especially penguins, known to be present in the area for much of the summer, it is probably not too unrealistic to accept the higher value for the present.

3.2 Consumption of Krill

Of the total estimated Krill consumption of 16×10^6 t, seabirds and seals breeding at South Georgia and in the rest of the Scotia Sea take 42% and 58%, respectively. Such a division is probably not typical of other sectors of the Southern Ocean, since the Scotia Sea has a very high ratio of land (with suitable breeding sites for seabirds) to sea, a significant portion of which is outside the zone of winter pack-ice and, hence, unsuitable for breeding Crabeater Seals.

As it is, Crabeater Seals (26%) and Chinstrap and Macaroni penguins (24% each) are the main consumers, followed by Dove Prions (8%), Antarctic Fur Seals (7%), and Antarctic Fulmars, *Fulmarus glacialoides,* and Adélie Penguins, *Pygoscelis adeliae,* (3 % each). Penguins account for 76% of the intake by seabirds (68% at South Georgia; 84% in the rest of the area), and Crabeater Seals 79% of that by seals. At South Georgia, Macaroni Penguins (57%), Dove Prions (20%) and Antarctic Fur Seals (16%) predominate, with only Whitechinned Petrels, *Procellaria aequinoctalis,* (2%), South Georgia Diving Petrels (2%) and Gentoo Penguins, *Pygoscelis papua,* (1%) contributing 1% or more. In the rest of the Scotia Sea, Crabeater Seals (45%) and Chinstrap Penguins (40%) are the main predators, followed by Antarctic Fulmars (6%), Adélie Penguins (5%) and Cape Pigeons, *Daption capense,* (2%).

3.2.1 Comparisons with Stock Estimates

Surveys of Krill, particularly using acoustic echo-integration techniques, are very recent and preliminary and so give only very rough indications of abundance. In February 1981 an acoustic survey of about 50% (300,000 nm^2; ca. 1.0×10^6 km^2) of the Scotia Sea gave a Krill biomass estimate of 12×10^6 t (BIOMASS 1982). If unsurveyed areas were similar to those covered then the total of 24×10^6 t is very low compared with the consumption by seabirds and seals of 16×10^6 t. However, there are several reasons why this comparison may not be realistic. Firstly, the survey assessed Krill stocks in late summer after considerable depredation had taken place. Secondly, the Krill supply may be regularly replenished by swarms coming into the Scotia Sea from the Bellingshausen and Weddell seas. Thirdly, the Krill in the top 10 m of the water were not surveyed by the equipment used. This stratum is of considerable importance to predators. Flying seabirds can only take Krill near the surface and nearly 50% of the dives of foraging Chinstrap Penguins do not exceed 10 m (Lishman and Croxall 1983); 20% of dives by Antarctic Fur Seals are shallower than 20 m (Croxall et al. 1985c). Also, Krill are not evenly distributed and it may be easy for large-scale grid surveys to miss locally rich areas.

3.2.2 Seasonal Variation in Predation

In both areas, Krill consumption by all seabirds combined rises to a peak during February and decreases rapidly thereafter (Figs. 1 and 2). This pattern is mainly attributable to increases in energy demand by the fast-growing chicks of Macaroni (Fig. 3) and Chinstrap (Fig. 4) Penguins, the principal Krill consumers in each of the two parts of the Scotia Sea. The rapid drop in requirements is due to breeding penguins coming ashore for their annual moult. In reality, levels of Krill consumption would be less drastically reduced then because the substantial population of newly fledged chicks will probably still be in the vicinity of the breeding colonies, before their densities and numbers are reduced by winter dispersal and mortality. There are other, more subtle, features of the pattern of Krill consumption of seabirds during the breeding season, in particular the offsetting of peak food demands by species which have marked ecological similarities (e.g., the two species of diving petrels and Blue Petrels, *Halobaena caerulea,* and Dove Prions) but there is not space to discuss these here.

Krill demand by Fur Seals also peaks between January and March, during their pup-rearing period (Doidge and Croxall 1985). The ice-breeding Crabeater Seal, the other main seal eating Krill, is likely to have its greatest food requirement by adults in September and November, before and after the 2–3 week breeding fast by both sexes in October, during which the pup is reared to independence.

3.2.3 Size and Status of Krill Taken

Knowledge of the length-frequency composition of Krill taken predators is important in evaluating the dynamics of the interactions between predators and their prey in the Southern Ocean. Detailed published information is largely lacking, except for the Baleen Whales (Mackintosh 1974). Data for penguins at islands near the southern edge of the Scotia Sea are summarized in Figs. 5 and 6. Although four species, three island groups and several seasons are involved, the basic pattern is fairly consistent, with Krill

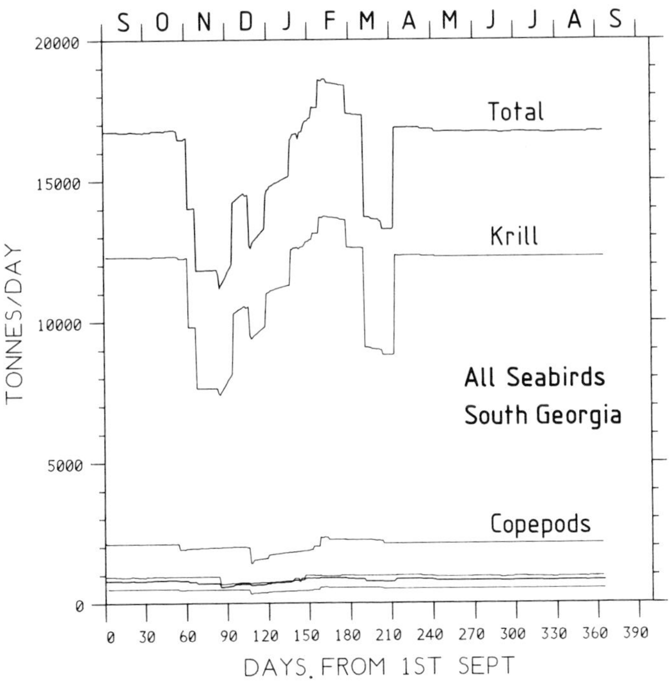

Fig. 1. Food consumption by all seabirds at South Georgia

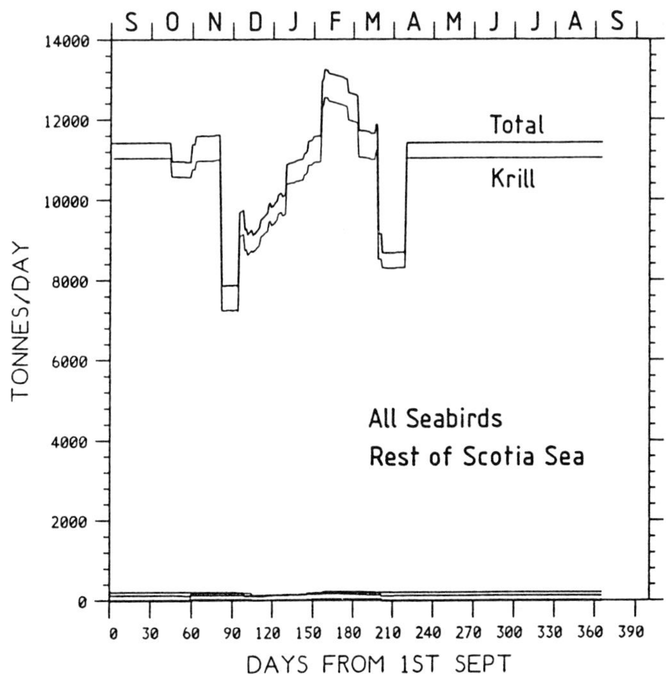

Fig. 2. Food consumption by all seabirds in the Scotia Sea, excluding South Georgia area

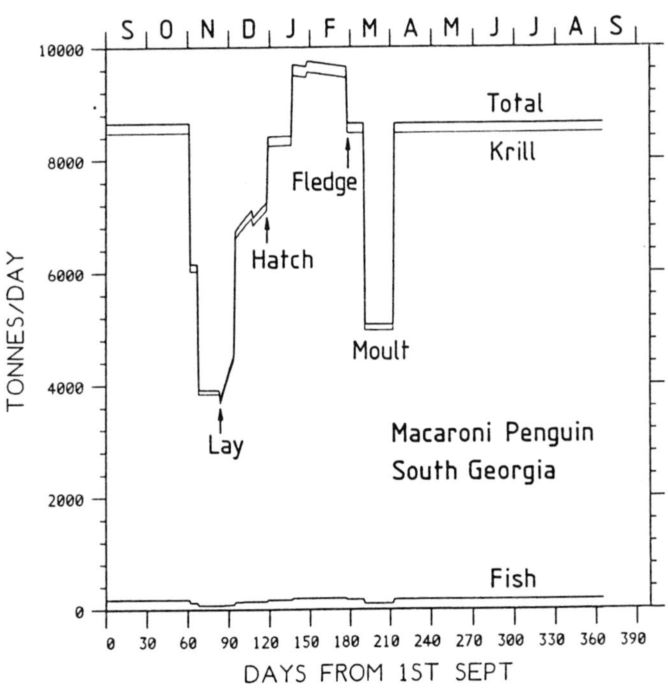

Fig. 3. Seasonal changes in Krill consumption by Macaroni Penguins at South Georgia

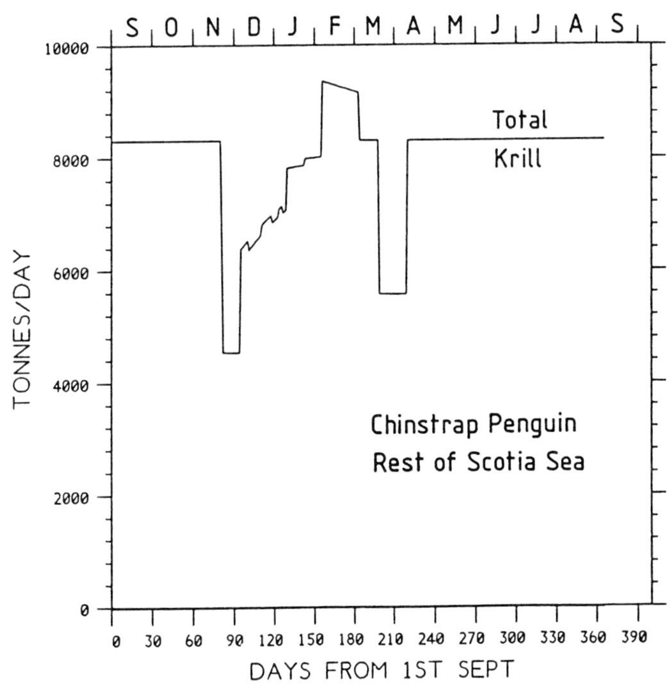

Fig. 4. Seasonal changes in Krill consumption by Chinstrap Penguins in the Scotia Sea, excluding the South Georgia area. Timing of events as in Fig. 3

of about 35–45 mm length being taken by all species. These are mainly sexually mature individuals (100%, 90%, and 95% for Gentoo, Chinstrap and Adélie Penguins, respectively (Volkman et al. 1980), and 91% in Crabeater Seals (J.L. Bengtson et al., unpubl.)). Within this pattern there are quite marked, albeit small-scale, variations between sites and within species and seasons. This is reinforced by a more detailed examination of data from Signy Island for Adélie and Chinstrap Penguins (Fig. 7). In 1980–81, Adélie Penguins took small Krill, absent from the Chinstrap Penguin samples, although both species took the larger size classes in similar proportions. In 1981–82, the range and

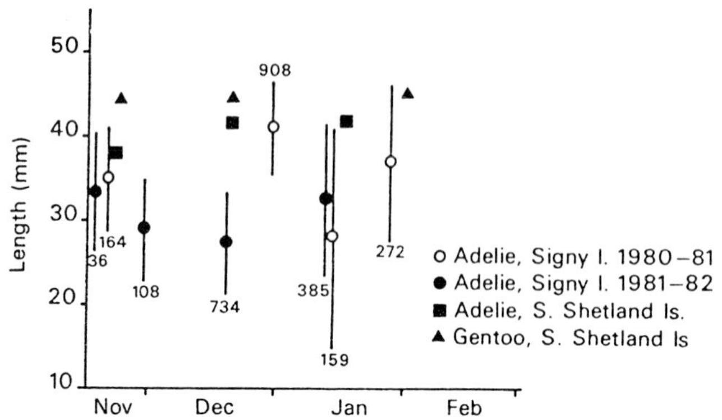

Fig. 5. Mean (± S.D.) length of Krill taken by Adélie and Gentoo Penguins in the southern Scotia Sea. (Data from Volkman et al. 1980); Lishman 1985)

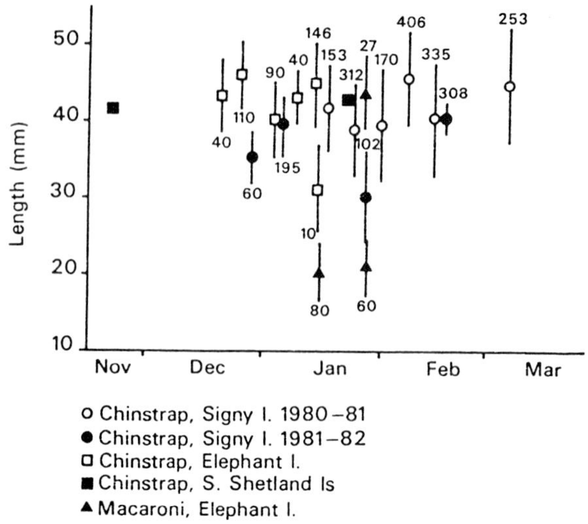

Fig. 6. Mean (± S.D.) length of Krill taken by Chinstrap and Macaroni Penguins in the southern Scotia Sea. (Data from Croxall and Furse 1980; Volkman et al. 1980; Lishman 1985)

size of prey taken were similar in both species but the proportions were considerably different with Adélie Penguins taking significantly more small Krill (Lishman 1985). It is not unexpected that Adélie Penguins might favour smaller prey, since in the Ross Sea their diet consists principally of *Euphausia crystallorophias*, ca. 20 mm in length (Emison 1968).

Interpreting differences in the lengths of Krill taken by predators should be done with great caution. Significant differences in mean lengths are often easily demonstrated (particularly because of the usually large sample sizes) but it is more difficult to assess their biological significance. This is partly because: Krill occurs in swarms and there is usually greater homogeneity within than between swarms; different water-masses tend to have Krill populations with small but distinctive morphometric differences; and, Krill

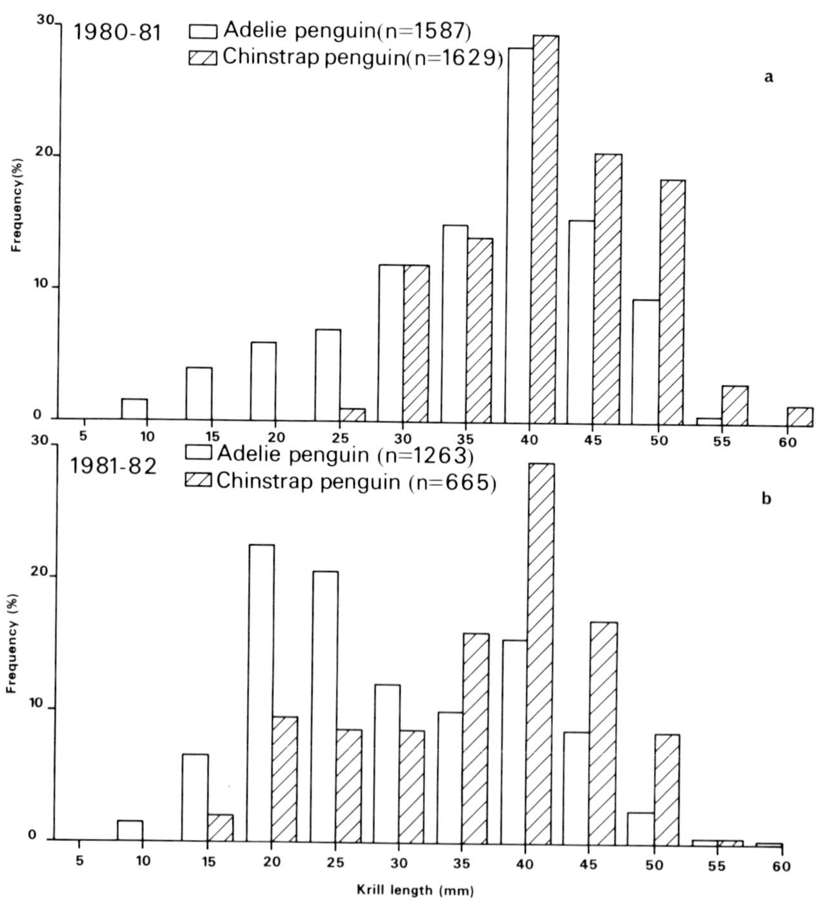

Fig. 7a, b. Length-frequency distribution of Krill taken by Adélie and Chinstrap Penguins at Signy Island, South Orkney Islands, for **a** 1980–81 and **b** 1981–82. (Data from Lishman 1983)

grow in length during the summer but populations in different areas do not necessarily do so at the same rate. Very comprehensive data are thus necessary to conclude that differences between seasons, sites, sexes and species reflect any more than differences in what predators encounter, because of the timing of predation or the location of foraging areas. Claims for selection of different-sized Krill by different sexes of penguins (Ainley and Emison 1972) cannot now be considered valid. Even the distinct tendency of Adélie Penguins to take smaller Krill than Chinstrap Penguins (and Gentoo Penguins to take larger prey than either (Volkman et al. 1980)) does not constitute evidence of active selection of prey of different sizes.

If we compare the size of Krill taken by these penguins with those caught by net-hauls in similar areas (Fig. 8), then at the South Orkney and South Shetland islands net-hauls show peaks at lengths (50 and 53 mm) some 10 mm greater than in the penguin diets. The detailed data of Stepnik (1982, Fig. 3) for 15 months from Admiralty Bay, King George Island, South Shetland Islands, are reasonably consistent with this picture, having a peak of the Krill size distribution in summer between 45 and 50 mm, about 3–10 mm greater than the mean length of Krill eaten by penguins breeding at exactly the same locality (Volkman et al. 1980). Krill data from Elephant Island (Fig. 8) and areas north of the South Orkney Islands (Nemoto et al. 1981, Fig. 12a) are more consistent with the data for penguins, although the largest Krill in the net-hauls are still unrepresented in the penguins' diets. It is not unexpected that some of the smaller prey taken by penguins should not show up in the net-hauls, since many of the nets used do not retain these smaller size classes. However, the relatively consistent pattern of the absence of the largest Krill from the penguins' diet suggests that the birds do not usually encounter such prey in foraging from their breeding sites, despite the fact that large Krill are known to occur in the vicinity of all the island groups.

Information on the size of Krill taken by six species of seabirds breeding at Bird Island, South Georgia (Fig. 9), shows a predominant pattern of consumption of very large (and sexually mature) Krill of ca. 45–55 mm in length. The diving petrels were the only birds which exclusively took small, immature Krill. Similar specimens in samples from Dove Prions usually occurred with large quantities of copepods and, like these, were probably caught by using the palatal lamellae of the mandible to filter small organisms from water ingested while hydroplaning (Prince 1980b). The only intraspecific pattern is the significant increase ($r = 0.87$, $P < 0.01$) with advancing time in the size of Krill taken by Blue Petrels. The tendency for prey length to decrease from January might simply reflect decreased availability of the largest Krill, due to post-reproductive mortality. The much greater variation in size of Krill taken by the prions and petrels undoubtedly reflects their foraging over much larger areas, encountering a greater diversity of Krill than do penguins.

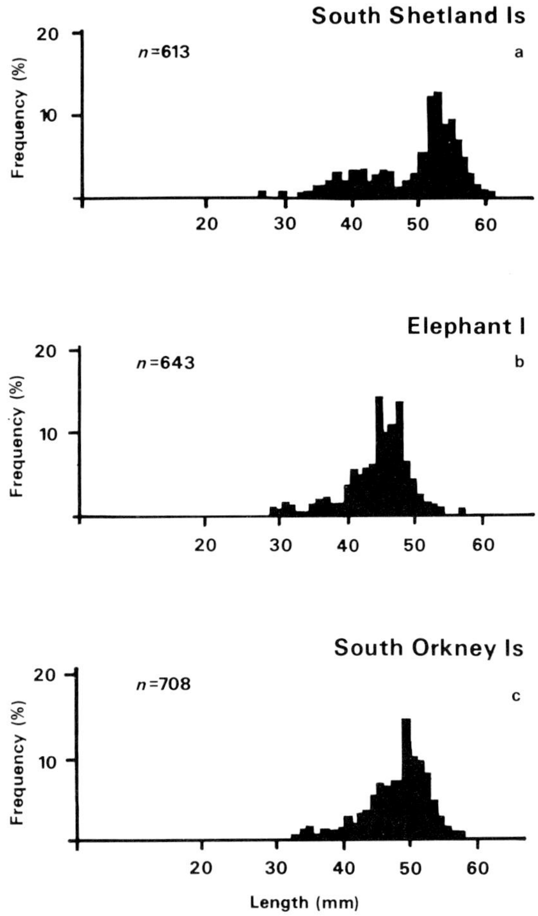

Fig. 8a–c. Krill length-frequency distributions for specimens caught by stern-deployed mid-water trawls, from three areas of the Scotia Sea in 1980–81. (Data from BIOMASS 1982)

In six seasons, Antarctic Fur Seals from Bird Island have consistently taken Krill of similar lengths and dimensions to those taken by seabirds, especially penguins (Croxall and Pilcher 1984). The size of Krill taken by seal and seabird predators at South Georgia is as consistent as in the Scotia Sea but there is a marked difference between the sizes taken in the two areas. Furthermore, at South Georgia there is much less overlap in Krill length between samples from predators and those from net-hauls. Three sets of Krill data from net-hauls made near South Georgia (Fig. 10) show peaks of frequency distribution at 25 mm (juveniles) and between 30–45 mm for adults. Even in the most extensive data set (Fig. 10b), representing over 50 net-hauls along the north coast of South Georgia, very few Krill over 50 mm in length were recorded. This sampling did not, however, include the foraging area of penguins and seals breeding at Bird Island, but the results of the few net-hauls available from this area (Croxall and Pilcher 1984) are very similar to the pattern shown in Fig. 10a.

Although it is standard practice to analyse Krill from net-hauls into maturity stages for each sex, such information is not available for specimens taken by predators and

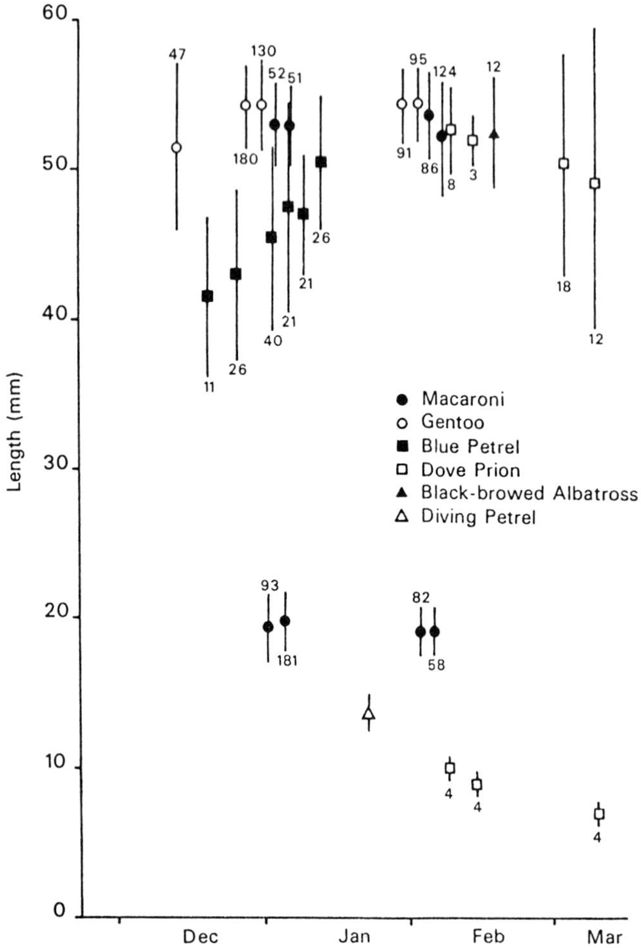

Fig. 9. Length of Krill (mean ± S.D.) taken by various seabirds at Bird Island, South Georgia. (Data from Payne and Prince 1979; Croxall and Prince 1980b; Prince 1980a, b)

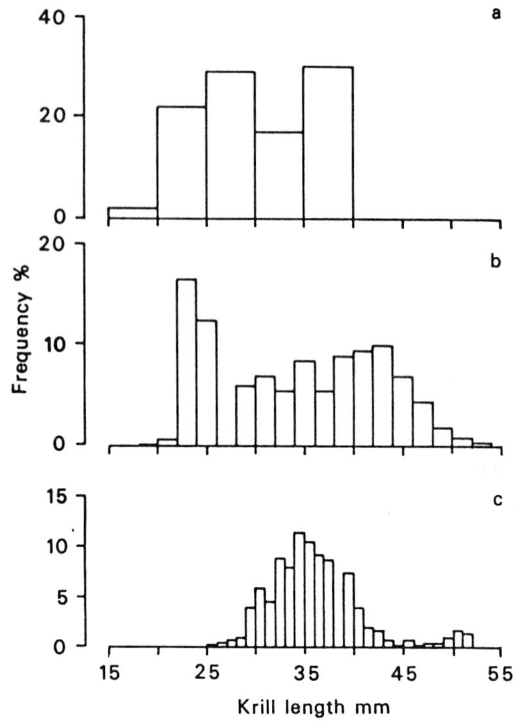

Fig. 10a–c. Length-frequency distributions of Krill caught by net-hauls around South Georgia. (Data for a from Nemoto et al. 1981; b Jazdzewski et al. 1978, n = 1472, c BIOMASS 1982, n = 300)

we can only compare data on the basis of overall composition by sex. Such sex ratios are available from only two areas: except for the Gentoo Penguin, there is a bias favouring male Krill in the southern Scotia Sea and a bias favouring female Krill at South Georgia, most notably in Antarctic Fur Seals (Table 2). If we compare sex ratios in net-hauls (Table 3), values for the southern Scotia Sea (except for one sample) have a consistent bias in favour of males, similar to that in samples from penguins and Crabeater Seals; the situation elsewhere is confusing. Comparison of data for different months and years is unsatisfactory, especially since sex ratios may vary appreciably with season. The data from Admiralty Bay (Fig. 11) show no clear pattern in change of overall sex ratio, despite major changes in population composition. Around South Georgia most Krill swarms in December and early January contained > 70% females; by late January, however, samples contained on average only a slight excess (ca. 55%) of females (Fig. 12). This pattern may reflect males and females coming together at the be-

Table 2. Sex ratios of Krill taken by penguins and seals

Species	No.	Male (%)	Female (%)	Ratio ♀/♂	Locality	Date	References
Adélie Penguin	5,250	54	46	0.85	S. Shetland Is.	November 1977–February 1978	Volkman et al. (1980)
Chinstrap Penguin	5,250	65	35	0.54	S. Shetland Is.	November 1977–February 1978	Volkman et al. (1980)
Gentoo Penguin	5,250	45	55	1.22	S. Shetland Is.	November 1977–February 1978	Volkman et al. (1980)
Gentoo Penguin	100	46	54	1.17	S. Georgia	January–February 1977	Croxall (unpubl.)
Macaroni Penguin	100	43	57	1.33	S. Georgia	January–February 1977	Croxall (unpubl.)
Antarctic Fur Seal	375	21	79	3.76	S. Georgia	January–March 1982	Croxall and Pilcher (1984)
Crabeater Seal	686	67	33	0.49	Bransfield Strait	November 1977	Bengtson et al. (unpubl.)

Table 3. Sex ratios of Krill taken by net-hauls

Net hauls	No. of Krill	Male[a] (%)	Female[a] (%)	Ratio ♀/♂	Locality[b]	Date	Reference
58	?	35	65	1.86	Scotia Sea	1972–73	Nemoto et al. (1981)
14	1,400[c]	61	39	0.64	S Ant Peninsula[c]	January–February 1976	Jazdzewski et al. (1978)
6	599	56	44	0.79	S Ant Peninsula	1980–81[d]	BIOMASS (1982)
11	685	47	53	1.13	S Bransfield St	1980–81	BIOMASS (1982)
?	613	68	32	0.47	S Shetland Is	1980–81	BIOMASS (1982)
12	1,200[c]	62	38	0.61	S Shetland Is[c]	February 1976	Jazdzewski et al. (1978)
?	300	67	33	0.49	King George Is	January 1978	Kittel (1980)
?	300	64	36	0.56	King George Is	February 1978	Kittel (1980)
?	300	74	26	0.35	King George Is	March 1978	Kittel (1980)
6	643	35	65	1.86	Elephant Is	1980–81	BIOMASS (1982)
12	1,969	48	52	1.08	Elephant Is	1980–81	BIOMASS (1982)
23	4,945	35	65	1.86	NE Ant Peninsula	1980–81	BIOMASS (1982)
5	500[c]	68	32	0.47	S Orkney Is	February 1976	Jazdzewski et al. (1978)
9	900[c]	47	53	1.13	C Scotia Sea[c]	March 1976	Jazdzewski et al. (1978)
6	1,135	46	54	1.17	C Scotia Sea	1980–81	BIOMASS (1982)
37	2,634	44	56	1.27	C Scotia Sea	1980–81	BIOMASS (1982)
7	1,274	49	51	1.04	S Sandwich Is	1980–81	BIOMASS (1982)
5	302	82	18	0.22	S Sandwich IS	1980–81	BIOMASS (1982)
34	1,440[c]	56	44	0.79	S Georgia[ce]	February–March 1976	Jazdzewski et al. (1978)
2	300	98	2	0.02	S Georgia	1980–81	BIOMASS (1982)
4	223	50	50	1.00	NW S Georgia	December 1981	Croxall and Pilcher (1984)

a Percentages excluding unsexed juveniles
b Mostly large areas of ocean; for actual extent see references, especially BIOMASS (1982) Fig. 1
c Hauls from "Profesor Siedlecki" only
d December–March, mainly January–February, for all BIOMASS operations
e Areas a and b only; virtually only juveniles in area c

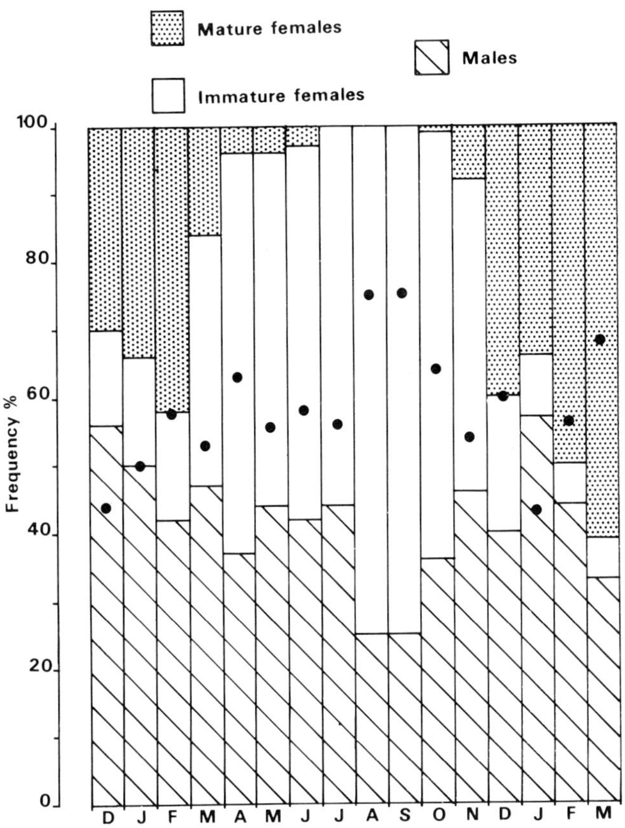

◂ **Fig. 11.** Proportion of male, immature and mature female Krill in monthy samples from Admiralty Bay, King George Island, South Shetland Islands. *Dots* denote sex ratio (% female Krill) in each sample. (After Stepnik 1982, Fig. 4)

ginning of mating, which is believed to occur mainly in January (Nemoto et al. 1981). Krill in Fur Seal samples, however, showed a high proportion of females (75%–83%) from early January to March (Croxall and Pilcher 1984).

3.2.4 Predator-Prey Interactions

Krill population dynamics are incompletely understood. There is particular uncertainty over the adult reproductive life of Krill lasts low many seasons and routine sampling cannot easily resolve this because conventional length-frequency analysis does not produce segregation into clear age classes (Mackintosh 1972).

From spring to autumn, seabird and seal predators mainly take adult (sexually mature) Krill, and this is particularly true of the small penguins and Fur and Crabeater seals which are by far the most important species. The main consumption by Crabeater Seals is likely to be of Krill which have either over-wintered as post-breeding adults, or of im-

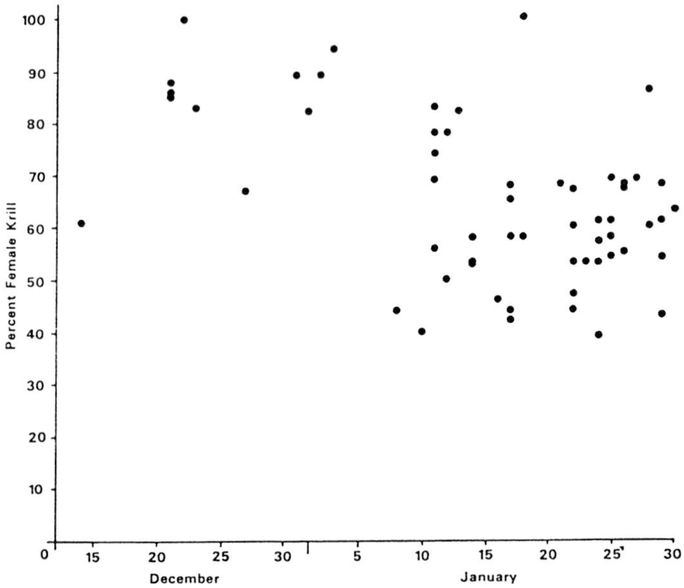

Fig. 12. Sex ratios of Krill in samples from swarms in the northern Scotia Sea, 1972–73. (Calculated from Nemoto et al. 1981)

matures which have already grown rapidly to reach sexual maturity. Penguins (and Fur Seals) build up to a peak consumption of Krill in February. This coincides with the time when there is active ovarian lipogenesis in female Krill, whose energy content may increase by some 60% (Clarke 1980, 1984). Male Krill may also show an increase in lipid content in late summer (Clarke, 1984). Catching mature Krill at this time of year probably represents a substantial energy bonus for predators. South Shetland Islands data (Fig. 11), probably reasonably typical of other parts of the Scotia Sea, show that mature females form a significant proportion of the Krill population for only a short period (December–March), imposing a fairly strict time contraint on their exploitation by predators.

The relatively short breeding phase of small-bodied penguins, such as Chinstrap and Macaroni (compared with procellariiform seabirds and Gentoo and larger penguins), gives them appreciable flexibility in timing of breeding, and it is conceivable that the period of peak energy demand for these penguins (late chick-rearing) is timed to coincide with the availability of a seasonally energy-rich supply of Krill.

The sex ratios in predator and net-hauls samples from the southern Scotia Sea provide no indication that there is any selection by predators for female Krill. Indeed, males predominate in both sets of samples, which is surprising, unless there is a large post-reproductive mortality in females Penguins and Fur Seals at South Georgia appear to take more female than male Krill; the Fur Seals to the extent that it is possible that they feed mainly in areas where swarms of female Krill predominate. Taking mature Krill during their breeding seaon (and following this up with a further substantial consumption of over-wintering survivors) may represent predation at particularly important stages of the life cycle of Krill, and thus an impact greater than would be deduced simply from comparison of food requirements and stock size. It is obviously important to try to determine the extent to which predation on female Krill takes place before they have spawned.

The predominance of immatures and the scarcity of gravid females (and eggs) from net-hauls around South Georgia have led to the conclusion that it is not a spawning ground for Krill (Marr 1962). However, since South Georgia penguins and Fur Seals mainly take mature Krill (including gravid females) of > 50 mm length, this idea needs reinvestigating. The discrepancies may be due to lack of net-haul samples from the feeding grounds of the main seal and penguin concentrations off northwestern South Georgia. Although we have good circumstantial evidence for the general area in which these predators feed (Croxall and Prince 1980a, b; Croxall et al. 1985b) we do not know the precise locations. Nor do we know whether predators exploit Krill in a manner different from nets, although it is recognized that large Krill are more likely than small ones to be able to avoid nets (Marr 1962). Fur Seals, and probably penguins, feed mainly at night on Krill swarms, most dives being shallower than 40 m (Croxall et al. 1985b; Kooyman et al., to be publ.). Energy demands and the number and durations of dives require that on average Fur Seals catch 60 Krill per dive (one every 2 s) and Chinstraps 16 per dive (one every 6 s). This argues for a relatively fast and efficient capture rate (especially if some dives are exploratory and not feeding) and it is possible that the escape responses of large Krill may be more effective at avoiding nets than avoiding fast-moving predators. However, large Krill are sampled perfectly adequately by identical nets in the southern Scotia Sea, where the largest Krill are caught in nets and the differences between predator and net-haul samples are much less.

4 Consumption of Cephalopods

Of the estimated 3.7×10^6 t of squid taken by seabird and seal predators in the Scotia Sea, 76% is consumed by Elephant Seals largely around South Georgia. Other species which take > 1% of all squid eaten are Whitechinned Petrels (10%), Antarctic Fulmars and Crabeater Seals (2.5% each) and King Penguins (1%). Sperm Whales, *Physeter catodon,* which take squid almost exclusively, are probably not of major importance in this sector. If only one-third of exploitable males migrate into the Southern Ocean and stay there for ca. 120 days in summer (Laws 1977b) and only animals from Whaling Division One move into the Scotia Sea, then, using the most recent stock estimates for this division (IWC 1978) only ca. 1,500 Sperm Whales will occur in the area. Using Laws' (1977b) data on food requirements, these Sperm Whales would take 162,000 t of squid, or about 4% of the consumption of seabirds and seals.

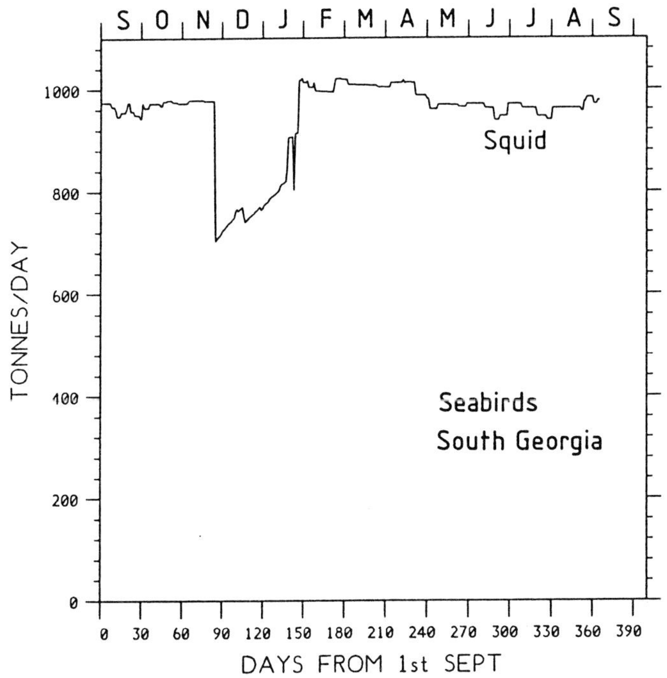

Fig. 13. Consumption of squid by seabirds breeding at South Georgia

About 94% of all squid consumed are taken by species breeding at South Georgia, with seals accounting for 88% and seabirds only 12%. In the rest of the Scotia Sea, Crabeater and Weddell Seals account for 48% and seabirds (mainly Antarctic Fulmar) 52%. There is no clear seasonal pattern to the consumption of squid by seabirds at South Georgia (Fig. 13), the "trough" being during the incubation period of Whitechinned Petrels which eat 78% of the squid taken by birds. The consumption by Elephant Seals, which far outweighs this, probably peaks either side (i.e., in October, December–January and March) of their breeding fast (November) and their moult fast (February).

4.1 Composition of Squid Diet of Predators

Only seven predator species have so far been studied quantitatively (by analysis of beak collections, and estimation of squid mass from standard relationships (MR Clarke 1980)) in the Scotia Sea (Appendix, see page 531); the main results are summarized in Fig. 14. Although most species of squid are taken by several predators, the main prey of each (except for the two mollymauks) is usually quite different. *Mesonychoteuthis* and *Taningia* are too large (and too deep?) to be taken by species other than Sperm Whales but other squid appear to be potentially available to most predators. The similarity between Sperm Whale and Wandering Albatross diet is remarkable, especially considering that the Whale samples were obtained in summer and the Wandering Albatross ones at the end of winter. The squid diets of the Lightmantled Sooty Albatross, *Phoebetria palpebrata,* and the two seals (Fig. 14) are broadly similar and rather distinct from the other species. This may reflect that the seals were sampled at Signy and Deception islands in the south of the Scotia Sea, an area in which Lightmantled Sooty Albatrosses are particularly common (Thomas 1982). Sperm Whales caught in the southern Scotia Sea show a slightly different diet composition (more *Gonatus* and *Moroteuthis*) consistent with this pattern, but the Sperm Whale diet is very similar in both areas.

We known very little of where and how predators catch squid. Sperm Whales have a diving capacity (depth attained and time spent submerged) much greater than even the largest seals. The abundance of inshore benthic octopods in the diet of Elephant and Weddell Seals suggest that both may feed near the sea floor, but as Weddell Seals dive readily to 350 m (Kooyman 1981) both species may still have a considerable foraging range over continental shelves. Although some large squid are probably scavenged (Clarke et al. 1981), albatrosses must take squid at the surface and probably at night. There is no evidence for their selection of bio-luminescent species (e.g., *Kondakovia* and *Moroteuthis* have no light organs) and they may well take squid associated with Krill swarms (the three *Kondakovia* caught at South Georgia at 0–50 m depth all had stomachs full of Krill (Filippova 1972)). Half the dives of King Penguins were deeper than 50 m (1 exceeding 240 m), a very different pattern from that of Krill-eating penguins. Also, only 10% of dives need result in prey capture to fulfil King Penguin energy requirements (Kooyman et al. 1982).

4.2 Comparisons with Net-Haul Data

There is little published information on squid caught by nets in the Antarctic, let alone in the Scotia Sea. Roper (1969) evaluating a collection of 600 specimens of 19 species from the Southern Ocean, concluded that the commonest species were *Bathyteuthis abyssicola* (39%), *Galiteuthis glacialis* (23%), *Brachioteuthis picta* (12%), *Gonatus antarcticus* (9%) and *Alluroteuthis antarcticus* (5%). *Bathyteuthis* has not been recorded above 500 m and is not represented in predator samples. *Brachioteuthis,* which is common in surface waters near South Georgia (67% of specimens reported on by Filippova (1972)), is a small squid (ca. 5 g) and this may account for its absence from predator samples from the area. The other species above are taken by five or six of the seven predators but contribute no more than 10% by weight to the diet of any one of them (*Galiteuthis* in Blackbrowed Albatross; *Gonatus* in Elephant Seal; *Alluroteuthis* in Lightmantled Sooty Albatross). Thus, the main species taken by birds and seals (and whales) are rare or absent from Roper's (1969) collection. However, Nemoto et al. (1984) have recently obtained significant net-haul catches of *Moroteuthis, Mesonychoteuthis* and *Kondakovia* (important species to many predators), permitting the first realistic comparisons with predator samples. There are

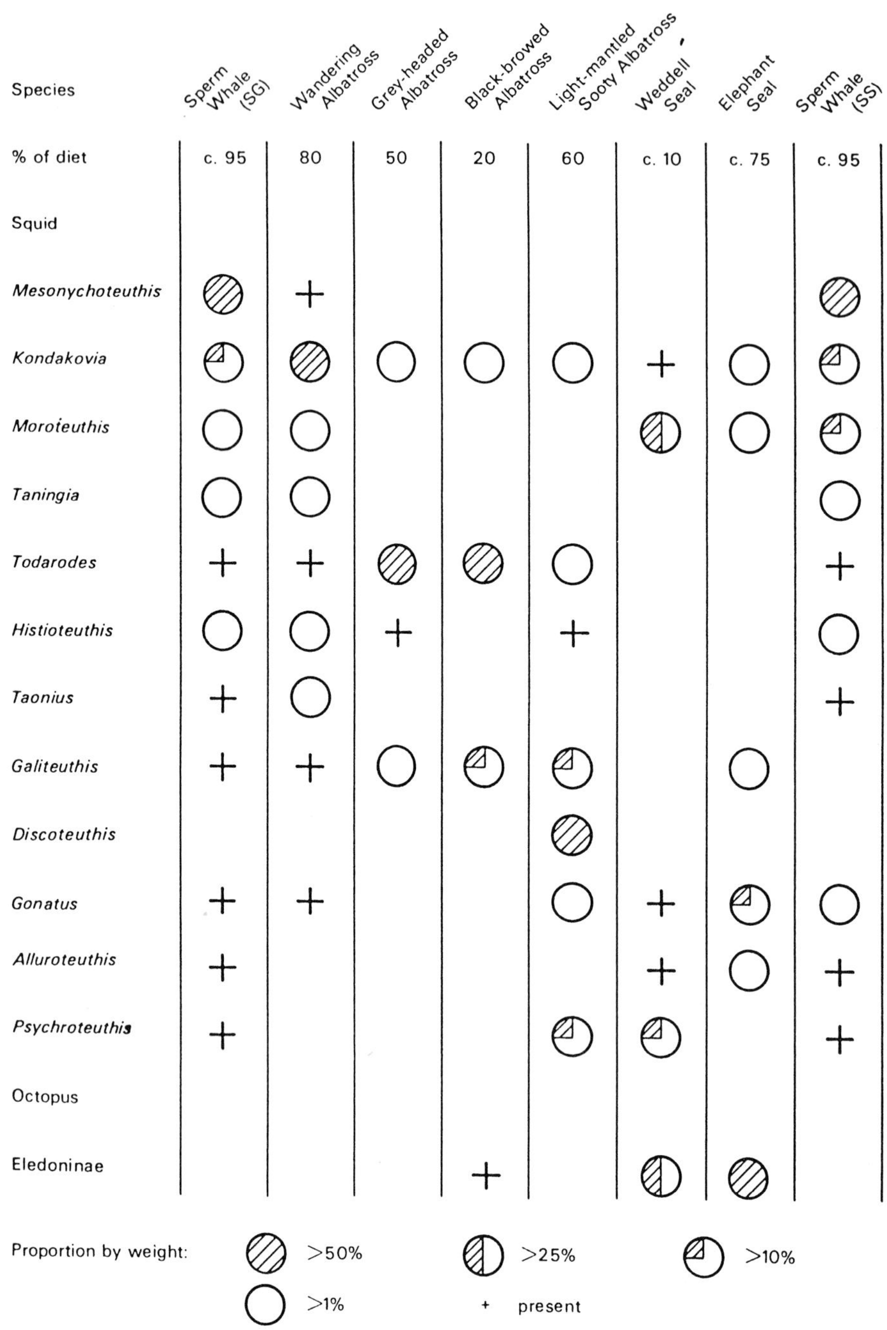

Fig. 14. Composition by weight of squid in the diet of albatrosses, seals and Sperm Whales in the Scotia Sea. Only squid comprising more than 1% by weight of the diet of any species are included. For sources see Appendix 1. *SG* South Georgia; *SS* rest of Scotia Sea

few data on such species for the Scotia Sea, although the distribution maps of Filippova and Yukhov (1979) show *Moroteuthis*, *Alluroteuthis* and *Psychroteuthis* to be fairly widespread, *Kondakovia* frequent near South Georgia and *Mesonychoteuthis* locally common around the South Orkney and South Shetland islands.

Disparity between samples from net-hauls and predators is not surprising given the difficulties in catching adult highly mobile squid with even the largest nets (Clarke 1977). However, predator samples presently provide a better indication of the range and abundance of Southern Ocean squid than samples from net-hauls, especially where meso- and epi-pelagic species (of greater relevance to any potential commercial exploitation) are concerned. Many more samples from seals (especially Elephant Seals) and petrels and *Aptenodytes* penguins are required before tentative assessments of the relative abundance of different squid species (in, say, the upper 500 m of the sea) can be made, let alone estimates of squid population sizes.

5 Consumption of Fish

On present estimates fish are the least important of the main resources taken by seabirds and seals. This may reflect the dominance of small (ca. 250 mm length) members of the Antarctic ichthyofauna and the absence of obligate pelagic schooling species (Evenson 1981). Of the 2 times, 10^6 t consumed, 77% is taken by seals (Elephant Seal 47%, Antarctic Fur Seal 15%, Crabeater and Weddell Seals both 7%) and the only seabirds taking more than 1% are White-chinned Petrels (10%), Macaroni Penguins (4%), Dove Prions (2%), Gentoo Penguins and Cape Pigeons (both 1.5%), and King Penguins (1%).

Species breeding at South Georgia take 82% of the total amount of fish (52% by Elephant Seals, 18% by Antarctic Fur Seals, the order of importance of other species being the same as above except for the absence of Crabeater and Weddell seals and Cape Pigeons). In the rest of the Scotia Sea seals take 82% (Crabeater and Weddell Seals 39% each, Leopard Seals 3%) and the main seabird predators are Cape Pigeons (8%), Gentoo Penguins (3%) and Imperial Cormorants, *Phalacrocorax atriceps,* and Snow Petrels, *Pagodroma nivea,* (2% each).

The peak of fish consumption of South Georgia seabirds (Fig. 15) in February is mainly attributable to the chick rearing demands of Whitechinned Petrels (and Macaroni Penguins, although fish form only 1% of their diet). However, the overwhelming importance of seals is such that the main demand, through Elephant Seals, is likely to be in October, December–January and March (see p. 527) which mostly are times of reduced consumption by seabirds. Antarctic Fur Seals take fish mainly in winter (Doidge and Croxall 1985).

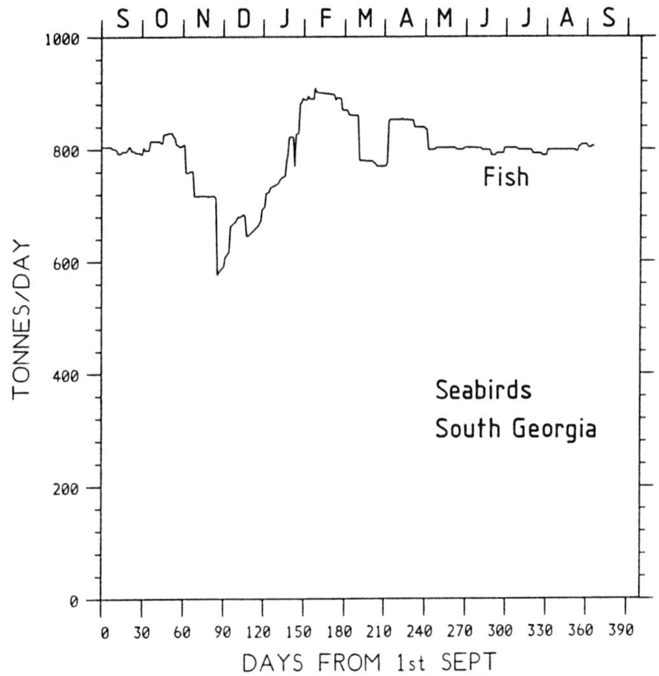

Fig. 15. Consumption of fish by seabirds breeding at South Georgia

In the Scotia Sea, Weddell Seals take *Notothenia* spp. and *Trematomus,* and they commonly take *Dissostichus* elsewhere (Øritsland 1977). *Champsocephalus gunnari* (76%), *N. rossii* (5%) and other nototheniid species (8%) are the main constituents by weight of the summer diet of Antarctic Fur Seals at South Georgia (North et al. 1983); fish are probably more significant in the winter diet but its composition then is unknown. The fish diet of Crabeater and Elephant seals is unknown, although the latter's might prove to be similar to that of Antarctic Fur Seals but perhaps with a greater proportion of benthic fish to match the quantity of octopus in its diet.

Imperial Cormorants and Gentoo Penguins, which feed largely inshore, mainly take juvenile fish which frequent such areas until reaching sexual maturity (Croxall and Prince 1980b; Croxall and North unpubl.). Antarctic Fur Seals, which are wider ranging (and catch the more pelagic *C. gunnari,* rather than *Notothenia* species), take mainly young adults (3–5 yrs-old) (North et al. 1983). Nothing is known about the age composition of the fish prey of other species.

There is little information on how and when predators catch fish. Flying birds presumably do so at night at the surface, and it is possible that many predators feed on the fish which are associated with Krill swarms. Of the fish mentioned above, *C. gunnari* and *Pseudochaenichthys* are classified as pelagic Krill-eaters (Everson 1981), whereas the *Notothenia* species are predominantly benthic-demersal, opportunistically feeding on Krill and other pelagic foods (Linkowski and Rembiszewski 1978; Targett 1981). *Pseudochaenichthys* and *C. gunnari* are associated with Krill chiefly at night (Rembiszewski et al. 1978).

5.1 Composition of Fish Diet of Predators

Knowledge of the diet of fish taken by most predators (including all the most important species) is very inadequate. Wide-ranging pelagic seabirds, such as Whitechinned Petrels, probably take mainly myctophids (Croxall and Prince 1980). Larger birds, such as albatrosses, also take *Pseudochaenichthys* (Channichthyidae) and *Notothenia* species (Nototheniidae) (Prince 1980a). Fish of these families are the main prey of diving seabirds (penguins, cormorants) which spend most time in inshore and continental shelf waters. At Signy Island, *Nototheniops nudifrons* (75%), *Notothenia rossii* (14%) and *Trematomus newnesi* (7%) comprise the bulk of the diet by weight of Imperial Cormorants (Shaw 1984). At South Georgia, however, *N. rossii* comprises about 50% of the diet, the remainder being other *Notothenia* species (North and Croxall, unpubl.). *Notothenia rossii, Patagonotothen larseni* and *Champsocephalus gunnari* are probably of almost equal importance in the diet of Gentoo Penguins (Croxall and Prince 1980b, unpubl.) at South Georgia. Southern Lampreys, *Geotria australis,* form 10% by weight of the diet of Greyheaded Albatrosses, *Diomedea chrysostoma;* this represents just over 1 million individuals taken per season (Prince 1980a), all being sub-adults presumably returning to spawn in South American streams (Potter et al. 1979).

5.2 Comparison with Stock Assessments

Despite extensive fishery operations in the Scotia Sea, few assessments of standing stocks have been made. Everson (1977) made a very preliminary estimate of 500,000 t for the South Georgia shelf. Fish consumption by seals and penguins, however, may total 1.35×10^6 t. Only if two-thirds of the foraging ranges of fish predators lie outside the shelf or if Elephant Seals only eat 5% (not 25%) fish, would values less than the standing stock be achieved. It is possible, therefore, that predators are more effective samplers of fish stocks than limited fishery surveys, and that the true stock size is appreciably larger than hitherto recognized.

Around South Georgia, *N. rossii* was the main commercial target but its share of the catch dropped from 98% in 1969 to 0.4% in 1980 (BIOMASS 1980). Burchett and Ricketts (1984) estimate the 1978–79 population biomass (after natural predation) to have been 117,000 t and the fishery catch 45,000 t. Seabirds (mainly Gentoo Penguins and Imperial Cormorants) take 11,000 t and Antarctic Fur Seals 15,000 t. If only 5% of the fish diet of Elephant Seals is *N. rossii,* then this adds a further 45,000 t, making a total of 71,000 t or 38% of an original stock of 188,000 t. This is probably a reasonably conservative estimate and serves to illustrate: the importance of seabirds and seals as predators; the need for more information on the fish taken by predators, especially Elephant Seals; the vulnerability of the system to overfishing, unless the standing stock is substantially under-estimated. Because seabirds and seals mainly take juvenile *N. rossii,* their impact is less significant than if mature adults were taken; but, where there is also enhanced mortality of adults (e.g., through the commercial fishery), the situation is potentially serious.

Champsocephalus gunnari has now replaced *N. rossii* as the main target of the commercial fishery with a big increase in catches from 1975–76 to 1977–78. Over this period, stock biomass around South Georgia is estimated to have decreased from 141,000 to 34,000 t (Kock 1981). If the winter and summer fish diets of Antarctic Fur Seals are similar, then, the species takes ca. 200,000 t of *C. gunnari* (73% in winter), representing 59% by weight of the 1975–76 total stock. Since both the seals and trawlers are chiefly exploiting adult fish, the combined predation pressure must be significant and suggests potential competition between seals and fisheries.

6 Conclusions

We are still at a very preliminary stage in the analysis of Southern Ocean predator-prey interactions, even for such a well-studied area as the Scotia Sea. Although this paper develops initiatives (Laws 1977b; Øritsland 1977) started at the last SCAR Symposium on Antarctic Biology, held in 1974, there are still many urgent requirements for data which must be satisfied before more realistic assessments can be made.

For estimating the food consumption of predators, we particularly need: (1) better estimates of breeding populations of Chinstrap Penguins at the South Sandwich Islands, of Macaroni Penguins at South Georgia and of small petrels, Crabeater and Weddell Seals throughout; (2) demographic information to estimate pre- and non-breeding populations; (3) quantitative dietary information for species in winter and for all seals, especially Elephant Seals (ideally this should include realistic estimates of minor constituents, since only 1% by weight of the diet of a large abundant predator can represent a very substantial consumption); and, (4) any data on seal bio-energetics and on energy costs of flight in seabirds.

In assessing the impact of predators on Krill, fish, and squid stocks, some of the main problems and requirements are: (1) studies of predator diets should in future record age, sex, size, and reproductive status of prey (relevant analytical techniques should exactly match those used with net-haul specimens); (2) discrepancies, especially at South Georgia, between length-frequencies in Krill in samples from nets and predators have important consequences (Krill biomass estimates using acoustic echo-integration techniques depend on calculations of target strength from net-haul samples. If these do not include the bulk of the length classes taken by predators in the same area then the estimates will be inaccurate and also misleading when assessing the biomass of Krill of actual relevance to the predators. Detailed Krill sampling, including acoustic work using upward-directed echosounders on a smaller geographical scale in areas of high concentrations of breeding penguins and seals would seem an appropriate first step towards resolving this problem); (3) squid populations are at present best sampled by predators but much more information is needed, particularly from seals and penguins; (4) fish stocks seem low compared with seal requirements, but the latter are estimated on the basis of minimal data on the fish diets of seals (studies at least of otoliths in stomach contents and/or faeces are urgently required); and, (5) data on activity budgets (including diving patterns) at sea are needed to define foraging ranges (and hence areas of effective impact of seabirds and seals (information on distribution in the non-breeding season can probably only come from the use of satellite telemetry and extensive quantitative surveys).

Despite these deficiencies, it is clear that seabirds and seals have an important impact on Krill and their interactions with Krill may have influenced the timing and nature of breeding cycles and population dynamics in both predators and prey. The impact of predators on squid populations cannot yet be assessed. Although fish are generally of lesser importance (than squid and Krill) to seals and seabirds, these predators may still have significant effects on prey stocks and important interactions with commercial fisheries.

Appendix. Composition by percentage numbers and weight[a] of the cephalod diet of albatrosses[b], seals[c] and Sperm Whales[d] in the Scotia Sea

Predator species	Sperm Whale (SG)		Wandering Albatross		Greyheaded Albatross		Blackbrowed Albatross		Lightmantled Sooty Albatross		Weddell Seal		Elephant Seal		Sperm Whale (SS)	
Proportion (% by wt.) squid in diet	ca 95		80		49		21		60		c 11		c 75		c 95	
Sample size (beaks)	1,640		527		322		100		191		366		50		5,406	
	No.	Wt	No.	Wt.	No.	Wt.	No.	Wt.	No.	Wt.	No.	Wt.	No.	Wt.	No.	Wt.
Prey species																
Squid																
Architeuthis sp	1.2	1.7														
Todarodes sagittatus	2.1	0.6	0.4	0.1	88.2	91.4	68.0	75.9	1.6	1.0					0.1	–
Kondakovia longimana	21.6	14.6	40.0	81.1	0.6	3.9	1.0	1.2	0.5	2.0	0.5	0.3	4.0	6.6	10.1	15.1
Moroteuthis knipovitchi	17.8	2.1	1.9	2.0							31.0	48.5	14.0	9.6	57.1	15.0
M. robsoni															2.5	1.8
Moroteuthis sp A	2.1	0.7														
Gonatus antarcticus	0.2	–	3.6	0.8					4.8	5.3	0.8	0.6	42.0	15.0	4.8	1.6
G. sp A			2.1	1.2												
Ancistrocheirus lesueuri	2.8	0.5	1.5	1.4			1.0	6.6							0.6	0.2
Octopoteuthis rugosa			1.5	0.3											0.4	–
Taningia dana	10.8	6.7	1.3	5.5											0.5	0.7
Lepidoteuthis grimaldi	0.9	0.3	0.9	0.8					1.1	c 1.0					0.3	0.3
Chiroteuthis veranyi	0.2	0.1	1.0	–	0.2	0.2	2.0	1.2							0.1	–
? *Mastigoteuthis* sp A	3.5	0.1	2.8	0.7	0.6	0.4										
? *Discoteuthis*									21.8	58.4						
Histioteuthis eltaninae	3.0	–	8.7	0.4	0.6	0.2			0.5	0.2						
H. atlantica			5.9	0.7												
H. miranda, H. dofleini	4.4	0.3														
H. sp A5									4.8	10.0					10.0	1.3
Psychroteuthis glacialis	0.6	0.1									28.7	23.3			0.4	0.1
sp aff *Psychroteuthis*													20.0	3.9		
Taonius pavo	1.0	–	16.8	3.4											0.3	–
Alluroteuthis antarcticus	0.3	–									2.2	0.5	2.0	0.2	1.2	0.1
Phasmatopsis cymoctypus	1.0	–													0.2	–
Galiteuthis glacialis	0.4	–	8.0	0.9	7.5	3.9	25.0	12.0	58.5	18.6			2.0	c 2.0		
Mesonychoteuthis hamiltoni	23.0	72.1	0.2	0.4											9.3	64.7
Brachioteuthis picta											1.4	0.3				
Octopus																
Alloposus mollis	3.0	0.2	1.7	1.0											0.6	0.1
Eledoninae							1.0	0.6			34.7	26.2	10.0	59.9		
Other (species)	II 2.6	0.1	(5) 1.2	0.5			I 1.0	0.2	I 0.5	0.4	(5) 1.0	0.2				

a Only species forming more than 1% by numbers or weight in the diet of any predator are included
b Data from Clarke et al. (1981), Clarke and Prince (1981), Croxall and Prince (1980a), Thomas (1982)
c Data from Clarke and McLeod (1982a, b)
d Numerical data from Clarke (1980) (Table 42 for South Georgia (SG); Table 43, whales 471–474 for rest of Scotia Sea (SS)); mean weights from Clarke (1980) (Table 10, using data from Durban for species not listed under Antarctic)

Acknowledgements. Many scientists have provided data used in this paper. We are grateful to them all but have particularly benefited from the specialist knowledge of A. Clarke, M.R. Clarke, R.W. Davis, G.L. Kooyman and A.W. North in our collaborations with them. We thank J.L. Bengtson, P. Shaw, and especially G.S. Lishman for allowing us to use their unpublished data. S.J. McInnes and A.C. Sylvester prepared the figures, E. Bailey typed the manuscript and C.M. Phillips obtained and checked numerous references.

References

Ainley DG; Emison WB (1972) Sexual size dimorphism in Adélie Penguins. Ibis 114:267–271

Beck JR; Brown DW (1972) The biology of Wilson's storm petrel, *Oceanites oceanicus* (Kuhl), at Signy Island, South Orkney Islands. Sci Rep Br Antarct Surv 69:54pp

BIOMASS (1980) Antarctic fish biology. Biomass Rep Ser 12:29pp

BIOMASS (1982) Post-Fibex data interpretation workshop. Biomass Rep Ser 20:38pp

Burchett MS; Rickets C (1984) The population dynamics of *Notothenia rossii* from South Georgia (Antarctica). Polar Biol 3:35–38

Clarke A (1980) The biochemical composition of krill, *Euphausia superba* (Dana), from South Georgia. J Exp Mar Biol Ecol 43:221–236

Clarke A (1984) Lipid content and composition of Antarctic krill *Euphausia superba* Dana. J Crust Biol Suppl 4:283–292

Clarke A; Prince PA (1980) Chemical composition and calorific value of food fed to mollymauk chicks *Diomedea melanophris* and *D. chrysostoma* at Bird Island, South Georgia. Ibis 122:488–494

Clarke MR (1977) Beaks, nets and numbers. Symp Zool Soc Lond 38:89–126

Clarke MR (1980) Cephalopoda in the diet of sperm whales of the Southern Hemisphere and their bearing on sperm whale biology. Discovery Rep 37:1–324

Clarke MR; MacLeod N (1982a) Cephalopods in the diet of elephant seals at Signy Island, South Orkney Islands. Bull Br Antarct Surv 57:27–31

Clarke MR; MacLeod N (1982b) Cephalopod remains in the stomachs of eight Weddell seals. Bull Br Antarct Surv 57:33–40

Clarke MR; Prince PA (1981) Cephalopod remains in regurgitations of blackbrowed and greyheaded albatrosses at South Georgia. Bull Br Antarct Surv 54:1–8

Clarke MR; Croxall JP; Prince PA (1981) Cephalopod remains in regurgitations of the wandering albatross at South Georgia. Bull Br Antarct Surv 54:9–22

Croxall JP (1982) Energy costs of incubation and moult in petrels and penguins. J Anim Ecol 51:177–194

Croxall JP (1984) Seabirds. In: Laws RM (ed) Antarctic Ecology. Academic Press, London New York, pp 533–619

Croxall JP; Furse JR (1980) Food of chinstrap penguins *Pygoscelis antarctica* and macaroni penguins *Eudyptes chrysolophus* at Elephant Island group, South Shetland Islands. Ibis 122:237–245

Croxall JP; Hunter I (1982) The distribution and abundance of burrowing seabirds (Procellariiformes) at Bird Island, South Georgia. II. South Georgia diving petrel *Pelecanoides georgicus.* Bull Br Antarct Surv 56:69–74

Croxall JP; Pilcher M (1984) Length-frequency distribution and sex ratio of Krill eaten by Antarctic Fur Seals *Arctocephalus gazella* at South Georgia. Bull Br Antarct Surv 63:117–125

Croxall JP; Prince PA (1980a) Food, feeding ecology and ecological segregation of seabirds at South Georgia. Biol J Linn Soc 14:103–131

Croxall JP; Prince PA (1980b) The food of gentoo penguins *Pygoscelis papua* and macaroni penguins *Eudyptes chrysolophus* at South Georgia. Ibis 122:245–253

Croxall JP; Prince PA (1982a) A preliminary assessment of the impact of seabirds on marine resources at South Georgia. Com Nat Franc Recherch Antarct 51:501–509

Croxall JP; Prince PA (1982b) Calorific content of squid (Mollusca: Cephalopoda). Bull Br Antarct Surv 55:27–31

Croxall JP; Prince PA; Hunter I; McInnes S; Copestake PG (1984). a) The seabirds of the Antarctic Peninsula, islands of the Scotia Sea and Antarctic Continent between 80°W and 20°W: their status and conservation. In: Croxall JP, Evans PGH, Schreiber RW (eds) The status and conservation of the world's seabirds. ICBP, Cambridge, pp 635–664

Croxall JP; Ricketts C; Prince PA (1984) b) Impact of seabirds on marine resources, especially krill, of South Georgia waters. In Whittow GC, Rahn A (eds) Seabird Energetics. Plenum, New York, pp 285–318

Croxall JP; Everson I; Kooyman GL; Ricketts C; Davis RW 1985 Fur seal diving behaviour in relation to vertical distribution of krill. J Anim Ecol 54

Davis RW; Kooyman GL; Croxall JP (1983) Water flux and estimated metabolism of free-ranging gentoo and macaroni penguins at South Georgia. Polar Biol 2:41–46

Doidge DW; Croxall JP (1985) Diet and energy budget of the Antarctic Fur Seal, *Arctocephalus gazella* at South Georgia. In: Siegfried WR, Condy PR, Laws RM (eds) Antarctic nutrient cycles and food webs (Proceedings of the 4th SCAR symposium on Antarctic biology). Springer, Berlin Heidelberg New York pp

Emison WB (1968) Feeding preferences of the Adélie penguin at Cape Crozier, Ross Island. In: Austin OL (ed) Antarctic Bird Studies (Antarctic Research Series 12) American Geophysical Union Washington DC, pp 191–212

Everson I (1977) The Living Resources of the Southern Ocean. GLO/SO/77/1 FAO, Rome, 156pp

Everson I (1981) Fish. In: Biological Investigations of Marine Antarctic Systems and Stocks Vol II. SCAR/SCOR, Cambridge, pp 79–97

Filippova JA (1972) New data on the squids (Cephalopoda: Oegopsida) from the Scotia Sea (Antarctic). Malacologia 11:391–406

Filippova JA; Yukhov VL (1979) A list of cephalopods and their distribution in the meso and bathypelagical of the Antarctic water. Antarktika 18:175–187 (In Russian)

Gilbert JR; Erickson AW (1977) Distribution and abundance of seals in the pack ice of the Pacific sector of the Southern Ocean. In: Llano GA (ed) Adaptations within Antarctic Ecosystems. Smithsonian Institution, Washington DÇ, pp 703–740

Griffiths DH; Barker PF (1971) Review of marine geophysical investigations in the Scotia Sea. In: Adie RJ (ed) Antarctic Geology and Geophysics. Oslo, Universitetsforlaget, pp 3–11

IWC (1978) Report of the scientific committee. Rep Int Whal Commun 28:82–83

Jazdzewski K; Dzik J; Porebski J; Rakusa-Suszczewski S; Witek Z; Wolnomiejski N (1978) Biological and populations studies on krill near South Shetland Islands, Scotia Sea and South Georgia in the summer 1976. Pol Arch Hydrobiol 25:607–631

Kendeigh SC; Dol'nik VR; Gavrilov VM (1977) Avian energetics. In: Pinowski J, Kendeigh SC (eds) Granivorous birds in ecosystems. Cambridge University Press, Cambridge, pp 127–204

Kittel W (1980) Populational studies on *Euphausia superba* Dana, 1852 (Euphausiacea, Crustacea) in waters of the Admiralty Bay during the Antarctic summer of 1978. Pol Arch Hydrobiol 27: 267–272

Kock K-H (1981) Fischereibiologische Untersuchungen an drei antarktischen Fischarten: *Champsocephalus gunnari* Lonnberg, 1904. *Chaenocephalus aceratus* (Lonnberg, 1905) und *Pseudochaenichthys georgianus* Norman, 1937 (Notothenioidei, Channichthyidae). Mitt Inst Seefisch BundForsch Anst Fisch 32: 226 pp

Kooyman GL (1981) Weddell Seal, consummate diver. Cambridge University Press, Cambridge, 135 pp

Kooyman GL; Davis RW; Croxall JP; Costa DP (1982) Diving depths and energy requirements of king penguins. Science 217:726 –627

Kooyman GL; Davis RW; Croxall JP (to be publ.) Diving behavior of the Antarctic fur seal *Arctocephalus gazella.* In: Kooyman GL, Gentry RL (eds) Fur Seals: Maternal Strategies on Land and at Sea. Princeton University Press, Princeton

Laws RM (1977a) The significance of vertebrates in the Antarctic marine ecosystem. In: Llano GA (ed) Adaptations within Antarctic Ecosystems. Smithsonian Institution, Washington DC, pp 411–438

Laws RM (1977b) Seals and whales of the Southern Ocean. Philos Trans R Soc Lond B279:81–96

Linkowski TB; Rembiszewski JM (1978) Ichthyological observations of the South Georgia coasts. Pol Arch Hydrobiol 25:697 –704

Lishman GS (1985) The food and feeding ecology of Adélie Penguins *Pygoscelis adeliae* and Chinstrap Penguins *P. antarctica* at Signy Island, South Orkney Islands. J Zool Lond 205

Lishman GS; Croxall JP (1983) Diving depths of chinstrap penguins. Bull Br Antarct Surv 61:21–25

Mackintosh NA (1972) Life cycle of Antarctic krill in relation to ice and water conditions. Discovery Rep 36:1–94

Mackintosh NA (1974) Sizes of krill eaten by whales in the Antarctic. Discovery Rep 36:157–178

Marr JWS (1962) The natural history and geography of the Antarctic krill (*Euphausia superba* Dana). Discovery Rep 32:33–464

McCann TS (1985) Size, status and demography of southern elephant seal *Mirounga leonina* populations. In: Ling JK, Bryden MM (eds) Sea Mammals of Southern Latitudes. South Australian Museum, Adelaide

Nemoto T; Doi T; Nasu K (1981) Biological characteristics of krill caught in the Southern Ocean. In: Biological Investigations of Marine Antarctic Systems and Stocks Vol II. SCAR/SCOR, Cambridge, pp 47–63

Nemoto T; Okiyama N; Takahashi M (1985) Aspects of the roles of squid in food chains of marine Antarctic ecosystems. In: Siegfried WR, Condy PR, Laws RM (eds) Antarctic nutrient cycles and food webs (Proceedings of the 4th SCAR symposium on Antarctic biology). Springer, Berlin Heidelberg New York pp

North AW; Croxall JP; Doidge DW (1983) Fish prey of the Antarctic fur seal *Arctocephalus gazella* At South Georgia. Bull Br Antarct Surv 61:27–37

Øritsland T (1977) Food consumption of seals in the Antarctic pack ice. In: Llano GA (ed) Adaptations within Antarctic Ecosystems. Smithsonian Institution, Washington DC, pp 749–768

Payne MR; Prince PA (1979) Identification and breeding biology of the diving petrels *Pelecanoides georgicus* and *P. urinatrix exsul* at South Georgia. NZ J Zool 6:299–318

Pinder RW (1966) The cape pigeon *Daption capensis* Linnaeus at Signy Island, South Orkney Islands. Bull Br Antarct Surv 8:19 –47

Potter IC; Prince PA; Croxall JP (1979) Data on the adult marine and migratory phases in the life cycle of the southern hemisphere lamprey *Geotria australis* Gray. Environ Biol Fishes 4:65–69

Prince PA (1980a) The food and feeding ecology of grey-headed albatross *Diomedea chysostoma* and black-browed albatross, *D. melanophris.* Ibis 122:476–488

Prince PA (1980b) The food and feeding ecology of the blue petrel *Halobaena caerulea* and dove prion *Pachyptila desolata.* J Zool Lond 190:59–76

Prince PA; Francis M (1984) Activity budgets of foraging grey-headed albatrosses. Condor 86:297–300

Rembiszewski JM; Krzeptowski M; Linkowski TB (1978) Fishes (Pisces) as by-catch in fisheries of krill *Euphausia superba* Dana (Euphausiacea, Crustacea). Pol Arch Hydrobiol 25:677–695

Roper CFE (1969) Systematics and zoogeography of the worldwide bathypelagic squid *Bathyteuthis* (Cephalopoda:Oegopsida). Bull US Nat Mus 291:1–208

Stepnik R (1982) All-year populational studies of Euphausiacea (Crustacea) in the Admiralty Bay (King George Island, South Shetland Islands, Antarctic). Pol Polar Res 3:49–68

Targett TE (1981) Trophic ecology and structure of coastal Antarctic fish communities. Mar Ecol Prog Ser 4:243–263

Thomas G (1982) The food and feeding ecology of the light-mantled sooty albatross at South Georgia. Emu 82:92–100

Thomas G; Croxall JP; Prince PA (1983) Breeding biology of the light-mantled sooty albatross at South Georgia. J Zool Lond 199:123–135

Volkman NH; Presler P; Trivelpiece W (1980) Diets of pygoscelid penguins at King George Island, Antarctica. Condor 82:373–378

Wiens JA (1984) Modelling the energy requirements of seabird populations. In Whittow GC, Rahn A (eds) Seabird Energetics. Plenum, New York

The Role of Giant Petrels in the Southern Ocean Ecosystem

S. HUNTER[1,2]

Summary. Giant petrels are the dominant scavenging seabird species in the Sub-Antarctic and maritime Antarctic ecosystems. They consume large amounts of penguin carrion, but also include significant numbers of seals, Antarctic Krill, *Euphausia superba,* and small petrels in their diets. Using results of detailed dietary studies at Bird Island, South Georgia, and qualitative data for other localities, notable variation in the quantitative impact of Giant petrels on their prey is revealed. Energy budgets of chicks are calculated from data on meal size and feeding frequency at all stages of growth. Differences in energy requirements between the Northern Giant Petrel, *Macronectes halli,* and the Southern Giant Petrel, *M. giganteus,* and between male and female chicks, are demonstrated. Published and unpublished data on numbers of giant petrels at known breeding sites are reviewed, and the present world breeding population is concluded to be ca. 8,600 pairs of *M. halli* and ca. 38,000 pairs of *M. giganteus.* Using survival and demographic data for South Georgia, total world non-breeding populations of ca. 26,000 and ca. 113,000 birds are calculated for *M. halli* and *M. giganteus,* respectively. The energy consumption of these populations in the breeding season is assessed and, taking into account geographical variation in diet, their impact on local ecosystems is evaluated. The approximations and assumptions of these estimates are discussed.

1 Introduction

The two sibling species of Giant petrels, *Macronectes halli* and *M. giganteus,* are the principal scavenging seabird species of the Southern Ocean, feeding chiefly on penguin and seal carrion (Conroy 1972; Johnstone 1977; Bonner and Hunter 1982; Hunter 1983). However, they also prey on other birds, Krill, *Euphausia superba,* squid and fish.

There are indications that diet differs between localities, due to differences in prey availability and/or different feeding habits. This paper aims to identify any such regional dietary differences during the breeding season, and to assess the impact on prey populations, by combining dietary, energetic and demographic data for Giant petrel populations.

1 British Antarctic Survey, NERC, High Cross, Madingley Road, Cambridge CB3 OET, United Kingdom

2 Present address: FitzPatrick Institute, University of Cape Town, Rondebosch 7700, South Africa

2 Methods and Data-Base

2.1 Diet

Bird Island, South Georgia, is the only locality for which both qualitative (frequency of occurrence) and quantitative (proportion by weight) dietary data are available for Giant Petrels (Table 1). The frequency of occurrence of dietary items at other localities (Table 2) was converted into proportions by weight, using the Bird Island Conversion factors (Table 3).

2.2 Demography

Breeding population sizes are summarized and combined into five geographical regions (Table 4), so as to ensure that each region contained a locality for which the composition of the species' diet was known. South Atlantic populations were put into Region 5, with the New Zealand populations, because both areas are north of the Antarctic Convergence.

The sizes of the pre-breeding and non-breeding populations were estimated from the number of breeding birds. The mean age of first breeding is ca. 10 years (British Antarctic Survey (BAS), unpubl. data), and the mean annual survival to that age is 0.88 (Hunter, to be publ.). About 35% of birds known to have bred previously do not breed in any one season (Hunter 1984). Table 5 summarizes the number of breeding, pre-breeding and non-breeding birds for each region. In addition, 30% of pairs that lay fail to incubate their eggs and a further 20% fail in rearing their chicks. After failure, these birds are treated as non-breeders and failed birds are taken as staying in the same region as the breeding adults.

2.3 Energy Requirements

Adult energy costs while brooding were taking as being equivalent to the existence energy requirements (EER) in non-passerines, given by Kendeigh et al. (1977) as

$$EER = a - bt \text{ (kcal d}^{-1}\text{)}$$

where a is the EER at 0°C, which is $4.142\ W^{0.544}$, b is a temperature coefficient ($0.2761\ W^{0.2818}$) and t is the

Antarctic Nutrient Cycles and Food Webs
(ed. by W.R. Siegfried, P.R. Condy, and R.M. Laws)

Table 1. Composition (%) of the diet of Giant Petrels at South Georgia and conversion factors for converting frequency of occurrence into percentage weight

Species	n	Food type						
		Seals and other mammals	Penguins	Other birds	Crustacea	Squid	Fish	Other
M. halli								
Frequency of occurrence, 0	165	13	54	17	34	12	10	4
Percentage weight, W	165	13	53	7	22	2	3	1
Conversion factor, W/0		1.00	0.98	0.41	0.65	0.17	0.30	0.25
M. giganteus								
Frequency of occurrence, 0	63	10	56	25	41	8	2	6
Percentage weight, W	63	6	62	9	21	1	1	1
Conversion factor, W/0		0.60	1.11	0.36	0.51	0.13	0.50	0.17

Table 2. Frequency of occurrence (%) of food types in the diet of Giant petrels at various localities

Locality	Species	n	Food type							Authority
			Seals and other mammals	Penguins	Other birds	Crustacea[a]	Squid	Fish	Other	
Bird Island	*M. halli*	165	13	54	17	34	12	10	4	Hunter (1983)
	M. giganteus	63	10	56	25	41	8	2	6	
Iles Crozet	Both	12	22	.... 94		3	13	3	3	Mougin (1975) including Voisin (1968)
Macquarie Island	*M. halli*	229	6[b]	87	20[c]	7	19	1	4	Johnstone (1977)
	M. giganteus	179	1[b]	96	12[c]	16	44	5	4	
Signy Island	*M. giganteus*	96	10	67	7	65	32	3	c 5[d]	Conroy (1972)
Terre Adélie	*M. giganteus*	18	28	.... 78		22	11	11	6	Mougin (1968)

a At Bird and Signy islands this includes Krill only; other Crustacea are in the "Other" column

b Johnstone also had an unidentified carrion category – presumably including mammal and avian remains. This occurred in 70% of *M. giganteus* and 36% of *M. halli*

c Johnstone divided "Other" remains into petrel feathers (given above) and petrel bones (5% in *M. giganteus* and 3% in *M. halli*)

d Conroy divided the "Other" category into other Crustacea (4%), vegetable material (7%) and other material (7%)

ambient temperature. For birds away from the nest energy requirements were calculated as 1.85 times EER (see Croxall and Prince 1982). Energy values were converted to kilo-Joules (1 kcal = 4.2 kJ). Croxall (1982) used empirical data to determine the relationship between body weight and daily energy consumption in incubating procellariiforms and his regression equation

$$y = 0.67\,x - 0.029$$

where y = $\log_{10}$ energy consumption (kcal d^{-1}) and x is $\log_{10}$ body weight (g), was used here to estimate daily energy requirements. Table 6 gives the weight and ambient temperature values used in these bio-energetic calculations.

2.4 Breeding

The duration of breeding events (Table 7), taken from Hunter (1984), shows that: males take a larger share

Table 3. Percentage biomass of food types consumed by Giant Petrels at various localities

Locality	Species	Food type						
		Seals	Penguins	Small birds	Crustacea	Squid	Fish	Other
Bird Island	*M. halli*	13	53	7	22	2	3	1
	M. giganteus	6	62	9	21	1	1	1
Signy Island	*M. giganteus*	5	61	2	27	4	1	1
Terre Adélie	*M. giganteus*	15	63	5	10	1	5	1
Iles Crozet	Both	15	65	15	1	2	1	1
Macquarie Island	*M. halli*	6	79	8	4	3	1	1
	M. giganteus	1	83	3	6	6	2	1

Table 4. Estimates of Giant petrel breeding populations

Region	Locality	No. pairs		Authority
		M. giganteus	*M. halli*	
1	South Georgia	ca. 6,000	ca. 2,500	Prince and Croxall (1983)
2	Antarctic Peninsula	1,125	0	Croxall et al. (1984)
	South Shetlands	5,315	0	Croxall et al. (1984)
	Elephant Island Group	870	0	Croxall et al. (1984)
	South Orkneys	8,755	0	Croxall et al. (1984)
	South Sandwich Is.	520+	0	Holdgate and Baker (1979)
	Bouvetøya	ca. 25	0	Fevolden and Sømme (1977)
	Total	ca. 16,610	0	
3	Frazier Islands	100–200	0	Cowan (1979)
	Hawker Island	25– 50	0	GW Johnstone (pers. comm.)
	Mawson Station area	10– 15	0	GW Johnstone (pers. comm.)
	Point Geologie	ca. 30	0	Mougin (1975)
	Total	ca. 305	0	
4	Iles Crozet	1,067	1,363	Jouventin et al. (1984)
	Iles Kerguelen	0	600?	Derenne et al. (1974)
	Marion Island	1,337	210	Williams et al. (1979)
	Prince Edward Island	410	ca. 100	Williams et al. (1979)
	Heard Island	ca. 4,000	0	Downes et al. (1959)
	McDonald Island	800–900	0	Johnstone (1982)
	Total	ca. 7,714	ca. 2,273	
5	Macquarie Island	4,000	1,000	Johnstone (1977)
	New Zealand Islands	0	2,820	CJR Robertson (pers. comm.)
	Isla de los Estados	30+	0	Imshaug (1972)
	Diego Ramirez	ca. 90	0	R Schlatter (pers. comm.)
	Falkland Islands	3,200	0	Croxall et al. (1984)
	Chubut, Argentina	ca. 100	0	PS Humphrey (pers. comm.)
	Gough Island	ca. 50[a]	0	Williams and Imber (1982)
	Total	ca. 7,470	3,820	
	Grand Totals	ca. 38,000	ca. 8,600	

a The specific status of these birds is in some doubt

Table 5. Estimate of the world population of Giant Petrels

Region	*M. halli*				*M. giganteus*			
	Breeding birds	Non-breeding birds	Pre-breeding birds	Chicks fledging	Breeding birds	Non-breeding birds	Pre-breeding birds	Chicks
1	5,000	1,750	5,718	1,250	12,000	4,200	13,722	3,000
2	0	0	0	0	33,220	11,627	37,987	8,305
3	0	0	0	0	610	214	698	153
4	4,546	1,592	5,200	1,137	15,428	5,400	17,642	3,857
5	7,640	2,674	8,733	1,910	14,940	5,229	17,084	3,735

Grand totals: *M. halli* 47,150
M. giganteus 208,737

Table 6. Weight of adult Giant Petrels and ambient temperatures for various localities

Region	Locality	Species	Sex	n	Adult weight (g) mean ± one s.d.	Authority	Ambient temperature[a] (°C)	Authority
1	South Georgia	*M. halli*	Males	56	4,902 ± 391	Hunter (1984)	4°	Richards and Tickell (1968)
			Females	43	3,724 ± 313			
		M. giganteus	Males	92	5,035 ± 406			
			Females	78	3,798 ± 357			
2	Signy Island	*M. giganteus*	Males	37	4,940 ± 410	Conroy (1972)	–1°	Holdgate et al. (1967)
			Females	37	3,850 ± 370			
3	Terre Adélie	*M. giganteus*	Males	5	4,864 ± 720	Voisin and Bester (1981)	–7°	Mougin (1975)
			Females	1	3,350			
4	Iles Crozet	*M. halli*	Males	18	4,711 ± 96	Voisin (1968)	6°	Mougin (1975)
			Females	12	3,571 ± 121			
		M. giganteus	Males	9	5,019 ± 76			
			Females	7	3,944 ± 224			
5	Macquarie Is.	*M. halli*	Males	57	4,790 ± 330	Johnstone (1977)	5°	Law and Burstall (1956)
			Females	49	3,580 ± 300			
		M. giganteus	Males	35	4,640 ± 440			
			Females	41	3,690 ± 340			

a Ambient temperatures are the average of the mean monthly temperatures for October–April

Table 7. Average time (days) spent by each sex on different breeding activities

Species	Incubation			Brood/Guard period			Remainder of fledging period	
	Males	Females	Total	Males	Females	Total	Male chicks	Female chicks
M. halli	33	27	60	18	13	31	82	78
M. giganteus	33	27	60	11	11	22	99	93

of incubation duties than females; there are differences between species and sexes in the brood-guard period; and, male chicks take longer to fledge than female chicks.

2.5 Food Consumption

Food consumption of chicks was estimated directly, using data on meal size and feeding frequency (Hunter 1983). This was combined with the dietary information from the five regions (Table 3) to estimate the composition of prey fed to chicks. Food consumption estimates for adults assumed a 75% assimilation efficiency. The energy values used in converting the energy requirements into prey estimates were 4.35 kJ g^{-1} for Krill, 3.98 for fish, 3.48 for squid (Croxall and Prince 1982), and 7.96 for penguin carrion (Burger 1981) which is also the value used for other birds. Fur Seal bulls have an energy value of ca. 10.47 kJ g^{-1}, though this varies according to an animal's amount of blubber (D.W. Doidge, pers. comm.). The "other" food category was given an arbitrary value of 4.00 kJ g^{-1}.

3 Results and Discussion

3.1 Diet

All dietary studies of Giant Petrels have shown that penguins are the most important prey of the two species throughout the Southern Ocean (Table 2–3). Penguins are either depredated as chicks (e.g., Warham 1962); Voisin 1968; Despin 1982), killed after being injured by seals (Bonner and Hunter 1982) or scavenged (e.g., Warham and Bell 1979; Voisin 1968; Johnstone 1977).

The main differences in diet between localities are in the importance of various secondary prey resources. Whereas penguins are abundant in all areas where Giant petrels breed, other food types vary in abundance and this can be reflected by their incidence in the diets of the two species. At localities south of the Antarctic Convergence, crustaceans (almost exclusively, *Euphausia superba*) can form up to 25% of the diets by weight, whereas at the Crozet and Macquarie islands, both north of Antarctic Convergence, they are mainly mysids which comprise only a small part of the diets of the two species (Johnstone 1977).

The only principal and consistent dietary separation between the two species is the greater utilization of seal carrion by *M. halli.* This occurs in *M. halli* mainly during the first 30 days of the chick period, when it is difficult to collect food samples from chicks (Hunter 1983). Thus the amount of seal carrion is considerably under-estimated. At Bird Island, more than 90% of the Giant petrels at Antarctic Fur Seal, *Arctocephalus gazella,* carcasses were male *M. halli,* and in this species chick-hatch coincides with the short period of seal carrion abundance (Hunter 1983). Johnstone (1977) suggested that the relatively greater tendency of *M. halli* to forage ashore may be related to the present or past high density of breeding seals at Sub-Antarctic islands occupied by the species.

Birds other than penguins are over-estimated in the diets of the two species, due to the retention of hard parts (heads and legs) in stomachs but, nevertheless, at certain sites (Bird Island and the Crozet Islands) they form a significant proportion of the diets. Small petrels, such as prions, *Pachyptila* spp., are probably killed at sea. Although burrowing petrels are abundant at Sub-Antarctic sites, they are replaced by larger, crevice-nesting species, such as Snow Petrels, *Pagodroma nivea,* farther south. These may be harder to catch and, thus, are less important in the diets of Giant Petrels at sites such as Signy Island.

Squid make up only a small fraction of the diets of the two species at all localities, although they are the second-most important prey (together with crustaceans) for *M. giganteus* at Macquarie Island. Fish are also unimportant as dietary items during the birds' breeding season, although they may constitute a larger part of both species' diets in winter (Hunter 1983).

3.2 Impact of Giant Petrels on Prey Resources

Male Giant Petrels require more energy than females, because they are larger (Table 8). The pre- and non-breeding portions of the population require 2–3 times the energy of the breeding birds (Table 8), partly because they outnumber the latter and partly because they do not spend time ashore fasting. The detailed impact of Giant Petrels on each prey type is given in Appendices 1 and 2; the main features being summarized in Table 9.

The average amount of flesh on a penguin is ca. 3 kg, although it varies markedly depending on species. On this basis, some 4.3×10^6 penguins are consumed annually by Giant Petrels during the breeding season, with totals of ca. 690,000 in Region 1 (from ca. 5.5 times 10^6 breeding pairs of penguins; Croxall et al. 1984) and ca. 1.5 times 10^6 birds taken in Region 2 (from ca. 7.2 times 10^6 breeding pairs of penguins; Croxall et al. 1984).

Estimating the number of small birds taken by Giant Petrels is more difficult. Such birds include many prions (ca. 140 g) but at certain sites representatives of larger species, such as the Whiteheaded Petrel, *Pterodroma lessoni,* (ca. 750 g) and Antarctic Fulmar, *Fulmarus glacialoides,* (ca. 800 g), are also taken. If a weight of 500 g per bird consumed is assumed, then, some 2.4×10^6 seabirds (excluding penguins) are taken by Giant Petrels in the Southern Ocean. Some 580,000 individuals (from ca. 30 $\times$ 10^6 breeding pairs of seabirds, Croxall et al. 1984) are taken in Region 1, and ca. 300,000 individuals (from ca. 9 million breeding pairs, Croxall et al. 1984) in Region 2.

This first estimate of the impact of Giant petrels in the Southern Ocean has involved the making of numerous assumptions and approximations. It was assumed that all Giant Petrels are restricted to waters around their sites of birth during the breeding season. Whereas this is reasonable

Table 8. Energy requirements (times 10^6 kJ) of adult Giant Petrel populations during the breeding season

Region	Species and sex	Breeders	Pre- and non-Breeders	Totals
1	*M. halli*			
	Males	756	2,067	2,823
	Females	684	1,776	2,460
	M. giganteus			
	Males	1,789	4,889	6,678
	Females	1,584	4,184	5,768
2	*M. giganteus*			
	Males	5,420	15,675	21,095
	Females	4,900	13,694	18,594
3	*M. giganteus*			
	Males	116	320	436
	Females	99	262	361
4	*M. halli*			
	Males	663	1,812	2,475
	Females	522	1,553	2,075
	M. giganteus			
	Males	2,431	5,851	8,282
	Females	2,199	5,116	7,315
5	*M. halli*			
	Males	1,132	3,095	4,227
	Females	885	2,634	3,519
	M. giganteus			
	Males	2,255	7,192	9,447
	Females	2,057	6,335	8,392

Table 9. Summary of total food consumption (t) by Giant Petrel populations

Region	Species	Food type							
		Seals	Penguins	Small birds	Crustacea	Squid	Fish	Other	Totals
1	*M. halli*	133.8	546.1	71.9	226.6	20.4	30.7	9.8	1,039.3
	M. giganteus	146.8	1,517.3	220.2	513.8	24.4	24.4	23.4	2,470.3
2	*M. giganteus*	369.6	4,510.0	147.8	1,996.2	295.7	73.8	70.5	7,463.6
3	*M. giganteus*	22.3	93.9	7.3	14.9	1.5	7.3	1.5	148.7
4	*M. halli*	121.4	526.2	121.4	8.0	16.0	8.0	7.7	808.7
	M. giganteus	415.7	1,801.8	415.7	27.6	55.4	27.6	26.4	2,770.2
5	*M. halli*	84.6	1,115.1	112.8	56.4	42.4	14.1	13.5	1,438.6
	M. giganteus	34.2	2,837.1	102.9	205.8	205.8	68.0	32.6	3,496.4

for failed breeders and many non-breeders, many younger immatures are in other regions or even outside the Southern Ocean altogether. In addition, the diet of these immature birds may be radically different; probably being more marine in nature since the birds are seen ashore infrequently.

Dietary data for Giant Petrels at many localities are meagre, and any major differences between localities within a region would markedly affect estimates of food consumption. This is most likely in Region 5, where the dietary information from Macquarie Island is unlikely to match closely that from the New Zealand and South Atlantic localities. Data from Bird Island have revealed significant differences between years in the prey taken by Giant Petrels (Hunter 1983), and ideally food sampling should be maintained over a number of years. Croxall and Prince (1982) have outlined the potential source of error in estimating the energy requirements of seabirds. Giant Petrels spend much of their foraging time sitting on the water or on land, and, consequently, the energy costs associated with foraging are probably over-estimated.

4 Conclusions

Despite potential errors, the important elements of the feeding impact of Giant Petrels in the Southern Ocean are probably as follows: throughout their ranges, both species consume significant numbers of penguins, although often they may not have killed them; in certain areas the level of predation on small petrels is high, although the effect is difficult to quantify. Seal carrion is an important food resource at certain sites, and may be an important factor in determining the abundance and distribution of the relatively early-breeding *M. halli;* Krill is probably important at most localities south of the Antarctic Convergence, as secondary prey, but nowhere is the impact likely to be very large compared with some other seabirds, such as certain penguins or Blackbrowed Albatrosses, *Diomedea melanophris,* (Croxall and Prince 1982); and, during the breeding season, squid and fish are unimportant components of the diets of Giant Petrels at most localities.

Appendix 1. Food consumption (t) of adult Giant Petrel populations, assuming 75% assimilation efficiency

Region	Species and sex	Food type							
		Seals	Penguins	Small birds	Crustacea	Squid	Fish	Other	Totals
1	*M. halli*								
	Males	66.6	271.5	35.8	112.7	10.2	15.3	4.9	517.0
	Females	58.0	236.6	31.2	98.2	8.9	13.3	4.2	450.4
	M. giganteus								
	Males	72.9	753.6	109.4	255.2	12.1	12.1	11.6	1,226.9
	Females	63.0	651.1	94.5	220.5	10.5	10.5	10.0	1,060.1
2	*M. giganteus*								
	Males	181.4	2,213.1	72.5	979.5	145.1	36.2	34.6	3,662.4
	Females	159.8	1,950.5	63.9	863.3	127.9	31.9	30.4	3,227.7
3	*M. giganteus*								
	Males	11.3	47.5	3.7	7.5	0.7	3.7	0.7	75.1
	Females	9.3	39.3	3.1	6.2	0.6	3.1	0.6	62.2
4	*M. halli*								
	Males	60.7	263.1	60.7	4.0	8.0	4.0	3.8	404.3
	Females	50.9	220.6	50.9	3.3	6.7	3.3	3.2	338.9
	M. giganteus								
	Males	203.1	880.4	203.1	13.5	27.0	13.5	12.9	1,353.5
	Females	179.4	777.7	179.4	11.9	23.9	11.9	11.4	1,195.6
5	*M. halli*								
	Males	42.6	561.2	56.8	28.4	21.3	7.1	6.8	724.2
	Females	35.4	467.1	47.3	23.6	17.7	5.9	5.6	602.6
	M. giganteus								
	Males	16.8	1,401.7	50.6	101.3	101.3	33.7	16.0	1,721.4
	Females	15.0	1,245.2	45.0	90.0	90.0	30.0	14.3	1,529.5

Appendix 2. Food consumption (t) for Giant petrel chick populations

Region	Species and sex	Food type							
		Seals	Penguins	Small birds	Crustacea	Squid	Fish	Other	Totals
1	*M. halli*								
	Males	5.1	21.0	2.7	8.7	0.7	1.1	0.4	39.7
	Females	4.1	17.0	2.2	7.0	0.6	1.0	0.3	32.2
	M. giganteus								
	Males	6.3	65.1	9.4	22.0	1.0	1.0	1.0	105.8
	Females	4.6	47.5	6.9	16.1	0.8	0.8	0.8	77.5
2	*M. giganteus*								
	Males	16.4	200.3	6.6	88.7	13.1	3.3	3.2	331.6
	Females	12.0	146.1	4.8	64.7	9.6	2.4	2.3	241.9
3	*M. giganteus*								
	Males	1.0	4.1	0.3	0.7	0.1	0.3	0.1	7.0
	Females	0.7	3.0	0.2	0.5	0.1	0.2	0.1	4.8
4	*M. halli*								
	Males	5.4	23.5	5.4	0.4	0.7	0.4	0.4	36.2
	Females	4.4	19.0	4.4	0.3	0.6	0.3	0.3	29.3
	M. giganteus								
	Males	19.2	83.1	19.2	1.3	2.6	1.3	1.2	127.9
	Females	14.0	60.6	14.0	0.9	1.9	0.9	0.9	93.2
5	*M. halli*								
	Males	3.6	47.9	4.8	2.4	1.9	0.6	0.6	61.8
	Females	3.0	38.9	3.9	2.0	1.5	0.5	0.5	50.3
	M. giganteus								
	Males	1.4	115.8	4.2	8.4	8.4	2.3	1.3	141.8
	Females	1.0	84.4	3.1	6.1	6.1	2.0	1.0	103.7

Acknowledgements. I thank Dr. J.P. Croxall for helpful comments during the preparation of this paper.

References

Bonner WN, Hunter S (1982) Predatory interactions between Antarctic Fur Seals, Macaroni Penguins and Giant Petrels. Br Antarct Surv Bull 56:75–79

Burger AE (1981) Time budgets, energy needs and kleptoparasitism in breeding Lesser Sheathbills (*Chionis minor*) Condor 83:106–112

Conroy JWH (1972) Ecological aspects of the biology of Giant petrel *Macronectes giganteus* (Gmelin) in the maritime Antarctic. Sci Rep Br Antarct Surv 75:74 pp

Cowan AN (1979) Ornithological studies at Casey, Antarctica, 1977–1978. Aust Bird Watcher 8:69–89

Croxall JP (1982) Energy costs of incubation and moult in petrels and penguins. J Anim Ecol 51:177–194

Croxall JP, Prince PA (1982) A preliminary assessment of the impact of seabirds on marine resources at South Georgia. CNFRA 51:501–509

Croxall JP, Prince PA, Hunter I, McInnes S, Copestake PG (1984) The seabirds of the Antarctic Peninsula, islands of the Scotia Sea and Antarctic Continent between 80°W and 20°W: their status and conservation. In: Croxall JP, Evans PGH, Schreiber RW (eds) The status and conservation of seabirds. ICBP Cambridge

Derenne P, Lufberry JX, Tollu B (1974) L'avifaune de l'Archipel Kerguelen. CNFRA 33:57–87

Despin B (1972) Note préliminaire sur Le Manchot Papou *Pygoscelis papua* de l'Ile de la Possession (Archipel Crozet). Oiseau Rev Fr Ornithol No special 42:69–83

Downes MC, Ealey EHM, Gynn AM, Young PS (1959) The birds of Heard Island. ANARE Rep Ser (B)51:135 pp

Fevolden SE, Sømme L (1977) Observations on birds and seals of Bouvetøya. Nor Polarinst Arbok 1976:367–371

Holdgate MW, Baker PE (1979) The South Sandwich Islands: 1 General description. Sci Rep Br Antarct Surv 91:76 pp

Holdgate MW, Allen SE, Chambers MJG (1967) A preliminary investigation of the soils of Signy Island, South Orkney Islands. Bull Br Antarct Surv 12:53–71

Hunter S (1983) The food and feeding ecology of the Giant petrels *Macronectes halli* and *M. giganteus* at South Georgia. J Zool (Lond) 200:522–538

Hunter S (1984) Breeding biology and population dynamics of Giant petrels *Macronectes* at South Georgia J Zool (Lond) 203: 441–460

Imshaug HA (1972) R/V Hero Cruise 71-5 to Isla de los Estados Antarct J US VII:42–44

Johnstone GW (1977) Comparative feeding ecology of the Giant petrels *Macronectes giganteus* (Gmelin) and *M. halli* (Mathews). In: Llano GA (ed) Adaptations within Antarctic ecosystems. Smithsonian Institute, Washington DC:647–688

Johnstone GW (1982) Zoology. In: Veenstra C, Manning J et al. Expedition to the Australian Territory of Heard and McDonald Islands, 1980. Technical Reports 31. Division of National Mapping, Canberra:33–39

Jouventin P, Mougin J-L, Stahl JC, Weimerskirch H (1984) The seabirds of the French Sub-Antarctic islands and Adélie Land, their status and conservation. In: Croxall JP, Evans PGH, Schreiber RW (eds). The status and conservation of seabirds. ICBP Cambridge

Kendeigh SC, Dol'nik VR, Gavrilov VM (1977) In: Pinowski J, Kendeigh SC (eds). Granivorous birds in ecosystems. Cambridge University, Cambridge:127–204

Law PG, Burstall T (1956) Macquarie Island. ANARE Interim Rep 14:48 pp

Mougin J-L (1968) Etude écologique de quatre espèces de pétrels antarctique. Oiseau Rev Fr Ornithol No special 38:1–52

Mougin J-L (1975) Ecologie comparée des Procellariidae Antarctiques et sub-Antarctiques. CNFRA 36:1–195

Prince PA, Croxall JP (1983) Birds of South Georgia: new records and re-evaluations. Br Antarct Surv Bull 59:15–27

Richards PA, Tickell WLN (1968) Comparison between the weather at Bird Island and King Edward Point, South Georgia. Bull Br Antarct Surv 15:63–69

Voisin J-F (1968) Les petrels geants (*Macronectes halli* et *M. giganteus*) de l'Ile de Possession. Oiseau Rev Fr Ornithol No special 38:95–122

Voisin J-F, Bester MN (1981) The specific status of Giant petrels *Macronectes* at Gough Island. In: Cooper, J (ed) Proceedings of the symposium on birds of the sea and shore, 1979. African Seabird Group, Cape Town:215–222

Warham J (1962) The biology of the giant petrel *Macronectes giganteus*. Auk 79:139–160

Warham J, Bell BD (1979) The birds of Antipodes Island, New Zealand, Notornis 26:121–169

Williams AJ, Imber MJ (1982) Ornithological observations at Gough Island in 1979, 1980 and 1981. S Afr J Antarct Res 12:40–45

Williams AJ, Siegfried WR, Burger AE, Berruti A (1979) The Prince Edward Islands: A sanctuary for seabirds in the Southern Ocean. Biol Conserv 15:59–71

Diet and Energy Budget of the Antarctic Fur Seal, *Arctocephalus gazella*, at South Georgia

D.W. DOIDGE and J.P. CROXALL[1]

Summary. The energy requirements and prey consumption of the rapidly expanding South Georgia population of Antarctic Fur Seals, *Arctocephalus gazella,* are estimated, using information on diet, population structure, activity patterns, and energy budgets and variations in these with sex and season. All quantitative information on diet is reviewed. Breeding female Fur Seals mainly take Antarctic Krill, *Euphausia superba.* This predominates also in the summer diet of males but fish (mainly *Champsocephalus gunnari*) and squid are also eaten. Fish is more important to males in winter. Life tables are used to determine the proportion of each age-class in the population, and especially to estimate the size of that portion of the immature stock which does not appear ashore. Energy costs of adult males are assessed from attendance data and age-specific weights. Female energy budgets are calculated from extensive data on the duration of periods ashore and foraging trips to sea together with information on activity budgets at sea. Pup energy consumption is calculated from body weight and growth data. A summer food consumption of 1.1×10^6 t is divided between juvenile males (39%), adult females (31%), adult males (22%) and juvenile females (8%). In winter, when females are absent, juvenile (67%) and adult males (33%) consume 0.4×10^6 t. Krill forms 69% of the annual consumption, fish 19% and squid 12%. The main peaks of demand are in October (all adults feeding at sea), January–February (rapidly growing pups being reared) and April–May (most of population, including weaners, feeding at sea). From data on diet and foraging ranges and depths, the main natural competitor of the Antarctic Fur Seal is the Macaroni Penguin, *Eudyptes chrysolophus;* significant competition with a developed Krill fishery would be expected. Continuing commercial exploitation of fish, especially *C. gunnari,* may influence male Fur Seal populations, particularly in winter.

1 Introduction

Antarctic Fur Seals, *Arctocephalus gazella,* at South Georgia, where the population has increased from near extinction last century to a present population of about 1×10^6 individuals (Payne 1977; Doidge et al. 1984a), must be an important local consumer of marine prey. With the present extensive exploitation of fish in South Georgia shelf waters and significant potential for developing an Antarctic Krill, *Euphausia superba,* fishery, it is timely to assess the role of Antarctic Fur Seals in more detail, especially in terms of potential interactions between Fur Seals, their prey and their competitors, both natural and commercial.

Previous assessments of food consumption of the Antarctic Fur Seal and other Antarctic seals (e.g., Laws 1977) have been derived simply by multiplying population estimates by the average daily food intake, assumed to represent a fixed proportion of the average seal's body weight. The composition of the diet has usually been taken (or modified) from the review of qualitative information by Øritsland (1977). In this paper we try to develop a more complex model, taking account of seasonal changes in energy requirements for the main categories of the Fur Seal population. Doing this highlights the many deficiences in our knowledge, even for what is probably now the best-known Antarctic seal. The main part of this paper, therefore, consists of a review of the data presently available, particularly on population structure and diet. This requires a brief introduction to the annual cycle of the species.

At South Georgia, female Antarctic Fur Seals haul-out on the breeding beaches during late November to mid-December (mean date 4 Dec.) and pup two days later (Doidge et al., to be publ.). They spend ca. 7 days with the pup (termed the perinatal attendance period) and then return to sea to feed. The feeding trips average four days and are separated by ca. two-day shore visits during which the pup is suckled. During the pup-rearing period about 65% of the female's time is spent at sea (Doidge et al., to be publ.). Pups wean and go to sea at age 112 days (pers. obs.). Most females return to shore for one or two visits after the pups have left.

Adult males establish territories from late October, the majority being ashore from late November to late December (McCann 1980). Small numbers of younger males continue to hold territories between January and April. Other immature animals make sporadic appearances from December to April, but their attendance patterns have not been well documented.

Little is known of the winter distribution of this species. Numbers of juvenile and adult males are seen ashore at South Georgia breeding sites throughout the winter (P.G. Copestake, pers. comm.). Females do not come ashore in winter and migration towards temperate waters, as seen in

1 British Antarctic Survey, NERC, High Cross, Madingley Road, Cambridge CB3 OET, United Kingdom

Antarctic Nutrient Cycles and Food Webs
(ed. by W.R. Siegfried, P.R. Condy, and R.M. Laws)

the Northern Fur Seal, *Callorhinus ursinus*, has been postulated; one juvenile from Bird Island was found on the Chilean coast (Payne 1979a).

2 Data Base and Methods

Prey consumption was modelled on a daily basis for each sex-age class using the following relationships

1. $PC = \sum_{ij} (D_{ij} \cdot PE_j \cdot ER_i / 0.75)$

where PC = prey consumption (biomass); D_{ij} = proportion of jth prey species in the diet of ith sex-age class; PE_j = energy content per unit mass of jth prey species; ER_i = daily energy requirement of the ith sex-age class; and 0.75 = conversion efficiency of prey energy to apparent digestible energy

2. $ER_i = \sum_{ij} (AC_ik \cdot T_{ik} \cdot N_i)$

where AC_{ik} = energy cost per unit time of the kth activity for the ith sex-age class; T_{ik} = time spent at kth activity by the ith sex-age class; and N_i = population size of the ith sex-age class.

Energy costs of the various activities were applied on a weight-specific basis to each sex-age class. Below we outline the data and methods used to estimate the above parameters.

2.1 Population Size, Structure and Body Weight

The Fur Seal population at South Georgia numbered ca. 369,000 in March 1976 (Payne 1979b), having increased by 16.8% per annum from 1958 to 1972 (Payne 1977). The occupation of new breeding beaches and increases at recently colonized sites suggest that the population is continuing to expand at a similar rate (Doidge et al. 1984a); if so, the population in March 1983 would number ca. 1.09×10^6 animals.

Herd structure, mean body weight and population biomass have been calculated by Payne (1979b) for males and females. Here we use his methods and data but sub-divide the herd further into adults, juveniles and weaners of each sex. The estimates of mean body weight, age-class size and biomass are given for each category in Table 1. Payne (1979b) used age-specific pregnancy and mortality rates of *C. ursinus* to estimate *A. gazella* female age-class size. The age structure of the male herd is assumed to be identical to that of the female herd to age six years, thereafter a 30% annual mortality is postulated (Payne 1979b). The mean age of males six years or older which die ashore is 8.8 years (unpubl. data), so such a mortality rate seems entirely realistic.

Eighty per cent of the adult female population is assumed to bear one pup annually. Pre-weaning pup mortality is fixed at 20% of pup production and decreases exponentially on a daily basis over the 112-day lactation period (Doidge et al., to be publ.).

2.2 Energy Requirements

No direct measurements of metabolism have been made on *A. gazella* so we rely heavily on studies conducted on *C. ursinus*, a species of similar size and habit (Miller 1978; Blix et al. 1979; Costa and Gentry, to be publ.).

The energy requirements used in the model are summarized in Table 2 and detailed below. Although we are dealing with the energy requirement for maintenance and growth rather than just metabolism, we have expressed these in units of standard metabolic rate (defined by Kleiber (1981) as SMR in $kJ\ d^{-1} = 16.73\ W^{0.75}$, where W is body weight in kg) in order to facilitate sex-age and species-specific comparisons. Doidge et al. (to be publ. b), using pup growth rates and estimates of body composition, calculated the energy consumption required by a female to rear a pup over a 115-day lactation period to be 1,787 MJ or 15.5 MJ d^{-1}. They assumed the conversion of prey to female stored energy was 70% efficient and, thus, in terms of female energy requirements, a pup would cost 10.9 MJ

Table 1. Mean body size, body mass, population size (March 1983) and biomass for the Antarctic Fur Seal *Arctocephalus gazella*

Category	Sex	Age class (years)	Mean weight (kg)	Population size no. × 10^3	%	Biomass (t)	%
Adult	F	3+	32.1	328	30	10,529	26
	M	7+	125.2	98	9	12,270	31
Juvenile	F	0–2	16.0	131	12	2,096	5
	M	0–6	37.8	306	28	11,567	29
Weaners	F	0	13.5	109	10	1,427	4
	M	0	17.0	109	10	1,853	5
Totals				1.09×10^6		39.8×10^3	

Table 2. Summary of metabolic rates used in Fur Seal model

Category and activity	Multiple of SMR[a]	Data source
Adult females		
Foraging for self and pup	5.6	Costa and Gentry (to be publ.)
Rearing pup	2.7	Doidge et al. (1984b)
Foraging for self	3.0	Difference between above
Lactating ashore	3.2	Costa and Gentry (to be publ.)
Adult males		
Average daily metabolic rate	3.1	Average of female rates for self
Juveniles		
Foraging at sea and growth	5.6	Female rate for self and pup
Ashore	3.2	Adult female rate
Weaners		
Foraging at sea and growth	5.6	Female rate for self and pup

a Standard metabolic rate (SMR) = 70 $W^{0.75}$ in kcal d^{-1}, weight in kg (Kleiber 1961)

d^{-1}. Applying *Callorhinus* weight-specific metabolic rates to *A. gazella,* Costa and Gentry (to be publ.) calculated the cost of foraging in a lactating female of 33.5 kg to be 23 MJ d^{-1} which is equivalent to 5.6 times SMR. For the same animal, the cost of rearing the pup is 2.7 times SMR. Thus, by substraction, the cost of foraging for self is 3.0 times SMR. Here, we assume that the metabolic rate of lactating females whilst ashore is 3.2 times SMR, as found for *C. ursinus* by Costa and Gentry (to be publ.).

Fasting territorial males lose between 1.57 and 1.99 kg d^{-1} during the breeding season (Payne 1979b). If this is exclusively fat, at 37.7 kJ g^{-1} fr. wt. (Diem and Lentner 1970), then this is equivalent to 5.4–6.8 times SMR for a 125 kg male. However, until the actual composition and energy content of the material lost is determined, these estimates remain uncertain. It is most unlikely that the energy content is < 20 kJ g^{-1}, based on data for phocid seals (M. Fedak, pers. comm.), which is equivalent to 2.9 –3.7 times SMR. Given these uncertainties, we use here the average rate (3.1 times SMR) of females foraging for self and when ashore, assuming that the lower rate expected for the larger males is compensated for by the considerable activity associated with maintaining a breeding territory. Food requirements are greater (on a weight-specific basis) for young growing animals. Thus we set juvenile and weaner metabolism at 5.6 times SMR, the cost for an adult female foraging for self and pup.

2.3 Activity Budgets

To simplify the energy consumption model, the activity patterns of all individuals of each category of the herd are assumed to change simultaneously. In reality there is obviously less synchrony than this; events such as the arrival of adult males and pupping taking place over ca. 3 weeks.

The winter distribution of adult females is not known. We presume they lead a pelagic existence, north of the Antarctic Convergence (i.e., outside the area we are considering) until the beginning of October. A typical female comes ashore at South Georgia on 4 December, gives birth two days later, and remains for a seven-day perinatal period (Doidge et al., to be publ.). The energy reserves required for this nine-day fast are considered to have been built up between the time of crossing the Convergence and arrival ashore. At the end of the perinatal attendance period she goes to sea and starts the feeding-shore attendance cycles which continue until the pup weans on 1 April. The female then remains ashore for three days and returns to sea to forage until late May when she is assumed to move north of the Convergence. The cost of pup rearing is thus accounted for within the female energy budget. The weaner class leaves the system at the same time as the adult females and thus feeds at sea between 1 April and late May.

No quantitative data exist for the proportion of the adult male herd at sea during the winter months. Since some males are seen ashore during this time, we assume 85% of the population is at sea at any one time. From November to January, the proportion of the adult male herd at sea is estimated using data on the length of tenure of territorial males (McCann 1980). Between 15 November and 24 December, 40% of the adult male herd is at sea while the balance (60%) is ashore holding territories. Adult males abandon their territories in late December; 95% are at sea between 25 December and 9 March.

We presume juveniles divide their time equally between land and sea. Juvenile females are assumed to be migratory; juvenile males are frequently seen in winter and their population is regarded as sedentary.

2.4 Diet

It is usually reported that breeding female Antarctic Fur Seals take largely Krill, but squid and fish are also taken (Bonner 1968; Payne 1977), although quantitative data and information on the diet of other elements of the population are lacking.

Three large samples of breeding females have been examined for stomach contents. Of 195 animals taken by Payne (1977) at Bird Island in 1971–73, the stomachs of 16 contained appreciable quantities of Krill. Indications of Krill presence were found in most of the specimens whose stomachs were not entirely empty, squid beaks were found in 15 (8%) stomachs and signs of fish in about 20 (M.R. Payne, unpubl.).

Adult females were collected at Schlieper Bay, South Georgia, during January–March 1982. Of the 97 stomachs, examined 54% were empty, 44% contained gravel or pebbles, 1% fish otoliths and 1% penguin feathers. Colonic contents were generally homogenous and in this analysis each colon was considered to contain only 1 food type. Krill occurred in 53% of the colons with exoskeleton parts visible in 35% and a further 18% were pink or orange, which is interpreted as being euphausiid in origin. Fish vertebrae and penguin feathers were each present in 1% of the samples. Three per cent of the colons were empty and 41% contained no discernible prey items.

At Bird Island, during January–March 1983, of 238 adult females five had stomachs with large quantities (ca. 400 g) of Krill; about 50% were empty. Most of the rest showed signs of Krill, and only very few had indications of fish or squid, including only one squid beak and no otoliths. Whole Krill from these and other samples from Bird Island covering six yrs and the months December–March, have been analyzed by Croxall and Pilcher (1984) who concluded that female Fur Seals consistently take only mature adult Krill. A typical length-frequency distribution, taken from the 1982–83 samples, is shown in Fig. 1.

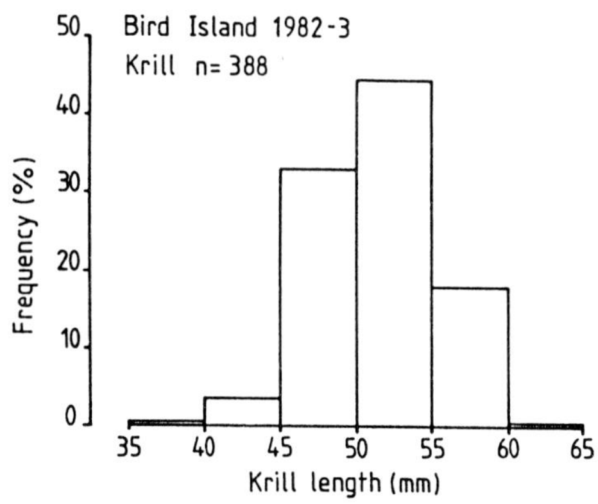

Fig. 1. Length-frequency distribution of Krill, *Euphausia superba*, from adult female Antarctic Fur Seal stomachs

Only four taxa of squid have been identified from beak collections: *Kondakovia longimana, Moroteuthis knipovitchi, ? Galiteuthis glacialis* and an unidentified ommastrephid, with *Kondakovia* estimated to form nearly half the squid diet by weight, and the ommastrephid one-third (MR Clarke,pers. comm.)

Fish otoliths from sub-adult male scat samples were analyzed by North et al. (1983) who found the pelagic icefish, *Champsocephalus gunnari,* to be the predominant prey both by number (55%) and weight (76%) of fish consumed. The *C. gunnari* taken were mainly young adults 3–5-yr-old, weighing on average 160 g. Lantern fish, *Gymnoscopelus nicholsi,* and the nototheniids *Notothenia rossii, N. gibberifrons, Patagonotothen larseni,* and *Pseudochaenichthys georgianus,* comprised the rest of the diet, summarized in Fig. 2. These fish are mainly epi-pelagic feeders on Krill and were probably taken from shoals associated with Krill. The fish contents of the scats represented about 10% of a Fur Seal's daily food requirement (North et al. 1983), so fish is possibly somewhat more important than its frequency of occurrence indicates. We have little information on winter diet, although fish were present in at least 50% of the scats of adult and immature males at Bird Island in August and September 1982 (Prince, in North et al. 1983).

Taking all these data into account we have assigned tentative compositions by weight to the diet of the various categories of Seals (Table 3). We have assumed that adult male Fur Seals take more fish and squid in summer than other classes of Fur Seals, because of their larger body size and hence deeper diving capacity which facilitates exploitation of more benthic resources, such as *Notothenia rossii* and octopods.

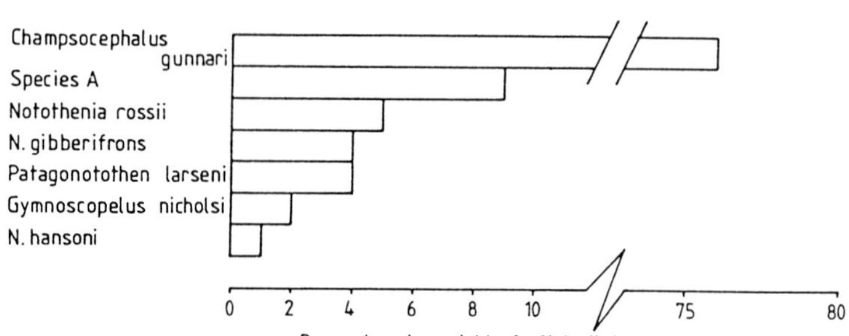

Fig. 2. Composition by weight of the fish diet of sub-adult male Fur Seals. (Data from North et al. 1983)

Table 3. Composition (% by weight) of Fur Seal diet used in model

Season	Category	Krill	Squid	Fish
Summer	Adult females	92	3	5
	Juveniles	92	3	5
	Weaners	92	3	5
	Adult males	75	10	15
Winter	Juvenile males	20	30	50
	Adult males	20	30	50

Penguins, especially Macaroni Penguins, *Eudyptes chrysolophus,* are often killed, but not necessarily eaten, by subadult males (Bonner and Hunter 1982) but we are unable to assess their contribution to the diet and have excluded them from this analysis.

Energy content of prey is taken as: Krill 4.35 kJ g^{-1}, fish 3.97 kJ g^{-1} and squid 3.47 kJ g^{-1} (Clarke and Prince 1980; Croxall and Prince 1982). Assimilation efficiency is assumed to be 75%, although this is somewhat lower than found by Miller (1978) in laboratory studies with *C. ursinus.*

3 Results

3.1 Energy Requirements and Food Consumption

The energy requirements and biomass of prey species consumed by each sex-age category are listed in Table 4. During the summer season (for which we have the best data), juvenile males consume 39% of the biomass, adult females take 31%, adult males 22% and juvenile females only 8% of the total. In winter, when only males are presumed to be present, juveniles account for 67% of the food taken. Over the entire year Krill accounts for 69% of the biomass taken by the herd, fish 19% and squid 12%.

The pattern of consumption is also of interest, being illustrated in Fig. 3. The biomass of prey consumed shows three main peaks during the year. The first is in October when adult males and females are all at sea, consumption falling as adult males begin to come ashore in November. From late December until early March the impact rises as the number of adult males at sea increases and females are rearing their rapidly growing pups. A small proportion of adult males starts to haul-out in March, thereby reducing prey consumption. A final peak in prey consumption occurs in April and May when the weaners go to sea, before the switch to winter levels occurs. Since we lack any data

Table 4. Energy requirement (MJ × 10^6) and prey consumption (t × 10^3) of the South Georgia Fur Seal herd, March 1983

	Season	Adult F	Adult M	Juvenile F	Juvenile M	Total
Energy required						
	Summer	1,125	765	307	1,429	3,627
	Winter	–	452	–	809	1,261
	Total	1,125	1,217	307	2,238	4,889
Prey consumption						
Krill	Summer	321	183	87	407	997
	Winter	–	31	–	56	87
	Total	321	213	87	463	1,084
Squid	Summer	11	24	3	13	51
	Winter	–	47	–	83	129
	Total	11	71	3	96	180
Fish	Summer	17	36	5	23	81
	Winter	–	77	–	139	216
	Total	17	113	5	162	297
Total	Summer	349	243	95	443	1,130
	Winter	–	155	–	277	432
	Total	349	398	95	740	1,562

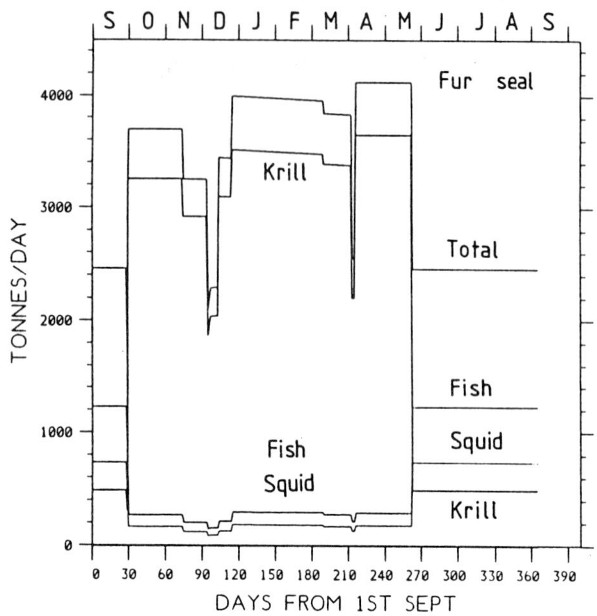

Fig. 3. Consumption of Krill, *Euphausia superba,* fish and squid by Fur Seals at South Georgia. Females are absent in winter (June to September)

on haul-out patterns in winter, the amount of prey taken by males remains constant until the next year's cycle starts.

In summer, the Fur Seal population consumes about 23% of its biomass daily, assuming an assimilation rate of 75%. Lavigne et al. (1982) suggested that in seals at least 90% of gross energy would be available as apparent digestable energy. At 90% efficiency, 19% of body weight would be required per day. This figure appears high but includes body growth and reproduction costs. Miller (1978) estimated that in *C. ursinus* juveniles needed to consume 14% of their body weight daily in order to meet existence metabolism, excluding the cost of growth. More recently, Costa and Gentry (to be published) calculated that 16% of body weight was needed daily in *C. ursinus.* The above estimates indicate that fur seals have higher energy demands than those (e.g., 7% of body weight daily) previously assumed (McAlister et al. 1976; Laws 1977; Payne 1979b).

4 Discussion

Our model is, of necessity, very preliminary. Its construction has made clear the data we lack for many of the key featues of the bio-energetics, activity patterns and budgets, and diet of *A. gazella.* Nevertheless, estimates of overall food consumption, the broad pattern of changes in this during the summer and the difference in the resource base (and consumption levels) in winter are probably not too unrealistic. The importance of refining this model is emphasized by Croxall et al. (1985) who indicate that Antarctic Fur Seals are the third most important consumers of Krill, after Macaroni Penguins and Dove Prions, *Pachyptila desolata,* in the South Georgia area. At present, however, there is no indication that food supply is currently limiting population size or survival of pups through reducing growth rates (Doidge et al. 1984b, to be publ.).

4.1 Interspecific Competition

At South Georgia, the animals whose foraging ranges and dietary habits suggest greatest potential competition with Antarctic Fur Seals are penguins and Imperial Cormorants, *Phalacrocorax atriceps* (Croxall and Prince 1980a).

Female Fur Seals that are rearing pups probably forage exclusively within the continental shelf. They feed mainly at night; dives at this time rarely exceeding 40 m and 26% of the dives being shallower than 20 m. Only 3% of all dives exceed 70 m (Croxall et al. 1985; Kooyman et al., to be publ.). This is entirely consistent with their predominantly Krill diet and suggests that feeding is timed to coincide with the vertical migration of Krill towards the surface at night. Except for the King Penguin, *Aptenodytes patagonicus,* which is a deep-diving species feeding primarily on squid and Krill (Kooyman et al. 1982), the other penguins at South Georgia mainly take mature Krill (Croxall and Prince 1980b). Diving data for Chinstrap Penguins, *Pygoscelis antarctica* (Lishman and Croxall 1983) and Gentoo Penguins, *P. papua* (Adams and Brown 1983) indicate that they are also shallow divers, 90% of Chinstrap Penguin dives being no deeper than 45 m and 40% being shallower than 10 m.

Chinstrap Penguins are rare at South Georgia, especially at the north-western end of the island (Prince and Payne 1979). Gentoo Penguins are widely distributed and abundant but their feeding activities are probably confined to are as much closer inshore than Fur Seals. Their theoretical maximum foraging radius of ca. 30 km (Croxall and Prince 1980a) compares with the distances to first and from last dives of Fur Seals of 62 km and 26 km, respectively (Croxall et al. 1985), assuming constant swimming in a straight line. Gentoo Penguins also take a significant proportion (33% by weight) of fish in their diet, mainly juvenile *Notothenia* spp. which frequent the inshore kelp beds (Croxall and Prince 1980b). Macaroni Penguins, however, have a theoretical foraging radius of ca. 115 km, indicating a foraging range essentially co-extensive with that of Fur Seals. Furthermore, both Fur Seals and Macaroni Penguins are observed leaving Bird Island on foraging trips heading in a north-west direction (P.A. Prince, unpubl.; J.L. Bengtson, unpubl.) and thus they may even share the same feeding grounds. Macaroni Penguins take mainly mature Krill but include a significant fraction (18% by weight) of juveniles in their diet (Croxall and Prince 1980b). Although Fur Seals appear to take mature adult Krill almost exclusively (Croxall and Pilcher 1984), the potential competition between Macaroni Penguins and Fur Seals would seem significant, especially since the diving behaviour of Macaroni Penguins is unlikely to be much different from the pygoscelid penguins and therefore similar to Fur Seals. Together, Macaroni Penguins and Fur Seals are estimated to account for 73% of the Krill consumption in the South Georgia region (Croxall et al. 1985).

The remaining diving bird, the Imperial Cormorant, feeds principally on fish in waters very close inshore. Its prey at South Georgia is chiefly the nototheniids, *N. rossii* and *N. gibberifrons*, and thus rather different from the essentially pelagic *Champsocephalus gunnari* taken by the more widely ranging Fur Seals. More significant competition for fish, especially in winter, may be provided by the Southern Elephant Seal, *Mirounga leonina*, which is probably the main piscivore in the region (Croxall et al. 1985). However, the nature of the fish portion of its diet is unknown, as are its foraging range and depth. It has the potential for diving very much deeper than Fur Seals and the octopods found in its diet at Deception and Signy islands (Clarke and MacLeod 1982) suggest benthic feeding during extended dives to the sea floor. Diving habits may thus be very different from Fur Seals, further reducing competition.

4.2 Competition with Commercial Fisheries

There is as yet, no substantial Krill fishery in the South Georgia area. Because Krill is by far the main element of the Fur Seal's summer diet, significant competition would probably result, particularly if commercial operations included the continental shelf off northwestern South Georgia. Recent exploitation of fish indicates the problems which may arise and be of consequence to Fur Seals.

If the 5% by weight of *N. rossii* in the fish diet of juvenile Fur Seals in summer is typical of the rest of the year, then the annual consumption would be ca. 15,000 t. In 1978–79 the total *N. rossii* biomass was 188,000 t, made up of an estimated depredated stock of 117,000 t (Burchett and Ricketts 1984) and a total predator consumption of ca. 71,000 (Croxall et al. 1985). The Fur Seal consumption would thus represent 8% of the 1978–79 stock, a year the fishery catch was 45,000 t (Burchett and Ricketts 1984). Since Fur Seals take mainly juvenile *N. rossii*, whereas the trawlers take mainly adults, direct competition with the fishery may be reduced. Also, *N. rossii* is relatively unimportant in the Fur Seal's diet, at least in summer.

The reduction in stock size of *N. rossii* age classes three yrs +, which are cannibalistic on younger fish (M.S. Burchett, pers. comm.), might even benefit both Fur Seal and fish stocks. This is similar to the interaction between *C. ursinus* and commercial fisheries in the Bering Sea where it has been suggested that the removal of adult cannibalistic Walleye, *Theragra chalcogramma*, lowers mortality among younger pollock stocks (Swartzman and Haar 1983). The benefit would remain, however, only as it was greater than the effects of reducing recruitment.

As the importance of *N. rossii* in the fishery decreased from 98% in 1969 to 0.4% in 1980 (Anon. 1980) so the importance of *C. gunnari* has increased. The stock biomass of this species, based on catch statistics, is estimated to have decreased from 141,000 t between 1975–76 and 1977–78 (Kock 1981). Since Fur Seals would take ca. 200,000 t of *C. gunnari* (if winter diet is the same as summer) and both Seals and trawlers exploit adult fish, then there is clearly the likelihood of extensive competition between the two.

There are obviously many potential sources of error in the assessments of both fishery stocks and Fur Seal consumption. Nevertheless, the evaluation above underlines the urgent need for realistic information on the fish component of Fur Seal diets and on Fur Seal population levels and activity patterns at South Georgia in winter.

Our model could be substantially refined, especially given better and more appropriate bio-energetic information, data on attendance patterns of male Fur Seals and, of course, knowledge of the at-sea distribution of males in summer and all classes in winter (and their diets at that time). Nevertheless, the model developed here indicates that Fur Seals may eat 1.6 times 10^6 t of food, including 1 times 10^6 t of Krill taken in summer alone. Their main natural competitors for Krill at South Georgia are Macaroni Penguins, but local commercial operations would also probably be in direct competition.

Acknowledgements. We thank M.R. Clarke and A.W. North for unpublished data on squid and fish respectively, S. McInnes who drew the figures. C. Ricketts for devising the computer model and M. Fedak for helpful discussions on pinniped energetics. W.N. Bonner, T.S. McCann, and D.B. Siniff commented on the manuscript.

References

Adams NJ; Brown CR (1983) Diving depths of the gentoo penguin *Pygoscelis papua.* Condor 85:503–504

Anon. (1980) Antarctic fish biology. Biomass Rep Ser 12:29 pp

Blix AS; Miller LK; Keyes MC; Gray HJ; Elsner R (1979) Newborn Northern fur seals (*Callorhinus ursinus*) – do they suffer from the cold? Am J Physiol 236:322–327

Bonner WN (1968) The fur seal of South Georgia. Sci Rep Br Antarct Surv 56:1–81

Bonner WN; Hunter S (1982) Predatory interactions between Antarctic fur seals, macaroni penguins and giant petrels. Bull Br Antarct Surv 56:75–79

Burchett MS; Ricketts C (1984) The population dynamics of *Notothenia rossii* from South Georgia (Antarctica). Polar Biol 3:35–38

Clarke A; Prince PA (1980) Chemical composition and calorific value of food fed to mollymauk chicks *Diomedea melanophris* and *D. chrysostoma* at Bird Island, South Georgia. Ibis 122: 488–494

Clarke MR; MacLeod N (1982) Cephalopods in the diet of elephant seals at Signy Island, South Orkney Islands. Bull Br Antarct Sur 57:27–31

Costa DP; Gentry RL (to be published) Free ranging and reproductive energetics of the Northern Fur Seal *Callorhinus ursinus.* In: Kooyman GL; Gentry RL (eds) Fur Seals: Maternal Strategies on Land and at Sea. Princeton University Press, Princeton

Croxall JP; Pilcher MN (1984) Characteristics of krill *Euphausia superba* eaten by Antarctic fur seals at South Georgia. Bull Br Antarct Surv 63:117–125

Croxall JP; Prince PA (1980a) Food, feeding ecology and ecological segregation of seabirds at South Georgia. Biol J Linn Soc 14:103–131

Croxall JP; Prince PA (1980b) The food of gentoo penguins *Pygoscelis papua* and macaroni penguins *Eudyptes chrysolophus* at South Georgia. Ibis 122:245–253

Croxall JP; Prince PA (1982) Calorific content of squid (Mollusca: Cephalopoda). Bull Br Antarct Surv 55:27–31

Croxall JP; Everson I; Kooyman GL; Ricketts C; Davis RW (1985) Fur seal diving activity relates to vertical distribution of krill. J Anim Ecol 54

Croxall JP; Prince PA; Ricketts C (1985) Relationships between prey life-cycles and the extent, nature and timing of seal and seabird predation in the Scotia Sea. In: Siegfried WR; Condy PR, Laws RM (eds) Antarctic nutrient cycles and food webs (Proceedings of the 4th SCAR symposium on Antarctic biology). Springer, Berlin Heidelberg New York

Diem K; Lentner C (eds) (1970) Documenta Geigy Scientific Tables. Geigy Pharmaceuticals, Macclesfield UK

Doidge DW; Croxall JP; Baker JR (1984) Density-dependent pup mortality in the Antarctic fur seal *Arctocephalus gazella* at South Georgia. J Zool Lond 202:449–460

Doidge DW; Croxall JP; Ricketts C (1984b) Growth rates of Antarctic fur seal *Arctocephalus gazella* pups at South Georgia J Zool Lond 203:87–93

Doidge DW; McCann TS; Croxall JP (to be published) Attendance behaviour of Antarctic fur seals *Arctocephalus gazella*. In: Kooyman GL; Gentry RL (eds) Fur Seals: Maternal Strategies on Land and at Sea. Princeton University Press, Princeton

Hamilton JE (1934) The southern sea lion, *Otaria byronia* (de Blainville). Discovery Rep 8:269–318

Kleiber M (1961) The Fire of Life. Wiley, London p 454

Kock K-H (1981) Fischereibiologische Untersuchungen an drei antarktischen Fischarten: *Champsocephalus gunnari* Lonnberg, 1904. *Chaenocephalus aceratus* (Lonnberg, 1905) und *Pseudochaenichthys georgianus* Norman, 1937 (Nototheniodei, Channichthyidae). Mitt Int Seefisch BundForschAnst Fisch Hamburg 32:226

Kooyman GL; Davis RW; Croxall JP; Costa DP (1982) Diving depths and energy requirements of king penguins. Science 217:726–727

Kooyman GL; Davis RW; Croxall JP (to be published) Diving behaviour of the Antarctic fur seal *Arctocephalus gazella*. In: Kooyman GL; Gentry RL (eds) Fur Seals: Maternal Strategies on Land and at Sea. Princeton University Press, Princeton

Lavigne DM; Barchard W; Innes S; Øritsland NA (1982) Pinniped bio-energetics. FAO Fish Ser(5) Vol 4:191–235

Laws RM (1977) Seals and whales of the Southern Ocean. Philos Trans R Soc Lond B279:81–96

Lishman GS; Croxall JP (1983) Diving depths of the Chinstrap penguin *Pycoscelis antarctica*. Bull Br Antarct 61:21–26

McAlister WB; Sanger GA; Perez MS (1976) Preliminary estimates of pinniped-finfish relationship in the Bering Sea. Background Paper 19th Meeting Fur Seal Commission, Moscow 1976. US Dept Commerce NOAA NMFS Seattle

McCann TS (1980) Territoriality and breeding behaviour of adult male Antarctic fur seals *Arctocephalus gazella*. J Zool Lond 192: 295–310

Miller LK (1978) Energetics of the northern fur seal in relation to climate and food resources of the Bering Sea. Final report to US Marine Mammal Commission, NTIS, US Dept. Commerce, Springfield, 27 pp

North AW; Croxall JP; Doidge DW (1983) Fish prey of the Antarctic fur seal *Arctocephalus gazella* at South Georgia. Bull Br Antarct Surv 61:27–38

Øritsland T (1977) Food consumption of seals in the Antarctic pack ice. In: Llano GA (ed) Adaptations within Antarctic ecosystems. Smithsonian Institution, Washington DC, pp 749–768

Payne MR (1977) Growth of a fur seal population. Philos Trans R Soc Lond B279:67–79

Payne MR (1979a) Fur seals *Arctocephalus tropicalis* and *A. gazella* crossing the Antarctic Convergence at South Georgia. Mammalia 43:93–98

Payne MR (1979b) Growth in the Antarctic fur seal *Arctocephalus gazella*. J Zool Lond 187:1–20

Prince PA; Payne MR (1979) Current status of birds at South Georgia. Bull Br Antarct Surv 48:103–118

Swartzman GL; Haar RT (1983) Interactions between fur seal populations and fisheries in the Bering Sea. Fish Bull Natn Ocean Atmos Admin 81:121–132

Cephalopod Prey of the Sub-Antarctic Fur Seal, *Arctocephalus tropicalis*, at Gough Island

M.N. BESTER[1] and P.A. LAYCOCK[2]

Summary. The Sub-Antarctic Fur Seal, *Arctocephalus tropicalis*, at Gough Island preys predominantly on cephalopods, but includes relatively small quantities of fish in its diet. Stomachs of Fur Seals (n = 220) culled on land were either empty (32%), contained only stones (8%) or contained almost exclusively prey remains resistant to digestion, such as cephalopod pens, 'beaks', eye balls, fish bones and otoliths. The pooled cephalopod lower beaks (n = 424) that could be identified (n = 337), showed that Ommastrephidae (52.5%), Histioteuthidae (25.2%), Onychoteuthidae (19.9%), Cranchiidae (2.1%) and Octopoteuthidae (0.3%) constituted the main prey items based on frequency of occurrence. Cephalopod mass estimates, from regression of lower rostral lengths against mass, approximated this relative arrangement of cephalopod families.

1 Introduction

The diet of the Sub-Antarctic Fur Seal, *Arctocephalus tropicalis,* at the Tristan da Cunha group of islands, including Gough Island (40° 20'S, 9° 54'W) is largely unknown. At its other breeding locations in the Southern Ocean, the species feeds on seasonally abundant Rockhopper Penguins, *Eudyptes chrysocome,* squid and fish (e.g., Snoek, *Thyrsites atun*) at Amsterdam Island (37° 50'S, 77° 35'E) (Tollu 1974). Unspecified squid, fish (Rand 1956) and perhaps small amounts of Krill (Euphausiacea) (Condy 1981) are taken at Marion Island (46° 54'S, 37° 45'E). The stomachs of vagrants (n = 7) stranded on the coast of South Africa contained almost exclusively squid "beaks" with those belonging to the families Histioteuthidae (tentative assignation, vide Ross 1979) and Loliginidae (*Loligo reynaudi*) being the most numerous (Shaughnessy and Ross 1980). The need for more detailed knowledge of the diets of seals in the Southern Ocean, as part of the BIOMASS (Biological Investigations of Marine Antarctic Systems and Stocks) programme, and the potential for competition between the local lobster, *Jasus tristani,* industry and the large, rapidly expanding, *A. tropicalis* population at Gough Island (Bester 1980), prompted the present investigation.

1 Mammal Research Institute, University of Pretoria, Pretoria 0002, South Africa

2 School of Environmental Studies, University of Cape Town, Rondebosch 7700, South Africa

2 Materials and Methods

Stomachs of specimens of all age categories of *A. tropicalis* (n = 220) were collected from animals culled on land between November 1977 and October 1978 at Gough Island. Stomach contents (if any) were removed and the various types of food remains separated in each sample. The cephalopod beaks were counted, after which they were pooled for later identification. Whole or part samples from stomach contents (n = 8) were stored in 10% formaldehyde when a substantial number of undigested soft specimens was present which could aid in further identification, and the wet mass and volume were recorded.

The lower beaks of squid were sorted into groups and identified as far as possible using descriptions by Clarke (1962, 1980) and Voss (1969). The designation histioteuthid type 8 and assignation to the family Histioteuthidae is tentative following Ross (1979). Lower rostral length (LRL) was measured following standardized procedures, and cephalopod mass was estimated from regressions of LRL against mass as developed for separate cephalopod families by Clarke (1962). The importance of each cephalopod family in the diet was obtained by expressing its frequency of occurrence as a percentage of the total number of lower beaks (n = 337) in the pooled sample, and using the predicted mass obtained by summing the products of the numbers multiplied by the mean mass of all species within each family, and expressed as a percentage of the total sum.

3 Results

A total of 719 intact beaks was isolated from the stomachs and examined. These included 424 lower beaks of which 337 could be used for identification and measurement of lower rostral lengths. Stomachs (n = 220) were usually empty (31.8%, n = 70), contained only stones (7.7%, n = 17) or almost exclusively prey remains resistant to digestion such as cephalopod pens, beaks, eye-balls, fish bones and otoliths. Stones were encountered in 22.3% (n = 49) of all stomachs, and as yet unidentified fish remains were found in only 2.3% (n = 3) of stomachs that contained

Antarctic Nutrient Cycles and Food Webs
(ed. by W.R. Siegfried, P.R. Condy, and R.M. Laws)

Table 1. Relative importance of squid in the diet of *A. tropicalis* at Gough Island, based on number of lower "beaks" and reconstituted mass of cephalopods in a pooled sample

Family	Species	Number	%	Mean LRL (mm)	Estimated mass Mean (kg)	Estimated mass Total (kg)	Estimated mass Total (%)	Family contribution % by number	Family contribution % by mass
Ommastrephidae	(?) *Todarodes* A	95	28.2	5.5	0.290	27.6	28.3	52.5	95.7
	Todarodes B	10	3.0	6.3	0.430	4.3	4.3		
	Todarodes (A or B)	17	5.0	6.6	0.480	8.2	8.4		
	Ommastrephid sp. A	50	14.8	8.3	0.950	47.5	48.7		
	Ommastrephid sp. B	5	1.5	10.1	1.160	5.8	5.9		
Histioteuthidae	Histioteuthid type 8	83	24.6	1.9	0.021	1.7	1.7	25.2	2.3
	(?) *Histioteuthis*	2	0.6	2.3	0.240	0.5	0.5		
Cranchiidae	*Galiteuthis armata*	1	0.3	2.8	0.030	< 0.1	< 0.1	2.1	0.4
	Bathothauma lyromma	2	0.6	3.4	0.045	0.1	0.1		
	Cranchiid sp.	4	1.2	1.5	0.079	0.3	0.3		
Onychoteuthidae	*Moroteuthis knipovitchi*	6	1.8	4.5	0.160	1.0	1.0	19.9	1.5
	(?) *Moroteuthis* sp.	61	18.1	1.6	0.008	0.6	0.5		
Octopoteuthidae	Octopoteuthid sp.	1	0.3	2.6	1.109	0.1	0.1	0.3	0.1

Table 2. Number and identity of cephalopod lower "beaks"[a] in the stomachs of 7 *A. tropicalis* at Gough Island, which also contained semi-digested squid flesh

Ref no. GFS	66	76	125	157	183	192	222
Sex	F	F	M	F	F	M	M
Age (years)	10	17+	7	8	9	18	4
Date collected	19 Feb.	24 Feb.	17 April	19 May	17 July	10 Aug.	5 Oct.
Histioteuthid type 8					2	2	2
(?) *Histioteuthis* sp.	1						
Ommastrephes (?) *bartrami*	4						1
(?) *Todarodes* sp.	4						
(?) *Symplectoteuthis* sp.	3		3				
Ommastrephid		2[b]				2	
Todarodes A					1		
Todarodes (?) *sagittatus*					1		
Onychoteuthid	2		1	1			
(?) *Moroteuthis* sp.			5				
Cranchiid			1				
Unspecified							1

a Not included in pooled sample on which Table 1 is based

b Part sample only, as a further 13 (upper and lower beaks) went into the pooled sample at least some prey remains (n = 131). Other food items included pieces of seaweed (n = 5), fragments of gastropod mollusc shell (n = 2), penguin (probably *E. chrysocome*) feathers (n = 1) and polyps of a colonial siphonophore (n = 1).

Table 1 shows that representatives of at least 10 different species, belonging to five different cephalopod families, were eaten by Fur Seals at Gough Island. The most important families by frequency of occurrence, in decreasing order, were the Ommastrephidae, Histioteuthidae and Onychoteuthidae. The remaining families, the Cranchiidae and Octopoteuthidae, together contributed only 2.4% in frequency of occurrence, and a meagre 0.5% by mass. The Ommastrephidae outstripped all other families in contribution by mass (95.7%), the Onychoteuthidae and the Histioteuthidae amounting to 1.5% and 2.3% respectively. Although *Todarodes* sp. and the histioteuthid type 8 were the most abundant, the relatively large-bodied ommastrephids "A" and "B" contributed the most in mass of flesh represented by beaks (54.6%). What appeared to be *Ommastrephes bartrami*, from GFS 22 (Table 2), was estimated to have a mass of approximately 4.5 kg. However, squid bodies were usually too far digested to allow definitive taxonomic identification, and reconstituted mass was far in excess of recorded wet wt.

A representative cross-section of cephalopod families was found together in single samples (Table 2), and in the case of certain ommastrephids (*Symplectoteuthis*) and onychoteuthids, buccal masses were still in situ. A mean of 5.6 ± 6.1 (S.D.) squid beaks (both upper and lower beaks), ranging from 1–32, was found in stomachs containing beaks (n = 129).

4 Discussion

Unlike immature Cape Fur Seals, *Arctocephalus pusillus*, in South Africa (Rand 1959), the Sub-Antarctic Fur Seals at Gough Island (this study) and Amsterdam Island (Tollu 1974) do not exploit the locally abundant lobsters (*Jasus* spp.). Nor do they take appreciable numbers of fish, although it has been assumed that fish are an important food item in their diet at the higher latitude of Marion Island (Rand 1956). In contrast to the Amsterdam Island population, members of *A. tropicalis* at Gough and Marion islands do not supplement their diet with penguins (Condy 1981; this study), although they may occasionally kill or injure these birds (Swales 1965; Bester 1977).

Pelagic squid constitute the most important dietary component of the Fur Seals at Gough Island, and the presence of the occasional semi-digested squid body, sometimes with an intact buccal mass, shows that at least some ommastrephids and onychoteuthids are taken in waters close to the island. *Todarodes sagittatus* (Ommastrephidae) is common in warm waters, but *Moroteuthis knipovitchi* (Onychoteuthidae) being an Antarctic species is a curious find at Gough Island which lies close to the Sub-Tropical Convergence and, therefore, in relatively warm water. *Moroteuthis knipovitchi* is furthermore well represented in the diet of a true Antarctic phocid, the Weddell Seal, *Leptonychotes weddelli*, at Deception Island (60° 57′S, 69° 38′W) and is of great numerical importance in the diet of Sperm Whales, *Physeter catodon*, near the South Orkney and South Shetland islands (Clarke 1980; Clarke and MacLeod 1982a). On the other hand, histioteuthids, absent in samples taken from Weddell Seals (Clarke and MacLeod 1982a) and Southern Elephant Seals, *Mirounga leonina* (Clarke and MacLeod 1982 b) at Signy Island (60° 43′S, 45° 38′W), are typical of warmer waters, being recorded from Sperm Whales off the coast of South Africa (Clarke 1980). Histioteuthids have also been observed in the stomachs of vagrants of *A. tropicalis*, hauled out on the South African coastline, in addition to the locally abundant *Loligo reynaudi* and *Sepia* spp.; *Loligo reynaudi* is commercially important and is caught, commonly in bottom trawls on the continental shelf of South Africa (Sea Fisheries Branch 1976). These differences in diet according to latitude in *A. tropicalis* and *P. catodon* probably reflect the availability of prey. Ashmole (1969) and Clarke et al. (1981) suggested, similarly, that the difference observed between squid samples taken from Wandering Albatrosses, *Diomedea exulans*, in New Zealand and South Georgia was a reflection of the latitudes in which the birds were feeding.

The diurnal haul-out pattern of *A. tropicalis* at Gough Island, where the number of animals peaks during the afternoon, following an increase in the rate of hauling out during the morning (Bester 1977), suggests that the Fur Seals feed from dusk to dawn in order to take advantage of vertical migrations of pelagic squid. This would account for the predominance of Ommastrephidae in their diet, since at least *T. sagittatus*, *O. bartrami* and *Symplectoteuthis* spp. regularly appear at the surface at night (Clarke 1966).

Acknowledgements. We are grateful to the South African Department of Transport for logistical and financial support, acting on the advice of SASCAR (South African Scientific Committee on Antarctic Research). We thank A. Seabrook for field assistance, and M.J. Imber and G.J.B. Ross for guidance in the identification of cephalopods.

References

Ashmole NP (1969) Body size, prey size, and ecological segregation in five sympatric tropical terns (Aves: Laridae). Syst Zool 17: 292–304

Bester MN (1977) Habitat selection, seasonal population changes, and behaviour of the Amsterdam Island fur seal *Arctocephalus tropicalis* on Gough Island. D.Sc. thesis, Univ Pretoria, South Africa (unpublished)

Bester MN (1980) Population increase in the Amsterdam Island fur seal *Arctocephalus tropicalis* at Gough Island. S Afr J Zool 15: 229–234

Clarke MR (1962) The identification of cephalopod beaks and the relationship between beak size and total weight. Bull Br Mus Nat Hist (Zool) 8:419–480

Clarke MR (1966) A review of the systematics and ecology of oceanic squids. Adv Mar Biol 4:91–300

Clarke MR (1980) Cephalopoda in the diet of sperm whales of the Southern Hemisphere and their bearing on sperm whale biology. Discovery Rep 37:1–324

Clarke MR; Croxall JP; Prince PA (1981) Cephalopod remains in regurgitations of the wandering albatross *Diomedea exulans* L. at South Georgia. Br Antarct Surv Bull 54:9–21

Clarke MR, MacLeod N (1982a) Cephalopod remains in the stomachs of eight Weddell seals. Br Antarct Surv Bull 87:33–40

Clarke MR, MacLeod N (1982b) Cephalopods in the diet of elephant seals at Signy Island, South Orkney Islands. Br Antarct Surv Bull 57:27–31

Condy PR (1981) Annual food consumption and seasonal fluctuations in biomass of seals at Marion Island. Mammalia 45:21–30

Rand RW (1956) Notes on the Marion Island fur seal. Proc Zool Soc Lond 126:65–82

Rand RW (1959) The Cape fur seal *Arctocephalus pusillus*). Distribution, abundance and feeding habits off the South Western coast of the Cape Province. Investl Rep Div Sea Fish S Afr 34: 1–75

Ross GJB (1979) The smaller cetaceans of the southeast coast of southern Africa. PhD thesis, Univ Port Elizabeth, South Africa (unpublished)

Sea Fisheries Branch (1976) Annual report for 1974: commercial fisheries; demersal fishery. Rep Sea Fish Brch S Afr 42:9–10

Shaughnessy PD; Ross GJB (1980) Records of the subantarctic fur seal (*Arctocephalus tropicalis*) from South Africa with notes on its biology and some observations of captive animals. Ann S Afr Mus 82: 71–89

Swales MK (1965) The seabirds of Gough Island. Ibis 107:17–42, 215–229

Tollu B (1974) L'otarie de l'île Amsterdam *Arctocephalus tropicalis tropicalis* (Gray 1872). These de doctorat de troiséme cycle, l'Universite Paris - VII Paris

Voss NA (1969) A monograph of the Cephalopoda of the North Atlantic. The family Histioteuthis. Bull Mar Sci 19:713–867

The Role of the Leopard Seal in the Tropho-Dynamics of the Antarctic Marine Ecosystem

D.B. SINIFF and S. STONE[1]

Summary. The Leopard Seal, *Hydrurga leptonys,* preys significantly on several species of vertebrates and invertebrates. Observations and specimens (stomach contents, reproductive tracts and skulls), collected during recent expeditions to the Antarctic, as well as data from past studies, provide insights into how this predator partitions pressure among trophic levels and species. These data and their analyses suggest hypotheses regarding the role of the Leopard Seal in influencing the Antarctic marine ecosystem. Variables, such as seasonal effects, sea-ice, and availability of prey species, cause changes in prey selection and influence predation rates. The impact that Leopard Seal predation has on the evolution of the life-history patterns of prey is considered, using data on observed foraging and escape behaviour. The seasonal distribution of prey items is hypothesized to influence the reproductive strategy of the Leopard Seal. Data from recent collections show that the breeding season of the Leopard Seal is much later in the austral summer than those of other Antarctic seals, and that delayed implantation probably does not exist in this species. The reason for these differences is hypothesized to be related to the ability of the Leopard Seal to utilize relatively many different prey species, resulting in a reproductive strategy that does not require the highly synchronized breeding and young-bearing periods characteristic of other seal species. Such considerations for the Leopard Seal are important in attempting to understand the dynamics of the Antarctic ecosystem, since this species may exert significant predatory pressure at several trophic levels.

1 Introduction

The Leopard Seal, *Hydrurga leptonyx,* is the largest of the four species of Antarctic seals and is well known for its predatory activities, because it preys regularly on warm-blooded animals as well as fish, cephalopods, and krill (Euphausiacea). It ranges between about 50° and 79°S, and periodically has been found on other southern continents (Bonner and Laws 1964). It is usually a solidary animal in the Antarctic pack-ice region (Erickson et al. 1971; Gilbert 1974). However, concentrations of immatures are seen regularly on certain Sub-Antarctic islands, since the animals migrate northward during the austral late-autumn and winter (King 1964; Rounsevell and Eberhard 1980).

The predatory behaviour of the Leopard Seal is commonly associated with penguin breeding colonies (Kemp and Nelson 1931; Csordas 1963; Penney and Lowry 1967; Muller-Schwarze 1971; Muller-Schwarze and Muller-Schwarze 1975). However, more recent studies in pack-ice (Tikhomirov 1974; Hofman et al. 1977; Øritsland 1977; Siniff et al. 1979; Siniff 1981; Stone and Meier 1981) have shown that several other prey species are taken commonly as well, and constitute important components of the species' diet. Both Hofman et al. (1977) and Øritsland (1977) noted that Krill was a major food item, and stressed the importance of the Antarctic Krill, *Euphausia superba,* in the functioning of the Antarctic marine ecosystem.

The reproductive cycle of the Leopard Seal is still not well understood, although recent findings have indicated that certain general characteristics of the species are similar to those of other species of Antarctic seals. DeMaster et al. (1979) reported that the Leopard Seal produces its young on the pack-ice during November but, whereas females and their pups were observed on ice floes, the males were absent and their role in pupping and weaning remains unknown. It has been suggested (Harrison 1969; Sinha and Erickson 1974) that mating occurs some time after weaning, but there has been disagreement over whether the species has the delayed implantation period that is common for other pinnipeds.

Information for this paper was taken from various sources, but relies heavily on data obtained during a 4-year period of research, and observations made by Stone (1983). We attempt to integrate all available data, to determine the likely life-history pattern of the Leopard Seal and to examine how changes throughout the species' annual cycle influence its impact on other species.

2 Materials and Methods

Four years of study were carried out north and west of the Antarctic Peninsula in an area bounded approximately by 62°–65°S and 58°–64°W. The R.V. Hero, of the U.S. National Science Foundation, was used to gain access to ice and seals. Cruises of R.V. Hero, lasting about 30 days, were carried out in October and November in 1977, 1978, and 1979. During these cruises, when the ship was moving through pack-ice, running tallies were maintained of all seals sighted in relation to ice density and other physical

1 Department of Ecology and Behavioral Biology, University of Minnesota, Minneapolis, MN 55455, USA

Antarctic Nutrient Cycles and Food Webs
(ed. by W.R. Siegfried, P.R. Condy, and R.M. Laws)

features. Primarily shore-based work was carried out near Palmer Station during January–March, 1981. Forty-seven female Leopard Seals were collected, to study the species' reproductive cycle. Standard length and axillary girth were measured, and reproductive tracts and other materials were preserved in 10% buffered formalin. When possible, the collected animal was weighed before dissection.

Additional information on reproduction was made available by Dr. J.L. Bengtson, from collections made by him at the Faraday Station of the British Antarctic Survey. Seven females, collected during 5 January–4 February, 1982, are included in our analysis, since they contributed important data that were not available previously. A total of 35 Leopard Seals was collected by Bengtson and, whereas only 7 were used in our reproductive studies, they all are included in our diet analysis.

A total of 94 Leopard Seals, of all age classes, was collected by means of shooting or over-doses of drugs; the sample included 64 females and 30 males. The stomach contents of these specimens were examined, and form the basis of our information on diet. Changes in diet throughout the annual cycle were determined from this information and from data in the literature for seasons in which we did not collect specimens.

3 Results

3.1 Reproductive Cycle

Our findings indicate that parturition occurs during late October to mid-November, as determined from females with near-term foetuses or nursing pups (Fig. 1). The earliest collections of specimens after the pupping season were obtained on 5 January, and they indicated a developing *corpus luteum* but not implanted foetus. Implanted foetuses were not found until mid-February. Thus, we have identified early January to mid-February as the period of delayed implantation. Implanted foetuses were found after mid-February when the *corpus luteum* had begun to increase in size, and the *corpus albicans* from the previous pregnancy had continued to regress (Fig. 1).

3.2 Seasonal Food Habits

Figure 2 shows that the diet of the Leopard Seal changes in relation to time of year. The species' consumption of Crabeater Seals, *Lobodon carcinophagus,* rose sharply in November, when newly weaned Crabeater pups became available (Siniff et al. 1979). After November, the proportion of Crabeater Seals continued to rise until early February in the Antarctic Peninsula area. A low level of Crabeater Seals in the diet is indicated from the collections of Tikhomirov (1974), but his specimens were taken in pelagic pack-ice where Crabeater Seal density may be relatively low. After February, the decrease in Crabeater Seals in the diet probably coincides with an increasing ability of growing, young animals to escape from predators.

Krill was an important food item in late September, and seemed to drop in importance as other prey items became available during the summer. Krill became important again in March, and probably remained so throughout the winter. Based on faecal samples, Hofman et al. (1977) showed that the occurrence of Krill peaked in December. Their data were not included in Fig. 2, because frequency of occurrence of prey items in faecal samples is not directly comparable with the results of analyses of stomach samples. It seems likely that the food taken by Leopard Seals during December, when breeding occurs, may reflect very opportunistic and infrequent foraging.

Penguins were an important food item when their young were fledging and entering the sea. Thus, penguins featured most prominently in the species' diet in mid-February, and

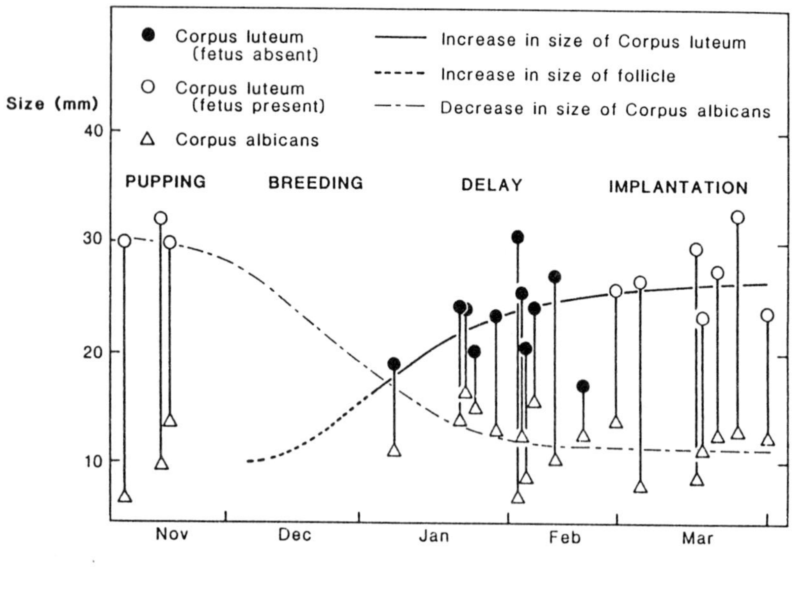

Fig. 1. The reproductive cycle of the Leopard Seal showing size of the *corpus luteum* and largest *corpus albicans* for individual specimens. Large dots: *corpus luteum* (foetus absent); small dots: *corpus albicans;* circles: *corpus luteum* (foetus present). Solid line: traces increase in size of *corpus luteum* (following increase in size of follicle) and broken line traces decrease in size of *corpus albicans*

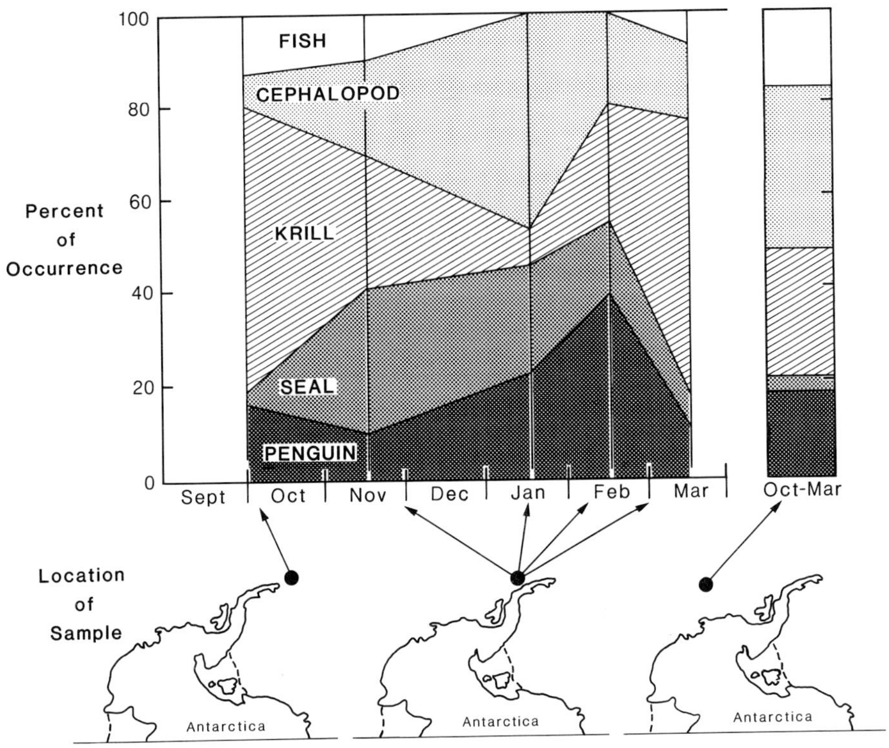

Fig. 2. Frequency of occurrence (%) of food items in the stomachs of Leopard Seals in relation to time of year. Sample sizes and sources of data, as follows: Sept.–Oct., n = 38 specimens with food in stomachs (from Øritsland 1977); pooled data for Oct.–March, n = 39 specimens with food (from Tikhomirov 1974) not broken down by month; n = 66 specimens with food (this study)

decreased rapidly as the season of fledging passed. Cephalopods were an important food item in January, but some were taken throughout the year. Cephalopods could be important as a mainstay in the diet, when prey such as seals and penguins are not readily available.

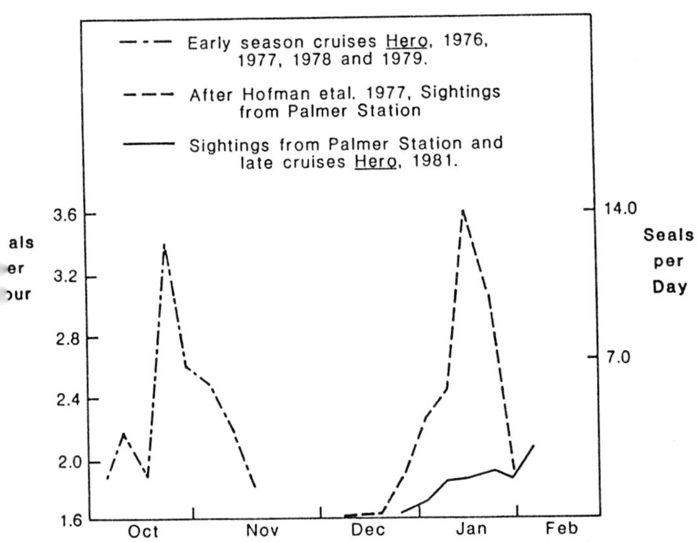

Fig. 3. Sightings of Leopard Seals in areas of pack-ice in the vicinity of the Antarctic Peninsula and off Palmer Station, Antarctica. *Broken line* on *left* is based on early-season cruises of R.V. Hero in 1976, 1977, 1978 and 1979. *Broken line* on *right* is based on data in Hofman et al. (1977) for Palmer Station. *Solid line* is based on late-season cruises of R.V. Hero in 1979

3.3 Dispersion of Leopard Seals

Figure 3 shows our sightings of Leopard Seals during early-season cruises, as well as sightings made near Palmer Station and those reported by Hofman et al. (1977). Sightings of Leopard Seals in the pack-ice region decreased markedly during mid-November to late December which coincides with the mating period. It seems likely that mating takes place in the water, since the number of Seals hauled out on ice floes was very low in the mid-November–late December period.

4 Discussion

In our representation of the reproductive cycle of the Leopard Seal the mating season is December to early January. Considering the solitary nature of this species, one would hypothesize that female Leopard Seals advertize their reproductive condition. Either vocalizations or chemical attractants from females in oestrus could be important cues for the attraction of males (Marlow 1967). Thus, the species' social organization might be similar to that of the Tiger, *Panthera tigris,* in which the female advertizes by roaring (Sunquist 1979), or the Manatee, *Trichenus manatus,* in which some type of chemical attractant seems likely (Bengtson 1981). Obviously, study of the Leopard Seal during December is necessary.

Most of the scarring of Crabeater Seals by Leopard Seals probably occurs in February and March when the immature animals are growing and increasing their ability to escape from attacks by Leopard Seals. Incidental sightings of newly-wounded Crabeater pups increased during February and March when seven out of nine immature animals were seen with fresh, raking wounds (Stone 1983). In contrast, a single, unscathed immature animal was seen prior to this period. Lindsey (1938) described a similar pattern for the Bay of Whales, in the Ross Sea, where young-of-the-year Crabeater Seals "commonly were seen" with fresh, raking gashes in February and March.

Our findings re-affirm the view that Leopard Seals are very catholic in their choice of food. Peaks in the incidence of various groups of prey in the diet of the Leopard Seal apparently occur when the prey are most available and vulnerable or when opportunities to take other prey items are few. This opportunism does not differ from the feeding behaviour of other pinnipeds (Geraci 1975), but it does seem extreme with respect to the size and taxonomic range of dietary items. Siniff (1981) suggested that the impact of Leopard Seal depredation on Crabeater Seals could be significant in certain areas of the Antarctic. Using the assumption that 20% of adult Leopard Seals eat at least one Crabeater pup per day, for a 90-day period from November to February, Siniff (op. cit.) suggested that about 2.4 Crabeater pups km^{-2} of pack-ice would be consumed. This predator pressure apparently varies geographically as well as seasonally. Condy (1976) found that 62.5% of all Crabeater Seals seen were scarred in the King Haakon VII Sea where the incidence of Leopard Seals is relatively very low. Siniff and Bengtson (1977) reported that 78% of all adult Crabeater Seals examined along the Antarctic Peninsula had scars, and believed that the scars had been caused by Leopard Seal attacks. Presumably many or most of the scars, unattributed as to cause, that Condy (1976) reported were also from Leopard Seals. Previous censuses in the King Haakon VII Sea yielded levels of scarring amounting to 63 and 25.2%. Although the frequency of scarring may be underestimated by shipboard censuses, these data suggest that predation by Leopard Seals on Crabeater Seals is lower in the King Haakon VII Sea than in the Antarctic Peninsula area.

Condy (1976) also reported scars of unattributed cause on 0, 0.8 and 3.2% of Ross Seals, *Ommatophoca rossii*, observed, respectively, in 3 years of censuses in the King Haakon VII Sea, the region of greatest known absolute density and relative species abundance of Ross Seals. We observed a Leopard Seal unsuccessfully try to capture a Ross Seal from an ice floe in November, and did not find Leopard Seal teeth marks on three Ross Seals which we handled. Nor have we observed scars or wounds attributable to Leopard Seals on Weddell Seals along the Antarctic Peninsula, although Weddell Seals have been reported in the diet of Leopard Seals (Øritsland 1977). Thus, preliminary observations indicate that the impact of Leopard Seal depredation on Ross and Weddell seals is likely to be insignificant.

Leopard Seal depredation has a low impact on adult penguin populations at rookeries in the spring and summer breeding season, as well as on fledging chicks leaving the rookeries (Penney and Lowry 1967; Muller-Schwarze and Muller-Schwarze 1975). In the pack-ice during January and February, Adélie Penguins, *Pygoscelis adeliae,* occur in densities varying from < 1 (Erickson et al. 1972) to almost 35 birds km^{-2} (Zink 1978). Ratios range from 0.08–438.1 penguins per Leopard Seal in census tracts. Densities of hauled-out Leopard Seals are fairly consistently around 0.03–0.07 animals km^2 (Condy 1976; Gilbert and Erickson 1977) in pack-ice throughout the Antarctic. Adélie Penguins, which perhaps constitute up to 90% of all Antarctic bird biomass (Croxall and Prince 1979), seem to be nearly as patchily distributed as a resource for Leopard Seals in pelagic pack-ice as are their rookeries along the coast. Even in Arthur Harbour, near the moderately sized rookeries (about 15,000 breeding pairs) at Palmer Station, Leopard Seals fed on Krill and Crabeater Seals, as well as on penguins throughout the fledging season. Penguins probably make up no more than 15–20% of the Leopard Seal's diet, by frequency of occurrence. If the Seals consume a similar proportion by weight of diet, then, a Leopard Seal eats about 1 penguin d^{-1}.

An Adélie Penguin weighs 2.7–6.5 kg (Prevost and Mougin 1970) and a Leopard Seal eats about 20–22 kg d^{-1} (Øritsland 1977; Stone 1983). Thus, if Leopard Seals are distributed in pack-ice at a density of roughly 0.12 animals km^2 (Gilbert and Erickson 1977) during January and February, they could consume 0.32–0.78 kg of Adélie Penguin km^2 d^{-1}. Leopard Seals do not seem to be preventing a current increase in Adélie Penguin numbers, if such an overall increase is indeed occurring (Croxall and Prince 1979). Penguin species may have a more significant impact on Leopard Seal populations as competitors for Krill than as prey themselves.

The annual combined biomass of prey consumed by Leopard Seals can be estimated, making several assumptions. We feel that a Leopard Seal fills its stomach once daily. The full stomachs that we obtained usually weighed 15–20 kg. We weighed 37 animals which were yearlings or older individuals, and they averaged 315 kg. Based on these data, Leopard Seals consume 5–6% of their body weight per day. Using 400,000 as a population estimate (see SCAR Group of Specialists on Seals, meeting report: 6–7 Sept., 1983, Pretoria, South Africa) and 335 feeding days per animal per annum, we estimate an annual consumption of 2.0–2.7 × 10^6 t (wet wt.) of prey by Leopard Seals in Antarctic.

Partitioning this biomass into food types becomes very speculative, since the species' diet changes seasonally (Fig. 2) and translation of diet proportion into mass is very uncertain. Some idea can be obtained by considering how individual items were found in the stomachs. For the occurrence of seal prey we assumed at least 1 full meal (20 kg) for each occurrence. Krill presence ranged from a few individuals to 16 kg, but it seems likely that at least 5–10 kg were taken during each feeding. An individual penguin weighs about 4 kg, for the species taken most frequently. Presence of fish and cephalopods probably represented a few individuals, in the range of 2–4 kg per meal. Using the general areas under the curves of Fig. 2, and assuming that the Leopard Seal's feeding is heavily biased in favour of Krill in winter, we suggest the biomass taken might be partitioned as 45% Krill, 35% seal, 10% penguin and 10 fish and cephalopods.

In considering seasonal changes in the diet of the Leopard Seal, it seems that winter could be a period of food stress in certain areas. During winter, Leopard Seals probably compete with Krill-feeding specialists such as Crabeater Seals and Adélie Penguins. The Leopard Seal is almost certainly not as efficient in foraging for Krill as these specialists and, thus, may be especially sensitive to any changes in Krill abundance that may be triggered by commercial exploitation. It has been noted (Øritsland 1970; Siniff 1981; Stone 1983) that the Leopard Seal has a relatively low pupping rate. It may be that the low rates observed are a reflection of food availability during winter and early spring, even though Øritsland's (1970) observations were made in 1964 when Krill abundance is thought to have been at its maximum (Bengtson and Laws 1985). Other evidence for food stress during winter might be the behaviour of immature and certain adult Leopard Seals which disperse to northern islands outside the pack-ice area (Gwynn 1953; Rounsevell and Eberhard 1980). Such movements might be due to competition for feeding areas, and could be a seasonal migration triggered by dominant individuals. If such migrations are a result of competition for food during winter, then it can be expected that the number, sex, and age of individuals migrating may change in response to human exploitation of Krill. Thus, monitoring Leopard Seals as agents for reflecting changes in Krill abundance may be an important activity in the future.

5 Conclusion

The Leopard Seal is an important predator which is a generalist in its feeding habits. Its reproductive cycle is displaced about 1 month later than that of the Crabeater Seal, and this may be an evolutionary response to food availability for lactating females and newly weaned pups. The Leopard Seal probably has a significant impact on Crabeater Seal populations in certain regions of the Antarctic, particularly the Antarctic Peninsula area. However, its impact on other prey species is probably not an important limiting factor. The Leopard Seal probably competes directly with Krill-feeding specialists during the winter, and, since it is relatively less efficient in its foraging abilities, the Leopard Seal may be the first of the Antarctic seal species to feel the impact of exploitation of Krill by man. Migration of immature Leopard Seals to northern islands during the winter may increase with an increase in Krill exploitation, and perhaps show changes in sex and age composition.

References

Bengtson JL (1981) Ecology of manatees (*Trichechus manatus*) in the St. Johns River, Florida. Ph.D. Thesis, University of Minnesota

Bengtson JL; Laws RM (1985) Trends in crabeater seal age of maturity: An insight into Antarctic marine interactions? In: Siegfried WR, Condy PR, Laws RM, (eds) Antarctic nutrient cycles and food webs (Proceedings of the 4th SCAR symposium on Antarctic biology). Springer, Berlin Heidelberg New York

Bonner WN; Laws RM (1964) Seals and sealing. In: Priestey R, Adie RJ, Robin G (eds) Antarctic Research: A review of Btitish scientific achievement in Antarctica. Butterworths, London

Condy PR (1976) Results of the third survey in the King Haakon VII Sea, Antarctica. S Afr J Antarct Res 6:2–9

Croxall JP; Prince PA (1979) Antarctic seabird and seal monitoring studies. Pol Rec 19(123):573–595

Csordas SE (1963) Leopard seals on Macquarie Island. Victorian Nat 79(12):358–362

DeMaster D; Thomas J; Stone S; Andriashek D (1979) Biological studies of seals in pack ice habitat. Antarct J US 14(5):179–180

Erickson AW; Siniff DB; Cline DR; Hofman RJ (1971) Distributional ecology of antarctic seals. In: Deacon G (ed) Symp. on antarctic ice and water masses. (19 Sept 1970, Tokyo) pp 55–76. SCAR Cambridge pp 55–76

Erickson AW; Gilbert JR; Petrides GA; Oehlenschlager RJ (1972) Populations of seal , whales and birds in the Bellingshausen and Amundsen Seas. Antarct J US 7(4):70–72

Geraci JR (1975) Pinniped nutrition. In: Ronald K, Mansfield AW (eds) Symposium on the biology of the seal (Guelph, Ontario, August, 1972). Rapp P-V Reun Cons Int Explor Mer 169:312 –32

Gilbert JR (1974) The biology and distribution of seals in Antarctic pack ice. PhD Thesis, Univ Idaho

Gilbert JR; Erickson AW (1977) Distribution and abundance of seals in the pack ice of the Pacific sector of the Southern Ocean. In: Llano GA (ed) Adaptations within Antarctic ecosystem. Proceedings of the 3rd SCAR Symposium on Antarct Biol. Gulf Pulb Co Houston, Smithsonian Inst Wash DC pp 703–740

Gwynn AM (1953) The status of the leopard seal at Heard Island and Macquarie Island, 1948–1950. ANARE Interim Report 3 (Publ #7)

Harrison RJ (1969) Reproduction and reproductive organs. In: Anderson HT (ed) The Biology of marine mammals. Academic Press, New York

Hofman RJ; Reichle RA; Siniff DB; Muller-Schwarze D (1977) The leopard seal *Hydrurga leptonyx*) at Palmer Station, Antarctica. In: Llano GA (ed) Adaptations within Antarctic ecosystems. Proceedings of the 3rd SCAR Symposium Antarct Biol. Gulf Publ Co Houston, Smithsonian Inst Wash DC pp 769–782

King JE (1964) Seals of the World. Br Mus (Nat Hist) London

Kemp S, Nelson AL (1931) The South Sandwich Islands. Discovery Rep 3:133–198, Seals pp 155–156

Lindsey AA (1938) Notes on the crabeater seal. J Mammal 19(4): 456–461

Marlow BJ (1967) Mating behaviour in the leopard seal, *Hydrurga leptonyx* (Mammalia:Phocidae), in captivity. Aust J Zool 15: 1–5

Muller-Schwarze D (1971) Behaviour of antarctic penguin and seals. In: Quam LO (ed) In research in the Antarctic. Publ No 93 AAAS Washington DC pp 259–276

Muller-Schwarze D, Muller-Schwarze C (1975) Relations leopard seals and Adélie penguins. In: Ronald K, Mansfield AW (eds) Symposium on the biology of the seal. Rapp R-V Reun Cons Int Explor Mer 169:312–323

Øritsland T (1970) Sealing and seal research in the South-west Atlantic Pack Ice, Sept–Oct 1964. In: Holdgate MW (ed) Antarctic ecology. Academic Press, London

Øritsland T (1977) Food consumption of seals in the Antarctic pack ice. In: Llano GA (ed) Adaptations within Antarctic ecosystems. Proceedings of the 3rd SCAR Symposium on Antarct Biol. Gulf, Houston for Smithsonian Inst Wash DC pp 749–768

Penney RL; Lowry G (1967) Leopard seal predation on Adélie penguins. Ecology 48:878–882

Prevost J; Mougin JL (1970) Guide des oiseaux et mammiferes des Terres Australes et Antarctique Francaise. Delachaux et Niestle, Paris

Rounsevell D; Eberhard I (1980) Leopard seals. *Hydrurga leptonyx* (Pinnipedia), at Macquarie Island from 1949 to 1979. Austr Wildl Res 7(3):403–415

Sinha AA, Erickson AW (1974) Ultrastructure of the placenta of Antarctic seals during the first third of pregnancy. Am J Anat 141(2):263–280

Siniff DB (1981) Seal population dynamics and ecology. J Roy Soc NZ 11(4):317–327

Siniff DB; Stirling I; Bengtson JL; Reichle RA (1979) Social and reproductive behaviour of crabeater seals (*Lobodon carcinophagus*) during the austral spring. Can J Zool 57:2243–2255

Stone HS (1983) Leopard seal reproductive and feeding ecology along the Antarctic Peninsula. MS Thesis University of Minnesota

Stone S; Meier T (1981) Summer leopard seal ecology along the Antarctic Peninsula. Antarct J US 16(5):151–152

Sunquist ME (1979) The movements and activities of tigers (*Panthera tigris tigris*) in Royal Chitawan National Park, Nepal. PhD Thesis, University of Minnesota

Tikhomirov EA (1974) Material on the food and feeding of Antarctic seals (in Russian). Issled Biol R Promisl Okeanogr 5:129–136

Zink RM (1978) Birds of the Weddell Sea. Antarct J US 13(4):142–145

Weddell Seal – Antarctic Cod Interactions in McMurdo Sound, Antarctica

J.W. TESTA[1], D.B. SINIFF[1], M.J. ROSS[1], and
J.D. WINTER[2]

Summary. The impact of Weddell Seals, *Leptonychotes weddellii,* on Antarctic marine fish communities appears to be influenced by the behaviour of the Seals, which concentrates predation temporally and spatially. Movements of prey species, possibly in response to the distribution of the Seals, may determine which members of the fish community are most affected by predation. The distribution and movements of the Antarctic Cod, *Dissostichus mawsoni,* were monitored under shore-fast ice near Ross Island, Antarctica, by set-line fishing. The distribution and the age of Weddell Seals were studied in the same area, in conjunction with long-term research dealing with the species' population dynamics. During the austral spring adult Weddell Seals concentrate in discrete breeding colonies near the shore, while sub-adults remain scattered in small groups at isolated breathing holes throughout the fast-ice areas. Experimental fishing success for Antarctic Cod was related to distance of fishing sites from Weddell Seal breeding colonies. Adult Seals from both sides of McMurdo Sound leave their colonies in mid-December and gradually concentrate along the edge of the permanent ice-shelf, when fishing success decreases rapidly. The diet of Weddell Seals in November, January and February consists mostly of small nototheniid fish (*Pleuragramma* and *Trematomus* spp.), decapod and amphipod crustaceans, and cephalopods. Although Weddell Seals apparently exclude Antarctic Cod from their preferred fast-ice areas, the Cod do not appear to be an important food resource for more than a limited time in December.

1 Introduction

The Weddell Seal, *Leptonychotes weddellii,* population in Antarctica, has been studied since 1962 (Smith 1966; Stirling 1969; Kaufman et al. 1975; Siniff et al. 1977; Siniff 1981). These studies concentrated on the species' social organization, behaviour and population parameters, and the physical environment and other factors potentially influencing changes in population size. Anecdotal accounts (Wohlschlag 1968; Calhaem and Christoffel 1969) of Weddell Seals eating large Antarctic Cod, *Dissostichus mawsoni,* led us to speculate that an important predator-prey relationship exists between Weddell Seals and Antarctic Cod in McMurdo Sound. Thus, in 1981, we initiated a study of the movements and distribution of the two species. Our purpose here is to show that the dispersion and movements of Weddell Seals and Antarctic Cod are not independent of each other, and to evaluate the possible importance of the predator-prey relationship affecting the two species.

1 Department of Ecology and Behavioral Biology, University of Minnesota, Minneapolis, MN 55455, USA
2 Environmental Resources Center, State University of New York at Fredonia, New York, USA

2 Materials and Methods

We used the successful technique described by Eastman and DeVries (1981) for capturing specimens and studying the distribution of Antarctic Cod under annual sea-ice. This method involves drilling a hole in the sea-ice, and setting a baited, stainless steel, fishing line with 20 hooks spaced at 10 m intervals from the bottom. During 24 October–7 December 1982, we drilled 17 ice-holes (370 mm diam.) in eastern McMurdo Sound (Fig. 1) into which set-lines, each bearing 20 hooks at 3 m apart and starting at 9 m from the bottom, were inserted. Two consecutive sets of approximately 12 h each were made at each site, with depths ranging from 225–607 m. Two sites at outlying depths (61 and 740 m, catch = 0) were later excluded from analysis. The effects of depth and distance from Weddell Seal colonies were analyzed. Catch-effort data, based on fishing in depths of 300–600 m within 8 km of McMurdo Station, were provided by Dr. A.L. DeVries (pers. comm.) for the period late September–late December in 1979 and in 1978 –1980. These data were analyzed at the University of Minnesota.

Fifteen and 17 scats (faecal droppings) were collected from two colonies of Seals, respectively, and another 17 scats from breathing holes away from the colonies on 23 and 30 November, 1982 (Fig. 2). The scats were < 1 month old, judged by their degree of desiccation and the Seal's occupancy of certain areas. Small scats which could have come from pups were not analyzed. Scats were soaked in detergent and water for several days before being washed through wire sieves (minimum pore size = 1 mm); their contents were preserved in 35% EtOH for identification later. Stomach contents from 36 Seals (31 males and 5 females), killed by members of the New Zealand Antarctic Research Programme (NZARP) during 22–27 January 1983 (Fig. 1), were washed similarly through 1 mm wire sieves and the remains stored in 10% formalin. All remains (mostly skeletal bones of fish and crustacean exoskeletons) were identified to broad taxonomic groups. Vertebrae of

Antarctic Nutrient Cycles and Food Webs
(ed. by W.R. Siegfried, P.R. Condy, and R.M. Laws)

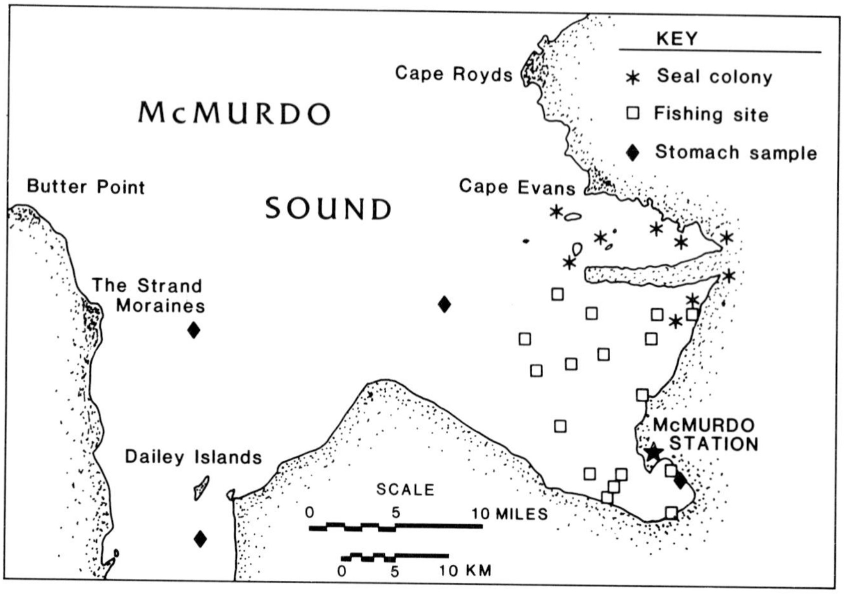

Fig. 1. McMurdo Sound, showing Weddell Seal colonies and sites at which seals were collected for examination of their stomach contents, and sites at which fishing for Antarctic Cod was carried out

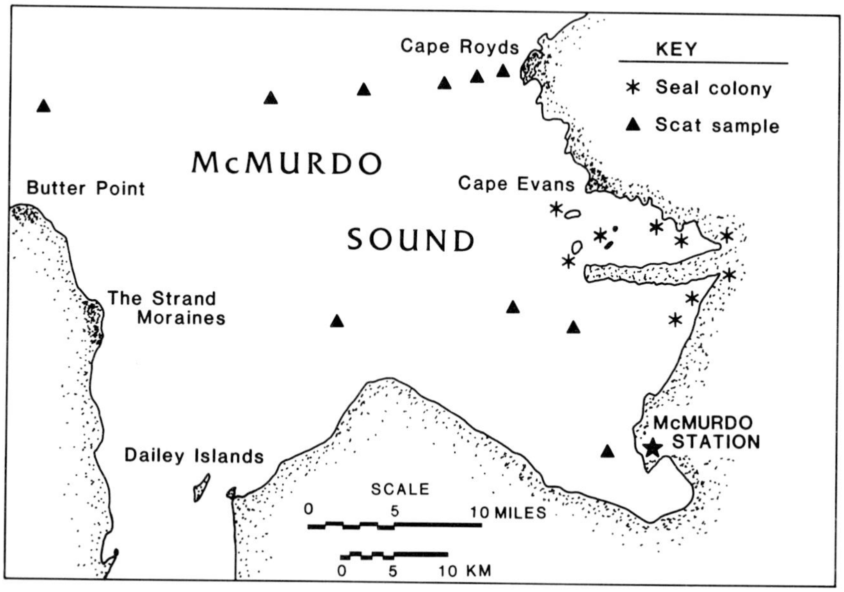

Fig. 2. McMurdo Sound, showing Weddell Seal colonies and sites at which scats of seals were collected

Pleuragramma antarcticum were distinctive enough in both scats and stomach contents to allow their reliable identification. The results of a cursory examination of the stomach contents of 52 Seals collected by members of NZARP during 15–16 February 1982 in the immediate vicinity of Scott Base (3 km from McMurdo Station and adjacent to the Ross Ice Shelf) are also presented.

3 Results

Fishing success at McMurdo Sound was relatively high between late October and late November (Fig. 3). A two-way analysis of variance indicated significant effects ($P < 0.001$) of both year and fishing period on catch per 24 h set. We cannot comment now on the importance of annual variation, but the peak in fishing success and, presumably, Antarctic Cod abundance in the area of deep water near McMurdo Station coincided with the congregation of adult Weddell Seals in breeding colonies 15–30 km away from our fishing sites. Seals dispersed from the colonies towards the permanent ice shelf (McMurdo Station) in early to mid-December, at the same time as fishing success decreased.

Fishing success in 1982 was evaluated by regression analysis, relative to depth of water and the distance from the nearest Seal colony. There was no discernible relationship between catch and depth, except general agreement with earlier work (Eastman and DeVries 1981) indicating that specimens of *D. mawsoni* are not caught at very shallow depths (< 300 m). However, because our shallow sites were

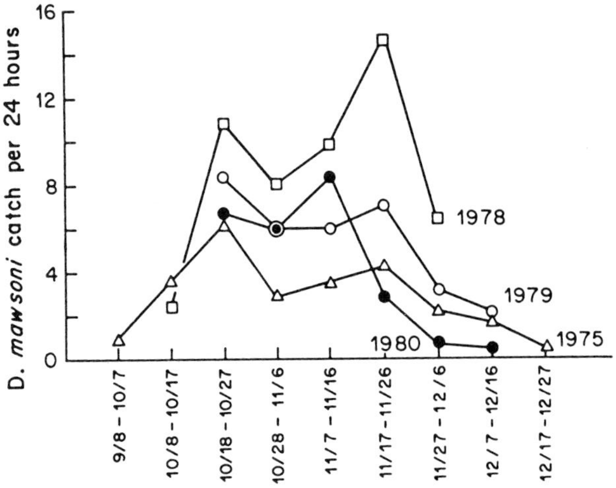

Fig. 3. Fishing success (catch/effort) for Antarctic Cod during September–December for 4 years in McMurdo Sound, based on data collected by A.L. DeVries

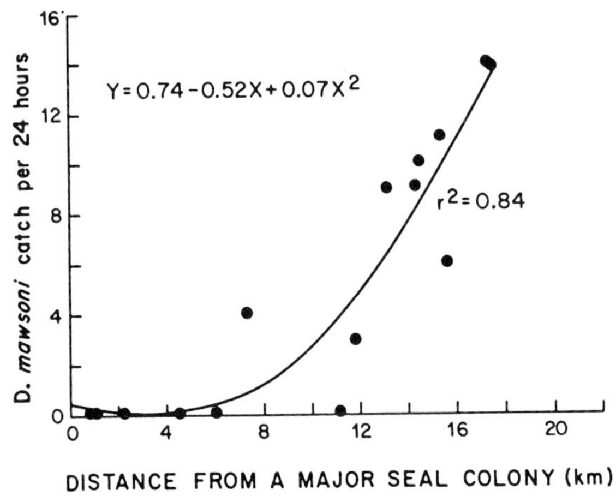

Fig. 4. Effect of concentrations of Weddell Seals on fishing success (catch/effort) for Antarctic Cod in McMurdo Sound

relatively close to Seal colonies, it is possible that our fishing success was more affected by distance from Seal colony than depth of water. Various analyses were attempted using the data in Fig. 4; the best fit regression equation (adjusted for degrees of freedom) being the second-order polynomial ($r^2 = 0.836$, 12 df).

Scat analysis for 1982 was intended to detect differences in diet between Weddell Seals in colonies and those dispersed in lower densities in other areas. Turtle Rock, located well within Erebus Bay and farthest removed from the deep water that underlies most of the fast-ice habitat in McMurdo Sound, was expected to have a different, possibly depauperate, prey community available to feeding Seals. Tent Island is the most peripheral of the Seal colonies, being nearest to the large expanse of ice covering the middle of McMurdo Sound, and the diet of animals there was expected to be intermediate between those at Turtle Rock and sites outside the colonies. The results are shown in Table 1. Eye lenses of fish were excluded from the analysis, since they were less common than vertebrae and gave lower resolution of dietary composition. No otoliths were found.

Multi-way Categorial analysis on the small sample of scats, using location, *P. antarcticum,* mysid shrimp, and amphipods as dimensions, detected no statistically significant associations ($P < 0.05$) between prey items or between prey and location with the log-linear model (Feinberg 1981). The dominance of *P. antarcticum* in the samples and the low occurrence of mysids and amphipods may have obscured a possible association of these items, which could have resulted through *P. antarcticum* feeding on the crustaceans (DeWitt and Hopkins 1977). There was a trend, however, towards a high occurrence of *P. antarcticum* in the diet of Seals outside colonies. This trend was slightly stronger (i.e. trend toward empty scats in the colonies) when prey types were lumped, but remained non-significant ($P = 0.18$). Interpretation of any differences in scat contents, as a reflection of dietary differences, is complicated

Table 1. Frequency of occurrence of items in scats of Weddell Seals, collected during November 1982 in McMurdo Sound

Prey item	Turle Rock	Tent Island	Outside sites
P. antarcticum	8	12	15
Other nototheniids	2	2	1
Cephalopoda	1	1	1
Mysidacea	1	5	5
Amphipoda	2	4	5
Copepoda[a]	4	3	1
Isopoda[a]	0	0	3
Helminth parasites*	12	10	12
Total scats examined	15	17	17

a Probably not a prey item

by potential effects of food and feeding (especially fasting) on rates of defecation, and immigration of Seals into colonies from areas of low density. In general there was a trend towards greater feeding away from colonies, but there were no statistically significant differences in the prey remains found in the three collections of scats made in late November; *P. antarcticum* being the major item present. In a total of 36 stomachs of adult Weddell Seals taken in 1983, eight contained no food, 26 contained *P. antarcticum,* four contained other nototheniids, and five contained cephalopod remains. All stomachs were heavily infested by helminth parasites. In terms of volume, crustaceans constituted an insignificant component of the stomach contents. Of the five stomachs containing cephalopod remains, one had a "crown" of octopod arms, and the rest contained only "beaks". Small sample size and the obvious similarity in the prey occurrence at all four sites precluded a categorial analysis of association between diet and site of collection.

The result of a cursory examination of 52 stomachs of Seals taken in 1982 was in general agreement with the preceding findings. Remains of small fish (including one whole bathydraconid) were found in 19 stomachs, cephalopods in two, fresh Mysidacea in one and *D. mawsoni* remains in at least one. A seal which had eaten a large specimen of *D. mawsoni* had also eaten 16 squid. Crustacea were a small part by volume of these contents. The person (JWT) who made these observations was unfamiliar with *P. antarcticum* at the time, so identification of this species was not possible.

4 Discussion

The Weddell Seal shows plasticity in its feeding behaviour. According to Bertram (1960), cephalopods and fish were represented equally in the stomachs of 367 Weddell Seals collected during the 2-year period of the British Graham Land Expedition. Clarke and MacLeod (1982) also found cephalopod remains in a small sample of Seals killed in the South Shetland Islands. At McMurdo Sound, however, *P. antarcticum* seems to be the most prevalent food item in both stomachs and scats of Weddell Seals collected in November, January and February (Dearborn 1965; this study), although anecdotal accounts suggest that *D. mawsoni* may be important to adult Seals in December (Calheam and Christoffel 1969). In spite of the apparent unimportance of Antarctic Cod in the Seal's diet, the distributions of the two species do not appear to be independent.

Our studies (unpubl. data) indicate that densities of Weddell Seals in the fast-ice areas, excluding pupping colonies, increase northward of the Ross Ice Shelf, with the proportion of sub-adults (1–3 year) increasing. Age-class segregation is present in all months of the summer at McMurdo (November–February), implying that the movement northward of adults to breeding colonies in late October is likely to affect the abundance of Antarctic Cod at fishing sites, as was observed. Furthermore, the apparent relationship between Cod catches and distance from Seal colonies (Fig. 4) is confounded by an increasing density of outlying Seals in the same area. If adult Seals depress Antarctic Cod catches in September and October and later in December they do so at a much lower density than that prevailing in the Seal colonies. We believe that relatively low densities of Seals can effectively exclude Cod from fast-ice areas, and that predation by Seals on Antarctic Cod is probably important only when Seals move into areas that they have not occupied for some time. The marked seasonal abundance of Antarctic Cod at McMurdo indicates that they quickly find those areas which are abandoned by Seals because of ice conditions or breeding behaviour. More extensive fishing in McMurdo Sound and data on the food habits of Seals in December are needed to verify this.

Direct predation is the obvious mechanism by which Antarctic Cod could be excluded from Seal areas. Detailed information on the availability of Cod and their movements into the fast-ice area of McMurdo Sound, concurrent with an assessment of predation on Cod by Seals, are necessary to demonstrate conclusively the importance of predation. Currently, we have only anecdotal accounts to support this mechanism. An alternative hypothesis is that Cod are capable of detecting the presence of Seals and avoiding Seal concentrations. It is known that fish in general are fairly sensitive to underwater sound (Tavolga 1971) and that Weddell Seals are extremely vocal in defence of underwater territories (Thomas and Kuechle 1982). A test of this hypothesis might be to subject captive Cod to experimental playbacks of Weddell Seal vocalizations or to monitor Cod acoustically under the ice during playback experiments in an area free of Seals.

The diets of Weddell Seals and Antarctic Cod in McMurdo Sound are very similar. Antarctic Cod collected in November had a relatively higher incidence of mysids in their stomachs but otherwise were similar in the occurrence of fish and cephalopods (Eastman 1985). The under-ice feeding niches of these two species, at the apex of the same trophic system, are partitioned by means of the feeding behaviour of the Weddell Seal at the expense of *D. mawsoni.* The effect of Weddell Seal predation on *P. antarcticum* abundance is not known, but could be severe in local areas (i.e., breeding colonies and late-summer aggregations of Seals, which occur as fast-ice breaks up). Year-round observations of the abundance of prey species and the diet of Seals, such as those underway at Australia's Davis Station (H. Burton, pers. comm.), are needed. In general, we believe that the wide geographical range and probably nomadic movements of *P. antarcticum* assist to minimize the impact of Weddell Seal predation, except in very restricted areas or seasons. One consequence of this is that while Weddell Seal populations might be limited by the availability of prey at certain times, food supply is not likely to be a density-dependent regulating factor affecting populations in the long term.

Studies of the abundance and life history of *P. antarcticum* are needed urgently in order to understand better the trophic web in the Ross Sea. Efforts to monitor trends in mid-water fish abundance will be important in areas where development occurs for the exploitation of resources, and also in pristine "control" areas, to assist in attempts to safeguard key trophic links, such as *P. antarcticum*, or sensitive vertebrate populations at higher trophic levels.

References

Bertrams GCL (1940) The biology of the Weddell and crabeater seals. Sci Rep Br Graham Land Exped 1:1–139

Calhaem I, Christoffel DA (1969) Some observations of the feeding habits of a Weddell seal, and measurements of its prey, *Dissostichus mawsoni*, at McMurdo Sound, Antarctica. NZ J Mar Freshwater Res 3(2):181–190

Clarke MR, MacLeod N (1982) Cephalopod remains in the stomachs of eight Weddell seals. Br Antarctic Surv Bull 57:33–40

Dearborn JH (1965) Food of Weddell seals at McMurdo Sound, Antarctica. J Mammal 46:37–43

DeWitt HH, Hopkins TL (1977) Aspects of the Diet of the Antarctic silverfish, *Pleuragramma antarcticum*. In: Llano GA (ed) Adaptations Within Antarctic Ecosystems. Gulf Publ Co, Houston, Texas pp 557–568

Eastman JT (1984) The evolution of neutrally buoyant notothenioid fishes: their specializations and potential interactions in the Antarctic marine food web. In: Siegfried WR, Condy PR, Laws RM (eds) Antarctic nutrient cycles and food webs (Proceeding of the 4th SCAR symposium on Antarctic biology). Springer, Berlin Heidelberg New York

Eastman JT (1985) The evolution of neutrally buoyant notothenioid fishes: their specializations and potential interactions in the Ant-

Feinberg SE (1981) The Analysis of Cross-Classified Categorial Data. MIT, Cambridge, Mass

Kaufman GW, Siniff DB, Reichle R (1975) Colony behaviour of Weddell Seals, *Leptonychotes weddellii*, at Hutton Cliffs, Antarctica. Rapp PV Reun Cons Int Explor Mer 169:228–246

Siniff DB, DeMaster DP, Hofman RJ, Eberhardt LL (1977) An analysis of the dynamics of a Weddell seal population. Ecol Monogr 47(3):319–335

Siniff DB (1981) Seal population dynamics and ecology. J R Soc NZ 11(4):317–327

Smith MSR (1966) Studies on the Weddell seal in McMurdo Sound, Antarctica. PhD Thesis, Univ of Canterbury, NZ pp 161

Stirling I (1969) Ecology of the Weddell Seal in McMurdo Sound, Antarct Ecol 50(4):573–586

Tavolga WN (1971) Sound Production and Detection. In: Hoar WS, Randall DJ (eds) Fish Physiology, vol 5, Academic Press, New York pp 135–205

Thomas JA, Kuechle VB (1982) Quantitative analysis of the underwater repertoire of the Weddell seal (*Leptonychotes weddellii*). J Acoust Soc Am 72(6):1730–1738

Wohlschlag DE (1968) Fishes beneath Antarctic ice. Aust Nat Hist 16:no 2

A Review of the Food Habits of the Small Cetaceans of the Antarctic and Sub-Antarctic

R.N.P. GOODALL and A.R. GALEAZZI[1]

Summary. This paper reviews the current state of knowledge of the food habits of the poorly known small cetaceans of the Southern Ocean and presents new information obtained through a study of stranded specimens in Tierra del Fuego. In the Antarctic and Sub-Antarctic, food data are available for only 10 of the 22 species present. Adding to this all the food records known from the South Temperate Zone south of 30° S, there is still no information at all for 5 species (*Mesoplodon layardii, M. hectori, M. bowdoini, Cephalorhynchus eutropia* and *Phocoena dioptrica*); that of 8 species is based on fewer than three samples, and more than 10 samples are available for only 2 species (*Orcinus orca* and *Cephalorhynchus commersonii.* Since these cetaceans are among the larger animals of the Southern Ocean, their role in the food web of the ecosystem cannot be determined until more information is available.

1 Introduction

Most accounts of food chains in the Southern Ocean recognize large cetaceans as major consumers of Krill (Euphausiacea), other zooplankton, and squid. However, few even mention the 22 species of smaller cetaceans which inhabit Antarctic and Sub-Antarctic waters. Although smaller than the great whales, these cetaceans are still among the larger animals of the Southern Ocean.

The large whales of the Antarctic have been well studied by the International Whaling Commission (IWC) and by the whaling nations themselves, so that a great fund of knowledge has accumulated on most aspects of their biology, including their food habits (Fraser 1964; Bengtson 1978). The smaller cetaceans, being commercially unimportant, have never been studied intensively. Although some of the smaller whales were taken at times by whalers (Godoy 1963), little has been published on those of the Southern Ocean. The dolphins and smaller beaked whales of the Sub-Antarctic are known mainly through opportunistic sightings and strandings.

The study of small cetaceans presents some unique problems. Foremost is the general public's emotional involvement with dolphins and whales, with strong feelings against their capture or killing. Being wholly aquatic, there are only three major ways we can learn about their food habits: by observing, from ship or shore, live animals feeding; by examining stomach contents of captured animals (which usually means killing them); or by studying the stomachs of stranded animals. These problems are aggravated in the far south because many of the species are difficult to observe at sea and there are few fishing activities which might take cetaceans incidentally in nets. There is little land, in relation to the water area, on which cetaceans might strand. Most of our information on the food habits of these animals comes from stranded specimens. Although extensive surveys of stranded cetaceans have been carried out in the northern hemisphere (Sheldrick 1979; Mead 1979), little has been done in the southern hemisphere where much information is lost due to the scarcity of available scientists. Strandings have been recorded, over the years, from the tips of all the southern continents, but systematic beach surveys have been made only recently.

In this paper we present a review of the literature dealing with the food habits of the southern small cetaceans and, in some instances, report new information from our field surveys in Tierra del Fuego, Argentina (Goodall 1978, unpubl.). We follow Brownell (1977) in our definition of a "small cetacean" and use Mitchell's (1975) nomenclature. The Minke Whale, *Balaenoptera acutorostrata,* is excluded from our review.

2 Definition of Geographical Zones

We define the distribution zones for the smaller cetaceans as: (I) the High Antarctic or Antarctic proper, south of 60°S; (II) the Low Antarctic or Antarctic Convergence zone, between the Convergence and 60°S; (III) the Sub-Antarctic, as delineated by Hedgpeth (1969) as the area from about 48°S to the Antarctic Convergence, but which may reach northward in the eastern South Atlantic and Indian Oceans and may be farther south in the South Pacific; and, (IV) the Southern Cool Temperate zone, north of 48°S. This area includes New Zealand (except for the islands to the south), southern Australia, both coasts of southern South America north of 48°S and southwestern Africa. In this paper, we are concerned with Zones I, II and III.

1 Centro Austral de Investigaciones Científicas (CADIC), 9410 Ushuaia, Tierra del Fuego, Argentina

Antarctic Nutrient Cycles and Food Webs (ed. by W.R. Siegfried, P.R. Condy, and R.M. Laws)

3 Distribution of Cetaceans

3.1 Zone I: High Antarctic

The U.S.Antarctic Conservation Act of 1978, which defines the Antarctic as the area south of 60°S, designated only five small cetaceans as native mammals (marked with * below). Fraser (1964) and Brownell (1974), who made the most complete summaries of Antarctic cetaceans, agreed with this designation. Aguayo and Torres (1967) reported an additional species, *Cephalorhynchus commersonii*, from south of 60°S, and Goodall et al. (to be publ.) provide information on two species which occur at times south of 60°S, and suggest that a species of *Mesoplodon* may reach the South Shetland Islands.
The species in Zone I are:

Arnoux's Beaked Whale	*Berardius arnuxii**
Southern Bottlenose Whale	*Hyperoodon planifrons**
Cuvier's Beaked Whale	*Ziphius cavirostris*
Killer Whale	*Orcinus orca**
Pilot Whale	*Globicephala melaena**
Hourglass Dolphin	*Lagenorhynchus cruciger**
Southern Right-Whale Dolphin	*Lissodelphis peronii*
Commerson's Dolphin	*Cephalorhynchus commersonii*

All of these species also occur in Zones II, III and IV, although *L. cruciger* in found mainly south of Zone IV.

3.2 Zone II: Low Antarctic

In addition to the original five Antarctic proper species, Fraser (1964) and Brownell (1974) recognized *L. peronii, C. commersonii, Lagenorhynchus obscurus* and *Phocoena dioptric* as occurring at or near the Antarctic Convergence. Two of these species have now been found in Zone I, and *Z. cavirostris* and an unidentified *Mesoplodon* have been recorded from the Convergence zone. *Lagenorhynchus obscurus* was included in this zone only because of a skull from Kerguelen Island, but Kerguelen should be considered as Sub-Antarctic (P Arnaud, pers. comm.), and the species probably should not be considered as Antarctic. Thus, Zone II includes the eight species from Zone I in addition to the following three species:

Dusky Dolphin (possibly)	*Lagenorhynchus obscurus*
Spectacled Porpoise	*Phocoena dioptrica*
Beaked Whale	*Mesoplodon* sp.

3.3 Zone III: Sub-Antarctic

Six species were recognized from the Sub-Antarctic islands south of New Zealand (Baker 1977), 11 from the Falklands (Malvinas) Islands (Hamilton 1952; Strange 1972) and nine from Tierra del Fuego (Brownell 1974; Aguayo 1975) before we began regular beach surveys (Goodall et al. to be publ. a). Two further species have been identified from museum specimens from the New Zealand Sub-Antarctic (GJB Ross, pers. comm.), bringing the Sub-Antarctic total to 22 (excluding the Minke Whale). These species, in addition to those found in Zones I and II, are:

Shepherd's Beaked Whale	*Tasmacetus shepherdi*
Blainville's Beaked Whale	*Mesoplodon densirostris*
Straptoothed Whale	*Mesoplodon layardii*
Hector's Beaked Whale	*Mesoplodon hectori*
Gray's Beaked Whale	*Mesoplodon grayi*
Andrews Beaked Whale	*Mesoplodon bowdoini*
Dusky Dolphin	*Lagenorhynchus obscurus*
Peale's Dolphin	*Lagenorhynchus australis*
Bottlenose Dolphin	*Tursiops truncatus*
Risso's Dolphin	*Grampus griseus*
Chilean Dolphin	*Cephalorhynchus eutropia*
Burmeister's Porpoise	*Phocoena spinipinnis*
Pygmy Right Whale	*Caperea marginata*

4 Food

Information is reported first according to cetacean families, and then by zones. Records from Zone IV (South Temperate) and the northern hemisphere are mentioned only to indicate foods taken in certain localities north of our study area. An animal feeding off South Africa will most probably not find the same prey as one feeding near the Antarctic Peninsula.

5 Balaenidae

The only baleen whale, apart from the Minke Whale, which is recognized as a small cetacean is the Pygmy Right Whale, *Caperea marginata*. This species is circumpolar in the southern hemisphere in Zones III and IV (Ross et al. 1975), occurring as far south as Tierra del Fuego (Goodall et al., to be publ.). It feeds on copepods, and its distribution resembles that of other cetaceans which feed on copepods (Ross et al. 1975).

6 Ziphiidae

The beaked whales are probably the least known cetacean family. In general, they are thought to be oceanic, deep-diving, and to feed on squid. They are seldom sighted, are known mainly from strandings, and the exterior form of at least one species is still unknown.

6.1. Tasmacetus shepherdi

Shepherd's Beaked Whale is known from only 12 specimens, from New Zealand, South America and, recently, Australia

(Mead and Payne 1975; Goodall 1978; Ling and Aitken 1981). It has not yet been found off South Africa or in the Antarctic; the southernmost record is at 55°S in the Beagle Channel. Stomach contents are known for one specimen in Zone IV, and included otoliths of *Merluccius hubbsi*, a serranid and a brotulid (forming the bulk), as well as a crab, *Peltarion spinulosum,* and one squid beak which may have been a remnant of the food of the fish (Mead and Payne 1975).

6.2. Berardius arnuxii

Arnoux's Beaked Whale occurs as far south as the ice edge. Although more than 30 specimens exist, and the species was sometimes taken by whalers, we could find only one original reference to stomach contents. According to Von Haast (1870), "about half a bushel of horny beaks . . . of the common sea-spider or octopus" were found in the stomach of a specimen from New Zealand (Zone IV). Stomachs of the northern hemisphere congeneric *B. bairdii* have been found to contain mainly deep-sea fish and squid, although other types of fish (mackerel, sardines, etc.) were represented as well (Nishiwaki and Oguro 1971).

6.3. Mesoplodon densirostris

This species is included in the Sub-Antarctic list on the basis of a specimen (National Museum of Victoria, Melbourne, no. C23596) from Macquarie Island (GJB Ross, pers. comm.). Normally an inhabitant of warm seas, about 40 specimens have been obtained in the southern hemisphere. Ross (1979, pers. comm.) found four stomachs to be empty, one contained 21 fish otoliths of the genera *Lampanyctus, Scopelogadus* and *Cepola?,* and one held single beaks of the squid *Todarodes sagittatus* and *Octopoteuthis* sp. *Mesoplodon densirostris* is known to feed on squid in the northern hemisphere (Leatherwood et al. 1976).

6.4. Mesoplodon layardii

Some 65 specimens or stranding observations are known for this species which is found off all the southern continents and major islands; the southernmost being Isla Navarino, Tierra del Fuego (Sielfeld 1980). It has not been found yet in the northern hemisphere. In spite of the number of specimens known, we could find no original information on the species' food (Ross 1979). Sclater (1900) stated that its food seemed to consist of cuttlefish and octopus, but did not report his source of information.

6.5. Mesoplodon hectori

Hector's Beaked Whale is known from 14 specimens, 10 of them in the southern hemisphere; the southernmost being from the south coast of Tierra del Fuego (Sielfeld 1980). No food records are known from the southern hemisphere. One California specimen had nematodes only in the stomach, whereas another had two lower beaks of a squid, *Octopoteuthis deletron*, and a piece of an unidentified invertebrate (Mead 1981).

6.6 Mesoplodon grayi

Some 53 stranding observations or specimens have been recorded from the tips of the southern continents, the southernmost being Tierra del Fuego. There is one record for the Netherlands in Europe. In spite of the large number of specimens known, we could find no published records of prey or feeding observations. Stomach contents were taken from an animal captured in the Indian Ocean, but were not published (Gambell et al. 1975). One pregnant female examined in Tierra del Fuego contained six large and four small otoliths, as yet unidentified.

6.7. Mesoplodon bowdoini

Some 11 specimens of this southern hemisphere beaked whale are known, mainly from New Zealand and Australia (GJB Ross pers. comm.) recently identified a museum specimen (Nat. Mus. of Victoria, Melbourne, C23595) of this species from Macquarie Island, which puts it in the Sub-Antarctic. No stomach contents have been recorded as yet.

6.8. Ziphius Cavirostris

More than 50 specimens of Cuvier's Beaked Whale have been found south of 30°S, the southernmost record being Tierra del Fuego (Goodall 1978). Goodall et al. (to be publ.) report sightings south of 60°S. We could find only two southern hemisphere food records. Ross (1979), in two animals from the southeastern coast of South Africa (Zone IV), found squid beaks belonging to at least eight oceanic families comprising about 15 species, otoliths of a species of *Antimora,* fragments of crustaceans, plastic bottle caps, pumice stone and a large seed. Nishiwaki and Oguro (1972) recorded deep-sea fish and squid in the diet of northern Pacific animals.

6.9. Hyperrodon planifrons

The Southern Bottlenose Whale ranges south to the edge of the ice pack. Hard to distinguish at sea from *Berardius*, most "beaked whale" sightings are thought to be of this species. At least three specimens and several catch records are known from Zones I and II and some 24 specimens from Zone III and IV (Goodall et al., to be publ.). Fraser (1945) reported eye lenses of cephalopods in stomachs of specimens captured off South Georgia, and Ash (1962) reproduced a photograph of "small squid" from the stomach of an animal killed in Zone II. This stomach also contained some Krill, *Euphausia superba.* In Zone III, two specimens in Tierra del Fuego had stomachs crammed with squid beaks (as yet undetermined). In Zone IV, a large number of squid beaks, possibly of *Polypus variolatus*, was found in the stomach of a stranded animal in South Australia (Hale 1931). Ross (1979) examined two specimens; one contained four cephalopod beaks. In the northern hemisphere, *H. ampullatus* is thought to feed at great depths, mainly on squid but also on a variety of fish, sea cucumbers, starfish and prawns; foreign objects, such as stones, fishing nets and plastic bags, have been found in stomachs (Benjaminsen and Christensen 1979).

7 Delphinidae

Dolphins are considered to feed mainly on fish, although certain taxa, such as *Grampus* and *Globicephala*, may feed primarily on squid. Gaskin (1982) reviewed the feeding habits of mostly northern hemisphere species, and emphasized that these cetaceans are "obviously opportunistic feeders and, as a result, studies in different geographical areas under different conditions can yield very divergent conclusions."

7.1 Orcinus orca

The Killer Whale occurs in Arctic and Antarctic seas (Brownell 1974; Jehl et al. 1980). Although common in the Antarctic, there is only one stranding observation from the region. Some 23 strandings have been found in Tierra del Fuego and some 54 in Zone IV. The Killer Whale is the only truly sarcophagic feeder among the Cetacea, although it is assumed to eat mainly fish. Smith et al. (1981) described Killer Whales in the Antarctic as top marine predators preying on large whales and seals. They described co-operative hunting behaviour for removing a seal from an ice floe. Killer Whales have been observed attacking whales, penguins and other marine birds (Goodall, unpubl.). Ash (1962) described Killer Whales eating the tongues of dead whales tied alongside boats. Shevchenko (1976) examined 49 stomachs of specimens caught between 50°S and the pack-ice: 5 were empty, 84% of the others contained remains of Minke Whales, 45% pinnipeds, 7% fish and only 2% contained squid.

At Kerguelen, in Zone III, Paulian (1953) noted wounds on pinnipeds, caused by Killer Whales. In Zone IV, Killer Whales have been seen feeding on whales off Australia (Mitchell and Baker 1980). Off the coast of Patagonia, they prey on whales, pinnipeds and penguins, but are also often seen together with these animals in "peaceful co-existence" (JC Lopez pers. comm.). Castello (1977) found the remains of a sting-ray (*Myliobatis* sp.) in the stomach of a Killer Whale from Brazil. Shevchenko (1976) found that while 11 out of 30 stomachs from 30–50°S were empty, 47% of the others contained dolphins, 42% fish, 20% pinnipeds and 16% toothed whales; two males contained remains of other Killer Whales. Tooth marks of Killer Whales were found on 53% of the Fin Whales, 24% of the Sei Whales and 6% of the Minke Whales taken in these waters by two whale catchers during 1962–1974, but no remains of large whales were found in Killer Whale stomachs (Shevchenko 1976). Fish are probably more important as prey in temperate waters.

Our knowledge of the diet of the Killer Whale in the southern hemisphere is biased, because it is based on two sources: observations of attacks on animals (usually near the surface), and on specimens caught by whalers who concentrate on adult males for higher yield, leaning the bias toward relatively large food items. Although more stomachs of Killer Whales have been examined than of any other southern small cetacean species, more work is needed. It would be interesting to compare the capture records with those of an equal number of stranded animals. Castello (1977) and Gaskin (1982) have reviewed the prey of Killer Whales in the northern hemisphere.

7.2 Globicephala melaena

Common throughout the cooler southern waters, the Pilot Whale strands singly or en masse. Published accounts of strandings south of 30°S embrace some 2,800 individuals, the largest group containing 340 animals. In spite of this, we could find no records of stomach content examinations. An indirect observation that squid and fish were abundant about the time a group stranded in Brazil was made by Pinero and Castello (1975). In Tierra del Fuego, we examined the stomachs of five specimens which contained mostly fish bones and otoliths, with some squid beaks.

7.3 Lagenorhynchus obscurus

The Dusky Dolphin's habitat is south temperate. There are specimens from the New Zealand Sub-Antarctic, but not from Tierra del Fuego or the Falkland (Malvinas) Islands. Fraser (1964) stated that a specimen from the Falkland Islands had cuttlefish in its stomach. He was probably quoting Hamilton, who supplied specimens to the British Museum. Hamilton (1952) later wrote that he had seen this species only at South Africa, and that he found cuttlefish in the stomachs of *L. australis* at the Falkland Islands. No Sub-Antarctic specimen has been examined for food remains. In the south-temperate zone, Gaskin (1872) reported a small squid, *Nototodarus sloanei,* and fish as prey. Würsig (1976) observed Dusky Dolphins co-operating in the herding of anchovy, *Engraulis anchoita.*

7.4 Lagenorhynchus cruciger

The Hourglass Dolphin inhabits waters from about 41° 30′S (Bastida and Lichtschein 1982) to 66°S (AR Galeazzi, pers.

obs.), and it is regarded as the only Antarctic dolphin. Only five specimens are known. There are two general descriptions of the species' stomach contents. Nichols (1908) reported "more or less digested small fish, which may have been mixed with other food matter" for the one specimen, and Ash (1962) stated that the other specimen "had been feeding on small squid".

7.5 Lagenorhynchus australis

Some 20 museum specimens exist, and we have found an additional 27 in Tierra del Fuego. This is an in-shore species, common on both coasts of South America. The only published food record for Peale's Dolphin is that of Hamilton (1952), who found "octopod molluscs" in an animal which had been diving among the kelp. A preliminary analysis was made of two stomachs taken from areas with few kelp beds in Tierra del Fuego. One contained cephalopods (probably *Loligo gahi*), hydroids and algae, and the other remains of fish, possibly *Eleginops* (R Bastida, pers. comm.).

7.6 Tursiops truncatus

Worldwide in distribution, the Bottlenosed Dolphin has been recorded for all the southern continents in the temperate zone. It has only recently been found in the Sub-Antarctic, based on a single stranding (Goodall et al., to be publ.). In Zone IV, off South Africa, Ross (1979), who recognized two forms, found this species feeding on two kinds of fish and squid, farther offshore than *T. aduncus* which appears to be an inshore opportunistic feeder on a variety of fish and squid. Marelli (1953) found fish skeletons in one specimen, and Mermoz (1977) found 949 otoliths of *Corvina, Micropogonias furnieri,* together with other fish, gastropods and crabs, which may have been the prey of the fish. R Bastida (pers. comm.) also found *Corvina* as well as other fish in the diet of *Tursiops* off Mar del Plata, Argentina. Leatherwood (1975) described a number of behaviour patterns of northern hemisphere *Tursiops* during feeding, and he identified a suite of prey species, as a series of fish (anchovies, menhaden, mullet, minnows), shrimp, eels, and discards from ships.

7.7 Grampus griseus

Known from temperate seas, Risso's Dolphin has only recently been found in the Sub-Antarctic (Sielfeld 1980; Goodall et al., to be publ.). Thirty-one specimens are known from south of 30°S. The stomachs of two of our specimens were empty. In Zone IV, Baker (1974) reported fish otoliths and a squid beak. Ross (1979) found stomachs of four specimens to be empty, and three others contained squid beaks representing nine families.

7.8 Lissodelphis peronii

The Southern Right-Whale Dolphin is known from about 18°S to 55°S (Baker 1981). There are some 56 stranding or capture records, about 50% deriving from our surveys in Tierra del Fuego. The species recently has been added to the list for the Antarctic proper on the basis of a skull found in the South Shetland Islands (Goodall et al., to be publ.). There are three food records. For the Sub-Antarctic, Bennett (1840) wrote that "Our crew never lost an opportunity of harpooning them, as we esteemed their flesh a delicacy". And, that these animals had their stomachs "distended by a vast number of calamars or flying-squid (*Loligo*)". For the South Temperate Zone, Torres and Aguayo (1979), in a male captured off Chile, found otoliths of five species of lantern fish (Myctophidae), mainly *Hygophum hanseni,* and four species of squid, mainly *Gonatus antarcticus.* Baker (1981) examined stomachs of two specimens, one of which contained mostly the fish *Macruronus novaezelandiae,* and both contained the Common Arrow Squid, *Nototodaris sloani.*

7.9 Cephalorhynchus eutropia

Known only from Chile, from Valparaiso (33°S) to near Cape Horn (about 56°S), some 30 cranial specimens have been reported by Goodall et al. (to be publ.). There are no stomach content records. Lesson and Garnot (1826) called it an "active destroyer of fish". Clarke et al. (1978) observed what we think was this species feeding on anchovies and sardines. Goodall et al. (to be publ.) reported the species feeding in in-shore tidal currents and felt that it may be a non-selective feeder.

7.10 Cephalorhynchus commersonii

Although this small dolphin has been sighted in the High Antarctic and the Convergence zones, no specimens are as yet known from these areas. Some 26 specimens were known in Zones III and IV before our study which now has recorded more than 200 specimens (Goodall et al., to be publ.). The only published stomach content record was that of Hamilton (1952), who found Krill and cuttlefish pens; this was repeated by all subsequent authors. In Tierra del Fuego, we have examined the stomachs of 31 incidentally captured specimens. Although not yet fully analyzed, 13 stomachs contained small squid beaks or pens, seven had fish vertebrate or otoliths and 13 had mysid shrimps. Also present were isopods, nematodes, tunicates and other invertebrates (R Bastida, pers. comm.) These findings are for animals feeding close inshore over a strongly tidal coast, and may not represent the food of animals feeding offshore. At Comodoro Rivadavia, just north of our arbitrary Sub-Antarctic Zone, Mermoz (1980) reported that the species fed on *Sprattus fueguensis,* silversides and anchovies.

8 Phocoenidae

Although the presence of this family in the southern hemisphere has been established for over a century, very little is known of the two species recorded.

8.1 Phocoena spinipinnis

Although known from both sides of South America in temperate waters, Burmeister's Porpoise was not found in the southern part of its range until recently (Goodall 1978). Some 50 beached or captured specimens are known, but we are aware of only one stomach record. A specimen from Uruguayan waters (Zone IV) contained remains of one Patagonian Hake, *Merluccius hubbsi,* one Porgy, *Pagrus sedecim* and one unidentified squid (Brownell and Praderi 1982). Sparse contents from three of our specimens (Zone III) contained remains of fish (probably clupeid), although one also contained a few mysid shrimps and euphausiids (R Bastida, pers. comm.).

8.2 Phocoena dioptrica

Eleven specimens of the Spectacled Porpoise were known before our survey recorded more than 120 beach strandings (Goodall and Cameron 1979; Goodall et al., to be publ.). No stomach contents are known for this species. The stomachs of two Tierra del Fuego specimens were empty except for one nematode.

9 Discussion

This review has outlined the current state of knowledge of the food habits of each of the southern cetacean species. There has been no attempt to list all the known prey species, or to describe the cetaceans' foraging or feeding behaviour. Food records exist for only three species in the area south of the Antarctic Convergence (bearing in mind that certain Russian reports which might contain data on cetacean food were unavailable to us). North of the Antarctic Convergence, within our Sub-Antarctic Zone, stomachs have been examined for nine species. In all four of our zones (including the South Temperate Zone IV) there is still no information at all for five species (*Mesoplodon layardii, M. hectori, M. bowdoini, Cephalorhynchus eutropia* and *Phocoena dioptrica*). Knowledge of the food habits of only two species (*O. orca* and *C. commersonii*) are based on sufficient data; those of eight species are based on fewer than three samples. Moreover, the information available may not reflect normal feeding. Most of our information comes from strandings; these animals may have been ill or feeding poorly (Ross 1979). However, the only source of data for most species is from strandings. Most of the records are from Zone IV which has more land and more scientists than the other zones, but which does not contain the entire ranges of many species. A species' food in a cold area may differ greatly from that in a temperate area.

In spite of all these limitations, our survey does indicate, in general terms, the food that certain small cetaceans tend to eat. It is perhaps surprising that several of the beaked whales, thought to feed on squid, eat fish as well. The diets of certain Delphinidae may be much more diverse than has been supposed to date.

It is clear that there is a dearth of information for 22 species of animals, living in the Southern Ocean, whose body-lengths range from 1.45 m in Commerson's Dolphin to almost 13 m in Arnoux's Beaked Whale. The importance of these 22 species in the Southern Ocean ecosystem probably has been under-rated. As predators, *Hyperoodon* and *Orcinus* are numerous enough to have a significant effect at their respective trophic levels, and this applies to the smaller cetaceans, as a group, as well. Dead cetaceans provide nourishment for a great variety of creatures. It is obvious that more studies are needed.

It has been pointed out that currently most of our information necessarily comes from strandings. Many cetaceans strand on beaches where there is no scientist to examine them. It is desirable that stranding "net-works" should be set up in coastal areas, and that scientists covering such net-works should be encouraged to obtain as much information as possible from each specimen.

Acknowledgements. Our studies in Tierra del Fuego have been partially supported by grants from the Committee for Research and Exploration of the National Geographic Society and from CADIC/CONICET (Centro Austral de Investigaciones Científicas/Consejo Nacional de Investigaciones Científicas y Tecnicas.) Dora Sandin assisted with research and typing. The manuscript was reviewed and many helpful suggestions offered by G.J.B. Ross, P.B. Best, and R. Bastida. We are grateful to G.J.B. Ross for permission to use his unpublished information on the occurrence of *Mesoplodon densirostris* and *M. bowdoini* in the Sub-Antarctic.

References

Aguayo A (1975) Progress report on small cetacean research in Chile. J Fish Res Board Can 32:1123–1143

Aguayo A; Torres D (1967) Observaciones sobre mamíferos marinos durante la Vigésima Comisión Antárctica Chilena. Rev Biol Mar Valparaiso 13(1):1–57

Ash C (1962) Whaler's Eye. The Macmillan Co, New York, 245 pp

Baker A (1974) Risso's dolphin in New Zealand waters, and the identity of "Pelorus Jack". Rec Dom Mus 8(16):267–276

Baker A (1977) Spectacled Porpoise, *Phocoena dioptrica,* new to the Subantarctic Pacific Ocean (Note). NZ J Mar Freshwater Res 11(2):401–406

Baker A (1981) The southern right whale dolphin *Lissodelphis peronii* (Lacépède) in Australasian waters. Nat Mus NZ Rec 2(4):17–24

Bastida R; Lichtschein V (1982) Avistajes de cetáceos realizados por buques balleneros on aguas Argentinas. Contribución INIDEP no. 421 (manuscript):16

Bengtson J (1978) Review of information regarding the conservation of living resources of the Antarctic marine ecosystem. US Marine Mammal Commission, Washington DC, 148 pp

Benjaminsen T; Christensen I (1979) The natural history of the bottlenose whale. *Hyperoodon ampullatus* (Forster). In: Winn H, Olla B (eds) Behaviour of marine animals, vol III:143–164

Bennett F (1940) Narrative of a whaling voyage round the globe, from the year 1833 to 1836. R Bentley, London. 395 pp

Brownell R (1974) Small odontocetes of the Antarctic. In: Bushnell V (ed) Antarctic map folio series, Folio 18, Am Geog Soc, New York, pp 13–19

Brownell R (1977) (ed) Report of the Sub-Committee on Small Cetaceans. London 7–9 June 1976. Rep Int Whal Commn 27: Annex L:474–484

Brownell R; Praderi R (1982) Status of Burmeister's Porpoise, *Phocoena spinipinnis*, in Southern South American waters. FAO Fish Series 5(4):91–96

Castello H (1977) Food of a Killer whale; eagle sting-ray, *Myliobatis*, found in the stomach of a stranded *Orcinus orca*. Sci Rep Whales Res Inst 29:107–111

Clarke R; Aguayo A; Basulto del Campo S (1978) Whale observation and whale marking off the coast of Chile in 1964. Sci Rep Whales Res Inst (30):117–177

Frazer F (1945) On a specimen of the Southern Bottlenosed Whale, *Hyperoodon planifrons*. Discovery Rep 23:19–36

Frazer F (1964) Whales and whaling. In: Priestley R, Adie R, Robin G (ed) Antarctic Research, Butterworths, London, pp 191–205

Gambell R, Best PB, Rice D (1975) Report on the International Indian Ocean whale marking cruise, 24 November 1973–3 February 1974. Rep Int Whale Commn 25:240–252

Gaskin D (1972) Whales, Dolphins and Seals (with special reference to the New Zealand Region). St Martins, New York, pp 107–138

Gaskin D (1982) The ecology of whales and dolphins. Heinemann, London:459 pp

Godoy J (1963) Caza Marítima; caza de ballenas. Fauna Silvestre. In: Serie Evaluación de los Recursos Naturales de la Argentina. CFI, vol VIII:124–132

Goodall R (1978) Report on the small cetaceans stranded on the coasts of Tierra del Fuego. Sci Rep Whales Res Inst 30:197–230

Goodall R; Cameron I (1979) *Phocoena dioptrica*, una nueva especie para aguas chilenas. Rev Mus Argent Cienc Nat Bernardino Rivadavia, Zool 12(11):143–152

Goodall R; Galeazzi A; Cameron I (to be publ. a) Distribution of the smaller cetaceans of the Antarctic and Subantarctic, Nat Geog Res

Goodall R; Norris K; Cameron I; Oporto J (to be published) On *Cephalorhynchus eutropia* (Gray 1849). IWC Special Issue on *Cephalorhynchus*

Haast J von (1870) Preliminary note of a Ziphoid whale, probably *Berardius arnuxii*, stranded on the 16th of December, 1868, on the sea-beach, near New Brighton Canterbury, New Zealand. Ann Mag Nat Hist 6, Ser 4:348–351

Hale H (1931) Beaked whales – *Hyperoodon planifrons* and *Mesopolodon layardii* – from South Australia. Res S Aust Mus 4:291–311

Hamilton J (1952) Cetacea of the Falkland Islands. Com Zool Museo Hist Nat Montevideo vol IV:1–6

Hedgpeth J (1969) Distribution of selected groups of marine invertebrates in waters south of 35°S latitude. In: Bushnell V, Hedgpeth J (eds) Antarctic Map Folio Series, Folio 11, American Geographical Society, New York p 44

Jehl J; Evans W; Awbrey F; Drieschmann W (1980) Distribution and geographic variation in the killer whale (*Orcinus orca*) population of the Antarctic and adjacent waters. Antarct J US, 1980 review vol XV:161–163

Leatherwood S (1975) Some observations of feeding behaviour of Bottle-nosed dolphins (*Tursiops truncatus*) in the northern Gulf of Mexico and (*Tursiops cf. T. gilli*) off Southern California, Baja California, and Nayarit, Mexico. Mar Fish Rev 37(9):10–16

Leatherwood S; Caldwell D; Winn H (1976) Whales, dolphins and porpoises of the western North Atlantic. A guide to their identification. NOAA Technical Report NMFS Circ-396, 176 p

Lesson R; Garnot P (1826) Observations générales sur quelques cétacés. In: Duperrey L Voyage Autour du Monde La Coquille Zoologie. Arthus Bertrand, Paris

Ling J; Aitken P (1981) Marine Mammals in South Australia. S A Year Book 1981, p 10

Marelli C (1953) Documentos Iconográficos Sobre cetáceos de las costas argentinas. Avales del Museo Nahuel Huapi 3:133–143

Mead J (1979) An analysis of cetacean strandings along the eastern coast of the United States. In: Geraci J, St. Aubin J (eds) The biology of marine mammals: insights through strandings. Mar Mamm Comm Rep 77/13:54–68

Mead J (1981) First records of *Mesoplodon hectori* (Ziphiidae) from the northern hemisphere and a description of the adult male. J Mammal 62(2):430–432

Mead J; Payne R (1975) A specimen of the Tasman beaked whale, *Tasmacetus shepherdi*, from Argentina. J Mammal 56(1):213–218

Mermoz J (1977) Sobre el veramiento de un delfín naríz de botella *Tursiops truncatus*, en la desembocadura del Rio de la Plata (Buenos Aires, Argentina). Physis Secc C 37(93):227–235

Mermoz J (1980) A brief report on the behaviour of Commerson's dolphin, *Cephalorhynchus commersonii*, in Patagonian shores. Sci Rep Whales Res Inst 32:149–153

Mitchell E (ed) (1975) Review of biology and fisheries for smaller cetaceans. Report of the Meeting on Smaller Cetaceans. J Fish Res Board Can 32:875–983

Mitchell E; Baker A (1980) Age of reputedly old killer whale, *Orcinus orca*, "Old Tom" from Eden, Twofold Bay, Australia. Rep Int Whal Commn (Special Issue):143–154

Nichols J (1908) Notes on two porpoises captured on a voyage into the Pacific Ocean. Bull Am Mus Hist 24:217–219

Nishiwaki M; Oguro N (1971) Baird's beaked whales caught on the coast of Japan in recent 10 years. Sci Rep Whales Res Inst 23: 111–122

Nishiwaki M; Oguro N (1972) Catch of Cuvier's beaked whale off Japan in recent years. Sci Rep Whales Res Int 24:35–42

Paulian P (1953) Pinnepédes, cétacés, oiseaux des Iles Kerguelen et Amsterdam, Mission Kerguelen 1951. Mem Inst Sci Madagascar, Ser A 8:111–234

Piñero M; Castello H (1975) Sobre "Ballenas piloto", *Globicephala melaena edwardi* (Cetacea, Delphinidae) varadas en la Isla Trinidad Provincia de Buenos Aires. Mus Argent Cienc Nat, Zool 12(2)

Ross G (1979) The smaller cetaceans of the southeast coast of Southern Africa. PhD thesis, University of Port Elizabeth p 415

Ross G; Best P; Donelly B (1975) New records of the pygmy right whale (*Caperea marginata*) from South Africa, with comments on distribution, migration, appearance and behaviour. J Fish Res Board Can 32:1005–1017

Sclater W (1900) Mammals of South Africa. Cetacea, vol II:177–214. Porter, London

Sheldrick M (1979) Cetacean strandings along the coasts of the British Isles 1913–1977. In: Geraci J, St. Aubin D (eds) Biology of Marine Mammals: insights through strandings. Nat Tech Inf Serv PB-293 890:35–53

Shevchenko V (1976) The nature of the interrelationships between killer whales and other cetaceans. Fisheries & Marine Ser Trans Series no 3839. Orig title: Kharakter vzaimootnoshenii kasatok i drugikh kitoobraznykh. Morsk. Mlekopitayushchie Chast' 2:174–174, 1975. 4 p typescript

Sielfeld W (1980) Mamíferos marinos en colecciones y museos de Chile. An Inst Patagonia, Punta Arenas, Chile 11:273–280

Smith T; Siniff D; Reichle R; Stone S (1981) Coordinated behaviour of killer whales, *Orcinus orca*, hunting a crabeater seal, *Lobodon carcinophagus*. Can J Zool 59(6):1185–1189

Strange I (1972) The Falkland Islands. David & Charles, Devon, p 256

Torres D, Aguayo A (1979) Habitos alimentarios de *Lissodelphis peronii* (Lacépède, 1804) en Chile central (Cetacea: Delphinidae). Rev Biol Mar Dep Oceanol Univ Chile 16(3):221–224

Würsig B (1976) Radio tracking of dusky porpoises (*Lagenorhynchus obscurus*) in the South Atlantic, a preliminary analysis. Sci Consult on Mar Mamm, Bergen, Norway, 31 Aug–9 Sept 1976, 21 pp

Energy Flow in the Southern Ocean Food Web

A. CLARKE[1]

Summary. Current knowledge of energy flow within the Southern Ocean food web is reviewed in the light of 6 recommendations for future work made by Knox (1970). The most recent estimate of total annual open ocean primary productivity is about 6,400 times 10^6 t, but the contribution from ice-associated algae is not known. Almost nothing is known about the quantitative importance of bacteria or the heterotrophic micro-zooplankton, although the few observations available suggest that the importance of this loop in the food web is likely to be as great as in other oceans. Knowledge of energy flow between the Antarctic Krill, *Euphausia superba*, and the higher predators is better, but by no means complete. Little is known about the role of dissolved or particulate organic carbon, or of fluxes to the benthic community, although many of the life-history features of the shallow water benthos indicate that this flux is important. Recent physiological and year-round autecological studies of benthic species indicate that winter metabolic rates are very low, and hence there is no need for the large lipid stores so characteristic of the plankton, and suggest that growth efficiencies are high and turnover rates low.

1 Introduction

In the conclusion to his discussion of the Antarctic marine ecosystem presented at the second SCAR Symposium on Antarctic Biology in 1968, Knox (1970) suggested six topics of research which he felt required urgent attention, and ended on a note of optimism that further symposia would be able to report significant advances. The six topics were:

1. the role of marine bacteria;
2. the role of organic detritus as a source of food;
3. the relationships between the biological communities of the sea-ice and open ocean;
4. quantitative investigations of food chains;
5. the estimation of turnover rates in Antarctic communities; and,
6. physiological studies of key marine organisms.

Significant advances have been made in certain of these fields, but in others we remain almost as lacking in information as we were in 1968. In this paper I will discuss briefly all six topics, but with particular emphasis on physiology and ecological energetics (topics 5 and 6).

1 British Antarctic Survey, NERC, High Cross, Madingley Road, Cambridge CB3 OET, United Kingdom

2 The Southern Ocean Food Web

The Southern Ocean food web has often been described as simple in comparison with other marine system, partly because of the numerical importance of a single species of primary consumer, the Antarctic Krill, *Euphausia superba*. It is also frequently suggested that the major route for energy in the Southern Ocean is the simple linear food chain diatoms-Krill-consumers, and a typical such food web has been described by Tranter (1982). In reality, the Antarctic marine ecosystem is more complex than this. For example, in some areas Krill are almost absent (see Maruyama et al. 1982), and organisms such as copepods or salps are the dominant macro-zooplankton herbivores. We are also witnessing a significant change in our view of marine food webs in general, notably in the realization of the importance of the microbial and micro-zooplankton communities. Any representative food web for the Southern Ocean must take all these features into account.

2.1 Primary Production

The organisms responsible for autotrophic (primary) production in the Southern Ocean are usually assumed to be the open-ocean phytoplankton. In so large an area it is very difficult to calculate an average, or even representative, figure for annual primary production, and various estimates have been collated by Everson (1977a). The most recent estimate, that of Holm-Hansen et al. (1977), is also the lowest. This value is 0.134 g m^2 d^{-1} (as carbon). Assuming a growing season of 150 days and a total area of 32 times 10^6 km^2, Holm-Hansen calculated the annual Southern Ocean phytoplankton production to be about 6,400 times 10^6 t (wet wt.). Although this is the best figure we have, there are many difficulties associated with using it as a basis for further calculation (e.g., of energy flow to *E. superba*). In essence, these difficulties are the unknown contributions from very small autotrophs and sea-ice phytoplankton, the significance of grazing by micro-zooplankton, and the relative importance of the various groups of macro-zooplankton herbivores in the Southern Ocean.

In Antarctic seas the organisms involved in primary production have long been known to include diatoms, dino-

Antarctic Nutrient Cycles and Food Webs
(ed. by W.R. Siegfried, P.R. Condy, and R.M. Laws)

flagellates and silicoflagellates. Until recently the larger diatoms have attracted most attention, but modern studies have shown that in tropical and temperate seas very small photosynthetic organisms are considerably more important that had previously been appreciated (Li et al. 1983). Since many of these organisms will pass through the filter used routinely for measurements of ^{14}C fixation, such estimates are likely to be low. The importance of these very small autotrophs in the Southern Ocean is not yet known, although at 24 stations in the Bellingshausen and Scotia seas, von Bröckel (1981) found that organisms passing through a 20 micron filter (nanoplankton) were responsible for about 70% of the total plant biomass and over 90% of the production (as measured by ^{14}C incorporation).

During the spring melt the total area of ice-free water south of the Antarctic Convergence increases from about 14 times 10^6 km^2 in September, to about 31 times 10^6 km^2 in February (Mackintosh 1972). In the course of this melt, plants contained within the sea-ice are released. The biomass liberated may be substantial, but virtually nothing is known of the quantitative importance of C fixation by sea-ice algae in situ, nor of the contribution by the bloom associated with the retreating ice. Both open-ocean and sea-ice primary production are very variable, and very high rates of fixation have been reported at some inshore sites, particularly during calm weather when turbulence is decreased (Whitaker 1982). Clearly we are not yet in a position to calculate an overall figure for Southern Ocean phototrophic production of any real value for further energy flow calculations.

2.2 The Importance of the Micro-Zooplankton

The immediate fate of fixed C in the Southern Ocean is still only poorly understood, but studies from other seas would suggest that a sizeable fraction enters the pool of dissolved organic C where it will be utilized by heterotrophic bacteria. The proportion of fixed C released by phytoplankton varies widely, and Joint and Morris (1982) have collated values from 1 to 69.7% from a variety of waters (none of them polar). Nutrient status is an important factor in governing this release, and modern techniques of estimating microbial biomass and production have shown that the marine microbial fauna is intimately linked to the phytoplankton biomass. Dissolved organic C is also utilized by heterotrophic micro-flagellates (see reviews by Azam et al. 1983; Linley et al. 1983). Very little is known about Southern Ocean dissolved organic C or bacteria, although open-ocean heterotrophic bacterial activity appears to be comparable with other areas of the world (Morita et al. 1977; Bölter and Dawson 1982), and Fuhrman and Azam (1980) reported an average bacterial cell number of 5.4 times 10^8 cells l^{-1} from the Antarctic, which is comparable with values from other seas (Williams 1981).

It is now becoming clear that "grazing" by micro-zooplankton is a process of major significance in many oceans. Laboratory studies have shown that organisms such as ciliate protozoans, micro-flagellates and pelagic rotifiers have the potential to remove a large fraction of daily plant production. Such grazing pressures have been demonstrated in the field (Caprulio and Carpenter 1980, 1983), and Linley et al. (1983) have speculated that between 20 and 60% of primary production passes to the microbial–micro-zooplankton community. Williams (1981) has suggested, from model calculations, that twice as much C reaches the next trophic level through micro-heterotrophs as goes through the larger herbivores.

Very little is known of the quantitative importance of the microbial–micro-zooplankton community in the Southern Ocean, but these organisms will likely prove to be significant grazers and important in nutrient vegetation. Tintinnids were reported by the Discovery investigations, but the quantitative importance of Protozoa in the Southern Ocean is largely unknown, although von Bröckel (1981) reported a mean protozoan biomass of 109 mg m^3 (as carbon). If grazing by micro-zooplankton is significant in Antarctic waters, then it is likely that current ^{14}C estimates of primary production will be in error on the low side. Current knowledge of this community in Antarctic waters has been summarized by Hewes et al. (1985).

2.3 Macro-Zooplankton

Our ability to make quantitative statements about the importance of the larger zooplankton in the Southern Ocean food web has shown little improvement since the summaries by Knox (1970) and Everson (1977a). Acoustic biomass estimates are becoming available for some areas (e.g., Hampton 1983), but energy requirements have been estimated for only one species, *E. superba.* Three such estimates (Chekunova and Rynkova 1974; Clarke and Morris 1983; Ikeda, to be publ.) have all suggested a value of about 5% body weight per day. It is not known to what extent this requirement is met by phytoplankton alone, or whether Krill also take the micro-zooplankton (ciliates, choanoflagellates, etc.) as food.

With no firm figure available for Krill biomass, and virtually no knowledge of all the biomass, production or food consumption of other planktonic herbivores, zooplankton carnivores, or the micro-zooplankton community, any attempt at an overall calculation of energy flow from producers to consumers is of little value. What data we do have suggest that euphausiids (other than *E. superba*), copepods, and salps will all prove to be quantitatively important in the Southern Ocean food web (see Table 5 in Knox 1970).

2.4 Higher Predators

Consumption of Krill and other zooplankton by vertebrate predators is the part of the Antarctic marine food web for which we have the best data, albeit only from selected areas. This topic has been discussed in detail by Croxall et al. (1984), and so will not be covered further here.

2.5 The Benthic Environment

The remaining major pathway for utilization of primary production is to the benthos. Although there are no published quantitative data on the magnitude of the energy flow to the benthos, several observations indicate that this is significant. Much of the shallow water benthos consists of sessile particulate or detritus feeders which must rely on settlement of food from above (Knox 1970; Dell 1972; Arnaud 1974, 1977). The importance of water column production to the underlying benthos is clearly demonstrated in McMurdo Sound. The east side of the Sound is fed by southerly currents (i.e. into the Sound from the Ross Sea); this is also an area of high primary productivity. In-faunal biomass is higher than almost anywhere else in the world, and the benthos is rich in particle feeders. By contrast, the west side of the Sound is fed by currents from beneath the Ross Ice Shelf. Here the infaunal biomass is very much lower (similar to that reported from bathyal communities), and there is only a sparse mobile epifauna (Dayton and Oliver 1977).

Plankton-benthos coupling is a major route for energy flow in many oceans of the world. Hargrave and Peer (1973) have demonstrated a strong correlation between benthic macro-faunal biomass and mean summer water column chlorophyll-a concentration, and Hargrave (1973) estimated that between 10 and 70% of primary production may reach the benthos. The upper range of these estimates is obviously incompatible with some estimates of the fraction of primary production passing to micro-zooplankton and herbivorous zooplankton. Joint and Morris (1982) consider that flux to the benthos is usually < 10% of primary production, and mainly via faecal pellets rather than direct sinking of cells.

There are very few published Antarctic data on C flux to the sediments (Table 1). The difference between the various estimates probably reflects the variations in the primary productivity of overlying waters. In all cases the measured fluxes decreased with depth, though not always smoothly. The sedimenting material recovered from traps in the Drake Passage included whole diatoms, cast crustacean moults and numerous faecal pellets. Faecal pellets, especially those from salps, may be a very important mechanism for transferring phytoplankton to the benthos, and Dearborn (pers. comm.) has suggested that the rich and diverse benthic community to the north of Elephant Island may be related to the persistent presence of a large concentration of Krill in the same area. The data in Table 2 are expressed on a per year basis, but much of the flux will be concentrated in summer only.

2.6 The Southern Ocean Food Web

It is clear from the above discussion that a simple food web for the Southern Ocean based on the linear chain diatoms – Krill – consumers is no longer tenable. In order to be realistic, any Antarctic marine food web must include herbivores other than Krill, a link to the benthos, and the dissolved organic carbon – microbial – micro-zooplankton communities. Such a food web is proposed in Fig. 1.

Table 1. Estimates of vertical carbon flux in the Southern Ocean. All data expressed as g mg^2 year^{-1}

Location and depth (m) of sediment trap	Total mass flux (carbon)	Organic carbon flux	Remarks and authority
Drake Passage (60° 54.6′S, 57° 06.0′W)			
Surface	471	36.5	Deployed in water of depth 3,625, 2 Dec. 1980–25 Jan. 1981
965	163	5.41	(523 d). Mean primary production 0.1 g m^2 d^{-1} (as carbon).
2,540	143	4.78	Wefer et al. (1982)
Southern Ocean, Indian Ocean sector (64° 50.5′S, 124° 36.7′E)			
50	–	67.16	Deployed 2–3 Jan. 1981. Fujita and Nishizawa (1982)
500	–	8.40	
1,000	–	5.84	
Southern Ocean, Australian sector (65° 01.1′S, 160° 03.4′E)			
50	–	31.03	Deployed 2–3 Feb. 1981. Fujita and Nishizawa (1982)
500	–	5.11	
1,000	–	1.46	

Table 2. A summary of patterns of lipid storage in polar marine invertebrates. (Data from Clarke 1983)

	Typical range of lipid contents (% fr. wt.)	Typical range of wax ester contents (% total lipid)	Taxa
Pelagic			
Herbivores	2 –20	15–55	*Calanoides acutus*
			Euphausia crystallorophias
Omnivores	1 – 7	< 1	*E. superba*
Carnivores	0.05– 3.5	generally <6 (but	Medusae
		amphipods and	*Tomopteris* sp.
		euchaetid copepods	*Vanadis antarctica*
		are exceptions)	Chaetognaths
			Themisto gaudichaudii
Benthic			
Herbivores	0.5 – 2	< 2	*Philine gibba*
			Gaimardia trapesina
			Lyothyrella uva
Deposit feeders	2	nd	*Limatula hodgsoni*
Carnivores	0.5 – 7	< 2	*Serolis cornuta*
			Chorismus antarcticus
			Chlanidota densesculpta
			Odontaster validus

Abbreviations: nd = no data

3 Physiological Studies

The physiology and biochemistry of Antarctic marine organisms have received considerable attention since the second SCAR Symposium on Antarctic Biology in 1968.

The coupling between the plankton and the benthos, discussed above, suggests that many of the biological features of the Antarctic benthos will be related to the intensely seasonal pattern of primary production in the overlying water column. In many cases this has been found to be so. Growth and vitellogenesis are slow and usually confined to summer, annual reproductive effort is reduced, and summer metabolic rates are low (Clarke 1983). Since the latter will have included contributions from the respiratory cost of growth and vitellogenesis, winter metabolic rates (and hence food requirements) will be very low indeed. Unfortunately there are very few measurements of winter metabolic rates, although Wohlschlag (1964) demonstrated a decreased oxygen consumption in the shallow water population of the demersal fish *Trematomus bernacchii* in winter. In a detailed study of the energetics of the carnivorous isopod *Serolis polita* – Luxmoore (1984) found a four fold decrease of population food intake in winter compared with the previous summer.

Very little is known about the availability of food for particulate feeders in winter, but this is usually assumed to be low. Certainly the vertical flux of particulate matter will decrease once primary production drops to winter levels, and as winter proceeds water column stability under sea-ice will accentuate this effect. Certain organisms (particularly amphipods, and likely also *E. superba*) may be able to "browse" the community which develops on the underside of sea-ice, and Whitaker (1981) has suggested that formation of particulate aggregates under sea-ice may delay turnover of summer production and provide food for some browsers and particulate feeders in winter.

One clear distinction which arises is the tendency for most planktonic herbivores and some carnivores to store large quantities of lipid, often as wax ester, whereas such stores are rarely, if ever, detected in benthic organisms (apart from those associated with vitellogenesis). These data are summarized in Table 2. The inference is that benthic organisms can survive winter without the need for a substantial energy reserve; either there is some food available, or winter energy demands are much lower (Clarke 1983). Since a low food availability will presumably affect planktonic herbivores and benthic particulate feeders alike, this difference in the pattern of lipid storage may reflect the costs of swimming activity. The extra metabolic cost of staying in the water column may be sufficient to require an over-wintering lipid store when feeding is impossible. Storage of energy as lipid has the additional

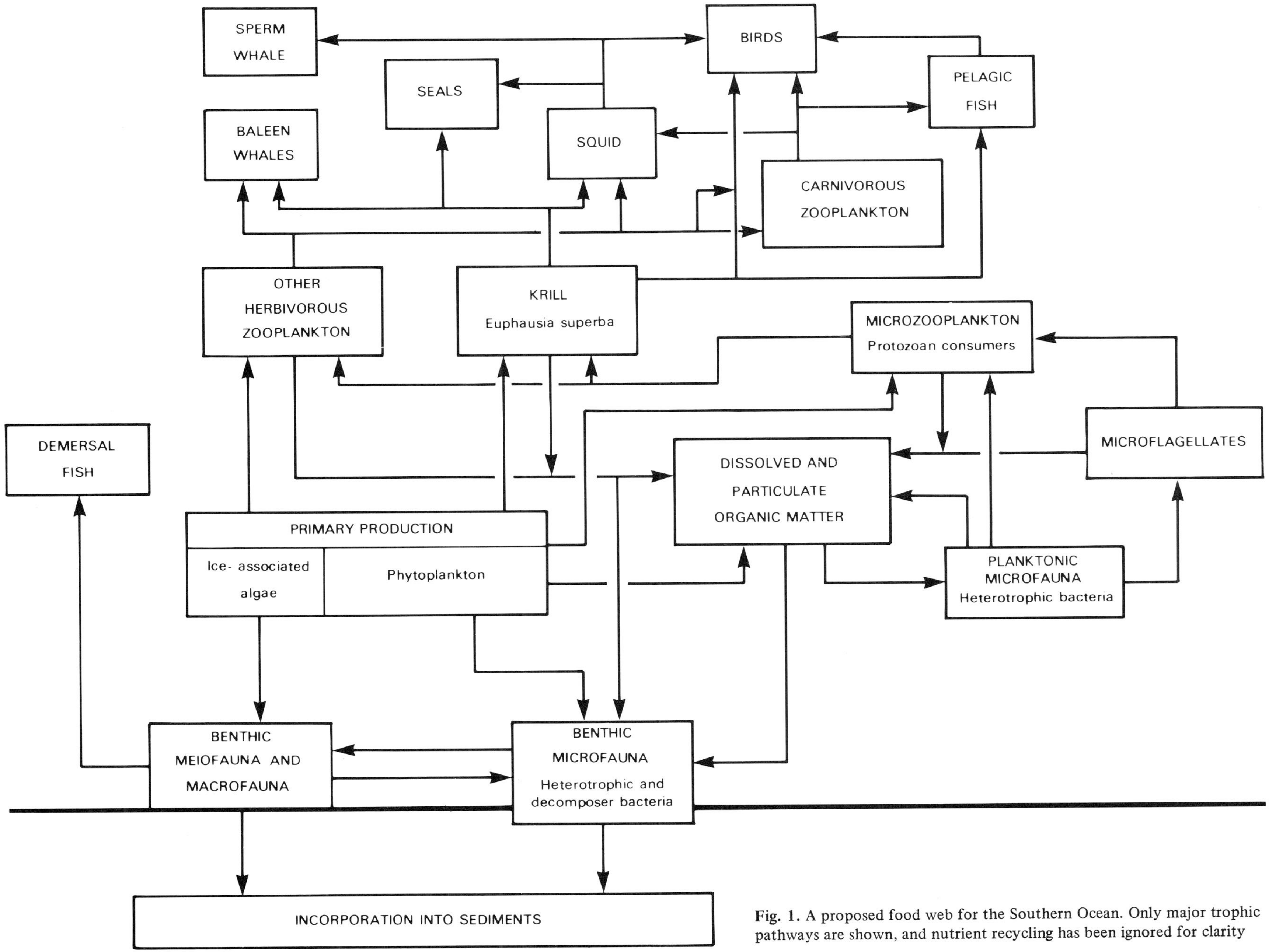

Fig. 1. A proposed food web for the Southern Ocean. Only major trophic pathways are shown, and nutrient recycling has been ignored for clarity

bonus of supplying lift, but has the disadvantage that not all of the summer food intake can be used immediately for growth or reproduction; some must be put aside for future use. It has been suggested that the use of wax ester, rather than triacylglycerol, as a lipid reserve reflects the need to make maximal use of a short-term food resource (Sargent 1976).

A lipid store also allows vitellogenesis to proceed during winter in the absence of active feeding, and this certainly occurs in *Euphausia crystallorophias* and possibly also the predatory copepod *Euchaeta antarctica* (Littlepage 1964). In the benthos, however, where egg production may take as long as 18 months, vitellogenesis must slow over winter. Differences in the timing of reproduction in relation to food supply may also explain the difference in the lipid biochemistry of marine and freshwater copepods (Clarke 1984).

Sessile particulate feeders are widespread and common components of the shallow water benthos in all oceans of the world. In most of these waters the availability of particulate food is sporadic and unpredictable, for plankton is generally variable in both time and space. Fixed suspension feeders must therefore typically be opportunistic (Levinton 1972), and the life-history patterns of these organisms frequently show features characteristic of such opportunists, such as massive production of planktonic larva. The distinguishing feature of primary production in polar waters, especially the Southern Ocean, is, however, its extreme predictability. This predictability is likely to be less marked in the open ocean where the spring bloom is intimately related to the break up of ice, but the overall pattern is still predictable in comparison with temperate waters. As a consequence, organisms can adjust their life-history strategies to exploit this predictable resource. Coupled with a very low basal metabolic rate this means that very efficient use can be made of the energy intake, and the Antarctic benthos is characterized by slow-growing organisms with a physiology finely tuned to the stable low temperature (Clarke 1980, 1983). Part of this suite of strategies involves the tendency to produce small numbers of large eggs, rather than massive release of very immature planktonic stages. These conclusions are based largely on studies of errant, scavenging benthos (whose food supply could be expected to be less seasonal than many), but the general paucity of planktonic larval forms in the Antarctic summer zooplankton indicates that these are general phenomena. Thus, this paucity of planktonic larval forms can be seen as part of a suite of adaptations, rather than as a direct response to an unfavourable open ocean environment for larval development.

3.1 Metabolic Cold Adaptation

In the last 10 years there have been several careful studies of the oxygen consumption of a variety of Antarctic marine invertebrates. These studies have cast doubt on the idea of

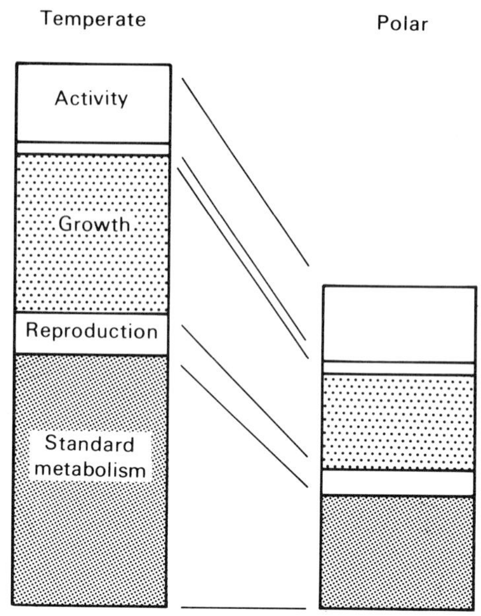

Fig. 2. A diagrammatic representation of the annual energy intake of a typical temperate-water marine invertebrate (*left-hand rectangle*), and a typical polar marine invertebrate (*right-hand rectangle*) showing the effect of adaptation to cold water as proposed by Clarke (1980). The relative sizes of the *boxes* representing metabolism, reproduction, growth and moulting (*unlabelled*) in the polar rectangle are based on data for the Antarctic isopod *Serolis polita* (Luxmoore 1985). The sizes of these boxes in the temperate-water species have been obtained from literature data for temperate-water species. The size of the box representing activity is arbitrary, and is the same in both temperate and polar species. (After Clarke 1980, 1983)

metabolic cold adaptation (i.e., that Antarctic marine invertebrates and fish have a higher resting metabolic rate than would be expected on the basis of studies of temperate organisms extrapolated to polar temperatures), and the concept is being abandoned (see Everson 1977b). An alternative picture of cold adaptation has been proposed, namely that in polar marine invertebrates resting metabolic rates are low, and both growth and annual and individual reproductive effort are reduced in comparison with related species from warmer waters (Clarke 1979, 1980, 1983). This alternative picture is summarized in Fig. 2.

3.2 Turnover Rates

Since basal metabolic rate is reduced in polar marine invertebrates, a relatively greater proportion of the energy intake should be available for growth and reproduction. This would suggest that the growth efficiency expressed as:

(energy used for growth and reproduction)/total energy intake

will be relatively high, especially during summer. In many areas benthic biomass is equal to, or even greater than, in other areas of the world, and since both growth and vitellogenesis are reduced, then the turnover ratio:

annual production/mean biomass

will be low. This, of course, is a straightforward consequence of a high biomass of relatively slow growing, long-lived species. Turnover ratio varies with age (since rapid growth occurs in the younger stages), and in species with a short period of annual recruitment both production and biomass measures will vary greatly from year to year depending on conditions. Gray (1981) has collated a wide range of production/biomass (P/B) ratios for temperate benthic macrofauna, from 0.1 to 5.5. Picken (1979) found that the P/B ratio of the gastropod *Laevilacunaria antarctica* was very variable, although overall the value was lower than those reported for other molluscs of similar life-span. Luxmooere (1985) has shown that when calculated over a complete life-span, the polar isopod *Serolis polita* had a relatively high growth efficiency and low turnover ratio.

4 Conclusions

This brief survey of the progress made in understanding the Antarctic marine ecosystem, since the review of Knox (1970) at the second SCAR Symposium on Antarctic biology, suggests that advances in at least three topics (the role of marine bacteria, the role of detritus, and the importance of the sea-ice communities) have been slow. Our knowledge of the energy flow within the higher levels of the food web has increased substantially, but we are hindered by a lack of reliable, quantitative data on the importance of the micro-zooplankton, heterotrophic bacteria, or C flux to the benthos. Several long-term autecological studies have now produced valuable physiological data which allow us to test predictions concerning growth efficiencies and turnover rates, but these are still few in number. The importance of these studies, and the paucity of data on the biology and physiology of particulate feeders in winter, emphasize the great value of year-round studies in Antarctic biology.

References

Arnaud PM (1974) Contribution a la bionomie marine benthique des regions antarctiques et subantarctiques. Tethys 6:465–656

Arnaud PM (1977) Adaptations within the Antarctic marine benthic ecosystem. In: Llano GA (ed) Adaptations within Antarctic ecosystems. The Smithsonian Institution, Washington DC, pp 135–157

Azam F; Fenchel T; Field JG; Gray JS; Meyer-Reil LA; Thingstad F (1983) The ecological role of water-column microbes in the sea. Mar Ecol Prog Ser 10:257–263

Bölter M; Dawson R (1982) Heterotrophic utilisation of biochemical compounds in Antarctic waters. Neth J Sea Res 16:315–332

Bröckel K von (1981) The importance of nanonplankton within the pelagic Antarctic ecosystem. Kieler Meeresforsch, Sonderh 5: 61–67

Caprulio GM; Carpenter EJ (1980) Grazing by 35 to 202 μm micro-zooplankton in Long Island Sound. Mar Biol (Berl) 56:319–326

Caprulio GM; Carpenter EJ (1983) Abundance, species composition and feeding impact of tintinnid micro-zooplankton in central Long Island Sound. Mar Ecol Prog Ser 10:277–288

Chenkunova VI; Rynkova TI (1974) Energy requirements of the Antarctic crustacean *Euphausia superba* Dana. Oceanology 14: 434–440

Clarke A (1979) On living in cold water: K-strategies in Antarctic benthos. Mar Biol (Berl)55:111–119

Clarke A (1980) A reappraisal of the concept of metabolic cold adaptation in polar marine invertebrates. Biol J Linn Soc 14: 77–92

Clarke A (1983) Life in cold water: the physiological ecology of polar marine ectotherms. Oceanogr Mar Biol Annu Rev 21:341–453

Clarke A (1984) Lipid content and composition of some Antarctic macrozooplankton. Bull Br Antarct Surv 63:57–70

Clarke A; Morris DJ (1983) Towards an energy budget for krill: the physiology and biochemistry of *Euphausia superba* Dana. Polar Biol 2:69–86

Croxall JP; Prince PA; Ricketts C (1985) Relationships between prey life-cycles and the extent, nature and timing of seal and seabird predation in the Scotia Sea. In: Siegfried WR, Condy PR, Laws RM (eds) Antarctic nutrient cycles and food webs (Proceedings of the 4th SCAR symposium on Antarctic biology). Springer, Berlin Heidelberg New York

Dayton PK; Oliver JS (1977) Antarctic soft-bottom benthos in oligotrophic and eutrophic environments. Science 197:55–58

Dell RK (1972) Antarctic benthos. Adv Mar Biol 10:1–216

Everson I (1977a) The living resources of the Southern Ocean. Rome, FAO United Nations Development Programme, Southern Ocean Fisheries Survey Programme GLO/SO/77/1:156 pp

Everson I (1977b) Antarctic marine secondary production and the phenomenon of cold adaptation. Philos Trans R Soc Lond B 279:55–66

Fuhrman JA; Azam F (1980) Bacterioplankton secondary production estimates for coastal waters of British Colombia, Antarctica and California. Appl Environ Microbiol 39:1085–1095

Fujita N; Nishizawa S (1982) Vertical flux of particulate matter in the Antarctic Ocean in summer 1981. Trans Tokyo Univ Fish 5: 43–52

Gray JS (1981) The ecology of marine sediments. Cambridge University Press, Cambridge, 185 pp

Hampton I (1983) Preliminary report on the FIBEX acoustic work to estimate the abundance of *Euphausia superba*. In: Nemoto T, Matsuda T (eds) Proceedings of the BIOMASS colloquium in 1982. Mem Nat Inst Polar Res, Tokyo, Special Issue No 27:165–175

Hargrave BT (1973) Coupling carbon flow through some pelagic and benthic communities. J Fish Res Board Can 30:1317–1326

Hargrave BT; Peer DL (1973) Comparison of benthic biomass with depth and primary production in some Canadian east coast inshore waters. ICES manuscript. 1973/E

Hewes C; Holm-Hansen; Sakshaug E (1985) Alternate carbon pathways at lower trophic levels in the Antarctic food web. In: Siegfried WR, Condy PR, Laws RM (eds) Antarctic nutrient cycles and food webs (Proceedings of the 4th SCAR symposium on Antarctic biology). Springer, Berlin Heidelberg New York

Holm-Hansen O; El-Sayed SZ; Franceschini GA; Cuhel RL (1977) Primary production and the factors controlling phytoplankton growth in the Southern Ocean. In: Llano GA (ed) Adaptations within Antarctic ecosystems. The Smithsonian Institution, Washington DC, pp 11–50

Ikeda T (to be published) Sequences in metabolic rates and elemental composition (C, N, P) during the development of *Euphausia superba* Dana, and estimated food requirement during its life span. J Crust Biol 4, Sp No

Joint IR; Morris RJ (1982) The role of bacteria in the turnover of organic matter in the sea. Oceanogr Mar Biol Annu Rev 20:65–118

Knox GA (1970) Antarctic marine ecosystems. In: Holdgate MW, (ed) Antarctic ecology, vol. 1. Academic Press, London, pp 69–96
Levinton J (1972) Stability and trophic structure in deposit-feeding and suspension-feeding communities. Am Nat 106:472–486
Li WKW; Subba Rao DV; Harrison WG; Smith JC; Cullen JJ; Irwin B; Platt T (1983) Autotrophic picoplankton in the tropical ocean. Science 219:292–295
Linley EAS; Newell RC; Lucas MI (1983) Quantitative relationsships between phytoplankton, bacteria and heterotrophic microflagellates in shelf waters. Mar Ecol Prog Ser 12:77–89
Littlepage JL (1964) Seasonal variation in lipid content of two Antarctic marine Crustacea. In: Carrick R, Holdgate MW, Prevost J (eds) Biologie Antarctique, Hermann, Paris. Actual Sci Indust 1312:463–470
Luxmoore (1985) The energy budget of a population of the Antarctic isopod *Serolis polita*. In: Siegfried WR, Condy PR, Laws RM (eds) Antarctic nutrient cycles and food webs (Proceedings of the 4th SCAR symposium on Antarctic biology. Springer, Berlin Heidelberg New York
Mackintosh NA (1972) Life cycle of Antarctic krill in relation to ice and water conditions. Discovery Rep 36:1–93
Maruyama T; Totoda H; Suzuki S (1982) Preliminary report on the biomass of macroplankton and micronekton collected with a Bongo net during the *Umitaka Maru* FIBEX cruise. Trans Tokyo Univ Fish 5:145–153
Morita RY; Griffiths RP; Hayasaka SS (1977) Heterotrophic activity of micro-organisms in Antarctic waters. In: Llano GA (ed) Adaptations within Antarctic ecosystems. The Smithsonian Institution, Washington DC, pp 99–113
Picken GB (1979) Growth, production and biomass of the Antarctic gastropod *Laevilacunaria antarctica* Martens 1885. J Exp Mar Biol Ecol 40:71–79
Sargent JR (1976) The structure, metabolism and function of lipids in marine organisms. In: Malins DC, Sargent JR (eds) Biochemical and biophysical perspectives in marine biology, vol. 3. Academic Press, London, pp 149–212
Tranter DJ (1982) Interlinking of physical and biological processes in the Antarctic Ocean. Oceanogr Mar Biol Annu Rev 20:11–35
Wefer G; Sness E; Balzer W; Liebezeit G; Muller PJ; Ungerer CA; Zenk W (1982) Fluxes of biogenic components from sediment trap deployment in circumpolar waters of the Drake Passage. Nature (Lond)299:145–147
Whitaker TM (1981) Observations on particulate material in seawater, during winter at Signy Island, South Orkney Islands. Bull Br Antarct Surv 53:261–262
Whitaker TM (1982) Primary production of phytoplankton off Signy Island, South Orkneys, the Antarctic. Proc R Soc Lond B Biol Sci 214:169–189
Williams PJ le B (1981) Incorporation of microheterotrophic processes into the classical paradigm of the planktonic food web. Kieler Meeresforsch, Sonderh 5:1–28
Wohlschlag DE (1964) Respiratory metabolism and ecological characteristics of some fishes in McMurdo Sound, Antarctica. In: Wells HW (ed) Biology of Antarctic Seas. Antarctic Research Series, vol. 1. American Geophysical Union, Washington, DC, pp 33–62

Part 4
Terrestrial Food Webs

Terrestrial Food Webs in the Sub-Antarctic: Island Effects

A.E. BURGER [1]

Summary. The tundra-like ecosystems of the Sub-Antarctic islands differ from those of the Arctic, in that they occur on small, very isolated landmasses. The effects of isolation, the climate and the paucity of vertebrate consumers on the food webs of the islands are reviewed. There are no native land mammals and few land birds, unlike in Arctic regions. The success of introduced mammals provides interesting insights into the functioning of Sub-Antarctic terrestrial ecosystems. The diets of land birds are reviewed. Data from Marion Island suggest that there is a comparatively large amount of food available to birds but few species exploit it, due to isolation and climatic constraints. Although there are few invertebrate species, they are key elements in food webs but their trophic roles are poorly understood. The few data available show some evidence of broad habitat use by invertebrates, possibly a response to low species diversity. The review highlights the lack of comprehensive studies of the dynamic structure and functioning of terrestrial food webs in the Sub-Antarctic.

1 Introduction

The structure and dynamics of certain tundra ecosystems are now fairly well documented, as a result of the International Biological Programme's (I.B.P.) focus on the tundra biome. Only 3 of the 23 I.B.P. tundra study sites were in the southern hemisphere: at Signy Island, South Georgia, and Macquarie Island (Bliss et al. 1981). Consequently, southern hemisphere tundra systems are poorly understood, relative to those in Arctic or northern alpine regions.

Sub-Antarctic terrestrial ecosystems have been difficult to categorize in a global scheme. In general appearance the habitats resemble Arctic or alpine tundra, but there are important differences, notably the lack of perma-frost. I have adopted the broad-scale approach used in the I.B.P. programme (Bliss et al. 1981), and have included the Sub-Antarctic among tundra-like regions where low temperatures, high precipitation and wind preclude the development of forests (Wielgolaski 1972; Smith 1977a). Smith (to be publ.) has produced a revised classification of Sub-Antarctic vegetation types.

Tundra-like habitat covers very little of the southern hemisphere and occurs mainly on islands, in alpine regions and in ice-free areas of the Antarctic continent. Thus, unlike the Arctic tundra which covers vast continental areas, the tundra of the southern hemisphere is insular; being surrounded by large areas of ocean, ice or temperate-zone vegetation. Island effects have been suggested as influencing many aspects of Sub-Antarctic life, including the low numbers of plant and animal species (Wace 1960; Gressitt 1970; Walton, to be publ.) and the low productivity of certain invertebrate (Holdgate 1977). One aim of this review is to elucidate some of the effects of insularity on terrestrial food webs in the Sub-Antarctic.

Another major difference between Arctic tundra and Sub-Antarctic systems is in the roles played by birds and mammals. This difference is of course largely due to the geographical isolation mentioned above. The I.B.P. studies revealed that, in the Arctic, a few species of birds and mammals played important roles in the flux of nutrients and energy and in determining the community structure (Wielgolaski 1975; Bliss 1977; Brown et al. 1980; Bliss et al. 1981). By contrast, the Sub-Antarctic islands support no endemic land mammals and very few land birds. The effects this has on terrestrial food webs are investigated in this paper.

Research into the structure and dynamics of terrestrial food webs is at a very early stage in the Sub-Antarctic. A qualitative description of the food web at Marion Island has been proposed (Smith 1977a) but the intensive, collaborative research which led to the quantitative models of food webs at Signy Island in the maritime Antarctic (Davis 1981; Block 1985) has not been undertaken at any Sub-Antarctic island.

The islands to be considered here are those included in Holdgate's (1977) Sub-Antarctic sub-region: South Georgia; Marion and Prince Edward islands; Iles Crozet; Iles Kerguelen; Heard Island; and, Macquarie Island. In general they are characterized by an oceanic climate with small daily and seasonal temperature range, mean temperatures above 0 °C, frequent high winds and high precipitation (Holdgate 1977). None of the islands has trees but all support extensive tundra-like vegetation dominated by low vascular plants and bryophytes. The organisms and food webs to be considered here are those supported by terrestrial primary productivity. The islands also have large populations of breeding seabirds and seals.

1 Biology Department and Newfoundland Institute of Cold Ocean Science, Memorial University, Corner Brook, Newfoundland A2H 6P9, Canada

Antarctic Nutrient Cycles and Food Webs
(ed. by W.R. Siegfried, P.R. Condy, and R.M. Laws)

2 Introduced Mammals

The Antarctic and Sub-Antarctic have no indigenous terrestrial mammals, but several species have been successfully introduced. Their history and biology were reviewed by Holdgate (1967) and Bonner (to be publ.). Many populations have not been studied adequately, notably those on the archipelagos of Kerguelen and Crozet. A review of the trophic roles of introduced mammals reveals some characteristics of the terrestrial ecosystems.

Herbivores have been very successful. At present there are Reindeer, *Rangifer tarandus*, on South Georgia and Kerguelen, feral sheep, *Ovis aries*, on Kerguelen and Iles Crozet, Corsican moufflon, *Ovis ammon musiman*, on Kerguelen and Rabbits, *Oryctolagus cuniculus*, on Kerguelen, Ile aux Cochons (Iles Crozet) and Macquarie Island (Bonner, to be publ.). In addition, horses, mules, donkeys, cattle, goats, and sheep were once kept as domestic stock at several islands but have now been removed (Bonner, to be publ.).

Herbivores do not appear to be limited by food on Sub-Antarctic islands, except in certain instances where their populations have increased until they caused overgrazing and habitat destruction (Pascal 1982; Bonner to be publ.). All of the introduced herbivores exhibit food preferences and in large areas of South Georgia, Iles Crozet, Kerguelen, and Macquarie Island they have eradicated or severely overgrazed preferred plants. In the absence of preferred species, the mammals have been able to eat other plants.

The introductions of herbivorous mammals have demonstrated interesting interactions within the terrestrial communities, in a similar manner to the experimental perturbations used to demonstrate the structure and dynamics of intertidal communities (Paine 1980). Something of the nature of interspecific competition between plants has been demonstrated. Prior to the arrival of mammals, under conditions of minimal herbivory, vegetation in productive areas tended to be dominated by species which were competitively dominant but susceptible to heavy grazing by mammals (Costin and Moore 1960; Smith 1982). Reindeer at South Georgia and rabbits at Macquarie Island have destroyed large areas of formerly dominant species, allowing other species to establish themselves more prominently. At heavily grazed areas on South Georgia, *Poa flabellata* and *Acaena magellanica* have been replaced by *P. annua* which is resistant to grazing and trampling (Smith 1982). In large areas of Kerguelen, rabbits eradicated *Pringlea antiscorbutica* and *Azorella selago* (the latter by the rabbits digging burrows) which have been replaced by *Acaena magellanica* (Holdgate 1967; Pascal 1982). The use of exclosures at South Georgia has shown that some of these changes were reversible (Smith 1982), but on many islands the changes are thought to be irreversible (Holdgate 1967; Bonner, to be publ.).

The effects of herbivorous mammals on entire ecosystems have not been studied, but plant species affected by mammals are often those important to invertebrate herbivores and detritivores (e.g., *Poa flabellata, Acaena magellanica,* and *Pringlea antiscorbutica*; Brown 1964; Smith and Walton 1975). Holdgate (1967) has suggested that the "unique endemic invertebrate fauna" formerly associated with *Pringlea* and *Azorella* vegetation at Kerguelen has been reduced in range and abundance by the actions of rabbits. The Sub-Antarctic food webs are largely detritus-based and large herbivores may have significant effects on such food webs by converting plant matter into faeces rather than litter. Exclosures at South Georgia have been used to investigate the effects of grazing by reindeer on certain insects (Vogel et al., to be publ.). Similar experiments could show the resilience of terrestrial communities to perturbations caused by introduced mammals and also provide data on natural interspecific interactions.

Omnivorous mammals accidentally introduced to Sub-Antarctic islands include Brown Rats, *Rattus norvegicus*, Black Rats, *R. rattus*, and House Mice, *Mus musculus*. Domestic pigs introduced to Iles Crozet and Kerguelen in the 19th century were omnivorous, but they have been eliminated there (Holdgate 1967).

The Brown Rats on South Georgia and the Black Rats on Macquarie Island are primarily vegetarian, preferring tillers of tussock grass, but both species also eat invertebrates, such as insect larvae, and carrion (Smith and Walton 1975; Pye and Bonner 1980; Jenkin et al. 1982). The diet of the Black Rat at Kerguelen and Iles Crozet is unknown (Pascal 1982). Early suggestions that rats killed large numbers of smaller seabirds (e.g. Luke 1917) are not supported by recent work at Macquarie Island (Jenkin et al. 1982) and South Georgia (Pye and Bonner 1980), although the South Georgia Pipit, *Anthus antarcticus*, and the Pintail, *Anas georgicus*, have been adversely affected by rats (Bonner, to be publ.). Rats have not caused other serious changes to the islands and damage to tussock grass is usually restricted to small areas (Bonner, to be publ.).

Mice feed at several trophic levels in the Sub-Antarctic. At Marion Island they prefer macro-invertebrates, particularly lepidopterous larvae, but also eat seeds and carrion (Gleeson and Van Rensburg, to be publ.). Seeds of the Macquarie Cabbage, *Stilbocarpa polaris*, are the preferred food of mice at Macquarie Island but invertebrates, such as spiders and lepidopterous larvae, are also important (Berry and Peters 1975; Jenkin et al. 1982). Mice on Ile aux Cochons (Iles Crozet) feed almost exclusively on plants (Derenne and Mougin 1976). Mice are not reported to have major effects on Sub-Antarctic ecosystems, although their populations are sometimes large (Bonner, to be publ.).

Of the surviving introduced mammals, only the House Cat, *Felis catus*, is a true carnivore. Cats occur on Marion Island, Ile aux Cochons (Iles Crozet), Kerguelen and Macquarie Island. They have had a serious impact on the populations of small petrels (Van Aarde 1980; Pascal 1982; Jenkin et al. 1982) but their effect on terrestrial food webs is not known. Their greatest impact might be to cause changes in the amounts of energy and nutrients supplied to the terrestrial systems by the petrels (Williams 1978). Cats kill very few of the birds which feed in the terrestrial food webs, such as Lesser Sheathbills, *Chionis minor*, (Van Aarde 1980;

Derenne and Mougin 1976) and Wekas, *Gallirallus australis*, (Jones 1977), but they have been implicated in the extinction of the Banded Rail, *Rallus philipensis*, and the Parakeet, *Cyanoramphus novaezelandiae erythrotis*, at Macquarie Island (Jenkin et al. 1982). In general, cats are only important top predators within terrestrial food webs when preying on other introduced mammals, and they are supported more frequently by marine birds.

There are remarkably few records of introduced mammals failing to colonize Sub-Antarctic islands. Cats and rabbits failed at South Georgia and this is probably due to the climate, which is colder than at the other Sub-Antarctic islands where one or both species have been successful (Holdgate 1967; Bonner, to be publ.). Mink, *Mustela vison*, failed to colonize Ile du Chat, at Kerguelen, in 1956, apparently because the founder population (1 male and 2 females) was too small (Bonner, to be publ.). Some introduced populations (e.g., dogs at Macquarie Island) have been exterminated by humans.

The successes of introduced mammals are clear indications that there were suitable food resources at most Sub-Antarctic islands which were unexploited or underexploited by the indigenous birds and invertebrates. Research on introduced mammals, currently focused on their primary effects on the fauna and flora and on eradication methods, could profitably include investigations into the effects of the mammals on whole communities.

3 Roles of Birds in Terrestrial Food Webs

Birds are a component of the terrestrial food web on all Sub-Antarctic islands. Their occurrence and diets are summarized in Table 1. The diets of most species are poorly known. Marine species, such as skuas, *Catharacta* spp. and giant petrels, *Macronectes* spp., which prey on other seabirds on the islands have been omitted, since they are not strictly

Table 1. Diets of birds using land-based food resources on Sub-Antarctic Islands

Species	Island	Food from tundra habitats	Food from the shoreline	Other food
South Georgia Pipit *Anthus antarcticus*	South Georgia	Insects; Collembola; other arthropods[a,b]. Plant seeds[c]	Beached marine crustaceans[a]	–
European Starling *Sturnus vulgaris*	Macquarie	Insects; earthworms; berries; (seeds?)[a]	Beached marine organisms[a]	–
Common Redpoll *Carduelis flammea*	Macquarie	Seeds of several plants; (insects?)[a,e]	–	–
Weka *Gallirallus australis*	Macquarie	Insects; earthworms; other arthropods[a,d,e]. Some plant matter[a,e]	Kelp fly larvae; beached fish and marine invertebrates[a,j]	Rodents; rabbit young; carrion; birds' eggs and chicks[a,d,e]
Wattled Sheathbill *Chionis alba*	South Georgia	Not studied	Not studied	Carrion; birds' eggs and chicks; kleptoparasitized food[a]
Lesser Sheathbill *Chionis minor*	Marion and Prince Edward	Insects; earthworms; other arthropods; snails[f]	Kelp fly larvae, beached marine animals; intertidal invertebrates; algae[f]	Carrion; birds' eggs and chicks; kleptoparasitized food; placentae; excreta[f]
Lesser Sheathbill *Chionis minor*	Îles Crozet, Kerguelen	Similar to Marion Island; not well-studied[f]	Similar to Marion Island; not well-studied[f]	Similar to Marion Island[f]
Lesser Sheathbill *Chionis minor*	Heard Island	Not studied	Similar to Marion Island; not well-studied[f]	Similar to Marion Island[f]
Kelp Gull *Larus dominicanus*	Marion and Prince Edward and others(?)	Lepidoptera larvae; earthworms; possibly other invertebrates[g]	Intertidal and shallow subtidal organisms[a]	Carrion; possibly birds eggs and chicks; seal placentae[a]
Kerguelen Tern *Sterna virgata*	Marion, Prince Edward, Îles Crozets, Kerguelen	Diptera; spiders; other invertebrates[a]	Not recorded	Marine organisms[a]
South Georgia Pintail *Anas georgica*	South Georgia	Nematodes; other invertebrates(?)[h]	Algae; crustaceans[a,h]	Marine and freshwater invertebrates; algae[h]; Carrion[i]
Speckled Teal *Anas flavirostris*	South Georgia	Insects and small arthropods[h]	Not recorded	Mainly plant material from freshwater ponds; Invertebrates[a]
Kerguelen Pintail *Anas eatoni*	Kerguelen, Îles Crozet	Seeds; roots; other plant matter; earthworms; insects[a]	Crustaceans; small fish etc. from tidal flats[a]	–
Grey Duck *Anas superciliosa*	Macquarie	Not recorded	Kelp fly larvae (and other foods?)[a]	Aquatic plants and animals?[a]

References: [a] Watson (1975); [b] Smith and Walton (1975); [c] Holdgate (1977); [d] Jenkin et al. (1982); [e] N.P. Brothers (in litt. 1980, 1983); [f] Burger (1981); [g] Burger (1978); [h] Weller (1975); [i] Parmelee (1980); [j] Jenkin (1975)

part of the terrestrial food webs. However, they do play a role in the introduction of energy and nutrients into the islands (Gillham 1961; Smith 1978a; Siegfried 1982).

The South Georgia Pipit is the only indigenous species which is a true land bird. Despite its unique status and the fact that it is threatened by rats (Pye and Bonner 1980), there has been no intensive field study of this bird. The European Starling, *Sturnus vulgaris*, and the Common Redpoll, *Carduelis flammea*, are Holarctic species which were introduced to New Zealand, and which have spread to several small islands to the south. Both species are now common and have persisted for more than 50 years at Macquarie Island (Jenkin et al. 1982).

Terrestrial foods are vital to the survival of the Pipit, Starling, and Redpoll. The Sheathbills and the Weka are primarily scavengers and predators at seal and seabird colonies, but also rely heavily on terrestrial food at certain times. Terrestrial foods appear to be a less important food for the Kelp Gull, *Larus dominicanus*, Kerguelen Tern, *Sterna virgata*, and the four species of dabbling ducks (Table 1). All of these species rely on marine or freshwater organisms for food. The diet of the recently established Mallard, *Anas platyrhynchos*, at Macquarie Island is unknown.

The macro-invertebrates appear to be the most important terrestrial food eaten by birds in the Sub-Antarctic (Table 1). Plant seeds are apparently abundant at some Sub-Antarctic localities (Dorne 1977; Tallowin 1977) but are not important food for birds, except for Common Redpolls at Macquarie Island. Plant foliage and roots are unimportant and apparently are eaten only by the Weka and Kerguelen Pintail, *Anas eatoni*. The Upland Goose, *Chloephhaga picta*, a grazing species, has been introduced twice to South Georgia but was extirpated (Watson 1975). All of the birds which use terrestrial foods are obliged to eat intertidal and marine species in some seasons. The South Georgia Pipit, for example, is forced by heavy snow to leave the tussock grass in winter to forage along the shoreline, and the birds apparently suffer heavy mortality at this time (Parmelee 1980).

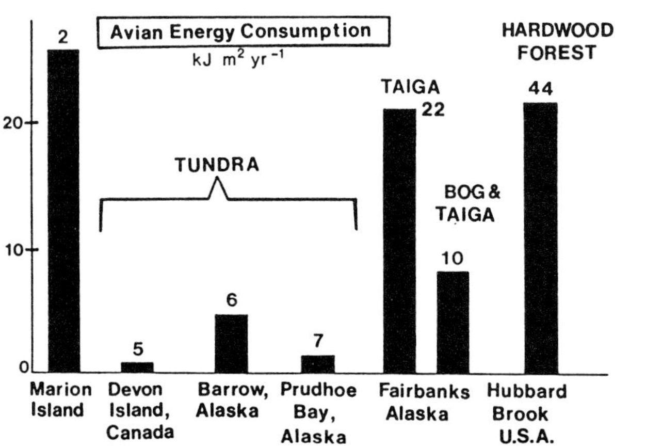

Fig. 1. Energy consumption by birds eating terrestrial invertebrates in the coastal lowlands at Marion Island (Burger 1978) compared to that of birds in northern tundra, taiga, and hardwood forests. *Numbers above columns* indicate the numbers of avian species involved. Data from Pattie (1977) and Norton et al. (1975): Truelove Lowland, Devon Island; MacLean (1980): Barrow; Norton et al. (1975): Prudhoe Bay; West and DeWolfe (1974): Fairbanks; Homes and Sturges (1973): Hubbard Brook, New Hampshire, USA

The roles of birds as consumers in Sub-Antarctic terrestrial ecosystems have received little attention, but should not be ignored. Predation by Lesser Sheathbills and Kelp Gulls on terrestrial invertebrates at Marion Island was important to both predators and prey (Burger 1978, 1981, 1982). The macro-invertebrates supported 33% of the 3,500 Sheathbills during the winter when the more preferred food in penguin and seal colonies was much less abundant. The Sheathbills exhibited acute habitat selectivity, foraging predominantly in those vegetation types in which invertebrates were most abundant and accessible (Burger 1982). The availability of invertebrates in winter probably reduced the Sheathbills' mortality rates, to the extent that the population appeared to be near or at the island's carrying capacity and limited by the availability of suitable nesting territories in penguin colonies (Burger 1979, 1981). The importance of invertebrates in the population dynamics of the Kelp Gull is unknown. Sheathbills and Kelp Gulls ate about 8% of the standing crop of invertebrates annually, within a 100 ha study site, and possibly a similar proportion over the whole of Marion Island (Burger 1978). Intensive foraging by Sheathbills in localized areas caused depletions of invertebrate populations.

In contrast to the Sub-Antarctic, the roles of birds in Arctic tundra have received considerable attention (e.g., Pattie 1977; MacLean 1980). Birds are conspicuous members of tundra food webs, although they are not always important in total energy or nutrient flux of the ecosystems. In Fig. 1 the annual energy consumption by Lesser Sheathbills and Kelp Gulls at a coastal site at Marion Island is compared with data from certain tundra, taiga, and forest areas. The site at Marion Island was representative of vegetated coastal lowland, which is more productive than the island's other habitats. Avian energy consumption within this area far exceeded that within tundra areas and was comparable to that of taiga and northern hardwood forests. This comparatively high intake at Marion Island is made possible by the high biomass of macro-invertebrates, which is several times greater than the biomass of invertebrates at tundra and bog sites used in the I.B.P. tundra programme (Fig. 2).

Although energy consumption by birds is high at Marion Island, virtually all of it goes to only two species, the Lesser Sheathbill and the Kelp Gull. A third species, the Kerguelen Tern eats negligible amounts of invertebrates at Marion Island (Burger 1978). By contrast, many more bird species are found regularly in Arctic tundra and taiga communities (Fig. 1) and often are members of relatively complex food webs (Pattie 1977). The paucity of bird species in terrestrial food webs in the Sub-Antarctic seems due to two primary factors: isolation and the climate.

Sub-Antarctic islands are among the most isolated in the world. In addition, the Antarctic continent is not a source region for terrestrial species and the environments at other

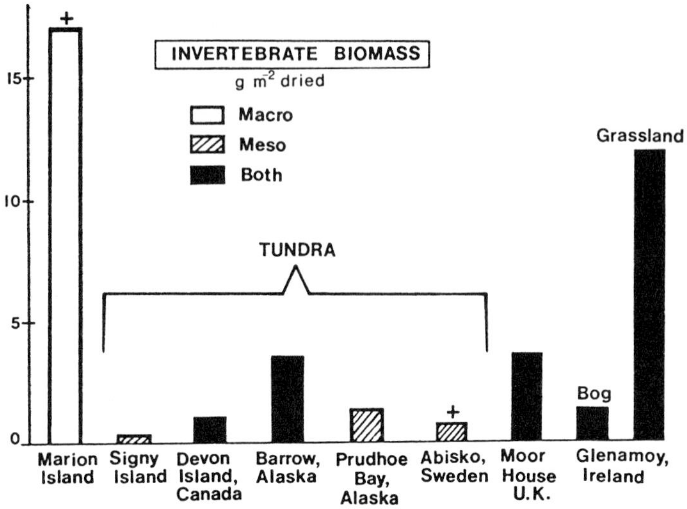

Fig. 2. The biomass of terrestrial macro-invertebrates (earthworms, molluscs, weevils, and spiders) at the coastal lowlands at Marion Island compared to that of macro- and meso-invertebrates at I.B.P. tundra biome study sites. Data from Burger (1978): Marion Island; Davis (1981): Signy Island; Ryan (1977): Truelove Lowland, Devon Island; MacLean (1980): Barrow; MacLean (1975): Prudhoe Bay; Rosswall et al. (1975): Abisko; Heal et al. (1975): Moor House; Moore et al. (1975): Glenamoy

potential source regions, the southern continents, are generally unlike those of the islands. Despite their isolation, Sub-Antarctic islands regularly receive vagrant land birds (Table 2). Most of these are not potential colonists, being migrants which breed in the northern hemisphere or birds from temperate continental regions not suited to the islands. Nevertheless, these records indicate that the isolation does not totally preclude colonization by land birds.

Abbott (1974) investigated factors affecting numbers of land bird species on Sub-Antarctic and cool-temperate islands. The number of plant species per island accounted for 71% of the variability in the number of bird species. Island area, the number of insect species and the distance from a mainland were also correlated with the number of avian species, but were less important. Abbott (1974) proposed that temperature was the ultimate factor determining numbers of plant species, and the plants in turn determined the numbers of land bird species by providing cover and food. Biotic factors on the islands were thus suggested to be the proximate factors affecting land birds.

Sub-Antarctic islands provide abundant food for land-based birds but most of it is available for limited seasonal periods (e.g., food from seal and seabird colonies) or is made unavailable by snow-cover (soil and litter invertebrates and seeds). Food from seals and seabirds is exploited successfully, in the absence of specialized terrestrial predator-scavengers, by seabirds (two species of giant petrels, two skuas and one gull). These species, unlike land birds, can switch to marine prey when seals and seabirds are unavailable.

Abbott (1974) suggested that the numbers of land birds on the southern islands might be determined by the accessibility of arthropods (and he should have included earthworms) to birds, rather than the numbers of arthropod species present. Snow-cover and frozen ground are known to affect the accessibility of available invertebrates to Pipits on South Georgia (Parmelee 1980) and Sheathbills on Marion Island (Burger 1979, 1981), thereby affecting their survival. Accessibility to seeds should be similarly affected. Thus, Abbott's (1974) proposal that the climate (temperature) was the ultimate factor affecting the number of land birds might also operate through the effects of climate on

Table 2. Numbers of land bird vagrants recorded at South Georgia and the Prince Edward Islands (Marion and Prince Edward)

	South Georgia[a]		Prince Edward Islands[b]	
	Species	Individuals	Species	Individuals
Palaearctic/Nearctic Migrants				
Waders (Charadriidae, Scolopacidae)	4	13	8	22
Swallows (Hirundinidae)	2	7	1	15
Others	1	1	5	7
Totals	7	21	14	44
Resident S. American/African Birds				
Egrets (Ardeidae)	3	56–66	2	9
Doves (Columbidae)	0	0	2	7
Others	1	1	2	2
Totals	4	57–67	6	18
Unidentified Species	0	0	5+	12
Overall Totals	11	78–88	25+	74
Years	1943, 1958, 1960, 1961, 1970–1982		1974–1983	

[a] Prince and Payne 1979; Prince and Croxall 1983
[b] Burger et al. 1980; Berruti and Schramm 1981; Newton et al. 1983

the accessibility of invertebrates, seeds and perhaps other terrestrial foods.

4 Invertebrates in Terrestrial Food Webs

Invertebrates are by far the most important heterotrophs within undisturbed terrestrial food webs in the Sub-Antarctic. Bacteria and other microbial groups, Protozoa, oligochaetes, Collembola, mites, spiders, and insects are major components of terrestrial communities, and nematodes, rotifers, molluscs, enchytraeids, and tardigrades are often present but not consistently common. The invertebrates occur in several habitats: intertidal zones; among piles of kelp and other jetsam on the shore; within colonies of seabirds and seals; and, in a fairly wide spectrum of terrestrial plant communities (Holdgate 1977; Sømme, to be publ.).

Research on these invertebrates has been centred on taxonomy and geographical distribution (e.g., Gressitt 1967, 1970; Dreux 1971; Deharveng 1981; Travé 1981), physiology (e.g., Block 1982; Vannier 1982), habitat use and life-histories (e.g., Brown 1964; Davies 1973; Travé 1977, 1982) and a few autecological studies (e.g., Paulian 1953; Smith 1977b; Davies 1982). Their diets and food webs are poorly known (Sømme, to be publ.). There are certainly insufficient data to construct quantitative food webs for any island.

Some effects of insularity are evident in the invertebrate communities, which influence their roles in terrestrial food webs. Gressitt (1970) found that the arthropod faunas of Sub-Antarctic islands were disharmonic, relative to Arctic tundra, and that several major groups were absent, possibly due to insularity. The numbers of free-living arthropods on the islands are consistently fewer than those at northern tundra sites at higher latitudes (Table 3), despite the high primary productivity and less extreme climate in the Sub-Antarctic. Both Gressitt (1970) and Abbott (1974) suggested that the physical and biotic environments of the islands

Table 3. Numbers of free-living arthropods on Sub-Antarctic islands (Gressitt 1970) and at I.P.B. Arctic, Sub-Arctic and sub-alpine tundra sites (Ryan 1981)

Sub-Antarctic islands		IBP Tundra sites	
South Georgia (55°S, 36°W)	96	Devon Island, Canada (75°N, 84°W)	167+
Marion Island (47°S, 37°E)	33	Point Barrow, Alaska (71°N, 156°W)	280+
Îles Crozet (46°S, 50°E)	81	Tareya, Taimyr, U.S.S.R. (72°N, 89°E)	395+
Îles Kerguelen (49°S, 65°E)	58	Kevo, Finland (70°N, 27°E)	452+
Heard Island (53°S, 73°E)	27	Stordalen mire (Abisko) Sweden (68°N, 19°E)	115+
Macquarie Island (54°S, 158°E)	95	Hardangervidda (Finse) Norway (60°N, 7°E)	296
		Hardangervidda (Stigstuv) Norway (60°N, 8°E)	260

Table 4. Habitat preferences of 30 species of surface-dwelling arthropods from Île de la Possession, Îles Crozet, January–April 1968 (from Davies 1973)

Habitat	Weevils	Spiders	Others	Total
Moorland only	0	1	5	6
Abundant in moorland and rare in fellfield	1	1	2	4
Total moorland species	1	2	7	10
Fellfield only	5	2	1	8
Abundant in fellfield only and rare in moorland	1	0	0	1
Total fellfield species	6	2	1	9
Moorland and fellfield	0	2	9	11

governed the invertebrate faunistic composition and not island area or distance from a source region as implied by the MacArthur and Wilson (1967) island-biogeography model. Hence, careful studies of the food webs and niches of the terrestrial invertebrates are needed to make meaningful conclusions about island biogeography in the Sub-Antarctic.

The effects of insularity on invertebrate niche dimensions in the Sub-Antarctic are poorly known, although Tréhen et al. (1985) showed that lumbricid earthworms and Diptera exhibit some habitat specialization at Ile de la Possession, Iles Crozet. Using somewhat broader habitat criteria, Davies (1973) showed that weevils and spiders on the same island were apparently habitat-specific, but that many other arthropods were not (Table 4). One-third of the species (11 of the 30) were considered to be regular inhabitants of both fellfield and moorland, habitats which differed greatly. Fellfield has well-drained, rocky soil of low organic content with a sparse cover (usually < 10%) of low cushion-plants, moss and grass tufts; moorland comprises poorly drained organic peat soils with almost complete vegetation cover including quite tall (0.5 m) plants (Davies 1973). A significant portion of the arthropod fauna thus exhibits very broad habitat tolerances.

Similar conclusions can be drawn from a study of macroinvertebrates at Marion Island (Burger 1978). Nineteen vegetation types were sampled on the coastal plain, which differed in plant species composition, physiognomy, the chemical composition of the plants, slope, aspect, soil water content, and soil organic content (Huntley 1971; Smith 1977c, 1978b,c, 1979; Burger 1978; Gremmen 1982). Accordingly, the biomass of invertebrates varied greatly (60-fold) between vegetation types, ranging from 0.7 to 46.9 g m^{-2} (dry wt.) (Burger 1978). Despite this, most invertebrate taxa were found within most vegetation types (Table 5). Furthermore, the ranking of the 19 vegetation types, with respect to biomass, was similar for all invertebrates. This was tested using a Friedman two-way analysis of variance (Siegel 1956:166). The null hypothesis, that the ranking of habitats by each organism was independent, was rejected (X_r^2 = 46.1, d.f. = 1, P < 0.001). Thus, for a wide

Table 5. Annual mean biomass (dry g m^{-2}) of terrestrial invertebrates in each of 19 vegetation types at Marion Island (from Burger 1978)[a]

Vegetation type[b]	Earthworms	Earthworm cocoons	Lepidoptera larvae	Weevil larvae and pupae	Weevil adults	Spiders	Snails	Slugs
Mires								
1	9.2	0.2	0.3	0.1	0.1	0	0	0
2	15.1	0.3	2.6	0.2	0.01	0.2	0	0.4
3	0.1	0	0.1	0.3	0.1	0.1	0	0
4	9.8	0.2	0.3	1.3	0.4	0.02	0	0
5	0.9	0.02	0.1	0.3	0.1	0.2	0	0
6	15.8	0.3	0.5	0.3	0.1	0.1	0.3	0
7	2.4	0.04	0.1	0.2	0.04	0.01	0	0
8	13.9	0.3	0.5	0.4	0.1	0.2	1.3	0
Slopes								
9	2.1	0.1	0.04	0.1	0.1	0.1	0.2	0.3
10	7.5	0.2	0.1	0.5	0.2	0.1	0.1	2.0
11	7.1	0.1	0.5	0.1	0	0.1	0.3	2.9
Saltspray Zone								
12	3.1	0.01	0.7	0.3	0.1	0	0	0.1
13	8.5	0.2	0.4	0.2	0.1	0.2	0	0
14	8.0	0.3	0.9	0.3	0.1	0.1	0	0
Manured Areas								
15	34.6	1.9	1.0	0.6	0.2	0.1	0	0
16	21.3	0.2	0.8	0.4	0.1	0.4	1.2	0.1
17	17.4	0.4	2.0	0.8	0.1	0.1	0	0
18	43.1	1.3	1.0	0.8	0.4	0.3	0	0.1
Fellfield								
19	6.1	0.1	0.2	0.5	0.1	0.01	0.2	0
Weighted mean[c]	14.6	0.4	0.6	0.4	0.1	0.1	0.3	0.2

[a] Lepidoptera adults and pupae, which occurred in only four vegetation types and which were inadequately sampled by the method used here have been omitted
[b] See Burger (1978) for descriptions of vegetation types
[c] Pooled data weighted in proportion to the % area of each vegetation type in the study area (Burger 1978)

range of vegetation types there were similar assemblages of macro-invertebrates which responded similarly to the differences between the habitats.

Gressitt (1970) also made generalized statements that animals in the Sub-Antarctic tend to occupy a wide variety of habitats. None of the studies mentioned here has had a primary objective of examining invertebrate niches and food webs in the light of modern island-biogeography theory. The Sub-Antarctic islands lend themselves well to comparative studies in this field.

Sub-Antarctic terrestrial food webs are primarily detritus-based (Holdgate 1977; Smith 1977a). Lumbricid earthworms have been found to be important elements in the food webs at Marion Island (Burger 1978), Iles Crozet (Tréhen et al. 1985) and possibly also Macquarie Island (Jenkin 1975) and Kerguelen (Bouché 1982). They are widespread but not abundant in the loamy soils and tussock peat at South Georgia (Smith and Walton 1975). At Marion Island, *Microscolex kerguelenarum* is spectacularly successful, comprising 68% of the total numbers and 87% of the total biomass of macro-invertebrates (Burger 1978), and occurring in all 19 vegetation types sampled (Table 5). This clearly is an ideal island colonist which, having overcome the considerable problems of reaching a remote island, has successfully exploited a wide range of different habitats. Earthworms are scarce or absent at most Arctic tundra sites (MacLean 1981; Ryan 1981), except for certain areas in the U.S.S.R (Rosswall and Heal 1975). Perma-frost, present in the Arctic tundra but not in the Sub-Antarctic, is perhaps a limiting factor for earthworms.

5 Arctic and Sub-Antarctic Tundra: Some Comparisons

Tundra communities in the Arctic and those on Sub-Antarctic islands have many differences. This is expected, given their disparate global positions, geological histories, surface areas, and floras and faunas. Major differences are the reduced climatic seasonality, lack of perma-frost, low summer temperatures, high rainfall and a longer growing season in the Sub-Antarctic. Both regions, however, have broadly similar vegetation structures dominated by low phanerogams, bryophytes and lichens (Rosswall and Heal 1975). Energy and nutrient flux occur mainly through detritus food webs in both regions (Holdgate 1977; MacLean 1980). Comparisons between these regions highlight some of the factors apparently important in structuring terrestrial food webs in the Sub-Antarctic.

Batzli et al. (1980, 1981) have reviewed the herbivore-based trophic systems in Arctic tundra. Mammals, and to a lesser extent birds, were the dominant herbivores. Homeotherms have an advantage, because they can remain active and productive at low temperatures and tend to have higher assimilation efficiencies than poikilotherms. Microtine rodents, ungulates and geese have substantial effects on the vegetation structure, diversity and productivity and are important in energy and nutrient flux at several Arctic localities. Invertebrate herbivores tend to be uncommon, appear to have little impact on the vegetation and are limited by the severe Arctic climate (Haukioja 1981). H. Remmert (pers. comm.) has proposed that herbivorous ectotherms are more susceptible to cold than other trophic groups.

The climate, particularly the lack of warmth and the prevailing strong winds, also limits herbivory by invertebrates in the Sub-Antarctic (Gressitt 1970). Insects are common above ground only in very sheltered places. Thus, the climate at the islands is important in shaping terrestrial food webs. Vertebrate herbivores, better able to tolerate the climate, are precluded by the geographical isolation of the islands, but have been very successful if introduced by humans.

One of the consequences of having few invertebrate and no native vertebrate herbivores is that plants at the Sub-Antarctic islands seem to be particularly susceptible to macroherbivory (Holdgate 1967). The plants seem to have been under no selection pressure to evolve effective chemical, morphological or life-history adaptations to reduce herbivory. Introduced mammals have thus had devastating effects on some islands. Plant-herbivore interactions and possible anti-herbivore adaptations have received scant attention in the Sub-Antarctic, but have been well studied in high northern latitudes and the proceedings of a special symposium on this topic have been published recently (see Oikos 40 (3), May 1983). Arctic and Sub-Arctic plants possess a wide range of adaptations to restrict herbivory (Haukioja 1981; McNaughton 1983; Tuomi et al. 1983).

Detritus-based food webs are more complex than grazing food webs in both the Arctic and the Sub-Antarctic. In low Arctic tundra, 93%–99% of the primary production becomes dead organic matter (DOM) and much accumulates as peat (MacLean 1980). The detritivores here are mainly enchytraeid worms, Collembola, mites and Diptera. They are preyed upon by few invertebrate carnivores, but support a large and diverse assemblage of migratory birds during the summer (MacLean 1980). In the Sub-Antarctic, virtually all of the plant productivity goes to DOM, some of which supports invertebrate detritivores (Holdgate 1977; Smith 1977a). The detritus food-webs appear to be truncated, relative to the Arctic, because so few bird species are involved. Thus, the isolation of the islands, by restricting colonization or migration of birds, affects the structure of these food webs.

6 Conclusions

The terrestrial ecosystems of the Sub-Antarctic have not been studied sufficiently to reveal the dynamics of their food webs. We also know little about the factors which affect the faunal communities and thus shape the food webs. The combined effects of isolation and the inclement climate appear to be important ultimate factors but the proximate factors remain unclear. Simple observations and experiments could do much to improve our understanding of these food webs. The diets, habitat use and life cycles of the dominant organisms, the invertebrates, are largely unknown. The intensive research at Signy Island in the maritime Antarctic, which led to the construction of quantitative food-web models (Davis 1981), should be emulated in the Sub-Antarctic.

Acknowledgments. I am grateful to Marthan Bester, Nigel Bonner, Nigel Brothers, David Parmelee, Ron Lewis Smith, Lauritz Sømme, and David Walton for sending me information or unpublished manuscripts. I thank Roy Siegfried, Patrick Condy, and other organizers of the 4th SCAR Symposium on Antarctic Biology for their assistance, and South African Airways for logistical support. Ms Elizabeth Behrens, librarian at Grenfell College, was very helpful in acquiring reference material. This paper was prepared while I was supported, in part, by a Canadian NSERC grant A0306.

References

Abbott I (1974) Numbers of plant, insect and land bird species on nineteen remote islands in the southern hemisphere. Biol J Linn Soc 6:143–152

Batzli GO, White RG, MacLean SF, Pitelka FA, Collier BD (1980) The herbivore-based trophic system. In: Brown J, Miller PC, Tieszen LL, Bunnell FL (eds) An Arctic ecosystem: the coastal tundra at Barrow, Alaska. IPB synthesis series 12. Dowden, Hutchinson and Ross, Stroudsburg, Pennsylvania, pp 335–410

Batzli GO, White RG, Bunnell FL (1981) Herbivory: a strategy of tundra consumers. In: Bliss LC, Heal OW, Moore JJ (eds) Tundra ecosystems: a comparative analysis. IBP series 25. Cambridge University Press, Cambridge, pp 359–375

Berruti A, Schramm M (1981) More nonmarine vagrant birds at the Prince Edward Islands. Cormorant 9:133–135

Berry RJ, Peters J (1975) Macquarie Island house mice: a genetic isolate on a sub-Antarctic island. J Zool (Lond) 176:375–389

Bliss LC (ed) (1977) Truelove Lowland, Devon Island, Canada: A high Arctic ecosystem. University of Alberta Press, Edmonton, Alberta, Canada

Bliss LC, Heal OW, Moore JJ (eds) (1981) Tundra ecosystems: A comparative analysis. IBP series 25. Cambridge University Press, Cambridge

Block W (1982) Respiration studies on some South Georgian Coleoptera. Colloque sur les Ecosystèmes Subantarctiques. 1981, Paimpont. CNFRA 51:183–192

Block W (1985) Arthropod interactions in an Antarctic Terrestrial community. In: Siegfried WR, Condy PR, Laws RM (eds) Antarctic nutrient cycles and food webs (Proceedings of the 4th SCAR symposium on Antarctic biology). Springer, Berlin Heidelberg New York

Bonner WN (to be publ.) Introduced Mammals. In: Laws RM (ed) Ecology of the Antarctic. Academic Press, London

Bouché MB (1982) Les Lombriciens (Oligochaeta) des terres Australes Françaises. Colloque sur les Ecosystèmes Subantarctiques. 1981, Paimpont. CNFRA. 51:175–179

Brown J, Miller PC, Tieszen LL, Bunnell FL (eds) (1980) An Arctic ecosystem: the coastal tundra at Barrow, Alaska. IBP synthesis series 12. Dowden, Hutchinson and Ross, Stroudsburg, Pennsylvania

Brown KG (1964) The insects of Heard Island. ANARE Sci Rep (B) 1:1–39

Burger AE (1978) Terrestrial invertebrates: a food resource for birds at Marion Island. S Afr J Antarct Res 8:87–99

Burger AE (1979) Breeding biology, moult and survival of Lesser Sheathbills *Chionis minor* at Marion Island. Ardea 67:1–14

Burger AE (1981) Food and foraging behaviour of Lesser Sheathbills at Marion Island. Ardea 69:167–180

Burger AE (1982) Foraging behaviour of Lesser Sheathbills *Chionis minor* exploiting invertebrates on a sub-Antarctic island. Oecologia (Berl) 52:236–245

Burger AE, Williams AJ, Sinclair JC (1980) Vagrants and the paucity of land bird species at the Prince Edward Islands. J Biogeog 7:305–310

Costin AB, Moore DM (1960) The effects of rabbit grazing on the grasslands of Macquarie Island. J Ecol 48:729–732

Davies L (1973) Observations on the distribution of surface-living land arthropods on the Subantarctic Ile de la Possession, Iles Crozet. J Nat Hist 7:241–253

Davies L (1982) Ecology of two *Amblystogenium* species (Carabidae) on Ile de la Possession, Iles Crozet. Colloque sur les ecosystémes Subantarctiques. 1981, Paimpont. CNFRA 51:167–172

Davis RC (1981) Structure and function of two Antarctic terrestrial moss communities. Ecol Monogr 5:125–143

Deharveng L (1981) Collemboles des Iles Subantarctiques de L'Océan Indien, Mission J. Travé 1972–1973. CNFRA 50:33–108

Derenne P, Mougin JL (1976) Donées ecologiques sur les Mammifères introduits de L'Ile aux Cochons, Archipel Crozet (46°06'S, 50° 14'E). Mammalia 40:21–53

Dorne A-J (1977) Analysis of the germination under laboratory and field conditions of seeds collected in the Kerguelen archipelago. In: Llano GA (ed) Adaptations within Antarctic ecosystems. Proc 3rd SCAR symp Antarct biol, pp 1003–1013

Dreux P (1971) Insecta. La fauna entomologiques des Iles Marion et du Prince Edouard. In: Van Zinderen Bakker EM, Winterbottom JM, Dyer RA (eds) Marion and Prince Edward Islands. Balkema, Cape Town, pp 335–343

Gillham M (1961) Modification of sub-Antarctic flora on Macquarie Island by sea birds and sea elephants. Proc R Soc Victoria 74:1–12

Gleeson JP, Van Rensburg PJJ (in press) Feeding ecology of the House Mouse *Mus musculus* on Marion Island. S Afr J Antarct Res 12

Gremmen NJM (1982) The vegetation of the Subantarctic islands Marion and Prince Edward. Junk, The Hague

Gressitt JL (1967) Introduction. Entomology of Antarctica. Antarct Res Ser 10:1–33

Gressitt JL (1970) Subantarctic entomology and biogeography. Pac Insects Monogr 23:295–374

Haukioja E (1981) Invertebrate herbivory at tundra sites. In: Bliss LC, Heal OW, Moore JJ (eds) Tundra ecosystems: a comparative analysis. IBP 25. Cambridge University Press, Cambridge, pp 547–555

Heal OW, Jones HE, Whittaker JB (1975) Moor House, UK. In: Rosswall T, Heal OW (eds) Structure and function of tundra ecosystems. Ecol Bull (Stockholm) 20:295–320

Holdgate MW (1967) The influence of introduced species on the ecosystems of temperate oceanic islands. IUCN Publ New Ser 9: 151–176

Holdgate MW (1977) Terrestrial ecosystems in the Antarctic. Philos Trans R Soc Lond B Biol Sci 279:5–25

Holmes RT, Sturges FW (1973) Annual energy expenditure by the avifauna of a northern hardwoods ecosystem. Oikos 24:24–29

Huntley BJ (1971) Vegetation. In: Van Zinderen Bakker EM, Winterbottom JM, Dyer RA (eds) Marion and Prince Edward Islands. Balkema, Cape Town, pp 98–160

Jenkin JF (1975) Macquarie Island, Subantarctic. In: Rosswall T, Heal OW (eds) Structure and function of tundra ecosystems. Ecol Bulls (Stockholm) 20:375–397

Jenkin JF, Johnstone GW, Copson GR (1982) Introduced animal and plant species on Macquarie Island. Colloque sur les Ecosystemes Subantarctiques. 1981, Paimpont, CNFRA 51:301–313

Jones E (1977) Ecology of the feral cat, *Felis catus* (L.) (Carnivora: Felidae) on Macquarie Island. Aust Wildl Res 4:249–262

Luke IA (1917) Faunal conditions in South Georgia islands. Science (NY) 45:502–503

MacArthur RH, Wilson EO (1967) The theory of island biogeography. Princeton University Press, Princeton

MacLean SF (1975) Ecology of tundra invertebrates at Prudhoe Bay, Alaska. In: Brown J (ed) Ecological investigations of the tundra biome in the Prudhoe Bay region, Alaska. Biol Pap Univ Alaska, Spec Rep 2, pp 114–123

MacLean SF (1980) The detritus-based trophic system. In: Brown J, Miller PC, Tieszen LL, Bunnell FL (eds) An Arctic ecosystem: The coastal tundra at Barrow, Alaska. US, IBP synthesis series 12. Dowden, Hutchinson and Ross, Stroudsburg, Pennsylvania, pp 411–457

MacLean SF (1981) Introduction: Invertebrates. In: Bliss LC, Heal OW, Moore JJ (eds) Tundra ecosystems: A comparative analysis. IBP 25. Cambridge University Press, Cambridge, pp 509–516

McNaughton SJ (1983) Compensatory plant growth as a response to herbivory. Oikos 40:329–336

Moore JJ, Dowding P, Healy B (1975) Glenamoy, Ireland. In: Rosswall T, Heal OW (eds) Structure and function of tundra ecosystems. Ecol Bulls (Stockholm) 20:321–343

Newton IP, Adams NJ, Brown CR, Enticott JW, Fugler SR (1983) Nonmarine vagrant birds at the Prince Edward Islands, June 1981 to May 1983. Cormorant 11:35–38

Norton DW, Ailes IW, Curatolo JA (1975) Ecological relationships of the inland tundra avifauna near Prudhoe Bay, Alaska. In: Brown J (ed) Ecological investigations of the Tundra Biome in the Prudhoe Bay region, Alaska. Biol Pap Univ Alaska Spec Rep 2:125–133

Paine RT (1980) Food webs: linkage, interaction strength and community infrastructure. J Anim Ecol 69:667–685

Parmelee DF (1980) Bird Island in Antarctic waters. University Minnesota Press, Minneapolis

Pascal M (1982) Les espèces Mammaliennees introduites dans L'Archipel des Kerguelen (Territoire des TAAF) bilan des recherches entreprises sur ces espèces. Colloque sur les Ecosystèmes Subantarctiques. 1981, Paimpont. CNFRA 51:269–280

Pattie DL (1977) Population levels and bioenergetics of arctic birds on Truelove Lowland. In: Bliss LC (ed) Truelove Lowland, Devon Island, Canada: a high Arctic ecosystem. University Alberta Press, Edmonton, pp 413–436

Paulian P (1953) Notes biologiques sur le *Pringleophaga kerguelensis* End. Rev Fr Entomol 20:45–47

Prince PA, Payne MR (1979) Current status of birds at South Georgia. Br Antarct Surv Bull 48:103–118

Prince PA, Croxall JP (1983) Birds of South Georgia: new records and re-evaluations of status. Br Antarct Surv Bull 59:15–27

Pye T, Bonner WN (1980) Feral Brown Rats *Rattus norvegicus* in South Georgia (South Atlantic Ocean). J Zool (Lond) 192:237–255

Rosswall T, Heal OW (eds) (1975) Structure and function of tundra ecosystems. Ecol Bull 20. Stockholm, Swedish Natural Science Research Council

Rosswall T, Flower-Ellis JGK, Johansson LG, Jonsson S, Ryden BE, Sonesson M (1975) Stordalen (Abisko), Sweden. In: Rosswall T,

Heal OW (eds) Structure and function of tundra ecosystems. Ecol Bull (Stockholm) 20:265–294

Ryan JK (1977) Synthesis of energy flows and population dynamics of Truelove Lowland invertebrates. In: Bliss LC (ed) Truelove Lowland, Devon Island, Canada: A high Arctic ecosystem. University Alberta Press, Edmonton, pp 325–346

Ryan JK (1981) Invertebrate faunas at IBP tundra sites. In: Bliss LC, Heal OW, Moore JJ (eds) Tundra ecosystems: A comparative analysis. IBP 25. Cambridge University Press, Cambridge, pp 517–539

Siegel S (1956) Nonparametric statistics for behavioral sciences. McGraw-Hill, New York

Siegfried WR (1982) The roles of birds in ecological processes affecting the functioning of the terrestrial ecosystem at sub-Antarctic Marion Island. Colloques sur les Ecosystèmes Subantarctiques, 1981, Paimpont. CNFRA 51:493–499

Smith RIL (1982) The impact of introduced reindeer on the vegetation of South Georgia. Colloques sur les Ecosystèmes Subantarctiques, 1981, Paimpont. CNFRA 51:531

Smith RIL (to be publ.) Terrestrial plant biology of the Sub-Antarctic and Antarctic. In: Laws RM (ed) Antarctic ecology. Academic Press, London

Smith RIL, Walton DWH (1975) South Georgia, Subantarctic. In: Rosswall T, Heal OW (eds) Structure and function of tundra ecosystems. Ecol Bull (Stockholm) 20:399–423

Smith VR (1977a) A qualitative description of energy flow and nutrient cycling in the Marion Island terrestrial ecosystem. Polar Rec 18:361–370

Smith VR (1977b) Notes on the feeding of *Ectemnorrhinus similis* Waterhouse (Curculionidae) adults on Marion Island. Oecologia (Berl) 29:269–273

Smith VR (1977c) Vegetation standing crop of the grey lava flows and of the eastern coastal plain on Marion Island. J S Afr Bot 43:105–114

Smith VR (1978a) Animal-plant-soil nutrient relationships on Marion Island (sub-Antarctic) Oecologia (Berl) 32:239–253

Smith VR (1978b) Plant ecology of Marion Island: a review. S Afr J Antarct Res 8:21–30

Smith VR (1978c) Soil chemistry of Marion Island (sub-Antarctic). S Afr J Sci 74:174–175

Smith VR (1979) The influence of seabird manuring on the phosphorus status of Marion Island (sub-Antarctic) soils. Oecologia (Berl) 41:123–126

Sømme L (to be publ.) Invertebrates. In: Bonner WN, Walton DWH (eds) Key environments – Antarctica. Pergamon, Oxford

Tallowin JRB (1977) The reproductive strategies of a Subantarctic grass *Festuca* contracta T. Kirk. In: Llano GA (ed) Adaptations within Antarctic ecosystems. Proc 3rd SCAR Symp Antarct Biol. Smithsonian Institution, Washington DC, pp 967–980

Travé J (1977) Recherches sur les Microarthropodes Terrestres de l'Archipel de Kerguelen-Etude Quantitive Globale. In: Llano GA (ed) Adaptations within Antarctic ecosystems. Proc 3rd SCAR Symp Antarct Biol Smithsonian Institution, Washington DC, pp 1035–1053

Travé J (1981) Biogéographie des Oribates (Acariens) subantarctiques. CNFRA 50:141–148

Travé J (1982) Structure des peuplements de Microarthropodes a Kerguelen essai de synthèse. Colloque sur les Ecosystèmes Subantarctiques, 1981, Paimpont. CNFRA 51:73–80

Tréhen P, Bouché M, Vernon Ph, Frénot Y (1985) Organization and dynamics of the Oligochaeta and Diptera populations on Possession Island. In: Siegfried WR, Condy PR, Laws RM (eds) Antarctic nutrient cycles and food webs (Proceedings of the 4th SCAR symposium on Antarctic biology). Springer, Berlin Heidelberg New York

Tuomi J, Salo J, Haukioja E, Niemlä P, Hakala T, Mannila R (1983) The existential game of individual self-maintaining units: selection and defence tactics of trees. Oikos 40:369–376

Van Aarde RJ (1980) The diet and feeding behavior of feral cats, *Felis catus*, at Marion Island. S Afr J Wildl Res 10:123–128

Vannier G (1982) Caractérisation des principaux stades de développment d'*Anatalanta aptera*. Eaton 1875 (Insecte Diptère des Iles Subantarctiques) par leur diagramme de transpiration. Colloque sur les Ecosystèmes Subantarctiques. 1981 Paimpont, CNFRA 51:193–207

Vogel M, Smith RIL, Remmert A (to be publ.) Introduced reindeer and their effects on the vegetation and the epigenetic invertebrate fauna of South Georgia (Sub-Antarctic). Oecologia (Berl)

Wace NM (1960) Botany of the southern oceanic islands. Proc R Soc Lond Ser B 152:475–490

Walton DWH (in press) The Sub-Antarctic islands. In: Bonner WN, Walton DWH (eds) Key environments – Antarctica. Pergamon Press, Oxford

Watson GE (1975) Birds of the Antarctic and sub-Antarctic. Amer Geophys Union, Washington DC

Weller MW (1975) Ecology and behaviour of the South Georgian pintail (*Anas g. georgica*). Ibis 117:217–231

West GC, DeWolfe BB (1974) Populations and energetics of taiga birds near Fairbanks, Alaska. Auk 91:757–775

Wielgolaski FE (1972) Vegetation types and plant biomass in tundra. Arctic Alpine Res 4:291–305

Wielgolaski FE (ed) (1975) Fennoscandian tundra ecosystems. Part 2, Animals and systems analysis. Springer, Berlin Heidelberg New York

Williams AJ (1978) Mineral and energy contributions of petrels (Procellariiformes) killed by cats to the Marion Island terrestrial ecosystem. S Afr J Antarct Res 8:49–53

Wind Transport of Electrostatically Charged Particles and Minute Organisms in Antarctica

W.S. BENNINGHOFF and A.S. BENNINGHOFF[1]

Summary. Dispersal of very small organisms and diaspores by wind is widely acknowledged, but the actual process and transport potential have been little studied. In the polar regions, and especially on the Antarctic continent, wind transport is enhanced by blown snow crystals and electric charges on the airborne particles. At several sites in Antarctica in 1977, we exposed pairs of Tauber sedimentation traps, grounded and not grounded, under different weather conditions and attendant electric potential gradients within 2 m of the ground. Relationships inferred from observation of the deposition of snow and dust particles in the grounded and ungrounded Tauber traps are reported. These relationships suggest that melting snow-banks in contact with moist or wet soil or bedrock will tend to accumulate airborne dust faster than melting snow overyling ice or frozen ground. Sea-ice, accreting from the underside, tends to incorporate sessile plankton, and, as the upper surface sublimes, the plankton residues may be dehydrated, lofted, charged, and transported by wind. Similarly, bursting bubbles on water surfaces inject micro-biota into the atmosphere as charged particles which probably have potential for long distance travel in the presence of snow being blown over ice surfaces. These and other examples of wind transport of charged particles are described and evaluated for their significance in Antarctic biogeography.

1 Introduction

From the time of earliest enquiry into polar terrestrial biology, wind transport of very small organisms and diaspores has been recognized as a leading process of dispersal (Ridley 1930; Rudolph and Benninghoff 1977). In the past several years we have been able to shed some additional light on this process through regard for the role of wind-blown snow and electrostatic charging of airborne particles (Benninghoff and Benninghoff 1978). We see prospects that similar roles of these forces at least extend, and perhaps in large part account for, dispersal of minute pelagic organisms over the seas surrounding the Antarctic continent. The particles of concern here range from 1.0 μm to approximately 500 μm. By the term diaspore we mean particles of any kind having potential for reproducing a whole living organism (e.g., a spore, an egg, or a fragment of lichen).

1 Matthaei Botanical Gardens, University of Michigan, Ann Arbor, MI 48105, USA

2 Rationale and Methods

In 1977 we undertook a reconnaissance of airborne particles of biological origin at several stations in Antarctica to learn what kinds of particles are being transported in the air and what sampling methods are effective. For sampling, we tried microscope slides coated with adhesive (Rudolph 1970), a Staplex Hi-volume sampler, Roto-rod samplers and Tauber sedimentation traps. The high-volume and Roto-rod samplers were not as effective as had been expected, for reasons we do not yet understand. We also recovered airborne dust by melting and filtering about 0.5 m^3 samples of freshly fallen or drifted snow. Further details of our methods have been published (Benninghoff and Benninghoff 1982) and only a summary is needed here. The Tauber traps yielded information of direct concern to the processes of wind transport.

The sedimentation trap, designed by Tauber (1974) of the Danish Geological Survey, has a special cap and aperture over a collecting jar. The cap is 150 mm in diameter, and the upper surface is cambered as an airfoil for smooth flow of air blowing over the surface. In the centre of the cap is an aperture 50 mm in diameter. Turbulence of air passing over that aperture results in a portion of the airborne particles falling into the collecting jar. We had become aware of the strong and fluctuating electric potential gradients in the lower atmosphere in polar regions, so at the suggestion of a colleague, Prof. A.D. Moore, we made the Tauber trap covers electrically conductive by means of special paint (TV Tube-Koat) and attached terminals for ground wires. The traps were then exposed in pairs, one grounded, the other not grounded. The main sampling site was at the Cosmic Ray Laboratory, mid-way between the American McMurdo Station and the New Zealand Scott Base. At that sampling site we operated an electric field mill to measure continuously the atmospheric potential gradient 2 m above the ground.

Atmospheric electrification in fair weather occurs mainly through ionization by cosmic rays, collision of particles (tribo-electrification), fracturing of particles (separation charging), or by water droplet separation (spray electrification) (Billings and Gussman 1976) as with bursting of bubbles in breaking waves on water. These are minor components within the global circuit between the solid earth (negative) and the ionosphere (positive). The potential gradient has an

Antarctic Nutrient Cycles and Food Webs
(ed. by W.R. Siegfried, P.R. Condy, and R.M. Laws)

average value of 130 V m^{-1} at sea level in fair weather, and it decreases with altitude, going to zero at or above an altitude of 50 km (the electrosphere or lower portion of the ionosphere) (Herman and Goldberg 1978). Using the electric field mill, we measured episodes of positive gradients as high as +5,000 V m^{-1} during blizzards and fair-weather blowing snow. Episodes of negative gradients as much as − 2,500 V m^{-1} occurred during falling snow, although sudden strong gusts of wind would for a fraction of 1 min change the polarity to positive. The record from this field mill at the Cosmic Ray Laboratory site in Antarctica resembles records from mountain top stations in middle latitudes, such as the one from the Zugspitze Observatory operated for several decades by Reiter (1982), as well as records from other observing stations in the Antarctic. The potential gradient is indicative of the predominant charges on particles in the atmosphere, and in the Antarctic environment the particles commonly include ice crystals (fine snow), as well as dust and minute bodies of biological origin.

3 Results

Snow and dust collected in the Tauber traps indicated that signs of charges on the particles agreed with the prevailing potential gradients in the lowermost atmosphere during the sampling period. For example, trap AT-4 (grounded) collected about 4 times as much water and particulate matter from blowing snow and dust as AT-3 (not grounded) during a blizzard with potential gradients greater than +1,000 V m^{-1}. In contrast, AT-17 (grounded) collected only half as much snow-water and sediment as AT-16 (not grounded) during a period of falling snow and negative potential gradients fluctuating from zero to − 800 V m^{-1}. Ice-fog conditions also produced steep positive potential gradients, but Tauber trap collections were minimal because there was no wind. These observations are summarized by the matrix of conditions and Tauber trap collections shown in Table 1.

Samples of newly fallen snow and blizzard drifted snow which were melted and filtered permitted inspection of the dust particles scavenged from the atmosphere. In the vicinities of stations, the most common organic particles were single cells and tissue fragments of conifer wood, readily identifiable by the bordered pits on the tracheid cells. Next most common were fibres, both natural and synthetic, from clothing. In both Tauber trap samples and surface-snow samples taken near McMurdo Sound we found fragments of filamentous micro-flora (especially one *Streptococcus*-like organism), minute fragments of presumed exoskeletal material from invertebrates, and numerous faecal pellets.

4 Discussion

Our results confirm the aerial transport of particles of biological origin together with snow, especially with wind-driven snow near the ground. The Tauber trap collections support the hypothesis that electrically grounded or earthed surfaces (moist soil, snow-flush areas, wet bedrock, sea surface) would tend to have more positively charged particles lodge on them than on neutral or positively charged surfaces. Many observers have speculated about melting snow-banks that seem to become increasingly dirty, and Warren Wilson (1958) described in detail the dirt patches accumulating on snowbanks in Spitzbergen. He could offer only wind transport as a responsible agent. It is probable that, at least during fair, windy weather, the grounded or negative polarity of melting snow-banks in effect augments the gravitational influence on positively charged particles being blown across the surface at low level. These thoughts may be extended to a consideration of the conservation effect for diaspores; a greater proportion would tend to lodge on sites with water which would permit germination and growth. In polar regions the dry air and strong winds commonly cause pools to dry up and ponds to lower their levels, so that dried surface sediments,

Table 1. Summary of inferred relationships between electric field gradients, weather conditions, and deposition of snow and dust particles in grounded and ungrounded Tauber traps (Benninghoff and Benninghoff 1982)

		Approximate range of electric field gradient (V m^{-1})					
		−1,500 to −1,000	−1,000 to −200	−200 to +200	+200 to +1,000	+1,000 to +2,000	+2,000
Weather condition		Low clouds, falling snow, little wind	Low clouds, lightly falling snow, little wind	Fair, calm	Fair or cloudy, blowing snow, moderate wind	Ice fog, calm	Fair or cloudy, blowing snow or blizzard, strong wind
Deposition of snow and dust in Tauber traps	Grounded	x	2x	3x	4x	2x	6x
	Ungrounded	2x	4x	3x	2x	x	2x

Aliquots of dust particles were counted, but snow particles were measured (as water equivalent) from Tauber traps under different conditions. The condition yielding the minimal amounts of dust and snow in the traps is given the symbol x, and the approximate relationships of amount caught under other conditions are indicated by multiples of the x amount algal mats, and other organic materials may be eroded by wind and blown snow and re-lofted for further transport.

We suggest that two other processes are probably important in the transport of airborne materials near the ground and, at least in part, are a result of electrostatic charging. One is the saltative or rebounding manner of travel of snow and dust particles being blown over a snow surface. The particles are both tumbling and rebounding, and they carry positive charges. The positive potential gradient probably exists down to within 1-2 mm of the snow surface. At some point slightly above the snow surface, or perhaps a short distance within the snow-pack, the potential gradient must regress to zero, but the level at which this occurs depends upon the conductivity of the snow-pack. We need to accomplish field measurements of the time constant for relaxation of the positive charge at the snow-cover surface, in order to test the hypothesis that particles carrying positive charges are to some extent repelled when they fall to within a few millimetres of the snow-pack surface during an episode of blowing snow.

Another process contributing to long-distance dispersal of biological particles is that of re-launching into the atmosphere. In the very dry air of polar regions and in the presence of frequent strong surface winds, sublimation of ice and snow is an important step in the hydrological cycle. Thus, dust particles encased in ice or snow undergoing ablation by sublimation eventually will be exposed to moving air, and to collisions with other airborne particles, and thereby to launching and further travel on the wind.

In scanning electron-microscope surveys of particles recovered from Tauber traps and snow samples on Ross Island, we found several fragments that appeared to be portions of chitinous appendages of small crustaceans or other arthropods and many faecal pellets of uniform shape and size (ca. 100 μm). Father Howard Brady, a micro-palaeontologist at Macquarie University, told us of sponge and other remains of sea-floor organisms which have been found on the surface of the Ross Ice Shelf. These materials were presumably frozen into the grounded base of the shelf and carried slowly upward as the uppermost layers of ice were removed by melting and sublimation. From the ice-shelf surface, the biological particles are released to the wind for transport widely over land and sea. Many "cryobiont" organisms can survive for years at sub-freezing temperatures; therefore, this process could be repeated for some diaspores, thus helping to account for the widespread distributions of micro-flora and micro-fauna species in polar regions. This process also contributes to the possibility of re-bedding, or multiple deposition, of micro-fossils, a disturbing hazard for the stratigraphic micro-palaeontologist.

Saltative travel over snow surfaces and re-launching from snow or ice by sublimation probably enhance the wind-driven exchanges of minute organisms and biological products between land and sea in Antarctic coastal areas. Some additional considerations apply to transport of particles at sea, especially where there is seasonal ice cover. Bursting air bubbles from breaking waves have been shown to be a major source of a positive electric potential gradient in the global atmosphere (Blanchard 1963) and are also the principal source of sea-spray (Wu 1981). The spray droplets are positively charged. The water of the droplets evaporates (relatively quickly for the small droplets, slowly for the large

Table 2. Inferred atmospheric potential gradients and conditions for aerial transport of diaspores over sea surfaces under different conditions of sea-ice. Ice status as defined by Armstrong et al. (1966)

State of sea surface	Relevant features of sea surface	Atmospheric potential gradient (V m^{-1})	Conditions for aerial transport of diaspores
Open	Wind waves, blown sea-spray from jet drops and droplets from bursting bubbles	~+150	Launching of microbionts and other particles from the sea to extent that breaking waves and wind strength permit
New ice (frazil, slush, pancake, etc.)	Few breaking waves, surface partly covered with ice; swells reduced	Mostly reduced below ~+50	Reduced launching and transport of particles from the sea
First-year ice	Continuous or near-continuous smooth ice cover up to 1 m thick; snow pack accumulates with sastrugi	Fluctuations similar to those over land	No launching and transport of particles from the sea. Best opportunity for transport from land to distant sea areas
Second-year ice	After second winter ice 2 m or more thick, standing higher out of water; surface relief with smooth hummocks snow drifts, pressure ridges; leads and polynas	Fluctuations similar to those over land; effects of water patches unknown	Some launching and transport from water patches and sublimation and wind erosion of ice ridges releasing particles from ice
Old ice	More than 2 years old, 3 m or more thick, smooth hummocks, ice nearly salt-free, locally free of snow and blue. Few water patches	Fluctuations similar to those over land	Some launching and transport by sublimation and wind erosion of ice ridges releasing particles from ice
Rotten ice	Ice honeycombed and dissected by melting; pools on ice, leads and polynas common; ice surface features subdued	Near that of open sea under calm conditions (~+150)	Launching and transport from water increasing, from sublimation decreasing

ones) and the positive charge is left on whatever solid particle (or particles) remains suspended in the atmosphere. In this manner, viruses (Baylor et al. 1977) and bacteria (Blanchard 1978) have been shown to be injected into the atmosphere as aerosols, and it is not unreasonable to expect other marine micro-biota to become airborne in this manner as well. By this mechanism the sea or adjacent land could be seeded with microbial organisms at some distance from the point of lofting. How large a particle would be lofted under ordinary circumstances needs to be determined.

In the Antarctic there is a very great seasonal change in the ice cover on the surrounding sea, which Gordon and Taylor (1975) accept as ca. 18×10^6 km^2, from a September maximum of 22×10^6 km^2 to a March minimum of 4×10^6 km^2. The average seasonal change affects an area larger than that of the Antarctic continent and its ice shelves. In Table 2 we present an estimate of the conditions for atmospheric transport of diaspores in relation to the different stages of development of the sea-ice.

As new sea-ice accumulates the breaking waves are suppressed, so the sea-spray source of charged droplets and particles is reduced, and the potential gradient in the atmosphere over the sea surface decreases. The new sea-ice is saline and less conductive electrically than seawater, but more conductive than freshwater ice or snow. With continuous ice cover snow can accumulate, the surface becomes less conductive as the snow-pack increases in thickness, and transport of particles by blowing snow may occur on this surface as on land. Thick multi-year ice and ice heaved up in pressure ridges tend to lose salinity and to some extent conductivity. Thereafter, such surfaces respond to sublimation, wind, and blowing snow much as terrestrial surfaces do, and minute organisms and diaspores that were once trapped in the sea-ice may be released and carried by wind and blowing snow for long distances.

As far as we know there have been no detailed measurements of electric potential gradients in the atmosphere over various kinds of sea-ice, leads, and polynyas, and in different weather conditions; nor have we found records of sampling for airborne particles of biological origin over sea-ice. The inferences in Table 2 are speculative, but they form the bases of hypotheses that deserve to be tested.

5 Conclusions

Wind transport of small diaspores and minute organisms has long been recognized in arid climates, but new evidence suggests that this transport in polar regions is enhanced by wind-blown snow crystals and the electric charges acquired by airborne particles. Experiments indicate that during fair weather positive potential gradients, wind-blown snow and dust (including particles of biological origin) tend to be deposited in measurably greater quantities on negatively charged (grounded or earthed) surfaces. This effect contributes to the processes whereby melting snow-banks become increasingly "dirty", and late snow-beds are frequently more eutrophic and richer in species than well-drained surfaces. Two effects of electrostatic charges on airborne particles are believed to occur but have not yet been confirmed by experiment: airborne particles from about 1 μm to 300 μm tend to remain dispersed in the air rather than clumped, because of their uniform polarity of charge during fair weather; and, positively charged particles being blown very near the surface of fresh snow-pack maintain a steep potential gradient at the surface of relatively dielectric snow, thus assisting the buoyant, saltative movement of blown snow and other particles near the ground.

Dust (including diaspores) is deposited together with falling snow and wind-blown snow. Re-launching of these dust particles into the atmosphere is made possible by sublimation of the enclosing snow (or firn or ice) and is assisted by positive charging of the particles as they are exposed to a fair weather positive potential gradient. Repeated deposition and re-launching of the same particle is conceivable, until mechanical erosion eventually grinds the particle into unidentifiable fragments.

Special and seasonally changing conditions for airborne particles exist over sea areas subject to annual ice cover. The positive potential gradient maintained by sea-spray from bursting bubbles on the open sea is replaced by a more nearly terrestrial regime of blowing snow and fluctuating potential gradients during ice-cover periods. Especially from old hummocky ice, sublimation, wind, and positive potential gradients may assist in launching biological particles that had been frozen into the ice. Thus, marine micro-biota that become incorporated into sea-ice are released for transport to distant areas.

On the Antarctic continent and over the adjacent seas covered by sea-ice, the geographical patterns of distribution of minute organisms and those with minute diaspores are determined by processes dominated by wind, blowing snow, sublimation of ice, and positive electrostatic charging of airborne particles. Broad, circumpolar distribution patterns are one consequence, and the establishment of species on remote and isolated habitats, such as nunataks, is another.

References

Armstrong T, Roberts B, Swithinbank C (1966) Illustrated glossary of snow and ice. Scott Polar Research Institute, Cambridge, England

Baylor ER, Baylor MB, Blanchard DC, Syzdek LD, Appel C (1977) Virus transfer from surf to wind. Science (Wash DC) 198:575–580

Benninghoff WS, Benninghoff AS (1978) Airborne particles and electric fields near the ground in Antarctica. Ant J US 13 (4):163–164

Benninghoff WS, Benninghoff AS (1982) Airborne biological particles and electric fields. Radio Science 17(5S):13S–15S

Billings CE, Gussman RA (1976) Dynamic behavior of aerosols. In: Dennis R (ed) Handbook on aerosols. Technical Information Center, US Department of Energy, Oak Ridge, Tennessee, pp 40–65

Blanchard DC (1963) The electrification of the atmosphere by particles from bubbles in the sea. In: Sears M (ed) Progress in oceanography, vol 1. Macmillan Pergamon, New York, pp 71–202

Blanchard DC (1978) Jet drop enrichment of bacteria, virus, and dissolved organic material. Pure Appl Geophys 116:302–308

Gordon AL, Taylor HW (1975) Seasonal change of Antarctic sea ice cover. Science (Wash DC) 187:346–347

Herman JR, Goldberg RA (1978) Sun, weather, and climate. National Aeronautics and Space Administration, SP-426, Washington DC, 360 pp

Reiter R (1982) Biological effects of atmospheric and man-made noise. In: Volland H (ed) Handbook of Atmospherics, vol 1 of 2 vols. CRC Press, Boca Raton, Florida, pp 291–306

Ridley HN (1930) The dispersal of plants throughout the world. L Reeve, Ashford, Kent. 744 pp

Rudolph ED (1970) Local dissemination of plant propagules in Antarctica. In: Holdgate MW (ed) Antarctic ecology, 2 vols. Academic Press, London, vol 2, pp 812–817

Rudolph ED, Benninghoff WS (1977) Competitive and adaptive responses of invading versus indigenous biotas in Antarctica – a plea for organized monitoring. In: Llano GA (ed) Adaptations within Antarctic ecosystems. Smithsonian Institution and Gulf Publishing Company, Houston, Texas, pp 1211–1225

Tauber H (1974) A static non-overload pollen collector. New Phytol 73:359–369

Warren Wilson J (1958) Dirt on snow patches. J Ecol 46:191–198

Wu Jin (1981) Evidence of sea spray produced by bursting bubbles. Science (Wash DC) 212:324–326

Succession in the Microflora of Leaves and Litter of Three Plant Species at Sub-Antarctic Macquarie Island

E.J. KERRY and G.M. WESTE [1]

Summary. Microbial ecology and succession were studied on leaves and litter of *Poa foliosa, Stilbocarpa polaris*, and *Pleurophyllum hookeri* which are the three dominant species of flowering plant of the Sub-Antarctic herbfields of Macquarie Island. Colonization was monitored on sample plots, with particular reference to the occurrence of filamentous fungi. The leaves and litter were examined by microscopic observation, and from cultures obtained after serial washing or surface sterilization. The patterns of microbial succession were similar for the three plant species and resembled those on most plants from temperate and tropical regions, except that on Macquarie Island most primary saprophytes became established on leaves before senescence. Despite similarity as to the types of micro-organisms and classes of filamentous fungi occurring at certain stages of the succession, there were obvious differences both between the three plant species and in comparison with plants from other parts of the world in the fungal species involved. The main differences referred to the fungal species which dominated the primary saprophytic microflora. These were Ascomycetes with Coelmycete anamorphs, which were often either host-specific or host-restricted, and included unidentified and probably new species of *Phoma* (3 spp.) and *Mycosphaerella* (1 sp.), and *Ascochyta stilbocarpae*. The *Phoma* spp. and *Mycosphaerella* sp. are restricted, as far as is known, to Macquarie Island. Other primary saprophytes, including *Mycosphaerella tassiana* and *Leptosphaeria eustoma* occur worldwide, but appear to favour cold climates.

1 Introduction

The nature and role of the microflora colonizing leaves and litter of many species of flowering plant have been investigated for a wide range of habitats, and the results have been the subject of a number of reviews (Last and Deighton 1965; Hudson 1968; Menna and Parle 1970; Bell 1974; Jensen 1974; Pugh 1974; Dickinson 1976). These reviews record the extensive work undertaken in temperate and tropical regions and at the same time demonstrate the paucity of comparable data from tundra regions.

Information on leaf and litter microflora of tundra plants has been assembled by Holding (1981) who reported on three major studies in tundra regions of northern Canada and Sweden and in Alaska (Dowding and Widden 1974; Flanagan and Scarborough 1974; Hayes and Rheinberg 1975). Results presented by Hayes and Rheinberg (1975) and Dowding and Widden (1974) suggest that fewer fungal species occur on the more limited range of flowering plant species found in the Arctic. Low species diversity is also a characteristic of Arctic tundra soils (Bunnell et al. 1980). The only detailed study of leaf and litter microflora in south polar regions was conducted on three plants at South Georgia in the Sub-Antarctic. The plants included two grasses, *Poa flabellata* and *Festuca contracta*, and one dwarf shrub, *Acaena magellanica* (Hurst and Pugh 1981; Hurst et al. 1983). A fungal succession was recorded on the leaves of each plant species as they senesced and dried out. A high percentage of sterile forms was recorded in isolates from both South Georgia (Hurst et al. 1983) and north polar sites (Dowding and Widden 1974), although Dowding and Widden found that the percentage varied according to the amount of time taken to induce fruiting.

Apart from the work at South Georgia, the only detailed information on the microflora of leaves and litter in south polar regions comes from annotated lists of the fungal species found on a variety of plants at Macquarie Island (Bunt 1965) and of the higher fungi of the Antarctic, Sub-Antarctic, and the Falkland Islands (Pegler et al. 1980). Jenkin (1972) showed that the early stages of leaf decomposition of *Poa foliosa* and *Pleurophyllum hookeri* on Macquarie Island proceeded at rates similar to those recorded for many plants in temperate regions. He noted that the decomposition of leaves of *Stilbocarpa polaris* was more rapid and resembled a bacterial rot. No other information is available on the agents of leaf decomposition on Macquarie Island, although Bunt (1965) postulated the involvement of the microfungi that he recorded.

The present investigation was conducted between 1975 and 1979 and involved a study of the microflora of leaves and litter of *Poa foliosa, Stilbocarpa polaris*, and *Pleurophyllum hookeri* which are the three dominant flowering plants on Macquarie Island. The study was concerned with identifying the active components of the microflora and their succession on leaves from emergence until soon after death, and on litter after 3 and 12 months on the ground. Filamentous fungi were isolated, examined and identified wherever possible and the occurrence of bacteria and yeasts on leaves and litter was also recorded. Limited experiments were conducted on the possible role of bacteria in the decomposition of *S. polaris* leaves.

1 School of Botany, University of Melbourne, Parkville, Victoria 3052, Australia

Antarctic Nutrient Cycles and Food Webs
(ed. by W.R. Siegfried, P.R. Condy, and R.M. Laws)

2 Materials and Methods

Macquarie Island is a small, isolated Sub-Antarctic island situated approximately half way between Australia and Antarctica, at a latitude of 55° S. Its climate is uniformly cold, wet and windy, with mean monthly temperatures ranging from 3.1 °C in July to 6.7 °C in January. A detailed description of the climate and biology of Macquarie Island is provided by Jenkin (1975).

The flora is impoverished, with only 45 species of vascular plants. Two of the plant species included in this study, *P. foliosa* and *S. polaris*, colonize much of the area near sea level and up to 200–300 m a.s.l. They often occur as codominants in wet tussock grassland associations (Taylor 1955). *Poa foliosa* is a tussock-forming grass. Tussocks grow up to 2 m in height, with a profuse growth of living shoots coming from a raised peat pedestal. The leaves are smooth and glabrous. Soon after death, they fall around the outside of the pedestal and remain attached for 1–2 years. *Stilbocarpa polaris* has large succulent leaves which grow on petioles up to 1 m in length. The leaf surface is uneven and sparsely hairy. Leaves decompose in most cases into a wet, slimy mass, but occasionally become dry and fragmented.

Pleurophyllum hookeri dominates the flora of many areas of higher altitude, and also occurs in peat bogs at lower altitudes (Taylor 1955). It only occurs with *P. foliosa* and *S. polaris* in ecotones between major plant formations. The leaves are blue-grey and leathery, with dense hairs, particularly on the lower surface. They form a rosette with youngest leaves innermost. As with *P. foliosa*, dead leaves remain attached around the outside for a long period.

The study sites were on north facing slopes at the northern end of the Macquarie Island plateau. *Poa foliosa* and *S. polaris* were studied at a site 45 m a.s.l. where they grew as codominants in a typical wet tussock grassland association. *Pleurophyllum hookeri* was studied in the nearest well developed stand of the species, 200 m to the southwest of the other site, at an altitude of 115 m a.s.l.

Four leaf age categories were studied: (1) Young (0 to 1 month) green, folded; (2) Mature (4–6 months) green, fully expanded; (3) Senescent *P. foliosa* (8–11 months), *S. polaris* and *P. hookeri* (6–9 months) yellow or yellow-green but not dry, or for *P. hookeri*, where no yellowing occurs, the outermost whorl of erect green leaves; and, (4) Dead (approx. 12 months) those most recently dead, dry and brown, but still attached. For *P. foliosa* these were the leaves lying around the outside of the tussock base and, for *P. hookeri*, those lying on top of the mass of fallen leaves. The least decomposed *S. polaris* leaves were selected.

Litter was studied by placing dead leaves of each plant species in separate "terylene" net bags with replicate leaves of each plant species in bags of 1- and 8-mm mesh diameter. The bags were secured in positions where the natural litter fell. Samples were withdrawn for examination after 3 and 12 months. Studies on leaf colonization were conducted in February or March, and in November in 1975 and 1976, and in March 1977. Dead leaves were placed in litter bags in November 1977 for studies in February 1978 (after 3 months) and November 1978 (after 12 months). The timing of collections varied by a few weeks in different years, because they were dependent on shipping time-tables. Samples were not available at other times of the year. Some experiments and observations were completed on the island. However, because of shortage of time and the limited laboratory facilities, some samples had to be taken to Melbourne for further investigation. All samples collected in February and March of all years and in November 1978, were processed in Melbourne. Specimens and cultures were transported to Melbourne in the ship's refrigerators at about 4 °C. The voyage took 5–7 days. This, together with unloading and quarantine formalities, meant that specimens to be processed in Melbourne were stored for approximately 14 days beforehand. The effects of this length of storage were investigated, and allowed for in the final analysis of data. Apart from such considerations, data from all collections were analyzed together regardless of time of year. Comparisons between the microflora of early and late summer samples were not made, because the late summer samples had been stored before processing.

2.1 Examination of Leaves and Litter

The following methods were used to obtain a comprehensive description of the microflora: (1) Microscopic examination, with identification and counts of different fungal species present. Twenty to 30 leaves and litter of each age category were examined in this way. (2) Serial washing. Thirty discs 5 mm in diameter (*S. polaris P. hookeri*) or squares 5 × 5 mm (*P. foliosa*) were cut with sterile cork borers or razor blades. The discs or squares were cut at random from the laminae except that large veins were avoided. The leaf pieces were washed, five of one type at the same time, in detergent (1% "Teepol") and then in 10 changes of sterile distilled water (Dickinson 1973). The pieces were blotted with sterile filter paper, then plated out, half onto "tomato-decoction agar" and half onto tap-water agar. The acidic tomato agar was selected to favour fungi, and, being a natural medium of low sugar content (10 g l^{-1}), had low fungal selectivity. The plates were incubated at 4 °C in the dark and observed at intervals for growth of colonies from the leaf pieces. Results were finalized when it was considered that no new colonies were forming (usually 2 weeks for tomato agar and up to 2 months for water agar). Litter was washed in a similar way except that 60 2-mm^2 pieces were used. The use of smaller pieces was considered necessary because high microbial populations usually occur on litter, and compete in culture. The chance of over-growth of slow, by faster growing colonies was reduced by using smaller pieces. (3) Surface sterilization. Leaf and litter pieces were prepared as described for washing. Mercuric chloride (0.1% w/v), was used for leaves and saturated calcium hypochorite was used to surface sterilize litter. The latter, a weak sterilant, was considered most suitable for litter, because other chemicals may penetrate the damaged cuticles of decomposing leaves. In each case, the pieces were rinsed in alcohol to aid wetting, placed in the sterilant for 1 min, and then rinsed in three changes of sterile distilled water. Plating, incubation and identification were as described for washed leves.

It was necessary in most cases to subculture colonies growing from leaf or litter pieces before identification was attempted. Particular care was taken to induce fruiting, to see whether sterile forms comprised a large part of the microflora, as recorded for other polar regions (Dowding and Widden 1974; Holding 1981; Hurst et al. 1983). In many cases this was achieved by keeping subcultures in the dark at 4 °C for periods of up to 3 months. Other treatments were used occasionally: tap-water agar prepared with sterilized pieces of leaves from the host plant; placing cultures under ultraviolet light for short periods; and, subjecting cultures to alternating periods of light and dark.

3 Results

Forty-one species of filamentous fungi were identified as colonizers of leaves and litter of the Macquarie Island plant species. These are listed in Appendix 1 (see page 604), together with plant species and leaf age from which each taxon was recorded. The fungal species belonged to 36 genera and included 7 *Penicillum* spp., 6 *Acremonium* spp., and 5 species of *Phoma*. Only 1 species, *Schizothyrioma stilbocarpae*, has been recorded previously (Pegler et al. 1980), although 1 other species, *Dasyscyphus* cf. *enzenspergerianus*, closely resembles *D. enzenspergerianus* (Dennis 1958). Four species, 1 of *Mycosphaerella* and 3 of *Phoma*, are considered to be new to science (Appendix 1). Certain fungi did not fruit despite the use of various methods to induce fruiting. Sterile forms comprised 5% of isolates from most types of samples. Exceptions were records of 10% sterile forms from surface sterilized leaves of *P. foliosa*, 17% from washed *S. polaris* leaves, 17% from washed 3-month-old litter of *P. hookeri*, and 34% from washed and 28% from surface sterilized 12-month-old *P. hookeri* litter. All sterile isolates from 12-month-old *P. hookeri* litter came from two very common sterile dematiaceous species.

Yeast species were not identified but white, and pale and bright pink and orange colonies grew in cultures, suggesting that a range of species was present. Both colourless and chromogenic bacteria were recorded.

A different range of taxa was recorded by the various methods used, the greatest differences being in the species recorded by microscopic examination (mostly Ascomycetes) and cultural isolation methods (washing and surface sterilization) (mostly Imperfect Fungi). In order to determine most accurately the patterns of microbial colonization on leaves as they aged and as litter decomposed, results obtained by the different methods were combined in schemes showing changes in frequency with leaf age of the different fungal species. Patterns recorded for the most common species are presented (Figs. 1-3). It was not possible to show actual frequencies in such schemes, because these were measured in different ways (microscopic or cultural isolation techniques). Instead, the taxa were determined as "very common", "common" or "rare" on leaves of each age category and on 3- or 12-month-old litter. The assessment was based on frequencies recorded by different methods in separate sampling seasons, with taxa ranked in order of frequency. Complete results and their assessment are given in Kerry (1982).

The schemes probably showed the anamorph and teleomorph stages of some Ascomycetes as separate species, but no attempt was made to pair species in the schemes because definite relationships have not been determined. No distinction was made in the schemes between the colonization of leaf surfaces and of internal tissues, although this information was obtained by surface sterilization.

A succession of different micro-organisms, including bacteria, yeasts and filamentous fungi, and of different species of filamentous fungi colonized the leaves of each plant species as they grew, senesced, and died and as litter decomposed (Figs. 1-3). The patterns of succession recorded for the three plant species were similar, particularly in the classes of filamentous fungi present at equivalent stages of the succession. However, different groups of species comprised the microflora of each plant species. The main features of the successions are summarized below.

3.1 Living Leaves

Bacteria were major colonizers of living leaves of *S. polaris*. Yeasts were dominant on living leaves of *P. foliosa* and *P. hookeri*, although they were not as common on the latter (Figs. 1-3). Certain Hyphomycete and Coelomycete species also colonized living leaves, but were usually less common than bacteria and yeasts. These included the Hyphomycetes *Aureobasidium pullulans* on *P. foliosa*, *Acremonium* spp. on *S. polaris* and *Botrytis cinerea* on *P. hookeri*. The Coelomycete species on living leaves were invariably species which became major colonizers of leaves after senescence. They occurred on living leaves as small inconspicuous colonies.

3.2 Senescent and Dead Leaves

Ascomycetes and Coelomycetes became dominant on leaves after senescence (i.e., they were the major primary saprophytes). A different group of species comprised the primary saprophytic microflora of each plant species, although some species occurred on more than one of the plant species (Figs. 1-3). A few species were dominant in this microflora on each plant species. These were, on *P. foliosa*, an unidentified *Phoma* species (sp. 1), *Camarosporium metableticum*, *Stagonospora ischmaemi*, *Mycosphaerella tassiana*

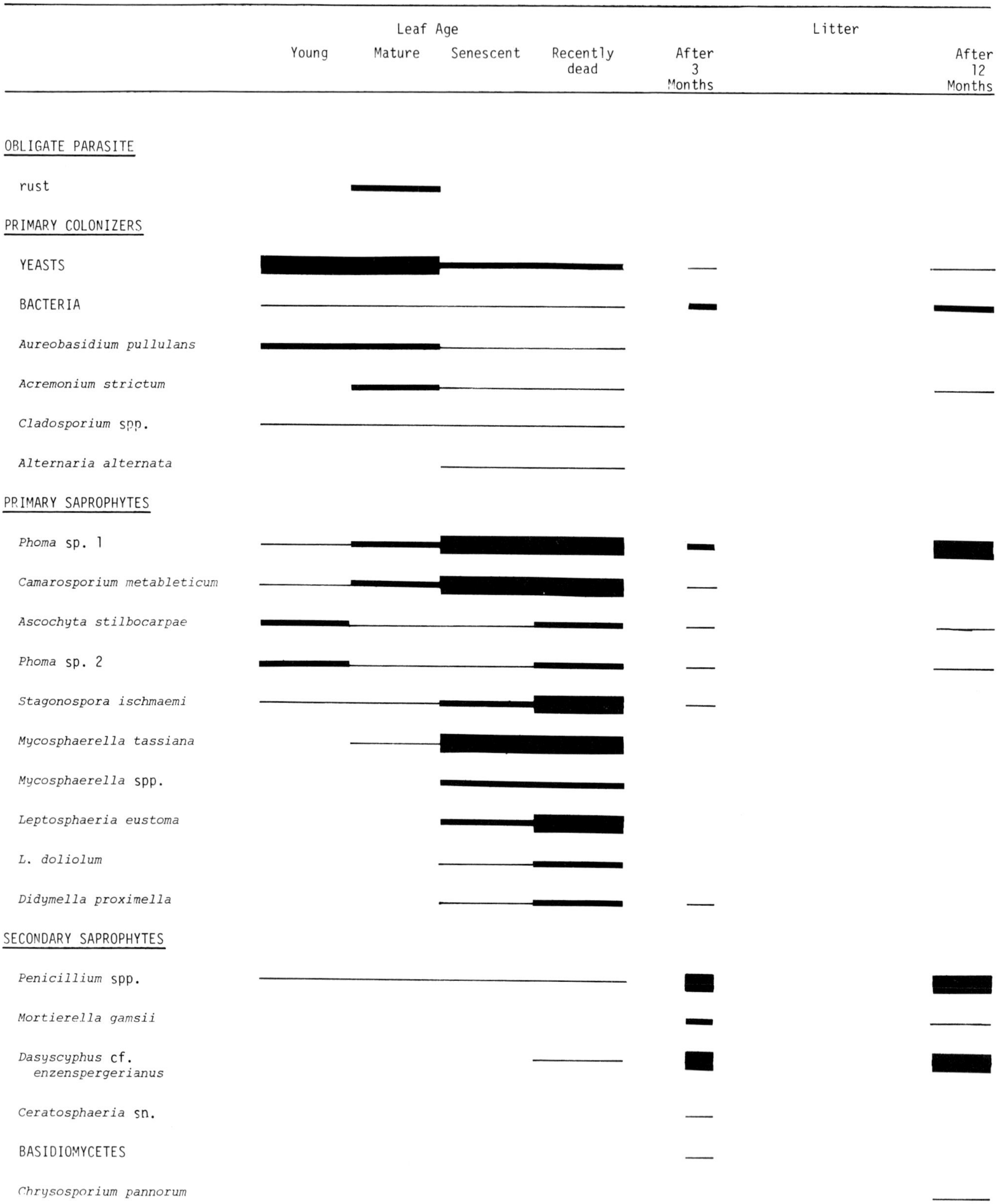

Fig. 1. Succession of microflora on leaves and litter of *Poa foliosa* from Macquarie Island. The decreasing thickness (three classes) of the *horizontal bars* indicates very common, common and rare taxa, respectively. Very common taxa were recorded frequently in most sampling seasons; common taxa were recorded frequently in one or two seasons or with moderate frequency in most seasons; and, rare taxa were recorded once, or rarely, by any or all methods. The *absence of a bar* indicates an absence of records. The data are based on direct observations and isolations from washed and surface sterilized leaves and litter. No distinction was made between internal and external colonization

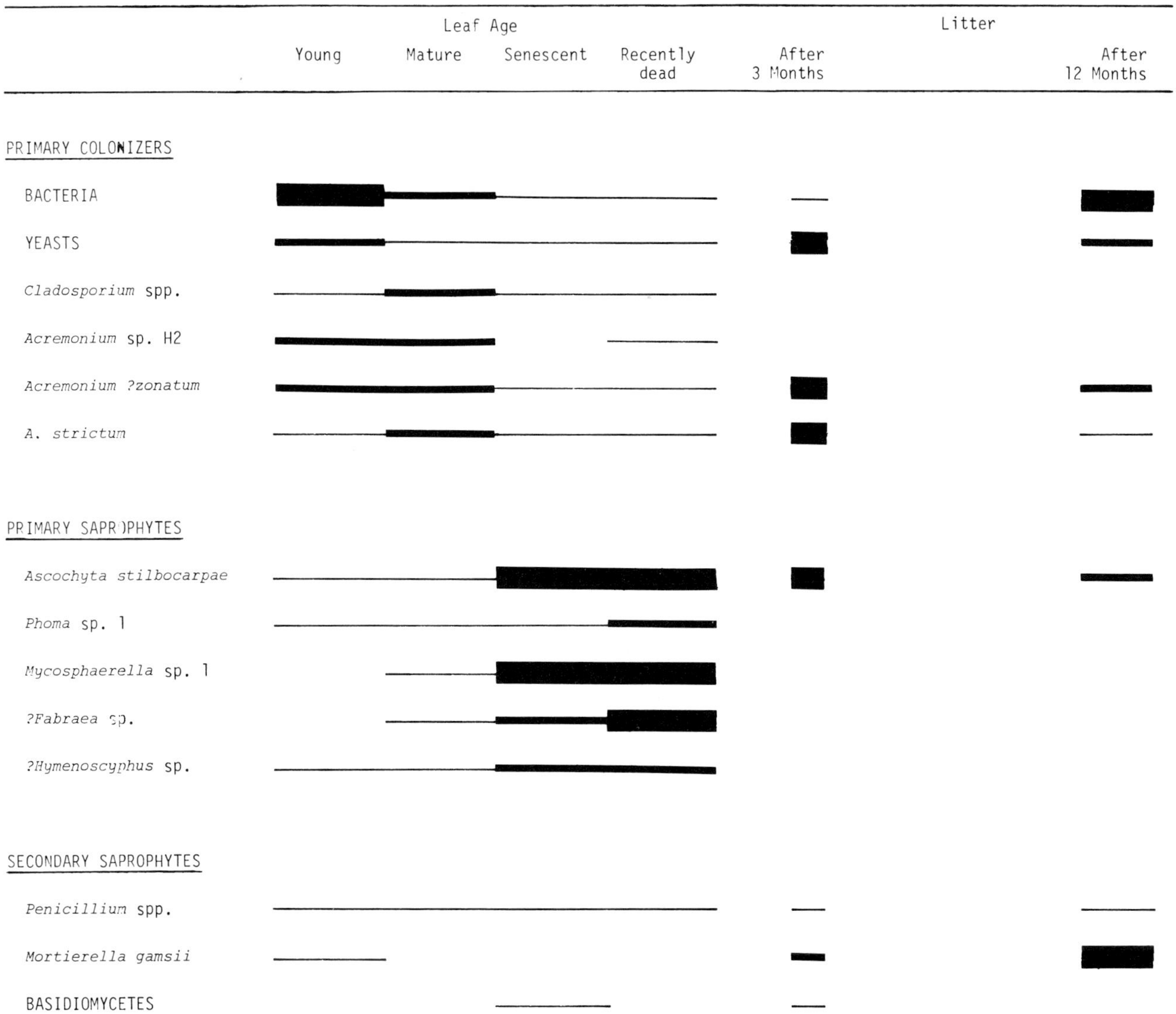

Fig. 2. Succession of microflora on leaves and litter of *Stilbocarpa polaris* from Macquarie Island. See caption to Fig. 1 for further particulars

and *Leptosphaeria* spp., especially *L. eustoma. Ascochyta stilbocarpae*, an unidentified *Mycosphaerella species* (sp. 1) and *?Fabraea* sp. were dominant on *S. polaris*, and *Phoma* spp. 1 and 3 and *"Rhodesiopsis gelatinosa"* were dominant on senescent and dead leaves of *P. hookeri*. Only one species, *Phoma* sp. 1, was a major primary colonizer of more than one of the plant species.

3.3 Litter

The litter of each plant species remained quite distinctive after both 3 and 12 months on the ground. Some fungi from the primary saprophytic microflora carried over onto the litter. These were not apparent on 3-month-old-litter, and the majority were less common on litter than on leaves. Exceptions were *Phoma* sp. 1 which was as common on litter as on leaves of *P. foliosa* (Fig. 1), and *Phoma exigua* which was more common on 12-month-old litter than on leaves of *P. hookeri* (Fig. 3). The carry-over of species from the primary saprophytic microflora meant that the specialized nature of this microflora carried over onto the litter.

As fungi from the leaf microflora decreased in number, they were succeeded in most cases by Basidiomycetes, members of the Mucorales (especially *Mortierella gamsii*) and *Penicillium* spp. (particularly *P. cyclopium*). These fungi occurred on the litter of each plant species, but varied in frequency (Figs. 1–3). In addition to these fungi, the Ascomycete *Dasyscyphus* cf. *enzenspergerianus* became a major colonizer of *P. foliosa* litter, and bacteria became major colonizers of 12-month-old litter of *S. polaris*. Bacteria were also common on young *S. polaris* leaves but may have been a different group of species from those on the litter.

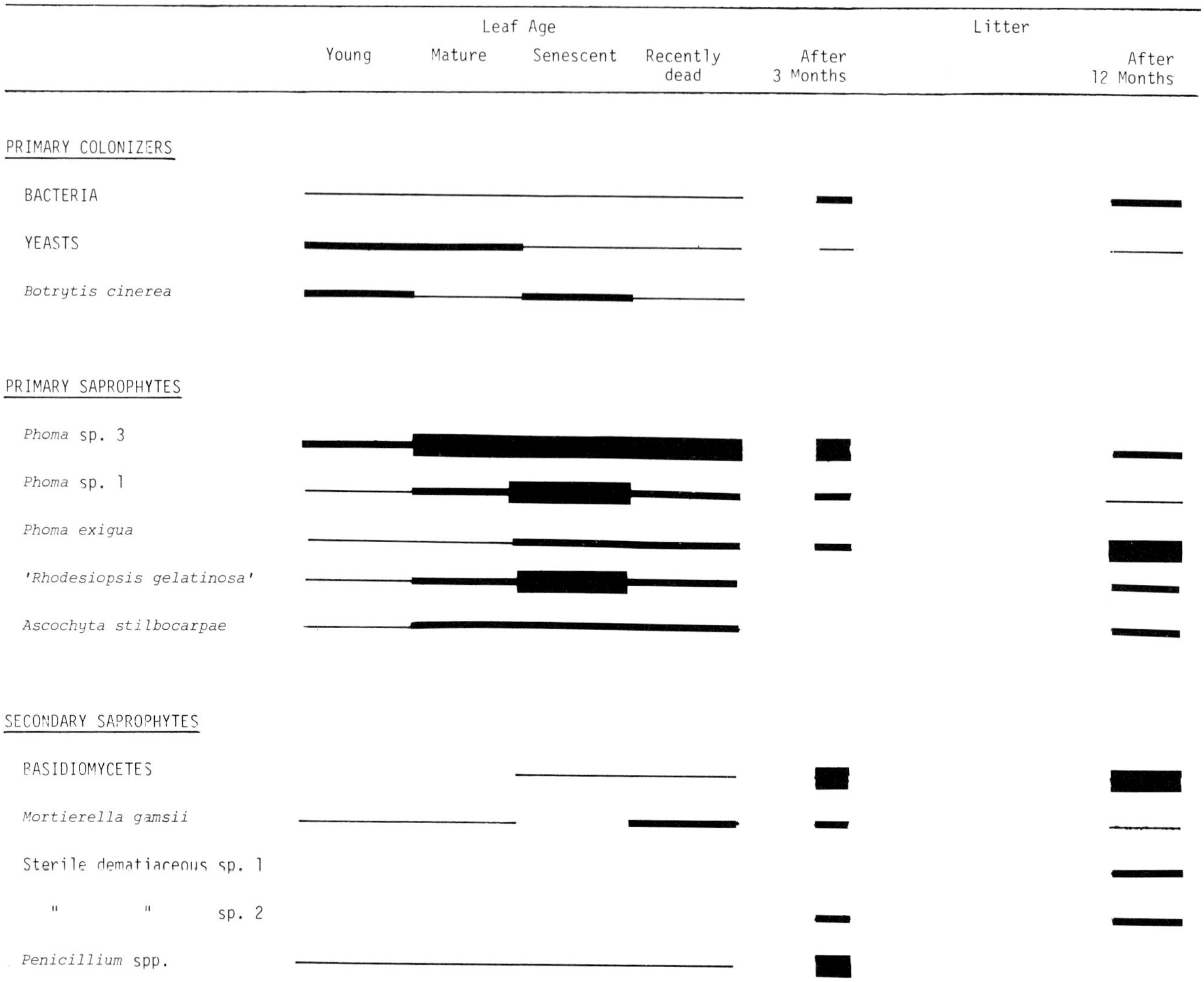

Fig. 3. Succession of microflora on leaves and litter of *Pleurophyllum hookeri* from Macquarie Island. See caption to Fig. 1 for further particulars

4 Discussion

The schemes of succession recorded on leaves of the Macquarie Island plants were similar to those which occur on the leaves of many plants from temperate and tropical regions. Bacteria and yeasts are major colonizers of living leaves of many plant species (Last and Deighton 1965; Menna and Parle 1970; Jensen 1974). One or other of these taxa were major colonizers of living leaves of each of the Macquarie Island plant species. Other features the Macquarie Island microflora had in common with other microflora were: (1) The primary saprophytic microflora was dominated by Ascomycetes and Imperfect Fungi, although the latter were mostly Coelomycetes and included few Hyphomycetes compared with northern hemisphere plants (Hudson 1968); (2) Some of the primary saprophytes carried over onto the litter and persisted for different periods of time (Kendrick and Burges 1962; Macauley and Thrower 1966; Ruscoe 1971); and, (3) Basidiomycetes, members of the Mucorales and *Penicillium* spp., colonized the litter, but varied in relative importance on the litter of each plant species (Hudson 1968).

The genera *Phoma, Mycosphaerella* and *Leptosphaeria* are common on leaves of the Macquarie Island plant species, as they are on leaves in warmer regions (Dickinson 1976). They have been recorded also from leaves and plant material in other south polar regions: *Mycosphaerella* and *Leptosphaeria* from leaves on South Georgia and Kerguelen (Pegler et al. 1980) and *Phoma* from leaves on South Georgia (Hurst and Pugh 1981) and plant material from the Mawson (Fletcher et al., to be publ.) and Casey regions of Antarctica (Kerry, unpubl.). These records suggest that all three genera are important as leaf colonizers in south polar regions.

The main differences between the leaf microflora of the Macquarie Island plants and those elsewhere occurred in the types of species which made up the primary saprophytic microflora. On most plants, the "common primary saprophytes" (Hudson 1968) included some or all of the same six species of Hyphomycetes: *Aureobasidium pullulans*, *Cladosporium herbarum*, *C. cladosporioides*, *Botrytis cinerea*, *Alternaria alternata*, and *Epicoccum purpurascens*. The "common primary saprophytes" occurred rarely on senescent and dead leaves of the Macquarie Island plant species, and the microflora of these were dominated by species of more restricted host range. Some were determined on present knowledge to be either host-specific or host-restricted. These included *Phoma* spp. 1 and 3, *Ascochyta stilbocarpae* and *Mycosphaerella* sp. 1 (Kerry 1982). *Mycosphaerella tassiana* and *Leptosphaeria eustoma* have a worldwide distribution including many cold regions. Pegler et al. (1980) commented that the range of *M. tassiana* indicated a preference for cold climates. The geographical distribution of *L. eustoma* indicates a similar preference. There was no indication from the study at South Georgia (Hurst et al. 1983) to indicate whether the "common primary saprophytes" or species of more restricted host range were dominant in the microflora, because few species were identified from dead leaves.

Most reports on the mycoflora of tundra regions indicate a high proportion of sterile forms; for example, 47% and 60% of fungi isolated from soil samples from MacRobertson and Enderby Lands, respectively (Fletcher et al., to be publ.), 41% from leaves and litter of South Georgia (Hurst et al. 1983), and > 50% of isolates from soil and litter from Barrow (Alaska) (Bunnell et al. 1980). In contrast, < 5% were sterile for most samples from Macquarie Island. It is possible that a longer period was allowed for the development of fruit in Macquarie Island cultures, and this may have reduced the records of sterile forms. The rate of occurrence of sterile fungi in the polar, compared with other, regions only may be assessed by the use of standardized methods for incubation of isolates.

The present results support the contention (Dickinson 1976) that the establishment of primary saprophytes on leaves before senescence is more common than thought previously. The constantly high humidity and the low levels of ultraviolet light (Jenkin 1975) may have made the surfaces of living leaves more favourable than those of plants from other environments for the development of fungi. The ability to behave parasitically is an important requirement for these fungi, and there are indications that a relationship exists between host restriction and this ability. This is another factor which may explain the colonization of leaves of the Macquarie Island plant species by primary saprophytes before senescence.

The results suggest that fungi are major agents of leaf decomposition on Macquarie Island, and may be more important than bacteria. However, the relative importance of these organisms cannot be determined finally because the methods favoured fungi. Plant decomposition is primarily fungal on South Georgia (D. Walton, pers. comm.). Bacterial biomass is greater than that of fungi in most arctic ecosystems (Bunnell et al. 1980), suggesting that bacteria are more important than fungi as decomposers of arctic plants.

The numbers of fungal species recorded from *P. foliosa*, *S. polaris*, and *P. hookeri* leaves, 38, 29, and 30 respectively, are in the same range as recorded for many plants in temperate regions (Pugh 1958; Bainbridge and Dickinson 1972; Dickinson 1976). Pugh (1958), working in England, recorded 30 species of fungi from dead leaves of *Carex paniculata*. This, like *P. foliosa*, is a tussock forming grass, and the number of species recorded may be compared directly with records of 31 species of fungi from dead *P. foliosa* leaves (Appendix 1). In contrast to these findings, five fungal species on average were recorded from each of five plant species studied at Abisko (northern Sweden) and four on average from each of two plant species at Barrow (Alaska) (Dowding and Widden 1974). Although relatively high species diversity was recorded for the Macquarie Island plants, very few of the species were dominant on the leaves of each plant species (Figs. 1–3).

A number of factors may have contributed to the high rates of initial leaf decomposition: the establishment of primary saprophytes on living leaves; the occurrence in the primary saprophytic microflora of host-specific or host-restricted species, which may be better able to grow on their substrates than species of wider host range; and, all the major leaf colonizers were capable of growth at 4 °C, and some had relatively high rates of growth at this temperature (Kerry 1982). Leaf colonizing fungi from South Georgia were able to utilize cellulose at 1 °C at from 40% to 70% of optimum rates (Hurst et al. 1983). Cellulase activity was not tested for the Macquarie Island species, but it is possible that high rates of activity at low temperatures are characteristic of Sub-Antarctic fungi. If so, this would be a major factor in the high rates of initial leaf decomposition.

The particularly high rates of decomposition recorded for *S. polaris* cannot be explained readily by any characteristics recorded for its leaf microflora. The type of breakdown was suggestive of bacterial rot caused by *Erwinia caratovora* or *Xanthomonas campestris*, but tests for these species proved negative. The rapid decomposition of *S. polaris* leaves was thought to be related to the succulent nature of the leaves (Bell 1974), and possibly also their chemical composition. High sugar and nitrogen contents were suggested as possible factors involved in the high rates of decomposition of *Acaena magellanica* on South Georgia (Hurst et al. 1983). The presence of *Mortierella* sp. which is a saccharophilic species was taken as an indication of the decomposition of leaf sugars. *Stilbocarpa polaris* decomposes in a similar way to *A. magellanica*. Its young leaves were colonized rarely by *Mortierella gamsii*, so that sugar decomposition may have occurred at times on those leaves. However, this does not explain the consistently high rates of decomposition for *S. polaris* leaves.

Appendix 1. Filamentous fungi recorded from leaves and litter of *Poa foliosa, Stilborcarpa polaris,* and *Pleurophyllum hookeri* from Macquarie Island. Plant species and leaf age categories from which taxa were recorded are shown on the right as Pf = *P. foliosa,* Sp = *S. polaris,* Ph = *P. hookeri,* Y = young, M = mature, S = senescent, D = dead, L = litter. Species new to science are shown as (o), and previous records from Macquarie Island are indicated as (+) for genera and (*) for species

Fungal Taxa	Plant species		
	Pf	Sp	Ph
Hyphomycetes		Leaf age	
Acremonium kiliense Grutz	YMSD	YMS	D
A. strictum W. Gams	YMSDL	YMSDL	YMSDL
A. terricola (Miller et al.) W. Gams	D		Y
A. ?zonatum (Sawada) W. Gams		YMSDL	
Acremonium sp. 1	Y	MS	
Acremonium sp. H2		YMD	D
Alternaria alternata (Fr.) Keissler	D		D
Aureobasidium pullulans (de Bary) Arnaud	YMSD	YMSD	
Botrytis cinerea Pers. ex Fr. Rabenh.	D		YMSD
Chalara sp.		SD	S
Chrysosporium pannorum (Link) Hughes	ML		Y
+*Cladosporium cladosporioides* (Fresen.) de Vries } *C. herbarum* (Pers.) Link	YMSD	YMSD	YMSD
?*Diploospora* sp.	YMSD	YS	Y
Epicoccum purpurascens Ehrenb. ex. Schlecht.	MD		YS
Fusidium sp.	SD		YMSD
Gliocladium sp.	S		
Graphium sp.	L		
+*Penicillium* brevi-compactum Dierckx			
P. corylophilum Dierckx			
P. cyclopium Westling			
P. decumbens Thom.			
P. frequentans Westling			
P. granulatum Bain.			
P. restrictum Gilman & Abbot			
Penicillium spp.	YMSDL	YMSDL	YMSDL
Stemphyllium botryosum Wallr.	Y		S
Trichoderma harizianum Rifae	D		
Trichoderma sp.			L
Volucrispora graminea Ingold, McDougall & Dann			S
Coelomycetes			
+*Ascochyta stilbocarpae* Syd.	YMSDL	YMSDL	YMSDL
+*Camarosporium metableticum* Trail.	YMSDL	MSD	YMSD
Colletotrichum gloeosporiodes Penz.		YMS	S
Phoma exigua Desm.	L		YMSDL
°*Phoma* sp. 1	YMSDL	YMSDL	YMSDL
°*Phoma* sp. 2	YMSDL	MS	
°*Phoma* sp. 3	L		YMSDL
Phoma ?herbarum Westend.		M	YD
"Rhodespiopsis gelatinosa"[a]	YSDL	MSD	YMSDL
Septoria sp.	D		YMSD
Stagonorpora ischaemi Sacc.	YMSDL	S	S
Tunicago sp.	YD	D	
Ascomycetes			
Ceratosphaeria sp.	L		
Dasyscyphus cf. *enzenspergerianus* (P. Henn.) Dennis	DL		L
Didymella proximella (Karst.) Sacc.	SDL		
?*Fabraea* sp.		MSD	
?*Hymenoscyphus* sp.		YMSD	
Leptosphaeria eustoma (Fuckel) Sacc.	YMSD		
L. doliolum (Pers. ex Fr.) Ces. & de Not.	SD		
Leptosphaeria sp.	SD	MS	
+*Mycosphaerella tassiana* (de Not.) Joh.	MS		
°*Mycosphaerella* sp. 1		MSD	
Mycosphaerella spp.	SD		
Niesslia exosporioides (Desm.) Wint.	SD		
+*Pleospora graminearum* Wehmeyer		D	
P. lutea Wehmeyer	MSD	Y	
P. vagans Neissl.	MSD		S
Pleospora sp.		S	
*_Schizothyrioma stilbocarpae_ Spooner		MS	
Zygomycetes			
Mortierella gamsii Mil'ko	MSL	YL	YMDL
Mortierella spp.	SL	L	DL
Mucor hiemalis Wehmer f hiemalis	D		MD
M. mucedo (L.) Fresenius	L	YD	SDL
Basidiomycetes			
Unknown species of rust	M		

[a] This species was named *Rhodesiopsis gelatinosa,* based on its occurrence on *Phormium tenax* in England (Sutton, pers. comm.). Since records from the present study indicate a wider host range, its taxonomy is under review. In the meanwhile it will be referred to as *"Rhodesiopsis gelatinosa"*. It is lodged in the herbarium at Commonwealth Mycological Institute as IMI No 210525

Acknowledgments. This research was supported by a University of Melbourne Postgraduate Scholarship. Support for field work was provided by the Antarctic Division, Australian Department of Science and Technology, and the Tasmanian National Parks and Wildlife Services.

References

Bainbridge A, Dickinson CH (1972) Effect of fungicides on the microflora of potato leaves. Trans Br Mycol Soc 59:31–41

Bell MK (1974) Decomposition of herbaceous litter. In: Dickinson CH, Pugh GJF (eds) Biology of plant litter decomposition, vol 1. Academic, New York, pp 37–67

Bunnell FL, Miller OK, Flanagan PW, Benoit RE (1980) The microflora: composition, biomass and environmental relations. In: Brown J, Miller PC, Tieszen LL, Bunnell FL (eds) An arctic ecosystem. US/IBP Synthesis 12. Dowden, Hutchinson & Ross Inc., Stroudsburg Pennsylvania, pp 255–290

Bunt JS (1965) Observations on the fungi of Macquarie Island. Austral Nat Antarctic Res Exp Sci. Report ser B (2) Botany No 78: 22pp

Dennis RWG (1958) Critical notes on some Australian Helotiales and Ostropales. Kew Bull 13:321–358

Dickinson CH (1967) Fungal colonization of *Pisum* leaves. Can J Bot 45:915–927

Dickinson CH (1973) Effects of Ethimerol and Zineb on phylloplane microflora of barley. Trans Br Mycol Soc 60:423–431

Dickinson CH (1976) Fungi on aerial surfaces of higher plants. In: Dickinson CH, Preece TF (eds) Microbiology of aerial plant surfaces. Academic, London New York, pp 293–324

Dowding P, Widden P (1974) Some relationships between fungi and their environment in tundra regions. In: Holding AJ, Heal OW, MacLean SF Jr, Flanagan PW (eds) Soil organisms and decomposition in tundra. Tundra Biome Steering Committee, Stockholm, pp 123–150

Flanagan PW, Scarborough AM (1974) Physiological groups of decomposer fungi on tundra plant remains. In: Holding AJ, Heal OW, MacLean SP Jr, Flanagan PW (eds) Soil organisms and decomposition in tundra. Tundra Biome Steering Committee, Stockholm, pp 159–181

Fletcher LD, Kerry EJ, Weste GM (to be published) Microfungi of MacRobertson and Enderby Lands, Antarctica. Polar Biol

Hayes AJ, Rheinberg P (1975) Microfungal populations of the Abisko area, north Sweden. In: Wielgolaski FE (ed) Fennoscanadian tundra ecosystems, part 1 (Ecological Studies 16). Springer, Berlin Heidelberg New York, pp 244–250

Holding AJ (1981) The microbiology of tundra. In: Bliss LC, Heal OW, Moore JJ (eds) Tundra ecosystems: a comparative analysis. Cambridge University Press, Cambridge, pp 561–585

Hudson HJ (1968) The ecology of fungi on plant remains above the soil. New Phytol 67:837–874

Hurst JL, Pugh GJF (1981) Fungi on plants on plant litter in South Georgia. In: Jouventin P, Massé L, Trehen P (eds) Colloque sur les ecosystèmes subantarctiques. CNFRA no. 51, Paimpont, pp 249–254

Hurst JL, Pugh GJF, Walton DWH (1983) Fungal succession and substrate utilization on the leaves of three South Georgia phanerogams. Br Antarct Surv Bull 58:89–100

Jenkin JF (1972) Studies of plant growth in a subantarctic environment. PhD thesis, University of Melbourne

Jenkin JF (1975) Macquarie Island, subantarctic. In: Rosswall T, Heal OW (eds) Structure and function of tundra ecosystems. Swedish National Science Research Council, Stockholm, pp 375–397

Jensen V (1974) Decomposition of angiosperm tree litter. In: Dickinson CH, Pugh GJF (eds) Biology of plant litter decomposition, vol 1. Academic, New York, pp 69–104

Kendrick WB, Burges A (1962) Biological aspects of the decay of Pinus silvestris leaf litter. Nova Hedwigia IV:313–342

Kerry E (1982) Succession in the microflora of leaves and litter of three plants from subantarctic Macquarie Island. PhD thesis, University of Melbourne

Last FT, Deighton FC (1965) The non-parasitic microflora on the surfaces of living leaves. Trans Br Mycol Soc 48:83–99

Macauley BJ, Thrower LB (1966) Succession of fungi in leaf litter of *Eucalyptus regnans*. Trans Br Mycol Soc 49:509–520

Menna ME di, Parle JN (1970) Moulds on leaves of perennial ryegrass and white clover. NZ J Agric Res 13:51–68

Pegler DN, Spooner BM, Smith RIL (1980) Higher fungi of Antarctica, the subantarctic zone and Falkland Islands. Kew Bull 35: 499–562

Pugh GJF (1958) Leaf litter fungi found on *Carex paniculata* L. Trans Br Mycol Soc 41:185–195

Pugh GJF (1974) Terrestrial fungi. In: Dickinson CH, Pugh GJF (eds) Biology of plant litter decomposition, vol II. Academic, New York, pp 303–336

Ruscoe QW (1971) Mycoflora of living and dead leaves of *Nothofagus truncata*. Trans Br Mycol Soc 56:463–474

Taylor BW (1955) The flora, vegetation and soils of Macquarie Island. Austral Nat Antarct Res Exp Sci. Report ser B (2) Botany:192 pp

Organization and Dynamics of Oligochaeta and Diptera on Possession Island

P. TRÉHEN[1], M. BOUCHÉ[2], Ph. VERNON[1], and
Y. FRENOT[1]

Summary. In terms of abundance and biomass, flies (Diptera) and earthworms (Acanthodrilidae) are principal organisms in the decomposition of organic matter. Their effects on decomposition processes, as compared to those of micro-organisms, vary in accordance with characteristics of the local substratum, altitude, distance from shore, origin (animal or plant) of litter, production periodicity of organic matter, etc. The organization and dynamics of these principal organisms in decomposer communities were investigated in the field and in the laboratory. The investigations included studies of their micro-distribution, demography, fecundity, growth, and trophic specificity. The co-existence of two species in the same biotope is examined. The low number of species and the nature of interspecific relationships indicates the fragility of these Sub-Antarctic decomposer communities.

1 Introduction

The organization of invertebrate populations at Possession Island is of special taxonomic interest, due to their high degree of endemism. Morphological, physiological, and behavioural adaptations to Sub-Antarctic conditions are evident. Current studies of some of these adaptations include alary polymorphism, duration of post-embryological development, longevity, fasting, reproductive behaviour and locomotor activity of Diptera (Vernon 1981; Tréhen 1982; Tréhen and Vernon 1982).

It is generally acknowledged that the diversity and abundance of many Sub-Antarctic invertebrate populations depend on seaweed deposits, which accumulate at high-water zones, and manuring by seabirds and seals. Smith (1978) demonstrated the inter-relationships between primary producers on Marion Island and exogenous manuring by animals. The invertebrate communities are organized in response to these factors, each mono-specific population revealing its own responses. Our investigations on Possession Island, where invertebrate species diversity is low, illustrate these interrelationships at various levels. Only two groups, Diptera and Oligochaeta, are considered in this study. The groups included similar numbers of species. These species show similar distribution patterns over the island but they occupy different niches. Three complementary aspects were covered by our investigations: the distribution of Diptera and Oligochaeta at Possession Island, focusing on the relationship between different species and local topographic, edaphic, and biological characteristics; animal-environment relationships, focusing on edaphic characteristics as principal factors in the micro-distribution and abundance of certain earthworm species and the role of manuring in the organization of necrosaprophagous Diptera communities; and, a trial working model of the *Durvillea*-Diptera-Oligochaeta ecosystem.

2 Material and Methods

Field work was carried out during 1980–1982. Live Diptera specimens were collected and counted by hand in sampling areas. Dispersal activity and behaviour were quantified by capturing the insects in pitfall traps distributed over these sampling areas. Earthworm communities were evaluated by manually examining ca. 4,000 cm^3 soil samples in the laboratory. The samples were removed from the upper layer of soil in the sampling areas. These areas were selected and delimited prior to quantitative estimation of population sizes. For the population estimates of *Anatalanta crozetensis*, individuals were marked with coloured metallic paints and a capture-recapture technique employed. Specimens of the sea-shore earthworm, *Phreodrilus crozetensis*, and its cocoons, were collected by washing fractions of soil taken from the sampling areas.

Dry weights (dry wt.) were obtained by drying specimens at 60 °C with their digestive tracts full. Experimental laboratory work was done at the Station Biologique de Paimpont (France), where animals were reared throughout the year and their reproductive cycles studied. Data on the distribution of the species were computed, using the map of Delettre et al. (1982). Factorial correspondence analysis was used to examine the species' specific affinities for environmental characteristics.

1 ERA 798 CNRS, Programme Biosol/TAAF, Station Biologique de Paimpont, 35380-F-Plélan-le-Grand, France
2 INRA/CEPE-CNRS, Route de Mende, BP 5051, F-34033-Montpellier Cedex, France

Antarctic Nutrient Cycles and Food Webs
(ed. by W.R. Siegfried, P.R. Condy, and R.M. Laws)

Table 1. Distribution and alar development of Diptera

Family	Species	Alar development	Distribution	Code (in Fig. 1a)
Helcomyzidae	*Paractora dreuxi*	Micropterous	Marion-Crozet	PAR
Ephydridae	*Amalopteryx maritima*	Stenopterous	Crozet-Kerguelen-Heard	AMA
Coelopidae	*Apetenus litoralis*	Micropterous	Marion-Crozet-Kerguelen	
	Listriomastax litorea	Macropterous and micropterous	Marion-Crozet-Kerguelen	LIS
Sphaeroceridae	*Siphlopteryx antarctica*	Sub-apterous	Crozet	SIP
	Anatalanta aptera	Apterous	Crozet-Kerguelen-Heard	APT
	Anatalanta crozetensis	Apterous	Crozet	CRO
	Dreuxiella spinosa	Apterous	Crozet	

3 Results

3.1 Distribution of Diptera and Oligochaeta

The brachycerous Diptera of the Crozet archipelago include eight species, of which five are known from other islands of the southern Indian Ocean (Table 1). Of these, this paper does not deal with *Dreuxiella spinosa* (only one suspected specimen collected which could have been an immature stage of *Anatalanta crozetensis*) and *Apetenus litoralis* (a strictly rock-dwelling species of the sea-shore, which together with the chironomid *Halirytus amphibius* (Diptera) and the curculionid *Palirheous eatoni* (Coleoptera) comprise an original seaweed-eating community). The following morphological features were common to the six other species found among organic debris: physogastry (mainly in the genus *Anatalanta*); reduced wing size with consequent loss of flight capability (except *Listriomastax*); and, wing polymorphism (e.g., *Anatalanta* is completely wingless) which does not seem to impair colonizing ability.

Four species of earthworms (Phreodrilidae and Acanthodrilidae), previously described by Michaelsen (1905), occur on Possession Island. Both morphological (size, colour, place of nephridial pores) and ecological features led us to distinguish more forms (Table 2). Taxonomic studies of these are in process. The presence of introduced Lumbricidae is common at most Sub-Antarctic islands. We have observed *Dendrobaena rubida tenuis* on Possession Island where it is a recent colonizer around the base which was established in 1962. Studies on the colonizing process and the response of the indigenous fauna to this are in process. This species and the other earthworms are all epigeous or epiendogeous (Bouché 1972). Only *P. crozetensis* is known from South Georgia and the southern part of Tierra del Fuego (Tetry 1947). The genus *Microscolex* (except for the two nomadic species, *M. dubius* and *M. phosphoreus*) is known only from the high latitudes of the southern hemisphere (Stephenson 1930). When and how the Sub-Antarctic islands were colonized by earthworms have been discussed by Benham (1903), Michaelsen (1905), Tetry (1947), and Bouché (1982).

Data from 60 Diptera and Lumbricidae samples were processed according to factorial analysis of correspondence (Figs. 1a and 1b). The main environmental factors taken into account were: elevation above sea-level (A), distance from the sea-shore (D), cover abundance (vegetation) (V), reference to general environment, such as shore type (L), grass type (H), fell-field type (F), type (E), and nature (O) of manuring, and soil moisture (U). The analysis showed that the organization and the distribution of communities depend on two principal factors: altitude and structure of vegetation.

Several types of organic manuring occur, despite a decreasing density of vertebrates as distance from the shore increases. This includes seaweed wracks on the shore, moulting areas of Elephant Seals, *Mirounga leonina*, and Gentoo Penguin, *Pygoscelis papua*, and King Penguin, *Aptenodytes patagonicus*, rookeries and moulting areas adjacent to the shore, nests of albatrosses in the grassland zones farther inland, and finally burrows of Whitechinned Petrel, *Procellaria aequinoctialis*, and Georgian Divingpetrel, *Pelecanoides georgicus*, in both the grassland and higher-altitude fell-field zones far inshore. Decaying seaweed, guano, seabird regurgitates and carcasses, and moulted skin and feathers constitute the main trophic substratum for Diptera and probably for some species of Lumbricidae as well (Figs. 1 and 2).

Table 2. Oligochaeta of Possession Island

Family	Species	Code (in Fig. 1b)
Phreodrilidae	*Phreodriius crozetensis*	PC
Acanthodrilidae	*Microscolex crozetensis*	C
	Microscolex luykeni var. *typique*	LT
	Microscolex enzenspergeri var. *typique*	ET
	Microscolex enzenspergeri var. *alba*	EA
	Microscolex enzenspergeri var. *nigra*	EN
	Microscolex enzenspergeri var. *bicolor*	EB
	Microscolex luykeni var. *halophila*	LH
	Microscolex luykeni var. *albinique*	LA
	Microscolex frenoti nov. sp.	F

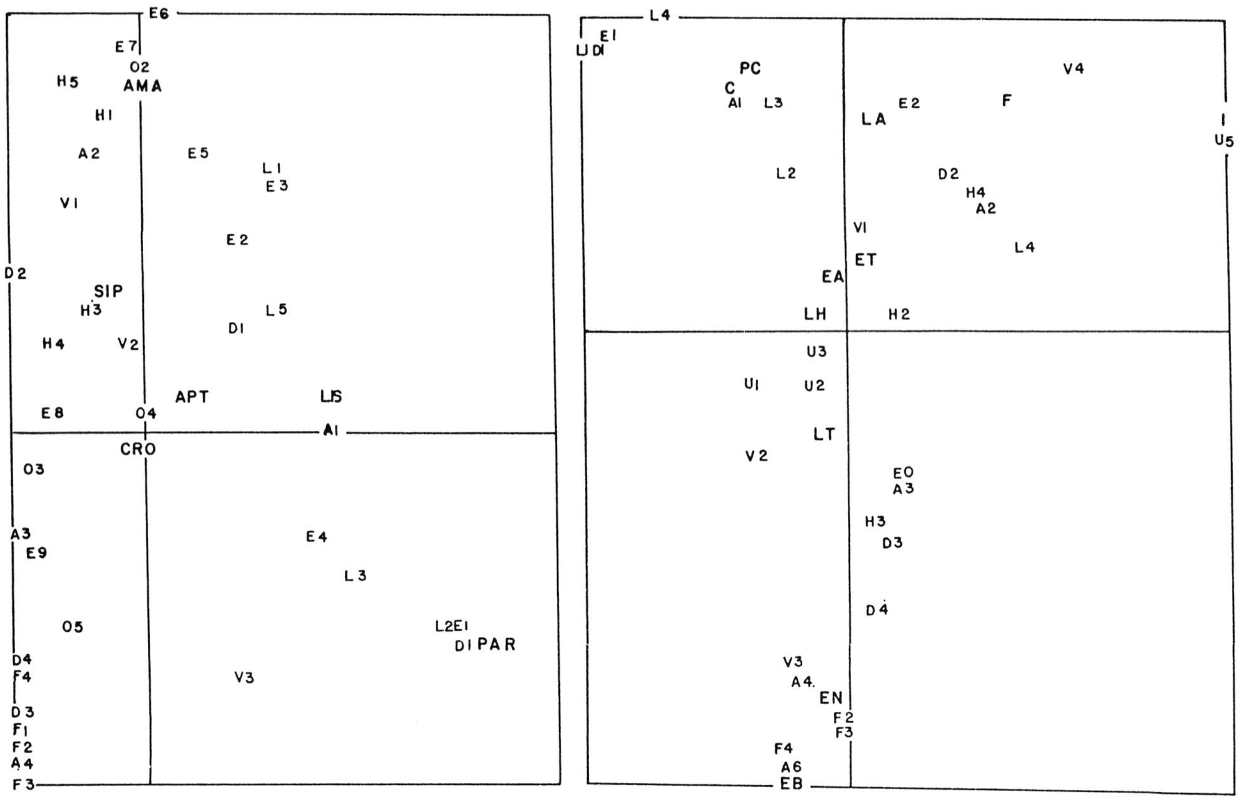

Fig. 1a,b. Graphical representation of Diptera-environment relationships (**a**) and Oligochaeta-environment relationships (**b**). Species codes are given in Tables 1 and 2 espectively. **a** *A1* to *A4* increasing elevation above sea-level; *D1* to *D4* increasing distance from the sea-shore; *V1* to *V3* decreasing cover abundance; *L1* to *L5* shore type; *H1* to *H5* grass type; *F1* to *F5* fell-field type; *E1* seaweed wracks; *E2* Elephant Seal wallows; *E3* King Penguin rookeries; *E4* Macaroni Penguin rookeries; *E5* Gentoo Penguin rookeries; *E6* King Penguin moulting areas; *E7* Albatross and Petrel nests; *E8* Petrel burrows; *E9* other types of organic enrichment; *O1* seaweeds; *O2* droppings; *O3* regurgitations; *O4* carcasses; *O5* other forms of organic enrichment. **b** *A1* to *A6* increasing elevation above sea-level; *D1* to *D4* increasing distance from sea-shore; *V1* to *V4* decreasing cover abundance; *L1* to *L4* shore type; *H2* to *H4* grass type; *F2* to *F4* fell-field type; *U1* to *U5* increasing soil moisture; *E0* no organic enrichment; *E1* rookeries; *E2* other forms of organic enrichment

4 Diptera

Paractora dreuxi is a shore-dwelling species common among pebbles and is observed exclusively at high-water marks in decaying seaweed. *Listriomastax litorea* may also occur on decaying seaweed, but it is more typical in the coastal rookeries of Gentoo Penguins, and the moulting areas of King Penguins. *Amalopteryx maritima* is found in the inland grass zone, wherever birds and mammals manure the phanerogamic vegetation. The Sphaeroceridae (*Siphlopteryx, Anatalanta*) show a wider distribution and occur wherever there is organic enrichment on Possession Island (Tréhen and Vernon 1982) and are the only Diptera able to survive on the fell-field above 400 m altitude, feeding on bird carcasses and regurgitates of Georgian Divingpetrels, and dwelling in petrel burrows. The two *Anatalanta* species occur from sea-level up to 600 m and are necrophagous, coprophagous, and phycophagous. It is therefore very difficult to discriminate between *A. aptera* and *A. crozetensis*, but they are, however, good biological indicators of the presence of any kind of manuring in Sub-Antarctic situations (Vernon 1981).

5 Oligochaeta

Two species are confined to the littoral zone. *Microscolex crozetensis* occurs in spindrift-influenced areas, on low, vegetation-covered coasts and in penguin rookeries, and also at higher altitudes (Fig. 2c) where spindrift occurs, which seems to indicate that salinity plays an important part in the distribution of this species. *Phreodrilus crozetensis* occurs on shingle beaches together with *Paractora dreuxi*, and both species feed at the high-water mark. It is also fairly common in rookeries, around wallowing areas of Elephant Seals, in grasslands surrounding the nests of albatrosses, and in the burrows of petrels. It is generally found wherever organic deposits can be observed and, like *Anatalanta*, *P. crozentensis* probably plays an important role in the decomposition of animal remains. *Microscolex e. nigra* and *M. e. bicolor* occur strictly in the fell-field zone, where the soil is mainly coarse and well-drained, and has a discontinuous vegetation cover. These two species are distinguished by their pigmentation pattern: *bicolor* has no ventral pigmentation and it occurs only above 300 m (Fig. 2b), usually in

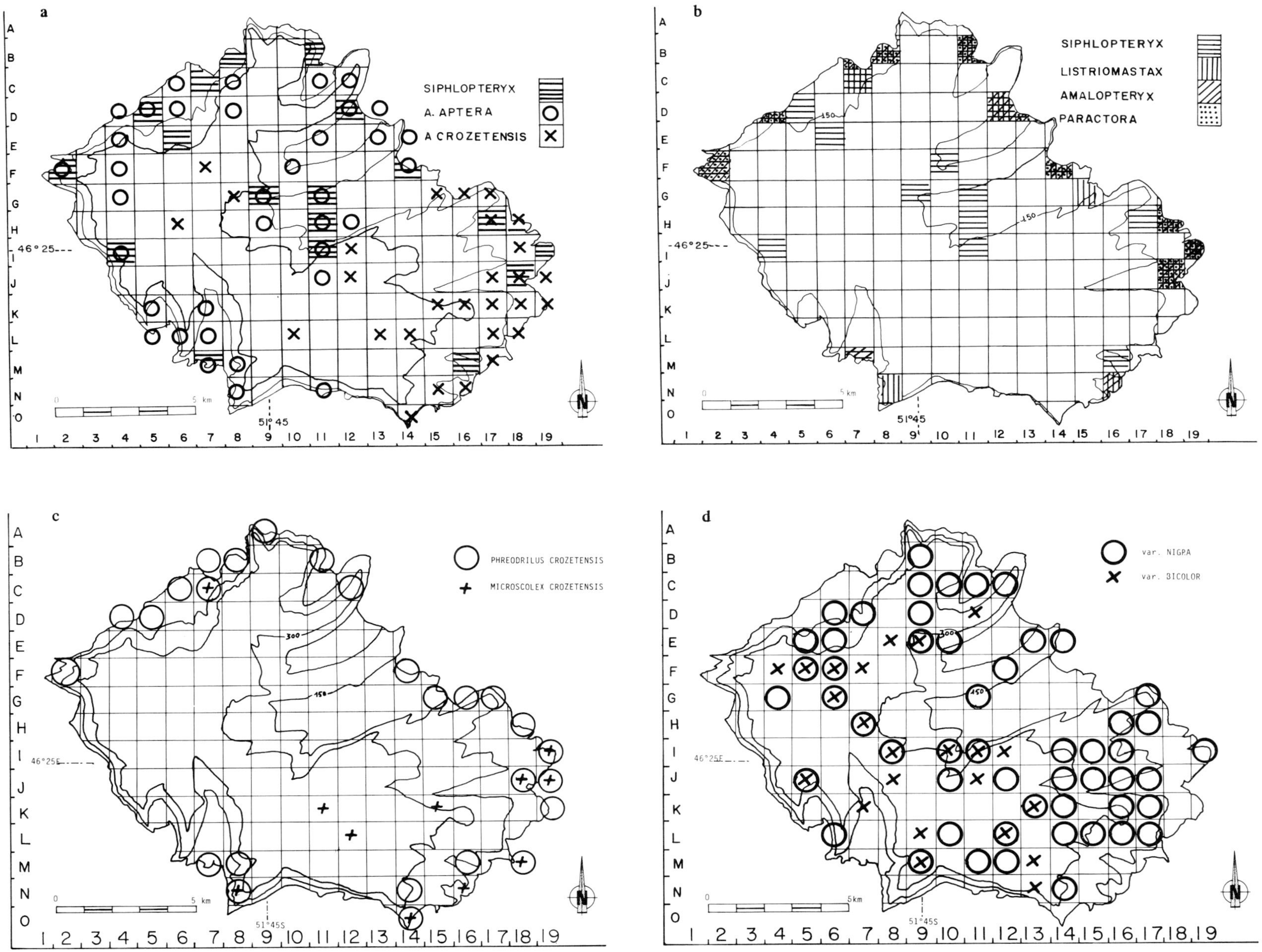

Fig. 2a–d. Distribution of Diptera brachycera and Oligochaeta on Possession Island. **a** Sphaeroceridae. **b** Other families including *Siphlopteryx antarctica*. **c** Sea-shore-dependent Oligochaeta. **d** Oligochaeta of the fell-field

Poa cookii tufts and rolling mosses, whereas *nigra* colonizes all types of fell-field. *Microscolex frenoti* is an aquatic spe cies, occurring among weeds and mosses in lakes and rivers or in flooded grassland areas.

It is more difficult to identify the inter-relationships between the five other species of earthworms (EA, ET, LT, LH, LA in the centre of Fig. 1b), based on the ecological factors considered here. Further study of the micro-distribution of the earthworms and of the structures of these communities is necessary.

5.1 Oligochaeta-Environment Inter-Relationships

During January 1983 observations were made along a transect established over the large, peaty bottom of a glacial valley (Vallée des Branloires), and divided into nine zones according to micro-topography, water-content of the soil, and vegetation (Fig. 3). Preliminary observations on one soil extraction (200 × 200 mm) per zone were made, except for Zone III where three samples were extracted.

Two main earthworm communities were observed: (a) Zones I to VI, where *M. enzenspergeri alba* and *M. luykeni halophila* were encountered and (b) Zones VII to IX, where only *M. e. typique* and *M. frenoti nov.* sp. were present. The pH of circulating water averaged 4.10 in the first community (Zones I–VI) and 5.10 in the second community (Zones VII to IX). *Agrostis magellanica* and several species of mosses form the vegetation of the first community, except in Zone III, where *Blechnum penna-marina* is dominant and earthworms are completely absent. Zone VII includes large boulders covered with *Acaena magellanica* and liverworts. The vegetation of Zone VIII (*A. magellanica* and *Juncus pusillus*) is scattered over a sandy substratum. The level of ground-water is almost flush with the soil surface in Zone I, but the water content of the top layer of the soil, where earthworms live, decreased progressively from Zones II to IV, and increased again from Zones V to IX. This factor seems important with regard to the density and distribution of earthworms.

5.2 Colonization of a Penguin Carcass by Diptera

Carcasses may be considered as temporary habitats. Their presence in a given area is random both in time and space. Three Diptera species (*Anatalanta crozetensis, Siphlopteryx antarctica*, and *Amalopteryx maritima*), respectively apterous, sub-apterous, and stenopterous, occur in Gentoo Penguin rookeries. The colonization patterns of these three species on the decaying carcass of a Gentoo Penguin chick (body mass = 2.1 kg) were followed from 18 February to 19 March, 1980. On 18 February, the carcass was placed under a 300 × 300 mm board. Pitfall traps (θ = 60 mm) filled with diluted teepol were set around the bait. Four traps were placed in a row radiating out in a north, south, east, and west direction from the carcass. The traps of each set were numbered from 1 to 4 according to their distance from the carcass (i.e., 10, 30, 100, and 300 cm). A control set of four traps was placed 20 m from the carcass of evaluate the density and activity of Diptera without a trophic substratum. Catches were collected at 5 day intervals during 1 month (R1–R6). The results showed that colonizing activity (Heydeman 1961, in Nentwig 1982) fluctuated greatly from one species to another. Compared with the first catch (R0), the third catch (R3) was 18.5 × as great for *A. crozetensis*, 1.8 × as great for *S. Antarctica*, and 2.0 × as great for *A. maritima*. The catch patterns for *S. antarctica* and *A. crozetensis* dif-

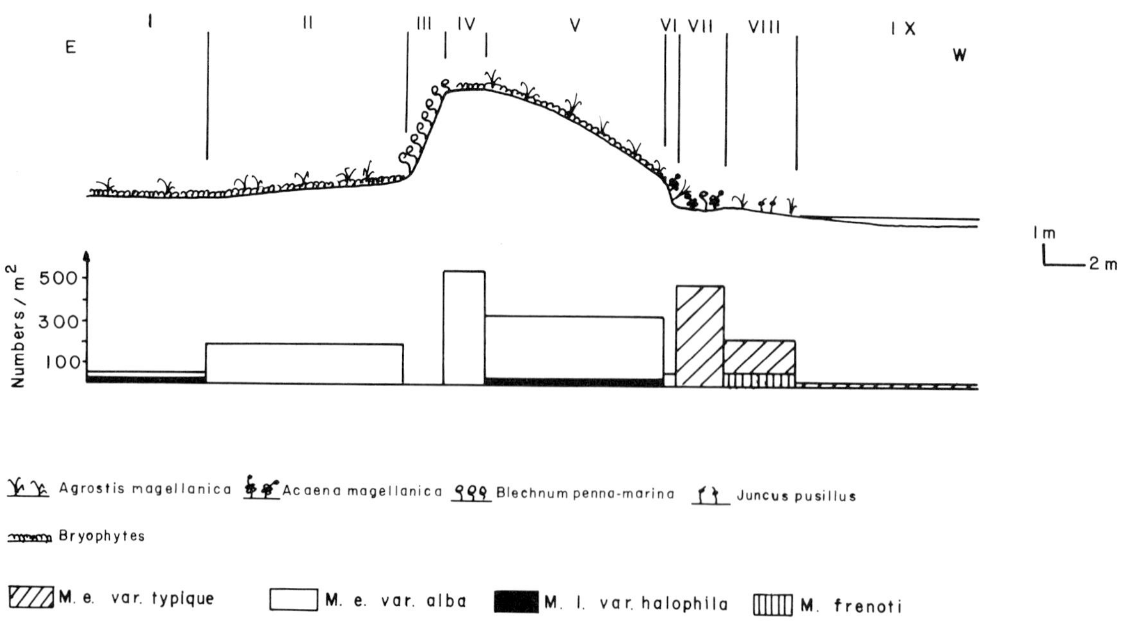

Fig. 3. Micro-distribution of Oligochaeta on a transect across Vallée des Branloires at Possession Island

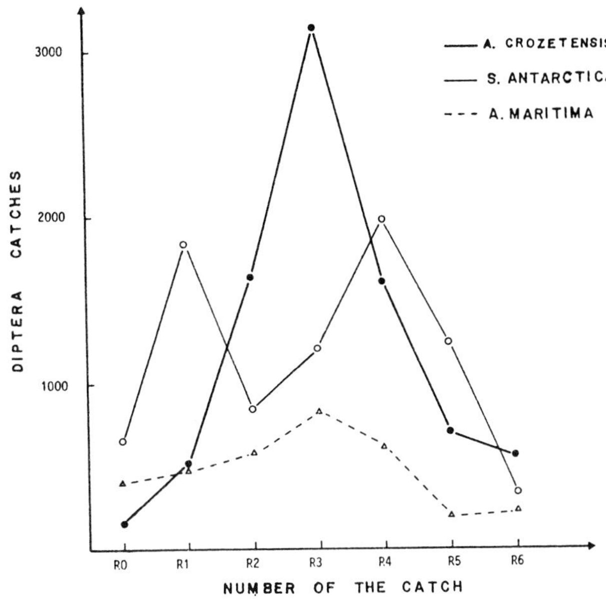

Fig. 4. Diptera catches at 5-day intervals (*R0* first catch; *R6* catch at 30 days after *R0*) in a penguin-manured study area

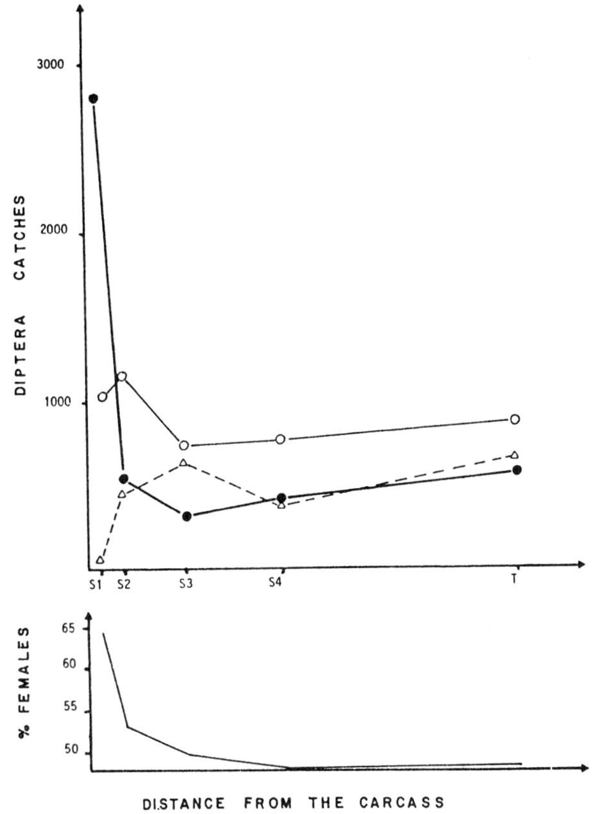

Fig. 5. Change in numbers of diptera caught at different distances from a decaying penguin carcass (*S1* 10 cm; *S2* 30 cm; *S3* 100 cm; *S4* 300 cm; *T* 20 m). *Top graph* represents total numbers, *bottom graph* represents proportion of females of *A. crozetensis*. See Fig. 4 for species' names

fered. The catch curve for adult *A. crozetensis* was symmetrical (Fig. 4), and the decrease in catches corresponded to an increased occurrence of larvae in the carcass. This kind of relationship between adults and larvae has also been found in other communities of necrophagous Diptera (Vernon 1983).

Some 9,200 Diptera were caught around the bait, but at different distances for each species. Approximately 65% of all catches (82% of all *A. crozetensis*, 59% of all *S. antarctica*, and only 39% of all *A. maritima*) were made within 300 mm of the bait (Fig. 5). Each catch peak was situated at a specific distance from the bait: S1 at 100 mm for *A. crozetensis*; S2 at 300 mm for *S. antarctica*; and S3 at 1.0 m for *A. maritima*. The latter species was never found on the carcass, and a comparison of capture frequencies showed that the attractive effect of the bait decreased for distances > 3 m, but may be different for males and females. *Anatalanta crozetensis* females were much more abundant near the bait, whereas the percentage of males increased at 1 m and at greater distances the sex-ratio tended towards parity (Fig. 5). The presence of the carcass clearly biased the spatial distribution of these species. It attracted *A. crozetensis* and *S. antarctica* and repulsed *A. maritima* during the first weeks. However, these biases may be modified by seasonal factors which influence the biological activity of these insects, as well as the probability of occurrence of vertebrate carcasses.

To examine some of these factors in greater detail, a population of *A. crozetensis* was monitored for 9 months, from February to October 1982. The study was conducted in a fenced area of 5 × 5 m, at an elevation of 126 m, in which live King Penguins were kept during February and March 1982 for physiological experiments. The pebbles used to secure the wire netting at ground level formed a suitable biotope for the insect community associated with the Penguins. This community included *A. crozetensis* and *S. antarctica* (Fig. 6). Insects among these pebbles were caught and marked with a different coloured metallic paint for each side of the enclosure where they were caught. Repeated counts were made throughout the period. The Penguins were released in April. Results for *A. crozetensis* are given in Table 3. Values are calculated from observed data according to Schumacher-Eschmeyer's formula (in Barbault 1981). The results indicate a significant decrease in numbers begining in May, and the percentage of females in the population increased during the winter months. The decrease in numbers in May is probably over-estimated, because some 335 flies (208 < N < 1,000) were attracted to a Wandering Albatross, *Diomedea exulans*, carcass 7 m away, and it is likely that many of these insects came from our enclosure.

The main factor influencing the decrease from May (Sub-Antarctic autumn) is probably related to the decrease in temperature during autumn and winter. Below 7 °C, insect emergence decreases and larval growth slows down (Tréhen 1982). The increased percentage of females during autumn and winter may be due either to a higher mortality or a higher mobility in the males. They were more widely dispersed from the shelter of the pebbles, whereas eggs, nymphs, and pupae were found only under the pebbles. Recaptures

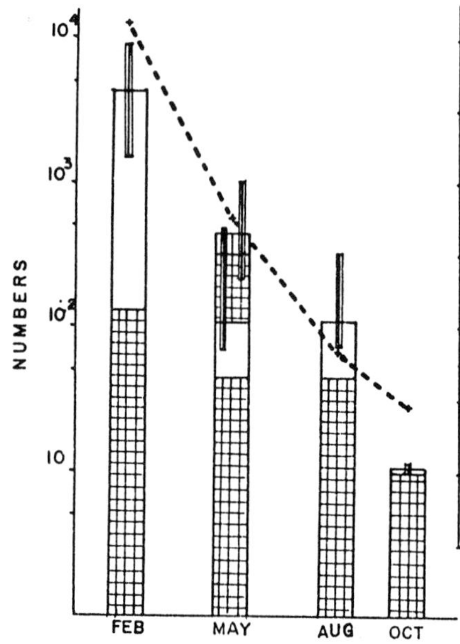

Fig. 6. Seasonal variation in numbers (*hatching* represents females, *clear* represents males) of *A. crozetensis* in relation to mean temperture (*broken line*)

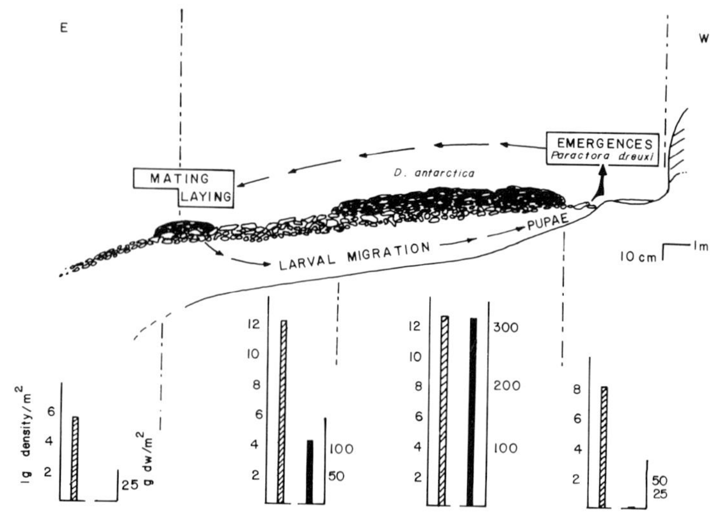

Fig. 7. Sea-weed (*Paractora* – *Phreodrilus*) transect at "Crique du Sphinx"; with earthworm density and biomass estimates at different sites on the beach

of marked individuals showed that 87% occurred at the site of marking. However, males were more frequent in catches made away from the marking sites, and formed 83% of these catches. It seems that during winter the more sedentary females are less susceptible to mortality than the more mobile males. It is possible that females move towards protected egg-laying sites while males, in search of sexual partners, disperse away from these sites where they are less protected from the elements and therefore suffer a higher mortality. Observations on *A. aptera* at Kerguelen (De Blignieres and Tréhen 1982) show that in this species males are also more mobile than females.

5.3 Model of the Durvillea-Diptera-Oligochaeta Ecosystem

The producer compartment in this model comprises seaweed wracks (*Durvillea antarctica*) decaying on pebbles just above the tidal zone on a beach called "Crique du Sphinx" (Fig. 7). The insect community, described previously by Tréhen and Vernon (1982), comprises only a small number of species. One of the Diptera, *Paractora dreuxi*, is a micropterous, mobile species able to colonize seaweed immediately after its deposition on the beach. Dead seaweed biomass can sometimes be very high (> 5 kg m^{-2}) but fluctuates very quickly and randomly. The biomass of *P. dreuxi*, the first colonizer, can reach 90 g (dry wt.) m^{-2} but also fluctuates in space and time with the fluctuations in seaweed.

The oligochaete *Phreodrilus crozetensis* does not colonize the actual seaweed deposits, but occurs in the products of its decay which flow seawards between the pebbles of the beach. Earthworm biomass is the highest ever recorded for terrestrial oligochaeta (density = 250,000 ind. m^2, biomass = 320 g (dry wt.) m^{-2}). Biomass fluctuates as the seaweed deposits fluctuate, but they are also controlled by the high reproductive potential of this sedentary earthworm species.

Table 3. Seasonal variation of numbers of individuals in a population of *Anatalanta crozetensis* on Possession Island

	February	May	August	October
Total number	1,457	119	122	18
Standard error 95% confidence except May: 90%	1 162 < N < 1 923	71 < N < 500	77 < N < 333	15 < N < 28
Number of females	N = 163 (11.18%)	45 (37.8%)	46 (37.7%)	12 (66%)
Standard error (95%)	66 < N < 260	26 < N < 166	35 < N < 83	10 < N < 16
T °C monthly average	8.2 °C	5.5 °C	3.7 °C	2.9 °C

Cocoon production is synchronized with arrivals of seaweed wrack, with a delay for sexual maturation. The earthworm *P. crozetensis* is a secondary colonizer and is probably also a prey species for the last instars of the fly *P. dreuxi*, and for other invertebrates (Acarina, Gamasida).

6 Conclusion

Possession Island has a poor dipteran fauna compared with continental communities. On the other hand, there is a relatively rich earthworm community on the island. This difference in species richness may be due to different colonizing strategies and to the fact that the oligochaetes occupy a more stable substratum.

Edaphic factors are dominant with regard to the distribution of Oligochaeta but altitude, vegetation type and different types of manuring play important roles in their distribution. The species present on the island occur in different biotopes ranging from the shore (*Microscolex crozetensis*) all the way up to the fell-field above 300 m (*M. enzenspergeri bicolor*). It is also shown that in the case of *M. frenoti nov.* sp., soil-water content and the pH of soil water strongly influence its occurrence.

Diptera communities on the island are generally unstable, with eight species having relatively specialized trophic forms: necrophagous, phycophagous, saprophagous, keratophagous etc. The sequential occurrence of *Anatalanta crozetensis* and *Amalopteryx maritima* on the same substratum (carcass) is influenced by these specializations as well as a difference in rate of larval development on the substratum. Males show a greater winter mortality than females. The latter lay eggs through winter and spring, so that larval development can occur in summer when temperatures and substratum availability are more favourable. Species composition of communities is determined mainly by the origin, form and periodicity of substrata.

One of the more important substrates is decaying seaweed which accumulates at high-water marks. *Paractora dreuxi* (Diptera) and *Phreodrilus crozetensis* (Oligochaeta) are the two main species associated with these. Their population sizes and dynamics are dependant on the random fluctuations of the seaweed deposits and each of these species has a specific strategy to cope with these fluctuations: *P. dreuxi* has a strong colonizing ability; *P. crozetensis* is strongly competitive. These two strategies are the two main factors influencing biological equilibrium between Oligochaeta and Diptera populations in terrestrial subantarctic regions.

References

Barbault R (1981) Ecologie des populations et des peuplements. Collection Biologie et Maitrise, Masson (ed), Paris:1–208

Benham WB (1903) The geographical distribution of earthworms and the paleogeography of the antarctic region. In: 9th Rep Meet Aust Assoc Adv Sci, Sect D, Hobart

Blignieres de X, Tréhen P (1982) Rythme circadien et phototaxie chez Anatalanta aptera Eaton, Diptère Sphaeroceridae des Iles Kerguelen. CNFRA 51:157–166

Bouché MB (1982) Les lombriciens (Oligochaeta) des Terres Australes et Antarctiques Francaises. CNFRA 51:175–180

Delettre Y, Tréhen P, Vernon Ph (1982) Premiers éléments pour une cartographie automatique de données écologiques à l'île Crozet. CNFRA 51:526–527

Michaelsen W (1905) Die Oligochaeten der Deutschen Südpolar-Expedition 1901–1903. Bd IX Zool 1:3–55

Nentwig W (1982) Epigeic spiders, their potential prey and competitors: relationship between size and frequency. Oecologia (Berl). 55:130–136

Smith VR (1978) Animal-Plant-Soil nutrient relationship on Marion Island (subantarctic). Oecologia (Berl) 32:239–253

Stephenson J (1930) The Oligochaeta. Clarendon Press, Oxford 1–978

Tetry A (1947) Croisière du Bougainville aux Iles Australes Françaises, XVIII-Oligochètes. Mem Mus Paris, NS 20:101–110

Tréhen P (1982) Cycles de développement et stratégies de la reproduction chez quelques Diptères des Iles Subantarctiques. CNFRA 51:149–156

Tréhen P, Vernon Ph (1982) Peuplement diptérologique d'une île subantarctique: La Possession (46°S, 51°E, Iles Crozet). Rev Ecol Biol Sol 19(1):105–120

Vernon Ph (1981) Peuplement Diptérologiques des substrats enrichis en milieu insulaire subantarctique (Iles Crozet). Etude des Sphaeroceridae du genre Anatalanta. Thèse de 3° cycle, Université de Rennes 1:1–115

Vernon Ph (1983) Décomposition en période estivale de cadavres de micro-mammiféres dans une formation pionniére de la lande armoricaine. Rev Ecol Biol Sol 20(3):317–347

Arthropod Interactions in an Antarctic Terrestrial Community

W. BLOCK[1]

Summary. A functional analysis of a low-diversity terrestrial arthropod community in the maritime Antarctic is made, especially in relation to the flux of energy and predation. Although net primary production of the main plants – mosses is 392–409 g m^2 yr^{-1} (dw), most arthropods feed on epiphytic algae, micro-flora and detritus. Invertebrates other than Protozoa contribute < 1% to total heterotroph respiration. Invertebrate carnivores utilize very little of primary consumer production (< 0.5%). The abundant collembolan herbivore *Cryptopygus antarcticus* alters its energy budget depending on food quality and temperature, but its grazing may have a significant impact only at algal dominated sites. The single arthropod predator *Gamasellus racovitzai* feeds on three species of arthropod, the principal one during summer being *C. antarcticus*, which forms the greatest proportion of available prey. It is non-selective with regard to prey when feeding during summer, and does not feed extensively in winter. It is unlikely that such a predator ever has a shortage of food in bryophyte habitats. It is concluded that: the principal grazing chain in such communities is ectothermic, being based entirely on invertebrates and regulated by algae and micro-flora production; Antarctic and temperate terrestrial systems differ in the pathways of energy flux with contrasting ecological efficiencies for both herbivores and carnivores; on current evidence the functional position of invertebrate predators such as *G. racovitzai* appears anomalous; and, several correlates of adversity or A-selection are found in Antarctic communities, which may preclude their recovery from perturbations.

1 Introduction

Notwithstanding the variation in tundra communities throughout the world, those of the Arctic and Antarctic regions have some features in common and many that are dissimilar. Their commonality resides in the limitation by their low temperature environments of the rates of most biological processes, the relatively short season for biological activity and the frequent occurrence of permafrost. The major differences are the rates of net primary production, especially the cryptogam component, their micro-floral biomasses, the amount of herbivory, the composition of the invertebrate community and utilization of the annual detritus input. In particular, Antarctic terrestrial habitats accommodate invertebrate communities of relatively low species diversity, and this is especially the case for arthropods, the majority of which are Acari and Collembola. Such a community composition with few interacting species, but often comprising large numbers of individuals (D.G. Goddard 1979; Block 1982), allows a functional analysis which is impossible at present elsewhere. Most other terrestrial communities are richer both in species and population numbers, and, hence, present too complex a suite of interactions for such analysis. Antarctic communities therefore provide an unrivalled opportunity for an attempt to understand some of the fundamental ecological processes governing species and energy interactions, and to test general ecological theory.

This paper is a first attempt at a functional analysis of a low-diversity terrestrial arthropod community at Signy Island in the maritime Antarctic (Holdgate 1977). As part of the analysis, the flux of energy through key species is examined, together with the role of predation in the community. Most of the basic data are derived for bryophyte-dominated communities (e.g., the Signy Island Reference Sites (Tilbrook 1973), but observations comparative with other terrestrial habitats such as fellfields are made where appropriate.

2 Community and Trophic Structure

In terrestrial communities of the maritime Antarctic, organic matter resides in three compartments; autotrophs (primary producers), heterotrophs (secondary producers) and dead organic matter. The primary producers are chiefly mosses, liverworts and lichens (consumed as dead material by heterotrophs) together with both uni- and multi-cellular algae (consumed as live material by heterotrophs). The secondary producers are comprised of saprotrophs (bacteria, yeasts, and filamentous fungi), primary consumers (protozoa, rotifers, tardigrades, nematodes, mites, and springtails), and secondary consumers (three carnivorous species: a nematode, *Coomansus gerlachei*, a tardigrade, *Macrobiotus furciger*, and a mite, *Gamasellus racovitzai*). The dead organic matter pool receives contributions from both auto- and heterotrophic components, but consists mainly of dead plant material.

The arthropods of such communities are all micro-forms (135–1,250 μm in body size), do not include higher insects and consist entirely of Acari (six to nine species) and Collembola (three to four species) (Tilbrook 1967; D.G. God-

1 British Antarctic Survey, NERC, High Cross, Madingley Road, Cambridge CB3 OET, United Kingdom

Antarctic Nutrient Cycles and Food Webs
(ed. by W.R. Siegfried, P.R. Condy, and R.M. Laws)

Table 1. The diet of terrestrial arthropods at Signy Island

Taxon	Algae	Mosses, liverworts lichens	Micro-flora	Dead organic matter	Micro-arthro-pods
Collembola					
Cryptopygus antarcticus	⊕	–	⊕	+	–
Parisotoma octo-oculata	+	–	⊕	⊕	–
Acari					
Alaskozetes antarcticus	⊕	⊕	+	+	–
Eupodes minutus	⊕	–	+	–	–
Ereynetes macquari-ensis	⊕	–	+	–	–
Stereotydeus villosus	⊕	–	–	–	–
Nanorchestes antarcticus	⊕	–	–	–	–
Tydeus tilbrooki	⊕	+	+	–	–
Gamasellus racovitzai	–	–	–	–	⊕

Note: (–) not observed in gut contents; (+) low proportion (< 40%) of gut contents; and, (⊕) high proportion (> 50%) of gut contents

dard 1979; Usher and Booth 1984). A generalized synopsis of the diet of the main species of arthropods is presented in Table 1 (after D.G. Goddard 1982; Burn 1984). Four principal food resources can be distinguished. The dominant foods for such micro-arthropods are algae, both micro- and macro-forms, which comprise a highly diverse and easily assimilated food resource. Broady (1979) recorded upwards of 162 species of algae from a range (122) of terrestrial habitats at Signy Island. A second-order resource is the micro-flora, with yeasts and fungi being the main items. The third food component is dead organic matter, and the fourth resource is entirely animal material as prey for the single arthropod predator. Mosses, liverworts and lichens are not consumed by the arthropods to any significant extent (< 0.1% of their net primary production being utilized by the primary consumers according to Davis (1981). Hence, under suitable environmental conditions, the majority of bryophyte and lichen production passes directly into the dead organic matter pool and accumulates as peat, a process which is characteristic of the maritime Antarctic. Thus algae, together with the micro-flora and detritus, provide the main fraction of the diet of many of the primary consumers, especially the micro-arthropods.

Live weight biomass of micro-algae growing epiphytically on a variety of mosses in summer ranged from ca. 4 to 30 g m^{-2} (Broady 1975), equivalent to a dry wt. biomass of 1.29–8.4 g m^{-2}. Net production estimates for algae were in the range 13–15 g m^{-2} yr^{-1} dry wt. (Davis 1981). These data contrast markedly with average values in the maritime Antarctic for total moss standing crop of these communities of 180–604 g m^{-2} dry wt. and a calculated net production of 392–409 g m^{-2} yr^{-1} dry wt.

3 Energy Flux

Few functional analyses of Antarctic terrestrial communities have been made (Smith 1977 in the Sub-Antarctic; Davis 1981 in the maritime Antarctic). Figure 1 shows a generalized energy flow diagram for the bryophyte-dominated ecosystems typical of the maritime Antarctic. Although the majority of the plant biomass is made up of mosses and liverworts (and sometimes lichens), the importance of algae, and to a lesser extent the micro-flora, in the functioning of such systems must be stressed. Davis (1981) estimated that in moss turf and carpet communities the consumption of algae by primary consumers (invertebrates) exceeded algal net production by between 8 and 77 times, respectively. In the same instance, the dietary demands of the invertebrates are largely transferred to the heterotrophic micro-flora with between 40 and 80% of micro-floral production being utilized in this way.

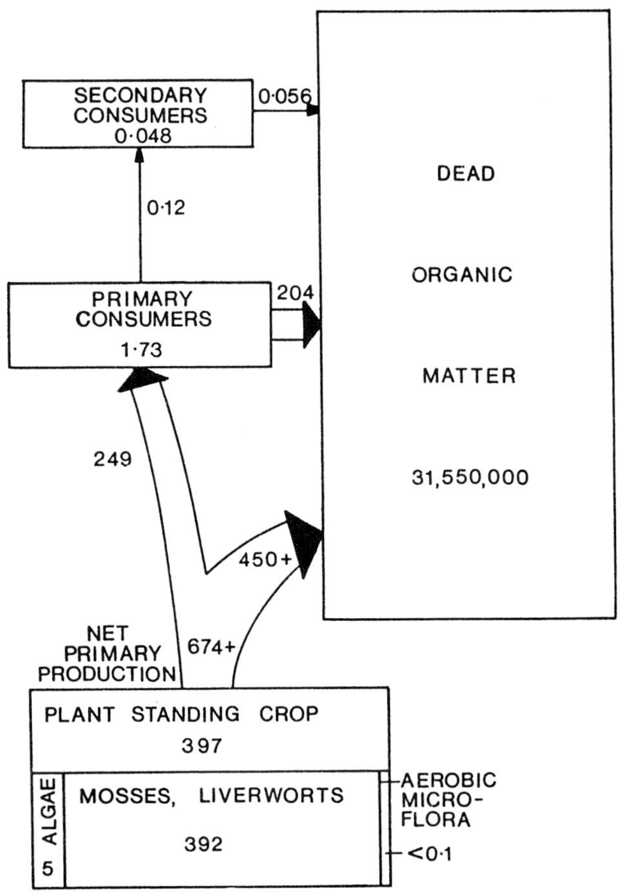

Fig. 1. Generalized diagram of energy flux in bryophyte-dominated ecosystems of the maritime Antarctic (After Davis 1981). Units are g m² yr^{-1} dry wt.

Since < 0.1% of the net primary production (bryophytes) is eaten by primary consumers, the consumption of algae by some invertebrate species is the only significant herbivory in such moss-dominated systems. The saprotrophs are functionally the most important heterotrophic organisms and may control decomposition processes to a large extent. Their dominant role is reflected in the proportion of total heterotrophic respiration: the micro-flora respire 81%-89% compared to 11%-19% by the Protozoa, and 0.4%-0.5% by the remainder of the invertebrates (Davis 1981). The reduced role of the invertebrates other than the Protozoa in such Antarctic ecosystems contrast with Arctic and temperate upland studies, where they contribute between 1% and 23% of the total heterotroph respiration (Whitfield 1977; Coulson and Whittaker 1978). However, Antarctic communities are greatly impoverished in respect of invertebrates compared to those of Arctic and upland systems, in which insect larvae, lumbricids, enchytraeids, etc., are often found. The reduced invertebrate component, together with low microbial activity, may contribute to the slow decomposition rates measured in Antarctic bryophyte communities (Davis 1980).

The supposed poor quality of bryophytes as food resources for many terrestrial invertebrates in conjunction with the overall limiting effects of their low temperature environment may account for their small part in energy flux. The low efficiency of utilization of primary consumer production by invertebrate carnivores is less easy to understand. Efficiencies of 0.3%-0.5% [(carnivore consumption/primary consumer production) × 100] have been calculated (Davis 1981), but these are very low compared with Arctic systems where efficiencies of 12%-33% have been estimated (Whitfield 1977; MacLean 1980).

It is appropriate to examine in detail the processes of herbivory and predation in Antarctic ecosystems in an attempt to clarify these anomalies. For this, attention will be confined to recent work on selected species of micro-arthropods at Signy Island.

4 Herbivory

The isotomid collembolan, *Cryptopygus antarcticus*, is very abundant in communities throughout the maritime Antarctic. Field populations in summer ranged from 19,000 (moss turf) through 39,000 (fellfield) to 180,000 (algal site) individuals m^{-2} at Signy Island (Burn 1984) with annual mean values of ca. 49,420 (moss turf) individuals m^{-2} (Block 1982). It feeds on unicellular green algae, dead moss material and fungal hyphae (in order of preference). Using a radioisotope C^{14} label, Burn (1984) measured feeding rates and individual growth rates on these three common field foods for two size classes of *C. antarcticus*. The interval between moults was similar on all foods for both sizes, and the dry weight increase per moult for the larger animals was similar on each food type. However, the growth rates of the smaller collembolans fed on fungi were significantly slower when compared to those fed on algae and homogenized moss peat.

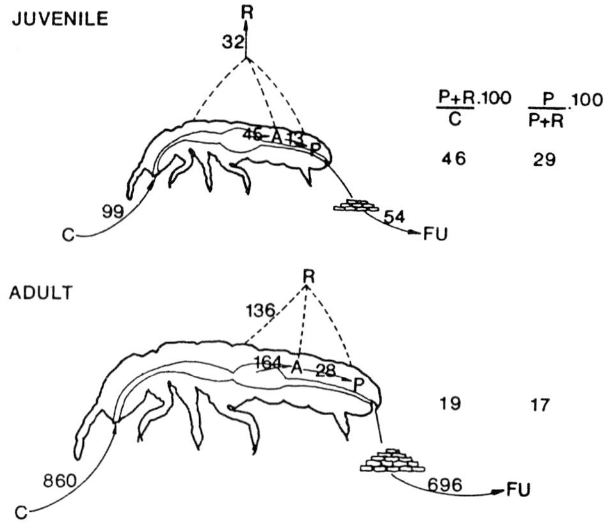

Fig. 2. A schematic representation of the energy budget for juvenile and adult individuals of *Cryptopygus antarcticus* when fed on algae at 5 °C, together with assimilation and production efficiencies. (After Burn 1984). *C* consumption, *A* assimilation, *P* production, *R* respiration, and *FU* rejecta. Units are J × 10^3 per individual per moult interval

A schematic representation of the energy budgets developed for both young and mature individuals of *C. antarcticus* is given in Fig. 2. At 5 °C, this species consumed 24%-50% of its body weight on moss peat, but only 7%-13% on algal food. Of particular note in the energy budgets are differences in assimilation [100(P+R)C]* and production [100P/(P+R)]* efficiencies between young and mature individuals. In the mature animal, both these efficiencies are < 50% those of the young instars. The range of both these energetic efficiencies is low compared to data for temperate species. Both age classes assimilated algae better than moss peat, although individuals fed more rapidly on the latter, which had a larger indigestible component, thereby compensating for a lowered assimilation efficiency and achieving similar rates of growth on both substrata. The energy balance of *C. antarcticus* was influenced further by temperature over the range 0°-10 °C (Fig. 3). The ingestion rates suggest cold adaptation by maintenance of a positive energy balance at low temperatures. In turn, feeding rates changed rapidly over 0°-5 °C, corresponding with previously reported changes in metabolic rate (Block and Tilbrook 1975).

In terms of population energetics, it was estimated that *C. antarcticus* would consume between 7 (moss turf) and 26 (algae) g m^2 yr^{-1} dry wt. (Burn 1984). This has an almost negligible effect on the net primary production of the bryophyte communities, but feeding of this species may have a more influential role in algae-dominated sites, where food

* P = production; R = respiration; C = consumption

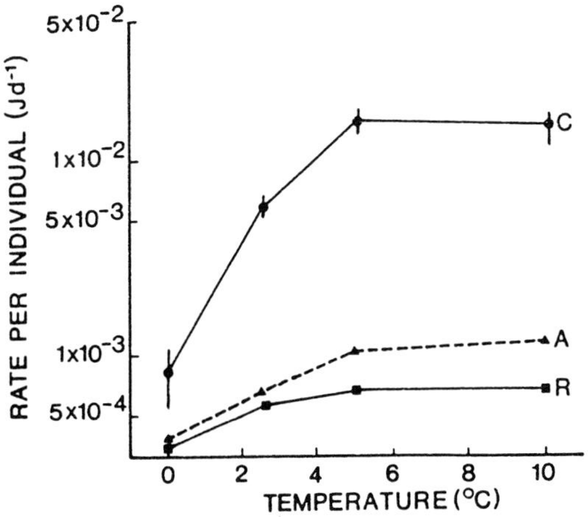

Fig. 3. Influence of temperature on the energy budget of juvenile *Cryptopygus antarcticus* fed on moss peat (Burn 1984). *C* consumption ± S.E.M., *A* assimilation, and *R* respiration

consumption may reach 135 g m^2 yr^{-1} dry wt. and plant production may be limited locally.

This detailed approach to the feeding ecology and energetics of an abundant species confirms the general low level of herbivory in such situations, and the importance of algae, when available, as a nutritious and comparatively readily assimilated, energy-rich substratum. The influence of food quality on individual energy budgets and growth rates is profound, and will be reflected in the population dynamics of *C. antarcticus* (Block 1982).

5 Predation

The sole arthropod predator in these simple communities is the mesostigmatid mite, *Gamasellus racovitzai*, which has been observed to feed on eight monospecific prey arthropods. Recent studies have enabled the identification of the three largest and most abundant prey items in extracts of individual predators by their characteristic esterase band patterns, using polyacrylamide gel electrophoresis (Lister, to be publ.). The data presented here are qualitative and based on the occurrence of prey traces. Food traces remain identifiable in the predator gut for up to 19 d at 5 °C.

Ignoring minor between-site differences, there are clear seasonal changes in the proportion of the predator population with detectable food or prey traces. Only about 18% of the population at Signy Island is non-feeding during summer (November–March), whereas no evidence of feeding was found in *G. racovitzai* throughout the winter (May–September). The proportions of two of the three prey species in *G. racovitzai* diet occur with the same frequency as their field abundance in the sites examined (Fig. 4). However, the mite *Alaskozetes antarcticus* appears to be underrepresented in the diet of the predator, probably as a result of its heavy sclerotization especially in the later nymph and adult stages making predation by a similar sized mite difficult. There is no evidence for active selection by *G. racovitzai* between potential prey species in the field.

There are no data available on predation rates of *G. racovitzai* under experimental or field conditions, but from preliminary observations at Signy Island it seems unlikely that starvation due to prey shortage would ever occur during summer. Comparison of population estimates for both the predator and its main prey species, *C. antarcticus*, shows that over the 2-yr study period, the prey population was, on average, over 100 times more numerous than the predator in a moss turf habitat. *Gamasellus racovitzai* is randomly distributed and generally found in the top 30 mm of the vertical profile, whereas *C. antarcticus* is strongly aggregated and is part of the green moss community in the upper 15 mm of the turf (D.G. Goddard 1979; Usher and Booth, to be published). It appears, therefore, that the predator has a super-abundance of potential food, and its extremely low utilization efficiency of herbivore production is brought about by other factors such as behavioural or physiological constraints.

6 Discussion

The absence of above-ground herbivores and the reduced invertebrate component of the fauna have profound effects

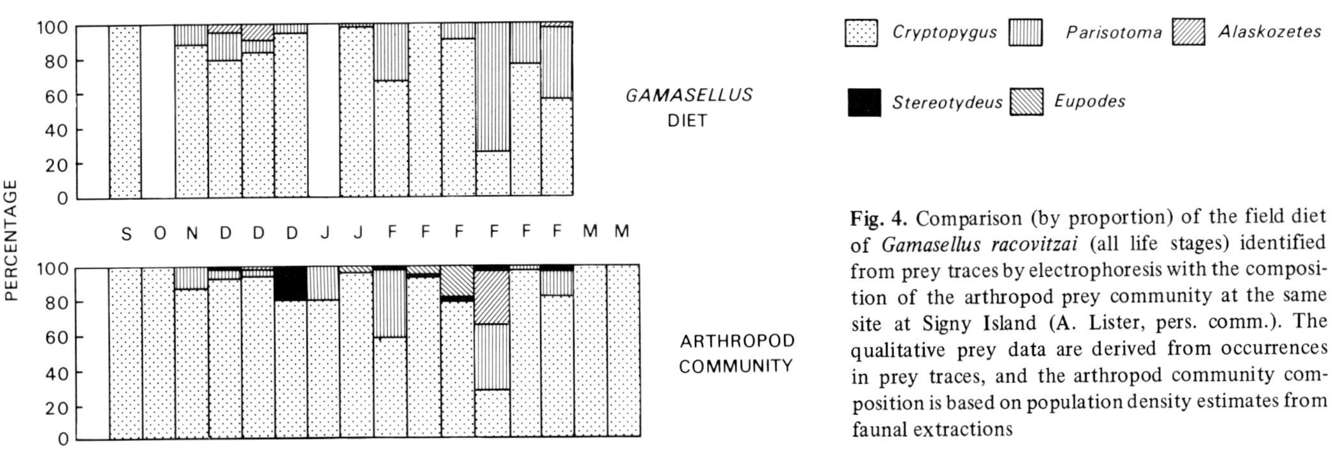

Fig. 4. Comparison (by proportion) of the field diet of *Gamasellus racovitzai* (all life stages) identified from prey traces by electrophoresis with the composition of the arthropod prey community at the same site at Signy Island (A. Lister, pers. comm.). The qualitative prey data are derived from occurrences in prey traces, and the arthropod community composition is based on population density estimates from faunal extractions

on both the structure and functioning of terrestrial ecosystems. It shifts the emphasis of energy flux onto smaller organisms in the below-ground-soil sub-system. This, in turn, greatly enhances the role of the micro-flora, especially in polar ecosystems, and the micro-flora-invertebrate interface is one that warrants further study under these conditions. Mammalian grazers tend to control net primary production and phytomass in communities where they occur. Invertebrates, on the other hand, appear less sensitive to annual variations in primary production, completing their life cycles when environmental conditions permit, and, hence, they may be more efficient at exploiting low temperature habitats. Thus, the principal grazing chain in the maritime Antarctic is ectothermic and invertebrate based.

The ecological efficiencies (Ricklefs 1973) for herbivores in the moss communities studied are 7%–9%, which lie within the range of temperate values (Davis 1981). However, the pathways of energy flux differ between Antarctic and temperate systems. Very little of the net primary production of mosses is eaten by primary consumers directly. Instead, the micro-flora assimilate dead organic matter and convert it to a form more readily metabolized by invertebrates. In this way, the soil fauna annually consume an amount equivalent to ca. 58% of the annual net primary production (after Davis 1981), which is very different from that which occurs in Arctic tundra where exploitation efficiencies are < 1.5% (Whitfield 1977). On the other hand, carnivore efficiency is slight, being 0.3%–0.5% of primary consumer production, which is in contrast to 15%–33% calculated for Arctic tundra. The major regulator of this type of herbivory in the maritime Antarctic is therefore the autotrophic micro-flora (mainly algae) which form one of the main herbivore food sources. Such a functional difference between northern and southern terrestrial systems was postulated by Holdgate (1977), on the basis that less invertebrate material was produced in the maritime Antarctic compared with the prediction of the model of Heal and MacLean (1975).

The position of invertebrate predators in the simple Antarctic communities is not clear. In the unique, single-predator arthropod community at Signy Island, *G. racovitzai* has a broad diversity of potential prey, but apparently exercises no selectivity and the proportions of the three main species in its diet fluctuate with their field abundance. These are characteristics more of a random rather than a prudent predator (Slobodkin 1974). There are no indications of the regulation of any of the herbivore populations by this predator under maritime Antarctic conditions. Invertebrate predators usually exhibit high basal metabolic rates, and therefore might be at a disadvantage in a low temperature environment compared to low metabolic rate herbivores (S.J. Goddard 1979). It is difficult to envisage how an extended life cycle could have evolved in such a predator without the interpolation of a diapause to conserve energy and withstand periods of stress. *Gamasellus racovitzai* has a high basal metabolism compared to the herbivorous and detritivorous mites of the maritime Antarctic (D.G. Goddard 1977); its life cycle is at least 2 yrs in duration, but no diapause has been found.

The idea of the habitat templet was advanced by Southwood (1977), in which habitat favourableness was assumed to be the converse of a stressful or adverse situation, and imparted a durational stability to the system through its predictability. To the concept of r- and K-selection was added adversity or A-selection, which favours the conservation of adaptations to environments which are consistently severe but nevertheless predictable. Several correlates of A-selection (Greenslade 1983) are found in Antarctic terrestrial habitats where relatively simple communities exist in harsh environments, and these include the poor migratory ability of species, extended life cycles and low reproductive rates. In the present analysis, the inability of the invertebrate herbivores to consume bryophytes and the way in which this is circumvented are related features. Furthermore, such habitats are species poor and many of the adaptations are to the physical rather than the biological environment. Thus, environmental stability and predictability allow the species to adapt closely, whilst environmental severity restricts competitors. Interspecific competition appears to be non-existent, and in dietary terms, generalists predominate over specialists. The Antarctic populations appear to remain well below habitat carrying capacity. However, the predicted low reproductive rates of certain species in such communities suggest a limited capacity for recovery after a decrease. The ability of these arthropod communities to adjust to physical perturbations or to counter the introduction of new species, whether accidentally by man (Block et al. 1984) or naturally by colonization, may not be adequate.

Acknowledgements. I appreciate the stimulus provided by the opportunity to study Antarctic systems, and the helpful criticism of many colleagues in the British Antarctic Survey in the preparation of this paper. I am indebted to Drs R.C. Davis, A.J. Burn, and A. Lister for the use of unpublished information, and to my wife May for skilful drawing.

References

Block W (1982) The Signy Island terrestrial reference sites: XIV. Population studies on the Collembola. Br Antarct Surv Bull No 55:33–49

Block W, Tilbrook PJ (1975) Respiration studies on the Antarctic collembolan *Cryptopygus antarcticus*. Oikos 26:15–25

Block W, Burn AJ, Richard K (1984) An insect introduction to the maritime Antarctic. Biol J Linn Soc 22:

Broady PA (1975) Studies on the terrestrial algae of Signy Island, South Orkney Islands. PhD thesis, University of Bristol, UK

Broady PA (1979) The terrestrial algae of Signy Island, South Orkney Islands. Br Antarct Surv Sci Rep No 98:1–117

Burn AJ (1984) Energy partitioning in the Antarctic collembolan *Cryptopygus antarcticus*. Ecol Entomol 9:11–21

Coulson JC, Whittaker JB (1978) Ecology of moorland animals. In: Heal OW, Perkins DF (eds) Production ecology of British moors and montane grasslands. Springer, Berlin Heidelberg New York, pp 52–93 [Ecological Studies 27]

Davis RC (1980) Peat respiration and decomposition in Antarctic terrestrial moss communities. Biol J Linn Soc 14:39–49

Davis RC (1981) Structure and function of two Antarctic terrestrial moss communities. Ecol Monogr 51:125–143

Goddard DG (1977) The Signy Island terrestrial reference sites: VI. Oxygen uptake of *Gamasellus racovitzai* (Trouessart) (Acari: Mesostigmata). Br Antarct Surv Bull No 45:1–11

Goddard DG (1979) The Signy Island terrestrial reference sites: XI. Population studies on the terrestrial Acari. Br Antarct Surv Bull No 48:71–92

Goddard DG (1982) Feeding biology of free-living Acari at Signy Island, South Orkney Islands. Br Antarct Surv Bull No 51:290–293

Goddard SJ (1979) The population metabolism and life history tactics of *Neobisium muscorum* (Leach) (Arachnida: Pseudoscorpiones). Oecologia (Berl) 42:91–105

Greenslade PJM (1983) Adversity selection and the habitat templet. Am Nat 122:352–365

Heal OW, MacLean SF (1975) Comparative productivity in ecosystems – secondary productivity. In: van Dobben WH, Lowe-McConnell RH (eds) Unifying concepts in ecology. Dr W Junk, The Hague, pp 89–108

Holdgate MW (1977) Terrestrial ecosystems in the Antarctic. Philos Trans R Soc Lond B 279:5–25

Lister A (to be publ.) Predation in an Antarctic micro-arthropod community. Proc VI Int Cong Acarology (Edinburgh)

MacLean SF (1980) The detritus-based trophic system. In: Brown J, Miller PC, Tieszen LL, Bunnell FL (eds) An Arctic ecosystem: the coastal tundra at Barrow, Alaska. Dowden, Hutchinson and Ross, Pennsylvania, pp 411–457 [US/IBP Synthesis Series 12]

Ricklefs RE (1973) Ecology. Thomas Nelson, London

Slobodkin LB (1974) Prudent predator does not require group selection. Am Nat 108:665–678

Smith VR (1977) A qualitative description of energy flow and nutrient cycling in the Marion Island terrestrial ecosystem. Pol Rec 18:361–370

Southwood TRE (1977) Habitat, the templet for ecological strategies? J Anim Ecol 46:337–365

Tilbrook PJ (1967) The terrestrial invertebrate fauna of the maritime Antarctic. Philos Trans R Soc Lond B 252:261–278

Tilbrook PJ (1973) The Signy Island terrestrial reference sites: I. An introduction. Br Antarct Surv Bull Nos 33/34:65–76

Usher MB, Booth RG (1984) Arthropod communities in a maritime Antarctic moss-turf habitat: three dimensional distribution of mites and Collembola. J Anim Ecol 53:427–442

Whitfield DWA (1977) Energy budgets and ecological efficiencies on Truelove Lowland. In: Bliss LC (ed) Truelove Lowland, Devon Island, Canada: a high Arctic ecosystem. University of Alberta Press, Edmonton, pp 607–620

The Feeding Ecology of a Decreasing Feral House Cat, *Felis catus*, Population at Marion Island

P.J.J. VAN RENSBURG [1]

Summary. A significant change has occurred in the diet of feral House Cats, *Felis catus*, at Marion Island, concomitant with an artificial reduction in the Cat population. Prior to population-reduction operations, burrowing petrels (Procellariidae) and the House Mouse, *Mus musculus*, constituted the main prey of the Cats. After control operations, the diet of the decreasing Cat population contained an increased proportion of House Mice, and the Kerguelen Petrel, *Pterodroma brevirostris*, no longer featured as a food item. The causes and effects of this change are discussed in terms of the flow of energy through the Cat population.

1 Introduction

Populations of feral House Cats, *Felis catus*, have become established on several Sub-Antarctic islands (Holdgate 1966) where they feed predominantly on burrowing petrels (Procellariidae and Pelecanoididae). The Cats are causing severe reductions and even the elimination of certain petrel species which breed on these islands (Jones 1977; Van Aarde 1980). Several studies have shown that feral Cats are opportunistic predators, and that differences in their diets mirror availability rather than selection of prey (Van Aarde 1980; Jones and Coman 1981; Apps 1983).

Van Aarde (1980) described the diet and feeding behaviour of feral Cats at Marion Island (46°54′S, 37°45′E) when the population was estimated to contain 2,139±290 individuals with a growth rate of 23.3% (Van Aarde 1979, 1983). Following the introduction of feline panleucopaenia (FPL) during 1977 (Van Aarde and Skinner 1981), the 1982 population was estimated at 615±105 Cats with the rate of decrease since 1977 estimated at 34.3% yr^{-1} (Van Rensburg). This paper describes changes in the diet of the Cats at Marion Island since 1975, and discusses these changes in terms of flow of energy through the Cat population.

2 Materials and Methods

Between July 1981 and January 1983, the remains of 989 prey items (mainly wings of burrowing petrels) of Cats and/or Antarctic Skuas, *Catharacta skua*, were collected each month in a study area of 0.43 km^2 described by Van Aarde (1980). These were grouped into three predator categories, i.e., either Cat, Skua or unidentified (Van Aarde 1980). In addition, 129 prey remains of Cats were collected elsewhere on the island. Prey remains were identified with the aid of reference material, based on size, shape and colour of wings and, where possible, of beaks and/or legs of birds (Van Aarde 1980). The stomach contents of 143 Cats, shot during May 1981, November 1981–April 1982, and September 1982–May 1983, were analyzed according to Van Aarde (1980). Bird remains were identified as described above. The remains of *Mus musculus* could be identified always with certainty.

The energy requirements of the Cat population were calculated, following Van Aarde (1980), by taking into account the sex ratio, age distribution, seasonality in breeding, population size and the minimum energy requirements of Cats for the various sex and reproductive classes according to Scott and Scott (1967). Predation rate was based on the energy requirements of the Cat population and the energy contents of the most important prey species (Van Aarde 1980).

3 Results

The frequencies of occurrence of items identified from prey remains collected during 1975 and 1982 are given in Table 1. Data from Van Aarde (1980) indicated that in 1975 there was a significant difference (χ^2 = 51.461, P < 0.005) in species composition between the total prey remains collected within and outside the study area, and those from the study area. This was not observed subsequently (χ^2 = 9.849, P > 0.1).

The Cats fed predominantly on Salvin's Prions, *Pachyptila salvini*, during 1975 and 1982, with no significant (χ^2 = 0.137, P > 0.5) change in remains from the study area being evident. To minimize possible observer differences, as a result of difficulties in distinguishing between the remains of Blue Petrels, *Halobaena caerulea*, and Salvin's Prions when wings were not measured, these two prey categories were combined into a single food item. This, however, still failed to reflect any difference in the frequency of occurrence (χ^2 = 0.726,

1 Mammal Research Institute, University of Pretoria, Pretoria 0002, South Africa
Present address: Transvaal Division of Conservation, Private Bag X209, Pretoria 0001, South Africa

Antarctic Nutrient Cycles and Food Webs
(ed. by W.R. Siegfried, P.R. Condy, and R.M. Laws)

Table 1. Frequency of occurrence of remains of prey items of *F. catus* in a particular study area and elsewhere in the field at Marion Island

Prey item	January–September 1975				January–September 1982		July–January 1981–1983	
	Study area[a]		Total[b]		Study area		Total	
	%	n	%	n	%	n	%	n
Pachyptila salvini	59.7	423	60.0	735	60.0	75	53.4	229
Halobaena caerulea	0	0	2.5	30	9.6	12	13.1	56
Pterodroma macroptera	16.6	118	10.0	123	12.0	15	9.1	39
Pterodroma mollis	9.3	66	9.4	115	15.2	19	13.5	58
Pterodroma brevirostris	7.3	52	12.8	157	1.6	2	4.7	20
Procellaria aequinoctialis	0.7	5	0.7	9	0.8	1	1.2	5
Fregetta spp.	0	0	0.1	1	0	0	0.2	1
Pelecanoides spp.	0.3	2	0.6	7	0	0	0	0
Unidentified petrels	5.2	37	3.0	37	0.8	1	4.7	20
Chionis minor	0.8	6	0.8	10	0	0	0	0
Pygoscelis papua	0	0	0	0	0	0	0.2	1
n		709		1,224		125		429

[a] Van Aarde (1977)
[b] Van Aarde (1980)

P > 0.1). The incidence of Greatwinged Petrels, *Pterodroma macroptera*, also did not change significantly (χ^2 = 2.036, P > 0.1) between 1975 and 1982, whereas Softplumaged Petrels, *Pterodroma mollis*, increased (χ^2 = 3.269, P < 0.1) and Kerguelen Petrels, *Pterodroma brevirostris*, decreased (χ^2 = 5.283, P < 0.025) significantly.

Salvin's Prions were the most important food item of Antarctic Skuas (Table 2), and their frequency of occurrence amongst prey remains did not change significantly between 1975 and 1982 (χ^2 = 0.338, P > 0.5). However, when grouped with Blue Petrels the frequency of occurrence increased significantly (χ^2 = 2.913, P < 0.1). Medium-sized petrels

Table 2. Frequency of occurrence of remains of prey items of *C. skua* in a particular study area at Marion Island. Data for 1975 taken from Van Aarde (1977)

Prey item	January–September 1975		January–September 1982	
	%	n	%	n
Pachyptila salvini	54.1	239	53.6	111
Halobaena caerulea	0	0	7.3	15
Pterodroma macroptera	1.1	5	5.3	11
Pterodroma mollis	26.5	117	12.6	26
Pterodroma brevirostris	10.4	46	0	0
Procellaria aqueinoctialis	0.9	4	7.7	16
Fregetta spp.	0	0	0.5	1
Unidentified petrels	6.8	30	12.6	26
Chionis minor	0.2	1	0	0
Phalacrocorax albiventer	0	0	0.5	1
n		442		207

(Softplumaged and Kerguelen petrels) decreased significantly (χ^2 = 9.611, P < 0.005 and χ^2 = 20.000, P < 0.005, respectively), whereas larger petrels (Greatwinged and Whitechinned, *Procellaria aequinoctialis*, petrels) increased significantly (χ^2 = 6.545, P < 0.025 and χ^2 = 23.810, P < 0.005, respectively) in the frequencies of occurrence of the prey remains of Antarctic Skuas between 1975 and 1982.

In 1982 the total petrel portion of the diets of Cats and Skuas, with the exception of White-chinned petrels (χ^2 = 8.743, P < 0.005), did not differ significantly.

The frequency of occurrence of prey items in the stomachs of Cats is summarized in Table 3. The incidence of stomachs containing Mice increased significantly (χ^2 = 18.837, P < 0.005), whereas those with petrels decreased significantly (χ^2 = 13.824, P < 0.005) between 1975 and 1982. A similar trend occurred in the frequency of occurrence of these items (χ^2 = 24.789 and 15.096, P < 0.005 for Mice and petrels, respectively). Mice, furthermore, contributed 16.6% and petrels 77.8% of the total weight of all stomach contents, whereas in 1975 they accounted for 4.4% and 87%, respectively, of the total volume of the stomach contents (Van Aarde 1977, 1980).

The energy requirements of the 615 Cats in 1982 (given 82.7% adults and a 1:1 sex ratio) was 337,640,913.2 kJ (Table 4). Thus, some 148,233 petrels would have to have been killed by the 1982 population of Cats to satisfy minimum requirements (Table 5), in comparison with 455,119 birds in 1975 (Van Aarde 1980). The contributions of energy made by the different prey items in the stomachs of the Cats in 1975 and 1982 are given in Table 6. Because of the high proportion of unidentified petrels in the stomachs, these items (1975, n = 124; 1982, n = 133) were assumed to occur in the same arrangement of species as was found in the collection of all prey remains (Table 1). For Mice, a mean body weight of 21.0 g (Berry et al. 1978) and an energy

Table 3. Frequency of occurrence of prey items in the stomachs of *F. catus* in 1975 and 1982 at Marion Island. No. stomachs = 116 and 143 in 1975 and 1982, respectively

Prey item	1975				1982			
	Number n	Stomach[a] %	Number n	Prey[b] %	Number n	Stomach %	Number n	Prey %
Plant material	43	37.1	–	–	7	4.9	–	–
Egg shells	6	5.2	–	–	0	0	–	–
Felis catus	1	0.9	1	0.6	1	0.7	1	0.4
Penguin	8	6.9	8	5.0	7	4.9	7	2.6
Mus musculus	19	16.4	27	16.9	59	41.3	130	48.5
Pachyptila salvini	35	30.2	39	24.4	21	14.7	29	10.7
Halobaena caerulea	1	0.9	2	1.3	3	2.1	3	1.1
Pterodroma macroptera	6	5.2	10	6.3	0	0	0	0
Pterodroma mollis	7	6.0	7	4.4	9	6.3	9	3.3
Pterodroma brevirostris	6	5.2	7	4.4	0	0	0	0
Procellaria aequinoctialis	6	5.2	7	4.4	4	2.8	5	1.8
Procellaria cinerea	0	0	0	0	2	1.4	3	1.1
Fregetta spp.	1	0.9	1	0.6	0	0	0	0
Unidentified petrels	51	44	51	31.9	81	56.6	84	31.0
Petrels (total)	113	97.4	124	77.5	120	83.9	133	49.1
n	116		160		143		271	

[a] Van Aarde (1980)
[b] Van Aarde (1977)

value of 7.1128 kJ g^{-1} (live weight) (Fleharty et al. 1973) were taken, and the energy contribution per unidentified petrel was based on the mean energy content of the identified petrels and their frequency of occurrence. According to Table 6, petrels contributed 92.98% and 90.91% in 1975 and 1982, respectively, and Mice 1.48% and 6.14% in 1975 and 1982, respectively, to the total energy requirements of the Cat population.

4 Discussion

Van Aarde (1980) suggested that the pattern of depredation of petrels by Cats at Marion Island reflected availability rather than selection of prey. All lack an anti-predator strategy to terrestrial mammalian predators and are therefore highly susceptible to cats. He, furthermore, ascribed the scarcity of the Common Divingpetrel, *Pelecanoides urinatrix*, at Marion Island to Cats of which some 200 were established on the island as early as 1965–1966.

In spite of the marked decrease in Cat density, Kerguelen Petrels are decreasing at Marion Island, and the increasing predation on Softplumaged Petrels since 1975 apparently is a function of the decreasing abundance of Kerguelen Petrels. Kerguelen Petrels probably disappeared from the study area after 1977. The increase in predation on Mice and the decrease in predation on petrels by Cats, and the change in the diet of Antarctic Skuas from medium-sized to larger and smaller petrels, probably reflect changes in the relative abundance of petrels. The current overlap of the petrel fraction in the diets of Cats and Skuas differs from the situation reported for 1975 (Van Aarde 1980) and again reflects changes in the relative abundance of prey species.

The overall decrease in the number of prey remains of Cats is ascribed to the reduction in the Cat population. A similar decrease in the number of prey remains of Skuas cannot be ascribed to a reduction in Skua numbers, since comparable data are not available. This decrease may, however, be ascribed to a decrease in petrel numbers which may either have affected Skua density or the Skua's diet, with Skuas possibly preying now more intensely on penguins.

Some of the shortcomings of an assessment of the impact of predation based on the annual energy requirements of the Cat population have been discussed by Van Aarde (1980). Furthermore, the digestive coefficients of the prey species are not known, and it is also not known how much of a prey animal is eaten by a Cat before it kills again. Van Aarde (1980) estimated that each Cat killed 213 birds yr^{-1}, whereas the data given here for 1982 indicate approximately 241 birds annually. This increase is due to a higher estimated annual mean energy requirement per Cat, resulting from a 1:1 sex ratio as opposed to a 1:0.55 adult sex ratio in 1975 (Van Aarde 1978). The average body weight of adult Cats was also higher in 1982 than in 1975 (Van Rensburg). Although the estimate of the number of prey killed, based on the prey items in the stomach contents (Table 6), does not show a large difference from that based on the collection of prey remains (Table 5), it highlights the marked increase of Mice in the diet of the Cats.

A detailed quantification of the impact of the Cat population on the Marion Island avifauna requires data on the

Table 4. Annual energy requirement of the *F. catus* population in 1982 at Marion Island. Weight specific energy requirement of each class of animal is based on Scott and Scott (1967)

Ecological age reproductive class[a]		kJ kg^{-1} [c]	Observed body weight	Days in specific ecological age or reproductive class[b]	kJ yr^{-1} individual	Number individuals/ specific class	kJ yr^{-1} class
Adult ♂		334.7	4.40 ± 0.70	365	537,528.2	254.5	136,806,926.9
Non-reproductive	Adult ♀	334.7	3.07 ± 0.59	135	138,716.42		
Lactating	Adult ♀	1,046.0	3.03 ± 0.50	100	316,938	254.5	168,012,841.3
Pregnant	Adult ♀	418.4	3.76 ± 0.69	130	204,513.92		
Sub-adult ♂	Growing cats	543.9	1.84 ± 0.35	230	230,178.48	53	16,346,221.84
Juvenile	Growing cats	836.8	1.10 ± 0.17	85	78,240.8		
Sub-adult ♀	Growing cats	543.9	1.84 ± 0.35	135	135,104.76	53	16,480,923.2
Adult ♀	Growing cats	334.7	3.07 ± 0.59	95	97,615.26		
							337,640,913.2

[a] Van Aarde (1978)
[b] Van Aarde (1980)
[c] Scott and Scott (1967)

Table 5. Contributions of birds to the annual energy requirement of the *F. catus* population in 1982 at Marion Island

Prey species	Frequency of prey species in the total prey remains collection[a]	Mean energy content/ individual (kJ)[b]	Energy contribution to the diet of Cats (%)	Energetic contribution to 94.87%[c] of the Cat populations energy requirements (kJ)	Number of birds killed
Pachyptila salvina	229	1,273.18	33.15	106,186,058.3	83,402
Halobaena caerulea	56	2,016.40	12.84	41,129,079.58	20,397
Pterodroma macroptera	39	4,002.61	17.75	56,856,788.36	14,205
Pterodroma mollis	58	3,130.11	20.64	66,114,034.46	21,122
Pterodroma brevirostris	20	3,552.03	8.08	25,881,850.7	7,286
Procellaria aequinoctialis	5	13,262.76	7.54	24,152,123.05	1,821
Total	407[d]				148,233

[a] Table 1
[b] Van Aarde (1980)
[c] 320,319,934.4 kJ
[d] 94.87% of total prey remains

Table 6. Contributions of prey of the annual energy requirement of the *F. catus* population in 1975 and 1982, respectively, at Marion Island

Prey species	Species contribution to total prey items in stomach contents[a] 1975	1982	Mean energy content per individual (kJ)	Energetic contribution to the diet of cats (%) 1975	1982	Energy contribution (kJ) 93.75% of the 1975 cat population's energy requirements	96.68% of the 1982 cat population's energy requirements	Number of prey killed to provide energy requirements 1975	1982
Mus musculus	27	130	149.37	1.48	6.33	14,342,779.13	20,742,176.05	96,022	138,864
Pachyptila salvini	75.07	71.16	1,273.18	35.15	29.54	340,640,790.50	96,796,821.56	267,551	76,028
Halobaena caerulea	3.06	17.40	2,016.4	2.27	11.44	21,998,708.23	37,486,649.92	10,910	18,591
Pterodroma macroptera	12.56	12.12	4,002.61	18.49	15.82	179,187,716.00	51,839,056.10	44,768	12,951
Pterodroma mollis	11.75	18.02	3,130.11	13.53	18.38	131,023,143.30	60,227,677.06	41,859	19.241
Pterodroma brevirostris	16.04	6.21	3,552.03	20.95	7.19	203,027,725.80	23,560,228.40	57,158	6,633
Procellaria aequinoctialis	0.92	1.55	13,262.76	4.49	6.70	43,512,863.42	21,954,593.92	3,281	1,655
Unidentified Petrels	4.60	6.53	2,160.60 (1975) 2,160.50 (1982)	3.65	4.60	35,372,372.27	15,073,303.29	16,372	6,977
Total	151[b]	263[c]						441,899 (Mice excluded)	142,076 (Mice excluded)

[a] Table 3
[b] 94.38% of prey items in stomachs
[c] 97.05% of prey items in stomachs

population dynamics of the prey species. Such information is currently not available for all the prey species. However, certain petrels, especially winter breeders such as the Great-winged and Grey, *Procellaria cinerea*, petrels, may well have been saved from local extinction through the introduction of FPL. Because it is likely that the lowered Cat population is still reducing the burrowing petrels, further measures to eliminate the Cats should be applied as soon as possible to prevent the possible extinction of the petrels. It is also important to reduce an increasing number of Cats which have developed immunity against FPL.

Acknowledgements. The South African Department of Transport provided logistical and financial support on the advice of the South African Scientific Committee for Antarctic Research. This project was carried out under the auspices of the Mammal Research Institute of the University of Pretoria. I thank Prof. J.D. Skinner, Dr. M.N. Bester, and Mr. R.J. van Aarde for their encouragement, support, and advice.

References

Apps PJ (1983) Behavioural ecology of the feral house cat (*Felis catus* Linnaeus) on Dassen Island, M.Sc. Thesis, University of Pretoria, South Africa

Berry RJ, Peters J, van Aarde RJ (1978) Sub-antarctic house mice: Colonization, survival and selection. J Zool (Lond) 184:127–141

Fleharty ED, Krause ME, Stinnett DP (1973) Body composition, energy content and lipid cycles of four species of rodents. J Mammal 54:426–438

Holdgate MW (1966) The influence of introduced species on the ecosystem of temperate oceanic islands. In: Towards a new relationship of man and nature in temperate lands. Proceedings of the 10th Technical Meeting IUCN, Lucerne, 1966, IUCN Publications New Series 9:151–176

Jones E (1977) Ecology of the feral cat, *Felis catus* (L), (Carnivora: Felidae) on Macquarie Island. Aust Wildl Res 4:249–262

Jones E, Coman BJ (1981) Ecology of the feral cat, *Felis catus* (L), in South-eastern Australia. I Diet. Aust Wildl Res 8:537–547

Scott PP, Scott MG (1967) Nutritive requirements for Carnivora. In: Conalty ML (ed) Husbandry and laboratory animals. Academic Press, London

Van Aarde RJ (1977) Voeding, habitatsvoorkeur en voortplanting van die wilde huiskat *Felis catus* (Linnaeus 1758) op Marioneiland M.Sc. Thesis, University of Pretoria, South Africa

Van Aarde RJ (1978) Reproduction and population ecology in the feral house cat *Felis catus* on Marion Island. Carniv Genet Newsl 3:288–316

Van Aarde RJ (1979) Distribution and density of the feral house cat *Felis catus* on Marion Island. S Afr J Antarct Res 9:14–19

Van Aarde RJ (1980) The diet and feeding behaviour of feral cats, *Felis catus* at Marion Island. S Afr J Wildl Res 10:123–128

Van Aarde RJ (1983) Demographic parameters of the feral cat *Felis catus* population at Marion Island. S Afr J Wildl Res 13:12–16

Van Aarde RJ, Skinner JD (1981) The feral cat population at Marion Island: Characteristics, colonization and control. Paimpont, Com nat fr Rech Antarctiques 51:281–288

Part 5

Interactions Between Marine, Freshwater, and Terrestrial Systems

Interactions Between Antarctic and Sub-Antarctic Marine, Freshwater and Terrestrial Organisms

J.C. HUREAU[1]

Summary. The various studies carried out since the beginning of the BIOMASS (Biological Investigations of Marine Antarctic Systems and Stocks) programme have revealed the complexity of interactions between Antarctic marine organisms, with krill (Euphausiacea) as a key group. On the one hand, the interactions between whales, seals, fish, birds, and krill, and, on the other hand, the interactions between krill and phytoplankton have been carefully described. However, several interactions have not been sufficiently stressed, more particularly those responsible for the impacts of terrestrial and freshwater ecosystems on marine organisms and *vice versa*. Certain recent studies have shown that the terrestrial micro-flora play an important role in Antarctica as well as in the Sub-Antarctic islands, due to a substantial influx of run-off water. Similarly, the terrestrial micro-fauna is linked with the coastal marine micro-fauna; coastal birds, such as terns and penguins, and seals, such as the Elephant Seal, *Mirounga leonina*, playing important roles in coastal ecosystems. It is necessary to emphasize the role played by seabirds which introduce significant quantities of nutrients into terrestrial ecosystems: it has been demonstrated that in one Sub-Antarctic island this amounts to about 56 kg ha^{-1} yr^{-1} of N alone. The role of seals which breed on land and feed at sea is also important, because they constitute a link between terrestrial micro-organisms and marine fauna. One type of interaction has not yet been studied in detail, and requires research: it is the role of parasites which need 2 or more hosts, invertebrate and vertebrate. Finally, there is the role of man whose presence on land and whose exploitation of whales, fish, krill, and other crustaceans can modify the natural dynamic equilibrium of the interactions between terrestrial, freshwater and marine ecosystems.

1 Introduction

Since the beginning of the BIOMASS (Biological Investigations of Marine Antarctic Systems and Stocks) programme, many studies have demonstrated the complexity of the interactions between the different elements of the marine Antarctic ecosystems. One group of species, the euphausiids or krill (Euphausiacea), has been considered as a key group within these ecosystems. Its inter-relationships with predators, such as whales, seals, birds, and fish, on the one hand, and with the phytoplankton, on the other hand, have been, or are being, carefully described and quantified. However, its role as a key group may have been excessively amplified to the detriment of other groups of zooplanktonic organisms, such as salps, other pelagic crustaceans and coelenterates or to the detriment of higher organisms, such as fish, birds, and seals, or also to the detriment of the benthic organisms whose role in the recycling of detritus should be better studied. An even greater neglect attends research dealing with organisms which interact between two or more of the three main groups of ecosystems (namely, marine, freshwater, and terrestrial) found in Antarctic and Sub-Antarctic regions. In this brief review, I highlight some of these inter-system interactions.

2 Inter-Ecosystem Interactions

Several important interactions between ecosystems have been stressed only in very recent years. Such studies have shown that it is now justified to define a "coastal ecosystem" as the link between land, freshwater and marine systems. Bellido (1982), in his study of the rocky shores of the Kerguelen archipelago, has shown that this biotope, known to be poor in macro-fauna (Arnaud 1974), has a rich micro-fauna dominated by insects and acari (mainly oribatids) which originate from a terrestrial fauna (Table 1). The marine micro-fauna (polychaete annelids, amphipods, and isopods) occurs only in the medio-littoral zone, and a terrestrial fauna is never found in the infra-littoral zone which is always submerged.

Another type of interaction which has been recently studied concerns the relationships between soil and fresh-

Table 1. Relative importance of the main groups of littoral invertebrates at Kerguelen Archipelago. (Based on data from Bellido 1982)

	Zone		
	Ad-littoral (%)	Supra-littoral (%)	Medio-littoral (%)
Tardigrads	–	–	1.5
Annelids	–	0.5	2.5
Amphipods	–	–	0.5
Isopods	–	–	0.5
Insects	58.4	7.1	10.7
Acarids	41.6	92.4	84.3

1 Muséum National d'Histoire Naturelle, Ichtyologie Générale et Appliquée 43 Rue Cuvier, F-75231 Paris Cedex 05, France

Antarctic Nutrient Cycles and Food Webs (ed. by W.R. Siegfried, P.R. Condy, and R.M. Laws)

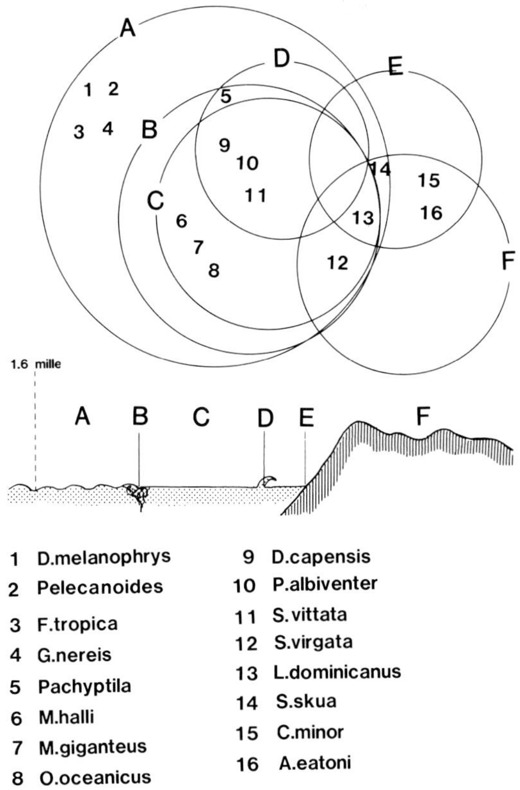

Fig. 1. Foraging zones of 16 avian species breeding at the Crozet Islands. *A* inshore waters; *B macrocystis* zone; *C* calm zone; *D* surf zone; *E* coast; *F* land. (After Jouventin et al. 1982)

water lake micro-flora (Ellis-Evans and Wynn-Williams 1984). The relationships between freshwater lakes and the marine micro- or macro-fauna are not yet well studied but may be maintained by the inter-change of fauna between lakes and coastal waters.

At this point, it is necessary to emphasize the major role played by seabirds which probably constitute the most important link between Antarctic and Sub-Antarctic marine, freshwater, and terrestrial ecosystems. The expression "coastal" seabirds is employed here according to the definition given by Jouventin et al. (1982). At the Crozet Islands, for example, this coastal avifauna includes 16 species (Fig. 1) which are linked to a number of foraging zones: land, coast, surf zone, calm zone, *Macrocystis* zone, and inshore waters which here are limited to ca. 3 km from the coast. These seabirds, and also penguins, are not only predators on terrestrial or marine animals but, as Siegfried (1982) has shown, they are significant participants in ecological processes affecting the functioning of terrestrial ecosystems. At Marion Island, for example, the biomass of breeding birds exceeds 9,000 tons in the coastal lowland of about 100 km^2. The amount of nutrients deposited in the terrestrial ecosystem, in the form of guano and other material, represents annually at least 33,000 t (4,000 t dry wt.), including a deposition of N alone amounting to 56 kg ha^{-1} yr^{-1}. Thus, the relatively high primary productivity of the vegetation on Sub-Antarctic islands is largely maintained by the permanent supply of nutrients transported by birds from the sea. The role of certain species, such as the Giant petrels, *Macronectes giganteus* and *M. halli*, Lesser Sheathbill, *Chionis minor*, Antarctic Skua, *Catharacta antarctica*, Kelp Gull, *Larus dominicanus*, and the terns, *Sterna virgata* and *S. vittata*, must be stressed because they are the main predators and scavengers in the terrestrial ecosystem and in the littoral zone. They re-distribute nutrients within the systems and, more particularly, transport them from the large coastal penguin colonies to the vegetated areas.

The impact of certain coastal seabirds (e.g., Lesser Sheathbill and Kelp Gull) on the terrestrial micro-fauna or on terrestrial macro-invertebrates of Sub-Antarctic islands cannot be neglected, since Siegfried (1982) estimated the amount consumed annually by these species at about 24 t (dry wt.) over an area of 100 km^2, representing about 25% and 12%, respectively, of the total annual energy requirements of these two species at Marion Island. Stahl and Weimerskirch (1982) have also shown the substantial predation of terns on the terrestrial micro-fauna at Kerguelen Island, where $> 30\%$ of the diet of the Kerguelen Tern, *Sterna virgata*, is based on insects, spiders and oligochaetes (Table 2), the rest of the prey being marine organisms, mainly fish and crustaceans. However, the main predatory impact of seabirds is directed at the marine fauna. At South Georgia, for example, Croxall and Prince (1982) estimated that about 1.7×10^6 t (fr.wt.) of food is consumed annually by breeding birds (mainly Macaroni Penguins, *Eudyptes chrysolophus*, and Dove Prions, *Pachyptila desolata*) with 88% of the food being Krill, *Euphausia superba*, 6% copepods, 3% fish and 3% squid. These amounts represent $1{,}556 \times 10^9$, 106×10^9, 48×10^9, and 42×10^9 kcal, respectively, using the calorific values of crustaceans, fish, and squid given by Clarke and Prince (1980) (Table 3). Croxall and Prince (1982), in taking into account the non-breeding populations of the various species and other correction factors, estimate that the total annual food consumption of South Georgia seabirds during the breeding season might be about 2.7×10^6 t (fr.wt.). More detailed and abundant data are required in order to quantify the relationships between Antarctic and Sub-Antarctic seabirds and their prey. A need for information on the diets of Krill- and fish-eating seabirds must be emphasized, because the probable responses of the avian communi-

Table 2. Diets of the Kerguelen Tern, *Sterna virgata*, and Antarctic Tern, *S. vittata*, at the Crozet Islands (Based on data from Stahl and Weimerskirch 1982)

	Kerguelen Tern (%)	Antarctic Tern (%)
Fish	49.3	92.5
Crustaceans	15.0	5.4
Molluscs	3.9	–
Insects Spiders	30.7	2.1
Earthworms	1.3	–
Holothurids	1.3	–

Table 3. Food (t × 10^3 fr.wt.) and corresponding energy (kcal × 10^9) consumed by breeding birds at South Georgia, based on data from Croxall and Prince (1982)

	Food	Energy
Krill	1,500	1,556
Copepods	102	106
Fish	51	48
Squid	51	42

ties to changes in the krill or fish populations might be used to indicate the state of the prey stocks.

Seals are also a major component of the coastal ecosystem. They feed at sea on various prey (e.g., Krill for the Crabeater Seal, *Lobodon carcinophagus*, fish and squid for the Elephant Seal, *Mirounga leonina*, and benthic organisms for the Weddell Seal, *Leptonychotes weddelli*), but they breed on land and on the ice. Elephant Seals, in the Sub-Antarctic islands, play a role similar to that of coastal seabirds in that they participate in the deposition of nutrients. According to Laws (1977), the diet of the Elephant Seal consists of fish (75%) and squid (25%). Pascal (1982) has shown that at the Kerguelen Islands the theoretical amount of fish consumed by Elephant Seals is much larger than the population of fish there, as indicated by the preliminary assessments made by Everson (1977) and Hureau (1979). This can be explained either by an under-estimate of the fish stocks (although I do not believe that this is the case) or by an over-estimate of the food requirements of the Elephant Seal population, or by a lack of knowledge of its feeding habits. There is a need for better assessments of the fish stock and the diet of the Elephant Seal. The Elephant Seal, similarly to seabirds, might be useful as an indicator of the state of the fish or squid stocks (Fig. 2). Along the coast of the Antarctic continent, the Crabeater Seal populations also might play a role as indicators of the state of Krill stocks.

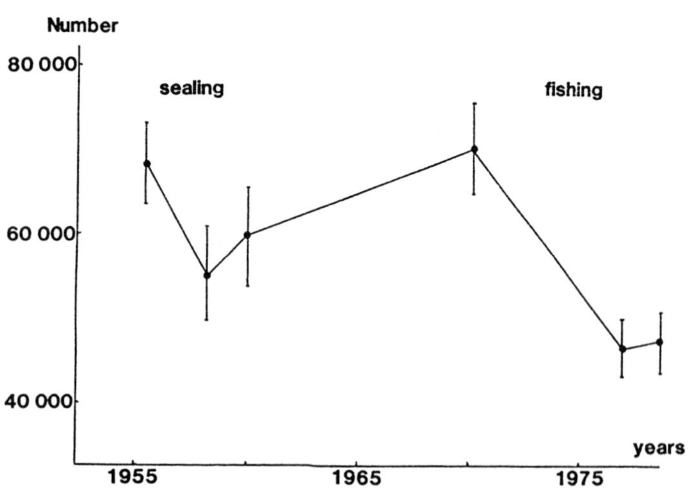

Fig. 2. Variation in the number of female Elephant Seals on the eastern coast of the Kerguelen archipelago between 1956 and 1979. (After Pascal 1982)

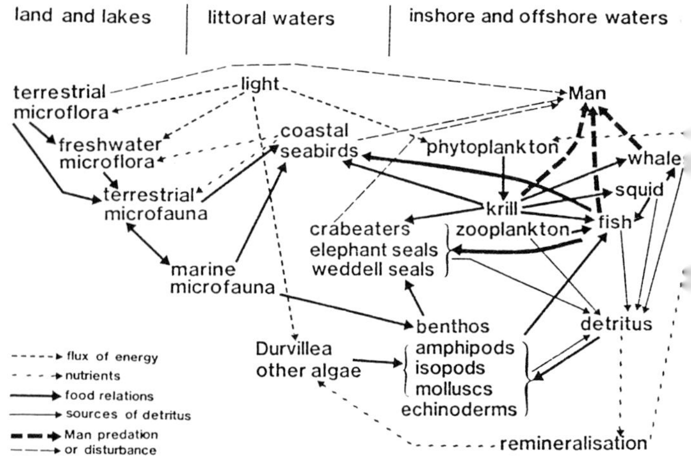

Fig. 3. The principal interactions between marine, freshwater, and terrestrial organisms in Antarctic and Sub-Antarctic regions

Apart from the micro-fauna, coastal birds and seals, there is another group which has not yet been studied carefully. These are the parasites which generally need two or more hosts which generally are an aquatic invertebrate (a mollusc or an annelid) and one or two vertebrates (fish, birds, or seals). These parasites might thus play a relatively important role within each ecosystem or between the various ecosystems, with regard to the dynamics of the host populations concerned.

3 Conclusion

The main interactions between terrestrial, freshwater, and marine organisms in the Antarctic and Sub-Antarctic regions are summarized in Fig. 3. The natural dynamic equilibria between the various organisms and their ecosystems involved in these interactions can, of course, be disturbed by man's injudicious exploitation of renewable and non-renewable natural resources, and through the introduction of alien biota and toxic pollutants into the ecosystems.

References

Arnaud PM (1974) Contribution à la bionomie marine benthique des régions antarctiques et subantarctiques. Théthys 6(3):467–553
Bellido A (1982) Les biocénoses du littoral rocheux aux îles Kerguelen. CNFRA 51:81–92
Clarke A, Prince PA (1980) Chemical composition and calorific value of food fed to mollymauk chicks *Diomedea melanophris* and *D. chrysostoma* at Bird Island, South Georgia, Ibis 122:488–494
Croxall JP, Prince PA (1982) A preliminary assessment of the impact of seabirds on marine resources at South Georgia. CNFRA 51: 501–509
Ellis-Evans JC, Wynn-Williams DD (1985) The interaction of soil micro-flora with lake micro-flora at Signy Island. In: Siegfried WR, Condy PR, Laws RM (eds) Antarctic nutrient cycles and food webs (Proceedings of the 4th SCAR symposium on Antarctic biology). Springer, Berlin Heidelberg New York

Everson I (1977) The living resources of the Southern Ocean. FAO. GLO/SO/77/1, Roma, 156 pp

Hureau JC (1979) La faune ichthyologique du secteur Indien de l'Océan Antarctique et estimation du stock de poisson autour des îles Kerguelen. Mém Mus Nat Hist Nat NS série C 43:235–247 247

Jouventin P, Mougin JL, Stahl JC, Weimerskirch H (1982) La ségrégation écologique entre les oiseaux des îles Crozet. Données préliminaires. CNFRA 51:457–467

Laws RM (1977) Seals and whales of the Southern Ocean. Philos Trans R Soc Lond B Biol Sci 279:81–96

Pascal M (1982) Evolution numérique de la population d'éléphants de mer (*Mirounga leonina* L.) de l'archipel des Kerguelen au cours des 30 dernières années. CNFRA 51:119–135

Siegfried WR (1982) The roles of birds in ecological processes affecting the functioning of the terrestrial ecosystem at sub-antarctic Marion Island. CNFRA 51:493–499

Stahl JC, Weimerskirch H (1982) La ségrégation écologique des deux espèces de sternes des îles Crozet. CNFRA 51:449–456

The Intertidal and Shallow Subtidal Food Web at Marion Island

W.O. BLANKLEY and J.R. GRINDLEY[1,2]

Summary. The diets of the seven main predators of intertidal and subtidal organisms at Marion Island are described, based on the food and feeding habits of the Kelp Gull, *Larus dominicanus*, Lesser Sheathbill, *Chionis minor*, Imperial Cormorant, *Phalacrocorax atriceps*, nototheniid fish, *Notothenia coriiceps*, and *N. magellanica*, Plunder Fish, *Harpagifer georgianus*, and the starfish, *Anasterias rupicola*. The biomass and standing crop energy values of the main prey species, and the energy requirements and biomass of the predators, were determined, to yield data for a quantitative account of major patterns of energy flow in the food web of the marine coastal ecosystem. The limpet, *Nacella delesserti*, has the highest biomass of all the prey species and is also the major food of *Larus dominicanus, Notothenia coriiceps*, and *Anasterias rupicola* which tend to exploit limpets of different sizes and from different habitats. *Nacella delesserti* maintains extremely high population densities despite the intense depredation. Connectivity in the food web is relatively weak, and there is little evidence for interspecific competition for common food resources. However, there are indications that intraspecific competition for grazing occurs in the *N. delesserti* population. The system, as a whole, is generally composed of opportunistic generalists which may be regarded as constituting a set of loosely interacting colonizers of a relatively young and isolated Sub-Antarctic island.

1 Introduction

Published information on the abundance and interactions of intertidal and subtidal organisms at Marion Island (46° 54′ S, 37° 45′ E) is relatively sparse (Siegfried et al. 1979). De Villiers (1976) reported qualitative descriptions of the species composition and the major zonation patterns of the Marion Island littoral community, but he did not study interactions between species. The diets of the seven most obvious predators of littoral organisms at the island have been described (Blankley 1981, 1982, 1984) and, in addition, the ecology of the limpet, *Nacella delesserti*, the most important prey species, has been studied (Blankley and Branch, unpubl.). This paper provides quantitative information on the biomass of the seven predators and their prey, and estimates of the energy requirements of the different predators in relation to the standing crop energy values of the major prey species. To the best of our knowledge, no studies of energy flow in the intertidal and subtidal zones of other Sub-Antarctic islands have been published.

1 School of Environmental Studies, University of Cape Town, Rondebosch 7700, South Africa

2 Present address: Oceanographic Research Institute, P.O. Box 10712, Marine Parade 4056, Durban, South Africa

2 Materials and Methods

The main study site was Transvaal Cove, a relatively sheltered rocky and boulder beach on the northeastern coast of Marion Island.

Estimates of the abundance of birds breeding at Marion Island are given by Burger (1978) and Siegfried et al. (1978). Data on the mean mass of each species were made available by A.J. Williams (pers. comm.). These data were used to calculate the biomass of the three species of birds considered here. The foraging areas available to the three species were estimated by means of field observations of the birds' activities, and the topography of the island's 77 km coastline. The abundance of fish was assessed from counts made along subtidal transects, by means of SCUBA diving.

Energy requirements of the bird species were calculated from the Existence Metabolism (E.M.) equation provided by Kendeigh et al. (1977), where E.M. (kcal d^{-1}) = $4.142\ W^{0.5444}$, and W = fresh mass (g) of the individual. Energy requirements of the predatory fish species were calculated by means of Winberg's (1956) equation $Q = 35.8\ W^{0.81}$ cal d^{-1}, where Q = the resting metabolic rate, and W = fresh mass (g) of the individual. All energy requirement values were converted to kJ $m^{-2}\ d^{-1}$.

The densities and biomass of prey and associated non-prey species on the shore were estimated from counts of animals in 118 0.1 m^2 quadrats randomly sampled in three zones during September 1979–May 1980. The spray and splash zones were treated as a single zone which was recognized as a belt extending up to 3 m above sea-level where wave and wind action, as opposed to tidal flux, wet the boulders, ledges, and cliffs. The intertidal zone was a narrow belt (tidal range = 0.7 m) whose upper limit was easily recognized in the field as the uppermost margin of pink encrusting *Lithothamnion* spp. The subtidal zone was sampled to a depth of 4 m. All organisms were scraped from the substratum and removed from each quadrat, and were either analysed as fresh material in the laboratory or preserved in 10% buffered formalin for later analysis.

Antarctic Nutrient Cycles and Food Webs
(ed. by W.R. Siegfried, P.R. Condy, and R.M. Laws)

Animals were sorted to species level and then counted, dabbed dry and wet-weighed to the nearest 0.01 g. Representative samples of at least 100 animals of large, and 1000 specimens of smaller, species were dried at 60 °C to constant mass for determination of dry mass values. Shells were removed from molluscan species to provide shell-free dry mass values.

The maximum population density of each species was obtained from the five quadrats in which the species occurred at the highest densities. The energy contents of the common species were determined by means of a Gentry diabatic microbomb calorimeter, from representative samples of dried powdered flesh. For the remaining species, all of which were rare or contributed little to the overall biomass, energy content was estimated from values given for related taxa by Field et al. (1980).

3 Results

The energy requirements of the seven main predators are summarized in Table 1. The Lesser Sheathbill, *Chionis minor*, population feeds highest on the shore and, although only 10% of its diet consists of marine organisms, it consumes seaweeds and small invertebrates amounting to 1.25 kJ m^{-2} d^{-1}. The starfish, *Anasterias rupicola*, population makes a major impact (3.9 kJ m^{-2} d^{-1}) on intertidal and subtidal organisms and preys mostly on the limpet, *Nacella delesserti*. The three species of fish occupy three different feeding niches in the intertidal and subtidal zones (Blankley 1982) and their total impact is in the region of 2.18 kJ m^{-2} d^{-1}. The Kelp Gull, *Larus dominicanus*, feeds on *Nacella delesserti* in the intertidal and shallow subtidal zones and also on the bivalve *Gaimardia trapesina* from the offshore *Macrocystis* canopy, and its population accounts for a total of 0.93 kJ m^{-2} d^{-1}. The Imperial Cormorant, *Phalacrocorax atriceps*, hunts fish and crustaceans in the inshore waters. Its impact is approximately 0.18 kJ m^{-2} d^{-1}.

Table 2 gives biomass and standing crop energy values of the main organisms in the intertidal and subtidal zones. The algal biomass is higher than animal biomass in each of the zones, and total biomass is greatest in the subtidal, and lowest in the spray and splash, zones. In the spray and splash zone, 25.9% of the total standing stock consists of non-prey species, whereas these species account for only 8% and 3% in the intertidal and subtidal zones, respectively. The amphipod, *Hyale grandicornis*, is the most important prey species in the spray and splash zone, although the small bivalve *Lasaea consanguinea* occurs at very high densities. *Nacella delesserti* forms over 90% of the standing stock of prey species in the intertidal and subtidal zones and occupies nearly all available space. It is also the most heavily exploited prey species (Table 3) and occurs in the diets of six of the seven predators. The polychaete, *Platynereis australis*, the isopod, *Dynamenella huttoni*, and various amphipod species are also widely taken as prey, and algae are an important source of food for the two species of nototheniid fish and also for the Lesser Sheathbill.

Figure 1 shows the main paths of energy flow in the system. The food web is obviously simplistic and connectance (i.e. number of connections amongst the com-

Table 1. Foraging areas, abundance, standing crop biomass and estimated energy requirements or seven main predators in the intertidal and subtidal zones of Marion Island

Predators	Foraging ares	Total number of individuals (birds) or density m^{-2}	Biomass (fresh) mass (m^{-2})	% diet biomass consisting of marine organisms	Energy requirements of population kJ m^{-2} d^{-1}
Aves					
Imperial Cormorant *Phalacrocorax atriceps*	Shallow and deep subtidal zones (1–15 m depth and 1–50 m off shore)	570	0.37 g m^{-2} (1,425 kg ÷ 38.5 × 10^5)	100	0.18
Kelp Gull *Larus dominicanus*	Intertidal and offshore *Macrocystis* surface canopy	923	1.80 g m^{-2} (831 kg ÷ 4.62 × 10^5)	66	0.93
Lesser Sheathbill *Chionis minor*	Spray-splash and intertidal zones	3,500	10.23 g m^{-2} (1,575 kg ÷ 1.54 × 10^5)	10	1.25
Pisces					
Notothenia coriiceps	Subtidal, benthic	?0.04 m^{-2}	?28.0 g m^{-2}	100	?1 21
Notothenia magellanica	Subtidal, midwater end benthic	?0.1 m^{-2}	?10.0 g m^{-2}	100	?0.61
Harpagifer georgianus	Intertidal and subtidal, benthic	?1.0 m^{-2}	?2.9 g m^{-2}	100	?0.36
Echinodermata					
Anasterias rupicola	Intertidal and subtidal, benthic	17.5 m^{-2}	63.6 g m^{-2}	100	3.9

Table 2. Density (no individuals/m²) and biomass (g dry mass/m²) of prey and associated (non-prey) organisms in the intertidal and subtidal zones of Marion Island

		Density m^{-2} (mean ± S.E.)	Maximum density (m^{-2})	Biomass ($g\ m^{-2}$)	kJ m^{-2}	Energy content kJ g^{-1} dry mass
Spray and splash zones						
Prey species						
Hyale grandicornis	Amphipod	463 ± 194	3,316	2.04	31.8	15.59
Lasaea consanguinea	Bivalve	1,037 ± 631	8,846	0.51	8.7	17.06[a]
Kerguelenella lateralis	Siphonariid	57 ± 2	458	0.28	4.5	16.07[a]
Insecta	Weevils, coleopteran larvae, kelp flies	49 ± 24	396	0.07	1.3	18.57[a]
Kidderia minuta	Bivalve	24 ± 14	220	0.01	0.2	16.52[a]
Hemiarthrum setulosum	Chiton	4 ± 3	57	0.01	0.1	10.00[a]
Totals				2.92	46.6	
Associated fauna						
Lumbricillus spp.	Oligochaetes	694 ± 420	5,712	0.77	13.6	17.66[a]
Planarians	Unidentified	38 ± 19	342	0.10	1.1	11.00[a]
Anatanais gracilis	Tanaid	41 ± 38	610	0.09	0.9	10.00[a]
Tigriopus angulatus	Copepod	353 ± 307	5,300	0.03	0.3	10.00[a]
Halozetes spp.	Mites	261 ± 2,552	23,378	0.01	0.1	10.00[a]
Others	13 species	–	–	0.02	0.3	15.00[a]
Totals				1.02	16.3	
Algae *(Porphyra, Bostrychia, Rama)* and lichens		–	–	77.0	1095.7	14.22[a]
Intertidal zone						
Prey species						
Nacella delesserti	Limpet	75 ± 9	380	62.23	1,292.0	20.76
Kidderia minuta	Bivalve	2,048 ± 816	16,394	2.70	44.6	16.52[a]
Platynereis australis	Polychaete	29 ± 12	248	1.30	25.7	19.76
Lasaea consanguinea	Bivalve	1,634 ± 1,146	16,534	0.80	13.6	17.00[a]
Dynamenella huttoni	Isopod	184 ± 99	1,398	1.08	12.0	11.11
Hyale hirtipalma	Amphipod	406 ± 225	2,973	0.63	9.8	15.56[a]
Hemiarthrum setulosum	Chiton	188 ± 126	1,553	0.47	6.6	14.04[a]
Exosphaeroma gigas	Isopod	46 ± 38	472	0.53	5.5	10.38
Jassa falcata	Amphipod	70 ± 37	632	0.14	2.2	16.00[a]
Shackletonia sp.	Amphipod	99 ± 62	1,014	0.09	1.3	14.44[a]
Antias bicornis	Isopod	588 ± 588	15,000	0.07	0.8	11.43[a]
Eophiliantidae	Amphipod	488 ± 290	4,728	0.04	0.6	15.00[a]
Eusiridae	Amphipod	9 ± 7	150	0.01	0.2	20.00[a]
Totals				70.09	1,414.9	
Associated fauna						
Romanchella spp.	Tube worms	531 ± 337	5,420	1.28	28.2	22.03[a]
Pseudocnus laevigatus	Holothuroid	8 ± 4	84	1.20	11.3	9.41[a]
Lumbricillus spp.	Oligochaetes	512 ± 389	4,396	0.57	10.0	17.54[a]
Nematodes	Unidentified	758 ± 744	7,632	0.08	1.2	15.00[a]
Others	16 species	–	–	6.03	72.4	12.00[a]
Totals				9.16	123.1	
Algae (*Rhodymenia, Ulva, Lithothamnion* etc.)	Not counting *Durvillea*	–	–	124.4	1,769.5	14.22[a]
Subtidal zone						
Prey species						
Nacella delesserti	Limpet	93 ± 11	463	115.60	2,400.1	20.76
Platynereis australis	Polychaete	114 ± 41	354	5.11	100.9	19.76
Hyale hirtipalma	Amphipod	2,183 ± 2,131	7,858	3.38	52.6	15.56[a]
Jassa falcata	Amphipod	1,383 ± 993	4,978	2.77	44.3	16.00[a]
Dynamenella huttoni	Isopod	298 ± 182	1,086	1.75	19.5	11.11
Gaimardia trapesina	Bivalve (juveniles)	156 ± 133	562	0.67	11.1	16.60

Table 2 (cont.)

		Density m^{-2} (mean ± S.E.)	Maximum density (m^{-2})	Biomass ($g\ m^{-2}$)	$kJ\ m^{-2}$	Energy content $kJ\ g^{-1}$ dry mass
Kidderia minuta	Bivalve	226 ± 178	812	0.30	4.9	16.33[a]
Eusiridae	Amphipod	44 ± 28	263	0.05	0.8	16.00[a]
Eophiliantidae	Amphipod	314 ± 204	1,810	0.03	0.4	13.33[a]
Lasaea consanguinea	Bivalve	26 ± 26	92	0.01	0.2	20.00[a]
Totals				129.67	2,634.8	
Associated fauna						
Pseudocnus laevigatus	Holothuroid	47 ± 12	54	3.76	35.3	9.41[a]
Polychaetes	Unidentified	19 ± 10	68	0.68	12.2	17.94[a]
Ophiuroidea	Brittle stars	18 ± 10	64	0.76	2.2	2.89[a]
Unidentified	Ascidian	433 ± 359	1,560	0.18	1.2	6.66[a]
Apseudes sp.	Tanaid	5 ± 2	18	0.03	0.3	10.00[a]
Tanystylum cavidorsum	Pycnogonid	5 ± 2	14	0.01	0.2	20.00[a]
Others	10 species			2.53	30.4	12.02[a]
Totals				7.95	81.8	
Algae (*Lithothamnion, Ulva,* mixed reds)				336.0	4,781.0	
Deep water and offshore						
Prey species						
Gaimardia trapesina (on *Macrocystis pyrifera*)	Bivalve	?	?2,000	?	?	16.60
Nauticaris marionis	Shrimp	?	?	?	?	
Cephalopoda	Squid	?	?	?	?	

[a] Estimates, based on values given for related taxa by Field et al. (1980)

ponent species) appears to be weak, since only the very strongest trophic links between species are shown (cf. Table 3 where minor feeding links are also quantified). *Nacella delesserti* stands out as a key species, since it provides more energy (4.25 kJ m^{-2} d^{-1}) for the predators than all the other prey species combined. This species is also the major source of energy in the diets of *Anasterias rupicola, Larus dominicanus*, and *Notothenia coriiceps*.

4 Discussion

It is difficult to provide a comprehensive quantitative account of all the interactions which occur at different levels in a fairly complex food web, even when the basic structure of the web is known (Cohen 1978). Food webs vary seasonally and geographically, and long-term broad-based studies are necessary before valid interpretations can be made (Paine 1980). Nevertheless, intense, short-term studies such as ours can provide useful information from which structures of communities can be described.

An important aspect ignored in many studies of community functioning is the influence of body size on predator-prey interactions, although this is a popular topic in one-to-one predator-prey studies (e.g., Brooks and Dodson 1965; Paine 1976; Griffiths and Seiderer 1980). At Marion Island, only relatively large members of *Anasterias rupicola* and *Notothenia corciiceps* are able to actually capture limpets (Blankley 1982; Blankley and Branch, unpubl.). Similar differences occur in the diets of small and large size classes of the three species of fish (Blankley 1982). Although adult *A. rupicola* are essentially monophagous on *N. delesserti*, the juveniles (< 30 mm) are only able to reach a size where they are capable of preying directly on limpets by initially feeding on smaller prey, such as amphipods and isopods, or by joining larger starfish for a meal of limpets (Blankley and Branch, unpubl.). Thus, food webs should not be regarded as simple systems consisting of one-to-one trophic links between species, but also as sequences of dynamic interactions between predators and prey which grow at different rates and alter the interactions between them as they do so.

Since the Marion Island ecosystem is younger than 300,000 year (McDougall 1971) it could be expected to accommodate a relatively impoverished biota and simple food webs. Macquarie Island is a far older system containing more complex assemblages of molluscs and echinoderms (Simpson 1976, 1982). About 64% of the Macquarie Island molluscan species are endemic (Briggs 1966), whereas there are no molluscs which are endemic to Marion Island. The kelp-bed communities of the Cape Peninsula, South Africa, are dominated by algae and filter feeders (Field et al. 1980),

Table 3. Relative composition (% fresh mass) of the marine portions of the diets of 7 predators of intertidal and subtidal organisms at Marion Island, based on data in Blankley (1981, 1984)

Prey	Predators						
	Harpagifer georgianus	*Notothenia magellanica*	*Notothenia coriiceps*	*Anasterias rupicola*	*Larus dominicanus*	*Chionis minor*	*Phalacrocorax atriceps*
Algae							
Ulva, Rhodymenia, Porphyra, Durvillea, Macrocystis etc.	2.4	35.5	57.6	b	b	90	b
Invertebrata							
Polychaeta							
Platynereis australis	10.4 (36)	13.3 (12.2)	1.9 (16)	1.9 (42)	b	1 (0.5)	b
Unidentified species	a	a	0.2 (7)	a	b		10 (1)
Gastropoda							
Nacella delesserti	0.5 (1)	3.1 (11)	27.1 (56)	90.2 (161)	45 (1–6)	1 (1)	b
Laevilitorina caliginosa	a	a	a	0.1 (4)	b	a	b
Kerguelenella lateralis	b	b	b	0.4 (20)	b	1 (1–5)	b
Polyplacophora							
Hemiarthrum setulosum	b	0.1 (8)	b	0.2 (10)	b	a	b
Bivalvia							
Lasaea consanguinea	1.6 (30)	0.3 (78)	b	0.1 (26)	b	a	b
Gaimardia trapesina	0.1 (1)	4.6 (103)	0.3 (18)	0.2 (3)	45 (6–10)	a	b
Kidderia minuta	a		b	0.1 (13)	b	a	b
Cephalopoda							
Unidentified squid	b	0.8 (1)	0.1 (1)	b	b	b	(3)
Isopoda							
Antias bicornis	9.7 (180)	b	b	a	b	b	b
Dynamenella huttoni	4.4 (34)	12.8 (338)	1.1 (112)	1.0 (40)	b	a	b
Exosphaeroma gigas	b	0.9 (14)	1.7 (17)	1.1 (16)	b	a	b
Amphipoda							
Jassa falcata	13.6 (187)	4.6 (1,336)	0.3 (114)	0.1 (15)	b	1 (2)	b
Eophiliantidae (unid.)	2.6 (41)	0.1 (61)	b	b	b	b	b
Shackletonia sp.	28.6 (415)	b	b	0.1 (5)	b	a	b
Hyale spp.	19.0 (146)	1.4 (202)	0.1 (13)	0.1 (12)	b	2 (20)	b
Eusiridae (unid.)	4.6 (41)	3.5 (312)	1.2 (292)	0.1 (3)	b	a	b
Pontogeneiella brevicornis	b	0.2 (18)	1.1 (61)	a	b	a	b
Natantia							
Nauticaris marionis	b	4.2 (59)	b	a	b	b	10 (10)
Brachyura							
Unidentified crab	b	1.5 (5)	a	a	b	b	b
Insecta							
Weevils, coleopteran larvae, kelp flies	0.2 (1)	a	b	a	a	1 (5)	b
Echinodermata							
Anasterias rupicola	b	a	a	a	2.0 (0.1)	1 (0.5)	b
Chordata							
Pisces							
Harpagifer georgianus	b	5.4 (9)	0.9 (3)	b	a	a	22 (4)
Notothenia magellanica		2.4 (1)	6.6 (2)	a	a	b	56 (1)
Unidentified fish	b	3.3 (3)	1.0 (3)	b	b	b	a
Other prey species	2.3	2.2	–	4.3	?8.0	2.0	?2.0

Figures in parentheses are estimates of numbers of individuals consumed daily (birds) or the total number of individuals counted in stomach content samples (fish and starfish)

[a] Slight predation occurs

[b] Never observed as prey

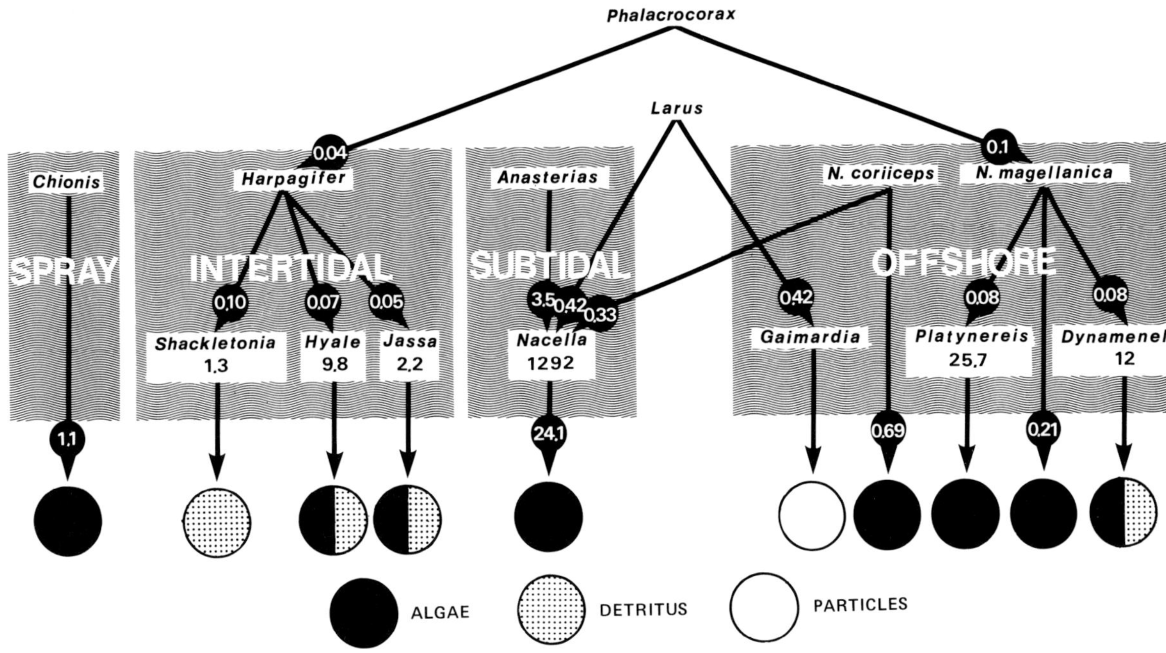

Fig. 1. The main predator-prey interactions in intertidal and shallow subtidal zones at Marion Island. *Arrows* direct predator impacts (kJ m^{-2} d^{-1}) which are shown by *white numerals* (on *black backgrounds*). Standing crop energy values (kJ m^{-2}) of prey are shown by *black numerals* (on *white backgrounds*)

and the rocky-shore biomass is dominated by algae in sheltered places and by filter feeders in exposed places (McQuaid 1980). At Marion Island, algae predominate in the biomass, and although filter feeders are very abundant they are mostly small bivalves accounting for a low biomass, compared to that of the dense populations of the herbivorous *N. delesserti*. However, although the production of the different species has not yet been determined, the many small species of amphipods and bivalves are likely to have higher rates of "turnover" than *N. delesserti*.

Molluscs and echinoderms have received most attention in studies of inshore marine organisms in the Antarctic and Sub-Antarctic, which is not surprising since limpets and starfish in particular appear to dominate the littoral systems of these regions (Simpson 1976; Brand 1980: Picken 1980). In our attempt to investigate all major members of an inshore community, we found that, although the species diversity of the diets of the seven predators was fairly high, only one or two species formed the major portion of the diet of any particular species. Brand (1980) found that food-web connectance was very high in the Antarctic shallow-water, benthic community of Arthur Harbour, due to the diverse diets of most organisms. He suggested that this reflects an extensive overlapping of niches, implying that competition for trophic resources is intense. He also suggested that the community remains stable by de-coupling processes which avoid trophic stress. Methods of de-coupling include switching of food preference to exploit relatively more abundant food items, changing feeding behaviour so that a different method of food procurement is used at different times, and lowering the rate of metabolism when food is scarce. Although such processes could well be operative in the Marion Island system, there appears to be an abundant supply of food in the system throughout the year. Algae are, however, more prominent in summer than in winter.

The large limpet population at Marion Island appears to make a heavy impact (24.1 kJ m^{-2} d^{-1}; Blankley and Branch, unpubl.) on algal sporeling settlement. However, although algal biomass dominates the inshore system, the availability of suitable algae which can be eaten by limpets could limit their growth. Circumstantial evidence for this view is based on surveys of an extremely crowded subtidal *N. delesserti* population at Prince Edward Island where the mean size of limpets is much smaller than those of a comparable, but less dense, population at Marion Island (Blankley, unpubl.).

Sub-Antarctic marine communities generally tend to follow patterns of organization described for systems in the Antarctic, although the fact that the Sub-Antarctic encompasses a number of discrete and often youthful islands introduces many problems when one is making generalizations for the Sub-Antarctic as a whole. Brand (1980) records the shallow-water benthic community of the Antarctic Peninsula as consisting predominantly of generalist species which have flexible repertoires of behaviour, allowing them to constantly adjust to changes in their relatively unstable physical and biological environment. Similarly, the animals studied by us are all generalists (with wide feeding niches), rather than specialists relying on a single specific feeding technique or particular species of prey. In view of the relative youth and paucity of endemic species in the Marion Island system (van Zinderen Bakker 1971), the shallow-water marine species may be regarded as constituting a set of loosely interacting, opportunistic colonizers, representing a late-pioneer stage in the evolution of a Sub-Antarctic community.

Acknowledgements. Special thanks are due to Prof. G.M. Branch for his advice, help and encouragement. Thanks to Peter Haxen for help in the field, and Chris Seebrechts for assistance in analyzing the biomass samples. This study was sponsored by the South African Scientific Committee for Antarctic Research.

References

Blankley WO (1981) Marine food of kelp gulls, lesser sheathbills and imperial cormorants at Marion Island (Subantarctic) Cormorant 9:77–84

Blankley WO (1982) Feeding ecology of three inshore fish species at Marion Island (Southern Ocean) S Afr J Zool 17(4):164–170

Blankley WO (1984) Ecology of the starfish *Anasterias rupicola* at Marion Island (Southern Ocean). Mar Ecol Prog Ser 18:131–137

Brand T (1980) Trophic interactions and community ecology of the shallow-water marine benthos along the Antarctic Peninsul. Ph.D. Thesis, University of California, Davis

Briggs JC (1966) Oceanic islands, endemism and marine paleotemperatures. Syst Zool 15:153–163

Brooks JL, Dodson SI (1965) Predation, body size and composition of plankton. Science (Wash DC) 150:28–35

Burger AE (1978) Terrestrial invertebrates: a food resource for birds at Marion Island. S Afr J Antarct Res 8:87–99

Cohen JE (1978) Food webs and niche space. Princeton University, Princeton, NJ, 189 pp

De Villiers AF (1976) Littoral ecology of Marion and Prince Edward islands (Southern Ocean). S Afr J Antarct Res, Suppl 1:1–40

Field JG, Griffiths CL, Griffiths RJ, Jarman N, Zoutendyk P, Velimirov B, Bowes A (1980) Variation in structure and biomass of kelp communities along the South-west Cape Coast. Trans R Soc S Afr 44(2):145–203

Griffiths CL, Seiderer JL (1980) Rock-lobsters and mussels – limitation and preferences in a predator-prey interaction. J Exp Mar Biol Ecol 44:95–109

Hureau JC (1970) Biologie comparee de quelques poissons Antarctiques (Nototheniidae). Bull Inst Oceanogr (Monaco) 68(1391): 1–244

Kendeigh SC, Dôlnik VR, Garilou VM (1977) Avian energetics. In: Pinowski J, Kendeigh SC (eds) Granivorous birds in ecosystems. Cambridge University, Cambridge, pp 127–204

McDougall I (1971) Geochronology in Marion and Prince Edward Islands. In: Van Zinderen Bakker EM, Winterbottom JM, Dyer RA (eds) Marion and Prince Edward Islands. A A Balkema, Cape Town, pp 396–408

McQuaid CD (1980) Spatial and temporal variations in rocky intertidal communities. PhD thesis, University of Cape Town

Paine RT (1976) Size-limited predation: an observational and experimental approach with the *Mytilus-Pisaster* interaction Ecology 57:858–873

Paine RT (1980) Food webs: linkage, interaction strength and community infrastructure. The third Tansley lecture. J Anim Ecol 49:667–685

Picken GB (1980) The distribution, growth and reproduction of the Antarctic limpet *Nacella (Patinigera) concinna* (Strebel 1908). J Exp Mar Biol Ecol 42:71–85

Siegfried WR, Williams AJ, Burger AE, Berruti A (1978) Mineral and energy contributions of eggs of selected species of seabirds to the Marion Island terrestrial ecosystem. S Afr J Antarct Res 8:74–86

Siegfried WR, Forbes PF, Condy PR (1979) Scientific research at the Prince Edward islands, 1847–1979; a bibliography. S Afr J Antarct Res 9:35–41

Simpson R (1976) Physical and biotic factors limiting the distribution of littoral molluscs on Macquarie Island (sub-Antarctic). J Exp Mar Biol Ecol 21:11–49

Simpson RD (1982) Reproduction and lipids in the sub-Antarctic limpet *Nacella (Patinigera) macquariensis* Finlay, 1927. J Exp Mar Biol Ecol 21:11–49

Targett TE (1981) Trophic ecology and structure of coastal Antarctic fish communities. Mar Ecol Prog Ser 4:243–263

Van Zinderen Bakker EM (1971) Introduction. In: Van Zinderen Bakker EM, Winterbottom JM, Dyer RA (eds) Marion and Prince Edward Islands; report on the South African biological and geological expedition 1965–1966. AA Balkema, Cape Town, pp 1–15

Winberg GG (1956) Rate of metabolism and food requirements of fish. (Engl. transl. from Russian: Fish Res Board Can Transl Ser No 194)

Durvillaea antarctica Production in Relation to Nutrient Cycling at Marion Island

P.G. HAXEN and J.R. GRINDLEY[1]

Summary. An account is presented of the production of *Durvillaea antarctica* at Marion Island, including its inter-relationships with higher trophic levels. Biomass was determined for shores with differing degrees of exposure. Data from aerial photography indicated that 48% of the coast is exposed and directly struck by swells, 33% is sheltered or parallel to swell direction, 15% is large boulder or block shore, and 2% is sandy or small-pebble beach. The total fresh mass of *D. antarctica*, approximately 3,300 t (ww), is situated with varying proportion along each shore type of the island. Calorific content varies with season and with plant organ. Recruitment generally takes place after August on sites having become available after autumn and winter storms. Production in a re-colonizing stand of kelp sporelings is equivalent to 172 × 10^3 kJ m^{-2}. Low erosion rate and high mortality is a feature of the *D. antarctica* population. The primary contributors to mortality are grazing by intertidal herbivores and occasional, violent storms. Holdfasts are weakened by boring isopods. Affected plants are torn free and cast onto beaches during storms, where they are attacked by grazing Amphipoda in the spray zone. Beyond this zone, decomposition takes place through bacterial action.

1 Introduction

Marion Island (46°54′S, 37°45′E) appeared as the result of volcanic activity approximately 276,000 years ago (McDougal 1971), so the biotic communities it supports are relatively young assemblages. Environmental conditions are characterized by cold water with little seasonal temperature variation, and inshore nutrient status is unknown. Frequent cloud cover allows on average only 3.5 h of sunshine daily.

The first biological expedition to visit the Island was that of the Challenger in 1873. Moseley (1879), the naturalist in the shore party, drew attention to the fact that "The rocks, about high-tide mark, are covered with a dense growth of the large brown seawed, *D'Urvillaea utilis* [= *D. antarctica*] which is of great assistance in breaking the surf". Further observations are included in the monograph on the Marion and Prince Edward Islands (Van Zinderen Bakker et al. 1971), and in De Villiers (1976), Grindley (1978), and Blankley and Grindley (1984).

Durvillaea antarctica occurs widely in the Sub-Antarctic region, Tierra del Fuego, Chile, New Zealand, and islands of the Southern Ocean including the Falkland Islands. By the early 20th century the general distribution of the species was known. More recent reports indicate that *D. antarctica* is present at Gough Island (Chamberlain 1965) and at Crozet and Kerguelen islands (Delepine 1964). *Durvillaea antarctica* is a large complex plant. At Marion Island it forms dense stands in the intertidal breaker-surge zone. It has a smooth conglomerate, almost hemispherical holdfast with stipes up to 45 mm in diameter, 300 mm long, and a large buoyant divided lamina up to 7.08 m long with smooth margins. These features are variable, since they may be modified environmentally. The palmate region is non-meristematic and is structured in such a way as to serve a shock-absorbing function between lamina and stipe. The most important factor modifying morphology is exposure to swell. Plants tend to become longer and less divided with shorter stipes in more sheltered conditions. The mature sporophyte of *D. antarctica* is dioecious with conceptacles densely scattered over both sides of the lamina, and similar in size and appearance. The germling develops from the fertilized zygote with no heteromorphic alternation in generations. Hay (1979) has provided a nomenclatural and taxonomic review of the genus *Durvillaea*.

2 Materials and Methods

In harvesting kelp plants from transects, large individuals were cut at the base of the stipe with a sharp diving knife, hatchet or spade and thrown up onto the shore out of wave reach. Smaller plants were placed directly into a diver's collecting net. When all plants in the transect had been cleared they were gathered up, about 25 kg per load, and carried to the laboratory for weighing, counting, and further analysis.

Holdfasts were levered off the substrata with a combination of hatchet, hammer, and diving knife. The holdfasts were usually concavely excavated by grazing herbivores, so that they tended to be strongly attached only by a narrow outer perimeter. After cutting the perimeter, the holdfast could be dislodged and was placed in a bucket as quickly as possible to minimize faunal loss. To calculate the percentage of available space occupied by the holdfasts, holdfast dimensions were noted before removal. In the laboratory, they were inspected for faunal content and measurements were

1 School of Environmental Studies, University of Cape Town, Rondebosch 7700, South Africa

Antarctic Nutrient Cycles and Food Webs
(ed. by W.R. Siegfried, P.R. Condy, and R.M. Laws)

made for each plant of the mass and length of both frond and stipe, stipe diameter, number of digits and state of fertility. Mass figures for stipes and fronds for each transect were summed later. Energy content of the plant and associated invertebrates was determined by bomb calorimetry.

De Villiers (1976) in his study of the littoral ecology of Marion Island recognized a number of different shore types. Unbroken shores are primarily made up of vertical or very steep cliffs and it is rare to find shores inclined at less that 50° to the horizontal. Boulder shores slope gently to the sea or exist as chaotically piled blocks of lava. Intermediate conditions between these extremes occur at some points. During the present study it was realized that both shore type and the angle at which the shore is presented to the prevailing swell direction have an important bearing on the biomass and population structure of kelp present at these localities. For this reason it was considered important to obtain biomass estimates for each of the shore categories. Using available aerial photographs, the shore of the island was categorized, on a percentage basis, into each of four types. Biomass was estimated within each shore type and then extrapolated on a proportional basis for the entire island's coastline. No kelp is attached to 1.72 km of sandy beach. Biomass values are expressed in two ways. Variation according to shore type is expressed as kg m^{-2} and km^{-1} of shore.

Durvillaea antarctica has a diffuse meristem, unlike the "moving belts of tissue" (Mann 1972) that characterize the Laminariales. This fact makes established hole-punching techniques for measuring growth considerably more difficult to implement. A further difficulty arises because the laminae are not tough and leathery. Since the cortex contains an air-filled honeycomb meristoderm, even slight holes are soon torn. Intense movement in the swell at a point of weakness leads to severing at that point. It was therefore necessary for conventional hole-punching methods (Mann 1972; Dieckmann 1980) to be modified in order to measure growth rate. Counts of hole numbers as well as increases in distance between holes had to be accurately monitored. Mean seasonal increments in total length and in stipe diameter were measured (Hay 1979). The method can be criticized in that account was not taken of erosion at frond tips. However, the rate of frond loss at the tips of *D. antarctica* is small. This was found from separate experiments in which tags placed 150 mm from the tips of fronds were still present up to 6 months later; the tips having grown up to 12 mm during this period. Tag disappearances were only noted after loss of a digit or loss of a whole plant.

Observations were made of grazing herbivores, and particularly the limpet, *Nacella delesserti*, to determine its role in cycling material produced by *Durvillaea*. Seasonal variations in the abundance of the burrowing isopod, *Limnoria antarctica*, were monitored to assess its role in *Durvillaea* mortality. Grazing experiments with Amphipoda, *Hyale grandicornis* and *Jassa falcata*, in wire cages were carried out to determine the rate of consumption of *Durvillaea* in the intertidal zone. Measurements of the rate of bacterial decay of *Durvillaea* when stranded beyond the spray zone were made.

3 Results

Data on biomass and energy of the frond, stipe, and holdfast portions are summarized in Table 1. Nine transects were studied, three located on each of the exposed, sheltered and boulder shores. Data from aerial photography indicated that 48% of the coast is exposed and directly struck by swells, 33% is sheltered or parallel to swell direction, 15% is large boulder or block shore, and 2% is sandy or small-pebble beach.

On exposed shores, fronds represent only 60%–68% of the standing-crop harvested. On more sheltered shores, the frond material amounted to 75%–82%. A reverse proportion is true of stipes. Boulder shores show more variability exemplified by the frond fraction lying between 60% and 74% with a mean of 67% (SD ± 7.12). The remaining percentage portion is stipe and holdfast material. The former shows an opposite pattern of variation. The figures presented in Table 2 represent the biomass for the length of each shore type. These figures have been used as a basis on which bio-

Table 1. Mean biomass and energy of *Durvillaea antarctica* fronds, stipes, and holdfasts on three different types of shore

Shore type	Means for each plant organ for each shore type	Mass harvested (g × 10^6)	Fresh mass (g × 10^7 km^{-1})	Dry mass (g × 10^6 km^{-1})	kJ km^{-1} (× 10^6)	Fresh mass (g × 10^4 m^{-2})	Dry mass (g × 10^3 m^{-2})	kJ × 10^4 m^{-2}
Exposed	Fronds	7.043	1.538	2.821	30.860	3.023	5.554	6.065
	Stipes	1.347	0.538	0.802	9.551	0.105	2.193	1.634
	Holdfasts	0.422	0.299	0.624	4.646	0.754	1.570	1.170
Sheltered	Fronds	10.450	4.334	7.949	86.960	12.790	23.460	25.660
	Stipes	1.554	0.702	1.047	12.460	2.110	3.144	3.744
	Holdfasts	1.020	0.430	0.896	6.672	1.269	2.645	1.970
Boulder	Fronds	7.742	3.235	5.934	64.920	1.895	3.475	3.802
	Stipes	2.549	1.115	1.637	19.500	0.636	0.948	1.129
	Holdfasts	0.950	0.475	0.989	7.371	0.271	0.565	0.421

Table 2. Total biomass and energy values of different organs of *D. antarctica* for three types of shore on Marion Island. No kelp is attached to 1.172 km of sandy beach

Shore category and plant organ		Shore length (km)	Fresh mass ($g \times 10^8$)	Dry mass ($g \times 10^8$)	kJ ($\times 10^8$)
Exposed	Fronds		5.880	1.078	11.800
	Stipes	38.23	2.058	0.307	3.651
	Holdfasts		1.145	0.238	1.776
Sheltered	Fronds		11.240	2.061	22.550
	Stipes	25.93	1.821	0.272	3.231
	Holdfasts		1.115	0.232	1.730
Boulder	Fronds		3.856	0.707	7.738
	Stipes	11.92	1.329	0.195	2.324
	Holdfasts		0.567	0.117	0.879

mass values, obtained from transects within each shore category, have been extrapolated to represent the biomass of *D. antarctica* on the island as a whole. The estimated total is 3,300 t (wetwt.).

A high mortality rate is characteristic of the *D. antarctica* population at Marion Island. Settlement of *D. antarctica* sporelings at the start of the growing season is up to 20,000 individuals m^{-2} but numbers decrease rapidly to between 100 and 200 m^{-2} after 1 year. Settlements of *D. antarctica* sporelings occur on space cleared in autumn and early winter through holdfast detachment or boulder rotation. Some of the plants which had settled in August exceeded 100 mm in length by October. At the end of the first year of growth, August 1979–July 1980, production for this period was 90.1 kg (wetwt.) m^2. The mean energy value for juvenile kelp was 11.1 kJ g^{-1}. Thus, production by a re-colonized stand of kelp sporelings over a period of 1 year is 150×10^3 kJ m^{-2}. Loss by erosion was assessed by an analysis of difference in mean weight of apparently intact plants and apparently eroded plants. In juvenile plants this amounted to 21% of the laminal portion of the plants unaccounted for in production estimation. The mean laminal portion of total biomass in juveniles is 71% and therefore this increases production estimates by 15.75% to 172×10^3 kJ m^{-2}. No attempt was made to measure loss as dissolved organic matter.

The limpet, *Nacella delesserti*, grazes on *D. antarctica* sporelings. When plants are small with holdfast diameters < about 30 mm, a limpet can destroy an entire plant. However, once plants become larger and holdfasts coalesce they were found to be immune to attack by *Nacella*. Once holdfasts become larger than about 100 mm they begin to be colonized by the burrowing isopod, *Limnoria antarctica*. Holdfasts eventually become extensively burrowed with narrow tunnels (approximately 4 mm diameter) proliferating in every direction. Most activity is close to the holdfast surface and small vents to the exterior are made consistently. These allow tunnels to be perfused with fresh oxygenated water. Up to 55 juveniles hatch in tunnels terminating in soft nutritious tissue. These burrow actively in all directions, forming an enlarged nursery.

Seasonal variations occur in the *Limnoria antarctica* populations. Mortality of *Durvillaea* results in death of individuals unable to escape from the dislodged holdfasts. Holdfasts weakened by boring are torn off in storms, providing space for re-colonization by juvenile sporelings. The breeding cycle of *L. antarctica* appears to be in synchrony with periods of rough seas, both of which reach peaks in February – March and August–October. A high frequency of the first size class of *L. antarctica* appears in May and in December, 3 months later in each case. February and September are the breeding peaks. Those holdfasts left have large populations of *Limnoria* ready for colonization.

Burrowing isopods weaken holdfasts, thus contributing to plant mortality in storms. Mature plants are then washed up onto the shore or into crevices between boulders. These plants are colonized by dense aggregations of amphipods, 438 ± 260 g kg^{-1} (wetwt.). The most important species contributing to kelp breakdown are *Hyale grandicornis* and *Jassa falcata*. Grazing pressure experiments showed that the kelp is consumed at a rate of 1.55 ± 0.3 mg wetwt. d^{-1} per amphipod. Amphipods in the intertidal zone and bacteria in higher regions were found to play a role in degrading kelp.

There are few inshore filter feeding organisms, but organic material released feeds the filter feeding pelecypods, *Kideria bicolor* and *Kideria oblonga* (Blankley and Grindley 1984). Kelp thrown beyond the spray zone is decomposed by a succession of bacteria but owing to low temperatures the process is slow, complete degradation taking from 2–4 months. The water temperature generally lies between 4 °C and 6 °C so bacterial colonization of particulate organic material is slow.

4 Discussion

Durvillaea antarctica is the dominant intertidal primary producer at Marion Island and is therefore of considerable significance to nutrient cycling at this island.

Littoral ecosystems are often largely dependent on the primary production of macro-algae. The kelp bed communities of the Cape Peninsula, South Africa, are dominated by the algae (Field et al. 1980). Rocky shores often have algae making up the greatest part of their biomass, particularly on sheltered shores. On Marion Island algae have the highest biomass on the shores (Blankley and Grindley 1985). The algae and particularly *Durvillaea* are extensively grazed by *Nacella delesserti* which comprise the greatest faunal biomass. *Limnoria antarctica* are also important consumers burrowing into holdfasts. Beach-cast specimens of *Durvillaea* further support dense populations of amphipods on the shore and communities of decomposers in and beyond the spray zone.

In its pattern of growth *Durvillaea antarctica* suffers little distal attrition and so produces little fine detritus. As a result there are relatively few in-shore filter feeding organisms, although the small pelecypod *Kideria bicolor*, may be present in considerable numbers. In comparison detritivores play a much more important role in some other kelp bed systems (Mann 1972; Field et al. 1980).

At Marion Island the intertidal herbivores, particularly *Nacella delesserti*, are most important, while violent storms cast large quantities of algae ashore. *Durvillaea* thus provides a massive input of primary production for direct consumption in the intertidal zone and on the shore. Similar patterns have been observed for other comparable islands by Bunt (1955), Chamberlain (1965), and Delépine (1964). The 3,300 t (wetwt.) standing stock of *Durvillaea antarctica* thus apparently plays a fundamental role in primary production and nutrient cycling in the near-shore and littoral ecosystem of Marion Island.

Acknowledgements. We thank SASCAR and the Department of Transport for making this study possible.

References

Blankley WO, Grindley JR (1985) The intertidal and shallow subtidal food web at Marion Island. In: Siegfried WR, Condy PR, Laws RM (eds) Antarctic nutrient cycles and food webs (Proceedings of the fourth SCAR symposium on Antarctic biology). Springer, Berlin Heidelberg New York

Bunt JS (1955) The importance of bacteria and other microorganisms in the sea water at Macquarie Island. Austr J mar Freshwater Res 6:60–65

Chamberlain YM (1965) Marine algae of Gough Island. Bull Br Mus (Nat Hist) Botany 3:175–232

Delépine R (1964) Le genera *Durvillaea* dans les Iles Australe Francais. Com Nat Franc Recherch Antarct 1:141–147

De Villiers AF (1976) Littoral ecology of Marion and Prince Edward Island (Southern Ocean). S Afr J Antarct Res Suppl 1:1–40

Dieckmann GS (1980) Aspects of the ecology of *Laminaria pallida* (Grev) J Ag off the Cape Peninsula. Bot Mar 23:579–585

Field JA, Griffiths CL, Linley EA, Carter RA, Zoutendyk P (1980) Upwelling in a nearshore marine ecosystem and its biological implications. Estuarine Coastal Mar Sci 2:133–150

Grindley JR (1978) Marine ecosystems of Marion Island. S Afr J Antarct Res 8:38–92

Hay CH (1979) A phytogeographical account of the Southern Bull-kelp seaweeds *Durvillaea* spp Bory 1826 (Durvilliales Petrov 1965) Proc int symp on marine biogeography in evolution in the southern hemisphere 1978, N Z DSIR Information Services 137 (2)

Mann K (1972) Ecological energetics of the seaweed zone in a marine bay on the Atlantic coast of Canada, II Productivity of the seaweeds. Mar Biol 14:199–209

McDougal I (1971) Geochronology. In: Marion and Prince Edward Islands. Report on the South African biological and geological expeditions 1965–66. Van Zinderen Bakker EM Sr, Winterbottom JM, Dyer RA (eds). AA Balkema, Cape Town

Moseley HN (1879) Notes by a naturalist on the 'Challenger', being an account of various observations made during the voyage of HMS 'Challenger' round the world in the years 1872–1876. MacMillan, London

Van Zinderen Bakker EM Sr (1971) In: Van Zinderen Bakker EM Sr, Winterbottom JM, Dyer RA (eds) Marion and Prince Edward Islands. Report on the South African biological and geological expeditions 1965–1966. AA Balkema, Cape Town

Impact of Fur Seals on the Terrestrial Environment at South Georgia

W.N. BONNER [1]

Summary. Antarctic Fur Seals, *Arctocephalus gazella*, have increased from a few tens of individuals at the beginning of this century to over 900,000 now, most occurring at the north-west end of South Georgia. At Bird Island in recent years some 150,000 seals have come ashore in their breeding season. Since 1960 profound changes have taken place in the vegetation. Fur Seals destroy tussock grass, *Poa flabellata*, which is the dominant plant cover, by trampling and lying on the tops of the tussocks. Meadows of *Deschampsia antarctica* and cryptogams are also destroyed by trampling and this can lead to local erosion. Destruction of the tussock deprives birds (the endemic Pipit, *Anthus antarcticus*, and Pintail, *Anas georgica*, and several burrowing petrels) of breeding habitat and exposes them to predation by Antarctic Skuas, *Catharacta antarctica*. The destruction of vegetation by Fur Seals is believed to be a new phenomenon associated with high density occupation by the recovering population. This may reach 4×10^6 by the year 2000, and the effects are likely to increase. The successful recovery of the Fur Seals may well be accompanied by significant decreases in several bird populations.

1 Introduction

Antarctic Fur Seals, *Arctocephalus gazella*, were numerous at South Georgia, South Atlantic Ocean, until nearly exterminated by skin hunters during the nineteenth century. A remnant survived, probably on inaccessible islands just off the north western extremity of South Georgia. They may have numbered only a few tens of animals (Bonner 1964; Payne 1977). This population underwent an explosive increase some time after the 1930's and now numbers ca. 900,000 individuals (including a relatively small number on other island groups of the Scotia Arc). The bulk of this population is still concentrated at the north western end of South Georgia. At Bird Island (Fig. 1), which is only some 5.5 km long by 1.5 km wide, and which has less than 5 km of shore-line accessible to seals, about 150,000 Fur Seals now come ashore annually during the breeding season from November to April. This period corresponds to the growing season of the plants on the island.

Profound changes have taken place in the vegetation cover of Bird Island since I first visited it in 1956, and these can be attributed to the physical impact of the Fur Seals. Because such a major increase in Fur Seal numbers was not foreseen at the time of earlier visits, systematic quantitative studies of the effect of the Fur Seals on the vegetation were not undertaken from the start. However, such studies, involving the erection of exclosures, will shortly be started at Bird Island. This paper describes the relevant behaviour of the Fur Seals and their effect on the vegetation, and its implications for other biota at Bird Island.

1 British Antarctic Survey, NERC, High Cross, Madingley Road, Cambridge CB3 OET, United Kingdom

2 Behaviour of Fur Seals

2.1 Breeding Seals

In uncrowded situations, Antarctic Fur Seals form breeding groups on rocky beaches which often incorporate small shingle patches merging into cliff talus (Bonner 1968). In such localities adult males establish territories which they defend against other males. In these territories groups of females arrive to produce their pups.

Breeding females arrive ashore in November or early December about 2 days before giving birth. Each female remains with her new-born pup for about 8 days before coming into oestrus and mating with the male in whose territory she is (Bonner 1968). The female then leaves the beach to start a series of feeding excursions to sea. On returning from a feeding excursion, the female will usually suckle her pup in an area away from male territories, which, in the crowded conditions now prevailing at Bird Island, means inland behind the beach. Characteristically at South Georgia, this inland ground is covered with a coarse tussock grass, *Poa flabellata*. The lactating female will usually choose a clump of tussock to lie on while feeding her pup (Fig. 2). The preference shown by the majority of lactating females to avoid the male territories at this stage frees them from the attentions of sexually active males and from being involved in disturbance resulting from agonistic encounters between males.

The females repeat alternate sequences of feeding at sea and returning to suckle their pups until early April when the pups are weaned. Soon after this, the females desert the beaches until the next breeding season (Bonner 1968).

Antarctic Nutrient Cycles and Food Webs
(ed. by W.R. Siegfried, P.R. Condy, and R.M. Laws)

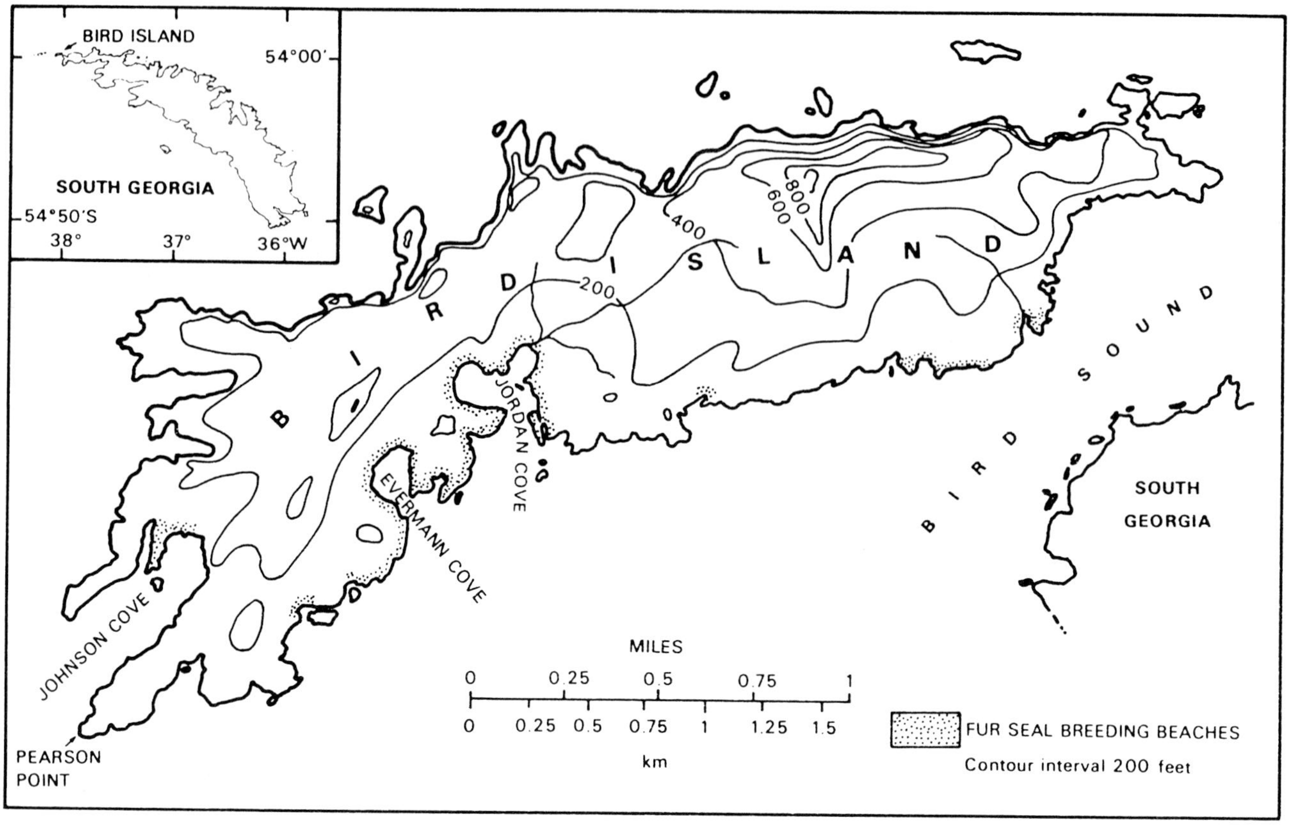

Fig. 1. Location of Bird Island, South Georgia

Fig. 2. Female Antarctic Fur Seal feeding pup. Note the typical attitude on a tussock clump

2.2 Non-Breeding Seals

Throughout the breeding season juveniles of both sexes, sub-adult and some senescent males, besides breeding animals, are ashore. The non-breeding animals in general avoid the breeding beaches and move inland over the tussock-covered slopes. Details of the duration of the stay ashore for different age and sex classes are not available, but individuals have been observed to spend at least 3 days ashore at a time (Bonner 1968). Like the lactating females, these animals, except for large males, show a preference for lying on top of tussock clumps. Seals of all classes tend to use relatively well-defined access routes to the inland and upland areas, often following water-courses.

3 Vegetation Cover at Bird Island

Greene (1964) has described the vegetation of South Georgia. Of the vegetated areas in the localities occupied by Fur Seals, tussock grassland is predominant. Hunter et al. (1982) determined by planimetry of low level aerial photographs that tussock grassland covered 190 ha (48%) of Bird Island. Greene (1964) described two growth forms of tussock, depending largely on the density of the plants. His categories are not always easily distinguished at Bird Island, but there is a general distinction between a more vigorous type, characteristic of areas with N levels enriched by bird and seal manure, where the grass grows in distinct clumps, and a less vigorous, closed community with the plants crowded together with little development of stools (Greene 1964; Plate Ib). Greene suggested that the latter predominated at Bird Island, but in fact the reverse is the case, probably because of the consistently high levels of manuring by birds (and then by seals).

The tussocks that fringe the beaches used by the Fur Seals are generally 0.5–0.6 m in diameter and about 0.4 m high, though they may exceed 1.0 m in both diameter and height. When undisturbed, the leaves grow out from the centre of the tussock. The spreading crown of living shoots is surrounded by a pendant fringe of attached dead leaves which cast a deep shade in which other plants cannot flourish. The dead and dying leaf bases, together with wind-blown inorganic material, form a peaty stool which slowly extends upwards. Deep gutters exist between neighbouring plants and at lower altitudes these are usually very wet. On higher or drier slopes and on less well-fertilized ground the tussock occurs as shorter plants with little stool development. Although pure associations are the rule, there may be an understorey of the suffruticose herb, *Acaena magellanica*, or mosses.

In places around the banks of streams or on raised beaches dense closed swards of Antarctic Hair Grass, *Deschampsia antarctica*, may occur. In some places, generally above an elevation of about 60 m, these are extensive enough to be referred to as meadows (Hunter et al. 1982). Shadier places, as on the banks of streams, may develop a sward with a high proportion of bryophytes. At the highest levels, or on scree or rock outcrops wherever they occur, there may be an open fellfield vegetation composed largely of *Colobanthus quitensis, Acaena tenera*, sparse tufts of the grass *Phleum alpinum* and various mosses, liverworts and lichens.

4 Damage to the Vegetation

When a Fur Seal lies on the crown of a tussock, the leaves and shoots are crushed (Fig. 3), but this damage, if not soon repeated, is only temporary. However, if a tussock is regularly used by the Fur Seals, the leaves in the centre of the crown are broken down, destruction of the shoot bases kills the growing points of new shoots, and the centre of the stool becomes a bare peaty mound, surrounded by a fringe of downwardly directed leaves (Fig. 4). Even in such a condition the grass can flower and set seed, but its vigour is reduced, and continued use by the Fur Seals leads to the death of the whole plant. The peaty stool becomes smooth and rounded and may be colonized by algae; it is then gradually eroded by the Fur Seals, although once all its leaves are dead it becomes less attractive as a resting place.

Fig. 3. Trampling damage to tussock caused by Fur Seals

Fig. 4. Severe erosion of tussock clumps

This type of damage is associated predominantly with the lactating females. It appears that they settle on the tops of the tussocks, as this allows them to avoid getting their fur clogged with the mud that generally covers the ground. The majority of lactating females do not go very far inland, so this extreme form of tussock damage is confined to a belt immediately behind the breeding beaches. Juvenile animals, particularly young and sub-adult males, however, travel far inland, occasionally reaching altitudes above 100 m. On the higher tussock grassland which, as noted above, does not show marked stool development, their effect is mainly to crush down the leaves. So far this damage has not been observed to result in the death of the plants.

The continual passage of Fur Seals from the sea to their inland resting sites causes severe impact on the vegetation along their routes. Because the seals tend to use stream-courses for routes, the areas of *Deschampsia* and cryptogam-dominated sward which lack resistance to trampling are soon destroyed. At Freshwater Inlet, a raised beach (about 80 $\times$ 100 m in extent, some 2 m above the existing sea level) was covered in the 1950's with scattered clumps of tussock and a continuous sward of *Deschampsia*, dissected by the tributaries of a small stream flowing between discrete banks of grass-topped beach deposits about 0.3 m high. This area was used as a main access corridor to the interior of the island by the Fur Seals. Signs of deterioration and erosion had appeared in 1967 (R.I.L. Smith, pers. comm.). By 1975 the *Deschampsia* sward had disappeared completely, save around the edges of the area, and its underlying peat and fine beach deposits had been washed out to sea to form a gravel fan at the mouth of the stream. The stream and its tributaries spread as broad ribbons of water with no defined banks. Although actual measurements are impossible now, it is estimated from observations of turf collars around some hold-down pickets inserted in 1958 that, about 0.15 m of material has been lost by erosion consequent on the destruction of the *Deschampsia* sward from the central and seaward parts of this area.

5 Effects on Other Biota

The direct effect on the vegetation is the most conspicuous, but not the only effect of the impact of the Fur Seals. Two endemic avian species, the Antarctic Pipit, *Anthus antarcticus*, and the South Georgia Pintail, *Anas georgica*, typically nest in tussock grass. The Pipit builds its nest of woven grass in the crown of tussock leaves, whereas the Pintail nests under the shelter of the overhanging leaves. Also, the Pipit appears to find much of its food on the ground between the tussocks where perimylopid beetles are abundant. The loss of tussock grass deprives both these birds of protected nest sites and, in the case of the Pipit, of foraging habitat as well.

Numerically far more important than the Pipit and Pintail are several species of burrowing petrels: the Dove Prion, *Pachyptila desolata*; the Blue Petrel, *Halobaena caerulea*; the Common Diving Petrel, *Pelecanoides urinatrix*; and the Whitechinned Petrel, *Procellaria aequinoctialis* (in order of abundance, Hunter et al. 1982). These all excavate burrows in the peaty soil beneath tussock grassland. The trampling of the Fur Seals can cause burrows to collapse, destroying nests, and flattens the tussock cover, leaving the smaller birds exposed to predation by Brown Skuas, *Catharacta lonnbergia*, when on the ground.

In many parts of mainland South Georgia these ground and burrow-nesting birds are already adversely affected by introduced Brown Rats, *Rattus norvegicus* (Pye and Bonner 1980), and the remaining rat-free offshore islands, of which Bird Island is a prime example, are important refuges for the birds. At Bird Island, some 60% of the tussock grassland is affected to some extent by Fur Seals (Croxall et al., to be publ.), and few areas of flat coastal grassland now afford shelter to breeding petrels; only steep slopes and rock ledges are left undisturbed.

The destruction of the tussock grassland profoundly alters the habitat of many invertebrates, including the perimylopid beetles mentioned above, other Coleoptera, Diptera, Collembola, and Acarina, as well as lower groups. However, no observations have been made on the abundance of these animals in areas affected or unaffected by the Fur Seals.

6 Discussion

The observed effects of the Fur Seals on the vegetation have, with very few and localized exceptions, arisen since 1960 and, thus, represent a recent phenomenon. It may be asked whether this is in fact no more than a return to the pre-exploitation conditions of the late eighteenth century, when the Fur Seal populations were in a pristine state, or whether it represents a completely new development.

The size of the pre-exploitation population is not known. From what records are available, principally the estimate by Weddell (1825) that by 1822 1.2×10^6 skins of Fur Seals had been taken at South Georgia and that the species was very nearly extinct, it seems unlikely to have been very much larger than the present-day population. It is known, from accounts of the areas used by the sealers and the occurrence of Fur Seal hairs in peat (R.I.L. Smith, pers. comm.), that the breeding animals were very much more widely distributed along the coast of the main island than they are today. This implies that there was probably a lower density of Fur Seals overall, with breeding beaches separated by stretches of coast where no breeding took place and non-breeding animals could have access to the shore without the need to go far inland. This is what happens today on parts of the South Georgia coast, as yet uncolonized by breeding Fur Seals.

One piece of evidence suggests strongly that the conditions seen at Bird Island are more extreme than they have been in the past several millenia: this is the destruction of the raised beach at Freshwater Inlet, referred to above. Clapperton (1971) described raised beaches on mainland South Georgia, and has suggested (pers. comm.) that the beach at Bird Island (which shows no evidence of glaciation) is likely to be of the order of 10,000 years old. The persistence of this feature throughout the pre-exploitation period, to be rapidly eroded in the course of 15 years after 1960, would indicate that recent Fur Seal densities are much higher than previously.

The effect of the Fur Seals on the terrestrial ecosystem is thus a new one. As no regeneration has been observed to take place in the devastated areas, the present situation is likely to get worse and the breeding environment of the Pipit, Pintail and the petrels will continue to deteriorate. There may even be significant population effects in the first two, if they are exposed to heightened predation from Antarctic Skuas on the ground.

The Fur Seal population is continuing to expand (though at Bird Island it has probably reached a peak). The ultimate level cannot be predicted, though Laws (1984) suggests a flattening of the curve of increase at a figure of 4×10^6 in the year 2000. With the decrease in abundance of krill-eating baleen whales (Laws 1984), the carrying capacity of the environment for Fur Seals (which also feed on Krill, *Euphausia superba*) may have increased significantly and the final figure reached may be very substantially above pre-exploitation values.

The recovery of a nearly exterminated population of a large mammal is generally seen as a desirable event. However, the explosive increase of Fur Seals at Bird Island, the resultant damage to the vegetation and the possible serious consequences for the birds (including two species which are much less abundant than the Fur Seals) is less desirable. This example demonstrates that, although components of an ecosystem may be resilient and capable of recovery from even extreme reduction, the recovered system may differ in important respects from the pristine condition.

The problem of local abundance of a protected species, and consequent damage to the environment is familiar enough (e.g., Jewell and Holt 1981). A common response to this in conservation practice today is management. This involves deciding arbitrarily what is the desired system and then taking active steps to achieve and maintain this objective. It would be possible to devise management policies for Fur Seals at South Georgia, which would control damage to the vegetation. Such a scheme might involve fencing off vulnerable areas, or culling Fur Seals, or a combination of both. No proposals for management have been made (though experimental enclosures may protect some areas). It remains to be seen what the outcome will be over the next two decades or so.

Acknowledgements. I am grateful to John Croxall, Bill Doidge, Dick Laws, Ron Lewis Smith, and Séamus McCann, who read the manuscript, or allowed me to use unpublished material, or both.

References

Bonner WN (1964) Population increase in the fur seal, *Arctocephalus tropicalis*, at South Georgia. In: Carrick R, Holdgate M, Prevost J (eds) Biologie Antarctique. 1962, Hermann, Paris

Bonner WN (1968) The fur seal of South Georgia. Br Antarct Surv 56:1–8

Clapperton CM (1971) Geomorphology of the Stromness Bay and Cumberland Bay area, South Georgia. Br Antarct Surv Sci Rep 70:1–25

Croxall JP, Prince PA, Hunter I, McInnes S, Copestake PG (to be publ.) The seabirds of the Antarctic Peninsula, islands of the Scotia Sea and Antarctic Continent between 80° W and 20° W: their status and conservation. In: Croxall JP, Evans PGH, Schreiber RV (eds) The status and conservation of seabirds. Cambridge University Press, Cambridge

Greene SW (1964) The vascular flora of South Georgia. Br Antarct Surv Sci Rep 45:1–58

Hunter I, Croxall JP, Prince PA (1982) The distribution and abundance of burrowing seabirds (Procellariiformes) at Bird island, South Georgia: 1. Introduction and methods. Br Antarct Surv Bull 56:49–67

Jewell PA, Holt S (eds) (1981) Problems in management of locally abundant wild mammals. Academic, London

Laws RM (1984) Seals. In: Laws RM (ed) Antarctic ecology. Academic, London, pp 621–715

Payne MR (1977) Growth of a fur seal population. Philos Trans R Soc London B Biol Sci 279:67–79

Pye T, Bonner WN (1980) Feral Brown rats, *Rattus norvegicus*, in South Georgia (South Atlantic Ocean). J Zool (Lond) 192:237–255

Weddell J (1825) A voyage towards the South Pole, performed in the year 1822–1824. Longman, Hurst, Tees, Orme, Brown and Green, London

Chlorinated Hydrocarbons in Seabird Eggs from the Southern Atlantic and Indian Oceans

B.D. GARDNER[1], W.R. SIEGFRIED[2], and
A.D. CONNELL[1]

Summary. The eggs of 24 species of seabirds breeding at Marion and Gough islands, and at islands off the west coast of South Africa, were analyzed for concentrations of chlorinated hydrocarbons (DDE, DDT, PCB, and dieldrin). The eggs of all 24 species contained organochlorines, but the levels of concentrations were far lower than those found for certain seabirds in North America and Europe. Residue levels in birds permanently resident in South African coastal waters were up to 10 times higher than those of related species breeding at Marion Island in the Southern Ocean. It appears that the incidence of chlorinated hydrocarbons is low in the sea in the vicinity of Marion Island. In general, species which spend most, if not all, of their lives south of the Sub-Tropical Convergence, and whose diet consists mainly of planktonic crustaceans, apparently are less susceptible to contamination by chlorinated hydrocarbons than species which range farther to the north, and which feed on squid, fish or birds. The eggs of species which regularly scavenge dead, and prey on moribund, birds and other animals near the tops of food chains contained relatively high concentrations of chlorinated hydrocarbons. Eggs collected after 1980 contained higher PCB levels than eggs of the same species collected in an earlier period.

1 Introduction

Chlorinated hydrocarbon residues occur commonly in coastal seabirds in the northern hemisphere, but there is a paucity of information on the degree of contamination of seabirds breeding in areas far from known sources of pollution (Ohlendorf et al. 1978), particularly in the southern Atlantic and Indian oceans, and the Antarctic and Sub-Antarctic regions (Tatton and Ruzicka 1967; Risebrough and Carmignani 1972; Johnston 1973; Bennington et al. 1975).

High levels of chlorinated hydrocarbon residues have been found in the fat-storage tissue of feral Cats, *Felis catus*, at Marion Island (46°54′S, 37°45′E) in the Indian Ocean (Gardner 1983). Since the Cats feed mainly on seabirds (Williams 1978), a study of the degree of contamination of the seabirds at Marion Island was warranted. Here we report on concentrations of pp′ DDT and its metabolite pp′ DDE, dieldrin and polychlorinated biphenyls (PCB) in the eggs of 24 species of seabirds, collected during 1979–1982 at Marion and Gough (40°20′S, 10°00′W) islands and at islands in Saldanha Bay (33°03′S, 17°58′E) on the west coast of South Africa. Tentative explanations are offered for differences found in the incidence of chlorinated hydrocarbon residues in eggs belonging to different populations of birds.

2 Methods

The eggs were wrapped in aluminium foil and stored frozen. "Whole eggs", without their shells, were analyzed, rather than the lipid fraction only, because some of the eggs contained embryos in various stages of development. In the laboratory, each egg was homogenized and 15 g of the homogenate was mixed with 30 g of anhydrous sodium sulphate and 50 ml hexane in an all-glass container, using a high-speed ultrasonic blender (Johnson and Starr 1972). The mixture was left to extract for a week. A 5 ml aliquot of the extract was passed through a clean-up column containing dry activated silica gel (Holden and Marsden 1969). The column was eluted with 10 ml hexane, followed by a further 30 ml of hexane. Both eluates were collected in separate containers. The first eluate contained the PCB, DDT, and DDE, whereas the second contained the dieldrin. Five microlitres of the eluate were injected into a gas chromatograph column. The gas chromatograph specifications were as follows: glass column 1.5 m × 3 mm O.D. containing mixed stationary phase 1.95% QF-1 + 1.5% OV-17 on Chromasorb WHP support at 220 °C; electron capture Ni^3 detector at 250 °C; and, for the mobile phase, high purity N at flow rate 40 ml min^{-1}. A second column containing OV-101 on Chromasorb WHP was used for confirmation. Further confirmation depended on exposing the eluate to ultraviolet light, causing DDE and dieldrin to decompose to new characteristic compounds (Glotfelty 1972).

Various sub-sets of the results of these analyses, particularly those for birds with taxonomic or ecological affinities and which are permanently resident in South African coastal waters or breed at Marion Island, were compared statistically to determine if there were significant (at least $P < 0.05$) geographical or diet-related differences in levels of chlorinated hydrocarbons. Since preliminary analyses indicated that there were significant differences between the variances in many of these comparisons, standard parametric tests

1 National Institute for Water Research, P.O. Box 17001, Congella 4013, South Africa
2 FitzPatrick Institute, University of Cape Town, Rondebosch 7700, South Africa

Antarctic Nutrient Cycles and Food Webs
(ed. by W.R. Siegfried, P.R. Condy, and R.M. Laws)

(e.g., t-tests and analysis of variance) were inappropriate. Therefore, the Mann-Whitney U test which does not require equivalency of variance (Sokal and Rohlf 1969) was employed throughout.

3 Results and Discussion

Although a few of the eggs were not fresh when collected, putrefaction does not significantly affect residue analysis for DDT and its metabolites (Mulhern and Reichel 1970) and organochlorine concentrations in the egg are reported to be similar to whole-body concentrations in the laying female (Keith and Gruchy 1972). However, some of the eggs which were not fresh could have been deserted ones, and contamination by chlorinated hydrocarbons might have had an influence in this regard. Hence, certain values for residue levels given here may represent extremes, and caution is required in extrapolating to larger populations. This problem can be overcome by completely randomizing the collecting of eggs.

Chlorinated hydrocarbon residues occurred in the eggs of all the species sampled (Tables 1-3), which accords with the general ubiquity of organochlorine contamination of aquatic birds (Stickel 1973). However, the concentrations of chlorinated hydrocarbons reported here tend to be far lower than those found in the eggs of certain seabirds in the northern hemisphere (Peakall 1975; Ohlendorf et al. 1982). Even the highest level (1,562 $\mu g\ kg^{-1}$) of DDE reported here, for the Antarctic Skua, *Catharacta antarctica* (Table 1), is $< 25\%$ of the mean DDE in eggs of the Great Skua, *Catharacta skua* (Furness and Hutton 1979) and $< 1\%$ of the level suspected of causing breeding failure in the Californian population of the Brown Pelican, *Pelecanus occidentalis* (Blus et al. 1974).

With the exception of dieldrin, the eggs of species breeding at islands in Saldanha Bay contained significantly higher levels of organochlorine residues than those of related species breeding at Marion Island. In the Kelp Gull, *Larus dominicanus*, eggs taken from islands in Saldanha Bay contained DDE levels significantly higher than eggs taken from Marion Island before 1980. Similarly, cormorants breeding at islands in Saldanha Bay had DDE concentrations 4-12 times higher than the Imperial Cormorant, *Phalacrocorax atriceps*, at Marion Island. DDE levels in the Jackass Penguin, *Spheniscus demersus*, were significantly (6-10 times) higher than in the three species of penguins at Marion Island; PCB concentrations in the Jackass Penguin were relatively high, as well. Thus, species which are permanently resident in areas close to the South African mainland presumably are exposed to a greater risk of organochlorine contamination than similar species breeding at remote localities in the Southern Ocean. Bennington et al. (1975) report a similar pattern of organochlorine contamination in seabirds off the coast of New Zealand, and the Snares and Auckland islands in the Sub-Antarctic.

The birds breeding at Marion and Gough islands can be divided roughly into groups according to their patterns of dispersion at sea. The Kelp Gull and the Imperial Cormorant are permanently resident at Marion Island where they forage mainly in the subtidal area (Siegfried 1978; Blankley 1981). Their eggs contained relatively low residues of DDE and PCB before 1980, as did those of the penguins *Aptenodytes patagonicus, Eudyptes chrysolophus,* and *E. chrysocome* (Table 1) which forage offshore, but not too distant from the island; the *Eudyptes* penguins probably feed within 100 km of the island during their breeding seasons (Williams and Siegfried 1980). The concentration of DDE in eggs of the Lesser Sheathbill, *Chionis minor*, which is the only landbird resident at Marion Island (Burger 1981), was significantly higher than the levels in the Kelp Gull, the Imperial Cormorant and the three species of penguins (Table 1).

The species which range relatively far from Marion and Gough islands can be divided as those which move into subtropical waters when not breeding, and those which tend to remain south of the Sub-Tropical Convergence. The latter group included the Fairy Prion, *Pachyptila turtur*, Broadbilled Prion, *P. vittata*, Blue Petrel, *Halobaena caerulea*, Kerguelen Petrel, *Pterodroma brevirostris*, and Softplumaged Petrel, *P. mollis*. DDE levels in these species were similar to those in the nearshore group, and only PCB was significantly higher. Species which are encountered more commonly north of the Sub-Tropical Convergence, and particularly in South African neritic waters, include the Greatwinged Petrel, *Pterodroma macroptera*, Whitechinned Petrel, *Procellaria aequinoctialis*, Yellownosed Albatross, *Diomedea chlororhynchos*, Wandering Albatross, *D. exulans*, and Antarctic Skua (Brooke 1981). The Great Shearwater, *Puffinus gravis*, ranges widely in the Atlantic as far north as the Arctic (Voous and Wattel 1963). DDE and PCB concentrations in most of these species were moderately high (Tables 1-2).

In addition to foraging dispersion, diet, and feeding behaviour are important considerations in interpreting chlorinated hydrocarbon levels in different species. The two *Eudyptes* penguins considered here feed mainly on planktonic crustaceans, whereas the King Penguin, *Aptenodytes patagonicus*, eats principally squid and fish (Croxall and Prince 1980). The significantly higher level of DDE in the King Penguin might reflect the fact that squid and fish are higher than crustaceans in the food chain of the Southern Ocean ecosystem. The prions and the Blue Petrel also feed predominantly on crustaceans (Croxall and Prince 1980) and, hence, as expected had significantly lower residues of DDE and DDT than the Kerguelen and Softplumaged petrels whose diets apparently include relatively more fish and squid (Watson 1975). All of these birds had far lower concentrations of chlorinated hydrocarbon residues than the Antarctic Skua and the Southern Giant Petrel, *Macronectes giganteus* (Table 1), which regularly scavenge dead, and prey on moribund, birds and other animals near the tops of food chains (Watson 1975), and ship-following species such as the Wan-

Table 1. Concentrations ($\mu g\ kg^{-1}$ for whole egg) of chlorinated hydrocarbons in eggs of seabirds at Marion Island

Species	DDE				DDT				PCB				Dieldrin			
	Mean[a]	S.D.	Range	No. eggs	Mean	S.D.	Range	No. eggs	Mean	S.D.	Range	No. eggs	Mean	S.D.	Range	No. eggs
King Penguin *Aptenodytes patagonicus*	56.1	32.7	27– 103	8	2.7	2.5	0– 8	8	29.4	20.3	0– 50	8	2.8	2.3	0 – 6	8
Rockhopper Penguin *Eudyptes chrysocome*	9.1	5.4	4– 20	8	2.9	1.0	1– 4	8	16.6	33.3	0– 96	8	0.9	1.7	0 – 5	8
Macaroni Penguin *Eudyptes chrysolophus*	14.7	4.2	9– 23	9	2.0	1.1	0– 3	9	5.4	10.9	0– 33	9	1.2	0.4	1 – 2	9
Wandering Albatross *Diomedea exulans*	227.3	119.5	84– 406	6	70.2	39.1	22–122	6	123.3	59.0	60– 211	6	1.3	1.2	0 – 3	6
Sooty Albatross *Phoebetria fusca*	83.0	22.6	58– 102	3	19.7	4.5	15– 24	3	75.3	37.2	50– 118	3	0.1	0.2	0 – 0.3	3
Southern Giant Petrel *Macronectes giganteus*	260.0	187.0	56– 424	3	11.0	9.5	5– 22	3	81.7	76.4	13– 164	3	1.7	1.2	1 – 3	3
Blue Petrel *Halobaena caerulea*	15.6	8.2	7– 32	9	8.3	8.0	2– 25	9	45.8	49.7	11– 167	9	0.6	1.1	0 – 3	9
Broadbilled Prion *Pachyptila vittata*	12.2	10.0	3– 40	19	2.5	2.4	0– 9	19	37.8	31.9	16– 137	19	0.6	1.2	0 – 5	19
Fairy Prion *Pachyptila turtur*	10.0			1	3.0			1	26.0			1	1.0			1
Whitechinned Petrel *Procellaria aequinoctialis*	127	124	14– 303	4	16.5	23.3	0– 51	4	54	46	0– 94	4	1.1	1.4	0 – 3	4
Greatwinged Petrel *Pterodroma macroptera*	71.1	37.1	33– 107	3	11.7	10.3	3– 23	3	68.7	38.3	25– 98	3	0.3	0.6	0 – 1	3
Kerguelen Petrel *Pterodroma brevirostris*	19.6	14.0	1– 58	13	5.7	6.0	0– 22	13	56.3	63.0	0– 184	13	0.4	0.6	0 – 2	13
Softplumaged Petrel *Pterodroma mollis*	61.7	42.2	15– 128	8	19.6	19.3	2– 57	8	28.9	28.8	0– 90	8	0.8	1.3	0 – 4	8
Imperial Cormorant *Phalacrocorax atriceps*	16	8.2	7– 23	3	4.3	3.5	1– 8	3	67.3	84.4	0– 162	3	0.3	0.6	0 – 1	3
Lesser Sheathbill *Chionis minor*	84	130	8– 348	6	2.2	4.3	0– 11	6	109	133	3– 369	6	1.5	0.9	0.4– 2.9	6
Antarctic Skua *Catharacta antarctica*	589	490	49–1,562	10	4.6	8.4	0– 26	10	1,453	1,250	54–3,171	10	15.9	26.8	0 –87	10
Kelp Gull *Larus dominicanus*	29.8	11.1	16– 45	10	10.2	19	0– 19	10	657	444	27–1,149	10	0.4	0.5	0 – 1.6	10

[a] Arithmetic mean

Table 2. Concentrations ($\mu g\ kg^{-1}$ for whole egg) of chlorinated hydrocarbons in the eggs of seabirds at Gough Island

Species		DDE	DDT	PCB	Dieldrin
Yellownosed Albatross	No. eggs	3	3	3	3
Diomedea chlororhynchos	Mean[a]	43	5.3	81.7	1.4
	SD	1.7	4.6	73.9	2.2
	Range	41 – 44	0 – 8	0 – 144	0 – 4
Sooty Albatross	No. eggs	4	4	4	4
Phoebetria fusca	Mean	152.8	33.5	202.3	0.7
	SD	23.3	3.9	85.0	0.2
	Range	132 –183	29 – 38	109 – 311	0.4– 0.8
Great Shearwater	No. eggs	3	3	3	3
Puffinus gravis	Mean	61.3	7.7	295.7	5.3
	SD	38.1	4.6	243.2	9.2
	Range	35 –105	5 – 13	100 – 568	0 –16
Kerguelen Petrel	No. eggs	1	1	1	1
Pterodroma brevirostris		17	0	0	0
Antarctic Skua	No. eggs	3	3	3	3
Catharacta antarctica	Mean	544	104	1,840	3.3
	SD	177	63	1,049	3.2
	Range	389 –737	65 –177	720 –2,800	1 – 7

[a] Arithmetic mean

Table 3. Concentrations ($\mu g\ kg^{-1}$ for whole egg) of chlorinated hydrocarbons in the eggs of seabirds at islands off the coast of South Africa

Species		DDE	DDT	PCB	Dieldrin
Jackass Penguin	No. eggs	4	4	4	4
Spheniscus demersus	Mean[a]	161.8	15.0	36.0	8.0
	SD	48.7	27.4	49.0	3.2
	Range	115 –220	4 – 56	0 – 108	4 –11
Whitebreasted Cormorant	No. eggs	3	3	3	3
Phalacrocorax carbo	Mean	53.7	15.3	183.7	0.7
	SD	18.5	3.8	58.5	1.2
	Range	42 – 75	11 – 18	130 – 246	0 – 2
Cape Cormorant	No. eggs	5	5	5	5
P. capensis	Mean	97.4	6.8	27.8	6.0
	SD	55.0	4.6	62.2	4.8
	Range	35 –176	0 – 12	0 – 139	2 –13
Bank Cormorant	No. eggs	7	7	7	7
P. neglectus	Mean	143.4	44.9	384.7	0.07
	SD	83.3	59.6	541.0	0.2
	Range	62 –297	0 – 164	0 –1,450	0 – 0.5
Crowned Cormorant	No. eggs	3	3	3	3
P. coronatus	Mean	76.3	12.3	30.3	0
	SD	19.1	4.7	31.0	0
	Range	62 – 98	7 – 16	0 – 62	0
Kelp Gull	No. eggs	8	8	8	8
Larus dominicanus	Mean	285.9	33.8	169.0	13.0
	SD	225.9	26.2	141.6	14.9
	Range	70 –798	0 – 84	0 – 428	0 –34

[a] Arithmetic mean

dering Albatross and the Whitechinned Petrel which feed on galley refuse.

Most of the eggs were collected during 1979 or between 1977 and 1979. Five King Penguin and six Antarctic Skua eggs were collected during 1981-1982, and five lesser Sheathbill and eight Kelp Gull eggs were collected during 1982 to 1983. The DDE level in the King Penguin eggs collected in 1981-1982 was significantly (2.5 times) higher than that found previously. The DDE levels in the Antarctic Skua and Kelp Gull eggs were about 1.5 times higher than in eggs collected previously, but these differences were not statistically significant. The PCB level in the more recent King Penguin eggs was significantly (ca. 7 times) higher than that of earlier samples. The PCB level in the Kelp Gull was significantly ($>$ 20 times) higher than the earlier level. The PCB level in the 1981-1982 Antarctic Skua eggs was about 2.5 times higher than the previous level, but the difference was not statistically significant. The PCB in one of the Kelp Gull eggs was $> 12{,}000\ \mu g\ kg^{-1}$. This result was not included in the statistical analysis, since it was more than five standard deviations outside the mean of 921 ± 158 of the other seven samples. Although there were insufficient samples of each species to demonstrate trends, the amount of PCB in the environment of Marion Island appears to be increasing. The relatively high levels of PCB in the Kelp Gull eggs suggests that the pollution has a local source.

Acknowledgements. Scientific research at Marion and Gough islands is carried out under the auspices of the South African Scientific Committee on Antarctic Research, and is assisted financially and logistically by the Antarctic Division of the Department of Transport. Permission to collect eggs of seabirds at Gough Island was granted by the Administrator of Tristan da Cunha. Permission to collect eggs at islands in Saldanha Bay was granted by the Sea Fisheries Research Institute. This paper is published with the permission of the Chief Director of the National Institute for Water Research, CSIR, Pretoria.

References

Bennington SL, Connors PG, Wolfram C, Risebrough RW (1975) Patterns of chlorinated hydrocarbon contamination in New Zealand sub-Antarctic and coastal marine birds. Environ Pollut 8: 135–147

Blankley WO (1981) Marine food of Kelp Gulls, Lesser Sheathbills and Imperial Cormorants at Marion Island (sub-Antarctic). Cormorant 9:77–84

Blus LJ, Burkett WS, Belisle AA, Prouty RM (1974) Organochlorine residues in Brown Pelican eggs: Relation to reproductive success. Environ Pollut 7:81–91

Brooke RK (1981) The place of South Africa in the world of seabirds and other marine animals. In: Cooper J (ed) Proc symp Birds of Sea and Shore. Cape Town, 1979. African Seabird Group, Cape Town

Burger AE (1981) Food and foraging behaviour of Lesser Sheathbills at Marion Island. Ardea 69:167–180

Croxall JP, Prince PA (1980) Food, feeding ecology and ecological segregation of seabirds at South Georgia. Biol J Linn Soc 14:103–131

Furness R, Hutton M (1979) Pollutant levels in the Great Skua Catharacta skua. Environ Pollut 19:261–268

Gardner BD (1983) Chlorinated pesticides in cats on Marion Island. S Afr J Sci 80:43–44

Glotfelty DE (1972) Identification of organochlorine pesticide residues by ultraviolet solid-phase photolysis. Anal Chem 44:1250–1254

Holden AV, Marsden K (1969) Single stage clean up animal tissue extracts for organochlorine residue analysis. J Chromatogr 44: 481–492

Johnson RE, Starr RI (1972) Ultra rapid extractions of insecticides from soil using new ultrasonic technique. J Econ Entomol 63: 165–168

Johnston DW (1973) Polychlorinated biphenyls in seabirds from Ascenscion Island, South Atlantic Ocean. Bull Environ Contam Toxicol 10:368–371

Keith JA, Gruchy IM (1972) Residue levels of chemical pollutants in North American birdlife. Proc Int Ornithol Congr 15:437–454

Mulhern BM, Reichel WL (1970) The effect of putrefaction of eggs upon residue analysis of DDT and metabolites. Bull Environ Contam Toxicol 5:222–225

Ohlendorf HM, Risebrough RW, Vermeer K (1978) Exposure of marine birds of environmental pollutants. US Wildl Res Rep 9, Washington DC

Ohlendorf HM, Bartonek JC, Divoky GJ, Klaas EE, Krynitsky AJ (1982) Organochlorine residues in eggs of Alaskan seabirds. US Wildl Serv Spl Sci Rep 245, Washington DC

Peakall DB (1975) PCBs and their environmental effects. CRC Critical Reviews in Environmental Control, pp 469–500

Risebrough RW, Carmignani GM (1972) Chlorinated hydrocarbons in Antarctic birds. In: Parker BC (ed) Conservation problems in Antarctica. Allen, Lawrence, Kansas

Siegfried WR (1978) Ornithological research at the Prince Edward islands: a review of progress. S Afr J Antarct Res 8:30–34

Sokal RR, Rohlf FJ (1969) Biometry. Plenum, London

Stickel LF (1973) Pesticide residues in birds and mammals. In: Edwards CA (ed) Environmental pollution by pesticides. Plenum, London

Tatton JO'G, Ruzicka JHA (1967) Organochlorine pesticides in Antarctica. Nature (Lond) 215:346–348

Voous KH, Wattel J (1963) Distribution and migration of the Great Shearwater. Ardea 51:143–157

Watson GE (1975) Birds of the Antarctic and Sub-Antarctic. American Geophysical Union, Washington DC

Williams AJ (1978) Mineral and energy contributions of petrels (Procellariiformes) killed by cats, to the Marion Island terrestrial ecosystem. S Afr J Antarct Res 8:49–53

Williams AJ, Siegfried WR (1980) Foraging ranges of Krill-eating penguins. Polar Record 20:159–175

Organochlorines and Heavy Metals in Seals and Birds from the Weddell Sea

R. SCHNEIDER[1], G. STEINHAGEN-SCHNEIDER[1], and
H.E. DRESCHER†

Summary. Nine Weddell Seals, *Leptonychotes weddelli*, 6 Crabeater Seals, *Lobodon carcinophagus*, 9 Adélie Penguins, *Pygoscelis adeliae*, 12 Emperor Penguins, *Aptenodytes forsteri*, and 7 South Polar skuas, *Catharacta maccormicki*, were collected along the ice-shelf of the Weddell Sea in the austral summer of 1980–1981 for a baseline investigation of concentrations of organochlorines and heavy metals in these animals. Muscle, liver, and kidney tissues were analysed for concentrations of cadmium, copper, and zinc. Adipose tissue was analysed for several organochlorines (polychlorinated biphenyls, DDT and its main analogues, and other insecticides). The results are reported and discussed.

1 Introduction

A unique feature of the Antarctic region is its remoteness from important sources of man-made toxic pollutants. However, in spite of its remoteness, the region is contaminated by chemical pollutants. For instance, residues of DDT, an exclusively man-made compound, occur in Antarctic animals (Sladen et al. 1966; George and Frear 1966; Tatton and Ruzicka 1967; Risebrough et al. 1968; Brewerton 1969).

Since deposits of certain minerals, coal, and oil have been found in Antarctica, rising ore and energy prices might soon lead to an intensification of exploration and make exploitation economically attractive in and around the continent (Holdgate and Tinker 1979). Such activities would dramatically increase the danger of heavy metal contamination but, since heavy metals are natural components of marine ecosystems, their proportions due to anthropogenic origin can only be determined if natural baseline concentrations are known (Anderlini et al. 1972). As part of a pilot project aimed at establishing the extent of organochlorine contamination and heavy metal baseline concentrations in the marine biota of the Weddell Sea, several specimens of seals and birds were collected during the German Filchner Shelf-Ice Expedition of 1980–1981.

2 Material and Methods

Totals of 9 Weddell Seals, *Leptonychotes weddelli*, 6 Crabeater Seals, *Lobodon carcinophagus*, 9 Adélie Penguins, *Pygoscelis adeliae*, 12 Emperor Penguins, *Aptenodytes forsteri*, and 7 South Polar Skuas, *Catharacta maccormicki*, were collected in Gould Bay and in Atka Bay during January and February 1981, respectively. The animals were immediately dissected on board the R.V. Polarsirkel, and samples of liver, kidney, and muscle were placed in aluminium foil (for organochlorine analyses) and polyethylene bags (for heavy metal analyses) and stored at −20 °C. Prior to analysis, the samples were freeze-dried and subsequently homogenized in an agate mortar mill.

For organochlorine analyses, aliquots of the homogenized samples were extracted for 6 h with n-hexane/acetone (85/15) in a Soxhlet apparatus. After solvent evaporation in tared flasks, the lipid contents were determined gravimetrically. A maximum of 110 mg of the extracted lipids was subjected to the clean-up procedure described by Holden and Marsden (1969), (i.e., column + chromatography on aluminium oxide and silica). The gaschromatograph employed was a Packard 438 equipped with a Chrompack on column injector, and a Chrompack 25 m × 0.32 mm i.d. CP Sil 19 CB fused silica-bonded phase temperature programmed capillary column. The instrument was coupled to a Hewlett-Packard HP 3380 integrator. For certain samples, the results were confirmed with a 25 m × 0.32 mm SGE 25 QC 3 column. The PCB concentrations were calculated as the mean of all peaks occurring in the samples, as well as in the technical mixture Clophen A 60. (BAYER AG) which contains mainly penta- to octa-chlorobiphenyls.

For heavy metal analyses, aliquots of the homogenates, as well as of NBS Standard Reference Material (SRM) 1566 (oyster tissue) applied to verify the accuracy of the analyses, were acid digested (3 ml HNO_3/1 ml $HClO_4$, both suprapur, MERCK) (Kruse 1980). Due to the different sensitivities of the AAS to different elements and to the different metal concentrations and tissue digestibilities, sample sizes varied as follows: muscles, kidneys, livers and SRM 200, 25, 50, and 25 mg for Cd and Cu, and 20, 10, 10, and 20 mg for Zn, respectively. A Perkin-Elmer 300 SG/HGA 76 B 1-beam flameless atomicabsorption spectrophotometer, equipped

1 Alfred-Wegener-Institut für Polarforschung, Columbus-Center, 2850 Bremerhaven, FRG

† H.E. Drescher was killed in a helicopter accident in June 1983

Antarctic Nutrient Cycles and Food Webs
(ed. by W.R. Siegfried, P.R. Condy, and R.M. Laws)

with a D_2 background corrector and an As-1 autosampler, was used. Three replicates of 20 μl per sample were injected. The standard analytical conditions outlined on the PE manual were used for all determinations. After preliminary SRM analyses, the best calibrations were obtained with external standards for Cu and Zn, and with a standard addition method for Cd.

3 Results and Discussion

3.1 Organochlorine Concentrations

Information on organochlorine concentrations in Antarctic animals is scarce; birds having been analysed more often than seals. The bulk of the information was obtained in the late 1960s and early 1970s, and deals with DDE and DDT. Data on PCBs are few, which probably reflects the relatively insensitive analytical instruments available at the time, rather than the absence of these compounds.

The DDT, DDE, and PCB levels determined by us (Table 1) indicate that concentrations of these pollutants in Antarctic animals are much lower than those in comparable North Sea species (e.g., Drescher et al. 1977; Harms et al. 1978). The relatively elevated levels in the South Polar Skua may be explained by this species' annual migrations which extend far north of the Antarctic Convergence (Risebrough 1977). No dieldrin was found in the seals. The high concentrations in the Emperor Penguins at Gould Bay might be related to the fact that the birds were starving at the time that they were sampled.

3.2 Heavy Metal Concentrations

The CD concentrations (Table 2) lie within the range of values given by other authors for Crabeater Seal liver and

Table 1. Organochlorines in seal blubber and bird abdominal adipose tissue from the Weddell Sea. Mean concentration and 1 ± S.D. in $\mu g\ g^{-1}$ (ppb) extractable lipids. Number of samples (n) in which a specific compound was measurable. Not detected = n.d.

Species and station	No. specimens	Lipid content (% wet wt.)	HCB	−BHC	Lindane	p,p'−DDE	p,p'−DDT	PCBs	Dieldrien
Crabeater Seal									
Gould Bay	2	85.6 ± 7.6	60 ± 11 (2)	9 − (1)	39 ± 11 (2)	46 ± 30 (2)	7 − (1)	326 ± 115 (2)	n.d.
Atka Bay	3	69.9 ±12.3	31 ± 2 (3)	32 ±35 (2)	103 ±129 (2)	65 ± 25 (3)	12 ± 3 (3)	340 ± 216 (3)	n.d.
Weddell Seal									
Gould Bay	2	71.6 ± 6.5	42 ± 11 (2)	20 ±18 (2)	13 − (1)	71 ± 18 (2)	37 ± 11 (2)	310 ± 30 (2)	n.d.
Atka Bay	3	71.4 ±17.8	33 ± 5 (3)	14 ± 1 (2)	20 ± 6 (2)	49 ± 14 (3)	31 ± 13 (3)	381 ± 15 (3)	n.d.
Emperor Penguin									
Gould Bay (starving chicks)	3	0.6 ± 0.4	898 ± 295 (3)	12 − (1)	118 ± 33 (2)	611 ± 110 (3)	39 ± 4 (3)	1,760 ±1,640 (3)	127 ± 30 (2)
Gould Bay	2	28.2 ± 6.8	1,180 ± 290 (2)	44 ± 1 (2)	26 ± 15 (2)	1,560 ± 398 (2)	71 ± 40 (2)	617 ± 260 (2)	72 ± 17 (2)
Atka Bay	1	61.1 −	467 −	− −	17 −	288 −	13 −	524 −	13 −
Adélie Penguin									
Atka Bay	9	79.7 ±11.5	232 ± 161 (9)	14 ± 3 (3)	73 ± 54 (9)	361 ± 767 (9)	38 ± 17 (6)	346 ± 135 (9)	21 ± 20 (7)
South Polar Skua									
Gould Bay	3	87.7 ± 0.8	568 ± 46 (3)	7 ± 1 (2)	28 ± 16 (2)	2,930 ± 380 (3)	783 ±415 (3)	4,170 ± 260 (3)	75 ± 45 (3)
Atka Bay	4	78.7 ± 5.4	1,330 ± 651 (4)	8 − (1)	17 ± 65 (4)	3,370 ± 580 (4)	953 ±557 (4)	4,160 ± 440 (4)	105 ± 59 (3)

Table 2. Cadmium in tissues of seals and birds from the Weddell Sea. Mean concentrations and 1 ± S.D. in $\mu g\ g^{-1}$ dry wt. Number of samples (n) analyzed

Species and station	No. specimens	Muscle		Liver		Kidney	
Crabeater Seal							
Gould Bay	2	0.20 ± 0.11	(2)	13.7 ± 5.9	(2)	98.7 ± 43.1	(2)
Atka Bay	4	0.17 ± 0.08	(3)	34.7 ± 15.8	(4)	139.4 ± 87.1	(4)
Weddell Seal							
Gould Bay	2	0.20 ± 0.13	(2)	14.9 ± 5.0	(2)	114.6 ± 43.8	(2)
Atka Bay	7	0.21 ± 0.12	(7)	22.5 ± 10.2	(7)	174.3 ± 67.3	(7)
Emperor Penguin							
Gould Bay	9	0.28 ± 0.37	(5)	5.7 ± 7.7	(8)	23.8 ± 39.3	(8)
Atka Bay	3	1.67 ± 1.03	(2)	48.3 ± 21.0	(3)	382.2 ± 199.2	(3)
Adélie Penguin							
Atka Bay	9	0.30 ± 0.14	(8)	15.2 ± 7.0	(9)	174.9 ± 88.1	(9)
South Polar Skua							
Gould Bay	3	0.32 ± 0.10	(3)	26.2 ± 2.3	(3)	178.6 ± 74.1	(3)
Atka Bay	4	0.16 ± 0.04	(3)	17.3 ± 5.7	(3)	132.4 ± 8.4	(3)

Table 3. Copper in tissues of seals and birds from the Weddell Sea. Mean concentrations and 1 ± S.D. in $\mu g\ g^{-1}$ dry wt. Number of samples (n) analyzed

Species and station	No. specimens	Muscle		Liver		Kidney	
Crabeater Seal							
Gould Bay	2	3.30 ± 0.99	(2)	49.8 ± 1.7	(2)	28.9 ± 4.7	(2)
Atka Bay	4	2.80 ± 0.28	(4)	79.6 ± 27.0	(4)	43.4 ± 13.6	(4)
Weddell Seal							
Gould Bay	2	3.00 ± 1.13	(2)	16.7 ± 0	(2)	21.9 ± 4.0	(2)
Atka Bay	7	2.80 ± 0.24	(7)	44.5 ± 21.3	(7)	35.8 ± 12.3	(7)
Emperor Penguin							
Gould Bay	9	3.70 ± 1.44	(7)	30.7 ± 14.5	(9)	21.2 ± 11.1	(8)
Atka Bay	3	3.70 ± 0.93	(3)	61.8 ± 36.5	(3)	24.8 ± 4.0	(3)
Adélie Penguin							
Atka Bay	9	7.23 ± 1.01	(8)	22.9 ± 6.9	(9)	15.9 ± 3.4	(9)
South Polar Skua							
Gould Bay	3	20.3 ± 1.20	(3)	15.4 ± 1.3	(3)	28.0 ± 8.8	(3)
Atka Bay	4	13.8 ± 0.95	(3)	20.9 ± 5.1	(3)	25.4 ± 5.7	(3)

Table 4. Zinc in tissues of seals and birds from the Weddell Sea. Mean concentrations and 1 ± S.D. in $\mu g\ g^{-1}$ dry wt. Number of samples (n) analyzed

Species and station	No. specimens	Muscle		Liver		Kidney	
Crabeater Seal							
Gould Bay	2	65.3 ± 1.7	(2)	76.8 ± 8.9	(2)	123.0 ± 58.5	(2)
Atka Bay	4	32.8 ± 11.3	(4)	143.6 ± 59.7	(4)	137.2 ± 30.7	(4)
Weddell Seal							
Gould Bay	2	71.4 ± 0.6	(2)	122.0 ± 94.6	(2)	251.8 ± 25.2	(2)
Atka Bay	7	60.4 ± 25.3	(7)	104.6 ± 41.0	(7)	142.0 ± 42.5	(7)
Emperor Penguin							
Gould Bay	9	108.3 ± 39.3	(8)	96.9 ± 70.8	(9)	137.6 ± 18.4	(8)
Atka Bay	3	57.2 ± 39.6	(3)	77.6 ± 11.6	(3)	177.8 ± 95.7	(3)
Adélie Penguin							
Atka Bay	9	44.7 ± 21.5	(9)	83.1 ± 10.8	(9)	203.5 ± 87.0	(9)
South Polar Skua							
Gould Bay	3	30.2 ± 9.1	(3)	62.4 ± 5.0	(3)	140.2 ± 32.7	(3)
Atka Bay	4	59.2 ± 11.9	(4)	83.2 ± 9.2	(3)	126.7 ± 56.0	(3)

kidneys (Schneppenheim[2] 1981), and Weddell Seal livers and muscle (Wagemann and Muir 1981). As expected, Cd concentrations were always highest in the kidney, followed by liver and muscle (Eisler 1981). Cadmium concentrations in all tissues of all the species analysed were similar, and fall within the ranges obtained for other species of marine birds and mammals (Eisler 1981).

The range of Cu concentrations (Table 3) in Crabeater Seal livers and muscles includes the value given by Wagemann and Muir (1981) for one animal. The Cu concentrations in bird livers were approximately the same as those in petrel livers from the Antarctic (Anderlini et al. 1972). In the petrels, however, Cu concentrations in muscles were much higher than in the livers analyzed. In both birds and mammals, livers generally contained the highest concentrations of Cu, followed by kidney and muscle tissues. Skua muscles contained relatively high Cu concentrations.

The range of Zn concentrations (Table 4) in Crabeater Seal livers and muscles also includes the value given by Wagemann and Muir (1981). Zinc concentrations in the tissues of the birds were similar to those found in petrels in the Antarctic (Anderlini et al. 1972). As in the case of Cd, Zn concentrations were in all cases highest in kidneys, followed by livers and muscles, and varied little between the species.

Since Cu and Zn are essential elements, their concentrations can be expected to be metabolically regulated in vertebrates (Anderlini et al. 1972), so that it is doubtful whether birds and mammals are suitable for monitoring Cu and Zn pollution. Although Cd is non-essential and toxic, the similarity of the concentrations in the tissues of all species analyzed indicates that Cd concentrations are probably also metabolically regulated. In future, our baseline study will therefore investigate potentially more suitable indicators, such as fish, benthic invertebrates and, where available, benthic macrophytes.

2 The data given by Schneppenheim (1981) should read ppm instead of ppb (R. Schneppenheim, pers. comm.)

Acknowledgements. We are grateful to Mrs. T. Koennecker for technical assistance.

References

Anderlini VC, Connors PG, Risebrough RW, Martin JH (1972) In: Proceedings of the Colloquium on Conservation Problems in Antarctica. Allan, Lawrence, Kansas, pp 49–61

Brewerton HV (1969) DDT in fats of antarctic animals. N Z J Sci 12:194–199

Drescher HE, Harms U, Huschenbeth E (1977) Organochlorines and heavy metals in the harbour seal *Phoca vitulina* from the German North Sea coast. Mar Biol 41:99–106

Eisler R (1981) Trace metal concentrations in marine organisms. Pergamon, New York, p 687

George JL, Frear DEH (1966) Pesticides in the Antarctic. J Appl Ecol 3 (suppl):155–167

Harms U, Drescher HE, Huschenbeth E (1978) Further data on heavy metals and organochlorines in marine mammals from German coastal waters. Meeresforschung 26:153–161

Holden AV, Marsden K (1969) Single-stage clean-up of animal tissue extracts for organochlorine residue analysis. J Chromatogr 44: 481–492

Holdgate MW, Tinker J (1979) Oil and other minerals in the Antarctic – The environmental implications of possible mineral exploration and exploitation in Antarctica. SCAR, p 51

Kruse R (1980) Vielfach-Bestimmung von Pb and Cd in Fischen durch elektrothermale AAS nach Naßveraschung in kommerziellen Teflon-Bechern. Z Lebensm Unters Forsch 171:261–264

Risebrough RW (1977) Transfer of organochlorine pollutants to Antarctica. In: Llano GA (ed) Adaptations within Antarctic ecosystems – Proceedings of the 3rd SCAR symposium on Antarctic biology. Gulf Publishing Company, Houston, Texas, pp 1203–1210

Risebrough RW, Reiche P, Peakall DB, Herman SG, Kirven MN (1968) Polychlorinated biphenyls in the global ecosystem. Nature 220: 1098–1102

Schneppenheim R (1981) Results of the biological investigations in the Weddell Sea during a site survey 1979/80. Polarforschung 51: 91–99

Sladen WJL, Menzie CM, Reichel WL (1966) DDT residues in Adélie penguins and a crabeater seal from Antarctica. Nature 210:670–673

Tatton JO'G, Ruzicka JHA (1967) Organochlorine pesticides in Antarctica. Nature 215:346–348

Wagemann R, Muir DCG (1981) Assessment of heavy metals and organochlorine concentrations in Marine Mammals of the Northern waters. ICES CM 1981/N:9

Metal Concentrations in Antarctic Zooplankton Species

H.F.-K.O. HENNIG[1], G.A. EAGLE[1], C.D. McQUAID[2], and L.H. RICKETT[3]

Summary. Trace metal analyses were carried out on 15 species of zooplankton, collected between New Zealand and McMurdo Sound, Antarctica. Three species, *Themisto gaudichaudi, Eukrohnia hamata*, and *Euphausia triacantha*, were widely distributed. An attempt is made to relate the concentrations of copper, iron, and zinc in these three species to different water-masses. Such relationships could be used to identify feeding grounds of predators which eat zooplankton.

1 Introduction

It is widely known that certain well-defined oceanic phenomena, such as fronts, convergences and divergences, are permanent features of the Southern Ocean. These features separate distinct water-masses which are usually characterized with the aid of temperature-salinity (T–S) diagrams. However, apart from variations in temperature and salinity, dramatic changes in chemical properties of the water can also occur across frontal zones. For example, there is normally a characteristic sharp change in silicate concentration across the Antarctic Polar Front (Lutjeharms et al. 1984), and significant differences in the trace metal concentrations of water-masses have also been found (e.g., Burton and Young 1980; Boyle et al. 1981). Boyle et al. suggested that metal concentrations were higher in cool nutrient-rich eastern boundary currents, and found that copper was distinctly higher in coastal water over the continental shelf. This was reported to be due to a nutrient trapping of diagenetically remobilized Cu from mildly reducing terrigenous sediments.

A number of water-masses have been identified in the Southern Ocean south of New Zealand (Burling 1961; Houtman 1967; Craig et al. 1981; Edwards and Emery 1982). In this region, the Circumpolar Current is deflected rather abruptly to the south, as a result of bottom topography possibly, or the southward curve of the Antarctic land-mass. The Antarctic Polar Frontal Zone is fairly wide (approx. 500 km) and incorporates meso-scale features (Edwards and Emery 1982). There is also evidence that individual plankters will remain within a particular water-body, and phytoplankton assemblages have been used to identify water-masses. We suggest, therefore, that zooplankters may accumulate trace metals in their bodies, to levels which are characteristic of particular water-masses. Thus, variations in trace metal content of zooplankton may be an indication of the water-mass from which the sample was derived.

An opportunity to make preliminary studies of trace metal concentrations in zooplankton in this area presented itself in the austral summer of 1980–1981, during the Transglobe Expedition led by Sir Ranulph Fiennes. Here we report on the possibility of using trace metal concentrations in zooplankters, as indicators of the water-masses from which they were derived. This could be useful in identifying the feeding grounds or feeding depths of animals which feed on zooplankton.

2 Materials and Methods

Sampling was carried out between Christchurch, New Zealand, and McMurdo Sound in Antarctica (Fig. 1). Altogether there were 23 stations, although samples specifically dealt with here were collected from 17 stations (8 on the outward journey and 9 on the return leg). These stations were reasonably close to those occupied during the GEOSECS cruise (Craig et al. 1981). Surface macro-plankton was collected from the M.V. Benjamin Bowring, using an N100 Discovery pattern plankton net of 1 m diam. Samples were taken 30 min after sunset on each day, until the ship moved into a zone of no sunset, when samples were taken at 22.00 hours local time. The net was towed just below the surface for 10 min at a speed of 7.5 km h^{-1}, during which a total of approximately 4,000 m^3 of water was filtered. Start and finish positions for each trawl were taken from the ship's satellite navigation system.

Specimens were preserved in analar grade formalin and subsequently sorted and identified to species level, using precleaned plastic instruments. Each species was then counted and dried for 48 h at 60 °C. Samples were stored in acid-washed glass vials for analysis later in the laboratory.

1 National Research Institute for Oceanology, P.O. Box 320, Stellenbosch 7600, South Africa

2 Zoology Department, University of Cape Town, Rondebosch 7700, South Africa

3 Applied Mathematics Department, University of Cape Town, Rondebosch 7700, South Africa

Antarctic Nutrient Cycles and Food Webs (ed. by W.R. Siegfried, P.R. Condy, and R.M. Laws)

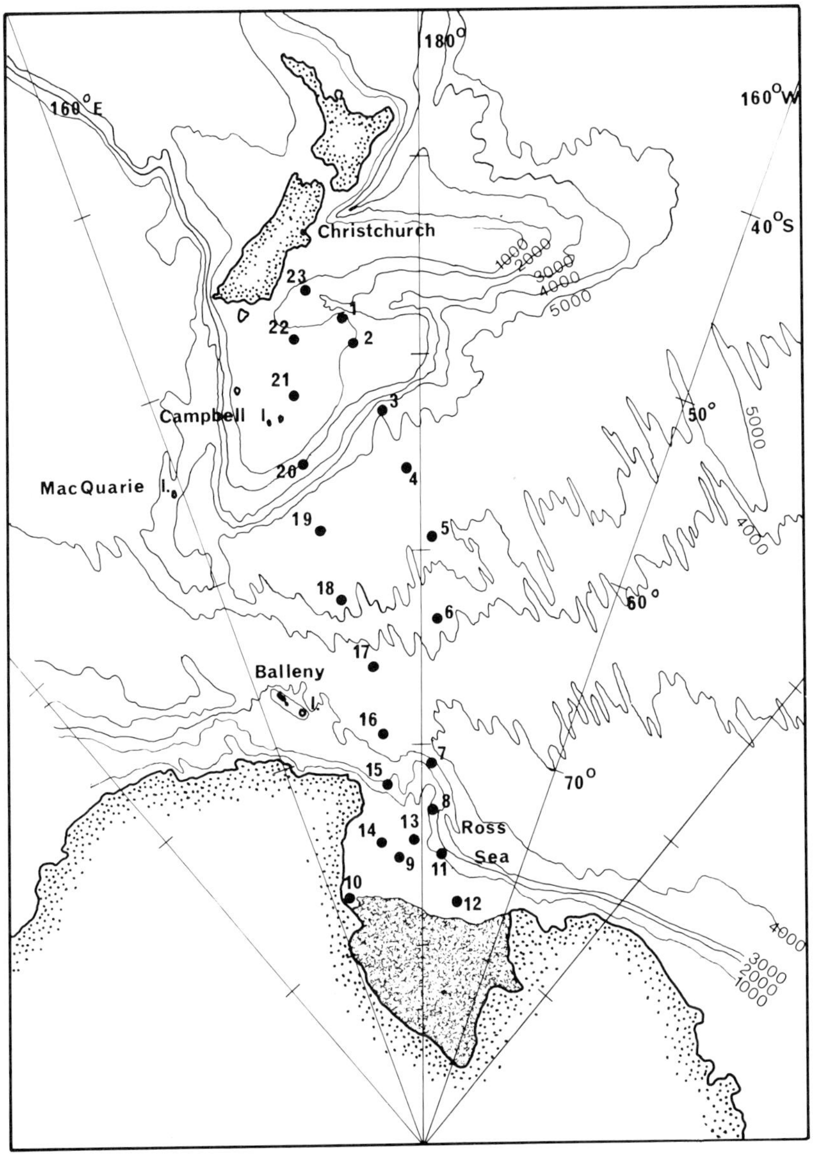

Fig. 1. Sampling positions of collected zooplankton between Christchurch, New Zealand, and McMurdo Sound, Antarctica

Salinity samples were collected from the surface, using a temperature bucket, and stored in sealed glass bottles at room temperature for later analysis at the University of British Columbia. Surface-water temperature was measured, using a mercury bulb thermometer.

In the laboratory, zooplankters were weighed (individually for large specimens, and grouped together for smaller ones) and digested in quartz-redistilled nitric acid, and then in a 4:1 mixture of nitric-perchloric acids. The residues were dissolved in 10% nitric acid and analyzed for Cd, Cu, Fe, Mn, Ni, Pb, Sr, and Zn by flame atomic-absorption, using standard conditions with background correction (Watling and Watling 1976). This method of analysis of biological material is standard in our laboratory and, in a recent inter-calibration exercise, we obtained results which were within 1 S.D. of the "true" values for all metals.

3 Results

Surface temperature decreased steadily from north to south (Fig. 2), and the Antarctic Polar Front was found from continuous surface temperature recordings to occur at approximately 62°20′S on the southbound journey (13 January) and 62°40′S on the northbound leg (20 February). The surface temperatures are given in more detail in Fig. 3.

Salinity also decreased towards the south, dropping from 34.77‰ near the ice edge at Stations 7 and 16 (Fig. 2) where the summer ice-melt had already begun. The decrease in salinity was, however, less regular than that of surface temperature and values were higher in February than at the same latitudes in January (Fig. 2). If these data are plotted as a temperature-salinity diagram (Fig. 4), and the station posi-

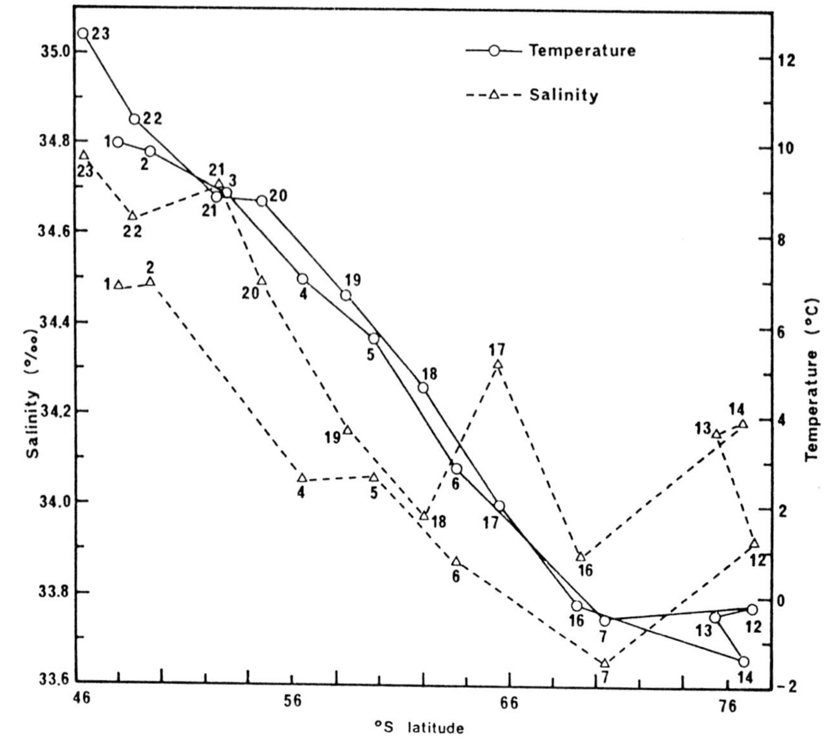

Fig. 2. Temperate and salinity between Christchurch and McMurdo Sound

tions considered, samples may be grouped into four sets representing different water-bodies as follows:

1. New Zealand continental shelf (Stations 1, 2, 20, 21, 22, 23)
2. Deep Oceanic basin (Stations 3, 4, 5, 6, 18, 19)
3. Antarctic continental shelf (Station 17)
4. Ross Sea (Stations 7, 8, 9, 10, 11, 12, 13, 14, 15, 16).

Fifteen of the zooplankton species collected were analyzed for concentrations of 8 metals (Tables 1+2). Many of the species were, however, found in only a few samples.

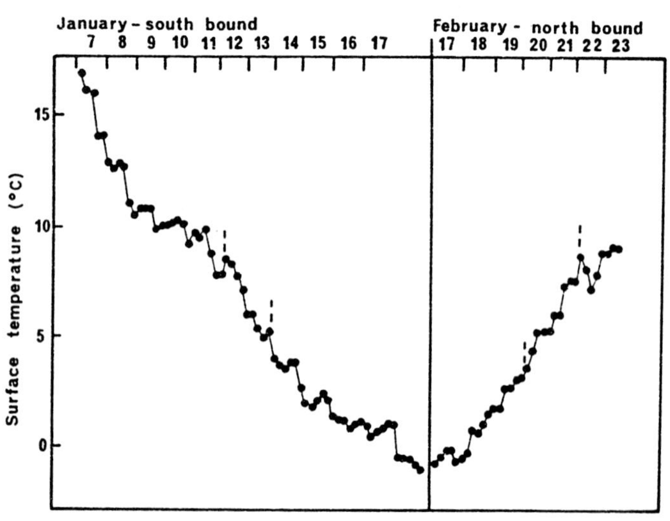

Fig. 3. Continuous surface temperature recording between Christchurch and McMurdo Sound from the M.V. Benjamin Bowring cruise

Thus, our account here is limited to concentrations of Cu, Fe, Zn, and Sr in three species which were found to be widely distributed and which occurred in a large number of the samples. These were the amphipod *Themisto gaudichaudii*, the chaetognath *Eukrohnia hamata*, and the euphausiid *Euphausia triacantha*.

Considerable interspecific differences in metal concentrations occurred. Although each species exhibited a particular pattern of metal content (Table 1), *T. gaudichaudii* generally contained greater concentrations than *E. hamata* or *E. triacantha*. There were striking differences between the stations representing different water-masses. Metal concentrations were generally highest at the New Zealand shelf stations and low above the oceanic basin. This pattern was clearest in the case of Cu, concentrations of which decreased from north to south with a marked drop between stations over the shelf and those over the basin (Table 1). The Cu concentrations in the shelf zooplankton were approximately 2 times those found farther south. The highest Fe concentrations occurred in animals from the Ross Sea. Zinc concentrations were also higher in animals obtained farther south, whereas Cd and Sr were only found in samples collected over the New Zealand shelf and in the oceanic basin.

There were also large differences in metal concentrations between samples taken at the same latitudes in early January and in late February. Salinities were higher at the stations sampled in February (Fig. 2). However, these stations were sampled about 2 months apart. The differences were particularly noticeable in comparisons of Stations 1 and 2 with Stations 20–23 (i.e. over the New Zealand shelf in January and in February) and between Stations 4 and 19 (i.e. over the oceanic basin). All three species followed a broadly

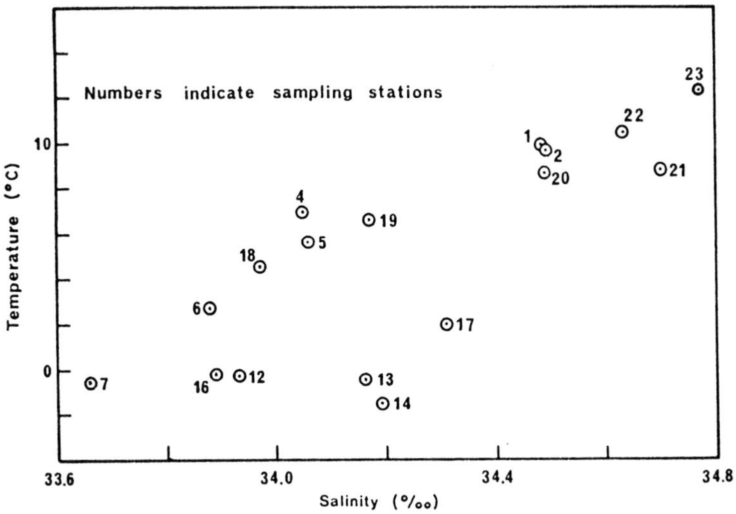

Fig. 4. Temperature-salinity diagram of stations with relevant zooplankton between Christchurch and McMurdo Sound

similar pattern of change in these comparisons. Zinc and especially Cu concentrations decreased, sometimes quite dramatically, in all three species, both over the shelf region and over the oceanic basin, whereas the Sr concentration generally increased. Values for Fe generally rose between January and February over the shelf, but decreased in the same period over deeper waters.

The salinity at Station 17 was much higher than at Stations 16 and 18 (south and north of Station 17, respectively), and the change clearly represents an intrusion between watermasses of the Ross Sea and the deep oceanic basin (Fig. 2). Similar dense water was reported by Craig et al. (1981), and Pickard and Emery (1982) found that this region has the densest water in the Southern Ocean. Of the three species considered, only *T. gaudichaudii* was found at Station 17, where it exhibited particularly low concentrations of all metals (Table 1).

4 Discussion

Very little chemical oceanographic information exists for the area of the Southern Ocean in which our study was made, the most comprehensive being that collected during the GEOSECS cruise (Craig et al. 1981). Available information on trace metals in zooplankton has been summarized in a review by Davies (1978). Almost without exception these data were obtained in the northern hemisphere, and there is no information of which we are aware on the trace metal content of marine zooplankton from the Southern Ocean. Therefore, although our data are relatively meagre, we nevertheless feel that some discussion of their significance is warranted.

Accumulation of metals appears to be associated with the nutrient cycle (Boyle et al. 1981). For example, Cu concentrations in surface waters correlate well with nitrate concen-

Table 1. Trace metal concentrations ($\mu g\ g^{-1}$ dry mass) in species of zooplankters

Species	Stations	Cu		Fe		Zn		Sr		Cd		Mn		Ni		Pb	
		$\bar{x}$	S.D.	$\bar{x}$	S.D.	$\bar{x}$	S.D.	$\bar{x}$	S.D.	$\bar{x}$	S.D.	$\bar{x}$	S.D.	$\bar{x}$	S.D.	$\bar{x}$	S.D.
Themisto	1, 2	875	462	283	54	1,183	1,081	n.d.		n.d.		n.d.		n.d.		n.d.	
gaudichaudii	3, 4, 5, 6	35	18	402	223	1,011	998	8	4	8	2	3	1	3	1	n.d.	
	7	79		514		1,996		n.d.		n.d.		n.d.		n.d.		n.d.	
	17	20		160		282		n.d.		n.d.		1		n.d.		n.d.	
	18, 19	34	26	233	139	529	238	66	76	118	81	10	8	n.d.		n.d.	
	20, 21, 22, 23	38	24	318	263	587	560	97	92	29	23	15	20	n.d.		n.d.	
Eukrohnia	2	115		96		402		39		n.d.		3		n.d.		n.d.	
hamata	4	118		85		172		19		8		3		1		n.d.	
	18, 19	34	40	171	161	593	713	19	4	17	17	7	3	n.d.		n.d.	
	20, 22, 23	36	18	186	52	485	65	18	17	14	9	2	2	1	1	n.d.	
Euphausia	4	168		121		310		2		11		3		n.d.		n.d.	
triacantha	16	23		250		694		n.d.		7		3		n.d.		n.d.	
	19	17		190		343		178		50		4		n.d.		n.d.	
	20, 21, 22, 23	31	28	252	208	402	110	58	45	9	6	10	12	n.d.		n.d.	

Table 2. Trace metal concentrations ($\mu g\ g^{-1}$ dry mass) in species of zooplankters

Species	Station No.	Cd		Cu		Fe		Mn		Ni		Pb		Zn		Sr	
		$\bar{x}$	S.D.	$\bar{x}$	S.D.	$\bar{x}$	S.D.	$\bar{x}$	S.D.	$\bar{x}$	S.D.	$\bar{x}$	S.D.	$\bar{x}$	S.D.	$\bar{x}$	S.D.
Cyprid larvae	1, 2	n.d.		109	94	59	19	n.d.		n.d.		n.d.		1,459	390	n.d.	
	3, 19	7	10	44	30	175	58	n.d.		n.d.		n.d.		416	115	n.d.	
	14	n.d.		73		1,812		n.d.		n.d.		n.d.		1,739		n.d.	
Sagitta gazellae	2	n.d.		175		66		n.d.		n.d.		n.d.		801		14	
	10	n.d.		n.d.		611		n.d.		n.d.		n.d.		96		n.d.	
Hyperiella dilatata	3, 4, 5, 6, 18, 19	231	416	140	107	610	836	15	38	n.d.		n.d.		1,657	2,796	188	496
Primo sp.	4, 5, 18, 19	31	29	38	35	194	121	4	4	n.d.		n.d.		687	589	49	84
Fish larvae	20	11	1	30	4	295	122	16	2	n.d.		n.d.		230	20	19	7
	4, 5, 19	27	33	38	47	145	109	2	2	n.d.		n.d.		466	380	28	33
Clione limacina antarctica	4, 5	2	2	133	94	993	725	8	11	n.d.		n.d.		898	349	n.d.	
	10, 14	n.d.		103	17	528	245	3	0	n.d.		n.d.		813	514	n.d.	
Salpa thompsoni	21, 22, 23	3	2	27	31	137	80	2	1	n.d.		n.d.		250	339	56	16
	6	2		4		78		2		n.d.		n.d.		33		22	
	17	0.2		2		18		1		n.d.		n.d.		20		9	
Cyllopus magellanicus	21, 22, 23	18	4	26	8	167	28	1	2	n.d.		n.d.		427	195	1	2
	6	7		35		119		n.d.		n.d.		n.d.		527		n.d.	
	17	n.d.		135		139		n.d.		n.d.		n.d.		1,495		n.d.	
Vibilia antarctica	21, 22, 23	n.d.		74	57	365	131	n.d.		n.d.		n.d.		753	210	n.d.	
	6	n.d.		12		201		n.d.		n.d.		n.d.		181		n.d.	
	17	n.d.		85		175		n.d.		n.d.		n.d.		1,121		n.d.	
Orchomene plebs	12, 13	n.d.		76	52	577	542	n.d.		n.d.		n.d.		306	48	59	83
Eusirus antarcticus	12	n.d.		n.d.		2,008		n.d.		n.d.		n.d.		236		n.d.	
Salpa maxima	22	3		34		170		1		n.d.		n.d.		431		41	

Where stations have been grouped the concentrations given are the means of all samples. n.d. = Not detected; S.D. = sample standard deviation

trations (Boyle and Edmond 1975). Metals are accumulated by zooplankton in three ways: by adsorption onto the animal's body surfaces; by absorption from water passing over the gills and other body tissues; and by assimilation from food and detrital particles. Elimination may occur by excretion, moulting, and egg-laying. Thus, the metal concentrations of any particular animal depend upon the degree of imbalance between the uptake and loss processes. Since certain of the loss processes are discontinuous (e.g., moulting), large variations in trace metal concentrations between individual animals can be expected (Davies 1978). This was found to be the case in our study, particularly with Zn which is known to be very closely associated with the moulting cycle in certain crustaceans (Eagle et al. 1983).

Our most striking result is the difference in metal concentrations associated with samples from different groups of stations (i.e. different water-masses). This was particularly noticeable for Cu, which has been described as having a high concentration in continental shelf water (Boyle et al. 1981). Direct comparison with other published information is difficult because of differences in species and region. Nevertheless, Martin and Knauer (1973) reported mean concentrations of 15.1 $\mu g\ g^{-1}$ Cu, 108 $\mu g\ g^{-1}$ Fe, 164 $\mu g\ g^{-1}$ Sr, and 71 $\mu g\ g^{-1}$ Zn in euphausiids from Monterey Bay, California. These concentrations are lower than ours for Cu, Fe, and Zn, whereas the Californian euphausiids contained more Sr. Additional data for euphausiids include, Cu and Zn concentrations between 8.5 and 42.2 $\mu g\ g^{-1}$ and between 50 and 131 $\mu g\ g^{-1}$, respectively, in *Euphausia pacifica* from the northeastern Pacific (Cutshall and Holton 1972) and mean Cu and Zn concentrations of 32.9 and 108 $\mu g\ g^{-1}$, respectively, in euphausiids from the Mediterranean (Fowler et al. 1976). Fujita (1972) reported Zn concentrations of 166 and 456 $\mu g\ g^{-1}$, respectively, in *Themisto oblivia* and *Themisto japonica* from the Sea of Japan. These data are of the same order of magnitude as ours for the north-bound leg of the cruise. However, certain samples collected at the beginning of the voyage, closer to New Zealand, had higher metal concentrations. For example, the Cu concentrations in *Themisto gaudichaudii* were particularly high at Stations 1 and 2, and for the other two species high Cu concentrations were found at Station 4, as well.

It is also clear from this limited data set, that meaningful comparisons can only be made between different samples of the same species. There were large differences in metal concentrations between different species, no doubt associated with their particular mode of uptake. These would mask any other trends.

Although metal concentrations in the water were not measured, the differences in the concentrations found in the zooplankton must, at least in part, be due to differences in the concentrations in the water. In describing the water-masses and fronts south of New Zealand, Houtman (1967) stated that this is one of the principal meandering regions of the Antarctic Polar Front, where apparent rates of change of over 4 km d^{-1} have been recorded. The rapidity and magnitude of change of trace metal concentrations (e.g., the change in Cu concentration in *T. gaudichaudii* between stations on the New Zealand continental shelf and the Deep Oceanic basin) leads us to speculate that the zooplankton samples came from populations occurring in different water-masses. Thus, it appears that zooplankters accumulate metals to a degree which is dependent on the water-mass in which they live. There seems to be little mixing of animal populations, each staying within its own water-mass, and we suggest that metal concentrations could be used to differentiate between zooplankton populations from different water-masses. This might be used to identify feeding areas of predators which eat zooplankton. Metal concentrations of zooplankton in the stomachs of krill-eating animals, for example, might indicate their feeding grounds. However, temporal changes should be considered, as indicated by the differences in metal concentrations between the two legs of the cruise which were separated by 2 months. These seasonal variations should be investigated in future studies.

Finally, our data show that trace metal concentrations in zooplankton from the Antarctic are not necessarily low. This has important implications for the use of metal levels in samples from remote areas, in establishing base-line levels for pollution studies. Natural enrichment of trace metals in unpolluted water can, and does, lead to relatively high levels of trace metals in animals which have not been subjected to pollution loads.

Acknowledgements. We thank Sir Ranulph Fiennes and members of the Transglobe Expedition who gave us the opportunity to carry out this research, and especially the crew of M.V. Benjamin Bowring for their help and patience. Thanks to Anton Bowring who patronised the research programme and to him and Terry Kenchington for driving the winch, under often harrowing conditions. We are also grateful to the Zoology Department of Canterbury University in Christchurch for laboratory facilities and the Zoology Department of the University of British Columbia for the use of their facilities and the salinity determinations. Anglo-American generously provided financial support. Financial support from the South African National Programme for Environmental Sciences is also gratefully acknowledged.

References

Boyle E, Edmond JM (1975) Copper in surface waters south of New Zealand. Nature (Lond) 253:107–109

Boyle EA, Huested SS, Jones SP (1981) On the distribution of copper, nickel and cadmium in the surface waters of the North Atlantic and North Pacific Ocean. J Geophys Res 86:8048–8066

Burling RW (1961) Hydrology of circumpolar waters south of New Zealand. NZ Dep Sci Ind Res Bull 143:pp 66

Burton JD, Young ML (1980) Trace metals in the shelf seas of the British Isles. North west European shelf seas: seabed motion. Elsevier Oceanography Series, pp 495–516

Craig H, Broecker WS, Spencer D (1981) GEOSECS Pacific Expedition, vol 4, Sections and profiles. Nat Sci Found Washington DC, pp 81–87

Cutshall N, Holton R (1972) Metal analyses in IDOE baseline samples. In: Goldberg ED (Chairman) Baseline studies of pollutants in the marine environment (Heavy metals, halogenated hydrocarbons and petroleum). Brookhaven Nat Lab Upton NY, pp 67–82

Davies AG (1978) Pollution studies with marine plankton. Part II. Heavy metals. Adv Mar Biol 15:381–508

Eagle GA, Airey DD, Connell AD, Fricke AH, Hennig HF-KO (1983) A comparative study using four different toxicity tests. Poster, 5th National Oceanographic Symposium Grahamstown, South Africa

Edwards RJ, Emery WJ (1982) Australasian Southern Ocean frontal structure during 1976–1977. Austr J Mar Freshwater Res 33(1): 3–22

Fowler SW, Oregioni B, La Rosa J (1976) Intercalibration measurements for non-nuclear pollutants: Trace metals in pelagic organisms from the Mediterranean Sea. Report of the International Laboratory of Marine Radioactivity, Monaco, pp 110–122

Fujita T (1972) The zinc content in marine plankton. Rec Oceanogr Works Jpn 11:73–79

Houtman TJ (1967) Water masses and fronts in the Southern Ocean south of New Zealand. NZ Dep Sci Ind Res Bull 174:pp 40

Lutjeharms JRE, Walters NM, Allanson BR (1984) Oceanic frontal systems and biological enhancement. In: Siegfried WR, Condy PR, Laws RM (eds) Antarctic nutrient cycles and food webs (Proceedings of the fourth SCAR symposium on Antarctic biology). Springer, Berlin Heidelberg New York

Martin JH, Knauer GA (1973) The elemental composition of plankton. Geochim Cosmochim Acta 37:1639–1653

Pickard GL, Emery WJ (1982) Descriptive physical oceanography. An introduction. 4th (SI) Enlarged Edition. Pergamon, Oxford, pp 249

Watling HR, Watling RJ (1976) Trace metals in *Choromytilus meridionalis*. Mar Pollut Bull 7:91–94

The Interaction of Soil and Lake Microflora at Signy Island

J.C. ELLIS-EVANS and D.D. WYNN-WILLIAMS[1]

Summary. Lakes on Signy are small and, during the summer ice-free period, receive a substantial influx of run-off from the surrounding catchment, greatly in excess of their total volume. Within the various habitats comprising each lake's catchment area, large populations of bacteria, yeasts and fungi exist. Transport of terrestrial micro-flora to the lakes in run-off water is demonstrated by the appearance of *Chromobacterium* and yeasts in lake surface waters at around the same time as their numbers increase in certain terrestrial habitats. However, close examination reveals substantial differences between lake and terrestrial populations, possibly the most obvious being the virtual absence of gram-positive bacteria from lake samples whilst comprising 20% of terrestrial populations. By means of numerical profiles, it has been possible to demonstrate that the composition of lake populations differs between lakes, and between lakes and their catchment populations. These differences can be explained in terms of catchment composition and nutrient availability. A study of temperature-specific growth rate characteristics indicates that successful freshwater bacteria can, in broad terms, be considered psychrophiles, whereas successful terrestrial bacteria are more psychrotolerant. A comparison of substrate affinity values indicates that freshwater bacteria are more efficient than terrestrial bacteria or yeasts in utilizing low concentrations of dissolved organic C at low temperatures. At higher temperatures, substrate affinity values are more equivalent. From the data presented, it is proposed that the two populations may be considered as a mixed-culture chemostat population. At high nutrient concentrations, a typical terrestrial population will exist and with fluctuations in temperature and nutrient level, a typical succession will also occur. But when nutrient levels drop suddenly for an extended period and temperature stabilizes at about 0 °C, a typical freshwater population emerges whose components had previously existed only in very low numbers.

1 Introduction

Freshwater lakes in the maritime Antarctic zone are usually small, relatively shallow systems covering a wide range of trophic states (Heywood et al. 1979). The catchment areas of such lakes vary in complexity from permanent ice cover to moss carpets overlying peat, usually frequented by seals and birds. Priddle and Heywood (1980), working at Signy Island, suggested that the complexity of a catchment is reflected in the trophic status of the lake within the catchment. Terrestrial studies by Baker (1970) demonstrated that large populations of bacteria, yeasts and moulds occur in the soils of Signy Island. More recently, Wynn-Williams (1980) established that large microbial populations are prevalent during the spring period when large volumes of snow-melt water flush through the catchment profile into each lake. This paper attempts to establish the significance of these seasonal inputs from the terrestrial systems to the lake microflora.

2 Methods

This work is based on sampling of enriched Heywood Lake and a moss carpet community dominated by *Calliergon* and *Drepanocladus*. The methodology for sampling and enumerating soil microbial populations has been fully described in Wynn-Williams (1979). Total viable bacterial counts were made on casein peptone starch agar (CPSA) amended with soil expressate and Actidione (Upjohn). Although the absolute limitations of viable counting are appreciated (Martin et al. 1982) they are justified for detecting relative changes in the opportunistic, fast-growing microbial populations subjected to freeze-thaw stress in moss peats. They also enable distinctive strains such as *Chromobacterium* to be recognized and monitored.

Yeasts and moulds were enumerated on Sabouraud Dextrose agar (SDA) amended with Aureomycin (Lederle). Lake bacteria were counted as colony forming units (CFU) on casein peptone starch agar (CPS) amended with Actidione only (Ellis-Evans 1981a), and yeasts and moulds were enumerated as described above. Isolates selected randomly, were purified by repeated sub-culturing and then maintained at 5 °C on semi-solid (0.3%) versions of CPS and SDA. The characterization tests employed standard methodology (Cowan and Steel 1974). Numerical profiles were obtained using ten characterization tests and the results were scored as outlined in Table 3. The tests essentially follow the scheme of Griffiths and Lovitt (1980) with the addition of the ecologically relevant test of ability to grow at 22 °C.

Temperature-growth studies were undertaken on several isolates using a temperature gradient incubator described by Baker (1974). He used full strength tryptone soya broth (Oxoid) as the growth medium but in the present study, 1/100 strength TSB in a modified version of the gradient

1 British Antarctic Survey, NERC, High Cross, Madingley Road, Cambridge CB 3 OET, United Kingdom

Antarctic Nutrient Cycles and Food Webs
(ed. by W.R. Siegfried, P.R. Condy, and R.M. Laws)

incubator was more appropriate (Ellis-Evans 1981b). Growth was expressed as change in optical density of cultures with time. Results were expressed as an Arrhenius plot ($\log_{10}$ specific growth rate vs. 1/K) to ascertain the cardinal temperatures for each isolate.

Substrate affinity was assessed using glucose limited cells harvested from a simple 1-1 chemostat system (Baker 1968). ^{14}C-labelled glucose was added at five concentrations to five separate sets of cell suspension and incubated at 2, 8, 15, and 22 °C for 5 h as described by Herbert and Bell (1977). At the end of incubation, samples were killed with Lugols iodine, filtered through 0.2 μm Nucleopore filters and counted in a liquid scintillation counter using Picofluor-15 (Packard) as the scintillation fluid. Counts were analyzed by Michaelis-Menten kinetics and a straight line plotted using the Lineweaver-Burk transformation. The intercept on the horizontal axis was expressed as K_s - a measure of substrate affinity. Herbert and Bell (1977) give further details.

3 Results and Discussion

3.1 Catchment-Lake Interactions During the Spring Melt

The data presented in Table 1 show the marked seasonal pattern present in both lake and terrestrial environments. Direct microscopy of the water samples indicated that total bacteria cell counts were 50-200 times higher than plate counts. Total cell counts followed a similar but less pronounced seasonal pattern than that described here for viable counts. Soil population counts for the later winter (pre-thaw) period (Table 1) indicate the potential for microbial development, since the peat profile is frozen solid at this time. Lake-water counts during the same period indicated that bacterial numbers were extremely low and yeasts undetectable. Phytoplankton at this time were inactive (Ellis-Evans 1981c) so readily assimilable dissolved organic C (DOC) was not available to the microbial population. At the time of the first major thaw the peat profile remained frozen and inflows comprised largely melt-water from overlying snow, as evidenced by the low bacterial counts. Peat bacterial numbers remained relatively constant until late spring but yeasts, which were characteristic of the moss phyllosphere, increased rapidly in spring. This was possibly due to utilization of DOC flushed from frost-damaged plant cells (Hurst 1982). Despite the low temperature prevailing (0 to + 1 °C), this phase of yeast growth was potentially exponential as shown by simulation in vitro of observed spring freeze-thaw cycles (Wynn-Williams 1982). This allowed the derivation of the regression $y = 0.023\,x + 2.207$ ($r^2 = 0.885$, $P < 0.001$), where y = ln viable yeast biomass (mg m^{-2}) and x = time elapsed after initial thaw (h).

Detection of exponential growth of yeasts in vivo was obscured, probably by a proportion being flushed from the moss surface and into the peat profile since they appeared concurrently in both inflow streams and lake water. The observed increase in bacterial numbers in lake water can be attributed to the increased algal production (and associated DOC release), as the melt removed snow cover allowing light to penetrate the ice cover of the lake.

During the 2 weeks following the onset of the thaw, the top 20 mm of the moss-peat profile had thawed completely and a series of samples showed that the peak bacterial population increased tenfold, whereas the yeast populations decreased markedly. This may be attributed partly to the larger yeast cells being flushed down and out of the peat profile, but also to a decrease in available DOC due to earlier yeast activity and decreased frequency and influence of major freeze-thaw cycles. The progressive increase in numbers of bacteria and yeasts in inflow streams after the disappearance of snow cover, confirmed by daily sampling, reflected the depth of penetration into the catchment profile of the melt-water which subsequently flowed into the lake. Samples taken 2 weeks later from the fully thawed peat profile revealed bacterial counts decreasing towards summer levels and yeasts stabilizing around 10^3 cells g^{-1} (wet wt.). Inflow bacterial counts had also stabilized but lake-water

Table 1. Fluctuations in counts of bacteria and yeasts in the catchment[a], inflow stream and water column[b] of Heywood Lake during spring thaw

Time	Microbial group[c]	Catchment	Water status	Inflow	Lake water	Comments
Pre-thaw	B	16,300 ± 3,200	Snow and inter-	Frozen	438 ± 86	Lake in winter conditions
	Y	800 ± 200	stitial ice		0	and peat frozen
Thaw	B	18,400 ± 3,900	Snow and surface	420 ± 66	6,040 ± 321	First thaw, peat profile still
	Y	10,700 ± 500	meltwater	200	100	frozen
11-17 days	B	351,300 ± 86,200	Peat surface water	25,490 ± 3,225	25,200 ± 628	Top 20 mm moss/peat
past thaw	Y	1,000 ± 250	flowing	3,900 ± 100	1,200 ± 200	profile thawed
29-34 days	B	148,700 ± 25,700	Peat profile water	20,000 ± 4,062	74,400 ± 10,677	100 mm peat profile
post thaw	Y	4,500 ± 700	flowing	2,200 ± 300	1,600 ± 200	thawed

[a] CFU ml^{-1} (lake water) Mean values ± 1 S.E. (n = 3)
[b] CFU g^{-1} FW (peat) Mean values ± 1 S.E. (n = 4)
[c] B = bacteria, Y = yeasts

Table 2. Fluctuations in counts of chromobacteria in the catchment[a], water column[b], and sediments of Heywood Lake

Season	Catchment 1976–77	Lake water 1976–77	1977–78	Lake sediment 1976–77	1977–78
Early winter	< 100	130 ± 28	85[c]	0	1,250 ± 114
Late winter	< 100	0	10	0	
Spring	< 100	0	50	900 ± 57	5,000 ± 518
Summer	522 ± 176	400 ± 29	75	5,000 ± 420	4,800 ± 384
Autumn	2,783 ± 2,420	1,350 ± 194	1,100 ± 207	12,000 ± 422	6,800 ± 190

[a] CFU g^{-1} FW (soil) mean ± 1 S.E. (n = 4)
[b] CFU ml^{-1} (lake water) mean ± 1 S.E. (n = 3)
[c] Errors are not computed on low counts

bacterial populations continued to increase in parallel with phytoplankton development as lake light levels improved.

Yeast counts in the inflow samples were consistently higher than in lake water. The lake-water samples (Table 1) were taken 1.5 m below the underside of the ice cover and the inflow stream is known to flow through the lake immediately under this ice cover. It is therefore significant that differences in yeast counts were nevertheless found between the two water-bodies. Mid-depth (3.0 m) lake-water samples consistently had 2–3 times lower yeast counts than the 1.5-m samples. Characterization of peat-water, inflow and lake-water yeast populations revealed the same species of yeast present in all three locations, which indicates that the peat was the source of the lake population. The trends outlined above were confirmed by detailed sampling the following year.

3.2 Catchment-Lake Interactions During the Summer Period

In addition to the influence exerted by the catchment on lake populations in spring (as indicated by the yeasts), the data (Table 2) demonstrate that peat populations were also contributing to the composition of the lake micro-flora during the ice-free summer period when the lakes are relatively homogeneous due to wind-induced mixing. Chromobacteria show distinctive purple coloration and they have been proposed as biological indicators of environmental status (Wynn-Williams 1983). Table 2 indicates that maximum counts of these bacteria in both soil and lake water occurred during late summer and early autumn. Wynn-Williams (1983) showed that the appearance of Antarctic strains of purple chromobacteria in soil coincides with pH rising above 4.6 which only occurred in late summer. The absence of chromobacteria in lake water until their simultaneous appearance in the catchment summer, despite favourable pH conditions in the lake year round (Ellis-Evans 1981c), indicates a link between the two environments via the inflows.

Similar microbial interaction studies of an oligotrophic lake (Moss Lake) showed its catchment to be largely rock and scree (fellfield) providing only snow-melt inflows (Ellis-Evans 1981a). Microbial populations were therefore lower than those detailed for Heywood Lake in Tables 1 and 2. Yeasts were present only in very low numbers in the catchment "soil" profile (c.f. counts from Antarctic fellfields; Wynn-Williams 1985) and were rarely detected in lake-water samples. The main input was therefore bacterial, but counts were low so although the pattern was repeated, the input of bacterial cells was less significant than autochthonous development. Chromobacteria were virtually absent from the water column all year.

3.3 Characteristics of Lake and Terrestrial Populations

Although substantial numbers of yeasts and bacteria are flushed from the catchments into the lakes, the terrestrial and aquatic environments differ in various ways (e.g., pH) so that introduced terrestrial strains may be inactivated or out-competed. The characteristics (Table 3) of pure bacterial isolates from terrestrial catchments and lake water were therefore examined in some detail. The numerical profiles for 310 isolates include data for a *Calliergon-Drepanocladus* dominated catchment and three lakes of differing trophic status (Table 4). Gneiss Lake is ultra-oligotrophic with a catchment mainly of permanent ice and snow. Moss Lake is oligotrophic with a fellfield catchment including some permanent ice. Heywood Lake is mesotrophic with a moss-carpeted catchment free of permanent ice cover. The range of aerobic heterotrophs was remarkably similar in all three lakes, but the dominant bacterial "sort" (a general group term with no preconceived taxonomic status – see Griffiths and Lovitt 1980) differed between lakes. Horizontal lines between columns indicate a degree of overlap of dominant "sorts" between ultra-oligotrophic and oligotrophic lakes; between oligotrophic and eutrophic lakes; but not between ultra-oligotrophic and eutrophic lake systems (Table 4).

Although fewer numerical profiles from catchment samples were available, slightly more "sorts" were recognized from this environment than from either of the lakes. A large number of the bacterial "sorts" from the catchment had no equivalent from the lakes, suggesting a marked difference in the composition of the lake and terrestrial populations. The relatively small terrestrial data set precluded the recognition of any dominant bacterial "sorts" as was apparent in the lakes.

Wynn-Williams (1980) has shown that single species can become numerically dominant in peat at different times of

Table 3. Characters used to obtain the numerical profiles to isolates

Score	First digit	Second digit	Third digit	Fourth digit
1	Gram-positive	Rod-shaped	Fermentative[a]	22 °C
2	Catalase	Oxidase	Oxidative[a]	
4	Motile	Proteolysis	NO_3 reduction	

[a] Fermentative or oxidative metabolism based on Hugh and Leifson test for glucose utilization

the year, and Ellis-Evans (1981b) has reported single species domination on a seasonal basis in a variety of Signy Island lakes. Numerical profiling must be applied to many more lake and catchment isolates before seasonal or topographical patterns can be established. However, Table 4 indicates that a restricted number of bacterial "sorts" were especially prevalent in the three lake systems, whereas greater variety occurred in catchment populations.

3.4 Temperature-Growth Characteristics of Lake and Terrestrial Isolates

Six lake and three terrestrial isolates, all of which commonly occurred in the respective environment, were studied in greater detail using the temperature gradient incubator. The cardinal temperatures of these isolates were ascertained from Arrhenius plots (Table 5). For comparison, data for other catchment isolates are included from Baker (1974) and Wynn-Williams (1983). All lake isolates had optimum temperatures < 20 °C, but maximum temperatures > 20 °C, indicating psychrotolerance. Of the three catchment isolates (C1, C2, C3) which were treated similarly, the two more typically terrestrial types (C1 and C2) showed slightly higher optimum and maximum temperatures than C3 *(Pseudomonas)*, which was indistinguishable from its aquatic equivalent. Data obtained by Baker (1974) are illustrated by isolates C5–C8 and here both cardinal temperatures are significantly higher than those of isolates C1–C4. This may be related to the use of full strength TSB by Baker, whereas 1/100 strength was employed in the present study. This highlights the problems of comparing temperature-growth characteristics on different media. Studies by Ellis-Evans (1981b) indicate that media such as full strength TSB give consistently lower counts and less species diversity than 1/100 strength TSB or 1/10 strength CPS and must therefore be interpreted with caution.

Figure 1 illustrates the low temperature portion of Arrhenius plots for isolates L1–L6 and C1–C2. The lake isolates gave straight line plots down to +0.9 °C, whereas plots for the two terrestrial isolates deviated from linearity below +4.0 °C. Harder and Veldkamp (1971) concluded that the slope of the Arrenhius plot for psychrophiles is linear to 0 °C and below, whilst psychrotrophs deviate from linearity below 4–5 °C and mesophiles tend to deviate at 10 °C or

Table 4. Distribution of bacterial "sorts" in three lake systems and a moss-peat catchment

Code	Gneiss Lake	Moss Lake	Heywood Lake	Moss-Peat Catchment
0101			1	
0111				1
0141			2	
0351				1
0700				1
0701	2	4	9	
0721		2	3	
1051				1
2301			2	
2341	1			3
2351				2
2701				2
2731				2
2741		2		
3011				1
3021			1	
3041				1
3101			1	
3111	1	1		1
3221			1	
3451				1
3571				1
3741				1
4151				1
4321		2		
4341	1	1		1
4360	1		5	
4361	1			1
4701				1
4721	1	1	1	
4741				1
4760	1			
4761	2	4	4	
6151			1	
6301	1	1	1	1
6311				1
6321				1
6341			1	
6351			1	
6360	1	1	1	
6361	44 common/dominant	22	7	
6701				5
6711		6	4	1
6720		2	1	
6721	4	16 common/dominant	13	1
6731				4
6741	6			
6751	3	18 common/dominant	15	2
6760	2		5	
6761	1	4	15	1
6770				1
6771				3
7101				2
7111				1
7141				2
7501				1
Totals	77	87	95	51

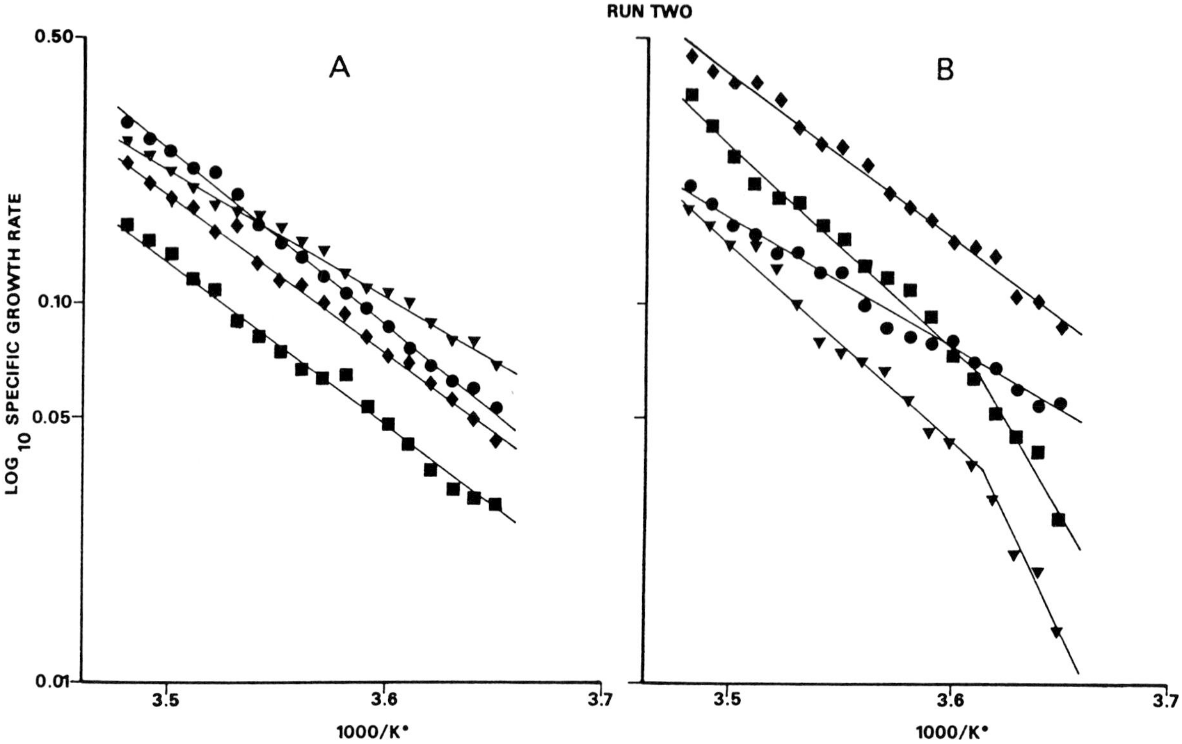

Fig. 1A,B. Arrhenius plots of log specific growth rate versus absolute temperature for isolates L1 ▾, L2 ◆, L3 ●, and L4 ■ (A) and L5 ◆, L6 ■, C1▾, and C2 ■ (B)

Table 5. Cardinal temperatures derived from Arrhenius plots for catchment and lake micro-organisms

Code[a]	Identity	Maximum Temp (°C)	Optimum Temp (°C)
L1	*Pseudomonas* sp.	23.5	19.4
L2	*Vibrio* sp.	22.7	14.4
L3	*Vibrio* sp.	21.4	16.7
L4	*Flavobacterium* sp.	24.1	17.8
L5	*Cytophaga* sp.	23.7	18.1
L6	*Chromobacterium fluviatile*	23.7	15.5
C1	*Corynebacterium* sp.	27.9	21.4
C2	*Candida* sp.	26.4	19.0
C3	*Pseudomonas* sp.	22.8	18.2
C4	*Chromobacterium lividum*	25.0	15.0
C5	*Cellulomonas* sp.	30.6	24.9
C6	*Pseudomonas* sp.	33.1	24.7
C7	Coryneform	31.7	28.9
C8	*Arthrobacter* sp.	29.6	23.1

All values for L1–L6 and C1–C4 are based on growth in 1/100 strength tryptone soya broth medium. Values for C5–C8 were based on growth in full strength TSB and were obtained by Baker (1974). C4 values were obtained by Wynn-Williams (1983) using casein peptone starch agar plates.

[a] L = lake isolate; C = catchment isolate

above. However, the cardinal temperatures (Table 5) indicate that the lake isolates are psychrotolerant. This evidence would suggest that the lake isolates fall between both definitions and illustrates the shortcomings of such artificial groupings. A few Antarctic aquatic psychrophiles have been isolated (Herbert and Bhakoo 1979), but unlike other permanently cold environments (e.g., marine), the maritime Antarctic lakes receive regular and substantial inputs of psychotolerant bacteria from the catchment. This raises the question of the existence of an indigenous lake microflora as opposed to a population selected from the terrestrial input.

3.5 Selection or an Indigenous Lake Microflora?

The existence of an indigenous lake micro-flora is supported by the predominance of substantial numbers of *Chromobacterium fluviatile* in Heywood Lake sediments throughout the year (Table 2). This species is not found in the lake water column or terrestrial catchments where the occurrence of *C. lividum* fluctuates seasonally (Wynn-Williams 1983). Similar restricted distribution of *C. fluviatile* has also been found in Moss Lake sediments (Ellis-Evans, unpubl.) and may therefore be a general trend. It is probable that such lake bacteria are present in catchment populations throughout the year, but in such low numbers that they are rarely detected by plate-culture methods. Data in Tables 2 and 5 indicate that *C. lividum* only occurs in detectable numbers in terrestrial samples during late summer-early autumn, though clearly propagules must be present throughout the spring-summer period. The occurrence of *C. lividum* is closely associated with environmental pH in the terrestrial environment (Wynn-Williams 1983). However, *C. lividum* only occurs in lake water when high numbers are also present in the terrestrial environment (Table 2). This suggests that the organism is

Table 6. Changes in substrate affinity with temperature in three Antarctic isolates

	K_s (μM glucose)				
	0 °C	2 °C	8 °C	14 °C	22 °C
L-3	–	9.2	13.4	14.9	27.1
C-1	–	34.0	25.5	12.1	15.4
AF-1[a]	5.0	–	6.5	24.0	no growth

[a] AF-1 data from Herbert and Bell (1977)

perhaps not actively growing in lake water, but is introduced by flushing.

Bacterioplankton must be able to survive relatively nutrient poor conditions as compared to many terrestrial environments. Readily assimilable nutrients produced by the algae are rapidly taken up by the lake bacteria, so maintaining very low DOC concentrations (Ellis-Evans 1985). Terrestrial bacteria washed into the lake must compete with the extant lake micro-flora for these low concentrations of DOC. This suggests a potential selection process and this was investigated using chemostat-cultured material. The results for isolates L3 and C1 (Table 6) are compared with data for psychrophilic *Vibrio*, AF-1 isolated from Heywood Lake by Herbert and Bell (1977). The findings are comparable with data from batch-cultured material representing the remaining 7 isolates used in this study (Ellis-Evans 1981b).

Pronounced differences were apparent between catchment and lake bacteria. The latter exhibit high substrate affinity at lake temperatures (L3, 9.2 μM glucose at 2 °C) and similarly at 8 and 14 °C. However, substrate affinity decreased significantly at 22 °C. In contrast, isolate C1 showed high substrate affinity at 14 and 22 °C and low affinity at 2 and 8 °C. This demonstrates that L3 is a more effective competitor for low substrate concentrations at lake temperatures. At higher temperatures, such as may occur in peat in summer, C1 was more efficient. The pattern of changing substrate affinity with temperatures shown by isolate AF-1 is similar to that of L3, but AF-1 is a true psychrophile with an optimum temperature of 10 °C and a maximum of <20 °C. Herbert and Bell (1977) have shown that, under C limitation, it has a maintenance energy coefficient of 0.02 at 0 °C compared to 0.10 at 20 °C with a relatively long generation time of ca. 23 h at 0 °C. They concluded that such slow growth by a psychrophile at low temperatures was due to high cell C production and low mineralization. The present studies of lake isolates revealed slightly shorter generation times than AF-1 (range 8–26 h at +0.2 °C) but otherwise followed the same strategy under C limitation. In a consequent experiment, isolates L3 and C1 were grown together in a chemostat under glucose limitation at different temperatures. As might be predicted from Table 6, L3 ultimately became dominant at low temperatures, whereas C1 dominated at high temperatures (+23 °C). Although this may not be true for other C substrates, it exemplifies the observed differences between terrestrial and freshwater populations. When C limitation was removed, C1 dominated L3 in the mixed culture chemostat experiments, a situation comparable to the terrestrial environment. The data (Fig. 1, Table 6) enable clear demarcation of two groups of bacteria: terrestrial opportunists which exploit pulses of DOC potentially released from cryptogamic cells damaged by freeze-thaw cycles, and oligotrophs which dominate the lake environment and compete successfully for regular low inputs of DOC in a dilute environment.

4 Conclusions

We suggest that the catchment-lake ecosystem behaves as a natural chemostat. In the presence of relatively high nutrient concentrations as found in the terrestrial environment, particularly during spring freeze-thaw cycles, a typical terrestrial population (as exemplified by isolate C1) exists. In the presence of free water in the catchment, fluctuations in temperature and nutrient level result in a succession of bacteria and yeasts appearing and disappearing from prominence. However, if the nutrient concentrations of the input are markedly reduced, the dilution rate is increased and the temperature maintained at a relatively constant low value (0 to +4 °C) then typical freshwater bacteria (such as L3 and AF-1) will emerge as the dominant groups and terrestrial forms diminish in importance. This is a simplification of a complex interaction of environmental variables and population characteristics. Nevertheless, we feel that this approach, coupled with field and supportive laboratory studies, offers a valuable means of assessing the significance of microbial interactions in vivo and furthering our understanding of the interactions between Antarctic ecosystems.

Acknowledgements. We gratefully acknowledge the support of the British Antarctic Survey and especially the help of the personnel at Signy Island base during 1975–1978. We thank Dr. W.C. Block for critical review of the manuscript.

References

Baker JH (1970) Quantitative study of yeasts and bacteria in a Signy Island peat. Brit Antarct Surv Bull 23:51–53

Baker JH (1974) The use of a temperature gradient incubator to investigate the temperature characteristics of some bacteria from Antarctic peat. Brit Antarct Surv Bull 39:49–59

Baker K (1968) Low cost continuous culture apparatus. Lab Pract 17:817–821

Cowan ST, Steel KJ (1974) Manual for the identification of medical bacteria (2nd edn) Cambridge University Press, 238 pp

Ellis-Evans JC (1981a) Freshwater microbiology in Antarctica I microbial numbers and activity in oligotrophic Moss Lake, Signy Island. Brit Antarct Surv Bull 54:85–104

Ellis-Evans JC (1981b) Freshwater microbiology at Signy Island, South Orkney Islands, Antarctica. Ph D thesis (CNAA) 282 pp

Ellis-Evans JC (1981c) Freshwater microbiology in Antarctica II Microbial numbers and activity in nutrient-enriched Heywood Lake, Signy Island. Brit Antarct Surv Bull 54:105–121

Ellis-Evans JC (1985) Interaction of bacteria and phytoplankton in nutrient cycling within eutrophic Heywood Lake, Signy Island. In: Siegfried WR, Condy PR, Laws RM (eds) Antarctic nutrient cycles and food webs (Proceedings of the 4th SCAR symposium on Antarctic biology). Springer, Berlin Heidelberg New York, pp

Griffiths AJ, Lovitt R (1980) Use of numerical profiles for studying bacterial diversity. Microb Ecol 6:35–44

Harder W, Veldkamp H (1971) Competition of marine psychrophilic bacteria at low temperature. Antonie Leeuwenhock J Microbiol 37:51–63

Herbert RA, Bell CR (1977) Growth characteristics of an obligately psychrophilic *Vibrio* sp. Arch Microbiol 113:215–220

Herbert RA, Bhakoo M (1979) Microbial growth at low temperatures. In: Russell AD, Fuller R (eds) Cold tolerant microbes in spoilage and the environment. Academic, London, pp 1–16

Heywood RB, Dartnall HJD, Priddle J (1979) The freshwater lakes of Signy Island, South Orkney Islands, Antarctica: data sheets. Brit Antarct Surv Data 3, 46 pp

Hurst JL (1982) Biology of fungi on plants in the Sub-Antarctic. Ph.D. thesis, University of Aston in Birmingham

Martin NJ, Siwasin J, Holding AJ (1982) The bacterial population of a blanket peat. J Appl Bacteriol 53:35–48

Priddle J, Heywood RB (1980) Evolution of Antarctic lake ecosystems. Biol J Linn Soc 14:51–66

Wynn-Williams DD (1979) Techniques for studying terrestrial microbial ecology in the Maritime Antarctic. In: Russell AD, Fuller R (eds) Cold tolerant organisms in spoilage and the environment. Academic, London, pp 67–82

Wynn-Williams DD (1980) Seasonal fluctuations in microbial activity in Antarctic moss peat. Biol J Linn Soc 14:11–18

Wynn-Williams DD (1982) Simulation of seasonal changes in microbial activity of Maritime Antarctic peat. Soil Biol Biochem 14: 1–12

Wynn-Williams DD (1983) Distribution and characteristics of *Chromobacterium* in the Maritime and Sub-Antarctic. Polar Biol 2: 101–108

Wynn-Williams DD (1985) Comparative microbiology of moss peat decomposition in the Scotia Arc and Antarctic Peninsula. In: Siegfried WR, Condy PR, Laws RM (eds) Antarctic nutrient cycles and food webs (Proceedings of the 4th SCAR symposium on Antarctic biology). Springer, Berlin Heidelberg New York

Trends in Crabeater Seal Age at Maturity: An Insight into Antarctic Marine Interactions

J.L. BENGTSON[1] and R.M. LAWS[2]

Summary. The age at which Crabeater Seals, *Lobodon carcinophagus*, reach sexual maturity is a parameter that might reflect changes in the abundance of Antarctic Krill, *Euphausia superba*. With accelerated body-growth, sexual maturity is reached at an earlier age. It was hypothesized that the decrease in Baleen Whale stocks would have resulted in an increase in Krill availability, and a subsequent decrease in the age of maturity of Crabeater Seals. As whaling decreased and populations of other consumers of Krill presumably increased in response to abundant prey, Krill availability to individuals would have decreased; age at maturity in Crabeater Seals would have been expected to rise as body-growth slowed. An apparent decrease in maturity, as evident from the transition zone of tooth cementum, was found to be unrelated to population trends. Instead, the decrease in maturity appears to be related to a differential mortality within cohorts; those individuals maturing earlier apparently die sooner. Hence, tooth back-calculation is an unsuitable method to assess past trends in sexual maturity. Counts of ovarian *corpora* indicated an increase in maturity during the predicted period (post 1962). Unfortunately, lack of ovarian samples from years prior to 1962 does not allow a complete evaluation during that time. However, the 1 sample available (reflecting individuals from the 1958–1962 cohorts) does fall within the range predicted. Hence, it appears that age of sexual maturity in Crabeater Seals fell and rose in response to a presumed changing availability of Krill. These past trends in maturity, reflecting changing trophic relationships within the marine community, indicate that the age of sexual maturity in Crabeater Seals might be a useful parameter to monitor in the future. Just as whaling altered Krill availability in the past, commercial Krill harvests may again affect the availability of Krill to predators. Monitoring the age at sexual maturity in Crabeater Seals may provide insight into the extent of these future perturbations on the Antarctic marine ecosystem.

1 Introduction

The Convention for the Conservation of Antarctic Marine Living Resources (CCAMLR) provides that the Antarctic marine ecosystem should be managed and conserved as a unit. Thus, consideration must be given to the conservation of particular species, such as the Antarctic Krill, *Euphausia superba*, as well as species dependent on Krill. However, obtaining data on the past and current status of all component groups within the ecosystem would be an exceedingly expensive, if not impossible, task. The use of ecosystem indices has therefore been suggested as a method whereby general ecological trends could be monitored.

In selecting suitable ecosystem indicators, it is necessary to find parameters which respond to perturbations of the ecosystem, particularly those reflecting changes in Krill populations. Investigating the interactions between Krill and its consumers represents one approach to this subject. Moreover, understanding the extent to which past commercial whaling and its presumed impact on Krill availability affected various parameters of the consumers of Krill would help identify useful indicators to evaluate impacts of commercial fisheries.

The Crabeater Seal, *Lobodon carcinophagus*, is a species which appears to be useful as an indicator of community interactions. Crabeater Seals eat Krill almost exclusively (Øritsland 1977; Bengtson 1982), live up to 40 years of age (Laws, unpubl.), have a circumpolar distribution (Erickson et al. 1971), and have a large standing stock estimated at between 15–30 × 10^6 individuals (Gilbert and Erickson 1977). Pending current re-evaluation of census correction factors, revised population estimates incorporating new data may well exceed 30 × 10^6 individuals (A.W. Erickson, pers. comm.). Laws (1977a) estimated that Crabeater Seals (assuming a population of 15 × 10^6) consume approximately 63 × 10^6 t of Krill annually, or about 1.5 times the quantity eaten by current baleen whale populations. Therefore, changes in Krill availability, whether increased by reduced competition from Baleen Whale populations or decreased by direct Krill harvests, might be reflected in growth, reproductive, and population parameters of Crabeater Seals.

The age at which Crabeater Seals reach sexual maturity is a parameter which might reflect changes in Krill availability. Attainment of sexual maturity in mammals is often related to attainment of a certain body size (Leslie et al. 1946; Sadleir 1969). Laws (1956) demonstrated that seals with high growth rates reached puberty at a younger age than slower-growing seals. Abundant food and relatively faster growth rates may therefore lead to attaining sexual maturity earlier. Decreased food availability would presumably have an opposite effect, "delaying" age at sexual maturity (Eberhardt 1977).

1 National Marine Mammal Laboratory, NMFS/NOAA, 7600 Sand Point Way N.E., Seattle, WA 98115, USA
2 British Antarctic Survey, NERC, High Cross, Madingley Road, Cambridge CB3 OET, United Kingdom

Antarctic Nutrient Cycles and Food Webs
(ed. by W.R. Siegfried, P.R. Condy, and R.M. Laws)

Laws (1977b) used tooth back-calculation (back-calculating age at maturity evidenced by the transition in rates of cementum deposition) from Crabeater Seals in the Marguerite Bay area to demonstrate a decrease in age at maturity from approximately 4.5 years in 1945 to 2.5 years in the late 1960s. If the observed trend occurred in response to numerical and trophic changes within the marine food chain caused by decreasing Whale populations, one might have expected a subsequent increase in age of maturity through the 1970s with the slowdown of whaling and enhanced competition between increasing populations of consumers of Krill. A small sample of Crabeater Seals was taken in 1977 near the South Shetland Islands to look for evidence of such an increase in age at maturity (Bengtson and Siniff 1981). Although a mean age of 3.8 years was found from ovarian examination, tooth back-calculation did not show a clear fall and rise in age of maturity. Here we review the data from these and other studies (Øritsland 1970) in the light of recent material from the Antarctic Peninsula and an expanded sample of Crabeater Seals from Marguerite Bay.

2 Methods

Crabeater Seals of both sexes and of all age classes were collected in 13 seasons for specimen material and dog food over an area extending from the southwestern coast of the Antarctic Peninsula northeast along the Scotia Arc to the South Orkney Islands. Øritsland (1970) collected 132 females and 86 males in 1964 from the area between Elephant Island and the South Orkney Islands. During 1966–1976, 753 females, 527 males, and 113 individuals of undetermined sex were collected from the Marguerite Bay area; 94 females were collected in Bransfield Strait and Admiralty Bay, King George Island, in 1977; and 261 females and 279 males were collected from the vicinity of the Argentine Islands, Bransfield Strait, Seymour Island, and James Ross Island in the 1981 to 1982 season. Standard measurements, teeth, and reproductive tracts were collected from most of these seals.

Mandibles from freshly collected seals were boiled 30 to 60 min prior to pulling all teeth, which were then stored in a 10% formalin solution. A longitudinal thin section (0.1 to 0.5 mm thick) was taken from the third post canine tooth (or alternative if this tooth was damaged) in each individual, using a high-speed diamond circular saw. Tooth wafers were cleared through serial solutions of ethanol and xylene and mounted on glass slides with an artificial mounting medium (D.P.X. or Permount). These mounted thin sections were examined under transmitted polarized light at 150–300 times magnification. Teeth from the 1977 collection were decalcified, thin-sectioned, and stained prior to microscopic examination (Bengtson and Siniff 1981).

The total age of each seal was estimated by counting the total number of cementum layers. Each main layer was assumed to represent 1 year. Replicate counts were made for all samples. For the 1966–1976 samples, two readers counted cementum layers, discussed discrepancies, and agreed on an appropriate score. For the 1977 and 1981–82 samples, the mean value of five blind replicate counts was used as the total age estimate.

Back-calculation of the age at sexual maturity (first ovulation or first sperm production) was accomplished by locating the transition zone in cementum layers. Early layers are broader and irregular, whereas layers laid down later in life are narrower and more regular (Hewer 1964; Laws 1977b; Bengtson and Siniff 1981). For the current analysis only seals 7 years or older were considered for tooth back-calculation. This value was obtained by assuming that all seals were mature and had a visible transition zone by 7 years of age. Previous work (Bengtson and Siniff 1981) indicated that all female seals 7 years or older had ovulated at least once. Inclusion of younger individuals which may not yet have matured would have biased the maturity estimates downward. Scoring of the transition zone was done with replicate counts as described for estimating total age. No tooth back-calculation was done on the 1964 sample. The variance between the readers of the two main data sets, 1966–76 (RML) and 1977–82 (JLB), was evaluated by having both authors read an identical subset of 100 teeth from the 1972–73 sample.

For the portion of this study that estimated sexual maturity by examining reproductive tracts, only material from females was considered. Ovaries were collected during August to March in 12 seasons between 1964–65 and 1981–82. Ovaries were preserved, prepared and evaluated in a similar way as those used by Øritsland (1970) and Bengtson and Siniff (1981). Ovaries were preserved in a 10% formalin solution, sectioned serially by hand to approximately 2 mm thickness, and macroscopically examined. *Corpora* and follicles were counted and measured. For this analysis, only female seals 2- to 7-years-old were considered (n = 217 pairs of ovaries). Assuming that 2 years is the youngest maturing age class and that *corpora* are macroscopically visible up to 5 years after formation, a *corpus* formed at 2 years should be visible 5 years later when that individual is 7-years old. An individual's age at maturity was calculated by subtracting the total number of *corpora* present in its ovaries from its total age estimated from cementum layers. Mean age at maturity was calculated separately for each collection year by calculating an arithmetic mean of all seals 2- to 7-years old in that sample.

3 Results

There was no difference between sexes for estimates of age at maturity calculated from transition zones in teeth (Laws 1977b). Therefore, males and females were lumped in all subsequent analyses of tooth back-calculation. When age at maturity was plotted against cohort year (cohort defined as all individuals born in a given year), the expected rise in maturity through the 1970s was not evident. Instead, curves

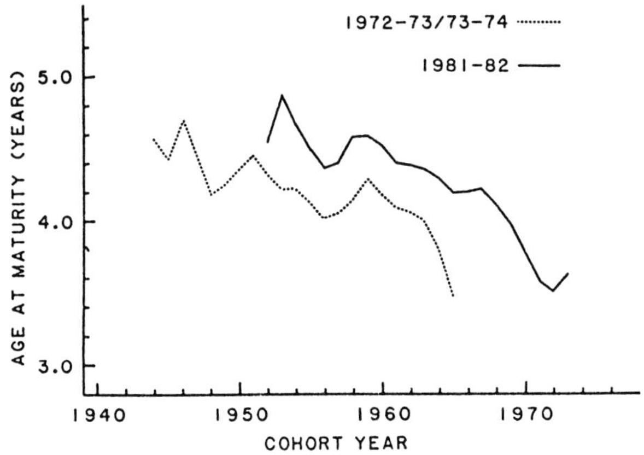

Fig. 1. Mean age of maturity in relation to cohort year (year of birth) of Crabeater Seals. Sample sizes for the two collections shown are: 1972–73/73–74 = 358, 1981–82 = 341

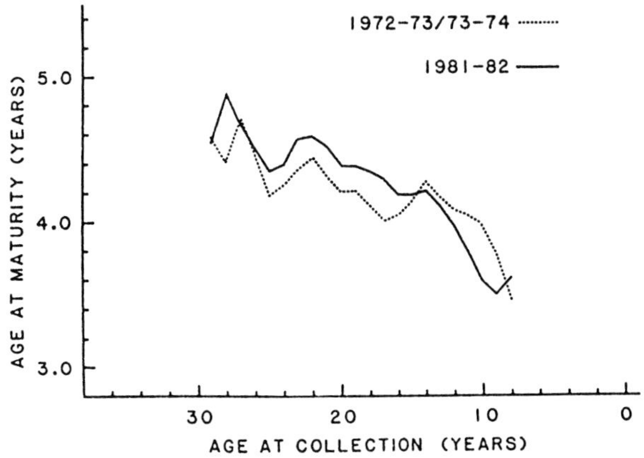

Fig. 2. Mean age at maturity in relation to age of Crabeater Seals at time of collection (year of death). Sample sizes are the same as in Fig. 1

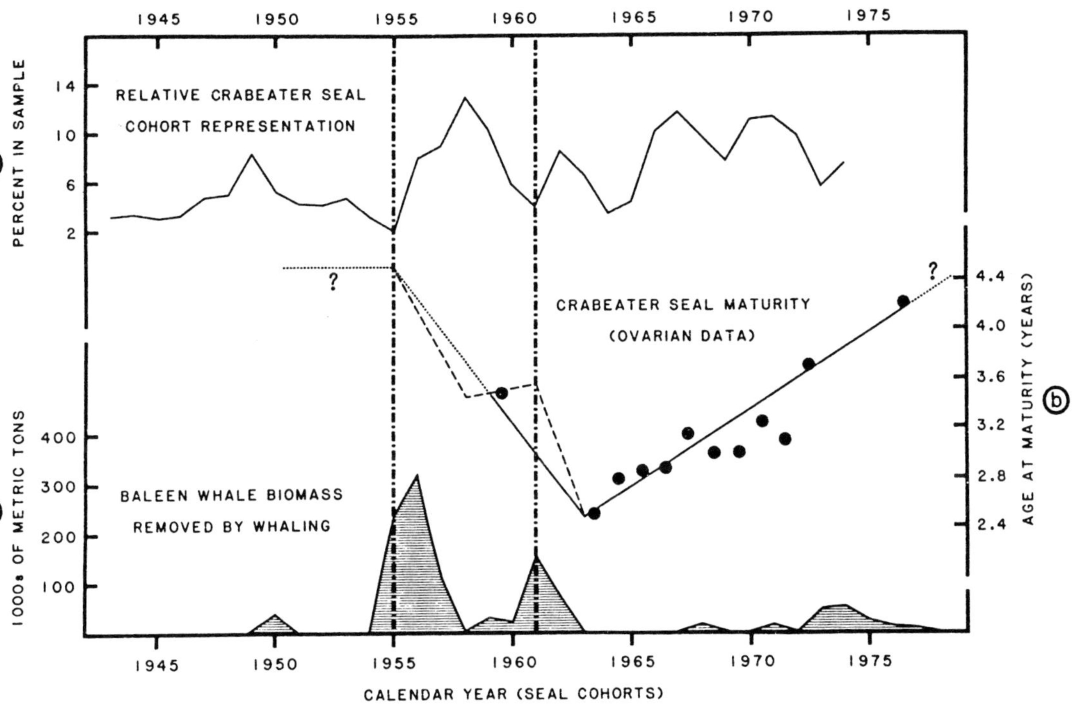

Fig. 3. a Relative representation of Crabeater Seal cohorts. The graph was calculated as follows: a subsample of each of 13 yearly samples was selected; only those seals 7- to 21-years-old at the time of collection were included; the age distribution of each of these subsamples was plotted independently against cohort year (the x-axis was the per cent of the subsample that was born in any given year); a composite graph of all 13 plots was made, with the separate plots overlapping cohort years included in their particular subsample; and, the mean value of the independent values for each cohort year was calculated and plotted as shown. b Age at maturity as estimated from ovarian material (earliest value estimated from the 1964–65 Norwegian sample (Øritsland 1970)). Mean values for 2- to 7-years-old animals from each collection year are plotted by mean cohort (4.5 years prior to collection year). *Dotted line* shows presumed trend of maturity, and *broken line* shows alternative trend of decrease. All lines fitted by eye. c Combined baleen whale biomass removed from Area I annually, based on the following average body mass (t): Blue (88), Fin (5), Humpback (27), Sei (18.5), and Minke (7)

from recent collections (e.g., 1981–82) had slopes parallel to past collections (e.g., 1972–73/73–74) (Fig. 1). Two potential explanations for the differences in these curves include variation in the tooth scoring technique between readers and differential survival rates related to the age of maturity. Analysis of variance between the scores of RML and JLB indicated no difference in scores of age at maturity ($P < 0.001$). A more likely explanation for the observed trends in age at maturity is an effect on the survivorship of cohorts related to the age at which any particular individual matures. Plotting the age at maturity vs. the age at collection (Fig. 2) illustrates that the age of maturity curves for animals between 7 and 30 years old, regardless of specific cohort year, are essentially the same. If seals that mature early tend to die relatively early, then these individuals would be under-represented in older cohorts. Conversely, seals from older cohorts which matured later would be the ones still present in the population at the time of sampling, producing elevated estimates for that cohort's mean age at maturity. If the age at maturity had no effect on the age at death, then there should be no change in the relative representation of the different ages at maturity between samples for any given cohort. This is not the case, as shown by an analysis of variance of the two samples considered. The majority of variation in age at maturity is accounted for by year of collection ($P < 0.001$).

An analysis of the relative strength of cohorts shows that in the 1950s and early 1960s, seals apparently prospered if born in the seasons immediately after major whaling activity (Fig. 3a). Conversely, as whaling slowed down, relative cohort strength decreased. However, the occurrence of two groups of strong cohorts starting in 1966 and 1970 does not appear directly related to whaling activities. The possibility of population cycles initiated by discrete blocks of heavy whaling is discussed later.

The estimates of mean age at maturity for each collection year as calculated by ovarian material show a drop in the age of maturity after the 1964 sample and a steady increase in the age of maturity through the 1960s and 1970s (Fig. 3b). Because each of the points shown in the figure was calculated identically, even if the true mean age at maturity is not reflected exactly by each estimate, the estimates should indicate a true population trend. Unfortunately, no estimate could be made from the 1966–67 sample because reproductive tracts were not collected.

4 Discussion

The cause of the transition zone in the cementum of teeth is not immediately apparent. It has been suggested that the changing rates in cementum deposition may be related to the additional energetic costs experienced by sexually mature individuals (Laws 1977b). Mature females (through pregnancy and lactation) and males (by fasting and defending females during pup-rearing) may utilize energy formerly directed to body (and tooth) growth and maintenance. However, perhaps the transition zone is related not to energetic demands, but to an overall physiological change at puberty. Variation in the rate of tooth formation near the time of puberty has been described in humans (Cherktov 1980).

Mean age at maturity estimated by tooth back-calculation is related to the age at which seals were collected. Possible causes for this trend include some aspect of tooth interpretation (e.g., systematic reader error) or a real interaction between age at maturity and longevity. In the first case, incorrect identification of ambiguous transition zones might bias estimates enough to artificially produce the observed trends. For example, in younger seals, with fewer layers laid down after maturity, the transition zone might be estimated at an earlier age. Conversely, older seals with more cementum layers might bias the reader to estimate the transition zone at a higher age. Reader error of this type has been proposed in cetacean studies where transition zones in earplugs are evaluated (Cooke and De la Mare 1983). Because the layers are much clearer and more regular in Crabeater Seal cementum than in whale earplugs, we do not believe that such a bias is present in our analysis of Crabeater Seal teeth. However, the possibility of reader error remains.

In the second case, longevity may be related to the age at which a Seal matures. If early maturing Seals die sooner than those maturing later, representation of different ages at maturity would vary within any one cohort when sampled 10, 20, or 30 years later. Individuals that had matured early would have lower representation in older cohorts, resulting in an upwards bias in mean age at maturity estimates. (This is the case in penguins, where individuals that breed at an early age may be subject to higher depredation (Ainley and DeMaster 1980). However, predation does not seem to be an important factor in adult Crabeater Seals). Seals with a relatively high metabolism may grow faster, mature earlier, and exhaust their physiological-energetic resources sooner. Experimental studies on mice have shown that longevity can be increased if metabolism and growth rates are experimentally lowered (McCay et al. 1935). Laws (1959) found that female seals of species with fast growth rates reached sexual maturity earlier and had shorter longevity than those species with slower growth rates.

The timing of changes in the age at maturity (estimated by ovarian material) and whaling activities strongly suggests that perturbations of one component of the marine ecosystem are reflected in other components. The Seals considered here were taken from an area that had two discrete peaks of whaling. No whaling was allowed in Antarctic Area I (as defined for whaling statistics) from 1939 to 1955. Area I was opened in 1955, and an estimated 70% of the Fin Whale, *Balaenoptera physalus*, stock was removed over the next 3 years (Breiwick 1977). From 1958 to 1960 there was a lull in Area I whaling, but in 1961 large catches resumed for 2 more years (Committee for Whaling Statistics). Some additional Fin Whales and Sei Whales, *Balaenoptera borealis*, were taken during the 1960s; the Minke Whale, *Balaenoptera acutorostrata*, fishery began in 1971 (Fig. 3c). The seals collected for this paper were taken from Area I and the western portion of Area II. Although Area II begins along the eastern coast of the Antarctic Peninsula, the principal whaling undertaken in Area II occurred farther east, nearer to South Georgia. Hence, from an ecological perspective, we consider the seals evaluated in this study to have been influenced by whaling in Area I alone.

There was an apparent drop in the female (ovarian) age at maturity from the 1959 to the 1963 cohort (Fig. 3b). Unfortunately, samples were not taken between the years these cohorts reached maturity or earlier, so our ability to estimate mean age of maturity in the interim years or earlier is limited. After 1963, there was a steady increase in female age at maturity through the 1976 cohort. This rise corresponds closely with the 1963 cessation of major whaling in

Area I. We suggest that the decrease and subsequent rise in age at maturity reflects an increasing availability of Krill as whales were being harvested; that abundant available food prompted faster growth and earlier maturity; and that with the conclusion of heavy whaling in 1963, subsequent seal cohorts grew slower and matured later as additional supplies of Krill were no longer being made available by further reductions in whale consumption.

Prior to whaling in 1955, the mean age at maturity in the population may have been stable, perhaps approximately 4–4.5 years of age. This seems a reasonable assumption in light of the fact that the mean age at maturity for the 1976 cohort was nearly 4.2 years, and the trend was still increasing. If one assumes that the age at maturity prior to 1955 was approximately 4.5, then once whaling began, mean age at maturity may have fallen rapidly through the 1950s and 1960s as indicated by the one available estimate for the 1959 cohort (Fig. 3b).

It is not known whether age at maturity is flexible enough to reflect short-term fluctuations in food availability. If so, then one would not have expected a linear function as age of maturity decreased during the two periods of heavy whaling in Area I. A possible pattern of decreasing age at maturity is proposed by the broken line in Fig. 3b and explained by the following scenario: With the onset of whaling, Krill availability increases rapidly, allowing earlier attainment of maturity. As whaling slows, competition increases, and age at maturity begins to rise until whaling once again causes age of maturity to fall. Future studies will need to test the degree to which maturity can change in response to short-term phenomena.

The strength of cohorts born in the whaling periods begun in 1955 and 1961 may be due to high pupping rates, high survival, abundant food, faster growth rates, or a combination of these factors. Abundant Krill, for example, could have stimulated faster growth rates, which in turn may have led to lower levels of depredation. Predation by Leopard Seals, *Hydrurga leptonyx*, has a significant impact on young Crabeater Seals (Siniff and Bengtson 1977), but this threat is thought to fall off dramatically with increasing age and body size (Siniff et al. 1979). The occurrence of weaker cohorts following the slowdown in whaling might have resulted if environmental conditions became less favourable and survival rate decreased.

The cause of the two peaks of strong cohorts starting in 1966 and 1970 is unclear. Why these cohorts showed relative strength in years when whaling was not undertaken in Area I remains to be determined. It is possible that the two groups of strong cohorts caused by whaling in the 1950s and 1960s may have had a shadow effect on the population in subsequent years. For example, Seals from "boom" cohorts would have entered the breeding population at approximately 3 to 4 years of age. At that time, the recruitment of their pups into the population would further accentuate an unusual age distribution. The behaviour of such reproductive pulses or cycles requires further study and modelling to determine their effect both on age distribution, as well as the analysis of population parameters themselves. It is not known what effect periodically strong age classes would have on a variable such as age at sexual maturity.

If the observed decrease and subsequent increase in age at maturity in Crabeater Seals is indeed related to changes in Krill availability resulting from whaling activity, one would expect to see similar changes in parameters of some Baleen Whale stocks. Much attention has been given to the evaluation of growth and reproductive parameters in southern Baleen Whales (Gambell 1973, 1975; Laws 1962; Lockyer 1972, 1977, 1979; Masaki 1977, 1979). These studies have indicated decreases in age at maturity and increases in pregnancy rates for various whale species. Unfortunately, few studies have evaluated trends since whaling decreased in the mid-1970s. This is due both to the time lag required to allow all individuals in recent cohorts to mature before calculating mean age of maturity and also due to the lack of specimen material once provided by whaling. Masaki (1979) evaluated age at maturity in the Minke Whale by four different methods, including examination of ovarian material and transition zone in earplugs. Using transition zones, he observed a decreasing age at maturity in all six statistical whaling areas. In Area I, he observed a decrease in maturity from approximately 13 years in 1945 to 5 years in 1970. From 1963 to 1970, a period when age at maturity in Crabeater Seals apparently increased (estimates from ovarian materials), he found a decrease in Minke Whale maturity from 8 to 5 years. His sample was insufficient to allow calculation of age at maturity from ovaries in Area I, but Areas III and IV showed no change in age at maturity. Back-calculating age at maturity using transition zones of Crabeater Seal teeth during the period falsely indicated a decreasing age at maturity, whereas ovarian material has demonstrated an actual increase. Perhaps similar problems are present in back-calculating age at maturity using whale earplugs. If so, it may be that the true age at maturity in Minke Whales (estimated from ovarian material) was fairly constant during that time.

The causes of the observed changes in age at maturity in Crabeater Seals are likely to be related to complex interactions within the marine ecosystem. Various functional and numerical responses probably occurred throughout the marine community in response to the massive removal of whale biomass from the Southern Ocean. Smaller whale populations probably resulted in less competition for food, contributing both to higher individual and population growth rates. It has been suggested that other groups such as penguins (Conroy and White 1973; Conroy 1975) have increased rapidly in numbers due to reduced whale competition and increased Krill availability. Moreover, Antarctic Fur Seal, *Arctocephalus gazella*, numbers have recovered faster than expected (Laws 1973, 1977b; Payne 1977). The true extent to which other consumers of Krill, including fish and cephalopods, might have been increasing during this period is unknown. Numerical increases in any of these groups presumably would have led to increased inter- and intraspecific competition for food and a subsequent slowing in seal growth rates and age of maturity. Lack of information on Krill population dynamics further complicates the difficult task of identifying which suites of factors are most important in affecting age at maturity in Crabeater Seals. Nevertheless, changes in Crabeater Seal age at maturity do give insight into general ecological trends in the Antarctic marine ecosystem by providing an index of whether conditions are relatively good or bad for growth, survival, and reproduction.

5 Conclusion

The information presented here suggests that the age of sexual maturity in Crabeater Seals decreased as a result of whaling in Area I and subsequently increased when whaling ceased. This trend may have been due to changing food availability and functional and numerical changes in consumers of Krill.

Modelling can serve a useful role in future evaluation of the interactions between different population parameters. Further attention should be directed to the significance of changes in relative cohort representation in various samples. How do periodically strong cohorts affect the population dynamics of Crabeater Seals in subsequent years? Do very strong cohorts affect survival, mean age of maturity, or competition for food? Does pulsed reproduction contribute to the formation of population cycles in Crabeater Seals? There is evidence for similar fluctuations in cohort strength in Leopard Seals (Rounsevell and Eberhard 1980). Further work should also examine the hypothesis that early maturation leads to a relatively earlier death, and the implications of this for life history strategies.

This study has demonstrated the value of one aspect of Crabeater Seals as an indicator species. It is important that sampling of Crabeater Seals should continue, for monitoring studies of the Antarctic marine ecosystem. Samples should be taken regularly from different areas to obtain growth, reproductive, and demographic data. Such a sampling programme would provide insight into how Crabeater Seal population dynamics and variations in reproductive success with time reflect ecological interactions within the Antarctic marine ecosystem.

Acknowledgements. We are grateful for the field assistance provided by L. Airey, M.J. Baker, B. Bostlemann, D. Burke, D.P. DeMaster, I. Everson, A. Morgan, D. Ragland, R.A. Reichle, P. Salino, D.B. Siniff, I. Stirling, T. Øritsland, B. Weaver, personnel at Rothera Station, and the assistance and support of the officers and crew of R.R.S. JOHN BISCOE, R.V. HERO, and U.S.G.C. GLACIER. Assistance in the laboratory by A. Baird, M. Fallows, and D. Regis is gratefully acknowledged. J. Beddington, I.L. Boyd, J.G. Cook, J.P. Croxall, W.K. de la Mare, R.C. Handley, and D. Lavigne participated in helpful discussions. D. Ricketts is thanked for help with statistical analysis. We are especially grateful to D.B. Siniff for his ideas on potential biases inherent in using tooth backcalculation to interpret past events. Financial and logistical support was provided by the British Antarctic Survey and the Division of Polar Programs, National Science Foundation (Grants # OPP 75-17719 and DPP 80-20087).

References

Ainley DG, DeMaster DP (1980) Survival and mortality in a population of adelie penguins. Ecology 61:522–530

Bengtson JL (1982) Reproductive ecology of crabeater and leopard seals. Antarct J US 17:185

Bengtson JL, Siniff DB (1981) Reproductive aspects of female crabeater seals *(Lobodon carcinophagus)* along the Antarctic Peninsula. Can J Zool 59:92–102

Breiwick JM (1977) Analysis of Antarctic fin whale stock in Area I. Rep Int Whal Comm 27:124–127

Chertkov S (1980) Tooth mineralization as an indicator of the pubertal growth spurt. Am J Orthod 77:79–91

Conroy JWH (1975) Recent increases in penguin populations in Antarctica and the subantarctic. In: Stonehouse B (ed) The biology of penguins. University Park Press, Baltimore, p 555, pp 321–336

Conroy JWH, White MG (1973) The breeding status of the king penguin *(Aptenodytes patagonica)*. Br Antarct Surv Bull 32:31–40

Cooke JG, De la Mare WK (1983) The effects of variability in age data on the estimation of biological parameters of minke whales *(Balaneoptera acutorostrata)*. Rep Int Whal Comm 33:333–338

Eberhardt LL (1977) Optimal policies for conservation of large mammals, with special reference to marine ecosystems. Environ Conserv 4:205–212

Erickson AW, Siniff DB, Cline DR, Hofmann RJ (1971) Distributional ecology of Antarctic seals. In: Deacon G (ed) Symposium on Antarctic Ice and Water Masses (Tokyo, 1970). Scientific Committee on Antarctic Research, Cambridge, pp 55–76

Gambell R (1973) Some effects of exploitation on reproduction in whales. J Reprod Fert Suppl 19:533–553

Gambell R (1975) Variations in reproduction parameters associated with whale stock sizes. Rep Int Whal Comm 25:182–189

Gilbert JR, Erickson AW (1977) Distribution and abundance of seals in the pack ice of the Pacific sector of the Southern Ocean. In: Llano GA (ed) Adaptations within Antarctic ecosystems: Proceedings of the Third SCAR Symposium on Antarctic Biology. Smithsonian Institution, Washington DC, p 1252, pp 703–740

Hewer HR (1964) The determination of age, sexual maturity, longevity and a life-table in the grey seal, *(Halichoerus grypus)*. Proc Zool Soc Lond 142:593–624

Laws RM (1956) Growth and sexual maturity in aquatic mammals. Nature 178:193–194

Laws RM (1959) Accelerated growth in seals, with special reference to the Phocidae. Nor Hvalfangst-T 9:425–452

Laws RM (1962) Some effects of whaling on the southern stocks of baleen whales. In: LeCren ED, Holdgate MW (eds) The exploitation of natural animal populations. Oxford

Laws RM (1973) Population increase of fur seals at South Georgia. Pol Rec 16:856–858

Laws RM (1977a) Seals and whales in the Southern Ocean. Philos Trans R Soc Lond B 279:81–96

Laws RM (1977b) The significance of vertebrates in the Antarctic marine ecosystem. In: Llano GA (ed) Adaptations within Antarctic ecosystems: Proceedings of the Third SCAR Symposium on Antarctic Biology. Smithsonian Institution, Washington DC, p 1252, pp 411–438

Leslie PW, Perry JS, Watson JS (1946) The determination of the median body weight at which female rats reach maturity. Proc Zool Soc Lond B 115:473–488

Lockyer C (1972) The age at sexual maturity of the southern fin whale *(Balaenoptera physalus)* using annual layer counts in the earplug. J Cons Perm Int Explor Mer 34:276–294

Lockyer C (1977) A preliminary study of variations in age at sexual maturity of the fin whale with year class in six areas of the Southern Hemisphere. Rep Int Whal Comm 27:141–147

Lockyer C (1979) Changes in a growth parameter associated with exploitation of southern fin and sei whales. Rep Int Whal Comm 29:375–395

Masaki Y (1977) Yearly change in the biological parameters of the Antarctic sei whale. Rep Int Whal Comm 28:421–429

Masaki Y (1979) Yearly change of the biological parameters for the Antarctic minke whale. Rep Int Whal Comm 29:375–395

McCay CM, Cromwell MF, Maynard LA (1935) The effect of retarded growth upon the length of life span and upon ultimate body size. J Nutr 10:63

Øritsland T (1970) Sealing and seal research in the southwest Atlantic pack ice, Sept–Oct, 1964. In: Holdgate MW (ed) Antarct Ecology vol 1. Academic, New York, p 604, pp 367–376

Øritsland T (1977) Food consumption of seals in the Antarctic pack ice. In: Llano GA (ed) Adaptations within Antarctic ecosystems: Proceedings of the 3rd SCAR Symposium on Antarctic Biology. Smithsonian Institution, Washinton DC, p 1252, pp 749–768

Payne MR (1977) Growth of a fur seal population. Philos Trans R Soc Lond B 279:67–79

Rounsevell D, Eberhard I (1980) Leopard seals, *Hydrurga leptonyx*, (Pinnipedia), at Macquarie Island from 1949 to 1979. Aust Wildl

Sadlier R (1969) The ecology of reproduction in wild and domestic mammals. Methuen, London

Siniff DB, Bengtson JL (1977) Observations and hypotheses concerning the interactions among crabeater seals, leopard seals and killer whales. J Mammal 58:414–416

Siniff DB, Stirling I, Bengton JL, Reichle RA (1979) Social and reproductive behaviour of crabeater seals *(Lobodon carcinophagus)* during the austral spring. Can J Zool 57:2243–2255

The Committee for Whaling Statistics (1955–1979) International whaling statistics. Compiled for the International Whaling Commission. Grondahl and Sons, Oslo

Summary and Conclusions

R.M. LAWS [1]

1 Introduction

This has been a very successful Symposium, the diverse contributions leading to many productive interactions between and within disciplines, but the absence of Russian ecologists is regretted and their presence would have resulted in a more complete picture of current knowledge. There were 170 participants from 13 nations and in addition to the five keynote addresses by invited speakers there were 106 papers, about half of them in the form of posters. No summary can be complete or unbiased and it would be invidious to single out individual authors by name, so I have not attempted to do this. However, I would like to acknowledge the five keynote speakers who have assisted me greatly with this final chapter by permitting me to incorporate into it their respective session summaries. Despite the title of the Symposium most of the contributions tend to deal with individual organisms or limited groups of organisms. I propose to start with the marine environment, its physics, chemistry, microbial components, algae, benthos, zooplankton, and higher consumers, as well as the interactions between them; then to discuss the ideas presented on the interactions between terrestrial and freshwater ecosystems; and, finally, the presentations on these ecosystems separately. Conclusions will be drawn as appropriate in the course of the discussion and also in a final section.

2 Marine Ecosystems

2.1 Physical Factors

The influence of physical environments on phytoplankton distribution, standing stock and primary productivity has been directly addressed by some contributors and mentioned in passing by others. Elevated values of surface chlorophyll-a are to a varying degree closely associated with physical fronts in the oceanic regime of the Southern Ocean, and a similar relationship was shown in bird density. A new front, the Sub-Antarctic Front, was reported south of South Africa. Very high chlorophyll levels were shown to occur near the Antarctic coast, with peaks also at the Antarctic and the Sub-Tropical convergences. High chlorophyll also correlates with islands but this so-called "island effect" has conflicting explanations: on the one hand wind-induced upwelling affected zooplankton species composition in the lee of Marion Island; on the other hand no upwelling was found on a different occasion, despite strong winds, and a true island effect was proposed. Further investigations are needed to resolve the problem.

The importance of physical mixing processes was stressed in controlling both primary production and marine nutrient cycles and it was emphasized that further attention should be given to the physical mechanisms in the open ocean. Investigations on the ability of internal waves to mix nutrients up into the euphotic zone, and hence increase its fertility, are needed, particularly in near-shore waters. Nimbus Coastal Zone Colour Scanner (CZCS) studies have been highly successful and in conjunction with surface observations showed that the borders of temperature gradients were correlated with peaks in chlorophyll. Sub-duction and shear fronts all contribute to the recorded distributions. The mathematical formulations need further testing in laboratory and field, and interdisciplinary studies are required.

It was formerly thought that wind-shear stress would produce upwelling at the ice edge and a productive zone, but recent work has shown no evidence of upwelling there and so alternative mechanisms have been sought. There is a need to follow the ice edge retreat over the summer months, because pack-ice expansion and retreat occurs over a vast area and influences the underlying water column. A correlation has been demonstrated between phytoplankton standing crop and time since the ice broke up; more chlorophyll is found away from the ice edge and chlorophyll-a levels are enhanced up to 200–250 km away from it, a reflection of the recent extent of ice cover. Wind moves the ice so there is an averaging effect over large areas, but sampling over the 24-h cycle indicates extreme patchiness at any one time.

1 British Antarctic Survey, NERC, High Cross, Madingley Road, Cambridge CB 3 OET, United Kingdom

Antarctic Nutrient Cycles and Food Webs
(ed. by W.R. Siegfried, P.R. Condy, and R.M. Laws)

2.2 Chemistry

Silicate has been long recognized as a most important nutrient in the Southern Ocean, and the very large gradients in silicate concentrations are a useful aid to physical oceanography in tracing water movements. Relatively large amounts of silicate are released from Antarctic abyssal sediments into the overlying water, and the evidence points to the release of other elements such as copper. Cu/Si ratios are useful in flux studies and regeneration processes. Trace elements can have a marked effect on primary production and their study should be encouraged because it may provide a clue as to why production is low in such nutrient-rich waters.

Elements such as copper, zinc, strontium, and iron in zooplankton differ sufficiently between samples to permit differentiation between water-masses and, if confirmed, such analyses will provide a further indicator of the geographical distribution and movements of zooplankton. The concentration of some elements in Weddell Sea seals underlines the need in all trace element work to understand the natural geochemical cycles of elements and to distinguish man-made effects from naturally high concentrations brought about by mineralization in rock bodies. Large amounts of phosphorus are transported from the ocean to penguin rookeries and many minerals are formed, involving specific bacteria in guano. It is interesting to note that fluoride-rich minerals occur – possibly linked to elevated fluoride in Krill through the penguins' diet.

Next, not only is microbial carbon input important, but marine bacteria were shown to supply much of the ammonia-N required by nanoplankton in the Ross Sea; they are often attached to particles rather than free-living. Ammonia accumulates over winter in sea-ice and may encourage phytoplankton blooms in spring.

2.3 Ice-Associated Algae

Although the likely importance of ice-associated algae in the Antarctic marine food web has been recognized since work in the 1960s at McMurdo Sound, it is only recently that quantitative studies have begun to demonstrate their importance.

The types of algae-bacteria – sea-ice assemblages will obviously vary with the type of ice (pack-ice, fast-ice, multi-year ice) and from area to area. Contributions to this Symposium have shown that: significant biomass is contained in sea-ice to be freed on melting; blooms, or at least increases in primary production, can occur as melting ice retreats; ice-associated communities are not simple; and, light is a major regulating factor because snow cover is critical.

It seems likely that "grazing" on these communities in the ice (by microbial consumers and larger herbivores) may prove to be important. Epiphytic bacteria on algae have specialized attachment processes and live both on healthy and senescing algae.

The potential for seeding the water column with algae released from such communities in and under the ice, different from the initial microflora of the water column, raises interesting chemical questions. Because salinity may vary from zero (in the case of melt-water), or at least from normal seawater salinity, to about 5 times seawater salinity in brine channels, the algae are obviously tolerant of large variations in ionic strength at exceptionally low temperatures for marine microbes. During brine formation nutrients and trace elements will be concentrated and it is important to study the effect of this on algal production. Also, high ionic strength differences may markedly alter the permeability of the cell membrane, altering the flow of chemicals in and out of the cell.

2.4 Phytoplankton and Microplankton

In the water column itself, the papers presented here have emphasized the growing awareness of the complexities of the lower parts of the food web. As a result, ideas on the structure of Antarctic marine food webs and the rates of nutrient cycling and energy transfer need to be revised. Notable are the significance of primary production by nanoplankton, and the possibly significant "grazing" by microplankton on these very small autotrophs. Two papers discuss grazing (or at least the potential for grazing) on the mucilagenous colonies of choanoflagellates, by protozoans, and suggested the possibility that consumption of these protozoans by Krill may represent an important route of energy flow. During discussion of this topic it was mentioned that choanoflagellate loricae had been found in *E. superba* faeces by Japanese workers. This is obviously an active area, and one whose results may change our picture of the Southern Ocean ecosystem.

Bacteria have been much neglected, with regard to: their interactions with other animals and role in carbon and nitrogen fluxes; their effects in nutrient regeneration processes; their role in other organisms as enzyme producers; and, as a food base for grazing organisms.

The detrital food chain is very important in the transformation of particles in transit in the water column. Microcosm studies on homogenized kelp showed bacterial growth rates to be similar to those in temperate regions – despite cold Antarctic water. The decomposer microflora increase the nitrogen-content and nutritious value of decaying macroalgae for epibenthic animals, and are important in mineralization processes.

2.5 Benthos

Benthic studies received relatively little attention at this Symposium, but studies of energy flow and trophodynamics were in evidence. Much of the short burst of primary production in the summer goes to detrital feeders in the benthos, and the role of macro-, micro-, and meiobenthos in recycling and remineralization of detritus demands attention. In general a low energy strategy of slow, seasonal growth, and low metabolism in winter at a high wintering biomass, contrasts with the high energy strategy characteristic of the pelagic ecosystem.

2.6 Zooplankton

Relatively few papers dealt directly with zooplankton energetics as such, although a number were concerned with biomass distribution, or feeding structures, or both. Those studies described, however, emphasize the current trend towards the investigation of rates or processes in zooplankton. Feeding studies in particular have revealed that the relationships between the grazer and the grazed are complex and will need careful energetic interpretation; we cannot easily discern the energetic costs to the Krill of its feeding processes, or their consequences to the phytoplankton and microbial communities.

However, Krill, *Euphausia superba*, is a fast swimmer with a high metabolic rate; it is calculated that 30% of its energy output is allocated to active feeding. Huge swarms comprising up to 2.1×10^6 t in 450 km^2 have been described, and the ammonia output by Krill is significant, compared with that by birds and mammals. Moulting represents a large loss of organic material and the growth processes of Krill are shown to be remarkably plastic. It is clear that in vivo and in vitro studies may show very different results. Growth can vary between − 15% and +20%, depending on experimental conditions. Moulting may be less frequent, or stop altogether in the winter. Krill feed on the most abundant food particles – over an enormous size range – from microorganisms to other Crustacea – and at high rates. Studies of enzyme systems are interesting and confirm their omnivorous habits. Also, Krill populations occupying different oceanic gyres are said to have different enzyme systems; this needs to be followed up as a possible technique for identifying the different Krill stocks.

2.7 Higher Levels

Consumption of Krill by higher predators received much attention at this Symposium, including extensive quantitative data on fish, birds, and seals and a valuable first such attempt for squid. Whales, however, received very little attention. There is increasing evidence of considerable overlap of Krill predator groups in feeding, their diurnal and seasonal cycles locked into the Krill cycle.

2.8 Squid

Data on squid distribution, life histories and ecology are very sparse and most information is derived from examination of bird, seal, and Sperm Whale, *Physeter catodon,* stomachs. What is for me an exciting paper in its implications considerably clarifies the problems inherent in the study of oceanic squid. Comparison of *Kondakovia* size frequencies from trawl hauls and Sperm Whale stomachs indicates a net avoidance problem or that Sperm Whales were feeding deeper than the nets were fished. (Similar differences between sizes caught by nets and predators were also noted for Krill.) In general squid size seems to increase with depth but we know that they have vertical migrations, the Krill feeders possibly diurnally and perhaps on a lunar cycle. There also appears to be a seasonal change in the squid depth range. Squid appear to select the same Krill size frequencies as baleen whales. It has been suggested that the larger squid feed on fish at depth.

2.9 Fish

Estimates of Antarctic fish taken by predators appear to be greater than estimates of the abundances of fish standing stocks. A similar discrepancy arises between Krill acoustic survey estimates and estimates of Krill consumed by predators. Clearly these need to be resolved.

I would single out two fish studies. First, it was shown that the commercially important species *Champsocephalus gunnari* feeds on Krill in surface waters at night and is quiescent on the bottom during the day. This pointed to the potentially great importance of zooplankters in the diet of so-called benthic fishes. Secondly, an elegant study of the evolution of neutrally buoyant fish indicated the dominant role of *Pleuragramma* in the Ross and Weddell seas and its importance in turn to the giant *Dissostichus*, also a midwater species. These fish are pelagic seasonally when there is an abundant pelagic food supply, but appear to stay on the bottom during the rest of the year – a parallel on a longer time scale with the diel pattern of *Champsocephalus*.

Areas I would identify as needing special investigation are the changes in fish diet with size, and the rate of winter feeding.

2.10 Birds

Birds are receiving much attention and at Bird Island, South Georgia, the combination of intensive terrestrial studies with telemetry to elucidate behaviour at sea (TDR's and activity recorders) is yielding exciting results on food consumption, feeding behaviour and range. These and other studies provide evidence for strong selection or preferences for Krill by predators.

Other studies are also contributing significantly to a more complete understanding of the place of birds in the ecosystem. The distributions of vertebrate biomass in the Ross Sea region, comprising 11% of the circumference of Antarctica, were mapped in relation to plankton blooms, oceanographic fronts and current ice conditions. Both birds and mammals were concentrated at the continental slope front and ice edges and apparently avoided the centre of an intense, persistent phytoplankton bloom over the continental shelf. Another study showed that frontal areas in the open ocean are characterized by elevated nutrient levels, biological production and seabird densities. Overall, however, open-ocean seabird biomass estimates are less than half those of neritic regions. Information is accumulating on foraging zones of seabirds and on their diet and energetics. A new method based on lipid and fatty acid contents in seabird stomach contents might make it possible to infer their diets. It was claimed that the proportions of sterols, wax esters and alkyldiacylglycerols in Antarctic seabirds suggested a diet dominated by squid and euphausiids; elsewhere stomachs high in triacylglycerols suggested a mainly fish diet. However, more and better data are needed to demonstrate that this is a viable quantitative method for characterizing predator diets. Petrel chicks can store food reserves to cope with temporary periods of low food availability and experimental studies on albatross species show that brooding birds need 40% more energy than non-breeders of the same size.

2.11 Seals

Studies on this group have advanced, again employing new telemetry techniques (TDR's, radio tagging). New information on the elusive Leopard Seal, *Hydrurga leptonyx*, indicates that as a generalist feeder it has to compete with the specialist Krill feeder, the Crabeater Seal, *Lobodon carcinophagus*, in winter. Its food requirement is estimated at 5% to 6% of body weight per day with Krill comprising 45% of the diet (mainly in winter), seals and penguins 45% (mainly in summer), and fish and squid 10%. Preliminary energy budgets for Antarctic Fur Seals, *Arctocephalus gazella*, have been calculated and indicate a daily food requirement of 16% of body weight.

A paper on Crabeater Seals, based on extensive collections in the Antarctic Peninsula region, has thrown some further light on the interactions between Krill and its predators. There appears to have been an increase in the seals' reproductive rate (as measured by age at maturity) in line with the decrease in whale stocks and the presumed increase in availability of Krill for other consumers. In the past two decades this has been followed by an increase in the age at sexual maturity (implying a reduced reproductive rate), which suggests increased competition for Krill as consumer stocks (penguins, pack-ice seals, Fur Seals, and presumably baleen whales) increased to fill the gap.

3 Estimates of Food Consumption and Ecological Interactions

The impact of modern techniques of data collection and analysis were evident in this field as one of the most striking developments since the last SCAR Biology Symposium. The British Offshore Biological Programme centred on South Georgia, the German programme around Elephant Island and the international BIOMASS studies centred on Prydz Bay, among others, have objectives which include the linking of Krill and its consumers.

Quantitative measures of food intake by predators may indicate overall energetic demands of those predators, but they do not always appear to be compatible with biomass estimates of their food resources obtained from acoustic surveys, or size frequency analyses from net hauls. As emphasized in several papers, it is also necessary to consider the rates of transfer involved, and the possible effects of competition when breeding communities are located close together (for example adjacent seal beaches and penguin rookeries).

In my opinion the estimates from predator consumption are likely to be more accurate, because they are integrated over the feeding ranges and over the year, thus minimizing the effect of the extreme patchiness of Krill on the results of an instantaneous survey. Also, the water-masses at South Georgia, for instance, have been likened to a conveyor belt bringing Krill into the region; the instantaneous acoustic estimate of Krill in that area is certainly an underestimate of Krill quantities available to predators over the year. Moreover it has been pointed out that Krill in the top 10 m of the water column, in which large quantities are to be expected by night, are underrepresented in acoustic surveys using downward-pointing sonar; towed, upward-pointing sonar should help to resolve this problem, and provide more survey data.

The relationships between prey life cycles and population structures and the timing, extent and nature of predation by seabirds and seals was reviewed. The timing of peak Krill demand may be influenced by Krill lipogenesis; lipid-rich gravid female Krill provide bonus energy for the predators. There is evidence of differential predation on male and female Krill, represented at South Georgia by differences between sex ratios and sizes of Krill in net hauls and in predator stomachs. Net hauls take mainly 25-30 mm Krill, most seabirds and fur seals 50-55 mm Krill. Clearly it is now important to study food abundance and predation concurrently.

The complex seasonal pattern of distribution, migrations and interactions of Weddell Seal, *Leptonychotes weddelli*, and Antarctic Cod, *Dissostichus mawsoni*, a major prey species, has been studied at McMurdo Sound; the Cod is taken up to December when it moves away, and is then replaced in the seals' diet by small nototheniid fish, cephalopods, and crustaceans.

4 Terrestrial and Inland-Water Ecosystems

The general state of research is that a relatively large amount of work is being done by a few national groups (mainly UK,

US, Australia, and South Africa) in very few localities or sites (especially Signy Island, Marion Island, South Georgia, the Vestfold Hills, and the Dry Valleys). A phase of surveys and descriptive studies has now been superseded by more intensive process-orientated research in which the value of year-round investigations is apparent.

Major advances have been made in such research, especially in microbiology and limnology. For many studies sampling periods were of short duration; therefore, it is difficult to interpret or model annual cycles meaningfully. Significant work has been undertaken on the Marion Island and Signy Island terrestrial ecosystems and currently in Signy Island lakes, including studies on sediments and catchment areas. Work is also continuing on a series of short-term field studies in the Dry Valleys' ice-covered meromictic lakes and on saline lakes there and in the Vestfold Hills.

At Signy Island attempts have been made to integrate terrestrial and freshwater studies by following the transfer of microbial indicator species from catchment areas to streams and lakes. Elsewhere, especially at Marion Island, inter-relationships between marine and terrestrial ecosystems are being studied through the influence of seabirds and seals on the input of nutrients.

In the drier climate of continental Antarctica, aerial distribution of microbes is affected by static charges on snow particles in blizzard conditions. Such charging may have considerable effects on the dispersal, accumulation and trapping efficiencies with respect to particles, spores and invertebrates.

There is much interest, in both terrestrial and inland-water systems, in the importance of nitrogen input, especially the role of cyanobacteria in the fixation of atmospheric nitrogen. Fixation is often inhibited or prevented by lack of utilizable energy sources (e.g., carbohydrates and phosphorus) or low temperatures or both. However, there is significant input of nitrogen in some Sub-Antarctic mire communities and Antarctic lake sediments.

Another important aspect of current studies is in anaerobic decomposition processes in wet terrestrial communities and in lake sediments. Methane accumulates during winter in sediments indicating significant activity at low light intensity and low temperatures and its production reflects strictly anaerobic decomposition by bacteria. Methanogenic bacteria isolated from both ecosystems are associated with nitrate respiration, sulphate respiration, fermentation, and cellulose breakdown.

5 Terrestrial Biology

One of the major differences between the Antarctic and Arctic terrestrial ecosystems is the absence of native vertebrate herbivores in the Antarctic. Not surprisingly, therefore, detritivore and microbial decomposition dominates consideration of energy and nutrient flow.

Terrestrial studies at Signy Island have shown that net primary production is poorly balanced by decomposition of dissolved organic carbon and cellulose by micro-organisms in moss peat. Such activity appears to be regulated primarily by substratum quality and moisture at a range of comparable sites along the Scotia Arc and Antarctic Peninsula. Signy Island appears to be representative of the maritime Antarctic in this respect.

In one study on Macquarie Island, succession of epiphyllous fungi from young, mature and senescent leaves was studied by direct observation, plate count and litter bag techniques. Some 44 species of filamentous fungi have been identified, several of which may be new to science and endemic to the Sub-Antarctic. Primary saprophytes colonize living leaves before they senesce. The microclimate of the litter layer (small temperature range, high relative humidity and low ultraviolet radiation) encourages a more diverse microflora there.

An interesting paper addressed the biology of Sub-Antarctic invertebrates on Iles Crozet. It concentrated on Diptera (8 spp.) and earthworms (10 spp.) in terms of their spatial distribution in the archipelago, the inter-relationships of invertebrates and the environment, and the dynamics of the supra-littoral system. The importance of kelp as an arthropod habitat was recognized. There is also a carrion habitat for a species of Diptera.

Work on arthropod interactions at Signy Island has shown that micro-algae are important, together with microflora, as food for the majority of species of mites and springtails, and that arthropods and other invertebrates play a relatively minor role in the energy flux in these communities. Herbivory accounts for $< 0.1\%$ of net primary production and the efficiency of utilization of herbivore production by carnivores is low (0.3–0.5%) compared with Arctic efficiencies of 15%–33%. Energy flow through the common herbivore *Cryptopygus* (Collembola) is influenced by the age of individuals, temperature, and food type. Predation by the mite *Gamasellus* is apparently non-selective, taking prey items in proportion to their occurrence in the arthropod community. Several correlates of adversity or A-selection have been noted in the maritime Antarctic where simple terrestrial communities exist in harsh environments.

Concerning terrestrial food chains on the Sub-Antarctic islands, it is clear that few studies extend beyond one or two trophic levels. Birds and mammals have a reduced role in Antarctic ecosystems compared with the Arctic, but introduced mammals have established vertebrate food chains easily, with no significant competition between species, although insularity may be important in the maintenance of the ecological balance. Invertebrates are a main food source for some birds; this correlates with the relatively high abundance of invertebrates, especially insects, at Marion Island compared to other Sub-Antarctic tundra sites. Some 40–50 species of macro-arthropods have been found at Marion Island. Future work in this region should concentrate on the dynamics of the food webs (on which current information is fragmentary), island effects and climatic constraints.

6 Inland Waters

Lakes are naturally closed ecosystems, and when sealed by surface ice they are completely closed boxes. Advantage is being taken of this to try to model whole-system energy, or nutrient flux, or both. The work reported is mainly from two lakes at Signy Island and the lakes of the Dry Valleys.

Ice-covered undisturbed lake water may become anaerobic in winter, but the water column is later re-aerated by wind mixing. Nutrient limitation is of vital importance in Antarctic lakes. Studies of nutrient cycling have shown that under anoxic conditions the release of elements such as phosphorus and nitrogen occurs from the underlying sediment, and they may be limiting factors according to the trophic status of the lake. These lakes are characterized by nutrient inflows and nutrient cycling. Turbulent mixing in mesotrophic lakes promotes the growth of shade-adapted phytoplankton under the ice in the spring.

The Dry Valley lakes have a low diversity of cosmopolitan algae and heterotrophic bacteria. Consumer animals are almost absent. Despite the high biomass of the benthic-algal-diatom-bacterial mat, these oligotrophic lakes show low productivity, low decomposition and negligible food chains. Their benthic mats had several forms, some resembling Pre-Cambrian stromatolites.

7 Conclusions

The earlier exploratory work in the Antarctic involving broad surveys has been succeeded by process-orientated research. Hypothesis-related research is likely to be most productive, and it is important to ask the right questions at the start of a research programme. It is also necessary to stress the importance of year-round observation or sampling in any studies of cycles – whether of abundance, production, decomposition, nutrients or energy. Thus, one doubts the value of a predictive model of an Antarctic lake based on a 6-week summer study.

There is a unique series of environments and organisms in the Antarctic which lend themselves to ecological and physiological studies. We need to exploit these opportunities and at the same time seek a better understanding of the environments. There is also a need to relate Antarctic biology to a wider field – not only to the Arctic, but also to work in temperate and tropical areas.

The importance of physical mixing processes in the control of primary production and marine nutrient cycles was stressed; also of upwelling mechanisms and effects and the need for further research on the physical, chemical, and biological processes at the pack-ice edge and of developments in the water column as the ice retreats. Investigations of trace elements and their effect on primary production are also needed.

Bacteria play a central role in chemical and nutrient cycling in all environments and much more work is required here. Benthic bacteria are important in mineralization processes. Many Antarctic bacteria do not relate closely to known taxonomic groups; physiological grouping is ecologically more meaningful than systematic nomenclature.

The well-known food chain of diatoms – Krill – large whales is only part of a more complex interacting food web. The nanoplankton is a neglected group now known to be a major food base for primary consumers. Alternative pathways, such as through bacteria and micro-zooplankton, or through predatory carnivorous zooplankton, such as chaetognaths, need to be elucidated.

There is a pressing need for more organic- and bio-chemical research in the Southern Ocean. Identification of wax esters and lipids in stomach contents may provide valuable information on diet. Organic studies are most promising and a more detailed approach is now called for. For example, instead of determining "polysaccharides" we should attempt to determine individual constituents. Many authors give concentrations of chlorophyll in their papers and it should be noted that chromatographic separation of chlorophyll pigments can, in many cases, give information on species' distribution and the growth status of a population. Many enzymes are only active if a specific metal is present; for example zinc activates carbonic anhydrase. Such element-enzyme effects need study, particularly in marine organisms. The temperature effect on enzyme activity also merits attention.

Benthic studies have been much neglected recently and because of their intrinsic interest, and the role of this group in nutrient cycling, they should now receive more attention.

For Krill the desirability of studies of physiology and biochemistry has been stressed; of great potential value is the possible use of enzyme systems for identifying unit stocks.

A major gap in marine studies is in winter, which may be the key to understanding Krill and other zooplankton population ecology; year-round investigations and studies within the pack-ice zone are urgently needed. Also year to year fluctuations in Krill larval abundance, and in the distribution of adult Krill are of considerable significance for its predators and point to the importance of long-term studies. A serious deficiency is the lack of quantitative historical data on Krill during the earlier episodes of seal and whale exploitation. In this connection it will be important to examine the historical data in the Discovery collections.

There is a need for accurate estimates of predation pressure on zooplankton, fish, and squid. In fish the nature of winter feeding should be established, in qualitative and quantitative terms, together with the dietary changes associated with growth in size.

The work on the role of seabirds in the ecosystem should be pursued on several fronts: improved estimates of numbers, trends, and population- and tropho-dynamics. The latter should include more studies to elucidate foraging ranges and behaviour and food consumption.

Investigations on the population dynamics of seals should continue in different geographical regions to establish how widespread is the change in reproductive rates as demonstrated for Crabeater Seals along the Antarctic Peninsula; these kinds of study could be extended to Leopard Seal, Weddell Seal, Elephant Seal, and Fur Seals. We also need more data on diet and energy budgets. A major gap is in our understanding of the current status of the baleen whales, which continue to be an important part of the pelagic ecosystem; with the decline and imminent cessation of whaling, knowledge of their continuing status and dynamics will be sparse.

An important aspect of nutrient cycling is that, if nutrients are rapidly turned over, the rate of primary production may be high, even though the nutrient concentration is low; simple nutrient-production correlations will not then be appropriate. On land, the weathering of rocks by lichens complements chemical weathering and emphasizes the close links between biological processes and geochemistry.

It is important to stress that despite much highly acclaimed research we still have a long way to go before we can generate models of ecosystem functioning, or more critically predictive models. Several presentations proposed models based on inadequate data sets, or data with great inherent variances or modified by unacceptable conversion factors. For future modelling attempts for terrestrial, inland-water and marine ecosystems it will be necessary to quantify elements in the food web in energy terms. The dynamics of terrestrial food webs, island effects and climatic constraints are all poorly understood.

By the time of the next SCAR Symposium on Antarctic Biology – perhaps in five years' time – we can expect that many of the questions posed at Wilderness will have been answered, but some will still be unresolved and no doubt more will be asked.

Subject Index

Further references to a group (e.g. Crustaceans) may be found under scientific or common names (e.g. *Lithodes murrayi*). '*Passim*' following a group of page numbers means that brief references to the subject are scattered throughout those pages.

A-selection 618, 680
Abisko, Sweden 206, 586, 587, 603
Abrams, R.W. 362, 466
Acaena magellanica 177, 211, 583, 597, 610, 643
  nutrient release 141, 143, 192–199
Acaena tenera 643
Acantheocidae 271
Acanthocyclus albatrossis 397, 398, 400, 401, 402
Acanthodrilidae 607
Acanthoeca spectabilis 273
Acanthoecopsis unguiculata 273
Acantholatris monodactylus 360
Acarina 613, 614–615, 626, 645
Acetate 144, 229, 230, 232, 254, 259
Acetylene reduction 141, 145, 186–191
Achnanthes brevipes 242
Acremonium 599, 601, 604
Acremonium kiliense 604
Acremonium strictum 600, 604
Acremonium terricola 604
Acremonium zonatum 601, 604
Actiniaria 385, 386, 387
Actinobolinidae 242
Adams, N.J. 460
Adelaide Island 204
Adenosine triphosphate (ATP) 116, 118, 125
Adenylates, total 286, 287, 292
Adipose tissue in notothenioid fish 434
Admiralty Bay, King George Island
  benthic macro-algae 104–107, 109–113, 115
  Crabeater Seals 670
  Krill 324, 326–328, 339, 343–344, 523, 524, 525
  penguins 156–161
Adversity-selection (A-selection) 618, 680
Aeromonas 126
Aethia 490
Aethia pusilla 490
Aethotaxis mitopteryx 431, 432, 433, 434
Aetidius armatus 45, 51
Agreed Measures for the Conservation of Antarctic Flora and Fauna 1
Agrostis 200
Agrostis magellanica 610
Agrostis stolonifera 177
Agulhas Current 11, 14, 16–18, 19
Ainley, D.B. 498
Alaskozetes antarcticus 615, 617
Albatrosses 221, 491
  Blackbrowed, *see Diomedea melanophris*
  flight energy requirements 467, 470
  Greyheaded, *see Diomedea chrysostoma*
  Lightmantled Sooty, *see Phoebetria palpebrata* nests 607, 608
  Shy 471
  Sooty, *see Phoebetria fusca*
  Wandering, *see Diomedea exulans*
  Yellownosed, *see Diomedea chlororhynchos*
Alcaligenes 125, 126
Alcids 491
Alciopidae 454
Algae (*see also* Ice-algae; Macro-algae; Phytoplankton)
  benthic mats 146, 150, 241, 243, 681
  blooms in Ross Sea 501, 502, 514
  'blue green', *see* Cyanobacteria
  consumption by vertebrates 570, 584
  freshwater lakes 146, 150, 241
  Lithodes murrayi consumption of 384, 385, 386, 387
  terrestrial 614, 615, 616, 617
Alkylglycerolethers 494, 495, 496, 497
Allanson, B.R. 11, 38
Alloposus mollis 531
Allotanais hirsutus 378–380
Alluroteuthis antarcticus 416–420, 527, 528, 531
Almirante Brown 206–210 *passim*
Alternaria alternata 600, 603, 604
Aluminium phosphate, amorphous 160, 165, 167
Amalopteryx maritima 607–611 *passim* 613
Amines, methylated 229
Amino acids 87, 144
Ammonia/ammonium
  in inland waters 146, 148, 150, 256, 257, 258
  marine: in Admiralty Bay 111, 112
    distribution of 23, 25, 28
    excretion by zooplankton 101–102, 324–329
    near receding ice edge 93–103, 677
    origin of 8, 9
    under fast-ice 53, 54
  in terrestrial ecosystems 144, 145, 159, 170, 222
Amos, A.F. 93
Amos Lake 148, 245, 253–259
Amphineura 409, 410, 411, 412
Amphipods 45, 389, 390, 410
  browsing on sea-ice communities 79, 576
  consumption of macro-algae 118, 120, 638, 639
  fish predation on 422–429, 431, 432, 453, 454, 456, 458
  in intertidal and subtidal zones 626, 631–635 *passim*
  Lithodes murrayi consumption of 385, 386, 387
  seal and seabird predation on 517, 518–519, 563
  squid predation on 417, 419
Amphiprora 80, 81
Amsterdam Island 551, 553
Amylase 116, 118, 120, 121, 255, 333–337
Anabaena 242
Anabaena cylindrica 242
Anaerobic microbial activity
  in lakes 147, 256, 259, 680
  in soil 188, 190
  in wet moss 144, 229–232, 680
Anas eatoni 584, 585
Anas flavirostris 584
Anas georgicus (South Georgia Pintail) 583, 584, 644, 645
Anas platyrhynchos 585
Anas supercilliosa 584
Anasterias rupicola 631, 633, 634, 635
Anatalanta aptera 607, 608, 609, 612
Anatalanta crozetensis 606–609 *passim* 610–612, 613
Anatanais gracilis 632
Anchoveta (*Engraulis ringens*) 269, 372
Anchovy 496, 570
  Cape (*Engraulis capensis*) 494
  South-West Atlantic (*Engraulis anchoita*) 449, 569
Ancistrocheirus lesueuri 531
Annelids 45, 424, 626
Anoxia in freshwater lakes 235–236, 237
Antarctic Bottom Water (AABW) 7, 12, 266, 267
  nutrients 23, 24, 27, 28
Antarctic Circumpolar Current 12, 656

Antarctic Continent 123, 145, 148–151
Antarctic Convergence (AC, or Polar Front, APF) 7, 124, 266, 267, 478
biological productivity 11, 12–16, 19, 24
Prince Edward Islands 47, 49–50, 355, 359
seabird distribution at 466, 467, 468, 469
south of New Zealand 656, 657, 661
Antarctic Divergence 7, 12, 13, 267, 478, 480
Antarctic Intermediate Water (AAIW) 24, 26, 27, 47
Antarctic Peninsula 268, 324, 449, 536, 670
freshwater lakes 146, 148
Leopard Seals 556, 557, 558
moss-peat decomposition 204–210
Antarctic Polar Front, *see* Antarctic Convergence
Antarctic Slope Front 501, 506, 509, 512
Antarctic Surface Water (ASW) 12, 54, 478
nutrients 23, 24, 25, 27, 28
Antarctic Treaty 1
Antarctomysis 431
Anthus antarcticus (South Georgia/Antarctic Pipit) 583, 584, 585, 586, 644, 645
Antias bicornis 632, 634
Antimora 568
Antiprod 22, 23
Anvers Island 60, 148, 324
Apetenus litoralis 607
Aphidae 175, 178
Apseudes 633
Aptenodytes forsteri (Emperor Penguin) 268, 445, 503, 652–655
Aptenodytes patagonicus (King Penguin)
foraging range 479, 484, 485
lipids in stomach contents 494, 495, 497
manuring by 607, 608, 611
organochlorine content of eggs 648, 649, 651
prey consumption 517, 518, 526, 527, 529, 548
Arachnida, *see* Spiders
Arbacia dufresnei 409, 410, 412
Architeuthis 531
Arctic
sea-ice micro-algae 85
tundra ecosystems 582, 585, 586, 587, 588–589, 597
Arctocephalus australis 411
Arctocephalus gazella (Antarctic Fur Seal) 3, 673
carcasses 538
diet and energy budget 543–549, 679
impact on terrestrial environment 228, 641–645
prey consumption 517, 519, 520, 523, 524, 525–526, 529, 530
Arctocephalus pusillus (Cape Fur Seal) 553
Arctocephalus tropicalis (Sub-Antarctic Fur Seal) 174, 551–553
Ardeidae 586
Argentine Islands 670
Argentine shelf 488
Arnaud, P.M. 381
Arthrobacter 666
Arthropods, terrestrial 587, 614–618, 680
consumption by birds 584, 586
Arthur Harbour 635
Ascidians 385, 386, 387, 409, 410, 633
Ascochyta stilbocarpae 600–604 *passim*
Ascomycetes 599, 602, 604
Ash content
bryophytes 201, 202
notothenioid fish 433
Assimilation efficiency 161, 365, 390–391, 395, 461, 616
Asterodon singularis 411
Asteroidea 384, 385, 386, 387
Atka Bay 652
Atka Ice Port 445, 446
Atlantic Ocean
South East 17
Southern 304–306, 647–651
Auckland Islands 648
Auklets 490
Least (*Aethiapusilla*) 490
Aulacomya ater 409, 410, 411
Aureobasidium pullulans 599, 600, 603, 604
Australia 568, 569
Australian National Antarctic Research Expeditions (ANARE) 452
Austrocidaris canaliculata 409, 410, 412
Avena sativa 225
Aytoella magellenica 360
Azorella selago 175, 177, 583

Bacteria 188, 587, 614, 677
benthic mats 241
in freshwater lakes 241, 250, 254–259, 261–264
gold chromogenic 204, 206, 207, 209
guano decomposition 158
ice 78–83, 677
ingestion by *Allotanais hirsutus* 378–380
leaf and litter decomposition 194, 198, 597, 599–603 *passim*
macro-algal degradation 104, 107, 115–121, 638, 639
marine: activity in a mussel-bed 128–134
ammonia-production 9, 677
choanoflagellate consumption of 273 275
creation of aerosols 595
decomposition of organic matter 9, 111, 112, 113
role in nutrient cycling 123–126, 368, 574
methanogenic 147, 229–232, 259, 680
moss peat decomposition 144, 207, 208, 209
nitrate reducing 125, 141, 257, 258
nitrifying 9, 144
nitrogen-fixing, *see* nitrogen fixation
phototrophic 125
soil and freshwater lake interactions 622–667
in soils 141, 186–191
sulphate reducing 144, 257–258, 259
taxonomy 125–126
Bacterio-plankton
in dry valley lakes 238, 241
in maritime lakes 147, 254, 261–264, 667
Balaenidae, *see* Balleen whales
Balaenoptera acutorostrata (Minke Whale) 569, 672, 673
biomass in Ross Sea 499, 505, 509–512
Krill consumption 366, 367
Balaenoptera borealis (Sea Whale) 508, 569, 672
Balaenoptera physalus (Fin Whale) 569, 672
Balanus 408, 409, 410
Baleen whales 70, 268, 375, 470, 567
and expansion of fur seal populations 645, 669, 671
sexual maturity 673
Barczuk, A. 163
Barnacles 409, 410
Barrett, J.A. 493
Barrow, Alaska 85, 152, 188, 205, 585, 586, 587, 603
Barrow Strait 85
Basidiomycetes 600, 601, 602, 604
Bathothauma lyromma 552
Bathydraconid fish 434, 564
Bathyteuthis abyssicola 527
Beagle Channel, Chile 407–413
Beauchêne Island 204, 210
Beaufort Sea 85
Bengtson, J.L. 669
Benguela Current 470, 493–497
Benjamin Bowring 656, 658
Benninghof, A.S. 592
Benninghof, W.S. 592
Benthic algal mats 146, 150, 241, 243, 681
Benthos, marine 2, 268, 677–678
bacteria 124
biological features 576–578
contribution of macro-algae 104, 107, 109, 113, 115
energy flow to 575
role of *Lithodes murrayi* 381–388
Berardius arnuxii 567, 568
Berardius bairdii 568
Bering Sea 76, 85, 466, 470, 487–491, 549
Berries 584
Berruti, A. 460
Bester, M.N. 551
–BHC 652
Bicarbonate 146
Bicosta spinifera 273, 274, 275
Biddulphia 312–322

Biggs, D.C. 93
Biological enhancement and oceanic frontal systems 11–19
BIOMASS (Biological Investigations of Marine Antarctic Systems and Stocks) 1, 2, 52, 266, 270, 339, 415, 551, 626
Biotic/sea-spray zone 200, 202
Bird Island 204, 210
  albatrosses 473–477
  Fur Seals 523, 546, 548, 641–645
  Giant Petrels 535, 536, 538, 540
  Krill 523, 524
Birds 2–3, 678–679
  biomass 70, 470, 487–491, 498–503, 514
  'coastal' 627
  consumption by giant petrels 535–541
  distribution in relation to prey availability 268, 466–471
  eggs 584, 647–651
  energy requirements 466–471
  enrichment of freshwater lakes 146, 148
  food consumption 487–491, 517–532, 627–628
  foraging zones 478–485, 491, 627
  Krill consumption 366, 367, 525–526
  lipids in diet of 493–497
  manuring by 140–141, 144, 145, 169–171, 606, 607
  organochlorines and heavy metals 647–651, 652–655
  terrestrial food webs, role in 582, 584–587, 589
  vagrant land 586
Bivalves
  consumption by *Lithodes murrayi* 384, 385, 386, 387
  littoral 632, 633, 634, 635
Blankley, W.O. 630
Blechnum penna-marina 610
Blepharidophyllum densifolium 200–202, 213
Block, W. 614
Blue Glacier 79
Boden, B. 38
Body size 395, 403
Bogs 141, 142, 211–220
Bölter, M. 109
Bonner, W.N. 641
Boreogadus saida 430
Bossiere Fjord 128–134
Bostrychia 632
Botrytis cinerea 194, 599, 602, 603, 604
Bouché, M. 606
Bouvet Island 13, 14, 445, 536
Bovichthys diacanthus 360
Bovichthys variegatus 432, 433
Brachidontes demissus 401
Brachiomonas submarina 242
Brachiopoda 384, 385, 386, 387, 424
Brachioteuthis picta 415–420, 527, 531
Brachythecium rutabulum 200–202
Brachythecium subplicatum 141, 200–202
Brachyurans 387
Branchinecta gaini 249
Bransfield Strait 670
  Krill 284, 304–306, 308–309, 324, 326–329
  nutrient cycling 109, 110, 113
  phytoplankton 60, 61, 62
Brown, C.R. 460
Bryophytes, *see* Mosses
Bryozoa 384, 385, 386, 387, 409
Buchholz, F. 339
Buoyancy of notothenioid fish 430–435
Burger, A.E. 582
Burton Island 498

C:Chl-*a* ratio 62, 81, 247
C:N ratio 111, 134, 312
  freshwater lakes 247, 250
  litter breakdown 219
  macro-algal degradation 115, 116–117, 118, 121
C:P ratio 247, 250
Cadmium (Cd) 7, 30, 34–36, 653–655, 659, 660
Cahet, G. 128
Calanoides acutus 454, 455, 576
  distribution 47, 309, 310
  feeding behaviour 311–322
Calanus propinquus 360, 454, 455
  distribution 47, 309, 310
  feeding behaviour 311–322
Calanus simillimus 45, 47, 358, 360
Calcite 180
Calcium (Ca)
  in guano 170
  in *Rostkovia magellanica* 141, 142, 211, 213–217, 218, 219
  in soils 143, 144
Calliacantha natans 273
Calliacantha simplex 273
Calliergidium austrostramineum 229
Calliergon 662, 664
Calliergon sarmentosum 211, 229
Callitriche antarcticae 174, 175
Callorhinus ursinus (Northern Fur Seal) 3, 544, 548, 549
Calonectris diomedea 471
Caloneis ventricosa var. *truncatula* 242
Calothrix gypsophila 242
Camarosporium metableticum 599, 600, 604
Campylonotus vagans 409
Candida 666
Cannibalism 353, 386
  fish 427, 431, 549
  squid 417, 418
Cape Armitage 79, 80, 84
Cape Bird 79
Cape Norwegia 308–310, 445, 446
Cape Peninsula 633–635, 639
Cape Royds 78, 79
Cape Town 13, 14, 285
Caperea marginata 567
Caprellida 384, 385, 386, 387, 424
Carassius 402
Carbohydrates
  in lakes 255
  marine 109–112, 331, 332
  terrestrial 143, 144
Carbon (*see also* C:Chl-*a* ratio; C:N ratio; C:P ratio)
  dissolved organic, *see* Dissolved organic carbon
  extracellular release 87
  in macro-algal debris 116–117, 118, 120, 121
  particulate organic, *see* Particulate organic carbon
  in penguin guano 158, 159
Carbon cycling
  anaerobic 144, 229
  dry valley lakes 150, 240, 241
  marine 9, 125
    at lower trophic levels 277–282
  maritime lakes 261, 262–263
Carbon dioxide 9, 144
  in freshwater lakes 146, 147
  for methanogenesis 144, 229, 259
  in moss peat decomposition 205–206
  phytoplankton activity and 262–263
Carbon flux
  to marine sediments 575
  to seabirds 466–471, 487–491
Carbonate 146, 236
Carcasses, colonization by Diptera 607, 608, 610–612, 613
Carcinus maenas 344
Carduelis flammea 584, 585
Carex paniculata 603
Carnivores
  invertebrate 614, 616, 617, 618
  mammalian 583
Carrion feeders 409, 410, 583, 584
Casey Station 152
Cassidinopsis emarginata 410
Castilla, J.C. 407
Cat, House (*Felis catus*) 583–584, 620–624, 647
Catharacta (skuas) 148, 151, 471, 627, 645
  organochlorines 648, 649, 650, 651
  pelagic distribution 479, 480, 484, 485, 648
  predation on petrels 620, 621
Catharacta maccormicki (South Polar Skua) 445, 471, 652–655
Cattle 583
Cavolina tridentata 358, 360
Cellana tramoserica 392
Cellulase 116, 121, 333–337, 603
Cellulomonas 666
Cellulose decomposition 192–199, 603
  in maritime lakes 255–256
  in moss communities 204, 205, 206, 209–210, 229, 232
Cephalopods (*see also* Squid) 268, 557, 559, 563, 564, 633, 634
Cephalorhynchus commersonii 567, 570, 571
Cephalorhynchus eutropia 567, 570, 571
Cephaloziella varians 229
Cepola 568
Ceratosphaeria 600, 604
Cetacea, small (*see also* Whales) 566–571
Chabrier Rock 157
Chaenocephalus aceratus 437–443

Chaenodraco wilsoni 453, 455
Chaetoceros 62, 284, 312, 314–322, 351
Chaetoceros radicans 40, 50
Chaetognaths 45, 58, 268, 576
fish consumption of 424, 450, 453, 454
squid predation on 417, 419
Chaetophoma 194
Chalara 604
Challenger 637
Champsocephalus esox 410, 411
Champsocephalus gunnari 434
diet 421–422, 425–429
Krill consumption 437–443, 678
predation on 423, 529, 546, 549
stock biomass 442, 530
Channichthyid fish 434, 437, 442
diet 421–429, 453, 457, 458
larvae 448, 449
predation on 431, 432, 529
Channichthys rhinoceratus 421, 422, 425, 429
Char Lake 250
Charadriidae 586
Charlotte Bay 324, 326–328
Chatham Islands 359
Cheirimedon femoratus 390
Chemical elements 6–7
Chemical pollutants 647–651, 652–655
Chile 637
Chionis alba 584
Chionis minor (Lesser Sheathbill) 583, 621
impact on intertidal zone 631, 634, 635
organochlorine content of eggs 648, 649, 651
role in terrestrial ecosystem 584, 585, 627
Chionodraco 448, 453, 456–457, 458
Chionodraco hamatus 437
Chionodraco rastrospinosus 437–443
Chiroteuthis veranyi 531
Chitin 158, 255
Chlamydomonas intermedia 242
Chlamydomonas subcaudata 238, 241, 242
Chlanidota densesculpta 576
Chloephaga picta 585
Chlorella 261, 263
Chlorella vulgaris 242
Chloride 146, 151
Chlorinated hydrocarbons 647–651, 652–655
Chlorococcum 242
Chloromonas alpina 241
Chlorophyll
in bryophytes 201, 202
near receding ice-edge 71–72, 73–75, 95, 99
Chlorophyll-a 676
and digestive enzymes of euphausiids 330–337
Krill distribution and 286, 287, 292, 300, 301, 302, 368
in maritime lakes 147, 149, 246–250 *passim* 254
in nanoplankton and microplankton 279–280
near Prince Edward Islands 41, 42, 43, 50
and oceanic frontal systems 11, 12–19
photosynthesis and 62, 64, 65, 68
satellite remote sensing 16–18
in sea-ice 78, 80, 81, 83, 90
under fast-ice 54, 57, 58
Choanoflagellates 271–275, 281, 677
Chorismus antarcticus 397–404, 576
Chorisodontium aciphyllum 144, 213, 228
Chromobacteria 664
Chromobacterium 204, 207, 209, 662
Chromobacterium fluviatile 666
Chromobacterium lividum 666
Chroomonas lacustris 238, 241, 242
Chrysosporium pannorum 194, 600, 604
Chthamalus 408, 410, 411
Chubut, Argentina 536
Ciliates 242, 280–281
Cirolana hardfordi 395
Cirolana imposita 392, 395
Cirripedia 384, 385, 386, 387
Cladonia rangeferina 213
Cladosporium 600, 601
Cladosporium cladosporioides 603, 604
Cladosporium herbarum 603, 604
Cladosporium sphaerospermum 194
Clarke, A. 573
Clasmotocolea vermicularis 200–202
Clausosocalanus laticeps 47
Clear Lake 272
Cleodora sulcata 360
Climate
maritime Antarctic 123
Sub-Antarctic islands 582, 586–587, 589
Clio 45
Clio pyramidata 454, 456
Clio pyrimidata antarctica 360
Clio pyrimidata sulcata 360
Clione 45
Clione antarctica 360
Clione limacina antarctica 359, 360, 660
Clostridia 144
Cloud cover in Southern Ocean 16
Clupea harengus (Herring) 449
Clupea harengus pallasi 430
Cnidaria 422, 423, 424, 426
Coastal ecosystems 104, 107, 115–121, 626–628
Coastal ponds 146
Coastal Zone Colour Scanner (CZCS) 1, 16, 18, 676
Coccolithus huxleyi 327
Cod
Antarctic, *see Dissostichus mawsoni*
Arctic 430
Atlantic (North Sea) 438, 440, 443
New Zealand Black 432, 433
Coelenterata 45
Coelomycetes 599, 602, 604
Coelopidae 607
Cold adaptation 397–404, 441, 578
Coleoptera 178, 632, 634, 645
Collembola 175, 178, 587, 645
consumption by birds 584, 589
diet 614–615
Colletotrichum gloeosporiodes 604
Collodictyon 242
Colobanthus quitensis 643
Columbidae 586
Compression feeding 375
Congelation ice 81, 84
Connell, A.D. 647
Continental Convergence 268
Continental Water Boundary 12–13, 466 467, 468
Convention for the Conservation of Antarctic Marine Living Resources (CCAMLR) 1, 49–50, 269, 669
Coomansus gerlachei 614
Cooper, J. 362
Copepodids 311–322
Copepods 368, 574
feeding behaviour 311–322, 354, 375
intertidal zone 632
lipids 496, 578
near Prince Edward Islands 42–47 *passim* 51
near receding ice-edge 100, 268
predation on 563, 567
by birds 517, 518–519, 523, 627, 628
by fish 422, 450, 453–458
under sea-ice 58, 79
vertical distribution 308–310
Copper (Cu)
Cu/Si ratios 32, 677
in sea water 7, 30, 32–33, 35, 36, 656
in seals and birds 654–655
in zooplankton 658–661, 677
Corethron 62, 268
Corethron criophilum 284, 312–322, 351, 352
Coriolis force 50
Cormorants 648
Bank 650
Cape 650
Crowned 650
Imperial, *see Phalacrocorax atriceps*
Whitebreasted 650
Corre, M.C. 330
Corvina 570
Corynebacterium 666
Coscinodiscus 86
Cosmasterias lurida 407–410 *passim* 412–413
Cotton strips 192, 193–194, 198, 207, 255–256
Cottoperca gobio 410, 411
Cotula plumosa 175, 177
Crabs 387, 388
King 381, 386–387
Stone 381–388
Cranchiidae 552, 553
Crassula moschata 175
Crepipatella dilatata 408, 409, 410
Crinoids 424
Crinolina aperta 273
Croxall, J.P. 516, 543

Crozet Islands 152, 381–388, 421–429, 627, 637
foraging zones of seabirds 478–485, 627
petrels 461, 535, 536, 537, 538
terrestrial food webs 582, 583, 584, 587, 588
Crustaceans 268, 389
cast moults 575
cold adaptation 397–404
consumption by birds 535–541, 584, 627, 631, 648
Cryodraco 448
Cryodraco antarcticus 434, 453, 455
Cryothenia peninsulae 432
Cryptomonads 281
Cryptophyceae 277
Cryptopygus 680
Cryptopygus antarcticus 615, 616–617
Crysophyceae 277
Ctenaria 422, 423, 424
Curculionidae 178
Cyanobacteria ("blue green algae"), 273, 680
in benthic algal mats 150
endolithic 145
epiphytic on mosses 140, 141, 143, 144, 147, 186, 191
free-living 141, 145, 186, 191
Cyanophytes 146
Cyanoramphus novaezelandiae erythrotis 584
Cycetra verrucosa 411
Cyllopus 454
Cyllopus magellanicus 360, 660
Cymbella 242
Cyprid larvae 660
Cytophaga 204, 206, 209, 666

Dallman Bay 324, 326–328
Daption capense (Cape Pigeon, Pintado Petrel) 471, 490, 496
energy expenditure 470
food consumption 360, 518, 520, 529
foraging range 480, 481, 484, 485, 627
Dasyscyphus cf. *enzenspergerianus* 599, 600, 601, 604
David, P. 46
Davis station 271
Dawson, R. 109
DDE 647–651, 653
DDT 647–651, 652–655, 653
Dead organic matter (DOM) 589, 614, 615, 618
Dead plant material 141, 192, 194–196, 211
Decapods 424
larvae 453, 454, 456
Decapterus longimanus 360
Deception Island 204, 324, 527, 549, 553
Decomposition (*see also* Leaf decomposition; Litter breakdown) 109–113
anaerobic 229–232, 259, 680
cellulose, *see* Cellulose decomposition
in dry valley lakes 243
of guano 158–160
of macro-algae 115–121, 639
in maritime lakes 147, 253–259
moss-peat 204–210, 229–232, 680
in *Rostkovia magellanica* 211–220
in terrestrial communities 139–140, 141, 142–143, 616
Delille, D. 128, 378
Delphinidae, *see* Dolphins
Dendrobaena rubida tenuis 607
Denitrifying bacteria 125, 141, 257, 258
Deschampsia antarctica 144, 225, 643, 644
Desmarestia 104
Desulfovibrio 125
Detritus feeders 269, 410, 575
Detritus-based food chains 409, 410, 588–589, 635, 677
role of macro-algae 104, 107, 115–121
Devon Island, Canada 152, 585, 586, 587
Diacylglycerides 494, 495
Diaphanoeca grandis 272, 273, 274, 275
Diaphanoeca multiannulata 272, 273, 275
Diaphanoeca pedicellata 273, 274, 275
Diaspores 592–595
Diatoms 272, 280, 573–574
ammonia uptake 327
dry valley lakes 238, 241
grazing by Krill and copepods 284, 312–322, 350–353
ice 81, 82, 85–86, 87
sedimentation 9, 575
silicate requirements 8, 28
Didymella proximella 600, 604
Dieckmann, G. 104, 115
Diego Ramirez 221, 536
Dieldrin 648–650, 653
Digestive enzymes of euphausiids 330–337
Dihydrohalite 240
Dimophyes arctica 358, 359, 360
Dinoflagellates 242, 272, 273, 573–574
grazing by Krill and copepods 314, 315, 316, 319
Dinophyceae 277
Diomedea cauta 471
Diomedea chlororhynchos (Yellownosed Albatross) 471
foraging range 479, 480, 483, 484, 485
organochlorines in eggs 648, 650
Diomedea chrysostoma (Greyheaded Albatross) 470, 471, 473–477, 490
food consumption 518, 528, 529, 531
foraging range 479, 481, 484, 485
Diomedea exulans (Wandering Albatross) 471, 611
energy requirements 467, 470
food consumption 518, 527, 528, 531, 553
foraging range 479, 483, 484, 485
lipid content of stomach 494, 495, 496
organochlorine content of eggs 648–651
Diomedea melanophris (Blackbrowed Albatross) 471, 473–477, 490, 540
energy requirements 470
food consumption 517, 518, 524, 527, 528, 531
foraging range 479, 481, 484, 485, 627
lipid content of stomach 494, 495, 496, 497
Diploospora 604
Diptera 178, 584, 587, 589, 645
biology of 606–613, 680
Discoteuthis 528, 531
Discovery 70, 324, 574, 681
Dissolved organic carbon (DOC) 574
in Admiralty Bay 109, 110, 111, 112
in decomposing moss-peat 205, 209
in maritime lakes 254–255, 263, 663, 667
Dissolved organic phosphorus (DOP) 22, 25, 28
Dissostichus 678
Dissostichus eleginoides 421, 424–425, 426, 429, 438
Dissostichus mawsoni (Antarctic Cod)
buoyancy 430, 431–432, 433, 434
diet 431–432, 445, 564
Weddell Seal interactions 432, 450, 561–565, 679
Distephanus 350
Distephanus speculum 313, 314, 315, 316, 319
Ditrichum strictum 141, 200–202
Ditylum brightwellii 373–376
Doidge, D.W. 543
Dolphins 566, 567, 569–570, 571
Bottlenose 567, 570
Chilean (*Cephalorhynchus eutropia*), 567, 570, 571
Commerson's (*Cephalorhynchus commersonii*) 567, 570, 571
Dusky 567, 569
Hourglass 567, 569–570
Peale's (*Lagenorhynchus australis*) 567, 570
Risso's (*Grampus griseus*) 567, 570
Southern Right Whale 567, 570
Donkeys 583
Doves 586
Drainage lines 200, 202
Drake Passage 11, 61, 62, 575
Drepanocladus 662, 664
Drepanocladus uncinatus 204–210, 211, 229
Drescher, H-E. 652
Dreuxiella spinosa 607
Dronning Maud Coast 366
Dry Valleys 145, 150, 680
lakes 152, 238–243, 681
Ducks
Common Eider 171
dabbling 584, 585
Grey 584
Duhamel, G. 421
Dunaliella tertolecta 373
Duncombe Rae, C. 38
Durvillea 634
Durvillea antarctica 612–613, 637–640

Dust particles 592–595
Dynamenella huttoni 631, 632, 634, 635

Eagle, G.A. 656
Earplugs, transition zone in 673
Earthworms 587, 606–613, 680
East Ongul Island 52, 53
East Wind Drift 7, 12, 267, 268
  phytoplankton 279, 280
  Pleuragramma larvae 449, 450
Eastman, J.T. 430
Echinoderms 268, 411, 424, 633, 634, 635
Echinoidea (sea-urchins)
  in *Macrocystis* belts 407, 410, 412–413
  predation on 384–387 *passim*
Ecklonia maxima 106, 118
Ecological efficiencies 618
Ecosystems (*see also* Coastal ecosystems; Freshwater ecosystems etc.)
  future development of research 151–153
  indicators 669
  interactions between 626–628
Eddies 39, 44, 102
Egesta 390
Egrets 586
El-Sayed, S.Z. 284
Electrona antarctica 440, 452, 454, 456, 457, 458
Electrona paucirastra 358, 360
Electrona subaspera 358, 360
Electrostatically charged particles, wind transport 592–595
Eledoniae 528, 531
Eleginops 570
Elephant Island 61, 163, 278, 536, 670
  fish 437–443, 458
  Krill 284, 304–306, 308, 324, 327, 343, 522, 523, 575
Ellis-Evans, J.C. 253, 261, 662
Eltanin 277
Embryonopsis halticella 178
Enchytraeids 587, 589
Enderby Land 151, 603
Endolithic cyanobacteria 145
Endolithic microbes 3
Energy budget
  Antarctic Fur Seal 543–549
  Cryptopygus antarcticus 616–617
  Notothenia neglecta 393–394
  Serolis polita 389–396
  Whitechinned Petrel chicks 460–464
Energy content
  bryophytes 201, 202
  Durvillaea antarctica 638–639
  of meals fed to chicks 461, 462, 475–476
Energy flow
  in intertidal and subtidal zones 630–636
  in Southern Ocean food web 573–579
  terrestrial food webs 588–589, 614–618
  to *Felis catus* 620–624
Energy requirements
  Anasterias rupicola 631
  Felis catus 620, 623
  fish 438, 441, 630, 631
  giant petrels 534–535, 538, 539, 540
  Krill 362–366
  seabirds 466–471, 487, 488, 489, 630, 631
Energy turnover rate 395, 578–579
Engraulis anchoita 449, 569
Engraulis capensis 494
Engraulis ringens (Anchoveta) 269, 372
Enhidra lutris 413
Enteromorpha 408
Enzymes
  cold adaptation 403
  digestive, of euphausiids 330–337
Eophiliantidae 632, 633, 634
Ephydridae 607
Epicoccum purpurascens 603, 604
Erebus 70
Ereynetes macquariensis 615
Erosion 173, 228, 644
Erwinia caratovora 603
Essential elements 6
Eucalanus longiceps 47
Euchaeta 45
Euchaeta antarctica 453, 454, 455, 578
Eudyptes chrysocome (Rockhopper Penguin) 360, 480, 485, 551, 648, 649
Eudyptes chrysolophus (Macaroni Penguin) 3, 608
  Antarctic Fur Seals and 546, 548–549
  food consumption 518, 520–526 *passim* 529, 530, 627
  foraging ranges 479, 480, 484, 485, 548
  organochlorine content of eggs 648, 649
Eukrohnia hamata 45, 358, 359, 360, 658–661
Euphausia crystallorophias 2, 58
  Adelie Penguin predation 367, 522
  Crabeater Seal predation 268
  fish predation on 268, 437, 450, 453–458 *passim*
  larvae 310
  lipid storage 576, 578
Euphausia frigida 454, 457
Euphausia longirostris 360
Euphausia lucens 45, 358, 359, 360
Euphausia pacifica 419, 660
Euphausia similis 45, 425
  var. *armata* 45, 358, 359, 360
Euphausia superba (Krill) 2, 58, 678
  ammonia excretion 324–329
  biomass estimates 36, 270, 294, 295–296, 301–302, 362, 366, 520, 530, 574, 679
  calorific content 517
  commercial fishing 549
  decomposition 125
  digestive enzymes 330–337
  eggs 308, 309
  feeding behaviour 284, 311–322, 372–377
  food requirements 278, 362–370
  growth and moulting 339–344
  larvae 270, 304–306, 308–310, 346–354
  lipids 335, 336, 526, 576, 679
  marine distribution 284–292, 294–303
  oxygen tension and 300–301, 303, 328–329
  predation on 366–367, 516–526, 530
    by albatrosses 473, 475, 476
    by cetaceans 366, 367, 569
    by Crabeater Seals 366, 367, 669, 673–674
    by fish 423, 432–433, 437–443, 450, 452–458
    by fur seals 546, 547, 548, 549
    by giant petrels 535, 538, 540
    by Leopard Seals 555, 556, 557, 559
    by penguins 161, 366–367
    by seabirds 366, 367, 470, 516–526, 627–628
    by seals 516–520, 523–526
    by squid 367, 417–418, 419, 420
  role in marine food webs 70, 266, 268, 269–270
  sex ratios 524–525, 526
  spawning 306
Euphausia triacantha 658–661
Euphausia vallentini 359, 360
Euphausiacea 45, 551
Euphausiids 58, 268, 574
  larvae 100
  lipids 496, 497
  metal concentrations 660
  moult and growth 339–344
  predation on 422–429, 452–458, 489
  Prince Edward Islands 43, 44
Eupodes minutus 615, 617
Euridyce pulchra 45
Eusiridae 632, 633, 634
Eusirus antarcticus 660
Excreta 584
Excretion, energy costs of 390, 391, 392, 394
Exosphaeroma gigas 409, 410, 422, 632, 634
Exposure, *Poa flabellata* growth and 222–223, 224, 227

Fabraea 601, 604
Faecal pellets 575, 593, 594
Faeces 142, 173, 175, 177
Fairbanks, Alaska 585
Falkland Islands 204, 221, 397, 536, 567, 569, 597, 637
Faraday 206, 207, 556
Farber-Lorda, J. 330
Fast-ice 52–58, 505, 564
Fatty acids 494, 495, 496
Fatty alcohols 494, 495, 496, 497
Feathers 40, 607
  penguin 161, 546, 553
  Stone Crab consumption of 384, 385, 386, 387
Feline panleucopenia (FPL) 620, 624

Felis catus (House Cat) 583–584, 620–624, 647
Fellfield 643, 664
invertebrates 587, 588, 608–610
micro-organisms 141, 204–210
moss production 200, 202
nutrients 139, 140
soil, *Poa flabellata* growth in 221–227
Fermentative processes 256–257
Festuca contracta 141, 192–199, 221, 225, 597
Festuca grassland 143
FIBEX (First International BIOMASS Experiment) 2, 271, 284, 285, 294–295, 331, 362, 363, 452
Filchner ice shelf 446
Filter feeders 410, 633, 635, 639, 640
Fish 2, 268, 678
biomass estimates 530
calorific content 517
commercial fishing 530, 549
eggs 423, 424
food consumption 360, 367, 386–387, 421–429, 437–443, 452–458
larvae 79, 268, 446, 448–449, 660
lipids 496
predation on
by birds 466, 469, 470, 471, 475, 476, 516–520, 528–530, 584, 627, 631, 648
by fish 422–429, 431, 454, 455, 456
by giant petrels 535–541
by seals 516–520, 528–530, 546–551 *passim* 557, 561–565, 628
by small cetaceans 568, 569, 570, 571
by squid 417, 418, 419
scales 384, 385, 386, 387, 433
Fissurella 409, 410, 411, 412
Fjaeldmark soil (*see also* Fellfield) 186–191
Flagellates (*see also* Micro-flagellates) 272, 273, 280
Flavobacterium 125, 126, 666
Flight, energy consumed by 467, 470
Fluorescence 66, 290–291, 292
Fluoride (F) 159, 677
Food chains
detritus-based, *see* Detritus-based food chains in dry valley lakes 238–243
Food consumption 679
Food webs
in intertidal and subtidal zones 631–635
in *Macrocystis pyrifera* belts 407–413
marine 8, 266–270, 470, 677
alternate pathways at lower trophic levels 277–282
energy flow 573–579
role of squid 415–420
terrestrial 582–589, 680
Foraminifera 45, 384, 385, 386, 387
Formate 144, 229, 230, 232
Fragilariopsis kerguelensis 28
Fragilariopsis nana 279, 280, 281
Fratercula corniculata 490
Frazier Islands 536
Frazil ice 81
Freeze-thaw cycles 139, 143, 144, 663
Fregetta 621, 622
Fregetta grallaria 471
Fregetta tropica (Blackbellied Storm-petrel) 471
food consumption 519
foraging range 479, 481, 483, 484, 485, 627
Frénot, Y. 606
Freshwater ecosystems (*see also* Lakes) 3, 679–680
future development of research 151–153
interactions with marine and terrestrial organisms 626–628
nutrient cycling 139, 145–151
Frittilaria 45
Frobisher Bay 85, 124
Fugler, S.R. 169
Fukuchi, M. 52
Fulmars, *see entries below*
Fulmarus glacialis (Northern Fulmar) 488, 490
Fulmarus glacialoides (Southern/Antarctic Fulmar) 471, 490, 518, 520, 526, 527, 538
Fungi 614, 615, 616
cellulolytic 194, 198
leaf and litter breakdown 219, 597–604, 680
peat decomposition 204, 207, 209–210
on rocks 184, 185
soil and freshwater lake interactions 662–667
in soils 141
sterile forms 602, 603
Furcilia lobosa 242
Fusidium 604
Fusitriton magellanicus 409, 410, 411, 412

Gadus morhua (Atlantic Cod) 438, 440, 443
Gaimardia trapesina 359, 360, 576
cold adaptation 397, 400, 401, 402
intertidal and subtidal food webs and 409, 410, 631–635 *passim*
ß-Galactosidase 333–337
Galatheida 384, 385, 386, 387
Galeazzi, A.R. 566
Galiteuthis glacialis 415–420, 527, 528, 531
Gallagher, J.B. 234
Gallirallus australis (Weka) 584, 585
Galliteuthis armata 552
Gamasellus 680
Gamasellus racovitzai 614, 615, 617–618
Gammarids 384, 387, 388, 424
Gammarus oceanicus 401
Gannet, Cape 471
Gardner, B.D. 647
Garrodia nereis (Greybacked Stormpetrel) 471, 483, 484, 485, 627
Gastric evacuation rates 441, 443
Gastropoda 45
Geese 589
Gelatinase 158
George, R.Y. 324
Geotria australis 475, 529
Glacier 71, 93, 94
Glenamoy, Ireland 586
Gliocladium 604
Globicephala melaena 567, 569
Globigerina 45
Globigerinoides 45
Glucose 128–134
Glutamate 128–134
Gneiss Lake 253–259, 664, 665
Gneiss Point 79
Goats 583
Goldfish 402
Gonatus antarcticus 527, 528, 531, 570
Goodall, R.N.P. 566
Goose, Upland 585
Gough Island 177, 221, 355–360, 536, 637
chlorinated hydrocarbons 647, 648, 650
Sub-antarctic Fur Seals 551–553
Grahamland 123
Grampus griseus 567, 570
Granite Harbour 71
Graphium 604
Grass, Antarctic Hair (*Deschampsia antarctica*) 643, 644
Gravel 384, 385, 386, 387
Grindley, J.R. 46, 630, 637
Growth
benthic organisms 576
efficiency 578
efficiency, net 394–395
energy costs 391, 392, 393
potential of sea water 8
Guano 607, 627
ammonia from 148, 159, 170
burrowing petrel chicks 169–171
mineralization and nutrient cycling 156–161, 677
phosphatized rock 159–160, 163
Guidi, L.D. 378
Gulf Stream 17
Gulls 491
Great Blackbacked 171
Herring 171
Kelp, *see Larus dominicanus*
Ringbilled 171
Gymnodraco acuticeps 431, 432
Gymnoscopelus braueri 453, 455, 457, 458
Gymnoscopelus nicholsi 453, 455, 457, 458, 546
Gyrodinium 314

Habitat templet 618
Hake
(*Merluccius capensis*) 494, 496
Patagonian (*Merluccius hubbsi*) 568, 571
Halicarcinus planatus 387, 409, 410, 422
Halirytus amphibius 607

Halobaena caerulea (Blue Petrel) 471, 644
cat predation on 620, 621, 622, 623
chick guano 169, 170, 171
food consumption 518, 520, 523, 524
foraging range 479, 480, 484, 485
organochlorine content of eggs 648, 649
Haloptilis 45
Haloptilus oxycephalus 51
Halozetes 178, 632
Hampton, I. 294, 362
Hantzschia amphioxys var. *maior* 242
Hardangervidda, Norway 205, 587
Harengula jaguana 449
Harmöthoe bervipalpa 360
Harpagifer georgianus 631, 634, 635
Harpagifer marionensis 360
Hawes, I. 245
Hawker Island 536
Haxen, P.G. 637
HCB 652
Heard Island 128, 139, 152, 422, 425, 536
terrestrial food webs 582, 584, 587
Helcomyzidae 607
Helminth parasites 563, 564
Hemiarthrum setulosum 632, 634
Hempel, G. 266
Hempel, I. 304, 308
Hennig, H.F.-K.O. 656
Henry, J. 362
Herbivores
invertebrate 616–617, 618, 680
mammals 583, 618
Hero 324, 555, 557
Herring, Pacific 430
(*Clupea harengus*) 449
Heterorhabdus 45
Heterorhabdus austrinus 51
Heterotrophy 9
in maritime lakes 250, 254
in a mussel-bed 128–134
in nanoplankton and micro-plankton 271, 272, 275, 280–282
sea-ice micro-algae and 79, 87
in a terrestrial community 616
Hewes, C.D. 277
Heywood Lake
decomposition processes 253–259
interaction of soil and lake micro-flora 662–667
nutrient cycling 147, 148, 149, 234–237, 261–264
phytoplankton population 245–251
Himantothallus grandifolius 104–107, 115–121
Hirundinidae 586
Histioteuthidae 551, 552, 553
Histioteuthis 528, 531, 552
Histioteuthis atlantica 531
Histioteuthis dofleini 531
Histioteuthis eltaninae 531
Histioteuthis miranda 531
Holm-Hansen, O. 6, 93, 277
Holothuroids 632, 633
Hope Bay 146
Horgan, I.E. 493
Horlick Mountains 123
Horses 583
Hoshiai, T. 89
Hubbard Brook, U.S.A. 585
Hubold, G. 445
Humic acids 146, 237
Hunt, G.L., Jr. 487
Hunter, S. 534
Hureau, J.C. 421, 626
Hyadesia 178
Hyale 634, 635
Hyale grandicornis 631, 632, 638, 639
Hyale hirtipalma 632
Hyalella azteca 380
Hydrobates pelagicus 471
Hydrogen 144, 229, 259
Hydroids 384–387 *passim* 411, 570
Hydroxylapatite 158, 159, 160, 163, 165
Hydrurga leptonyx (Leopard Seal) 3, 555–559, 673, 674, 679
food consumption 517, 519, 529, 556–557, 558–559
in Ross Sea 499, 504, 505, 507
Hygophum hanseni 570
Hymenoscyphus 601, 604
Hyperia 454
Hyperiella dilata 660
Hyperiid amphipods 454
Hyperoodon ampullatus 569
Hyperoodon planifrons (Southern Bottle-nose Whale) 567, 568–569, 571
Hyphomycetes 599, 602, 603, 604

Ice (*see also* Fast-ice; Pack-ice; Sea-ice)
bacteria 78–83, 677
congelation 81, 84
dry valley lakes 238, 240
frazil 81
maritime lakes 147, 245, 246, 253
nucleation, biologically mediated 79
shelves (*see also* Ross Ice Shelf), *Pleuragramma antarcticum* and 446, 449, 450
and wind dispersal of diaspores 594, 595
zonation in Southern Ocean 267–269
Ice-algae 268–269, 677
autumnal proliferation 89–92
ice-edge phytoplankton bloom and 76
influences of light 78–83, 84–87
Ice-edge
ammonium distribution 93–103
phytoplankton biomass near receding 70–77, 676
whale distribution and 509, 512
Ice-platelet layer 84–87
Idotea baltica 395
Idotea metallica 360
Indian Ocean 23, 47, 607
Indian Ocean Central Water (IOCW) 23, 24, 27
Inland waters (*see also* Freshwater ecosystems; Lakes) 681
Insects
consumption by birds 584, 627
intertidal zone 626, 632, 634
in terrestrial communities 587, 589
International Biological Programme (IBP) 211
Tundra Biome 138, 192, 582
International Whaling Commission (IWC) 366, 566
Intertidal zone 584, 587
food webs 630–636
role of *Durvillaea antarctica* 637–640
Invertebrates (*see also* Zooplankton)
intertidal and subtidal 584, 626, 631–635
terrestrial 583, 680
consumption by birds 584–587, 680
Elephant Seal influences 175, 178
food webs 587–588, 680
Iron (Fe)
in freshwater lakes 234–237
in *Rostkovia magellanica* 141, 213–215, 218, 219
in soil and plants 141
zooplankton 658–661, 677
Isla de los Estados 536
Island effects 38, 44, 676
Isochrysis galbana 373–376
Isopods 45, 387, 389
consumption by small cetaceans 570
fish predation on 422, 423, 424, 426
intertidal and subtidal zones 626, 632, 633, 634
Lithodes murrayi predation on 384, 385, 386, 387
Weddell Seal predation 563
Isotomidae 178

James Ross Island 146, 670
Jamesionelletum coloratae 186
Japetella diaphana 419
Jassa falcata 632, 634, 635, 638, 639
Jasus 553
Jasus lalandei 382
Jasus tristani 360, 551
Jenny Island 206
Jones Sound 85
Jouventin, P. 478
Juncus pusillus 610
Juncus scheuchzerioides 211

K-selection 403, 618
Kattegat 339
Kelp (*see also* Macro-algae; Seaweed) 587
communities 104, 407–413, 633–635, 639
flies 632, 634
fly larvae 584
invertebrates 587
nutrient cycling 637–640
Kerguelen Islands 125, 146, 152, 536, 567, 637
Allotanais hirsutus 378–380
fish 421–429
marine nutrients 22, 23, 25, 26
mussel-bed 128–134
terrestrial communities 582, 583, 584, 587, 602
Kerguelen-Gaussberg Ridge 128, 267
Kerguelenella lateralis 632, 634
'Kerguelenian biotic province' 359

Kerry, E.J. 597
Kevo, Finland 587
Keystone species 412, 413
Kidderia bicolor 639, 640
Kidderia minuta 632, 633, 634
Kidderia oblonga 639
King Edward VII Peninsula 500
King George Island 104–107, 109–113, 156–161, 163–168, 324, 339
King Haakon VII Sea 558
Kita-no-seto Strait 52, 89, 90, 91
Kittiwakes (*Rissa* spp.) 490, 491
Kock, K-H. 437
Kondakovia 527, 528, 678
Kondakovia longimana 415–420, 531
Kottmeier, S. 78
Krefftichthys anderssoni 454, 455–456, 457, 458
Krill, *see Euphausia superba*
Krogh's Curve 402

Labidiaster annulatus 411
Lactate dehydrogenase 256, 257
Laevilacunaria antarctica 579
Laevilitorina caliginosa 634
Lagenorhynchus australis 567, 570
Lagenorhynchus cruciger 567, 569–570
Lagenorhynchus obscurus 567, 569
Lake Bonney 148–150, 238
Lake Bonney-East 239–243
Lake Chad 239–243
Lake Fryxell 238–243
Lake Hoare 239–243
Lake Joyce 239–243
Lake Miers 239–243
Lake Vanda 148, 239–243
Lakes (*see also specific lakes*) 681
  decomposition processes 253–259
  Dry Valley 152, 238–243, 681
  eutrophic 148
  mesotrophic 148
  nutrient cycling 145–151, 234–237, 261–264
  oligotrophic 147–148
  phytoplankton production 245–251
  saline 145–146
  soil and lake micro-flora interactions 626–627, 662–667
Lamellibranchs 45
Laminaria 116
Laminaria digitata 106
Laminaria longicruris 106
Laminaria pallida 106, 107
Laminaria saccharina 118
Laminarinase 333–337
Lampanyctus 568
Lamprey, Southern 475, 529
Lantern Fish, *see Gymnoscopelus nicholsi*
Laridae 491
Larus argentatus 171
Larus delawarensis 171
Larus dominicanus (Kelp Gull)
  foraging range 479, 484, 485, 627
  organochlorines 648, 649, 650, 651
  role in intertidal zone 584, 627, 631, 633, 634, 635
  terrestrial diet 584, 585, 627
Larus marinus 171
Lasaea consanguinea 631, 632, 633, 634
Laws, R.M. 1, 669, 676
Lawson, G.J. 211
Laycock, P.A. 551
Le Jehan, S. 22
Lead (Pb) 7, 659, 660
Leaf decomposition 194–196, 198
  in *Rostkovia magellanica* 141, 142, 211–220 *passim*
  succession of micro-flora 597–604, 680
Lecanora 185
Lecidea 183
Lecidella bullata 180–185
Lepidoptera 178
  larvae 584, 588
Lepidoteuthis grimaldi 531
Leptonychotes weddelli (Weddell Seal) 558
  Dissostichus mawsoni interactions 432, 561–565, 679
  food consumption 268, 517, 519, 527–531 *passim* 553, 628
  organochlorines and heavy metals 652–655
  Pleuragramma consumption 431, 445, 450
  in Ross Sea 499, 504, 505, 506
Leptosomia simplex 115–121
Leptosphaeria 601, 602, 604
Leptosphaeria doliolum 600, 604
Leptosphaeria eustoma 600, 601, 603, 604
Lessonia flavicans 408, 410, 412
Lessonia vadosa 408, 410, 412
Leucophosphite 159, 160, 163, 163–164, 165
Lewis Smith, R.I. 138, 221
Lichens 144, 186, 213, 614, 615, 632, 643
  ion leakage from 144
  weathering of rocks 180–185
Life-cycles, prey 516–532, 679
Life-history strategies 403–404, 578, 618
Light
  dependence of marine phytoplankton photosynthesis 60–68
  growth of sea-ice microbial communities and 78–83
  penetration of sea-ice 78–79, 85, 86, 91, 92
  and photosynthesis in maritime lakes 247, 249
  physiology of sea-ice micro-algae and 84–87
  transmission in dry valley lakes 240, 243
Ligia dilatata 395
Ligia oceania 391
Limacina 45, 450
Limacina helicina 454, 455
Limatula hodgsoni 576
Limnoria antarctica 638, 639
Lindane 652
Linneus 411
Lipids 255
  'compound X' 494, 495, 496
  in diet of seabirds 463, 493–497
  in Krill 335, 336, 526, 576, 679
  marine particulate 331, 332
  storage: by marine invertebrates 576–578
    in notothenioid fish 434
Lipofuscins 2
Lissodelphis peronii 567, 570
Listriomastax litorea 607, 608, 609
Lithodes 426
Lithodes antarctica 386, 411
Lithodes murrayi 381–388
Lithothamnion 630, 632, 633
Litter breakdown 141, 143, 192, 193, 196–198
  succession of micro-flora 597–604
Litterbags 193, 211, 213, 217, 218, 219–220, 598
Littoral ecology (*see also* Intertidal zone) 630
  role of *Durvillaea antarctica* 637–640
  shore types 638
Liverworts 213, 610, 614, 615, 643
Lobodon carcinophagus (Crabeater Seal) 3, 628, 679
  food consumption 268, 517, 519, 520, 524–530 *passim*
  organochlorines and heavy metals 652–655
  predation by Leopard Seals 556, 557–558, 559
  predation on Krill 366, 367, 520, 524, 525–526
  in Ross Sea 499, 504, 505–506, 508
  trends in age at maturity 669–674
Lobster, Rock 382
Loliginidae 551
Loligo 494
Loligo gahi 570
Loligo reynaudi 551, 553
Loxechinus albus 407, 409, 410, 411, 412
Lumbricillus 632
Lutjeharms, J.R.E. 11
Lutra felina 411
Lützow-Holm Bay 52–58, 273
Luxmoore, R.A. 389
Lycopodio magellanicae 186
Lycopodium annotinium 211
Lyngbya 242
Lyngbya martensiana 241, 242
Lyothyrella uva 576

McDonald Island 536
McGrath-Grossi, S. 78
McMurdo Sound 52, 54, 71, 430, 431, 561–565, 575
  heavy metals 656, 657
  sea-ice micro-algae 78–83, 84–87
McQuaid, C.D. 656
Macquarie Cabbage (*Stilbocarpa polaris*) 583, 597–604
Macquarie Island 138, 152, 173, 177, 225, 568
  freshwater lakes 146–147

Macquarie Island
giant petrels 535, 536, 537, 538, 540
leaf and litter micro-flora 597–604
littoral community 633
terrestrial food webs 582, 583, 584, 585, 587
'Macro' elements 6
Macro-algae (*see also* Kelp; Seaweed)
Beagle Channel 407–413
Marion Island 631, 632, 633–635, 639
role in Antarctic coastal waters 104–107, 109–113, 115–121
Macro-molecules 206–207, 255–256
MacRobertson Land 603
Macrobiotus furciger 614
Macrocystis 631, 634
Macrocystis pyrifera 407–413, 633
Macronectes 470
Macronectes giganteus (Southern Giant Petrel) 471, 490, 518, 534–541
foraging range 479, 480, 484, 485, 627
organochlorine content of eggs 648, 649
phospholipids 496
Macronectes halli (Northern Giant Petrel) 471, 518, 534–541
foraging range 479, 484, 485, 627
Macruronus novae-zelandiae 570
'Magellanic biotic province' 359
Magellanica (Neothyris) venosa 409, 410, 411
Magnesium (Mg)
in guano 170
in plants 141
Rostkovia magellanica 213–215, 218, 219
in soil 140, 144
Malate dehydrogenase 256, 257
Mallard 585
Malvinas, *see* Falkland Islands
Mammals 411, 589, 679
age at sexual maturity 669
biomass in Ross Sea 498–514
herbivorous 583, 618
introduced 583–584, 680
in terrestrial food webs 582, 583–584, 589
Manatee (*Trichenus manatus*) 557
Manganese (Mn)
in freshwater lakes 234–237
in *Rostkovia magellanica* 141, 213–215, 219
in sea water 7
in zooplankton 659, 660
Manuring, *see under* Birds *and* Seals
Marchant, H.J. 271
Margarella violacea 409, 410, 412
Marguerite Bay 670
Marie Byrd Land 500
Marine ecosystems 1–2, 266–267, 676–679
indicators 669
interactions with freshwater and terrestrial 626–628
nutrient cycling 6–10
Marion Dufresne 46, 381, 478
Marion Island 3, 30, 285, 422, 680
bryophytes 200–202
cats 620–624
chlorinated hydrocarbons 647–651
freshwater lakes 146
intertidal and subtidal zones 630–636, 637–640
nutrient upwelling 38–45, 46–51, 676
petrels 169–171, 460–464, 536
seals 173–178, 551, 553
soils 186–191
terrestrial food webs 582–588 *passim*
terrestrial nutrient cycling 140–141
Maritime Antarctic 123, 163, 662
Marschall, H.P. 346
Mastigoteuthis 531
Mawson Station 52, 536
Maxwell, J.G.H. 397
Mayzaud, P. 330
Mediterranean Sea 660
Medusae 45, 576
Meganyctiphanes norvegica 339–444
Merluccius capensis (Hake) 494, 496
Merluccius hubbsi 568, 571
Mesonychoteuthis 527, 528
Mesonychoteuthis hamiltoni 418, 531
Mesoplodon 567
Mesoplodon bowdoini 567, 568, 571
Mesoplodon densirostris 567, 568
Mesoplodon grayi 567, 568
Mesoplodon hectori 567, 568, 571
Mesoplodon layardii 567, 568, 571
Metabolic rates
benthic organisms 576, 578
cold adaptation and 397–404, 578
fur seals 544–545
of large birds 476
and life-history strategies 403–404
terrestrial invertebrate predators 618
Metals
heavy, in seals and birds 652, 653–655
in Southern Ocean 30–36
in zooplankton 656–661
Meteor 60, 61, 308, 339
Methanogenesis 144, 147, 229–232, 259, 680
Methanol 230, 232
Methionine 144, 230, 232
Methylated amines 144
Metridia curticauda, feeding behaviour 311–322
Metridia gerlachei 45, 47, 454, 455
feeding behaviour 311–322
vertical distribution 309, 310
Mica-schist 180, 181–183, 184
'Micro' elements 6
Micro-flagellates (μ-flagellates) 574
grazing by Krill and copepods 312–322
heterotrophic 79, 271, 272, 275, 574
Micro-organisms 3
guano decomposition 158
moss-peat decomposition 204–210
Microbes, endolithic 3
Microbial organisms, marine 8, 9
Microcalanus 45
Microcoleus paludosus var. *acuminatus* 242
Microflora
interactions between soil and lake 627–628, 662–667
in moss-peat 207–209
succession in leaf and litter decomposition 597–604
in a terrestrial community 615, 616, 618
wind transport 593
Micropogonius furnieri 570
Microscolex crozetensis 607, 608, 609, 613
Microscolex dubius 607
Microscolex enzenspergeri
var. *alba* 607, 610
var. *bicolor* 607, 608, 609, 613
var. *nigra* 607, 608–610
var. *typique* 607, 610
Microscolex frenoti 607, 610, 613
Microscolex kerguelenarum 588
Microscolex luykeni
var. *albinique* 607
var. *halophila* 607, 610
var. *typique* 607
Microscolex phosphoreus 607
Microthyrium fuegianum 219
Miller, D.G.M. 355, 362
Mink (*Mustela vison*) 584
Minyulite 159, 160, 164, 165, 166, 167
Miquel, J.C. 381
Mires 140, 141, 588
acetylene reduction 186–191
moss production 200, 202
Rostkovia magellanica 141, 142, 192–199, 212
Mirounga leonina (Southern Elephant Seal) 3, 553
diving habits 549
food consumption 517, 519, 526–531 *passim* 628
moulting areas 174–177, 607
in Ross Sea 499, 505, 507
terrestrial influences 173–178
wallows 186, 188, 190, 221, 225, 608
Mites 587, 589, 614, 632
Moe, R. 78
Molluscs 45, 454, 455, 587, 633, 635
Molodezhnaya 151
Monoacylglycerides 494, 495
Monodus coccomyxa 242
Monosaccharides 109, 110, 112
Monteiro, P.M.S. 30
Monterey Bay, California 660
Montio fontanae 175
Moor House, U.K. 586
Moorland 587
Moraxella 125
Moroteuthis 527, 528, 531, 552
Moroteuthis knipovitchi 415–420, 531, 552, 553
Moroteuthis robsoni 531
Mortierella 204, 209, 210, 603, 604
Mortierella gamsii 600, 601, 602, 603, 604

Morus capensis 471
Moss Lake 148, 245
decomposition processes 253–259
microflora 664, 665, 666
Moss-peat
Cryptopygus antarcticus consumption 616, 617
decomposition 204–210, 680
microflora of lake catchments 663–666
Mosses (bryophytes) 610, 614, 643
cyanobacteria 140, 141, 143, 144
epiphytic algae 615
methanogenesis and anaerobic microbial activity 144, 229–232
nitrogen cycling 144–145
primary consumers 615, 616, 617
production 200–202
Moufflon, Corsican (*Ovis ammon musiman*) 583
Mougin, J.L. 478
Moulting
of euphausiids 339–344
(exuvia) energy costs 391, 392, 394
Mouse, House, *see Mus musculus*
Mucor 204, 209–210
Mucor hiemalis 604
Mucor mucedo 604
Mucorales 601, 602
Mules 583
Munida 411
Munida subrugosa 409, 410, 411
Murres (*Uria* spp.) 490
Mus musculus (House Mouse) 583, 620, 621, 622, 623
Muscles in notothenioid fish 434
Mussel (*Mytilus edulis*) 401
Mussel-beds 128–134
Mustela vison (Mink) 584
Mycosphaerella 599–604 *passim*
Mycosphaerella tassiana 599, 600, 603, 604
Myctophid fish 268, 425, 426, 427, 449, 454, 455, 457, 529
Myliobatis (Sting-ray) 569
Myrcha, A. 156
Myriophyllum elatinoides 146
Mysids 431
consumption by giant petrels 538
consumption by small cetaceans 570, 571
fish predation on 424, 426, 427, 431
Weddell Seal predation 563, 564
Mytilids 408
Mytilus chilensis 411
Mytilus edulis 401
Mytilus edulis kerguelensis 128

N:P ratio 24, 25, 28, 40, 247, 250
Nacella concinna 397, 400, 401, 402
Nacella delesserti 630–635 *passim* 638, 639, 640
Nacella magellanica 409, 410, 411
Nacella mytilina 409, 410, 412
Nanoplankton 9, 271, 574, 677
food webs and 277–282
N assimilation 8
Nanorchestes antarcticus 615
Narwhal Island 85
Nauticaris marionis 360, 633, 634
Natotodarus sloanii 569, 570
Navicula cryptocephala 242
Navicula fragilarioides 242
Navicula gausii 242
Navicula glaciei 393
Navicula litos 242
Navicula muticopsis 242
Navicula seminulum 242
Navicula shackletoni 242
Nelson, D.M. 70
Nematocarcinus 453, 454
Nematodes 570, 571, 587, 614
in dry valley lakes 241, 242
intertidal and subtidal zones 632
predation on 384–387 *passim* 584
Nematoscellis atlantica 360
Nemastoscellis megalops 45
Nemichthys scolopaceus 358, 359, 360
Nemoto, T. 415
Neopagetopsis ionah 434
Nereis callaoana 360
Net growth efficiency 394–395
New Harbor 79
New Zealand 536, 553, 567, 568, 569, 637, 648
Nickel (Ni) 7, 659, 660
Nidamental gland 420
Niesslia exosporioides 604
Nitrate (*see also* N:P ratio)
in freshwater lakes 146
from guano decomposition 159
reduction 125, 141, 257, 258
in sea water 7–8, 22–28, 61, 98–99
in Admiralty Bay 110, 111, 112, 113
at oceanic frontal systems 12, 13, 14, 15
copper levels and 659–660
near Prince Edward Islands 40, 41, 42, 48–49
under ice 53, 54
Nitrifying bacteria 9, 144
Nitrite 23, 24, 25, 26, 27
in Admiralty Bay 111, 112
near Prince Edward Islands 40, 41, 42
under ice 53, 54
Nitrogen
deposition by birds 140–141, 158, 159, 170, 627
in macro-algal debris 116–117, 118, 120, 121
radiolabelled isotopes 8, 9
Rostkovia magellanica 141, 213–217, 218, 219–220
Nitrogen cycling 124
in Antarctic waters 8, 28
dry valley lakes 240, 241
in freshwater lakes 146–151 *passim* 235, 236, 237, 248, 249, 250, 261, 263
in moss communities 144–145
in soils 140–141, 143
terrestrial 141, 145
Nitrogen fixation 680
in inland waters 125, 147, 150, 243
in terrestrial ecosystems 141, 143, 144, 145, 186, 191
Nitzschia 62, 314, 319, 352
Nitzschia acicularis 380
Nitzschia angustata 242
Nitzschia cyclindrus 87
Nitzschia frustulum 242
Nitzschia stellata 80, 81
Nitzschia sublinearis 242
North Atlantic Deep Water 23, 31
North Polar Basin 123
Northwind 498
Nostoc commune 145, 147, 150, 242
Nostoc longistaffi 242
Nostoc muscorum 144
Notocrangon antarcticus 397, 398, 400, 401, 402
Notolepsis coatsi 452, 453, 455, 457, 458
Notopogon notopogon 360
Notothenia 410, 434, 529, 548
Notothenia angustata 432, 433
Notothenia angustifrons 438
Notothenia coriiceps 631, 633, 634, 635
Notothenia gibberifrons 437–443, 546, 549
Notothenia larseni 421, 422, 423–424, 425, 437–442, 529
Notothenia macrocephala 360
Notothenia magellanica 631, 634, 635
Notothenia marionensis 359, 360
Notothenia neglecta 393–394, 438
Notothenia nudifrons 529
Notothenia nybelini 437
see also N. larseni
Notothenia rossii 437–443, 529, 530, 546, 549
Notothenia rossii marmorata 422, 423, 434
Notothenia rossii rossii 421–423, 425, 429
Notothenia squamifrons 421, 423, 424, 425, 429
Nototheniid fish 457
diet 421–429
larvae 448–449
in *Macrocystis* belts 410, 411
Weddell Seal predation 563, 564
Notothenioid fish 437–443
benthic 430, 432–433
biomass estimates 438–439, 442
evolution of buoyancy in 430–435
gastric evacuation rates 441, 443
Krill consumption 437, 440–442, 443
pelagic 433, 434, 445
Nutrient cycling 682
in dry valley lakes 238–243
Durvillaea antarctica production and 637–640
in freshwater ecosystems 3, 139, 145–151, 681
in freshwater lakes 234–237
future development of research 151–153
marine 6, 7–10
in Admiralty Bay 109–113
in maritime lakes 248–250, 251, 261–264

Nutrient cycling
  role of bacteria 123–126
  role of penguins 156–161
  in terrestrial ecosystems 3, 138–145, 192–199, 211–220
Nutrients
  from Elephant Seals 175, 177
  marine 6–7, 22–28
    from benthic macro-algae 109–113
    near Prince Edward Islands 40, 41, 42, 48–49
    and oceanic frontal systems 12–16
    phytoplankton relationships 61–62
    trace metals and 31–36
    under sea-ice 53, 54
    upwelling 7, 17–18, 38, 50–51, 76, 470
  mires 188
  Poa flabellata growth and 221–228
  in terretrial ecosystems 140, 141, 142, 196
Nyctiphanes australis 346

Oates Piedmont Glacier 71
Oats (*Avena sativa*) 225
Obelia geniculata 360
Oceanic frontal systems 11–12, 676
  and biological enhancement 11–19
  bird distribution and 467, 482, 485, 679
  horizontal shearing zone 19
  metal concentrations 656–661
  satellite remote sensing 16–18
  seal distribution and 514
  vertical convergence zone 19
  and whale distribution 509, 512, 514
Oceanites oceanicus (Wilson's Stormpetrel) 470, 471, 517
  food consumption 519
  foraging range 480, 481, 483, 484, 627
Oceanodroma leucorhoa 464, 471
Ochromonas 242
Ochromonas minuscula 242
Octopods 527, 546, 549
Octopoteuthidae 552, 553
Octopoteuthis 568
Octopoteuthis deletron 568
Octopoteuthis rugosa 531
Octopus 411
Octopus 426, 429, 528, 531
Odontaster validus 411, 576
Ohtsuka, H. 52
Oikopleura 45
Oithona 45
Oithona frigida 47
Okiyama, M. 415
Oligochaetes 178, 587, 606–613, 627, 632
Ommastrephes bartrami 552, 553
Ommastrephidae 552, 553
Ommatophora rossi (Ross Seal) 499, 504, 505, 506–507, 509, 558
Oncaea 45
Ongul Islands 151
Ongul Strait 52
Onychoteuthidae 552, 553
Ophiuroidea 384–387 *passim* 424, 633
Oradera ocellata 360
Orchomene (Orchomenella) 431
Orchomene plebs 660
Orcinus orca (Killer Whale) 432, 450, 567, 571
  feeding habits 569
  in Ross Sea 499, 504, 505, 512–513
Organochlorines 647–651, 652–655
Oribatei 175, 178
Oribatids 626
Ornithogenic phosphates 160–161, 163–168
Ornithogenic soils 156, 158, 159, 160, 165
Orren, M.J. 30
Oryctolagus cuniculus (rabbit) 583, 584
Oscillatoria 242
Oscillatoria limosa 242
Oscillatoria priestleyi 242
Oscillatoria retzii 242
Ostracods 43, 44, 45, 58, 454
Otaria flavescens 411
Ovarian *corpora* 670
Ovis ammon musiman (Corsican moufflon) 583
Ovis aries 583
Oxygen
  and ammonia excretion by Krill 328–329
  consumption: in marine invertebrates 391, 392, 397–404, 578
    moss peat decomposition 205–206
    petrel chicks 461
  dissolved: Krill distribution and 300–301, 303
    off Prince Edward Islands 48
    under sea-ice 53, 54
  in maritime lakes 147, 253–254
  supersaturation in dry valley lakes 240–241, 243

Pachyptila (prions) 360, 467, 470, 471, 538, 627
Pachyptila belcheri (Thinbilled Prion) 479, 484, 485
Pachyptila desolata (Dove Prion) 490, 644
  food consumption 517, 518, 520, 523, 524, 529, 548, 627
  foraging range 479, 482, 483, 484, 485
Pachyptila salvini (Salvin's Prion)
  cat predation on 620, 621, 622, 623
  foraging range 479, 482, 483, 484, 485
Pachyptila turtur (Fairy Prion) 484, 485, 648, 649
Pachyptila vittata (Broadbilled Prion) 463, 648, 649
Pacific Ocean 660
Pacific Sub-Arctic zone 470
Pack-ice
  Adélie Penguins 558
  ammonium and phytoplankton biomass under 95–99, 102
  concentration and distribution of mammals 504, 506–509 *passim* 512, 514
  Leopard Seals 555, 557
  permanent 268–269
  in Ross Sea 501–502
  seasonal 268
Pagetopsis 448–449
Pagetopsis macropterus 431, 453, 455
Pagodroma nivea (Snow Petrel) 70, 471, 518, 529, 538
Pagothenia borchgrevinki 431, 432
Pagrus sedecim 571
Pagurida 384, 385, 386, 387
Pagurus forceps 409, 410
Palaemonetes vulgaris 401
Palinurichthys antarcticus 358, 359, 360
Palirheous eatoni 607
Palmer Peninsula 85
Palmer Station 278, 556, 557, 558
Palmisano, A.C. 78, 84
Panagis, K. 173
Panthera tigris (Tiger) 557
Paractora dreuxi 607, 608, 609, 612, 613
Paraeuthria plumbea 409, 410
Parakeet (*Cyanoramphus novaezelandiae erythrotis*) 584
Paralabidocera antarctica 58
Paraliparis devriesi 431
Paralithodes camtschatica 381, 386–387
Paralomis aculeata 387, 388
Paralomis granulosa 409, 410, 411
Parasites 628
  helminth 563, 564
Parisotoma octooculata 615, 617
Parker, B.C. 238
Parker, L. 38
Parodiochloa flabellata, see Poa flabellata
Particles, wind transport of 592–595
Particulate feeders (*see also* Filter feeders) 575, 576, 578
Particulate organic carbon (POC) 81, 110, 111, 112, 113
Particulate organic matter 9, 109, 273–275, 331–333
Parvicorbicula 275
Parvicorbicula circularis 273
Parvicorbicula socialis 273
Patagonotothen larseni 546
Patella aspersa 401
Patella vulgata 392, 401
Patigera kerguelensis 422
Patiriella fimbriata 411
Peat (*see also* Moss-peat) 615
Pectins 255
Pelagadroma marina 471
Pelagobia longicirrata 453
Pelecanoides (divingpetrels) 471, 523, 524, 621, 627
Pelecanoides georgicus (South Georgia Divingpetrel) 607, 608
  food consumption 519, 520
  foraging range 479, 484, 485
Pelecanoides urinatrix (Common Divingpetrel) 622, 644
  food consumption 517, 519, 520
  foraging range 479, 480, 484, 485
Pelecanoididae 620
Pelecanus occidentalis 648
Pelecypods 45, 639, 640

Pelican, Brown 648
Peltarion spinulosum 568
Penaeus merguiensius 403
Penguins 2–3, 70, 221, 269, 622
Adélie, *see Pygoscelis adeliae*
Chinstrap, *see Pygoscelis antarctica*
chlorinated hydrocarbons 648–651
competition with fur seals 548–549
Diptera colonization of a carcass 610–612
Emperor (*Aptenodytes forsteri*) 268, 445, 503, 652–655
feathers 161, 546, 553
food consumption 360, 366–367
Gentoo, *see Pygoscelis papua*
giant petrel consumption of 535–541
Jackass (*Spheniscus demersus*) 161, 494, 495, 648, 650
King, *see Aptenodytes patagonicus*
Macaroni, *see Eudyptes chrysolophus*
ornithogenic soils 163–168
Rockhopper, *see Eudyptes chrysocome*
role in nutrient cycling 144, 156–161, 627
seal predation on 546, 553, 555, 556–557, 558, 559
squid predation 367
Penicillium 599, 600, 601, 602, 604
Penicillium brevi-compactum 604
Penicillium corylophilum 604
Penicillium cyclopium 601, 604
Penicillium decumbens 604
Penicillium frequentans 604
Penicillium granulatum 604
Penicillium restrictum 604
Perimylopid beetles 644, 645
Perisinusoidal cells 434
Peritrich 242
Perma-frost 588
Perumytilus purpuratus 409, 410, 411
Peruvian upwelling area 470
Petrels 221
Antarctic (*Thalassoica antarctica*) 445, 471, 503
Atlantic 471
Blue, *see Halobaena caerulea*
burrowing: impact of fur seals on 644, 645
soil enrichment by 141, 169–171, 607, 608
diving, *see Pelecanoides* spp.
giant, *see Macronectes* spp.
Greatwinged, *see Pterodroma macroptera*
Grey, *see Procellaria cinerea*
Kerguelen, *see Pterodroma brevirostris*
Pintado, *see Daption capense*
predation by House Cats 583, 620
Snow (*Pagodroma nivea*) 70, 471, 518, 529, 538
Softplumaged, *see Pterodroma mollis*
Whitechinned, *see Procellaria aequinoctialis*
Whiteheaded, *see Pterodroma lessoni*
pH
inland waters 146, 237, 664
methanogenesis and 230, 231–232
in *Rostkovia magellanica* bogs 211–213, 219
run-off water from penguin rookeries 159
of soil water 610, 613, 666
under fast-ice 53, 54
Phaeocystis 62
Phaeocystis pouchetii 272, 275
Phaeodactylum tricornutum 373–376
Phaeopigments 90, 286, 292
Phalacrocorax albiventer 621, 627
Phalacrocorax atriceps (Imperial Cormorant) 635
competition with fur seals 548, 549
food consumption 519, 529, 530, 634
foraging range 479, 484, 485, 631
organochlorines 648, 649
role in shallow subtidal food webs 631, 634, 635
Phalacrocorax capensis 650
Phalacrocorax carbo 650
Phalacrocorax coronatus 650
Phalacrocorax neglectus 650
Phalaropus fulicarus (Grey Phalarope) 471
Phasmatopsis cymoctypus 531
Philine gibba 576
Phleum alpinum 225, 643
Phocoena dioptrica 567, 571
Phocoena spinipinnis 567, 571
Phocoenidae 571
Phoebetria fusca (Sooty Albatross) 470, 471
foraging range 479, 480, 483, 484, 485
organochlorine content of eggs 649, 650
Phoebetria palpebrata (Lightmantled Sooty Albatross) 471
food consumption 518, 527, 528, 531
foraging range 479, 484, 485
Pholidoeuthis boschmai 416–420
Phoma 599, 600, 601, 602, 603, 604
Phoma exiqua 601, 602, 604
Phoma herbarum 604
Phormidium frigidum 241, 242
Phormium tenax 604
Phoronima sedentaria 360
Phosphatase 158
Phosphate
in freshwater lakes 146, 151
ornithogenic 160–161, 163–168, 170
in sea water 8, 12, 22–28, 125
in Admiralty Bay 110, 111, 112, 113
near Prince Edward Islands 40, 41, 42, 48, 49
trace metals and 31, 33, 34–36
under fast-ice 53, 54
Phosphatized rock 159, 163, 165
Phospholipids 494, 495, 497
Phosphorus
anomalies in Antarctic waters 23–28
dissolved organic, *see* Dissolved organic phosphorus
in *Rostkovia magellanica* 141, 142, 213–217, 218, 219–220
Phosphorus cycling
dry valley lakes 149–150, 240, 241, 243
in maritime lakes 146, 147, 148
iron and manganese and 235, 236, 237
phytoplankton growth and 147, 246, 248–249, 250, 261, 263, 264
in soils 140–141, 143, 144
Photosynthesis
by sea-ice micro-algae 85, 86–87, 574
in maritime lakes 246–247, 248, 249, 251, 263, 264
in sea water, light-dependence of 60–68
Phototrophic bacteria 125
Phreodrilidae 607
Phreodrilus crozetensis 606, 607, 608, 609, 612–613
Phronimella elongata 45
Phycolimnoria 412
Physeter catodon (Sperm Whale) 367, 553
size of squid consumed by 415, 416, 418–419, 678
squid consumption in Scotia Sea 526, 527, 528, 531
Physiological studies 576–579
Phytoplankton (*see also* Algae; Chlorophyll-*a*; Primary production) 1–2, 573–574, 677
ammonia uptake 8, 327–328
extra-cellular products 150, 273–275
freshwater 146
dry valley lakes 241, 242
maritime lakes 147–148, 245–251, 261–264, 663, 664
lipids 494, 496
marine: at oceanic fronts 14–16
bacteria and 124–125
copepod feeding behaviour and 311–322
food webs and 277–282
Krill and 284–292, 311–322, 330–337, 362–370, 372–377
light-dependence of photosynthesis 60–68
near Prince Edward Islands 40–43, 50
near receding ice-edge 70–77, 93, 95–99, 101–102, 268, 574
nutrient upwelling and 7
spatial variability 284–292
Picton Channel 407, 408
Pietr, S.J. 156
Pigeon, Cape, *see Daption capense*
Pigs 583
Pinnipeds 569
Pintail
Kerguelen 584, 585
South Georgia (*Anas georgicus*) 583, 584, 644, 645
Pipit, Antarctic/South Georgia (*Anthus antarcticus*) 583, 584, 585, 586, 644, 645
Placentae 584

Planarians 632
Plankton (*see also* Bacterio-plankton; Nanoplankton; Phytoplankton; Zooplankton)
in dry valley lakes 241, 242
-eating birds, pelagic distribution 466, 467, 469, 470–471
ichthyoplankton 445–446, 449
macro-plankton 355–360
micro-plankton 8, 279–282, 677
pico-plankton 281, 282
Plants (*see also* Leaf decomposition; Litter breakdown; Vegetation)
dead material 141, 192, 194–196, 211
ion leakage from 144
matter eaten by birds 584, 585
resistance to grazing 589
Platynereis australis 631, 632, 634, 635
Pleospora 604
Pleospora graminearum 604
Pleospora lutea 604
Pleospora vagans 604
Pleuragramma antarcticum 268, 452, 678
buoyancy 431, 432, 433, 434
early life-history 445–450
eye-size 448
food consumption 438, 455, 456, 457, 458
Weddell Seal predation 562, 563–565
Pleurasiga 273, 275
Pleurasiga orculaeformis 273
Pleurasiga reynoldsii 273
Pleurasiga sphyrelata 273
Pleuromamma robusta 45
Pleurophyllum hookeri 597–604
Poa annua 175, 177, 583
Poa cookii 141, 174, 175, 610
Poa flabellata 143, 583, 597
growth and production 221–228
impact of fur seals 641–645
litter breakdown 141, 192–199
Poa foliosa 225–226, 597–604
Point Demay 157
Point Llano 157, 160–161
Point Patelnia 157
Point Thomas 109, 110, 157, 160
Point Uchatka 157
Pointe Geologie 536
Polar Star 278
Polarsirkel 445, 652
Pollutants 647–651, 652–655
Polychaetes 45, 390
fish predation on 450, 453, 454, 455
Lithodes murrayi consumption of 384, 385, 386, 387
role in coastal ecosystems 626, 632, 633, 634
Polychlorinated biphenyls (PCBs) 647–651, 653
Polyplacophora 384, 385, 386, 387
Polypus variolatus 569
Polysaccharidaes 109, 110, 112
particulate 331–332
Polytrichum alpestre 144, 192–199, 204–210, 213, 221, 228
Ponds 146, 151
Pontogeneiella brevicornis 634
Porania antarctica 409, 410, 411
Porgy (*Pagrus sedecim*) 571
Porifera 384, 385, 386, 387, 409
Porphyra 632, 634
Porphyrosiphon notarisii 242
Porpoise
Burmeister's (*Phocoena spinipinnis*) 567, 571
Spectacled 567, 571
Posidonia oceanica 118
Possession Island 587, 606–613
Potamogeton 146
Potassium (K)
in guano 170
in plants 140–141, 143
Poa flabellata growth and 222
in *Rostkovia magellanica* 213–217, 218, 219, 220
in soils 141, 144
Prawns, Banana 403
Predator-prey relationships 516–517, 525–526, 530, 633–635
Predators
estimates of food consumption 679
in *Macrocystis pyrifera* belt communities 407–413
marine vertebrate 574–575, 678–679
terrestrial invertebrate 614, 617, 618
Prey life-cycles 516–532, 679
Priapulido 409, 410
Pribilof Islands, Bering Sea 487–491
Primary consumers, terrestrial 614, 615
Primary production
coastal, role of *Durvillaea antarctica* 637–640
freshwater 148, 150
marine 23, 124, 266, 267, 277, 573–574
consumed by birds 469, 470
copper distribution and 33
flux to benthos 575
near Prince Edward Islands 40–43, 50
in permanent pack-ice zone 268
sea-ice microbial communities 81, 82–83
and sea-surface fronts 11–19
seabird distribution and 482, 485
terrestrial 140, 614–615, 616
Primno 660
Primno macropa 45, 453, 454
Prince, P.A. 473, 516
Prince Edward Island 38–45, 46, 536, 582, 584, 586, 635
Prince Edward Islands 38–45, 46–51, 355–360
Princess Elizabeth Land 145
Pringlea antiscorbutica 583
Pringleophaga marioni 178
Prions, *see Pachyptila* spp.
Procellaria aequinoctialis (Whitechinned Petrel) 471, 607, 644
cat predation on 621, 622, 623
chick energy balance 460–464
chick guano 169, 170, 171
energy requirements 470
food consumption 519, 520, 526, 527, 529
foraging range 479, 481, 482, 483, 484, 485
lipid content of stomach 494, 495, 496
organochlorine content of eggs 648, 649, 651
Procellaria cinerea (Grey Petrel) 471, 624
chick guano 169, 170, 171
foraging range 479, 480, 484, 485
Procellariidae 169–171, 620
Procellariiformes 491
Production
efficiency 394, 616
energy costs of 389, 390, 394, 395
Production/biomass (P/B) ratios 366, 395, 579
Prosobranchia 384, 385, 386, 387
Proteases 116, 118, 120, 121, 158, 256, 257
Proteinase 255
Proteins
content of Krill 335, 336
particulate 331, 332
Protomyctophum anderssoni 457
see also Krefftichthys anderssoni
Protoperidinium 314
Protozoa
freshwater 3, 241
marine 45, 79, 280–282, 574
terrestrial 587, 614, 616
Prudhoe Bay, Alaska 585, 586
Prydz Bay 267, 271–275, 450, 452, 458
Pseudechinus magellanicus 409, 410, 411, 412
Pseudochaenichthys 529
Pseudochaenichthys georgianus 434, 437–443, 546
Pseudocnus laevigatus 632, 633
Pseudocyphellaria freycinetii 213
Pseudomonas 125, 126, 665, 666
Psychrophile bacteria 123–124
Psychroteuthis 528, 531
Psychroteuthis glacialis 531
Pterodroma brevirostris (Kerguelen Petrel) 471
cat predation on 621, 622, 623
foraging range 479, 480, 482, 484, 485
organochlorine content of eggs 648, 649, 650
Pterodroma incerta 471
Pterodroma lessonii (Whiteheaded Petrel) 471, 479, 482, 484, 485, 538
Pterodroma macroptera (Greatwinged Petrel) 471, 648, 649
cat predation on 621, 622, 623, 624
chick guano 169, 170, 171
foraging range 479, 480, 481, 482, 484, 485
lipids in stomach contents 494, 495, 496
Pterodroma mollis (Softplumaged Petrel) 471, 648, 649
cat predation on 621, 622, 623
chick guano 169, 170, 171
foraging range 479, 481, 484, 485
Pteropods 45, 100–101, 368, 456

Puffinus 490
Puffinus assimilis 471
Puffinus gravis (Great Shearwater) 471, 648, 650
Puffinus griseus 471
Puffinus puffinus 471
Puffinus tenuirostris 489
Pycnogonids 411, 633
Pygoscelis adeliae (Adélie Penguin) 2, 268, 445, 490
food consumption 518, 520, 521–523, 524
Krill consumption 366–367, 521–523
organochlorines and heavy metals 652–655
predation by seals 558, 559
rookeries and nutrient cycling 148, 156, 158, 160, 161
in Ross Sea 499, 503, 505
Pygoscelis antarctica (Chinstrap Penguin) 490, 548
food consumption 518, 520, 521–523, 524, 526, 530
rookeries and nutrient cycling 156, 157, 158, 161
Pygoscelis papua (Gentoo Penguin) 445, 485, 621
carcass colonization by Diptera 610
competition with fur seals 548
food consumption 518, 520, 521–523, 524, 529, 530
rookeries and nutrient cycling 156, 157, 161, 607, 608

Quartzite 180
Queen Maud Land 285
Queen Maud Rise 267
Quetin, L.B. 372

R- and K-selection 403, 618
Rabbits 583, 584
Radiolaria 45
Rainwater 144, 146
Rallus philipensis (Banded Rail) 584
Ralph, R. 397
Rama 632
Rangifer tarandus (Reindeer) 141, 228, 583
Ranunculetosum biternatis 177, 186
Rats 585
Rattus norvegicus (Brown Rat) 584, 644
Rattus rattus (Black Rat) 583
Redox potential 230, 231
Redpoll, Common 584, 585
Refuge Islands 204, 206
Regurgitates 607, 608
Reichardt, W. 104, 115
Reindeer (*Rangifer tarandus*) 141, 228, 583
Reproduction, energy costs of 391
Respiration
energy costs of 390, 391, 392, 393, 394, 395
peat 205–206
Rhincalanus gigas 45, 360, 453
distribution 47, 309, 310
feeding behaviour 311–322
Rhincalanus masutus 45
Rhizocarpum geographicum 185
Rhizophila dearboni 402
Rhizosolenia 62, 313, 316
Rhizosolenia curvata 40
"Rhodesiopsis gelatinosa" 601, 602, 604
Rhodomonas palustris 261, 263
Rhodymenia 632, 634
Rickett, L.H. 656
Ricketts, C. 516
Rissa 490, 491
Rocks
micro-organisms 145
phosphatized 159, 163, 165
weathering by lichens 180–185
Rodents 584, 589
Romanchella 632
Ronne ice shelf 446
Roots 584, 585
Ross, M.J. 561
Ross, R.M. 372
Ross Bank 102
Ross Ice Shelf 54, 500–501, 594
ammonium distribution 93–103
phytoplankton biomass 71, 75, 76, 79
Project (RISP) 2
Ross Island 94
Ross Sea 54, 267, 268, 368
ammonium flux 93–103, 324, 328
biomass of birds and mammals 498–514
fish 431, 445, 449, 452
metal concentrations 658
phytoplankton biomass 70–77
sea-ice microbial community 78–83
Rostkovia magellanica 141–142, 143, 192–199, 211–220, 221
Rothera Point 204, 206, 207, 208, 209
Rotifers 241, 242, 574, 587, 614
Roux, J.P. 478
RSA 46
Russell, S. 200
Rust (fungal) 600, 604

S A Agulhas 13, 14, 38, 284–285, 290, 291, 292, 294, 295, 299, 362, 363, 466
Saepicula leadbeateri 273
Sagitta 43, 44
Sagitta gazellae 45, 360, 417, 660
Sagitta lyra 45
Sagitta maxima 45, 359, 360
Sagitta planktonis 360
Sagitta tasmanica
St. George Island, Bering Sea 487–491
St. Lawrence Estuary, Canada 170–171
St. Lawrence Island, Bering Sea 487–491
St. Matthew Island, Bering Sea 487–491
St. Paul Island, Bering Sea 487–491
Sakshaug, E. 277
Saldanha Bay 647, 648
Salinity 657, 658
dry valley lakes 240, 243
near Ross Ice Shelf 76, 95, 96
off Prince Edward Islands 39, 40, 48
under sea-ice 53, 54, 55, 84, 87, 677
Salpa 45
Salpa fusiformis 360
Salpa maxima 660
Salpa thompsoni 360, 660
Salps 268, 368, 574, 575
fish predation on 422, 423, 424, 426, 429
Salt-spray, *see* Sea-spray
Saprophytes 599–603, 680
Saprotrophs 614, 616
Sardine, Scaled (*Harengula jaguana*) 449
Satellite remote sensing 1, 16–18
Scaphocalanus magnus 45
Scavengers 648
Schirmacher Oasis 151
Schizothrix antarctica 242
Schizothyrioma stilbocarpae 599, 604
Schnack, S.B. 311
Schneider, R. 652
Schramm, W. 109
Scientific Committee on Antarctic Research (SCAR) 1, 266
Scientific Committee on Oceanic Research (SCOR) 415
Scina latifrons 45
Scina rattrayi kielhacki 45
Scolecithricella glacialis 45, 47
Scolopacidae 586
Scopelogadus 568
Scotia Arc 204, 204–210, 267, 389, 437, 670
Scotia Ridge 279, 280
Scotia Sea
birds 489, 491
fish 427–428, 452, 458
Krill 270, 304–306, 308–310, 324, 325, 343
phytoplankton 60, 62, 277–278, 280
predator-prey relationships 516–532
Sea of Japan 660
Sea-ice (*see also* Fast-ice; Ice; Pack-ice)
algae 84–87, 89–92, 573, 574, 677
ammonium content 102
Antarctic Cod under 561, 564
browsers under 576
light penetration 78–79, 85, 86, 91, 92
microbial communities (SIMCOs) 78–83, 84
Sea-spray
and aerosol creation 594–595
inland waters and 146
and splash zone 630, 631, 632
terrestrial invertebrates and 588
vegetation and 174–175, 177, 202
Sea-urchins, *see* Echinoidea
Seals 3, 70, 679
Antarctic Fur, *see Arctocephalus gazella*
biomass in Ross Sea 498–502, 503–508, 509, 510, 514
Cape Fur (*Arctocephalus pusillus*) 553
Crabeater, *see Lobodon carcinophagus*
in diet of giant petrels 535–541
Killer Whale predation on 569
Krill predation 366
Leopard, *see Hydrurga leptonyx*
manuring by 141, 146, 147, 221, 606, 607, 628

Seals Northern Fur (*Callorhinus ursinus*) 3, 544, 548, 549
 organochlorines and heavy metals 652–655
 predator-prey relationships 516–532
 Ross, *see Ommatophora rossi*
 Southern Elephant, *see Mirounga leonina*
 squid, predation on 415
 Sub-Antarctic Fur (*Arctocephalus tropicalis*) 174, 551–553
 Weddell, *see Leptonychotes weddelli*
Seaweed (*see also* Macro-algae; Kelp) 606, 607, 631
 decaying 608, 612–613
Secondary consumers, terrestrial 614, 615
Secondary production
 marine 124
 terrestrial 614–615
Sediments
 freshwater 255, 256
 marine 9, 125, 575
 mussel-bed 128–134
Seeds 584–587 *passim*
Seepage slope 221
Semicossiphus pulchrum 410
Sepia 553
Septoria 604
Serolis 422
Serolis cornuta 576
Serolis pagenstecheri 397, 399–400, 401, 402
Serolis polita 389–396, 576, 578, 579
Serpula vermicularis 358, 360
Serpulids 409
Serranids 568
Sexual maturity in mammals 669–674
Seymour Island 670
Shackletonia 632, 634, 635
Shag Island 157
Shearwaters 490
 Cory's 471
 Great 471
 Great (*Puffinus gravis*) 471, 648, 650
 Little 471
 Manx 471
 Shorttailed 488–489
 Sooty 471
Sheathbills 585, 586
 Lesser, *see Chionis minor*
 Wattled 584
Sheep 583
Shore types 638
Si:N ratio 24, 28
Siegfried, W.R. 647
Sigma-t 39–42 *passim* 53, 96, 100, 101
Signy Island 3, 138, 152, 680
 coastal waters 124, 125, 389
 giant petrels 535, 536, 537
 Krill taken by penguins 521, 522
 lakes 146, 147–148, 234–237, 245–251, 253–259, 261–264, 662–667, 681
 seals 527, 549, 553
 terrestrial studies 3, 144–145, 180, 204–210, 229–232, 582, 586, 614–618, 680
Silicate/silica (*see also* Si:N ratio)
 in freshwater lakes 235, 236, 237
 marine 8–9, 22–28, 677
  near Prince Edward Islands 40, 41, 42, 49
  oceanic frontal systems and 12, 13, 15
  trace elements and 31–34, 35, 36
  under fast-ice 53, 54
 phytoplankton and 61–62, 286, 287, 289, 292
Silicoflagellates 8, 574
Silverfish, *see Pleuragramma antarcticum*
Silversides 570
Simmons, G.M., Jr. 238
Siniff, D.B. 555, 561
Siphlopteryx antarctica 607, 608, 609, 610–611
Siphonoectes dellavallei 379–380
Siphonophora 45
Skeleton of notothenioid fish 433
Skuas 148, 151
 Antarctic, *see Catharacta antarctica*
 Arctic 471
 Great, *see Catharacta skua*
 Longtailed 471
 Pomarine 471
 South Polar (*Catharacta maccormicki*) 445, 471, 652–655
 Subantarctic 471
Slopes 200, 588
Slugs 178, 588
Smetacek, V. 60
Smith, V.R. 186
Smith, W.O. 70
Snails 584, 588
Snares Islands 648
Snoek (*Thyrsites atun*) 360, 551
Snow
 depth, and growth of sea-ice microbes 78–81, 91, 92
 light penetration 78–79, 91, 253
 melt water 662, 663
 and wind transport of airborne particles 592, 593, 594, 595
Sodium (*see also* Salinity)
 in freshwater lakes 146, 151
 in guano 170
 in plants 141
 Rostkovia magellanica 141, 213–215, 219
 in soils 140, 144
Soils 3
 acetylene reduction 186–191
 Elephant Seal influences 173, 175, 176
 interactions with lake micro-flora 626–627, 662–667
 nutrient cycling 140–141, 142–143, 144
 ornithogenic 156, 158, 159, 160, 165
 Poa flabellata growth and 221–228
 water content 198, 201, 202, 610, 613
Solar activity 151
Somateria mollissima 171
Sombre Lake 147, 149, 261, 263
 decomposition processes 253–259
 deep water anoxia 234–237
 phytoplankton 245–251
South Africa 551, 553, 568, 570, 633, 639, 647, 648, 650
South African Antarctic base (SANAE) 13, 14
South Georgia 3, 138, 270, 397, 582–587 *passim* 607, 680
 fish 437–443, 445, 446, 458
 lakes 152
 leaf and litter micro-flora 597, 602, 603
 plant decomposition 192–199, 206–210 *passim* 211–220
 Poa flabellata growth 221–228
 seabirds 463, 490, 491, 517–531, 537, 553, 627–628
 seals 517–531, 543–549, 641–645
 terrestrial nutrient cycling 139, 141–144, 211–220
South Orkney Islands 85, 278–280 *passim* 304–306, 308–310, 437–443, 487–491, 536, 670
South Sandwich Islands 146, 458, 536
South Shetland Islands 146, 152, 339, 415, 437–443, 522, 536, 570, 670
Southern Ocean
 African sector 12–19, 30, 285, 466–471
 Australian sector 575
 Crozet Islands sector 478–485
 food web 573–579
 frontal systems 11–12
 Indian Ocean sector 284–285, 362, 421, 575, 647–651
 satellite remote sensing 16–18
 south of New Zealand 656–661
 zonation 267–269
Soyer, J. 378
Spartina 118
Sphaeroceridae 607, 608, 609
Sphaerophyra 242
Spheniscidae 70
Spheniscus demersus (Jackass Penguin) 161, 494, 495, 648, 650
Spiders (Arachnida) 178, 584, 587, 588
Spinocalanus 45
Sponges 268
Spongiobranchia australis 360
Sprattus fueguensis 570
Springtails, *see* Collembola
Square Lake 146
Squid 268, 633, 634, 648, 678
 calorific content 517
 food consumption 367, 417–418, 419
 lipids 494, 496, 497
 predation on 516–520, 526–528, 530, 531
  by birds 466–471 *passim* 475, 476, 535–541 *passim* 627–628
  by fish 425, 426, 429, 431, 432
  by seals 546, 547, 548, 551–553, 628
  by small cetaceans 567–571 *passim*
 role in marine food webs 415–420

Stagonospora ischmaemi 599, 600, 604
Stahl, J.C. 478
Stanley, Falkland Islands 204, 210
Staphylinidae 178
Starling, European (*Sturnus vulgaris*) 584, 585
Stauroneis anceps 242
Steinhagen-Schneider, G. 652
Stemphyllium botryosum 604
Stephanoeca diplocostata paucicostata 273
Stercorarius longicaudus 471
Stercorarius parasiticus 471
Stercorarius pomarinus 471
Stereotydeus villosus 615, 617
Sterna hirundu 471
Sterna paradisaea 471
Sterna virgata (Kerguelen Tern) 479, 584, 585, 627
Sterna vittata (Antarctic Tern) 471, 479, 484, 485, 519, 627
Sterol esters 494, 495, 496, 497
Stilbocarpa polaris (Macquarie Cabbage) 583, 597–604
Sting-ray (*Myliobatis*) 569
Stomach oils 494–496
Stombidium 242
Stone, S. 555
Stordalen, Sweden 188
Stormpetrel
  Blackbellied, *see Fregetta tropica*
  European 471
  Greybacked (*Garrodia nereis*) 471, 483, 484, 485, 627
  Greyrumped 471
  Leach's (*Oceanodroma leucorhoa*) 464, 471
  Whitebellied 471
  Whitefaced 471
  Wilson's, *see Oceanites oceanicus*
Strand Moraines 79
Streptococcus 593
Strontium (Sr) 658–661, 677
Struvite 158, 159, 160, 163, 164, 165, 166
Sturnus vulgaris 584, 585
Styella melicae 409, 410
Stylocheiron elongatum 45
Sub-Antarctic Front 11–16, 466–469 *passim* 676
Sub-Antarctic Intermediate Water 12
Sub-Antarctic Islands, food webs 582–589
Sub-Antarctic Surface Water 11, 12, 14
Sub-Tropical Convergence (STC) 7, 11, 12, 19, 23, 478
  distribution of pelagic seabirds at 466, 467, 468, 469, 471, 480
  nutrients 12–15
  satellite remote sensing 16–17
Sub-Tropical Front 355, 359
Sub-Tropical Surface Water 11, 12, 14
Subtidal zones 630–636
Sugar content of leaves 603
Sullivan, C.W. 78, 84
Sulphate reduction 144, 257–258, 259
Sulphide 144, 235–236, 256, 258
Sulphur cycling 124, 125
Sunspot maxima 151
Swallows 586
Symplectoteuthis 552, 553
Synechococcus aeruginosa 242
Synedra reinboldii 280
Syowa Station 52, 54, 58, 85, 89, 91

Tachyeres pleneres 411
Takahashi, M. 415
Tanaidacea 378, 385, 386, 387, 632, 633
Tanimura, A. 52
Taninigia 527, 528
Taninigia dana 531
Tanner, A.C. 123
Tanystylum cavidorsum 633
Taonius 528
Taonius pavo 531
Taranakite 159, 160, 164, 165
Tardigrades 241, 242, 587, 614, 626
Tareya, Taimyr, U.S.S.R. 587
Tasmacetus shepherdi 567–568
Tatur, A. 156, 163
Taylor Valley 238
Teal, Speckled 584
Tegula 409, 412
Temperature (*see also* Cold adaptation)
  bacterial activity and 126, 665–666, 667
  cardinal 665, 666
  cellulose decomposition and 196–198
  dependence of photosynthesis 65–66
  in dry valley lakes 240, 243
  energy balance of *Cryptopygus antarcticus* and 616, 617
  in maritime lakes 245, 248
  methanogenesis and 230, 231–232
  mineralization of guano and 158
  moss peat decomposition 205–206
  in petrel burrows 461
  sea water 657, 658
    ammonium distribution and 101
    at oceanic fronts 12–16
    and Krill distribution 299–300, 301, 302
    in Lützow-Holm Bay 53, 54–55
    near Prince Edward Islands 39, 47–48
    near Ross Ice Shelf 95, 96, 97
    phytoplankton and 72, 74, 75, 76
  soil 189
Temperature-salinity diagram 657–658, 659
Terns
  Antarctic (*Sterna vittata*) 471, 479, 484, 485, 519, 627
  Arctic 471
  Common 471
  Kerguelen (*Sterna virgata*) 479, 584, 585, 627
Terre Adélie 273, 535, 536, 537
Terrestrial biology 680
Terrestrial ecosystems 3, 679–680
  arthropod interactions 614–618
  future development of research 151–153
  influences of seals 173–178, 641–645
  interactions with marine and freshwater 626–628
  nutrient cycling 138–145
Terror 70
Testa, J.W. 561
Tetracystis 242
Thalassiosira 62, 268, 351
Thalassiosira eccentrica 373–376
Thalassoica antarctica (Antarctic Petrel) 445, 471, 503
Thaliacea 45
Thawing of frozen cells (*see also* Freeze-thaw cycles) 140
Themisto gaudichaudii 45, 360, 576
  fish consumption of 427, 428, 453, 454
  metal concentrations 658–661
  squid consumption of 417, 418
Themisto japonica 419, 660
Themisto oblivia 660
Theragra chalcogramma (Walleye) 549
Thorakomonas feldmannii 242
Thornfish, New Zealand 432, 433
Thyrsites atun (Snoek) 360, 551
Thysanoessa 310, 489
Thysanoessa macrura 45, 58, 359, 360
  digestive enzymes 330–337
  fish predation on 453, 454, 456, 457, 458
  squid predation on 417, 419
Thysanoessa vicina 58, 358, 359, 360
Tierra del Fuego 221, 398, 566–571, 607, 637
Tiger (*Panthera tigris*) 557
Tigriopus angulatus 632
Tillea moschata 174–175, 177
Tilzer, M.M. 60
Tintinnids 272, 280, 574
Todarodes 528, 552, 553
Todarodes pacificus 419
Todarodes sagittatus 475, 531, 552, 553, 568
Tomopteris 45, 576
Tooth back-calculation 670, 672
Tortula robusta 211, 221
Trace metals 30–36, 656, 677
Trachipterus trachipterus 360
Tréguer, P. 22
Tréhen, P. 606
Trematomus 448, 452–455, 455
Trematomus bernacchii 432, 433, 576
Trematomus centronotus 432
Trematomus hansoni 432
Trematomus loennbergii 431, 432
Trematomus newnesi 529
Trematomus nicolai 432
Trematomus scotti 455, 456, 457
Triacyclglycerides 494, 495, 496, 497
Triacylglycerol 578
Trichenus manatus (Manatee) 557
Trichoderma 604
Trichoderma harizianum 604
Triglycerides 434
Tristan da Cunha Islands 359, 551
Trophon 409, 410, 411
Trypsin 333–337
Tullbergia antarctica 178

Tundra ecosystems 138, 192, 582–589
Tunicago 604
Tunicates 45, 570
Turnover ratio 395, 578–579
Tursiops aduncus 570
Tursiops truncatus 567, 570
Turtle Rock 79
Tydeus tilbrooki 615

Ulva 408, 632, 633, 634
Upwelling of nutrients 7, 17–18, 38, 50–51, 76, 470
Urea
  degradation in maritime lakes 257
  in sea water 22, 23, 25, 26, 111–112
Uria 490
Uric acid 158
Urine, Elephant Seal 173, 175, 177

Van Rensburg, P.J.J. 620
Vanadis antarctica 576
Van't Hoff rule 397, 402
Variscite 160, 165
Vegetation (*see also* Plants)
  Bird Island 643
  classification 192
  grazing resistance 583
  impact of seals 173, 177, 641–645
  invertebrate distribution and 587–588
  salt-spray and 173, 174–175, 177, 202
  Sub-Antarctic islands 140–141, 582, 627
  tundra 588
Vellella vellella 360
Vernon, Ph. 606
Vestfold Hills 145, 680
Vibilia antarctica 45, 660
Vibilia armata 45
Vibrio 125, 126, 666, 667
Vibrio fluvialis 126
Victoria Land 71, 145, 238–243, 500, 501
Viruses 595
Viscera in notothenioid fish 434
Vitellogenesis 576, 578
Vivianite 160
Volucrispora graminea 604
Von Bodungen, B. 60
Vulcan 9

Waders 586
Walleye (*Theragra chalcogramma*) 549
Walters, N.M. 11
Walter Herwig 346
Walton, D.W.H. 180, 192
Warm Deep Current 47
Warm Deep Water (WDW) 23, 24, 26, 27, 28, 266
Water
  content of soil 198, 201, 202, 610, 613
  ground-water 150
  melt-water: entering lakes 148, 150, 151, 662, 663
    sea-ice 76, 81
  in moss-peat 205
  plant-tissue, in bryophytes 201, 202
  run-off, from penguin rookeries 156, 158, 159
Waves, internal 102
Wax esters 434, 494, 495, 496, 497, 576, 578
Weber, L.H. 284
Weddell Drift 267
Weddell Gyre 268, 449
Weddell Sea 81, 267, 268, 270
  choanoflagellates 271, 273
  fish 431, 445–450
  Krill 304–306, 308, 324, 325, 343
  organochlorines and heavy metals 652–655
  phytoplankton 70, 75, 368
  plankton 277–278, 279–281
  sea-ice micro-algae 85
Weddell-Scotia Confluence 267
Weevils 587, 588, 632, 634
Weimerskirch, H. 478
Weka (*Gallirallus australis*) 584, 585
West Wind Drift 7–8, 38, 44, 50, 267, 268
Weste, G.M. 597
Whales 3, 269, 445, 566
  Andrew's Beaked (*Mesoplodon bowdoini*) 567, 568, 571
  Arnoux's Beaked (*Berardius arnuxii*) 567, 568
  baleen, *see* Baleen whales
  beaked 566, 567–569, 571
  biomass in Ross Sea 498–502, 504, 505, 508–514
  Blainville's Beaked (*Mesoplodon densirostris*) 567, 568
  Cuvier's Beaked 567, 568
  Fin (*Balaenoptera physalus*) 569, 672
  Gray's Beaked (*Mesoplodon grayi*) 567, 568
  Hector's Beaked (*Mesoplodon hectori*) 567, 568, 571
  Killer, *see Orcinus orca*
  Krill consumption 366, 375
  Minke, *see Balaenoptera acutorostrata*
  Pilot 567, 569
  Pygmy Right (*Caperea marginata*) 567
  Sei (*Balaenoptera borealis*) 508, 569, 672
  Shepherd's Beaked 567–568
  Southern Bottlenose (*Hyperoodon planifrons*) 567, 568–569, 571
  Sperm, *see Physeter catodon*
  stranding on beaches 571
  Straptoothed (*Mesoplodon layardii*) 567, 568, 571
Whaling 671, 672, 673
Wilhelm II Land 445
Williams, R. 452
Wind
  currents induced by 44, 50
  mixing induced by 253
  nutrient upwelling and 38, 46, 50, 51, 76
  plant growth and 227
  transport of airborne particles 592–595
Windmill Islands 152
Winter, J.D. 561
Wynn-Williams, D.D. 204, 229, 662

Xanthomonas campestris 603
Xanthoria 184–185

Yarrington, M.R. 229
Yeasts 185, 614, 615
Yeasts
  in dry valley lakes 241
  in freshwater lakes 150
  leaf and litter decomposition 597, 599, 600, 601, 602
  moss peat decomposition 207, 208, 209, 210
  phyllosphere 204
  soil and freshwater lake interactions 662–667
Yoshino-Maru 415

Zielinski, K. 104
Zinc (Zn) 677
  in sea water 7, 30, 33–34, 35, 36
  in seals and seabirds 654–655
  in zooplankton 658–661
Ziphiidae (beaked whales) 566, 567–569, 571
Ziphius cavirostris 567, 568
Zooplankton 2, 678
  ammonia excretion 8, 99–100, 325–326
  bird predation on 481–482, 483–485
  distribution in relation to ammonium 99–102
  lipids 494, 496
  macro-zooplankton 355–360, 573, 574
  in marine food webs 278–282
  in maritime lakes 249–250
  metal concentrations 656–661, 677
  micro-zooplankton 279, 281, 282, 368, 573, 574
  near Prince Edward Islands 43–44, 45, 46–47, 50–51
  predation by fish 421–429
  under sea-ice 58
  zonation 268